Homogeneous Catalysis with Metal Complexes

Homogeneous Catalysis with Metal Complexes

Kinetic Aspects and Mechanisms

O. N. TEMKIN
M.V. Lomonosov Moscow State University of Fine Chemical Technology

Translated from the Russian
by P. P. Pozdeev

WILEY
A John Wiley & Sons, Ltd., Publication

This edition first published 2012

An earlier version of this work was published in the Russian language by Академкнига under the title Гомогенный металлокомплексный катализ: *Кинетические аспекты.*

Registered office
John Wiley & Sons Ltd, The Atrium, Southern Gate, Chichester, West Sussex, PO19 8SQ, United Kingdom

For details of our global editorial offices, for customer services and for information about how to apply for permission to reuse the copyright material in this book please see our website at www.wiley.com.

A catalogue record for this book is available from the British Library.
ISBN: 9780470666999

Set in 10/12pt Times by Laserwords Private Limited, Chennai, India
Printed and bound in Malaysia by Vivar Printing Sdn Bhd

1 2012

To My Wife and Friend,
Raisa Vasil'evna Basova,
with Gratitude for Everything

Contents

Notations and Abbreviations

Notations:

A	reaction affinity
a	activity of molecules or ions
an	acceptor number of a solvent molecule
$\boldsymbol{B}_m$	stoichiometric matrix of reaction mechanism
$\boldsymbol{B}_N$	stoichiometric submatrix of reaction participants
$\boldsymbol{B}_p$	stoichiometric matrix of overall route equations
$\boldsymbol{B}_{\mathrm{X}}$	stoichiometric submatrix of intermediates
$C^{\pm}$	cyclic characteristic of a kinetic graph
C_{A}	concentration of substance A
C_{pn}	weight of nth cycle on pth route
C_{X}	concentration of substance X
D_i	weight of the graph root determinant in ith vertex
D_{ik}	weight of kth rooted tree in ith vertex
dn	donor number of a solvent molecule
D_{pn}	weight of subgraph of nth cycle on pth route
E°	standard electrode potential
ee	enantiomer excess
F_i^*	complexity of ith intermediate (ith vertex)
F_{M}	complexity of metal catalyst
F_{S}	complexity of substrate
$\boldsymbol{G}$	matrix of stoichiometric numbers of steps
G	Gibbs energy
$\boldsymbol{H}$	atomic (molecular) matrix
H_0	Hammet acidity function
h_0	Hammet acidity
I	number of intermediates; ionic strength; induction factor; indicator ratio
K	reaction equilibrium constant
k_j	reaction rate constant
n	average coordination number; Bjerrum formation function
N	number of reaction participants (reactants and products)
N_I	number of linearly independent intermediates
P	number of linearly independent routes (basis set of routes); pressure
$Q_{\max}$	maximum number of linearly independent reactions (in stoichiometric basis set)
Q_r	number of linearly independent overall route equations
r	reaction rate

$r^{(p)}$	reaction rate over pth route
$r^{\pm}$	reaction rates in forward (+) and reverse (−) direction
S	number of steps in reaction mechanism
W_j	rate on jth steps
[X]	concentration of substance X
α	gas distribution coefficient between gaseous phase and solution
α_i	polarizability of ith substance (or solvent)
β^*_{ij}	stoichiometric coefficient of ith intermediate at jth step
β_{12}	exchange integral
χ	degree of conversion
ε	small parameter; concentration fluctuation; dielectric permittivity
ε_i	extinction coefficient
γ	kinetic chain length; activity coefficient
γ_i	activity coefficient (molar)
η	viscosity small parameter
μ	chemical potential; molecularity of step; dipole moment
σ	surface tension
$\nu_j^{(p)}$	stoichiometric number of jth step in pth route
ξ	extent of reaction; chemical variable
ω_j	weight of jth step

Abbreviations:

AA	acrylic acid
AC	acetone
Ac	acetyl
acac	acetylacetonate (anion)
acacH	acetylacetone
AcO	acetate anion
AcOH	acetic acid
ADMA	acetaldehyde dimethyl acetal (1,1-dimethoxyethane)
ADN	adipodinitrile
ADP	adenosine diphosphate
Alk	alkyl
All	allyl
AN	acetonitrile
AP	acetophenone (methyl phenyl ketone)
ATP	adenosine triphosphate
BA	butyl acrylate
BDM	dibutyl dihydromuconate
BEE	2-bromoethyl ethyl ether (1-bromo-2-ethoxyethane)
BG	bipartite graph (nonlinear mechanism)
BG_X	subnetwork of intermediates X in bipartite graph
bipy	2,2′-bipiridyl
BME	2-bromoethyl methyl ether (1-bromo-2-methoxyethane)

$[bmim]^+$	1-bityl-3-methylimidazolium cation
BP	benzoyl peroxide
BQ	*para*-benzoquinone
Bu	butyl
Bu^t	*tert*-butyl
Bz	benzoyl
CB	chlorobutadiene
CH	cyclohexene
CHA	cyclohexanecarboxylic acid
CME	2-chloroethyl methyl ether (1-chloro-2-methoxyethane)
CSTR	continuous-flow stirred tank reactor
Cod	1,5-cyclooctadiene
Cot	cyclooctatetraene
Cp	cyclooctapentadienyl
CQ	chloranil (tetrachloro-*p*-benzoquinone)
CVA	2-chlorovinylacetylene (2-chloro-1-but-3-ine)
Cy	cyclooctene, cyclohexyl
DA	diacetylene
dba	dibenzilydeneacetone
DCA	dichloroacetylene
DCE	1,2-dichloroethylene
DIOP	2,2-dimethyl-4,5-bis(diphenylphosphinomethyl)-1,3-dioxolan
DMC	dimethyl carbonate
DMDA	dimethyldiacetylene
DMF	dimethylformamide
DMO	dimethyl oxalate
DMSO	dimethyl sulfoxide
dppe	1,2-bis(diphenylphosphin)ethane (diphenylphosphinoethane)
dppm	1,2-bis(diphenylphosphin)methane (diphenylphosphinomethane)
DQ	duroquinone (tetramethyl-*p*-benzoquinone)
dtc	diethylthiocarbamate
DVA	1,2-divinylacetylene
EA	ethyl acetate
$[emim]^+$	1-ethyl -3-methylimidazolium cation
Et	ethyl
GA	glycolate anion
GBL	γ-butyrolactone
Hal	halogenide (halide)
HE	halide ethers
HFB	hexafluorobenzene
HMPTA	hexamethylphosphoric triamide
HP	hydroperoxide
HPA	heteropolyacid
IL	ionic liquid
Im	imidazole

IP	ion pair
KF	kinetic function
KG	kinetic graph (subnetwork of intermedites in linear mechanism)
KIE	kinetic isotope effect
LTI	labeled topology identificator
MCA	monochloroacetylene
MCS	multicomponent system
MDA	methyldiacetylene
Me	methyl
Mes	2,4,6-trimethylphenyl
MMA	methylmethacrylate
MO	molecular orbital
MP	methyl propionate
MSA	methanesulfonic acid
MSS	multiplicity of steady states
NADH	nicotinamide adenine dinucleotide
NBD	norbornadiene
NC	nanocluster
NMP	N-methylpyrrolidone
NQ	*para*-naphthoquinone
Nu	nucleophile
Oc	octene
OEP	octaethylporphyrinato anion
Ol	olefine
Ox	oxidant
PA	phenylacetylene
Pc	phthalocyaninato anion
p-cymol	*para*-cymol (*para*-methylisopropylbenzene)
PE	quasi-equilibrium (pre-equilibrium) approximation
PEG	polyethylene glycol
PFCS	polyfunctional catalytic system
Ph	phenyl
Phen	1,10-phenanthroline
Pipy	piperidine
Pr	propyl
Py	pyridine
Q	quinone
Red	reductant
RN	reaction network
SA	succinic anhydride
salen	N, N′-ethylene-bis(salicylideniminato) anion
Salophen	N, N′-phenylene-bis(salicylideniminato) anion
SS	quasi-steady-state approximation
SSM	simple submechanism
TCE	trichlotoethylene; tetrachloroethane
TEM	transmission electron microscopy

THF	tetrahydrofuran
TI	transition-state-topology identificator
tmeda	1,2-bis(dimethylamino)ethane (tatramethylethylenediamine)
TOF	catalyst turnover frequency
TON	catalyst turnover number
TPP	tetraphenylporphyrinato anion
TS	transition state
TsOH	*para*-toluenesulfonic acid
Tyr	thyrosine
VA	vinylacetylene monomer; vinyl acetate
VC	vinyl chloride (1-chloroethylene)
VD	2-vinyl-1,4-dioxane
VDC	vinylidene chloride (1,1-dichloroethylene)
XPS	X-ray photoelectron spectroscopy
XRD	X-ray diffraction
Ep	epoxide
Tol	toluene
X_i	intermediate species (intermediates)

Preface to English Edition

In recent years, many books have appeared that are devoted to catalysis – a central, unifying concept in chemistry, including the rapidly developing homogeneous catalysis with metal complexes. This type of catalysis has demonstrated impressive achievements in synthetic organic chemistry and commercial chemical technology.

A reader might be surprised that yet another monograph on catalysis is offered to his or her attention. However, an unbounded field of knowledge such as catalytic chemistry can be considered in various aspects. The present book is aimed at providing a notion about the state-of-the-art in the theory of mechanisms of catalytic reactions in solutions, the state and possibilities of the kinetic method of investigation of the mechanisms of reactions involving metal complexes, a relationship between mechanistic hypotheses and existing kinetic models, and the kinetic models and mechanisms of many homogeneous catalytic processes employed in synthetic and commercial chemistry. Considerable attention in this monograph is devoted to the development of a rational strategy for kinetic models design, to the introduction of new concepts, and to an analysis of problems that are encountered in catalysis with metal complexes (including catalysis with nanoclusters and colloidal particles, homogeneity and heterogeneity of active catalysts, polyfunctional homogeneous catalytic systems, mechanisms of formation and decay of the active centers, chain mechanisms in the catalytic chemistry, allowance for a nonideal character of reaction media, etc.).

The Russian school of chemical kinetics has gained the respect of the world scientific community. It will suffice to mention only works by the Nobel Prize winner N.N. Semenov and other well known scientists such as N.M. Emmanuel', M.I. Temkin, S.Z. Roginskii, G.K. Boreskov, I.I. Moiseev, A.E. Shilov, A.M. Zhabotinskii, S.L. Kiperman, G.S. Yablonskii and A.Ya. Rozovskii. On the other hand, for many reasons, the works of Russian scientists in chemical kinetics (as well as in other fields) only began to regularly appear in international editions in the last 20–25 years. This monograph, generalizing the main results obtained in the field of kinetics and mechanisms of homogeneous catalytic reactions involving metal complexes for the last 60 years, naturally also presents the most interesting investigations performed both in the former USSR and modern Russia.

The author is pleased to know that this book is now available to the whole catalytic community, rather than to Russian-speaking readers only, and is highly grateful to John Wiley & Sons for deciding to publish the English translation of his most recent monograph. This book is a combination of a scientific monograph and a handbook, and the author hopes that it will be useful to specialists as well as to advanced students, graduates, and postgraduates of universities and higher technology colleges by providing a deeper insight into catalytic chemistry, theory of reaction mechanisms, and chemical kinetics of homogeneous catalytic processes.

It is a great pleasure for the author to express his gratitude to Dr P. P. Pozdeev, the translator, for his kind consent to translate this huge monograph from the Russian, creative approach to this work, and fruitful cooperation in all stages of translation manuscript preparation.

O. N. Temkin

Preface

This book is an attempt to summarize the results of an approximately 60 year-long period in which the kinetic method was applied to investigations of the mechanisms of homogeneous catalytic reactions catalyzed by metal complexes. This period of time simultaneously featured both the establishment of homogeneous catalysis with metal complexes as one of the most important directions in catalytic chemistry and the development of a kinetic method and the corresponding approach to studying reaction mechanisms and constructing kinetic models of catalytic processes.

The theory of mechanisms of homogeneous catalytic reactions, including an analysis of the results of investigations of the structure of intermediates and the possible ways to their formation and transformation, noticeably outstrips possibilities of the experimental identification of the mechanisms of particular catalytic processes. Indeed, at the beginning of the 1950s, it was difficult to formulate even a single non-contradictory hypothesis concerning a possible mechanism of one or another catalytic reaction, whereas now the topical problem is how to perform discrimination of numerous probable, theoretically justified hypotheses. Both possibilities and limitations of the kinetic method have become evident. Consideration of the entire set of questions related to the kinetic aspects of homogeneous catalysis with metal complexes is the subject of this monograph.

Despite the great significance of catalysis with metal complexes in both commercial chemistry and organic synthesis, peculiarities and problems in the kinetics of homogeneous catalytic reactions in solutions of metal complexes – in contrast to the kinetics of gas-phase, enzymatic, and heterogeneous catalytic and topochemical reactions – are inadequately reflected in both basic monographs and teaching handbooks.

An analysis of the available literature, including monographs on separate types of catalytic reactions, and his own half-a-century experience led the author to the conclusion that writing a special monograph on the kinetics of catalytic reactions with metal complexes is expedient. This book presents a generalization of the results of studying the kinetics of various homogeneous catalytic (and, in some cases, noncatalytic) processes, which have been obtained since the beginning of 1950s for the reactions in solutions of both transition and nontransition metal complexes.

Traditionally, the kinetics of complex chemical reactions was developed within the framework of adjacent disciplines such as physical chemistry, chemical physics, and biophysics. Modern chemical kinetics can also be considered as a direction in the rapidly developing faculty of mathematical chemistry. Indeed, the direction in mathematical chemistry devoted to the kinetics analyzes the structure and dynamic properties of some special types of differential and algebraic equations. The first issue of a special international journal devoted to mathematical chemistry (*Journal of Mathematical Chemistry*) was published in 1987.

It is conventional that chemists employ the kinetic method for studying the reaction mechanisms, whereas mathematicians are engaged to solve the inverse problem of chemical kinetics (i.e., to estimate rate constants and parameters of kinetic models and assess the possibility of their identification), analyze the dynamical behavior of the system of differential equations, etc. The optimum situation would naturally imply a collaboration of chemists and mathematicians but, as the author's experience shows, their effective cooperation is hardly possible for many reasons. In this context, one of the author's goals is to concisely present in this book, intended mostly for chemists, the main mathematical approaches, ideas and problems that are important to understand when setting a kinetic experiment, discriminating hypotheses, and interpreting kinetic data. It is hoped that this book will also suggest interesting research objects to specialists engaged in numerical simulations and mathematical chemistry.

The monograph considers the potential, achievements, and limitations of applying the kinetic approach to the identification of mechanisms of complex reactions and dwells on the issues of a rational strategy in constructing theoretically justified kinetic models. The kinetics of reactions in systems with associates and polynuclear complexes of metals is considered in detail. Factors that account for a multi-route character and a relationship between the topological structure of mechanisms and features of kinetic models are analyzed. The problem of kinetic and thermodynamic conjugation in complex reaction kinetics is discussed. Information on the basic principles and specific features of the dynamic behavior of nonlinear kinetic models (including mechanisms of oscillatory reactions) is presented and the thermodynamic, chemical, and mathematical principles of nonlinear dynamics are considered. In addition to data on the mechanisms of well-known processes such as the Belousov–Zhabotinskii reactions, the book presents the results of studying the oscillatory reactions of oxidative carbonylation of alkynes in solutions of palladium complexes, which were discovered in the Department of Chemistry and Technology of Basic Organic Synthesis at the M.V. Lomonosov Moscow State University of Fine Chemical Technology, in the Laboratory of Kinetics and Catalysis headed by the author. The monograph briefly considers existing notions about the influence of a reaction medium and a nonideal character of the solutions of electrolytes and metal complexes in aqueous and nonaqueous media on the kinetics of reactions and the equilibrium of complex formation processes. Approaches to the elimination or allowance for these effects in setting kinetic experiments for the discrimination of hypotheses are discussed.

All sections of this book contain the results of original investigations that have not been considered previously in scientific monographs or teaching handbooks. In Chapters 1, 6, and 8, significant emphasis is placed on teaching aspects, whereas Chapters 2–5 and 7 mostly tend towards a scientific research character, although they can also serve as an additional teaching guide for advanced students, graduates, postgraduates and young specialists engaged in catalysis with metal complexes, complex reaction kinetics, and the theory of mechanisms of catalytic reactions.

O. N. Temkin

Acknowledgments

For my interest in catalysis with metal complexes, catalytic chemistry of alkynes, and kinetic methods of investigation, I am greatly indebted to Professor R. M. Flid – my teacher and friend for 20 years (1954–1974). Professor Flid was a student of Professor M.I. Usanovich (academician of the Kazakh Academy of Sciences) and Professor M.Ya. Kagan, and my meetings with Professor Usanovich for 18 years were an important school that significantly influenced my chemical outlook (see collection of memories *Vospominaniya o Professore R.M. Flide* [Remembering Professor R.M. Flid], Ekonomika, Moscow, 2006).

I am pleased to heartily express my acknowledgement to Professor I.I. Moiseev, academician of the Russian Academy of Sciences, for highly fruitful and stimulating contacts over more than 50 years.

Writing this book would be impossible without many years of collaboration and fruitful discussions with my co-workers, colleagues, and postgraduate students whose results are also reflected in this monograph. These are Professor G.K. Shestakov, S.M. Brailovskii (Cand.Sci), Professor L.G. Bruk, Professor O.L. Kaliya, Professor D.G. Bonchev (Bulgaria), A.V. Zeigarnik (Cand.Sci), Dr R.E. Val'des-Peres (USA), Professor D. Kamenski (Bulgaria), Professor M.G. Mys'kiv (Ukraine) and Candidates of Sci. N.F. Alekseeva, S.M. Airyan, I.V. Bozhko, V.S. Vartanyan, N.Yu. Vsesvyatskaya (Kozlova), E.G. Gel'perina, S.N. Gorodskii, G.V. Emel'yanova, I.A. Esikova, L.N. Zhir-Lebed', L.A. Zakharova, T.T. Zung, A.S. Zakieva (Abdullaeva), L.A. Il'ina,, A.P. Kozlova, A.V. Kulik, A.E. Kuz'min, O.V. Marshakha, N.G. Mekhryakova, L.Ya. Mesh, F.B.O. Nazarov, A.N. Nyrkova, D.I. Otaraku, I.V. Oshanina, S.A. Panova, A.B. Svetlova (Pshenichnikova), L.V. Reshetnikova (Mel'nikova), L.A. Sil'chenko, M. Skumov, T.G. Sukhova, G.F. Tikhonov, I.V. Trofimova, A.A. Khorkin, Kh.Kh. Man', M.S. Shlapak, G.M. Shulyakovskii, L.V. Shchel'tsyn, and L. Elefteriu A.I. and PhD A.I. Kozlov.

I would like to express my deep gratitude to all of them.

I would like to thank S.M. Brailovskii for kindly permitting the use (in Chapters 1–3) of some materials from our manuscript, written in cooperation (1974) but yet unpublished, devoted to the kinetics of reactions in catalysis with metal complexes and to B.M. Mykhalichko and M.G. Mys'kin for kindly permitting the use of materials from our joint review on polynuclear copper(I) complexes.

I would like to gratefully mention E.D. German, my first supervised diploma student and friend, for that work marked the beginning of potentiometric investigations *in situ* for catalytic reactions in solutions of Cu(I, II), Ag(I), Hg(II), and Pd(I, II) complexes.

I am also heartily grateful to L.G. Bruk – my student, colleague and friend – for permanent and highly fruitful discussions of all aspects related to catalysis with metal complexes and the entire content of this monograph.

I would like to thank Yu.A. Pisarenko. N.B. Librovich, I.S. Kislina, and A.V. Zeigarnik for kindly reading separate chapters of this book and making highly valuable remarks, and to A.V. Kulik and A.P. Ivanov for their help in preparing the manuscript.

I would like to express my gratitude to D.K. Novikova, editor of the book, for her informal, constructive, and kind cooperation in carrying out a huge work on the manuscript preparation.

Finally, I am grateful to the Russian Foundation for Basic Research for financial support to publishing this book.

O. N. Temkin

Acknowledgments

About the Author

O. N. Temkin, the author of the monograph *Homogeneous Catalysis with Metal Complexes: Kinetic Aspects and Mechanisms*, is a well-known specialist in the field of chemical kinetics, catalysis with metal complexes, chemistry of alkynes, and mathematical chemistry. He is a professor of the M.V. Lomonosov Moscow State Institute of Fine Chemical Technology (now the M.V. Lomonosov Moscow State University of Fine Chemical Technology).

Professor Temkin is the co-author of three monographs and two chapters in monographs:

- O. N. Temkin and R. M. Flid, *Kataliticheskie prevrashcheniya atsetilenovykh soedinenii v rastvorakh kompleksov metallov* (Catalytic Transformations of Alkynes in Solutions of Metal Complexes), Nauka, Moscow, 1968.
- O. N. Temkin, G. K. Shestakov and Yu. A. Treger, *Atsetilen: khimiya, mekhanizmy reaktsii, tekhnologiya* (Acetylene: Chemistry, Reaction Mechanisms, Technology), Khimiya, Moscow, 1991.
- O. N. Temkin, A. V. Zeigarnik and D. G. Bonchev, *Chemical Reaction Networks. A Graph-Theoretical Approach*, CRC Press, Boca Raton, Fl., USA, 1996.
- O. N. Temkin and D. Bonchev, in *Mathematical Chemistry Series*, Vol. 2. *Chemical Graph Theory. Reactivity and Kinetics*, D. Bonchev and D. H. Rouvray (Eds), Abacus Press–Gordon & Breach Sci. Publ., Philadelphia, 1992.
- O. N. Temkin, A. V. Zeigarnik and D. G. Bonchev, *Understanding Chemical Reactivity*, Vol. 9, *Graph Theoretical Approaches to Chemical Reactivity*, D. Bonchev and O. Mekenyan (Eds), Kluwer Academic Publishers, Dordrecht, 1994.

Professor Temkin is the author and co-author of more than 380 scientific publications, including reviews and a chapter (O. Temkin and L. Bruk, Oxidative Carbonylation: Homogeneous) in the *Encyclopedia of Catalysis*, I. Horvath (Ed.), John Wiley & Sons, 2003, Vol. 5, pp. 394–424. He is the author of article "Homogeneous Catalysis" in the new *Big Russian Encyclopedia* (Rossiiskaya Entsiklopediya, Moscow, 2007), Vol. 7. His works have been reported in many international scientific journals, including *J. Mol. Catal., Organometallics, J. Phys. Chem., J. Chem. Soc. Chem. Commun., Inorg. Chim. Acta, Langmuir, J. Chem. Ed., React. Kinet. Catal. Lett., J. Chem. Inf. Comput. Sci., Math. Chem., J. Comput. Chem., J. Mol. Structure (Theochem)*, and in a number of Russian journals.

Professor Temkin designed for the first time kinetic models of numerous catalytic reactions of alkynes in superconcentrated Cu(I) chloride complex solutions, kinetic models of oxidation, oxidative carbonylation, and chlorination reactions of olefins, dienes, alkynes, and alcohols. He established catalytic systems for the anti-Markovnikov addition of water and hydrogen chloride molecules to alkynes, developed the catalytic chemistry of Pd(I) complexes, discovered a new type of oscillatory reaction in Pd-catalyzed carbonylation of alkynes and many other catalytic systems, and reactions catalyzed by metal complexes.

Introduction

The second half of the 20th century was marked in catalytic chemistry by an extensive development and effective use of catalysts based on metal complexes for homogeneous processes in chemical industry [1–7]. Since then, metal complex catalysts have become an important tool in modern synthetic organic chemistry [8–12], while their investigation provided a basis for ideas concerning the mechanisms of reactions in heterogeneous catalysis [13–16]. Academician I.I. Moiseev pointed out that "catalysis with metal complexes is an integral part of the scientific-technological revolution in the 20th century" [17].

At the end of the 19th century – i.e., much later than when K.S. Kirchhoff carried out his works on the acid catalysis of potato starch hydrolysis (1811) – three new directions were established in homogeneous catalysis [18, 19]. The use of aluminum complexes as catalysts for the alkylation and acylation of aromatic compounds by Ch. Friedel and J.M. Crafts in 1887 initiated the development of homogeneous electrophilic aprotic catalysis ($AlCl_3$, $SnCl_4$, $SbCl_5$, $FeCl_3$, etc.). A boost to investigations into the reduction–oxidation (redox) reactions with the participation of H_2O_2 and ROOH catalyzed by metal complexes was provided in 1894 by the work of G.J. Fenton on the oxidation of tartaric acid by dihydroperoxide in Fe(II) salt solution (now known as the Fenton reaction and reagent).

In continuation of the works performed by E. Linnemann, K.M. Zaitsev, and G.N. Glinskii in 1866–1867 [20], which discovered the hydrolysis of propenyl bromide and vinyl bromide in the presence of mercury acetate via the following scheme:

$$CH_2{=}CHBr + H_2O \xrightarrow{Hg(II)} CH_3CHO + HBr,$$

M.G. Kucherov (Kutscheroff) in 1881 suggested that the formation of carbonyl compounds in this reaction is due to the elimination of HBr from alkenyl bromide and the addition of H_2O to the intermediate alkyne. Although Kucherov's hypothesis was incorrect, the idea that Hg(II) salts catalyze the hydration of alkynes was successfully confirmed. This discovery, having drawn considerable interest from commercial chemistry, together with the aforementioned works by Linnemann, Zaitsev, and Glinskii, can be considered as triggering the development of the catalysis of organic reactions by complexes of post-transition (Cu, Ag, Au, Hg) and transition metals, involving the formation of organometallic intermediate compounds.[1] Subsequently, it was demonstrated that Hg(II) complexes possess the properties of typical aprotic acids and exhibit some features characteristic of platinum-group metals. The method of acetaldehyde synthesis was patented in 1910 (one year before the death of Kucherov), while the first commercial production

[1]It should be noted that 2011 marks the 100th anniversary of the death of M.G. Kucherov, the well-known Russian scientist, and the founder of the catalysis with metal complexes.

Homogeneous Catalysis with Metal Complexes: Kinetic Aspects and Mechanisms, First Edition. O. N. Temkin.

of acetaldehyde via the Kucherov reaction was launched in Germany and Canada in 1916. At the same time, it was found by F. Klatte in 1913 that mercury salts in solution are capable of catalyzing the addition of acetic acid to acetylene with the formation of vinyl acetate and ethylidene diacetate, and the addition of HCl to acetylene with the formation of vinyl chloride [21].

In 1929–1931, the American chemist, J. Nieuwland, discovered that copper halides catalyze the reactions of acetylene dimerization and trimerization [21]. The dimerization reaction scheme

$$2C_2H_2 \xrightarrow{Cu(I)} CH_2{=}CHC{\equiv}CH$$

provided a basis for the commercial synthesis of chloroprene as

$$C_4H_4 + HCl \xrightarrow{Cu(I)} CH_2{=}C(Cl)CH{=}CH_2,$$

and the related production of synthetic rubber as Neoprene (United States) and Sovprene (USSR). A significant contribution to investigations of this process and the development of chloroprene-based synthetic rubber technology was made by W. Carothers (USA) and A.L. Klebanskii (USSR). Copper(I) complexes also proved to be catalytically active in the hydrochlorination and hydration of acetylene. A "non-mercury" catalyst ($CuCl–ZnCl_2–H_2O$) of acetylene hydration (proposed in 1958) was comparable with the Kucherov catalyst in the activity, but significantly exceeded it with respect to stability [22]. In 1929, Nieuwland discovered the reaction of oxidative chlorination of acetylene to *trans*-1,2-dichloroethylene in the $CuCl–CuCl_2–H_2O$ system. Somewhat later (in 1939), P. Kurtz established in Germany that the Nieuwland catalyst ($CuCl–MCl–HCl–H_2O$) was also active in the hydrocyanation of acetylene according to the following scheme:

$$C_2H_2 + HCN \xrightarrow{Cu(I)} CH_2{=}CHCN,$$

which was used in 1942 by IG Farbenindustrie for the production of acrylonitrile.

Copper(I) complexes in solution were the first homogeneous catalysts of H_2 activation in the reactions of reduction of inorganic oxidizers and hydrogenation of *p*-benzoquinone (originally proposed by M. Calvin in 1938) [23]. In 1939, M. Iguchi discovered that rhodium complexes exhibited catalytic activity in the reaction of fumaric acid hydrogenation.

In the course of investigation of the hydrocarbon synthesis via the Fischer–Tropsch process on a $CoTh/SiO_2$ heterogeneous catalyst, O. Roelen in 1938 revealed a very interesting reaction of hydroformylation of olefins:

$$C_2H_4 + CO + H_2 \rightarrow CH_3CH_2CHO.$$

It should be noted that Roelen at that time believed that the mechanism of this reaction included the formation of $HCo(CO)_4$.[2] In 1951, D.M. Rudkovskii with co-workers showed that this process (oxosynthesis) is homogeneous and can be catalyzed by cobalt carbonyl complexes [24]. Somewhat later, this was confirmed by I. Wender, G. Sternberg and M. Orchin [23]. As a result, $Co_2(CO)_8$ was introduced as the new type of catalyst into commercial catalysis with metal complexes.

Investigations of the reactions of carbon oxide with olefins, acetylene, and alcohols were carried out by W. Reppe from 1938 to 1945 (the results had become known only after World War II) and continued from 1951 to 1957 by the BASF company. Reppe with co-workers discovered several important catalytic reactions, including the carbonylation of methanol in solutions of Co, Ni, and Fe complexes,

$$CH_3OH + CO \rightarrow CH_3COOH,$$

[2]According to communication from M. Orchin, *Accounts Chem. Res.*, **14** (9), 259 (1981).

the hydrocarboxylation and hydrocarbalcoxylation of acetylene and olefins in solutions of Ni(0) and Ni(II) complexes,

$$C_2H_2 + CO + ROH \rightarrow CH_2{=}CHCOOR \ (R{-}H, \text{alkyl});$$

$$C_2H_4 + CO + ROH \rightarrow CH_3CH_2COOR \ (R{-}H, \text{alkyl}),$$

and the cyclotrimerization (C_6H_6) and cyclotetramerization (C_8H_8) of acetylene (and alkynes) in nickel cyanide solutions in tetrahydrofuran. Thus, nickel and iron complexes were introduced into homogeneous catalysis with metal complexes. Reppe with co-workers also developed an original heterogeneous liquid-phase synthesis of 2-butyn-1,4-diol:

$$C_2H_2 + 2CH_2O \rightarrow HOCH_2C{\equiv}CCH_2OH,$$

which was catalyzed by an organometallic compound (Cu_2C_2) supported on silica (Cu_2C_2/SiO_2 obtained from CuO_2/SiO_2). All these reactions (known as Reppe's chemistry) became commercial processes. In particular, in 1948 Rohm & Haas Co. (USA) launched the synthesis of acrylic acid esters in $Ni(CO)_4$ solutions, and BASF (FRG) in 1952 developed the production of propionic acid.

On the whole, twelve commercial processes using metal complexes were used in commercial chemistry in the middle of the 20th century. It is also interesting to note that it was Reppe who originally (in 1948) introduced phosphine ligands into practical catalysis with metal complexes. In particular, he established that $Ni(CO)_2(PPh_3)_2$ exhibited very high activity in the oligomerization of acetylenes and $NiBr_2(PPh_3)_2$ – in the synthesis of acrylates from acetylene.

Thus, at the end of the *first stage* in the development of catalysis with metal complexes, which terminated after World War II (at the end of the 1940s), investigations of the reactions catalyzed by metal complexes in solution were carried out mostly by industrial companies. The discoveries of Kucherov, Klatte, Nieuwland, Roelen, and especially Reppe had drawn the interest of chemical companies to the homogeneous catalysis with metal complexes, and this interest considerably increased after the new discoveries made in the 1950s to the 1970s. However, the academic community did not consider the development of new catalysts and processes as an independent direction in catalytic chemistry. In the first half of the 20th century, scientific research laboratories studied the catalysis of redox type reactions (e.g., reactions of hydrogen peroxide in solutions of metal ions and complexes, catalysis of the oxidation of metal ions and organic compounds) and modeled the catalase and peroxidase functions of enzymes. In particular, Shpitalskii [25] developed the first theory of intermediate compounds in homogeneous catalysis with metal complexes, which was based on the results of studying the kinetics of the catalytic decomposition of H_2O_2. The proposed theory was used by F. Haber and J. Weiss in 1934 to develop a mechanism of this reaction in the Fenton system. Extensive research was also devoted to reactions in which metal ions and complexes played the role of electrophilic catalysts (Lewis acids) [23], including the hydrolysis of esters, amides, and phosphoric acid ethers, and reactions of decarboxylation, transamination, epoxide polymerization, aldole condensation, bromination, etc. Investigations of that period were summarized Langenbeck's 1948 monograph [26] (with additional author's comments and a bibliography extended to 1959 in the Russian edition of 1961), reviews of Baxendale [27] and Weiss [28], and monographs by Hein [29], Reppe [30], and Copenhaver and Bigelow [31]. At the First International Congress on Catalysis that was held in 1956 [32], five reports were devoted to the homogeneous catalysis with metal complexes in reactions of hydrogenation and reduction, oxidation, hydroformylation, isomerization, and hydrocyanation of olefins.

The role of π- and σ-organometallic intermediates in the catalysis of organic reactions by metal complexes was not understood at the first stage of development of the catalysis with metal complexes. Although Kucherov tried to isolate and study the proposed organomercuric compounds in the reaction of acetylene hydration, Nieuwland and Klebanskii considered *trans*-β-chlorovinylmercury chloride (ClHgCH=CHCl,

Biginelli complex) formed from mercuric chloride and acetylene (as was known from 1898 [22]) and the product of acetylene insertion into the Hg–OH bond as intermediates in the reactions of acetylene hydrochlorination and hydration (see, e.g., [33]), and the first organocopper products of the interaction of C_2H_2 and CuCl were originally synthesized as long ago as 1900 [22]. Apparently, the role of organometallic intermediates was most adequately recognized by Yu.S. Zal'kind and B.M. Fundyler in 1936, when it was shown that the stoichiometric oxidation of Cu(I) acetylides (known as the Glaser reaction since 1869 [22])

$$2RC{\equiv}CCu + Ox \rightarrow RC{\equiv}CC{\equiv}CR + Red + 2H^+,$$

$$(Ox\text{–}O_2, Fe^{3+}, Cu^{2+})$$

could be implemented for the catalytic oxidative coupling of alkynes by combining the reactions of acetylide formation and oxidation in a common system (Glaser–Zal'kind reaction) as

$$2RC{\equiv}CH + {}^1\!/_2\,O_2 \xrightarrow{Cu(I)} RC{\equiv}CC{\equiv}CR + H_2O.$$

The situation in the homogeneous catalysis with metal complexes dramatically changed as a result of a series of discoveries made and investigations performed by academic scientists in cooperation with industrial companies in the period from the beginning of 1950s up to 1961–1963. This period can be considered as the *second stage* in the development of the catalysis with metal complexes. The following investigations performed at that time in the fields of organometallic, coordination, and catalytic chemistry were of key importance for the subsequent rapid progress in the catalysis with metal complexes.

(i) Immediately after the synthesis of ferrocene $(C_5H_5)_2Fe$ in 1952, J. Wilkinson, R. Woodword and E. Fischer explained the structure of this organometallic compound, and then (in 1955) Fischer synthesized a no less remarkable complex of $(C_6H_6)_2Cr$. These events marked the onset of the systematic development of the organoelemental chemistry of transition metals and the theory of the structure of π complexes of transition metals. For these works, G. Wilkinson and E. Fischer received the Nobel Prize for Chemistry in 1973.

(ii) In 1953–1955, K. Ziegler and G. Natta proposed the heterogeneous (and then, homogeneous) organometallic catalysts for the stereoregular polymerization of α-olefins and dienes ($TiCl_4$–$AlEt_3$, $TiCl_3$–AlR_3, $(C_5H_5)_2TiCl_2$–$Al(C_2H_5)_3$, etc.). In 1963, these scientists were awarded the Nobel Prize for creating a new approach to the chemistry and technology of polymers. The results of investigations into the polymerization of olefins and dienes, which were performed at the second stage of development of the catalysis with metal complexes, were summarized by Gaylord and Mark [34].

Investigations of the NiX_2–AlR_3 system, which led to the discovery of Ziegler's catalysts (or the so-called "nickel effect"), provided a basis for the development of the commercial technologies of olefin dimerization and diene cyclooligomerization (G. Wilke) and the creation of the catalytic chemistry of Ni(0) complexes and bis-π-allyl Ni(II) complexes.

(iii) In 1959–1960, J. Smidt and W. Hafner with coworkers in Germany as well as I.I. Moiseev, M.N. Vargaftik, and Ya.K. Syrkin in the USSR reported on the discovery of a new type of catalytic reactions for olefin oxidation in solutions of Pd(II) complexes, in particular,[3]

$$C_2H_4 + {}^1\!/_2\,O_2 \xrightarrow{PdCl_2\text{–}CuCl_2} CH_3CHO.$$

[3]The first patent application that described a gas-phase reaction on carbon-supported $PdCl_2$-$CuCl_2$ catalyst was submitted in January 1957 (see R. Jira, *Angew. Chem. Int. Ed.*, **48**, 9034 (2009)).

In 1960, Russian chemists developed a new method for the synthesis of vinyl acetate (Moiseev's reaction) [17, 19, 35]:

$$C_2H_4 + {}^1\!/\!_2\,O_2 + CH_3COOH \xrightarrow{Pd(II)} CH_2{=}CHOCOCH_3 + H_2O.$$

The same year, Wacker Chemie and Hoecht AG (FRG) launched the first plants for the commercial production of acetaldehyde via ethylene oxidation (Wacker process). Investigations into the reactions of olefin oxidation provided a basis for the development of a new direction in the catalysis with metal complexes, namely, the use of palladium complexes for the catalysis of numerous reactions of oxidation, carbonylation, cyclization, isomerization, dimerization, and polymerization of olefins, etc.

(iv) In 1957, the so-called Speier catalyst (H_2PtCl_6–isopropanol) was discovered for the hydrosilylation of olefins and alkynes according to the following scheme:

$$RCH{=}CH_2 + HSiR_3 \rightarrow RCH_2CH_2SiR_3.$$

(v) Investigations into the stereospecific polymerization of acetylene, which were initiated by G. Natta in 1956–1958, were continued in 1960 by Luttinger [36] and Green et al. [37] using Ni(II), Co(II) and some other transition metal complexes. The results of these studies provided a basis for H. Shirakawa to obtain polyacetylene films in 1971. These films were used to create various electroconductive organic materials, for which H. Shirakawa, A. Heeger, and A. MacDiarmid received the Nobel Prize for Chemistry in 2000.

(vi) Since the beginning of the 1950s, systematic investigations were devoted to the kinetics and mechanisms of both known and newly-discovered reactions catalyzed by metal complexes in solutions and the kinetics of their separate stages involving organometallic and complex compounds, including the following issues:

- Kinetics and mechanisms of olefin hydroformylation in solutions of $Co_2(CO)_8$ and $HCo(CO)_4$ (G. Natta, 1952; G. Sternberg, I. Wender, and M. Orchin, 1953–1959; D. Breslow and R. Heck, 1960–1963) [23, 24].
- Mechanism of CO insertion into Mn–R bonds (T. Coffild, 1957).
- Kinetics and mechanisms of the reduction of inorganic substrates and the hydrogenation of olefins by molecular hydrogen in solutions of Cu(I, II), Ag(I), and Ru(II, III) complexes (J. Halpern, 1955–1963) [23, 32].
- Kinetics and mechanisms of the oxidative dehydrocondensation of alkynes, including their oxidative cyclization in both aqueous and nonaqueous solutions of Cu(I, II) complexes (W. Reppe, 1955; J. Baxendale, 1955; G. Eglinton, 1956; F. Sondheimer, 1957).
- Kinetics and mechanisms of olefin oxidation in $PdCl_2$ solutions (I.I. Moiseev with co-workers, 1959–1963).
- Kinetics and mechanisms of the addition of HX molecules (X = OH, Cl, Br, I) to acetylene in solutions of Hg(II) and Cu(I) complexes (R.M. Flid, 1952–1963; D. V. Sokol'skii, 1955) and in solutions of Ru(III) complexes (X = OH) (J. Halpern, 1961) [22].

At the second stage, an important role was played by the works of Sternberg, Wender, Orchin, Heck, Breslow, Moiseev, and Halpern in understanding of the essence of the catalysis with metal complexes as a phenomenon related to the transformation of molecules in the coordination sphere of metal complexes. These investigations disclosed almost all typical stages of mechanisms involved in the catalysis with metal complexes. Separate stages in the oxidative addition of HCl (J. Wilkinson, L. Vaska), H_2(L. Vaska, $IrCl(CO)L_2$), and C_3H_5Cl (R. Heck, E. Fischer, $Ni(CO)_3L$) molecules to metal complexes were revealed in 1955–1961. It is noteworthy that Moiseev put forward the first experimentally and theoretically justified kinetic model of a process involving metal complexes and admitted for the first time the participation of organopalladium compounds as intermediates in catalytic reactions. This hypothesis, successfully confirmed by subsequent investigations, led to the development of new directions in synthetic chemistry. In particular, the reaction of oxidative esterification of ethylene is a basis of the modern commercial synthesis of vinyl acetate.

It should be noted that the second stage in the development of the catalysis with metal complexes was accompanied by commercial implementation of the processes proposed previously (at the first stage) and modified (improved) due to the improved knowledge. For example, BASF (FRG) in 1956 launched the production of acrylic acid based on acetylene carbonylation under pressure in $NiBr_2$ solutions containing $CuBr_2$ additives in tetrahydrofuran (with an annual production of about 90 thousand tons).

Despite significant achievements in the catalysis with metal complexes, catalytic chemistry was developing until the beginning of 1960s as the science of heterogeneous catalysis (except for the homogeneous acid catalysis). Indeed, in both industrial and academic laboratories, the heterogeneous catalysis occupied main positions and the syntheses of HNO_3, H_2SO_4, NH_3, methanol, ethanol, and butadiene (according to Lebedev), Fischer–Tropsch process as well as cracking, reforming, and hydrocracking of oil hydrocarbons and the hydrogenation and oxidation of organic compounds were predominantly discussed at all conferences. In the international *Journal of Catalysis* edited since 1962 and intended to consider both heterogeneous and homogeneous processes, almost all of about four hundred articles published in the first five years were devoted to heterogeneous catalysis [38].

At the same time, it was the second stage in development of the catalysis with metal complexes that played a decisive role in the establishment of this direction as an independent branch of catalytic chemistry and in the joint "autocatalytic" development of coordination chemistry, the chemistry of organometallic compounds, and catalysis with metal complexes.

The *third stage* in the development of catalysis with metal complexes, which may be conditionally restricted to a period of 1962 to 1972, also featured very important events in synthetic, theoretical, and industrial chemistry.

In 1964–1965, several research groups have synthesized a $RhCl(PPh_3)_3$ complex that was called the Wilkinson catalyst and proved to be highly active in the hydrogenation of olefins and alkynes and in the hydroformylation olefins. Thus, in view of Ir(I) complexes introduced by L. Vaska in 1961–1965, two new metals (Ir and Rh) have appeared in the catalysis with metal complexes.

The possibility of N_2 reduction to ammonia in the presence of typical Ziegler's catalysts was originally demonstrated in 1964 by M.E. Volpin and V.G. Shur. The Volpin–Shur reaction triggered the search for homogeneous catalytic systems of N_2 fixations. Already in 1965, A.E. Shilov, A.K. Shilova, and Yu.G. Borod'ko obtained a $Ru(III)N_2$ complex from molecular nitrogen and then in 1970 A.E. Shilov with co-workers discovered the reaction of nitrogen reduction to hydrazine by strong reducers (Mo and V complexes) in protonic media.

In 1964, R. Banks and G. Bailey – chemists working for the Phillips Petroleum Co. – apparently for the first time clearly demonstrated that acyclic olefins could feature a new, remarkable reaction of metathesis on a $Mo(CO)_6/Al_2O_3$ catalyst. At the same time, the investigation by G. Natta and G. Dall'Asta of the polymerization of cycloolefins with the cycle opening and the retention of double bonds (which was later also shown to proceed via the "metathesis" mechanism [6, 12]) marked the beginning of systematic studies of this process, although it was apparently first reported in 1956 for cyclopentene by the DuPont company. In 1967, N. Calderon discovered homogeneous systems of the Ziegler type (e.g., WCl_6–$EtAlCl_2$–EtOH) for the metathesis of olefins and the polymerization of cycloolefins. For studying the mechanism of this unusual reaction and developing catalysts (Mo, W, Ru) that allowed this process to become a convenient laboratory and commercial technology of valuable and otherwise hardly accessible organic products, Y. Chauvin, R. Shrock and R. Grubbs were awarded the 2005 Nobel Prize in Chemistry.

In 1965, N. Indictor and W. Brill described the process of olefin epoxidation by hydroperoxides in solutions of Cr, V, and Mo complexes as the development of the well-known Prilezhaev reaction of olefin epoxidation by peroxy acids [39]:

$$RCH{=}CH_2 + R'OOH \longrightarrow \underbrace{RCH{-}CH_2}_{\diagdown O \diagup} + R'OH,$$

and in 1967 the Halcon and Arco companies implemented the commercial technology of propylene epoxidation by ethylbenzene hydroperoxide in solutions of Mo(VI) complexes. One year before that, BASF launched the commercial production of acetic acid via carbonylation of methanol on cobalt catalysts (according to Reppe).

In the period from 1965–1972, the main types of catalytic reactions that involved the formation of C–C bonds with the participation of Pd and Ni complexes were discovered, which led to a revolution in the synthetic organic chemistry and small-scale commercial synthesis. These reactions were as follows.

- Allylation of CH acids (allylic substitution):

$$CH_2{=}CHCH_2X + RH \rightarrow CH_2{=}CHCH_2R + HX$$

  (J. Tsuji, 1965 [11, 40]).
- Oxidative coupling of arenes and oxidative arylation of olefins:

$$2ArH + Pd(OAc)_2 \rightarrow Ar{-}Ar + Pd^0 + 2AcOH$$

  (Van Helden, 1965 [8]);

$$ArH + CH_2{=}CHR + Pd(OAc)_2 \rightarrow ArCH{=}CHR + Pd^0 + 2AcOH$$

  (Moritani–Fujiwara reaction, 1967 [8, 10, 40]); and

$$ArHgOAc + CH_2{=}CHPh + Pd(OAc)_2 \rightarrow ArCH{=}CHPh + Pd^0 + Hg(OAc)_2 + AcOH$$

  (R. Heck, 1968 [10, 41]). The description of these stoichiometric reactions and identification of the product of oxidative addition of ArI to PdL_4 ($L = PPh_3$) as

$$ArI + PdL_4 \rightarrow Pd(I)L_2 + 2L$$

  (P. Fitton et al., 1968 [42]) showed the possibility of the catalytic arylation of olefins (Heck reaction).
- Heck reaction [10, 11, 43, 44]:

$$RX + CH_2 = CHR' + B \rightarrow RCH = CHR' + BH^+X^-$$

  where
  (i) $RX = PhBr$, $R' = H$, PdL_4; O.N. Temkin, O.L. Kaliya, et al., 1970 (from PhBr via $PhPdBr(L)_2$; in addition, vinyl chloride was carbonylated to acrylate in an alcohol solution of PdL_4 [45]);
  (ii) $RX = PhI$, $R' = COOMe$, Pd/C; T. Mizoroki, K. Mori, and A. Ozaki., *Bull. Chem. Soc. Jpn.*, **44**, 581(1971), see also review [40]; or
  (iii) $RX = PhI$, $R' = Ph$, $Pd(OAc)_2$; R. Heck and J. Nolley, 1972 [46].
- Cross-coupling of RHal molecules with organometallic compounds of non-transition metals (catalyzed by Ni(II) complexes):

$$RX + R'MgX \rightarrow R{-}R + MgX_2$$

  (K. Tamao, K.Sumitani, and M.Kumada, *J. Am. Chem. Soc.*, **94**, 4374 (1972), see also reviews [40, 47]).

In the next stage of development of the catalysis with metal complexes, more effective Pd complexes were discovered for catalysis of the cross-coupling of organometallic compounds of Mg, Al, Zn (E. Negishi), Sn (M. Kosugi, J. Stille, T. Migita), and B (A. Suzuki, N. Miyaura) with RX [44, 47, 48]. In 2010, R.F. Heck, A. Suzuki, and E. Negishi were awarded the Nobel Prize in Chemistry for their investigations devoted to the palladium-catalyzed cross-coupling in organic synthesis, which have important implications, particularly in medicine and electronics.

In 1968, it was reported that the Monsanto company (USA) developed the synthesis of acetic acid from methanol under mild conditions in solutions of Rh(I) iodide complexes (J. Roth) and by 1970 the first commercial technology of acetic acid production using homogeneous rhodium catalysts was implemented in practice [3].

In 1969, A.E. Shilov with co-workers demonstrated for the first time the possibility of activating alkanes (CH_4) by Pt(II) complexes in aqueous acetic acid solutions, and somewhat later they showed the ability of Pt(II) to catalyze the oxidation of alkanes by Pt(IV) complexes. These findings gave impact to the development of the homogeneous catalysis with metal complexes in the chemistry of alkanes [8, 48, 49].

From 1969–1972, K. Oliver and D. Fenton (researchers from Union Oil of California) discovered the following reactions of oxidative carbonylation of olefins and alcohols in solutions of palladium and Cu(II) complexes [50]:

$$C_2H_4 + CO + CH_3OH + 0.5O_2 \rightarrow CH_2{=}CHCOOCH_3 + H_2O;$$

$$2ROH + CO + 0.5O_2 \rightarrow (RO)_2CO + H_2O;$$

$$2ROH + 2CO + 0.5O_2 \rightarrow (COOR)_2 + H_2O.$$

The stoichiometric synthesis of acrylates and succinates was described previously [40]. The commercial synthesis of oxalates and carbonates was implemented in solutions of Pd(II) and Cu(I)/Cu(II) complexes, respectively.

In 1968, W. Knowles demonstrated the possibility of the enantioselective hydrogenation of atropic acid in solutions of the Wilkinson complex containing chiral monophosphine ligands [Horner's phosphine, (Ph)(Me)(Pr)P] and in 1971, H. Kagan synthesized a chiral diphosphine ligand (DIOP) with a rather high enantioselectivity (*ee*, 63 %). These results inspired an extensive development of the so-called asymmetric catalysis. The first commercial process of this kind was implemented by the Monsanto company in the middle of 1970s for the synthesis of L-DOPA, a drug for the treatment of Parkinson's disease. For their investigations in the field of asymmetric catalysis of various reactions, W. Knowles, R. Noyori and K. Sharpless were awarded 2001 Nobel Prize in Chemistry.

In 1971, the DuPont company implemented the commercial production of adiponitrile using the reaction

$$C_2H_4 + 2HCN \rightarrow NC(CH_2)_4CN,$$

which was carried under mild conditions with high selectivity in $Ni[P(OR)_3]_4$ and $ZnCl_2$ complex solutions.

Thus, at the third stage of development of the catalysis with metal complexes, extensive investigations into all aspects of this field were carried out in academic laboratories, and the first generalizations of accumulated facts and formulated notions appeared concerning the mechanisms of reactions involving metal complexes on the whole and on the particular reaction types [22, 23, 35, 51–65]. During this period of time, it was established that some reactions could be catalyzed by Pd(I) complexes [35] and a chain mechanism of oxosynthesis in $HCo(CO)_4$ solutions was proposed by V. Yu. Gankin [54]. The so-called "alcoholate" mechanism of acetylene hydroalcoxycarbonylation involving a PdCOOR intermediate was kinetically justified [66], approaches to studying the kinetics of reaction in solutions of polynuclear metal complexes were developed, and the main features of the kinetics of complex reactions in solutions of metal complexes were described [67].

The next, *fourth stage* in development of the catalysis with metal complexes, which began in about 1972 and lasts to the present, features significant changes in the character of both the innovative activity of industrial companies and the direction of research in academic laboratories. According to Parshall et al. [68], these changes are related to the following factors.

(i) A certain level of development was achieved in the industry employing the catalysis with metal complexes. There was a decrease in the number of new large-scale products (especially polymers and monomers). The existing technologies were optimized and became quite perfect, so that only a change in the nature of raw materials could lead to major changes.
(ii) The first "oil shocks" (1973 and 1979) stimulated extensive search for new sources of raw materials necessary for the synthesis of currently used products. This led, in particular, to the development of C_1 chemistry and the chemistry of natural compounds.
(iii) There appeared a need for applying advanced technologies to low-scale production of expensive chemicals used in semiconductor technology, pharmacy, medicine (diagnostics), agrochemistry (ferromones, growth accelerators, etc.) and home chemistry. Already at the end of the 1980s, high-selectivity catalysis with metal complexes was well adapted to meet challenges for the synthesis of complex organic molecules, such as the aforementioned reactions of C–C bond formation (including cycle formation), oxidation, asymmetric catalysis, metathesis of olefins, etc.
(iv) Researchers began to devote more attention to the ecological aspects of the methods used in the chemical synthesis and technology of various products, which led to the development of "green chemistry" and a tendency to increase the selectivity of processes, in particular, atomic selectivity (atom economy) [69].

The above trends, pointed out by Parshall et al. [68], were ongoing after 1985. New types of catalytic systems were created and new approaches to conducting homogeneous reactions were developed, which provided more-or-less successful solutions to the problems of recirculating the metal complex catalysts and further increasing the selectivity of reactions.

In 1976–1977, first generalizations appeared concerning the use of cluster metal complexes (with metal–metal bonds) in homogeneous catalysis [70, 71]. There was rapid progress in this direction of research, especially with respect to Pd clusters [35, 72–76], including the synthesis and characterization of giant palladium clusters such as Pd_{561} (with an idealized formula of $Pd_{561}(Phen)_{60}(OAc)_{180}$) [75, 76].

Investigations of the catalysts containing $CuX-CuX_2$ and $PdCl_2-CuCl_2$ moieties and their activity in oxidation processes initiated the development of homogeneous catalysis with multicomponent polyfunctional catalytic systems [22, 35, 77]. These systems are considered in detail in Chapter 4.

In view of the increasing interest of industrial companies in the catalysis with metal complexes, much effort since the beginning of 1970s has been devoted to solving problems of the recirculation of used metal catalysts and the continuous separation of products from the catalytic systems. The main technologically acceptable approaches to solving these problems consist of the immobilization of metal complexes in one phase of a two-phase system or on the surface of a solid carrier occurring in a liquid solvent. This goal can be achieved using thermomorphous ligands and/or solvents that change their phase state depending on the temperature [78–80], water-soluble ligands and metal complexes [81], or ionic liquids [80, 82]. One of the first systems based on ionic liquids was studied in 1972 by G.W. Parshall. This was a homogeneous $Et_4N(SnCl_3)$ melt containing $PtCl_2$, which was capable of dissolving olefins in the reactions of hydrogenation and hydroformylation and converting into a two-phase system on cooling below 78°C [83]. Methods for the immobilization (heterogenization) of metal complexes on the surface of supports have been developed since the middle of 1970s [84–87]. Two-phase liquid systems under conditions of micellar catalysis are also effectively used in the catalysis with metal complexes [88, 89]. Very useful directions for the intensification of processes in two-phase systems were related to the development of interphase transfer catalysts [90, 91] and the use of methods of the supramolecular chemistry [92–94]. Significant contribution to the development of "aqueous" organometallic catalysis was provided by the investigations of I.P. Beletskaya with co-workers [88, 93].

Although more than a century has passed since N.A. Shilov published his monograph *On Conjugate Oxidation Reactions* in 1905 [95], the conjugated processes (making possible the realization of thermodynamically hindered reactions and involved in many biochemical processes ensuring the functioning of cells in nature) have been employed in the catalysis with metal complexes only during the last 20–30 years (see Chapters 3 and 4). Note that the so-called cascade (tandem or domino) type reactions [96] extensively developed in recent years are kinetically conjugated processes (see Section 4.6.3).

Investigations performed at the second and third stages in the development of the catalysis with metal complexes discovered new reactions and opened ways to major new processes. For example, the handbooks cited in [44] provide information on 84 types of organic reactions that are catalyzed by only palladium complexes. Some of the new reactions used in the catalysis with metal complexes are as follows:

- Catalytic methods that are widely employed for the formation of C–C bonds include the Heck reaction [40, 43], cross-coupling [40, 47], and ArX carbonylation, in particular, the double carbonylation ($ArCOCOOR$, $ArCOCONR_2$) [10]. The investigations by Shvartsberg et al. [97, 98] ($CuI{-}K_2CO_3$) and L. Cassar, R. Heck, and K. Sonogashira ($PdCl_2L_2$, $PdCl_2L_2{-}CuI$, see [40]) made possible the reactions of acetylene condensation with all types of organic halide compounds, including the previous reactions with allyl chlorides (P. Kurtz, 1954) and alkynyl bromides (P. Cadio, W. Chodkiewicz, 1955) [22]:

$$RX + HC{\equiv}CR' + B \rightarrow RC{\equiv}CR' + BH^+X^-,$$

  where R = alkyl, aryl, hetaryl, alkenyl, allyl, ethynyl, or acyl.
- Numerous new reactions of oxidative carbonylation [99], which are still not implemented in commercial processes:

$$RC{\equiv}CH + CO + 0.5O_2 + MeOH \rightarrow RC{\equiv}CCOOMe + H_2O$$

  (with $CuCl{-}PdCl_2{-}CuCl_2$);

$$RC{\equiv}CH + 2CO + {}^1\!/_2O_2 \longrightarrow$$

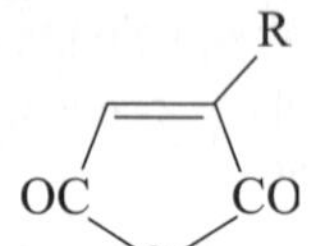

  (with Pd(I, II));

$$C_2H_2 + CO + HCl + 0.5O_2 \rightarrow trans\text{-}ClCH{=}CHCOOH$$

  (with $HgCl_2{-}PdCl_2{-}FeCl_3$);

$$ArH + CO + 0.5O_2 \rightarrow ArCOOH$$

  (with PdX_2, RhX_3);

$$CH_4 + CO + Ox \rightarrow CH_3COOH + Red$$

  (with $PdX_2{-}CuX_2$, RhX_3);

$$PhCH{=}CH_2 + CO + 0.5O_2 + ROH \rightarrow PhCH{=}CHCOOR + H_2O$$

  (with $PdCl_2{-}CuCl_2{-}Mn(OAc)_2$) [100].

- Reactions with the participation of CO, including the reductive carbonylation [101]:

$$ArNO_2 + 3CO \rightarrow ArN{=}C{=}O + 2CO_2,$$

$$ArNO_2 + 3CO + MeOH \rightarrow ArNHCOOMe + 2CO_2;$$

copolymerization of CO and olefins in solutions of Pd complexes with the formation of polyketones [102]

$$CH_3CH_2\text{–}(\text{–}COC_2H_4\text{–})_n\text{–}COOCH_3;$$

carbonylation of acetylene to succinic anhydride (in Pd^I solution) [103]:

$$C_2H_2 + 2CO + H_2O \longrightarrow$$

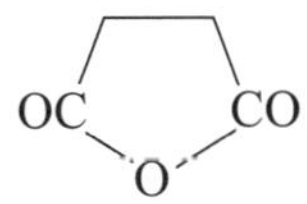

(40 °C, 1 bar);

CO oxidation conjugated with the dehydration of acetic acid [104]:

$$CO + 2CH_3COOH + {}^1\!/_2\,O_2 \xrightarrow{Pd(OAc)_2} CO_2 + (CH_3CO)_2O + H_2O$$

and with the oxidation of water [105, 106]:

$$CO + H_2O + O_2 \rightarrow CO_2 + H_2O_2$$

($PdBr_2$, Pd(II) complexes with phenanthroline).

- Regioselective addition reactions of HX molecules with acidic H atom to alkenes and alkynes:

$$C_3H_6 + CF_3COOH \rightarrow CH_3CH_2CH_2OCOCF_3$$

(Pd–Mo clusters [107]);

$$CH_3C{\equiv}CH + H_2O \rightarrow CH_3CH_2CHO$$

(selectivity, 70 %; Cu(I)–RSH–HCl–H_2O [108, 109]);

$$CH_3C{\equiv}CH + H_2O \rightarrow CH_3CH_2CHO$$

(selectivity, >99 %; Ru(II) [110]);

$$RC{\equiv}CH + R'COOH \rightarrow RC(H) = CHOCOR'$$

(selectivity, >90–98 %, Ru(II) [111]);

$$CH_3C{\equiv}CH + HCl \rightarrow ClCH = CHCH_3$$

(selectivity, 50 %, $RuCl_3$–DMF [112]).
Catalysts were found and conditions were established for obtaining the products of alkene and alkyne hydroamination of the Markovnikov and *anti*-Markovnikov types [113].

- Dehydrocondensation of alcohols:

$$2RCH_2OH \rightarrow RHC{=}CHR + 2H_2O$$

($Cp_4Mo_4Pd_4$ clusters [114]).

- New reactions with participation of hydrogen peroxide in $H_2O_2{-}V^V{-}CF_3COOH$ system [19, 115, 116]:

$$3H_2O_2 \rightarrow O_3 + 3H_2O, \tag{1}$$

$$H_2O_2 + N_2 \rightarrow N_2O + H_2O, \tag{2}$$

  and the formation of O_2 singlet oxygen and O^+ radical cation species.
- Asymmetric catalysis of hydrogenation, hydrosilylation, hydroamination and some other reactions of addition to C=C and C=O groups, as well as epoxidation, dimerization and cyclopropanation of olefins, etc. [9].
- New syntheses with the participation of acetylene compounds and metal complexes (including those mentioned above) as considered in [117]. Two examples are offered by
  (i) metathesis of alkynes (Mo complexes):

$$2RC{\equiv}CR' \rightleftharpoons RC{\equiv}CR + RC'{\equiv}CR';$$

  and
  (ii) co-cyclization of acetylene with RCN (Co complexes):

$$2C_2H_2 + CH_3CN \longrightarrow$$

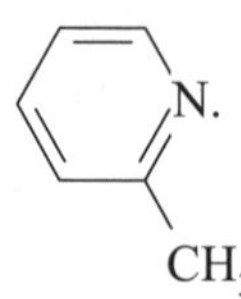

In the last quarter of the 20th century, the industry widely implemented the reactions of oxosynthesis with the participation of rhodium catalysts [7, 118] and the dimerization of olefins on Wilke's catalysts (IFP Dimersol process). In 1988, about 20 enterprises employed these methods. Hüls and Shell developed the synthesis of cyclododecatriene on Ziegler's catalysts and higher α-olefins on Ni(II) complexes (SHOP). In particular, the traditional process of alkyne carbonylation was successfully carried out on palladium complexes with high rate and selectivity for methylacetylene [119]. As a result, Shell developed an efficient technology of methyl methacrylate synthesis. BP Chemicals improved the Monsanto process of acetic acid synthesis from methanol by replacing Rh with Ir (Cativa process) [120]. Other commercially implemented processes included the metathesis polymerization of cycloolefins, production of polyketones, synthesis of acetic anhydride from methyl acetate, and the production of diethyloxalate and dimethylcarbonate [118].

The knowledge gained over the last 25 years in both theory and practice of the catalysis with metal complexes has been summarized in numerous monographs, handbooks, encyclopedias, and textbooks. Some of the editions devoted to the chemistry and technology of this catalysis and the achievements of organometallic chemistry of transition metals have been already mentioned above. Modern notions about the reaction mechanisms and applied aspects of the catalysis with metal complexes are presented in the comprehensive *Encyclopedia of Catalysis* (I.T. Horvath (Ed), Wiley, Hoboken, 2003, Vols. 1–5)) and in the comprehensive contributed monograph [5]. During the fourth period in the development of the catalysis with metal complexes, an important contribution to the development of this field was played, in addition to the aforementioned publications [2–4, 6, 8–12, 24, 39, 44, 48, 49, 81, 86, 87, 75, 117, 118], by monographs [121–138] and the handbooks and lectures [139–148].

As with all other fields of chemistry, a basis for the theory of the mechanisms of reactions catalyzed by metal complexes and the theory of reactivity of these complexes is provided by chemical kinetics, naturally, together with the chemical and physical methods of investigations. Chemical kinetics, being a science about the rates of chemical reactions and the dynamic behavior of chemical systems, provides knowledge that is common to various fields of chemistry and is an important tool for discriminating between various hypotheses concerning the mechanisms of reactions. However, in view of the unavoidable differentiation of these fields, features of the phase states of different chemical media (gases, liquids, solids), reaction types (homogeneous versus heterogeneous), and the chemical specificity of various objects (protonic acids and bases, metal complexes, enzymes, metals, etc.), the theory of chemical kinetics as a method of investigation has been developed independently in particular fields of chemistry, using specific terminologies and different approaches – but essentially solving the same tasks in the description of kinetics of chemical processes. Achievements in the development and use of the kinetic method in various fields of chemistry have been generalized in numerous monographs and teaching handbooks, e.g., for acid–base processes [149–151], gas-phase reactions [152–156], radical-chain processes in the liquid phase [52, 157, 158] (see also references in [158, 159]), enzyme reactions, [160–166], heterogeneous catalytic processes [167–174], and topochemical reactions [175].

In current monographs and textbooks, kinetic features of the catalytic reactions involving metal complexes on the whole are not described and generalized to the same extent as, e.g., those of enzymatic reactions. On the other hand, the kinetics of some particular types of reactions or the reactions involving certain types of substrates have been addressed in a large number of monographs. Among these, it is necessary to mention first the monographs by Basolo and Pearson [23] and Wilkins [176] devoted mostly to the mechanisms of inorganic reactions in solutions of metal complexes, books by Moiseev [35] and Henry [132] on the kinetics and mechanisms of olefin oxidation, and books on the homogeneous hydrogenation [121], oxidation and activation of alkanes [48, 49], and acetylene chemistry [117]. Concise presentation of the feature of reaction kinetics in the catalysis with metal complexes is given in textbooks [142, 147, 148] and reviews [177].

This book analyzes the most significant features in kinetics of the catalytic reactions in solutions of metal complexes, considers the state-of-the-art including achievements and current problems in this field of chemical kinetics, and discusses general problems in the strategy of investigations of the mechanisms of complex reactions and the role of the kinetic method in these investigations. The opinion that no one statement for a reaction mechanism cannot be accepted unless it can be demonstrated that it agrees with the observed kinetics [178] is still valid at the beginning of the 21st century. The author's wish to support this statement based on his almost half-a-century experience in studying complex reaction kinetics was among the stimuli for writing this book. In an analysis of the kinetic aspects of reactions in the catalysis with metal complexes, we do not always discuss alternative hypotheses and consider possible details of mechanisms, structures of transition states, and proofs of the existence of proposed intermediates. The main attention is devoted to elucidating the composition of intermediates and checking the correspondence of the proposed reaction schemes to the results of kinetic experiments and possibilities of the kinetic method.

In order to facilitate understanding of the content without recourse to additional sources, the book includes small sections that briefly present some important achievements in the theory of complex reactions kinetics, including the theory of routes, quasi-steady-state and quasi-equilibrium approximations, kinetic applications of the graph theory methods, and the selection rules for elementary stages. This monograph also considers questions pertaining to the formation of catalytically active centers from precursors, the influence of medium on the rates of catalytic processes, and some aspects of the theory of nonlinear dynamical systems.

References

1. I. I. Moiseev, *Khim. Prom.*, No. 4, 27 (1985); *Kinet. Katal.*, **42** (1), 5–29 (2001).
2. A. Mortreux and F. Petit (Eds.), *Industrial Aspects of Homogeneous Catalysis*, D. Reidel Publ., Dordrecht, 1988.

3. G. W. Parshall and D. I. Ittel, *Homogeneous Catalysis*, 2nd ed., John Wiley & Sons, New York, 1992.
4. G. Fink, R. Muelhaupt and H. H. Brintzinger (Eds.) *Ziegler Catalysts: Recent Scientific Innovations and Technical Improvements*, Springer, Berlin, 1995.
5. B. Cornils and W. A. Herrmann (Eds.), *Applied Homogeneous Catalysis with Organometallic Compounds*, Vols. 1 and 2, VCH, Weinheim, 1996.
6. K. J. Ivin and J. C. Mol, *Olefin Metathesis and Metathesis Polymerization*, Academic Press, 1997.
7. P. W. N. M. Van Leeuwen, *Homogeneous Catalysis: Understanding the Art*, Kluwer Academic Publishers, 2004.
8. G. B. Shul'pin, *Organicheskie reaktsii, kataliziruemye kompleksami metallov* (Organic Reactions Catalyzed by Metal Complexes), Nauka, Moscow, 1988.
9. R. Noyori, *Asymmetric Catalysis in Organic Synthesis*, John Wiley & Sons, New York, 1994.
10. M. Beller and C. Bolm (Eds.), *Transition Metals for Organic Synthesis*, Wiley–VCH, New York/Weinheim, 1998.
11. J. Tsuji, *Transition Metal Reagents and Catalysts*, John Wiley & Sons, Chichester, 2000.
12. R. H. Grubbs (Ed.), *Handbook of Metathesis*, Vols. 1–3, Wiley–VCH, New York/Weinheim, 2003.
13. O. V. Krylov and V. A. Matyshak, *Promezhutochnye soedineniya v geterogennom katalize* (Intermediates in Heterogeneous Catalysis), Nauka, Moscow, 1996.
14. F. Zaera, *Chem. Rev.*, **95**, 2651 (1995).
15. B. E. Bent, *Chem. Rev.*, **96**, 1361 (1996).
16. O. N. Temkin, A. V. Zeigarnik, A. V. Kuz'min, L. G. Bruk and E. V. Slivinskii, *Izv. Akad. Nauk, Ser. Khim.*, 1, 1–34 (2002).
17. I. I. Moiseev, My Way in Catalysis with Metal Complexes, in *Iz istorii kataliza* (From the History of Catalysis), V. D. Kal'ner (Ed.), Kalvis, Moscow, 2005, p. 361.
18. O. N. Temkin, *Vvedenie v metallokompleksnyi kataliz. Kataliz i koordinatsionnaya khimiya* (Introduction to Catalysis with Metal Complexes: Catalysis and Coordination Chemistry), MITKhT, Moscow, 1980.
19. M. N. Vargaftik, A. E. Gekhman and I. I. Moiseev, Catalysis with Metal Complexes in the N. S. Kurnakov Institute of General and Inorganic Chemistry), in *Iz istorii kataliza* (From the History of Catalysis), V. D. Kal'ner (Ed.), Kalvis, Moscow, 2005, p. 413.
20. V. I. Esafov, *Mikhail Grigor'evich Kucherov* (M. G. Kucherov), Nauka, Moscow, 1972.
21. J. A. Nieuwland and R. R. Vogt, *The Chemistry of Acetylene*, Am. Chem. Soc., New York, 1945; J. Nieuwland and R. Vogt, *Khimiya Atsetilena* (The Chemistry of Acetylene), A. D. Petrov (Ed.), GIIL, Moscow, 1947, Additional Chapters Vl and Vll.
22. O. N. Temkin and R. M. Flid, *Kataliticheskie prevrashcheniya atsetilenovykh soedinenii v rastvorakh kompleksov metallov*, (Catalytic Transformations of Alkynes in Solutions of Metal Complexes), Nauka, Moscow, 1968.
23. F. Basolo and R. Pearson, *Mechanisms of Inorganic Reactions: A Study of Metal Complexes in Solutions*, 2nd ed., John Wiley & Sons, 1967.
24. V. Yu. Gankin and G. S. Gurevich, *Tekhnologiya oksosinteza* (Technology of Oxosynthesis), Khimiya, Leningrad, 1981.
25. E. N. Shpitalskii, *Z. Phys. Chem.*, **122**, 257 (1926).
26. W. Langenbeck, *Die Organischen Katalysatoren und ihre Beziehungen zu den Fermenten*, Springer-Verlag, 1949; W. Langenbeck, in *Organicheskie katalizatory i ikh otnoshenie k fermentam* (Organic Catalysts and Their Relation with Enzymes), G. Vovchenko (Ed.), Inostr. Lit., Moscow, 1961 [additional bibliography from 1949 up to 1959 and author's comments].
27. J. Baxendale, in *Advances in Catalysis*, P. H. Emmett and H. S. Taylor (Eds.), Acad. Press, 1952, Vol. 4, pp. 31–86.
28. J. Weiss, in *Advances in Catalysis*, P. H. Emmett and H. S. Taylor (Eds.), Acad. Press, 1952, Vol. 4, pp. 343–365.
29. F. Hein, *Chemische Koordinationslehre*, Leipzig, Hirzel Verlag, Leipzig, 1950.
30. W. Reppe, *Neue Entwiclungen auf dem Gebiete der Chemie des Acetylenes und Kohlenoxyds*, Springer-Verlag, Berlin/Göttingen/Heidelberg, 1949.
31. J. W. Copenhaver and M. M. Bigelow, *Acetylene and Carbon Monoxide Chemistry*, Reinhold, New York, 1949.

32. A. Farkas (Ed.), *Proceedings of the 1st Int. Congress on Catalysis*, Acad. Press, New York, 1956; D. D. Eley, W. G. Frankenburg and V. I. Komarevsky (Eds.), *Advances in Catalysis*, Acad. Press New York, 1957.
33. G. F. Hennion, R. R. Vogt and J. A. Nieuwland, *J. Org. Chem.*, **1**, 159 (1937).
34. N. Gaylord and H. F. Mark, *Linear and Stereoregular Addition Polymers*, Int. Publ., 1959.
35. I. I. Moiseev, *π-Kompleksy v zhidkofaznom okislenii olefinov* (π-Complexes in Liquid-Phase Oxidation of Olefins), Nauka, Moscow, 1970.
36. L. B. Luttinger, *Chem.* & Ind., **36**, 1135 (1960).
37. M. L. H. Green, M. Nehme and G. Wilkinson, *Chem. & Ind.*, **36**, 1136 (1960).
38. *Catalysis in France: an Adventure* (SCF Publication at the 13th Int. Congress on Catalysis), Société Français de Chimie, Paris, 2004, p. 21.
39. G. A. Tolstikov, *Reaktsii gidroperekisnogo okisleniya* (Reactions of Hydroperoxide Oxidation), Nauka, Moscow, 1976.
40. J. Tsuji, *New J. Chem.*, **24**, 127 (2000).
41. R. Heck, *J. Am. Chem. Soc.*, **90**, 5518, 5526, 5531 (1968).
42. P. Fitton, M. P. Johnson and J. E. McKeon, *J. Chem. Soc. Chem. Comm.*, 6 (1968).
43. I. P. Beletskaya and A. V. Cheprakov, *Chem. Rev.*, **100**(8), 3009–3066 (2000).
44. J.-L. Malleron, J.-C. Fiad and J.-Y. Legros, *Handbook of Palladium Catalyzed Organic Reactions*, Acad. Press, San Diego, 1997; E. Negishi (Ed.), *Handbook of Organopalladium Chemistry for Organic Synthesis*, Wiley-Interscience, New York, 2002, Vols. 1 and 2.
45. O. N. Temkin, O. L. Kaliya, G. K. Shestakov, S. M. Braylovskii, R. M. Flid and A. P. Aseeva, *Kinet. Katal.*, **11**(6), 1592 (1970).
46. R. Heck and J. P. Nolley, *J. Org. Chem.*, **37**, 2320 (1972).
47. I. P. Beletskaya and N. A. Bumagin, *Zh. Obshch. Khim.*, **32**(12), 1771 (1996).
48. A. E. Shilov and G. B. Shul'pin, *Aktivatsiya i kataliticheskie reaktsii uglevodorodov* (Activation and Catalytic Reactions of Hydrocarbons), Nauka, Moscow, 1995.
49. E. S. Rudakov, *Reaktsii alkanov* (Reactions of Alkanes), Naukova Dumka, Kiev, 1985.
50. K. L. Oliver, D. M. Fenton, and G. Biale, *Hydrocarbon Process.*, **51**, 95 (1972).
51. O. A. Chaltykian, *Kuprokataliz* (Cuprocatalysis), Aipetrat, Yerevan, 1963; *Copper-Catalytic Reactions*, A. E. Stubbs (transl.), Consultants Bureau, New York, 1966.
52. N. M. Emanuel, E. T. Denisov and Z. K. Maizus, *Tsepnye reaktsii okisleniya uglevodorodov v zhidkoi faze* (Chain Reactions of Liquid-Phase Oxidation of Hydrocarbons), Nauka, Moscow, 1965.
53. C. W. Bird, *Transition Metals Intermediates in Organic Synthesis*, Logos Press, London, 1967.
54. D. M. Rudkovskii (Ed.), *Karbonilirovanie nenasyshchennykh uglevodorodov* (Carbonylation of Unsaturated Hydrocarbons), Khimiya, Leningrad, 1968.
55. M. Orchin, in *Advances in Catalysis*, D. D. Eley, H. Pines, and P. B. Weisz (Eds.), Acad. Press, 1966, Vol. 16, pp. 1–47.
56. A. J. Chalk and J. F. Harrod, in *Advances in Organometallic Chemistry*, F. Stone and R. West (Eds.), Acad. Press, 1968, Vol. 6, p. 119.
57. B. A. Dolgoplosk, K. L. Makovetskii, E. I. Tinyakova and O. K. Sharaev, *Polimerizatsiya dienov pod vliyaniem π-allil'nykh kompleksov* (Polymerization of Dienes under the Influence of π-Allyl Complexes), Nauka, Moscow, 1968.
58. N. N. Korneev, A. F. Popov and B. A. Krentsel, *Kompleksnye metalloorganicheskie katalizatory* (Complex Organometallic Catalysts), Khimiya, Leningrad, 1969.
59. R. Ugo (Ed.), *Aspects of Homogeneous Catalysis*, Carlo Manfredi, Milan, 1970, Vol. 1.
60. J. P. Candlin, K. A. Taylor and D. T. Thompson, *Reactions of Transition-Metal Complexes*, Elsevier Publ., Amsterdam, 1968.
61. I. Wender and P. Pino (Eds.), *Organic Synthesis via Metal Carbonyls*, John Wiley & Sons, New York, 1968.
62. J. Falbe, *Synthesen mit Kohlenmonoxyd*, Springer Verlag, 1967; J. Falbe, *Syntezy na osnove okisi ugleroda* (Syntheses Based on Carbon Oxide), N. S. Imyanitov (Ed.), Khimiya, Leningrad, 1971 [additional comments].
63. P. W. Jolly and G. Wilke, *The Organic Chemistry of Nikel*, Acad. Press, New York, 1971, Vol. 1.
64. P. M. Maitlis, *Organic Chemistry of Palladium*, Acad. Press, New York, 1971, Vols. 1 and 2.

65. D. V. Sokol'skii and Ya. A. Dorfman, *Kataliz ligandami v vodnykh rastvorakh* (Catalysis with Ligands in Aqueous Solutions), Nauka, Alma-Ata, 1972.
66. O. N. Temkin, O. L. Kaliya, R. M. Flid and N. G. Mekhryakova, *Dokl. Akad. Nauk SSSR*, **199**, 1321 (1971).
67. O. N. Temkin, R. M. Flid and G. K. Shestakov, Specific Features of the Kinetics of Complicated Reactions in Metal Complex Solutions, in *Gomogennyi kataliz* (Homogeneous Catalysis: A Collection of Papers Presented at the 1st Coordination Meeting on Homogeneous Catalysis, Frunze, 1969), Ilim, Frunze, 1970, pp. 90–112.
68. G. W. Parshall, W. A. Nugent, D. M.-T. Chan and W. Tam, *Pure Appl. Chem.*, **57**(12), 1809 (1985).
69. (a) R. Sheldon, *Chem. Technol.*, **39** (1994); (b) R. A. Sheldon, I. Arends and U. Hanefeld, Green Chemistry and Catalysis, Wiley–VCH, Weinheim, 2007.
70. E. Muetterties, *Bull. Soc. Chim. Belg.*, **85**, 451 (1976).
71. A. K. Smith and J. M. Basset, *J. Mol. Catal.*, **2**, 229 (1977).
72. I. I. Moiseev, Palladium Clusters in Catalysis, in *Chemistry Reviews (Sov. Sci. Rev. Sec. B)*, M. E. Volpin (Ed.), Harwood Acad. Publ., New York, 1982, Vol. 4, pp. 1–35.
73. O. N. Temkin and L. G. Bruk, *Usp. Khim.*, **52**, 206 (1983).
74. I. I. Moiseev, Catalysis by Complexes with Pd–Pd bonds, *Itogi Nauki Tekh.: Kinet. Katal. (VINITI)*, **13**, 147 (1984).
75. I. I., Moiseev and M. N. Vargaftik, Catalysis with Giant Palladium Clusters, in *Perspectives in Catalysis*, J. A. Thomas and K. I. Zamaraev (Eds), Blackwell, Oxford, 1992, pp. 91–123.
76. I. I. Moiseev and M. N. Vargaftik, Catalysis with Palladium Clusters, in *Catalysis by Di- and Polynuclear Metal Cluster Complexes*, R. D. Adams and F. A. Cotton (Eds), Wiley–VCH, New York, 1998, pp. 395–442.
77. O. N. Temkin, O. L. Kaliya, G. K. Shestakov and R. M. Flid, *Dokl. Akad. Nauk SSSR*, **190**, 398 (1970).
78. I. T. Horvath and J. Rabai, *Science*, **266**, 72 (1994).
79. I. T. Horvath, *Acc. Chem. Res.*, **31**, 641 (1998).
80. H. W. Bohnen and B. Cornils, *Adv. Catal.*, **42**, 1–64 (2002).
81. B. Cornils and W. Herrmann (Eds.), *Aqueous Phase Organometallic Catalysis: Concepts and Applications*, John Wiley & Sons, Weinheim, 1998.
82. P. Wasserscheid and W. Keim, *Angew. Chem. Int. Ed.*, **39**, 3772 (2000).
83. G. N. Parshall, *J. Am. Chem. Soc.*, **91**, 8716 (1972).
84. E. M. Cernia and M. Graziani, *J. Appl. Polym. Sci.*, **18**, 2725 (1974).
85. Yu. I. Ermakov, *Zh. Vses. Khim. o-va im. D.I. Mendeleeva*, **22**, 532 (1977).
86. G. V. Lisichkin and A. Ya. Yuffa, *Geterogennye metallokompleksnye katalizatory* (Heterogeneous Metal Complex Catalysts), Khimiya, Moscow, 1981.
87. Yu. I. Ermakov, V. A. Zakharov and B. N. Kuznetsov, *Zakreplennye kompleksy na okisnykh nositelyakh v katalize* (Immobilized Complexes on Oxide Carriers in Catalysis), Nauka, Novosibirsk, 1980.
88. I. P. Beletskaya and A. V. Cheprakov, Aqueous Palladium Catalysis, in *Handbook of Organopalladium Chemistry for Organic Synthesis*, E. Negishi (Ed.), 2002, Vol. 2, pp. 2119–2140.
89. M. Niyaz Khan, *Micellar Catalysis*, CRC–Taylor & Francis, Boca Raton, Fl, 2007.
90. W. P. Weber and G. W. Gokel, *Phase Transfer Catalysis in Organic Syntheses*, Springer-Verlag, Berlin, 1977.
91. E. V. Dehmlow and S. S. Dehmlow, *Phase Transfer Catalysis*, Verlag Chemie, 1983.
92. H.-J. Schneider and A. Yatsimirsky, *Principles and Methods in Supramolecular Chemistry*, John Wiley & Sons, New York, 1996.
93. I. P. Beletskaya and A. V. Chuchuryukin, *Usp. Khim.*, **69**, 699 (2000).
94. E. A. Karakhanov, A. L. Maksimov and E. A. Runova, *Usp. Khim.*, **74** (1), 104 (2005).
95. N. A. Shilov, *O sopryazhennykh reaktsiyakh okisleniya* (On Conjugate Oxidation Reactions), Mamontov Publishers, Moscow, 1905.
96. L. F. Tietz, *Chem. Rev.*, **96** (1), 115 (1996).
97. M. S. Shvartsberg, A. A. Moroz, and I. L. Kotlyarevskii, *Izv. Akad. Nauk SSSR, Ser. Khim.*, 981 (1972).
98. E. B. Merkushev and M. S. Shvartsberg, *Iodistye organicheskie soedineniya i sintezy na ikh osnove* (Iodine-Containing Organic Compound and Related Syntheses), TGPI, Tomsk, 1978.
99. O. N. Temkin and L. G. Bruk, Oxidative Carbonylation: Homogeneous, in *Encyclopedia of Catalysis*, I. T. Horvath (Ed.), John Wiley & Sons Ltd, New York, 2003, Vol. 5, pp. 394–424.

100. A. R. El'man, O. V. Boldyreva, E. V. Slivinskii, and S. M. Loktev, *Izv. Akad. Nauk, Ser. Khim.*, 552 (1992).
101. S. Cenini and F. Ragaini, *Catalytic Carbonylation of Organic Nitro Compounds*, Kluwer Academic Publishers, Dordrecht, 1997.
102. E. Drent and P. H. M. Budzelaar. *Chem. Rev.*, **96**, 663 (1996).
103. L. G. Bruk, I. V. Oshanina, A. P. Kozlova, E. V. Vorontsov and O. N. Temkin, *J. Mol. Catal.*, **104**, 9 (1995).
104. I. I. Moiseev, M. N. Vargaftik, O. I. Gentosh, et al., *Dokl. Akad. Nauk SSSR*, **237**, 645 (1977).
105. V. N. Zudin, V. N. Likholobov and Yu. I Ermakov, *Kinet. Katal.*, **18**, 921 (1977); **20**, 1559 (1979).
106. L. G. Bruk, A. S. Abdullaeva, A. V. Vyrodov, I. V. Oshanina and O. N. Temkin, *Proceedings of the 14th Int. Symp. on Homogeneous Catalysis* (Munich, Germany, 2004), Book of Abstracts, p. 141.
107. I. I. Moiseev and A. E. Gekhman, *Kinet. Katal.*, **21**, 284 (1980).
108. G. K. Shestakov, O. N. Temkin, N. Yu. Vsesvyatskaya and A. M. Stepanov, *Zh. Org. Khim.*, **15**(2), 248 (1979).
109. N. Yu. Vsesvyatskaya, D. V. Tsyganov, A. M. Bykov, et al., *Zh. Org. Khim.*, **21**(10), 2041 (1985).
110. M. Tokunaga and Y. Wakatsuki, *Angew. Chem. Int. Ed.*, **37**, 2867 (1998).
111. H. Doucet, B. Martin-Vaca, C. Bruneau and P. Dixneuf, *J. Org. Chem.*, **60**, 7247 (1995).
112. O. N. Temkin, O. Yu. Antonova, S. A. Kezin, N. G. Akhmedov and Yu. A. Ustynyuk, *Proceedings of the 12th Int. Symp. on Homogeneous Catalysis* (Stockholm, 2000), Book of Abstracts, p. 224.
113. T. E. Mueller, Hydroamination: Homogeneous, in *Encyclopedia of Catalysis*, I. T. Horvath (Ed.), John Wiley & Sons, New York, 2003, Vol. 3, pp. 429–518.
114. T. A. Stromnova, I. A. Busygina, S. B. Katser, et al., *Izv. Akad. Nauk, Ser. Khim.*, **6**, 1435 (1987).
115. N. I. Moiseeva, A. E. Gekhman and I. I. Moiseev, *J. Mol. Catal. A: Chem.*, **117**, 39 (1997).
116. A. E. Gekhman, I. P. Stoljarov, N. I. Moiseeva and I. I. Moiseev, *Compt. Rend. Chim.*, **7**, 833 (2004).
117. O. N. Temkin, G. K. Shestakov and Yu. A. Treger, *Atsetilen: khimiya, mekhanizmy reaktsii, tekhnologiya* (Acetylene: Chemistry, Reaction Mechanisms, Technology), Khimiya, Moscow, 1991.
118. K. Weissermel and H.-J. Arpe, *Industrial Organic Chemistry*, Wiley–VCH, New York, 1997.
119. E. Drent, D. Arnoldy and P. H. M. Budzelaar, *J. Organomet. Chem.*, **455**, 247 (1993); **475**, 57 (1994).
120. J. H. Jones, *Platinum Metals Rev.*, **44**, 94 (2000).
121. B. R. James, *Homogeneous Hydrogenation*, John Wiley & Sons, New York, 1973.
122. P. W. Jolly and G. Wilke, *The Organic Chemistry of Nikel*, Academic Press, New York, 1975, Vol. 2.
123. N. M. Chirkov, P. E. Matkovskii and F. S. D'yachkovskii, *Polimerizatsiya na kompleksnykh metalloorganicheskikh katalizatorakh* (Polymerization on Complex Organometallic Catalysts), Khimiya, Moscow, 1976.
124. E. L. Muetterties (Ed.), *Transition Metal Hydrides*, Marcel Dekker, New York, 1971.
125. D. V. Sokol'skii and N. F. Noskova, *Katalizatory tipa Tsiglera–Natta v reaktsii gidrirovaniya* (Ziegler–Natta Type Catalysts in Hydrogenation Reaction), Nauka, Alma-Ata, 1977.
126. J. K. Kochi, *Organometallic Mechanisms and Catalysis*, Academic Press, New York, 1978.
127. J. P. Collman and L. S. Hegedus, *Principles and Application of Organotransition Metal Chemistry*, Oxford University Press, Oxford, 1980.
128. G. W. Parshall, *Homogeneous Catalysis*, John Wiley & Sons, New York, 1980.
129. W. Keim (Ed.), *Catalysis in C1 Chemistry*, D. Reidel Publ., Dordrecht, 1983; *Kataliz v C1 Khimii* (Catalysis in C1 Chemistry), W. Keim and I. I. Moiseev (Eds.), Khimiya, Leningrad, 1987 [additional comments].
130. R. A. Sheldon, *Chemicals from Synthesis Gas*, D. Reidel Publ., Dordrecht, 1983; R. A. Sheldon, *Khimicheskie produkty na osnove sintez-gaza* (Chemicals from Synthesis Gas), S. M. Loktev (Ed.), Khimiya, Moscow, 1987 (additional chapter).
131. F. K. Shmidt, *Kataliz kompleksami metallov pervogo perekhodnogo ryada reaktsii gidrirovaniya i dimerizatsii* (Catalysis with First Transition Series Metal Complexes in Hydrogenation and Dimerization Reactions), Irkutsk,Izd-vo IrGU, 1986.
132. P. Henry, *Palladium Catalyzed Oxidation of Hydrocarbons*, D. Reidel Publ., Dordrecht, 1980.
133. J. Tsuji, *Organic Syntheses by Means of Palladium Compounds*, Springer-Verlag, Berlin, 1980.
134. R. F. Heck, *Palladium Reagents in Organic Syntheses*, Academic Press, New York, 1985.
135. A. Muellen and J. Falbe (Eds.), *New Syntheses with Carbon Monoxide*, Springer, Berlin, 1980.
136. G. Wilkinson, E. G. A. Stone, and E. W. Abel (Eds.), *Comprehensive Organometallic Chemistry*, Pergamon, Oxford, 1982.

137. H. M. Colquhoun, D. J. Thompson and M. V Twigg, *Carbonylation*, Plenum Press, New York, 1991.
138. I. T. Horvath and F. Joo, *Aqueous Organometallic Chemistry and Catalysis* (NATO ASI Series), Kluwer, Dordrecht, 1995; F. Joo, *Aqueous Organometallic Catalysis*, Kluwer Academic Publishers, Dordrecht, 2001.
139. H. Pracejus, *Koordinations-Chemische Katalyse Organischer Reactionen*, Verlag T. Steinkopff, Dresden, 1977.
140. G. Henrichi-Olive and S. Olive, *Coordination and Catalysis*, Verlag-Chemie, Weinheim, 1977.
141. O. N. Temkin, *Vvedenie v metallokompleksnyi kataliz* (Introduction into Catalysis with Metal Complexes), MITKhT, Moscow, 1980.
142. O. N. Temkin, *Khimiya i tekhnologiya metallokompleksnogo kataliza: khimicheskie katalizatory i fermenty* (Chemistry and Technology of Catalysis with Metal Complexes: Chemical Catalysts and Enzymes), MITKhT, Moscow, 1980.
143. A. Nakamura and M. Tsutsui, *Principles and Application of Homogeneous Catalysis*, John Wiley & Sons, New York, 1980.
144. G. Masters, *Homogeneous Transition Metal Catalysis*, Chapman and Hall, London, 1981.
145. R. Taube, Homogene Katalyse, Academie-Verlag, Berlin, 1988.
146. K. Kurtev, E. Balbolov and O. Tomkin, *Metalokompleksen kataliz* (Catalysis with Metal Complexes), VKHTI, Burgas, Bolgariya, 1989.
147. R. A. van Santen, J. A. Moulijn, P. W. N. van Leeuwen and B. A. Averill (Eds.), *Catalysis: An Integrated Approach*, 2nd ed., Elsevier, Amsterdam, 1999.
148. I. Chorkendorff and J. W. Niemantsverdriet, *Concepts of Modern Catalysis and Kinetics*, Wiley–VCH, 2003.
149. L. P. Hammett, *Physical Organic Chemistry*, 2nd ed., McGraw-Hill, NewYork, 1970.
150. R. P. Bell, *The Proton in Chemistry*, 2nd ed., Chapman and Hall, London, 1973; R. Bell, *Proton v Khimii* (The Proton in Chemistry), R. R. Dogonadze (Ed.), Mir, Moscow, 1977 [additional chapter by E. D. German and R. R. Dogonadze].
151. W. P. Jencks, *Catalysis in Chemistry and Biochemistry*, McGraw-Hill, New York, 1969.
152. N. N. Semenov, *O nekotorykh problemakh khimicheskoi kinetiki i reaktsionnoi sposobnosti* (On Some Problems in Chemical Kinetics and Reactivity), Izd-vo AN SSSR, Moscow, 1954.
153. N. N. Semenov, *Tsepnye reaktsii* (Chain Reactions), 2nd ed., Nauka, Moscow, 1985.
154. V. N. Kondrat'ev and E. E. Nikitin, *Kinetika i mekhanizm gazofaznykh reaktsii* (Kinetics and Mechanisms of Gas Phase Reactions), Nauka, Moscow, 1974.
155. S. W. Benson, *The Foundations of Chemical Kinetics*, McGraw-Hill, New York, 1960.
156. O. V. Krylov and V. S. Arutyunov, *Okislitel'nye prevrashcheniya metana* (Oxidative Transformations of Methane), Nauka, Moscow, 1999.
157. Kh. S. Bagdasar'yan, *Teoriya radikal'noi polimerizatsii* (Theory of Radical Polymerization), Nauka, Moscow, 1966.
158. E. T. Denisov, O. M. Sarkisov and G. I. Likhtenshtein, *Khimicheskaya kinetika* (Chemical Kinetics), Khimiya, Moscow, 2000.
159. N. M. Emanuel', G. E. Zaikov and V. A. Kritsman, *Tsepnye reaktsii: Istoricheskii aspekt* (Chain Reactions: Historical Aspect), Nauka, Moscow, 1989.
160. M. Dixon and E. C. Webb, *Enzymes*, 2nd ed., Academic Press, New York, 1964.
161. C. Walter, *Enzyme Kinetics*, Ronald Press, New York, 1966.
162. A. Cornish-Bowden, *Principles of Enzyme Kinetics*, Butterworths, London, 1976.
163. I. H. Segel, *Enzyme Kinetics*, John Wiley & Sons, New York, 1975.
164. I. V. Berezin and A. A. Klesov, *Prakticheskii kurs khimicheskoi i fermentativnoi kinetiki* (Practical Course on Chemical and Basic Kinetics), MGU, Moscow, 1976.
165. S. D. Varfolomeev and S. V. Zaitsev, *Kineticheskie metody v biokhimicheskikh issledovaniyakh* (Kinetic Methods in Biochemical Investigations), MGU, Moscow, 1982.
166. S. D. Varfolomeev and K. G. Gurevich, *Biokinetika: Prakticheskii kurs* (Biokinetics: Practical Course), Fair-press, Moscow, 1999.
167. S. L. Kiperman, *Vvedenie v kinetiku geterogennykh kataliticheskikh reaktsii* (Introduction to Kinetics of Heterogeneous Catalytic Reactions), Nauka, Moscow, 1964.

168. I. I. Ioffe and L. M. Pis'men, *Inzhenernaya khimiya geterogennogo kataliza* (Chemical Engineering of Heterogeneous Catalysis), Khimiya, Leningrad, 1972.
169. M. I. Temkin, in: *Advances in Catalysis*, D. D. Eley, H. Pines, and P. B. Weiss (Eds), Academic Press, New York, 1979, Vol. 28, pp. 173–291.
170. Yu. S. Snagovskii and G. M. Ostrovskii, *Modelirovanie kinetiki geterogennykh kataliticheskikh protsessov* (Modeling the Kinetics of Heterogeneous Catalytic Processes), Khimiya, Moscow, 1976.
171. S. L. Kiperman, *Osnovy khimicheskoi kinetiki v geterogennom katalize* (Principles of Chemical Kinetics in Heterogeneous Catalysis), Khimiya, Moscow, 1979.
172. G. S. Yablonskii, V. N. Bykov and A. N. Gorban', *Kineticheskie modeli kataliticheskikh reaktsii* (Kinetic Models of Catalytic Reactions), Nauka, Novosibirsk, 1983.
173. A. Ya. Rozovskii, *Geterogennye khimicheskie reaktsii* (Heterogeneous Chemical Reactions), Nauka, Moscow, 1980.
174. D. Murzin and T. Salmi, *Catalytic Kinetics*, Elsevier, New York, 2005.
175. A. Ya. Rozovskii, *Kinetika topokhimicheskikh reaktsii* (Kinetics of Topochemical Reactions), Khimiya, Moscow, 1974.
176. R. G. Wilkins, *Kinetics and Mechanism of Reaction of Transition Complexes*, VCH, Weinheim, 1991.
177. C. A. Tolman and J. W. Faller, Mechanistic Studies of Catalytic Reactions Using Spectroscopic and Kinetic Techniques, in *Homogeneous Catalysis with Metal Phosphine Complexes*, L. H. Pignolet (Ed.), New York, Plenum Press, 1983, pp. 13–110.
178. J. C. Bond and R. H. Cunningham, *J. Catal*., **166**, 172 (1997).

1

State-of-the-Art in the Theory of Kinetics of Complex Reactions

Principles of the theory of kinetics of complex reactions are based on the investigations of Ostwald, Van't Hoff, Bodenstein, Shpitalskii, Michaelis, Semenov, Hinshelwood, Christiansen, and many others. After the works by Horiuti [1, 2] and Temkin [3, 4], a fairly consistent theory applicable to complex steady-state and quasi-steady-state reactions of any type has been formed in the second half of the 20th century, although refinements have been introduced until quite recently [5]. For about a century, one of the most important questions under discussion was the choice of criteria for the use of quasi-steady-state and quasi-equilibrium approximations in studying reaction kinetics in closed systems (very popular in catalysis with metal complexes and enzymatic catalysis), but now a certain clarity in this issue has been achieved. Significant influence on the understanding of features in the kinetics of complex, multi-route reactions was produced by methods of graph theory, which have been applied to the representation of the mechanisms of complex reactions, their classification, derivation of kinetic equations, and analysis of the dynamic behavior of reacting systems [6, 7]. Considerable progress has been achieved in understanding the nature of an elementary act (R. Bell, M.J. Evans, M. Polanyi, H. Eyring, K. Fukui, R. Hoffmann, H. Taube, J. Polanyi et al.) and in formulating the selection rules for elementary steps and the criteria of the elementarity of steps in the reaction mechanism [6]. Said issues will be briefly discussed in this chapter. Nonlinear mechanisms and related features in the dynamic behavior of kinetic models will be considered in more detail in Chapter 6.

1.1 Main concepts of the Horiuti–Temkin theory of steady-state reactions

The kinetics of reactions involving metal complexes and reactions in catalysis with metal complexes are studied in open, semi-open, and closed reactors operating under stationary, quasi-stationary, or nonstationary conditions. Most popular are the gas-phase-open (semi-open) reactors for gas–liquid systems and the closed reactors for various liquid-phase and gas–liquid reactions, in which various processes are carried out under steady-state and quasi-steady-state conditions, respectively. These regimes are described by the Horiuti–Temkin theory of steady-state reactions, an important part of which is the analysis of the stoichiometry of mechanisms and an important point is the concept of a reaction route. Therefore, it is natural to begin this presentation by considering the main principles of this theory.

Homogeneous Catalysis with Metal Complexes: Kinetic Aspects and Mechanisms, First Edition. O. N. Temkin.

1.1.1 Reaction mechanism: Stoichiometry and routes

Reaction mechanism is a system of elementary steps (consecutive, parallel-consecutive), within the framework of which a given set of initial reactants is converted into reaction products, that is, a certain overall (brutto) reaction equation is implemented. In a broader sense, the concept of "reaction mechanism" involves, besides the set of steps, the necessary notions concerning the structures of all intermediate species and transition states, estimations of rate constants for each step (i.e., of the reactivity of substances participating in this step) and, eventually, estimations of the effect of parameters (C, P, T) on the dynamic behavior of the given reacting system.

Thus, the reaction mechanism concept is dual, including two components [8]:

(i) a set of steps (usually called the mechanistic reaction scheme), which determines a *structure* of the mechanism – that is, the interplay of reactants and intermediate species – thus providing a *topological* component of the reaction mechanism;

(ii) notions about the structures and reactivities of all reactants and intermediates, which determine the *chemical* (or physicochemical) component of the reaction mechanism.

As will be shown below (see Chapter 7), there is no hierarchical subordination between the two components and they are equally necessary for establishing the reaction mechanism.

In an analysis of mechanisms, it is commonly accepted to consider reactants and products (referred to jointly as the *reaction participants* that enter into the overall equations) and intermediate species (intermediates). A strict definition of intermediate species was formulated based on topological characteristics of a reaction network [6]. According to this, let us define an *intermediate* as a compound that is formed in one step (or in several steps) and disappears as a result of participation in some other steps. Intermediate species (X_i) can be either *observable*, that is, involved in the material balance of reactants and products (entering into overall equations), or *unobservable*, that is, not making any significant (i.e., exceeding experimental error) contribution to the material balance. The character of unobservable (or Bodenstein type) intermediate species allows them to be excluded from the stoichiometry of the overall equation of the process. Evidently, overall equations not involving intermediates can be obtained only for steady-state or quasi-steady-state processes, that is, in cases where $\frac{dC_{X_i}}{dt} = 0$ or $\frac{dC_{X_i}}{dt} \cong 0$, respectively. If the rates of variation of the concentrations of intermediates are nonvanishing, the corresponding X_i components will enter into the overall equations.

There are *linear* and *nonlinear* steps and, accordingly, linear and nonlinear mechanisms. By definition, the equations of linear steps involve no more than one intermediate compound to the left and right of the arrow, and the linear mechanisms consist of only linear steps.

In cases where the set of initial reactants and products (i.e., reaction participants) is known, it is possible to determine the maximum number $Q_{\max}$ of linearly independent overall equations of chemical reactions. For this purpose, we have to write a molecular (or atomic) matrix $\boldsymbol{H}$ of dimension $N \times l$ (where N is the number of reaction participants and l is the number of elements and/or atomic groups entering into reactants and products, which remain unchanged during the reaction). According to the so-called Gibbs stoichiometric rule [5, 9], we have

$$Q_{\max} = N - \text{rank}\boldsymbol{H}, \tag{1.1}$$

where rank$\boldsymbol{H}$ is the matrix rank that characterizes the number of linearly independent elements and/or atomic groups (i.e., the number of linearly independent stoichiometric conservation laws).

Not all of the overall equations possible for the given set of reaction participants are represented by reactions in a particular reacting system. Therefore, in the general case, the number Q of linearly independent equations is $Q \leq Q_{\max}$. The corresponding stoichiometric rule was originally formulated in

1946 by Brinkly [10] as follows:

$$\text{rank}\,\boldsymbol{B} \leq N - \text{rank}\boldsymbol{H},$$

where $\boldsymbol{B}$ (in what follows, $\boldsymbol{B}_r$) is called the stoichiometric matrix of overall equations. The rank of the stoichiometric matrix $\boldsymbol{B}_m$ of the reaction mechanism obeys the following relation:

$$\text{rank}\boldsymbol{B}_m \leq N + I - \text{rank}\boldsymbol{H},$$

where I is the number of unobservable intermediates including a catalyst (active center).

In order to describe the kinetics of a complex reaction proceeding in a steady-state or quasi-steady-state regime, it is important to establish a set of overall equations *determined by the reaction mechanism*, the number of which (Q_r) can be smaller than or equal to Q_{max} [11]. In a chemical experiment, the overall equations are established using the results of an analysis of the material balance of the process. The necessity of writing the set of overall equations based on the reaction mechanism rather than on the material balance alone is demonstrated by the following example.[1]

Example 1.1

► Consider a process with the material balance described by Eq. (1.2) and the reaction scheme described by Eqs (1.3)–(1.5):[2]

$$\text{A} \rightleftharpoons \text{B}, \tag{1.2}$$

$$\text{A} + \text{M} \underset{k_{-1}}{\overset{k_1}{\rightleftharpoons}} \text{MA}, \tag{1.3}$$

$$\text{MA} + \text{A} \underset{k_{-2}}{\overset{k_2}{\rightleftharpoons}} \text{MB} + \text{B}, \tag{1.4}$$

$$\text{MB} \underset{k_{-3}}{\overset{k_3}{\rightleftharpoons}} \text{M} + \text{B}, \tag{1.5}$$

where M is the catalyst and MA and MB are intermediates. Upon summing the equations of steps (for steady-state or quasi-steady-state regimes), the catalyst and intermediates vanish and we arrive at the following overall equation:

$$2\text{A} \rightleftharpoons 2\text{B}. \tag{1.6}$$

From the standpoint of stoichiometry and material balance, Eqs (1.2) and (1.6) are equivalent, whereas from the standpoint of kinetics, the rate of converting reactant A into B is determined by the overall Eq. (1.6). It is this rate r, defined as a difference between the rates of processes in the forward (r^+) and reverse (r^-) directions ($r = r^+ - r^-$), that corresponds to the mechanism described by Eqs (1.3)–(1.5). Assuming that [A], [B] $\gg$ $[\text{M}]_\Sigma$ and $[\text{M}]_\Sigma \gg$ [MA], [MB] ($[\text{M}]_\Sigma \cong [\text{M}]$), we obtain the following formula for the rate of a steady-state or quasi-steady-state regime:

$$r = r^+ - r^- = -\frac{1}{2}\frac{d[\text{A}]}{dt} = \frac{1}{2}\frac{d[\text{B}]}{dt} = \frac{(k_1k_2k_3[\text{A}]^2 \quad k_{-1}k_{-2}k_{-3}[\text{B}]^2)[M]_\Sigma}{k_2k_3[\text{A}] + k_{-1}k_3 + k_{-1}k_{-2}[\text{B}]}. \tag{1.7}$$

[1]Example proposed to the author by G.S. Yablonsky. See also A.A. Frost and R.G. Pearson, *Kinetics and Mechanism*, 2nd ed., Wiley & Sons, New York, 1961.

[2]Mechanisms of close types are inherent in the skeletal isomerization of n-butane [12, 13] and isobutane [14].

In the case of equilibrium ($r^+ = r^-$), formula (1.7) yields the following equilibrium constant for reaction (1.6):

$$K = [B_{eq}]^2/[A_{eq}]^2.$$

The rate of the forward reaction can be expressed via that of the reverse reaction using the following de Donde relation:

$$RT \ln r^-/r^+ = \Delta\mu \ (\Delta\mu = -A), \tag{1.8}$$

where A is the chemical affinity and $\Delta\mu$ is the change in the chemical potential according to the overall reaction equation. An expression for $\Delta\mu$ should also be written using an equation that follows from the reaction mechanism, in the given case, Eq. (1.6). Note that relation (1.8) is valid only for single-route reactions (see below). ■

The general algorithm of finding overall equations for the reaction mechanisms of any complexity is called (Horiuti) the method of routes [1–7, 9, 15] and is a constituent part of the Horiuti–Temkin theory of steady-state reactions [1–4]. This algorithm determines a relationship between the stoichiometry of a mechanism and that of the overall reaction equation. The mathematical principles of this theory are presented in more detail elsewhere [5, 9]. Here, we dwell on the basic concepts and corollaries of the theory.

The *reaction route* is defined as a sequence of steps entering into the complex reaction mechanism, such that, upon summing the equations of these steps multiplied by the so-called stoichiometric numbers ν_j of steps, the obtained overall equation contains no intermediates that play important roles as participants of the mechanism of the given complex reaction.

The term 'reaction route' also refers to a vector with components representing the *stoichiometric numbers* ν_j of steps. For example, the reaction route vector for a mechanism described by Eqs (1.3–1.5) is a set of three components: $\nu_{1.3} = 1$, $\nu_{1.4} = 1$, $\nu_{1.5} = 1$ or $\boldsymbol{\nu}^{(1)} = (1, 1, 1)$. A different set of stoichiometric numbers, $\boldsymbol{\nu}^{(1)} = (0.5, 0.5, 0.5)$, is provided by the equation A = B, but the above example shows that this equation contradicts the kinetics of the steady-state process under consideration.

The condition for writing the overall equation without intermediates is formally determined by the following set of equations:

$$f_i = \sum_j \beta^*_{ij} \cdot \nu_j^{(p)} = 0, \tag{1.9}$$

where β^*_{ij} is the stoichiometric coefficient of the ith intermediate at the jth step of the reaction mechanism and $\nu_j^{(p)}$ is the stoichiometric number of the jth step in the pth route. Function f_i is written for each set of linearly independent intermediates. Note that, in generally, the *stoichiometric coefficient* β^*_{ij} in Eq. (1.9) (and any other β_{ij} value) is the difference of the stoichiometric coefficients for the same substance that enters into the reaction equation on the right and left sides. For a jth step described by the equation

$$\sum_i \alpha_{ij} A_{ij} = \sum_i \gamma_{ij} A_{ij},$$

we have $\beta_{ij} = \gamma_{ij} - \alpha_{ij}$. In particular, for a reaction considered in Section 6.3,

$$A + X_i \xrightarrow{j\text{th step}} 2X_i + B,$$

this coefficient is $\beta^*_{ij} = 2 - 1 = 1$.

The system of Eqs (1.9) can also be written in a matrix form as follows:

$$\overline{\boldsymbol{B}}_{\mathrm{X}}^{\mathrm{T}} G = 0, \tag{1.10}$$

where $\overline{B}_{\mathrm{X}}^{\mathrm{T}}$ is the transposed matrix of stoichiometric coefficients of the linearly independent intermediates and $\boldsymbol{G}$ is the $S \times \nu^{(p)}$ matrix of the stoichiometric numbers of steps, where S is the number of steps. Note that matrix $\boldsymbol{B}_{\mathrm{X}}$ is a block of the matrix $\boldsymbol{B}_{\mathrm{m}}$ of stoichiometric coefficients of the reaction mechanism. The second block in this matrix represents the matrix $\boldsymbol{B}_N$ of stoichiometric coefficients of the reaction participants (reactants and products) with a dimension of $S \times N$ (where S is the number of steps and N is the total number of participants):

$$\boldsymbol{B}_{\mathrm{m}} = ||\boldsymbol{B}_N \vdots \boldsymbol{B}_{\mathrm{X}}||. \tag{1.11}$$

In order to determine the stoichiometric numbers of steps, Eq. (1.10) can be rewritten as follows:

$$\boldsymbol{B}_{\mathrm{X}}^{\mathrm{T}} \boldsymbol{\nu}_j^{(p)} = 0, \tag{1.12}$$

where $\boldsymbol{\nu}_j^{(p)}$ is the vector of stoichiometric numbers of jth step in the pth route. Since the number of unknowns (equal to the number of steps S) in the system of Eqs (1.12) is greater than the number of equations (equal to the number of linearly independent intermediates N_{X}), the number of linearly independent solutions of system (1.12) or the number of linearly independent routes P is

$$S - N_{\mathrm{X}} = P. \tag{1.13}$$

Example 1.2

▶ Consider the following mechanistic reaction scheme:

$$\begin{aligned} &(1)\quad \mathrm{A} \underset{k_{-1}}{\overset{k_1}{\rightleftharpoons}} \mathrm{X}, \\ &(2)\quad 2\mathrm{X} \underset{k_{-2}}{\overset{k_2}{\rightleftharpoons}} \mathrm{P}, \end{aligned} \tag{1.14}$$

which involves one independent intermediate X (so that $N_{\mathrm{X}} = 1$), two steps ($S = 2$), and a single route ($P = 2 - 1 = 1$). The matrix of stoichiometric coefficients of intermediates $\boldsymbol{B}_{\mathrm{X}}$ can be written as the row-vector $\boldsymbol{\beta}_{ij}^{*} = [1\ -2]$. Using the condition that $\boldsymbol{B}_{\mathrm{X}}^{\mathrm{T}} \boldsymbol{\nu}_j = 0$ and multiplying the row-vector $\boldsymbol{\beta}_{ij}^{*}$ and column-vector $\boldsymbol{\nu} = [\nu_1\ \nu_2]$, we obtain the following equation:

$$\nu_1 - 2\nu_2 = 0, \tag{1.15}$$

which has a single linearly independent solution. Setting $\nu_1 = 1$, we obtain $\nu_2 = 0.5$; for $\nu_1 = 2$ we have $\nu_2 = 1$ and so on. In order to exclude X from the overall equation, let us sum reaction steps (1) and (2) multiplied by $\boldsymbol{\nu}^{(1)} = [1\ 0.5]$ or $\boldsymbol{\nu}^{(2)} = [2\ 1]$. This yields the following overall equations for routes $\boldsymbol{\nu}^{(1)}$ and $\boldsymbol{\nu}^{(2)}$:

$$\begin{aligned} &\boldsymbol{\nu}^{(1)} \quad \mathrm{A} = 1/2\ \mathrm{P}; \\ &\boldsymbol{\nu}^{(2)} \quad 2\mathrm{A} = \mathrm{P}. \end{aligned}$$

Evidently, a change in the Gibbs energy $\Delta\mu^{(p)}$ on route $\boldsymbol{\nu}^{(p)}$ is as follows:

$$\Delta\mu^{(p)} = \sum \nu_j^{(p)} \Delta\mu_j. \tag{1.16}$$

According to Eq. (1.8), we have:

$$\frac{r_p^-}{r_p^+} = \prod \left(\frac{W_j^-}{W_j^+} \right)^{\nu_j^{(p)}}, \tag{1.17}$$

where $W_j^{\pm}$ are the rates of the elementary step in the forward and reverse directions. This yields

$$\frac{r_1^-}{r_1^+} = \left(\frac{W_1^-}{W_1^+} \right) \left(\frac{W_2^-}{W_2^+} \right)^{1/2} \tag{1.18}$$

for route $\boldsymbol{\nu}^{(1)}$ and

$$\frac{r_2^-}{r_2^+} = \left(\frac{W_1^-}{W_1^+} \right)^2 \left(\frac{W_2^-}{W_2^+} \right) \tag{1.19}$$

for route $\boldsymbol{\nu}^{(2)}$.

Let us assume step (1) of mechanism (1.14) to be the rate controlling (limiting) and step (2) to be a quasi-equilibrium step ($W_2^-/W_2^+ \simeq 1$). Then, considering an equilibrium brutto process ($r_1^- = r_1^+$) and using Eq. (1.18), we obtain the following equilibrium constant for the overall equation according to route $\boldsymbol{\nu}^{(1)}$:

$$K^{(1)} = K_1 K_2^{1/2} = \frac{[P_{eq}]^{1/2}}{[A_{eq}]},$$

while Eq. (1.19) for $r_2^- = r_2^+$ yields the following equilibrium constant for route $\boldsymbol{\nu}^{(2)}$:

$$K^{(2)} = K_1^2 K_2 = \frac{[P_{eq}]}{[A_{eq}]^2}.$$

Analogous equations for $K^{(1)}$ and $K^{(2)}$ can be obtained assuming that step (2) is the limiting stage. Thus, once the route is set [Eqs (1.18) and (1.19)], we obtain the overall equations corresponding to this route irrespective of which step is assumed to be limiting.

For the kinetic equations obtained in the experiment, the overall equations are not chosen arbitrarily. For example, in the case of mechanism (1.14) with $r^+ \propto [A]$ (rate-controlling step (1)), the overall equilibrium equation corresponds to $\boldsymbol{\nu}^{(1)}(\nu_1 = 1)$, while for $r^+ \propto [A]^2$ (rate-controlling step (2)) the overall equation refers to $\boldsymbol{\nu}^{(2)}$ ($\nu_2 = 1$). Thus, once the limiting step is experimentally established or set *a priori*, the overall equations would correspond to the routes in which the stoichiometric numbers of limiting steps are equal to unity. Therefore, for determining r^- from the known r^+ (and vice versa), one should use the overall equations that correspond to the actual kinetics. Thus, the choice of a route in the presence of a limiting step is determined by the reaction kinetics. ■

For any single-route nonlinear mechanism with a rate limiting step, Eq. (1.17) yields the following formula:

$$\frac{W_l^-}{W_l^+} = \left(\frac{r_p^-}{r_p^+} \right)^{\frac{1}{\nu_l}},$$

where l refers to the limiting step. Using this expression and Eq. (1.8), it is possible to obtain a relationship between the ratio of the rates of forward and reverse reactions and the thermodynamic potential ($\Delta\mu$), that is, determine the equilibrium constant of the overall equation of the given route in the following form:

$$W_l = W_l^+ \left[1 - \left(\frac{\prod C_i^{\beta_i}}{K^{(p)}} \right)^{\frac{1}{\nu_l}} \right].$$

Such relationships were originally obtained by Horiuti [2] and Boreskov (see [7]).[3]

For mechanism (1.14) and route $\boldsymbol{\nu}^{(2)}$ with a limiting second step ($\nu_2 = 1$), we obtain:

$$W_2 = W_2^+ \left[1 - \left(\frac{[P]/[A]^2}{K^{(2)}} \right) \right].$$

Using the Boreskov equation [7]

$$W_l = W_l^+ \left[1 - \left(\frac{\prod C_{i\,\min}^{\beta_i}}{K_{\min}} \right)^m \right],$$

it is possible to determine the so-called *multiplicity coefficient* (m) of the limiting step by comparing the experimental ratio (W/W^+) to the quantity $\frac{\prod C_{i\,\min}^{\beta_i}}{K_{\min}}$, where $\prod C_{i\,\min}^{\beta_i}$ and $K_{\min}$ are the concentration product and equilibrium constant, respectively, for the overall equation with minimum integer coefficients with respect to the key reactant. In this case, we have $m = 1/\nu_l$, where ν_l is the stoichiometric number of the route with minimum integer coefficients. In particular, for mechanism (1.14) with a limiting second step, the multiplicity coefficient according to Boreskov is $m = 2$, since the stoichiometric number for the second step in overall equation (with minimum integer coefficients with respect to reactant A) is 0.5 (route $\boldsymbol{\nu}^{(1)}$).

In the general case, the Horiuti stoichiometric rule (1.13) is written as follows [4]:

$$P = S - I + W, \tag{1.20}$$

where I is the total number of intermediates and W is the number of linearly independent stoichiometric conservation laws (or the number of linear relations between intermediates); evidently, $N_X = I - W = \mathrm{rank}\boldsymbol{B}_X$. For catalytic reactions with the same type of catalysts (or active centers), we have $W = 1$, which implies that there is a single stoichiometric conservation law – a material balance with respect to the catalyst. The case of two catalysts participating in the reaction mechanism corresponds to $W = 2$. If intermediates of different natures are involved in the reaction, then the determination of P requires finding the rank of matrix $\boldsymbol{B}_X$ and using Eq. (1.13).

Thus, in order to find the overall equations of routes, it is necessary to have a basis set of linearly independent routes (vectors of stoichiometric numbers), that is, to know the number P of these routes. Various vector sets $\boldsymbol{\nu}^{(p)}$ (columns of the matrix $\boldsymbol{G}$ of stoichiometric numbers) lead to different sets of overall route equations, which are equally adequate in describing the mechanism of a complex reaction. The structure of solutions for Eqs (1.19) or (1.10) was considered in [6, 16], according to which there are three possible variants.

Variant 1: $S = N_X$, $P = 0$

In this case, the system of equations has a single, trivial solution with $\boldsymbol{\nu}_j = 0$ (all stoichiometric numbers are zero). This variant corresponds to the case of autocatalytic (catalyst is a product) or autoinhibitory (catalyst is a reactant) reactions.

[3] These relationships (the Horiuti–Boreskov problem) were considered in more detail by M.Z. Lazman and G.S. Yablonsky, *Adv. Chem. Eng.*, **34**, 47 (2008) and by N. Guy and G.S. Yablonsky, *Kinetics of Chemical Reactions: Decoding Complexity*, Wiley-VCH, Weinheim, 2011.

Example 1.3

▶ According to a three-step mechanism of the oxidative chlorination of alkynes [17],

$$\begin{aligned}&(1)\quad RC{\equiv}CH + CuCl \rightleftharpoons X_1 + HCl,\\&(2)\quad X_1 + 2CuCl_2 \rightleftharpoons X_2 + CuCl,\\&(3)\quad X_2 \longrightarrow RC{\equiv}CCl + 2CuCl,\end{aligned} \tag{1.21}$$

the corresponding matrix $\boldsymbol{B}_X$ includes three intermediates (because, formally, CuCl is also an intermediate):

$$\boldsymbol{B}_X = \begin{matrix}(1)\\(2)\\(3)\end{matrix}\overset{\begin{matrix}CuCl & X_1 & X_2\end{matrix}}{\begin{bmatrix}-1 & 1 & 0\\1 & -1 & 1\\2 & 0 & -1\end{bmatrix}};\ \mathrm{rank}\ \boldsymbol{B}_X = N_X = 3\ (\text{i. e.}, S = N_X).$$

If we remove the column of coefficients for CuCl from matrix $\boldsymbol{B}_X$ and introduce this column into the matrix of participants ($\boldsymbol{B}_N$), then the rank of the new stoichiometric matrix of intermediates is rank $\boldsymbol{B}'_X = 2$ and $P = S - N'_X = 1$. Thus we obtain two equations for three unknown quantities:

$$\nu_1 - \nu_2 = 0; \quad \nu_2 - \nu_3 = 0.$$

Setting $\nu_3 = 1$, we obtain $\nu_1 = 1$ and $\nu_2 = 1$, so that $\boldsymbol{\nu} = (1\ 1\ 1)$. The overall equation (1.22) for a single route is as follows:

$$RC{\equiv}CH + 2CuCl_2 \longrightarrow RC{\equiv}CCl + HCl + 2CuCl. \tag{1.22}$$

■

Variant 2: $S > N_X$ (and there are steps that do not enter into any route)

Solutions of Eqs (1.10) are nonzero and there always are some routes, but one or several steps have $\nu_j = 0$ in all vectors (routes) and, hence, these steps do not enter into any route. This situation also arises in autocatalytic or autoinhibitory reactions, provided that one of reactants or products is an intermediate and enters into matrix $\boldsymbol{B}_X$ (for an example, see [16]), as well as in the presence of equilibrium steps (buffer step) involving the formation of compounds unaccounted for in overall equations (dead end), which do not directly participate in steps of the mechanism but enter into the material balance with respect to the catalyst. Steps that do not enter into any route form a *defect* of the mechanism [6, 18], and this feature is used in the classification of mechanisms [18]. The typical example is a nonbranching chain process with a reversible step of chain initiation in the absence of other steps of chain termination.

Variant 3: $S > N_I$ (and each step is involved in some route)

Using matrices $\boldsymbol{B}_N$ and $\boldsymbol{G}$, we find matrix $\boldsymbol{B}_r$ of stoichiometric coefficients of the overall equations of routes and, thus, write the overall equations as

$$\boldsymbol{B}_r^T = \boldsymbol{B}_N^T G, \tag{1.23}$$

where the dimension of matrix $\boldsymbol{B}_r$ is $P \times N$ and, as was noted above, rank$\boldsymbol{B}_r = Q_r \leq Q_{max}$ [11]. The number of the linearly independent overall equations of routes can also be determined from the following relation:

$$Q_r = \mathrm{rank}\boldsymbol{B}_r = \mathrm{rank}\boldsymbol{B}_m - N_X. \tag{1.24}$$

A comparison of the number $Q_{\max}$ of equations, number Q_r of the overall route equations, and number P of routes shows that $Q_{\max} \geq Q_r$, $Q_r \leq P$, and $Q_{\max} \geq (\leq) P$. It should be noted that, since the considerations in [19, 20] did not allow for the (rather rare) possibility of $Q_{\max} > Q_r$, it was ascertained that $Q_{\max} \leq P$.

Example 1.4

▶ The reaction of carbon with water vapor leads to the formation of three products (CO, CO_2, and H_2):

$$C_{\text{solid}} + H_2O \longrightarrow CO, CO_2, H_2.$$

According to the Gibbs stoichiometric rule (1.1), the number of independent reactions in this system is $Q_{\max} = 2$, which implies that only two overall equations can be independent on the given set of reaction participants. An answer to the question as to which are these equations is provided by an analysis of the reaction mechanism. Let us consider the following simplified mechanism:

$$\begin{aligned} &(1) \quad C_{\text{solid}} + H_2O + z \rightleftharpoons zCO + H_2, \\ &(2) \quad zCO \rightleftharpoons z + CO, \\ &(3) \quad zCO + CO \rightleftharpoons z + CO_2 + C_{TB}, \end{aligned} \tag{1.25}$$

where z is an active center on the carbon surface. According to this mechanism, $\text{rank}\boldsymbol{B}_X = 1$ $(N_X = 1)$ and $P = S - N_X = 2$, which implies that the number of independent routes is $P = 2$. Multiplying the row-vector of stoichiometric coefficients $\beta^*_{ij} = [1 \quad -1 \quad -1]$ for the intermediate zCO and column-vector $\boldsymbol{\nu}^{(p)} = [\nu_1 \quad \nu_2 \quad \nu_3]$ and equating the product to zero, we obtain an equation with three unknowns:

$$\nu_1 = \nu_2 + \nu_3, \tag{1.26}$$

which has an infinite number of solutions, two of which are linearly independent (two solutions, two routes). Let us consider various sets of solutions defined by setting certain ν_2 and ν_3 values, for example, $\nu_2 = 1$, $\nu_3 = 0$ and $\nu_2 = 0$, $\nu_3 = 1$, which correspond to $\boldsymbol{\nu}^{(1)} = (1\ 1\ 0)$ and $\boldsymbol{\nu}^{(2)} = (1\ 0\ 1)$, respectively. The other possible sets of ν_j can be obtained by summing and subtracting $\boldsymbol{\nu}^{(1)}$ and $\boldsymbol{\nu}^{(2)}$:

$$\boldsymbol{\nu}^{(3)} = (2\ 1\ 1) = \boldsymbol{\nu}^{(1)} + \boldsymbol{\nu}^{(2)}; \quad \boldsymbol{\nu}^{(4)} = (0\ -1\ 1) = \boldsymbol{\nu}^{(2)} - \boldsymbol{\nu}^{(1)}.$$

Upon writing the corresponding matrix $\boldsymbol{G}$ as

$$\boldsymbol{G} = \begin{matrix} (1) \\ (2) \\ (3) \end{matrix} \overset{\begin{matrix} \boldsymbol{\nu}^{(1)} & \boldsymbol{\nu}^{(2)} & \boldsymbol{\nu}^{(3)} & \boldsymbol{\nu}^{(4)} \end{matrix}}{\begin{bmatrix} 1 & 1 & 2 & 0 \\ 1 & 0 & 1 & -1 \\ 0 & 1 & 1 & 1 \end{bmatrix}},$$

and using Eq. (1.23), we obtain the stoichiometric matrix $\boldsymbol{B}_r$ of overall equations and write these equations as follows:

$$\begin{aligned} &\mathbf{v}^{(1)} \quad C_{\text{solid}} + H_2O = CO + H_2; \\ &\mathbf{v}^{(2)} \quad CO + H_2O = CO_2 + H_2; \\ &\mathbf{v}^{(3)} \quad C_{\text{solid}} + 2H_2O = CO_2 + 2H_2; \\ &\mathbf{v}^{(4)} \quad 2CO = CO_2 + C_{\text{solid}}. \end{aligned}$$

All these equations follow from the proposed reaction mechanism, but only two of them are linearly independent (since $Q_r = \text{rank}\boldsymbol{B}_r = 2$). In this example, we have the case of $P = 2, Q_r = 2$ and $Q_{max} = 2$.

Although the standard values $\Delta G_0^{(p)}$ (standard chemical potentials) of the overall equations are different, no one of these equations is preferred from the standpoint of kinetics as long as all the steps are reversible. All steps of the mechanism in both forward and reverse directions will determine the reaction rate for each overall equation (see below).

If irreversible steps appear, we can formally select the routes with negative ν_j that refer to these irreversible steps, but it is more expedient from both chemical and thermodynamic standpoints to choose the routes so that their directions would correspond to the positive direction of irreversible steps. In this case, the overall equations will not proceed in thermodynamically forbidden directions (see Section 1.4). Using these overall equations, it is possible to calculate the yields of reaction product and selectivities of their formation. For example, if steps (2) and (3) in mechanism (1.25) under the given process conditions is irreversible, then route $\boldsymbol{\nu}^{(4)}$ will be thermodynamically forbidden in one direction and kinetically prohibited in both directions. ■

Example 1.5

► The reaction of ethylene hydrogenation at $T = 25\,°\text{C}$ is thermodynamically allowed in the forward direction. A simplified mechanism of this reaction

$$C_2H_4 + H_2 \longrightarrow C_2H_6$$

in solutions of transition metal complexes can be described by the following scheme:

$$\begin{aligned}
&(1)\quad M + H_2 \rightleftharpoons M(H)_2,\\
&(2)\quad M(H)_2 + C_2H_4 \rightleftharpoons HMC_2H_5,\\
&(3)\quad HMC_2H_5 \longrightarrow M + C_2H_6,\\
&(4)\quad HMC_2H_5 + H_2 \longrightarrow M(H)_2 + C_2H_6,
\end{aligned} \tag{1.27}$$

for which we can distinguish three intermediates (including the catalyst) and write one law of conservation as $[M]_\Sigma = [M] + [M(H)_2] + [HMC_2H_5]$. The rank of the stoichiometric matrix of intermediates is $\text{rank}\boldsymbol{B}_X = 2$ ($N_X = 2$) and, hence, $P = S - N_X = 4 - 2 = 2$. On the given set of reactants and products (H_2, C_2H_4, C_2H_6), we have a single linearly independent overall equation (since $Q_{max} = 1$, $P > Q_{max}$). Upon multiplying matrix $\boldsymbol{B}_X^T$ (with two linearly independent rows)

$$\boldsymbol{B}_X^T = \begin{matrix} \\ M \\ M(H)_2 \end{matrix} \begin{matrix} (1) & (2) & (3) & (4) \\ \end{matrix} \begin{bmatrix} -1 & 0 & 1 & 0 \\ 1 & -1 & 0 & 1 \end{bmatrix}$$

by column-vector $\boldsymbol{\nu} = (\nu_1, \nu_2, \nu_3, \nu_4)$, we obtain two equations with four unknowns:

$$\nu_1 = \nu_3; \quad \nu_2 = \nu_1 + \nu_4. \tag{1.28}$$

Rewriting these equations as

$$\nu_1 = \nu_3; \quad \nu_2 = \nu_3 + \nu_4, \tag{1.29}$$

and setting the values of ν_3 and ν_4, let us calculate the values of ν_1 and ν_2. As a result, we obtain a basis set of routes, for example, in the form of matrix $\boldsymbol{G}_1$, and the corresponding overall equations of routes:

$$\boldsymbol{G}_1 = \begin{matrix} \boldsymbol{\nu}^{(1)} & \boldsymbol{\nu}^{(2)} \\ \end{matrix} \begin{bmatrix} 1 & 0 \\ 1 & 1 \\ 1 & 0 \\ 0 & 1 \end{bmatrix} \Rightarrow \begin{matrix} \boldsymbol{\nu}^{(1)} & C_2H_4 + H_2 = C_2H_6; \\ \boldsymbol{\nu}^{(2)} & C_2H_4 + H_2 = C_2H_6. \end{matrix}$$

The overall equations for both routes are the same, and each route yields the reaction product. Evidently, only one equation is linearly independent. Indeed, rank $\boldsymbol{B}_r = Q_r = 1$ and, in the given case, $Q_r = Q_{max}$. By selecting the proper sets of stoichiometric numbers, we can reduce the system of overall equations to the basis set of such equations, the number of which is Q_r. In this case, some $(P - Q_r)$ of the overall equations will refer to "empty" routes (with the overall equation $0 = 0$), the reaction rate over which is the rate of the cyclic conversion of intermediates that leads to neither reactant consumption nor product formation. This set of routes is called the *stoichiometric basis set* of routes.

Let route $\boldsymbol{\nu}^{(3)}$ be characterized by set $\boldsymbol{\nu}^{(3)}(1 \quad 0 \quad 1 \quad -1) = \boldsymbol{\nu}^{(1)} - \boldsymbol{\nu}^{(2)}$, so that

$$\boldsymbol{G}_2 = \begin{matrix} \boldsymbol{\nu}^{(1)} & \boldsymbol{\nu}^{(3)} \\ \end{matrix} \begin{bmatrix} 1 & 1 \\ 1 & 0 \\ 1 & 1 \\ 0 & -1 \end{bmatrix} . \quad \begin{matrix} \boldsymbol{\nu}^{(1)} & C_2H_4 + H_2 = C_2H_6; \\ \boldsymbol{\nu}^{(2)} & 0 = 0 \end{matrix}$$

Since step (4) is irreversible ($W_4^- = 0$), the reaction rate over this route is negative. The number of empty routes in this case is $P^{(0)} = S - \text{rank}\boldsymbol{B}_m$. ■

Let us briefly consider alternative approaches to an analysis of the stoichiometry of reaction mechanisms. Corio [21, 22] analyzed the stoichiometry of mechanisms using the Gibbs rule in the form of Eq. (1.1), which was written in the following form:

$$K = \rho + \pi + I = s + m,$$

where ρ is the number of reactants, π is the number of products $(\rho + \pi = N), I$ is the total number of intermediates, $K = N + I, s$ is the number of linearly independent reactions (rank$\boldsymbol{B}_m$), and m is the number of linearly independent stoichiometric conservation laws (m = rank$\boldsymbol{H}$, where $\boldsymbol{H}$ is the atomic matrix for the total number K of substances [in Eq. (1.20) this number is denoted W]). Barbara and Corio [23] also suggested a method for finding overall equations, but without allowing for the case that Q_{max} can be greater than Q_r.

Interesting observations concerning the number of linear conservation laws were reported by Korzukhin [24]. Since rank$\boldsymbol{B}_m \leq (N + I) - \text{rank}\,\boldsymbol{H}$, the total number L of these laws can be greater than the total number of the stoichiometric conservation laws (m = rank$\boldsymbol{H}$), provided that $L = (N + I) - \text{rank}\boldsymbol{B}_m > \text{rank}\boldsymbol{H}$. Then, $L - m$ is the number of the linear kinetic laws of conservation, the number of which for catalytic reactions can be determined from the following equation [9, 11]:

$$L - m = I - \text{rank}\boldsymbol{B}_X - K^*, \qquad (1.30)$$

where K^* is the number of catalysts (or the types of active centers).

1.1.2 Kinetics: Reaction rates with respect to substances and over routes

Any study of kinetics is aimed at finding the type and constants of a kinetic model, which has the form of a system of algebraic, algebraic-differential, or differential equations that describe the dependence of the net rates of consumption or production of participants (r_N, reaction rate with respect to a substance) or the reaction rates for the overall equations of routes (r_p, reaction rate over a route) as functions of the concentrations of participants and the temperature, or the dependences of the concentrations of participants and intermediates on the time and the temperature. The system of equations that quantitatively describes the entire set of chemical and physical stages, including the formation of a catalyst, is necessary for an analysis of the kinetic behavior of a process under non-steady-state conditions [25].

The theory of steady-state reactions employs the concepts of a *run over step* and *run over route*. According to [20], the *number of runs over step* is defined as the difference between the numbers of elementary acts (reactions) in the forward and reverse directions per unit time. Then, the rate of a given step $W_j = W_j^+ - W_j^-$ is the number of runs over step per unit volume of the reaction space. The *run over route* implies that the number of runs over every step in the given route is equal to the stoichiometric number $\nu_j^{(p)}$ of this step in the route. All runs over routes can be represented as linear combinations of runs over routes of the basis set. Then, the reaction rate over a basis-set route (or the rate for the overall equation of this route) is defined as the number of runs over this route per unit volume of the reaction space, provided that the runs over all steps are distributed over routes of the given basis set.

Now let us derive a relationship between the rates of steps and the rates over basis-set routes r_p, which is very important in the theory of steady-state reactions. The rate with respect to a substance is naturally related to the rates over basis-set routes via matrix $\boldsymbol{B}_p$ as follows [19]:

$$\boldsymbol{r}_N = \boldsymbol{B}_p^{\mathrm{T}} \boldsymbol{r}_p \tag{1.31}$$

$\left(\text{or } \boldsymbol{r}_N = \sum_p \beta_{ij} \boldsymbol{r}_p\right)$, where $\boldsymbol{r}_N$ and $\boldsymbol{r}_p$ are the column-vectors of reaction rates with respect to substances and over routes, respectively. On the other hand, the r_N values are related to the rates of steps by the following system of equations:

$$\boldsymbol{r}_N = \boldsymbol{B}_N^{\mathrm{T}} \boldsymbol{W}_j, \tag{1.32}$$

where $\boldsymbol{W}_j$ is the column-vectors of the step rates. The condition of steady-state (or quasi-steady-state) reactions in open and closed systems can be written as

$$\boldsymbol{r}_{\mathrm{X}} = \boldsymbol{B}_{\mathrm{X}}^{\mathrm{T}} \boldsymbol{W}_j = 0, \tag{1.33}$$

where $\boldsymbol{r}_{\mathrm{X}}$ is the column-vector of reaction rates with respect to unobservable intermediates. Then, equating expressions (1.31) and (1.32) and multiplying the obtained equation by $\boldsymbol{B}_N$ from the left, we arrive at the so-called *condition of the steady state of steps* (M.I. Temkin [4, 9]):

$$\boldsymbol{W}_j = \boldsymbol{G}\boldsymbol{r}_p, \tag{1.34}$$

or

$$W_j = W_j^+ - W_j^- = \sum_p \nu_j^{(p)} \boldsymbol{r}_p \quad (j = 1, 2, 3, \ldots, S). \tag{1.35}$$

This system of equations is equivalent to Eq. (1.33) and presents an alternative form of the Bodenstein principle ($\frac{dC_{\mathrm{X}}}{dt} = 0$, $\boldsymbol{r}_{\mathrm{X}} = 0$, where $\boldsymbol{C}_{\mathrm{X}}$ and $\boldsymbol{r}_{\mathrm{X}}$ are the vectors of concentrations and reaction rates, respectively; the Bodenstein condition is considered; see below). Upon solving system (1.35) that consists of S equations with $S = P + N_I$ unknowns, we obtain P reaction rates r_p over basis-set routes and N_I unknown concentrations of intermediates. The r_p values can be determined using the Temkin algorithm [9, 26, 27] or the method of graphs (for linear mechanisms, see below). It is also possible to determine $\boldsymbol{C}_{X_i}$ using condition (1.33) and solving the corresponding system of algebraic equations, and then to substitute these C_{X_i} values into Eqs (1.35) for W_j of steps with required r_p values. In the case of a large number of routes at small N_I, it is convenient to use the condition of $\boldsymbol{r}_X = 0$, while in the case of a large set of $X_i(N_I)$ and a small number of routes, it is more convenient to use system (1.35), since a solution of the set of linear equations using the Cramer rule directly yields the required r_p values.

Example 1.6

► Let us derive equations for the rates over basis-set routes and substances for the reaction used in Example 1.4. Consider routes $\boldsymbol{\nu}^{(1)}$ and $\boldsymbol{\nu}^{(2)}$ with the following matrix $\boldsymbol{G}_1$ of stoichiometric numbers:

$$\begin{matrix} & \boldsymbol{\nu}^{(1)} & \boldsymbol{\nu}^{(2)} \end{matrix}$$
$$\boldsymbol{G}_1 = \begin{bmatrix} 1 & 1 \\ 1 & 0 \\ 0 & 1 \end{bmatrix} \Rightarrow \begin{matrix} \boldsymbol{\nu}^{(1)} & C_{solid} + H_2 = CO + H_2; \\ \boldsymbol{\nu}^{(2)} & CO + H_2O = CO_2 + H_2. \end{matrix}$$

Using condition (1.34) of the steady state of steps, we obtain the following system of equations:

$$\begin{bmatrix} W_1 \\ W_2 \\ W_3 \end{bmatrix} = \begin{bmatrix} 1 & 1 \\ 1 & 0 \\ 0 & 1 \end{bmatrix} \times \begin{bmatrix} r_1 \\ r_2 \end{bmatrix}. \tag{1.36}$$

Taking the product in (1.36) we obtain the following set of three equations ($S = 3$) with three unknowns ($[zCO]$, r_1, r_2):

$$\begin{aligned} W_1 &= W_1^+ - W_1^- = r_1 + r_2; \\ W_2 &= W_2^+ - W_2^- = r_1; \\ W_3 &= W_3^+ - W_3^- = r_2, \end{aligned} \tag{1.37}$$

from which it follows that the rates r_1 and r_2 over the basis set routes are W_2 and W_3, respectively, and that step (1) ($W_1 = W_2 + W_3$) proceeds via both routes. Let us introduce the *step weight* ω_j (or ω_{-j}), which is defined as the ratio of the step rate in one direction to the concentration $[X_i]$ of an intermediate product involved in this step:

$$\omega_j = \frac{W_j^+}{[X_{ij}]}; \quad \omega_{-j} = \frac{W_j^-}{[X_{i+1,j}]}.$$

Taking into accout the material balance equation $[z]_\Sigma = [z] + [zCO]$, assuming that the relative concentrations obey the relation $[z]_\Sigma = 1$, and solving system (1.37), we obtain the following expressions:

$$r_1 = W_2 = \frac{\omega_2(\omega_1 + \omega_{-3}) - \omega_{-2}(\omega_{-1} + \omega_2)}{D}; \tag{1.38}$$

$$r_2 = W_3 = \frac{\omega_3(\omega_1 + \omega_{-2}) - \omega_{-3}(\omega_{-1} + \omega_2)}{D}; \tag{1.39}$$

$$D = \omega_{-1} + \omega_2 + \omega_3 + \omega_1 + \omega_{-2} + \omega_{-3}. \tag{1.40}$$

As can be seen, the rate over the basis-set route contains the rates of all steps of the given mechanism (including the rates of reverse steps) not entering in W_2 and W_3 in the numerators of expressions (1.38) and (1.39). The rates r_N for substances can be found using relation (1.31), written as

$$\begin{bmatrix} r_C \\ r_{H_2O} \\ r_{CO} \\ r_{H_2} \\ r_{CO_2} \end{bmatrix} = \overset{\quad\boldsymbol{\nu}^{(1)} \quad \boldsymbol{\nu}^{(2)}}{\begin{bmatrix} -1 & 0 \\ -1 & -1 \\ 1 & -1 \\ 1 & 1 \\ 0 & 1 \end{bmatrix}} \times \begin{bmatrix} r_1 \\ r_2 \end{bmatrix},$$

which yields $-r_C = r_1$; $-r_{H_2O} = r_1 + r_2$; $-r_{CO} = r_1 - r_2$; $-r_{H_2} = r_1 + r_2$; $-r_{CO_2} = r_2$.

In order to describe the reaction, it is sufficient to select two key substances (i.e., linearly independent columns in matrix $\boldsymbol{B}_N$), for example, CO_2 and H_2. For matrix $\boldsymbol{G}_2$ with routes $\boldsymbol{\nu}^{(1)}$ and $\boldsymbol{\nu}^{(4)}$ we have

$$\boldsymbol{G}_2 = \overset{\quad\boldsymbol{\nu}^{(1)} \quad \boldsymbol{\nu}^{(4)}}{\begin{bmatrix} 1 & 0 \\ 1 & -1 \\ 0 & 1 \end{bmatrix}} \Rightarrow \begin{matrix} \boldsymbol{\nu}^{(1)} & C_{solid} + H_2O = CO + H_2, & r_1'; \\ \boldsymbol{\nu}^{(4)} & 2CO = CO_2 + C_{solid}, & r_4. \end{matrix}$$

In this case, we obtain

$$W_1 = W_1^+ - W_1^- = r_1';$$

$$W_2 = W_2^+ - W_2^- = r' - r_4;$$

$$W_3 = W_3^+ - W_3^- = r_4;$$

$$r_1' = W_1 = \frac{\omega_1(\omega_2 + \omega_3) - \omega_{-1}(\omega_{-2} + \omega_{-3})}{D}; \tag{1.41}$$

$$r_4 = W_3 = \frac{\omega_3(\omega_1 + \omega_{-2}) - \omega_{-3}(\omega_{-1} + \omega_2)}{D}. \tag{1.42}$$

A comparison of expressions (1.38), (1.39) and (1.41), (1.42) shows that: (i) the structure of equations does not change in the case of using a route with a negative stoichiometric number; and (ii) similar to system (1.37) we have $r_2 = r_4$ and $W_1 = W_2 + W_3$. The expression for r_1' (at which ν_j are positive) contains the rates of all reverse steps. Then, we have

$$\begin{bmatrix} r_C \\ r_{H_2O} \\ r_{CO} \\ r_{H_2} \\ r_{CO_2} \end{bmatrix} = \overset{\quad\boldsymbol{\nu}^{(1)} \quad \boldsymbol{\nu}^{(4)}}{\begin{bmatrix} -1 & 1 \\ -1 & 0 \\ 1 & -2 \\ 1 & 0 \\ 0 & 1 \end{bmatrix}} \times \begin{bmatrix} r_1' \\ r_4 \end{bmatrix}, \tag{1.43}$$

which yields $r_C = r_4 - r_1'$; $-r_{H_2O} = r_1'$; $r_{CO} = r_1' - 2r_4$; $-r_{H_2} = r_1'$; $r_{CO_2} = r_4$. Let us compare the rates of formation, e.g., for hydrogen (r_{H_2}) via the two sets of routes. For the first set we have $r_{H_2} = r_1 + r_2$, while for the second set $r_{H_2} = r_1'$. Adding equations (1.38) and (1.39) yields formula (1.41), which shows that the rates for substances are invariant with respect to the sets of routes. It is important to bear in mind that both an expression for the rate and a value of the rate over a given route have sense only provided that the entire basis set of routes is defined.

For better understanding the reasons for which we can include steps with negative ν_i (i.e., reactions proceeding in thermodynamically prohibited directions) into the route, let us also consider the sets of routes $\mathbf{v}^{(1)}$ and $\mathbf{v}^{(5)}$:

$$G_3 = \begin{array}{c} \begin{matrix} \boldsymbol{\nu}^{(1)} & \boldsymbol{\nu}^{(5)} \end{matrix} \\ \begin{bmatrix} 1 & 0 \\ 1 & 1 \\ 0 & -1 \end{bmatrix} \end{array} \Rightarrow \begin{matrix} \boldsymbol{\nu}^{(1)} & C_{solid} + H_2O = CO + H_2; \\ \boldsymbol{\nu}^{(5)} & C_{solid} + CO_2 = 2CO. \end{matrix}$$

In this case, we have

$$W_1 = W_1^+ - W_1^- = r_1'';$$
$$W_2 = W_2^+ - W_2^- = r_1 + r_5;$$
$$W_3 = W_3^+ - W_3^- = -r_5,$$

which yields

$$r_5 = -W_3 = \frac{\omega_{-3}(\omega_2 + \omega_{-1}) - \omega_3(\omega_{-2} + \omega_1)}{D}. \tag{1.44}$$

Thus, the use of a negative rate over the route also leads to invariant rates with respect to substances. For example, according to matrix G_3, the rate of CO_2 formation is $r_{CO_2} = -r_5 = r_4$ (cf. equation (1.42) obtained using basis set G_2). ■

In the case of a reversible multi-route reaction, the rate over the route can be presented as the difference between rates in the forward and reverse directions: $\boldsymbol{r}_p = \boldsymbol{r}_p^+ - \boldsymbol{r}_p^-$ [for example, $r_2 = r_2^+ - r_2^-$ (1.39)], but the ratio of rates $\boldsymbol{r}_p^+/\boldsymbol{r}_p^-$ in this case is no longer related to the chemical affinity A_p on route $\boldsymbol{\nu}^{(p)}$, since A_p includes only the chemical affinities of steps entering into the given route,

$$A_p = \sum A_j^{(p)}$$

and does not include the chemical affinities of other steps that enter into the kinetic equation of rate over the route. Therefore, for a reversible multi-route reaction, we have

$$RT \ln(\boldsymbol{r}_p^+/\boldsymbol{r}_p^-) \neq A_p.$$

In Example 1.6 above, the rate of the first step was

$$W_1 = r_1 + r_2,$$

so that the fraction of the affinity of the first step was distributed between two routes, whereas the equation of affinities for the second route,

$$A_\Sigma^{(2)} = A_1 + A_3$$

does not reflect the condition $W_1 = W_2 + W_3$. In this case, the relation $r_2^+/r_2^- = 1$ in equilibrium (see Eq. 1.39) does not lead to an expression for the equilibrium constant of the overall equation of the second route. The equilibrium of one route leads automatically to the equilibrium of all other routes. Indeed, for the equilibrium of step 3 ($W_3 = 0$) in Example 1.4, the rate over route $\boldsymbol{\nu}^{(2)}$, is zero and $\omega_3(\omega_1 + \omega_{-2}) = \omega_{-3}(\omega_{-1} + \omega_2)$. However, the equality $r_2 = 0$ is only possible for $W_1 = W_2 = 0$, but this implies that the overall equation of route $\boldsymbol{\nu}^{(1)}$ is also at equilibrium, as well as the overall equation of all other derivative routes.

Example 1.7

► Let us obtain the kinetic equations of rates over the basis-set routes and substances for Example 1.5 [reaction scheme (1.27)]. For the first basis set of routes with matrix $\boldsymbol{G}_1$, we have

$$W_1 = W_1^+ - W_1^- = r_1;$$
$$W_2 = W_2^+ - W_2^- = r_1 + r_2;$$
$$W_3^+ = r_1;$$
$$W_4^+ = r_2;$$
$$r_1 = W_3^+ = \frac{[M]_\Sigma \omega_1 \omega_2 \omega_3}{D}; \tag{1.45}$$
$$r_2 = W_4^+ = \frac{[M]_\Sigma \omega_1 \omega_2 \omega_4}{D}; \tag{1.46}$$
$$D = \omega_{-1}\omega_3 + \omega_2\omega_3 + \omega_{-1}\omega_{-2} + \omega_1\omega_3 + \omega_1\omega_{-2} + \omega_1\omega_4 + \omega_1\omega_2. \tag{1.47}$$

Then, evidently, for the overall equations of routes $\mathrm{v}^{(1)}$ and $\mathrm{v}^{(2)}$, we have $\boldsymbol{r}_{C_2H_6} = \boldsymbol{r}_1 + \boldsymbol{r}_2$, and

$$r_{C_2H_6} = \frac{[\mathrm{M}]_\Sigma \omega_1 \omega_2 (\omega_3 + \omega_4)}{D}. \tag{1.48}$$

By the same token, for the basis set of routes with matrix $\boldsymbol{G}_2$, we have

$$W_1 = r_1' + r_3;$$
$$W_2 = r_1';$$
$$W_3^+ = r_1' + r_3;$$
$$W_4^+ = -r_3.$$

Ethane is formed via the first route at a rate of

$$r_1' = W_2 = \frac{[\mathrm{M}]_\Sigma \omega_1 \omega_2 (\omega_3 + \omega_4)}{D}. \tag{1.49}$$

Thus, $\boldsymbol{r}_{C_2H_6} = \boldsymbol{r}_1 + \boldsymbol{r}_2 = \boldsymbol{r}_1'$, and we again infer that the rates for substances are invariant with respect to the selection of routes, since

$$W_3^+ + W_4^+ = r_1 + r_2 \text{ (for } G_1\text{)};$$
$$W_3^+ + W_4^+ = r_1' \text{ (for } G_2\text{)}.$$

■

As is known, for a cyclic route with a zero overall equation (empty route), such as that for the mechanism

$$\mathrm{A} \rightleftharpoons \mathrm{B} \rightleftharpoons \mathrm{C} \rightleftharpoons \mathrm{A},$$

the chemical affinity (A_p) is zero and the equilibrium constant for the overall equation of this three-step reaction is $K = K_1K_2K_3 = 1$ [15]. This circumstance imposes certain relationships between rate constants of elementary steps, which can be used to estimate the constants for kinetic models in multi-route reactions with empty routes as was originally pointed out by Avetisov et al. [28]. Let us consider mechanism (1.27) analogous to that in Example 1.5, but with all stages being reversible. The equations of rates over routes $\boldsymbol{\nu}^{(2)}$ (matrix $\boldsymbol{G}_1$) and $\boldsymbol{\nu}^{(3)}$ (matrix $\boldsymbol{G}_2$ with an empty route) are

$$r_2 = W_4 = W_4^+ - W_4^- \quad \text{and} \quad r_3 = -W_4 = W_4^- - W_4^+,$$

only differing in the sign and containing a term $(\omega_1\omega_3\omega_{-4} - \omega_{-1}\omega_{-3}\omega_4)$ in the numerator that refers to the empty route:

$$r_3 = \frac{(\omega_1\omega_3\omega_{-4} - \omega_{-1}\omega_{-3}\omega_4) + (\omega_1 + \omega_{-3})(\omega_{-2}\omega_{-4} - \omega_2\omega_4)}{D}.$$

From thermodynamic considerations, we infer that $k_1k_3k_{-4} = k_{-1}k_{-3}k_4$ and the concentrations of products and reactants for the empty route cancel each other so that

$$\omega_1\omega_3\omega_{-4} = \omega_{-1}\omega_{-3}\omega_4$$

and

$$r_3 = -r_2 = \frac{(\omega_1 + \omega_{-3})(\omega_{-2}\omega_{-4} - \omega_2\omega_4)}{D}.$$

Thus, while the chemical affinity for the empty route $\boldsymbol{\nu}^{(3)}$ is zero ($A_3 = 0$), the rate over route $\boldsymbol{\nu}^{(3)}$ is not zero and the empty route is not an equilibrium one.

If a mechanism involves reversible steps that do not enter into any route – i.e., steps with zero stoichiometric numbers ν_j ("defect" of the mechanism) – then, according to equation (1.34), the rates of these steps are zero and, hence, they are equilibrium steps. For example, if we add a "dead-end" step to the mechanism according to scheme (1.27),

$$(5) \quad M + C_2H_4 \rightleftharpoons MC_2H_4,$$

for which $\boldsymbol{\nu}_5^{(1)} = 0$ and $\boldsymbol{\nu}_5^{(2)} = 0$, then

$$W_5^+ - W_5^- = \nu_5^{(1)} r_1 + \nu_5^{(2)} r_2 = 0$$

and, hence, $W_5^+ = W_5^-$.

The rate of an n-step reversible single-route reaction with a linear mechanism according to Temkin [3, 27] is expressed as follows:

$$r = \frac{W_1^+ \cdot W_2^+ \cdot \ldots \cdot W_n^+ - W_1^- \cdot W_2^- \cdot \ldots \cdot W_n^-}{\nu_1 \cdot W_2^+ \cdot \ldots \cdot W_n^+ + W_1^- \cdot \nu_2 \cdot \ldots \cdot W_n^+ + W_1^- \cdot W_2^- \cdot \nu_3 \cdot \ldots \cdot W_n^+ + W_1^- \cdot W_2^- \cdot \ldots \cdot W_{(n-1)}^- \cdot \nu_n}. \tag{1.50}$$

All stoichiometric numbers for linear single-route mechanisms are $\nu_j = 1$, but their position in terms of the denominator helps the memorizing of the algorithm of writing the denominator in formula (1.50). For a catalytic process, this formula (written in terms of step weights) has the following form:

$$r = \frac{[\mathrm{M}](\omega_1 \cdot \omega_2 \cdot \ldots \cdot \omega_n - \omega_{-1} \cdot \omega_{-2} \cdot \ldots \cdot \omega_{-n})}{\omega_2 \cdot \ldots \cdot \omega_n + \omega_{-1} \cdot \ldots \cdot \omega_n + \omega_{-1} \cdot \omega_{-2} \cdot \ldots \cdot \omega_n + \omega_{-1} \cdot \omega_{-2} \cdot \ldots \cdot \omega_{-(n-1)}}. \tag{1.51}$$

where $[M]$ is the concentration of active centers (active complexes).

Thus, the theory of steady-state and quasi-steady-state reactions employs three equivalent conditions of a steady-state process:

- The condition of zero total rate of variation of the concentration of X_i intermediate:

$$\boldsymbol{r}_x = \boldsymbol{B}_X^T \boldsymbol{W}_j = 0 \quad \text{and} \quad \frac{d\boldsymbol{C}_X}{dt} = 0,$$

  (in both open and closed systems);
- The Horiuti condition (1.10) of excluding intermediates from overall equations:

$$\boldsymbol{B}_X^T \boldsymbol{G} = 0,$$

  which is valid only provided that the total rate of X_i accumulation (see the first condition above) is zero;
- The condition of the steady state of steps:

$$W_j^+ - W_j^- = \sum_p \nu_j^p r_p.$$

Now let us perform the stoichiometric and kinetic analysis of a more complicated, nonlinear mechanism of a catalytic reaction that involves two active catalysts (types of active centers).

Example 1.8

▶ The reaction of acetylene hydrochlorination

$$C_2H_2 + HCl \longrightarrow CH_2{=}CHCl \tag{1.52}$$

is catalyzed by Cu(I) and Hg(II) chloride complexes (for simplicity, CuCl and $HgCl_2$) in an $HCl–H_2O$ solution and probably proceeds according to the following scheme [29]:

$$\begin{aligned}
&(1) \quad C_2H_2 + HgCl_2 \rightleftharpoons X_1;\\
&(2) \quad X_1 + HCl \longrightarrow CH_2{=}CHCl + HgCl_2;\\
&(3) \quad C_2H_2 + CuCl \rightleftharpoons X_2;\\
&(4) \quad X_2 + HCl \longrightarrow CH_2{=}CHCl + CuCl;\\
&(5) \quad X_1 + CuCl \longrightarrow X_2 + HgCl_2.
\end{aligned} \tag{1.53}$$

The corresponding matrix $\boldsymbol{B}_m$ consists of two blocks ($\boldsymbol{B}_N$ and $\boldsymbol{B}_X$):

$$\boldsymbol{B}_m = \begin{array}{c} \begin{array}{ccccccc} C_2H_2 & HCl & CH_2{=}CHCl & HgCl_2 & X_1 & CuCl & X_2 \end{array} \\ \left[\begin{array}{ccccccc} -1 & 0 & 0 & -1 & 1 & 0 & 0 \\ 0 & -1 & 1 & 1 & -1 & 0 & 0 \\ -1 & 0 & 0 & 0 & 0 & -1 & 1 \\ 0 & -1 & 1 & 0 & 0 & 1 & -1 \\ 0 & 0 & 0 & 1 & -1 & -1 & 1 \end{array}\right] \\ \boldsymbol{B}_N \qquad\qquad\qquad \boldsymbol{B}_x \end{array}.$$

If C_2H_2 and HCl groups are selected to be the groups retained during the reaction under consideration, the rank of the atomic (molecular) matrix $\boldsymbol{H}_{\text{part}}$ of reaction participants (reactants and products) is $\text{rank}\boldsymbol{H}_{\text{part}} = 2$ and, hence, $Q_{\max} = N - \text{rank}\boldsymbol{H}_{\text{part}} = 3 - 2 = 1$, i.e., there is a single overall equation [reaction (1.52)]. The reaction involves two catalysts ($W = 2$), obeys two material balances ($[\text{CuCl}]_\Sigma = [\text{CuCl}] + [X_2]$ and $[\text{HgCl}_2]_\Sigma = [\text{HgCl}_2] + [X_1]$) and, hence, there are only two linearly independent columns in matrix $\boldsymbol{B}_\text{X}$, so that rank $\boldsymbol{B}_\text{X} = N_I = 2$. From this it follows that the number of linearly independent routes (i.e., the rank of the matrix of stoichiometric numbers G) is $P = S - N_I = 5 - 2 = 3$ or $P = S - I + W = 5 - 4 + 2 = 3$ [see Eqs (1.13) and (1.14)]. The total number of conservation equations ($L = N + I - \text{rank}\boldsymbol{B}_\text{m} = 7 - 3 = 4$) in this case is equal to the number of stoichiometric conservation laws, since $\text{rank}H_{\text{part}} = 2$ and $\text{rank}\boldsymbol{B}_\text{X} = 2$.

Let us find the corresponding matrix $\boldsymbol{G}$. For this purpose, let us select and use two linearly independent columns for $\mathbf{X}_1$ and $\mathbf{X}_2$ as matrix $\boldsymbol{B}_\text{X}$ and write the condition $\boldsymbol{B}_\text{X}^T \boldsymbol{\nu}_j = 0$ as follows:

$$\begin{matrix}\mathbf{X}_1\\ \mathbf{X}_2\end{matrix}\begin{bmatrix}1 & -1 & 0 & 0 & -1\\ 0 & 0 & 1 & -1 & 1\end{bmatrix} \times \begin{vmatrix}\nu_1\\ \nu_2\\ \nu_3\\ \nu_4\\ \nu_5\end{vmatrix} = 0. \tag{1.54}$$

Equation (1.54) yields a set of two equations with five unknowns,

$$\begin{matrix}\boldsymbol{\nu}_2 + \boldsymbol{\nu}_5 = \boldsymbol{\nu}_1\\ \boldsymbol{\nu}_5 = \boldsymbol{\nu}_4 - \boldsymbol{\nu}_3\end{matrix} \quad \text{or} \quad \begin{matrix}\boldsymbol{\nu}_2 = \boldsymbol{\nu}_1 - \boldsymbol{\nu}_4 + \boldsymbol{\nu}_3\\ \boldsymbol{\nu}_5 = \boldsymbol{\nu}_4 - \boldsymbol{\nu}_3,\end{matrix} \tag{1.55}$$

which has three linearly independent solutions (i.e., three sets of the stoichiometric numbers of routes). Setting the sequences of whole numbers and zeros for ν_1, ν_3, and ν_4, we obtain the corresponding values of ν_2 and ν_5. Choosing the basis-set routes, for example, $\boldsymbol{\nu}^{(1)}$, $\boldsymbol{\nu}^{(2)}$ and $\boldsymbol{\nu}^{(3)}$, we write the matrix $\boldsymbol{G}$ as

$$G = \begin{matrix}\boldsymbol{\nu}^{(1)} & \boldsymbol{\nu}^{(2)} & \boldsymbol{\nu}^{(3)} & \boldsymbol{\nu}^{(4)} & \boldsymbol{\nu}^{(5)}\end{matrix}\\ \begin{bmatrix}1 & 0 & 1 & 1 & 1\\ 1 & 0 & 0 & 0 & 1\\ 0 & 1 & 0 & -1 & 1\\ 0 & 1 & 1 & 0 & 1\\ 0 & 0 & 1 & 1 & 0\end{bmatrix}, \tag{1.56}$$

and determine routes $\boldsymbol{\nu}^{(4)}$ and $\boldsymbol{\nu}^{(5)}$ as the difference $(\boldsymbol{\nu}^{(3)} - \boldsymbol{\nu}^{(2)})$ and sum $(\boldsymbol{\nu}^{(1)} + \boldsymbol{\nu}^{(2)})$, respectively, of the given sets of stoichiometric numbers.

The overall equations for the first three routes (basis $\boldsymbol{G}_1$), that is, matrrix $\boldsymbol{B}_p$, can be found using Eq. (1.23). From this matrix $\boldsymbol{B}_p$, it follows that the first three independent routes correspond to three identical equations:

$$\boldsymbol{B}_p = \begin{matrix}\boldsymbol{\nu}^{(1)}\\ \boldsymbol{\nu}^{(2)}\\ \boldsymbol{\nu}^{(3)}\end{matrix}\overset{\begin{matrix}C_2H_2 & HCl & CH_2{=}CHCl\end{matrix}}{\begin{bmatrix}-1 & -1 & 1\\ -1 & -1 & 1\\ -1 & -1 & 1\end{bmatrix}} \Rightarrow \begin{matrix}\boldsymbol{\nu}^{(1)} & C_2H_2 + HCl & \rightarrow & CH_2{=}CHCl;\\ \boldsymbol{\nu}^{(2)} & C_2H_2 + HCl & \rightarrow & CH_2{=}CHCl;\\ \boldsymbol{\nu}^{(3)} & C_2H_2 + HCl & \rightarrow & CH_2{=}CHCl,\end{matrix}$$

so that $\text{rank}\boldsymbol{B}_p = Q_p = 1$. Equation (1.24) also yields $Q_p = \text{rank}\boldsymbol{B}_m - N_I = 3 - 2 = 1$, since mechanism (1.53) has only three linearly independent steps.

By the same token, using routes $\boldsymbol{\nu}^{(1)}$, $\boldsymbol{\nu}^{(2)}$ and $\boldsymbol{\nu}^{(4)}$ and the corresponding matrix $\boldsymbol{G}_2$, we arrive at matrix $\boldsymbol{B}'_p$ with an "empty" third route,

$$\boldsymbol{B}'_p = \begin{matrix} \\ \boldsymbol{\nu}^{(1)} \\ \boldsymbol{\nu}^{(2)} \\ \boldsymbol{\nu}^{(4)} \end{matrix} \begin{matrix} \mathrm{C_2H_2} & \mathrm{HCl} & \mathrm{CH_2{=}CHCl} \\ \end{matrix} \begin{bmatrix} -1 & -1 & 1 \\ -1 & -1 & 1 \\ 0 & 0 & 0 \end{bmatrix} \Rightarrow \boldsymbol{\nu}^{(4)} \quad 0 = 0,$$

in which case the rate over the empty route is nonzero.

The basis set of routes can be reduced to a stoichiometric basis set, that is, a set with two empty routes, by adopting negative stoichiometric numbers for irreversible steps (4) and (5). Selecting matrix $\boldsymbol{G}_3$ as

$$G_3 = \begin{matrix} (1) \\ (2) \\ (3) \\ (4) \\ (5) \end{matrix} \overset{\begin{matrix} \boldsymbol{\nu}^{(1)} & \boldsymbol{\nu}^{(4)} & \boldsymbol{\nu}^{(6)} \end{matrix}}{\begin{bmatrix} 1 & 1 & 0 \\ 1 & 0 & 1 \\ 0 & -1 & 0 \\ 0 & 0 & -1 \\ 0 & 1 & -1 \end{bmatrix}} \Rightarrow \begin{matrix} \boldsymbol{\nu}^{(1)} & \mathrm{C_2H_2 + HCl \longrightarrow CH_2{=}CHCl;} \\ \boldsymbol{\nu}^{(4)} & 0 = 0; \\ \boldsymbol{\nu}^{(6)} & 0 = 0, \end{matrix}$$

and writing the sets of equations (1.34) of the steady state of steps for matrices $\boldsymbol{G}_1$, $\boldsymbol{G}_2$, and $\boldsymbol{G}_2$,

$$\begin{array}{lll} \boldsymbol{G}_1 & \boldsymbol{G}_2 & \boldsymbol{G}_3 \\ W_1 = r_1 + r_3 & W_1 = r_1 + r_4 & W_1 = r_1 + r_4 \\ W_2 = r_1 & W_2 = r_1 & W_2 = r_1 + r_6 \\ W_3 = r_2 & W_3 = r_2 - r_4 & W_3 = -r_4 \\ W_4 = r_2 + r_3 & W_4 = r_2 & W_4 = -r_6 \\ W_5 = r_3 & W_5 = r_4 & W_5 = r_4 - r_6, \end{array} \tag{1.57}$$

we can express the rate of vinyl chloride formation (r_{VC}) via basis set routes using matrix $\boldsymbol{B}_p$ ($\boldsymbol{r}_N = \boldsymbol{B}_p^{\mathrm{T}} \boldsymbol{r}_p$) as follows:

$$\boldsymbol{G}_1: \quad r_{\mathrm{VC}} = r_1 + r_2 + r_3 = W_2 + W_4 \quad (\text{or } W_1 + W_3);$$

$$\boldsymbol{G}_2: \quad r_{\mathrm{VC}} = r_1 + r_2 = W_2 + W_4 \quad (\text{or } W_1 + W_3);$$

$$\boldsymbol{G}_3: \quad r_{\mathrm{VC}} = r_1 = W_2 + W_4 \quad (\text{or } W_1 + W_3).$$

Evidently, the r_{VC} value must be invariant with respect to the basis set of routes, so that in all cases $r_{\mathrm{VC}} = W_2 + W_4$ (or $W_1 + W_3$). An expression for r_{VC} can be obtained by solving sets of Eqs (1.57) for unknowns $[\mathrm{X_1}]$, $[\mathrm{X_2}]$, and r_p. In the latter case, the set of equations (for $\boldsymbol{G}_3$) is readily reduced to a set of three equations with three unknowns (r_1, $[\mathrm{X_1}]$, $[\mathrm{X_2}]$):

$$W_1 + W_3 = r_1; \quad W_2 + W_4 = r_1; \quad W_3 + W_5 = W_4,$$

from which the r_1 value is readily determined by the Cramer method. Assuming that the step rates obey the relations

$$[\mathrm{CuCl}]_\Sigma \gg [\mathrm{X_2}], \quad [\mathrm{HgCl_2}]_\Sigma \gg [\mathrm{X_1}], \quad [\mathrm{CuCl}]_\Sigma \cong [\mathrm{CuCl}], \quad [\mathrm{HgCl_2}]_\Sigma \cong [\mathrm{HgCl_2}],$$

we eventually obtain

$$r_1 = \frac{k_1k_2[\mathrm{HgCl_2}]_\Sigma[\mathrm{HCl}]P_{\mathrm{C_2H_2}}}{k_{-1}+k_2[\mathrm{HCl}]+k_5[\mathrm{CuCl}]_\Sigma} + \frac{k_3k_4[\mathrm{CuCl}]_\Sigma[\mathrm{HCl}]P_{\mathrm{C_2H_2}}}{k_{-3}+k_4[\mathrm{HCl}]} + \frac{k_1k_4k_5[\mathrm{HgCl_2}]_\Sigma[\mathrm{CuCl}]_\Sigma[\mathrm{HCl}]P_{\mathrm{C_2H_2}}}{[k_{-3}+k_4[\mathrm{HCl}]](k_{-1}+k_2[\mathrm{HCl}]+k_5[\mathrm{CuCl}]_\Sigma)}, \tag{1.58}$$

where $P_{\mathrm{C_2H_2}}$ is the partial pressure of acetylene.

It is also possible to use the relation $r_x = \boldsymbol{B}_{\mathrm{X}}^T\boldsymbol{W}_j = 0$ (Bodenstein condition) and determine $[\mathrm{X_1}]$ and $[\mathrm{X_2}]$ by solving a system of two equations for two linearly independent concentrations of intermediates X_i. Substituting these $[\mathrm{X_1}]$ and $[\mathrm{X_2}]$ values into equations for the rates of steps (2) and (4), in which vinyl chloride is formed, we eventually obtain

$$r_{\mathrm{VC}} = W_2 + W_4 = k_2[\mathrm{HCl}][\mathrm{X_1}] + k_4[\mathrm{HCl}][\mathrm{X_2}].$$

The values of $[\mathrm{X_1}]$ and $[\mathrm{X_2}]$ determined in this way are as follows:

$$[\mathrm{X_1}] = \frac{k_1[\mathrm{HgCl}]_\Sigma P_{\mathrm{C_2H_2}}}{k_{-1}+k_2[\mathrm{HCl}]+k_5[\mathrm{CuCl}]_\Sigma}; \tag{1.59}$$

$$[\mathrm{X_2}] = \frac{k_3[\mathrm{CuCl}]_\Sigma P_{\mathrm{C_2H_2}}}{k_{-3}+k_4[\mathrm{HCl}]} + \frac{k_1k_5\mathrm{CuCl}_\Sigma[\mathrm{HgCl_2}]_\Sigma P_{\mathrm{C_2H_2}}}{(k_{-3}+k_4[\mathrm{HCl}])(k_{-1}+k_2[\mathrm{HCl}]+k_5[\mathrm{CuCl}]_\Sigma)}. \tag{1.60}$$

As a result, we eventually also obtain Eq. (1.58) for r_{VC}. The empty route $\boldsymbol{N}^{(6)}$ in the stoichiometric basis set of routes includes the irreversible step (4) with a negative stoichiometric number. Note that the sets of non-empty basis-set routes (overall equations of vinyl chloride formation) in all variants are written in accordance with the thermodynamically allowed directions of steps.

Assuming that steps (1) and (3) are quasi-equilibrium ($W_{-1} \gg W_2, W_5$; $W_{-3} \gg W_4, W_5$), the rate of vinyl chloride formation is described by the following equation with two terms in the right-hand side:

$$r_{\mathrm{VC}} = k_2K_1[\mathrm{HgCl_2}]_\Sigma[\mathrm{HCl}]P_{\mathrm{C_2H_2}} + k_4K_3[\mathrm{CuCl}]_\Sigma[\mathrm{HCl}]P_{\mathrm{C_2H_2}}, \tag{1.61}$$

in which a synergistic effect caused by step (5) disappears and the rates of reactions on two independent centers (catalysts) are merely added. ■

In concluding this section, it is expedient to emphasize the following important features of the method of routes:

(i) An elementary step in one direction is characterized by the rates (r_i) with respect to all substances – participants of the step and by the reaction rate $W_j^{\pm}$ ($r_i = \sum \beta_i W_j^{\pm}$) that is invariant with respect to these substances. The reaction rate constant of the step refers to the latter invariant rate W_j. Similarly, for a multistep process proceeding under steady-state and quasi-steady-state conditions, it is necessary to use the rates with respect to substances (r_N) and rates for overall equations (r_p), which also characterize the whole process in the case of a single-route mechanism.

(ii) The method of routes can also be used to find the overall equations proceeding from the process mechanism, with the rates for substances being invariant with respect to the selection of the basis set of routes (see Examples 1.7 and 1.9).

(iii) The theory of routes shows that a reaction mechanism plays the determining role, while the overall equations merely follow from the mechanism. For a completely reversible process, the question as to what is the overall equation according to which product is formed is incorrect. The issue to be discussed is what a sequence of steps is, a run over which ensures the formation of this product (as described by one of the overall equations).

(iv) Formally speaking, all the reaction routes and overall equations are equivalent. At the same time, for the overall equations to have a clear chemical sense, to not contradict thermodynamically-allowed directions, and to obey the observed kinetics, it is expedient to take these considerations into account when selecting a basis set of routes. "Reasonable" overall equations are determined with allowance for the mechanism, kinetics (see Examples 1.1 and 1.2), and thermodynamic constraints (Example 1.3).

Features of the kinetics of multi-route reactions, which are related to the possibility of different conjugation of routes, will be considered in Chapter 3, which is specifically devoted to multi-route catalytic reactions (see Section 3.4).

1.1.3 Kinetic polynomial

Since the system of equations (1.34) for the condition of the steady state of steps in the general case has no solutions for nonlinear mechanisms, a new representation of this condition supplemented by the equations of material balance with respect to reaction intermediates X_i was proposed in the form of so-called *kinetic polynomial* [30–33] (see also M.Z. Lazman and G.S. Yablonsky, *Adv. Chem. Eng*., **34**, 47 (2008)).

Considering an n-step single-route catalytic reaction,

$$W_j^+ - W_j^- = \sum_j \nu_j r, \quad j = 1, 2, 3, \ldots, n;$$

$$\sum X_i = \text{const} \quad ([M]_\Sigma \text{ for solutions; 1 for surfaces}),$$

of the weights of and excluding $[X_i]$, we obtain a polynomial with respect to r, with coefficients representing polynomials of step weights (ω_j):

$$B_0 + B_1 r + B_2 r^2 + \cdots + B_k r^k = 0 \; (N = 0, 1, \ldots, k). \tag{1.62}$$

This polynomial can be used to describe the dependence of the observed reaction rate on the parameters measured and to solve the inverse problem of kinetics [34], i.e., to restore the polynomial coefficients and find the reaction rate constants k_i of steps, since the structure of coefficients $\boldsymbol{B}_N$ for the known mechanism is determined. Let us consider some examples.

Example 1.9

► For the reaction mechanism (1.14) with a set of $\boldsymbol{\nu} = (2, 1)$, we can write

$$\begin{aligned} W_1^+ - \omega_{-1}[X] &= 2r; \\ \omega_2[X]^2 - W_2^- &= r. \end{aligned} \tag{1.63}$$

Upon excluding [X], we obtain the following polynomial:

$$B_0 - B_1 r + B_2 r^2 = 0, \tag{1.64}$$

where $B_0 = k_2 k_1^2 [A]^2 - k_{-2} k_{-1}^2 [P]$; $B_1 = 4k_1 k_2 [A] + k_{-1}^2$; and $B_2 = 4k_2$. For $r = 0$, we have $B_0 = 0$ and $K_1^2 K_2 = [P_{eq}]/[A_{eq}]^2$, which is evidence for the validity of the requirements of thermodynamics in writing the polynomial. ■

Example 1.10

► For the reaction scheme

$$\begin{aligned} A + 2M &\rightleftharpoons 2X; \\ X + B &\longrightarrow M + P \end{aligned} \tag{1.65}$$

with the overall equation

$$A + 2B = 2P,$$

a system of equations (1.34) can be written as follows:

$$\begin{aligned} \omega_1[M]^2 - \omega_{-1}[X]^2 &= r, \\ \omega_2[X] &= 2r, \\ [M] + [X] &= [M]_\Sigma. \end{aligned} \tag{1.66}$$

Expressing [X] from the second equation and substituting it into the first one, we obtain the following polynomial:

$$B_0 - B_1 r + B_2 r^2 = 0, \tag{1.67}$$

where

$$B_0 = k_1 k_2^2 [M]_\Sigma^2 [A][B]^2; \quad B_1 = k_2^2 [B]^2 + 4k_1 k_2 [A][B][M]_\Sigma; \quad B_2 = 4k_1[A] - 4k_{-1}.$$

The dependence of the polynomial coefficients on $[M]_\Sigma$ and the steady-state and quasi-steady-state concentrations $[A]$, $[B]$, allows the task of estimating the rate constants to be solved. In the case of multi-route reactions, the kinetic polynomials can be written for the rates r_p over routes. ■

Example 1.11

► For the mechanism

$$\begin{aligned} A + 2Z_1 &\rightleftharpoons 2Z_2, \\ Z_2 &\rightleftharpoons Z_1 + P_1, \\ 2Z_2 &\rightleftharpoons 2Z_1 + P_2 \end{aligned}$$

with the matrix of stoichiometric numbers

$$\boldsymbol{G} = \begin{array}{c} \begin{array}{cc} \boldsymbol{\nu}^{(1)} & \boldsymbol{\nu}^{(2)} \end{array} \\ \left[\begin{array}{cc} 1 & 1 \\ 2 & 0 \\ 0 & 1 \end{array}\right] \end{array} \Rightarrow \begin{array}{ll} \boldsymbol{\nu}^{(1)} & A = 2P_1; \\ \boldsymbol{\nu}^{(2)} & A = P_2 \end{array}$$

and the system of equations (1.34) written as

$$\begin{aligned} &(1) \quad \omega_1[Z_1]^2 - \omega_{-1}[Z_2]^2 = r_1 + r_2; \\ &(2) \quad \omega_2[Z_2] - \omega_{-2}[Z_1] = 2r_1; \\ &(3) \quad \omega_3[Z_2]^2 - \omega_{-3}[Z_1]^2 = r_2; \\ &(4) \quad [Z_1] + [Z_2] = 1, \end{aligned} \tag{1.68}$$

we obtain the following polynomial with respect to r_1:

$$B_0 + B_1 r_1 + B_2 r_1^2 = 0,$$

where

$$B_0 = (\omega_{-1} + \omega_3)\omega_{-2}^2 - (\omega_1 + \omega_{-3})\omega_2^2;$$
$$B_1 = 4[(\omega_{-1} + \omega_3)\omega_{-2} - (\omega_1 + \omega_{-3})\omega_2] - (\omega_2 + \omega_{-2})^2;$$
$$B_2 = 4(\omega_3 - \omega_{-3} + \omega_{-1} - \omega_1).$$

The polynomial form can also be used to represent the dependences of concentrations $[X_i]$ and the relaxation time (used for an analysis of the relaxation regimes) on the reaction parameters [35]. ■

1.1.4 Determining the number of independent parameters in a kinetic model. The problem of identifiability of parameters

A necessary stage in any kinetic investigation is evaluating the parameters (constants) of an adopted kinetic model from experimental data, which is referred to as the *inverse problem of chemical kinetics*.

For some objective and subjective reasons, the solution of the inverse kinetic problem is a rather ambiguous procedure. Indeed, there appears to be a possibility of equally well describing the experimental data by different sets of parameters within the same mechanism [5, 36–40].

Objective factors that account for the appearance of ambiguity in solving the inverse kinetic problem are: (i) nonlinearity of adopted models with respect to their parameters (for steady-state and quasi-steady-state systems); (ii) rigidity of the sets of differential equations (for non-steady-state processes) [39]; and (iii) the deficiency of experimental data that are related to the presence of both fast stages and active intermediates whose concentrations cannot be measured [40].

Subjective factors include poor quality of experiments (large errors), absence of statistically justified experimental planning (i.e., unpredictable informativity), and inadequate choice of a functional determining the proximity of the experimentally measured and calculated values of the observable variables [39, 40].

The first question is to what extent the structure of a given model allows its parameters to be unambiguously estimated *a priori* for conditions of the proposed type of experiment, proceeding from a hypothetical mechanism? If the unambiguous determination of parameters is impossible, then it is necessary to establish which rate constants or complexes of these constants can be determined independently and to understand reasons of the non-identifiability of parameters. Possible approaches to the problem of the aperiodic identifiability of parameters are considered in [41–47].

Methodology of an analysis of the aprioric identifiability of parameters is based on the concept of a matrix of the sensitivity of a response function η (typically, the concentration C or the reaction rate r) with respect to the required rate constants k_i, that is, the Jacoby matrix, the partial derivatives of η with respect to k_i:

$$\boldsymbol{J}_{m\times S} = \left\{\frac{\partial \eta}{\partial k_i}\right\},$$

where m is the number of response functions and S is the number of rate constants.

There are two types of non-identifiability of parameters of a model [42, 43]. The *local non-identifiability* implies that (i) only some functions of the model parameters (constants) rather than all these parameters can be determined using an adopted experimental scheme, and (ii) the number of elements in a complete set of the independent parametric functions (basis set) is smaller than the number of parameters. This type corresponds to the matrix J of an incomplete rank and the corresponding model is called an incomplete-rank model (IRM). The *global non-identifiability* means that (i) we deal with a complete-rank model (CRM) such that all parameters of the model can, in principle, be determined, but (ii) the problem of determining the parameters has no single solution. Both types of non-identifiability of parameters of a model are related to the *invariance* of a measured response with respect to some transformations of the model parameters – continuous and discrete transformations for the local and global non-identifiability, respectively. Let us consider some examples of the analysis of models for the non-identifiability of parameters.

Example 1.12

► Consider a simple scheme of a two-step enzymatic or catalytic reaction, which is known as the Michaelis–Menten scheme and will be repeatedly used in what follows:

$$\mathrm{E+S} \underset{k_{-1}}{\overset{k_1}{\rightleftharpoons}} \mathrm{X_1} \xrightarrow{k_2} \mathrm{E+P}, \qquad (1.69)$$

where E is an enzyme [45]. Under quasi-steady-state conditions, the rate of the S → P conversion can be expressed as follows:

$$r = \frac{d[\mathrm{P}]}{dt} = \frac{k_1 k_2 [\mathrm{E}]_0 [\mathrm{S}]}{k_1[\mathrm{S}] + k_{-1} + k_2} = \frac{k_2[\mathrm{E}]_0[\mathrm{S}]}{[\mathrm{S}] + \dfrac{k_{-1}+k_2}{k_1}}. \qquad (1.70)$$

Apparently, the parameters to be determined in this simple case are k_2 and $K_m = \frac{k_{-1}+k_2}{k_1}$, or $\gamma_1 = \frac{k_{-1}+k_2}{k_1 k_2}$ and $\gamma_2 = \frac{1}{k_2}$. These two independent quantities result from an analysis of the non-identifiability of parameters of the given model. In this case, we deal with the local non-identifiability of an IRM type, where only two of the three columns of the $\boldsymbol{J}$ matrix are linearly independent. If the scheme of mechanism (1.69) is complicated to include steps corresponding to the conversion of intermediates,

$$\mathrm{E+S} \underset{k_{-1}}{\overset{k_1}{\rightleftharpoons}} \mathrm{X_1} \xrightarrow{k_2} \mathrm{X_2} \longrightarrow \ldots \xrightarrow{k_n} X_n \xrightarrow{k_{n+1}} \mathrm{E+P}, \qquad (1.71)$$

the number of independent parametric functions will remain unchanged:

$$\gamma_1 = \frac{k_{-1}+k_2}{k_1 k_2}; \quad \gamma_2 = \left(\frac{1}{k_2} + \ldots + \frac{1}{k_{n+1}}\right).$$

Indeed, any complication of the mechanism without involving additional measurements (only [S] or [P]) does not increase the informativity of kinetic data. ■

Example 1.13

► Let us return to Example 1.1, where the measured variables were [A] and [B] and the six elementary reactions included three pairs of the same type: (3) and (−5), (−3) and (5), (2) and (−2). This circumstance makes the system locally non-identifiable. If the equilibrium constant K of the overall reaction

$$2\mathrm{A} = 2\mathrm{B} \quad (K = K_1 K_2 K_3)$$

is known, we can obtain three linearly independent solutions (parametric functions) defined as $\gamma_1 = k_1$, $\gamma_2 = k_2 K_1$, and $\gamma_3 = k_3 K_1 K_2$. ■

The reaction schemes in Examples 1.12 and 1.13 involved steps of the same type, the weights of which had an identical form ($\omega_{-1} = k_{-1}$, $\omega_2 = k_2$, $\omega_n = k_n$ in Example 1.12). As a result, like terms and the coefficients representing sums of the products of rate constants appear in the concentration polynomial in the denominator of the equation of steady-state kinetics. These complexes of rate constants (parametric functions) are non-Arrhenius complexes of the constants in various powers [48]. If the number of Arrhenius complexes is equal to or greater than the number of parameters to be determined, then the problem of determining k_i becomes solvable, whereas if the number of these complexes is smaller than the number of k_i, there appears a local non-identifiability of parameters.

Example 1.14

► An interesting example was proposed by Yablonsky et al. [48] which is related to an analysis of a four-step mechanism of the heterogeneous catalytic conversion of methane:

$$\begin{aligned}
&(1)\quad CH_4 + Z \rightleftharpoons ZCH_2 + H_2;\\
&(2)\quad H_2O + ZCH_2 \rightleftharpoons ZCHOH + H_2;\\
&(3)\quad ZCHOH \rightleftharpoons ZCO + H_2;\\
&(4)\quad ZCO \rightleftharpoons Z + CO
\end{aligned} \tag{1.72}$$

with an overall equation of

$$CH_4 + H_2O \rightleftharpoons CO + 3H_2.$$

The conversion rate according to this overall equation can be expressed as follows:

$$r = \frac{k_1k_2k_3k_4[CH_4][H_2O] - k_{-1}k_{-2}k_{-3}k_{-4}[CO][H_2]^3}{D}, \tag{1.73}$$

where

$$\begin{aligned}
D = {} & K_1[H_2O] + K_2[CH_4] + K_3[H_2] + K_4[CH_4][H_2O] + K_5[CO][H_2]\\
& + K_6[H_2O][CO] + K_7[CH_4][H_2] + K_8[H_2]^2 + K_9[CO][H_2]^2 + K_{10}[CH_4][H_2]^2\\
& + K_{11}[H_2O][H_2][CO] + K_{12}[CH_4][H_2O][H_2] + K_{13}[H_2]^3;
\end{aligned}$$

$$K_1 = k_2k_3k_4;\quad K_2 = k_3k_4k_1;\quad K_3 = k_3k_4k_{-1};\quad K_4 = k_1k_2(k_3 + k_4);\; K_5 = k_3k_{-4}k_{-1};$$

$$K_6 = k_2k_3k_{-4};\quad K_7 = k_4k_1k_{-2};\quad K_8 = k_4k_{-1}k_{-2};\quad K_9 = k_{-4}(k_{-2}k_{-3} + k_{-1}k_{-3} + k_{-1}k_{-2});$$

$$K_{10} = k_1k_{-2}k_{-3};\quad K_{11} = k_2k_{-3}k_{-4};\quad K_{12} = k_1k_2k_{-3};\quad K_{13} = k_{-1}k_{-2}k_{-3}.$$

This scheme contains two similar reactions of desorption ($\omega_3 = k_3, \omega_4 = k_4$) and three similar reactions of H_2 with X_i ($\omega_{-1} = k_{-1}P_{H_2}, \omega_{-2} = k_{-2}P_{H_2}, \omega_{-3} = k_{-3}P_{H_2}$). Accordingly, the polynomial D includes two parameters (K_4 and K_9) of a non-Arrhenius type. Among the eleven Arrhenius complexes, we can separate eight independent parameters ($K_3, K_5, K_6, K_7, K_8, K_{10}, K_{11}, K_{12}$) according to the number of elementary reactions. The other three complexes can be represented as linear combinations of these independent parametric functions: $K_1 = K_3K_6/K_5$; $K_2 = K_3K_7/K_8$; $K_{13} = K_8K_{10}/K_7$. Complexes in the numerator are determined via those in the denominator as

$$\begin{aligned}
K^{\pm} &= (K_3K_5K_6K_7K_8K_{10}K_{11}K_{12})^{1/3} = k_1k_2k_3k_4k_{-1}k_{-2}k_{-3}k_{-4};\\
K^{+} &= K_3K_6K_7K_{12}/K^{\pm} = k_1k_2k_3k_4;\\
K^{-} &= K_5K_8K_{10}K_{11}/K^{\pm} = k_{-1}k_{-2}k_{-3}k_{-4}.
\end{aligned}$$

Note that, if all the 13 complexes were determined as independent, the solution would be ambiguous and deprived of a physicochemical sense, since the constants K_1, K_2, K_4, K_9, and K_{13} are combinations of the independent values. In complicated cases, the independent complexes should be determined using the approach described above. As will be shown below, the total number of parameters to be determined in linear mechanisms can also be evaluated using methods of the graph theory (see Section 1.3). ■

Example 1.15

► Let us consider the case of a global non-identifiability which is encountered in the investigation of non-steady-state processes [46, 47]. An analysis of the mechanism of a non-steady-state consecutive reaction

$$\mathrm{A} \xrightarrow{k_1} \mathrm{X}_1 \xrightarrow{k_2} \mathrm{X}_2 \xrightarrow{k_3} \mathrm{B} \tag{1.74}$$

shows that we deal with a CRM case in which all three constants k_i can be determined, but the solution is not single-valued.The concentrations of A and B do not vary with the time upon interchanging k_2 and k_3. If the concentration of X_2 is additionally measured, then we can also find a single solution for k_2 and k_3. An analysis showed that it is not just a symmetry in the arrangement of points in the space of parametric functions (k_2 and k_3) that accounts for the global non-identifiability [46]. There are some other discrete transformations (besides the permutation) for which the parametric functions are invariant. ■

Example 1.16

► The typical scheme of a parallel-consecutive mechanism with similar steps [46]

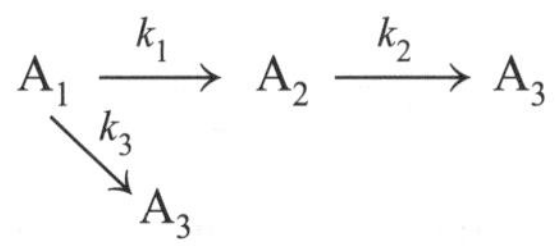

(1.75)

in which only one parameter $[A_3] = f(t)$ is measured, corresponds to a locally identifiable model with three independent parametric functions:

$$\gamma_1 = k_3, \quad \gamma_2 = k_1 + k_2 + k_3, \quad \gamma_3 = k_2(k_1 + k_3).$$

However, it was established that there is a linear discrete transformation, which provides an additional solution for the constants k_1^*, k_2^* and k_3^*:

$$k_1^* = k_2 - k_3, \quad k_2^* = k_1 + k_3, \quad k_3^* = k_3;$$
$$\gamma_1 = k_3^*, \quad \gamma_2 = k_1^* + k_2^* + k_3^*, \quad \gamma_3 = k_2^*(k_1^* + k_3^*).$$

■

It is believed that *a priori* analysis for the identifiability of parameters in kinetic models will become a necessary part of any kinetic investigation, the more so that the corresponding numerical methods of analysis are already available [49].

1.2 Quasi-steady-state and quasi-equilibrium approximations in chemical kinetics

An important method of describing the kinetics of homogeneous reactions in closed and semi-open systems is stationary concentrations, or the *quasi-steady-state approximation*, which allows kinetic models analogous to those for steady-state processes to be formulated in the case of nonstationary concentrations of reactants and products [7, 15, 50–54]. This approximation was originally used by Chapman in 1913 and then developed by Bodenstein and Semenov [50]. Known as the Bodenstein quasi-steady-state principle or condition, this approximation is now widely used in investigations of the reaction kinetics in homogeneous acid catalysis, nucleophilic and enzymatic catalysis, and homogeneous catalysis with metal complexes, as well as in polymerization, gas- and liquid-phase oxidation, chlorination, and other processes. However, this approximation is still actively discussed by physicists, chemists, and mathematicians. Following the pioneering work by Frank-Kamenetskii [55], significant contributions to deeper insights into the essence of the quasi-steady state were made by specialists in the enzymatic catalysis (see, e.g., [6]) and in many other fields [53, 57, 58] (see also monographs [7, 15] and references cited therein). A detailed analysis of this issue using a computational experiment was recently given in [59].

Under steady-state conditions (open systems) with unobservable intermediates X_i (which either do not leave a reactor that is closed for both a catalyst and these intermediates or do not enter into the equations of material balance for the catalyst and intermediates), the system obeys relations $\overline{\boldsymbol{B}}_X^T G = 0$ (1.10) and $\boldsymbol{B}_X^T \boldsymbol{W}_j = 0$ (1.33), which allows C_{X_i} to be determined for linear mechanisms (i.e., those with steps involving no more than one X_i on both left and right sides) and the expressions for r_N and r_p to be obtained. For example, in the case of a mechanism

$$\mathrm{A} \underset{k_{-1}}{\overset{k_1}{\rightleftharpoons}} \mathrm{X} \xrightarrow{k_2} \mathrm{P} \tag{1.76}$$

with the overall equation A = P for $dC_X/dt = 0$ we have $k_1 C_A - k_{-1} C_X - k_2 C_X = 0$ and

$$C_X = \frac{k_1 C_A}{k_{-1} + k_2}. \tag{1.77}$$

In this case, the rate of process (1.76) is determined by the following equation:

$$r = |r_A| = r_P = k_2 C_X = \frac{k_1 k_2 C_A}{k_{-1} + k_2}. \tag{1.78}$$

What are the experimental and theoretical criteria for the conditions $dC_{X_i}/dt \cong 0$ and $\boldsymbol{B}_X^T \boldsymbol{G} = 0$ to be applicable to a closed system, i.e., for the quasi-steady-state approximation to be valid? Is it possible to obtain equations for dC_A/dt and dC_P/dt in a closed system that would be analogous to Eq. (1.78)? Evidently, the condition that intermediates X_i are unobservable is ensured provided that $C_X \ll C_A, C_P$ [for mechanism (1.76)]. Then, the balance equation $C_{0A} = C_A + C_X + C_P$ yields $C_{0A} - C_A \cong C_P$ and, hence,

$$\left|\frac{dC_A}{dt}\right| \cong \left|\frac{dC_P}{dt}\right|, \quad k_2 C_X \cong k_1 C_A - k_{-1} C_X,$$

which implies that $dC_X/dt \cong 0$. Thus, the condition that

$$\sum C_{X_i} \ll C_A$$

is both *necessary and sufficient* for the validity of the Bodenstein principle ($dC_{X_i}/dt = 0$) or the condition of quasi-steady state in a closed system. The approximate equality

$$\left|\frac{dC_A}{dt}\right| \cong \frac{dC_P}{dt}$$

also implies that

$$\frac{dC_X}{dt} \ll \left|\frac{dC_A}{dt}\right|.$$

Differentiation of Eq. (1.77) yields the following relation:

$$\frac{dC_X}{dt} = \frac{k_1}{k_{-1} + k_2}\frac{dC_A}{dt}, \tag{1.79}$$

from which it is seen that the inequality

$$\frac{dC_X}{dt} \ll \left|\frac{dC_A}{dt}\right|$$

is possible for

$$\frac{k_1}{k_{-1} + k_2} = \varepsilon_1 \ll 1.$$

It is very important to emphasize that, in a multistep reaction, the condition $dC_{X_i}/dt \cong 0$ must be obeyed for each X_i and over a long interval of time, since in certain intervals the equality of $|r_A|$ and r_P can be provided merely by the summation of C_{X_i} derivatives with opposite signs:

$$\sum \frac{dC_{X_i}}{dt} = 0.$$

Equation (1.79) also shows that, on attaining a quasi-steady state (at $t > t_{max}$ for C_X) the rate of decrease in C_X is proportional to the rate of decrease in C_A. Under these conditions ($C_X \ll C_A, \varepsilon_1 \ll 1$), it is possible to obtain a system of equations that describes the process (1.76) proceeding in a quasi-steady-state (SS) regime [57, 58]:

$$\frac{dC_P^{SS}}{dt} = k_{eff}^{SS} C_A^{SS} \tag{1.80}$$

$$-\frac{dC_X^{SS}}{dt} = k_{eff}^{SS} C_X^{SS}; \tag{1.81}$$

$$-\frac{dC_A^{SS}}{dt} = k_{eff}^{SS} C_A^{SS}. \tag{1.82}$$

Evidently, the quasi-steady-state approximation is only applicable for $t > t_{max}$ (maximum on the $C_X = f(t)$ curve), since Eqs (1.77), (1.79) and (1.81) are not valid in the initial interval of time ($t \to 0$), where C_X vanishes rather than tends to $C_{0X} = \varepsilon C_{0A}$ [see Eq. (1.77)].

The above relationships indicate that the inequalities $C_X \ll C_A$ and $\varepsilon_1 \ll 1$ can serve, respectively, the experimental and theoretical criteria for the validity of the quasi-steady-state approximation. Now we will consider in more detail various criteria that were discussed in the literature.

1.2.1 Theoretical criteria of quasi-steady-state intermediate concentrations and quasi-equilibrium steps

Once the rate constants of steps are known, it is possible to analyze the behavior of intermediates and the applicability of the quasi-steady-state approximation using various criteria.

1.2.1.1 *Frank-Kamenetskii criteria* [55]

In order to understand the physical essence of the quasi-steady-state approximation, it is important to know one of the criteria proposed by Frank-Kamenetskii [55], which follows from Eq. (1.81) (for more detail, see [15]). According to this criterion of a quasi-steady state (SS) condition, it is necessary that the lifetime of an intermediate X (or the time to its disappearance), which can be expressed as $\tau = \frac{1}{k_{-1}+k_2}$, would be much shorter than the characteristic time τ_X of the variation of $C_X (\tau_X = 1/k_{eff}^{SS})$. The criterion $\tau \ll \tau_X$ or $k_{eff}^{SS} \ll (k_{-1} + k_2)$ is valid for $k_1 \ll (k_{-1} + k_2)$. The condition $\tau \ll \tau_X$ is analogous to the relation

$$\frac{d \ln C_X^{SS}}{dt} \ll k_{-1} + k_2 \tag{1.83}$$

[see Eq. (1.81)], since

$$\frac{d \ln C_X^{SS}}{dt} = k_{eff}^{SS}.$$

Variables C_X for the intermediates with short lifetimes are called *fast*, in contrast to *slow* variables C_A and C_P. Thus, the inequality $k_1 \ll (k_{-1} + k_2)$ that leads to the condition $\tau \ll \tau_X$ in the case under consideration is also the *necessary and sufficient* condition for the validity of the quasi-steady-state approximation.

1.2.1.2 *Wong's criterion* [56]

Let us transform the equation for dC_X/dt as written for mechanism (1.76),

$$\frac{dC_X}{dt} = k_1 C_A - (k_{-1} + k_2) C_X, \tag{1.84}$$

and write it as

$$\delta_1 = \frac{dC_X/dt}{k_{-1} + k_2} = \frac{k_1 C_A}{k_{-1} + k_2} - C_X, \tag{1.85}$$

or

$$\delta_1 + C_X = \frac{k_1 C_A}{k_{-1} + k_2}. \tag{1.86}$$

The value of δ_1 is what determines the error of the quasi-steady-state approximation (or the method of stationary concentrations). The *necessary and sufficient* condition for this approximation can be written as $\delta_1 \ll C_X$ (Wong's criterion), which also implies the validity of the Frank-Kamenetskii criterion written as

$$\frac{dC_X/dt}{k_{-1} + k_2} \ll C_X \quad \text{or} \quad \frac{d \ln C_X}{dt} \ll k_{-1} + k_2.$$

Another important conclusion following from Wong's criterion is that a negligibly small value (in comparison to C_X) is the product of dC_x/dt and a small parameter:

$$\delta_1 = \frac{dC_X}{dt} \frac{1}{k_{-1} + k_2} \ll 1 \quad (\delta_1 \to 0).$$

Alternatively, Wong's criterion can be written in terms of the relative rate Δ as

$$\Delta = \frac{dC_X/dt}{(k_{-1} + k_2) C_X} \ll 1. \tag{1.87}$$

The rate of variation of the C_X value can also be compared to the rate of formation of intermediate X (k_1C_A) [56]. The relative rate Δ_1 must obey the condition

$$\Delta_1 = \frac{dC_X/dt}{k_1C_A} \ll 1. \tag{1.88}$$

The value of Δ_1 tends to zero for $dC_X/dt \ll k_1C_A$, i.e., under conditions where the rate of variation of the C_X value is much smaller than the rate of formation of intermediate X. The Δ_1 value also provides an estimate for the error of determination of the rate of product P formation in the quasi-steady-state approximation:[4]

$$\Delta_1 = \frac{\left.\frac{dC_P}{dt}\right|_{SS} - \frac{dC_P}{dt}}{\left.\frac{dC_P}{dt}\right|_{SS}}. \tag{1.89}$$

Now let us turn to a more complicated mechanism of a two-step catalytic reaction that was considered above (see Section 1.1):

$$A + E \underset{k_{-1}}{\overset{k_1}{\rightleftharpoons}} EA, \tag{1.90}$$

$$EA \xrightarrow{k_2} P + E. \tag{1.91}$$

This mechanism is known as the Michaelis–Menten scheme since it was studied in detail by these researchers in the enzymatic catalysis [52, 56]. Here, E denotes a homogeneous catalyst such as an enzyme, metal complex, HA acid, etc. (in the heterogeneous catalysis, E is an active surface center), EA is an intermediate compound X ($C_{EA} = C_X$). Writing a differential equation for C_X as

$$\frac{dC_X}{dt} = k_1C_EC_A - (k_{-1} + k_2)C_X, \tag{1.92}$$

taking into account the material balance for the catalyst ($C_{0E} = C_E + C_X$), and assuming that $C_{0A} \gg C_{0E}$, we obtain an expression analogous to (1.85):

$$\delta_2 = \frac{dC_X/dt}{k_1C_A + k_{-1} + k_2} = \frac{k_1C_{0E}C_A}{k_1C_A + k_{-1} + k_2} - C_X. \tag{1.93}$$

For $dC_X/dt = 0$ (Bodenstein principle) or, more strictly, for $\delta_2 \ll C_X$ (Wong's criterion), this yields

$$C_X^{SS} = \frac{k_1C_{0E}C_A^{SS}}{k_1C_A^{SS} + k_{-1} + k_2}; \tag{1.94}$$

$$\frac{dC_P}{dt} = k_2C_X^{SS} = \frac{k_1k_2C_{0E}C_A^{SS}}{k_1C_A^{SS} + k_{-1} + k_2}. \tag{1.95}$$

The latter equation is more frequently written in the following form:

$$\frac{dC_P}{dt} = \frac{k_2C_{0E}C_A}{C_A + k_m}, \tag{1.96}$$

[4]Equation (1.89) was also derived by S.M. Brailovskii and A.F. Fel'dman (S.M. Fel'dman, *Diploma*, Lomonosov Institute of Fine Chemical Technology, Moscow, 1970) in the general case with an arbitrary number of intermediates X_i. The derivation of equation for Δ_1 in the case of a single intermediate X in the same study was more rigorous than in [53].

where $k_m = \frac{k_{-1}+k_2}{k_1}$ is the Michaelis constant. In this case, the lifetime of intermediate X is given by the following formula:

$$\tau = \frac{1}{k_1 C_{0A} + k_{-1} + k_2}. \tag{1.97}$$

A detailed analysis of conditions for the validity of the quasi-steady-state approximation and an estimation of the error of this approximation were performed by Wong [56]. In particular, it was shown that, for $C_{0A} \gg C_{0E}$, the error in description of the reaction rate by Eq. (1.96) tends to zero. The relative rate is

$$\Delta_2 = \frac{dC_X/dt}{k_1 C_{0E} C_A} \ll 1 \tag{1.98}$$

also tends to zero in this case for $dC_X/dt \ll k_1 C_{0E} C_A$.

1.2.1.3 *Bodenstein's criterion*

The equation $dC_X/dt = 0$ provides a *sufficient condition* for attaining a quasi-steady state, since it ensures the vanishing of all errors $(\delta_1, \delta_2, \Delta, \Delta_1, \Delta_2)$. On the other hand, it is a very rough criterion, since C_X is a finite but variable quantity and the equation

$$C_X = \frac{k_1 C_A}{k_{-1} + k_2}$$

yields relation (1.79). In addition, Eqs (1.81) and (1.82) show that

$$\frac{d \ln C_A^{SS}}{dt} = \frac{d \ln C_X^{SS}}{dt} = k_{eff}^{SS}. \tag{1.99}$$

The proximity of dC_A/dt and dC_P/dt values indicates that $dC_X/dt \ll dC_A/dt$ rather than $dC_X/dt = 0$. Therefore, the inaccurate (or even erroneous) approach (adopting the condition $dC_X/dt = 0$) leads to a correct and useful result (see [7, p. 146]) The Bodenstein criterion (condition) is useful for interpreting the quasi-steady-state approximation as the requirement of excluding the time as an explicit variable that controls the concentrations of intermediates in a quasi-steady state. The admission of the equality $dC_{X_i}/dt = 0$ in the mathematical description converts some differential equations into algebraic relations, thus making the set of initial differential equations degenerate.

1.2.1.4 *Eyring's criterion* [58]

This criterion (for brevity, referred to below as Eyring's D_E criterion) was analyzed in [58, 60]. It is based on a comparison of the rates of formation and consumption of intermediates. In particular, for the mechanism

$$A \xrightarrow{k_1} X \xrightarrow{k_2} P$$

the ratio of these rates is

$$D_E = \frac{k_2 C_X}{k_1 C_A}, \tag{1.100}$$

and for mechanism (1.76) we have

$$D_E = \frac{(k_{-1} + k_2)}{k_1} \frac{C_X}{C_A}, \tag{1.101}$$

where C_X and C_A are determined by the exact solution of the corresponding systems of differential equations. The accuracy of the quasi-steady-state approximation is determined by the degree of deviation of the D_E value from unity. In the case of $k_{-1} \gg k_2$, expression (1.101) converts into the approximate equation

$$D_E = \frac{C_X}{C_A K_1},$$

which characterizes the deviation of the C_X/C_A ratio from an equilibrium value and, hence, the degree of the system's deviation from a quasi-equilibrium state of the first step. The closer D_E to unity, the more equilibrium is the regime under consideration.

1.2.1.5 Rozovskii criterion [53]

For an analysis of factors determining the adequacy and accuracy of the quasi-steady-state approximation and for the estimation of k_i values that are necessary to ensure the required accuracy, it is suggested to use the ratio

$$D_R = \frac{C_X}{C_X^{SS}}, \tag{1.102}$$

where C_X is the value provided by an analytical solution of the system of differential equations and C_X^{SS} is the concentration of X in a quasi-steady state. An analysis of the conditions under which $D_R \to 1$ (or the D_R ratio differs from unity by a preset small value) was performed in [53]. It should be noted that, in writing expressions for C_X^{SS} in the case of mechanism (1.76), it would be more expedient to use the formula

$$C_X^{SS} = \frac{k_1 C_{0A}}{k_1 + k_{-1} + k_2} \exp\left(-k_{\text{eff}}^{SS} t\right), \tag{1.103}$$

since the exact solution $C_A = f(t)$ of the system of differential equations is not equal to C_A^{SS} expressed as

$$C_A^{SS} = \frac{(k_{-1} + k_2) C_{0A}}{k_1 + k_{-1} + k_2} \exp\left(-k_{\text{eff}}^{SS} t\right).$$

In the case of using a strict solution for C_A and C_X in the indicated mechanism, the D_R criterion is identical to D_E since

$$D_R = \frac{C_X}{C_X^{SS}} = \frac{C_X}{\dfrac{k_1}{k_{-1} + k_2} C_A} = D_E.$$

Similar to D_E, the D_R criterion for $k_{-1} \gg k_2$ characterizes the approach to a quasi-equilibrium regime. If the concentration C_X^{SS} in Eq. (1.102) is replaced by formula (1.103), we obtain a modified Rozovskii criterion $D'_R \left(D'_R = \frac{C_X}{C_{X(R)}^{SS}} \right)$, which differs from $D_R = D_E$.

1.2.1.6 Interrelation between quasi-steady-state approximation and numerical solutions of the complete system of differential equations

It has been demonstrated that, during a numerical solution of the system of differential kinetic equations by the method of successive approximations (iterative procedure) [56, 61], it is possible to improve the quasi-steady-state description by introducing additional terms. This approach reveals that the quasi-steady-state solution (or solutions) represents the first term in an approximate expansion that can be used for a numerical solution (in the region of establishment of a quasi-steady-state regime for $t > t_{max}$). Rudakov [57] convincingly demonstrated that the classical Bodenstein method based on the condition of $dC_{X_i}/dt = 0$ yields an approximate solution for k_{eff}^{SS} [see Eqs (1.80–1.82)] that is equivalent to the first term in the expansion of the exact solution (at $t > t_{max}$) for k_{eff} into a series in powers of a small parameter $\varepsilon_1 = \frac{k_1}{k_{-1}+k_2}$, and the C_A, C_X and C_P values as given by the following equations:

$$-\frac{dC_A}{dt} = k_{eff}C_A; \tag{1.104}$$

$$-\frac{dC_X}{dt} = k_{eff}C_X; \tag{1.105}$$

$$-\frac{dC_P}{dt} = k_{eff}(C_{0A} - C_P); \tag{1.106}$$

$$k_{eff} = \frac{k_1 + k_{-1} + k_2}{2} - \left[\frac{(k_1 + k_{-1} + k_2)^2}{4} - k_1 k_2\right]^{1/2}. \tag{1.107}$$

A strict mathematical analysis of the problem of quasi-steady-state was originally performed by Sayasov and Vasil'eva [62] based on the theory of singular perturbed differential equations [63]. According to this approach, the quasi-steady-state hypothesis is justified based on the presence of a small coefficient (parameter) at some derivatives in the system of differential equations (see also [7, 64]). For the reaction under consideration in a closed reactor, the role of this small parameter is played by the above ε_1 value. For example, Eq. (1.84) can be transformed into

$$\varepsilon_1 \frac{dC_X}{dt} = k_1(\varepsilon_1 C_A - C_X),$$

which leads for $\varepsilon_1 \ll 1$ to the following equation:

$$\varepsilon_1 C_A = C_X.$$

In the general case, consider a system of differential equations that can be separated into a set of slow motions $\dot{X}_i$ for the concentrations of reactants and products and a set of fast motions $\varepsilon\dot{y}$ for intermediates:

$$\begin{aligned} \dot{X}_i &= f_i(x, y), \quad i = 1, \ldots, p; \\ \varepsilon\dot{y} &= g_i(x, y), \quad i = 1, \ldots, q. \end{aligned} \tag{1.108}$$

If this system transforms for $\varepsilon \to 0$ into a degenerate system such that

$$\begin{aligned} \dot{X}_i &= f_i(x, y); \\ O &= g_i(x, y), \end{aligned} \tag{1.109}$$

then, according to the Tikhonov theorem [63], a solution of system (1.108) tends to the solution of system (1.109) as $\varepsilon \to 0$, provided that the corresponding solution for the set of fast motions $\varepsilon y = g_i(x, y)$ is robust and the initial conditions occur in the region of attraction of this solution. This theorem is valid for $t \gg \tau$, where $\tau \sim |\varepsilon \ln \varepsilon|$.

1.2.1.7 Small parameters as criteria of the Quasi-steady-state and Quasi-equilibrium approximations

Let us first consider the conditions for attaining the regime of quasi-equilibrium. In the literature, a quasi-equilibrium stage is frequently called pre-equilibrium (PE), that is, a stage preceding the limiting step [65]. This is not a strict term, since in a reversible reaction with one slow (rate-limiting) step, all steps both before and after the limiting step will be fast and quasi-equilibrium [4, 27]. Moreover, the quasi-equilibrium steps can also be involved in multi-route reactions in the absence of a limiting step. For the sake of retaining traditional notation, quantities using the PE index will refer to the quasi-equilibrium steps.

Steps involving the formation of intermediates X_i can be fast and quasi-equilibrium (PE) for both observable and unobservable intermediates. For the reaction mechanism (1.76), the ratio $C_X/C_A = \alpha$ tends to $K_1 = k_1/k_{-1}$, and the quasi-equilibrium concentration of X is

$$C_X^{PE} = \frac{K_1}{1+K_1}(C_{0A} - C_X^{PE}), \tag{1.110}$$

so that

$$\frac{dC_P^{PE}}{dt} = k_2 C_X^{PE} = \frac{k_2 K_1}{1+K_1}(C_{0A} - C_P^{PE}) = k_{eff}^{PE}(C_{0A} - C_P^{PE}).$$

It can be readily shown [59] that, under quasi-equilibrium conditions at the first step, the rate of consumption of intermediate X (at $t > t_{max}$ for C_X) is

$$-\frac{dC_X^{PE}}{dt} = k_{eff}^{PE} C_X^{PE}. \tag{1.111}$$

Since we have $C_{0A} - C_P^{PE} \cong C_X^{PE}$ for large K_1 ($K_1 > 10$), it is evident that,

$$\frac{dC_P^{PE}}{dt} \cong \left|\frac{dC_X^{PE}}{dt}\right|. \tag{1.112}$$

This is a manifestation of the difference between a quasi-equilibrium regime with observable large $C_X (\alpha > 10)$ and a quasi-steady-state regime. For $K_1 \gg 1$, we obtain $k_{eff}^{PE} = k_2$; while for $K_1 \ll 1$ (unobservable intermediates X_i), we have $k_{eff}^{PE} = K_1 k_2$, and relation (1.112) is no longer valid. Analogous expressions for k_{eff} are obtained from Eqs (1.78) and (1.80) for $k_{-1} \gg k_2$. Under these conditions, both quasi-steady-state and quasi-equilibrium regimes can exist. The equality of rates of the first step in the forward and reverse directions essentially implies that we admit the quasi-steady-state approximation with respect to the reactant, that is, $dC_A/dt \cong 0$ [57]. By analogy with the criterion of quasi-steady-state, one can readily obtain an expression for the quasi-equilibrium criterion δ_{PE} as

$$\delta_{PE} = \frac{-dC_A/dt}{k_1 C_A} = \frac{k_2}{k_1 + k_{-1} + k_2}. \tag{1.113}$$

Evidently, we have $\delta_{PE} \ll 1$ for

$$\frac{k_2}{k_{-1} + k_1} = \varepsilon_2 \ll 1.$$

The ratio of constants at which $\varepsilon_2 \ll 1$ provides for a quasi-equilibrium regime. In cases where $\varepsilon_1 \ll 1$ and $\varepsilon_2 \ll 1$, both quasi-steady-state and quasi-equilibrium conditions are satisfied and, hence, the C_X and C_A concentrations correspond to a quasi-steady and quasi-equilibrium state.

Here, it is expedient to consider the problem of reversible fast and slow steps. Under steady-state and quasi-steady-state conditions, the rate of each step is equal to the rate of a single-route reaction (see the condition of the steady state of steps in Section 1.1):

$$W_j^+ - W_j^- = r, \tag{1.114}$$

or (for a single-route reaction with a nonlinear mechanism)

$$W_j^+ - W_j^- = \nu_j r. \tag{1.115}$$

In order to determine the slow and fast steps, the rate of the jth step in the forward (or reverse) direction is compared to the value of r (or $\nu_j r$). If $W_j^{\pm} \gg r$, this step is called fast and quasi-equilibrium so that $W_j^+ \cong W_j^-$. If W_j^+ is comparable with r [see Eq. (1.114)], the step is considered as slow (for an irreversible step, $W_j^+ = r$). If there is a single slow step in the mechanism, it is called the *rate-controlling* or *limiting* step. The first step is limiting in irreversible reaction provided that $W_1^+ = r$.

The rate of a two-step reaction (1.76) under the quasi-steady-state conditions is given by the formula

$$r = \frac{W_1^+ W_2^+}{W_1^- + W_2^+}. \tag{1.116}$$

If this is supplemented by the condition of quasi-equilibrium ($W_1^+ \gg W_2^+, W_2^+ = r$), then $W_1^- \cong W_1^+$ and

$$r = \frac{W_1^+ W_2^+}{W_1^+ + W_2^+}, \tag{1.117}$$

which implies that

$$\frac{1}{r} = \frac{1}{W_1^+} + \frac{1}{W_2^+}, \tag{1.118}$$

or in the general case (quasi-equilibrium for all X_i),

$$\frac{1}{r} = \sum_j \frac{1}{W_j^+}, \tag{1.119}$$

where r is the rate of a single-route reaction (for the overall equation) with a linear mechanism.

An analysis of the relations between constants k_1, k_{-1}, and k_2 that result in $\varepsilon_1 \ll 1$ and $\varepsilon_2 \ll 1$, led to a conclusion [65] that, at a significant (not less than tenfold) difference in k_i, there are only six variants (Nos. I–VI) of relations between constants and four cases of the validity of each or both of the quasi-steady-state ($\varepsilon_1 \ll 1$) and quasi-equilibrium ($\varepsilon_2 \ll 1$) conditions (Table 1.1).

As can be seen, *strong conditions* determining the implementation of a quasi-steady-state regime are $k_2 \gg k_1, k_{-1}$ (I, II) and $k_{-1} \gg k_1, k_2$ (V, VI), which make $\varepsilon_1 \ll 1$ in the former case the result of a rapid conversion of X, and in the latter case due to a quite small $K_1 = k_1/k_{-1}$. There are three variants (I, II, V) that coincide with the "first condition" ($k_2 \gg k_1$) of quasi-steady state according to Rozovskii [53], while variant VI corresponds to the "second condition" ($k_{-1} \gg k_2$) supplemented by the condition $k_{-1} \gg k_1$. Note that writing the second condition as $k_1, k_{-1} \gg k_2$ [53] is incorrect, since the equality of k_{-1} and

Table 1.1 Relationship between rate constants and regimes of process (1.76)

Variant No.	Relations between k_i	$\varepsilon_1 = \dfrac{k_1}{k_{-1}+k_2}$	$\varepsilon_2 = \dfrac{k_2}{k_{-1}+k_1}$	Regime	Limiting step
I	$k_2 \gg k_1 \gg k_{-1}$	$\ll 1$	$\gg 1$	SS	1
II	$k_2 \gg k_{-1} \gg k_1$	$\ll 1$	$\gg 1$	SS	1
III	$k_1 \gg k_2 \gg k_{-1}$	$\gg 1$	$\ll 1$	PE*	2
IV	$k_1 \gg k_{-1} \gg k_2$	$\gg 1$	$\ll 1$	PE	2
V	$k_{-1} \gg k_2 \gg k_1$	$\ll 1$	$\ll 1$	SS, PE	2
VI	$k_{-1} \gg k_1 \gg k_2$	$\ll 1$	$\ll 1$	SS, PE	2

Note: *Conclusion made in [65]; SS = (quasi)-steady state; PE = quasi-equilibrium (pre-equilibrium).

Table 1.2 Relationship between rate constants and regimes

Variant No.	Close parameters	Relations between k_i	ε_1	ε_2	Regime	Limiting steps
VII (see I)	$k_1 \approx k_{-1}$	$k_2 \gg k_1 \approx k_{-1}$	$\ll 1$	$\gg 1$	SS	1
VIII (see IV)	$k_1 \approx k_{-1}$	$k_2 \ll k_1 \approx k_{-1}$	≈ 1	$\ll 1$	PE	2
IX (see II)	$k_2 \approx k_{-1}$	$k_1 \ll k_{-1} \approx k_2$	$\ll 1$	≈ 1	SS	1
X (see III)	$k_2 \approx k_{-1}$	$k_1 \gg k_{-1} \approx k_2$	$\gg 1$	$\ll 1$	PE*	2
XI (see V, VI)	$k_1 \approx k_2$	$k_{-1} \gg k_1 \approx k_2$	$\ll 1$	$\ll 1$	SS, PE	2

*For $k_1/k_{-1} \geq 100$.

k_1 makes the quasi-steady-state regime ($\varepsilon_1 \approx 1$) impossible (see below). The quasi-steady-state regime is possible only provided that $k_{-1} \gg k_1$. This is the *only possible variant of a quasi-steady-state regime* for $k_1 \gg k_2$. Variant IV is also a non-steady-state regime despite the fact that k_1, $k_{-1} \gg k_2$: this is a quasi-equilibrium regime with the second limiting stage. These conclusions were fully confirmed by the results of computer experiments [59]. Variant III is also a quasi-equilibrium regime ($\varepsilon_2 \ll 1$), but only at a large difference between all constants. At a tenfold difference between constants, the quasi-equilibrium regime in this variant is established later than in other cases, but with a higher yield of product P at the final step. This variant in fact reduces to a process with the irreversible first step under the conditions of almost complete and rapid conversion of A into X, that is, to virtually a single-step conversion reaction (see below).

The aforementioned six special cases are expediently supplemented by the variants of implementation of the quasi-steady-state and quasi-equilibrium regimes under conditions of equal (close) constants.[5] The case where all constants are close ($k_1 \approx k_{-1} \approx k_2$) does not obey the criteria of quasi-steady-state and quasi-equilibrium conditions. The equality of constants in pairs adds five cases (Table 1.2).

In the case of close k_2 and k_{-1} values (as well as in variant III), the regime is not strictly quasi-equilibrium (the ratio $C_X/C_A = \alpha$ is not constant during the process). The equality of k_1 and k_2 (variant XI) ensures (in contrast to the statement made in [53]) the quasi-steady state under quasi-equilibrium conditions ($D_E, D_P \cong 1, \Delta \cong 0$) [59].

Thus, the quasi-steady-state regime is attained in the following cases:

- for $k_2 \gg k_1$ (variants I, II, V, VII, VIII, IX);
- for $k_1 \gg k_2$ (variant VI);
- for $k_1 \approx k_2$ (variant XI).

[5]The expediency of this analysis was pointed out by V.F. Shvets.

The quasi-equilibrium approximation is valid in the following cases:

- for $k_2 \gg k_1$ (variant V);
- for $k_1 \gg k_2$ (variants IV, VI, VIII and III, X with large values of k_1/k_{-1});
- for $k_1 \approx k_2$ (XI).

Let us consider in more detail a difference between variants of the quasi-equilibrium approximation (see Tables 1.1 and 1.2) that leads to the appearance of k_2 in the kinetic equation and makes the second step limited in mechanism (1.76). Three variants (V, VI and XI) are characterized by small equilibrium constants ($K_1 \ll 1$), quasi-equilibrium first stage, quasi-equilibrium low concentrations of X, second limiting step, and the kinetic equation written in the following form:

$$r = \frac{k_2 K_1 [A_0 - P]}{1 + K_1} \cong k_2 K_1 [A_0 - P]. \tag{1.120}$$

Note that, for the indicated variants, the quasi-equilibrium regime is also quasi-steady-state with respect to X.

In two cases (regimes IV and VIII), the quasi-equilibrium concentrations of X are large and the reaction rate is described by the following equations:

$$r = k_2 [A_0 - P] \tag{1.121}$$

(for variant IV, $K_1 \gg 1$, $[A_0 - P] \cong [X]$) and

$$r = \frac{k_2 K_1 [A_0 - P]}{1 + K_1} \tag{1.122}$$

(for variant VIII, $K_1 \cong 1, [A_0 - P] \cong [A] + [X]$).

Variants III and X lead formally to the same result as does variant IV, but the concentrations of X are not quasi-equilibrium. Variant III with $k_1 \gg k_2 \gg k_{-1}$ ($K_1 \gg 1$) represents a regime with a fast irreversible first step proceeding under conditions $[A_0 - P] \cong [X]$, followed by the rapid transition to a single-step reaction with the rate given by Eq. (1.121). Variant X with $k_1 \gg k_2 \approx k_{-1}$ is closer to equilibrium, step (1) is reversible, but the concentration of X is also not quasi-equilibrium. In this case, $r = k_2[A]$ for $[A] = [A_0 - P] - [X]$.

For a catalytic reaction proceeding under quasi-steady-state conditions ($[A] \gg [M]_\Sigma$, where M is a catalyst), according to a two-step scheme

$$A + M \underset{k_{-1}}{\overset{k_1}{\rightleftharpoons}} X \xrightarrow{k_2} M + P$$

it is possible to separate three cases (with all constants being different).

I. Fast reversible step (1) with $K_1 \ll 1$ ($k_{-1} \gg k_2 \gg k_1[A]$ and $k_{-1} \gg k_1[A] \gg k_2$).

Here, the limiting step is (2) with

$$r = k_2 K_1 [A][M]_\Sigma;\ [M]_\Sigma \cong [M]. \tag{1.123}$$

II. Fast reversible step (1) with $k_1[A] \gg k_{-1} \gg k_2$ or fast irreversible step (1) with $k_1[A] \gg k_2 \gg k_{-1}$ and $K_1 \gg 1$.

Here, the limiting step is (2) with

$$r = k_2[M]_\Sigma;\ [M]_\Sigma \cong [X]. \qquad (1.124)$$

III. Fast step (2) and almost irreversible step (1) with $K_1 \gg 1 (k_2 \gg k_1[A] \gg k_{-1})$ or $K_1 \ll 1 (k_2 \gg k_{-1} \gg k_1[A])$.

In this case, the limiting step is (1) and

$$r = k_1[A][M]_\Sigma;\ [M]_\Sigma = [M].$$

Thus, the quasi-equilibrium step (1) as well as a fast irreversible step (1) with $K_1 \gg 1$ (case II) lead formally to the same kinetic equation (1.124).

The results presented in Tables 1.1 and 1.2 can be generalized by determining the regions where the reaction proceeds in a quasi-steady-state regime and/or under quasi-equilibrium conditions when constants are varied for preset ε_1 and ε_2 values.[6] Equations for ε_1 and ε_2 yield the following relations:

$$\frac{k_{-1}}{k_1} = \frac{1}{\varepsilon_1} - \frac{k_2}{k_1} \text{ and } \frac{k_{-1}}{k_1} = \frac{k_2}{k_1}\frac{1}{\varepsilon_2} - 1.$$

For preset ε_1 and ε_2 values, these equations determine straight lines that bound the regions of validity of the quasi-steady-state and/or quasi-equilibrium approximations. For example, with $\varepsilon_1 = 0.1$ and $\varepsilon_2 = 0.1$ [where $\varepsilon_1 \leq 0.1$ is the condition for quasi-steady-state (SS) regimes and $\varepsilon_2 \leq 0.1$ is the quasi-equilibrium (PE) condition], we obtain two straight lines (Figure 1.1) in the coordinates of $y = k_{-1}/k_1$ and $x = k_2/k_1$.

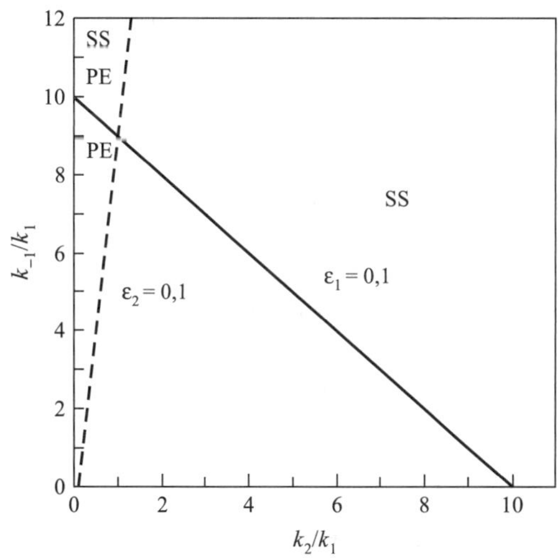

Figure 1.1 Domains of quasi-steady-state (SS) and quasi-equilibrium (pre-equilibrium, PE) regimes for mechanism (1.76)

[6] This analysis was proposed by K.Yu. Odintsov.

In Figure 1.1, a region above the solid line corresponds to quasi-steady states ($\varepsilon_1 \leq 0.1$), while a region on the left from dashed line corresponds to quasi-equilibrium states. A region between these lines above their intersection point corresponds to the validity of both quasi-steady-state (SS) and quasi-equilibrium (PE) approximations.

In [66], the quasi-steady-state condition was considered using the conventional Bodenstein criterion of $dC_{X_i}/dt = 0$, but it was ascertained that the concentrations of intermediates X_i differ only very slightly from the equilibrium values. However, this statement in the general case is incorrect. Small deviations from the equilibrium values of $[X_i]$ take place only in the case of simultaneous validity of the quasi-steady-state and quasi-equilibrium approximations. The condition of quasi-steady-state regimes can be valid far from the equilibrium – and this situation is frequently encountered in practice ($W_j^+ - W_j^- = r$). In this case, Eq. (1.119) is inapplicable and it cannot be considered as a basic result of the quasi-steady-state approximation as stated in [66]. Fully reversible chemical reactions that take place near the equilibrium are very rarely investigated. Note also that a system can occur close to the equilibrium, while not obeying quasi-steady-state criterion, for example, in the case of reaction (1.76) with $K_1 \gg 1$ and large $C_X (C_X/C_A = \alpha > 1)$.

1.2.2 Experimental criteria of applicability of quasi-steady-state approximation in various systems

The theoretical criteria of quasi-steady and quasi-equilibrium states provide deeper insight into the nature of these approximations and, if the model constants are known, allow the validity of approximations used for the processing of data to be assessed. On the other hand, for the correct planning and description of a kinetic experiment, it is also important to know the experimental criteria of applicability of the quasi-steady-state approximation.

- The above analysis shows that the *necessary and sufficient condition* of validity of the quasi-steady-state approximation is expressed by the relations $C_X \ll C_A, C_P$ or

$$C_X/C_A = \alpha \ll 1. \tag{1.125}$$

Equations (1.80)–(1.82) also imply the proximity of the rates of product formation and reactant consumption (for the indicated reaction scheme),

$$\left|\frac{dC_A}{dt}\right| \cong \frac{dC_P}{dt}, \tag{1.126}$$

as well as the condition $dC_X/dt \cong 0$ and the validity of criterion $\varepsilon_1 \ll 1 \left(\varepsilon_1 = \frac{k_1}{k_{-1}+k_2}\right)$.

Condition (1.125) can be experimentally verified by an analysis of the material balance of the process under consideration:

$$C_{0A} \cong C_A + C_P. \tag{1.127}$$

Violation of the material balance as expressed by an equation of the (1.127) type may be indicative of either the presence of unidentified side products or the deviation from condition (1.125). It should be borne in mind that the employed analytical schemes must ensure the determination of C_A without decomposing X in cases where the formation of X is a reversible step.

As was demonstrated in Section 1.1, the exclusion of concentrations C_{X_i} (of Bodenstein's intermediates) from the material balance and overall equations (containing only the observable substances) is ensured by the condition of the steady state of steps and the Horiuti stoichiometric rule. Relation (1.126) can also be quite readily checked by experiment. If this condition is valid over a sufficiently large time interval (e.g., corresponding to 70–90 % of the product yield recalculated for C_{0A}) with a preset accuracy, so that the error (*delta*) defined as

$$delta = \frac{\left|\frac{dC_A}{dt}\right| - \frac{dC_P}{dt}}{\left|\frac{dC_A}{dt}\right|} \quad (1.128)$$

does not exceed the experimental uncertainty (e.g., 5–10 %), then it is possible to speak of the applicability of the quasi-steady-state approximation. It should be noted that criterion (1.128) is applicable to any type of reaction mechanisms.

- In the case of *homogeneous catalytic reactions* [e.g., such as described by equations (1.90) and (1.91)], concentration C_X cannot exceed C_{0E}. For this reason, such systems have to be studied for $C_{0A} \gg C_{0E}$. According to Wong's estimation [56], the error Δ_2 of the quasi-steady-state approximation for $C_{0A}/C_{0E} > 15$ does not exceed 1 %. For $C_{0A}/C_{0E} < 10$, the applicability of this approximation is doubtful (probably only being valid for very large k_2/k_1 ratios) and the reaction kinetics should be described by differential equations.
- In the case of *heterogeneous catalysis* in closed (and open – see, e.g., [7]) systems, the amount of reactants in the gas phase is usually much greater than the amount (number of moles) of intermediates (C_X^s) occurring on the surface of a catalyst and, hence, a regime with $dC_X^s/dt = 0$ is also attained. In this case, the role of a small parameter is played by the ratio

$$\frac{N_{tot}^s}{N_{tot}^g} = \varepsilon < 1,$$

where N_{tot}^s and N_{tot}^g are the total amounts (number of moles) of intermediates on the catalyst surface and in the gas phase within the reactor volume.

The analysis of *enzymatic reactions* provided additional experimental criterion of the applicability of the quasi-steady-state approximation. According to Eq. (1.95), there must be a linear relationship between the reaction rate and the initial concentration of an enzyme. These relations are generally typical of the enzymatic catalysis [52, 56]. However, on replacing C_X^{SS} in Eq. (1.95) by C_X expressed from Eq. (1.93), we obtain the following equation for the reaction rate:

$$\frac{dC_P}{dt} = \frac{k_1 k_2 C_{0E} C_A - k_2 \dfrac{dC_X}{dt}}{k_1 C_A + k_{-1} + k_2}, \quad (1.129)$$

which indicates that the reaction rate in this case is not a linear function of C_{0E}. Although the homogeneous catalysis encounters many other factors that also lead to nonlinearity of the relation $r = f(C_{0E})$, this circumstance should also be borne in mind when considering the problem of applicability of the quasi-steady-state approximation. *A linear dependence of the reaction rate on the catalyst concentration is evidence in favor of the validity of the quasi-steady-state approximation within the limits of experimental errors.*

Kinetic investigations are frequently performed using the *approximation of initial reaction rates* (r_i for $t \to 0$). These rates are determined from the temporal variation of concentrations C_A and C_P and then it is assumed that the obtained values of $r_{0A} = \left|\frac{dC_A}{dt}\right|_{t\to 0}$ and $r_{0P} = \left|\frac{dC_P}{dt}\right|_{t\to 0}$ correspond to a quasi-steady-state. Since the C_X value passes through a maximum, and the quasi-steady state is attained for $t > t_{max}$ (upon a certain transient regime), the extrapolation of dC_P/dt to $t = 0$ can lead to an error in determining r_{0P}, which increases with decreasing C_{0A}/C_{0E} and C_A/C_E ratios. For $C_{0A}/C_{0E} < 10$, the method of initial reaction rates is not recommended for use. For the Michaelis–Menten scheme, the relative error δ_C due to the neglect of the transient regime depends on the lifetime τ of intermediate X ($\delta_C = \exp(-t/\tau)$ [56]). A rational method of determining the initial reaction rates r_{0P} for simple mechanisms was proposed by Boeker [67].

1.3 Methods of graph theory in chemical kinetics and in theory of complex reaction mechanisms

The application of the main concepts and methods of graph theory to chemical kinetics and theory of chemical reaction mechanisms in the past five decades constituted a special branch of mathematical chemistry [48, 68, 69] (see also review [6]). Graph-theoretical investigations enable specialists to carry out the following:

(i) construct graphs for reaction mechanisms of any type [48, 68–70] (see also reviews [6, 7]) and reveal the structural (topological) basis of the concept of "reaction mechanism" [8, 72, 73];
(ii) classify, encode, and enumerate the mechanisms of complex reactions [6, 74–79];
(iii) develop a nomenclature of reaction mechanisms [6, 18, 72, 73, 80–83];
(iv) obtain kinetic equations for linear mechanisms under steady-state and quasi-steady-state conditions [6, 7, 15, 70, 76, 84–104];
(v) analyze the identifiability of parameters (and determine the number of independent parameters) for an adopted kinetic model) [98];
(vi) study and predict the dynamic behavior and critical phenomena in a system under consideration [7, 98, 105–109];
(vii) assess the complexity of mechanisms [6, 78, 110–112];
(viii) formulate mechanistic hypotheses on graphs [6, 113, 114].

The possibility of applying graphs to the study of mechanisms of complex reactions was originally pointed out by Balandin [80, 115]. Balandin was also the first to apply the elements of topology to reaction mechanisms and proposed the first classification of mechanisms based on their topology [80]. Next, Christiansen [81] used graphs to classify the reaction mechanisms and King and Altman [84] gave a graphical interpretation of the Cramer method for solving systems of linear algebraic equations and used it to derive the kinetic equations of enzymatic processes.

The active use of graphs in chemical and enzymatic kinetics was initiated by Temkin [70] in his work devoted to planar cyclic graphs. Nonplanar graphs were originally used by Balaban et al. [116] for enumerating the possible reactions of isomerization of pentasubstituted ethylcation. At the same time, Volkenstein and Goldstein [86–88] proposed a method (known as the Volkenstein–Goldstein algorithm) for deriving kinetic equations using Temkin's kinetic graphs (KGs), which was later justified [94] and used to obtain a number of important results for the kinetics of steady-state reactions as described in monographs [6, 7, 98].

1.3.1 Linear mechanisms

As noted above, the class of *linear mechanisms* contains those for which the reactions of all steps involve no more than one intermediate substance on both the left and right sides of the corresponding equations. The rates of these steps are either independent of or linearly depend on the concentrations of intermediates. Linear mechanisms are naturally described by Temkin's KGs, the vertices of which represent intermediates and the edges (arcs, links) connecting these vertices correspond to reaction steps. For example, a two-route

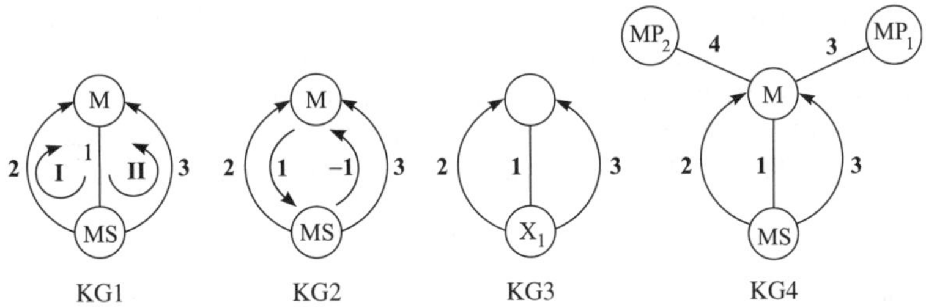

Figure 1.2 Variants of kinetic graphs (KGs) for linear mechanisms

mechanism of the catalytic reaction

$$\begin{aligned}&(1)\quad M+S \rightleftharpoons MS;\\&(2)\quad MS \longrightarrow M+P_1;\\&(3)\quad MS \longrightarrow M+P_2\end{aligned}\qquad(1.130)$$

(where M is a catalyst) can be presented in the form of KG1 (Figure 1.2), in which directed edges (with arrows) correspond to irreversible reaction steps and undirected edges denote reversible steps. A undirected edge can alternatively be depicted by two oppositely directed edges (variant KG2, Figure 1.2), but – for the sake of simplicity – it is more convenient to use variant KG1 [and bear in mind the reversibility of the corresponding edge (in this case, step 1)]. In the case of noncatalytic reactions, Temkin proposed to use the concept of *null-species* (0-species) representing a hypothetical intermediate with a concentration equal to unity ($[X_0] = 1$). Thus, cyclic graphs can be used to represent linear mechanisms of arbitrary complex reactions.

For example, the mechanism

$$\begin{aligned}&(1)\quad A+B \rightleftharpoons X_1;\\&(2)\quad X_1+C \longrightarrow P_1;\\&(3)\quad X_1+D \longrightarrow P_2\end{aligned}\qquad(1.131)$$

and the equivalent scheme

$$\begin{aligned}&(1)\quad X_0+A+B \rightleftharpoons X_1;\\&(2)\quad X_1+C \longrightarrow P_1+X_0;\\&(3)\quad X_1+D \longrightarrow P_2+X_0\end{aligned}\qquad(1.132)$$

can be depicted by a graph (KG3, Figure 1.2) with an empty vertex representing null-species X_0.

The mechanisms of both catalytic and noncatalytic reactions can include steps involving the formation of compounds of catalysts and/or intermediates with reactants, products, ligands and other components of the reaction medium, which do not enter into the stoichiometry of overall equations but contribute to the material balance with respect to the catalysts and reactants (see Section 1.1). For example, mechanism (1.130) can be supplemented by the reactions

$$\begin{aligned}&(4)\quad M+P_1 \rightleftharpoons MP_1;\\&(5)\quad M+P_2 \rightleftharpoons MP_2.\end{aligned}$$

Such steps and the corresponding compounds are depicted on the graphs by "pendant" vertices, since compounds MP_1 and MP_2 do not belong to intermediates (variant KG4, Figure 1.2).

Using the KGs of linear mechanisms, it is possible to establish the number of linearly independent routes, since the basis set of routes corresponds to the number of linearly independent (simple) cycles of the graph as determined by the so-called cyclomatic number (Euler's characteristic) F defined as [70]

$$F = q - \rho + C, \tag{1.133}$$

where $q = S$ is the number of edges (reaction steps), $\rho = I$ is the number of vertices (intermediates) (plus an empty vertex with the null-species, where necessary), and C is the number of graph components ($C = 1$ for KGs). Note that Eq. (1.133) is equivalent to Eq. (1.13). In the case of planar graphs (i.e., graphs that can be arranged on a plane), the number of simple cycles is equal to the number of KG faces (e.g., two for KG1 in Figure 1.2).

Possible ways of connecting cycles in KGs (topological characteristics) were used as a basis for the classification of linear multi-route mechanisms, their encoding, and enumeration [6, 74–79]. For example, in cases where the number of linearly independent routes is $P = F = 2$, there are only three topologically different ways of connecting cycles (routes) and, accordingly, three classes of mechanisms: class A, whereby cycles are connected by a bridge consisting of one or several steps; class B, in which cycles are shearing a vertex (common intermediate); and class C, in which cycles have one or several common steps (see Figure 1.3, which shows examples with minimum numbers of vertices).

For $P > 2$, there appears one more way of connecting cycles – class Z, in which unadjacent cycles are connected via other cycles [6, 78]. In the case of three-route reactions, it is possible to distinguish 15 topologically different classes of graphs based on classes A, B, C and $Z(n - \mathrm{A}^i\mathrm{B}^j\mathrm{C}^k\mathrm{Z}^v)$:

$$3-\mathrm{A}^2, 4-\mathrm{A}^2\mathrm{Z}, 3-\mathrm{ABZ}, 4-\mathrm{ACZ}, 2-\mathrm{B}^2\mathrm{Z}, 3-\mathrm{BCZ}, 4-\mathrm{C}^2\mathrm{Z},$$
$$2-\mathrm{A}^2\mathrm{B}, 3-\mathrm{A}^2\mathrm{C}, 2-\mathrm{B}^2\mathrm{C}, 2-\mathrm{BC}^2, 4-\mathrm{C}^3, 4-\mathrm{A}^3, 1-\mathrm{B}^3, 3-\mathrm{BC}^3,$$

where n is the number of vertices with degrees above 2. Tables of graphs with $P = 2, 3$, and 4 are presented in [6, 78]. Using this classification, a code has been developed that ensures single-valued identification of KGs.

Now let us introduce some important definitions and consider algorithms developed in order to derive kinetic equations for linear mechanisms based on methods of graph theory. A *graph's cycle* is any sequence

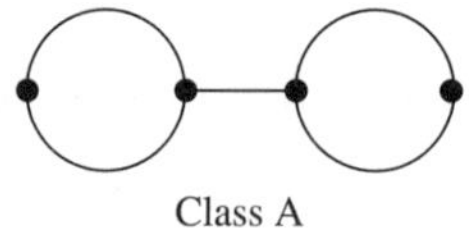

Class A

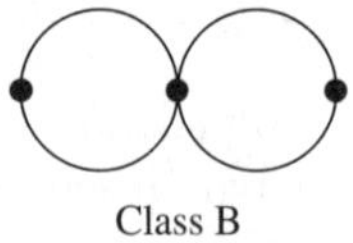

Class B

Class C

Figure 1.3 Classes of two-route mechanisms

of directed edges that originate from and terminate in the same vertex. A KG cycle corresponds to a cyclic conversion of intermediates. The *weight* of a cycle is defined as the product of the weights of edges (elementary reactions) entering into this cycle:

$$C^+ = \prod_j \omega_j; \quad C^- = \prod_j \omega_{-j}. \tag{1.134}$$

It should be recalled that the edge weight (i.e., step weight) is equal to the rate of the jth step in one direction divided by the concentration of the ith intermediate participating in this step

$$\omega_{\pm j} = W_j^{\pm}/[X_{ij}]. \tag{1.135}$$

Note that, for $[X_i] = [X_0] = 1$, we have $\omega_{\pm j} = W_j^{\pm}$. For KG1 and KG2 (Figure 1.2), the weights of cycles, including steps 1, 2 and 1, 3 of mechanism (1.130), are as follows:

$$C_1^+ = \omega_1\omega_2; \quad C_2^+ = \omega_1\omega_3.$$

The directions of cycles on KGs are chosen in accordance with the directions of routes which, in turn, are determined by the directions of steps and the vector of stoichiometric numbers. The directions of two cycles for two routes on KG1 (Figure 1.2) is indicated by arrows.

A *tree* is any sequence of edges on a graph which contains no cycles. The *maximum tree* (or frame) is a sequence of edges that passes through all vertices (and contains no cycles). A *rooted tree* or a tree with root at the ith vertex (frame of the ith vertex) is a maximum tree with all edges directed toward ith vertex. For graph KG5 (Figure 1.4) of a two-route catalytic reaction, the possible rooted trees are presented in Figure 1.5.

Let us define the *weight* D_{ik} of the kth rooted tree at the ith vertex as a product of the weights of edges entering into this tree:

$$D_{ik} = \prod_j \omega_j \ (j \in \{i, k\}). \tag{1.136}$$

A *root determinant* D_i of the ith vertex is a sum of the weights of rooted trees (sum of frame weights) at the ith vertex:

$$D_i = \sum D_{ik}. \tag{1.137}$$

Several methods have been proposed for determining D_i (and all D_{ik}) values [90, 101, 117–120]. A simple algorithm [117, 118] consists of the following operations. First, D_i^* is determined as a product of

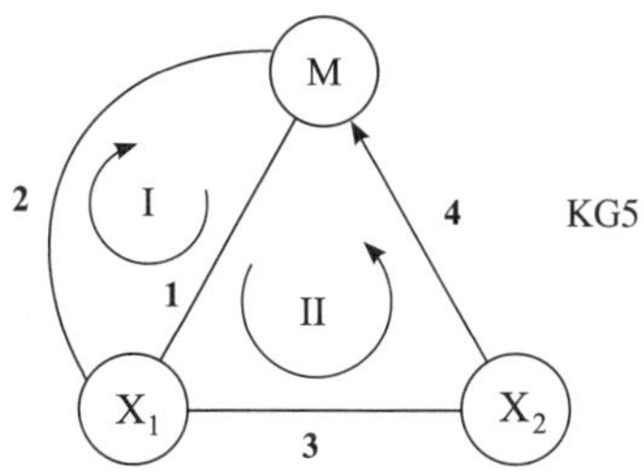

Figure 1.4 Kinetic graph for a two-route catalytic reaction

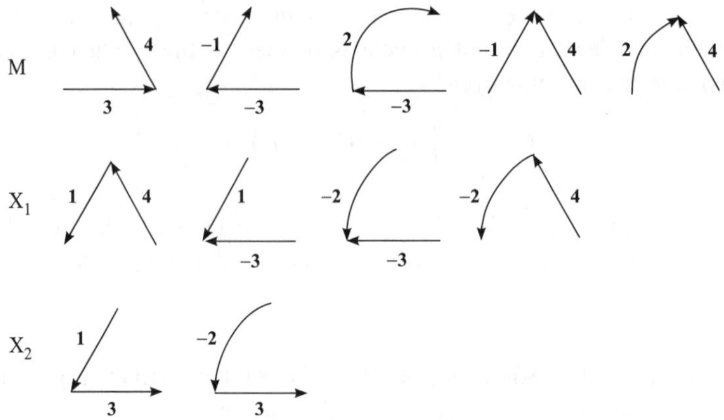

Figure 1.5 Sets of rooted trees at vertices M, X_1 and X_2 of KG5 (Figure 1.4)

the weights of edges originating from all vertices except the ith one. For example, for vertex M in KG5 (Figure 1.4) we have

$$D^*_{M} = (\omega_{-1} + \omega_{-2} + \omega_3)(\omega_{-3} + \omega_4)$$
$$= \omega_{-1}\omega_{-3} + \omega_2\omega_{-3} + \omega_3\omega_{-3} + \omega_{-1}\omega_4 + \omega_{-2}\omega_4 + \omega_3\omega_4.$$

Then, a product of the weights of edges that form a cycle (including the product of weights of the forward and reverse steps such as $\omega_3\omega_{-3}$) is excluded from D^*_{M}, which yields

$$D_{M} = \omega_3\omega_4 + \omega_{-1}\omega_{-3} + \omega_2\omega_{-3} + \omega_{-1}\omega_4 + \omega_2\omega_4;$$
$$D^*_{X_1} = (\omega_1 + \omega_{-2})(\omega_{-3} + \omega_4) = \omega_1\omega_{-3} + \omega_1\omega_4 + \omega_{-2}\omega_{-3} + \omega_{-2}\omega_4;$$
$$D^*_{X_1} = D_{X_1};$$
$$D^*_{X_2} = (\omega_1 + \omega_{-2})(\omega_{-1} + \omega_2 + \omega_3) = \omega_1\omega_{-1} + \omega_{-1}\omega_{-2} + \omega_1\omega_2 + \omega_2\omega_{-2} + \omega_{-2}\omega_3 + \omega_1\omega_3.$$

Finally, cycle products $\omega_1\omega_2$, $\omega_1\omega_{-1}$ and $\omega_2\omega_{-2}$, $\omega_{-1}\omega_{-2}$ are removed to leave

$$D_{X_2} = \omega_{-2}\omega_3 + \omega_1\omega_3.$$

A general method for deriving an equation of the rate over route (for an overall equation of this route) for steady-state and quasi-steady-state reactions reduces to obtaining expressions for the concentrations of intermediates X_i by solving a system of linear algebraic equations $\boldsymbol{B}^T_{X}\boldsymbol{W}_j = 0$ for linearly independent $[X_i]$ (see Section 1.1). This system of equations is solved using the Cramer rule as

$$[X_i] = \frac{\Delta_{X_i}}{\Delta}, \tag{1.138}$$

where Δ is the determinant of the system of linear algebraic equations (written for the coefficients at the unknowns) and Δ_{X_i} is a determinant in which the column of coefficients at X_i is replaced by a column of constant free terms.

The diagram method was originally used for determining Δ_{X_i} and Δ values by King and Altman [84]. Later, S.J. Mason (see [101, 102]) formulated a general rule that allowed graphs to be used for solving the problems related to linear laws of the $y = ax$ type and the systems of Kirchhoff equations in the theory of electric circuits (where x is the current, a is the resistance, and y is the potential difference). In the application to the kinetics of chemical reactions with linear mechanisms, x is the concentration of the ith intermediate, a is the step weight ($\omega_j^{\pm}$), and y is the step rate ($W_j^{\pm}$). The essence of this rule is expressed by the following relation:

$$\frac{[\mathrm{X}_i]}{[\mathrm{X}_j]} = \frac{D_i}{D_j}. \tag{1.139}$$

Volkenstein and Goldstein [86–88] employed an analogy with this rule and used the graphs method to derive kinetic equations for the rate of enzymatic reactions. Yablonsky et al. [7, 94] proved Eq. (1.139) and showed its relationship to the Cramer rule. Upon expressing Δ_{X_i} and Δ in terms of step weights and relating these quantities for a catalytic reaction as $\Delta_{\mathrm{x}} = \Delta^*_{\mathrm{X}_i}$ [M] (i.e., explicitly introducing the concentration [M] of a catalyst in KG5), we obtain

$$[\mathrm{X}_i] = [\mathrm{M}]\frac{\Lambda^*_{\mathrm{X}_i}}{\Delta} = [\mathrm{M}]\frac{D_i}{D_{\mathrm{M}}}, \tag{1.140}$$

where $D_i = \Delta^*_{\mathrm{X}_i}$ and $D_{\mathrm{M}} = \Delta$.

Using Eqs (1.138) and (1.140), it is possible to obtain the following relations:

$$\frac{[\mathrm{X}_i]}{[\mathrm{X}_j]} = \frac{\Delta_{\mathrm{X}_i}}{\Delta_{\mathrm{X}_j}} = \frac{\Delta^*_{\mathrm{X}_i}}{\Delta^*_{\mathrm{X}_j}} = \frac{D_i}{D_j}. \tag{1.141}$$

For noncatalytic reactions, the concentration of intermediate X_i is written using the concentration of a null-species at the empty vertex:

$$[\mathrm{X}_i] = [\mathrm{X}_0]\frac{D_i}{D_0} = \frac{D_i}{D_0}. \tag{1.142}$$

If all $[\mathrm{X}_i]$ in a catalytic reaction are related to [M], we obtain the following expression for the total concentration of the catalyst:

$$[\mathrm{M}]_{\Sigma} = [\mathrm{M}]\left(1 + \frac{\sum\limits_{i \neq \mathrm{M}} D_i}{D_{\mathrm{M}}}\right) = [\mathrm{M}]\frac{\sum\limits_{i} D_i}{D_{\mathrm{M}}}, \tag{1.143}$$

where

$$[\mathrm{M}] = \frac{[\mathrm{M}]_{\Sigma} D_{\mathrm{M}}}{\sum\limits_{i} D_i}. \tag{1.144}$$

Equations (1.143) and (1.144) imply that

$$[\mathrm{X}_i] = \frac{[\mathrm{M}]_{\Sigma} D_i}{\sum\limits_{i} D_i}. \tag{1.145}$$

For heterogeneous processes with all $[\mathrm{X}_i]$ normalized to $[\mathrm{X}]_{\Sigma}$ and expressed via the surface fractions Θ_{X_i}, we obtain

$$\Theta_{\mathrm{X}_i} = \frac{D_i}{\sum\limits_{i} D_i}. \tag{1.146}$$

Allowance for pendant vertices in the material balance with respect to a catalyst can be made by finding the root determinants for these vertices and including them into $\sum_i D_i$, so that $[M]_\Sigma$ would also refer to the substances occurring in the pendant vertices. Since the graph edges incident to pendant vertices for steady-state and quasi-steady-state reactions represent equilibrium steps, the second approach consists of introducing a supplementary function F_i, called the *complexity* of an intermediate (or any vertex of a cyclic graph), which is defined as follows [6, 68]:

$$F_i = \frac{|X_i|}{[X_i]} = 1 + \frac{\sum_S [X_S]}{[X_i]} = 1 + \sum_S \frac{\omega_S}{\omega_{-S}}, \tag{1.147}$$

where $[X_i] + \sum_S [X_S] = |X'_i|$, $[X_S]$ is the concentration of a compound occupying a pendant vertex linked to the graph by step S, and ω_S and ω_{-S} are the weights of the edge (step) incident to the pendant vertex and directed from X_i to X_S. Evidently, the ω_S/ω_{-S} ratio includes the equilibrium constant K_S and the concentrations of reaction participants that enter into ω_S and ω_{-S}. For vertex M in KG4 (Figure 1.2), we obtain

$$F_M = 1 + K_3[P_1] + K_4[P_2].$$

Formula (1.145) can be modified by taking into account that $|X'_i|' = [X_i]F_i$, which yields

$$[X_i] = \frac{[M]_\Sigma D_i}{\sum_i F_i D_i} \tag{1.148}$$

1.3.1.1 *Algorithms of deriving kinetic equations for steady-state and quasi-steady-state processes*

Relations (1.139) and (1.148) lead to the following expression for the rate of any step in the reaction mechanism:

$$W_j = W_j^+ - W_j^- = \omega_j [X]_i - \omega_{-j} [X]_{i+1}.$$

For a catalytic reaction, we have

$$W_j = \left[\omega_j \frac{D_i}{D_M} - \omega_{-j} \frac{D_{i+1}}{D_M}\right][M] \tag{1.149}$$

$$W_j = \frac{[\omega_j D_i - \omega_{-j} D_{i+1}][M]_\Sigma}{\sum_i F_i D_i}. \tag{1.150}$$

Then, using the condition of the steady-state of steps as expressed by Eqs (1.34) and (1.35), one can readily establish a relationship between the step rate and the rate over a route and thus find r_p [91].

Another algorithm was proposed by Volkenstein and Goldstein [86–88] and modified by Yablonsky et al. [7, 48]. According to this, a step belonging to one of the routes (with $W_j = r_p$) is selected on the graph of a multi-route reaction and the rate of this step is expressed as

$$W_j = r_p = \frac{[M]_\Sigma \sum_{n=1}^{K} (C_{pn}^+ - C_{pn}^-) D_{pn}}{\sum_i D_i} \tag{1.151}$$

(or in terms of $\sum F_i D_i$ in the case of a graph with pendant vertices), where $C^{\pm}_{pn}$ is the weight of the nth cycle over route P involving the jth step; D_{pn} is the determinant of a subgraph obtained upon the contraction of the nth cycle over route P to one vertex with a root in the vertex obtained upon the contraction; and K is the number of cycles passing through the jth step. If the rate over route P is described by a combination of steps rates W_j, then Eq. (1.151) is written for all of these steps.

Example 1.17

► Consider graph KG5 (Figure 1.4) for a two-route reaction with two simple cycles and let these cycles be the basis set. The first route includes steps (1) and (2), while the second route includes steps (1), (3) and (4). It is evident from the graph that $W_2 = r_1$ and $W_4 = r_2$. It is also natural that $W_3 = r_2$, but it is expedient to select irreversible step (4) in order to simplify the considerations. According to the second algorithm based on relation (1.151), we begin with writing the weights of cycles C_{pn} as

$$C^+_{11} = \omega_1\omega_2; \quad C^-_{11} = \omega_{-1}\omega_{-2}; \quad C^+_{12} = 0(\omega_{-4} = 0); \quad C^-_{12} = \omega_{-2}\omega_3\omega_4;$$
$$C^+_{21} = \omega_1\omega_3\omega_4; \quad C^-_{21} = 0; \quad C^+_{22} = \omega_{-2}\omega_3\omega_4; \quad C^-_{22} = 0.$$

Then we write expressions for the D_{pn} values of subgraphs as $D_{11} = \omega_{-3} + \omega_4$ (a sum of the weights of trees entering into the vertex obtained upon the contraction of cycle 11), $D_{12} = 1$ (one vertex corresponds to $D_{pn} = 1$), $D_{21} = 1$, and $D_{22} = 1$. Finally, using the values of D_M, D_{X_1} and D_{X_2} determined as described above, we can write the following expressions for r_1 and r_2:

$$r_1 = \frac{[M]_\Sigma\{(\omega_1\omega_2 - \omega_{-1}\omega_{-2})(\omega_{-3} + \omega_4) - \omega_{-2}\omega_3\omega_4\}}{\sum D_i}; \tag{1.152}$$

$$r_2 = \frac{[M]_\Sigma\{\omega_1\omega_3\omega_4 + \omega_{-2}\omega_3\omega_4\}}{\sum D_i}; \tag{1.153}$$

$$\sum D_i = D_M + D_{X_1} + D_{X_2} = \omega_3\omega_4 + \omega_{-1}\omega_{-3} + \omega_2\omega_{-3} + \omega_{-1}\omega_4$$
$$+ \omega_2\omega_4 + \omega_1\omega_{-3} + \omega_1\omega_4 + \omega_{-2}\omega_{-3} + \omega_{-2}\omega_4 + \omega_{-2}\omega_3 + \omega_1\omega_3.$$

■

For a single-route reaction, the step rate is $W_j = \nu_S r$. In the case of a linear mechanism, we have all $\nu_S = 1$ and, accordingly,

$$r = \frac{[M]_\Sigma(C^+ - C^-)}{\sum D_i}. \tag{1.154}$$

It is useful to know that, as pointed out in [7], the cyclic characteristic $C = C^+ - C^-$ in this case corresponds to the law of mass action written for an overall equation of the single-route reaction that is considered as the elementary step. For example, in the mechanism (1.3)–(1.5) with an overall equation of 2A = 2B, the cyclic characteristic is

$$C = C^+ - C^- = k_1k_2k_3[A]^2 - k_{-1}k_{-2}k_{-3}[B]^2 = k^+[A]^2 - k^-[B]^2.$$

Denominators (denoted by D in Section 1.1) in the kinetic equations (1.38), (1.39), (1.45) and the like represent the sums of weights of the corresponding trees ($D = \sum D_i$). In formulas (1.50) and (1.51), denominators D are equal to the determinants of vertices D_M and D_0, respectively.

The method of graphs is conveniently applied to single-route reactions with a limiting step (i.e., with all other steps obeying the quasi-equilibrium condition). In this case, the weight of a graph edge corresponding to a quasi-equilibrium step (similar to the case of a graph with edges incident to pendant vertices) is given by the ratio ω_j/ω_{-j}. For example, the step

$$\mathrm{A} + \mathrm{X}_i \underset{k_{-j}}{\overset{k_j}{\rightleftharpoons}} \mathrm{X}_{i+1} + \mathrm{P} \tag{1.155}$$

is characterized by the following ratio:

$$\frac{\omega_j}{\omega_{-j}} = \frac{k_j[\mathrm{A}]}{k_{-j}[\mathrm{P}]} = K_j\frac{[\mathrm{A}]}{[\mathrm{P}]}.$$

Let us denote the product of edge weights on a path from the ith to kth vertex by D_{ik} and call this value the determinant of the kth vertex. By definition, determinants of an empty vertex (null-species) and a vertex with the initial catalytically active center are assigned the values $D_{ii} = 1$. Consider an example of the mechanism of an enzymatic reaction with two pendant vertices (graph KG6, Figure 1.6), which involves inhibitor E and has a limiting step (3) [100]:

$$\begin{array}{ccccccc} \mathrm{E} & \underset{1}{\overset{\mathrm{A}}{\rightleftharpoons}} & \mathrm{EA} & \underset{2}{\overset{\mathrm{B}}{\rightleftharpoons}} & \mathrm{EAB} & \xrightarrow[3]{} & \mathrm{E} + \mathrm{P}; \\ \mathrm{I}\,\updownarrow\,5 & & \mathrm{I}\,\updownarrow\,4 & & & & \\ \mathrm{EI} & & \mathrm{EAI} & & & & \end{array} \tag{1.156}$$

According to Eq. (1.150), the rate of the reaction of product P formation can be expressed as

$$r = \frac{[\mathrm{E}]_0\omega_3 D_{\mathrm{EAB}}}{\sum D_i F_i} = \frac{[\mathrm{E}]_0\omega_3 D_{\mathrm{EAB}}}{D_{\mathrm{EE}}F_{\mathrm{E}} + D_{\mathrm{EA}}F_{\mathrm{EA}} + D_{\mathrm{EAB}}}. \tag{1.157}$$

Determinants of the vertices in this graph are as follows:

$$D_{\mathrm{E}} = D_{\mathrm{E}\to\mathrm{E}} = 1(D_{\mathrm{E}} = D_{\mathrm{EE}}); \qquad D_{\mathrm{EA}} = D_{\mathrm{E}\to\mathrm{EA}} = K_1[\mathrm{A}];$$

$$D_{\mathrm{EAB}} = D_{\mathrm{E}\to\mathrm{EA}\to\mathrm{EAB}} = K_1K_2[\mathrm{A}][\mathrm{B}]; \qquad D_{\mathrm{EI}} = D_{\mathrm{E}\to\mathrm{EI}} = K_4[\mathrm{I}];$$

$$D_{\mathrm{EAI}} = D_{\mathrm{E}\to\mathrm{EA}\to\mathrm{EAI}} = K_1K_5[\mathrm{A}][\mathrm{I}]; \qquad F_{\mathrm{E}} = 1 + K_4[\mathrm{I}]; \quad F_{\mathrm{EA}} = 1 + K_5[\mathrm{I}],$$

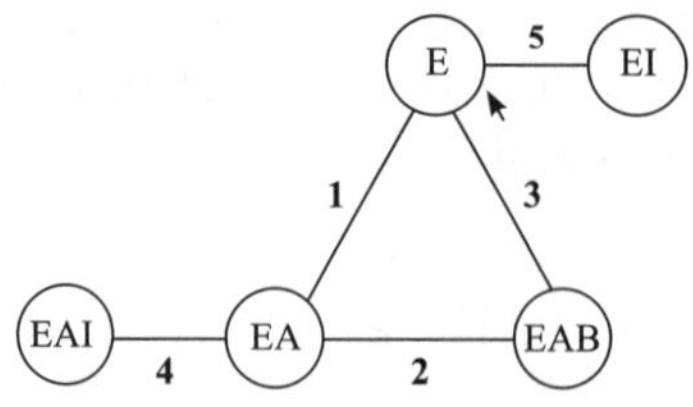

KG6

Figure 1.6 Kinetic graph for scheme (1.156)

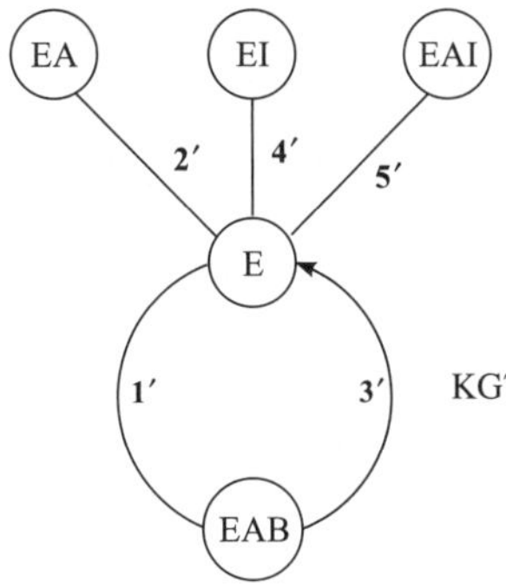

Figure 1.7 Kinetic graph for scheme (1.159)

where K_i are the equilibrium constants of formation of complexes with catalyst E. Substituting these D_{ik} into formula (1.157) yields

$$r = \frac{k_3 K_1 K_2 [A][B][E]_0}{1 + K_4[I] + (1 + K_5[I])K_1[A] + K_1 K_2 [A][B]}. \tag{1.158}$$

Evidently, in the presence of a limiting stage, the kinetic equation can be derived without using F_i. In this case, a sum in the denominator of formula (1.157) is replaced by the sum of D_{ik} for all vertices of the given graph ($\sum D_{ik}$), including the pendant vertices:

$$\sum D_{ik} = \Sigma D_i F_i.$$

Taking into account the quasi-equilibrium character of all steps except (3), the three-step mechanism (1.156) can be reduced to a two-step mechanism

$$A + B + E \underset{1'}{\rightleftharpoons} EAB \underset{3'}{\longrightarrow} P + E \tag{1.159}$$

represented by a graph with three pendant vertices corresponding to the formation of compounds EA, EI and EAI from E (KG7, Figure 1.7). In this case, we can obtain an equation that is equivalent to (1.158) with different K_i that are related to the constants entering into (1.158).

Methods that can be used to simplify the reaction schemes with quasi-equilibrium steps were proposed in [89, 100]. Graphs were also used to find relationships between the equilibrium constants and complexity

$$F = \frac{[M]_\Sigma}{[M]} = \sum D_i$$

during the analysis of complicated systems of equilibrium complex formation reactions [122] and to describe the kinetics of non-steady-state reactions [123, 88]. However, the latter approach, based on the Laplace–Carson transform, did not receive further attention–probably because of a considerable difficulty in the transition from reaction rate images to originals in the case of graphs with more than three vertices.

1.3.2 Nonlinear mechanisms

In the graphs of linear mechanisms, the vertices of two intermediates (including a null-species) are connected by an edge (reaction step). Relations between elements of a set, in which the elements appear in pairs, are called *binary* relations. Any edge on a graph determines a pair of vertices. However, if a mechanism is nonlinear, binary relations are not valid and, for example, a step of the type

$$X_1 + X_2 \longrightarrow 2X_3 \qquad (1.160)$$

cannot be presented on a KG according to the above approach. In this situation, Temkin [27] suggested using supplementary edges (dashed lines on KGs), but the choice of main and supplementary edges was arbitrary [71]. There are many alternative variants of describing nonlinear mechanisms (see, e.g., [6, 71]), but the author believes that the most convenient approach to the representation of any mechanisms is that employing bipartite graphs (BGs). BGs were originally used by Balandin [80], Volpert [105], and Clarke [106, 107].

The BG of a reaction mechanism is constructed as follows. Each substance is represented by a vertex of type w and each elementary reaction (step) is represented by a vertex of type u. Reversible steps are represented by two mutually reciprocal elementary reactions and, accordingly, depicted by two vertices u_j. Thus, the set of vertices is subdivided into two subsets, each containing the vertices of one (w or u) type. Vertices of one subset are non-adjacent, while vertices w_i and u_j are adjacent.

The *multiplicity* (m) of an edge of a BG is defined as the stoichiometric coefficient of the ith substance in the jth reaction step. If vertices of the w type are depicted by open symbols (circles with indicated substance) and the u type vertices are depicted by black symbols (e.g., squares), then an elementary reaction (step) can be represented by a graph (Figure 1.8) with all directed edges (also called segments). Segments $[X_1, u]$ and $[X_2, u]$ reflect the participation of species X_1 and X_2 in step u on the left of the arrow, while segment $[u, X_3]$ corresponds to the formation of substance X_3 via reaction u. The multiplicity of an edge, which is equal to the stoichiometric coefficient of a substance (e.g., $m = 2$ for X_3) can be indicated either by writing the stoichiometric coefficient at the corresponding arrow

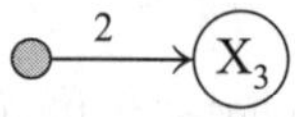

or by using the corresponding number of arrows [124]:

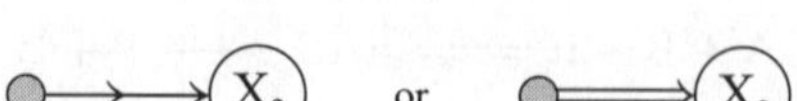

A reversible reaction is represented by two vertices (u_j and u_{-j}) and, hence, depicted by two graphs with opposite directions of arrows. The cycles on BGs can be either directed or undirected. The directed cycles (with the arrows in all segments pointing in the same direction) are called circuits. All cycles (including circuits) of a BG contain even numbers of segments.

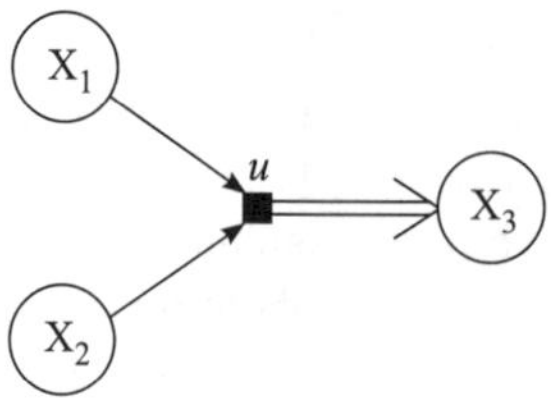

Figure 1.8 Elementary step (1.160)

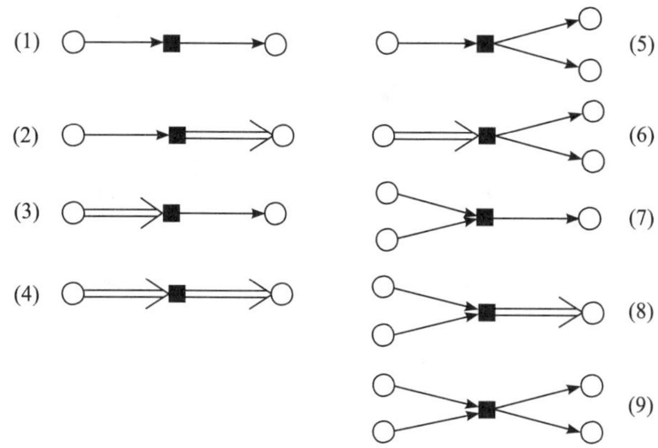

Figure 1.9 Graphical representation for nine types of elementary steps

Assuming that elementary reactions are predominantly mono- and bimolecular (see Section 1.4), we can separate nine types of elementary reactions that are distinguished by their stoichiometry [6]:

$$\begin{array}{llllll} (1) & A \to C; & (4) & 2A \to 2C; & (7) & A+B \to C; \\ (2) & A \to 2C; & (5) & A \to C+D; & (8) & A+B \to 2C; \\ (3) & 2A \to C; & (6) & 2A \to C+D; & (9) & A+B \to C+D. \end{array}$$

Graphical representations of these reactions are shown in Figure 1.9. For example, reaction (7) involving intermediates X_i can be depicted as follows:

$$\begin{array}{llll} (10) & A+B \to X_i; & (13) & X_i+X_j \to C; \\ (11) & A+X_i \to X_j; & (14) & X_i+X_j \to X_k. \\ (12) & A+X_i \to C; & & \end{array}$$

Figure 1.10 shows a BG for the typical mechanism of a catalytic reaction with irreversible steps:

$$\begin{aligned} A+M &\xrightarrow{(1)} MA; \\ B+M &\xrightarrow{(2)} MB; \\ MA+MB &\xrightarrow{(3)} 2M+P. \end{aligned} \tag{1.161}$$

BG is essentially the reaction network of a given mechanism, which includes all substances and reaction steps and reflects all relations between substances. The mutual one-to-one correspondence between stoichiometric matrices $\boldsymbol{B}_{\mathrm{m}}$ and BGs was proved in [6, 125]. Sakamoto et al. [124] developed an algorithm of writing a set of differential equations for a given BG. By analogy with a Temkin's KG, it is possible to construct a BG that images a given mechanism in the space of intermediates and is equivalent to the $\boldsymbol{B}_{\mathrm{X}}$ matrix of this mechanism [6, 125].

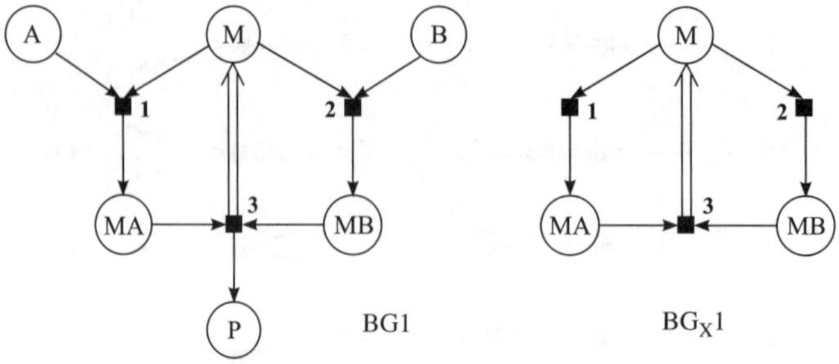

Figure 1.10 Bipartite graphs (BGs) for mechanism (1.161)

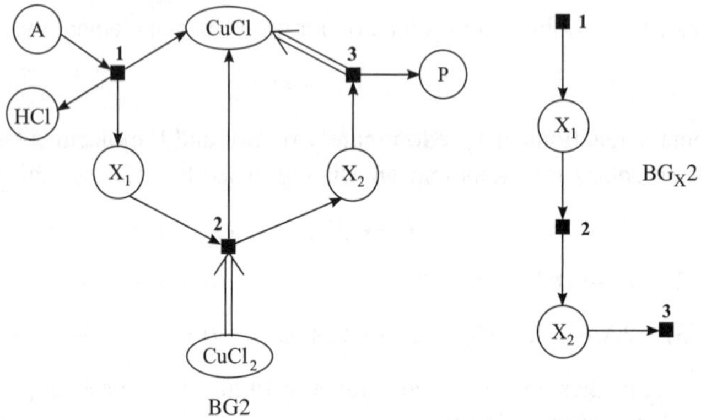

Figure 1.11 Bipartite graphs (BGs) for mechanism (1.21)

Let us remove from graph BG1 the vertices corresponding to the substances (A, B, P) that enter into overall equations and also remove the corresponding incident edges (segments ([A, 1], [B, 2], [3, P]) (Figure 1.10, left). The obtained graph BG$_X$1 (Figure 1.10, right) is called a bipartite network or subnetwork of intermediates. Note that vertices A and B are characterized by zero in-degrees, while vertex P has zero out-degree. Let us consider some other examples of BGs. Figure 1.11 shows graphs BG2 and BG$_X$2 for mechanism (1.21) with three irreversible steps, where A is RC≡CH and P is RC≡CCl. Upon removal of the terminal vertices of reaction participants (A, $CuCl_2$, HCl and P), we obtain graph BG$_X$ of the subnetwork of intermediates. Removal of the CuCl vertex that corresponds to a dual species of intermediate and product (see Example 1.3) yields graph BG$_X$2.

Figures 1.12 and 1.13 show graphs BG3 and BG$_X$3 for the mechanism (1.53) of vinyl chloride synthesis with the participation of two catalysts. As can be seen, graphs representing a subnetwork of intermediates reveal a topological structure of mechanisms and their features that can serve as criteria for classification. Indeed, BG$_X$1 is a cyclic graph with two circuits and contains no pendant vertices of the u type (catalytic reaction), whereas BG$_X$2 is a noncyclic graph with a linear topology and pendant vertices u_1 and u_3 (noncatalytic reaction). The mechanisms of chain reactions can also differ by the presence of circuits and pendant vertices of type u (mixed topology).

A useful concept for the classification of mechanisms on the topological basis is provided by a *simple submechanism of the network of intermediates* [6, 7, 18, 72, 73]. A submechanism of a reaction mechanism

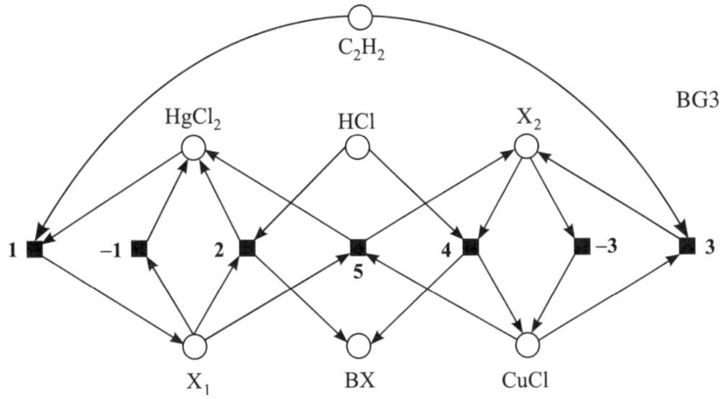

Figure 1.12 Bipartite graph (BG) for mechanism (1.53)

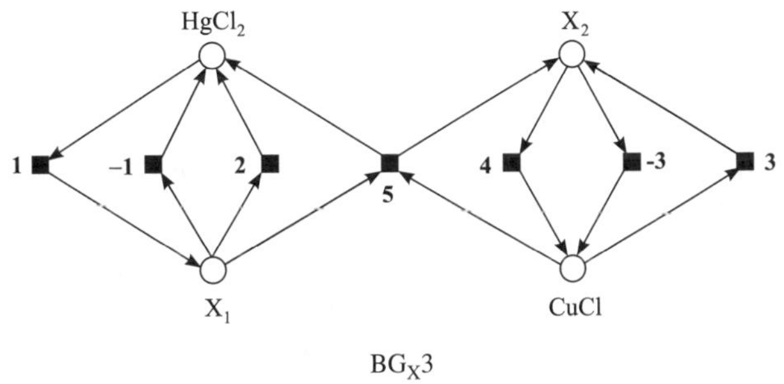

Figure 1.13 Bipartite network of intermediates for BG3 of mechanism (1.53)

is defined as a subset of the steps of this mechanism, for which a vector ν_j can be found that has no zero components. Submechanisms correspond to subgraphs $BG_X{}^*$ of graph BG_X. The set of submechanisms (subgraphs $BG_X{}^*$) is obtained as a result of decomposition of the reaction mechanism on the space of intermediates. Since the number of subgraphs is finite, the number of submechanisms (in contrast to the number of routes) corresponding to a fraction of subgraphs is also limited (finite). Submechanisms that cannot be further subdivided into two (or more) simpler ones are called simple submechanisms (SSMs). The set of such SSMs is an important characteristic of both BGs and KGs. Special software (GERM) for finding SSMs was developed in [125]. Figure 1.14 shows the subgraphs for four SSMs of graph BG_X3 for the aforementioned mechanism (1.53). As can be seen, all these SSMs contain circuits. SSM1–SSM3 represent routes of the vinyl chloride formation and SSM4 is an empty route. The elementary reaction steps (u_1, u_2), (u_3, u_4), (u_{-1}, u_4, u_5), and (u_1, u_{-3}, u_5) enter into SSM1–SSM4 with the vectors of stoichiometric numbers containing only positive components.

The number of steps (S_{SSM}) in an SSM is given by formula

$$S_{SSM} = \mathrm{rank}\boldsymbol{B}_{X(SSM)} - 1,$$

where $B_{X(SSM)}$ is the matrix of stoichiometric coefficients for the intermediates entering into the SSM [125]. The topological characteristics of SSMs and their relationships in BG_X constitute a basis for the classification of reaction mechanisms [16, 18, 72, 73].

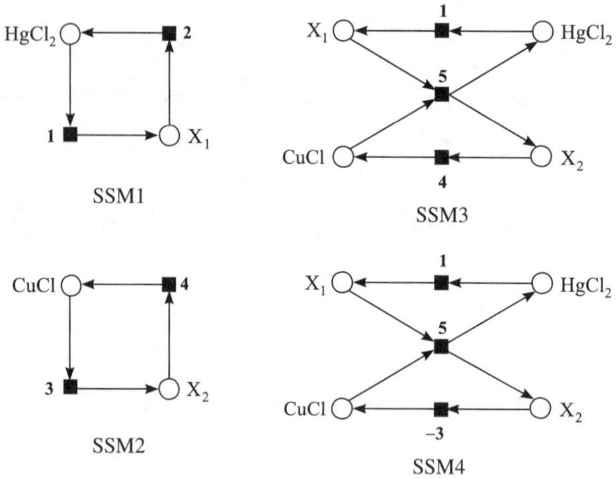

Figure 1.14 Simple submechanisms (SSMs) of the network of intermediates for bipartite subgraph (BG_X3) of mechanism (1.53)

1.3.3 Other fields of application of kinetic and bipartite graphs in chemical kinetics and in theory of complex reaction mechanisms

In the context of an analysis of the identifiability of parameters of kinetic models (see Section 1.1), Yablonsky et al. [48] proposed an algorithm for determining the number of independent parametric functions for linear mechanisms on KGs. This algorithm is based on determining the "color" of KG trees. According to this, *colored steps* are defined as those involving one of the observed substances – initial reactants or products if the corresponding reactions are reversible (on KGs, the weights of these steps include the concentrations of reaction products). *Uncolored steps* are those in which the observed substances do not interact with intermediates ($X_i \rightarrow X_k$, $X_i \rightarrow X_k + P$). The uniformity of the weights of steps (like colors of the graph edges) result in some trees being similar (having the same sets of colors). The tree color is characterized by a set of figures indicating the number of edges with the same color. An analysis of the sets of these values (assigned according to certain rules) allows one to determine the numbers of trees of different colors among real trees of the KG – that is, the number of independent parametric functions. Using this approach, interesting laws have been established that relate the coloration of KG edges to the structure of a kinetic equation for single-route mechanisms, i.e., to the form of terms (ΣD_i) in the denominator of the kinetic equation [48].

Let us consider the application of the graph-theoretical approaches to assessing the complexity of reaction mechanisms. The notion of complexity in this context belongs to intuitively determined properties. In its application to the theory of complex reaction mechanisms, it would be expedient to speak of the complexity of a kinetic equation (and the whole kinetic model), which would determine the probability of identifying the model from experimental data and the possibility of evaluating the model parameters. The complexity of a model is intuitively taken into account in the stage of discrimination between a set of hypothetical mechanisms, the simplest of which are considered first. Evidently, the complexity of a kinetic model in this sense depends on the numbers of intermediates, colored and uncolored steps, routes, and pendant vertices and on the ways of connecting routes (i.e., on the class of the mechanism).

It was suggested to characterize the complexity of linear mechanisms (as the complexity of a kinetic model of a steady-state multi-route reaction [6, 78, 110, 111]) in terms of the complexity index $\mathcal{E}$ (kinetic index). This value is determined by counting the numbers of weights (or rate constants) of steps in the fractional rational expressions for the rates over routes, obtained using the algorithm of Yablonsky et al. [7, 48] (Eq. (1.151)). According to this, the complexity index is calculated using the following formula:

$$\mathcal{E} = P(I-1)T_p^d + I\sum_{p=1}^{P} T_p^n, \tag{1.162}$$

where P is the number of linearly independent routes (KG cycles), I is the number of intermediates (graph vertices), T_p^d is the number of trees contributing to the denominator (ΣD_i) of the kinetic equation, and T_p^n is the number of trees in the specific subgraphs contributing to the numerator (D_{pk}) of the kinetic equation. The second term in expression (1.162) can be calculated using the following relation:

$$I\sum_{p=1}^{P} T_p^n = 2I\sum_{p=1}^{P}\sum_{k=0}^{k_{\max}} D_{pk}, \tag{1.163}$$

where k is the number of a cycle passing through the step, the rate of which is equal to the rate over route $\nu^{(p)}$. Expressions for calculating the T_p^d and T_p^n values without the derivation of kinetic equations are given in [6, 111].

An increment $\Delta\mathcal{E}$ in the complexity index due to the appearance of pendant vertices in the KG of a mechanism can be evaluated as

$$\Delta\mathcal{E} = P(I-1)\sum_{i=1}^{I} T_i P_i, \tag{1.164}$$

where T_i is the number of trees at the ith vertex of the initial KG without pendant vertices and P_i is the number of pendant vertices at the ith vertex of the KG.

A detailed analysis of the effect of the numbers of routes and vertices, the ways of connecting routes, and the class of a mechanism on the $\mathcal{E}$ value was performed in [6, 8, 78, 111]. The influence of the topological structure of a mechanism on the complexity of a kinetic model was clearly illustrated [8] in the case of three-route mechanisms with the KGs depicted in Figure 1.15, which are characterized by the same numbers of vertices ($I = 4$) and steps ($S = 6$). The reaction rate over route I of the catalytic reaction mechanism of class B^3 according to formula (1.151) is

$$r^{(1)} = \frac{[\mathcal{K}]_\Sigma (C_{10}^{+} - C_{10}^{-}) D_{10}}{\sum D_i}. \tag{1.165}$$

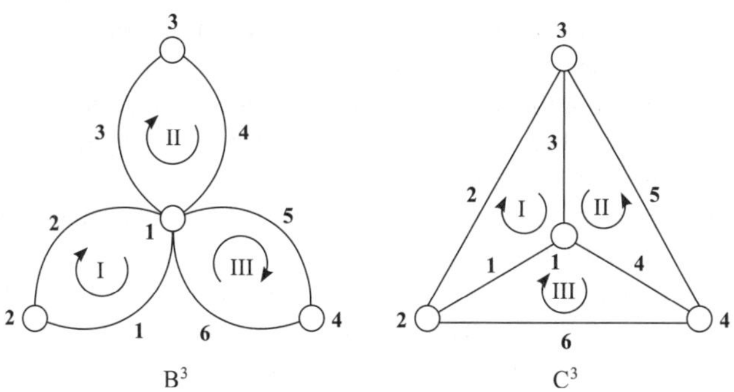

Figure 1.15 Kinetic graphs for three-route mechanisms of classes B^3 and C^3

In the more general form, this expression can be written as follows [7]:

$$r^{(1)} = \frac{[\mathcal{K}]_{\Sigma} W_{\mathrm{I}} P_{\mathrm{I}}}{\sum D_i}, \tag{1.166}$$

where W_{I} is the cyclic characteristic of route I ($W_{\mathrm{I}} = C_{10}^{+} - C_{10}^{-}$) and P_{I} is the conjugation parameter (root determinant of cycle I). For the given mechanism, $D_{10} = P_{\mathrm{I}}$, $[\mathcal{K}]_{\Sigma}$ is the total catalyst concentration, $W_{\mathrm{I}} = \omega_1\omega_2 - \omega_{-1}\omega_{-2}$, and $P_I = (\omega_4 + \omega_{-3})(\omega_6 + \omega_{-5})$.

The value of index $\mathcal{E}^*$ calculated for one route using formula (1.166) is $\mathcal{E}^* = 128$. If the concentration $[\mathcal{K}]$ of the free catalyst is known, the equation for r_1 can be written in a simpler form as

$$r^{(1)} = \frac{[\mathcal{K}]W_{\mathrm{I}}}{D_1^{\mathrm{I}}} = \frac{[\mathcal{K}](\omega_1\omega_2 - \omega_{-1}\omega_{-2})}{\omega_2 + \omega_{-1}}, \tag{1.167}$$

where $D_1^{(1)}$ is the root determinant of vertex 1 (calculated for cycle I only). In this case, the equation includes only the weights of route I, so that $\mathcal{E}^{\bullet\prime} = 6$ (the complexity index for the first route kinetic equation with the free catalyst concentration $[\mathcal{K}]$).

Now let us turn to the mechanism of class C^3. Owing to a strong conjugation of routes (due to the presence of common steps), formula (1.151) takes the following form:

$$r_1 = \frac{[\mathcal{K}]_{\Sigma}(W_{\mathrm{I}}P_{\mathrm{I}} + C^*)}{\sum D_i}, \tag{1.168}$$

where C^* is the conjugation factor calculated as

$$C^* = \sum_{l \neq \mathrm{I}} W_l P_l$$

(the sum is taken over all cycles except I, which include step 2 belonging to route I only) and P_l is the value determined for a subgraph with root in cycle l. In this case, $\mathcal{E}^* = 256$. It is important to note that, in strongly conjugated mechanisms (including classes C and Z), even the use of an equation with the free catalyst concentration

$$r_1 = \frac{[\mathcal{K}](W_{\mathrm{I}}P_{\mathrm{I}} + C^*)}{D_1} \tag{1.169}$$

leads to $\mathcal{E}' = 112$ (cf. $\mathcal{E}' = 6$ for the above mechanism of class B^3). For the complete kinetic model, the complexity index calculated using formula (1.162) for the mechanism of class C^3 is $\mathcal{E} = \mathcal{E}^* \times 3 = 768$.

The index of complexity of a mechanism, which is applicable to the mechanisms of any type (represented by both KGs and BGs) was constructed based on calculations of the information content of the stoichiometric matrix of a given mechanism and stoichiometric matrices of SSMs [112]. The information content index I_{mech} takes into account that the complexity of a mechanism increases with growing numbers of steps, and intermediates, the number, dimension, and degree of conjugation of SSMs, and the variety of the types of nonlinear steps. Therefore, it can be expected that the so defined I_{mech} value would also be correlated with the complexity of a kinetic model. Indeed, for the mechanisms of classes B^3 and C^3 (Figure 1.15), $I_{\mathrm{mech}} = 589.14$ and 1870.7, respectively [112].

The application of bipartite and other types of graphs to an analysis of the dynamic behavior of reacting systems (multiplicity of steady states, self-sustained oscillation regimes, etc.) will be considered in Chapter 6. Chapter 7, which is devoted to the problems of selecting a rational strategy in the investigations of reaction mechanisms, will also consider the use of KGs and BGs in formulating hypotheses of the mechanisms of catalytic reactions [6, 113, 114].

1.4 Elementary steps – Selection rules

Notions concerning the nature of elementary steps of a chemical reaction and the meaning of a concept of the "reaction mechanism" were formed during the accumulation and generalization of chemical knowledge and the development of a theory of elementary reactions. Steps which had been considered elementary at a certain stage of the development of chemistry proved later to be complex, consisting of still simpler reactions. Evidently, this process will continue, as the methods of detecting short-lived reactive species become more and more perfect. On the other hand, it should be borne in mind that very fast processes involving such species, which occur beyond the limiting step in single-route reactions, may no longer influence the kinetic laws of the overall process and, hence, can be ignored in constructing an adequate kinetic model.

In any case, in writing a reaction mechanism or formulating a set of hypothetical mechanisms, it is necessary to take into account the main theoretical concepts (postulates, laws, principles) and heuristic rules that can elucidate the nature of a particular reaction, i.e., to assess the level of confidence in classifying one or another reaction into elementary steps. Naturally, in assessing this probability, it is important to predict the possible kinetic behavior of the proposed elementary step and estimate whether the rate of this step is acceptable for the overall reaction. A quantitative answer to this question is provided by laborious quantum-chemical calculations, which are especially difficult for catalytic reactions with heavy metal atoms contained in intermediate species. Nevertheless, these calculations are now actively used in the catalysis with metal complexes and even in heterogeneous catalysis (see Chapter 7). In addition to model calculations, quantitative estimations can also be provided by experimental data on the kinetics of elementary steps, but the amount of such information in the catalysis with metal complexes is still very small. For this reason, considerable help in this respect is offered by the simple heuristic rules and generalizations that shed light on the expected kinetic characteristics of elementary steps. If a proposed mechanism involves a stage that appears to be a block of elementary steps, this possibility has also to be substantiated.

There are numerous monographs and handbooks that summarize the following principal theoretical notions concerning the elementary steps of chemical reactions:

- theory of collisions in gases [58, 126–130];
- theory of collisions in liquids [130–132];
- theory of activated complex (or transition-state theory) [50, 58, 127, 130, 133];
- quantum-mechanical theory of reactions in polar liquids [131, 134–136];
- quantum-mechanical selection rules for elementary steps [137–139];
- topological selection rules for elementary steps [6, 77].

It is expedient to begin considering problems related to the elementarity of steps and the main rules of selecting these steps with an analysis of the main concepts of chemical kinetics.

1.4.1 Main postulates, laws, and principles

A very important concept, and one of the main parameters in chemical kinetics, is the concentration of molecules or other particles involved in a reaction. The concentration C_i is defined as the number of molecules (particles) or moles of the ith species per unit volume. The number of particles must be sufficiently large so that it would be possible to treat the concentration, using the laws of statistics, as an averaged quantity that is the same in any part of the reaction volume. The statistical character of the concentration and the statistical independence of all particles are also closely related to the fact that a large number of both identical and different particles randomly colliding with each other are involved in the mutual energy exchange. The resulting distribution of particles with respect to their internal degrees of freedom (energies) and translational velocities (Maxwell–Boltzmann distribution, MBD) leads to the establishment of a statistical (thermodynamic) equilibrium between excited states. These systems are characterized by a single value of the temperature as a parameter determining this distribution. The time of establishment of the thermal equilibrium is shorter than the characteristic times of excited particle formation and disappearance.

The Boltzman law of the distribution of molecules (particles) with respect to their energies determines the relative fraction N_i/N of these particles occurring in the ith state with energy E_i:

$$\frac{N_i}{N} = \frac{\exp(-E_i/kT)}{Q}, \tag{1.170}$$

where Q is the sum over states (statistical sum, partition function) defined as

$$Q = \sum_i \exp(-E_i/kT)$$

and T is the absolute temperature. The Maxwell law of the distribution of molecules with respect to their velocities is expressed as follows:

$$\frac{dN(v)}{N} = A\exp\left(-\frac{mv^2}{2kT}\right)v^2\,dv, \tag{1.171}$$

$$f(v) = A\exp\left(-\frac{mv^2}{2kT}\right), \tag{1.172}$$

where v is the velocity of a molecule, m is its mass, A is a constant factor at $T = \text{const}$, and $f(v)$ is the velocity distribution function.

If chemical reactions do not violate the thermal equilibrium, that is, the reactions proceed for a time that exceeds the time necessary for the MBD to establish, the reaction kinetics is called *equilibrium* [130, 127, 140]. These are the reactions in gases at $T < 1000–1200\,°\text{C}$, in solutions, and on solid surfaces. In the liquid phase, the efficiency of MBD establishment increases due to the collisions between reaction participants and intermediates with solvent molecules. The kinetics of all the reactions considered in this book refers to the region of equilibrium (in the above sense) kinetics. The science of nonequilibrium kinetics studies the photochemical and plasmachemical reactions and the combustion and explosion processes, in which definite quantum states of the reactants and reaction products are taken into consideration [140], whereby the molecules in several energy states can simultaneously participate in both the chemical reactions and physical energy transfer processes.

The interaction between several particles (reactant molecules) with the formation of species called reaction products, which proceeds by surmounting a single potential barrier, is referred to as the *elementary act* of chemical reaction. Being multiply repeated with a large number of molecules, these acts lead to the formation of a measurable amount of products that is called *elementary reaction* or *elementary step*. The stoichiometry of the elementary act and step are naturally the same, for example,

$$\text{A} + \text{B} \longrightarrow \text{P}. \tag{1.173}$$

The number of participants of the elementary step (i.e., the sum of the stoichiometric coefficients of reactants) is called the *molecularity* (μ) of the elementary step. It is a common practice to distinguish between the mono-, bi-, and trimolecular steps. Since any elementary step takes place as a result of collisions between reactants and molecules, there arises the natural question as to what collision should be considered an elementary act of the chemical reaction. The average energy of all collisions in the gas is $1.5kT$. It is commonly accepted that, when the energy of atomic collisions exceeds the average energy of thermal motion ($E > 2kT$), a chemical compound is formed that is characterized by a minimum on the potential energy surface. Thus, if compounds (AB) with $\Delta E > 2kT$ (~1200 cal/mole) do not appear on the passage from reactants (A, B) to products (P) along the reaction variable trajectory on the potential energy surface, we can speak of the elementary act (1.173) of the product P formation with a single potential barrier along the reaction coordinate. A reaction that is reverse to (1.173) also proceeds along the reaction trajectory via the same saddle point (barrier), but in the reverse direction (starting from P). The elementary reaction act or step is alternatively called the consistent reaction.

In some cases, bimolecular reactions of the type (1.173) or

$$\mathrm{A} + \mathrm{B} \longrightarrow \mathrm{C} + \mathrm{D} \tag{1.174}$$

are considered as reactions proceeding via the formation of a long-lived excited state [130]

$$\mathrm{A} + \mathrm{B} \rightleftharpoons \mathrm{AB}^* \rightarrow \mathrm{P}_i$$

with a minimum on the potential energy surface, which corresponds to an AB* associate. Using this approach, it is possible to use methods of the theory of monomolecular reactions (AB* conversion) for calculating the rate constants. However, in this case, we are not dealing with an elementary step, since the reaction proceeds via two potential barriers and is represented by a block of two elementary steps.

A change (Δn_i) in the amount (number of moles) of the ith participant of a step (or the overall reaction equation, see Section 1.1) as a result of the chemical reaction is determined by the rate of formation or consumption (variation of the concentration n_i) of this species, the reaction time t, and volume V, so that $\Delta n_i = f(r_i, t, V)$. Therefore, the reaction rate r_i for this substance is given by the second-order derivative of n_i with respect to V and t [19]:

$$r_i = \frac{\partial^2 n_i}{\partial V \partial t}. \tag{1.175}$$

If r_i is the same over the entire volume and $V = \text{const}$ (closed system), then $\Delta n_i \propto V$ and the differentiation with respect to V can be replaced by the division as

$$r_i = \frac{dn_i}{Vdt} = \frac{dC_i}{dt}. \tag{1.176}$$

If the reaction proceeds in the open system in a steady-state regime, then $\Delta n_i \propto t$ and the differentiation with respect to time can be replaced by the division as:

$$r_i = \frac{dn_i}{tdV} = \frac{dF_i}{dV}, \tag{1.177}$$

where F_i is the molar flux (n_i/t). Should both these conditions be satisfied (as in gradient-free, continuous-flow ideally stirred-tank reactor, CSTR), then

$$r_i = \frac{n_i}{tV} \text{ or } r_i = \frac{\Delta F_i}{V}. \tag{1.178}$$

The rate of the jth step W_j at $V = \text{const}$ is a derivative of the extent of reaction (or chemical variable) ξ_j with respect to the time per unit volume:

$$W_j = \frac{d\xi_j}{Vdt}, \tag{1.179}$$

where $\xi_j = \Delta n_i/\beta_i$, from which it follows that

$$W_j = \frac{dC_i}{dt} \cdot \frac{1}{\beta_i} = \frac{1}{\beta_i} r_i. \tag{1.180}$$

Let us consider the assumptions, laws, and heuristic rules underlying the theory of elementary reaction steps. The most important laws and postulates in the chemical kinetics are as follows [127, 140].

- **Conservation of mass and elements in the course of chemical reaction.** These laws are expressed as

$$\boldsymbol{\beta}_i \times \boldsymbol{M}_i = 0; \tag{1.181}$$

$$\boldsymbol{\beta}_i \times \boldsymbol{H} = 0, \tag{1.182}$$

  where $\boldsymbol{\beta}_i$ is the row-vector of stoichiometric coefficients of the reaction participants, $\boldsymbol{M}_i$ is the column-vector of their molecular masses, and $\boldsymbol{H}$ is the atomic (molecular) matrix (see Section 1.1). The mass defect that arises due to the energy evolution or absorption in the course of chemical reactions is negligibly small and can be ignored.
- **Law of energy conservation.**
- **Law of mass action.** According to this law (Guldberg–Waage, Van't Hoff), the rate of an elementary step in a certain direction is proportional to a product of the concentrations of reactants in powers corresponding to the stoichiometric coefficients in the step rate equation

$$W^+ = k^+ \prod C_i^{\beta_i}. \tag{1.183}$$

  For example, in the case of reaction (1.173), we have

$$W^+ = k^+ C_A C_B,$$

  where k+ is the rate constant (specific rate at $\prod C_i^{\beta_i} = 1$). The sum of β_i corresponds to the molecularity of reaction, while the exponent β_i is called the reaction order. In the case of elementary steps, the total order coincides with the molecularity. Other models proposed for description of the elementary step rate (e.g., de Donde's model, see [7, 127]) did not receive further development in the chemical kinetics.
- **Necessity of collisions between particles (molecules) for chemical transformation.** Even in the case of monomolecular reactions, all theories [129] are based on the Lindemann–Christiansen scheme, according to which a molecule of reactant A in the reaction A → B gains an energy that is necessary for its conversion into B as a result of collisions with any molecules M (inert gas, product, or reactant):

$$A + M \underset{k_{-1}}{\overset{k_1}{\rightleftharpoons}} A^* + M;$$

$$A^* \xrightarrow{K_2} B.$$

  In the case of a quasi-steady-state process with respect to [A*], the rate of this conversion is

$$r = \frac{k_1 k_2 [A][M]}{k_{-1}[M] + k_2}. \tag{1.184}$$

  For large concentrations (high pressures) of M, we have $k_{-1}[M] \gg k_2$, so that

$$r = \frac{k_1 k_2}{k_{-1}}[A].$$

At low concentrations of M, the first step with second-order kinetics in the forward direction becomes the limiting step. Since the probability of collisions of two molecules ($z_0^B = 10^{-10} - 10^{-11} cm^3/s$) is significantly greater than that of triple collisions ($z_0^T = 10^{-33} - 10^{-35} cm^6/s$), the probability of trimolecular reactions is low. In known examples, the trimolecular reactions predominantly take place due to combinations (blocks) of bimolecular elementary steps.

Simple considerations based on the theory of gas phase collisions lead to the first, very important, selection rule that is applicable to both gas and liquid phases, according to which the molecularity of elementary steps does not exceed 2 ($\mu \leq 2$).

Particles in the gas phase exhibit collisions over the entire volume, while molecules in the liquid collide only within a free volume of a cage of solvent molecules. In the first sphere of this cage, a molecule of reactant A is surrounded by 10–12 solvent molecules. In order to collide under these conditions, molecules A and B must initially diffuse from individual solvent cages A_{sol} and B_{sol} to a common cage $(AB)_{sol}$, that is, to form a so-called *diffusion pair*. Since the free volume V_f of a cage amounts to 0.2–2 % of the corresponding solvent volume [130], the frequency of collisions within the cage is greater than in the volume of a gas phase. The experimental ratio k_{liq}/k_{gas} of the rate constants for the same bimolecular reaction in nonpolar solvents amounts to 10–150. Apparently, the probability of collision for three molecules in the cage will not be significantly greater than in the gas phase, whereas the probability of formation of a diffusion *triad* in one cage is much smaller than that for a diffusion pair.

In the reactions of complex molecules such as enzymes, the molecularities of separate steps also do not exceed 2. However, in the case of enzymatic catalysis, the active site of an enzyme can admit multicenter binding and the simultaneous participation of a greater number (e.g., three or four) of active groups in the reaction. Thus, while the reaction with respect to the enzyme–substrate (ES) complex is, for example, bimolecular ($ES + H_2O$), the cavity of the active site can feature a multicenter process. A strong drop in the entropy of activation in this case is compensated by an increase in the entropy due to a change in the tertiary structure of a protein and its dehydration as a result of the reconstruction caused by the ES complex formation.

- **Principle of microscopic reversibility.** This principle stipulates the reversibility of any elementary act or microscopic process on a molecular level. Upon attaining equilibrium, the sum of elementary acts in the forward direction (macroscopic process) is compensated by the sum of elementary acts in the reverse direction. It is a macroscopic description of large molecular ensembles that takes into account the MBD and statistical thermodynamic characteristics (such as ΔH, ΔS) and, accordingly, leads to irreversibility. A microscopically reversible process in a macroscopic system can be kinetically irreversible. In a macrosystem of reversible steps, each step (reaction) independently attains equilibrium when the chemical potential $\Delta\mu$ (or the chemical affinity A) ceases to change:

$$\frac{dG}{d\xi} = \Delta\mu = -A = 0\left(\Delta\mu = \sum \beta_i\mu_i\right).$$

  The condition $\Delta\mu = -A = 0$ implies the equality of rates of the forward and reverse reactions: $W^+ = W^-$.
- **Principle of detailed balance.** This principle determines a statistical relationship between the rate constants of an elementary process in the forward (k^+) and reverse (k^-) directions under the MBD conditions as an equilibrium constant of this process ($k^+/k^- = K$). The principle of detailed balance follows from the principle of microscopic reversibility and the equality of step rates ($W^+ = W^-$) at the point of chemical equilibrium ($dG/d\xi = 0$). Thus, the principle of detailed balance is a macroscopic manifestation of the principle of microscopic reversibility. Indeed, proceeding from the latter principle and assuming that $W^+ = W^-$ at equilibrium, we arrive at the principle of detailed balance – and vice

versa: on accepting that $k^+/k^- = K$ is an equilibrium constant, we obtain the equality of the rates $W^+ = W^-$ at equilibrium. For example, consider reversible reaction (1.173), where

$$\frac{W^+}{W^-} = \frac{k^+}{k^-} \cdot \frac{C_A C_B}{C_P}; \tag{1.185}$$

$$RT \ln \frac{W^+}{W^-} = RT \ln \frac{k^+}{k^-} - RT \ln \frac{C_P}{C_A C_B}. \tag{1.186}$$

Assuming that $k^+/k^- = K$, we obtain

$$RT \ln \frac{W^+}{W^-} = -\Delta G^\circ - RT \ln \frac{C_P}{C_A C_B},$$

or

$$\ln \frac{W^+}{W^-} = -\frac{\Delta G}{RT}. \tag{1.187}$$

Equation (1.187) implies that, for $\Delta G = dG/d\xi = 0$, we have $W^+ = W^-$ (here, $\Delta G = \Delta\mu$). The above considerations are valid for the ideal gases and ideal solutions. The principle of microscopic reversibility leads to some other important conclusions [15, 51] concerning mechanisms of complex reactions. In particular, it follows that a complex reaction in the forward and reverse directions passes via the same elementary steps and intermediates.

- **Arrhenius law.** This law in the form of the corresponding equation describes the basic property of an elementary step under the MBD conditions – the exponential dependence of the rate constant on the temperature:

$$k = A \exp(-E_{act}/RT). \tag{1.188}$$

This equation was experimentally established by Hood in 1885. Theoretically, this type of dependence was predicted by van't Hoff (1883) based on the equilibrium thermodynamics and then confirmed and substantiated by Arrhenius (1889) within the framework of statistical physics [7]. The classical mechanism of pair collisions between molecules treated as elastic balls leads to the following expression:

$$k = A(T) \exp(-E_{act}/RT). \tag{1.189}$$

Taking into account the necessary orientation of molecules, a kinetic equation should be written in a simple form as follows:

$$k = pZ_0 \exp(-E_{act}/RT), \tag{1.190}$$

where p is a sterical factor and Z_0 is the total number of collisions.

The activation energy E_{act} of an elementary step entering in Eqs (1.188–1.190) represents a difference between the average energy of interacting particles and that of all particles, while the exponential factor characterizes the fraction of "active" collisions – that is, collisions with the energy E equal to or exceeding the barrier energy. This expression follows from the fact that the main contribution to E_{act} in the gas phase is due to the translational motion of A and B molecules. In the liquid phase, the character of molecular motions becomes oscillatory. If the total system energy (kinetic and potential) is described by $2S$ terms (for S independent harmonic oscillators), then the fraction of active particles increases. In particular, a simple approximation [130, 131]

$$\frac{N_{(E \geq E_i)}}{N} = \exp\left(-\frac{E_i}{kT}\right) \frac{[E_i/(kT)]^{S-1}}{(S-1)!} \tag{1.191}$$

yields the following formula for a bimolecular reaction:

$$k^{\mathrm{bm}} = Z_0 \exp\left(-\frac{E_i}{RT}\right) \frac{(E_i/RT)^{(S-1)}}{(S-1)!}. \tag{1.192}$$

Restriction of the expansion in Eq. (1.191) to the first two terms yields an expression analogous to that for the translational motion:

$$k^{\mathrm{bm}} = Z_0 \left(\frac{E}{RT}\right) \exp\left(-\frac{E}{RT}\right). \tag{1.193}$$

In this case, the activation energy in the Arrhenius law is $E_{\mathrm{act}} = E - 0.5RT$.

An expression for k^+ can be obtained within the framework of the transition state (activated complex) theory. According to this theory, the rate W^+ of elementary step (1.173) is

$$W^+ = C^{\neq}_{\mathrm{AB}}/\tau, \tag{1.194}$$

where $C^{\neq}_{\mathrm{AB}}$ is the concentration of activated complexes $(\mathrm{AB})^{\neq}$ at the barrier top and τ is the time of passage over the barrier, which obeys the relation

$$1/\tau = kT/h$$

(kT/h is the frequency of transitions over the barrier). The state of the activated complexes occurring on top of the barrier (i.e., in the transition state) corresponds to their equilibrium with the initial reactants, so that $C^{\neq}_{\mathrm{AB}}$ can be expressed in terms of the thermodynamic equilibrium constant $K^{\neq}_a$ for the formation of activated complex (postulate) under the MBD conditions for all particles (including the activated complex $(\mathrm{AB})^{\neq}$):

$$K^{\neq}_a = \frac{a^{\neq}_{\mathrm{AB}}}{a_{\mathrm{A}} a_{\mathrm{B}}} = \frac{C^{\neq}_{\mathrm{AB}}}{C_{\mathrm{A}} C_{\mathrm{B}}} \cdot \frac{\gamma^{\neq}_{\mathrm{AB}}}{\gamma_{\mathrm{A}} \gamma_{\mathrm{B}}}, \tag{1.195}$$

where a is the activity and γ is the activity coefficient. This relation yields

$$C^{\neq}_{\mathrm{AB}} = K^{\neq}_a \frac{\gamma^{\neq}_{\mathrm{AB}}}{\gamma_{\mathrm{A}} \gamma_{\mathrm{B}}} C_{\mathrm{A}} C_{\mathrm{B}} \tag{1.196}$$

and

$$W^+ = \frac{kT}{h} K^{\neq}_a \frac{\gamma^{\neq}_{\mathrm{AB}}}{\gamma_{\mathrm{A}} \gamma_{\mathrm{B}}} C_{\mathrm{A}} C_{\mathrm{B}}; \tag{1.197}$$

$$k^+ = \frac{kT}{h} K^{\neq}_a \frac{\gamma^{\neq}_{\mathrm{AB}}}{\gamma_{\mathrm{A}} \gamma_{\mathrm{B}}}. \tag{1.198}$$

In an ideal system (gas, solution), all the activity coefficients are $\gamma_i = 1$ and, hence, $K^{\neq}_a$ is equal to the concentration constant $K^{\neq}_C$. Since the transition frequency kT/h has a dimensionality of [s^{-1}], while $K^{\neq}_a$ is dimensionless, the dimensionality of the reaction rate constant is determined by that of the activity factor

$F = \gamma_A \gamma_B / \gamma_{AB}^{\neq}$. In the given case, F has a dimensionality of [L · mole^{-1}] (same as γ_i). This dimensionality is retained in the case where all activity coefficients are $\gamma_i = 1$.

In order to improve the accuracy of description, a transmission coefficient χ is introduced into Eq. (1.198) so as to specify the fraction of activated complexes that overcome the barrier:

$$k^+ = \chi \frac{kT}{h} K_a^{\neq} F. \tag{1.199}$$

For $F = 1$, we have

$$k^+ = \chi \frac{kT}{h} \exp\left(\frac{-\Delta G^{\circ\neq}}{RT}\right) = \chi \frac{kT}{h} \exp\left(\frac{\Delta S^{\circ\neq}}{R}\right) \exp\left(\frac{-\Delta H^{\circ\neq}}{RT}\right), \tag{1.200}$$

where $\Delta G^{\circ\neq}$ ($\Delta\mu^{\circ\neq}$) is a change in the standard free energy (or the chemical potential) upon the formation of the transition state and $\Delta S^{\circ\neq}$ and $\Delta H^{\circ\neq}$ are the changes in the standard entropy and enthalpy, respectively, of activation of the process

$$A + B \rightleftharpoons (AB)^{\neq}.$$

The value of $\Delta H^{\circ\neq}$ is related to E_{act} in the Arrhenius equation (1.188) as $\Delta H^{\circ\neq} = E_{act} - RT$. Then

$$k^+ = \chi \frac{kTe}{h} \exp\left(\frac{\Delta S^{\circ\neq}}{R}\right) \exp \frac{-E_{act}}{RT}, \tag{1.201}$$

where $\Delta S^{\circ\neq}$ is the standard entropy of activation at V = const. Equation (1.201) describes the rate constant of a liquid-phase reaction. In the reactions involving light quantum particles (electron, proton), the transition to products in a reacting system can take place even before the energy reaches the barrer height – that is, by means of the tunneling effect, in which case $\chi > 1$. The preexponential factor A in the Arrhenius equation (1.188) is

$$A = \chi \frac{kTe}{h} \exp\left(\frac{\Delta S^{\circ\neq}}{R}\right); \tag{1.202}$$

$$\Delta S^{\circ\neq} = R \ln \frac{Ah}{\chi ekT}. \tag{1.203}$$

Despite some inconsistency [50, 127, 140], the transition state theory has proved to be very useful for the creation of a semiempirical theory of chemical reactivity and is still in progress [51, 58, 141]. Quite recently (1999), A. Zewail was awarded the Nobel Prize in Chemistry for developing the method of femtosecond laser spectroscopy and studying the transition states of chemical reactions. Using this technique with characteristic times of 10–100 fs (1 femtosecond = 10^{-15} s) it is possible to measure the characteristics of transition states in some reactions with lifetimes on the order of $\sim 10^{-13}$ s [142].

The law of mass action (1.183) is written in terms of the product of volume concentrations C_i. This circumstance is related to the fact that, in the main models of elementary steps (employing the theory of collisions and the transition state theory), the volume of a system plays a determining role, since the frequency of collisions and the number of activated complexes passing over the barrier per unit time depend on the system volume. Within the framework of the transition state theory, the reaction rate can also be expressed in terms of the activities of reactants as in Eq. (1.197). Accordingly,

$$W^+ = k_0^+ \frac{a_A a_B}{\gamma_{AB}^{\neq}}, \tag{1.204}$$

where k_0 is the rate constant that is independent of the properties of a medium. This writing agrees with the principle of microscopic reversibility, since the equation $W^+ = W^-$ allows the thermodynamic equilibrium constant to be obtained only using activities (or only for an ideal solution, where $a_i = C_i$) as

$$W^- = k_0^- \frac{a_P}{\gamma_P^{\neq}}, \tag{1.205}$$

because the transition state for the forward and reverse reactions is the same and $\gamma_{AB}^{\neq} = \gamma_P^{\neq} = \gamma^{\neq}$, so that

$$\frac{k_0^+}{k_0^-} = \frac{a_P}{a_A a_B} = K_a. \tag{1.206}$$

It should be noted that, during any variation of the composition of medium (solution), it is easier to maintain the value of activity $\gamma^{\neq}$ than the value of factor F (because of a very low concentration of activated complexes). Therefore, it can be expedient to use Eq. (1.204), especially in the case of organic reactions where determination of the activities of components of the medium (reaction participants) is a relatively simple, well developed routine procedure. The use of activities can also be useful in the case of ionic systems (see Chapters 2 and 8).

- **Principle of least motion.** The most important postulates of classical mechanics – the principle of least action (Euler, Maupertuis) and the variational principles (Lagrange, Hamilton) [143] – undoubtedly strongly influenced the first attempts at formulating the general approach to chemical reactivity [144]. In 1850, Kekule suggested that molecules subject to least structural changes in the course of a chemical reaction would react at a greater rate, and in 1924 Mueller and Paitral formulated a close principle of the least deformation of molecules at the elementary reaction step. Eventually, in 1938 Rice and Teller formulated the principle of least motion. According to this, an elementary step proceeds fast, with low activation energy E_{act}, provided that this step is characterized by
  (i) the least motion of nuclei (i.e., a minimum change in the coordinates of these nuclei); and
  (ii) the least motion of electrons (for which rearrangement of the electron shells does not lead to a change in the valence state).

  If a system exhibits only an insignificant change in the coordinates of the nuclei, this implies that the potential energy terms of the reactants and products are close. In general, it is true that the closer the terms, the lower the barrier (E_{act}). However, this is not always the case. The pathway of least motion may not coincide with the pathway of minimum energy [144]. Indeed, consider the system

  $$A + BC \longrightarrow AB + C. \tag{1.207}$$

  For the least motion, the approach of reactant A to group B is accompanied by the removal of group C (Figure 1.16, line *1*). In some cases, it is energetically more favorable to attach A to BC with subsequent rupture of the B–C bond (curve *2*), whereas in other cases it is better first to rupture the B–C bond and then add A to B (curve *3*). The principle of least motion for electrons is more universal. In the case of, e.g., monomolecular reactions, this condition implies that the electron structure of the reactive center of a reactant molecule must be close to the electron structure of the activated complex, so that the barrier height (or $\Delta G^{o\neq}$) would possibly be small.
- **Principle of independence of chemical reactions.** The elementary steps proceeding simultaneously in a reacting system must obey a postulate that these chemical reactions are independent of each other. According to this principle, introduced by W. Ostwald, all chemical reactions, including their steps, proceed independently (see, e.g., [51]). Relations between independent reactions are only established

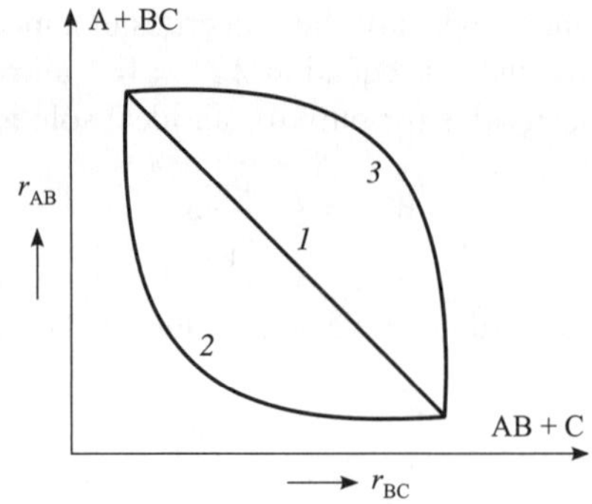

Figure 1.16 Pathways of least motion and minimum energy on potential energy surface; r_{AB} (r_{BC}) is the distance between groups A and B (B and C)

on the level of a material balance, through a change in the concentrations, and the appearance of new substances. According to the same principle, the forward and reverse elementary reactions also proceed independently and this allows a chemical equilibrium to be established (see also the principle of microscopic reversibility).

1.4.2 Energy selection rules for elementary steps

According to the principles and laws considered above, the possibility for a given reaction to occur as an elementary (consistent) process is determined either by the fraction of active species with energies E equal to or greater than the activation energy ($E \geq E_{act}$), or by the value of free energy $\Delta G^{o\neq}$ that is dependent on the barrier on the potential energy surface ($\Delta H^{o\neq}$) and the shape of this surface in the region of a saddle point ($\Delta S^{o\neq}$). There are different reasons for which the $\Delta G^{o\neq}$ and $\Delta H^{o\neq}$ values may vary. First, we will consider the energy selection rules, that is, the rules in which the thermodynamic characteristics of reactions are manifested.

- **Thermodynamic limitations on elementary reaction steps.** The main restriction in the case of an endothermal reaction is related to the value of the standard change in the enthalpy ΔH^{o}, which determines the lower limit of E_{act}. Indeed, in most cases, $E_{act} \geq \Delta H^{o}$ because maxima on the curves of potential energy and free energy are situated at the same value of the reaction coordinate [131]. In cases where the ΔH^{o}_{298} value is large (e.g., $\Delta H^{o}_{298} = 40\,000$ cal/mole) and the preexponential factor for a bimolecular reaction of the S_N2 type has a typical value of $A \sim 10^{11}\text{M}^{-1}\,\text{s}^{-1}$ (A usually varies within $10^{10} \div 10^{12}\,\text{M}^{-1}\,\text{s}^{-1}$), the rate constant k at $T = 298$ K will amount to

$$k = 10^{11} \exp\left(-\frac{40\,000}{RT}\right) \cong 5 \cdot 10^{-19}\,\text{M}^{-1}\text{s}^{-1}.$$

In order to ensure that the step rate (W_j) would be no less that 0.01 M/h (value acceptable in practice), the product of reactant concentrations in reaction of type (1.173) must be about

$$C_A C_B = \frac{W_j}{k} = \frac{0.01}{3600 \cdot 5 \cdot 10^{-19}} = 5.6 \cdot 10^{12}\,\text{M}^2,$$

which is unrealistic. For the same rate and $C_A C_B = 10^{-4}$ M^2, we have $k = 2.8 \cdot 10^{-2}$ M^{-1} s^{-1}, which yields $E_{act} \cong 17\,000$ cal/mole.

Thus, depending on the reaction temperature and the expected rate of a steady-state or quasi-steady-state process, we can set limitations on the ΔH°_{298} value in selecting a candidate reaction for the elementary step. In the case of a multi-route reaction, the choice of the rate can also take into account that the step rate can be equal to a sum (or difference) of the rates over routes:

$$\left(W_j = \sum_p \nu_j^{(p)} r_p \right).$$

In evaluating ΔH° for an elementary step, it is necessary to use data that refer to the states of reaction participants in which they actually occur during the reaction. For example, the exothermal and exoergonic reaction

$$Pd^{2+}_{aq} + H_{2(gas)} \rightarrow Pd^{0}_{solid} + 2H^{+}; \tag{1.208}$$

with $\Delta G^{\circ}_{298} = -45.5\,\text{kcal/mole}$ evidently cannot be elementary, since it results in the formation of a solid palladium phase. An elementary step could lead to the formation of an atom ($Pd^{0}_{(aq)}$) in solution:

$$Pd^{2+}_{aq} + H_{2(gas)} \longrightarrow Pd^{0}_{aq} + 2H^{+}; \tag{1.209}$$

with $\Delta G^{\circ}_{298} = +38.5\,\text{kcal/mole}$ (hydration of Pd° does not significantly contribute to the enthalpy of reaction), but the enthalpy of reaction (1.209) is greater than that of reaction (1.208) by 83–90 kcal/mole and, hence, this reaction also cannot be the elementary step.

Use of the ΔG° value as a thermodynamic criterion for the elementary step is not as strict as it is for assessing the possibility of implementing a given brutto process (overall reaction equation). In the latter case, the reaction conditions (P, T) are selected by estimating ΔG_T^P, K_{eq}, and the equilibrium product yield (which in all cases should be large). In the case of an elementary step of intermediate X formation via the reaction

$$A \underset{}{\overset{K_1}{\rightleftharpoons}} X \xrightarrow{k_2} P \tag{1.210}$$

an admissible concentration of X must provide for a positive affinity ($A > 0$, $\Delta\mu < 0$) on the first step at a sufficiently high rate of product P formation on the second step that also depends on k_2 (for more detail, see Section 3.4):

$$W_1 > 0 \text{ at } \frac{W_1^+}{W_1^-} = \frac{K_1}{\prod C_i^{\beta_i}} = \varphi > 1.$$

At the step of formation of intermediate X, we have

$$\varphi = \frac{K_1}{[X]/[A]} > 1. \tag{1.211}$$

For a quasi-equilibrium first step, $K_1 \cong [X]/[A]$. Assuming that $K_1 = 10^{-3}$($\Delta G^{\circ}_{298} = 4090$ cal/mol) and $[A]_0 = 1M$, we have $[X] = 0.999 \cdot 10^{-3}M$. Then, for $W_2 \geq 0.01$ M/h, we can estimate k_2 as

$$k_2 \geq \frac{W_2}{[X]} = \frac{0.01}{3600 \cdot 0.999 \cdot 10^{-3}} = 2.8 \cdot 10^{-3}\,\text{s}^{-1}.$$

and find E_{act} for the monomolecular reaction (with $A = 10^{13} - 10^{15}\ \text{s}^{-1}$) at $T = 298\,\text{K}$:

$$k_2 = 5 \cdot 10^{13} e^{-\frac{E_{act}}{RT}} \Rightarrow E_{act} \leq 22\,000\ \text{cal/mol}.$$

This value of E_{act} on the second step allows the proposed first step to be considered as an allowed elementary reaction despite the positive value of ΔG^{o}_{298}.

In the above example, we dealt with a kinetic conjugation of two elementary steps via a common intermediate X, which allows the product P to be synthesized despite a thermodynamically (energetically) unfavorable first step. Speaking of the kinetic conjugation [145] (see Section 3.4), one should distinguish between the two cases:

(i) $\Delta G^{o} > 0$ for the first step, [X] is small, and the affinity is $A \sim 0$. The appearance of step X → P leads to a decrease in [X], increase in A ($A > 0$), and growth in the difference of $W_1^+ - W_1^-$ (in the limit, up to W_1^+ that is a maximum possible rate of the process). An increase in the rate as a result of the conjugation of two steps is ensured, in fact, by making the first step irreversible.

(ii) $\Delta G > 0$ ($A < 0$) for the first step and $[X]/[A] > K$, the reaction goes leftward ([X] is above the equilibrium level). Here, the conjugation of the first and second steps (X → P) leads to a decrease in [X], so that $\varphi = \frac{K}{[X]/[A]}$ becomes greater than unity and the process direction switches to rightward (forward direction) and at its limit attains the maximum possible rate W_1^+. In this case, the conjugation changes the direction of the first step.

Now let us refine some notions in the context of the issues considered above. The rate of any elementary step is the difference between the rates of the forward and reverse reactions: $W_j = W_j^+ - W_j^-$. The kinetic reversibility of a step is a macroscopic manifestation of the principle of microscopic reversibility. On the other hand, a kinetically reversible reaction is thermodynamically irreversible, since a spontaneous chemical reaction proceeds only in one direction: toward a decrease in the affinity and increase in the entropy until attaining equilibrium ($A_j = 0$, $dS_j/dt = 0$, closed system).

A kinetically irreversible step can be represented by an elementary step in which the ratio $K_{eq}/\prod C_i^{\beta_i} = \varphi$ is $\gg 1 (W_j^+/W_j^- \gg 1)$ during the entire process up to reaching the degree of reactant conversion $\alpha = 0.999$. For example, if $K_{eq} \cong 10^{15}$, then we have $\varphi \gg 1$ over the entire process (for the conversion varying from $\alpha = 0$ to $\alpha = 0.999$) and the ratio W_j^+/W_j^- varies within $10^{12} - 10^7$ (decreasing, but still remaining very large). In other words, processes in which

$$W_j^+ - W_j^- \cong W_j^+$$

can be considered kinetically irreversible.

- **Empirical dependences of E_{act} and $\Delta G^{o\neq}$ on thermodynamic characteristics of steps.** Based on the results of experimental observations (J. Brønsted, R. Bell, M. Polanyi) and theoretical investigations (R. Bell, M. Evans, M. Polanyi), the so-called *linear free-energy relationships* were formulated, which are also referred to as the Bell–Evans–Polanyi (BEP) rule [51, 144, 146]. This rule reflects relationships between the kinetic and thermodynamic characteristics in a series of analogous elementary reactions:

$$\Delta G^{o\neq} = \beta + \alpha \Delta G^{o}; \tag{1.212}$$

$$E_{act} = \beta + \Delta H^{o}. \tag{1.213}$$

Let us consider a series of the curves of potential energy (representing cross sections of the potential energy surface along the reaction coordinate) for reactants and products of the reaction

$$A_i + BC \longrightarrow A_iB + C, \tag{1.214}$$

which yields a diagram presented in Figure 1.17. Since the energy of the BC bond rupture (E_{BC}) is independent of A_i, we can fit all curves for different A_i at point 2. By the same token, the curves of

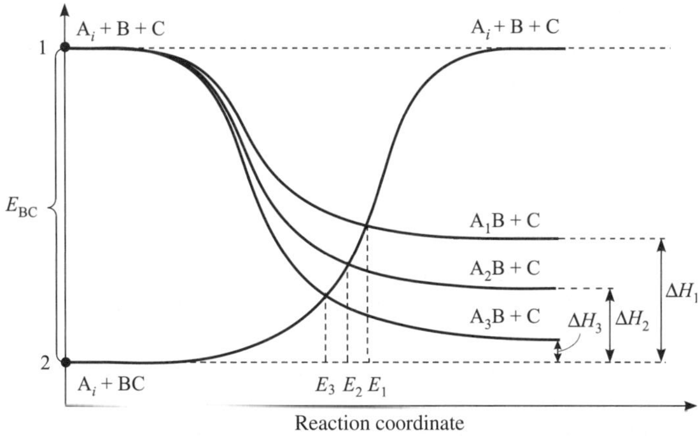

Figure 1.17 Diagram illustrating the Bell–Evans–Polanyi (BEP) rule

A_iB formation from A_i, B and C can be constructed from point 1. On varying group A_i in the given series of analogous reactions and assuming that the energy terms of $A_iB + C$ do not intersect (rule of non-crossing, or repulsion of terms), it is natural to expect a certain relationship between ΔH_i and E_i since an increase in the endothermicity of reaction (1.214) leads to an increase in E_{act}. Polanyi suggested to approximate this relationship using linear equation (1.213) in a certain interval of ΔH_i, and N.N. Semenov determined coefficients of this equation for several elementary steps with some radicals and molecules (Polanyi–Semenov equation [50, 130]; see Chapter 7):

$$E_{exo} = 11.5 - 0.25|\Delta H^{o}|; \tag{1.215}$$

$$E_{endo} = 11.5 + 0.75|\Delta H^{o}|. \tag{1.216}$$

According to the BEP rule, the elementary steps can be selected more accurately by using E_{act} rather than ΔH^{o} as the upper limit.

A question of principal importance is what the form of function $E_i = f(\Delta H_i)$ should be. As long ago as the 1930s, Polanyi understood that the linear free-energy relationships provide only a rough linear approximation (valid in a narrow interval of thermodynamic quantities ΔG^{o} and ΔH^{o}) of more complicated functions such as, for example, a quadratic relation

$$E = \frac{(A + \Delta H^{o})^2}{4A}. \tag{1.217}$$

An equation of this type was later obtained for the reactions with electron transfer (R. Marcus, 1956) and proton transfer (R. Marcus, 1968) and then refined and theoretically justified by Levich, Dogonadze, and Kuznetsov [134–137]. Parabolic and more complicated power equations for the calculation of E_{act} of radical reaction steps using ΔH^{o} values are presented in [130] (for more detail, see Chapter 7).

In developing a theory of the electron transfer in polar solvents, Marcus introduced important concepts of the internal and external barriers by separating $\Delta G^{o\neq}$ into two components as

$$\Delta G^{o\neq} = \Delta G^{\neq}_{int} + \Delta G^{\neq}_{ext}. \tag{1.218}$$

Reactions in polar media are accompanied by changes not only in the internal coordinates of atoms in reacting species, but also in the coordinates of molecules of the solvent (in the general case, of the medium) along the reaction coordinate, since the solvent (medium) exhibits reorganization around the transition state [131]. Accordingly, the A value in Eq. (1.217) for the reactions with electron transfer is related to the energy of solvent reorganization.

1.4.3 Quantum-chemical selection rules for elementary steps

Any thermodynamically allowed reaction in which the nuclei exhibit insignificant displacement (i.e., the minima of energy terms are close), the electron states change but little (in accordance with the principle of least motion), and the molecularity μ does not exceed 2 and has a chance to be a consistent process, that is, an elementary reaction step. However, for the $\Delta G^{o\neq}$ value to be small and the reaction to proceed at a measurable rate, two conditions must be fulfilled that follow from the quantum-chemical theory. According to these, an effective interaction of two particles with a sufficiently low barrier can take place only provided that their overlapping molecular orbitals (MOs) possess the same symmetry and that the energies of these MOs are close. For example, as will be shown below, the bimolecular reaction

$$H_2 + I_2 = 2HI \tag{1.219}$$

with small changes in the nuclear coordinates and electron shell structures cannot be an elementary step because it is prohibited with respect to the symmetry of frontier MOs. Below we will consider some quantum-chemical approaches to the problem of chemical reactivity in more detail.

- **Perturbation theory in the approximation of frontier MOs [137, 139]** According to the BEP rule, the knowledge of the energy states of the initial and final products allows the kinetic characteristics of elementary steps (probabilities of elementary acts) to be evaluated. Using the MO perturbation method and operating only with frontier MOs (highest occupied MO, HOMO; lowest unoccupied MO, LUMO) and charges on atoms, molecules, and atomic orbitals (AOs) in MOs, it is possible in some cases to predict the probability, direction, and efficiency of interaction between two reactants.

The more effective the interaction, the lower the activation energy (E_{act}) and the higher the probability of a consistent elementary act.

If the energies of the frontier orbitals χ_1 and χ_2 are close, then their interaction energy is determined by the resonance (exchange) integral β_{12} defined as

$$\beta_{12} = \int_0^\tau \chi_1 \hat{H} \chi_2 dv, \tag{1.220}$$

where $\hat{H}$ is the Hamiltonian of the system and v is the volume element in which the orbitals overlap.

In this case, the value of β_{12} also determines the splitting of new MOs (Ψ_1 and Ψ_2) and the stabilization energy ($\varepsilon = \beta_{12}$):

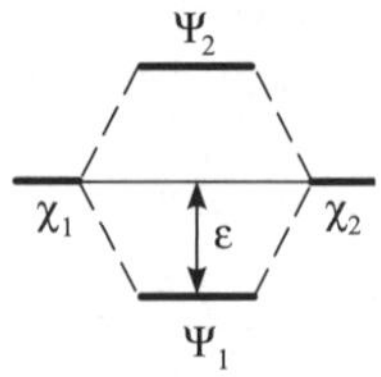

In contrast, if the energies of χ_1 and χ_2 differ significantly, then the ε value depends, besides β_{12}, on the difference of the frontier orbital energies as

$$\varepsilon = \frac{\beta_{12}^2}{E_1 - E_2}, \tag{1.221}$$

where E_1 and E_2 are the energies of the lowest and highest MOs, respectively:

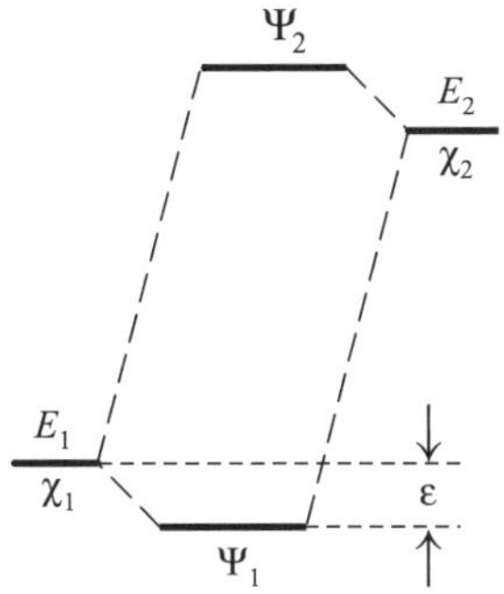

The greater the ε value, the higher the stability of the resulting adduct and the lower the activation energy E_{act} of its formation. The method of MO perturbation does not provide estimation of the transition state energy and E_{act}, and it only makes possible the calculation of a difference between the total electron energy E of the reacting system and the energies of initial reactants E_{0S} and E_{0T} (small perturbation):

$$\Delta E = E - E_{0S} - E_{0T}, \tag{1.222}$$

which is only valid for the initial interval of the reaction coordinate. Despite the fact that this approximation is only valid for large distances between reactants, where the MOs exhibit no mixing and there is no intermolecular repulsion (so that we can speak about pure MOs of the initial reactants), this approach allows the most probable reaction pathway to be assessed.

The perturbation energy ΔE for the interaction of reactants S and T (both being molecules or active centers in molecules) is calculated using the following equation:

$$\Delta E = -\frac{q_S q_T}{R_{ST}\varepsilon} + \gamma_{ST} \sum_{m\text{OMO}} \sum_{n\text{UMO}} \left(\frac{C_{Sm}^2 C_{Tn}^2 \beta_{ST}^2}{E_m - E_n} \right). \tag{1.223}$$

which simplifies in the presence of only two frontier MOs (e.g., electron donor and acceptor) to

$$\Delta E = -\frac{q_S q_T}{R_{ST}\varepsilon} + \gamma_{ST} \left(\frac{C_S^2 C_T^2 \beta_{ST}^2}{E_1 - E_2} \right). \tag{1.224}$$

Here, q_S and q_T are the effective charges on the S and T centers; R_{ST} is the distance between these centers; ε is the dielectric permittivity of the medium; γ_{ST} is the coefficient that takes into account the electron population of χ_1 and χ_2 orbitals; and C_S and C_T are the coefficients at the wave functions of the AOs of S and T centers, respectively, in the wave functions of the frontier MOs χ_1 and χ_2; β_{ST} is the exchange integral; and E_1 and E_2 are the energies of χ_1 and χ_2 orbitals, respectively.

The first term (first-order perturbation) in these equations reflects the Coulomb interaction energy, while the second term (second-order perturbation) describes the energy due to the orbital overlap. Various cases of different electron populations on the χ_1 and χ_2 orbitals of reacting particles and the corresponding γ_{ST} values are presented below:

Numbers of electrons on frontier orbitals χ_1 and χ_2	γ_{ST}
$2+2, 0+0$	0 (no overlap)
$2+1, 1+0$	1
$2+0, 1+1$	2 (maximum overlap)

If the difference $E_1 - E_2$ in the denominator of the second term of Eq. (1.224) is small, the overlap population is $\gamma_{ST} = 1$ or 2, and the orbitals possess the same symmetry ($\beta_{12} > 0$), then the orbital geometry is convenient for the overlap (coefficients C_S and C_T exhibit large values in the same regions of space) and the second term is significantly greater than the first, so that we can speak of an *orbital-controlled reaction*.

In cases where the $E_1 - E_2$ value is large, the second term becomes small even for large C_S and C_T. If the q_S and q_T values are also large, then we deal with the case of a *charge-controlled reaction*. Using these simple estimations by the MO–LCAO methods, it is possible to determine (without calculating the potential energy surface) in what direction (via which centers) a reaction would proceed and whether a high rate at the selected elementary step can be expected. Naturally, all the above considerations concerning the estimation of ΔE refer only to the elementary steps.

Now let us consider a donor molecule – e.g., SCN^-, the typical nucleophile – reacting with an acceptor that possesses a LUMO. If the HOMO of the donor (χ_1) is close to the LUMO (χ_2) of the acceptor, then the reaction is orbital controlled. This reaction involves the acceptor molecule and that center of the nucleophile (donor) which possesses a maximum charge density (C_i^2) on the frontier orbitals. In particular, the HOMO (Ψ_2) of the SCN^- written in the form of an LCAO is as follows:

$$\Psi_2 = 0.74\varphi_S + 0.33\varphi_C - 0.59\varphi_N;$$

$$C_S^2 = 0.55, \quad C_N^2 = 0.35 \Rightarrow C_S^2 > C_N^2.$$

Accordingly, the reaction with an acceptor will proceed at the S atom:

$$SCN^- + Ag^+ \longrightarrow AgSCN. \qquad (1.225)$$

If the acceptor orbital energy is much higher than the donor orbital energy and the difference $E_1 - E_2$ is large, the reaction is controlled by the Coulomb interaction between charges. In this case, the first (Coulomb's) term in Eq. (1.224) is greater for that center of the nucleophile, which possesses a higher q_i. For calculating q_S and q_N, it is necessary to take into account the coefficient C_i^2 for these centers on all orbitals, that is, to use both Ψ_2 and LOMO (Ψ_1) of the nucleophile:

$$\Psi_1 = 0.33\varphi_S + 0.59\varphi_C + 0.74\varphi_N.$$

The effective charges on the interacting centers are calculated using the following formula:

$$q_i = 1 - 2\left(\sum C_i^2\right),$$

which yields

$$q_S = 1 - 2(0.33^2 + 0.74^2) = -0.313,$$

$$q_N = 1 - 2(0.59^2 + 0.74^2) = -0.7914.$$

Thus, in the $[\overset{-0.31}{\mathrm{S}}{-}\mathrm{C}{\equiv}\overset{-0.79}{\mathrm{N}}]^-$ anion, the effective (negative) charge on the N atom is greater than that on the S atom: $|q_N| > |q_S|$ (and $q_C \approx 0.1$). Therefore, under the charge control conditions, the nucleophile SCN^- will interact with the acceptor via nitrogen atom as

$$SCN^- + R^+ \longrightarrow RNCS.$$

The MO perturbation method provided a theoretical explanation for several empirical rules and generalizations. In 1958, Ahrland, Chatt and Davies [147] (see also [148]) proposed a classification of metal complexes by separating metal ions into two groups. The first group (a) includes the ions of metal (in most frequently encountered degrees of oxidation) that form the most stable complexes with ligands containing donor atoms of N, O, and F. The second group (b) contains metal ions that form the most stable complexes with ligands containing donor atoms of the third and subsequent periods (P, S, Cl, Br, I). For example, the stability of halide complexes varies depending on the metal in the following order:

$$Zn^{2+}[\text{group(a)}] : F^- \gg Cl^- > Br^- > I^-;$$

$$Hg^{2+}[\text{group(b)}] : I^- > Br^- > Cl^- \gg F^-.$$

Thus, on the passage from Zn^{2+} to Hg^{2+}, the order of stability is reversed as compared to that in the "normal" series for Zn^{2+} (consistent with simple electrostatic considerations). A quantitative measure of the "reversed stability" is provided by the value of $(1/n)\lg\beta_n$ [149] for the process of complete replacement of n chloride ions by bromide ions in an aqueous solution:

$$MCl_n^{z-} + nBr^- \overset{\beta_n}{\rightleftharpoons} MBr_n^{z-} + nCl^-. \tag{1.226}$$

It turns out [149] that, in addition to metals occurring in the so-called "Ahrland–Chatt triangle", some other metals also belong to the group with reversed order of stability. Evidently, for ions of the first group, the stability is determined by the charge factor, whereas for the second group of ions, it is the orbital factor. Analogous explanations were obtained for the rule of interaction between hard and soft acids (Pearson, 1963; see [137]), and the Kornblum rule in organic chemistry [146]. In terms of Pearson's theory, the interaction of hard species (acid and bases, acceptors and donors) corresponds to the charge control, while the interaction of soft species is orbital controlled. The degree of hardness and softness of an acceptor (A) and donor (D) can be assessed using various criteria [139, 150]. In particular, the values of orbital electronegativities E_n(A) and E_m(D) (in electronvolts, eV) according to Klopman [139] are as follows:

Acceptors: (acids)	Al^{3+}	Mg^{2+}	Cr^{3+}	Fe^{2+}	H^+	Na^+	Cu^{2+}	Zn^{2+}	Cu^+	Hg^{2+}
	6	2.42	2.06	0.69	0.42	0.0	−0.55	−1.0	−2.3	−4.6
Donors: (bases)	F^-	H_2O	OH^-	Br^-	CN^-	SH^-	I^-	H^-		
	−12.18	−10.7	−10.45	−9.2	−8.78	−8.59	−8.31	−7.37		

In these series, the hardest acid is Al^{3+}, the hardest base is F^-, the softest acid is Hg^{2+}, and the softest base is H^-.

- **Orbital symmetry and selection rules**. The use of the symmetry of orbitals or electron states for determining allowed and forbidden elementary reaction steps is described in detail elsewhere [137–139, 144]. Here we will only consider the main concepts and rules of this approach. The requirement of identical orbital symmetry for their effective overlap (i.e., for the interaction of components without high energy barriers) is also taken into account when calculating the exchange integrals β_{ST} for the MO perturbation method. Even simple comparison of the symmetry of MOs in interacting molecules allows us, in some cases, to judge the possibility of a consistent elementary act. For example, consider reaction (1.219), for which a consistent character implies that an elementary act must involve the rupture and formation of two bonds. Therefore, the process must pass via a four-center cyclic transition state as follows:

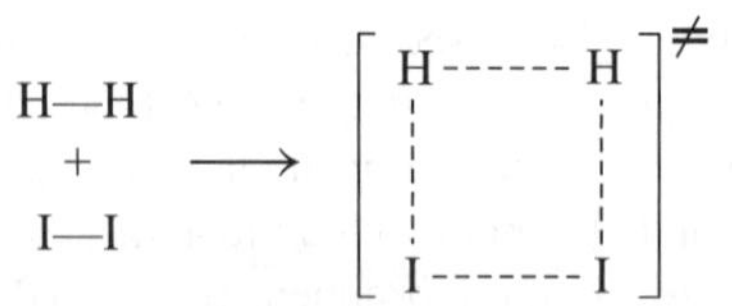

(1.227)

Figure 1.18 shows the σ–MOs of H_2 and I_2 molecules. If we use a single element of symmetry, as represented by the P_{intra} plane passing through the center of the molecule and perpendicular to the H–H and I–I bonds, and the symmetry operation of mirror reflection in this plane, then the σ-MOs of both molecules are symmetric (S) with respect to this operation, whereas their σ^*-MOs are antisymmetric (A). Thus, consistent (synchronous) overlap within the framework of a cyclic transition state (1.227) between the HOMO of H_2 (S) and LUMO of I_2 (A) as well as between LUMO of H_2 (A) and HOMO of I_2 (S) is impossible. Thus, reaction (1.219) with transition state (1.227) cannot be an elementary step.

An analogous simple analysis of the reaction

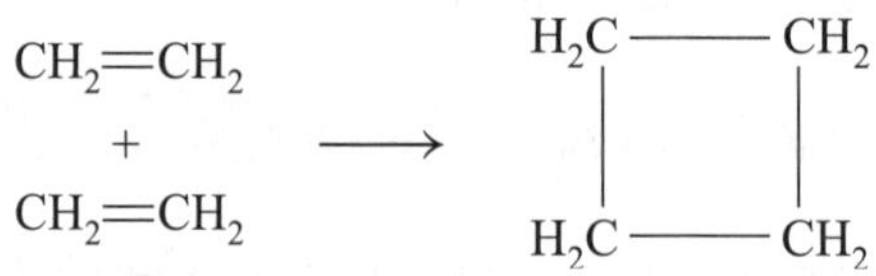

(1.228)

(2+2 cycloaddition) shows that the occupied and vacant MOs of the two molecules possess different symmetries (Figure 1.19). It is evident that, for this symmetry analysis, it is sufficient to use only a diagram of the *p* orbitals (Figure 1.19b). It should be recalled that the shaded and open lobes of the orbitals are indicative of the opposite (±) signs of the angular parts of their wave functions.

The general rules of selecting elementary steps with respect to the symmetry of orbitals in the reacting system were formulated by Woodword and Hoffmann [138] for the elementary reactions passing via cyclic transition states (pericyclic reactions):

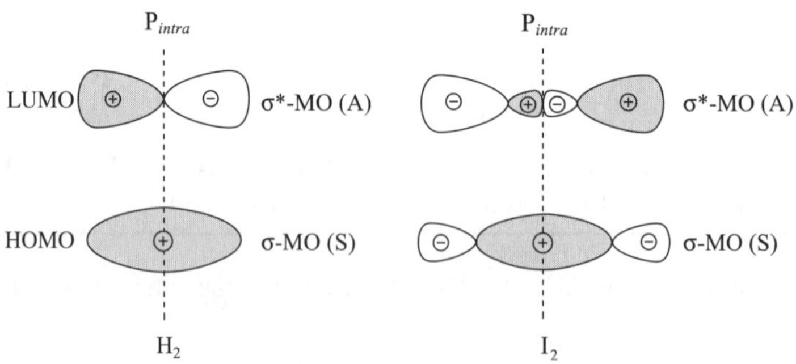

Figure 1.18 Frontier σ MOs formed by s (H_2) and p (I_2) orbitals

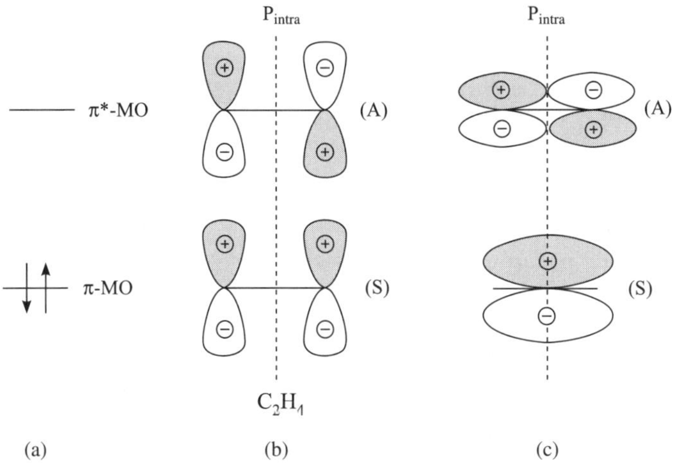

Figure 1.19 Diagrams of (a) energies, (b) *p* orbitals, and (c) frontier π and π^* MOs of C_2H_4

These rules are as follows:

If the occupied bonding MOs of reactants are symmetry-correlated (possess identical symmetry) with occupied bonding MOs of the reaction products, the reaction will proceed as consistent thermally (can be an elementary step). During this reaction, the symmetry of interacting orbitals is conserved along the reaction trajectory on the potential energy surface. Should this correlation be absent, a consistent reaction is possible only in a photochemical regime [138].

The class of pericyclic reactions includes consistent reactions of σ bond exchange (σ bond metathesis) such as

$$\begin{aligned} &H_2 + I_2 \rightleftharpoons 2HI, \\ &M{-}R + H_2 \rightleftharpoons MH + RH; \end{aligned} \tag{1.229}$$

consistent reactions of addition of HX and X–Y molecules to multiple bonds

$$\begin{aligned} &Cl_2 + C_2H_4 \longrightarrow ClCH_2CH_2Cl, \\ &HCl + C_2H_4 \longrightarrow CH_3CH_2Cl, \end{aligned} \tag{1.230}$$

reactions of cycle formation and decay [see also (1.228)]

(1.231)

electrocyclic reactions

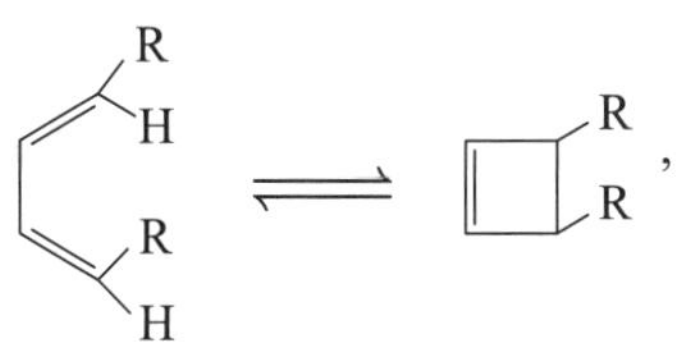

(1.232)

(in which intramolecular σ bonds are formed or ruptured), and so-called sigmatropic rearrangements

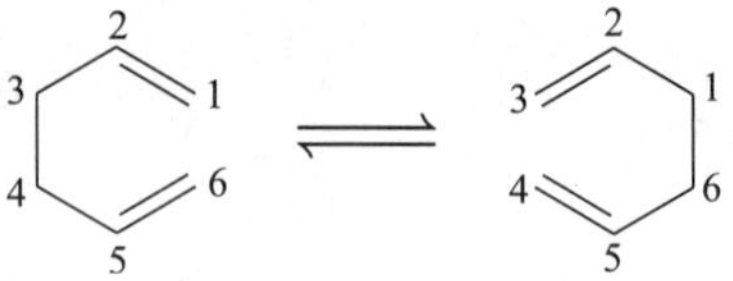

(1.233)

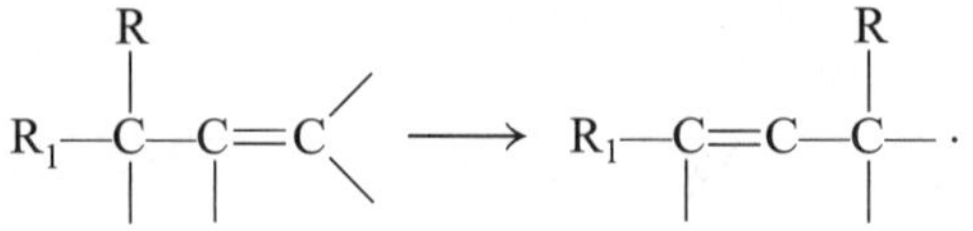

(1.234)

The reactions are also classified with respect to (i) the number of AOs of the reactants (or molecular fragments) involved in the formation of a transition state, and (ii) the types of bonds linking the reaction centers. For example, the aforementioned reactions are denoted as $[2\sigma + 2\sigma]$ (1.229), $[2\sigma + 2\pi]$ (1.230), $[2\pi + 2\pi]$ (1.228), $[2\pi + 4\pi]$ (1.231), $[2\pi + 2\pi]$ (1.232), $[2\sigma + 4\pi]$ (1.233), and $[2\sigma + 2\pi]$ (1.234).

Let us analyze the symmetry of the orbitals of reactants and products for the $[2\pi + 2\pi]$ type reaction (1.228) by considering a pair of reacting ethylene molecules and the corresponding MOs (Figure 1.20) [1.244]. For this purpose, we will use two planes – the local symmetry plane P_{intra}, which is common for both molecules, and the plane P_{inter} that is lying between the two ethylene molecules and is parallel to their planes. Two pairs of the π-MOs of ethylene molecules form two occupied MOs representing the combinations of $\pi + \pi'$ and $\pi - \pi'$. Both these combinations are symmetric with respect to the reflection in P_{intra}, while for the reflection in P_{inter}, the former combination is symmetric and the latter is

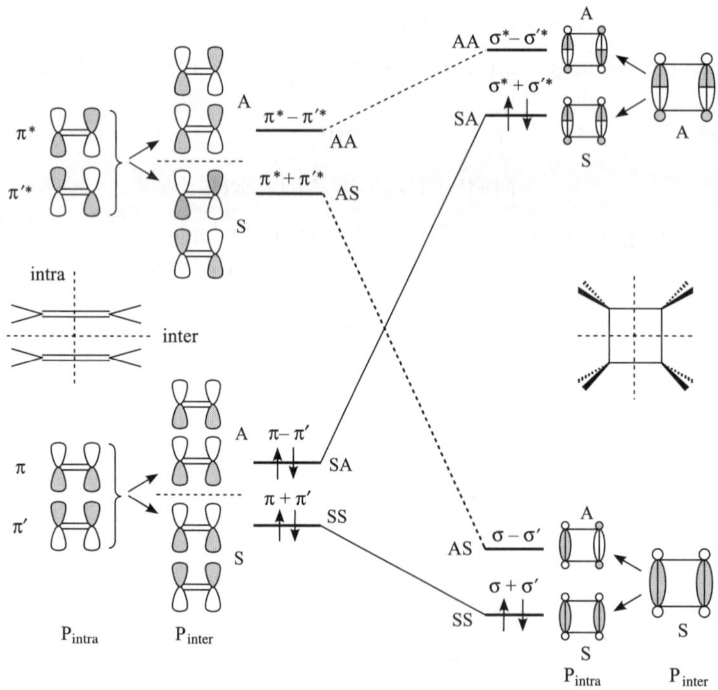

Figure 1.20 Woodward–Hoffmann correlation diagram for concerted cyclodimerization of C_2H_4 (Reprinted from [14] with kind permission from Springer Science+Business Media © 1998)

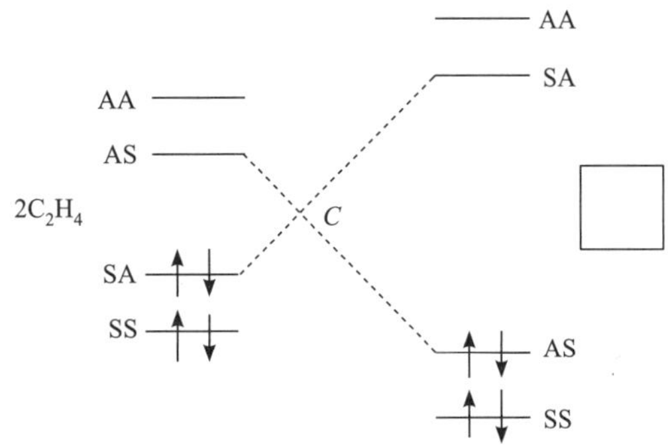

Figure 1.21 Alternative variant of correlation diagram in Figure 1.20

antisymmetric. The general symmetry of the two π-MOs is SS and SA. For the π^*-MOs, we also obtain two combinations ($\pi^* + \pi^{*'}$ and $\pi^* - \pi^{*'}$), the symmetry of which is illustrated in Figure 1.20. Note that the antisymmetric character of the $\pi - \pi'$ orbital relative to the P_{inter} plane implies the presence of a nodal plane between molecules.

In the product (cyclobutane), we consider σ-MOs of the two newly-formed C–C bonds. The two σ-MOs are symmetric with respect to reflection in the P_{inter} plane, while the two σ^*-MOs are antisymmetric in this respect. Thus, the possible combinations of these bonds yield four MOs with the symmetries SS, AS, SA, and AA. As can be seen from Figure 1.20, two ethylene molecules in the ground state (SS, SA) are symmetry-correlated with a cyclobutane molecule in the doubly excited state (SS, SA). The pair of electrons from an orbital of the SA symmetry has to surmount a high energy barrier in order to reach the SA level of cyclobutane, which implies that this reaction is forbidden with respect to orbital symmetry. In order to ensure the formation of a cyclobutane molecule in the ground state, the energy of a pair of electrons from MOs of SA symmetry must be increased so as to reach point C, where the energies of SA and AS orbitals intersect and the electron pair can be transferred to the orbital of AS symmetry (Figure 1.21). The magnitude of this barrier is evaluated as the energy of two bonding π electrons, which is too large for a thermoactivated reaction (270 kJ/mole) [144]. In the case of a photochemical excitation, the π^*(AS) orbital is populated by electrons (two-electron excitation is also possible) and the symmetries of the occupied MOs of the reactants and products are retained in the course of reaction and, hence, the reaction proceeds as an elementary act.

It was demonstrated by the method of laser spectroscopy [142] that the thermal cycloaddition reaction (1.228) proceeds via two symmetry-allowed steps with the formation of an intermediate biradical (with a lifetime of 700 fs):

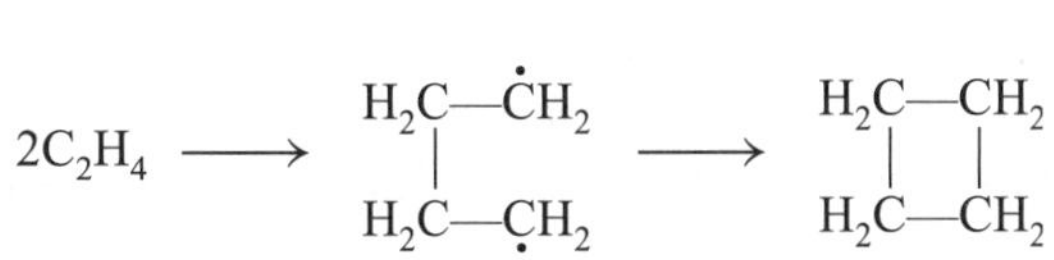

(1.235)

In contrast to the 2+2 cycloaddition, the 2+4 addition (Diels–Alder reaction) can proceed consistently as an elementary step via a hexa-center cyclic transition state. Even without an analysis of the combinations of reactants and products, the analysis of reactants for their MO symmetry with respect to the P_{intra} plane shows that the HOMO of ethylene (π-MO) can overlap with the LUMO of butadiene (π_1^*- MO), while the LUMO of ethylene (π^*-MO) can overlap with the HOMO of butadiene (π_2-MO) (Figure 1.22).

Based on the rule of conservation of the orbital symmetry, it is also possible to predict the stereochemistry of the reaction products [137, 138]. For example, a thermal process according to scheme (1.232) yields

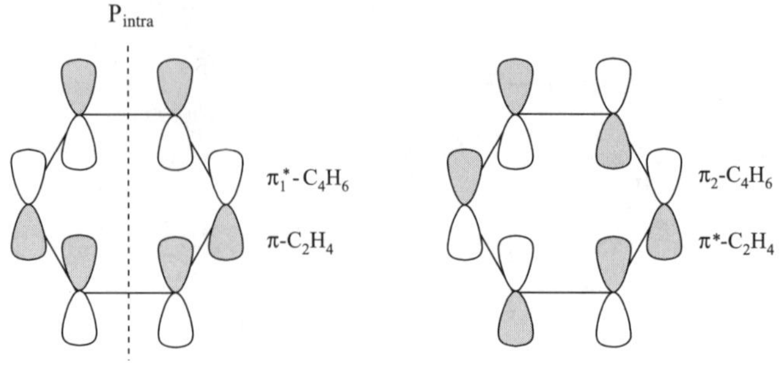

Figure 1.22 MO overlap in reactions of C_2H_4 and C_4H_6 with the formation of six-center transition states

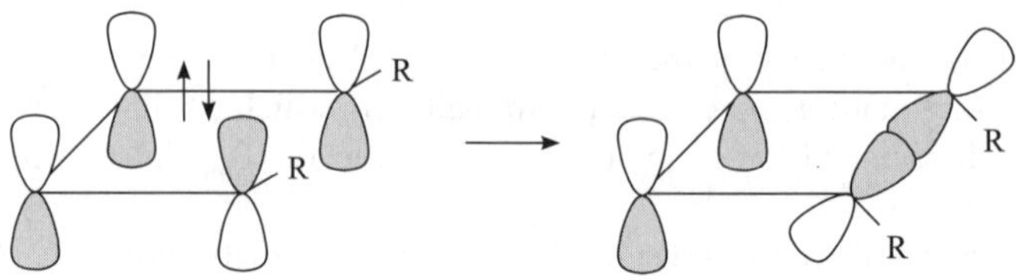

Figure 1.23 Exchange interaction of frontier MOs of reactant in $[2\pi + 2\pi]$ cyclization process

an isomer with groups R in the *cis* position relative to the ring plane, whereas the photochemical process yields a product with groups R in the *trans* position. This conclusion can be derived from an analysis of the symmetry of frontier MOs (Figure 1.23). Indeed, for the overlap of an occupied π-MO of one π bond with a free (antibonding) MO of the second π bond, it is necessary to rotate both bonds in the same direction (conrotator), e.g., clockwise to obtain a *cis* isomer. In an excited state of the diene molecule, the overlap of an occupied π^*-MO with the second bond takes place upon the counter-rotation of bonds (disrotator), which leads to the formation of a *trans* isomer (photochemical cyclization).

An important reason for the appearance of prohibitions with respect to the symmetry of orbitals is a change in the position of a nodal plane (or the shape of the nodal plane) [144]. For example, a horizontal nodal plane (P_{inter}) in the SA combination in ethylene dimer passes to a vertical nodal plane (P_{intra}) in the AS combination in cyclobutane (Figure 1.21). In other words, the upper bonding electron pair must rearrange so that the horizontal nodal plane becomes the vertical plane, i.e., this pair must receive a kinetic energy sufficient to reach point C.

Correlation diagrams for various diatomic and polyatomic molecules are presented in [137]. It should also be emphasized that a symmetry-forbidden reaction is also prohibited in the reverse direction.

Upon the appearance of substituents in olefin molecules in the $[2\pi + 2\pi]$ cycle formation process

$$\begin{matrix} H_2C{=}CH_2 \\ + \\ H_2C{=}CHCH_3 \end{matrix} \longrightarrow \begin{matrix} H_2C - CH_2 \\ | \quad\quad | \\ H_2C - CHCH_3 \end{matrix} \tag{1.236}$$

the P_{intra} and P_{inter} planes cease to be the planes of symmetry, but if the MO energies are weakly disturbed by the introduction of substituents, the main conclusions of the Woodward–Hoffmann theory remain valid. However, if the asymmetry increases due to the introduction of both donor (X) and acceptor (Y)

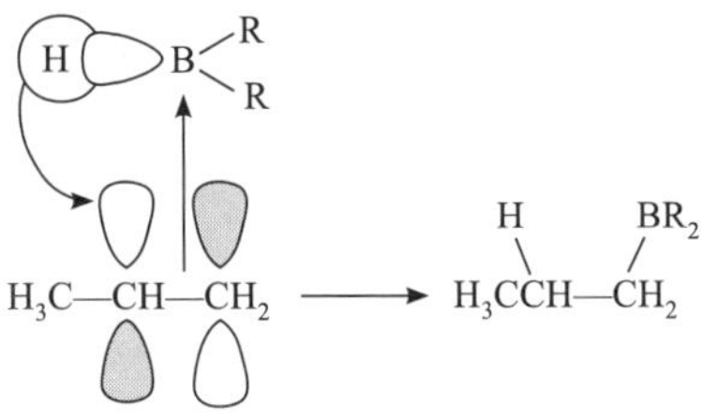

Figure 1.24 Addition of B–H bond to olefin in π complex of $R_2BH{\cdot}C_3H_6$

substituents (X–CH=CH–Y), the reaction no longer belongs to the class of usual pericyclic reactions and may become an allowed thermal process [144].

The symmetry also decreases during the reaction of the addition of HX molecules to multiple bonds. In this case, the overlap is nonzero (as it is in the reactions of H–H and Cl–Cl addition), but the barrier is still high [137]. Apparently, this accounts for the fact that no examples of consistent reactions were reported for X = OH, Cl, Br, I, SR, NR_2, etc. For studying the reactions of this type, it is expedient to use the MO perturbation method together with an analysis of the symmetry of frontier MOs.

The replacement of H by a metal (metal complex) in the HX molecule substantially changes the situation in reactions of M–X fragment addition to multiple bonds (X = H, R, Hal, OH, OR, etc.). Consistent addition of M–X in the case of *nontransition metals* is still of low probability with respect to the symmetry of frontier MOs. These systems most probably feature pre-coordination due to the free orbitals of a metal and π-MOs of an unsaturated molecule with the subsequent interaction between the M–X bond and one of the p orbitals of the antibonding MO and with an almost synchronous addition of the M–X fragment to the multiple bond. This mechanism is operative, e.g., during the addition of B–H bond in R_2B–H to olefins (Figure 1.24) [139]. Note that these processes are not considered in the Woodward–Hoffmann theory.

In the case of *transition metals*, the addition of M–X fragments to multiple bonds as well as the metathesis of σ bonds (M–X + H–H) are allowed as consistent processes (Figure 1.25). The participation of a d orbital in the formation of σ bond M–X allows two electrons from the σ-MO to overlap with π^*-MO of an olefin or σ^*-MO of an X–Y molecule.

The reaction of [$2\pi+2\pi$] cycle formation also becomes allowed provided that a multiple bond in one of the reactants is formed with the participation of a transition metal (Figure 1.26), i.e., a carbene (or carbyne) metal complex reacts with an olefin (or alkyne) [1561]:

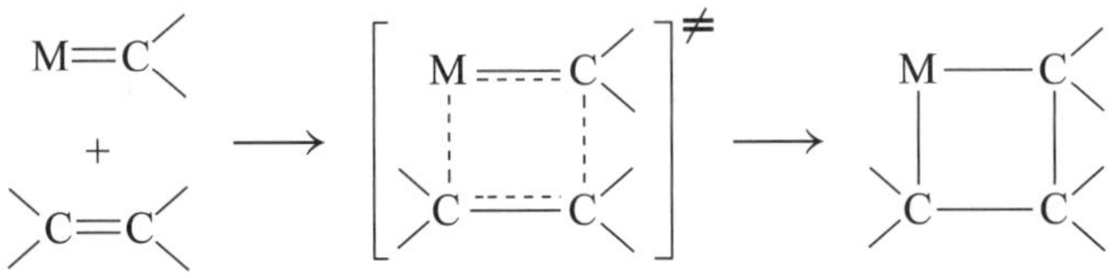

(1.237)

In this case, the d orbital involved in the formation of a d_π–p_π bond in the carbene metal complex can overlap with a bent π bond of the olefin molecule, in which substituents deviate from the olefin plane similar to the case of metal π complexes. The reactions of M–X or M=C < addition to multiple bonds most probably involve the pre-coordination of olefin (alkyne) to the metal with the formation of a π complex, followed by the consistent addition to the multiple bond via a cyclic transition state.

The possibility of reduction (or removal) of the prohibition with respect to symmetry (that is, lowering the corresponding barrier) due to the coordination of a metal to a substrate was considered for some time (in the 1970s) as a very attractive idea for explaining the nature of the catalytic effect [152, 153]. Indeed, theoretical analysis showed that the participation of both free and occupied d-AOs of a metal can actually

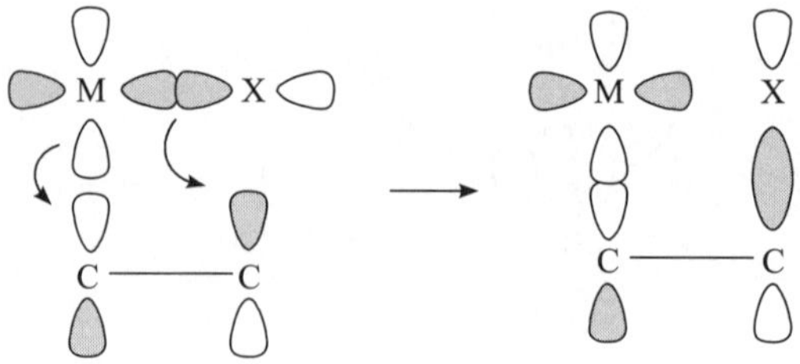

Figure 1.25 Addition of M–X bond to olefin in the case of transition metal M

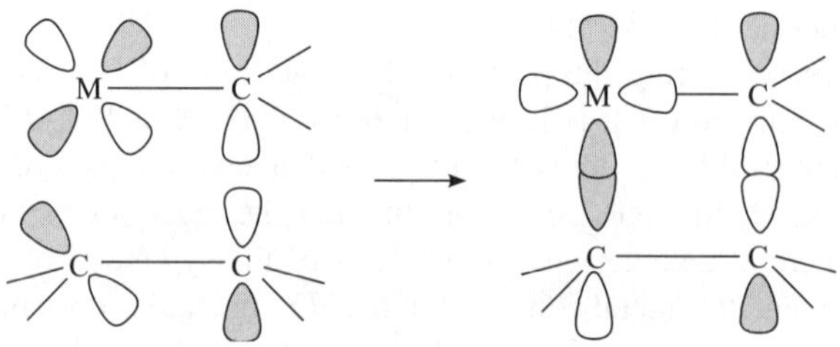

Figure 1.26 Reaction of $[2\pi + 2\pi]$ cycle formation with participation of metal carbene complex

reduce the prohibition in the case of a consistent $[2\pi + 2\pi]$ cycloaddition of olefins

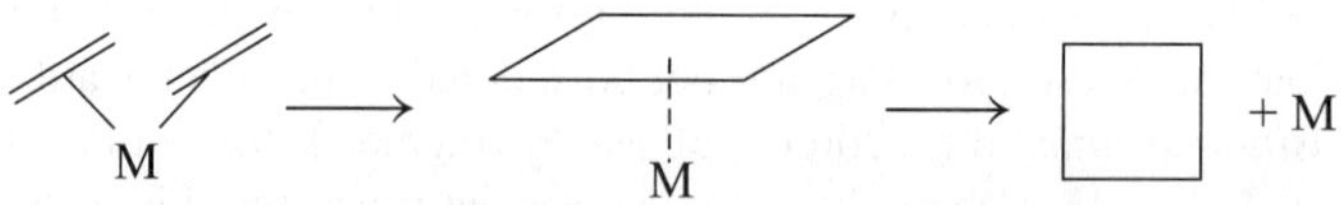

(1.238)

and in the reaction of H_2 addition to a metal-coordinated olefin (Figure 1.27). In the latter case, the antibonding π^*-MO of the olefin occupied by d electrons of the metal overlaps with σ^*-MOs of H_2. However, these ideas have not been confirmed in experiment: there is no one proved case of addition of nondissociated H_2 molecules to multiple bonds, and the cyclodimerization of olefins proceeds via a symmetry-allowed formation of metallacyclobutane intermediates as a result of the consistent oxidative addition of two olefin molecules to the metal (Figure 1.28).

Thus, the role of a metal (metal complex) consists in making possible the appearance of new intermediates, the formation and transformation of which proceed as consistent, symmetry-allowed processes with relatively low activation energies.

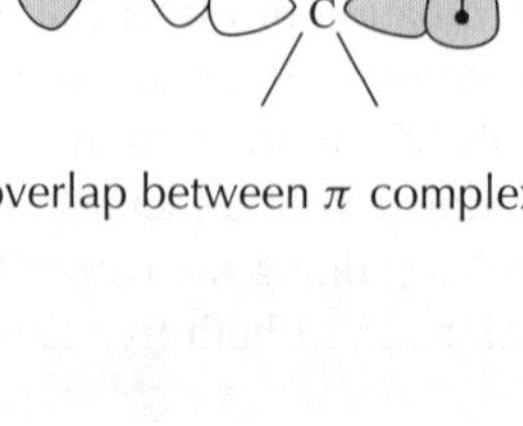

Figure 1.27 MO overlap between π complex and H_2 molecule

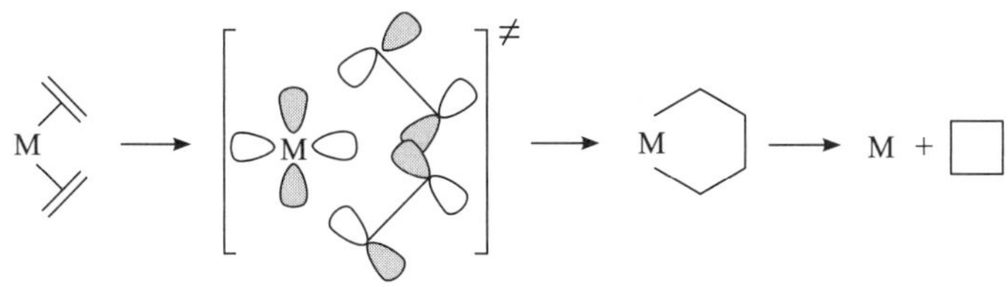

Figure 1.28 Catalysis of ethylene cyclodimerization reaction

- **Isolobal analogy**. This heuristic principle, which combines the main postulates of quantum chemistry (symmetry rules, energy characteristics of MOs) and the principle of least motion was formulated by Hoffmann [154] (see also [155–157]):

  Molecules and/or molecular fragments must readily (with low barriers) react, provided that these species possess the same symmetry of frontier (or hybrid) orbitals, same number of frontier orbitals, same number of electrons on these orbitals, close energies, and similar geometry (spatial arrangement) of interacting orbitals.

  Said fragments are called isolobal (from the Greek "$\lambda o \beta o \sigma$" for lobe). Since the isolobal analogy (or principle) does not specify any particular nature of the fragment, this principle is essentially a rule of reactivity that combines all fields of chemistry, including organic, inorganic, element-organic, and organometallic.

  For example, groups such as $\bullet CH_3$ (with unpaired electron on the sp^3 hybrid orbital) and $\bullet Mn(CO)_5$ (with unpaired electron on the d^2sp^3 hybrid orbital) are isolobal. This implies that they can react in any combination: $CH_3{-}CH_3$, $CH_3{-}Mn(CO)_5$, and $Mn(CO)_5{-}Mn(CO)_5$. If some fragments are isolobal (which is indicated by a special sign as in the example below), this by no means implies that the reaction product will always exist in the form predicted based on the interaction of fragments. The structure can be somewhat modified for sterical or thermodynamic reasons. Nevertheless, the isolobal analogy provides a good prognostic principle that offers a convenient tool for synthetic chemists. It is assumed that isolobal fragments can react in a consistent elementary act, but their interaction most frequently is a multistep process.

  Another example of isolobal fragments is as follows:

  $$\vdots CH \text{ (carbyne)} \overset{\leftarrow\;\rightarrow}{\circ} \vdots Co(CO)_3$$

  Pair interactions of these fragments lead to the following compounds: $HC{\equiv}CH$, $HC{\equiv}Co(CO)_3$, and $(CO)_3Co{\equiv}Co(CO)_3$. The latter compound, $Co_2(CO)_6$, exists but is highly reactive and readily adds another 2CO via the triple bond $Co{\equiv}Co$ and stabilizes in the form of $Co_2(CO)_8$ with a structure of

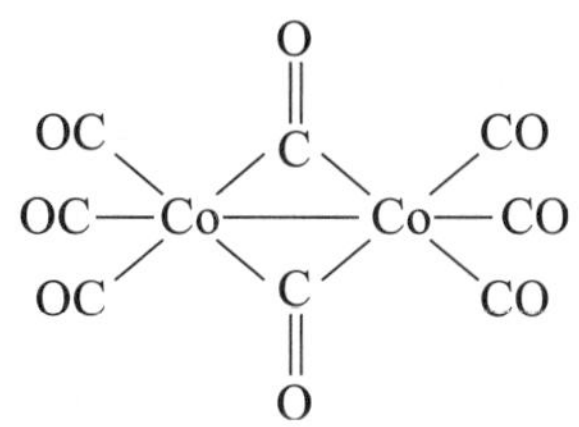

All the formed molecules can interact with the like species and with each other. In particular, fragments L⁝ may combine in four to yield one molecule

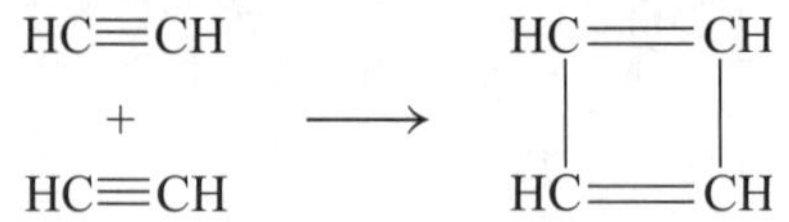

(1.239)

The cyclobutadene molecule is unstable (although it can exist as a reactive intermediate) and exhibits isomerization into a more stable form of tetrahedron:

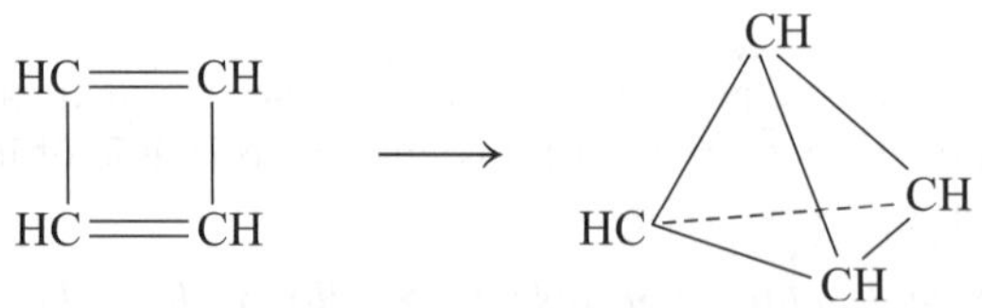

(1.240)

Examples (1.239)–(1.240) formally illustrate the pathway of combining isolobal fragments. In other words, each isolobal fragment L⁝ interacts with 3L⁝ fragments. All other possible molecules also exist in the form of tetrahedra:

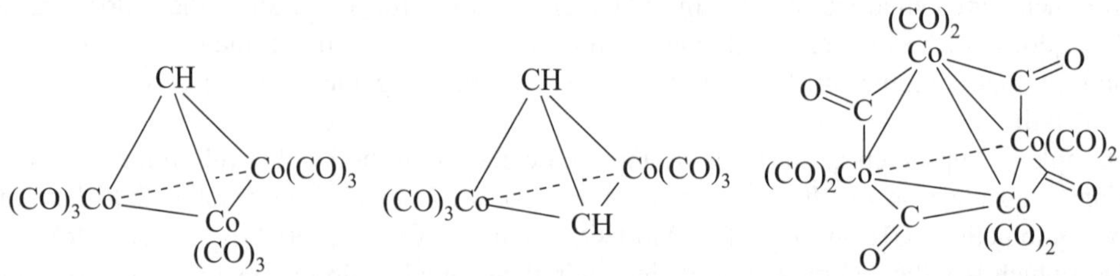

There are tables of isolobal fragments [155, 158]. Evidently, neither the formation of molecules from three or four fragments, nor the interaction of two acetylene molecules (1.239) in the course of a thermal reaction represents an elementary act. These reactions are essentially the blocks of elementary steps.

Sometimes, fragments that are isolobal with respect to almost all parameters possess strongly different orbital energies. Therefore, a simple analysis of the symmetry and electron configurations does not guarantee that the selected fragments would readily interact with the formation of anticipated products.

- **Selection rules related to the conservation or insignificant variation of electron configurations.** A strong perturbation of the electron configuration leads, in accordance to the principle of least motion, to significant values of E_{act} in the following cases [137, 145, 157]:
  (i) interaction proceeds via the orbitals of reactants that lead to the formation of products in excited states;
  (ii) orbitals of the ruptured and formed bonds are uncorrelated with respect to symmetry;
  (iii) absence of the low-lying excited states for reactants with symmetry-forbidden ground states;
  (iv) overlap of orbitals is weak or absent ($\beta_{ST} \cong 0$, see MO perturbation method).

The pathway of least motion for elementary steps is a pathway that creates or retains the maximum number of symmetry elements on the passage from reactants to products [137]. Let us consider several examples and rules, in which the principle of least motion is manifested in the conservation of electron configuration.

According to the electron transfer reaction

$$\mathrm{Co(Phen)_3}^{2+} + \mathrm{Co^*(Phen)_3}^{3+} \longrightarrow \mathrm{Co(Phen)_3}^{3+} + \mathrm{Co^*(Phen)_3}^{2+} \tag{1.241}$$

with $\Delta G^{o}_{298} = 0$ (symmetric reaction), Co^{2+} (d^7) converts into Co^{3+} (d^6). The electron configuration (valence shell) in the course of this transfer $\left(t_{2g}^6 e_g^1 \rightarrow t_{2g}^6 e_g^0\right)$ remains unchanged, since six electrons on the triply degenerate bonding level $\left(t_{2g}^6\right)$ are retained and a single electron is removed from the antibonding e_g level. The second-order rate constant for reaction (1.241) amounts to $1.1\ \mathrm{M^{-1} \cdot s^{-1}}$. Because phenanthroline (Phen) belongs to strong ligands, the maximum number of $7d$ electrons is paired (spin-paired state). In the case of a weak NH_3 ligand, the situation dramatically changes. Fragment $\mathrm{Co(NH_3)}_n^{2+}$ ($n = 4, 5, 6$) occurs in a spin-unpaired (high-spin) $t_{2g}^5 e_g^2$ state:

↑ ↑ e_g

⇅ ⇅ ↑ t_{2g}

A strong complex of $\mathrm{Co(NH_3)_6}^{3+}$ (which is $\sim 10^{30}$ times stronger than $\mathrm{Co(NH_3)_6}^{2+}$ occurs in a spin-paired $t_{2g}^6 e_g^0$ state like the complex with Phen ligand. Therefore, the $t_{2g}^5 e_g^2 \rightarrow t_{2g}^6 e_g^0$ electron transfer process is accompanied by a significant reconstruction of the valence shell and, accordingly, $k = 10^{-9}\ \mathrm{M^{-1} \cdot s^{-1}}$.

The degree of Co^{2+} conversion into Co^{3+}, about 50 %, is reached with Phen ligand for 1 s, while for NH_3 ligand it would take $\sim$30 years. Evidently, a step (formally, elementary) with this rate can be excluded from the set of elementary steps to be considered in the analysis of reaction mechanisms.

Reactions of radical species (atoms) with other radicals, radicals with molecules, and stable molecules in the triplet state (such as O_2) with other molecules obey the *Wigner–Witmer spin conservation rule* [137].

The conservation of a valence electron shell in the course of reactions of transition metal complexes is related to the *Tolman's 16/18 electron rule* [137]. It was noticed long ago (N. Sidgwick, 1929) that, in stable complex compounds, the total number of electrons surrounding the metal atom is equal to that in the electron shell of the closest inert gas atom. This number was called the effective atomic number (EAN). For d metals, the number of electrons in the valence shell of an atom bound to ligands is 18 ($d^{10}s^2p^6$), and this shell is considered stable. For example, in the case of $Ni(CO)_4$, we have $Ni^0(d^{10})$ and CO (two-electron ligand), which yields $n = 10 + 8 = 18\ e^-$.

In coordination compounds of transition metals, the number of electrons n introduced by each metal atom to the valence electron shell is equal to the number of electrons in its d^5sp^3 shell, that is, corresponds to the given group of elements. For example:

$V(CO)_6$	group V	$n = 5 + (2 \cdot 6) = 17\ e^-$;
$Cr(CO)_6$	group VI	$n = 6 + (2 \cdot 6) = 18\ e^-$
$Mn_2(CO)_{10}$	group VII	$n = 7 + (2 \cdot 5) + 1$ (from Mn) $= 18\ e^-$

By analogy with Mn, the formation of C–C bonds from $2CH_3\bullet$ supplements the valence shell of C atom to octet. Because of sterical hindrances, $V(CO)_6$ exists as a paramagnetic complex and forms a stable $\mathrm{V(CO)_6}^-$ anion (18 e^-), while $Co_2(CO)_8$ contains two bridging CO groups and appears as $Co_2(CO)_2$ (μ-CO)$_2$. It is commonly accepted that each bridging carbonyl introduces 1 e^- to the shell of Co, so that we have $9\ e^- + (2 \cdot 3)\ e^- + 1\ e^-$ (from Co) $+ (1 \cdot 2)\ e^-$ (from μ-CO) $= 18\ e^-$. The number n of

Table 1.3 Numbers of electrons introduced by typical ligands [158]

Ligand	Ligand type	CM*	IM*
Cl, OH, NR_2 Me, Ph, R, H	X	1	2
CO, NH_3, PR_3	L	2	2
C_2H_4	L	2	2
H_2	L	2	2
η^3- C_3H_5	LX	3	4
η^3- CH_3COO	LX	3	4
η^4- C_4H_6	L_2	4	4
η^5- C_5H_5	L_2X	5	6
η^6- C_6H_6	L_3X	6	6

Note: *IM = ion model; CM = covalent model.

electrons in the valence shell can be calculated using ion models (IMs) and covalent models (CMs) [158]. For example:

$$HMn(CO)_5,\ \ CM: n = 7 + (2 \cdot 5) + 1\ (\text{from H}) = 18\ e^-;$$

$$H^-Mn^+(CO)_5,\ \ IM: n = 6 + (2 \cdot 5) + 2\ (\text{from } H^-) = 18\ e^-.$$

For using IM or CM, it is necessary to come to an agreement on the number of electrons introduced by particular ligands (Table 1.3).

For example, in a simple case of $MX_aL_b^{c+}$, the number of introduced electrons according to the CM is

$$n = N + a + 2b - c,$$

while according to the IM this number is

$$n = N - (a + c) + 2a + 2b = N + a + 2b - c,$$

where $a + c$ is the metal oxidation number and N is the number of its valence electrons (or the group number in the periodic table).

In organometallic chemistry, there are many exceptions to the 18 electron rule, some examples being offered by Me_2NbCl_3 ($n = 10$), WMe_6 ($n = 12$), and $NiCp_2$ ($n = 20$). The EAN rule works satisfactorily for small but strong ligands (H, CO, CN). At the ends of transition metal series (Groups 8–11) the tendency to form 16-electron complexes is manifested on the passage from left to right. Nevertheless, the generalization over a large body of experimental material allowed Tolman to formulate the following rule:

Intermediates formed in the reactions of complex and organometallic compounds usually possess 16- or 18-electron shells. These intermediates usually exist in significant amounts.

Thus, in reactions involving *d* metals that can be considered as elementary steps, the number of valence electrons must typically change by two (18 → 16 → 18 and so on) [159]. Complexes with 16 electrons in the valence shell are naturally more reactive with respect to ligand substitution, since they admit substitution according to the associative mechanism:

$$\underset{16\ e^-}{ML_4} + :L^* \rightleftharpoons \underset{18\ e^-}{ML_4L^*} \rightleftharpoons \underset{16\ e^-}{ML_3L^*} + L \tag{1.242}$$

For example, $Rh(Acac)(C_2H_4)_2$ (16 e^-) exchanges ethylene ($^{13}C_2H_4$) by the associative mechanism at a rate constant above $10^4\ s^{-1}$ ($T = 25\,°C$, $P = 1$ bar), while $(C_5H_5)Rh(C_2H_4)_2$ (18 e^-) exchanges ethylene by the dissociative mechanism at a rate constant of $\sim 4 \cdot 10^{-10}\ s^{-1}$ [159].

Among the three possible mechanisms of CO insertion into the CH_3–Mn bond, the least probable variant, according to the Tolman rule, is (1):

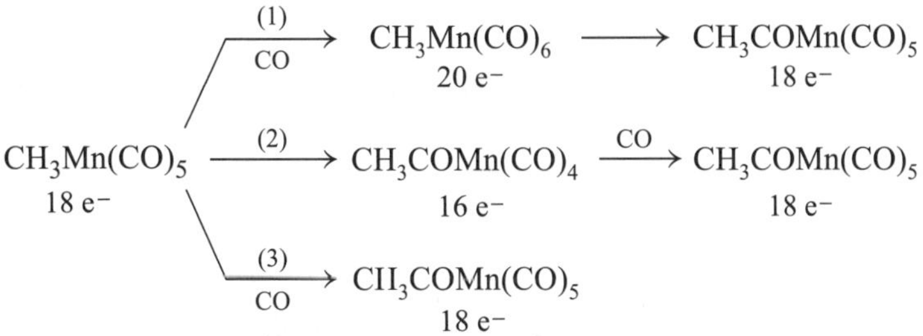

(1.243)

In the presence of π-allyl, π-cyclopentadienyl, and π-indenyl ligands, the associative mechanism for 18-electron complexes is sometimes possible without transition to 20-electron shells, which is due to a change in the type of η-ligand coordination (on the passage from η^5 to η^3 type and from η^3 to η^1 type):

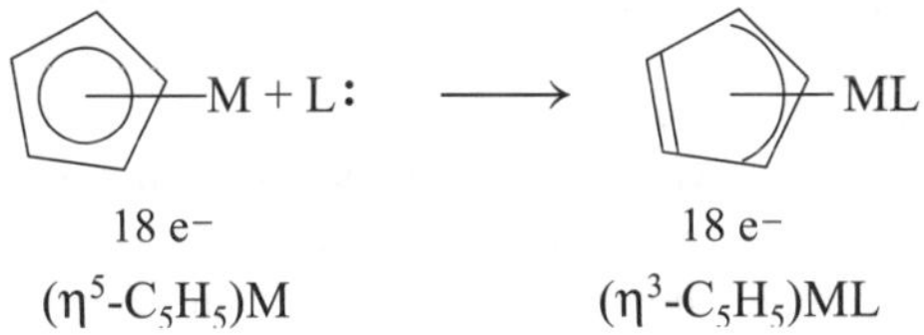

(1.244)

The rule of 18-electron-shell stability is also applicable to metal clusters [160], although a more general rule for metal clusters is provided by calculating the Wade "magic" numbers of stable valence shells [160].

For selecting a ligand to be replaced by another one in substitution reactions and evaluating the relative rate of ligand L_1 substitution by L_2 in the case of flat square and octahedral complexes of d metals, it is also possible to use the Chernyaev law of *trans*-influence and the Grinberg–Kushkin law of *cis*-influence [161–166]. The former law, established by I.I. Chernyaev in 1926 for Pt(II) complexes, says that certain ligands X accelerate the substitution of ligand L_1 in *trans* position relative to X (*trans*-coordinate X–Pt–L_1). The order of efficiency of a given ligand X is retained (to the first approximation) irrespective of the nature of ligand L_1. This effect exhibited a kinetic character, and the series of *trans*-activity were constructed with respect to the rate of ligand L_1 substitution. The *trans*-influence of ligands X decreases in the following series:

$$CN^- \sim CO \sim C_2H_4 \sim NO > SC(NH_2)_2, R_2S,\ R_3P,\ NO_2{}^-,\ I^-,\ SCN^- >$$

$$Br^- > CL^- > OH^- > RNH_2 > NH_3 > H_2O.$$

A strong *trans*-influence is known for ligands H^- ($\sim R_2S$), CH_3^- ($\sim SC(NH_2)_2$), and $C_6H_5^-$($\sim NO_2^-$). Investigations showed that the *trans*-influence is also manifested by changes in the equilibrium thermodynamic properties and geometric parameters of M–L_1 bonds. For this reason, the term *trans*-influence usually implies the effect of ligand X on the equilibrium characteristics of M–L_1 bonds, while the term *trans*-effect refers to the influence of ligand X on the kinetic properties of ligand L_1 in the

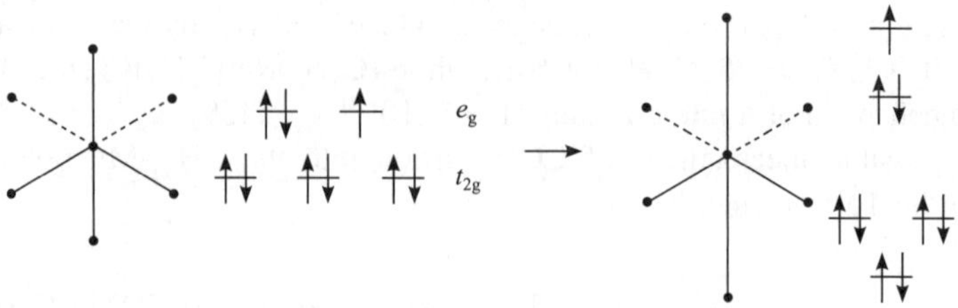

Figure 1.29 Distortion of $Cu(H_2O)_6{}^{2+}$ configuration caused by the Jahn–Teller (JT) effect

substitution reaction. Quantum-chemical models have been developed that explain the role of σ- and π-donor and π-acceptor properties of ligands X and L_1 in the manifestations of both *trans*-influence and *trans*-effect [164, 165].

Another phenomenon related to the properties of electron configurations of metal complexes is known as the Jahn–Teller effect (or the structure distortion due to the Jahn–Teller effect) [146]. This effect helps in selecting the type of elementary step or even the direction of transformation of metal complexes. The Jahn–Teller theorem is as follows:

If the ground-state electron configuration in a nonlinear molecule is degenerate, then the molecule exhibits distortion that removes the degeneracy and increases the stability.

For example, in octahedral aqua-complexes of Cu^{2+} and Cr^{2+} (with d^9 and d^4 configurations), the degeneracy of t_{2g} and e_g levels is removed as a result of the elongation of axial bonds between metal ions and H_2O molecules (Figure 1.29). Here, two axial water molecules are less strongly bound and more readily replaced by ligands (water molecules or anions) than water molecules in the equatorial plane, the substitution taking place within the framework of the dissociative exchange mechanism.

1.4.4 Topological selection rules for elementary steps

In the analysis of ethylene cyclodimerization reaction, it was pointed out that the transition states in elementary steps can possess various structures (depending on the number of reacting centers and the mode of their interaction) and, hence, different topologies. In particular, two ethylene molecules cannot form a (symmetry-forbidden) cyclic transition state A via a thermal reaction, but they can form a linear four-center transition state B:

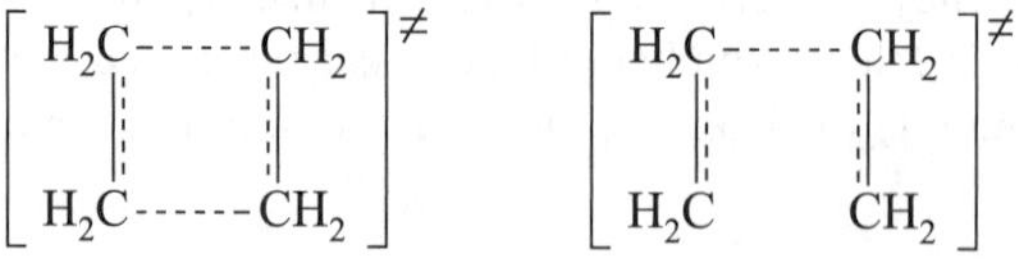

A B

A transition state formed during the Diels–Alder reaction is a hexa-center cycle. This circumstance poses some questions as to which structures of transition states are realized in the elementary steps studied and what is the possible topology of the redistribution of bonds and unshared electron pairs in the course of consistent reactions.

Here, by the *topology of bond redistribution*, we imply the structure of a graph that is obtained upon superposition of the ruptured and created bonds. Any chemical reaction can be represented by a graph from which atoms and atomic groups not involved in changing bonds are removed. For example, the Diels–Alder reaction can be represented by the following graphical equation that describes the redistribution of bonds:

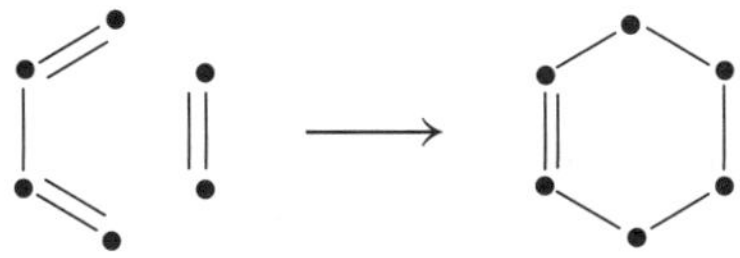

G_1 G_2

(1.245)

This equation type is refered to as the *symbolic* [167] or the *base reaction* [168–171]. Then it is possible to remove bonds that remain unchanged, which yields an equation that reflects the *reaction type* [6]:

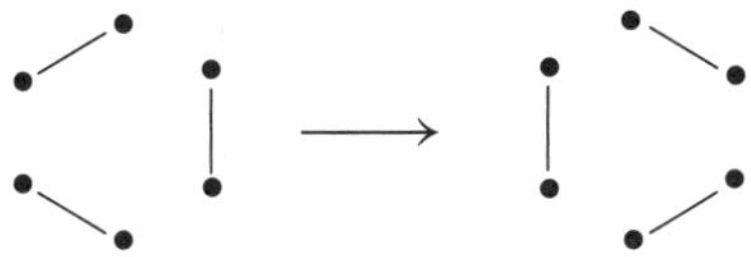

G_3 G_4

(1.246)

The superposition of graphs G_3 and G_4 yields the following graph G_5:

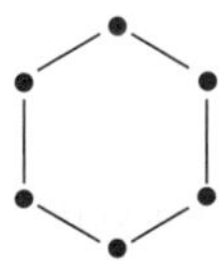

Γ_5

G_5

which is called the *topology identifier* (TI) [167] or the *topology of reaction category* [168]. The TI characterizes the topology of bond redistribution and presents a simplified topology of the transition state (in the given case, a hexagonal cyclic topology). If the simultaneous rupture and/or formation of multiple bonds is expected, then graphs G_3, G_4, and G_5 contain multiple edges that form additional cycles in the TI.

It is important to note [6] that the *reactions, which can be classified on sufficient grounds as elementary steps, virtually never have mixed (linear-cyclic) or more complicated (polycyclic) TIs*. This conclusion is based on an analysis of databases available for the elementary steps of reactions involving metal complexes and organometallic compounds [6] and on the results [170] obtained from an analysis of the ORAC and REACS databases for about 80 000 organic reactions. It should be noted that the processes with complex topologies referred to as "coarctate" reactions with coarctate transition states [169, 170, 172] apparently do not represent elementary steps. Indeed, these reactions, for example,

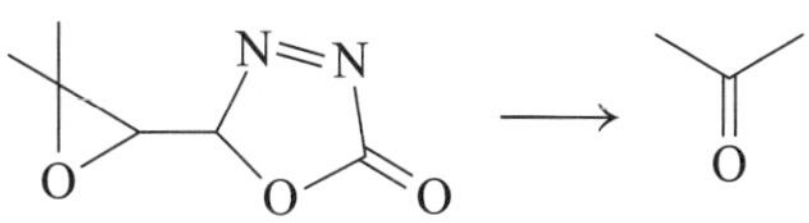

$+ HC{\equiv}CH + N{\equiv}N + CO_2,$ (1.247)

with a TI of the following type (G_6)

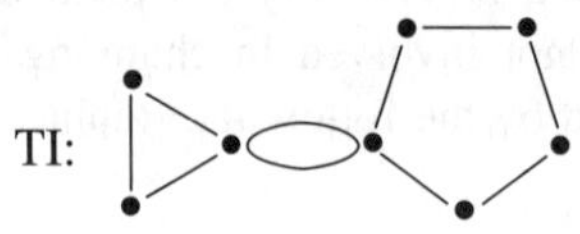

G_6

involve a change in the coordinates of a large number of nuclei, for which the synchronization of many oscillations (in this case, of nine bonds with various directions, which is necessary for the fragmentation of four molecules) seems be hardly probable. The given TI (G_6) is a complex polycyclic graph. The molecularity of the reverse reaction ($\mu = 4$) is also indicative of a low probability that reaction (1.247) represents an elementary step, since the reverse reaction cannot proceed via the same transition state.

As was noted above, the molecularity μ of a reverse step estimated with allowance for the basic principle of microscopic reversibility must not exceed 2, but this circumstance was ignored during the analysis of coarctate reactions. According to R. Ponec (see, e.g., [6]), the quality of quantum-chemical calculations of the potential energy surfaces used to justify the elementarity of coarctate reactions was insufficient for distinguishing between multistep and consistent mechanisms.

Let us consider the application of the above-described procedure to an analysis of transition states for the reaction of olefin epoxidation in the reaction with a molybdenum peroxy complex [173]:

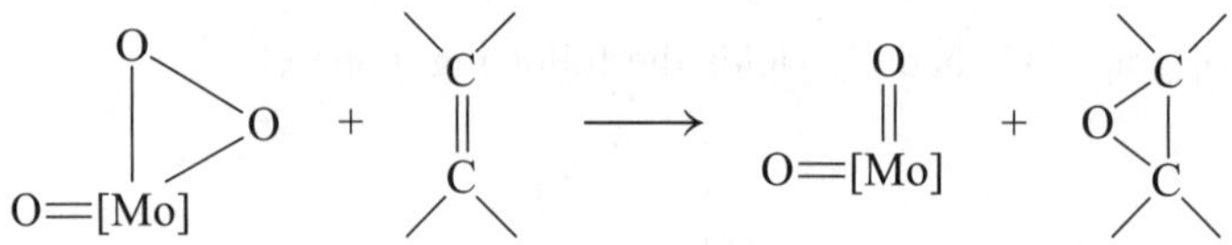

(1.248)

Assuming that this is an elementary step that proceeds via a transition state C according to the scheme

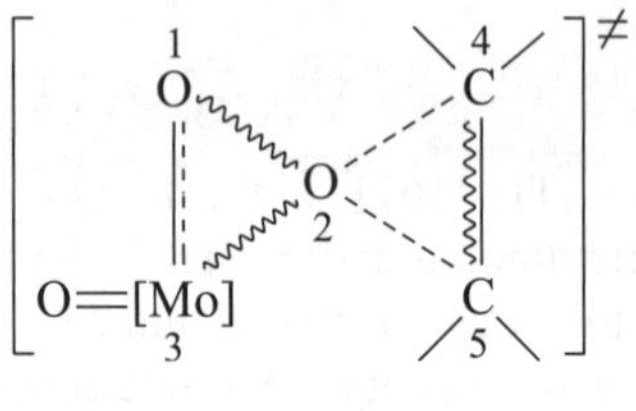

we obtain a TI with the following complex (bicylic) structure:

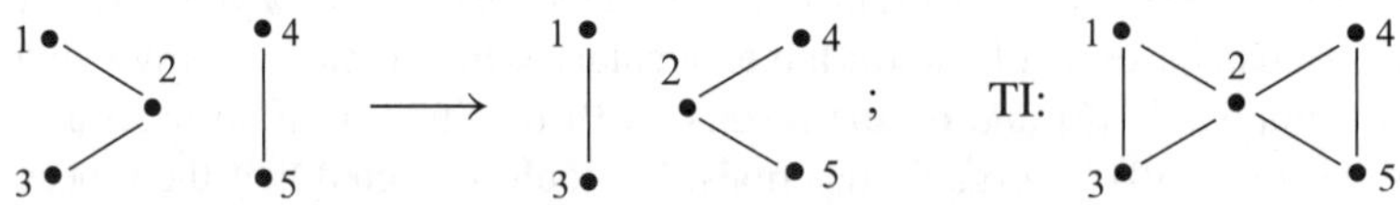

(1.249)

Then, it can be suggested that this is not an elementary reaction. Based on an analysis of reaction kinetics [174, 175], the following two-step mechanism was proposed:

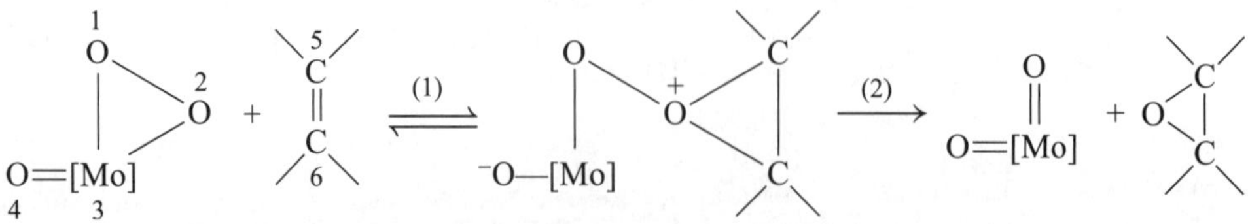

(1.250)

Let us consider the topology of transition states for the two steps:

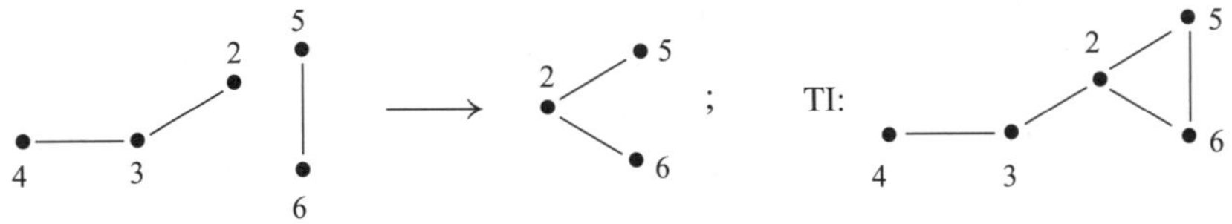

(1.251)

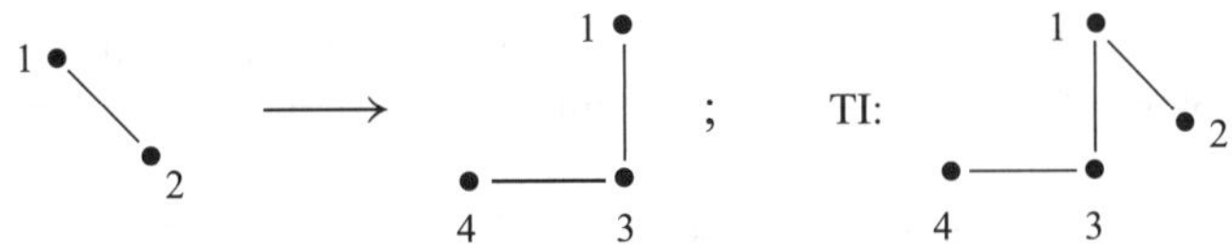

(1.252)

As can be seen, the first step (1.251) exhibits a mixed (linear and cyclic) topology. It can be suggested that there is one more elementary step, at which the M–O bond in the peroxy complex exhibits pre-rupture as

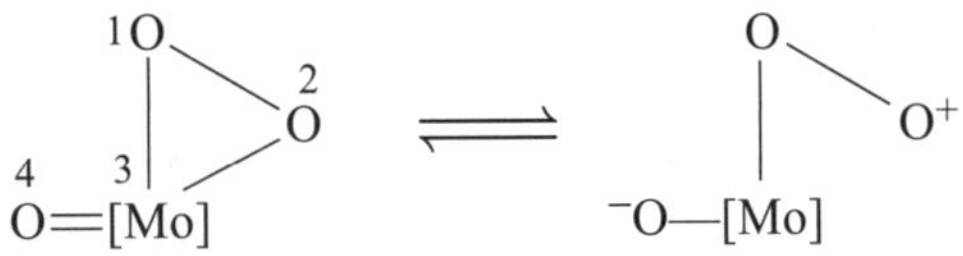

(1.253)

which has a linear TI structure (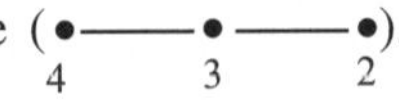). The resulting electrophilic center ($-O^+$) is added to olefin with a cyclic TI:

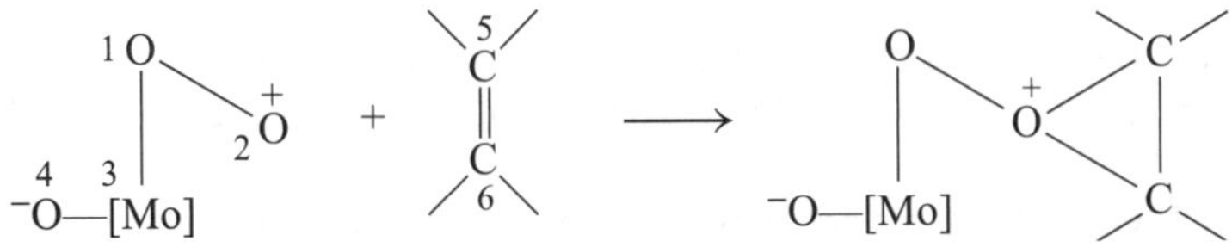

(1.254)

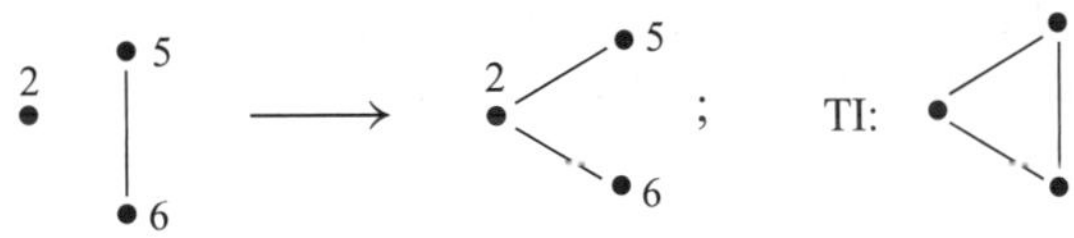

(1.255)

The intermediate compound (1.254) is converted into the reaction product in the second step (1.250) via a linear transition state (1.252).

Thus, an analysis of the topological structure (as represented by the TI) can be used to determine the possible elementary steps, that is, to find reactions which, from the standpoint of topology, have a high probability to be elementary steps according to the following rule:

Reactions whose transition states do not exhibit simple topology (cyclic or linear) are not considered to be elementary steps.

If a reaction involves simultaneous (synchronous) rupture and/or formation of multiple bonds, these bonds are represented on the graphs by double (or triple) edges, for example:

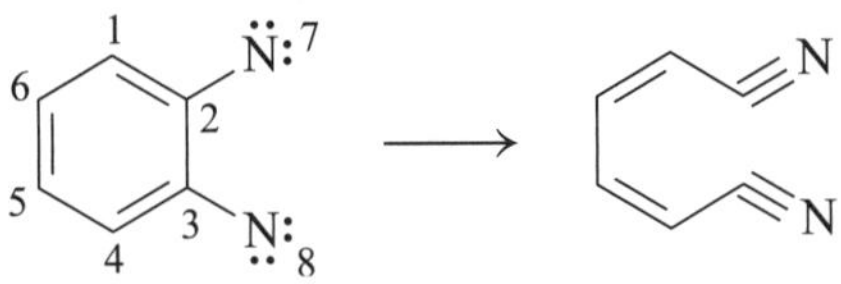

(1.256)

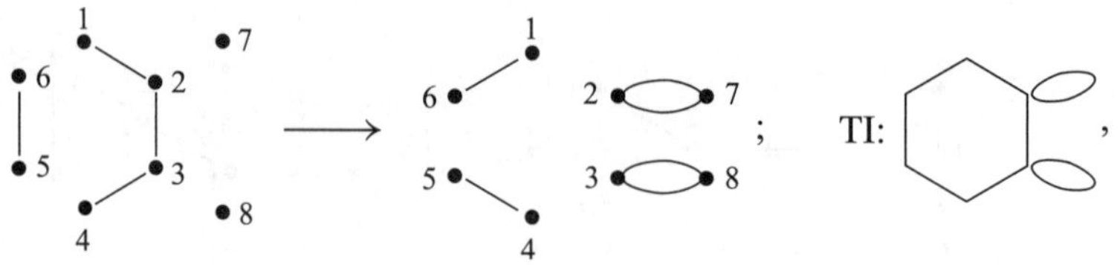

(1.257)

According to the above notions, reaction (1.256) should not be considered as a consistent elementary step. Some other topological criteria of the elementarity of steps have been formulated for pericyclic reactions [176].

The set of topological principles can include a heuristic rule concerning the number of bonds ruptured and formed in the course of an elementary reaction. An analysis of elementary steps in this respect [6, 77] showed that, in an overwhelming majority of cases, the difference of the numbers of ruptured and formed bonds does not exceed one:

$$|\Delta q| \leq 1$$

(irrespective of the sign of ΔH^0 for the elementary step). According to this rule, reaction (1.256) is not an elementary step since $|\Delta q| = 2$.

Based on the above brief review, it is possible to formulate a *set of rules for determining thermodynamically allowed elementary thermal reaction steps*. To the first approximation, reactions be considered elementary, provided that:

- molecularity of the forward and reverse reactions does not exceed two ($\mu \leq 2$);
- number of the ruptured bond does not differ from the number of the formed bond by more than one (bond compensation rule, $|\Delta q| \leq 1$);
- occupied frontier MOs of reactants are symmetry-correlated with those of the products (frontier MOs of the donor and acceptor must possess the same symmetry);
- topology of bond redistribution (topological structure of the transition state) is either linear or cyclic.

The principle of least motion can be considered as an additional rule, the expediency of which was demonstrated in Section 1.4.

Let us return to considering reaction (1.219) between hydrogen and iodine. The molecularity of a proposed consistent elementary step is $\mu = 2$, the bond compensation index is $|\Delta q| = 0$, and the topological structure of the transition state is cyclic. The only criterion for the impossibility of this elementary act, which turns out to be decisive in this case, is prohibition with respect to the symmetry. However, reaction (1.219) proceeds in accordance with the kinetic equation

$$W^+ = k^+[H_2][I_2] \tag{1.258}$$

and $E_{act} \cong 40$ kcal/mole (at $T = 500$–600 K). Then we have to consider the two possible variants with the dissociation of I_2 molecule.

(i) Chain mechanism:

$$I_2 \overset{K_0}{\rightleftharpoons} 2I; \tag{1.259}$$

$$I + H_2 \xrightarrow{k_1} HI + H; \tag{1.260}$$

$$H + I_2 \xrightarrow{k_2} HI + I. \tag{1.261}$$

The activation energy of this process according to the equation

$$W^+ = k_1 K_0^{1/2}[H_2][I_2]^{1/2} \tag{1.262}$$

amounts to ~ 54 kcal/mole and the reaction order with respect to $[I_2]$ differs from that observed in experiment. The other variants of radical chain termination are low probable, since $[I]/[H] \cong 10^{12}$ at $T = 600\,K$.

(ii) Radical mechanism, which includes the IH_2 complex formation and its subsequent decomposition at the limiting step via a symmetry-allowed transition state $|I \cdots H \cdots H \cdots I|^{\neq}$:

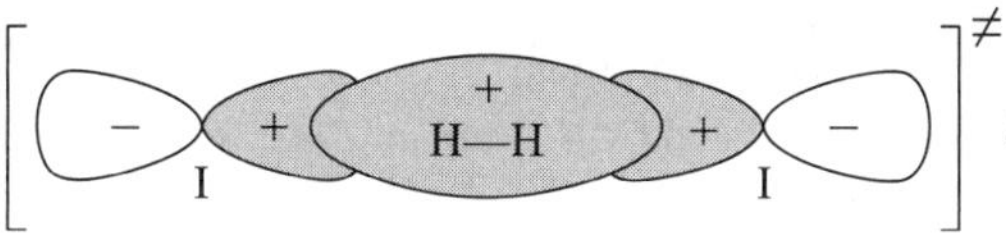

;

$$I + H_2 \underset{}{\overset{K_1}{\rightleftharpoons}} IH_2;$$

$$IH_2 + I \xrightarrow{k_2} 2HI; \tag{1.263}$$

$$W^+ = k_2 K_0 K_1 [H_2][I_2]; \tag{1.264}$$

$$E_{act} = E_2 + \Delta H_0^o + \Delta H_1^o. \tag{1.265}$$

Since the experimental activation energy is $E_{act} \cong 40$ kcal/mole [177] and $\Delta H_0^o = 35.5$ kcal/mole, we obtain $E_2 + \Delta H_1^o \cong 4.5$ kcal/mole. At a small estimated value of ΔH_1^o (about −0,4 kcal/mole), we have $E_2 \sim 5$ kcal/mole.[7] Estimating the values of preexponential factors for the rate constants and equilibrium constant of steps (1.259), (1.263), and (1.264) in the second variant of the proposed mechanism, we eventually obtain the observed rate constant $k_{obs} = k_2 K_0 K_1 = 2 \cdot 10^{-4}$ L·mole^{-1}·s^{-1} at $T = 600\,K$, which is close to the experimental value of k^+ in Eq. (1.258).

References

1. J. Horiuti and M. Ikusima, *Proc. Imp. Acad.* (Tokyo), **15**, 39 (1939).
2. J. Horiuti, *Adv. Catal.*, **9**, 339 (1957).
3. O. N. Temkin, *Dokl. Akad. Nauk SSSR*, **152**, 156 (1963).
4. M. I. Temkin, *Adv. Catal.*, **28**, 173 (1979).
5. V. G. Gorskii, T. N. Shvetsova-Shilovskaya, and V. A. Petrunin, *Fenomenologicheskaya i statsionarnaya kinetika khimicheskikh reaktsii* (Phenomenological and Steady-State Kinetics of Chemical Reactions), Oikumena, Smolensk, 2002.

[7]Results of ΔH_1^o calculations performed by E.D. German using the DFT method with RPBE functional and 3-21G** basis set for various configurations of the complex $|I \cdots H \cdots H|$.

6. O. N. Temkin, A. V. Zeigarnik, and D. G. Bonchev, *Chemical Reaction Networks: A Graph-Theoretical Approach*, CRC Press, Boca Raton, FL, 1996.
7. G. S. Yablonsky, V. N. Bykov, and A. N. Gorban, *Kineticheskie modeli kataliticheskikh reaktsii* (Kinetic Models of Catalytic Reactions), Nauka, Novosibirsk, 1983.
8. O. N. Temkin, L. G. Bruk, and D. Bonchev, *Teor. Eksp. Khim.*, **24**(3), 282 (1988).
9. Yu. S. Snagovskii and G. M. Ostrovskii, *Modelirovanie kinetiki geterogennykh kataliticheskikh protsessov* (Modeling Heterogeneous Process Kinetics), Khimiya, Moscow, 1976.
10. S. R. Brinkley, *J. Chem. Phys.*, **14**(9), 563 (1946).
11. A. G. Zyskin, Yu. S. Snagovskii, G. M. Ostrovskii, in *Khimicheskaya hinetika v katalize: Teoreticheskie problemy kinetiki* (Chemical Kinetics in Catalysis: Theoretical Problems of Kinetics), S. L. Kiperman (Ed.), Akad. Nauk SSSR, Chernogolovka, 1985, p. 56.
12. C. Bearez, F. Avendano, F. Chevalier, and M. Guisnet, *Bull. Soc. Chim. Fr.*, **3**, 346 (1985).
13. M. R. Guisnet, *Acc. Chem. Res.*, **23**, 392 (1990).
14. K. B. Fogash, Z. Hong, and J. A. Dumesic, *Catal. Lett.*, **56**, 85 (1998).
15. S. L. Kiperman, *Osnovy khimicheskoi kinetiki v geterogennom katalize* (Principles of Chemical Kinetics in Heterogeneous Catalysis), Khimiya, Moscow, 1979.
16. A. V. Zeigarnik, *Kinet. Katal.*, **36**(5), 653 (1995).
17. L. V. Shchel'tsyn, S. M. Brailovskii, and O. N. Temkin, *Kinet. Katal.*, **31**(6), 13–16 (1990).
18. A. V. Zeigarnik and O. N. Temkin, *Kinet. Katal.*, **40**(4), 538 (1999).
19. M. I. Temkin, *Kinet. Katal.*, **22**(3), 551 (1981).
20. M. I. Temkin, *Zh. Vses. Khim. O-va im. D.I. Mendeleeva*, **20**(1), 7 (1975).
21. P. L. Corio and B. G. Johnson, *J. Phys. Chem.*, **95**, 4166 (1991).
22. P. L. Corio, *J. Phys. Chem.*, **97**(46), 12100 (1993).
23. T. M. Barbara and P. L. Corio, *J. Chem. Educ.*, **57**, 243 (1980).
24. M. D. Korzukhin, *Zh. Fiz. Khim.*, **46**, 1845 (1972).
25. M. G. Slin'ko and A. K. Avetisov, *Kinet. Katal.*, **19**(1), 18 (1978).
26. M. I. Temkin, *Dokl. Akad. Nauk SSSR*, **152**, 156 (1963).
27. M. I. Temkin, Kinetics of Steady-State Complex Reactions, in *Mekhanizmy slozhnykh reaktsii. Trudy IV Mezhdunarodnogo kongressa po katalizu* (Mechanisms of Complex Reactions. Proceedings of the 4th International Congress on Catalysis), Nauka, Moscow, 1970, p. 57.
28. A. K. Avetisov, D. Yu. Murzin, and V. L. Kuchaev, *AIChE J.*, **52**(12), 4273 (2006); D. Yu. Murzin, and A. K. Avetisov, *Chem. Eng. Sci.*, **62**, 6492 (2007).
29. O. N. Temkin, R. M. Flid, and G. Shestakov, Specific Features of the Kinetics of Complex Catalytic Reactions in Solutions of Metal Complexes, in *Gomogennyi kataliz* (Homogeneous Catalysis: A Collection of Papers Presented at the 1st Coordination Meeting on Homogeneous Catalysis, Frunze, 1969), Ilim, Frunze, 1970, pp. 90–112.
30. G. S. Yablonsky, M. Z. Lazman, and V. I. Bykov, *Dokl. Akad. Nauk SSSR*, **269**(1), 166 (1983).
31. M. Z. Lazman, G. S. Yablonsky, and V. I. Bykov, *Khim. Fiz.*, **2**(2), 239 (1983); **2**(3), 413 (1983).
32. M. Z. Lazman, S. I., Spivak, and G. S. Yablonsky, *Khim. Fiz.*, **4**(4), 479 (1985).
33. V. I. Bykov, A. M. Kytmanov, M. Z., Lazman, and G. S. Yablonsky, in *Khimicheskaya hinetika v katalize: Teoreticheskie problemy kinetiki* (Chemical Kinetics in Catalysis: Theoretical Problems of Catalysis), S. L. Kiperman (Ed.), Akad. Nauk SSSR, Chernogolovka, 1985, p. 69.
34. G. S. Yablonsky, M. Z. Lazman, V. I. Bykov, in *Termodinamika neobratimykh protsessov i ee primenenie* (Thermodynamics of Irreversible Processes and Its Applications), Chernovtsy State University, Chernovtsy, 1984, Vol. 2, p. 311.
35. N. I. Kol'tsov and B. V. Alekseev, *Zh. Fiz. Khim.*, **63**(6), 1623 (1989).
36. V. N. Pisarenko and A. G. Pogorelov, *Planirovanie kineticheskikh eksperimentov* (Planning of Chemical Experiments), Nauka, Moscow, 1969.
37. *Primenenie vychislitel'noi matematiki v khimicheskoi i fizicheskoi kinetike* (Application of Computational Mathematics in Chemical Kinetics), L. S. Polak (Ed.), Nauka, Moscow, 1969.

38. L. S. Polak, M. Ya. Goldenberg, and A. A. Levitskii, *Vychislitel'nye metody v khimicheskoi kinetike* (Computational Mathematics in Chemical Kinetics), Nauka, Moscow, 1984.
39. V. G. Gorskii, *Planirovanie kineticheskogo eksperimenta* (Chemical Experiment Planning), Nauka, Moscow, 1984.
40. E. F. Brin, *Usp. Khim*., **54**(3), 428 (1987).
41. M. V. Klibanov, S. I. Spivak, V. I. Timoshenko, and M. G. Slinko, *Dokl. Akad. Nauk SSSR*, **208**(5), 1387 (1973).
42. S. I. Spivak and V. G. Gorskii, *Dokl. Akad. Nauk SSSR*, **257**(2), 412 (1981).
43. V. G. Gorskii and S. I. Spivak, *Zavod. Lab*., **47**(10), 39 (1981).
44. V. G. Gorskii and S. I. Spivak, in *Matematicheskie metody khimicheskoi termodinamiki* (Mathematical Methods of Chemical Thermodynamics), Nauka, Novosibirsk, 1982, p. 139.
45. S. I. Spivak and S. D. Varfolomeev, *Zh. Fiz. Khim*., **59**(12), 3040 (1985).
46. V. G. Gorskii and S. I. Spivak, *Zh. Strukt. Khim*., **29**(6), 119 (1988).
47. S. I. Spivak and V. G. Gorskii, *Khim. Fiz*., **1**(2), 237 (1982).
48. G. S. Yablonsky, V. A. Evstigneev, and V. I. Bykov, Graphs in Chemical Kinetics, in *Primenenie teorii grafov v khimii* (Application of Graph Theory in Chemistry), N. S. Zefirov and S. I. Kuchanov (Eds.), Nauka, Novosibirsk, 1988, p. 70.
49. V. G. Gorskii, E. A. Katsman, F. D. Klebanova, and A. A. Grigor'ev, *Teor. Eksp. Khim*., 23(2), 191 (1987).
50. N. N. Semenov, *O nekotorykh problemakh khimicheskoi kinetiki i reaktsionnoi sposobnosti* (Some Problems in Chemical Kinetics and Reactivity), Izdat. AN SSSR, Moscow, 1954; N. N. Semenov, *Some Problems in Chemical Kinetics and Reactivity*, Princeton University Press, Princeton (NJ), 1958, Vol I; 1959, Vol. II.
51. L. Hammett, *Physical Organic Chemistry: Reaction Rates, Equilibria and Mechanisms*, McGraw-Hill, New York, 1970.
52. S. D. Varfolomeev and S. V. Zaitsev, *Kineticheskie metody v biokhimicheskikh issledovaniyakh* (Kinetic Methods in Biochemical Investigations), MGU (Moscow State University), Moscow, 1982.
53. A. Ya. Rozovskii, *Geterogennye khimicheskie reaktsii* (Heterogeneous Chemical Reactions), Nauka, Moscow, 1980.
54. R. Schmid and V. N. Sapunov, *Non-Formal Kinetics*, Verlag-Chemie, Weinheim, 1982 (*Neformal'naya Kinetika*, Mir, Moscow, 1985, p. 41).
55. D. A. Frank-Kamenetskii, *Zh. Fiz. Khim*., **14**, 6 (1940).
56. J. Tze-Fei Wong, *J. Am. Chem. Soc*., **87**, 1788 (1965).
57. E. S. Rudakov, *Teor. Eksp. Khim*., **10**, 241 (1974).
58. G. Eyring, S. H. Lin, and S. M. Lin, *Basic Chemical Kinetics*, John Wiley & Sons, New York, 1983.
59. O. N. Temkin, K. Yu. Odintsov, and L. G. Bruk, *Priblizheniya kvazistatsionarnosti i kvaziravnovesiya v khimicheskoi kinetike: uchebnoe posobie* (Quasi-Steady-State and Pre-Equilibrium Approximations: A Teaching Aid), MITKhT (Moscow Institute of Fine Chemical Technology), Moscow, 2001.
60. L. Volk, W. Richardson, K. H. Lau, et al, *J. Chem. Educ*., 54, 95 (1977).
61. J. O. Hirschfelder, *J. Chem. Phys*., **26**, 271 (1957).
62. Yu. S. Sayasov and A. B. Vasil'eva, *Zh. Fiz. Khim*., **29**, 802 (1955).
63. A. N. Tikhonov, Systems of Differential Equations with Small Parameters at Derivatives, *Mat. Sbornik*, **31** (**73**), No. 3, 575 (1952).
64. V. M. Vasil'ev, A. I. Vol'pert, and S. I. Khudyaev, *Zh. Vychisl. Mat. Mat. Fiz*., **13**, 683 (1973).
65. K. J. Laidler, *J. Chem. Educ*., **65**, 250 (1988).
66. R. A. Van Santen and J. W. Niemantsverdriet, *Chemical Kinetics and Catalysis*, Plenum Press, New York/London, 1995.
67. E. A. Boeker, *Biochem. J*., **203**(1), 117 (1982).
68. *Mathematical Chemistry Series*, Vol. 2. *Chemical Graph Theory: Reactivity and Kinetics*, D. Bonchev and D. H. Rouvray (Eds.), Abacus Press/Gordon & Breach Sci. Publ., Philadelphia, 1992.
69. *Understanding Chemical Reactivity*, Vol. 9, *Graph-Theoretical Approaches to Chemical Reactivity*, D. Bonchev and O. Mekenyan (Eds.), Kluwer Acad. Publ., Dordrecht, 1994.

70. M. I. Temkin, *Dokl. Akad. Nauk SSSR*, **165**(3), 615 (1965).
71. A. V. Zeigarnik and O. N. Temkin, *Kinet. Katal.*, **35**(5), 691 (1994).
72. O. N. Temkin, A. V. Zeigarnik, and D. G. Bonchev, *J. Chem. Inform. Comput. Sci.*, **35**, 729 (1995).
73. O. N. Temkin, A. V. Zeigarnik, and D. G. Bonchev, *J. Chem. Inform. Comput. Sci.*, **36**, 973 (1996).
74. D. Bonchev, O. N. Temkin, and D. Kamenski, *React. Kinet. Catal. Lett.*, **15**(1), 113 (1980).
75. D. Bonchev, D. Kamenski, and O. N. Temkin, *J. Comput. Chem.*, **3**(1), 95 (1982).
76. O. N. Temkin and D. Bonchev, *J. Chem. Educ.*, **65**, 53 (1988).
77. O. N. Temkin, A. V. Zeigarnik, and D. G. Bonchev, in *Understanding Chemical Reactivity*, Vol. 9, *Graph-Theoretical Approaches to Chemical Reactivity*, D. Bonchev and O. Mekenyan (Eds.), Kluwer Acad. Publ., Dordrecht, 1994, p. 244.
78. E. V. Gordeeva, A. V. Bonchev, D. Kamenski, and O. N. Temkin, *J. Chem. Inform. Comput. Sci.*, **34**, 436 (1994).
79. A. V. Zeigarnik, *Kinet. Katal.*, **35**(5), 711 (1994).
80. A. A. Balandin, *Mul'tipletnaya teoriya kataliza*, Part 3, *Teoriya gidrogenizatsii. Klassifikatsiya kataliticheskikh reaktsii. Strukturnaya algebra v khimii* (Multiple Theory of Catalysis, Part 3: Theory of Hydrogenation. Classification of Catalytic Reactions. Structural Algebra in Chemistry), Izdat. MGU, Moscow, 1970.
81. J. A. Christiansen, *Adv. Catal.*, **5**, 311 (1953).
82. K. B. Yatsimirskii, *Teor. Eksp. Khim.*, **8**(1), 17 (1972).
83. A. V. Zeigarnik and O. N. Temkin, *Kinet. Katal.*, **40**(4), 550 (1999).
84. E. L. King and C. A. Altman, *J. Phys. Chem.*, **60**(10), 1375 (1956).
85. M. I. Temkin, *Vsesoyuznaya konferentsiya po khimicheskim reaktoram* (Proceedings of the All-Union Conference on Chemical Reactors), Nauka, Novosibirsk, 1966, Vol. 4, p. 628.
86. M. V. Volkenstein and B. N. Goldstein, *Biokhimiya*, **31**(3), 541 (1966); *Dokl. Akad. Nauk SSSR*, **170**(4), 963 (1966).
87. M. V. Volkenstein and B. N. Goldstein, *Biochim. Biophys. Acta*, **115**, 471 (1966).
88. M. V. Volkenstein, *Enzyme Physics*, Plenum Press, New York, 1969 (*Fizika fermentov*, Nauka, Moscow, 1967, p. 155).
89. B. N. Goldstein, Yu. B. Magarshak, and M. V. Volkenstein, *Dokl. Akad. Nauk SSSR*, **191**(5), 1172 (1970).
90. H. J. Fromm, *Biochem. Biophys. Res. Commun.*, **40**(3), 692 (1970).
91. M. V. Klibanov, M. G. Slin'ko, S. I. Spivak, and V. I. Timoshenko, *Upravlyaemye sistemy*. (Controllable Systems), Institute of Mathematics/Institute of Catalysis, Siberian Branch, Acad. Sci. USSR, 1970, Issue No. 7, p. 64.
92. K. B. Yatsimirskii, *Int. Chem. Eng.*, **15**(1), 7 (1975).
93. G. S. Yablonsky and V. I. Bykov, *Kinet. Katal.*, **18**(6), 1561 (1977); *Dokl. Akad. Nauk SSSR*, **238**(3), 645 (1978).
94. V. A. Evstigneev, G. S. Yablonsky, and V. I. Bykov, *Dokl. Akad. Nauk SSSR*, **245**(4), 871 (1979).
95. V. A. Evstigneev and G. S. Yablonsky, *Kinet. Katal.*, **20**(6), 1549 (1979).
96. G. S. Yablonsky, V. A. Evstigneev, A. S. Noskov, and V. I. Bykov, *Kinet. Katal.*, **22**(3), 738 (1981).
97. A. S. Noskov and G. S. Yablonsky, *Kinet. Katal.*, **23**(1), 191 (1982).
98. G. S. Yablonsky and V. A. Evstigneev, *Teor. Eksp. Khim.*, **15**(1), 41 (1979).
99. O. N. Temkin and D. G. Bonchev, *J. Chem. Educ.*, **69**(7), 544 (1992).
100. I. V. Berezin and A. A. Klesov, *Prakticheskii kurs khimicheskoi i fermentativnoi kinetiki*. (Practical Handbook of Chemical and Enzymatic Kinetics), Izdat. MGU, Moscow, 1976.
101. L. A. Petrov, Application of Graph Theory to the Study of the Kinetics of Heterogeneous Catalytic Reactions, in *Mathematical Chemistry Series*, Vol. 2. *Chemical Graph Theory: Reactivity and Kinetics*, D. Bonchev and D. H. Rouvray (Eds.), Abacus Press/Gordon & Breach Sci. Publ., Philadelphia, 1992, pp. 1–52.
102. O. N. Temkin, *Kinetika kataliticheskikh reaktsii v rastvorakh kompleksov metallov* (Kinetics of Catalytic Reactions in Metal Complex Solutions), MITKhT (Moscow Institute of Fine Chemical Technology), Moscow, 1980.
103. K. Kurtev, E. Balbolov, and O. Tomkin, *Metalokompleksen kataliz* (Catalysis with Metal Complexes), VKHTI, Burgas (Bulgaria), 1989.

104. O. N. Temkin, *Teoriya mekhanizmov slozhnykh reaktsii i kataliz. Konspekt lektsii* (Theory of Mechanisms of Complex Reactions and Catalysis: A Synopsis of Lectures), MITKhT (Moscow Institute of Fine Chemical Technology), Moscow, 2003.
105. A. I. Vol'pert, Differential Equations on Graphs, *Mat. Sbornik*, **88**(130), No. 4(8), 578 (1972).
106. B. L. Clarke, *J. Chem. Phys.*, **60**(4), 1481, 1493 (1974); **64**(10), 4165 (1976).
107. B. L. Clarke, *Adv. Chem. Phys.*, **43**, 7 (1980).
108. A. N. Ivanova, *Kinet. Katal.*, **20**(4), 1019, 1024 (1979).
109. A. N. Ivanova and B. L. Tarnopol'skii, *Kinet. Katal.*, **20**(6), 1514 (1979).
110. D. Bonchev, O. N. Temkin, and D. Kamenski, *React. Kinet. Catal. Lett.*, **15**(1), 119 (1980).
111. D. Bonchev, D. Kamenski, and O. N. Temkin, *J. Math. Chem.*, **1**, 345 (1987).
112. A. V. Zeigarnik and O. N. Temkin, *Kinet. Katal.*, **37**(3), 372 (1996).
113. D. Kamenski, O. N. Temkin, and D. Bonchev, *Appl. Catalysis A: General*, **88**, 1 (1992).
114. O. N. Temkin, Kh. I. Abanto-Chavez, and K. B. Khoang, *Kinet. Katal.*, **41**(5), 701 (2000).
115. A. A. Balandin, *Izv. Akad. Nauk SSSR, Ser. Khim.*, No. 1, 35 (1943).
116. A. T. Balaban, D. Farcasiu, and R. Banica, *Rev. Roum. Chim.*, **11**(10), 1205 (1966); A. Balaban, Reaction Graphs, in: *Understanding Chemical Reactivity*, Vol. 9, *Graph-Theoretical Approaches to Chemical Reactivity*, D. Bonchev and O. Mekenyan (Eds), Kluwer Acad. Publ., Dordrecht, 1994, pp. 137–180.
117. B. A. Orsi, *Biochim. Biophys. Acta*, **258**, 4 (1972).
118. L. G. Bruk and O. N. Temkin, *Abstract of Papers. The 4th All-Union Conf. "Mathematical Methods in Chemistry"* (Yerevan, 1982), p. 230.
119. K.-C. Chou and S. Forsen, *Chem. Scripta*, **18**, 82 (1981).
120. K.-C. Chou, *Biophys. Chem.*, **35**, 1 (1990).
121. S. Mason and H. Zimmermann, *Electric Circuits, Signals and Systems*, New York, 1960.
122. K. B. Jacimirskij, *Z. Chem.*, **13**(6), 201 (1973).
123. M. V. Volkenshtein, B. N. Goldshtein, and V. E. Stefanov, *Molek. Biol.*, No. 1, 52 (1967).
124. A. Sakamoto, H. Kawakami, and K. Yoshikawa, *Chem. Phys. Lett.*, **146**(5), 444 (1988).
125. A. V. Zeigarnik and O. N. Temkin, *Kinet. Katal.*, **35**(5), 702 (1994).
126. V. N. Kondrat'ev and E. E. Nikitin, *Kinetika i mekhanizm gazofaznykh reaktsii* (Kinetics and Mechanism of Gas-Phase Reactions), Nauka, Moscow, 1974.
127. V. I. Dimitrov, *Prostaya kinetika* (Simple Kinetics), Nauka, Novosibirsk, 1982.
128. I. A. Semiokhin, B. V. Strakhov, and A. I. Osipov, *Kinetika khimicheskikh reaktsii* (Kinetics of Chemical Reactions), Izdat. MGU, Moscow, 1995.
129. P. Robinson and K. Holbrook, *Unimolecular Reactions*, John Wiley & Sons, London, 1972.
130. E. T. Denisov, O. M. Sarkisov, and G. I. Likhtenshtein, *Khimicheskaya kinetika* (Chemical Kinetics), Khimiya, Moscow, 2000; E. T. Denisov, O. M. Sarkisov, and G. I. Likhtenshtein, *Chemical Kinetics: Fundamentals and New Developments*, Elsevier Science Ltd., Amsterdam, 2003.
131. S. G. Entelis and R. P. Tiger, *Kinetika reaktsii v zhidkoi faze* (Kinetics of Liquid-Phase Reactions), Khimiya, Moscow, 1973; S. G. Entelis and R. P. Tiger, *Reaction Kinetics in the Liquid Phase*, John Wiley & Sons, New York, 1976.
132. E. A. Moelwyn-Hughes, *The Kinetics of Reactions in Solutions*, Oxford University Press, Oxford, 1947.
133. S. Glasstone, K. Leidler, and H. Eyring, *The Theory of Rate Processes*, McGraw-Hill, New York, 1941.
134. E. D. German and R. R. Dogonadze, The Quantum-Mechanical Theory of the Kinetics of Proton Transfer Reactions, appendix to Russian translation of the book: R. Bell, *Proton in Chemistry*, Chapman & Hall, London, 1973; *Proton v khimii*, Mir, Moscow, 1977, p. 350.
135. R. R. Dogonadze and A. M. Kuznetsov, Kinetics of Chemical Reactions in Polar Media, in *Itogi Nauki i Tekhniki, Ser. Fiz. Khim. Kinetika* (Advances in Science and Technology, Ser. Physical Chemistry: Kinetics), VINITI, Moscow, 1973, Vol. 2.
136. R. R. Dogonadze, *Kvantovaya teoriya khimicheskikh reaktsii v polyarnoi zhidkosti* (Quantum Theory of Chemical Reactions in Polar Liquids), Znanie, Moscow, 1973.

137. R. Pearson, *Symmetry Rules for Chemical Reactions*, Academic Press, New York, 1976.
138. R. Woodward and R. Hoffmann, *The Conservation of Orbital Symmetry*, Springer, Weinheim, 1970.
139. G. Klopman (Ed.) *Chemical Reactivity and Reaction Paths*, John Wiley & Sons, New York, 1974.
140. L. S. Polak, *Neravnovesnaya khimicheskaya kinetika i ee primenenie* (Non-Equilibrium Chemical Kinetics and its Applications), Nauka, Moscow, 1979.
141. D. G. Truhlar, B. C. Garrett, and S. J. Klippenstein, *J. Phys. Chem.*, **100**, 12771 (1996).
142. S. Pedersen, J. L. Herek, and A. H. Zewail, *Science*, **266**, 1359 (1994).
143. L. S. Polak, *Variatsionnye printsipy mekhaniki* (Variation Principles of Mechanics), Fizmatlit, Moscow 1960.
144. L. Salem, *Electrons in Chemical Reactions*, John Wiley & Sons, New York, 1982.
145. M. J. Boudart, *Phys. Chem.*, **87** (15), 2786 (1983).
146. A. S. Dneprovskii and T. I. Temnikova, *Teoreticheskie osnovy organicheskoi khimii* (Theoretical Foundations of Organic Chemistry), Khimiya, Leningrad, 1979.
147. S. Ahrland, J. Chatt, and N. R. Davies, *Quart. Rev. Chem. Soc. (London)*, 12, 265 (1958).
148. V. Yu. Kukushkin and Yu. N. Kukushkin, *Teoriya i praktika sinteza koordinatsionnykh soedinenii* (Synthesis of Coordination Compounds: Theory and Practice), Nauka, Leningrad, 1990.
149. V. I. Belevantsev and B. I. Peshevitskii, *Issledovanie slozhnykh ravnovesii v rastvorakh* (Investigation of Complex Equilibria in Solutions), Nauka, Novosibirsk, 1978.
150. K. B. Yatsimirskii, *Teor. Eksp. Khim.*, **6**(4), 462 (1970).
151. K. Tatsumi and M. Tsutsui, in *Fundamental Research in Homogeneous Catalysis*, M. Tsutsui (Ed.), Plenum Press, New York, 1978, Vol. 3, p. 55.
152. F. D. Mango, The Removal of Orbital Symmetry Restrictions, in *Dynamic Chemistry* (*Topics in Current Chemistry*, Vol. 45), Springer, Berlin, 1974, p. 39.
153. V. I. Labunskaya, A. D. Shebaldova, and M. L. Khidekel', *Usp. Khim.*, **43**, 3 (1974).
154. R. Hoffmann, *Angew. Chem.*, **21**, 711 (1982).
155. F. G. A. Stone, *Angew. Chem.*, **23**, 89 (1984).
156. *Khmicheskaya entsiklopediya* (Chemical Encyclopedia), Sov. Entsiklopediya, Vol. 2, p. 362.
157. Ya. A. Dorfman, *Zhidkofaznyi kataliz* (Liquid-Phase Catalysis), Nauka, Alma-Ata, 1981.
158. R. H. Crabtree, *Organometallic Chemistry of the Transition Metals*, John Wiley & Sons, New York, 1988.
159. J. Collman, L. Hegedus, J. Norton, and R. Finke, *Principles and Applications of Organotransition Metal Chemistry (Graduate Textbook)*, University Science Books, Mill Valley, 1987.
160. S. P. Gubin, *Khimiya klasterov* (Chemistry of Clusters), Nauka, Moscow, 1987.
161. A. A. Grinberg, *Vvedenie v khimiyu kompleksnykh soedinenii* (Introduction to the Chemistry of Complex Compounds), Khimiya, Moscow, 1966; A. A. Grinberg, *An Introduction to the Chemistry of Complex Compounds* [Translated from the second edition (1951) by J. Rovtar Leach, D. H. Busch and R. F. Trimble, Jr., Eds.], Pergamon, London/Addison-Wesley, Reading (MA), 1962.
162. K. Langford and G. Gray, *Ligand Substitution Processes*, Benjamin, New York, 1966.
163. M. Tobe, *Inorganic Reaction Mechanisms*, Thomas Nelson, London, 1972.
164. E. M. Shustorovich, *Khimicheskaya svyaz' v koordinatsionnykh soedineniyakh* (Chemical Bonding in Coordination Compounds), Znanie, Moscow, 1975.
165. V. I. Nefedov and M. M. Gofman, Mutual Influence of Ligands in Inorganic Compounds, in *Itogi Nauki i Tekhniki, Ser. Neorg. Khim*. (Advances in Science and Technology, Ser. Inorganic Chemistry), VINITI, Moscow, 1978, Vol. 6.
166. O. N. Temkin, *Vvedenie v metallokompleksnyi kataliz. Kataliz i Koordinatsionnaya khimiya* (Introduction to Catalysis with Metal Complexes: Catalysis and Coordination Chemistry), MITKhT (Moscow Institute of Fine Chemical Technology), Moscow, 1980.
167. N. S. Zefirov and S. S. Tratch, *Anal. Chim. Acta*, **235**, 115 (1990).
168. R. Herges, *Tetrahedron Comput. Methodol*., **1**, 15 (1998).
169. R. Herges, *J. Chem. Inform. Comput. Sci*., **30**, 377 (1990).
170. R. Herges, *J. Chem. Inform. Comput. Sci*., **34**, 91 (1994).
171. J. Baner, R. Herges, E. Fontain, and I. Ugi, *Chimia*, **39**, 43 (1985).

172. R. Herges, *Angew. Chem. Int. Ed.*, **33**, 255 (1994).
173. H. Mimoun, I. Seree de Roch, and L. Sajus, *Tetrahedron*, **26**, 37 (1970).
174. A. A. Akhrem, T. A. Timoschtschuk, and D. I. Meteliza, *Tetrahedron*, **30**, 3165 (1974).
175. V. I. Metelitsa, *Modelirovanie okislitel'no-vosstanovitel'nykh fermentov* (Modeling of Redox Enzymes), Nauka i Tekhnika, Minsk, 1984.
176. R. Ponec, Similarity Model in Theory of Pericyclic Reactions, in *Molecular Similarity II*, P. J. Artymiuk, P. C. Mishra, and A. Kumar (Eds.), Springer-Verlag, Berlin, 1995 [*Topics in Current Chemistry*, **174**, 1 (1995)].
177. S. Benson, *The Foundations of Chemical Kinetics*, McGraw-Hill, New York, 1960.

2

Complexity Functions of Catalysts and Reactants in Reactions Involving Metal Complexes

An important feature of the catalysis with metal complexes (as well as noncatalytic reactions involving such complexes), which differs these reactions from those of the enzymatic, acid-base, electrophilic, and nucleophilic catalysis, is the presence of a large set of species, including various forms of complex compounds. This multispecies character is the first factor that hinders the verification of hypotheses concerning reaction mechanisms and the determination of the form of a kinetic equation. In this context, there arises the need of having a large volume of pre-kinetic information about the material balance with respect to the catalyst and reactants.

The necessity of taking into account the concentration of the intermediate complex catalyst–substrate in the description of kinetic relations was originally noticed by A. Brown (1902) and V. Henry (1903) in their investigations of enzymatic reactions [1]. In 1913, Henry, Michaelis, and Menten [1, 2] proceeded from the assumption concerning the quasi-equilibrium step of formation of an intermediate complex according to the following scheme:

$$E + A \underset{k_{-1}}{\overset{k_1}{\rightleftharpoons}} EA \xrightarrow{k_2} P + E, \tag{2.1}$$

(also considered in Sections 1.1 and 1.2), according to which

$$r = k_2 K_1 [E][A]. \tag{2.2}$$

Note that Eq. (2.2) involves the concentrations of catalyst E and substrate A that are not bound in a complex (EA). For the sake of simplicity, these quantities will be called *free* in contrast to the total (analytical) concentrations $[E]_0$ (or $[E]_\Sigma$) in solution. For $[A]_0 \gg [E]_0$ the initial condition is as follows:

$$[E]_0 = [E] + [EA] = [E](1 + K_1[A]_0). \tag{2.3}$$

Homogeneous Catalysis with Metal Complexes: Kinetic Aspects and Mechanisms, First Edition. O. N. Temkin.

Using formula (2.2) and condition (2.3), we obtain the following relation for the initial rate:

$$r_0 = \frac{k_2 K_1 [E]_0 [A]_0}{1 + K_1 [A]_0}, \tag{2.4}$$

which can be rewritten as

$$r_0 = \frac{k_2 [E]_0 [A]_0}{1/K_1 + [A]_0} = \frac{v_{max} [A]_0}{k_m + [A]_0}, \tag{2.5}$$

where $v_{max} = k_2 [E]_0$ and $k_m = 1/K_1$.

In 1925, G. Briggs and J. Haldane [1, 3] obtained an analogous relation under the assumption of a quasi-steady state condition with respect to [EA], in which case $k_m = \frac{k_{-1}+k_2}{k_1}$. In the enzymatic catalysis, the latter quantity is called the Michaelis constant. In the catalysis with metal complexes, different variants of the participation of metal complexes in homogeneous reactions under the conditions of quasi-equilibrium (or pre-equilibrium) with allowance for the material balance were originally analyzed by Schpitalsky in 1926 [4].

The first attempt to measure the "free" catalyst concentration [M] and use it instead of $[M]_0$ in the kinetic equation was undertaken by Flid and Moiseev [5, 6] in investigations of the hydration of acetylene in the $HgSO_4$–H_2SO_4–$Fe_2(SO_4)_3$ system by the potentiometric method. They obtained an equation that related the first-order rate constant (k_{obs}) to the electrode potential (E_{Hg}) of the Hg^{2+}/Hg^0 pair, i.e., to a quantity proportional to the activity of Hg^{2+} and, hence, of a catalytically active mercury(II) complex:

$$\lg k_{obs} = a + bE_{Hg}. \tag{2.6}$$

This approach was subsequently applied to other reactions of acetylene [7, 8] and further developed in investigations of the reactions involved in the catalysis with metal complexes (see below).

Let us consider in more detail various aspects of using the material balance in reactions involving mononuclear and polynuclear complexes and using the free concentrations of active components in reactions involving metal complexes.

2.1 Mononuclear metal complexes

In coordination chemistry, the equilibria in solutions of metal complexes are described in terms of the step (K_n) and total (β_n) equilibrium constants of complex formation (stability constants) and various material-balance functions [9–14]. In the case of binary complexes ML_n with homogeneous ligands, the total equilibrium constant is defined as

$$M + nL \rightleftharpoons ML_n, \qquad \beta_n = \frac{[ML_n]}{[M][L]^n},$$

and the individual step equilibrium constants as

$$M + L \rightleftharpoons ML, \qquad K_1 = \frac{[ML]}{[M][L]};$$

$$ML + L \rightleftharpoons ML_2, \qquad K_2 = \frac{[ML_2]}{[ML][L]};$$

$$ML_{n-1} + L \rightleftharpoons ML_n, \qquad K_n = \frac{[ML_n]}{[ML_{n-1}][L]},$$

so that we evidently have $\beta_1 = K_1$ and $\beta_n = \prod_{n=1}^{n} K_n$.

For roughly estimating the step equilibrium constants, it is convenient to use statistical considerations (under assumption of the equivalence of coordination sites and with neglect of the mutual influence of ligands). In particular, for the reaction

$$ML_n + L \overset{K_{n+1}}{\rightleftharpoons} ML_{n+1}, \tag{2.7}$$

the step equilibrium constant for the formation of complex ML_{n+1} will include a statistical factor f_{n+1} defined as

$$f_{n+1} = \frac{N-n}{n+1},$$

where N is the maximum coordination number, and the ratio of sequential step constants is determined by the following relation:

$$\frac{K_n}{K_{n+1}} = \frac{f_n}{f_{n+1}} = \frac{(N-n+1)(n+1)}{n(N-n)} = f. \tag{2.8}$$

For example, in the case of $N = 4$:

$$\frac{K_1}{K_2} : \frac{K_2}{K_3} : \frac{K_3}{K_4} = \frac{8}{3} : \frac{9}{4} : \frac{8}{3}.$$

According to Bjerrum [9], the value of K_n/K_{n+1} written as

$$T_{n/n+1} = \lg K_n/K_{n+1} = S_{n/n+1} + L_{n/n+1} \tag{2.9}$$

is called the total stepwise effect. Here, $S_{n/n+1}$ is the statistical factor ($\lg f$) and $L_{n/n+1}$ is the ligand factor that takes into account the nature and degree of the mutual influence of added ligands. The $L_{n/n+1}$ value characterizes a kind of "cooperative effect," which is positive for $L_{n/n+1} > 0$ (whereby the preceding ligand with number n facilitates the attachment of $(n+1)$th ligand) and negative for $L_{n/n+1} < 0$.

In some cases, it was demonstrated [9, 13, 14] that a correction of the step constants for the statistical factor $(\lg K_n - \lg f_n) = \lg K'_n$ leads to the disappearance of stepwise effects to within ± 0.1–0.2 log units, so that K'_n becomes a geometric mean value for the total equilibrium constant β_n $\left(\lg K' \cong \frac{1}{n} \lg \beta_n\right)$ (see Section 1.4).

The influence of the nature of a ligand on the stepwise effects can also be described using the so-called scattering factor χ via the following relation:

$$\frac{K_n}{K_{n+1}} = f\chi^2. \tag{2.10}$$

If this factor (determined experimentally from the shape of the "formation curves" is constant for the entire system, one can obtain the following expressions for the relationship between K'_n, K_n, β_n, and χ [13]:

$$K_n = \frac{(N-n+1)}{n} K' \chi^{N+1-2n}; \tag{2.11}$$

$$\beta_n = C_N^n (K')^n \chi^{n(N-n)}, \tag{2.12}$$

where the binomial coefficient is defined as

$$C_N^n = \frac{N!}{n!(N-n)!}.$$

In the case of mixed complexes, the number of variants of the coordination-saturated complexes is given by the following formula [13]:

$$l = \frac{(N+Z-1)(N+Z-2)\cdot\ldots\cdot(N+1)}{(Z-1)!}, \tag{2.13}$$

where Z is the number of various ligand types. For example, if the total number of ligands is $N = 6$ and the number of ligand species is $Z = 3$, the number of possible compounds is $l = 28$.

The analytical (total) concentrations of the complex-forming metal (C_M, $[M]_\Sigma$) and ligand (C_L) under equilibrium conditions are related via the material balance equation to the total equilibrium constant β_n and free concentrations of the ligand ([L]) and metal ions ([M]) (or other metal-containing species) as:

$$C_M = [M] + \sum_{i=1}^{n} \beta_i [M][L]^i; \tag{2.14}$$

$$C_L = [L] + \sum_{i=1}^{n} i\beta_i [M][L]^i. \tag{2.15}$$

Various functions obtained on the basis of these equations can be used for determining the compositions of complexes and the corresponding β_n values.

If the free concentration of a metal (in the form of ions or some other species) is known, it is possible to use the function F_M defined as

$$F_M = \frac{C_M}{[M]} = 1 + \sum_{i=1}^{n} \beta_i [L]^i, \tag{2.16}$$

which depends on [L] only. According to Yatsimirskii [10], this function is called the *complexity* of a metal, and various derivatives of F_M are called complexity functions.[1] This value characterizes the degree of metal binding by the ligands.

In cases where the ligand concentration [L] is measured, it is possible to use the Bjerrum formation function $\tilde{n}$ (representing the average coordination number) that is defined as

$$\tilde{n} = \frac{C_L - [L]}{C_M} = \frac{\sum_{i=1}^{n} n\beta_i [L]^i}{1 + \sum_{i=1}^{n} \beta_i [L]^i}. \tag{2.17}$$

As can be seen, the F_M and $\tilde{n}$ values are related by a simple formula as

$$\frac{d \lg F_M}{d \lg [L]} = \tilde{n}. \tag{2.18}$$

Since $\tilde{n} = f([L])$, the fact that $\tilde{n}$ vs. [L] curves are independent of C_M indicates that only mononuclear metal complexes exist in the system. A molar fraction of the nth complex is calculated as

$$\alpha_n = \frac{[ML_n]}{C_M} = \frac{\beta_n [L]^n}{F_M}. \tag{2.19}$$

[1] The F_M value is also frequently called the complexity function.

An analysis of the α values as dependent on n and the relations between step equilibrium constants for a large number of complexes is given in [14].

As will be shown below, functions such as the complexity of a metal (F_M) or substrate (F_S) are directly related to the kinetics of reactions with the participation of metal complexes and other catalysts (enzymes, acids, etc.). Indeed, the denominator in Eq. (2.4) represents the complexity of enzyme E (for $F_A = 1$):

$$F_E = 1 + K_1[A]_0. \tag{2.20}$$

Let us consider the kinetics of a two-substrate reaction A + B = P catalyzed by metal complexes under the conditions where before the reaction $n = 2$

$$C_M([M]_\Sigma) = \sum_0^2 [ML_n].$$

For a two-step mechanism involving an active form of metal catalyst species M with concentration [M]

$$A + M \underset{k_{-1}}{\overset{k_1}{\rightleftharpoons}} X_1 \underset{B}{\overset{k_2}{\rightleftharpoons}} P + M, \tag{2.21}$$

the kinetic equation under steady-state or quasi-steady-state conditions can be written as

$$r = \frac{k_1 k_2 [M][A][B]}{k_{-1} + k_2[B]}, \tag{2.22}$$

and the metal complexity in this case is expressed as

$$F_M = 1 + \beta_1[L] + \beta_2[L]^2 + \frac{[X_1]}{[M]}. \tag{2.23}$$

Taking into account that, under steady-state or quasi-steady state conditions,

$$[X_1] = \frac{k_1[A][M]}{k_{-1} + k_2[B]},$$

we eventually obtain

$$F_M = 1 + \beta_1[L] + \beta_2[L]^2 + \frac{k_1[A]}{k_{-1} + k_2[B]}. \tag{2.24}$$

In contrast to the value F_M of metal complexity in the case of equilibrium, here we deal with the *kinetic complexity of a metal-catalyst*, which becomes equilibrium ($\tilde{F}_M$) at $k_{-1} \gg k_2[B]$. The material balance with respect to the substrate A is described by the following equation:

$$[A]_0 = [A] + [X_1] + [P] \tag{2.25}$$

Denoting by $[A]_t$ the concentration of reactant A not converted into product P, we can express it as

$$[A]_t = [A]_0 - [P] = [A] + [X_1]$$

and the kinetic complexity F_A as

$$F_A = \frac{[A]_t}{[A]} = 1 + \frac{k_1[M]}{k_{-1} + k_2[B]}. \tag{2.26}$$

Since substrate B is not involved in the process of complex formation, we have $F_B = 1$ and $[B]_t = [B]_0 - [P] = [B]$. Taking into account Eqs (2.24) and (2.26), the kinetic equation (2.22) can be written in the general form as

$$r = \frac{k_1 k_2 [A]_t [B]_t C_M}{(k_{-1} + k_2 [B]_t) F_M F_A}. \tag{2.27}$$

A kinetic equation in this form was probably used for the first time by Moiseev [15]. Since the expressions for F_M and F_A contain the unknown quantities [M] and [A], this approach must generally simplify writing with the passage to a single variable F_i.

Solving the system of Eqs (2.24) and (2.26) with respect to [M] and [A] is possible only in simple cases (in particular, when there is only one metal complex ML_n in solution), but even in these cases the kinetic equations are rather complicated. For this reason, it is commonly accepted to work under conditions of $[A]_0 \gg C_M$, $F_A \cong 1$, and $F_i \cong 1$ (for other substrates). In this case, the condition of quasi-steady state in closed systems (see Section 1.2) is better obeyed. As a result, using Eq. (2.27) and omitting for simplicity the subscript "t" at the substrate concentrations, we obtain

$$r = \frac{k_1 k_2 [A][B] C_M}{(k_{-1} + k_2 [B]) \left[(1 + \beta_1 [L] + \beta_2 [L]^2) + \dfrac{k_1 [A]}{k_{-1} + k_2 [B]} \right]}. \tag{2.28}$$

This equation can be written in a simpler form by taking product in the denominator as

$$r = \frac{k_1 k_2 [A][B] C_M}{(k_{-1} + k_2 [B])(1 + \beta_1 [L] + \beta_2 [L]^2) + k_1 [A]}. \tag{2.29}$$

Let us consider the structure of Eq. (2.28). The expression in parentheses in the denominator reflects the stepwise character of the process and the absence of a limiting step. The expression in square brackets describes the kinetic complexity function of the catalyst under steady-state conditions (F_{SS}), including the complexity of the metal $F_M^* = 1 + \beta_1[L] + \beta_2[L]^2$ in the initial form (prior to the onset of the catalytic reaction) and the complexity related to the formation of intermediate X_1. For $k_{-1} \gg k_2[B]$, that is, in the quasi-equilibrium approximation for the first step, the denominator in Eqs (2.28) and (2.29) becomes the equilibrium complexity ($\tilde{F}_M$), and Eq. (2.29) takes the following form:

$$r = \frac{k_2 K_1 [A][B] C_M}{1 + \beta_1 [L] + \beta_2 [L]^2 + K_1 [A]}. \tag{2.30}$$

In order to obtain equations of this type, it is convenient to use the method of graphs (see Section 1.3) in both steady-state (Eq. (2.28)) and quasi-equilibrium (Eq. (2.30)) approximations. It should be recalled that, in terms of the graph theory used in the kinetic aspect, we have

$$F_M = \frac{\sum F_i^* D_i}{D_i}, \tag{2.31}$$

where F_i^* are the complexities of vertices (with allowance for the pendant vertices) and D_i are the root (base) determinants of the vertices involved in the cycles of the graph for the given mechanism. In Eq. (2.29), the determinant of vertex M is $D_M = (k_{-1} + k_2[B])$ and the complexity of this vertex is $F_M^* = 1 + \beta_1[L] + \beta_2[L]^2$ ($k_1[A]$ is the determinant of vertex X_1). Note that D_M is the denominator in Eq. (2.22) written for the free concentration of M. If the denominator in Eq. (2.28) under the quasi-steady state conditions is denoted D_{SS}, the kinetic complexity can be written as $F_{SS} = D_{SS}/D_M$. In the particular case under consideration, we have

$$F_{SS} = F_M^* + D_X/D_M. \tag{2.32}$$

In the case of quasi-equilibrium reverse steps, the total metal concentration can be expressed via equilibrium concentrations $[\tilde{M}]$ and $[\tilde{L}]$, that is, via the equilibrium complexity $\tilde{F}_M$: $C_M = [\tilde{M}]\tilde{F}_M$. Then, substituting this quantity into the expression for $F_{SS} = D_{SS}/D_M = C_M/[M]$, we obtain

$$\frac{D_{SS}}{D_M} = \frac{[\tilde{M}]\tilde{F}_M}{[M]}. \tag{2.33}$$

Evidently, on approaching equilibrium we have $[\tilde{M}]/[M] \to 1$, which implies that

$$\lim_{[M]\to[\tilde{M}]} \frac{D_{SS}}{D_M} = \tilde{F}_M. \tag{2.34}$$

This relationship was strictly proved in a different form and for a more general case by Pryakhin [16–18]. It was shown that, in the case of a reversible catalytic reaction with a linear mechanism, the polynomial in the denominator of the kinetic equation (D_{SS} in our notation) is divided without residue by (i.e., is a multiple of) the polynomial representing the equilibrium complexity of the catalyst ($\tilde{F}_M$), provided that all concentrations of reactants and products in polynomial D_{SS} are replaced by the equilibrium values (for the equilibrium overall reaction). Using this law, called the *multiplicity relationship*, it is possible to check for the meaning of constants in the D_{SS} polynomial and establish how these constants (complexes of constants of the elementary steps) are related to the equilibrium constants measured during investigation of the steady-state or quasi-steady-state kinetics.

Example 2.1

► Let us perform the kinetic analysis of a three-step reaction with the overall equation

$$S_1 + S_2 \rightleftharpoons P \tag{2.35}$$

and the mechanistic reaction scheme

$$S_1 + ML \underset{(1)}{\rightleftharpoons} X_1 \overset{S_2}{\underset{(2)}{\rightleftharpoons}} X_2 \underset{(3)}{\rightleftharpoons} ML + P \tag{2.36}$$

supplemented by several additional equilibria with the formation of complexes not representing intermediates in the catalytic process:

$$ML + P \underset{4}{\rightleftharpoons} MLP \quad (K_4); \tag{2.37}$$

$$ML + L \underset{5}{\rightleftharpoons} ML_2 \quad (K_5); \tag{2.38}$$

$$X_2 + S_1 \underset{6}{\rightleftharpoons} X_1S_1 \quad (K_6); \tag{2.39}$$

$$X_1 + L \underset{7}{\rightleftharpoons} X_1L \quad (K_7). \tag{2.40}$$

The mechanism of this reaction is represented in Figure 2.1 by graph G_1 with pendant vertices. The rate of a single-route reaction under steady-state conditions is determined by the following overall equation (see Section 1.1):

$$r = \frac{C_M(C^+ - C^-)}{\sum D_i F_i^*} = \frac{C_M(C^+ - C^-)}{D_{ML}F_{ML}^* + D_{X_1}F_{X_1}^* + D_{X_2}F_{X_2}^*}, \tag{2.41}$$

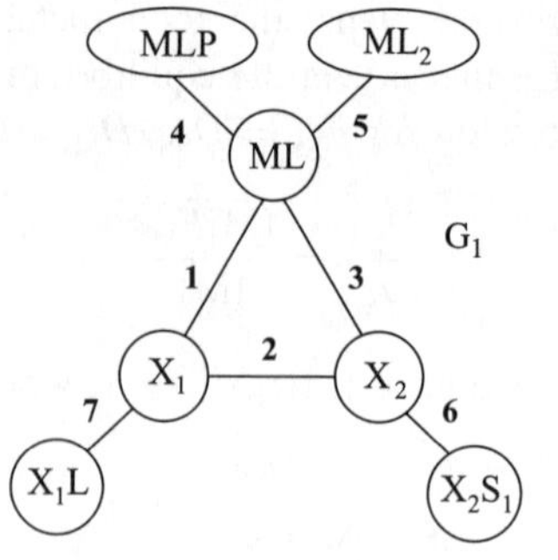

Figure 2.1 Graph G_1 of reaction mechanism (2.35)–(2.40)

where

$$C^+ = k_1k_2k_3[S_1][S_2], \qquad C^- = k_{-1}k_{-2}k_{-3}[P];$$
$$D_{ML} = k_2k_3[S_2] + k_{-1}k_3 + k_{-1}k_{-2}; \qquad D_{X_1} = (k_1k_3 + k_1k_{-2})[S_1] + k_{-2}k_{-3}[P];$$
$$F^*_{ML} = 1 + K_4[P] + K_5[L]; \qquad F^*_{X_1} = 1 + K_7[L]; \quad F^*_{X_2} = 1 + K_6[S_1].$$

The kinetic complexity $F_{SS} = D_{SS}/D_{ML}$ in this case is

$$F_{SS} = F^*_{ML} + \frac{D_{X_1}}{D_{ML}}F^*_{X_1} + \frac{D_{X_2}}{D_{ML}}F^*_{X_2}. \tag{2.42}$$

If we consider the conditions of equilibrium's overall reaction, substitute the equilibrium values of $[\tilde{S}_1]$, $[\tilde{S}_2]$ and $[\tilde{P}]$ into Eq. (2.42), and express the product concentration as $[\tilde{P}] = K_{eq}[\tilde{S}_1][\tilde{S}_2]$, where K_{eq} refers to the overall reaction ($K_{eq} = k_1k_2k_3/k_{-1}k_{-2}k_{-3}$), then polynomial F_{SS} given by Eq. (2.42) will become the polynomial $\tilde{F}_M$:

$$\tilde{F}_M = (1 + K_4[\tilde{P}] + K_5[\tilde{L}]) + K_1[\tilde{S}_1](1 + K_7[\tilde{L}]) + K_1K_2[\tilde{S}_1][\tilde{S}_2](1 + K_6[\tilde{S}_2]). \tag{2.43}$$

In other words, D_{SS} under these conditions is divided by $\tilde{F}_M$ without residue in accordance with the multiplicity relation. ■

Returning to scheme (2.21) and Eq. (2.29) with allowance for the fact that

$$F^*_{ML} = 1 + \sum_{i=1}^{N} \beta_i[L]^i,$$

where N is the maximum coordination number of metal M with respect to ligand L, and ML_k is considered to be the active complex, the reaction rate is determined by the following equation:

$$r = \frac{k_1k_2\beta_k C_M[L]^k[A][B]}{(k_{-1} + k_2[B])\left(1 + \sum_{i=1}^{N}\beta_i[L]^i\right) + k_1[A]}.$$

At $N > k$, the rate of this reaction will exhibit a maximum with respect to [L]. Examples of catalytic processes involving several complexes ML_n will be considered below.

The question arises whether the polynomial form of complexity F_M of the type given by Eq. (2.16) is the only form that can be used for writing the material balance in chemical kinetics. In other words, can polynomials of the type represented by formulas (2.16) and (2.17) be transformed to simpler expressions under definite conditions? The question was posed in the investigations of coordination compounds and, as a result, new approximate approaches to the description of complexity F_M and the formation function $\tilde{n}$ were found [19]. This problem is also important in the kinetics of reactions involving metal complexes, since the descriptions of kinetics in these investigations encounter unusual complexity functions.

For example, in studying the kinetics of the oxidation of alcohols by potassium periodate in an alkaline medium, which is catalyzed by Rh(III) aqua complexes [20], the following simple reaction mechanism was proposed:

$$\begin{aligned} \mathrm{Rh(H_2O)_6}^{3+} + \mathrm{ROH} &\overset{K_2}{\rightleftharpoons} X_1; \\ X_1 &\xrightarrow[\text{Slow}]{k} \mathrm{Rh(I)} + \text{Products}; \\ \mathrm{Rh(I)} + \mathrm{IO_4}^- &\xrightarrow{\text{Fast}} \mathrm{Rh(III)} + \mathrm{IO_3}^-; \\ \mathrm{Rh(H_2O)_6}^{3+} + \mathrm{OH}^- &\overset{K_1}{\rightleftharpoons} \mathrm{Rh(H_2O)_5OH}^{2+} \end{aligned} \tag{2.44}$$

Under the conditions of $[\mathrm{ROH}] \gg [\mathrm{OH}^-] \gg [\mathrm{Rh(III)}]_0 = [\mathrm{Rh(III)}]_t$, the corresponding kinetic equation is written as

$$-\frac{d[\mathrm{IO_4}^-]}{dt} = \frac{kK_2[\mathrm{ROH}][\mathrm{Rh(III)}]_t}{(1 + K_2[\mathrm{ROH}])(1 + K_1[\mathrm{OH}^-])}, \tag{2.45}$$

where the denominator represents the complexity of rhodium(III) written in the form of

$$F_{\mathrm{Rh}} = (1 + K_2[\mathrm{ROH}])(1 + K_1[\mathrm{OH}^-]). \tag{2.46}$$

It should be noted that no explanation was given in [20] for using this form of F_{Rh} in the kinetic equation instead of the function

$$F_{\mathrm{Rh}} = 1 + K_1[\mathrm{OH}^-] + K_2[\mathrm{ROH}] \tag{2.47}$$

that naturally follows from the reaction mechanism (2.44). No comparison of the two forms of complexity function was performed and it remains unclear how the function (2.46) was derived from the proposed reaction scheme.

An even odder kinetic equation was proposed [21] for the reaction of diacetone alcohol (DAA) oxidation with cerium(IV) catalyzed by Ir(III) complexes in sulfuric acid solutions under the conditions of ($[\mathrm{DAA}] > [\mathrm{Ce(IV)}]_0 \gg [\mathrm{Ir(III)}]_0$). Here, a very simple reaction scheme was also adopted:

$$\mathrm{DAA} + \mathrm{Ir(III)} \overset{K}{\rightleftharpoons} \text{Complex}$$

$$\text{Complex} + \mathrm{Ce(IV)} \xrightarrow[\text{Slow}]{k} \text{Products} + \mathrm{Ir(III)} + \mathrm{Ce(III)}$$

and the rate of cerium(IV) consumption was described by the equation

$$-\frac{d[\mathrm{Ce^{IV}}]}{dt} = \frac{kK[\mathrm{Ce^{IV}}][\mathrm{Ir(III)}][\mathrm{DAA}]}{1 + K[\mathrm{Ir(III)}] + K[\mathrm{DAA}]}, \tag{2.48}$$

where the denominator that must evidently express the complexity of only Ir(III) (since $F_{\mathrm{DAA}} \cong 1$ and $F_{\mathrm{Ce}} = 1$), nevertheless includes [Ir(III)].

Let us consider some possible simplified approaches to description of the complexity F_{M} and the formation function $\tilde{n}$.

If we accept that the step equilibrium constants for the series of ML_i complexes differ only by statistical factors, then the distribution of these complexes will obey the binomial law. This observation was apparently first made in 1963 by Hsu (Xu) and Li [19]. Writing the complexity as

$$F_{\mathrm{M}} = 1 + \sum_{i=1}^{N} \beta_i [\mathrm{L}]^i \tag{2.49}$$

and accepting that

$$\frac{K_i}{K_1} = \frac{N - i + 1}{iN},$$

where N is the maximum coordination number, we obtain the following expression for coefficients:

$$\beta_i = K_1^i \prod_{i=1}^{N} \frac{N - i + 1}{iN}. \tag{2.50}$$

Substituting expression (2.50) into formula (2.49) yields

$$F_{\mathrm{M}} = 1 + K_1[\mathrm{L}]\frac{N}{N} + K_1^2[\mathrm{L}]^2\frac{N(N-1)}{2!N^2} + \cdots + K_1^N[\mathrm{L}]^N\frac{N!}{N!N^N}. \tag{2.51}$$

Note that Eq. (2.51) is essentially a polynomial form of the Newton binomial:

$$F_{\mathrm{M}} = \left(1 + \frac{K_1}{N}[\mathrm{L}]\right)^N = (1 + a[\mathrm{L}])^N. \tag{2.52}$$

Thus, under certain assumptions, the metal complexity can be represented in a binomial form. Then, taking into account that $\partial \lg F / \partial \lg [\mathrm{L}] = \tilde{n}$ according to Eq. (2.18), and using the derivative of F_{M} written in the form of (2.52) we eventually obtain the following expression for the formation function:

$$\tilde{n} = \frac{Na[\mathrm{L}]}{1 + a[\mathrm{L}]}, \tag{2.53}$$

which has the form of an equation of the Langmuir-type adsorption isotherm.

The same Langmuir-type dependence can be obtained in a different way. Indeed, accepting again that all coordination sites are accessible for the ligands and assuming that all step constants K_n are the same and equal to K_1, we have

$$\tilde{n} = \frac{C_{\mathrm{L}} - [\mathrm{L}]}{C_{\mathrm{M}}} = \frac{\sum_{i=1}^{N} iK_1^i[\mathrm{L}]^i}{1 + \sum_{i=1}^{N} K_1^i[\mathrm{L}]^i}.$$

Upon dividing $\tilde{n}$ (i.e., the average coordination number) by N (the maximum number of sites accessible for coordination), we obtain $\tilde{n}/N = \alpha$, which is the average fraction of all sites occupied by ligand L in solution, so that $(1-\alpha)$ is the fraction of vacant coordination sites. Then, considering conditions of the equilibrium between the coordination and dissociation (adsorption and desorption) processes and assuming identity (homogeneity) of all coordination sites, we can express the equilibrium constant K_L of this process as

$$K_L = \frac{\alpha}{(1-\alpha)[L]}, \tag{2.54}$$

which yields

$$\alpha = \frac{K_L[L]}{1+K_L[L]} = \frac{C_L - [L]}{C_M N} = \frac{\tilde{n}}{N}. \tag{2.55}$$

From a comparison of expressions (2.55) and (2.53), it follows that $K_L = a = K_1/N$.

A satisfactory validity of the equation for $\tilde{n}$ in the form of the Langmuir-type adsorption isotherm (2.53) was demonstrated for a large body of experimental data (about 200 metal complex systems) [19]. However, in some cases the formation function was better described by a relation analogous to the Freundlich isotherm:

$$\tilde{n} = a'[L]^b.$$

Now let us proceed to the case of two ligands (substrates), A and B, which corresponds to the following expression:

$$C_M = [M] + \sum_{i=1}^{l*}\sum_{j=1}^{n*}[MA_iB_j], \tag{2.56}$$

where l^*, n^* are the maximum coordination numbers for ligands A and B, respectively, and $N = l^* + n^*$ is the maximum coordination number of metal M.

We assume that the coordination sites are specific for ligands A and B and the corresponding step equilibrium constants K_A and K_B are different, but still independent of the complex formation step number. In addition, we assume that the step constant for one ligand does not depend on the occupancy of coordination sites specific of the other ligand, for example:

$$MA + A \rightleftharpoons MA_2 \quad (K_A);$$
$$MA_2 + A \rightleftharpoons MA_3 \quad (K_A);$$
$$MB_2 + A \rightleftharpoons MB_2A \quad (K_A);$$

which implies complete absence of the mutual influence (L effect) and/or the cooperative effects.

Under the above assumptions, it can be readily shown that

$$\frac{C_M}{[M]} = F_M = \left(1 + \sum_{i=1}^{l*} K_A^i[A]^i\right)\left(1 + \sum_{j=1}^{n*} K_B^j[B]^j\right). \tag{2.57}$$

If the step equilibrium constants for the ligands of one type differ by the statistical coefficients, then the complexity F_M for octahedral complexes of the MA_2B_4 type ($l^* = 2$, $n^* = 4$) can be expressed as

$$F_M = (1 + 2K_A[A] + K_A^2[A]^2)(1 + 4K_B[B] + 6K_B^2[B]^2 + 4K_B^3[B]^3 + K_B^4[B]^4),$$

or

$$F_M = (1 + K_A[A])^2(1 + K_B[B])^4, \tag{2.58}$$

that is, as a product of two binomials.

If the step equilibrium constants for each of the ligands depend (but not very strongly) on the complex formation step (so that the ligand factor is not large), then polynomial (2.56) can be approximated by the following product:

$$F_M = \left(1 + \sum_{i=1}^{l*} \beta_i [A]^i\right)\left(1 + \sum_{j=1}^{n*} \beta_j [B]^j\right). \tag{2.59}$$

Functions of this type must be more probable for large clusters, that is, for complexes with large numbers of coordination sites that are specific for particular ligands.

Let us return to the example described by Eqs (2.45) and (2.46) [20]. Evidently, if the assumptions made in deriving Eqs (2.57) and (2.58) are valid for $l = 1$ and $n = 1$ with allowance for the formation of complex $Rh(H_2O)_4(OH)(ROH)^{2+}$, then the complexity of type (2.46) can take place. However, it would be difficult to justify these assumptions for the system under consideration. Indeed, if Eq. (2.45) is valid, then the reaction order with respect to the alcohol must not depend on $[OH^-]$, but this was not checked in [20]. Therefore, the use of the complexity function in the form of a product of polynomials in that study should be considered an accident. There are not even hypothetical arguments in favor of writing the complexity of Ir(III) in the form of Eq. (2.48). This equation was obviously used by mistake, as well as the erroneous equation written in [21] for the perchlorate system under the conditions of formation of $(Ce^{IV} \cdot DAA)$ and $(Ir(III) \cdot DAA)$ complexes.

Thus, the above analysis demonstrates that, under certain conditions, the formation of ML_i complexes admits writing F_M in a binomial form of Eq. (2.52) and $\tilde{n}$ in the Langmuir form of Eq. (2.53). In the case of two ligands that form complexes of the MA_lB_n type, the metal complexity can be expressed as a product of polynomials (2.57) or binomials (2.58) of various powers for the different ligands (substrates) only under very special assumptions. In each particular case, this unconventional description of F_M requires special consideration.

Equations (2.28), (2.29) and (2.41), as well as other examples, clearly demonstrate that, for an adequate description of reaction kinetics, it is necessary to have information on the "pre-catalytic" equilibria in the system (preferably knowing the β_n values and the number of complexes) and the [L] values in the course of catalysis.

In order to determine the reaction orders with respect to C_M and the substrate concentrations, a necessary condition is that the free ligand concentration [L] would remain constant during variation of the concentrations of complexes formed between metal (catalyst) and reactants. To obey this condition, it is necessary either to be able to determine [L] or, if the corresponding β_n values are known, to select the region of concentrations so as to ensure that $C_L \cong [L] = \text{const}$. The knowledge of [L] is also necessary for determining the dependence of the reaction rate on the ligand concentration, that is, for establishing the composition of a catalytically active form of the complex (see Chapter 3). If the determination of [L] is impossible (measuring this value in the course of catalysis is an extremely difficult problem), then it is conventional to use the regions of concentrations where $C_L \gg C_M$. Then, the ligand concentration will be virtually constant while C_M varies, and the free metal complex concentration will be proportional to its total concentration ($[ML_n] \propto C_M$). In cases where very strong complexes ML_n (with the stability constants of ML species exceeding 10^6–$10^8\ M^{-1}$) are formed and the step equilibrium constants differ by no more than three to four orders of magnitude, the complex composition is determined by the C_L/C_M ratio and is practically independent of the metal concentration. In these systems, the dependence of the reaction rate on the concentration of ligand L frequently exhibits a maximum and the $(C_L/C_M)_{max}$ value corresponds to the composition of a catalytically active complex.

Should the determination of [L] be possible, then the concentrations of substrates ($[A]_0$ and $[B]_0$ in Eq. (2.28)) can be varied so as to maintain constant concentrations of all complexes [8, 22, 23].

An analysis of the structure of an expression for the kinetic complexity of a metal-catalyst under quasi-steady-state conditions (F_{SS}) reveals four variants of this structure (the last representing the most general case):

1. The simplest variant of F_{SS} is encountered in systems that involve relatively strong, coordination-saturated complexes ML_n (typically, in non-aqueous solvents). In these systems, the step of interaction with a substrate is preceded by the step of dissociation of the complex, whereby one or several ligands are replaced by the solvent:
$$ML_n \rightleftharpoons ML_{n-1} + L. \tag{2.60}$$
The reaction can proceed either without the formation of intermediates X_i or under the condition that $\Sigma X_i \ll [M]_0$. Then, only two complexes will contribute to the material balance:
$$C_M = [M]_0 = [ML_{n-1}] + [ML_n]. \tag{2.61}$$
2. In this case, a set of complexes ML_n exist in solution, but no intermediates are formed in significant amounts and, hence, no other species enter into the material balance:
$$C_M = \sum_{i=0}^{N} [ML_i], \tag{2.62}$$
where N is the maximum coordination number in the system. The pendant vertices are formed only by ligands.
3. If the substrate–catalyst complexes and other intermediate species involved in the steps of a reaction mechanism (i.e., species entering into the kinetic or bipartite graphs in the space of intermediates) are formed in significant amounts, then
$$C_M = \sum_{i=1}^{n} [ML_i] + \sum_{j=1}^{k} X_j, \tag{2.63}$$
where k is the number of intermediates. In the enzymatic catalysis, this variant is known as the *competitive inhibition by the ligands (inhibitors).*
4. The process of complex formation with ML_n involves the reaction products P and substrates S_i that form "dead-end" complexes (inactive for the overall process) of the ML_xS_i and ML_yP types, and the formation of complexes with X_j involves ligands L, S_i, and P.

The latter variant of the structure of complexity function F_{SS} includes the appearance of various pendant vertices on the kinetic graph of the reaction mechanism, which contribute to the material balance. As was demonstrated above, in this case we can use the functions F_i^* that represent complexities of the vertices of the kinetic graph.

2.1.1 Complexity functions: variants I and II

Now we will analyze features of the manifestation of the metal complexity in reaction kinetics in cases where the structure of this function is determined only by the initial forms of complexes. Let us consider the sequence of steps

$$MX_nL \underset{k_{-1}}{\overset{k_1}{\rightleftharpoons}} MX_n + L, \tag{2.64}$$

$$MX_n + A \xrightarrow{k_2} P \tag{2.65}$$

where P are reaction products and metal complexes. This scheme is frequently encountered in both stoichiometric (noncatalytic) with respect to metal complexes and catalytic reactions. Step (2.64) in most cases consists of the replacement of ligand L by a molecule of solvent S:

$$MX_nL + S \rightleftharpoons MX_nS + L. \tag{2.66}$$

In cases of concentrated solutions of reactants, complexes, or inert electrolytes, the kinetic equation should be derived taking into account the solvent concentration.

In the general case of a quasi-steady-state process, a kinetic equation for the reaction scheme (2.64)–(2.65) has the following form:

$$r = \frac{k_1k_2C_M[A]}{(k_{-1}[L] + k_2[A])\left[1 + \dfrac{k_1}{k_{-1}[L] + k_2[A]}\right]}; \tag{2.67}$$

$$F_{SS} = \frac{C_M}{[MX_nL]} = 1 + \frac{k_1}{k_{-1}[L] + k_2[A]} \tag{2.68}$$

or

$$r = \frac{k_1k_2C_M[A]}{k_1 + k_{-1}[L] + k_2[A]}, \tag{2.69}$$

where the denominator is what was denoted above by D_{SS}.

If step (2.65) is rate-controlling, so that

$$r = k_2K_1[MX_nL][A][L]^{-1} \tag{2.70}$$

and $F_M = 1 + K_1/[L]$, then we obtain

$$r = \frac{k_2K_1C_M[A]}{K_1 + [L]}. \tag{2.71}$$

Equation (2.70) corresponds to cases where the ligand concentration [L] can be directly measured and $F_M \cong 1$.

Example 2.2

► The rate of the reductive elimination reaction

$$2HCo(CO)_4 \longrightarrow Co_2(CO)_8 + H_2, \tag{2.72}$$

which proceeds according to the scheme

$$HCo(CO)_4 \overset{K_1}{\rightleftharpoons} HCo(CO)_3 + CO; \tag{2.73}$$

$$HCo(CO)_3 + HCo(CO)_4 \xrightarrow{k_2} H_2Co_2(CO)_7 \xrightarrow{CO} Co_2(CO)_8 + H_2, \tag{2.74}$$

is described by the following equation [24]:

$$r = \frac{d[H_2]}{dt} = \frac{k[HCo(CO)_4]^2}{p_{CO}}, \tag{2.75}$$

where $k = k_2K_1$. ■

If the concentration [L] is not determined and ligand L is not added to the solution in excess, then for $F_M \cong 1$ we have $[MX_n] = [L]$, $[L] = \sqrt{K_1C_M}$, and Eq. (2.70) yields

$$r = k_2K_1^{1/2}C_M^{1/2}[A]. \tag{2.76}$$

Example 2.3

▶ The rate of protodemetallation of bis(triphenylphosphine)phenylpalladium bromide (I) by anhydrous HBr in dioxane (10–45 °C) [25] according to the reaction

$$L_2Pd(Ph)Br + HBr \longrightarrow L_2PdBr_2 + C_6H_6 \tag{2.77}$$

is described by the equation

$$r = k[I]_0^{1/2}[HBr] \tag{2.78}$$

if the ligand ($L = PPh_3$) is not added, and by the equation

$$r = k'[I]_0[HBr][L]^{-1} \tag{2.79}$$

if [L] is varied from $0.08 \cdot 10^{-3}$ to $2.2 \cdot 10^{-3}$ M at $[I]_0 = 1 \cdot 10^{-3}$ M. Both these equations correspond to the following mechanistic scheme:

$$L_2Pd(Ph)Br + S \overset{K_1}{\rightleftharpoons} LPd(Ph)Br(S) + L; \tag{2.80}$$

$$LPd(Ph)Br(S) + HBr \xrightarrow{k_2} L(S)PdBr_2 + PhH; \tag{2.81}$$

$$L(S)PdBr_2 + L \xrightarrow{\text{Fast}} L_2PdBr_2 + S, \tag{2.82}$$

where at $F_{Pd} \cong 1$ (i.e., at small K_1) we have $k = k_2K_1^{1/2}$ and $k' = k_2K_1$ ($K_{1(25\,°C)} = 2 \cdot 10^{-5}$ M). ■

Example 2.4

▶ Fractional reaction order with respect to a metal complex is also obtained in the catalytic reaction [26]

$$Me_3SiAr + \tfrac{1}{2}H_2O \xrightarrow[DMF]{Li_2PdCl_4} \tfrac{1}{2}Me_3SiOSiMe_3 + ArH, \tag{2.83}$$

which is probably related to the dissociation of the initial catalyst at 80–100 °C in 2 % aqueous dimethylformamide (DMF) according to the scheme

$$PdCl_4^{2-} + L \overset{K_1}{\rightleftharpoons} PdCl_3L^- + Cl^-; \tag{2.84}$$

$$PdCl_3L^- + Me_3SiAr \xrightarrow{k_2} X \longrightarrow \cdots \xrightarrow[Cl^-]{H_2O} P + PdCl_4^{2-} \tag{2.85}$$

where L = H_2O, DMF and P denotes products. The rate of decrease in the concentration of silane is described by the following equation:

$$r = k[Me_3SiAr]C_{Pd}^{1/2}. \tag{2.86}$$

In the case of LiCl addition, the equation is

$$r = k'[Me_3SiAr]C_{Pd}[L][Cl^-]^{-1}, \tag{2.87}$$

where $k = k_2K_1^{1/2}[L]^{1/2}$ and $k' = k_2K_1$. ■

It should be emphasized that, even in these reactions with relatively simple mechanisms, the experimental kinetic equation cannot be considered as a proof of the proposed mechanistic schemes (hypotheses). Independent information concerning the prevailing forms of complexes in solution is necessary.

Example 2.5

▶ Investigations of the kinetics of oxidative addition of CH_3I to Pt(0) complex [27]

$$PtL_2(C_2H_4) + CH_3I \longrightarrow CH_3PtL_2I + C_2H_4 \tag{2.88}$$

led to the following equation of the type of Eq. (2.71):

$$-\frac{d[Pt(0)]}{dt} = \frac{kC_{Pt}[CH_3I]}{K + [C_2H_4]}, \tag{2.89}$$

In the presence of excess C_2H_4 ($[C_2H_4] > K$), this equation led to a relation of the minus first order with respect to ethylene, which was in complete agreement with schemes (2.64)–(2.65) and

$$PtL_2(C_2H_4) \overset{K}{\rightleftharpoons} PtL_2 + C_2H_4; \tag{2.90}$$

$$PtL_2 + CH_3I \xrightarrow{k_2} CH_3PtL_2I. \tag{2.91}$$

However, the results of spectrophotometric and ^{1}H NMR measurements [28, 29] showed that a prevailing form of Pt(0) complexes in the presence of ethylene is $PtL_2(C_2H_4)_2$. This circumstance led to a new hypothesis concerning the mechanism of reaction (2.88) with the kinetic equation (2.97) that is formally analogous to Eq. (2.89):

$$PtL_2(C_2H_4) + CH_3I \xrightarrow{k} CH_3PtL_2I(C_2H_4); \tag{2.92}$$

$$CH_3PtL_2I(C_2H_4) \xrightarrow{\text{Fast}} CH_3PtL_2I + C_2H_4; \tag{2.93}$$

$$PtL_2(C_2H_4) + C_2H_4 \overset{K_3}{\rightleftharpoons} PtL_2(C_2H_4)_2; \tag{2.94}$$

$$r = k[\mathrm{PtL_2(C_2H_4)}][\mathrm{CH_3I}]; \tag{2.95}$$

$$F_{\mathrm{Pt}} = \frac{C_{\mathrm{M}}}{[\mathrm{PtL_2(C_2H_4)}]} = 1 + K_3[\mathrm{C_2H_4}]; \tag{2.96}$$

$$r = -\frac{d[\mathrm{Pt(0)}]}{dt} = \frac{kC_{\mathrm{M}}[\mathrm{CH_3I}]}{1 + K_3[\mathrm{C_2H_4}]}. \tag{2.97}$$

Here, it should be noted that, if the kinetics of reaction (2.88) were studied under the conditions of inert gas supply, this would allow the scheme (2.90)–(2.91) to be rejected because a shift of the equilibrium of reaction (2.90) must reduce the reaction order with respect to [CH_3I] down to zero. Monitoring of the amount of evolved ethylene could also provide a ground for correctly judging on the mechanistic scheme.

Note also that the process according to the sequence of steps (2.64)–(2.65) can involve no rate-controlling step according to Eq. (2.67) under the conditions of $F_{\mathrm{SS}} \cong 1$ or $F_{\mathrm{SS}} > 1$. ■

Example 2.6

► The oxidative addition reaction [30]

$$\mathrm{HSiR_3 + HIr(CO)L_3 \longrightarrow H_2Ir(SiR_3)(CO)L_2 + L},$$

is described by the following kinetic equation with a fractional order with respect to $HSiR_3$ and the inhibition by ligand (PPh_3):

$$r = \frac{k_1 k_2 C_{\mathrm{Ir}}[\mathrm{R_3SiH}]}{k_{-1}[\mathrm{L}] + k_2[\mathrm{R_3SiH}]} \tag{2.98}$$

(see Eqs (2.64)–(2.65) for $F_{\mathrm{SS}} \cong 1$). ■

Example 2.7

► A similar equation was obtained for the oxidative addition of H_2 to the Wilkinson complex $RhCl(PPh_3)_3$ [31]. In the presence of excess phosphine, the reaction proceeds via two routes, which involve $RhClL_3$ and $RhClL_2$. For $F_{\mathrm{Rh}} \cong 1$ ($C_{\mathrm{Rh}} \cong [\mathrm{RhClL_3}]$), we have

$$-\frac{d[\mathrm{Rh(I)}]}{dt} = \left[k_1 + \frac{k_2 k_3}{k_{-2}[\mathrm{L}] + k_3[\mathrm{H_2}]}\right] C_{\mathrm{Rh}}[\mathrm{H_2}] \tag{2.99}$$

in accordance with the following scheme:

$$\mathrm{RhClL_3 + H_2 \xrightarrow{k_1} RhCl(H)_2L_3}; \tag{2.100}$$

$$\mathrm{RhClL_3 \underset{k_{-2}}{\overset{k_2}{\rightleftharpoons}} RhClL_2 + L}; \tag{2.101}$$

$$\mathrm{RhClL_2 + H_2 \xrightarrow{k_3} RhCl(H)_2L_2}; \tag{2.102}$$

$$\mathrm{RhCl(H)_2L_2 + L \overset{\text{Fast}}{\rightleftharpoons} RhCl(H)_2L_3}. \tag{2.103}$$

Without additions of PPh_3, complex $RhClL_2$ readily forms dimer $Rh_2Cl_2L_4$ and this dimer reacts with H_2. In this case, the kinetic equation is

$$-\frac{d[\mathrm{Rh(I)}]}{dt} = 2k_4[\mathrm{Rh_2Cl_2L_4}][\mathrm{H_2}]. \tag{2.104}$$

Results of estimation of the rate constants led to a conclusion that the activity of complex $RhClL_2$ (with respect to H_2) is at least 10^4 times as large as that of both $RhClL_3$ and the dimer. ■

In reversible reactions, the terms of the leaving ligand [e.g., L in step (2.64)] in the kinetic equation may be canceled. This circumstance can be used for the discrimination of hypotheses.

Example 2.8

► The reductive elimination reaction

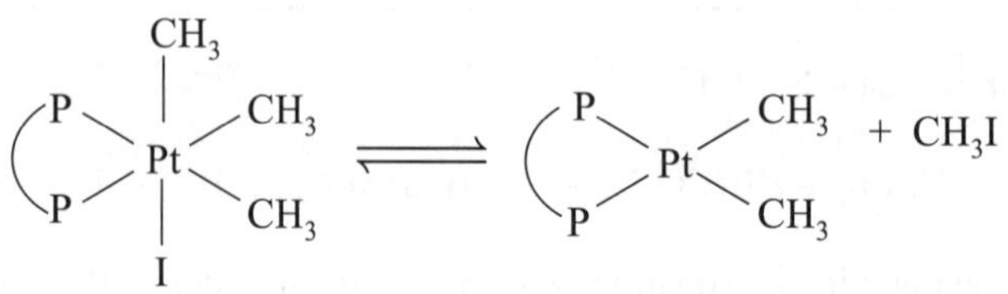

II III (2.105)

is an intermolecular process involving I^- from the solution (acetone-d_6, P–P = dppe, 70–100 °C) [32]. Two hypothetical mechanisms were considered for the nucleophilic substitution at C atom of the methyl group. According to mechanism A:

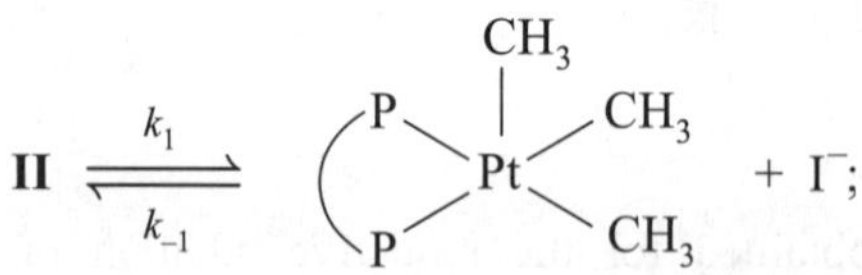

IV

$$\mathbf{IV} + \mathrm{I}^- \underset{k_{-2}}{\overset{k_2}{\rightleftharpoons}} \mathbf{III} + \mathrm{CH_3I};$$

$$-\frac{d[\mathbf{II}]}{dt} = \frac{k_1k_2[\mathbf{II}][\mathrm{I}^-] - k_{-1}k_{-2}[\mathbf{III}][\mathrm{CH_3I}][\mathrm{I}^-]}{k_{-1}[\mathrm{I}^-] + k_2[[\mathrm{I}^-]} = \frac{k_1k_2[\mathbf{II}] - k_{-1}k_{-2}[\mathbf{III}][\mathrm{CH_3I}]}{k_{-1} + k_2}. \tag{2.106}$$

According to mechanism B:

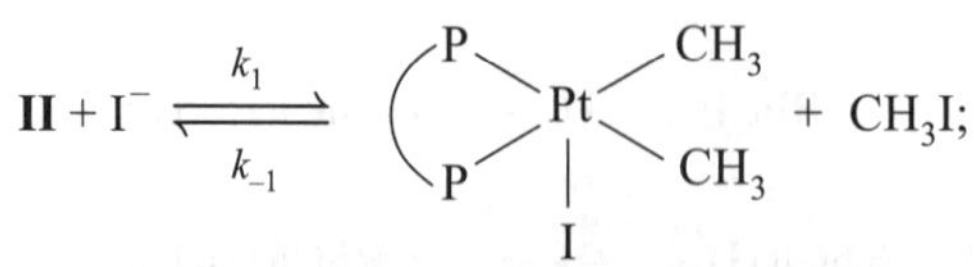

V

$$\mathbf{V} \underset{k_{-2}}{\overset{k_2}{\rightleftharpoons}} \mathbf{III} + I^-;$$

$$-\frac{d[\mathbf{II}]}{dt} = \frac{\{k_1k_2[\mathbf{II}] - k_{-1}k_{-2}[\mathbf{III}][CH_3I]\}[I^-]}{k_{-1}[CH_3I] + k_2}. \tag{2.107}$$

It was experimentally established that the rate of reaction (2.105) was independent of $[I^-]$, which allowed scheme B to be rejected. ■

The above examples demonstrate that monitoring of the ligand concentration [L] is among the most important prerequisites for the correct description of reaction kinetics and the formulation of a well-justified mechanistic scheme. Naturally, the [L] value is implied that corresponds to the quasi-steady-state or quasi-equilibrium conditions which cannot be separated into parts, such as $[Cl^-]_{SS}$ and $[Cl^-]_{add}$ as was done in the example presented below [33].

Example 2.9

► Investigation of the hydroxymercuration of olefins (Ol) [33] under irreversible conditions

$$HgCl_2 + Ol + H_2O \longrightarrow ClHgCH_2CH(R)OH + H^+ + Cl^-. \tag{2.108}$$

led to the following equation:

$$r = \frac{a[HgCl_2][Ol]}{b + c[HCl]}, \tag{2.109}$$

which reflected retardation of the process by Cl^- ions (the influence of hydrogen ion was ignored).

This equation was interpreted based on a scheme that had been described in the literature, but without an analysis and allowance for the complexity of mercury. A quasi-steady-state approximation was used and it was assumed that reaction (2.108) proceeds without a rate-controlling step, although this process most probably involves a limiting step of the nucleophilic addition of water to the cationic π complex of mercury(II). In this case, the scheme adopted in [33]

$$HgCl_2 \overset{K_1}{\rightleftharpoons} HgCl^+ + Cl^-;$$

$$HgCl^+ + Ol \overset{K_2}{\rightleftharpoons} HgCl(Ol)^+;$$

$$HgCl(Ol)^+ + H_2O \overset{k_3}{\longrightarrow} ClHgCH_2HC(R)OH + H^+$$

yields the following equation:

$$r = k_3K_2K_1\frac{[HgCl_2][Ol]}{[Cl^-]}, \tag{2.110}$$

where $[Cl^-]$ is the concentration of Cl^- ions in solution. Judging from the first order with respect to [Ol], the complexity of mercury(II) coordinated to olefin is insignificant and, with allowance for

$$F_{Hg} = \frac{[HgCl_2]_\Sigma}{[HgCl_2]} = 1 + \frac{K_1}{[Cl^-]},$$

we arrive at the equation

$$r = \frac{k_3 K_2 K_1 [\mathrm{HgCl_2}]_\Sigma [\mathrm{Ol}]}{[\mathrm{Cl^-}] + K_1}, \tag{2.111}$$

which is comparable with the experimental relation (2.109). At a large concentration of introduced Cl^- ions, the reaction order with respect to $[Cl^-]$ must be minus first, but this was not found in [33]. ■

In Example 2.7, a two-route mechanism was considered for one reaction that could proceed via different species of complexes occurring in equilibrium. The situation, whereby a fraction or all of a set of complexes (e.g., $\sum_{i=0}^{n} \mathrm{ML}_i$) catalyze the same reaction under conditions of one slow step for each complex, is frequently encountered in homogeneous catalysis. In this case (in a simple one-substrate variant), we have

$$r = \sum r_i = \left(\sum_{i=1}^{n} k_i [\mathrm{ML}_i] \right) [\mathrm{A}] = k_{\mathrm{obs}} [\mathrm{A}], \tag{2.112}$$

where k_i are the partial rate constants, [A] is the substrate concentration, and

$$k_{\mathrm{obs}} = \sum_{i=1}^{n} k_i [\mathrm{ML}_i]$$

is the observed first-order rate constant. The k_{obs} value can be expressed via the total metal concentration $[\mathrm{M}]_\Sigma$ (C_M), metal complexity F_M, and k_i as follows:

$$k_{\mathrm{obs}} = k_{\mathrm{eff}} [\mathrm{M}]_\Sigma,$$

where

$$k_{\mathrm{eff}} = \frac{k_0 + k_1 \beta_1 [\mathrm{L}] + k_2 \beta_2 [\mathrm{L}]^2 + \cdots + k_i \beta_i [\mathrm{L}]^i}{1 + \sum_{i=1}^{n} \beta_i [\mathrm{L}]^i}. \tag{2.113}$$

Example 2.10

► In the study of a catalytic reaction of H_2 oxidation by Fe(III), potassium bichromate, or benzoquinone that is catalyzed by palladium(II) chloride or bromide under homogeneous conditions (in the absence of palladium black) the reaction rate was determined by the distribution method based on the variation of H_2 concentration in the gas phase [34].

At a constant concentration of Pd(II), the rate of $[H_2]$ decrease is described by the following equation of the first order with respect to $[H_2]$:

$$-\frac{d[\mathrm{H_2}]}{dt} = k_{\mathrm{obs}} [\mathrm{H_2}], \tag{2.114}$$

where k_{obs} is independent of the oxidant concentration and type (Cr^{VI}, Fe^{III}, benzoquinone). The rate constant k of a homogeneous reaction of dissolved hydrogen with PdX_i^{2-i} is related to k_{obs} via the following equation:

$$k = k_{\mathrm{obs}} (1 + \lambda \alpha), \tag{2.115}$$

where $\lambda = V_g/V_{sol}$, $\alpha = [H_2]_g/[H_2]_{sol}$; V_g and V_{sol} are the volumes of gas and solution in the reactor, respectively; and α is the coefficient of H_2 distribution between the gas phase and solution. It was established that, in the system of Pd(II)–Cr(VI)–HCl–$HClO_4$ at 80 °C, [HCl] + [$HClO_4$] = 1.0 M, and the intervals of concentrations [Pd(II)] = (3.7–41.3) × 10^{-3} M and [Cl^-] = 0.001–0.8 M, the value of $k_{eff} = k/[Pd]_\Sigma$ exhibits a maximum with respect to the chloride ion content (the maximum is attained at [Cl^-] = 0.01 M). The partial constants k_i in the equation $k = \sum k_i[PdCl_i^{2-i}]$ or Eq. (2.113) were calculated using the values of β_i from [35, 36]. As a result, the following k_i [$M^{-1} \cdot s^{-1}$] were determined for all the possible species:

$$Pd^{2+}(0.12 \pm 0.003);\ PdCl^+(3.4 \pm 0,4);\ PdCl_2(5.2 \pm 0.3);$$
$$PdCl_3{}^-(2.34 \pm 0.07);\ PdCl_4{}^{2-}(\sim 0).$$

Thus, $PdCl^+$, $PdCl_2$ and $PdCl_3{}^-$ complexes exhibit maximum activity with respect to H_2 activation. A slow step in the process of H_2 oxidation is probably the reaction

$$PdCl_i^{(2-i)} + H_2 \xrightarrow{k_i} HPdCl_i^{(2-i-1)} + H^+ \quad (i = 1, 2, 3). \tag{2.116}$$

In the analogous bromide system Pd(II)–Fe(III)–HBr–$HClO_4$, the value of k_{eff} exhibits a sharp maximum with respect to [Br^-] (the maximum is attained at [Br^-] = 0.015–0.02 M), and the k_{eff}^{max} value in the bromide system is about five times as large as the corresponding value in the chloride system. However, the partial constants in the bromide system were not determined because of the lack of data on the constants βi of bromide complex formation. Nevertheless, the shape of the curve of k_{eff} vs. [Br^-] led to a conclusion on the activity of $PdBr^+$, $PdBr_2$ and $PdBr_3{}^-$ complexes in H_2 activation. ■

In view of the discovery of the reaction of exchange between alkanes and D_2O in solutions of Pt(II) complexes [37–39] and the oxidation of arenes [40] and alkanes [39, 41, 42] in Na_2PtCl_4–H_2PtCl_6 systems, extensive investigations were devoted to determining the kinetics of these reactions and elucidating the contributions of various $PtCl_i^{2-i^i}$ complexes to the total reaction rate.

Example 2.11

▶ An original investigation [43] established that the deuterium exchange rate via reaction

$$RH + D_2O \longrightarrow RD + HOD, \tag{2.117}$$

in a D_2O–CH_3COOD solution at 100 °C, as determined using the kinetic curves of RD accumulation in a closed system, is described by the following equation:

$$-\frac{d[RH]}{dt} = k_{obs}[RH][Pt(II)]_\Sigma^n, \tag{2.118}$$

where $n < 1$. The rate of reaction (2.117) (RH = C_2H_6) is independent of [$HClO_4$] varied in a range of 0.2–1.0 M. The first-order rate constant $k_1 = -\frac{d\ln[RH]}{dt}$ decreases with increasing [Cl^-] in a 10^{-3}–0.3 M interval.

Gol'dshleger et al. [44, 45] studied the exchange kinetics in more detail for the reaction with cyclohexane in a D_2O–CH_3COOD system (1 : 1 molar ratio) in a temperature interval of 80–112 °C. Under the conditions of existence of $PtCl_2$ ([$PtCl_2$] = 0.004 M, [$HClO_4$] = 0.1 M, 100 °C), it was established that the rate of deuterium exchange is first-order with respect to $[PtCl_2]$, while the second-order rate constant is $k_{PtCl_2} = (8.2 \pm 0.7) \cdot 10^{-3}\,M^{-1} \cdot s^{-1}$. The variation of $[Cl^-]$ was considered for the equilibria

$$PtCl_4{}^{2-} \overset{K_1}{\rightleftharpoons} PtCl_3{}^- + Cl^-;$$

$$PtCl_3{}^- \overset{K_2}{\rightleftharpoons} PtCl_2 + Cl^-.$$

Using the equation

$$k_{\text{eff}} = \frac{W_0}{[PtCl_2]_\Sigma} = \frac{k_2 + k_3 K_2^{-1}[Cl^-] + k_4 K_1{}^{-1} K_2{}^{-1}[Cl^-]^2}{1 + K_2{}^{-1}[Cl^-] + K_1{}^{-1}K_2{}^{-1}[Cl^-]^2}, \tag{2.119}$$

the equilibrium constants K_1 and K_2 were determined and the partial rate constants k_2 ($PtCl_2$), $k_3(PtCl_3{}^-)$ and $k_4(PtCl_4{}^{2-})$ were evaluated. For the reaction with cyclohexane at 100 °C, it was found that $k_2 : k_3 : k_4 = 100 : 12 : 1$. ■

Example 2.12

► The kinetics of oxidation and deuterium exchange for alkanes, acetic acid, and benzene in K_2PtCl_4–H_2PtCl_6 systems in D_2O–CH_3COOD and $CF_3COOH(D)$ were studied in [46–50]. The partial rate constants were measured for the oxidation of some hydrocarbons in the presence of various forms of Pt(II) [47]. In the reaction of cyclohexane oxidation, the partial rate constants are as follows:

Complex form	Pt^{2+}	$PtCl^+$	$PtCl_2$	$PtCl_3{}^-$	$PtCl_4{}^{2-}$
k_i, $M^{-1} \cdot s^{-1}$	0	5.3 ± 0.2	9.5 ± 2.4	12 ± 0.3	0

■

Example 2.13

► In contrast to the reactions of deuterium exchange and RH oxidation, in which the activity of $PtCl_4{}^{2-}$ complexes is minimum or absent, the methylation of K_2PtCl_4 by dimethyl sulfate [51, 52]

$$Me_2SO_4 + Pt(II) \xrightarrow{k_{Pt}} MePt(IV) + MeSO_4{}^- \tag{2.120}$$

increases when chloride ions are introduced into the system. The values of

$$r_0 = \frac{d[MePt(IV)]}{dt}$$

exhibit linear growth with an increase in the initial *equilibrium* concentration of $PtCl_4{}^{2-}$ species. At small values of $[Cl^-]$, when 98 % of platinum atoms enter into the aqueous chloride complexes $PtCl_i(H_2O)_{4-i}^{2-i}$ ($i = 1$–3), the methylation of Pt(II) is virtually absent. Allowance for the parallel reaction of hydrolysis

$$Me_2SO_4 + H_2O \xrightarrow{k_S} MeOH + HMeSO_4 \tag{2.121}$$

was made according to the following equation:

$$\frac{d[MePt(IV)]}{dt} = k_{Pt}[PtCl_4{}^{2-}][Me_2SO_4]_0 \exp(-k_S t) \tag{2.122}$$

with $[Me_2SO_4] \gg [K_2PtCl_4]_\Sigma$. ■

As was pointed out in [52], it is important to take into account the kinetics of attaining equilibrium in the Pt(II)–[Cl^-] system because Pt(II) chloride complexes are inert in the reactions of ligand substitution. In these inert systems, the calculation of partial constants k_i using the k_{eff} values determined from the current concentrations of reactants may be incorrect because of a difference in the reactivity of various species and a change in their ratio in the course of experiments. In addition, the concentrations of complexes can be nonequilibrium and, hence, not described by equations of the (2.119) type.

Example 2.14

► Shologon et al. [52–54] used spectrophotometric techniques to study the kinetics of alkylation of Pt(II) complexes by ethylene oxide (EO) in aqueous acid solutions via the reaction

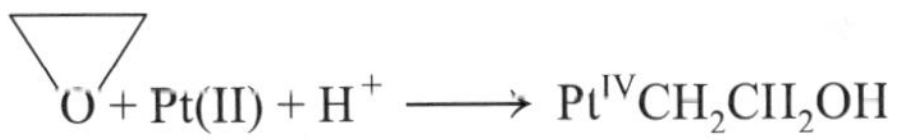

under the conditions of EO excess relative to Pt(II) in two systems. In the first system (EO–$HClO_4$–H_2O–$PtCl_4{}^{2-}$), the experiments were performed using freshly prepared K_2PtCl_4 solutions, that is, in the presence of $PtCl_4{}^{2-}$ species. In the second system (E–$HClO_4$–H_2–$PtCl_i(H_2O)_{4-i}^{2-i}$), the experiments were performed with K_2PtCl_4 solutions that were allowed to stand for one month so as to reach the equilibrium initial ratio of $PtCl_4{}^{2-}$, $PtCl_3{}^-$ and $PtCl_2$ species in the medium. In the former case, the reaction rate at a constant ionic strength of [NaCl] + [$NaClO_4$] = 0.25 M and [$HClO_4$] = 0.001–0.01 M is described by the following equation:

$$\frac{d[\mathrm{Pt(IV)R}]}{dt} = (k_0 + k_1[\mathrm{H}^+])[\mathrm{PtCl_4}{}^{2-}][\mathrm{EO}],$$

according to which the reaction of $PtCl_4{}^{2-}$ with protonated EO predominates. In order to evaluate the contributions from various routes involving different $PtCl_i(H_2O)_{4-i}^{2-i}$ species to the total reaction rate, the accumulation of Pt(IV) was studied in the second system at [EO] = const, [$HClO_4$] = const, fixed ionic strength (0.25 M), and variable concentrations of Cl^- ions. The values of k_i were calculated using the initial rates of variation of the optical density D and the initial concentrations of Pt(II) complexes according to the following equation:

$$\left(\frac{dD}{dt}\right)_0 = \sum_2^4 \varepsilon_i k_i [\mathrm{EO}][\mathrm{PtCl}_i\ (\mathrm{H_2O})_{4-i}^{2-i}]_0, \qquad (2.123)$$

where ε_i is the coefficient of extinction for the Pt(IV) alkyl complex formed from the ith species. These coefficients were determined from the relation for the total optical density

$$D_\infty = \sum_2^4 \varepsilon_i [\mathrm{PtCl}_i(\mathrm{H_2O})_{4-i}^{2-i}]_0, \qquad (2.124)$$

assuming that a redistribution of the Pt(II) and Pt(IV) complex species during the ligand exchange (Cl^- and H_2O) can be ignored. The values of ε_2, ε_3, and ε_4 were found to be 1020 ± 120, 1570 ± 150 and $5350 \pm 210\,M^{-1} \cdot cm^{-1}$, respectively. The value of ε_4 determined in independent experiments with $PtCl_4{}^{2-}$ solutions was $5300\,M^{-1} \cdot cm^{-1}$. The corresponding values of rate constants k_i determined from these experiments were $k_2 = (0.5 \pm 0.4) \times 10^{-3}\,M^{-1} \cdot s^{-1}$, $k_3 = (2.1 \pm 0.4) \times 10^{-3}\,M^{-1} \cdot s^{-1}$ and $k_4 = (2.7 \pm 0.2) \times 10^{-3}\,M^{-1} \cdot s^{-1}$.

Processes in neutral solutions of Pt(II) complexes lead to the parallel formation of $Pt^{IV}CH_2CH_2OH$ and a platinaoxetan complex (VI) [52–54]:

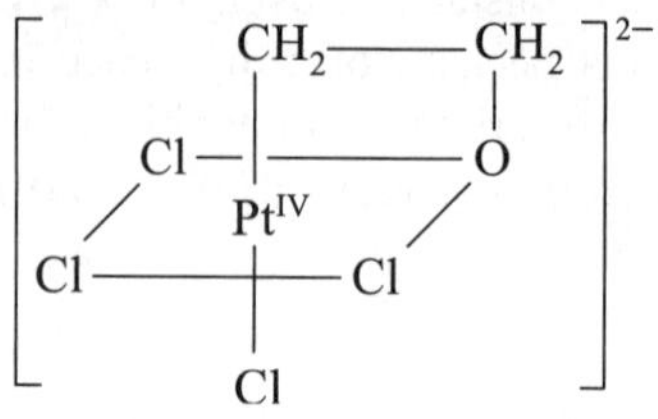

(VI)

It was established that active species in the synthesis of Pt^{IV}–ethanol and Pt^{IV}–oxetan complexes are $PtCl_2(H_2O)_2$ and $PtCl_4^{2-}$, respectively, with the corresponding rate constants of $k_2 = 0.125\,M^{-1}\cdot s^{-1}$ and $k_4 = 0.0024\,M^{-1}\cdot s^{-1}$ at 50 °C. ■

For the catalysis involving various species of metal complexes in the absence of any significant contribution of intermediates to the complexity, equations of the (2.113) type can lead to various dependences of k_{eff} on [L], the character of which is determined by the values of k_i and K_i. There are well-known examples of both catalytic and stoichiometric reactions involving identical complexes of various metals, which are characterized by significantly different forms of the $k_{eff} = f([L])$ function.

Example 2.15

► The kinetics of homogeneous reactions of the oxidation of metal phthalocyaninates (Pc′M) by benzoyl peroxide (BP) in nitrobenzene

$$Pc'M + BP \longrightarrow \text{Products} \tag{2.125}$$

and the oxidation of diphenylpicrylhydrazine (DPPH) catalyzed by metal phthalocyaninates with the formation of a stable diphenylpicrylhydrazyl radical (D•)

$$2D + BP \xrightarrow{Pc'M} 2D\bullet + 2PhCOOH, \tag{2.126}$$

where D = DPPH and Pc′ = tetra-4-(*tert*-butyl)phthalocyaninate anion was investigated in [55–57]. The rate of reaction (2.125) is described for M = Zn, Cu by the following equation:

$$r = k_{obs}[Pc'M][BP]. \tag{2.127}$$

The observed rate constant k_{obs} according to the proposed scheme

$$Pc'M + BP \underset{k_{-1}}{\overset{k_1}{\rightleftharpoons}} X \tag{2.128}$$

$$X \xrightarrow{k_2} \text{Products}$$

is defined as

$$k_{obs} = \frac{k_1 k_2}{k_{-1} + k_2}. \tag{2.129}$$

Equation (2.127) is also valid for initial rates in the case of catalysis with iron phthalocyaninate dipyridinate.

The rate of reaction (2.126) was determined by measuring the accumulation of radical D•. In this case, the scheme (2.128) is supplemented by the following steps:

$$\begin{aligned} \mathrm{X} + \mathrm{D} &\xrightarrow{k_3} \mathrm{D\bullet} + \mathrm{Pc'M} + \mathrm{PhCOOH} + \mathrm{PhCOO\bullet}; \\ \mathrm{PhCOO\bullet} + \mathrm{D} &\xrightarrow{k_4} \mathrm{PhCOOH} + \mathrm{D\bullet}. \end{aligned} \tag{2.130}$$

For the adopted schemes (2.128) and (2.130) [reaction (2.126)] with quasi-steady-state approximation, the accumulation of the radical species obeys the following equation:

$$\frac{d[\mathrm{D\bullet}]}{dt} = \frac{2k_1 k_3 [Pc'\mathrm{M}][\mathrm{BP}][\mathrm{D}]}{k_{-1} + k_2 + k_3[\mathrm{D}]}. \tag{2.131}$$

For $k_3[\mathrm{D}] \gg k_{-1} + k_2$, this yields

$$\frac{d[\mathrm{D\bullet}]}{dt} = 2k_1[\mathrm{Pc'M}][\mathrm{BP}] = k_1^*[\mathrm{Pc'M}][\mathrm{BP}]. \tag{2.132}$$

The values of k_1^* and k_1 were experimentally determined and studied as functions of [L] in various Pc′M–L systems (L = Py).

Initial complex Pc′FePy$_2$ [55, 57]. This complex occurs in equilibrium with Pc′Fe monopyridinate:

$$\mathrm{Pc'FeL} + \mathrm{L} \overset{K}{\rightleftharpoons} \mathrm{Pc'FeL_2}.$$

For the adopted two-route reaction scheme

$$\mathrm{Pc'FeL} + \mathrm{BP} \xrightarrow{k'_1} \mathrm{X_1};$$

$$\mathrm{Pc'FeL_2} + \mathrm{BP} \xrightarrow{k'_2} \mathrm{X_2},$$

the effective rate constant k_1^* in Eq. (2.132) can be expressed as

$$k_1^* = \frac{k'_1 + k'_2 K[\mathrm{L}]}{1 + K[\mathrm{L}]}. \tag{2.133}$$

For $k'_2 \gg k'_1$, an increase in [L] leads to a sharp decrease in the k_1^* value (Figure. 2.2). At $K[\mathrm{L}] \gg 1$ and $k'_2 K[\mathrm{L}] \gg k'_1$, this value ceases to depend on [L] in the interval of 0.1–6.3 M and becomes equal to $k'_2 (k_1^* = k'_2)$. The experimental values of rate constants at 23 °C are as follows:

$$k'_1 = 25\,\mathrm{M^{-1}\cdot s^{-1}}; \quad k'_2 = 0.14\,\mathrm{M^{-1}\cdot s^{-1}}; \quad K = 23\cdot 10^3\,\mathrm{M^{-1}}.$$

The results show evidence for the two possible mechanisms of BP activation, which can proceed via the BP coordination to metal (sixth axial coordinate) or via the BP coordination to macro-ring of the phthalocyaninate anion.

System Pc′Cu–Py [56]. The shape of the dependences of k_{obs} and k_1^* on [L] in this system (Figure 2.3) indicates that all the three possible species (Pc′Cu, Pc′CuL, and Pc′CuL$_2$) participate in the process, so that

$$k_{obs}(k_1^*) = \frac{k'_1 + k'_2 K_1[\mathrm{L}] + k'_3 K_2[\mathrm{L}]^2}{1 + K_1[\mathrm{L}] + K_2[\mathrm{L}]^2}, \tag{2.134}$$

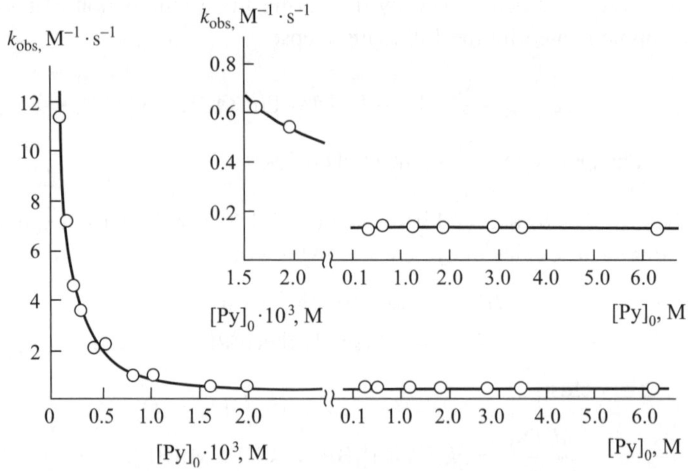

Figure 2.2 Dependence of the observed rate constant k_{obs} ($k_{obs} = k_1^*$) of the catalytic oxidation of DPPH on $[Py]_0$ (reproduced with permission from N. I. Bundina et al. © 1978 [55])

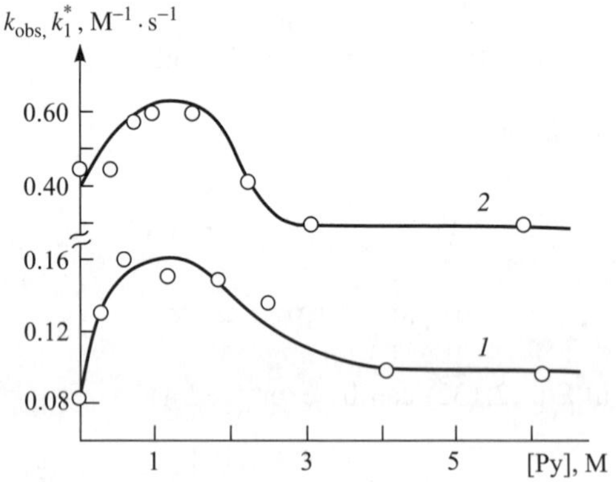

Figure 2.3 Plots of the rate constant for PcCu versus [Py]: (1) k_{obs} at $[PcCu] = 1.0 \cdot 10^{-5}$ M, $[BP] = 2.5 \cdot 10^{-2}$ M; (2) k_1^* at $[PcCu] = 2.0 \cdot 10^{-5}$ M, $[BP] = 2.5 \cdot 10^{-2}$ M, $[DPPH] = (1\text{–}5) \cdot 10^{-3}$ M (reproduced with permission from Yu. M. Golub et al. © 1979 [56])

where the values of k_1' and k_3' differ only slightly. It was suggested that the BP coordination to Pc′ macro-ring also takes place in copper phthalocyaninates.

System Pc′Zn–Py [57]. In this system, Pc′ZnPy complex is formed already at small pyridine concentrations ($[Py] = 1 \cdot 10^{-5}\text{–}1 \cdot 10^{-4}$ M), while the interval of $[Py] = 1 \cdot 10^{-4}\text{–}1 \cdot 10^{-3}$ M features a coexistence of complexes Pc′ZnPy and Pc′ZnPy$_2$. Accordingly, the dependence of k_{obs} and k_1^* on [L] are described by Eq. (2.133). The values of k_1'(Pc′Zn), k_2'(Pc′ZnPy), and k_3'(Pc′ZnPy$_2$) amount to 15.3, 2.4, and 0.12 $M^{-1} \cdot c^{-1}$, respectively. The equilibrium constant K for the transition from Pc′ZnL to Pc′ZnL$_2$ is 5680 M^{-1}. It was established that, in the system with Pc′Zn, the oxidation of Pc′Zn can also proceed via an alternative route, either without the formation of intermediate complexes X or with the formation of a complex X* that neither oxidizes DPPH nor participates in the catalytic process. This feature accounts for the fact that $k_{obs} > k_1^*$ only in the system with Pc′Zn.

System Pc′Ni–Py [57]. In the case of Pc′Ni (as well as Pc′Pd and Pc′Pt [57]) the rate of noncatalytic reaction (2.125) is described by an equation of second order in [Pc′Ni], while in the presence of Py the process kinetics is described by Eq. (2.127), which is the first-order equation in [Pc′Ni]. For this reason, the form of functions $k = r/[BP] = f([L])$ and $k_{obs} = f([L])$ depends on the total concentration $[Pc'Ni]_\Sigma$. The description is additionally complicated, since the reaction orders with respect to nickel complexes are different for the three routes: second for Pc′Ni and first for Pc′NiL and $Pc'NiL_2$. In this case, the value of k_{obs} in Eq. (2.127) represents a sum of two functions of [L], $k_{obs} = k'_{obs} + k''_{obs}$, where

$$k'_{obs} = \frac{k'_1[Pc'Ni]_\Sigma}{(1 + K_1[L] + K_2[L]^2)^2}; \tag{2.135}$$

$$k''_{obs} = \frac{k'_2K_1[L] + k'_3K_2[L]^2}{1 + K_1[L] + K_2[L]^2}. \tag{2.136}$$

The value of k'_{obs} sharply drops with increasing [L] (being dependent on $[Pc'Ni]_\Sigma$), whereas k''_{obs} can increase with [L] for $F_{PC'Ni} \gg 1$ and a sufficiently small value of $K_2[L]$ ($K_2[L] < K_1$). The sum of the two components accounts for a variety of dependences of the reaction rate on [L], which are observed in experiments at different $[Pc'Ni]_\Sigma$ [56, 57]. ■

Example 2.16

▶ The process of cyclopalladation of N,N-dimethylbenzylamines, by analogy with reactions of the oxidative addition of H_2 to $RhClL_3$ (see Example 2.7) and reaction (2.126) for $Pc'FeL_2$, proceeds with the participation of two complexes [58]:

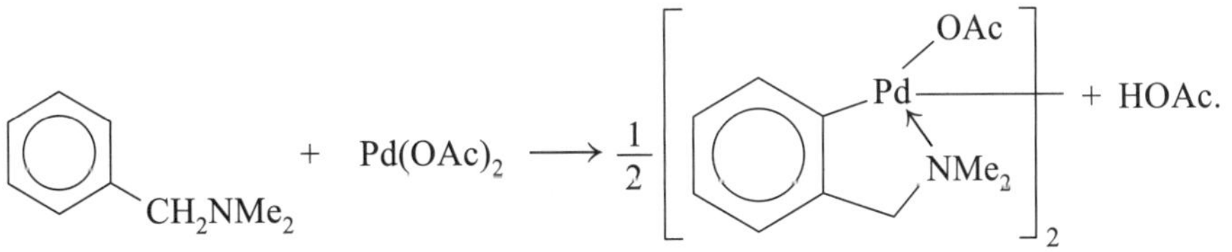

Trimer $Pd_3(OAc)_6$ reacts with benzylamine (L) in chloroform and converts into $PdL_2(OAc)_2$ complex, which serves the initial compound in the reaction of cyclopalladation. The observed rate constant of the first-order reaction with respect to $[Pd(II)]_\Sigma$ decreases with increasing concentration of benzylamine as described by the equation

$$k_{obs} = \frac{k_1K + k_2[L]}{K + [L]}, \tag{2.137}$$

which was obtained for the following reaction scheme [58, 59]:

$$\begin{array}{ccc} PdL_2(OAc)_2 & \overset{K_0}{\rightleftharpoons} & trans\text{-}PdL(OAc)_2 + L \\ \downarrow {\scriptstyle -HOAc,\ k_2} & & \updownarrow {\scriptstyle K'} \\ PdLL'(OAc) & & cis\text{-}PdL(OAc)_2 \\ \downarrow {\scriptstyle -L} & & \downarrow {\scriptstyle -HOAc,\ k_1} \\ [PdL'(OAc)]_2 & \longleftarrow & PdL'(OAc) \end{array} \tag{2.138}$$

(where L′ is the chelate ligand N–C in the product of cyclopalladation). According to this scheme, the constant K in Eq. (2.137) represents K_0K'. Substituting $K = 1/K^*$, we obtain an equation that is analogous in form to Eq. (2.133) with $k'_1 = k_1K'$:

$$k_{obs} = \frac{k_1K' + k_2K^*[L]}{1 + K^*[L]}.$$

The values of k_1 and k_2 for N,N-dimethylbenzylamine in chloroform at 25 °C are equal $4.3 \cdot 10^{-3}\ s^{-1}$ and $3.1 \cdot 10^{-5}\ s^{-1}$, respectively, and $K \cong 1.35 \cdot 10^{-2}$ M. ■

Sometimes, the values of step equilibrium constants in investigations of the kinetics of complex formation reactions are estimated based on the assumption that these constants only differ by their statistical coefficients (see Section 2.1). In these cases, for a sequence of reactions written as

$$ML_{n-1} + L \overset{K_n}{\rightleftharpoons} ML_n \qquad (n = 1, 2, \ldots, N),$$

e.g., for $N = 4$, we have

$$F_M = 1 + K_1[L] + K_1K_2[L]^2 + K_1K_2K_3[L]^3 + K_1K_2K_3K_4[L]^4. \qquad (2.139)$$

With allowance for the statistics $\left(\frac{K_n}{K_1} = \frac{N-n+1}{nN}\right)$, Eq. (2.139) yields

$$F^*_M = 1 + K_1[L] + 0.375K_1^2[L]^2 + 0.063K_1^3[L]^3 + 0.0039K_1^4[L]^4. \qquad (2.140)$$

If some reaction involves only one of a set of complexes (e.g., $PtCl_2(H_2O)_2$ from the set of $PtCl_i(H_2O)_{4-i}^{(2-i)+}$), then

$$k_{eff} = \frac{k_2K_1^2[L]^2 0.375}{F^*_M}. \qquad (2.141)$$

In this case, k_{eff} passes through a maximum with respect to [L], which corresponds to a maximum fraction α of the corresponding complex:

$$\alpha = \frac{0.375K_1^2[L]^2}{F^*_M}. \qquad (2.142)$$

Evidently, the condition of $d\alpha/d[L] = 0$ allows the value of $K_1[L]_{max}$ to be determined. In particular, for Eq. (2.142), we have $K_1[L]_{max} \cong 3.25$. Upon determining $k_{eff} = f([L])[L]_{max}$ from the experimental data, it is possible to estimate K_1. If a reaction involves several complexes ML_n with the corresponding partial rate constants k_n, this approach can only be used provided that all these partial constants have the same value, which is usually unlikely. A "statistical" approach to the estimation of K_1 was used in [60].

Example 2.17

▶ Investigation [60] of the kinetics of hydrocarbon (RH) oxidation in acetic acid (AcOH) in the presence of cobalt bromide catalysts showed that the rate constant k_4 of the reaction

$$RO_2^{\bullet} + Co(II) \xrightarrow{H^+} ROOH + Co(III) \qquad (2.143)$$

(or the parameter $k_4/\sqrt{2k_6}$, where k_6 is the rate constant of the $RO_2^{\bullet}$ – $RO_2^{\bullet}$ interaction) exhibits an extremal dependence on the concentration of acetic acid in the toluene – AcOH or xylene

(0.32 M) – benzene – AcOH systems under conditions where the reaction rate is independent of the concentration of the oxidized RH. In order to explain the form of this function, it was suggested that $Co(OAc)_2$ forms four complexes with AcOH, namely $Co(OAc)_2(AcOH)_n$ ($n = 1 \div 4$), three of which ($n = 1, 2, 3$) are active in the process of RH oxidation (the formation of cobalt–bromide complexes was ignored). In this case, the fraction (α) of three active complexes with allowance for the statistical factors is determined by the following relation:

$$\alpha = \frac{[\mathrm{Co(II)}]_{\mathrm{active}}}{[\mathrm{Co(II)}]_{\Sigma}} = \frac{K_1[\mathrm{L}] + 0.375K_1^2[\mathrm{L}]^2 + 0.063K_1^3[\mathrm{L}]^3}{F^*_{\mathrm{Co}}}, \tag{2.144}$$

where L = AcOH and F^*_{Co} is given by formula (2.140). The condition $d\alpha/d[\mathrm{L}] = 0$ yields $K_1[\mathrm{L}]_{\mathrm{max}} = 3.8$. Using this relation and the value of $[\mathrm{L}]_{\mathrm{max}}$ on the plots of the rate constant versus [AcOH], it is possible to estimate K_1. Since $[\mathrm{L}]_{\mathrm{max}}$ in a kinetic experiment also depends on the partial rate constants k_1, k_2, and k_3, the obtained K_1 value can be correct (provided all other assumptions are valid) only in the case of $k_1 \cong k_2 \cong k_3$, but this condition was not discussed in [60]. ■

2.1.2 Complexity functions: variants III and IV

The formation of intermediate products during the interaction of metal complexes with reactants in both noncatalytic and catalytic reactions is one of the most frequently encountered factors responsible for the appearance of complexity functions (F_{M}, F_{S}) in kinetic equations. Let us consider some examples that illustrate the main features of kinetics in both simple and complicated cases.

Example 2.18

► The reaction of sodium glycolate (GA^-) oxidation by hexachloroiridate(IV) anion [61]

$$\mathrm{CH_2(OH)COO^-} + 2\mathrm{IrCl_6}^{2-} \longrightarrow \mathrm{HCHO} + \mathrm{CO_2} + 2\mathrm{IrCl_6}^{3-} + \mathrm{H^+} \tag{2.145}$$

is stoichiometric with respect to the metal complex and can be described by an equation that is first-order in $[\mathrm{IrCl_6}^{2-}]$ and variable-order in $[\mathrm{GA^-}]$ (Michaelis–Menten equation). The proposed reaction scheme

$$\begin{aligned} &\mathrm{CH_2(OH)COO^-} + \mathrm{IrCl_6}^{2-} \overset{K}{\rightleftharpoons} \mathrm{X}; \\ &\mathrm{X} \overset{K}{\longrightarrow} \mathrm{IrCl_6}^{3-} + {\bullet}\mathrm{CH_2OH} + \mathrm{CO_2}; \\ &{\cdot}\mathrm{CH_2OH} + \mathrm{IrCl_6}^{2-} \overset{\text{Fast}}{\rightleftharpoons} \mathrm{CH_2O} + \mathrm{H^+} + \mathrm{IrCl_6}^{3-} \end{aligned} \tag{2.146}$$

for $[\mathrm{GA^-}] \gg [\mathrm{Ir(IV)}]_{\Sigma}$ and pH = const yields the following equation:

$$-\frac{d[\mathrm{IrCl_6}^{2-}]}{dt} = \frac{2kK[\mathrm{IrCl_6}^{2-}]_{\Sigma}[\mathrm{GA}]}{1 + K[\mathrm{GA^-}]}, \tag{2.147}$$

which is consistent with the obtained experimental results. The formation of free radicals was also confirmed in experiment. A more complicated equation is has been derived with allowance for the equilibrium dissociation of glycolic acid (introducing the dependence e on H^+). An increase in the observed pseudo-first-order reaction rate constant ($k_{\mathrm{obs}} = -d\ln[\mathrm{IrCl_6}^{2-}]/dt$) with increasing [NaCl] and [KCl] was considered in [61] as indicative of a positive salt effect (influence of the ionic strength of solution) and the absence of Cl^- replacement by GA^- anion during the formation of intermediate X. ■

The kinetics of decomposition of alkylhydroperoxides (ROOH) was studied in much detail in the context of investigations of the oxidation of hydrocarbons [62] and the epoxidation of olefins [63, 64].

Example 2.19

► The rate of ROOH decomposition catalyzed by copper(II) stearate ($CuSt_2$) [65], is described by the following equation:

$$r = \frac{k_1[\mathrm{ROOH}]_\Sigma^2[\mathrm{CuSt}_2]_\Sigma}{1 + K_1[\mathrm{ROOH}]_\Sigma^2}. \tag{2.148}$$

For the adopted sequence of steps

$$\begin{aligned} &\mathrm{CuSt}_2 + 2\mathrm{ROOH} \overset{K_1}{\rightleftharpoons} \mathrm{CuSt}_2(\mathrm{ROOH})_2; \\ &\mathrm{CuSt}_2(\mathrm{ROOH})_2 \xrightarrow{k} \mathrm{CuSt}_2 + \Sigma\mathrm{P}_i, \end{aligned} \tag{2.149}$$

(where P_i are the products of the free-radical decomposition of ROOH), the complexity is defined as $F_{\mathrm{Cu}} = 1 + K_1[\mathrm{ROOH}]_\Sigma^2$ for a quasi-equilibrium first step, $F_{\mathrm{ROOH}} \cong 1$, and $k_1 = kK_1$. The formation of Cu(II) complexes with ROOH was confirmed by the NMR data. ■

Example 2.20

► Investigations of the kinetics of ethylbenzene hydroperoxide (HP) decomposition showed that, in the case of catalysis with Mo(V) resinate (representing a mixed salt of abietic acid isomers, $C_{20}H_{31}O_2$) [66, 67] that converts into Mo(VI) in the presence of ROOH, the initial reaction rate is described by the following equation:

$$r_0 = \frac{kK_{\mathrm{HP}}[\mathrm{HP}]_0[\mathrm{Mo}]_\Sigma}{1 + K_{\mathrm{HP}}[\mathrm{HP}]_0}. \tag{2.150}$$

This kinetic equation describes experimental data in agreement with the following reaction scheme:

$$\begin{aligned} &\mathrm{Mo(VI)} + \mathrm{ROOH} \overset{K_{\mathrm{HP}}}{\rightleftharpoons} \mathrm{Mo(VI)} \cdot \mathrm{ROOH}; \\ &\mathrm{Mo(VI)} \cdot \mathrm{ROOH} \xrightarrow{k_1} \mathrm{MPC} + \mathrm{Mo(VI)}; \\ &\mathrm{Mo(VI)} \cdot \mathrm{ROOH} \xrightarrow{k_2} \mathrm{AP} + \mathrm{H_2O} + \mathrm{Mo(VI)}, \end{aligned} \tag{2.151}$$

where $k = k_1 + k_2$, MPC is methyl phenyl carbinol, and AP is acetophenone.

In the course of experiments, the reaction rate decreases compared to that calculated using Eq. (2.150), which is related to the formation of inactive complexes between Mo(VI) and reaction products of MPC and H_2O. The equilibrium constants for the formation of these complexes (represented by pendant vertices on the corresponding kinetic graph) determined from the initial reaction rates are as follows: $K_{\mathrm{HP}} = 0.416\,\mathrm{M}^{-1}$, $K_{\mathrm{MPC}} = 1.04\,\mathrm{M}^{-1}$, $K_{\mathrm{H_2O}} = 7.25\,\mathrm{M}^{-1}$ (100 °C). Accordingly, we have,

$$r = \frac{kK_{\mathrm{HP}}[\mathrm{HP}][\mathrm{Mo}]_\Sigma}{1 + K_{\mathrm{HP}}[\mathrm{HP}] + \sum K_i[\mathrm{In}_i]}, \tag{2.152}$$

where $[\mathrm{In}_i]$ is the concentration of products that inhibit the process. ■

In some investigations [68, 69], the inhibition by products is treated as a result of the formation of a complex species between the catalyst and a generalized inhibitor (In), the concentration of which is determined as the difference $[ROOH]_0 - [ROOH]_t$. However, this is only possible provided that the concentrations of all inhibitors are the same, which is not the case in ROOH decomposition processes [68], especially where the free-radical decay mechanisms are operative [68, 69]. Results of these investigations are considered in more detail in Section. 2.2.

During the epoxidation of olefins by hydroperoxide in solutions of molybdenum complexes, the active catalysts are Mo(VI) diolate complexes that are formed from olefin oxides [70]. In these systems, the reaction is highly selective with respect to the hydroperoxide because the rate of ROOH decomposition in the presence of diolate complexes (e.g., Mo(VI) dioxo-bis-1,2-propanediolate [70]) is significantly lower than that in Mo resinate solutions [71]. Therefore, the main inhibitors of epoxidation in these systems are olefin oxides and alcohols.

Example 2.21

► Gavrilenko, Evzerikhin and Moiseev [66, 67, 71–74] studied the kinetics of propylene epoxidation by ethylbenzene hydroperoxide (HP)

$$C_3H_6 + ROOH \xrightarrow{Mo(VI)} C_3H_6O + ROH, \tag{2.153}$$

catalyzed by Mo(VI) in the closed ideal mixing reactor, and determined the solubility of propylene in ethylbenzene as dependent on the pressure P and temperature T [72]. The experimental kinetic equation

$$r = \frac{k_3 K_{HP}[HP]C_3H_6[Mo]_\Sigma}{F_{Mo}} \tag{2.154}$$

corresponds to the following rate-limiting step:

$$Mo(VI) \cdot ROOH + C_3H_6 \xrightarrow{k_3} C_3H_6O + ROH, \tag{2.155}$$

The complexity of molybdenum(VI) is determined by the following equation:

$$F_{Mo} = 1 + K_{HP}[HP] + K_{MPC}[MPC] + K_{PO}[PO] + K_{H_2O}[H_2O],$$

where MPC = ROH is methyl phenyl carbinol and PO is propylene oxide. Thus, three substances appear as inhibitors in this process. ■

A set of hypotheses for the mechanism of the epoxidation reaction with allowance for 14 possible intermediates and inactive metal complexes with participants of reaction (2.153) was proposed [75] based on the analysis of kinetic graphs for one- and two-route reactions.

Now let us consider specific features of the kinetics of reactions involving various forms of the initial metal complex and intermediate compounds from the example of a reaction of olefin oxidation in aqueous solutions of $PdCl_i^{2-i}$ complexes [15, 76]. In the case of ethylene oxidation to acetaldehyde, this reaction is known as the Wacker process, from the name of the company that developed a commercial technology for the production of acetaldehyde:

$$C_2H_4 + {}^1\!/_2\,O_2 \xrightarrow{PdCl_2-CuCl_2} CH_3CHO. \tag{2.156}$$

Investigations of the kinetics of this process with another oxidant, *para*-benzoquinone (Q), revealed the formation of intermediate binuclear π complexes of palladium(II), which were absent in the initial solution in the studied interval of palladium(II) chloride and chloride ion concentrations [15].

Example 2.22

▶ The kinetics of a reaction

$$C_2H_4 + Q + H_2O \longrightarrow C_2H_4O + QH_2, \quad (2.157)$$

was studied in a system containing Na_2PdCl_4 and electrolytes with a constant sum of concentrations:

$$[NaCl] + [LiClO_4] + [HCl] + [HClO_4] = I,$$

where the ionic strength I was varied within 0.5–3.0 M [15, 77–82], or in an analogous system with NaCl and $NaClO_4$ [76, 83, 84]. The process was carried out and studied in a gas-phase-free syringe reactor with dissolved olefin [77–79], where the reaction was monitored using a potentiometric technique for determining the ratio of quinone and hydroquinone concentrations ($[Q]/[QH_2]$) [80–82]. Alternatively, the reaction was studied using a volumetric method in a two-phase system, by measuring the absorption of olefin in the reactor with a stirrer [83, 84]. It was established that the rate of acetaldehyde formation is independent of [Q] and, hence, a process that determines the oxidation kinetics is the oxidation of ethylene by tetrachloropalladate anion:

$$C_2H_4 + PdCl_4^{2-} + H_2O \longrightarrow C_2H_4O + Pd(0) + 2H^+ + 4Cl^-. \quad (2.158)$$

In the interval of palladium and NaCl concentrations studied, Pd(II) occurred predominantly in the form of $PdCl_4^{2-}$, while the concentration of $PdCl_3^-$ was below 5 % of $[Pd^{II}]_\Sigma$ for $[Pd^{II}]_\Sigma < 0{,}02$ and $[NaCl] = 0.5$ M. Therefore, for $[Pd^{II}]_\Sigma < 0.02$ M, it can be assumed that $[PdCl_4^{2-}] \cong [Pd^{II}]_\Sigma$.

Investigation of the equilibrium formation of ethylene π complexes with Pd(II) [15] showed the existence of two equilibria:

$$C_2H_4 + PdCl_4^{2-} \overset{K_1}{\rightleftharpoons} PdCl_3(C_2H_4)^- + Cl^-; \quad (2.159)$$

$$C_2H_4 + PdCl_4^{2-} \overset{K_2}{\rightleftharpoons} PdCl_2(C_2H_4)(H_2O) + 2Cl^-. \quad (2.160)$$

In the experiments with ethylene solutions in the absence of a gaseous phase, we have

$$[C_2H_4]_\Sigma = [C_2H_4] + K_1\frac{[C_2H_4][PdCl_4^{2-}]}{[Cl^-]} + K_2\frac{[C_2H_4][PdCl_4^{2-}]}{[Cl^-]^2}$$

$$F_{C_2H_4} = \frac{[C_2H_4]_\Sigma}{[C_2H_4]} = 1 + K[PdCl_4^{2-}], \quad (2.161)$$

where

$$K = \frac{K_1}{[Cl^-]} + \frac{K_2}{[Cl^-]^2}.$$

In the system with the gaseous phase at a constant pressure, $[C_2H_4]$ does not change with time (it remains equal to the concentration of ethylene that is physically dissolved under the given conditions), and enters into the kinetic equation as a constant value. The complexity of palladium F_{Pd} in both systems is determined by the following equation:

$$F_{Pd} = \frac{[PdCl_4^{2-}]_\Sigma}{[PdCl_4^{2-}]} = 1 + \frac{K_1[C_2H_4]}{[Cl^-]} + \frac{K_2[C_2H_4]}{[Cl^-]^2}. \quad (2.162)$$

In the range of low concentrations of $PdCl_2$ (within 0.02 M) and low concentrations of ethylene (within $1 \cdot 10^{-4}$–$3 \cdot 10^{-3}$ M), the rate of reaction (2.157) is given by the following equation:

$$r = k_{\mathrm{I}} \frac{[\mathrm{PdCl_4}^{2-}][\mathrm{C_2H_4}]}{[\mathrm{H^+}][\mathrm{Cl^-}]^2} = k_{\mathrm{I}} \frac{[\mathrm{Pd^{II}}]_\Sigma [\mathrm{C_2H_4}]_\Sigma}{[\mathrm{H^+}][\mathrm{Cl^-}]^2}, \tag{2.163}$$

which is valid for $F_{\mathrm{C_2H_4}} \cong 1$ and $F_{\mathrm{Pd}} \cong 1$. This equation was obtained by all researchers studying reactions (2.157) and (2.158) for ethylene [83] and other lower olefins [84]. In the region of $PdCl_2$ concentrations within 0.02–0.2 M at $[\mathrm{Pd^{II}}]_\Sigma > [\mathrm{C_2H_4}]_\Sigma$, it is necessary also to take into account the complexity with respect to olefin ($F_{\mathrm{C_2H_4}} = 1 + K[\mathrm{PdCl_4}^{2-}]_\Sigma$). In this region of concentrations, the observed second-order rate constant

$$k_{\mathrm{obs}} = \frac{r}{[\mathrm{C_2H_4}]_\Sigma [\mathrm{PdCl_4}^{2-}]_\Sigma} \tag{2.164}$$

was found to depend on $[\mathrm{Pd^{II}}]_\Sigma$ in accordance with the following equation [15, 80, 81]:

$$k_{\mathrm{obs}} = \frac{A + B[\mathrm{PdCl_4}^{2-}]_\Sigma}{1 + K[\mathrm{PdCl_4}^{2-}]_\Sigma}, \tag{2.165}$$

where $A = k_{\mathrm{I}}/[\mathrm{H^+}][\mathrm{Cl^-}]^2$ and $B = k_{\mathrm{II}}/[\mathrm{H^+}][\mathrm{Cl^-}]^3$. This leads to the following equation for a two-route reaction with two slow steps:

$$r = k_{\mathrm{I}}' \frac{[\mathrm{PdCl_4}^{2-}][\mathrm{C_2H_4}]}{[\mathrm{H^+}][\mathrm{Cl^-}]^2} + k_{\mathrm{II}} \frac{[\mathrm{PdCl_4}^{2-}]^2[\mathrm{C_2H_4}]}{[\mathrm{H^+}][\mathrm{Cl^-}]^3}. \tag{2.166}$$

A mechanism of this reaction proposed by Moiseev [15] includes equilibria (2.159), (2.160) and the step

$$\mathrm{PdCl_2(C_2H_4)(H_2O) + H_2O \overset{K_3}{\rightleftharpoons} PdCl_2(OH)(C_2H_4)^- + H_3O^+}, \tag{2.167}$$

which are common for both routes, and the following steps for the first route:

$$\mathrm{PdCl_2(OH)(C_2H_4)^- \xrightarrow[k_4]{\text{Slow}} Cl_2PdCII_2CII_2OH^-}; \tag{2.168}$$

$$\mathrm{Cl_2PdCH_2CH_2OH^- + H_2O \xrightarrow{\text{Fast}} CH_3CHO + H_3O^+ + Cl^- + PdCl^-} \tag{2.169}$$

and the second route:

$$\mathrm{PdCl_2(OH)(C_2H_4)^- + PdCl_4^{2-} \overset{K_4}{\rightleftharpoons} Pd_2Cl_5(OH)(C_2H_4)^{2-} + Cl^-}; \tag{2.170}$$

$$\mathrm{Pd_2Cl_5(OH)(C_2H_4)^{2-} \xrightarrow[k_5]{\text{Slow}} Pd_2Cl_5(C_2H_4OH)^{2-}}; \tag{2.171}$$

$$\mathrm{Pd_2Cl_5(C_2H_4OH)^{2-} + H_2O \xrightarrow{\text{Fast}} CH_3CHO + Pd_2Cl_4^{2-} + H_3O^+ + Cl^-}. \tag{2.172}$$

Kinetic equation (2.166) implies that the second molecule of $PdCl_4^{2-}$ is involved in a reversible step rather than participates as an oxidant in the step of decay of a σ palladoethanol complex. Step (2.172) leads to the formation of a complex of Pd(I) rather than Pd(0). Other questions under discussion relate to the formation of σ palladoethanol via reaction (2.168) and the mechanism of heterolytic demetallation of a σ organometallic compound via reactions (2.169) and (2.172) [15, 76].

The absence of a significant kinetic isotope effect ($k_H/k_D = 1.07$ for the replacement of C_2H_4 by C_2D_4 [83]) is consistent with the proposed limiting stage of intraspheric *syn*-addition of a hydroxy group to ethylene (2.168), while the facts of CH_3CHO formation from C_2H_4 and D_2O [85] and CD_3CDO formation from C_2D_4 and H_2O [83] show evidence for the intramolecular transfer of hydride ion at the step of σ palladoethanol demetallation (2.169).

Investigations of the stereochemistry of 1,2-dideutero-2-chloroethanol, which is a product of the 1,2-oxidation of 1,2-dideuteroethene [86], and the products of 1,2-dimethyl-1,4-cyclohexadiene oxidation [87] under conditions that are strongly different from those for the Wacker process ($[Cl^-] > 3$ M, $[CuCl_2] > 2.5$ M [87], acetone–water (10 mass %), $KHCO_3$, $PdCl_2(CH_3CN)_2$ [87]) led researchers to the conclusion that the *anti*-addition of water molecules to the π complex $PdCl_2(H_2O)(C_2H_4)$ is possible via the following quasi-equilibrium (according to the kinetic data) reaction:

$$PdCl_2(C_2H_4)(H_2O) + H_2O \rightleftharpoons Cl_2Pd(C_2H_4OH)(H_2O)^- + H^+, \tag{2.173}$$

followed by a slow step of σ-palladoethanol equation [87] as

$$Cl_2Pd(C_2H_4OH)(H_2O)^- + H_2O \longrightarrow ClPd(C_2H_4OH)(H_2O)_2 + Cl^-. \tag{2.174}$$

The possibility of reaction (2.173) proceeding as a slow irreversible process at higher temperatures, which had been originally proposed by Moiseev [15], was confirmed for another catalytic system containing a very large concentration of chloride ions, $PdCl_2$ ($3 \cdot 10^{-3}$ M) – $HgCl_2$ (2 M) – $FeCl_3$ (1 M) and $[HCl + LiCl] =$ 5 M, which was used in the process of CO oxidation [88]. When [HCl] is varied from 0.1 to 2 M under conditions of an almost constant activity of water at 70 and 90 °C, the rate of ethylene oxidation in a gradient-free flow reactor is described by the following equation:

$$r = \frac{A}{[HCl]} + B \tag{2.175}$$

with a significant contribution from an [HCl]-independent route involving reaction (2.173) as the irreversible limiting step.

$$Cl_2Pd(C_2H_4)(OH_2) + H_2O \xrightarrow[H_2O]{Slow} Cl_2Pd(CH_2CH_2OH)(H_2O)^- + H_3O^+.$$

In later works by P. Henry and co-workers,[2] an important conclusion was made based on the results of kinetic, isotopic, and stereochemical investigations of the oxidation and isomerization of $CD_2{=}CHCH_2OH$ and mono- and disubstituted allyl alcohols. It was suggested that, under conditions of the Wacker process ($[Cl^-] = 0.1$–0.8 M, low concentration of $CuCl_2$), a limiting step is the *syn*-addition of the OH nucleophile (bound to palladium) to the olefin, whereas under the conditions of a large concentration of chloride ions and the predominant occurrence of π-$Pd(C_2H_4)Cl_3^-$ complex in solution, the *anti*-addition of water to the π complex takes place. Unfortunately, the kinetic isotope effect was not studied under the conditions described above. The possibility of the *anti*-addition of water to a $PdCl_2(OH)(C_2H_4)^-$ hydroxo complex was only considered in quantum-chemical investigations (see the aforementioned review). A mechanism of this kind was considered for the reaction of ethylene oxidation in PdX_2–CuX_2–MeOH systems (see Example 4.1). ■

The existence of a reversible step of the type described by Eq. (2.173) was considered within the framework of a mechanism proposed for one of the two routes of styrene oxidation in the following example.

[2] See, e.g., the recent review by J.A. Keith and P.M. Henry, The Mechanism of the Wacker Reaction: A Tale of Two Hydroxypalladations, *Angew. Chem. Int. Ed.*, **48**, 9038 (2009).

Example 2.23

► The formation of phenylacetic aldehyde (PA) in addition to acetophenone (AP) during the oxidation of styrene by palladium chloride [90] posed questions concerning the possible mechanism of this product formation and the influence of $[H_3O^+]$ and $[Cl^-]$ on the value of $n = [PA]/[AP]$. At a constant ionic strength of solution (3.0 M) in the $LiClO_4$–LiCl–HCl–$HClO_4$ system, the rate of styrene oxidation by *para*-benzoquinone at 25 °C in the intervals of $[H_3O^+] = 0.1 - 1.0\,M$, $[Cl^-] = 0.1 - 1.0\,M$, and $[PdCl_4{}^{2-}] = 0.02 - 0.2M$ is described by the following equation of a two-route process [89]:

$$r = k_2[PdCl_4{}^{2-}]_\Sigma[C_8H_8]_\Sigma, \tag{2.176}$$

where

$$k_2 = \frac{k_I}{[II_3O^+][Cl^-]^2} + \frac{k_{II}}{[Cl^-](1 + p[H_3O^+])}.$$

Under the conditions of $F_{Pd} \cong 1$ and $F_{C_8H_8} \cong 1$, the first term of Eq. (2.176) corresponds to Eq. (2.163), while the second term is indicative of the participation of a π-$PdCl_3(C_8H_8)^-$ complex and the liberation of H_3O^+ species during the formation of σ organopalladium intermediate in the following quasi-steady-state process:

$$PdCl_4{}^{2-} + C_8H_8 \overset{K}{\rightleftharpoons} PdCl_3(C_8H_8)^- + Cl^-; \tag{2.177}$$

$$PdCl_3(C_8H_8)^- + H_2O \underset{\beta_1}{\overset{\alpha_1}{\rightleftharpoons}} [Cl_3Pd\text{-}C_8H_8\overset{+}{O}H_2]^-; \tag{2.178}$$

$$[Cl_3Pd\text{-}C_8H_8OH_2]^- + H_2O^+ \underset{b_2}{\overset{a_2}{\rightleftharpoons}} Cl_3Pd(C_8H_8OH)^{2-} + H_3O^+; \tag{2.179}$$

$$Cl_3Pd(C_8H_8OH)^{2-} \xrightarrow{\alpha_3} C_8H_8O + HCl + Cl^- + PdCl^-. \tag{2.180}$$

The rate over the second route according to scheme (2.177)–(2.180) is described by the following equation that also corresponds to the overall Eq. (2.176):

$$r = \frac{k_{II}[PdCl_4{}^{2-}]_\Sigma[C_8H_8]_\Sigma}{[Cl^-](1 + p[H_3O^+])}, \tag{2.181}$$

where

$$k_{II} = K\frac{\alpha_1\alpha_2}{\alpha_2 + \beta_2}; \quad p = \frac{\beta_1\beta_2}{\alpha_3(\alpha_2 + \beta_1)}.$$

An analogous equation is obtained if step (2.178) is quasi-equilibrium, while reactions (2.178) and (2.179) are combined into a single slow step with a rate comparable to that of step (2.180).

It is interesting to note that the ratio of rates over the routes, r_{II}/r_I, does not coincide with the value of $n = [PA]/[AP]$. Therefore, both routes contribute to the formation of PA and AP. It can be suggested that, in the case of the *trans*-addition of a water molecule to the π complex via reaction (2.178), the anti-Markovnikov orientation (favoring the PA formation) is predominant. This assumption is consistent with the fact that both n and r_{II}/r_I increase with $[Cl^-]$ at a weak dependence of n and r_{II}/r_I on $[H_3O^+]$. ■

Kinetic equations of greater-than-first orders with respect to a metal complex are frequently encountered in the description of reactions of various types both with and without the formation of intermediate dimer complexes. Now we will consider some characteristic examples.

Example 2.24

► The second order with respect to [Pd(II)] was observed for the oxidation of thiophene [91–95] and arenes [96]. The rate of thiophene oxidation to 2-thienol (and products of its conversion) was studied using the distribution method in a two-phase system, by monitoring a decrease in the thiophene concentration [TH] in the gas phase under the conditions of equilibrium between phases, and in a syringe reactor (free of a gaseous phase) [93] by measuring the loss of thiophene from solution. The reactions were carried out in perchloric and sulfuric acid solutions under conditions of the existence of $Pd(H_2O)_4{}^{2+}$. For $[Pd^{II}]_\Sigma \gg [TH]_0$, the first-order rate constant $k = k_{obs}(1 + \alpha\lambda)$ [see Eq. (2.115)] is described by the following equation:

$$k = \frac{C_1[Pd^{II}]_\Sigma^2}{C_2 + [Pd^{II}]_\Sigma}. \tag{2.182}$$

It was established that the oxidation rate is independent of the addition of an oxidant (Fe(III), $NaNO_3$, etc.). The value of C_2 exhibits linear growth with increasing $[H_3O^+]$, so that $C_2 = C_3[H_3O^+]$ (at $[HClO_4] + [LiClO_4] = 2\,M$, $[Pd^{II}]_\Sigma = 0.01\,M$). It can be suggested that the appearance of [Pd(II)] in the denominator of Eq. (2.182) is related to the participation of Pd(II) in the quasi-steady-state process in the absence of a limiting step:

$$Pd^{2+} + TH \overset{K}{\rightleftharpoons} Pd^{2+}(TH);$$

$$Pd^{2+}(TH) + H_2O \underset{k_{-1}}{\overset{k_1}{\rightleftharpoons}} \sigma\text{-}PdT^+ + H_3O^+; \tag{2.183}$$

$\sigma\text{-}PdT^+ + Pd^{2+} \xrightarrow[-Pd_2^{2+}]{k_2}$ 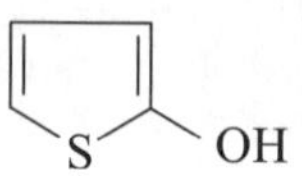 $\xrightarrow{Pd^{2+}}$ Oxidation products ;

according to which

$$r = \frac{k_1 k_2 K [Pd(II)]_\Sigma^2 [TH]}{k_{-1}[H_3O^+] + k_2[Pd(II)]_\Sigma}. \tag{2.184}$$

The step of the nucleophilic substitution of Pd in σ PdT by a molecule of water (or OH group from ~PdOH) probably involves the formation of an intermediate binuclear palladium(II) complex, while the role of the second Pd(II) particle as the oxidant reduces to favoring the nucleophilic substitution process. It is interesting to note that the oxidation process at $[TH]_0 \gg [Pd^{II}]_\Sigma$ leads to the products of oxidative dimerization of TH (2,2-dithienyl, 2,3′-dithienyl), and the total rate of formation of these products is described by the following equation [92]:

$$r = \frac{C_3[Pd^{II}]_\Sigma[TH]^2}{C_4[H_3O^+] + [TH]}. \tag{2.185}$$

Investigation of the oxidation of thiophene in the $PdCl_2$–HCl–H_2O system [95] in the interval of concentrations $[TH] = (0.2 - 1) \times 10^{-3}$ M, $[Pd^{II}]_\Sigma = 0.05$–0.22 M, $[H_3O^+] = 0.05$–0.24 M, and $[Cl^-] = 0.3$–1 M at 70 °C and an ionic strength of 1–2 M, i.e., under the conditions where $PdCl_4^{2-}$ predominantly occurs in solution (with a two- to fourfold excess over $PdCl_3^-$), led to a conclusion that the process of thienol formation under these conditions has a limiting step and can be described by the following relation:

$$r = k_{eff} \frac{[Pd^{II}]_\Sigma^2 [TH]}{[H_3O^+][Cl^-]^3}, \tag{2.186}$$

which is analogous to the term in Eq. (2.166) for the second route in the oxidation of ethylene to acetaldehyde. According to one hypothesis explaining Eq. (2.186), the limiting step is related to the conversion of a σ $Pd_2Cl_5(T)^{2-}$ complex (see scheme (2.183)):

$$\sigma\ Pd_2Cl_5(T)^{2-} + H_2O \longrightarrow \text{Oxidation products.} \tag{2.187}$$

One difference between the mechanisms of reactions in the oxygen-free system (2.183) and a low-chloride system [94] on the one hand, and a system with high Cl^- concentrations [95] on the other, consists of the participation of a second Pd(II) complex in the latter case in a quasi-steady-state step preceding the limiting step [95]. ■

Example 2.25

► The kinetics of olefin (1-octene) epoxidation by hydrogen peroxide, catalyzed with Pt(II) complexes was studied in aqueous tetrahydrofuran solutions for the catalysis with $P_2Pt(CF_3)OH$ (complex A) and in the two-phase H_2O–H_2O_2–CH_2Cl_2–1-octene system for the catalysis with $P_2Pt(CF_3)^+BF_4^-$ (complex B), where P_2 is *cis*-1,2-bis(diphenylphosphino)ethane [97]. Both systems were characterized by second-order kinetics with respect to $[Pt^{II}]_\Sigma$ and exhibited sharp maxima with respect to $[H_3O^+]$. In the first system, the Pt(II) complex occurs predominantly in the form of ≅PtOOH ($[Pt^{II}]_\Sigma$ = [≅PtOOH]), and the step leading to epoxide involves two species, ≅PtOOH and Pt^+ (1 octene):

$$PtOOH + Pt(1\text{-octene})^+ \longrightarrow \text{Epoxide} + Pt^+ + PtOH. \tag{2.188}$$

This reaction is a limiting step in the second system catalyzed by complex B. The reaction involving equimolar amounts of complexes $P_2Pt(CF_3)(1\text{-octene})BF_4$ and $P_2Pt(CF_3)OOH$ (10^{-2} M in CH_2Cl_2) leads to the quantitative formation of 1,2-epoxyoctane and a mixture of complexes A and B.

Kinetic investigations performed using the IR and ^{19}F NMR spectroscopy techniques allowed the following mechanism to be proposed for the second system (cation complex B – Pt^+, Ol = 1-octene):

$$Pt^+ + Ol \overset{K_1}{\rightleftharpoons} Pt(Ol)^+;$$

$$Pt^+ + H_2O \overset{K_2}{\rightleftharpoons} Pt(H_2O)^+;$$

$$Pt^+ + H_2O_2 \overset{K_3}{\rightleftharpoons} Pt(H_2O_2)^+;$$

$$Pt^+ + H_2O + H_2O_2 \overset{K_4}{\rightleftharpoons} Pt(OOH) + H_3O^+;$$

$$\mathrm{Pt(OOH) + Pt(Ol)^+ \xrightarrow{k_5} Epoxide + Pt^+ + PtOH;}$$

$$\mathrm{PtOH + H_2O_2 \xrightarrow{Fast} Pt(OOH) + H_2O,}$$

for which

$$F_{\mathrm{Pt}} = 1 + K_1[\mathrm{Ol}] + K_2[\mathrm{H_2O}] + K_3[\mathrm{H_2O_2}] + \frac{K_4[\mathrm{H_2O_2}][\mathrm{H_2O}]}{[\mathrm{H_3O^+}]}; \tag{2.189}$$

$$r = \frac{k_5K_4K_1[\mathrm{Pt^{II}}]^2_\Sigma[\mathrm{Ol}][\mathrm{H_2O_2}][\mathrm{H_2O}]}{(F_{\mathrm{Pt}})^2[\mathrm{H_3O^+}]} \tag{2.190}$$

at $F_{\mathrm{Ol}} \cong 1$. Equations (2.189)–(2.190) show that the dependence of rate r on $[\mathrm{H_3O^+}]$ exhibits an extremum, where

$$r = \frac{\alpha[\mathrm{H_3O^+}]}{(1 + \beta[\mathrm{H_3O^+}])^2}. \tag{2.191}$$

■

Example 2.26

► In the context of a discovery that the reaction of C_2H_2 hydroiodation at 353 K is catalyzed by Pt(IV) complexes [98],

$$\mathrm{C_2H_2 + HI \xrightarrow{Pt^{IV}} CH_2{=}CHI,}$$

the kinetics of the formation of intermediate σ-organoplatinum compounds in the Pt(IV)–NaI–$HClO_4$ system was studied at 313 and 333 K, i.e., under conditions where the rate of vinyl iodide formation is insignificant [99]. The process was studied by the distribution method in a catalytic "duck"-type shaken reactor. At 313 K, the kinetics corresponds to the establishment of equilibrium in the reaction

$$\mathrm{C_2H_2 + Pt^{IV} + I^- \underset{k_{-1}}{\overset{k_1}{\rightleftharpoons}} \underset{(X)}{Pt^{IV}CH{=}CHI}.} \tag{2.192}$$

The observed first-order rate constant with respect to $[C_2H_2]$ at $F_{\mathrm{Pt}} \cong 1$ obeyed the relation

$$k_{\mathrm{obs}} = k_{-1} + \frac{k_1[\mathrm{Pt^{IV}}]_\Sigma}{1 + \alpha_0\lambda}, \tag{2.193}$$

and the material balance with respect to acetylene was described by the following equation:

$$\frac{\varphi_0 - \varphi_\infty}{\varphi_\infty} = \frac{K_1}{1 + \alpha_0\lambda}[\mathrm{Pt^{IV}}]_\Sigma, \tag{2.194}$$

where φ_0 and φ_∞ are the relative concentrations of acetylene before (at $\tau = 0$) and after attaining the equilibrium in reaction (2.192), respectively; $\alpha_0 = [C_2H_2]_{g,0}/[C_2H_2]_{s,0}$; and $\lambda = V_g/V_s$.

At 333 K, the kinetics becomes more complicated because product X is converted via reaction with Pt(IV) into product Y (apparently, Pt(IV)–CHI–CHI–Pt(IV)):

$$\mathrm{X + Pt^{IV} + I^- \underset{k_{-2}}{\overset{k_2}{\rightleftharpoons}} Y.} \tag{2.195}$$

For a quasi-equilibrium first step (2.192), the observed rate constant obeys the following equation:

$$k_{obs} = k_{-2} + \frac{k_2 K_1 [\mathrm{Pt^{IV}}]_\Sigma^2}{1 + \alpha_0 \lambda + K_1 [\mathrm{Pt^{IV}}]_\Sigma}, \tag{2.196}$$

which is consistent with the experimental data. With allowance for the equilibrium (2.195), the material balance equation for C_2H_2 takes the following form:

$$\frac{\varphi_0 - \varphi_\infty}{\varphi_\infty} = \frac{K_1 K_2 [\mathrm{Pt^{IV}}]_\Sigma^2}{1 + \alpha_0 \lambda + K_1 [\mathrm{Pt^{IV}}]_\Sigma}. \tag{2.197}$$

An iodovinyl Pt(IV) derivative of type X containing two σ iodovinyl groups was obtained for methylpropargyl ether Pt(CH=CI–CH$_2$OMe)$_2$(I)$_2$ [100]. Diiodovinyl compounds are also considered as possible intermediates in reactions of the oxidative coupling of alkynes [101]:

$$\mathrm{I_2 + 2C_2H_2 \xrightarrow{Pt^{IV}-I^-} ICH{=}CH{-}CH{=}CHI.} \tag{2.198}$$

E,E-diiodo-1,3-butadiene

■

In determining the dependence of the reaction rate on the concentration of a complex of the metal catalyst in cases where the structure of the metal complexity function (F_M) probably belongs to variants III and IV, it is important to pay attention to a possible change in the concentration of ligands in solution during the variation of $[ML_n]_\Sigma$. Such a change in [L] takes place, for example, in the course of hydrogenation of olefins and alkynes in solutions of the phosphine complexes of Rh(I) [29,102–106], Ru(II) [106, 107] and Ir(I) [106, 108].

Example 2.27

▶ In the first investigations devoted to the kinetics of the hydrogenation of olefins in benzene solutions of $RhClL_3$ ($L = PR_3$), the reaction rate was monitored by measuring a change in the partial pressure of hydrogen (P_{H_2}) at a constant volume and constant values of λ and α ($\lambda = V_g/V_s$, $\alpha = [H_2]_g/[H_2]_s$) [102–106]. It was established that the reaction rate at 25 °C passes through a maximum with respect to the ratio $[L]_\Sigma/[Rh]_\Sigma$ (between 2:1 and 3:1). The reaction orders with respect to H_2 and olefin change from 1 to zero with increasing reactant concentration, while the reaction order with respect to $[Rh]_\Sigma$ is below unity. The dependence on $[L]_\Sigma$ was not studied. The proposed scheme of a two-route mechanism

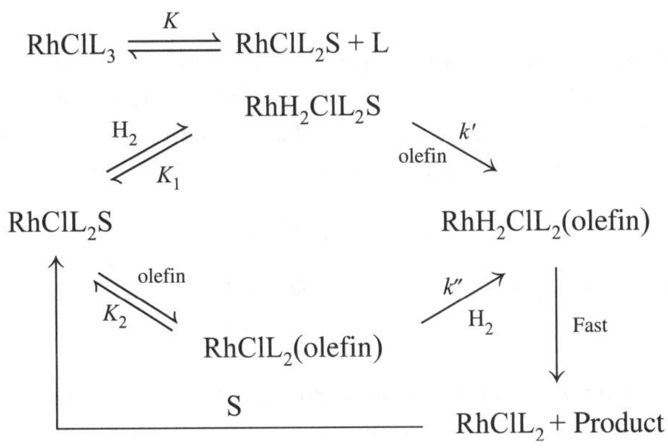

(2.199)

corresponds to the following kinetic equation:

$$r = \frac{(k'K_1 + k''K_2)[H_2][\text{olefin}][Rh]_\Sigma}{1 + \frac{[L]}{K} + K_1[H_2] + K_2[\text{olefin}]}. \qquad (2.200)$$

The denominator in Eq. 2.200 is

$$F_{Rh} = \frac{[Rh]_\Sigma}{[RhClL_2S]}.$$

A fractional (below 1) reaction order with respect to $[Rh]_\Sigma$ is related to the growth in [L] with increasing $[Rh]_\Sigma$. In these investigations, the results were processed and constants evaluated under assumptions that $k'K_1 \gg k''K_2$ and $K_1[H_2] + K_2[\text{olefin}] \gg 1 + [L]/K$. The first assumption proved to be incorrect [105] and the latter assumption contradicted the experimental results reported in [102–104], so that the obtained values of k' and activation parameters had no physical sense. It was shown later that, in the absence of added L, the order with respect to $[Rh]_\Sigma$ is ≅0.5 [107]. The first order in $[Rh]_\Sigma$ and almost minus first in [L] were observed at an added ligand concentration of [L] ~ 0.01 M and $[Rh]_\Sigma \cong 2 \cdot 10^{-3}$ M [109, 110]. The development of kinetic investigations in this system was reviewed in [29]. ■

Example 2.28

► It was established that $RuCl_2L_3$ (L = PPh_3) complex in polar solvents (ethanol, dimethylacetamide) is rapidly reduced by hydrogen to $HRuClL_3$ (and HCl) [106, 107]. According to experimental data and the proposed scheme

$$HRuClL_3 \overset{K_1}{\rightleftharpoons} HRuClL_2 + L;$$

$$HRuClL_2 + \text{Olefin} \overset{K_2}{\rightleftharpoons} RuClL_2(\text{alkyl}); \qquad (2.201)$$

$$RuClL_2(\text{alkyl}) + H_2 \xrightarrow{k_3} HRuClL_2 + \text{Product}$$

the rate of maleic acid hydrogenation in dimethylacetamide in the presence of this hydride complex is described by the following equation:

$$r = -\frac{d[H_2]}{dt} = \frac{k_3K_2[Ru]_\Sigma[\text{olefin}][H_2]}{1 + K_2[\text{olefin}] + \frac{[L]}{K_1}}. \qquad (2.202)$$

At low olefin concentrations, the reaction rate is proportional to $[L]^{-1}$. If the ligand is not added and the [olefin]/[L] ratio is small, the reaction order with respect to $[Ru]_\Sigma$ is below unity. If the ligand is in excess (i.e., [L] ceases to change with increasing $[Ru]_\Sigma$) and there is large excess of olefin relative to [L], the reaction becomes first-order with respect to $[Ru]_\Sigma$. In the latter case, the reaction rate can be expressed as

$$r = k_3[Ru]_\Sigma[H_2], \qquad (2.203)$$

and all ruthenium occurs in the alkyl form. The values of coefficients K_1 and K_2 at 35 °C are 0.03 M and 80 M^{-1}, respectively. ■

In most of the examples considered above, the structure of the catalyst complexity function was justified by chemical, spectroscopic, and kinetic data. In complicated cases, a useful approach consists of formulating hypotheses concerning the structure of complexity and planning experiments for their discrimination. In the absence of physicochemical information on the processes of complex formation and the structure of intermediates, the investigation of kinetics can reveal the presence of complexity and suggest the possible structure of F_M.

Example 2.29

► Investigation of the kinetics of butylacrylate synthesis [111] in the $PdBr_2$–$P(OPh)_3$–HBr–n-BuOH system

$$C_2H_2 + CO + ROH \longrightarrow CH_2{=}CHCOOR, \tag{2.204}$$

which was carried out in a gas-flow ideal mixing reactor at low degrees of ROH conversion, showed that the reaction rate at $P_{CO,C_2H_2} = \text{const}$ is a linear function of $[H^+]^{-1}$. The dependence of the reaction rate on P_{CO} and $P_{C_2H_2}$ at $[PdBr_2]_\Sigma = 0.046\,M$, $[LiBr] = 0.5\,M$, and $[P(OPh)_3]_\Sigma = 0.12\,M$ was described by the following equation:

$$r = \frac{k_{obs} P_{CO} P_{C_2H_2}}{[H^+](1 + K P_{CO} P_{C_2H_2} + K_1 P_{C_2H_2})}. \tag{2.205}$$

An analysis of the electronic absorption spectra and the dependence $r = f([P(OPh)_3]_\Sigma/[PdBr_2]_\Sigma)$ showed that the active complex is *cis*-$PdBr_2[P(OPh)_3]_2$ [112, 113] (see also Section 5.3.1). The form of Eq. (2.205) suggests that the limiting step involves an intermediate compound containing a $[Pd(CO)(C_2H_2)(OR)]$ fragment and allows the following alternative hypotheses to be considered concerning the structure of complexity function F_{Pd}.

1. The complexity of Pd (term in parentheses in the denominator of Eq. (2.205)) does not contain a contribution due to intermediates. Then, the main contribution to F_{Pd} is due to Pd(II)(CO) and $Pd(II)(C_2H_2)(CO)$ complexes that are not involved in the elementary steps of the process (i.e., correspond to pendant vertices in the kinetic graph). Apparently, the formation of intermediates in the case of *cis*-PdL_2Br_2 is not related to the rupture of the Pd–L bond and corresponds to the following scheme:

$$\begin{aligned} &L_2PdX_2 + ROH \rightleftharpoons L_2PdX(OR) + HX;\\ &L_2PdX(OR) + CO \rightleftharpoons L_2PdX(COOR);\\ &L_2PdX(COOR) + C_2H_2 \longrightarrow L_2PdX(CH{=}CHCOOR) \xrightarrow[HX]{\text{Fast}}\\ &\qquad \longrightarrow L_2PdX_2 + CH_2{=}CHCOOR, \end{aligned} \tag{2.206}$$

   where $L = P(OPh)_3$ and $X = Br$. Note that the $L_2PdX(C_2H_2)^+X^-$ and $LPdX(CO)(C_2H_2)^+X^-$ complexes enter into the material balance and the concentration of intermediates is small compared to $[Pd^{II}]_\Sigma$.
2. Complexes $L_2PdX(C_2H_2)^+X^-$ and $LPdX(CO)(C_2H_2)^+X^-$ are involved in elementary steps of the process and explicitly contribute to F_{Pd}:

$$\begin{aligned} &L_2PdX_2 + C_2H_2 \overset{K_1}{\rightleftharpoons} L_2PdX(C_2H_2)^+X^-;\\ &L_2PdX(C_2H_2)^+X^- + CO \overset{K_2}{\rightleftharpoons} LPdX(C_2H_2)(CO)^+X^- + L;\\ &LPdX(C_2H_2)(CO)^+X^- + ROH \rightleftharpoons LPdX(C_2H_2)(COOR) + HX; \end{aligned} \tag{2.207}$$

$$\mathrm{LPdX(C_2H_2)(COOR)} \xrightarrow[\mathrm{L}]{\mathrm{Slow}} \mathrm{L_2PdX(CH{=}CHCOOR)} \xrightarrow[\mathrm{HX}]{\mathrm{Fast}} \longrightarrow \mathrm{L_2PdX_2 + CH_2{=}CHCOOR}.$$

In this case, the reaction rate can be expressed as

$$r = \frac{k_{\mathrm{H}} P_{\mathrm{CO}} P_{\mathrm{C_2H_2}} [\mathrm{Pd}]_\Sigma}{[\mathrm{HX}][\mathrm{L}]\left(1 + K_1 P_{\mathrm{C_2H_2}} + K_1 K_2 \frac{P_{\mathrm{CO}} P_{\mathrm{C_2H_2}}}{[\mathrm{L}]}\right)}, \tag{2.208}$$

provided that $[\mathrm{LPdX(C_2H_2)(COOR)}] \ll [\mathrm{Pd}]_\Sigma$.

In order to discriminate between the two hypotheses, it is necessary to perform additional investigations into the dependence $r = f([\mathrm{L}])$, although the second hypothesis *a priori* seems to be less probable. ■

Example 2.30

► Investigation [114] of the kinetics of oxidative decomposition of $(\eta^3\text{-}\mathrm{C_3H_5PdCl})_2$ via the reaction with *para*-benzoquinone (Q)

$$\tfrac{1}{2}(\eta^3\text{-}\mathrm{C_3H_5PdCl})_2 + 2\mathrm{HCl} + 2\mathrm{Cl^-} + \mathrm{Q} \longrightarrow \mathrm{C_3H_5Cl} + \mathrm{QH_2} + \mathrm{PdCl_4}^{2-} \tag{2.209}$$

showed that the reaction rate at an ionic strength of 3.7 M, $[\mathrm{H_3O^+}] = 0.1$ M, $[\mathrm{Cl^-}] = 3.7$ M and a temperature of 25 °C was proportional to the concentrations of the monomer π allyl complex $\mathrm{C_3H_5PdCl_2}^-$ (π_X) and *para*-benzoquinone (Q):

$$r = k_{\mathrm{II}}[\pi_\mathrm{X}][\mathrm{Q}], \tag{2.210}$$

where $k_{\mathrm{II}} = f([\mathrm{H_3O^+}], [\mathrm{Cl^-}])$ and can be written as

$$k_{\mathrm{II}} = \frac{[\mathrm{Cl^-}][\mathrm{H_3O^+}]}{A + B[\mathrm{Cl^-}][\mathrm{H_3O^+}]}. \tag{2.211}$$

The above equations indicate that the intermediate compounds containing Q, π_X, and $\mathrm{H^+}$ do not contribute to the complexity, $F_{\mathrm{Pd}} \cong 1$, and the denominator of Eq. (2.211) reflects the absence of a limiting step under quasi-steady-state conditions [114]. The mechanistic scheme of reaction (2.209) written according to these considerations is as follows:

$$\mathrm{(C_3H_5PdCl)_2 + 2Cl^-} \longrightarrow \underset{(\pi\mathrm{X})}{2\mathrm{C_3H_5PdCl_2}^-};$$

$$\pi_\mathrm{X} + \mathrm{Q} \underset{k_{-1}}{\overset{k_1}{\rightleftharpoons}} \pi_\mathrm{X}\mathrm{Q}\ (K_1);$$

$$\pi_\mathrm{X}\mathrm{Q} + \mathrm{H_3O^-} \underset{k_{-2}}{\overset{k_2}{\rightleftharpoons}} \pi_\mathrm{X}\mathrm{QH^+} + \mathrm{H_2O}\ (K_2); \tag{2.212}$$

$$\pi_\mathrm{X}\mathrm{QH^+} + \mathrm{Cl^-} \xrightarrow{k_3} \mathrm{X} \xrightarrow{\mathrm{H_3O^+,2Cl^-}} \mathrm{C_3H_5Cl} + \mathrm{PdCl_4}^{2-} + \mathrm{QH_2}.$$

It is possible that the nucleophilic attack by Cl^- leads to the formation of C_3H_5Cl in the elementary act and X is a Pd(II) hydroquinolate with the structure of

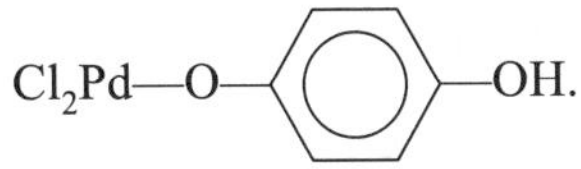

Then, the reaction rate in the quasi-steady-state approximation can be expressed as

$$r = \frac{k_1k_2k_3[\pi_X][Q][H_3O^+][Cl^-]}{k_2k_3[H_3O^+][Cl^-] + k_{-1}k_3[Cl^-] + k_{-1}k_{-2}}, \tag{2.213}$$

and for $k_2[H_3O^+] > k_{-1}$ we obtain the required Eqs (2.210) and (2.211) with $A = 1/K_1K_2k_3$ and $B = 1/k_1$.

Assuming that the second step ($\pi_XQ + H_3O^+$) is quasi-equilibrium, we also arrive at the experimental kinetic equations. ■

In some cases considered above, the description of kinetics involved the substrate complexity F_S (see Examples 2.22 and 2.26). Very complicated kinetic equations are obtained in cases where both F_M and F_S differ from unity.

Example 2.31

▶ It was established that hydrogen peroxide in alkaline solutions (pH 8–12) at 25 °C in the presence of MoO_4^{2-} catalyst exhibits decomposition with the formation of singlet ($'\Delta_g$) oxygen species [115, 116]:

$$2H_2O_2 \longrightarrow 2H_2O + {}^1O_2. \tag{2.214}$$

The observed decomposition kinetics was explained by the formation and conversion of MoO_6^{2-} diperoxo complex via the following reactions:

$$2H_2O_2 + MoO_4^{2-} \overset{K_4'}{\rightleftharpoons} MoO_6^{2-} + 2H_2O; \tag{2.215}$$

$$MoO_6^{2-} \overset{k_5}{\longrightarrow} MoO_4^{2-} + {}^1O_2, \tag{2.216}$$

with a rate of

$$r_{^1O_2} = k_5K_4'[H_2O_2]^2[MoO_4^{2-}]. \tag{2.217}$$

At pH 7, a protonated diperoxo complex is formed as

$$MoO_6^{2-} + H^+ \overset{K_6'}{\rightleftharpoons} HMoO_6^-, \tag{2.218}$$

and in strongly basic media H_2O_2 can dissociate as

$$H_2O_2 \overset{K_7}{\rightleftharpoons} H^+ + HO_2^-, \tag{2.219}$$

with a rightward shift of this equilibrium. Therefore, the description of kinetic features (in particular, an extremum in the reaction rate with respect to pH), requires taking into account two material balances (at $[H_2O_2]_\Sigma \gg [Mo]_\Sigma$):

$$[Mo]_\Sigma = MoO_4^{2-} + MoO_6^{2-} + HMoO_6^-;$$

$$F_{Mo} = \frac{[Mo]_{\Sigma}}{[MoO_4{}^{2-}]} = 1 + K_4'[H_2O_2]^2 + K_4'K_6'[H^+][H_2O_2]^2;$$

$$[H_2O_2]_{\Sigma} = [H_2O_2] + [HO_2{}^-];$$

$$F_{H_2O_2} = \frac{[H_2O_2]_{\Sigma}}{[H_2O_2]} = 1 + \frac{K_7}{[H^+]}.$$

Substituting F_{Mo} and $F_{H_2O_2}$ into Eq. (2.217) leads to the following kinetic equation:

$$r = \frac{k_5K_4'[Mo]_{\Sigma}[H_2O_2]_{\Sigma}^2}{\left(1 + \frac{K_7}{[H^+]}\right)^2 + K_4'[H_2O_2]_{\Sigma}^2(1 + K_6'[H^+])}, \tag{2.220}$$

which describes both the aforementioned extremal dependence on pH and the experimental concentration dependence of $r = f([H_2O_2]_{\Sigma})$. ■

Example 2.32

► A more complicated equation was obtained for the rate of triplet oxygen species formation via the reaction

$$2H_2O_2 \longrightarrow 2H_2O + O_2$$

in a catalytic system of Ni^{2+}–monoethanolamine (MEA) [117]. Under the conditions of [MEA] $\gg$ $[Ni^{2+}]$, the complex of Ni^{2+} with MEA has a limiting composition of $Ni(MEA)_3{}^{2+}$ and the reaction proceeds in a quasi-steady-state regime. The experimental dependences of the reaction rate on $[H_2O_2]_{\Sigma}$, $[Ni]_{\Sigma}$, $[MEA]_0$, and $[H^+]$ were explained using the following simple mechanistic reaction scheme:

$$\begin{aligned} &Ni(MEA)_3{}^{2+} + HO_2{}^- \underset{k_{-1}}{\overset{k_1}{\rightleftharpoons}} \underset{(X_1)}{Ni(MEA)_3(HO_2)^+}; \\ &X_1 + H_2O_2 \underset{k_{-2}}{\overset{k_2}{\rightleftharpoons}} X_2; \\ &X_2 \xrightarrow{k_3} Ni(MEA)_3{}^{2+} + H_2O + O_2 + OH^-. \end{aligned} \tag{2.221}$$

With allowance for the material balance with respect to Ni^{2+} and H_2O_2,

$$[Ni^{2+}]_{\Sigma} = [Ni(MEA)_3{}^{2+}] + [X_1] + [X_2];$$

$$[H_2O_2]_{\Sigma} = [H_2O_2] + [H_2O_2 \cdot MEA]$$

and reactions

$$H_2O_2 + MEA \overset{K_5}{\rightleftharpoons} H_2O_2 \cdot MEA; \tag{2.222}$$

$$OH^- + H_2O_2 \overset{K_4}{\rightleftharpoons} H_2O + HO_2{}^- \tag{2.223}$$

at $[HO_2{}^-] \ll [H_2O_2]_{\Sigma}$ and $[X_2] \ll [H_2O_2]_{\Sigma}$, the proposed scheme yields an equation that coincides with the equation for the initial rate of O_2 evolution (at $[H_2O_2]_{\Sigma} = [H_2O_2]_0$), which was obtained upon processing of the kinetic data:

$$r_0 = \frac{k_1'[Ni^{2+}]_\Sigma[H_2O_2]_0^2}{[H_2O_2]_0^2 + (k_2' + k_3'[MEA])[H_2O_2]_0 + k_4'[H_2O_2]_0[H^+] + k_5'[MEA]^2[H^+]}. \qquad (2.224)$$

■

2.1.3 General problems and recommendations

Let us consider some general problems encountered in investigations of the kinetics of reactions involving metal complexes.

2.1.3.1 Possible factors responsible for the deviation from unity of the reaction order (n) with respect to the concentration of a metal complex (n < 1, n > 1)

In the case of mononuclear complexes, there are several possible reasons for the appearance of $n \neq 1$:

- presence of nonlinear steps in the forward and/or reverse directions in the reaction mechanism (X_i + $X_j \rightarrow \ldots$), where one or more X_i contain a metal ($n < 1$, $n > 1$);
- use of complexity functions for both metal and substrate, where F_M and F_S contain contributions due to one or more common complexes ($n < 1$);
- use of F_S at $F_M = 1$ ($n < 1$).

Nonlinear steps were encountered in Examples 2.2 and 2.3 (L–intermediate), 2.4 (Cl^-–intermediate), 2.24, and 2.25. Other variants of the appearance on nonlinearity are also known, in which only one intermediate contains a metal. For example, when a catalyst reacts with a species formed from substrate S in a quasi-steady-state step (rather than directly with the substrate):

$$SH \underset{k_{-1}}{\overset{k_1}{\rightleftharpoons}} S^- + H^+;$$

$$ML_n + S^- \underset{k_{-2}}{\overset{k_2}{\rightleftharpoons}} X_1;$$

$$X_1 + B \xrightarrow{k_3(H^+)} P + Ml_n.$$

For $F_M \cong 1$, this yields

$$r = \frac{k_1k_2k_3[SH][B][ML_n]}{k_2k_3[ML_n][B] + k_{-1}k_3[H^+][B] + k_{-1}k_{-2}[H^+]} \qquad (2.225)$$

and the resulting reaction order with respect to $[ML_n]$ is below unity. An analogous form of the dependence of r on $[ML_n]$ is obtained when a catalyst is separated not in the last step of the mechanism, e.g., as in the following scheme [118]:

$$R'OH + SnCl_4 \underset{k_{-1}}{\overset{k_1}{\rightleftharpoons}} X_1;$$

$$X_1 + RNCO \underset{k_{-2}}{\overset{k_2}{\rightleftharpoons}} X_2 + SnCl_4;$$

$$X_2 \xrightarrow{k_3} RNHCOOR'.$$

For $F_{SnCL_4} \cong 1$, this scheme corresponds to

$$r = \frac{k_1k_2k_3[ROH][RNCO][SnCl_4]_\Sigma}{k_2k_3[RNCO] + k_{-1}k_3 + k_{-1}k_{-2}[SnCl_4]_\Sigma}. \tag{2.226}$$

In cases where two (or more) catalysts are involved in polyfunctional catalytic systems (see Chapter 4), nonlinear steps can also appear as a result of the interaction of additional catalysts with metal-containing intermediates.

In an autocatalytic reaction (where one of the products plays the role of a catalyst), the presence of nonlinear steps can result in fractional reaction orders with respect to reactant and catalyst. For example, a very complicated mechanism was observed [119] for a relatively simple process of reduction of $CF_3COOCH_2CF_3$ ester (B) by hydride complex IrH_5L_2 (A) via reaction

$$IrH_5L_2 + CF_3COOCH_2CF_3 \longrightarrow \underset{(X_1)}{IrH_2L_2(OCH_2CF_3)} + CF_3CH_2OH, \tag{2.227}$$

where L = Pd(i-Pr)$_3$. It was found that complex X_1 catalyzed this reaction and, in the case where B, X_1, and ROH (CF_3CH_2OH) were in excess, the observed kinetics obeyed the relations

$$-\frac{d[A]}{dt} = k_{obs}[A]^{1/2}; \tag{2.228}$$

$$k_{obs} = k^*[X_1]^{1/2}[B][ROH]^{-1/2}. \tag{2.229}$$

The following reaction scheme corresponding to Eqs (2.228) and (2.229) was proposed and experimentally confirmed:

$$A + X_1 \overset{K_1}{\rightleftharpoons} 2X_2 + ROH, \tag{2.230}$$

$$X_2 + B \xrightarrow{k_2} X_1 + X_3;$$

$$X_3 + X_2 \xrightarrow{\text{Fast}} X_1,$$

where $X_2 = IrH_3L_2$, $X_3 = CF_3CHO$, and $k^* = K^{1/2}k_2$.

2.1.3.2 *Comparison of the reaction orders with respect to the initial and current (time-dependent) concentrations of reactants*

In the case of a catalyst complexity function according to variant IV, where the reaction is inhibited by products, the reaction orders with respect to reactants as determined using the initial rates and concentrations can differ from the values obtained for the current rates and concentrations of the integral temporal dependences $C_i = f(t)$ for the kinetics studied under quasi-steady-state conditions. For example, for the reaction

$$A \xrightarrow{M} P$$

at $F_A \cong 1$ and $F_P \cong 1$, the scheme

$$M + A \overset{K_1}{\rightleftharpoons} MA;$$

$$MA \overset{k_2}{\rightleftharpoons} P + M;$$

$$M + P \overset{K_2}{\rightleftharpoons} MP$$

with a limiting second step yields

$$r = \frac{k_2 K_1 [\mathrm{A}][\mathrm{M}]_\Sigma}{1 + K_1[\mathrm{A}] + K_2[\mathrm{P}]}. \tag{2.231}$$

For $[\mathrm{A}] \cong [\mathrm{A}]_0 - [\mathrm{P}]$, we obtain the following kinetic equation:

$$r = \frac{k_2 K_1 [\mathrm{A}][\mathrm{M}]_\Sigma}{1 + K_2[\mathrm{A}]_0 - (K_2 - K_1)[\mathrm{A}]}, \tag{2.232}$$

according to which there are three possible cases depending on the relation between coefficients K_2 and K_1:

(i) $K_2 \cong K_1$. The reaction order with respect to [A] as determined from the dependence $r_0 = f([\mathrm{A}]_0)$ will vary from 1 to 0, while the integral plot of [A] versus time will correspond to the first order.
(ii) $K_2 > K_1$. The reaction order determined for the initial rates varies as in the preceding case, while the "temporal" order will exceed unity.
(iii) $K_2 < K_1$. Both the "temporal" order with respect to [A] and that determined for the initial rates will vary, but with different meaning of observed constants in the numerator and denominator of Eq. (2.232).

In concluding this section, it should be emphasized that, when constructing kinetic models for reactions with mononuclear metal complexes and describing results of kinetic investigations, it is *necessary* to analyze the material balance for both metal complexes and substrates and indicate all assumptions that were used in obtaining expressions for F_M, F_S and the whole kinetic model.

2.2 Polynuclear complexes in homogeneous catalytic and noncatalytic reactions

Various catalytic and noncatalytic processes and reactions involve all the possible kinds of polynuclear species:

- simple associates (e.g., $(\mathrm{LiR})_n$, $(\mathrm{FeCl_3})_n$, $\mathrm{Pd_3(OCOR)_6}$);
- polynuclear complexes (e.g., $\mathrm{Pd_2X_6}^{2-}$, $\mathrm{Cu}_m\mathrm{X}_n^{(n-m)-}$, $[\mathrm{Rh(CO)_2Cl}]_2$, $[\mathrm{Pd(PPh_3)Br_2}]_2$);
- cluster complexes ($\mathrm{M}_n(\mathrm{CO})_m$, $\mathrm{M}_n\mathrm{L}_m$, $\mathrm{Cr_2(OCOR)_4}$, $\mathrm{Mo_2(OR)_6}$, $\mathrm{Pd_2(CO)_2Cl_4}^{2-}$, $\mathrm{Rh_6(CO)_{16}}$, and big and giant clusters).

Situations where complexes of different kinds are simultaneously present may also be encountered.

In the absence of pre-kinetic information concerning the presence of polynuclear species in solution, it is very difficult to distinguish mechanisms, e.g., those involving associates from mechanisms with the sequential formation of intermediates that are polynuclear with respect to a catalyst. For example, consider two mechanisms in which the steps of product (P) formation are limiting:

$$\text{(A)}\quad \mathrm{S} + \mathrm{M} \overset{K_1}{\rightleftharpoons} \mathrm{SM} \underset{K_2}{\overset{\mathrm{M}}{\rightleftharpoons}} \mathrm{SM_2} \overset{k_\mathrm{A}}{\rightleftharpoons} \mathrm{P} + 2\mathrm{M};$$

$$\text{(B)}\quad 2\mathrm{M} \overset{K_\mathrm{d}}{\rightleftharpoons} \mathrm{M_2};\quad \mathrm{M_2} + \mathrm{S} \overset{K_1}{\rightleftharpoons} \mathrm{SM_2} \overset{k_\mathrm{B}}{\rightleftharpoons} \mathrm{P} + 2\mathrm{M}.$$

For scheme A: $r_{(A)} = k_A K_1 K_2[S][M]^2$ and the material balance is as follows:

$$[M]_\Sigma = [M] + [SM] + 2[SM_2] = [M] + K_1[S][M] + 2K_1K_2[S][M]^2.$$

For scheme B: $r_{(B)} = k_B K_1 K_d[S][M]^2$ and the material balance is as follows:

$$[M]_\Sigma = [M] + 2[M_2] + 2[SM_2] = [M] + 2K_d[M]^2 + 2K_dK_1[S][M]^2.$$

In general, after solving the quadratic equations of material balance with respect to [M] and substituting the obtained expressions for [M] into the kinetic equations for schemes A and B, we obtain complicated and hardly distinguishable functions $r_{(P)} = f([M]_\Sigma)$ with a reaction order with respect to $[M]_\Sigma$ that varies between 1 and 2, depending on the intervals of concentrations $[M]_\Sigma$ and the equilibrium constants for the two schemes. In particular cases, it is possible to use dependences $r_{(P)} = f([S])$ for making less ambiguous conclusions.

For example, if polynuclear species are predominantly formed ($2[SM_2] \cong [M]_\Sigma$ in the schemes of both mechanisms A and B), then we obtain identical equations for $r_{(P)}$ of the first order with respect to $[M]_\Sigma$, which are independent of [S]. However, if $2[M_2] \cong [M]_\Sigma$ in scheme B, then, for the first-order reaction with respect to $[M]_\Sigma$, the rate will also linearly depend on [S] (unlike the case with scheme A). When the concentrations of dimeric metal species are small in both systems (i.e., $[M]_\Sigma \cong [M] + [SM]$ in scheme A and $[M]_\Sigma \cong [M]$ in scheme B), then the reaction rates will differ in their dependence on [S] in a large interval of its variation:

$$\text{(A)} \quad r_{(P)} \propto \frac{[M]_\Sigma^2[S]}{(1 + K_1[S])^2};$$

$$\text{(B)} \quad r_{(P)} \propto [M]_\Sigma^2[S].$$

This simple example shows how important the information on the forms of complexes in solution can be when a kinetic approach is used for the discrimination of hypotheses about the mechanisms. Some possible ways to discriminate between hypotheses in the case of $[M]_\Sigma \cong n[M_n]$ are considered below.

2.2.1 Systems with formation of associates

Let us begin with the consideration with a simple case of the formation of associates (dimer, n-mer, oligomer) from monomers M and ligands L in solution according to schemes

$$n\text{ML} \xrightleftharpoons{\beta_n} (\text{ML})_n$$

or

$$n\text{M} \xrightleftharpoons{\beta_n} \text{M}_n.$$

In order to reveal the possible reaction orders with respect to $[M_\Sigma]$, we should take into account the material balance

$$[M]_\Sigma = [M] + n[M_n] \tag{2.233}$$

and analyze the role of monomeric and oligomeric complex species in the reaction under consideration.

If monomers are active in the reaction and the reaction rate is proportional to [M] ($r \propto$ [M]), then there are two possible limiting cases:

$$\text{(a) } [\mathrm{M}] \ll [\mathrm{M}]_\Sigma, \quad [\mathrm{M}_n] \cong \frac{[\mathrm{M}]_\Sigma}{n};$$
$$\text{(b) } [\mathrm{M}] \gg n[\mathrm{M}_n], \quad [\mathrm{M}] \cong [\mathrm{M}]_\Sigma.$$

In the former case, we have

$$[\mathrm{M}]^n = \frac{[\mathrm{M}_n]}{\beta_n} \cong \frac{[\mathrm{M}]_\Sigma}{n\beta_n}, \quad [\mathrm{M}] = \frac{[\mathrm{M}]_\Sigma^{1/n}}{(n\beta_n)^{1/n}},$$

and, hence,

$$r \propto [\mathrm{M}]_\Sigma^{1/n}.$$

In the latter case, $r \propto [\mathrm{M}]_\Sigma$, so that the reaction is of the first order with respect to $[\mathrm{M}]_\Sigma$.

Writing Eq. (2.233) with allowance for the equilibrium association leads to the following relation:

$$[\mathrm{M}]^n + \frac{[\mathrm{M}]}{n\beta_n} - \frac{[\mathrm{M}]_\Sigma}{n\beta_n} = 0, \tag{2.234}$$

which determines the form of the function $r_0 = f([\mathrm{M}]_\Sigma)$ in the general case. Evidently, the reaction order with respect to $[\mathrm{M}]_\Sigma$ can vary in the interval from $1/n$ to 1.

If the reaction is dominated by the activity of associates M_n, then the corresponding limiting cases are

$$\text{(a)} \quad [\mathrm{M}_n] \cong \frac{[\mathrm{M}]_\Sigma}{n},$$

so that $r \propto [\mathrm{M}]_\Sigma$, and

$$\text{(b)} \quad [\mathrm{M}] \cong [\mathrm{M}]_\Sigma; \quad [\mathrm{M}_n] \cong \beta_n[\mathrm{M}]^n \cong \beta_n[\mathrm{M}]_\Sigma^n,$$

in which case $r \propto [\mathrm{M}]_\Sigma^n$ and, hence, the reaction order with respect to $[\mathrm{M}]_\Sigma$ varies from 1 to n.

The role of association in the appearance of fractional orders with respect to the total concentration of an initiator (LiR) or the total concentration of active centers of the molecular chain growth was thoroughly studied in the process of anionic polymerization of styrene, butadiene, and isoprene in hydrocarbon solvents [120, 121]. Under the conditions of "living" polymerization, the concentration of active centers is equal to the initial concentration C_0 of the initiator (or the "living" polymer) occurring in the associated state. Under conditions where the polymer chain grows on the monomeric species of active center (LiR) whose concentration is small compared to C_0, the polymer chain growth rate is described by the following equation:

$$-\frac{d[\mathrm{m}]}{dt} = K[\mathrm{m}][C_0]^{1/n}, \tag{2.235}$$

where [m] is the monomer molecule concentration and $n = 2$, 4, or 6 depending on the system.

A more complicated case involving several types of associates was considered in the investigation of a reaction of poly(butadienyl)lithium with butadiene in aliphatic hydrocarbons [122]. The reaction order with respect to $[\mathrm{LiR}]_\Sigma$ varies from 0.5 to 1 in the range of LiR concentrations of 10^{-5}–1 M. Let us briefly analyze this case. A material balance with respect to the initiator is written based on a sum of the concentrations of dimers and tetramers

$$C_0 = 2C_2 + 4C_4, \tag{2.236}$$

which assumes that the concentrations of monomers (C_1) and trimers (C_3) are small compared to C_0 (for the sake of simplicity, the corresponding substances are denoted by the same letters). Denoting the equilibrium constants of formation of C_2 and C_4 species by β_2 and β_4, respectively,

$$2C_1 \overset{\beta_2}{\rightleftharpoons} C_2; \quad 4C_1 \overset{\beta_4}{\rightleftharpoons} C_4,$$

we can rewrite Eq. (2.236) as

$$2\beta_2 C_1^2 + 4\beta_4 C_1^4 = C_0. \tag{2.237}$$

Substituting the variable $C_1^2 = x$ and solving the obtained quadratic equation, we obtain

$$C_1 \cong \sqrt{\frac{a}{\beta_2}}\left(\sqrt{1+\frac{C_0}{a}}-1\right)^{1/2}, \tag{2.238}$$

where $a = \beta_2^2/4\beta_4$.

If the observed first-order rate constant for the chain propagation

$$k_1 = -\frac{d\,\ln[\mathrm{m}]}{dt}$$

is proportional to the concentration C_1 of active centers in the monomer form,

$$k_1 = k_{p1}C_1,$$

then we can consider two limiting cases.

(i) For small values of C_0/a in Eq. (2.238), we have $(1 + C_0/a)^{1/2} \approx (1 + C_0/2a)$ and $C_2 \gg 2C_4$, so that $k_1 \cong k_{p1}\sqrt{a/\beta_2}(C_0/2a)^{1/2}$, i.e., the reaction order with respect to C_0 is ½ and the active centers exist predominantly in the form of dimers C_2.

(ii) For $C_0/a \gg 1$ ($C_2 \ll 2C_4$), we have $k_1 \cong k_{p1}\sqrt{a/\beta_2}(C_0/2a)^{1/4}$, so that the reaction order with respect to C_0 is 1/4 and a predominant form of LiR in solution is tetramer C_4. Assuming that both monomers C_1 and tetramers C_4 participate in steps of the chain propagation, we can write

$$k_1 = k_{p1}C_1 + k_{p4}C_4. \tag{2.239}$$

Then, for large C_0 we have $C_0 = 4C_4$ (because $C_1 \ll C_0$) and $k_1 = k_{p1}C_1 = k_{p4}(C_0/4)$, which implies that the reaction is of the first order with respect to C_0, in agreement with experiment [122].

The behavior of other associates is similar to that of $(LiR)_n$. For example, palladium(II) acetate is used in both catalytic and stoichiometric conversions of olefins and aromatics [15, 76, 123]. In aromatic hydrocarbons below 40 °C (e.g., benzene) [124, 125] and in glacial acetic acid [126, 127] Pd(II) acetate occurs predominantly in the form of trimers $Pd_3(OAc)_6$, like many other Pd(II) carboxylates [128]. For the depolymerization of these associates in acetic acid,

$$Pd_3(OAc)_6 \overset{K_D}{\rightleftharpoons} 3Pd(OAc)_2, \tag{2.240}$$

the equilibrium constant at 75 °C amounts to $(2.7 \pm 0.6) \times 10^{-10}$ M^2 [127].

The kinetics of reactions involving tri-, di-, and monomeric Pd(II) complexes is considered in detail in Section 2.2.2. Prior to proceeding with this analysis, let us briefly mention some cases of catalysis with monomeric and trimeric forms of Pd(II) acetate.

Example 2.33

► For the oxidative dimerization of styrene

$$2C_6H_5CH{=}CH_2 + Pd(II) \longrightarrow (C_6H_5CH{=}CH)_2 + Pd(0) \tag{2.241}$$

in the interval of Pd(II) concentrations within (0.5–5.0) × 10^{-4} M that corresponds to the predominant existence of $Pd_3(OAc)_6$, the reaction rate is proportional to $[Pd]_\Sigma$ [127]:

$$r = k_{obs}[Pd]_\Sigma = k_1[Pd_3(OAc)_6],$$

where $k_1 = 3k_{obs}$ (at a constant concentration of styrene). In more (approximately tenfold) dilute solutions, the depolymerization of associates becomes pronounced, the trimer fraction decreases, and the dependence on $[Pd]_\Sigma$ ceases to be linear. Using the reaction rate and k_1 values, it is possible to determine the equilibrium concentration of trimers, their relative fraction α (defined as $\alpha = 3[Pd_3]/[Pd]_\Sigma$), and the equilibrium constant of the depolymerization process [127]:

$$K_D = \frac{3[Pd]_\Sigma^2(1-\alpha)^3}{\alpha}.$$

■

Example 2.34

► For the oxidative coupling of arenes in an aromatic substrate medium [129]

$$2ArH + Tl(OCOCF_3)_3 \xrightarrow{Pd(II)} Ar{-}Ar + TlOCOCF_3 + 2CF_3COOH, \tag{2.242}$$

which is catalyzed by Pd(II) acetate, the rate of reaction at 70 °C under the conditions of existence of $Pd_3(OAc)_6$ and $Pd(OAc)_2$ is described by the following equation:

$$\frac{d[Ar_2]}{dt} = k_{obs}[Tl(OCOCF_3)_3][Pd]_\Sigma^{1/3}. \tag{2.243}$$

A mechanistic reaction scheme that corresponds to this equation includes the steps of ArH metallation by trifluoroacetate Tl(III), transmetallation of $ArTl(OCOCF_3)_2$ by monomeric Pd(II) acetate, and the subsequent redox decomposition of ArPd(OAc) under the condition of complete Tl(III) binding by arene (and a small contribution of ArPd(OAc) to the material balance with respect to Pd):

$$ArH + TlX_3 \xrightarrow{k_A} ArTlX_2 + HX;$$

$$ArTlX_2 + Pd(OAc)_2 \xrightarrow{k_B} ArPdOAc + TlX_2(OAc);$$

$$2ArPdOAc \xrightarrow{Fast} Ar{-}Ar + Pd(OAc)_2 + Pd(0);$$

$$Pd(0) + Tl(III) \longrightarrow Pd(II) + Tl(I);$$

$$Pd_3(OAc)_6 \rightleftharpoons 3Pd(OAc)_2.$$

The kinetics of $ArTlX_2$ conversion into Ar_2 is described by an analogous kinetic equation:

$$\frac{d[Ar_2]}{dt} = k'_{obs}[ArTl(OCOCF_3)_2][Pd]_\Sigma^{1/3}. \tag{2.244}$$

■

A general approach to the description of equilibria involving a large number of arbitrary associates in solution was developed in the 1940s. In particular, a simple method for the description of associated systems was based on investigations of the process of phenol association in CCl_4 [130, 131]. Let $K_{C_{n-1}}$ denote the step dissociation constants of phenol associates (n-mers). Then, the sequential equilibria

$$(\mathrm{PhOH})_2 \rightleftharpoons 2\mathrm{PhOH} \quad (K_{C_1});$$

$$(\mathrm{PhOH})_n \rightleftharpoons (\mathrm{PhOH})_{n-1} + \mathrm{PhOH} \quad (K_{C_{n-1}}).$$

can be characterized by the general degree of dissociation $\alpha = b/C_B$ (where C_B is the analytical concentration of PhOH and b is the equilibrium concentration of a monomer form) and the fraction α_n of monomer molecules used for the formation of n-mers:

$$\alpha_n = n[(\mathrm{PhOH})_n]/C_B. \tag{2.245}$$

Using the definitions of α, α_n and the step equilibrium constants for the dissociation of associates, one can obtain the following relation:

$$\alpha_n = n\alpha \frac{(\alpha C_B)^{n-1}}{K_{C_1} K_{C_2} \ldots K_{C_{n-1}}}. \tag{2.246}$$

Assuming that stepwise dissociation constants are independent of the association degree n ($K_{C_1} = K_{C_2} = \ldots = K_{C_{n-1}} = K_{C_n} = K_C$), it is possible to rewrite this relation as

$$\alpha_n = \alpha n \left(\frac{\alpha C_B}{K_C}\right)^{n-1}. \tag{2.247}$$

Taking into account that $\sum_1^n \alpha_n = 1$, we obtain the following equation:

$$\alpha \sum_1^n n \left(\frac{\alpha C_B}{K_C}\right)^{n-1} = 1.$$

Substituting the variable $\alpha C_B/K_C = x$, this equation can be rewritten as

$$\alpha \sum_1^n n x^{n-1} = 1. \tag{2.248}$$

Assuming that $x < 1$ and n is sufficiently large, the term $\sum_1^n n x^{n-1}$ can be considered as a function of the sum of an infinite converging series

$$S = 1 + x + x^2 + \cdots + x^n = \frac{1}{1-x}.$$

Indeed, the series $\sum_1^n n x^{n-1}$ is a derivative of S with respect to x and, accordingly,

$$\sum_1^n n x^{n-1} = \frac{dS}{dx} = \frac{1}{(1-x)^2}. \tag{2.249}$$

Using the relation $\frac{\alpha}{(1-x)^2} = 1$, we obtain the following expression for the dissociation constant K_C:

$$K_C = \frac{\alpha C_B}{1 - \sqrt{\alpha}} \tag{2.250}$$

or an equation for the parameter b ($b = f(C_B)$):

$$C_B = \frac{b}{(1 - K_q b)^2}, \tag{2.251}$$

where $K_q = 1/K_C$ is the equilibrium step constant of associate formation. Note that Eq. (2.251) can also be derived directly from the equation of material balance in the case where all the sequential step constants K_q are identical:

$$C_B = b + \sum_2^n n K_q^{n-1} b^n = b\left(1 + \sum_2^n n K_q^{n-1} b^{n-1}\right). \tag{2.252}$$

Indeed, substituting $K_q b = y$, we obtain the relation $1 + \sum_2^n n y^{n-1} = \frac{1}{(1-y)^2}$ (provided that $y < 1$) and then arrive at Eq. (2.251).

The results of investigations of the association of phenols and alcohols [131] show that, for phenol in CCl_4, the dissociation constant K_1 for dimers differs from the other step constants: $K_1 = 0.72$ and $K_C = 0.34$. As a result, Eq. (2.247) takes a different form:

$$\alpha_n = n\alpha \frac{K_C}{K_1}\left[\frac{\alpha C_B}{K_C}\right]^{n-1}. \tag{2.253}$$

Equation (2.251) was used [12] to determine the constants of formation K_q of the associates of metal complexes with allowance for the fact that the equilibrium constant of dimer formation (β_{20}) differs from the other constants (K_q):

$$\frac{(C_B - b)K_q}{\beta_{20}} = \sum_2^n n K_q^{n-1} b^n;$$

$$\frac{(C_B - b)K_q}{\beta_{20} b} = \sum_2^n n (K_q b)^{n-1} = \sum_2^n n y^{n-1}.$$

Taking into account that

$$\sum_2^n n y^{n-1} = \frac{1}{(1-y)^2} - 1 = \frac{y(2-y)}{(1-y)^2}, \tag{2.254}$$

we obtain the following equation:

$$\frac{(C_B - b)K_q}{\beta_{20} b} = \frac{K_q b(2 - K_q b)}{(1 - K_q b)^2}, \tag{2.255}$$

from which one can readily calculate the monomer fraction in the sum of associates contributing to C_B:

$$C_B = b + b\left[\frac{K_q b(2 - K_q b)}{(1 - K_q b)^2}\right]. \tag{2.256}$$

Example 2.35

► Akhmed'yanova et al. [68, 132–134] studied the reactions of olefin epoxidation by hydroperoxides (HPs) and HP decomposition in solutions of a commercial molybdenum catalyst (CMC) and revealed nonlinear dependences of the initial rates of olefin epoxidation and HP decomposition on the initial CMC concentration. The reaction order with respect to $[Mo]_{\cong}$ changed from unity to nearly zero. Although CMC actually contains a mixture of Mo complexes [132] with the metal in various oxidation states, it was believed that, under the conditions of epoxidation reaction, all catalytic species converted into molybdenum diolate under the action of olefin oxide and HP. It was suggested that the nonlinear dependences of the initial reaction rates on $[Mo]_{\Sigma}$ in all cases were related to the formation of catalytically inactive oligomeric forms of $[Mo]_n$, i.e., inactive associates of Mo(VI) diolate complexes.

Then, it was assumed that the constants of oligomerization in all steps have the same value (K_{ol}), and the aforementioned approach [12, 130, 131] was applied for the first time to kinetic investigations for determining the fraction of Mo that enters into the material balance in the form of oligomers under the assumption that $K_{ol}[Mo] = y < 1$. The material balance also included the Mo–HP complex with the equilibrium constant K'_S. As a result, the material balance is expressed as follows (symbols $[Mo]_n$ and [Mo] refer to the oligomeric and monomeric diolate complexes of Mo(IV)) [134]:

$$[Mo]_{\Sigma} = [Mo] + K'_S[Mo][HP] + [Mo]\left\{\sum_{2}^{n} n(K_{ol}[Mo])^{n-1}\right\}$$

$$= [Mo] + K'_S[Mo][HP] + [Mo]\left\{\frac{K_{ol}[Mo](2 - K_{ol}[Mo])}{(1 - K_{ol}[Mo])^2}\right\} \quad (2.257)$$

(for $\sum_{2}^{n} n(K_{ol}[Mo])^{n-1} = \sum_{2}^{n} ny^{n-1}$ and $y < 1$).

Upon solving this equation with respect to [Mo] and ignoring terms with $[Mo]^n$ at $n \geq 2$, one can obtain an interesting hyperbolic equation that suggests the possibility that [Mo] is independent of $[Mo]_{\Sigma}$ at sufficiently high concentrations of molybdenum:

$$[Mo] = \frac{[Mo]_{\Sigma}}{1 + K'_S[HP] + 2K_{ol}[Mo]_{\Sigma}}. \quad (2.258)$$

This approach, albeit using rather deliberate assumptions (e.g., that K_{ol} is the same for all steps and $K_{ol}[Mo] < 1$) that obviously require some justification, is useful for the description of complex systems featuring the formation of polynuclear complexes-associates $(M)_n$. Naturally, the free catalyst concentration ([Mo]) also depends on the equilibrium formation of intermediates and inactive polynuclear complexes of the catalyst with reactants and products (corresponding to pendant vertices on the kinetic graphs), which must be taken into account when writing the material balance. Since it would be incorrect to use an independent, "classical" complexity function of mononuclear complexes under the conditions where oligomers do exist, the proposed approach should employ the common function

$$F_M = \frac{[Mo]_{\Sigma}}{[Mo]} = 1 + \sum_i K_i S_i + \frac{K_{ol}[Mo](2 - K_{ol}[Mo])}{(1 - K_{ol}[Mo])^2}, \quad (2.259)$$

where K_i are the equilibrium constants of formation of monomeric Mo complexes with substrates and products S_i that enter into the material balance. ■

Example 2.36

► An analogous approach was used in the investigation of HP decomposition in cobalt complex solutions [69], where oligomeric complexes were described within the framework of a general material balance equation and common complexity function [of the type of Eq. (2.259)]. However, with a unified system of equilibrium and quasi-equilibrium reactions, the equilibrium complexity also includes the nonequilibrium processes of depolymerization of oligomers under the action of substrates and reaction products. However, the process

$$M_{n+1} + S \underset{k_1^*}{\overset{k_{-1}^*}{\rightleftharpoons}} M_n + MS \tag{2.260}$$

has to be equilibrium provided that the equilibrium of reactions

$$M_n + M \overset{K_{ol}}{\rightleftharpoons} M_{n+1}; \tag{2.261}$$

$$M + S \overset{K_S'}{\rightleftharpoons} MS. \tag{2.262}$$

is attained. ■

Example 2.37

► The decomposition of associates under the action of substrates and the corresponding effect of the substrate concentration on the observed reaction orders with respect to the catalyst were revealed in the study of the kinetics of dichlorobutene (DCB) isomerization catalyzed by dimeric Fe_2Cl_6 (D) complexes in thoroughly dehydrated ($\cong$0.0002 mass % H_2O) chloroorganic solvents [135, 136]. The reaction order with respect to $[Fe]_\Sigma$ changed from 2 at $[DCB]_0 < 0.1$ M to 4 at $[DCB]_0 > 1$ M. At a low substrate concentration and $[DCB]_0 \cong [Fe]_\Sigma$, the rate of the isomerization reaction

$$\text{3,4-DCB} \rightleftharpoons \text{1,4-DCB}$$

was described by the following equation:

$$\frac{d[\text{1,4-DCB}]}{dt} = [D]^2(k_1[\text{3,4 DCB}] - k_{-1}[\text{1,4-DCB}]), \tag{2.263}$$

where $[D] = {}^1/_2[Fe]_\Sigma$. According to this equation, a transition state of the limiting step involves four Fe atoms. At an increased concentration of DCB, both isomers interact with dimers and decompose them with the formation of a complex between the $FeCl_3$ monomer (M) and DCB (M · DCB):

$$D + 2DCB \overset{K_{eq}}{\rightleftharpoons} 2(M \cdot DCB). \tag{2.264}$$

Using the material balance with respect to iron written as

$$[D]_0 = [D] + 1/2[M \cdot DCB],$$

and an expression for the concentration of M · DCB complexes

$$[M \cdot DCB] = [DCB]_0\sqrt{K_{eq}[D]},$$

and making some simplifying assumptions, we eventually obtain the following relation:

$$[D] \cong \frac{4[D]_0^2}{K_{eq}[DCB]_0^2} = \frac{[Fe]_\Sigma^2}{K_{eq}[DCB]_0^2}. \tag{2.265}$$

Substituting this expression into the kinetic equation (2.263), we arrive at the following:

$$\frac{d[1,4\text{-DCB}]}{dt} = \frac{[\text{Fe}]_\Sigma^4}{K_{\text{eq}}^2[\text{DCB}]_0^4}(k_1[3,4\text{-DCB}] - k_{-1}[1,4\text{-DCB}]). \tag{2.266}$$

In both cases, the isomerization process involves four Fe atoms and a polynuclear intermediate is formed that consists of two dimers. However, at high concentrations of the substrate and under the conditions favoring the formation of low-active monomeric forms, the concentration of dimers D sharply decreases and the reaction rate begins to depend on $[\text{Fe}]_\Sigma$ in the fourth power. The observed first-order rate constant exhibits a sharp drop with increasing $[\text{DCB}]_0$, while the isomerization rate passes through a sharp maximum with respect to $[\text{DCB}]_0$ in the region of low substrate concentrations. High reaction orders with respect to $[\text{FeCl}_3]_\Sigma$ ($n = 5$) were observed in the reactions of hydrochlorination of vinylidene chloride [137, 138] and ethylene [139] in methylene chloride. It was suggested that this behavior was related to the participation of polynuclear iron complexes, the amount of which was small as a result of the formation of low-active $\text{FeCl}_3 \cdot \text{HCl}$ species. ■

Interesting kinetic laws were observed in multi-electron redox processes that involved polynuclear complexes (associates) acting as reactants or catalysts, where mechanisms of the A and B type (see pp. 167) under certain conditions could be distinguished. Multi-electron redox reactions involving the transfer of two or more electrons in the case of one-electron oxidant and reductants are in fact noncomplementary reactions [140], the mechanisms of which include (because of the bimolecular character of elementary steps, see Section 1.4) several sequential stages of electron transfer or several consecutive steps that lead to the formation of polynuclear species containing several molecules of oxidant (reductant). A polynuclear (with respect to the oxidant or reductant) intermediate formally switches a noncomplementary process to multi-electron complementary process [141]. In some cases, it is the step-by-step formation of either the polynuclear complexes of an oxidant (reductant) capable of multi-electron one-step processes or the polynuclear transition states, which allow the thermodynamic and kinetic constraints to be surmounted [141].

An even more favorable situation from the standpoint of kinetics is provided by the existence of a metal-oxidant or metal-reductant species in the form of polynuclear complexes (associates) M_n, because the reaction with a substrate in this case becomes a bimolecular (or monomolecular) process. The presence of polynuclear metal complexes in redox enzymes and their models [142, 143] is probably related to the aforementioned factors. Let us consider some examples of noncomplementary reactions.

Example 2.38

► The one-electron oxidant Ce^{4+} produces oxidation of Cr(III) to Cr(VI) [140]:

$$3\text{Ce}^{4+} + \text{Cr}^{3+} + 4\text{H}_2\text{O} \longrightarrow 3\text{Ce}^{3+} + \text{HCrO}_4{}^- + 7\text{H}^+$$

The rate of this reaction is described by the following equation:

$$r = \frac{k[\text{Ce(IV)}]^2[\text{Cr(III)}]}{[\text{Ce(III)}]}, \tag{2.267}$$

which corresponds to the following mechanistic scheme:

$$\text{Ce(IV)} + \text{Cr(III)} \underset{}{\overset{\text{Fast}}{\rightleftharpoons}} \text{Ce(III)} + \text{Cr(IV)};$$

$$\text{Ce(IV)} + \text{Cr(IV)} \xrightarrow{\text{Slow}} \text{Ce(III)} + \text{Cr(V)};$$

$$\text{Ce(IV)} + \text{Cr(V)} \xrightarrow{\text{Fast}} \text{Ce(III)} + \text{Cr(VI)}.$$

■

Example 2.39

▶ The noncomplementary reaction

$$3H_2 + 2Cr(VI) \xrightarrow{Cu(II)} 2Cr(III) + 6H^+$$

is catalyzed by the Cu^{2+} ion that activates H_2, and the reduction of Cu^{2+} by hydrogen is also a noncomplementary process [144]:

$$H_2 + 2Cu^{2+} \longrightarrow 2Cu^+ + 2H^+. \quad (2.268)$$

The resulting Cu^+ ion acts as a reductant for Cr(VI) in three consecutive steps as described by the following stoichiometric equation:

$$3Cu^+ + Cr(VI) \longrightarrow 3Cu^{2+} + Cr(III). \quad (2.269)$$

The rate of the overall process is determined by reaction (2.268), the kinetics of which indicates that two Cu^{2+} ions participate in a transition state during the formation of two Cu^+ ions [144]:

$$Cu^{2+} + H_2 \underset{k_{-1}}{\overset{k_1}{\rightleftharpoons}} CuH^+ + H^+;$$

$$CuH^+ + Cu^{2+} \xrightarrow{k_2} 2Cu^+ + H^+;$$

$$-\frac{d[H_2]}{dt} = \frac{k_1k_2[Cu^{2+}]^2[H_2]}{k_2[Cu^{2+}] + k_{-1}[H^+]}. \quad (2.270)$$

Thus, the binuclear transition state is involved at the second step. In this case, the available kinetic information rejects the possibility that $(Cu^{2+})_2$ dimers participate in the reaction. ■

The reactions of four-electron oxidation of water by strong one-electron oxidants

$$4A^+ + 2H_2O \longrightarrow 4A + O_2 + 4H^+ \quad (2.271)$$

[A^+ = Co(III), Mn(IV)] and the reactions of N_2 reduction by one-electron reductants

$$4A + N_2 + 4H^+ \longrightarrow 4A^+ + NH_2NH_2; \quad (2.272)$$

$$6A + N_2 + 6H^+ \longrightarrow 6A^+ + 2NH_3 \quad (2.273)$$

[A = V(II), Mo(III) etc.], involve polynuclear complexes of the $(A^+)_n$ or $(A)_n$ type [143, 145].

As was noted above, if the reaction involves M_n (e.g., $(A^+)_n$) complexes with a concentration that is small compared to $[M]_\Sigma$, than the reaction orders with respect to the current ($[A^+]_t$) or initial ($[A^+]_0$) analytical concentrations of the reactant correspond to the composition (i.e., n value) of a complex involved in the limiting (or the last irreversible) step of the process. Here, the main problem is to distinguish this mechanism from that involving consecutive addition of oxidant species A^+ with the formation of the last intermediate. In the case of the process of DCB isomerization catalyzed with Fe_2Cl_6 dimers (Example 2.37) [135, 136], a tetrameric intermediate (with a fourth order of reaction with respect to Fe_Σ or the second order with respect to Fe_2Cl_6) was formed.

A quite different situation takes place in the systems where the oxidant (A^+) or reductant (A) occurs in solution *only* in the form of polynuclear complexes. The reaction kinetics in this case was analyzed in much detail by Dzhabiev et al. [145–150]. The main results of this analysis are as follows.

Consider a two-electron process such as

$$2Co^{3+} + 2H_2O \longrightarrow H_2O_2 + 2Co^{2+} + 2H^+, \tag{2.274}$$

where the oxidant ($Co^{3+} \equiv A^+$) occurs in the form of dimers $(A^+)_2$, and the reaction proceeds with product P formation via a bimolecular process:

$$(A^+)_2 + S \xrightarrow{k_2} 2A + P. \tag{2.275}$$

As was shown above, the variation of the initial and current reaction rates in this case is described by first-order equations with respect to the initial ($[A^+]_0 = a$) and current($[A^+]_t$) concentrations of the oxidant. If the reductant (A) also occur in the form of dimers $(A)_2$ that interact with $(A^+)_2$ so as to form inactive dimers (A^+A), e.g., via the election transfer reaction

$$(A^+)_2 + (A)_2 \overset{K}{\rightleftharpoons} 2(A^+A), \tag{2.276}$$

the process becomes self-inhibited (i.e., negative feedback arises [145]). In this case, the reaction remains first-order with respect to the initial rate $[A^+]_0$, while the reaction order with respect to $[A^+]_t$ at certain (not very large) values of K increases to 2, that is, corresponds to the composition of the active associate $(A^+)_2$. In the given system:

$$[A^+]_t = 2[A_2{}^+] + [A^+A]; \tag{2.277}$$

$$[A^+A] = \sqrt{K[A_2{}^+][A_2]}. \tag{2.278}$$

Here and below, the writing of concentrations is simplified by denoting $[(A^+)_2] = [A_2{}^+]$, $[(A_2)] = [A_2]$, and $[(A^+A)] = [A^+A]$.

If the equilibrium constants of formation of the three forms of associates from elements A^+ and A are close, it is possible to calculate the concentrations of these associates and the constant K of equilibrium (2.276) for a statistically equiprobable distribution of A^+ and A over the dimeric forms as follows [145, 148]:

$$[A_i^+A_{2-i}] = \frac{a}{2}C_2^i x^i (1-x)^{2-i}, \tag{2.279}$$

where $a = [A^+]_0$, x is the molar fraction of $[A^+]_t (x = [A^+]_t/a)$, and C_2^i is the number of combinations of two elements, i at a time (binomial coefficient) such that $C_n^n = 1$, $C_n^0 = 1$, and $C_2^1 = 2$.

Formula (2.279) yields the concentrations

$$[A_2{}^+] = \frac{a}{2}x^2; \quad [A_2] = a(1-x)^2; \quad [A^+A] = \frac{a}{2}x(1-x),$$

and that K for reaction (2.276) is 4. If the rate of H_2O_2 formation via reaction (2.274) is determined by step (2.275), we obtain (for $[S]_0 \gg a$) the following equation:

$$-\frac{1}{2}\frac{d[A^+]_t}{dt} = k_2[A_2{}^+][S] = k_1[A_2{}^+]. \tag{2.280}$$

Substituting $[A^+]_t$ and $[A_2{}^+]$ expressed via molar fraction x into Eq. (2.280), we obtain:

$$-\frac{dx}{dt} = k_1 x^2. \tag{2.281}$$

Expressing $[A_2^+]_t$ via $[A^+]_t$, we eventually obtain

$$r = -\frac{d[A^+]_t}{dt} = \frac{k_1}{a}[A^+]_t^2. \quad (2.282)$$

Thus, for $K = 4$, the reaction rate as a function of the time is described by a second-order equation with respect to the current concentration $[A^+]_t$, although all reactants exist in the form of dimers. The rate (2.282) of the reaction of oxidant consumption can be expressed via the degree of conversion $\alpha = 1 - x$, which yields

$$r = k_1 a(1 - \alpha)^2. \quad (2.283)$$

$$r_0 = k_1 a. \quad (2.284)$$

Therefore, using a dimensionless equation of the type

$$\varphi = r/r_0 = (1 - \alpha)^2, \quad (2.285)$$

plotted in the coordinates of φ (varying from 1 to 0) versus α (from 0 to 1) (Figure 2.4), it is possible to estimate the reaction order by determining the derivative $d\varphi/d\alpha$ at $\alpha \to 0$.

Using the material balance equation (2.277) written as

$$xa = 2[A_2^+] + [A^+A], \quad (2.286)$$

$$(1 - x)a = 2[A_2] + [A^+A], \quad (2.287)$$

and Eq. (2.278), it is possible to determine $[A_2^+]$, in the general case (for arbitrary K) by solving the following quadratic equation:

$$4(4 - K)[A_2^+]^2 + (2aKx - 8ax - aK)[A_2^+] + 2a^2x^2 = 0. \quad (2.288)$$

In particular, for $K = 4$ we have

$$[A_2^+] = -\frac{2a^2x^2}{8ax - 8ax - 4a} = \frac{a}{2}x^2$$

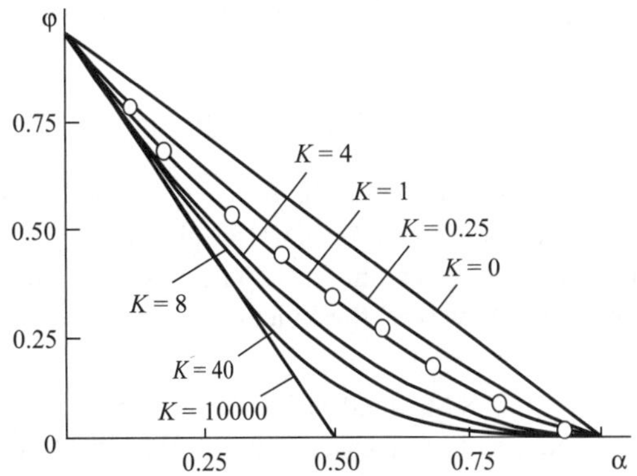

Figure 2.4 Phase trajectories on plane for the process described by Eqs. (2.280) and (2.289). Experimental points correspond to the kinetics of reaction (2.274) [149] (reproduced with permission from editorial board of *Izv. Acad. Nauk, Ser. Khim.* journal © 2001 [145])

and Eq. (2.288) yields

$$[A_2^+]_t = \frac{a}{2} \cdot \left\{ \frac{2x(4-K) + K - [K^2 + 4Kx(1-x)(4-K)]^{\frac{1}{2}}}{2(4-K)} \right\}. \tag{2.289}$$

For $K = 0$ (see Figure 2.4), we have $[A_2^+]_t = ax/2$ and the reaction is first-order with respect to $[A^+]_t$. At very large K, every mole of A in the reduced form binds (inactivates) one mole of oxidant A^+, so that the kinetics is also described by the first-order equation with respect to $[A^+]_t$. This is similar to the case of a bimolecular step (2.275), but in that case ($K \gg 1$) four moles of A^+ (for a maximum of $\alpha = 0.5$) were consumed per mole of product P. For all curves of φ versus α, which are lying between the boundary straight lines in Figure 2.4, the derivative $\left.\frac{d\varphi}{d\alpha}\right|_{\alpha\to 0}$ determines the reaction order with respect to $[A^+]_t$ and, accordingly, the composition of an associate that is involved in the product formation.

Let us check for this in the case of reactions involving dimers considered above. Denoting the expression in brackets of Eq. (2.289) by L, and the ratio L/x^2 by f_1, we obtain

$$[A_2^+]_t = \frac{a}{2} L \frac{x^2}{x^2} = \frac{ax^2}{2} \frac{L}{x^2} = \frac{ax^2}{2} f_1, \tag{2.290}$$

so that $[A_2^+]_t = [A_2^+]_{t,K=4} \cdot f_1$, where coefficient f_1 describes the extent of deviation of the $[A_2^+]_t$ value from the current concentration $[A_2^+]_{t,K=4}$ of dimers. Since the initial concentration of dimers ($x \to 1$) is always $[A_2^+]_0 = a/2$, we have

$$\varphi = x_2^2 f_1 = (1-\alpha)^2 f_1. \tag{2.291}$$

Taking into account that $\lim_{\alpha\to 0} f_1 = 1$, we obtain $\left|\frac{d\varphi}{d\alpha}\right|_{\alpha\to 0} = 2$.

The reactions of water oxidation by Mn(IV) ions were studied under the conditions of existence of Mn_2^{IV} dimers (D), Mn_4^{IV} tetramers (T), and even Mn_6^{IV} hexamers (H) [145–149]. If a solution predominantly contains dimers ($a \cong 2D$), then it is evident that the concentration of tetramers according to equilibrium

$$2D \underset{}{\overset{K_T}{\rightleftharpoons}} T \tag{2.292}$$

is approximately $a^2K_T/4$, i.e., proportional to a^2. Under these conditions, the initial rate of Mn(IV) consumption is described by the following equation:

$$r_0 = k_1[Mn^{IV}]_0 + k_2[Mn^{IV}]_0^2, \tag{2.293}$$

which corresponds to a two-route process involving the dimer ($[D]_0 = a/2$) and tetramer ($[T]_0 = a^2K_T/4$), where a is the initial analytical concentration of Mn(IV). Analogous equations describe the oxidation of water by the ions of other transition metals [148], in particular, by Co(III) ions in a sulfate system. [151]. In the case of Co(III) and perchloric acid, however, the kinetics is more adequately described by an equation including a term with $[Co(III)]_t^{3/2}$ [152].

At a concentration of $[Mn(IV)]_0 > 0.038$ M in 9.4 M H_2SO_4 at 355 K, the system predominantly features tetramers, and all manganese is contained in five associates: Mn_4^{IV}, $Mn_3^{IV}Mn^{III}$, $Mn_2^{IV}Mn_2^{III}$, $Mn_1^{IV}Mn_3^{III}$ and Mn_4^{III}. If Mn^{4+} and Mn^{3+} ions are statistically distributed over tetramers, we have

$$[Mn_i^{IV}Mn_{4-i}^{III}] = \frac{a}{4} C_4^i x^i (1-x)^{4-i}, \tag{2.294}$$

where $C_4^1 = 4$, $C_4^2 = 6$, $C_4^3 = 4$, $C_4^4 = 1$, $C_4^0 = 1$ and the concentration of active tetramers is

$$[\mathrm{Mn_4^{IV}}]_t = \frac{a}{4}x^4, \qquad (2.295)$$

where x is the molar fraction of $\mathrm{Mn^{IV}}$ and $x = [\mathrm{Mn^{IV}}]/a$.

In the case where only one active tetramer $\mathrm{Mn_4^{IV}}$ exhibits oxidative conversion,

$$4\mathrm{Mn(IV)} + 2\mathrm{H_2O} \longrightarrow 4\mathrm{Mn(II)} + \mathrm{O_2} + 4\mathrm{H^+}, \qquad (2.296)$$

the rate of Mn(IV) consumption is described by the following equation:

$$-\frac{d[\mathrm{Mn^{IV}}]}{dt} = 4k_4[\mathrm{T}] = 4k_4\left(\frac{a}{4}x^4\right) = \frac{k_4}{a^3}[\mathrm{Mn^{IV}}]_t^4. \qquad (2.297)$$

This, by analogy with the conversions of dimers $(\mathrm{A^+})_2$ considered above, implies that the reaction rate as a function of time t is described by a fourth-order equation with respect to the current concentration $[\mathrm{Mn^{IV}}]_t$ and the first-order equation with respect to the initial concentration $[\mathrm{Mn^{IV}}]_0 = a$ (see phase trajectories in [145] and integral plots of $[\mathrm{Mn^{IV}}]^{-3}$ versus t in [148]).

It is interesting to note that, under conditions where tetramers predominate, the reaction is first-order with respect to initial Mn(IV) concentrations for the initial rates (see Eq. (2.297)), while in the case of the predominant existence of dimers (statistic distribution) at a small content of tetramers, the reaction over the tetramer route is second-order with respect to the initial concentrations of Mn(IV) ($[\mathrm{T}] \propto a^2$, see Eq. (2.293)).

At a temperature of 341 K and an initial concentration of $[\mathrm{Mn(IV)}]_0 = 0.046\,\mathrm{M}$, the kinetic curves are rectified in the coordinates of an equation of the sixth order with respect to $[\mathrm{Mn(IV)}]_t$ ($[\mathrm{Mn(IV)}]^{-5}$ versus t) [148], which was explained by the participation of hexamers in a six-election oxidation of water to $\mathrm{O_3}$. Simple linear anamorphoses of the kinetic curves in cases of tetramers and hexamers showed evidence that the distribution of associates for $\mathrm{Mn^{4+}}$ and $\mathrm{Mn^{3+}}$ ions was close to statistical, as was confirmed by the coincidence of the experimental points [148] with the curve of a phase trajectory [145] calculated for the statistically probable binomial distribution of $\mathrm{Mn^{4+}}$ and $\mathrm{Mn^{3+}}$ in $\mathrm{Mn_4}$ associates. In the case of tetramers, the problem is also solvable in the general (rather than only statistical) case, which allows the composition of active associates to be determined.

An analysis of the results of investigations of the kinetics of $\mathrm{N_2}$ and $\mathrm{H^+}$ reduction by vanadium(II) in the model systems $\mathrm{VCl_2}$ – 3,5-di-*tert*-butylpyrocatechol – $\mathrm{NaOCH_3}$ – MeOH (I) and $\mathrm{VCl_2}$ – pyrocatechol – $\mathrm{NaOCH_3}$ – MeOH (II) [150] showed that the formation of $\mathrm{H_2}$ (system I) and $\mathrm{NH_2NH_2}$ (system II) involves $(\mathrm{V^{2+}})_4$ tetramer complexes that appear as kinetically independent species. In the case of ammonia formation, the kinetic analysis led to a conclusion that $(\mathrm{V^{2+}})_8$ octamers could probably also participate in the reaction

$$(\mathrm{V^{2+}})_8 + \mathrm{N_2} + 8\mathrm{H^+} \longrightarrow (\mathrm{V^{3+}})_8 + 2\mathrm{NH_3} + \mathrm{H_2},$$

the order of which with respect to $[\mathrm{V^{2+}}]$ was close to 8.

It should be borne in mind that the presence of monomers (in addition to a set of associates $\mathrm{M_2}$, $\mathrm{M_3}$, $\mathrm{M_4}$ etc.) and a nonstatistical distribution of $\mathrm{A^+}$ and A over these associates significantly complicate both the determination of rate constants of elementary steps and the analysis of hypotheses concerning the participation of various associates in the steps of electron transfer as manifested by the integral concentration versus time curves or their linear anamorphoses. Therefore, an analysis of the φ versus α curves can provide useful information in addition to the analysis of usual kinetic curves and initial reaction rates [145].

A general approach to description of the kinetics of catalytic reactions under the conditions of formation of homo- and heteroassociates of reagents and catalyst was developed in the course of investigation of the reaction between RNCO isocyanates and ROH alcohols in nonpolar media (heptane), catalyzed by organotin compounds [153]. In particular, it was established that the limiting step of the process

$$\underset{(A)}{n\text{-BuOH}} + \underset{(B)}{\text{RNCO}} \xrightarrow{Q} \underset{(P)}{\text{RNHCOOBu}} \qquad (2.298)$$

(where $Q = Bu_2SnX_2$) involves an alcohol dimer (A_2) as

$$A_2 + B + Q \overset{K}{\rightleftharpoons} A_2BQ \xrightarrow{k_{lim}} P,$$

and the material balance with respect to catalyst $[Q]_0$ and alcohol $[A]_0$ includes heteroassociates (AQ and A_4Q_4) and homoassociates (A_2 and A_4):

$$[Q]_0 = [Q] + K_{11}[A][Q] + 4K_{44}[A]^4[Q]^4;$$
$$[A]_0 = [A] + 2K_1[A]^2 + 4K_4[A]^4.$$

A mathematical apparatus was proposed that allowed, for a given $[B]_0$ and $[Q]_0 \ll [A]_0$, to use the dependences of the observed first-order reaction constant with respect to [B] ($k_{obs} = r/[B]$) on $[Q]_0$ (at $[A]_0 = \text{const}$) and on $[A]_0$ (for $[Q]_0 = \text{const}$) for determining the compositions of intermediates in the limiting step and the compositions of A_pQ_s associates. The alcohol homoassociates were studied by the IR spectrophotometry and it was established [154–157] that, for all possible types of tin-containing catalysts $R_{4-x}SnX_x$ ($x = 0, 1, 2, 3, 4$), the intermediate compound represents an A_2BQ complex, while the compositions of A_pQ_s associates depend on the nature of Sn(IV) compounds:

for Bu_4Sn: AQ, A_2Q_2;

for Bu_3SnCl: AQ, A_4Q_4;

for Br_4Sn: A_8Q_2.

A solution of the inverse problem of kinetics for a set of equations including an equation for the reaction rate and the equations of material balance with respect to A, B, and Q (with various initial concentrations of reactants and catalysts) allows the equilibrium constants and the Kk_{lim} product to be determined.

2.2.2 Systems with mononuclear and polynuclear complexes of various types

In an analysis of the kinetic features of systems containing associates, which was presented in Section 2.2.1, it was pointed out that a unified material balance equation should be used. The task of obtaining this equation is the most difficult part of constructing kinetic models for these systems. This requirement naturally refers to all systems containing polynuclear complexes.

Example 2.40

► The study of the reaction of oxidative decarboxylation of acetic acid by Ce(IV) in aqueous H_2SO_4 solutions in the presence of catalysts – chloride complexes of Rh(III), Ru(III) and Ir(III) at very low concentrations (10^{-4}–10^{-5} M) – showed evidence for variable reaction orders (below unity) with respect to CH_3COOH (S) and metal M(III) [158]. Under assumptions that intermediate acetate complexes are formed as

$$\text{M(III)} + CH_3COOH \overset{K}{\rightleftharpoons} \text{M}(OCOCH_3)^- + H^+$$

and polynuclear M(III) complexes are present in solution according to a (rather low probable) equilibrium

$$mM(III) \overset{K_{eq}}{\rightleftharpoons} M(III)_m,$$

the classical complexity function was defined as

$$F_M = \frac{[M]_\Sigma}{[M]} = 1 + \frac{K[S]}{[H_3O^+]} + mK_p[M]^{m-1}, \tag{2.299}$$

and used to write the following kinetic equation for the processing and interpretation of experimental data:

$$-\frac{d\ln[Ce^{IV}]}{dt} = k_{obs} = \frac{kK[M]_\Sigma[S]}{[H_3O^+] + K[S] + mK_p[M]^{m-1}[H_3O^+]}. \tag{2.300}$$

However, since the concentration $[M] = f([M]_\Sigma, [S], [H_3O^+])$ was unknown, the processing of results (constructing linear anamorphoses) [158] based on Eq. (2.300) made no sense. ■

Erroneous results are also obtained when two complexity functions are used in a system involving mononuclear and polynuclear complexes (see also Section 2.2.1).

Example 2.41

► In the study of the kinetics of cyclohexene (CH) hydrogenation in a Ziegler type system $Cp_2TiCl_2-LiC_4H_9$ [159, 160], the experimental kinetic equation

$$r = \frac{k'[CH][P_{H_2}]^{3/2}\sqrt{[Ti]_\Sigma}}{1 + K^*[CH]}, \tag{2.301}$$

was compared to the following mechanistic scheme:

$$Cp_2TiCl_2 + LiC_4H_9 \longrightarrow Cp_2Ti\text{-}TiCp_2;$$

$$Cp_2Ti\text{-}TiCp_2 + H_2 \overset{K_1}{\rightleftharpoons} (Cp_2TiH)_2;$$

$$(Cp_2TiH)_2 \overset{K_2}{\rightleftharpoons} 2Cp_2TiH;$$

$$Cp_2TiH + CH \overset{K_3}{\rightleftharpoons} \pi\text{-complex}$$

$$\pi\text{-complex} + H_2 \overset{k}{\longrightarrow} Cp_2TiH + C_6H_{12}.$$

This comparison suggests that the denominator in Eq. (2.301) represents the complexity of a monomer form of Cp_2TiH bound to CH (balance with respect to the monomeric form), while the square root in the numerator represents the balance with respect to dimers under the assumption that $2[(Cp_2Ti)_2] \cong [Ti]_\Sigma$ and $K^* = K_3$. The experimental relation (2.301) also indicates that the reaction orders with respect to $[Ti]_\Sigma$ and P_{H_2} are independent of [CH], while the order with respect to [CH] is independent of the concentrations of titanium and hydrogen. However, if the common material balance including dimers, monomers, and π complexes is written as

$$[Ti]_\Sigma = 2[(Cp_2Ti)_2] + 2[Cp_2TiH]_2 + [Cp_2TiH] + [Cp_2TiH \cdot CH], \tag{2.302}$$

then Eq. (2.301) cannot be obtained. When $[\mathrm{Ti}]_\Sigma \cong 2[(\mathrm{Cp_2Ti})_2]$, then $[\mathrm{Cp_2TiH}] = \sqrt{K_1K_2\frac{[\mathrm{Ti}]_\Sigma}{2}}$, the concentration of a monomer form is very small (so that the denominator in the kinetic equation is unity), and the contribution of π complexes to the material balance is insignificant. However, if the concentrations of the monomeric form and π complex are comparable with $[\mathrm{Ti}]_\Sigma$, then [CH] appears in the denominator, but the reaction is first-order with respect to $[\mathrm{Ti}]_\Sigma$. Under the assumption that Eq. (2.302) includes only contributions due to the initial dimer and π complex, the reaction order with respect to $[\mathrm{Ti}]_\Sigma$ depends on the concentration of cyclohexene and hydrogen and the order with respect to CH depends on the concentrations of titanium and hydrogen. Thus, neither the use of any assumptions within the framework of the proposed scheme nor the use of Eq. (2.302) without additional assumptions can lead to Eq. (2.301). This experimental equation apparently only provides a mathematical approximation of partial dependences that do not lead to the proposed reaction scheme.

A relation of the type of Eq. (2.301) with independent fractional orders with respect to olefin, hydrogen, and catalyst can probably be obtained as a consequence of the quasi-steady-state character at FM $\cong$ 1, but with a different reaction scheme. ■

Example 2.42

► The interaction of $\mathrm{RuCl_2(PPh_3)_3}$ with H_2 in dimethylacetamide (80 °C, 1 bar H_2) leads to the reduction of Ru(II) to Ru(I), which is capable of catalyzing the hydrogenation of olefins, e.g., fumaric and maleic acids [161, 162]. The experimental data are described by an equation:

$$-\frac{d[\mathrm{H_2}]}{dt} = \frac{k_1k_2K'[\mathrm{H_2}][\mathrm{Ol}][\mathrm{Ru^I}]_\Sigma^{0.5}}{k_{-1} + k_2[\mathrm{Ol}]}, \tag{2.303}$$

which is consistent with the following scheme under conditions of quasi-steady-state process ($[\mathrm{Ru}]_\Sigma \cong 2[\mathrm{Ru^I}]_2$):

$$\begin{aligned} &\{\mathrm{Ru(I)}\}_2 \overset{K}{\rightleftharpoons} 2\mathrm{Ru(I)};\\ &\mathrm{Ru(I)} + \mathrm{H_2} \underset{k_{-1}}{\overset{k_1}{\rightleftharpoons}} \mathrm{Ru(III)(H)_2};\\ &\mathrm{Ru(III)(H)_2} + \mathrm{Ol} \xrightarrow{k_2} \cdots \longrightarrow \mathrm{Ru(I)} + \mathrm{P}. \end{aligned} \tag{2.304}$$

■

Example 2.43

► Investigation of the kinetics of olefin arylation by benzene or naphthalene in glacial acetic acid at 70 °C in the presence of $\mathrm{Pd_3(OAc)_6}$ (Motitani–Fujiwara reaction)

$$\mathrm{ArH + RCH{=}CH_2 + Pd(OAc)_2 \longrightarrow ArCH{=}CHR + Pd^0 + 2HOAc} \tag{2.305}$$

led to the following kinetic equation [163, 164]:

$$r = \frac{k_1[\mathrm{Pd^{II}}]_\Sigma^{0.5}[\mathrm{ArH}]}{K_1 + [\mathrm{ArH}]}. \tag{2.306}$$

It was suggested that palladium acetate trimers dissociate into dimer and monomer, after which the latter forms a π complex with ArH:

$$[\mathrm{Pd^{II}}]_\Sigma = 3[\mathrm{Pd_3(OAc)_6}] + 2[\mathrm{Pd_2(OAc)_4}] + [\mathrm{Pd(OAc)_2}] + [\mathrm{Pd(OAc)_2(ArH)}]. \tag{2.307}$$

If the concentration of π complex is small, $[Pd_2(OAc)_4] \cong [Pd(OAc)_2]$, and $3[Pd_3(OAc)_6] \cong [Pd^{II}]_\Sigma$, then we obtain $[Pd(OAc)_2] \propto [Pd^{II}]_\Sigma^{0.5}$. However, π complex of palladium(II) with arene does not enter into the complexity function and the reaction is first-order with respect to [ArH] (cf. Eq. (2.306)). ■

The interaction of palladium acetate with sodium acetate (and other alkali metal acetates) leads to the following equilibria [76, 126]:

$$2Pd_3(OAc)_6 + 6NaOAc \overset{K_2}{\rightleftharpoons} 3Na_2Pd_2(OAc)_6; \tag{2.308}$$

$$Na_2Pd_2(OAc)_6 + 2NaOAc \overset{K_3}{\rightleftharpoons} 2Na_2Pd(OAc)_4 \tag{2.309}$$

with the following equilibrium constants for the reaction in glacial acetic acid at 25 °C: $K_2 = 7.77 \cdot 10^4\ M^{-5}$ and $K_3 = 7.6 \cdot 10^{-4}\ M^{-1}$. The study of these equilibria revealed a circumstance that is very important for the kinetic investigation, according to which bridging acetate groups in both trimers and dimers are virtually inert to the substitution with other ligands, so that the establishment of equilibria (2.308) and (2.309) takes several hours [76, 126].

The equilibrium in reaction (2.240) of trimer depolymerization at 75 °C is attained within 1–2 h [127]. In the presence of NaOAc at a concentration of 0.15–0.5 M, the molar fraction of dimeric complexes for [Pd(II)] in the interval from $0.2 \cdot 10^{-2}$ to $2.0 \cdot 10^{-2}$ M and [NaOAc] amounts to ~0.91.

Example 2.44

▶ Under conditions of the predominant formation of dimers, the rate of reaction (2.305) is proportional to the first power of $[Pd(II)]_\Sigma$ and is described by the equation

$$r = \frac{k_2 K_2 [Na_2Pd_2(OAc)_6]_t [C_6H_6]}{K_2[C_6H_6] + [NaOAc]}, \tag{2.310}$$

which agrees with the following scheme [163, 164]:

$$\begin{aligned}
&(1)\quad Na_2Pd_2(OAc)_6 + C_6H_6 \overset{K_2}{\rightleftharpoons} NaPd_2(OAc)_5(C_6H_6) + NaOAc;\\
&(2)\quad NaPd_2(OAc)_5(C_6H_6) \xrightarrow[NaOAc]{k_2} Na_2Pd_2(OAc)_5(C_6H_5) + HOAc;\\
&(3)\quad Na_2Pd_2(OAc)_5(C_6H_5) + C_6H_5CH{=}CH_2 \xrightarrow{\text{Fast}}\\
&\longrightarrow C_6H_5CH{=}CHC_6H_5 + Pd^0 + NaOAc + HOAc + {}^1\!/_2 Na_2Pd_2(OAc)_6.
\end{aligned} \tag{2.311}$$

The complexity function of the dimeric complex is expressed as follows:

$$F_{Pd_2} = 1 + K_2\frac{[C_6H_6]}{[NaOAc]}. \tag{2.312}$$

In a consecutive-parallel process of the arylation of ethylene by benzene to styrene and the conversion of styrene into stilbene [165], the products (styrene and stilbene) are formed virtually simultaneously, since already at small concentrations of styrene the rate of stilbene formation differs but little from the rate of styrene formation. This is caused by close values of the concentrations of ethylene and styrene, while the rate of benzene consumption is determined by the concentrations of benzene and $[Pd_2(II)]_2$. ■

Example 2.45

► The kinetics of styrene arylation by phenylmercury acetate under the conditions of existence of trimers (at [NaOAc] = 0) and dimeric complexes of palladium is described by first-order equations with respect to $[\mathrm{Pd(II)}]_\Sigma$, while the dependence on the olefin concentration is more complicated [164]:

$$r_0 = (k_3 + k_2'[\mathrm{Ol}])[\mathrm{Pd^{II}}]_\Sigma[\mathrm{PhHgOAc}]_0 \ \text{(for [NaOAc] = 0)};$$

$$r_0 = (k_4 + k_4'[\mathrm{Ol}])\frac{[\mathrm{Pd(II)}]_\Sigma[\mathrm{PhHgOAc}]_0}{[\mathrm{NaOAc}]} \ \text{(for [NaOAc]} \neq 0).$$

It was found [164] that the arylation of olefins according to the reaction

$$\mathrm{PhCH{=}CH_2 + PhHgOAc + Pd(OAc)_2} \longrightarrow \mathrm{PhCH{=}CHPh + Hg(OAc)_2 + Pd^0 + AcOH}$$

under conditions of the zero-order process with respect to olefin proceeds much faster than the oxidative coupling of arenes with the formation of Ar_2 (in the absence of olefin under otherwise identical conditions). An interesting hypothesis was proposed that explained this fact by the participation of different σ-aryl intermediates in these reactions according to the following scheme:

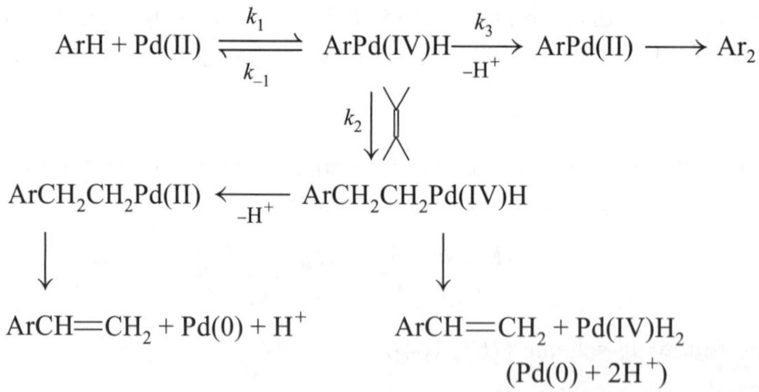

(2.313)

In the presence of olefin, we have $k_2[\mathrm{Ol}] \gg k_{-1}$ and the reaction rate is determined by the first step (k_1). In the absence of olefin, the rate of ArPd(II) complex formation is determined by the value of k_1k_3/k_{-1}. For $k_3/k_{-1} \cong 10^{-4}$, the ratio of the rates of diphenyl and styrene formation is close to that observed in experiment. ■

As was noted above (see Section 2.2.1), the rate of the oxidative dimerization of styrene in the $Pd_3(OAc)_6$–CH_3COOH system is described by an equation of the first order with respect to $[\mathrm{Pd^{II}}]_\Sigma$. The reaction rate sharply drops upon the addition of NaOAc and is close to zero at [NaOAc] = 1.2 M (for $[\mathrm{Pd^{II}}] = 1.5 \cdot 10^{-4}$ M, 25 °C) at a $Na_2Pd(OAc)_4$ concentration of ~82 % [166].

Since the material balance in the presence of trimers, dimers, and π complexes of dimers with olefins in reaction (2.241) is rather complicated, it was suggested [166] to use a special approach to determining the contributions of the catalysis by trimers and dimeric acetate complexes to the total rate of this reaction. According to this, the concentrations of Pd(II) and NaOAc in each experiment are varied with allowance for the equilibrium constants of reactions (2.308) and (2.309) so as to ensure a constant concentration of one of these complexes in two series of experiments. The observed reaction rate is described by the following equation:

$$r_0 = k_2[\mathrm{Na_2Pd_2(OAc)_6}] + k_3[\mathrm{Pd_3(OAc)_6}]. \qquad (2.314)$$

The values of k_2 and k_3 in the series of experiments with a constant concentration of dimers [about $(21.1-21.9) \times 10^{-5}$ M] at a variable concentration of trimers [within $(1.11-35.1) \times 10^{-5}$ M] and in the series with a constant concentration of trimers [about $(15.6-16.4) \times 10^{-5}$ M] at a variable concentration of dimers [within $(1.25-49.0) \times 10^{-5}$ M] were rather close, so that their average values were $k_2 = (4.5 \pm 1.0) \times 10^{-6}\,\text{s}^{-1}$ and $k_3 = (2.67 \pm 0.17) \times 10^{-5}\,\text{s}^{-1}$.

In the presence of olefin, the dimeric species occur predominantly in the form of π complex. For this reason, k_2 is independent of the concentration of styrene (St), while k_3 is a linear function of [St]. Accordingly, Eq. (2.314) should be rewritten with a sum of two terms that correspond to the mono- and bimolecular routes:

$$r_0 = k_2[\text{Na}_2\text{Pd}_2(\text{OAc})_5\text{St}] + k_3'[\text{Pd}_3(\text{OAc})_6][\text{St}]. \tag{2.315}$$

In contrast to the oxidative dimerization (2.241) of olefins, the exchange of carboxylate groups in vinyl esters according to the reaction

$$\text{AcO}^- + \text{CH}_2\text{=CHOCOC}_2\text{H}_5 \rightleftharpoons \text{CH}_2\text{=CHOAc} + \text{C}_2\text{H}_5\text{COO}^-, \tag{2.316}$$

catalyzed by Pd(II) under the same conditions [123, 167] showed a higher activity of dimers without a change in the reaction order with respect to [St] on the passage from dimers to trimers:

$$r = k_d[\text{Na}_2\text{Pd}_2(\text{OAc})_6][\text{St}] + k_t[\text{Pd}_3(\text{OAc})_6][\text{St}]. \tag{2.317}$$

It is interesting to note that, in the absence of NaOAc, trimer is decomposed to Pd(OAc)_2, the π complexes of which is virtually do not participate in the oxidation reactions [166]. A change (ΔA_{360}) in the optical density of π complexes in solution during the variation of $[\text{Pd}^{\text{II}}]$ is proportional to the cubic root of $[\text{Pd}^{\text{II}}]$: ($\Delta A = a[\text{Pd}^{\text{II}}]_\Sigma^{1/3}$).

In benzene, the oxidative dimerization of styrene is catalyzed by $\text{Pd}_3(\text{OAc})_6$ in a temperature interval of 40–80 °C. In the presence of oxygen, the process is partly catalyzed by Pd(II) [168]. A quasi-steady state of the process (up to about 40 % conversion) is described by the following kinetic equation:

$$r = \frac{(k_1 + k_2[\text{St}])[\text{Pd}^{\text{II}}]_0[\text{St}]}{K + [\text{St}]}. \tag{2.318}$$

Example 2.46

► The importance of "pre-catalytic" information about the state of a catalytic system and the equilibrium constants of the formation of various forms of the catalyst and intermediates, which is important to obtain an adequate interpretation of the experimental data and the construction of a kinetic model, can be illustrated by the results of investigations of the kinetics of oxidative acetoxylation of olefins (Moiseev reaction) in the Pd(OAc)_2–NaOAc–HOAc–p-benzoquinone (Q) system [169–172]:

$$\text{C}_2\text{H}_4 + \text{HOAc} + \text{Q} \longrightarrow \text{CH}_2\text{=CHOAc} + \text{QH}_2. \tag{2.319}$$

After the discovery of this reaction [15, 173], it was also the first acetate system for which the kinetics and equilibria of the formation of π complexes were studied. It was established that, in the presence of excess sodium acetate, the reaction is described by an equation of the first order with respect to $[\text{Pd(II)}]_\Sigma$ (at $[\text{Pd(II)}]_\Sigma/[\text{C}_2\text{H}_4]_\Sigma \cong 5$) and with respect to $[\text{C}_2\text{H}_4]_\Sigma$ (in excess of ethylene):

$$r = k_2[\text{Pd(II)}]_\Sigma[\text{C}_2\text{H}_4]_\Sigma. \tag{2.320}$$

The most important information for the interpretation of experimental data on the kinetics and equilibria of the formation of π complexes was related to the dependence of k_2 on $[\text{NaOAc}]_\Sigma$ in the interval of concentrations within 0.02–1.0 M, which exhibited an extremal shape with a sharp maximum at 0.2 M and a nearly quadratic inhibition by [NaOAc] at concentrations in the region of 0.2–1.0 M. The initial growth in the reaction rate until reaching the maximum was interpreted as being due to decomposition of the inactive palladium(II) associates with the formation of active monomeric acetate complexes:

$$\frac{1}{n}[Pd(OAc)_2]_n + 2NaOAc \rightleftharpoons Na_2Pd(OAc)_4. \tag{2.321}$$

This interpretation provided a basis for investigations of the equilibrium π complex formation and for the description of reaction kinetics [169–172]. According to these notions, the main intermediate is a π complex formed via the following reaction [172]:

$$C_2H_4 + Na_2Pd(OAc)_4 + HOAc \overset{K}{\rightleftharpoons} \underset{\pi\text{-complex}}{(C_2H_4)Pd(OAc)_2 \cdot HOAc} + 2NaOAc \tag{2.322}$$

Under the condition of $F_{C_2H_4} \cong 1$, the complexity of Pd(II) according to reaction (2.322) is expressed as follows:

$$F_{Pd} = 1 + K_1 \frac{[C_2H_4]_\Sigma}{[NaOAc]^2_\Sigma}, \tag{2.323}$$

where $[Pd]_\Sigma = [Na_2Pd(OAc)_4] + [\pi\,\text{complex}]$ and $K_1 = K[HOAc]$. In this case, the reaction rate can be expressed either via the total concentrations of participants and the F_{Pd} value as

$$r = \frac{k_{obs}[Pd^{II}]_\Sigma [C_2H_4]_\Sigma}{[NaOAc]^2_\Sigma \left(1 + K_1 \dfrac{[C_2H_4]_\Sigma}{[NaOAc]^2_\Sigma}\right)}, \tag{2.324}$$

or via the quasi-equilibrium concentrations of all participants as

$$r = \frac{kK_1[Na_2Pd(OAc)_4][C_2H_4]}{[NaOAc]^2}, \tag{2.325}$$

where k is the rate constant of the limiting step of conversion of the π complex (as a result of the acetoxypalladation of ethylene) into a σ organometallic intermediate compound whose decay leads to the formation of vinyl acetate.

An alternative interpretation of the kinetics of vinyl acetate synthesis [76, 123] was proposed in 1974 based on the results of the investigations of equilibria [reactions (2.308) and (2.309)] in the $Pd(OAc)_2$–NaOAc–HOAc system. According to the data obtained by Henry et al. [123, 126], the maximum on the k_2 versus [NaOAc] curve is attained at a concentration of sodium acetate that corresponds to the maximum concentration of $Na_2Pd_2(OAc)_6$ dimeric complexes. In this case, active species in the synthesis of vinyl acetate are represented by the dimeric complexes rather than by $Na_2Pd(OAc)_4$. However, a decrease observed in the reaction rate with the further growth in [NaOAc] proved to be more rapid than could be expected in the case of the active dimer conversion into an inactive monomeric acetate complex. Calculations [123] showed that the experimental results obtained in [169–173] could be described by the following equation:

$$r = \frac{k[Na_2Pd_2(OAc)_6][C_2H_4]_\Sigma}{[NaOAc]_\Sigma}. \tag{2.326}$$

In the investigations of kinetics of the exchange reaction (2.316) with a close mechanism [123, 167], the most active species also represented dimeric complexes (more active than trimer), while the $Na_2Pd(OAc)_4$

complex was virtually inactive. It is interesting to note that the rate constants k_d and k_t in the corresponding kinetic equation (2.317) are independent of [NaOAc]. This is explained by the fact that, since the formation of a π complex from dimeric species of Pd(II) requires a free coordination site and the reaction of vinyl propionate acetoxypalladation is a step in the process under consideration, the concentration of sodium acetate (involved in extra-spheric attack on the π complex) in the numerator and denominator of expression (2.317) is canceled. For the reaction scheme with dimers

$$Na_2Pd_2(OAc)_6 + CH_2{=}CHX \overset{K}{\rightleftharpoons} NaPd_2(OAc)_5(CH_2{=}CHX) + NaOAc;$$

$$NaPd_2(OAc)_5(CH_2{=}CHX) + NaOAc \xrightarrow{k_{lim}} NaPd_2(OAc)_5\left(CH_2CH\begin{matrix}\diagup OAc\\ \diagdown X\end{matrix}\right)^- Na^+;$$

we obtain

$$r = k_{lim}K[Na_2Pd_2(OAc)_6][CH_2{=}CH{-}X], \qquad (2.327)$$

where $k_{lim}K = k_d$. A comparison of Eqs (2.326) and (2.327) under the assumption that the mechanisms of vinyl propionate and ethylene acetoxypalladation on the dimeric acetate complex are the same leads to a conclusion that a limiting step in the ethylene oxidation is the decay (rather than the formation) of σ,β-acetoxyethylpalladium, which is inhibited by NaOAc [123] [see Eq. (2.326)]. According to this approach, a general scheme of the mechanism of reaction (2.319) can be represented by the following sequence of steps [123]:

$$Na_2Pd_2(OAc)_6 + C_2H_4 \overset{K_1}{\rightleftharpoons} NaPd_2(OAc)_5(C_2H_4) + NaOAc;$$

$$NaPd_2(OAc)_5(C_2H_4) + NaOAc \overset{K_2}{\rightleftharpoons} Na_2Pd_2(OAc)_5(CH_2CH_2OAc);$$

$$Na_2Pd_2(OAc)_5(CH_2CH_2OAc \overset{K_D}{\rightleftharpoons} NaPd_2(OAc)_4(CH_2CH_2OAc) + NaOAc;$$

$$NaPd_2OAc_4(CH_2CH_2OAc) \xrightarrow{K_{lim}} \underset{(I)}{NaPd_2(OAc)_4(CH_2{=}CHOAc)(H)};$$

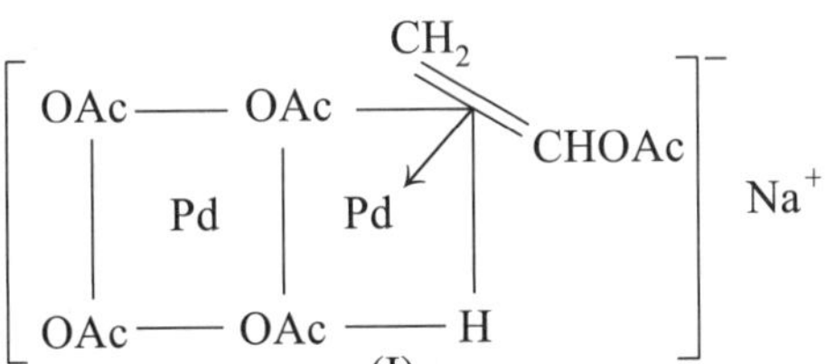

(I)

The higher activity of trimer species according to Eq. (2.314) as compared to the activity of $Pd(OAc)_2$ and $Na_2Pd_2(OAc)_6$ dimeric anion complexes in the reactions of oxidative dimerization of olefins [166] was related to the mechanism of electrophilic addition of Pd_3 at the double bond with the formation of a carbenium ion:

$$R_1R_2{=}CH_2 + Pd_3(OAc)_6 \longrightarrow R_1R_2C^+{-}CH_2Pd_3(OAc)_6^-$$
$$\longrightarrow R_1R_2C{=}CHPd_3(OAc)_5 + HOAc. \qquad (2.328)$$

The reaction of the oxidative coupling of olefins actually proceeds faster in the presence of two alkyl and one aryl, or two aryl groups in the olefin molecule, which stabilize the carbocation. In this context, it is also interesting to mention the direct mercuration of 1,1-diphenylethylene by mercury trifluoroacetate [174, 175]:

$$Ph_2C{=}CH_2 + Hg(OCOCF_3)_2 \longrightarrow Ph_2C{=}CHHg(OCOCF_3) + CF_3COOH. \qquad (2.329)$$

The above data and examples show that the kinetics and mechanism of the Moiseev reaction (oxidative acetoxylation or oxidative esterification of olefins) are worthy of further investigation. Quite recently, a study of the oxidative carboxylation of cyclohexene and linear olefins provided justification for a new scheme involving the steps of carbenium ion formation [176, 177]. ■

A formal kinetic description of catalytic reactions involving metal clusters (complexes with M–M bonds) does not differ from the description of reactions catalyzed by other polynuclear complexes, in particular, associates. If a cluster does not decay into low-molecular components during the reaction, then its kinetic behavior is essentially the same as that of mononuclear complexes. A characteristic example is offered by the catalysis with rhodacarborane clusters [178, 179].

Example 2.47

▸ A study of the kinetics of hydrogenation and isomerization of olefins in solutions of closo-rhodacarborane clusters, in combination with NMR measurements and the use of deuterium-labeled complexes, D_2 and D_20, allowed Behnken et al. [178, 179] to establish some specific features in the functioning of carborane cluster ligands. In particular, the equilibrium of closo–$HRh^{III}(PPh_3)_2(C_2B_8H_{10})$ (**I**) and exo–nido $Rh^I(PPh_3)_2{}^+(C_2B_8H_{11})^-$ (**II**) structures was observed (Figure 2.5), where the initial hydride ligand (HRh^{III} bond) did not participate in the steps of hydrogenation and isomerization, while the carboranyl ligand remained bound to Rh(I) via one or two B–H–Rh bridges so that the Rh atom "traveled" over the cluster surface.

The hydrogenation of 3-methyl-3-phenyl-butene (A) with a catalyst precursor (**I**) in tetrahydrofuran (THF) at 41 °C is described by a simple kinetic equation, in which intermediate compounds **1–5** (Figure 2.6) do not significantly contribute to the material balance ($F_{Rh} \cong 1$):

$$-\frac{d[H_2]}{dt} = \frac{[Rh^{III}]_\Sigma[A][H_2]}{[PPh_3](m + c[H_2])}. \qquad (2.330)$$

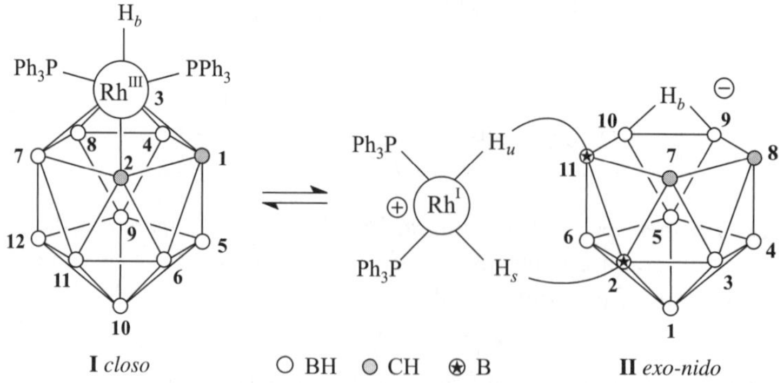

Figure 2.5 Reductive elimination with transfer of bridging hydrogen atom H_b in complex II; H_u and H_s are hydrogen atoms bound to coordination-unsaturated and saturated boron atoms, respectively

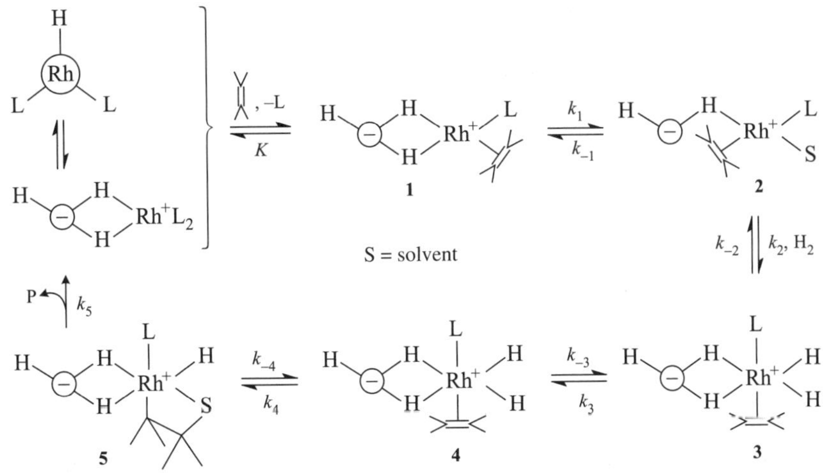

Figure 2.6 Mechanistic scheme of hydrogenation of 3-methyl-3-phenyl-1-butene and butyl acrylate (Reprinted with permission from [179]. © 1979 American Chemical Society)

This equation is consistent with the mechanism of a quasi-steady-state process, in which the first step is the quasi-steady-state formation of π complex **1** (with carboranyl ligand). The reaction kinetics was studied by varying or monitoring the concentration of PPh_3 (L).

In the reaction of butyl acrylate (C) hydrogenation with the same catalyst precursor **I**, the character of the material balance changes, since the interaction of **I** with butyl acrylate during a long induction period leads to the formation of a dead-end complex **VI** as a result of the acrylate insertion into the Rh(III)–H bond:

$$\mathbf{I} + \mathrm{C} \underset{}{\overset{K_{\mathrm{VI}}}{\rightleftharpoons}} \mathbf{VI} + \mathrm{L}; \tag{2.331}$$

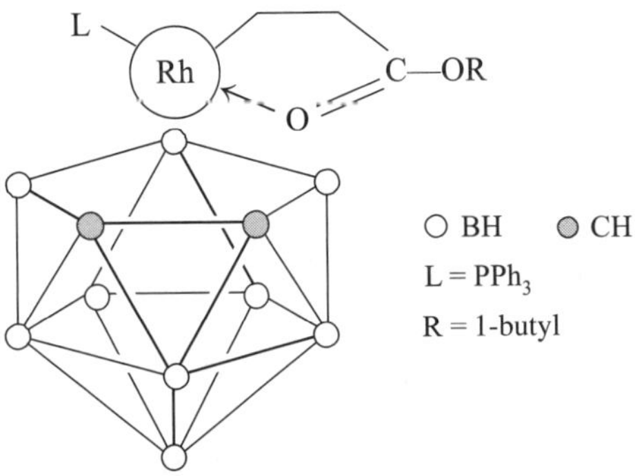

VI

It should be emphasized that the concentration of L here is determined by reaction (2.331) rather than by the first step in the hydrogenation scheme

$$\mathbf{I} + \mathrm{C} \overset{K}{\rightleftharpoons} \mathbf{1} + \mathrm{L}; \tag{2.332}$$

$$\mathbf{1} \underset{k_{-1}}{\overset{k_1}{\rightleftharpoons}} \mathbf{2} \underset{k_2}{\overset{H_2}{\longrightarrow}} \cdots \longrightarrow \mathrm{P} + \mathbf{I}. \tag{2.333}$$

At constant P_{H_2}, the reaction rate is described by the following equation:

$$-\frac{d[H_2]}{dt} = k[C]^{1/2}[Rh]_{\Sigma}^{1/2}. \tag{2.334}$$

According to reaction schemes (2.332) and (2.333), we have

$$-\frac{d[H_2]}{dt} = \frac{k_1 k_2 K[\mathbf{I}][C]}{[L](k_{-1} + k_2[H_2])}. \tag{2.335}$$

For $[H_2] = \text{const}$, $[\mathbf{VI}] \cong [L]$; $[L] = \sqrt{[\mathbf{I}][C]K_{VI}}$, and $[\mathbf{I}] \cong [Rh]_{\Sigma}$, this equation reduces to

$$-\frac{d[H_2]}{dt} = k_{obs}[C]^{1/2}[Rh]_{\Sigma}^{1/2},$$

i.e., is equivalent to Eq. (2.334).

However, the processing of experimental data showed that neglecting the contribution from complex **VI** to the material balance with respect to $[Rh]_{\Sigma}$ impairs the description of results. For this reason, once the concentration of VI was known from NMR data, the results were processed and constants evaluated using an equation that was corrected as follows:

$$-\frac{d[H_2]}{dt} = k'_{obs}\frac{\{[Rh]_{\Sigma} - [\mathbf{VI}]\}[C]}{[L]} \tag{2.336}$$

at $[L] \cong [VI]$. For strictly writing the kinetic equation, it is necessary to express [I] from the relation

$$[Rh]_{\Sigma} = [\mathbf{I}] + \sqrt{K_{VI}[C][\mathbf{I}]} \tag{2.337}$$

and substitute it into Eq. (2.335) that describes a complicated nonlinear character of the reaction kinetics, which is only related to the nature of a single step (2.331). ■

In the case of very large clusters, especially for giant clusters of the types $Pd_{561}Phen_{60}(OAc)_{180}$ (A) or $Pd_{561}Phen_{80}(PF_6)_6$ (B) [180, 181], the number of sites accessible for the coordination of substrates on the surface of a metal polyhedron sharply increases and these sites exhibit specificity with respect to different substrates. In these cases, the fractions of occupied coordination sites can be described by dependences of the Langmuir adsorption isotherm type, and the complexity can represent a product of polynomials.

Example 2.48

► An interesting kinetic equation was obtained [181, 182] for the catalyzed reaction

$$C_3H_6 + {}^1\!/_2O_2 + AcOH \longrightarrow CH_2{=}CHCH_2OAc + H_2O \tag{2.338}$$

in solutions of giant clusters A and B in AcOH. Under conditions of the constant catalytic activity at $P_{C_3H_6} \geq 0.07$ MPa and 60 °C, the rate of reaction (2.338) is described by the following equation:

$$r_0 = \frac{k[Pd]_{\Sigma}[C_3H_6][O_2][AcOH]}{(K_1 + [O_2])(K_2 + [AcOH])(K_3 + [C_3H_6])}. \tag{2.339}$$

A high reaction rate is achieved for the simultaneous coordination of three substrate molecules. Formally, Eq. (2.339) is analogous to the equations of heterogeneous kinetics for quasi-homogeneous surfaces, where the active centers of each kind obey the Langmuir adsorption laws. It is also possible that the denominator in

Eq. (2.339) represents one approximation of the complexity function expressed as a product of polynomials [see Eq. (2.59)].

The kinetics of reactions involving nanoclusters is undoubtedly worthy of further investigation, as well as the formal kinetic approaches to the study of these boundary systems between homo- and heterogeneous catalysis (microheterogeneous catalysis). Analogous kinetic features will probably take place for the homogeneous catalysis with fullerenes. ■

Additional difficulties in the description of reaction kinetics are also encountered in the case of polynuclear catalytic systems, where the formation and fragmentation of clusters enter as stages into the evolution of a catalytic system from the active to inactive state and vice versa, rather than represent equilibrium steps. These issues are considered in more detail in Chapter 5.

2.3 Catalysis with polynuclear copper(I) halide complexes in superconcentrated solutions

It has been shown in the preceding sections that the task of constructing a kinetic model for processes in solutions of polynuclear complexes, with a view to determining the reaction orders with respect to reactants, assessing the composition of active complexes, and evaluating the partial rate constants, is very complicated and has no simple answer. Once the compositions of complexes in solution and the equilibrium constants of transitions between complexes are known and the structure of the complexity function F_M is relatively simple, then the concentrations of complexes that enter into this function can be calculated and the contributions of each complex to the total reaction rate can be estimated (see Section 2.2.2). However, in order to provide for the adequacy of a model, it was found useful to maintain the concentration of one (of the two present) active complexes constant for determining the partial rate constants even in these relatively simple cases [166, 167].

Another approach to the study of kinetics in systems with polynuclear complexes is based on maintaining a constant concentration of all metal complexes in solution by variation of the concentrations of reactants and other participants of a reaction under consideration. A convenient model for demonstrating this approach is offered by the $CuCl-MCl-H_2O$ and $CuCl-MCl-HCl-H_2O$ ($M = Na, K, NH_4$) systems that were proposed by Nieuwland in 1929 as industrial catalysts (Nieuwland catalyst) for the dimerization of acetylene (Nieuwland process)

$$2C_2H_2 \longrightarrow CH_2{=}CHC{\equiv}CH \qquad (2.340)$$

and hydrochlorination of acetylene compounds [7, 8, 23, 183–185]. The main feature of these systems is the presence of a large number of polynuclear copper complexes $Cu_mCl_n^{(n-m)-}$, which exhibit catalytic properties in superconcentrated solutions containing (e.g., for $M = NH_4$) from 20 up to 56 mass % CuCl and from 20 to 30 % mass % NH_4Cl.

In the last quarter of the 20th century, investigations into the synthesis and crystalline structure of copper-containing complexes (binary chlorides as well as π and σ organometallic compounds) were performed [23, 186–189], which confirmed the existence of polynuclear Cu(I) complexes and stimulated new discussion [23] about the results obtained in the chemistry of acetylene and its derivatives [8, 185].

The aforementioned systems catalyze numerous reactions of alkynes [8, 23, 185], 1,3 and 1,2-dienes [8, 184, 190, 191], hydrolysis of allyl chlorides [184], and some other processes. The ecologically safe and low-waste synthesis of 2-chloro-1,3-butadiene (a monomer for the production of rubbers such as neoprene and chloroprene) from vinyl acetylene is still economically competitive.

2.3.1 Copper(I) chloride complexes in solution and in crystalline state

Systems involving Cu(I) chloride and aqueous solutions of alkali metal chlorides (MCl), ammonium chloride, or amine hydrochloride proved to be very interesting, so that investigations of the state of copper in these solutions and the mechanisms of related catalytic reactions, which began as long ago as the end of the 19th century, are continued to the present. The study of solubility [184] and other physicochemical properties of concentrated aqueous MCl and CuCl solutions and the isolation of the products of acetylene interaction with CuCl that contain 1, 2, 3, and even 6 molecules of CuCl per C_2H_2 molecule [7, 192, 193] showed evidence that polynuclear $(Cu_mCl_n)^{(n-m)-}$ complexes were present in these systems.

2.3.1.1 Physicochemical properties of concentrated MCl–CuCl–H_2O solutions

Numerous data on the solubility of CuCl in aqueous MCl solutions are summarized in [184]. One of the most thoroughly studied systems in this respect is NH_4Cl–CuCl–H_2O [194, 195], which is characterized by a sharp increase in the solubility of CuCl with increasing concentration of NH_4Cl and decreasing molar fraction of water (Figures 2.7 and 2.8). The ratio of [CuCl] : [NH_4Cl] at 80 and 100 °C is significantly greater than unity, which is indicative of the formation of polynuclear Cu(I)–chloride complexes. Additional evidence is provided by calculations of the Bjerrum formation function $\tilde{n}$ (or the Sillen function Z) based on the results of potentiometric determination of the equilibrium ligand concentration [Cl^-]:

$$\tilde{n} = \frac{[Cl^-]_\Sigma - [Cl^-]}{[CuCl]_\Sigma}.$$

At a total NH_4Cl and NH_4NO_3 concentration of $10\,\tilde{m}$ ($\tilde{m}$ is the molality defined as the number of moles of a dissolved substance per 1000 g of solvent), the $\tilde{n}$ value varies from 2.80 to 1.73 when [CuCl] is changed from 0.5 to $10\,\tilde{m}$ and [NH_4Cl] is changed from 1.5 to $10\,\tilde{m}$ (80 °C). The dependence of $\tilde{n}$ on [CuCl] is also indicative of the formation of polynuclear copper complexes [195].

The process of CuCl dissolution in aqueous NH_4Cl solutions is weakly endothermal (with ΔH° changing from 16.7 to 4.2 kJ · mole^{-1} when the molality of NH_4Cl is increased from 1.5 to $15\,\tilde{m}$) and proceeds

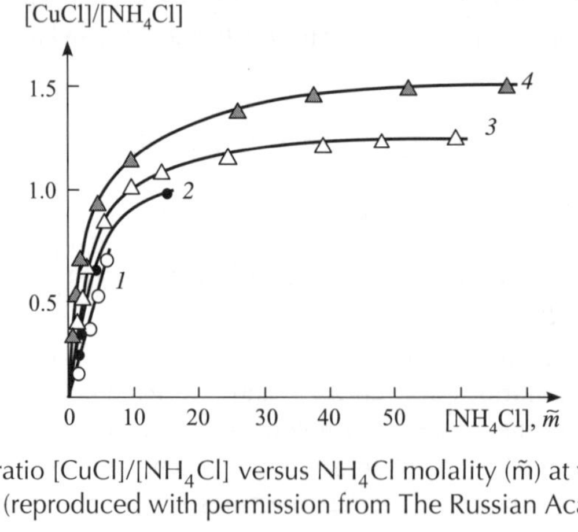

Figure 2.7 Plot of the molar ratio [CuCl]/[NH_4Cl] versus NH_4Cl molality ($\tilde{m}$) at various temperatures: (1) 25 °C; (2) 50 °C; (3) 80 °C; (4) 100 °C (reproduced with permission from The Russian Academy of Sciences and Turpion Ltd © 2000 [23])

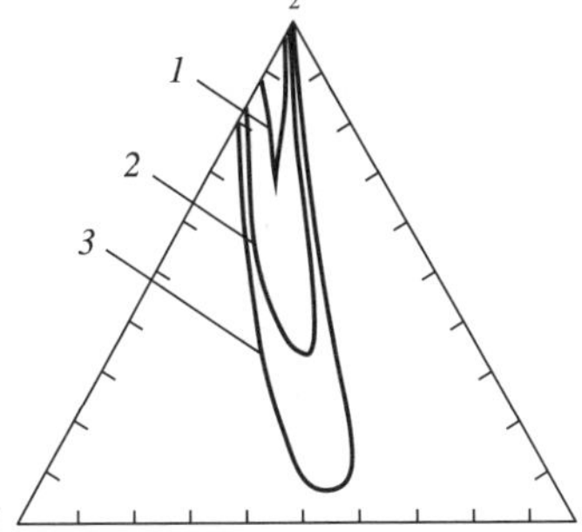

Figure 2.8 CuCl solubility triangle (molar fractions): (1) 25 °C; (2) 80 °C; (3) 100 °C (reproduced with permission from The Russian Academy of Sciences and Turpion Ltd © 2000 [23])

with increasing entropy (ΔS^{o}_{298} changes from 36 to 45 J · mole^{-1} · K^{-1}) [194]. The dissolution of CuCl is accompanied by the dehydration of chloride ions, which leads to an increase in the vapor pressure (P_{H_2O}) and activity (a_{H_2O}) of water [196]:

$$m\mathrm{CuCl_{solid}} + p\mathrm{Cl(H_2O)}_q^- \rightleftharpoons [\mathrm{Cu}_m\mathrm{Cl}_{m+p}(\mathrm{H_2O})_{pq-x}]^{p-} + x\mathrm{H_2O}.$$

In the range of [CuCl] from 0 to 10 $\tilde{m}$ (at [NH_4Cl] = 8.84 $\tilde{m}$), the activity a_{H_2O} increases from 0.74 to 0.83 (at 55 °C) and 0.88 (at 85 °C).

The study of the electronic absorption spectra (with the peaks at λ_{max} = 200, 235, and 273–400 nm) of concentrated aqueous solutions of copper(I) chloride and NH_4Cl [197] showed that there are two groups of complexes, since the optical density of solutions in the region of 250–290 nm increased and that in the region of 340–370 nm decreased with increasing concentration of NH_4Cl. The absorption band observed in the region of 340–370 nm was a sum of the bands due to different polynuclear complexes.

2.3.1.2 Equilibrium complex formation in MCl–CuCl–H_2O and NH_4Cl–CuCl–H_2O catalytic systems

The equilibrium formation of Cu(I) chloride complexes in relatively dilute solutions were studied in detail by Ahrland and Rawsthorne [198]. The concentration equilibrium constants ($\overline{\beta}_{m,n}$) of $(Cu_mCl_n)^{(n-m)-}$ complexes in the $NaClO_4$–NaCl–HCl–CuCl system at an ionic strength of I = 5 M and [HCl] = 0.1 M (25 °C) were determined by two methods. A potentiometric technique was used at [NaCl] = 0.2–0.88 M, [CuCl] ≤ 0.012 M, and [NaCl] = 0.2–4.268 M, [CuCl] ≤ 0.047 M, while the solubility method was used at [NaCl] ≤ 5.0 M and [CuCl] ≤ 1.6 M. In the latter case, it was established that binuclear complexes were formed and the corresponding $\overline{\beta}_{2,4}$ values for $Cu_2Cl_4{}^{2-}$ were evaluated (Table 2.1). A close value of pK_S = 6.73 was reported in [204].

The composition of Cu(I) chloride complexes formed in concentrated CuCl and NH_4Cl solutions employed in the catalytic reactions of alkynes was determined in the investigations of equilibria in the NH_4Cl–CuCl–H_2O system at 25, 50, and 80 °C for CuCl concentrations in a range from 0.03 to 10 $\tilde{m}$ and various constant concentrations of the background electrolyte NH_4Cl + NH_4NO_3 (6.5, 10, 14 $\tilde{m}$) [195, 202, 203, 205]. The activities of Cu^+ (a_{Cu^+}) and Cl^- (a_{Cl^-}) in these investigations were calculated from the results of measurements of the potentials of copper and silver chloride electrodes (relative to standard

Table 2.1 Stability constants for some $(Cu_mCl_n)^{(n-m)-}$ complexes (reproduced with permission from editorial board of *Usp. Khim.* journal © 2000 [23])

Temperature, °C	Ionic strength of solution, M	$\lg\beta_{1,2}$	$\lg\beta_{1,3}$	$\lg\beta_{1,4}$	$\lg\beta_{2,4}$	pK_S	References
25	0	5.31				6.50	[199, 200][a]
20	0	5.30	5.70				[201][a]
25	5.0	6.00	5.99	<4.7	13 ± 0.2	7.38	[198][a]
25	6.5	6.04	5.98	5.6	12.2	6.73	[202][b]
25	14.0	6.30	6.08	5.7	18.8		[203][b]

Notes: K_S is the solubility product of $CuCl_{solid}$; [a]Stability constants $\overline{\beta}_{m,n}$; [b]Stability constants $\beta_{m,n}$ according to Eq. (2.343) for a constant sum of $NH_4Cl + NH_4NO_3$ concentrations (molalities, $\tilde{m}$).

hydrogen electrode) in the following chains with charge transfer:

$$\text{(I) Cu} \mid Cu^+, Cl^- \vdots H^+(a_{H^+} = 1),\ H_2(1\ \text{atm}) \mid \text{Pt};$$

$$\text{(II) Ag} \mid AgCl_s \mid Cl^- \vdots H^+(a_{H^+} = 1),\ H_2(1\ \text{atm}) \mid \text{Pt}.$$

Other standard reference electrodes (with the potentials defined relative to the standard hydrogen electrode) were also used.

Under the standard conditions ($P_{H_2} = 1$ atm, $a_{H^+} = 1$, $a_{Cu^0} = 1$, $a_{AgCl_s} = 1$, $a_{Ag^0} = 1$), $E_{Pt} = 0$ and the electromotive forces (*EMF*) in chains (I) and (II) are given by the following formulas:

$$EMF_I = E_{Cu} = E^o_{Cu|Cu^+} + E_j + \frac{RT}{F}\ln a_{Cu^+}; \tag{2.341}$$

$$EMF_{II} = E_{AgCl} = E^o_{Ag|AgCl} + E_j - \frac{RT}{F}\ln a_{Cl^-}. \tag{2.342}$$

The activities a_{Cu^+} and a_{Cl^-} calculated using the Nernst equations (2.341) and (2.342) are directly proportional to the thermodynamic activities of Cu^+ and Cl^- ions (for $a_{H^+} = 1$ in the reference electrode) and are adopted as equal to these with neglect of the diffusion potential E_j ($E_j = 0$). Under the conditions of E_j = const [202, 203, 205], the proportionality coefficient (still close to unity) remains constant in the entire range of CuCl and NH_4Cl concentrations, which is just what ensures reliable determination of the composition of complexes. Using the difference of EMFs in the two chains, it is possible to determine the product of activities $a_{Cu^+}a_{Cl^-}$ with neglect of the diffusion potential. In the presence of $CuCl_s$, the value of $a_{Cu^+}a_{Cl^-} = K_S$ determines the thermodynamic solubility product. The K_S values determined from a_{Cl^-} and a_{Cu^+} at a cationic background value of 6.5 $\tilde{m}$ and nearly zero ionic strength are close to each other (see Table 2.1). It was established that the activity coefficient of Cl^- ions(γ_{Cl^-}) remains constant in the entire range of NH_4Cl and CuCl concentrations, provided that the concentration of $NH_4{}^+$ ions is constant, and it was suggested that the $\gamma_{(Cu_mCl_n)^{(n-m)-}}$ value under these conditions also remains constant.

Determination of the ions' activities (from the data of potentiometric measurements), which are proportional to the concentrations at constant values of ionic strength, ion medium, or cationic background (Chapter 8), also makes it possible to monitor the ion concentration and maintain a constant concentration of complexes during the catalytic reaction. This possibility is used for determining the compositions of catalytically active complexes (see below) and evaluating their stability constants.

The compositions of polynuclear complexes were determined and their stability constants evaluated using the Sillen–McKay–Hedström methods [12, 206, 207]. The stability constants $\beta_{m,n}$ determined from the a_{Cu^+} and a_{Cl^-} values differ from the thermodynamic constants $\beta^*_{m,n}$ by a factor equal to the coefficient of activity ($\gamma_{m,n}$):

$$\beta_{m,n} = \beta^*_{m,n}/\gamma_{m,n}, \tag{2.343}$$

so that

$$[CuCl]_\Sigma = \sum_{m=1}\sum_{n=2} m\beta_{m,n}a^m_{Cu^+}a^n_{Cl^-} \tag{2.344}$$

$$[Cl^-]_\Sigma = [Cl^-] + \sum_{m=1}\sum_{n=2} n\beta_{m,n}a^m_{Cu^+}a^n_{Cl^-}. \tag{2.345}$$

The values of $\beta_{m,n}$ for 19 complexes of the $(Cu_mCl_n)^{(n-m)-}$ type are listed in Table 2.2. It should be borne in mind that, when this method is used to calculate the $\beta_{m,n}$ values for complexes with large m (4, 5) and n (7, 8, 9), the error of determination can be about one order of magnitude. Therefore, it is possible to

Table 2.2 Stability constants for reaction (2.346) at various temperatures and NH_4^+ concentrations (molalities, m̃) (reproduced with permission from AIPC Nauka © 1969 [203])

Complex anion	$\beta_{m,n}$			
	25 °C		50 °C	80 °C
	$[NH_4^+] = 6.5$	$[NH_4^+] = 14.0$	$[NH_4^+] = 10.0$	$[NH_4^+] = 10.0$
$CuCl_2^-$	$1.1 \cdot 10^6$	$2.0 \cdot 10^6$	$7.0 \cdot 10^5$	$4.8 \cdot 10^5$
$CuCl_3^{2-}$	$9.6 \cdot 10^5$	$1.2 \cdot 10^6$	$2.8 \cdot 10^5$	$1.0 \cdot 10^5$
$CuCl_4^{3-}$	$4.0 \cdot 10^5$	$5.0 \cdot 10^5$	$7.3 \cdot 10^4$	$2.0 \cdot 10^4$
$Cu_2Cl_3^-$	$2.0 \cdot 10^{12}$	$7.5 \cdot 10^{12}$	$1.6 \cdot 10^{12}$	$1.3 \cdot 10^{11}$
$Cu_2Cl_4^{2-}$	$1.5 \cdot 10^{12}$	$6.4 \cdot 10^{12}$	$7.0 \cdot 10^{11}$	$5.0 \cdot 10^{10}$
$Cu_2Cl_5^{3-}$	$1.0 \cdot 10^{12}$	$5.5 \cdot 10^{12}$	$4.0 \cdot 10^{11}$	$2.8 \cdot 10^{10}$
$Cu_2Cl_6^{4-}$	$4.3 \cdot 10^{11}$	$1.5 \cdot 10^{12}$	$2.0 \cdot 10^{11}$	$1.0 \cdot 10^{10}$
$Cu_3Cl_4^-$	$3.6 \cdot 10^{18}$	$1.7 \cdot 10^{19}$	$1.5 \cdot 10^{18}$	$3.0 \cdot 10^{16}$
$Cu_3Cl_5^{2-}$	$4.0 \cdot 10^{18}$	$2.3 \cdot 10^{19}$	$1.1 \cdot 10^{18}$	$3.0 \cdot 10^{16}$
$Cu_3Cl_6^{3-}$	$2.8 \cdot 10^{18}$	$1.7 \cdot 10^{19}$	$5.0 \cdot 10^{17}$	$7.0 \cdot 10^{15}$
$Cu_3Cl_7^{4-}$	$1.8 \cdot 10^{18}$	$1.5 \cdot 10^{19}$	$3.0 \cdot 10^{17}$	–
$Cu_4Cl_5^-$	$2.0 \cdot 10^{25}$	$2.4 \cdot 10^{26}$	$3.0 \cdot 10^{24}$	$6.0 \cdot 10^{21}$
$Cu_4Cl_6^{2-}$	$2.0 \cdot 10^{25}$	$1.2 \cdot 10^{26}$	$1.8 \cdot 10^{24}$	$5.0 \cdot 10^{21}$
$Cu_4Cl_7^{3-}$	$5.0 \cdot 10^{24}$	$8.0 \cdot 10^{25}$	$1.0 \cdot 10^{24}$	–
$Cu_4Cl_8^{4-}$	$5.0 \cdot 10^{24}$	–	–	–
$Cu_5Cl_6^-$	$1.0 \cdot 10^{32}$	$8.0 \cdot 10^{32}$	–	–
$Cu_5Cl_7^{2-}$	$5.0 \cdot 10^{31}$	$8.0 \cdot 10^{32}$		
$Cu_5Cl_8^{3-}$	$3.0 \cdot 10^{31}$	–		
$Cu_5Cl_9^{4-}$	$2.0 \cdot 10^{31}$			

establish with confidence only that singly, doubly, or triply charged anions containing one to four copper atoms are present in solution. The deviation of experimental and calculated CuCl concentrations at all temperatures does not exceed 10 % [195]. As can be seen from Table 2.2, an increase in the temperature leads to the disappearance of multiply charged complexes ($n - m = 4$) and those with $m = 5$, and the growth in $[NH_4^+]$ does not really influence the composition of complexes. The possibility of determining a greater number of complexes using the Sillen–McKay–Hedström methods is explained primarily by the independent determination of a_{Cu^+} and a_{Cl^-} values. Calculations of the activity (and concentration) of Cl^- by the Hedström equation using the values of a_{Cu^+}, $[CuCl]_\Sigma$, and $[Cl^-]_\Sigma$ showed that the calculated and measured values of a_{Cl^-} coincide only for CuCl concentrations up to about 0.5 $\tilde{m}$. At still greater concentrations, the difference increases as a result of large errors related to the graphical method of solution of the Hedström equation [207].

An analysis of the distribution of $(Cu_mCl_n)^{(n-m)-}$ complexes in the NH_4Cl–CuCl–H_2O system (Figure 2.9) shows that polynuclear complexes appear in solution in significant amounts beginning with CuCl concentrations of about 3–4 $\tilde{m}$. The concentrations of $(Cu_mCl_n)^{(n-m)-}$ complexes with $m = 2$, 3, and 4 reaches ~30 % of the sum of concentrations of all complexes. Using the values of $\beta_{m,n}$ at I = 6.5 M (see Table 2.2) and the values of a_{Cu^+} and a_{Cl^-} obtained for the maximum concentrations of NaCl and CuCl [198], it is possible to show that the molar fraction of complexes with $m \geq 2$ is below 5 %. For this reason, these complexes could not be experimentally detected in [198]. It is interesting to note that different approaches led to close values of the stability constants $\overline{\beta}_{m,n}$ and $\beta_{m,n}$ [202, 203] for $CuCl_2^-$ and $CuCl_3^{2-}$ (see Table 2.1).

Since the activity coefficients of various chlorides (KCl, NaCl, NH_4Cl) in aqueous solutions are almost independent of the temperature, it could be assumed that $d\gamma_{m,n}/dT \cong 0$. This assumption was used [205] to evaluate the enthalpy ΔH° of reactions

$$mCu^+ + nCl^- \rightleftharpoons (Cu_mCl_n)^{(n-m)-}. \tag{2.346}$$

These estimations are presented in Table 2.3.

Taking into account that $\beta_{m,n}$ values vary in a regular way with the temperature ($\lg\beta_{m,n}$ is a linear function of $1/T$), it was possible to use a linear interpolation of the data in Table 2.2 so as to obtain the values of $\beta_{m,n}$ at 25 °C for $[NH_4^+] = 10\,\tilde{m}$. The addition of every next CuCl molecule to the $(Cu_mCl_n)^{(n-m)-}$

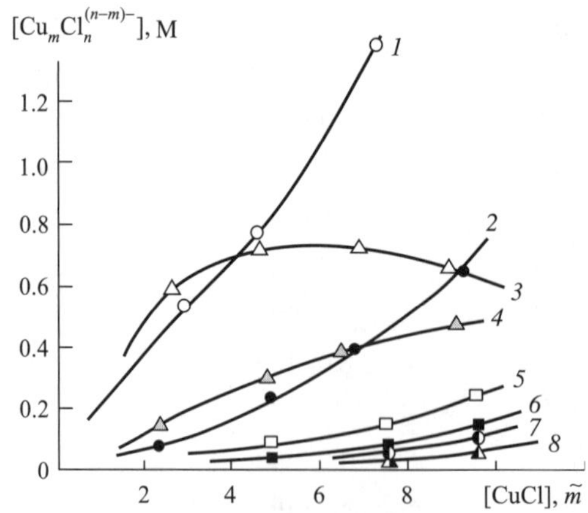

Figure 2.9 Distribution of $(Cu_mCl_n)^{(n-m)-}$ complexes in CuCl–NH_4Cl–H_2O system as dependent on CuCl concentration (80 °C, $[NH_4^+] = 10\,\tilde{m}$): (1) $CuCl_2^-$; (2) $Cu_2Cl_3^-$; (3) $CuCl_3^{2-}$; (4) $Cu_2Cl_4^{2-}$; (5) $Cu_3Cl_5^{2-}$; (6) $CuCl_3^{4-}$; (7) $Cu_4Cl_6^{2-}$; (8) $Cu_4Cl_5^-$ (reproduced with permission from The Russian Academy of Sciences and Turpion Ltd © 2000 [23])

Table 2.3 Enthalpy of reaction (2.346) (reproduced with permission from editorial board of *Usp. Khim.* journal © 2000 [23])

$(Cu_mCl_n)^{(n-m)-}$	ΔH^o_{298}, kJ · mole^{-1}	$(Cu_mCl_n)^{(n-m)-}$	ΔH^o_{298}, kJ · mole^{-1}
$CuCl_2^-$	-17.1 ± 2	$Cu_2Cl_6^{4-}$	-70.6 ± 4
$CuCl_3^{2-}$	-36.3 ± 2	$Cu_3Cl_4^-$	-89.5 ± 6
$CuCl_4^{3-}$	-48.5 ± 2	$Cu_3Cl_5^{2-}$	-93.6 ± 6
$Cu_2Cl_3^-$	-55.6 ± 4	$Cu_3Cl_6^{3-}$	-111.5 ± 6
$Cu_2Cl_4^{2-}$	-66.9 ± 4	$Cu_4Cl_5^-$	-139.6 ± 10
$Cu_2Cl_5^{3-}$	-73.6 ± 4	$Cu_4Cl_6^{2-}$	-147.1 ± 10

Note: $[NH_4^+] = 10\,\tilde{m}$.

complex leads to a decrease in ΔH^0 on the average by 30–42 kJ·mole^{-1}. According to these estimations, the formation of polynuclear complexes from those involving a few nuclei also proceeds with a decrease in the enthalpy. However, because of a large decrease in the entropy, the equilibrium exhibits a significant rightward shift at relatively low temperatures (below 50–60 °C). The increase in $\beta_{m,n}$ values with the concentration of a cationic background can be related both to a decrease in $\gamma_{m,n}$ and the formation of extraspheric complexes with NH_4^+ ion. The degree of influence of the anion complexes with $n-m \geq 2$ on the ionic strength of a solution was evaluated [8] using the Moiseev–Flid equation [6, 208] that describes the dependence of the HCl acidity function ($-H_0$) on the ionic strength of the solution (see also the review by Paul and Long [209]):

$$-H_0 = \lg C_{H_3O^+} + LI - \lg a_{H_2O},$$

where L is a constant that depends on the natures of the salt and solvent ($L = 0.051\,\tilde{m}^{-1} \equiv 0.063\,M^{-1}$ at 40–80 °C and [HCl] $= 0.18 - 0.8\,\tilde{m}$)

The results of measurements [210] showed that, at [HCl] $= 0.16\,\tilde{m}$ and 80 °C, the H_0 value remains almost unchanged in a large interval of CuCl concentrations (Figure 2.10) [8, 211]. With allowance for a change in a_{H_2O}, the ionic strength increases on the passage from NH_4Cl–HCl–H_2O to NH_4Cl–CuCl–HCl–H_2O system by no more than 0.5 for $[NH_4Cl] = 12\,\tilde{m}$. From this it follows that multiply charged ions virtually do not influence the ionic strength because of both their low concentrations (as calculated using data in Table 2.2) and because of the interaction with outer-sphere NH_4^+ ions.

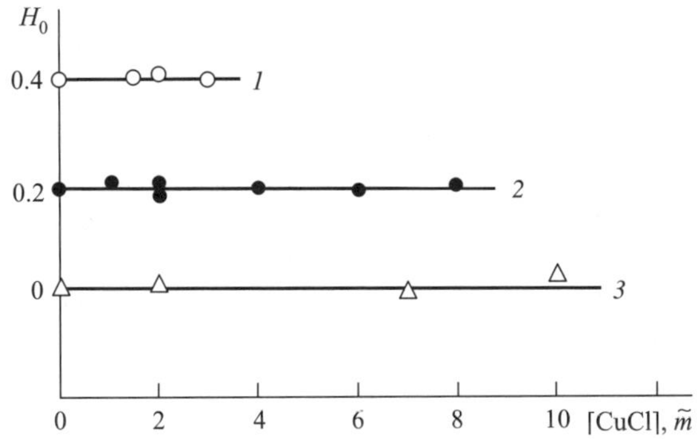

Figure 2.10 Effect of CuCl concentration on acidity function H_0 at 80 °C, [HCl] $= 0.16\,\tilde{m}$, and various NH_4Cl molalities ($\tilde{m}$): (1) 5; (2) 9; (3) 12 (reproduced with permission from The Russian Academy of Sciences and Turpion Ltd © 2000 [23])

Thus, $\beta_{m,n}$ values (with allowance for the coordination of water) should be described by the following relation [203]:

$$\beta_{m,n} = \sum_p \sum_q \frac{\beta^*_{m,n}}{\gamma_{m,n,p,q}} a^p_{NH_4^+} a^q_{H_2O}. \tag{2.347}$$

The existence of a variety of polynuclear complexes and their participation in catalytic reactions were confirmed by subsequent investigations devoted to the ethinyl complexes of copper, synthesis and isolation of crystalline complexes, and kinetics of reactions catalyzed by copper (I) complexes.

2.3.1.3 Crystal chemistry of $(Cu_mX_n)^{(n-m)-}$ anionic complexes

For better understanding of the mechanisms of catalytic processes and factors accounting for a high catalytic activity of concentrated solutions containing polynuclear complexes of copper, it is very useful to analyze the information available on the compositions and crystalline structures of complexes isolated from copper halide systems and compare the compositions of isolated solid $(Cu_mX_n)^{(n-m)-}$ anionic complexes to those found in solution. Table 2.4 presents some characteristics of $(Cu_mX_n)^{(n-m)-}$ (X = Cl, Br) complex anions in compounds with various cations, the crystalline structures of which were established. A detailed analysis of the structures of these complexes can be found in review [23] and references to original investigations cited therein. It should be noted that index m in solid complexes varies within a rather broad interval of $m = 1$–6, and it was also reported on the isolation of a complex with $m = 15$. Analogous complexes are formed in iodide systems ($Cu_2I_3^-$, $Cu_2I_4^{2-}$, $Cu_2I_5^{3-}$, $Cu_2I_6^{4-}$, $Cu_3I_4^-$, $Cu_4I_6^{2-}$, $Cu_4I_8^{4-}$, $Cu_5I_7^{2-}$, $Cu_6I_{11}^{5-}$, $Cu_8I_{13}^{5-}$) [212], where complexes with greater charges than those found for X = Cl and Br were reported as well.

Let us consider the main specific features inherent in the structure of these complexes (see Table 2.4). All types of $(Cu_mX_n)^{(n-m)-}$ anions encountered in copper(I) chloride and bromide complexes can be divided into four groups, which are characterized by the island, chain (ribbon), layer, and framework patterns.

A crystallochemical analysis [23] led to a conclusion that copper atoms in the $(Cu_mX_n)^{(n-m)-}$ (X = Cl, Br) anion complexes most frequently occur at the center of a tetrahedron (coordination number 4). The vertices of the tetrahedron are usually not equidistant with respect to the center. The X–Cu bond length depends on the coordination number of atom X, that is, on the degree of association in the coordination polyhedron of Cu(I). Atoms X that are characterized by greater coordination numbers occur at greater distances from the central atom.

Copper(I) halide complexes can also involve Cu atoms in a planar trigonal environment (coordination number 3) and a less frequently encountered angular or linear coordination (coordination number 2). At the same time, no one of the structurally characterized copper(I) halide complexes reported so far had the shape of an isolated $[CuX_4^{3-}]$ tetrahedron; apparently, a discrete $[CuX_3^{2-}]$ anion with a planar trigonal coordination of the metal atom also does not exist.

In $(Cu_mX_n)^{(n-m)-}$ anions involving two or more Cu atoms, the coordination polyhedra can be differently conjugated so as to have common vertices, edges, and faces. The most frequently encountered combined mode of association consists in simultaneously sharing vertices and edges. For example, the tetranuclear $[Cu_4X_5^-]$ anion forms $(Cu_4X_5^-)_m$ chain fragments, which is a rather unusual association for Cu(I) coordination polyhedra. In compound **10** (see Table 2.4), these chain fragments exist in the form of Cu(I) coordination tetrahedra that share edges and faces. In some cases, combined conjugation leads to the formation of $(Cu_mX_n)^{(n-m)-}$ anions with zeolite-like structures, the numerous cavities of which admit the incorporation (via the host–guest scheme) of small molecules (HC≡CH, H_2O, CuC≡CH, etc.) and even complexes (e.g., Cu(II) complexes in compounds **15a**, **15b**, and **16**).

The entire variety of structural fragments encountered in copper(I) chloride complexes described above exhibits a stable trend of forming more complicated structures with increasing number of nuclei in the

Table 2.4 Structural characteristics of $(Cu_mX_n)^{(n-m)-}$ anions in copper(I) halide complexes (reproduced with permission from editorial board of *Usp. Khim.* journal © 2000 [23])

Compound	Anion	Structure type	Coordination number for Cu atoms	Interatomic distances				
				Cl–Cu[a]	μ_2-Cl–Cu	μ_3-Cl–Cu	μ_4-Cl–Cu	Cu… Cu
$(PyH)CuCl_2$ (**1**)	CuX_2^-	Chain	4, 3	2.269(4), 2.269(4)	2.353(4), 2.433(4)			2.803(5)
$KCuBr_2 \cdot {}^1/_2H_2O$ (**2a**)[b]	CuX_2^-	Layered	4		2.44(1), 2.545(8)			2.891(8)
$NH_4CuBr_2 \cdot {}^1/_2H_2O$ (**2b**)	CuX_2^-	Layered	4		2.44(1), 2.545(8)			2.891(8)
K_2CuCl_3 (**3a**)	CuX_3^{2-}	Chain	4	2.34(2), 2.35(2)	2.44(2), 2.48(2)			4.08(1)
K_2CuBr_3(**3b**)	CuX_3^{2-}	Chain	4	2.34(2), 2.35(2)	2.44(2), 2.48(2)			4.08(1)
$(NH_4)_2CuBr_3$ (**3c**)	CuX_3^{2-}	Chain	4	2.34(2), 2.35(2)	2.44(2), 2.48(2)			4.08(1)
$(NH_4)_2CuCl_3$ (**4**)	CuX_3^{2-}	≫	4	2.34(2), 2.35(2)	2.44(2), 2.48(2)			4.08(1)
$CsCu_2Cl_3$ (**5**)	$Cu_2X_3^-$	Chain (II)	4		2.37(1), 2.37(1)		2.45(1), 2.45(1)	2.81(1)
$(ImH)Cu_2Cl_3$ (**6**)[c]	$Cu_2X_3^-$	Chain (I)	4, 3		2.283(4), 2.366(4)	2.375(3), 2.513(4)		2.748(3)
$Cs_3Cu_2Cl_5$ (**7**)	$Cu_2X_5^{3-}$	Chain	4	2.262(5), 2.262(5)			2.34(1), 2.881(8)	2.911(8)
$(Me_4N)_3Cu_2Br_5$ (**8**)[b]	$Cu_2X_5^{3-}$	Island	3	2.38(1), 2.38(1)	2.40(1), 2.40(1)			–
$(NH_4)_2Cu_3Cl_5 \cdot {}^1/_3H_2O \cdot {}^1/_{15}CuCl$ (**9a**)	$Cu_3X_5^{2-}$	Framework	4, 3 + 1		2.25(1), 2.30(1)	2.30(1), 2.71(1)	2.40(1), 2.68(1)	2.98(1)
$(NH_4)_2Cu_3Cl_5 \cdot {}^4/_9H_2O \cdot {}^1/_9CuCl$ (**9b**)[d]	$Cu_3X_5^{2-}$	Framework	4, 3 + 1			2.30(1), 2.71(1)		2.98(1)
$(NH_4)_2Cu_3Cl_5 \cdot {}^4/_9H_2O \cdot {}^4/_9CuCl$ (**9c**)[d]	$Cu_3X_5^{2-}$	Framework	4, 3 + 1			2.30(1), 2.71(1)		2.98(1)
$Rb_9Cu_{13}Cl_{22} \cdot (H_3O^+ Cl^-)$ (**9d**)	$Cu_3X_5^{2-}$	Framework	4, 3 + 1					2.98(1)
$NH_4Cu_4Cl_3(I, Cl)_2$ (**10**)	$Cu_4X_5^-$	Framework	4	2.29(2), 2.39(2)			2.60(2), 2.74(2)	2.48(2)

(*continued overleaf*)

Table 2.4 *(continued)*

Compound	Anion	Structure type	Coordination number for Cu atoms	Interatomic distances				
				Cl–Cu[a]	μ_2-Cl–Cu	μ_3-Cl–Cu	μ_4-Cl–Cu	Cu... Cu
$(C_6H_5NH_3)_3Cu_4Cl_7$ (**11**)	$Cu_4X_7^{3-}$	Chain	4, 3	2.549(7), 2.549(7)	2.240(8), 2.41(1)	2.56(1), 2.631(8)		2.801(9)
$(PyH)_2Cu_5Cl_7 \cdot {}^1/_5\ H_2O$ (**12**)	$Cu_5X_7^{2-}$	Framework	4, 3	2.31(2), 2.31(2)	2.24(2), 2.64(2)	2.48(2), 2.63(2)		2.74(2)
$Rb_4Cu_5Cl_9$ (**13**)	$Cu_5X_9^{4-}$	Chain (Ribbon)	4		2.196(5), 2.578(5)	2.299(6), 2.652(5)		3.092(4)
$(MeEt_3N)_3Cu_6Br_9$ (**14**)[b]	$Cu_6X_9^{3-}$	Island	4, 2		2.386(3), 2.528(2)		2.697(3), 2.697(3)	2.650(3)
$Rb_{11}Cu_{15}Cl_{16}Br_6$-$(Cu^{II}Cl_6)\cdot$ CuC≡CH (**15a**)	$Cu_{15}X_{24}^{9-}$	Layered	4, 2		2.287(7), 2.340(7)	2.305(7), 2.63(1)		3.006(6)
$Rb_{11}Cu_{15}Cl_{16}Br_6$-$(Cu^{II}Cl_6)$ (**15b**)	$Cu_{15}X_{24}^{9-}$	Layered	4, 2		2.287(7), 2.340(7)	2.305(7), 2.63(1)		3.006(6)
$\{Rb_{11}[Cu_2(H_2O)_2]\}$-$Cu_{15}Cl_{24}(Cu^{II}Cl_6)$ (**16**)	$Cu_{15}X_{24}^{9-}$	Layered	4, 2		2.287(7), 2.340(7)	2.305(7), 2.63(1)		3.006(6)

Notes: For Cu–Cl bonds, first and second values correspond to minimum and maximum lengths, respectively; for Cu–Cu bonds, data refer to shortest lengths. Complexes **2a, 2b, 3a–3C** and **4, 9a–3d**, as well as **15a, 15b** and **16** are isostructural. [a]Distance from Cu to nonbridging Cl; [b]Cu–Br bond lengths; [c]ImH = imidazolium cation; [d]acetylene is absorbed and does not occupy definite crystallographic positions.

$(Cu_mCl_n)^{(n-m)-}$ anions. In particular, an increase in m is accompanied by the growing ability of chlorine atoms to play the role of bridging ligands [23]. The established regularities agree well with the results of studying equilibria in the CuCl–MCl systems (see Table 2.2). Among the $(Cu_mCl_n)^{(n-m)-}$ ($m = 1$–5, $n = 2$–7) complexes present in a concentrated solution of NH_4Cl and CuCl, the dominating fraction represents complexes with $m = 2$, 3, and 4. The ability of the coordination polyhedra of Cu(I) atoms to form complicated $(Cu_mCl_n)^{(n-m)-}$ associates with $m \geq 2$, 3, and 4 is probably among the main factors that account for the high catalytic activity of concentrated MCl–CuCl solutions.

Complicated associates in MCl–CuCl solutions are formed from associates of the layer, chain, island, and framework types as a result of rupture of the μ_n-Cl bonds, e.g., by water molecules and/or Cl^- anions, which occupy the coordination vacancies in $(Cu_mCl_n)^{(n-m)-}$ complexes in solution. In addition, NH_4^+ cations are capable of forming crystalline complexes with highly involved structures of the zeolite type such as compounds **9a–9c** (Table 2.4). Their appearance is caused by the tetrahedral shape of ammonium cations, which interact with Cl^- via the formation of hydrogen bonds of the (N)H$\cdots$Cl type [213]. "Necklace" structures of compounds **1, 6, 11**, and **12** involving planar PyH^+, ImH^+ and $C_6H_5NH_3^+$ cations are also formed due to the (N)H$\cdots$Cl bonds. The replacement of NH_4Cl by KCl leads to anionic complexes **2a**, **3a**, and **3b**, in which K_p^+ ions cannot form directed bonds with Cl^-. For this reason, the MCl–CuCl system exhibits no complicated combinations of Cu(I) coordination polyhedra.

In addition, NH_4^+ cations and protonated amines are capable of generating complicated structural forms of $(Cu_mCl_n)^{(n-m)-}$ anions due to the spatially oriented, directed interaction $M^+ \cdots Cl^-$. Interactions of this type, which lead to the formation of extraspheric complexes in solution [see Eq. (2.347)], are probably among the factors that account for the higher catalytic activity of systems with NH_4^+ and amine hydrochlorides in reactions of alkynes as compared to the activity of the KCl–CuCl–H_2O system.

2.3.2 Kinetics of catalytic reactions of alkynes in concentrated NH_4Cl–CuCl aqueous solutions at constant complexity functions F_{Cu} and F_{Cl}

If a solution contains polynuclear catalytic metal complexes and no excess chloride ions relative to the metal, then the concentration of (one or several) free catalytically active complex species cannot be expressed via $[CuCl]_\Sigma$ (so as to take into account the material balance with respect to the catalyst in the kinetic equation). Therefore, the only reasonable approach to constructing theoretically justified, correct kinetic models and elucidating mechanisms of reactions in these systems consists of measuring the concentrations (or activities) of free reactants and solution components (Cu^+, Cl^-, H_3O^+) under steady-state catalytic reaction conditions and controlling these parameters of the catalyst components (Cu^+, Cl^-) so as to keep them constant during variation of the partial pressures or concentrations of reactants, that is, in maintaining constant ratios of $[CuCl]_\Sigma$: $[Cu^+]$ (F_{Cu} = const) and $[Cl^-]_\Sigma$: $[Cl^-]$ (F_{Cl} = const) and, hence, a constant value of $[(Cu_mCl_n)^{(n-m)-}]$. In other words, as the catalyst is bound by the reactant, it is necessary to add the initial catalyst components (CuCl and, if necessary, MCl) to the solution. Then, using a potentiometric technique (at a constant cationic background), it is possible to determine the values that are directly proportional to the activities (and concentrations) of Cu^+, Cl^-, and H_3O^+ ions (see Section 2.3.1) [195, 211, 214].

For determining $[H_3O^+]$ (or $a_{H_3O^+}$), it is possible to use, besides pH measurements, the Hammett indicator method [210, 211] based on the measurement of acidity h_0. In the aforementioned systems, this function is related to $a_{H_3O^+}$ by the following equation:

$$a_{H_3O^+} = bh_0a_{H_2O},$$

where b is a constant coefficient (for a constant cationic background). The kinetic dependences obtained for $[H_3O^+]$ (as pH or $D_{BH^+} = f[(H_3O^+)]$, where D_{BH^+} is the optical density of a solution of a protonated form of the indicator) and $a_{H_3O^+}$ have similar shapes. At a constant high cationic background and an insignificant change (not exceeding 10 %) in the activity of water, the activities of gaseous reactants (C_2H_2, C_3H_4, C_4H_4, HCN, C_5H_6, etc.) are proportional to their partial pressures over the solution.

This approach was used in the investigations of 12 reactions of acetylene compounds in 15 catalytic systems, which are listed in Table 2.5. The kinetic equations obtained for these reactions at constant cationic background and the a_{Cu^+} and a_{Cl^-} values are presented in Table 2.6. It was established that the form of a kinetic equation is independent of the CuCl concentration that only influences the observable constants k_{ij} (where subscript i refers to the reaction number and j indicates the catalytic system) in the kinetic equations (Table 2.6). In Table 2.6, the values of H_3O^+ activity and h_0 acidity in the kinetic equations are replaced by the proportional values of [HCl] and $[H_3O^+]$ determined under steady-state conditions. Let us consider the kinetics of some reactions in more detail.

2.3.2.1 *Reaction of acetylene dimerization (8, Table 2.5)*

The reaction was studied in a CuCl (6.6 $\tilde{m}$)–NH_4Cl (8.84 $\tilde{m}$) system at 80 °C [211]. The reaction rate was measured under steady-state conditions in a gas-flow gradient-free reactor [227]. The activities a_{Cu^+} and a_{Cl^-} were measured in situ (in the reactor) and the h_0 value was determined in the samples of solution taken from the reactor with a heated syringe and analyzed by spectrophotometry in a thermostatted cell (the indicator was pyrocatechol sulfophthalein ($\lambda_{max\,BH^+}$ = 558 mm, pK_a = −0.24 at 80 °C)).

Table 2.5 Catalytic reactions in $CuCl–NH_4Cl–H_2O$ system (reproduced with permission from AIPC Nauka © 1990 [185])

Catalytic reaction	Catalytic system	[CuCl], $\tilde{m}$	References
$C_2H_2 + HCl \rightarrow CH_2{=}CHCl$ (**1**)	*a*	5–12	[215]
$C_3H_4 + HCl \rightarrow MeC(Cl){=}CH_2$ (**2**)	*b*	0.6–12	[216]
$C_3H_4 + HCl \rightarrow MeCH{=}CHCl$ (**3**)	*c*	4–12	[216]
$C_4H_4 + HCl \rightarrow CH_2{=}CHC(Cl){=}CH_2$ (**4**)	*d*	0.5–8	[217]
$C_2H_2 + H_2O \rightarrow MeCHO$ (**5**)	*e* *f*[a]	5–12, 5–12	[215] [218]
$C_3H_4 + H_2O \rightarrow Me_2CO$ (**6**)	*g* *h*[b]	0–12, 4–12	[216] [219]
$C_3H_4 + H_2O \rightarrow MeCH_2CHO$ (7)	*i*[b]	4–12	[219]
$2C_2H_2 \rightarrow HC{\equiv}CCH{=}CH_2$ (**8**)	*j*	5–12	[211, 220–223]
$3C_2H_2 \rightarrow CH_2{=}CHC{\equiv}CCH{=}CH_2$ (**9**)	*k*	5–12	[224]
$3C_2H_2 \rightarrow CH_2{=}CHCH{=}CHC{\equiv}CH$ (**10**)	*l*	5–12	[225]
$C_2H_2 + HCN \rightarrow CH_2{=}CHCN$ (**11**)	*m*	5–10	[226]
C_2H_2 + 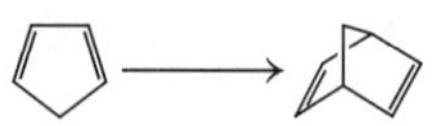 (**12**)	*n* *o*	7–12, 7–21	[8] [8]

Notes: For reactions (1)–(11), $[NH_4Cl] = 12\tilde{m}$; for reaction (12), $[NH_4Cl] = 21\tilde{m}$. [a]Besides NH_4Cl is present Cu_2S_{solid}; [b]Besides NH_4Cl is present RSH.

The reaction rate was considered as a function of the following interrelated parameters:

$$r = f(P_{C_2H_2}, a_{Cu^+}, a_{Cl^-}, a_{H_3O^+}, T).$$

At the first stage, the values of [HCl] and $P_{C_2H_2}$ were varied while the activities a_{Cu^+} and a_{Cl^-} were kept constant by decreasing $[NH_4Cl]$ (ultimately from 8.84 to 8.54 $\tilde{m}$) with an increase in HCl or by increasing [CuCl] with an increase in $P_{C_2H_2}$.

The rate of vinylacetylene (VA) formation decreased upon adding either HCl or NH_4OH [211]. The results of measurements of the potential E_{Cu} of a copper electrode showed that the extremal dependence of the rate correlates with the variation of E_{Cu} (i.e., a_{Cu^+}) (Figure 2.11). For $a_{Cu^+} \cong$ const (E_{Cu} = 99 ± 0.2 mV) and constant $P_{C_2H_2} = 350 \pm 10$ Torr, the reaction rate exhibited a hyperbolic decrease with increasing [HCl] and increased with $[NH_4OH]$. The amount of H_3O^+ ions evolved during the interaction of C_2H_2 with CuCl solutions was strictly measured for the first time (Figure 2.12). The gain in $[H_3O^+]$ at an initial [HCl] concentration of 0.1 $\tilde{m}$ amounted to 0.05 $\tilde{m}$. The results indicated that soluble ethynyl (acetylenide) complexes containing $CuC{\equiv}CH$ groups were formed at one step of the process.

The reaction order with respect to $P_{C_2H_2}$ under all conditions was greater than unity and varied depending on the interval of the initial solution acidity [220]. The rate of VA formation was described by the following equation:

$$r_{VA} = \frac{k^*_{obs} P^2_{C_2H_2}}{k' P_{C_2H_2} + h_0} = \frac{k_{obs} P^2_{C_2H_2}}{k P_{C_2H_2} + [H_3O^+]}, \quad (2.348)$$

Table 2.6 Kinetic equations for reactions of acetylene compounds in $CuCl–NH_4Cl–H_2O$ system system (reproduced with permission from AIPC Nauka © 1990 [185])

Kinetic equation	References
$r_{1a} = k_{1a}P_{C_2H_2}[HCl]$	[215]
$r_{2b} = k_{2b}P_{C_3H_4}[HCl] + kp_{C_3H_4}a_{Cl^-}$	[216]
$r_{3c} = k_{3c}P_{C_3H_4}[HCl]$	[216]
$r_{4d} = k_{4d}P_{C_4H_4}[HCl] + kp_{C_4H_4}[HCl](a^2_{Cl^-}/a_{H_2O})$	[217]
$r_{5e} = k_{5e}P_{C_2H_2}[H_3O^+]/(1 + k[H_3O^+])$	[215]
$r_{5f} = k_{5f}P_{C_2H_2}[H_3O^+] + kp_{C_2H_2}$	[218]
$r_{6g} = k_{6g}P_{C_3H_4}$	[216]
$r_{6h} = k_{6h}P_{C_3H_4}$	[219]
$r_{7i} = k_{7i}P_{C_3H_4}$	[219]
$r_{8j} = k_{8j}P^2_{C_2H_2}/(kp_{C_2H_2} + [H_3O^+])$	[211,220–223]
$r_{9k} = k_{9k}p_{C_2H_2}p_{C_4H_4}/(kp_{C_2H_2} + [H_3O^+])$	[224]
$r_{10l} = k_{10l}p_{C_2H_2}[H_3O^+]/(kp_{C_2H_2} + [H_3O^+])$	[225]
$r_{11m} = k_{11m}p_{C_2H_2}p_{HCN}/[H_3O^+]$	[226]
$r_{12n(o)} = k_{12n(o)}p_{C_2H_2}p_{C_5H_6}/[H_3O^+]$	[8]

Notes: In subscript at reaction rate and rate constant in kinetic equation, digit refers to reaction number and letter indicates catalytic system (see Table 2.5).

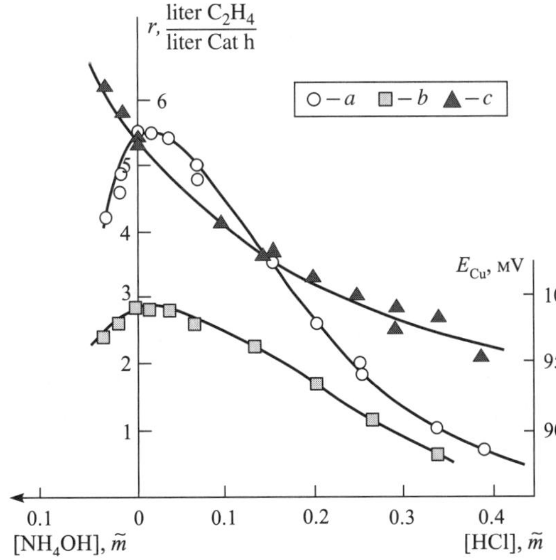

Figure 2.11 Effect of initial concentrations of HCl and NH_4OH on the (*a*, *b*) dimerization rate *r* and copper electrode potential E_{Cu}, respectively, in one series of experiments and (c) dimerization rate r in another series of experiments with E_{Cu} = const (reproduced with permission from editorial board of *Kinet. Katal.* journal © 1967 [211])

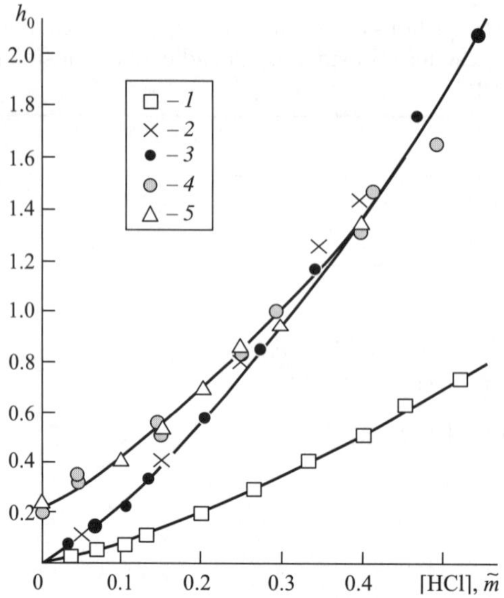

Figure 2.12 Plots of acidity h_0 versus initial concentrations of HCl in copper complex solutions at 80 °C: (1) pure hydrochloric acid; (2) acidity of complex solution with [CuCl] = 6.66 $\tilde{m}$ and $[NH_4Cl]$ = 8.84 $\tilde{m}$ prior to acetylene passage; (3) acidity calculated according to Moiseev–Flid equation [6, 208] (see p. 199); (4) acidity of complex solution after acetylene passage; (5) acidity of solution under conditions of steady-state kinetic experiments (reproduced with permission from editorial board of *Kinet. Katal.* journal © 1967 [211])

which was valid for the CuCl concentration varied within 5–11 $\tilde{m}$ (in six series of experiments with a_{Cu^+} and a_{Cl^-} maintained constant in each series). Equation (2.348) corresponds to a two-step mechanism without a limiting step ([Cu] is the concentration of copper chloride complexes). However, in the proposed reaction scheme

$$\text{(1)}\quad [Cu] + C_2H_2 \underset{k_{-1}}{\overset{k_1}{\rightleftharpoons}} [Cu]C_2H^- + H_3O^+;$$

$$\text{(2)}\quad [Cu]C_2H^- + C_2H_2 \overset{k_2}{\rightleftharpoons} X_1;$$

$$\text{(3)}\quad X_1 + H_3O^+ \overset{k_3}{\rightleftharpoons} [Cu] + C_4H_4,$$

steps (1) and (2) that follow from the observed kinetics are supplemented by step (3), which is necessary for obtaining the overall stoichiometric equation of catalytic reaction (**8**) according to Table 2.5.

It was established that constant k in the denominator of Eq. (2.348) was independent of $[CuCl]_\Sigma$ (and of a_{Cu^+}), but depends on a_{Cl^-} ($k = k_0/a_{Cl^-}$). This circumstance leads to the following formula:

$$r_{VA} = \frac{k'_{obs}P_{C_2H_2}}{k_0 P_{C_2H_2} + [H_3O^+][Cl^-]}, \tag{2.349}$$

according to which Cl^- is evolved in the same step as H_3O^+ (i.e., either in reaction (1) or at the subsequent quasi-equilibrium step). Information on the composition of intermediate X_1 (and, to some extent, on its nature) was obtained from the study of kinetics of the synthesis of VA and 2-chlorovinylacetylene (CVA) in systems containing small additives ($\sim 10^{-3}$ M) of $CuCl_2$ [228, 229]). In the presence of $CuCl_2$, the conversion of X_1 proceeds via two parallel routes, namely, the above step (3) and the following step:

$$(4)\ X_1 + CuCl_2 \xrightarrow{k_4} X_2 \xrightarrow{CuCl_2} CH_2{=}C(Cl){-}C{\equiv}CH + 2CuCl.$$

The rate of the oxidative reaction of CVA formation was studied under the conditions of a steady-state electrolytic oxidation of Cu(I) to Cu(II). The overall equation of the route catalyzed by Cu(I) and Cu(II) is as follows:

$$2C_2H_2 + HCl \longrightarrow C_4H_3Cl + H_2.$$

The activities of Cu^+ and Cl^- were maintained constant, while the activity of Cu^{2+} and the proportional concentration $[Cu^{2+}]$ were determined from the measurements of E_{Pt} and E_{Cu}. Under these conditions, the sum of rates r^*_{VA} (under the conditions of CVA synthesis) and r_{CVA} is $r_{VA} = r^*_{VA} + r_{CVA}$ (which is the rate of formation of intermediate X_1, Eq. (2.348)), while the ratio of these rates is described by the following equation:

$$\frac{r^*_{VA}}{r_{CVA}} = K' \frac{[H_3O^+]}{[CuCl_2]}, \tag{2.350}$$

where $K' = k_3/k_4$. The resulting equation for the rate of VA formation involves the concentration of H_3O^+ ions that enter step (3) after the irreversible step (2):

$$r^*_{VA} = \frac{k_{obs} K' P^2_{C_2H_2} [H_3O^+]}{(kP_{C_2H_2} + [H_3O^+])\ (K'[H_3O^+] + [CuCl_2])}. \tag{2.351}$$

This factor accounts for the presence of a maximum in the function $r^*_{VA} = f([H_3O^+])$. It was suggested that intermediate X_1 has a composition of $[Cu]C_4H_3$ and is formed as a result of the insertion of acetylene into Cu–C≡CH bond, that is, represents a β-ethynylvinyl derivative of copper (CuCH=CHC≡CH). The reactions of this intermediate consequently with two $CuCl_2$ molecules yield CVA.

An analogous process takes place during the interaction of CuCH=CHCl with two molecules of $CuCl_2$, which leads to the formation of $CH_2{=}CCl_2$ (with the parallel acidolysis to $CH_2{=}CHCl$) [230, 231]. The formation of CVA from X_1 (common intermediate for both vinyl- and chlorovinylacetylenes) is confirmed by the fact that CVA is not formed by conversion of VA, diacetylene, vinylidene chloride, or vinyl chloride in the indicated systems under given conditions [228, 229].

The results of investigations into the effect of $HgCl_2$ additives on the VA synthesis led to a conclusion that an increase in acidity of the $CuCl{-}NH_4Cl{-}H_2O$ system upon $HgCl_2$ addition is related to an exchange reaction that leads to the formation of a stable (with respect to aqueous HCl) organomercuric derivative (ClHgCH=CHC≡CH), from which VA can be isolated only under the action of concentrated HCl. Thus, Eq. (2.349) describes the dimerization process under consideration and $k'_{obs} = f(a_{Cu^+}, a_{Cl^-})$.

2.3.2.2 *Reaction of acetylene hydrocyanation (11, Table 2.5)*

Reaction (**11**) according to Table 2.5 is carried out in the $CuCl{-}NH_4Cl{-}HCl{-}H_2O$ system at greater values of [HCl] than those used in reaction (**8**) considered above. This is necessary to avoid the formation of polycyanide complexes and large amounts of CuCN. In a solution of 30 mass % CuCl and 15 mass % NH_4Cl, the rate of reaction (**11**) exhibits a maximum with respect to P_{HCN} and [HCl] (Figure 2.13) [226, 232].

The reaction rate was considered as a function of the following parameters:

$$r_{11m} = f(P_{C_2H_2}, P_{HCN}, a_{H_3O^+}),$$

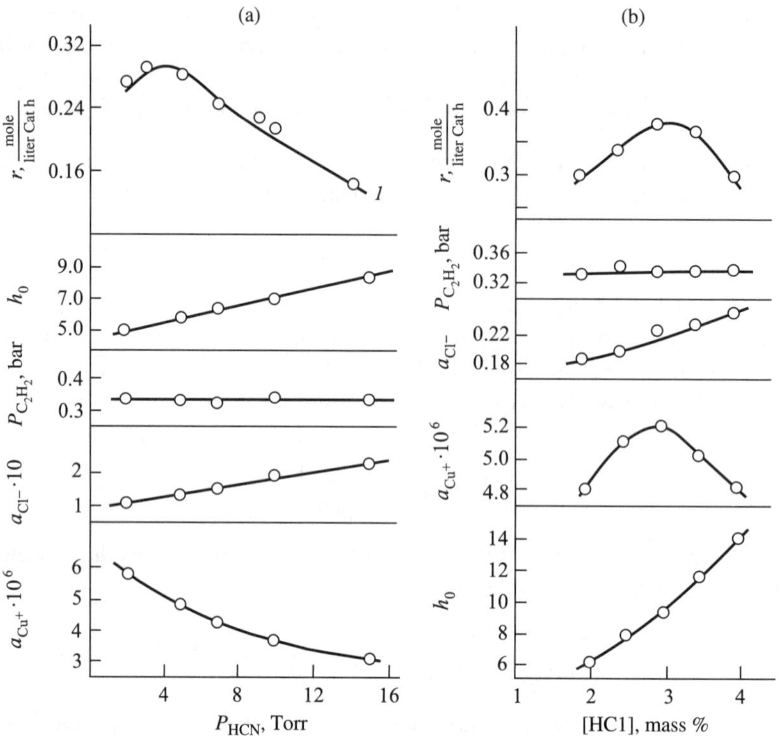

Figure 2.13 Plots of the rate r (points, experiment; curve 1, calculation using Eq. (2.352)) and some other parameters of acetylene hydrocyanation versus (a) P_{HCN} and (b) [HCl] (reproduced with permission from editorial board of *Kinet. Katal.* journal © 1973 [226])

and the form of this equation was determined by studying the reaction kinetics at constant activities a_{Cu^+} and a_{Cl^-} using an approach analogous to that described above, whereby the sum of NH_4Cl and HCl concentrations was maintained constant in series of experiments with variable partial pressures of acetylene and hydrogen cyanide. Under these conditions (with a cationic background of ~9), the reaction rate was described by the following equation:

$$r_{11m} = \frac{k_{11m} P_{C_2H_2} P_{HCN}}{[H_3O^+]}. \tag{2.352}$$

The first order with respect to P_{HCN} and minus first order with respect to $[H_3O^+]$ were obtained in the intervals of [HCl] and P_{HCN}, where the reaction rate exhibits extremal variation under conditions of variable a_{Cu^+} and a_{Cl^-} (see Figure 2.13). The extrema in r_{11m} as a function of P_{HCN} and [HCl] are described by Eq. (2.352) with allowance for the average orders with respect to $a_{Cu^+} (m = 4.2)$ and $a_{Cl^-} (n = 4.4)$ ([226], Section 2.3.3).

Although the presence of a limiting step does not allow one to conclude that this step involves free CN^- ions or complexes containing ~CuCN fragments, the entire combination of experimental data gives grounds to suggest that the reaction mechanism is analogous to that of acetylene dimerization. Since the concentration of free cyanide ions in acid CuCl solutions cannot exceed 10^{-9}–10^{-10} M, the attack of free CN^- on π coordinated acetylene is of a low probability and, hence, the reaction scheme should be

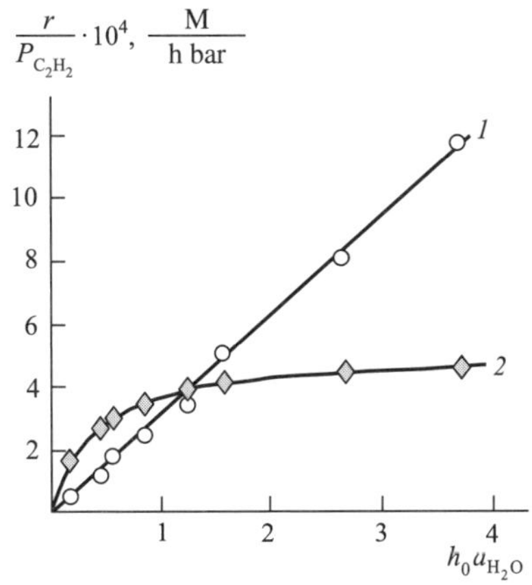

Figure 2.14 Plots of the rates of reactions (curve 1) 1 and (curve 2) 5 (see Table 2.5) versus a_{H_2O} (reproduced with permission from AIPC Nauka © 1970 [215])

described by the following sequence of steps:

$$
\begin{aligned}
&(1)\quad [Cu] + HCN \rightleftharpoons CuCN + H_3O^+ + Cl^-;\\
&(2)\quad [Cu]CN + C_2H_2 \longrightarrow X_2;\\
&(3)\quad X_2 + H_3O^+ + Cl^- \longrightarrow CH_2{=}CHCN + [Cu].
\end{aligned}
$$

2.3.2.3 *Synthesis of bicyclo[2.2.1]hepta-2,5-diene from acetylene and cyclopentadiene (12, Table 2.5)*

The kinetics of reaction (**12**) according to Table 2.5 was studied at a temperature of 80 °C, a cationic background of 15.3, and [CuCl] = 15.8 $\tilde{m}$ under conditions of a_{Cu^+}, a_{Cl^-} = *const* [233, 214]. The concentration of H_3O^+ was determined using calibration plots of $pH = A - B\lg[H_3O^+]$. The rate of the Diels–Alder reaction of [2 + 4]-cycloaddition obeys the relation

$$r_{12n} = \frac{k_{12n} P_{C_2H_2} P_{C_5H_6}}{[H_3O^+]}, \tag{2.353}$$

which corresponds to the following mechanistic scheme [23]:

$$
\begin{aligned}
&(1)\ [Cu] + C_2H_2 \rightleftharpoons [Cu](C{\equiv}CH) + H_3O^+ + Cl^-;\\
&(2)\ [Cu](C{\equiv}CH) + C_5H_6 \longrightarrow X_3;\\
&(3)\ X_3 + H_3O^+ + Cl^- \longrightarrow C_7H_8 + [Cu].
\end{aligned}
$$

where $X_3 = C_7H_7Cu$.

2.3.2.4 *Reactions of acetylene hydrochlorination (1) and hydration (5, Table 2.5)*

The kinetics of parallel reactions (**1**) and (**5**) according to Table 2.5, was studied [215] in a solution of $[NH_4Cl] + [HCl] = 12\,\tilde{m}$ and $[CuCl] = 11\,\tilde{m}$. The steady-state concentration of H_3O^+ (or activity $a_{H_3O^+}$)

was replaced by the proportional value of $h_0 a_{H_2O}$. A series of experiments yielded the following kinetic relations (Figure 2.14)

$$r_{1a} = k_{obs} P_{C_2H_2} h_0 a_{H_2O} = k_{1a} P_{C_2H_2} [HCl], \tag{2.354}$$

$$r_{5e} = k_{obs} \frac{P_{C_2H_2} h_0 a_{H_2O}}{1 + K_a h_0 P_{H_2O}} = k_{5e} \frac{P_{C_2H_2} [H_3O^+]}{1 + k[H_3O^+]}. \tag{2.355}$$

It was established that Eq. (2.355) was invariant with respect to variation of the initial concentration of CuCl and the constant $K_a = 2.67$ was independent of a_{Cu^+} and a_{Cl^-}. Equation (2.354) was attributed to the following mechanistic scheme with limiting step (2):

$$\begin{aligned}&(1)\ [Cu] + C_2H_2 + Cl^- \rightleftharpoons X_4;\\&(2)\ X_4 + H_3O^+ \longrightarrow CH_2{=}CHCl + [Cu].\end{aligned}$$

The kinetic equation (2.355) of acetylene hydration can be assigned to the following scheme without a limiting step:

$$\begin{aligned}&(1)\ [Cu] + C_2H_2 \rightleftharpoons X_5 + H_2O;\\&(2)\ X_5 + H_2O \rightleftharpoons X_6;\\&(3)\ X_6 + H_3O^+ \longrightarrow CH_2{=}CH^+OH_2 + [Cu];\\&(4)\ CH_2{=}CH^+OH_2 + H_2O \rightleftharpoons CH_3CHO + H_3O^+,\end{aligned} \tag{2.356}$$

where the rate of step (3) is comparable with the rate of reverse step (2).

2.3.2.5 *Reactions of methylacetylene hydrochlorination (2, 3) and hydration (6, Table 2.5)*

Chloride solutions of Cu(I) complexes can also catalyze reactions of the formation of 2-chloropropene (2-CP), *cis*-1-chloropropene (1-CP), acetone (A), and (at $[CuCl] > 10\,\tilde{m}$) propyonaldehyde (PA), which represent the products of addition to alkyne according to (2-CP, A) and against (1-CP, PA) the Markovnikov rule. The reaction kinetics was studied for $[NH_4Cl] = 9\,\tilde{m}$, $[CuCl] = 12\,\tilde{m}$ and a temperature of 80 °C. At $a_{Cu^+}, a_{Cl^-} = \text{const}$ and $[HCl] \propto [a_{H_3O^+}]$, the reaction rates obeyed the following equations:

$$r_{2,3} = k_{2,3} P_{C_3H_4} [HCl]; \tag{2.357}$$

$$r_{6g} = k_{6g} P_{C_3H_4} + k_6 P_{C_3H_4} [HCl]. \tag{2.358}$$

The second term in Eq. (2.358) corresponds to an acid-catalyzed route of methylacetylene hydration to acetone, while the first term refers to the cuprocatalytic hydration of methylacetylene, a special feature of which is that it does not depend on $[H_3O^+]$. The mechanism of methylacetylene hydration is described by a scheme with the limiting second step, which is analogous to the scheme of acetylene hydrochlorination. The catalytic formation of acetone is analogous to the process represented by scheme (2.356), but also with the second limiting step that is irreversible in the case of C_3H_4.

2.3.3 Determination of compositions of catalytically active copper(I) complexes in various reactions

The values of observed constants k_{ij} determined as described in the preceding section are complicated functions of a_{Cu^+} and a_{Cl^-} and, hence, depend on the partial rate constants $k_{eff(m,n)}$ and concentrations of the corresponding complexes. The dependence of k_{ij} on $[CuCl]_\Sigma$ is illustrated in Figure 2.15 [185]. Despite the

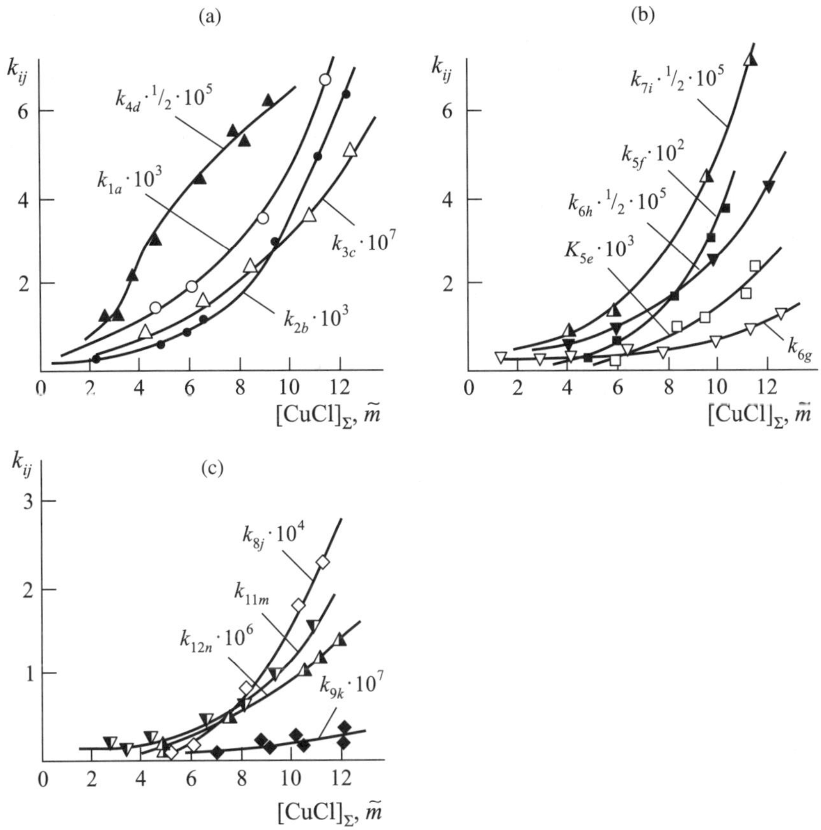

Figure 2.15 Plots of observed reaction rate constants k_{ij} versus CuCl concentration (reproduced with permission from AIPC Nauka © 1970 [219])

fact that k_{ij} in all cases increases with $[CuCl]_\Sigma$, the character of $k_{ij} = f[CuCl]_\Sigma$ is significantly different for various reactions (see Table 2.5). For the reactions of acetylene dimerization (**8**) and hydrocyanation (**11**), methylacetylene hydration (**6h** and **7i**), acetylene hydration (**5f**), and the reaction of cyclopentadiene (**12**), a sharp growth in k_{ij} begins only at $[CuCl]_\Sigma \geq 5\,\tilde{m}$, i.e., in the region of concentration where polynuclear complexes ($Cu_3Cl_4{}^-$, $Cu_3Cl_5{}^{2-}$, $Cu_4Cl_5{}^-$, $Cu_4Cl_6{}^{2-}$, etc.) appear in significant amounts. Only a reaction with HCl (**4d**) has the observed order with respect to $[CuCl]_\Sigma$ below unity. However, the same shape of the dependence of k_{ij} on $[CuCl]_\Sigma$ is observed for a $Cu_2Cl_4{}^{2-}$ complex (see Figure 2.9).

In simple cases where the Cl^- content in the initial complex is the same as that in the transition state at the limiting step, or the subtraction of Cl^- from $Cu_mCl_n^{(n-m)-}$ complex takes place not as a quasi-equilibrium step, we obtain the following equation:

$$k_{ij} = \sum_m \sum_n k_{\text{eff}(m,\,n)}[Cu_mCl_n^{(n-m)-}]. \tag{2.359}$$

This expression of the observed constant implies that, in the process without a limiting step, not all constants of the forward and reverse steps depend on the composition of the active complex. For example, the values of k_0 [Eq. (2.349)] and k [Eq. (2.355)] are independent of a_{Cu^+} and a_{Cl^-}. As a result, the summation of the rates over routes originating from different Cu(I) complexes yields an expression with a denominator that is the same for all routes.

Determining $k_{\mathrm{eff}(m,n)}$ from Eq. (2.359) is only possible provided that all complexes present in the system are taken into account, which implies that a set of equations of the (2.359) type has to be solved. However, taking into account the large number of different complex species present in solution, large errors (10–15 %) in the determination of k_{ij}, and uncertainty in defining a complete set of complexes, one can hardly expect that the calculated coefficients in the formulated set of equations would reflect the true contributions of particular complexes to the process rate.

In this context, it would be of interest to estimate the average composition of complexes that give maximum contributions to k_{ij} due to both high values of $k_{\mathrm{eff}(m,n)}$ and large concentrations of $\mathrm{Cu}_m\mathrm{Cl}_n^{(n-m)-}$. According to Eq. (2.359), we have

$$k_{ij} = \sum_m \sum_n k_{eff\,(m,\,n)} \beta_{m,\,n} a_{\mathrm{Cu}^+}^m a_{\mathrm{Cl}^-}^n = \sum_m \sum_n k_{m,\,n} a_{\mathrm{Cu}^+}^m a_{\mathrm{Cl}^-}^n.$$

An increase in $[\mathrm{CuCl}]_\Sigma$ leads to a decrease in a_{Cl^-} because a_{Cu^+} and a_{Cl^-} are interrelated [see Eqs (2.344) and 2.345)]. Therefore, independent determination of the m and n values is impossible and it is necessary to use special planning of experiment with controlled variation of $[\mathrm{NH_4Cl}]_\Sigma$ and $[\mathrm{CuCl}]_\Sigma$.

Different variants of estimating the m and n values were considered that required the verification of relations

$$k_{ij} = k_{\mathrm{eff}(i,\,j)} a_{\mathrm{Cu}^+}^m a_{\mathrm{Cl}^-}^n \tag{2.360}$$

for various preset values of $p = n - m$ (0, 1, 2, ...), e.g., in the form of equation

$$\lg k_{ij} - p \lg a_{\mathrm{Cl}} = m \lg a_{\mathrm{Cu}^+} a_{\mathrm{Cl}^-}. \tag{2.361}$$

In particular, it was established for the reaction of acetylene dimerization that m does not strongly vary for different p values [221]. A minimum sum of square deviations was obtained for $p = 1$ and $m = 3.93$ ($n = 4.93$). Then, different p and m were tried (with a step of 0.02 in both p and m), the form of an equation that related the activity of water to the rate of reaction was chosen, and the value of variance for the rates of dimerization was evaluated [223]. In this case, σ_{min}^2 was obtained for $m = 3.5$ and $n = 4.5$. The value of k'_{obs} in Eq. (2.349) was best described by the following two-term expression:

$$k'_{\mathrm{obs}} = k_{eff\,(3,\,4)} a_{\mathrm{Cu}^+}^3 a_{\mathrm{Cl}^-}^4 + k_{eff\,(4,\,5)} a_{\mathrm{Cu}^+}^4 a_{\mathrm{Cl}^-}^5, \tag{2.362}$$

although a contribution due to the first term is relatively small (Figure 2.16). An analogous analysis of the composition of complexes in the reaction of acetylene hydrocyanation (**11**, Table 2.5) led to the values of $m = 4.2$ and $n = 4.4$ for the transition state in the limiting step [226].

An alternative approach to estimating the interval of possible m and n values for the complexes that contribute to k_{ij} or the initial active $\mathrm{Cu}_m\mathrm{Cl}_y^{(y-m)}$ complexes (with $y \neq n$ for the subtraction or addition of Cl^- at particular reaction steps) with allowance for the interrelation between a_{Cu^+} and a_{Cl^-} was developed in [8, 23, 185]. It was demonstrated [185] that the a_{Cu^+} and a_{Cl^-} values determined by the potentiometric method in the entire range of a_{Cu^+} ($a_{\mathrm{Cu}^+} = 10^{-8}$–$10^{-6}$) for the catalytic systems studied obey the following empirical relationship:

$$a_{\mathrm{Cl}^-} = a_{\mathrm{Cl}^-,\,0} / b a_{\mathrm{Cu}^+}^g. \tag{2.363}$$

The values of parameters b and g have to be experimentally determined for each particular kinetic equation presented in Table 2.6. Using Eqs (2.359) and (2.363), we obtain the following expression:

$$k_{ij} = \sum_m \sum_n k_{m,\,n} a_{\mathrm{Cu}^+}^m a_{\mathrm{Cl}^-}^n = \sum_m \sum_n \frac{k_{m,\,n} a_{\mathrm{Cl}^-,0}^n}{b^n} a_{\mathrm{Cu}^+}^{m-ng} = \sum_k k_k a_{\mathrm{Cu}^+}^{l_k}. \tag{2.364}$$

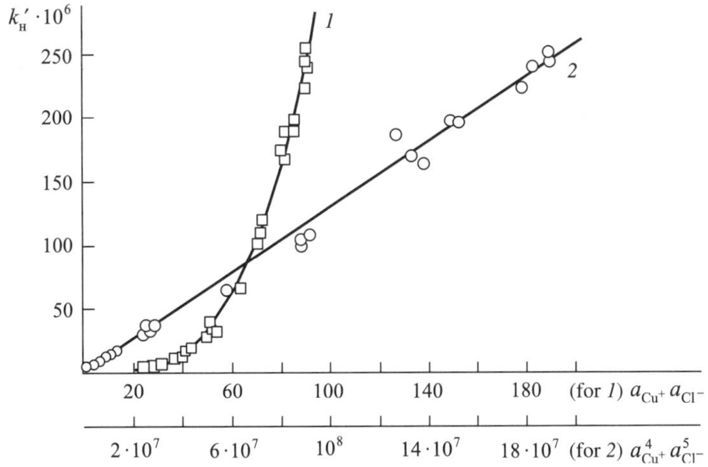

Figure 2.16 Plots of observed reaction rate constant k'_{obs} versus activity of copper and chlorine ions: (1) $a_{Cu^+} a_{Cl^-}$; (2) $a^4_{Cu^+} a^5_{Cl^-}$

The average observed reaction order ($\hat{l}$) with respect to a_{Cu^+} can be determined by approximating Eq. (2.364) as follows:

$$k_{ij} = k_{eff\,(i,j)} a^{\hat{l}}_{Cu^+}, \tag{2.365}$$

where $\hat{l} = \hat{m} - \hat{n}g$ is the weighed mean that takes into account l_k for each complex and the relative contribution (Θ_k) to the total reaction rate due to the participation of the kth complex, so that

$$\frac{d \ln k_{ij}}{d \ln a_{Cu^+}} = \hat{l} = \sum l_k \Theta_k. \tag{2.366}$$

According to Eq. (2.366), the so defined $\hat{n}$ and $\hat{m}$ also represent weighed mean values. In the general case, $\hat{l}$ is variable because Θ_k can vary depending on $[CuCl]_\Sigma$. The constancy of $\hat{l}$ in the indicated interval of CuCl concentrations for all processes studied is indicative of the proximity of m (n) values in the complexes that give the main contribution to k_{ij}. The values of m and n for the $(Cu_mCl_n)^{(n-m)-}$ complexes were estimated using the following equation:

$$\hat{m} = \frac{\hat{l}}{1-g} + \frac{\hat{q}g}{1-g} = \hat{m}_{min} + \frac{\hat{q}g}{1-g}, \tag{2.367}$$

where $\hat{q} = \hat{n} - \hat{m}$. It was assumed that multiply charged complexes that are formed in small amounts in concentrated CuCl solutions do not participate in the reactions involving acetylene compounds. This assumption restricted the analysis of the composition of $Cu_mCl_y^{(y-m)-}$ complexes to the cases where $y - m = 1, 2$. Another assumption was that, during the formation of a π complex, acetylene replaces water molecules so that the charge of the complex is retained. Then, the variable $\hat{q}$ in Eq. (2.367) can only take the values of 0, 1, 2, and 3, depending on the presence of reaction steps with the removal of one Cl^- ion ($y = n + 1$, $\hat{q} = 0, 1$), the intermediate conversion steps involving a single Cl^- ion ($y = n - 1$, $\hat{q} = 2, 3$), and the mechanisms with $y = n(\hat{q} = 1, 2)$. The corresponding values of $\hat{m}$ vary from $\hat{m}_{min}(\hat{q} = 0)$ to $\hat{m}_{min} + 3g/(1 - g)$. With these assumptions, a physical sense is only inherent in the cases of $\hat{m} \geq 1$ and $\hat{n} \geq \hat{m}$. Data on the values of observed reaction orders with respect to $a_{Cu^+}(\hat{l})$ are presented in Table 2.7.

Table 2.7 Parameters of Eq. (2.367) and probable compositions of initial $(Cu_mCl_y)^{(y-m)-}$ complexes (reproduced with permission from AIPC Nauka © 1990 [185])

k_{ij}	$\hat{\imath}$	G	$\hat{q}=0$	$\hat{q}=1$		$\hat{q}=2$		$\hat{q}=3$
			$\hat{m}_{min}$	$\hat{m}$	$(Cu_mCl_y)^{(y-m)-}$	$\hat{m}$	$(Cu_mCl_y)^{(y-m)-}$	$\hat{m}$
k_{1a}	0.73	0.29	1.03			1.85	$CuCl_2^-$, $Cu_2Cl_3^-$	2.25
k_{2b}	0.88	0.33	1.31			2.30	$Cu_2Cl_3^-$, $Cu_3Cl_4^-$	2.79
k_{3c}	1.10	0.33	1.64			2.65	$Cu_2Cl_3^-$, $Cu_3Cl_4^-$	3.12
k_{4d}	0.71	0.33	1.06			2.26	$Cu_2Cl_3^-$, $Cu_3Cl_4^-$	2.54
k_{5e}	0.63	0.32		1.49	$CuCl_2^-$, $Cu_2Cl_3^-$	1.96	$Cu_2Cl_4^{2-}$	
k_{5f}	1.16	0.29	1.64	2.04	$Cu_2Cl_3^-$	2.45	$Cu_2Cl_4^{2-}$, $Cu_3Cl_5^{2-}$	
k_{6g}	0.67	0.33	1.00	1.49	$CuCl_2^-$, $Cu_2Cl_3^-$	1.99	$Cu_2Cl_4^{2-}$	
k_{6h}	0.77	0.47	1.45	2.34	$Cu_2Cl_3^-$, $Cu_3Cl_4^-$	3.23	$Cu_3Cl_5^{2-}$, $Cu_4Cl_6^{2-}$	
k_{7i}	0.99	0.47	1.87	2.75	$Cu_2Cl_3^-$, $Cu_3Cl_4^-$	3.64	$Cu_3Cl_5^{2-}$, $Cu_4Cl_6^{2-}$	
k_{8j}	2.60	0.32	3.83	4.30	$Cu_4Cl_5^-$, $Cu_5Cl_6^-$	4.76	$Cu_4Cl_6^{2-}$, $Cu_5Cl_7^{2-}$	
k_{9k}	0.13	0.33	–	–	–	1.18	$CuCl_3^{2-}$, $Cu_2Cl_4^{2-}$	
k_{10l}	0.64	0.33	–	1.45	$Cu_2Cl_3^-$, $CuCl_2^-$	1.94	$CuCl_2^-$, $Cu_2Cl_3^-$ $Cu_2Cl_4^{2-}$	
k_{11}	2.21	0.39	3.62	4.26	$Cu_4Cl_6^{2-}$, $Cu_5Cl_7^{2-}$			
k_{12n}	1.52	0.60	3.80	5.30	$Cu_5Cl_7^{2-}$, $Cu_6Cl_8^{2-}$			
k_{12o}	2.02	0.60	5.05	6.55	$Cu_7Cl_9^{2-}$			

Notes: For $\hat{q}=0$, reactions (11) and (12) involve $(Cu_mCl_y)^{(y-m)-}$ complexes with $m=3$ and 4 (catalytic systems ***m***, ***n***) and $m=3$ (catalytic system ***o***).

For example, the reactions of acetylene hydrocyanation (**11**, Table 2.5) and diene condensation (**12**), which contain limiting steps, involve $(Cu_mCl_nCN)^{(n-m+1)-}$ and $[Cu_mCl_n(C{\equiv}CH)]^{(n-m+1)-}$ complexes, respectively, formed as a result of the substitution of Cl^- ions by cyanide ions and $HC{\equiv}C^-$ ions:

$$(Cu_mCl_y)^{(y-m)-} + HC{\equiv}CH \rightleftharpoons [(Cu_mCl_n(C{\equiv}CH)]^{(n-m+1)-} + H_3O^+ + Cl^-.$$

Accordingly, these reactions are characterized by $y=n+1$ (for $\hat{q}=0,1$) and $\hat{m}=3.6$–6.5. In the reactions (**8**)–(**10**), which contain no limiting steps, the value of $\hat{q}$ refers to the composition of a transition state ($y=n$). Reactions (**8**), (**11**), and (**12**) predominantly involve polynuclear complexes with $m=4$–7 (see Table 2.7).

Using the obtained average values of m and n, it was possible to describe the extremal temperature dependences of the reaction rates for acetylene dimerization [222] and hydrocyanation [226], as well as the dependences of the dimerization rate on the nature of chlorides and the hydrocyanation rate on the value of the ionic strength of the medium. In the case of acetylene dimerization, metal chlorides were introduced into the standard system ([CuCl] $=6.6\ \hat{m}$, [NH_4Cl] $=8.74\ \hat{m}$) at an amount of 0.3–1.5 $\hat{m}$ [223]. These experiments revealed three groups of chlorides that differently influence the rate of acetylene dimerization (Figure 2.17):

(i) chlorides that increase the reaction rate ($ZnCl_2$, $CdCl_2$, $BiCl_3$);

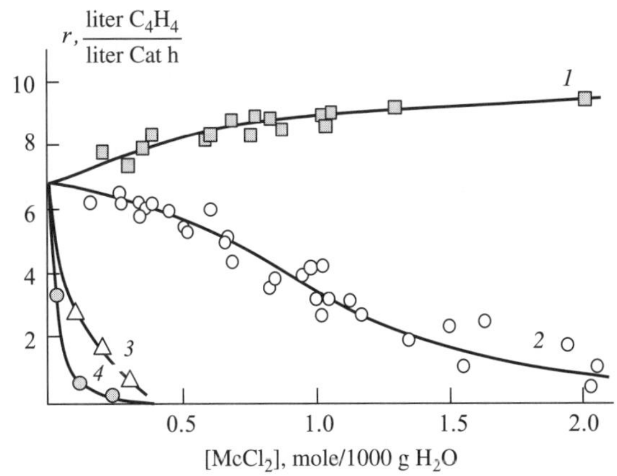

Figure 2.17 Plots of acetylene dimerization rate r versus chloride concentration: (1) $ZnCl_2$, $CdCl_2$, $BiCl_3$; (2) $BeCl_2$, $MgCl_2$, $CaCl_2$, $SrCl_2$, $BaCl_2$, $FeCl_2$, $CoCl_2$, $MnCl_2$, $2/3CrCl_3$, NH_4Cl; (3) $HgCl_2$; (4) $SnCl_2$ (reproduced with permission from editorial board of *Kinet. Katal.* journal © 1970 [223])

(ii) chlorides that decrease the reaction rate (NH_4Cl, $BeCl_2$, $MgCl_2$, $CaCl_2$, $SrCl_2$, $BaCl_2$, $MnCl_2$, $FeCl_2$, $CoCl_2$, $NiCl_2$, $CrCl_3$);

(iii) chlorides that sharply decrease the reaction rate ($SnCl_2$, $HgCl_2$, $CuCl_2$).

A change in the reaction rate upon adding chlorides of the first two groups was related to their influence on some variable parameters (a_{Cu^+}, a_{Cl^-}, h_0 and a_{H_2O}) in the kinetic equation (2.369) [221]:

$$r_{VA} = \frac{k_{eff} a^4_{Cu^+} a^5_{Cl^-} P^2_{C_2H_2} a_{H_2O}}{h_0 a^2_{H_2O} a_{Cl^-} + K_a P_{C_2H_2}}. \tag{2.368}$$

A change in the value of the ionic strength of the medium ($[NH_4Cl] = 7.5 - 12\,\tilde{m}$) affects neither the values of k_{eff} and K_a (i.e., the activity coefficients of transition states) nor the acetylene activity $a_{C_2H_2}$ (Figure 2.18). For example, the steady-state acidity h_0 decreased in the indicated concentration range from 0.3 to 0.01, while the product $a^4_{Cu^+} a^5_{Cl^-}$ decreased by a factor of about 25. An increase in the concentration

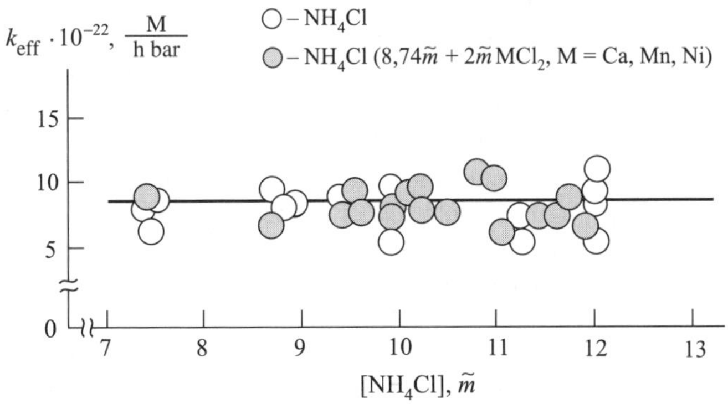

Figure 2.18 Plot of k_{eff} defined by Eq. (2.368) versus concentrations of NH_4Cl and other chlorides at 80 °C and [CuCl] = 6.56 m̃ [214]

of other chlorides of the second group also did not influence the values of k_{eff}. A close value of k_{eff} was also obtained for $ZnCl_2$. A significant increase in the reaction rate by the addition of $ZnCl_2$, $CdCl_2$ and $BiCl_3$ was related to the binding of Cl^- ions and to an increase in a_{Cu^+}. In these series of experiments, the activity product $a^4_{Cu^+}a^5_{Cl^-}$ exhibited an approximately twofold increase.

A sharp inhibition of the process upon the introduction of $CuCl_2$ and $HgCl_2$ (chlorides of the third group) into the solution was related to a conversion of the intermediate σ organocopper compounds. In particular, $CuCl_2$ reacts with the intermediate acetylenide [Cu]C_2H [or $Cu_4Cl_4(C{\equiv}CH)^-$] with the formation of diacetylene, and with intermediate X_2 ([Cu]CH=CHC≡CH) with the formation CVA (see steps (1)–(4) of VA and CVA formation); $HgCl_2$ converts σ organocopper compound X_2 into a stable (with respect to acidolysis) organomercuric derivative ClHgCH=CHC≡CH [223], while $SnCl_2$ exhibits hydrolysis that strongly increases h_0 and decreases a_{Cu^+}.

The lack of the influence of the ionic strength of the medium on k_{eff} (within $[NH_4Cl] + [HCl] = 6.4 - 12\hat{m}$ at $[HCl] = 0.5 - 2\hat{m}$) was also demonstrated in the case of reaction (**11**, Table 2.5) [226].

The participation of polynuclear Cu(I) complexes in the dimerization of acetylene according to reaction (**8**, Table 2.5) was confirmed by the results of a spectrophotometric investigation of the dependence of the reaction rate on the concentration of so-called "yellow" complexes ($Cu_4Cl_4(C{\equiv}CH)^-$ and $Cu_4Cl_5(C{\equiv}CH)^{2-}$ and the dependence of their composition on $a_{Cu^+}, a_{Cl^-}, P_{C_2H_2}$ and $[H_3O^+]$ [234]. Additional evidence is provided by the isolation of crystalline copper chloride, ethynyl, and π acetylene complexes formed in the Nieuwland catalytic system.

2.3.4 Studying π and σ complexes of copper(I) with alkynes in crystalline state and in solution

The composition and structure of $M_{(n-m)}Cu_mCl_n$, complexes formed in concentrated aqueous solutions of MCl and CuCl, as well as complexes isolated from solutions in the crystalline form (see Section 2.3.1), provide notions about the state of the initial catalytic system [23]. During the interaction of the MCl–CuCl–H_2O system with acetylene (e.g., under the conditions of oligomerization in the Nieuwland catalytic system), the molecules of acetylene form π and σ organocopper compounds with $(Cu_mCl_n)^{(n-m)-}$ anions. These anions form complexes, first with acetylene and then with the products of its conversion (mono- and disubstituted derivatives) [7, 8, 184].

The π complexes formed at the first stage are converted, depending on the pH of the reaction medium, into compounds of the mono- and diacetylenide types [234]. In the initial, weakly acidic medium, the organometallic compounds predominantly occur in the form of acetylenide (ethynyl) derivatives, while a decrease in pH leads to their reversible conversion into π complexes representing the main products of reactions between alkynes and Cu(I). This section is devoted to π- and σ-organocopper compounds that are formed during the interaction of acetylene compounds with concentrated aqueous solutions containing MCl and CuCl. An analysis of data on the structural features of polynuclear Cu(I) ethynyl complexes is necessary for deeper insight into their role in catalytic processes (provided that the main structural features of the active fragments of crystalline compounds are retained in solution). This analysis allows one to judge the efficiency of π interaction in π complexes of copper (I), estimate the degree of activation of the triple bond, and trace the possible pathways of conversion of the π complexes into acetylenide (ethynyl) complexes (Tables 2.8 and 2.9). It should be noted that the analysis is restricted to the main structural features of compounds under consideration.

Let us consider the main conclusions of a detailed analysis of these issues [23]. For the sake of simplicity, the type (mode) of coordination of the C≡C group (η^1 or η^2) will be referred to as the σ or π type, respectively (or as σ,π type if both modes of coordination take place).

Table 2.8 Specific features of coordination of C≡C groups in structures of π complexes of Cu(I) chlorides with acetylene and its derivatives (reproduced with permission from editorial board of *Usp. Khim.* journal © 2000 [23])

Complex	Type of coordination of C≡C group	Cu–C* bond length, Å	Difference of Cu–C bond lengths, Å	C≡C bond length, Å	Ref.
Cu_2Cl_2(HC≡CH) (17)	μ-η^2,η^2	2.01(8)	0.00(8)	1.18(14)	[235]
		1.76(8)	0.00(8)		
		2.12(7)	0.11(7)	1.26(13)	
		1.99(9)	0.07(9)		
		2.16(9)	0.00(9)	0.94(24)[a]	
		2.29(9)	0.00(9)		
Cu_3Cl_3(HC≡CH) (**18**)	μ-η^2,η^2	2.01(3)	0.08(3)	1.12(4)[a]	[235]
		2.01(3)	0.08(3)		
		1.97(2)	0.02(2)	1.32(5)	
		1.97(2)	0.02(2)		
Cu_6Cl_6(HC≡CH) (**19**)[b]	μ-η^2,η^2	–	–	–	[187]
KCu_3Cl_4(HC≡CH) (**20**)	μ-η^2,η^2	1.940(4)	0.011(4)	1.26(1)	[235]
		1.940(4)	0.011(4)		
		1.947(4)	0.004(4)	1.28(1)	
		1.947(4)	0.004(4)		
$NH_4Cu_3Cl_4$(HC≡CH) (**21a**)	μ-η^2,η^2	1.94(1)	0.00(1)	1.32(2)	[236]
		1.94(1)	0.00(1)		
		1.93(1)	0.03(1)	1.25(1)	
		1.93(1)	0.03(1)		
$RbCu_3Cl_4$(HC≡CH) (**21b**)	μ-η^2,η^2	2.03(3)	0.00(3)	1.33(3)	[237]
		1.93(3)	0.00(3)		
		1.89(2)	0 04(2)	1.27(3)	
		1.89(2)	0 04(2)		
K$Cu_8Cl_9(HC≡CH)_4\cdot{}^1/_2HCu_2Cl_3\cdot H_2O$ (**22**)	μ-η^2,η^2	1.948(4)	0.000(4)	1.23(1)	[235]
		1.948(4)	0.000(4)		
$NH_4Cu_8Cl_9(HC≡CH)_4 \cdot 2/5$ $\{[Cu(H_2O)_2]\cdot\cdot[CuCl_2]\cdot H_2O\}$ (**23**)	μ-η^2,η^2	1.957(5)	0.000(5)	1.233(9)	[238]
		1.957(5)	0.000(5)		
CuCl(HC≡CPh) (**24**)	η^2	1.941(4)	0.067(4)	0.213(5)	[239]

(continued overleaf)

Table 2.8 *(continued)*

Complex	Type of coordination of C≡C group	Cu–C* bond length, Å	Difference of Cu–C bond lengths, Å	C≡C bond length, Å	Ref.
CuCl(HC≡CCH$_2$OH) (**25**)	η^2	1.92(1)	0.012(9)	1.20(1)	[239]
Cu_2Cl_2(HC≡CCH$_2$OH) (**26**)	μ-η^2,η^2	1.94(1)	0.00(1)	1.26(2)	[240]
		1.98(1)	0.02(1)		
		1.95(1)	0.03(1)	1.24(2)	
		1.99(1)	0.04(1)		
		1.91(1)	0.01(1)	1.25(2)	
		2.10(1)	0.08(1)		
Cu_7Cl_7(HC≡CCH=CH$_2$)$_3$ (**27**)	μ-η^2,η^2	1.98(1)	0.03(1)	1.26(2)	[241]
		1.96(2)	0 02(2)		
Cu_2Cl_2(CH$_2$=CHC≡CCH=CH$_2$) (**28**)	η^2	1.961(4)[b]	0.050(4)	1.361(6)	[242]
Cu_3Cl_3(CH$_2$=CHC≡CCH=CH$_2$) (**29**)	η^2	1.951(7)[c]	0.034(6)	1.34(1)	[242]
		1.962(7)[c]	0.008(6)	1.36(1)	
		1.910(6)	0.002(5)	1.225(8)	

Notes: Cu–C* is the distance from Cu atom to the middle of multiple bond; [a]positions of C atoms in acetylene fragment are not refined because of low stability of complexes; [b]first stage of X-ray diffraction analysis; [c]Cu(I) atom coordinates C=C bonds; numbering of complexes according to [23].

Table 2.9 Type of coordination of C≡C bonds in Cu(I) ethynyl complexes (reproduced with permission from editorial board of *Usp. Khim.* journal © 2000 [23])

Complex	Type of coordination of C≡C group	Cu–C* bond length, Å	Difference of Cu–C bond lengths, Å	C≡C bond length, Å	Ref.
$(ImH)_4Cu_9Cl_{11}(C{\equiv}CH)_2$ (**30**)	μ-η^1,η^1		–	1.21(2)	[243]
$(PyH)_4Cu_9Cl_{12}(C{\equiv}CH)$ (**31**)	μ-η^1,η^1		–	1.19(7)	[244]
$(NH_4)_8Cu_{29}Cl_{29}(C{\equiv}C)_4 \cdot 7H_2O$ (**32**)	μ_3-η^1,η^1,η^1	2.18(3)	0.34(3)	–	[245]
	μ-η^2,η^2	2.01(2)	0.01(2)	1.27(5)[b]	
	μ_3-η^1,η^1,η^1	2.01(2)	0.01(2)	–	
	μ-η^1,η^1	2.06(2)	0.06(2)	1.23(5)	
	μ_3-η^2,η^2,η^2	2.06(2)	0.06(2)	–	
	μ_3-η^1,η^1,η^1		–	–	
$NH_4Cu_4Cl_4(C{\equiv}CCH_2OH) \cdot H_2O$ (**33**)[a]	μ-η^1,η^1 μ-η^2,η^2	–	–	–	[188]
$(NH_4)_2Cu_4Cl_5(C{\equiv}CCH_2OH) \cdot H_2O$ (**33a**)	μ-η^1,η^1	1.986(8)	0.028(8)	1.29(1)	[246]
	μ-η^2,η^2	2.005(8)	0.028(8)		
$(C_6H_5NH_3)_2Cu_4Cl_5$-$(C{\equiv}CCH_2OH)$ (**34**)	μ-η^1,η^1	2.00(2)	0.04(2)	1.25(4)	[247]
	μ-η^2,η^2	2.01(2)	0.03(3)		
$Cs_2Cu_5Cl_6(C{\equiv}CCH_2OH)$ (**35**)	μ-η^1,η^1	1.94(3)	0.08(2)	1.22(3)	[248]
	μ-η^2,η^2	1.98(2)	0.03(2)		
$Rb_3Cu_7Cl_9(C{\equiv}CCH_2OH) \cdot H_2O$ (**36**)	μ_3-η^1,η^1,η^1	2.09(2)	0.00(2)	1.23(3)	[249]
	μ-η^2,η^2	1.97(2)	0.02(2)		
$(PyH)_2Cu_8Cl_8(C{\equiv}CCH_2OH)_2 \cdot H_2O$ (**37**)	μ_3-η^1,η^1,η^1	1.974(5)	0.024(5)	1.258(6)	[244]
	μ-η^2,η^2	1.994(5)	0.009(5)		
	μ-η^1,η^1	1.989(5)	0.127(5)	1.267(6)	
	μ-η^2,η^2	2.020(5)	0.009(5)		
$(C_6H_5NH_3)_3Cu_8Cl_{10}(C{\equiv}CCH_2OH) \cdot 2H_2O$ (**38**)	μ-η^1,η^1	1.97(1)	0.04(1)	1.25(2)	[250]
	μ-η^2,η^2	1.98(1)	0.02(1)		
$HO(CH_2)_2NH_3Cu_9Cl_8(C{\equiv}CCH_2OH)_2 \cdot (H_2O)_2$ (**39**)	μ_3-η^1,η^1,η^1	1.988(6)	0.012(7)	1.259(8)	[251]
	μ-η^2,η^2	1.965(6)	0.069(6)		
$Cu_{25}Cl_{17}(C{\equiv}CCH{=}CH_2)_8 \cdot 14H_2O$ (**40**)	μ-η^1,η^1	1.93(4)	0.03(3)	1.21(5)	[241]
	μ-η^2,η^2	2.00(3)	0.03(3)		

Notes: [a]First stage of X-ray diffraction analysis; [b]C≡C group is coordinated to three Cu(I) atoms according to η^2 type; numbering of complexes according to [23].

2.3.4.1 π Complexes of acetylene and alkynes

In the structures of all the known π complexes of Cu(I), acetylene always appears as a bridging π ligand that is coordinated simultaneously by two metal atoms (Figure 2.19). In the π complexes, acetylene is isolobal with the μ_2-Cl ligand. An analogous behavior is known for some alkynes (compounds **26, 27**, Table 2.8). However, in some other alkynes and alkenynes (e.g., in divinyl acetylene derivatives), the C≡C bond in π complexes is coordinated (similar to olefin bonds) by a single copper atom (compounds **24, 25, 28, 29**). Note that the C≡C bond exhibits a greater ability to π interaction with Cu(I) than does the C=C bond.

According to the Dewar–Chatt–Duncanson concept (see, e.g., [8, 252, 253]), the formation of π complexes involves the appearance of two three-center molecular orbitals (MOs), which is related to the overlap between the vacant sp^3 hybrid orbital of Cu(I) and π-MO of acetylene (donor–acceptor M ← L bond) and the overlap between one occupied *d* orbital of the metal and π^*-MO of acetylene (electron-donating M → L bond). It is also possible that two *d* orbitals of Cu(I) overlap with $\pi^*_\perp$- and $\pi^*_\parallel$-MO of acetylene, since the *cis*-distortion of acetylene in the π complex removes the degeneracy of the bonding and antibonding MOs (acetylene becomes the best donor and acceptor of electrons). Both these three-center bonds (M ← L, M → L) contribute to an increase in the strength of the Cu–π ligand bond, but they differently influence the properties of the π ligand. For example, in π complexes of Cu(I), the occupation of antibonding π^*-MO of acetylene by elections from the metal (electron-donating component) leads to a change in C≡C–R bond angles (*cis*-distortion) and an increase in the C≡C bond length. On the other hand, the electron transfer from the π^*-MO of acetylene to one of the sp^3 hybrid orbitals of Cu(I) (donor–acceptor component) accounts for a decrease in the electron density on the C≡C bond, which leads to the appearance of a positive charge (δ^+) on carbon atoms, so that this π coordinated acetylene group becomes a soft electrophile. The electron density transfer from the ligand to Cu(I) also results in a decrease in the distance from Cu(I) to the middle of the C≡C bond (Cu–C* distance). Thus, specific features of the activation of acetylene in the π complexes of Cu(I) are determined primarily by the donor–acceptor (M ← L) interaction [254–256].

The formation of Cu(I)–(C≡C) (or Cu(I)–(C=C)) bonds in Cu(I) chloride π complexes leads to a deformation of the tetrahedral environment of the metal atom (which is typical of Cu(I) compounds with

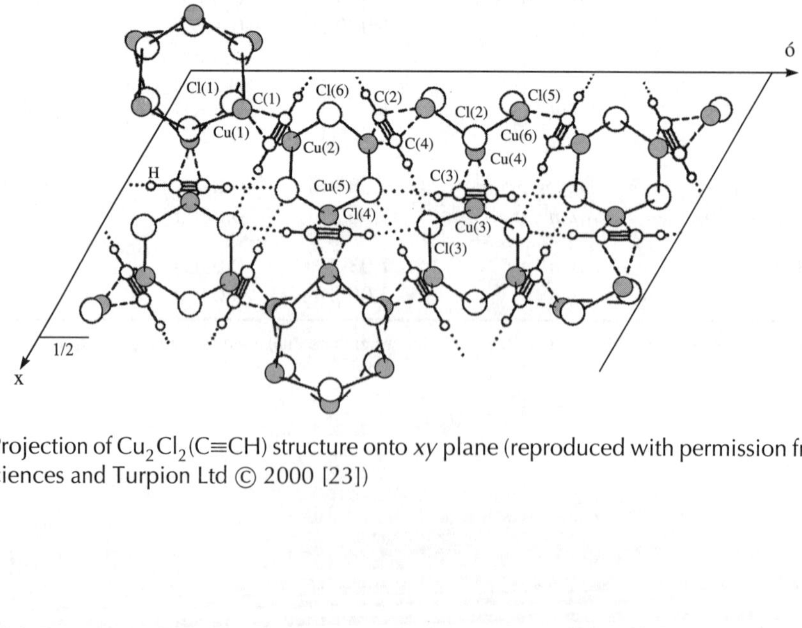

Figure 2.19 Projection of $Cu_2Cl_2(C{\equiv}CH)$ structure onto *xy* plane (reproduced with permission from The Russian Academy of Sciences and Turpion Ltd © 2000 [23])

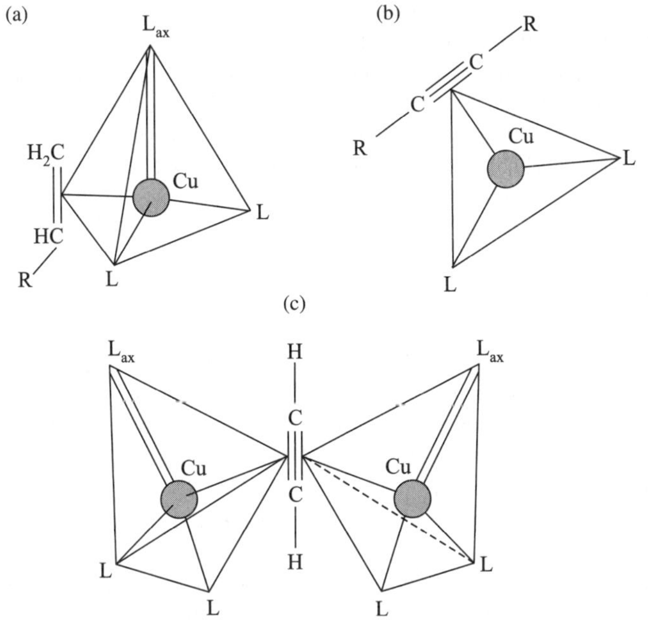

Figure 2.20 Types of coordination environment of Cu(I) in π complexes of CuCl and CuBr: (a) trigonal pyramid typical of Cu(I) atom π coordinated to C=C bond; (b) planar trigonal environment of Cu(I) atom in π complexes with disubstituted acetylene derivatives; (c) bridging π coordination of C≡C bond by Cu(I) atoms in π complexes of C_2H_2 and its monosubstituted derivatives (reproduced with permission from The Russian Academy of Sciences and Turpion Ltd © 2000 [23])

σ ligands) into a trigonal pyramid or even a planar triangle [257, 258]. The interaction of Cu(I) with a multiple C–C bond is accompanied by a decrease in the copper atom–ligand bond lengths at the pyramid base and by an increase in the Cu–L_{ax} bond length (i.e., Cu(I) atom sinks down to the base of the polyhedron). The extent of this deformation increases with the efficiency of the electron-donating component of the π bond. Owing to a synergism of the Cu(I) $\leftarrow$ L and Cu(I) $\rightarrow$ L components, this deformation is accompanied by a decrease in the Cu–C* distance [23].

The coordination of disubstituted acetylene molecules at Cu(I) produces a planar-trigonal environment. In the case of olefins, the coordination environment of Cu(I) has a trigonal pyramid geometry. The formation of bridging bonds in the π complexes of acetylene and substituted alkynes is also related to Cu(I) in a trigonal pyramid environment (Figure 2.20).

Deformation of the coordination polyhedra of Cu(I) atoms, which change from an almost regular tetrahedron ($R_{Cu-L_{ax}} = 2.416$ Å) to a trigonal pyramid ($R_{Cu-L_{ax}} = 2.6$–3.05 Å) and then to a planar-trigonal geometry ($R_{Cu-L_{ax}} = 3.738$ Å) is accompanied by a decrease in the Cu–$C^*_{C\equiv C}$ distance, which is evidence of an increase in the efficiency of π bonding in this sequence [23].

In structures **17–23** (Table 2.8), the Cu–C* distances are significantly greater than those in the case of monosubstituted alkynes and are close to Cu–C* (C=C) distances. The average Cu–C* distance in molecular complexes **17** and **18** (1.76–2.29 Å) is significantly greater than in anionic compounds **20–23** (1.89–1.957 Å). The latter compounds are characterized by an almost planar-trigonal environment of Cu(I) atoms. In most of the structurally characterized Cu(I) π complexes, the end carbon atom in terminal alkynes is somewhat closer to the Cu atom. The formation of an asymmetric Cu(I)(C≡C) π-node is mostly caused by the formation of hydrogen bonds (H-bonds) of the ≡C–H ···Cl type [23].

During the formation of π complexes with Cu(I) and some other metals, the acidity of C–H bonds in alkynes increases. The polarization of $C^{\delta-}$–$H^{\delta+}$ bonds makes the π complex a strong acid that is capable, depending on the pH of the solution, to dissociate with the formation of acetylenide compounds. The tendency to dissociation is related not only to the polarization of the C–H bond, but also to a high stability of the product of coordination of the $RC{\equiv}C^-$ anions to Cu(I) [7, 8]. At the same time, the polarization of the C–H bond in a π complex favors the formation of an H-bond with a chloride ion or a coordinated chloride ligand. For example, in a strongly acid medium, acetylene and its monosubstituted derivatives are bound with CuCl into π complexes, in which H-bonds of the ≡C–H ···Cl type produce a stabilizing effect on the complex structure [259, 260]. H-bonds of this type are characterized by an energy that is comparable with the energy of π interaction ($E_{C-H\cdots Cl} \sim 20\,kJ/mole$; $E_{Cu(I)-C^*_{C\equiv C}} \sim 50\,kJ/mole$). In the structures of π complexes with acetylene, the presence of H-bonds of the ≡C–H ···Cl type leads to an additional polarization of Cu–Cl bonds, which favors enhanced π bonding in the Cu(I)–(C≡C)–Cu(I) chain [187] that, in turn, influences the geometry of the π-coordinated C≡C bond. Moreover, there is a certain relationship between the elongation of a coordinated C≡C bond and the length of the ≡C–H ···Cl contact: extension of the triple bond is accompanied by weakening of the corresponding H-bond [23].

During catalytic reactions in concentrated solutions of anionic complexes, an important role is played by NH_4^+, K^+, and Na^+ cations, which influence the activity of the catalytic system. It is interesting to trace how the cation nature is manifested in the features and characteristics of hydrogen bond formation, e.g., in the case of complexes **20** ($KCu_3Cl_4[C_2H_2]$) and **21a** ($NH_4Cu_3Cl_4[C_2H_2]$) [23]. In compound **20**, hydrogen atoms of two crystallographically independent C_2H_2 molecules form H-bonds of almost the same length [2.85(8) and 2.80(9) Å] with chlorine atoms. In the structure of compound **21a**, these H-bonds are significantly different (2.98(12) and 2.55(12) Å), which makes the π-coordinated triple bond in one C_2H_2 molecule significantly longer (1.32(2) Å) as compared to that (1.25(1) Å) in the other molecule. In complex **20**, the parameters of both π-coordinated triple bonds are almost the same ([~1.28(1) Å) and correspond to a mean of the two extreme values observed for the C≡C bond length in the structure of compound **21a**. In addition, the ammonium cation in compound **21a** forms H-bonds of the (N)H ···Cl type (2.20–2.87 Å), which favor the formation of a lattice that is different from that of complex **20**. Thus, the anionic complexes can feature a competition of the $M \cdots Cl^{-\delta}$ and $\equiv C{-}H^{+\delta} \cdots Cl^{-\delta}$ interactions.

In the case of soluble anionic π complexes, the formation of H-bonds is also possible, provided that the bridging and/or terminal chlorine atoms occupy appropriate positions in the coordination polyhedrons. Weak H-bonds can also be formed between NH_4^+ ion and terminal Cl atoms [see Eq. 2.347)]. In addition, H-bonds can undoubtedly be formed with water molecules:

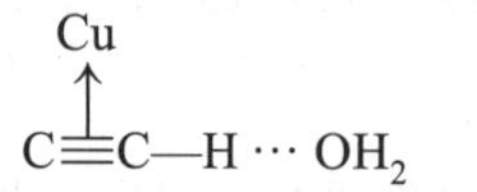

2.3.4.2 *Acetylenide (ethynyl) complexes of Cu(I)*

As noted above, the formation of ethynyl compounds (i.e., those containing at least one M–C ≡ bond) in protophilic media (or in the presence of bases) is a characteristic feature of the reactions of acetylene and monosubstituted alkynes with Cu(I) complexes. The ≡C–H bonds in π complexes of Cu(I) are highly polarized, while the Cu–C≡ bonds are very strong, so that Cu^+ ions react with C_2H_2 with the formation of CuC≡CH even in strongly acidic media, e.g., in the Cu_2SO_4–H_2SO_4 system with [H_2SO_4] up to 20–25 %. This is accompanied by the formation of vinylacetylene in small amounts [221]:

$$Cu(C_2H_2) + H_2O \rightleftharpoons CuC_2H + H_3O^+.$$

Spectrophotometric and potentiometric investigation [234] of the formation of "yellow" complexes during the interaction of C_2H_2 with solutions of polynuclear $(Cu_mCl_n)^{(n-m)-}$ anions at 25–80 °C showed that, depending on the acidity of medium, the following equilibria take place in the system:

$$(Cu_m)(C_2H_2) \underset{-H^+}{\rightleftharpoons} (Cu_m)C_2H \underset{-H^+}{\overset{(Cu_p)}{\rightleftharpoons}} (Cu_{m+p})C_2.$$

By increasing the acidity, it is possible to pass from a symmetric $(Cu_{m+p})C_2{}^{2-}$ acetylenide to asymmetric $(Cu_m)(C_2H^-)$ compound and then again to π complex $(Cu_m)(C_2H_2)$.

Vestin [261] thoroughly studied the formation of solid acetylenide complexes containing Cu_2C_2 in the $CuCl–KCl–H_2O$ system and, in particular, determined the equilibrium constant ($K = a^2_{HCl}/P_{C_2H_2} = 4$ at 25 °C) for the formation of a "violet" complex **41** (Table 2.9)

$$C_2H_2 + (x+2)CuCl_{solid} \overset{K}{\rightleftharpoons} \underset{\mathbf{41}}{C_2Cu_2(CuCl)_{x\ solid}} + 2HCl,$$

established the composition of a "yellow" complex **42**

$$C_2H_2 + 8CuCl_{solid} + 2K^+ \rightleftharpoons \underset{\mathbf{42}}{K_2C_2(CuCl)_{8\ solid}} + 2H^+,$$

and concluded that the solution contained $[C_2(CuCl)_n]^{2-}$ anions. Hui and James [107] reported that asymmetric ethynyl complexes were formed in solution at [HCl] = 0.05–0.6 $\hat{m}$ and an acidity of $h_0 = 0.2–5$ (pH 1.1–0.0) with predominant compositions of $[Cu_4Cl_4(C{\equiv}CH)]^-$ and $[Cu_4Cl_5(C{\equiv}CH)]^{2-}$.

Recent syntheses and structural characterization of ethynyl complexes and complex acetylenides of Cu(I) chloride (see Table 2.9) convincingly demonstrated that asymmetric polynuclear ethynyl complexes with compositions $[Cu_mCl_n(C{\equiv}CR)]^{(n-m+1)-}$ actually exist [23]. Some of these investigations revealed structures that were not previously reported for nonhalide complexes [262–281].

In the case of acetylene, crystalline (insoluble) acetylenide complexes are represented by $(ImH)_4Cu_9Cl_{11}(C{\equiv}CH)_2$ (**30**) [243] and $(PyH)_4Cu_9Cl_{12}(C{\equiv}CH)$ (**31**) [244]. In these compounds, C≡CH represent bridging groups that are coordinated to copper via one α-C atom (coordination type: μ-η^1, η^1–C≡CH, or σ,σ) (Figure 2.21), the C≡C bond lengths are 1.21(2) and 1.19(7) Å, and the Cu–C bond lengths amount to 1.97(1) and 1.92(3) Å. Complex **30** contains $Cu_4Cl_5(C_2H)^{2-}$ anion. Analogous ligands are frequently encountered in ethynyl complexes of the $Cu_2(C{\equiv}CR)_2L_4$ type ($L = PR_3$) [262]. Acetylenide complex $(NH_4)_8Cu_{29}Cl_{29}(C{\equiv}C)_4 \cdot 7H_2O$ (**32**), contains $C_2{}^{2-}$ anion which, unlike C_2H^- anion in compound **30**, is coordinated in both η^1 and η^2 modes (σ,π type) and is bridging between eight copper atoms [245]. This structure contains two crystallographically independent $C_2{}^{2-}$ units, each representing a bridging σ,π,σ-type ligand bound simultaneously to eight metal atoms. One should also note the modes of coordination of Cu(I) atoms to two independent $C_2{}^{2-}$ moieties. Each of the two carbon atoms in the C(3)≡C(4) group forms σ bonds with three copper atoms (Figure 2.22a). Thus, the C(3)≡C(4) group is coordinated in μ_3-$\eta^1,\eta^1,\eta^1,\mu_2$-$\eta^2,\eta^2,\mu_3$-$\eta^1,\eta^1,\eta^1$ modes, i.e., represents a σ,π,σ-type ligand. The influence of six η^1-coordinated and two η^2-coordinated Cu(I) atoms (bound to each $C_2{}^{2-}$ group) on the C≡C bond is so strong that the η^2 site geometry changes and the Cu–C*–Cu angle reaches 106°. However, the C(1) ≡C(2) bond is still capable of coordinating a third copper atom (Figure 2.22, *b*).

The ability of unusual coordination to Cu(I) atoms is apparently a characteristic property of $C_2{}^{2-}$ anion. Indeed, the structure of a $[Cu_4(Ph_2PCH_2PPh_2)_4(C{\equiv}C)](BF_4)_2 \cdot 4Me_2CO$ complex [282] contains

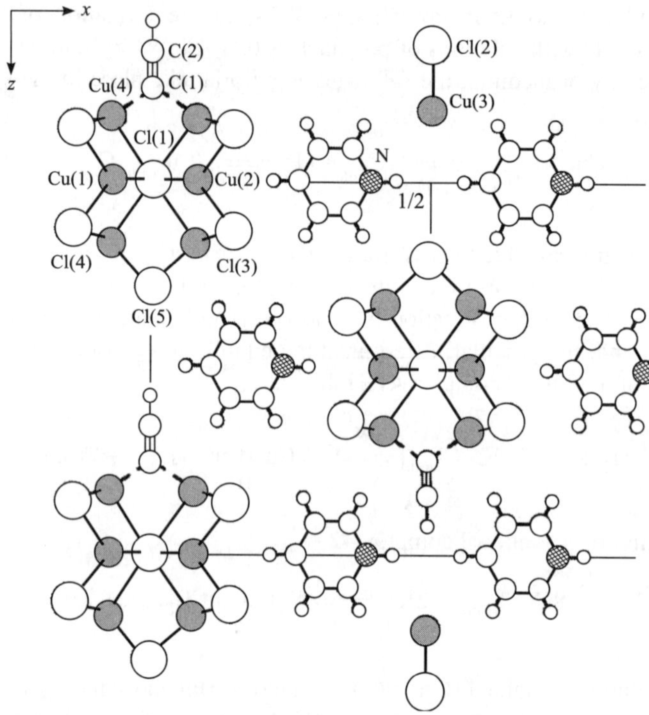

Figure 2.21 Projection of $(PyH)_4Cu_9Cl_{12}(C{\equiv}CH)$ (31) structure (one layer) onto *xz* plane (reproduced with permission from The Russian Academy of Sciences and Turpion Ltd © 2000 [23])

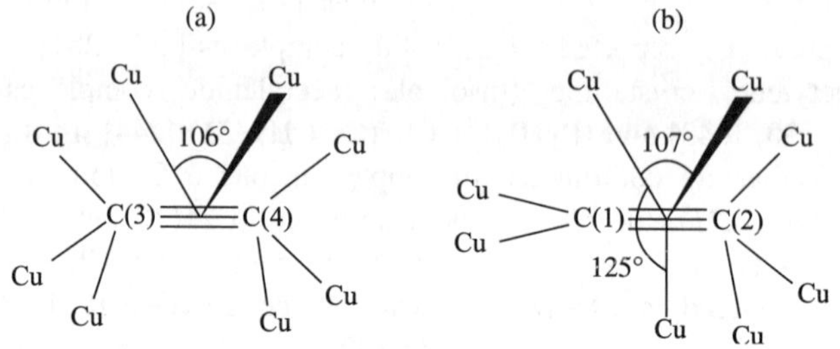

Figure 2.22 π,σ-coordination nodes in the structure of $(NH_4)_8Cu_{29}Cl_{29}(C{\equiv}C)_4 \cdot 7H_2O$ (32) complex (reproduced with permission from The Russian Academy of Sciences and Turpion Ltd © 2000 [23])

an absolutely planar σ,π,π,σ-type coordinated fragment

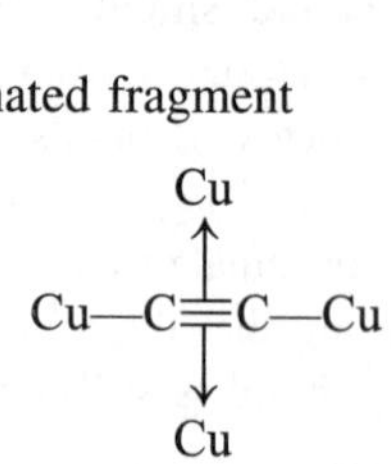

in which the Cu–C*–Cu angle is 180°. The structure of a "yellow" complex studied by Vestin [261] also probably contained fragments (a) and (b) depicted in Figure 2.22. The σ,π-type binding of several copper atoms by the C_2^{2-} group leads to the formation of clusters with a shortest Cu · · ·Cu distance of 2.452(6) Å. The coordination of an ethynyl ligand to three Cu(I) atoms (μ_3-η^1,η^1,η^1-C≡C) forming a three-nuclear cluster is also rather typical of ethynyl complexes [262]. The issue of the nature of η^2-bonds of C≡C with three metal atoms is worthy of special analysis.

A large variety of structural motifs was also found in ethynyl complexes containing RC≡C groups. The efficiency of η^2-type Cu(I) binding to RC≡C depends on the η^1-type binding mode. An analysis of the geometric characteristics of Cu(I) binding (η^2-C≡CR) showed [23] that, in polynuclear Cu(I) ethynyl complexes, one of the two metal atoms in the π-site (μ-η^2,η^2-C≡CR) is somewhat closer than another to the middle of the C≡C bond. This situation is similar to the case in π complexes of CuCl with $HC{\equiv}CCH_2OH$ and $HC{\equiv}C{-}CH{=}CH_2$ in compounds **26, 27** (Table 2.8). Let us consider the stereochemistry of [Cu_4(C≡CR)]] and [Cu_5(C≡CR)] fragments encountered in the structures of σ,π complexes of Cu(I). There are four main types of such "clusters" [23], the structural distinctions of which are related to the different orientations of two mutually perpendicular π-MOs of the ethynyl group of the σ,π-coordinated C≡CR ligand relative to the plane P passing through the C≡C bond (Figure 2.23).

In the structures of complexes containing type I clusters, the plane of one π-MO (e.g., $\pi_{||}$) of the C≡C group coincides with plane P, so that a π-coordinated Cu(2) atom and the C≡C bond with two σ-bound atoms (Cu(3) and Cu(4)) occur in the same plane, whereas the normal mn to this plane coincides with the axis of $\pi_\perp$-MO (axis Cu(1)–C*) (Figure 2.23a). In contrast to this, both π-MOs ($\pi_{||}$ and $\pi_\perp$) of the C≡C group in the clusters of types II–IV are oriented relative to normal mn (as well as relative to plane P) at an angle of 45°. In the clusters of types III and IV (Figures 2.23c and 2.23d) plane P is also drawn perpendicular to the normal mn so as to indicate the spatial orientation of π-coordinated Cu(I) atoms, which is the same as in type II cluster (Figure 2.23b). The main difference between cluster types II, III, and IV is related to the orientation of σ-coordinated Cu atoms. In clusters of type II, the σ-coordinated Cu(3) and Cu(4) atoms are arranged in plane P (Figure 2.23b). These σ-coordinated Cu(3) and Cu(4) atoms rotate by 90° about the C≡C axis, so as to fall in plane Q containing normal mn, yields cluster

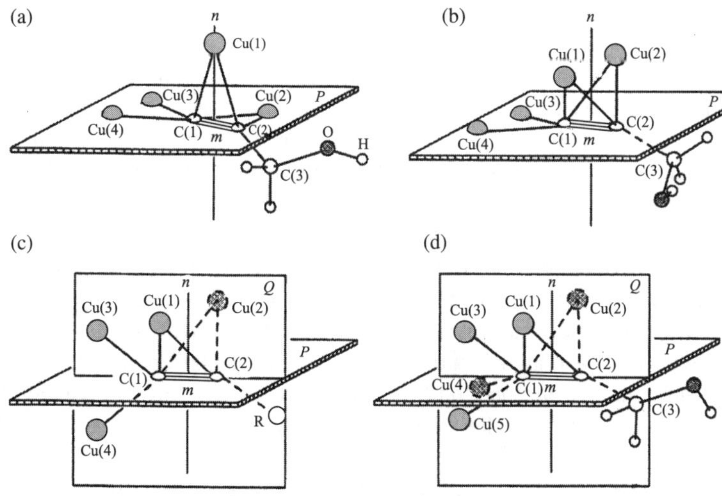

Figure 2.23 Types of clusters encountered in structures of acetylenide(ethynyl) copper chloride complexes: (a) type I cluster [Cu_4(C≡CCH_2OH)]; (b) type II cluster [Cu_4(C≡CCH_2OH)]; (c) type III cluster [Cu_4(C≡CR), R = CH=CH_2, CH_2OH]; (d) type IV cluster [Cu_4(C≡CCH_2OH)], mn is the normal to P plane (reproduced with permission from The Russian Academy of Sciences and Turpion Ltd © 2000 [23])

of type III (Figure 2.23c). Finally, replacing one σ-coordinated Cu(4) atom by two situated out of plane Q (Cu(4) behind and Cu(5) in front of this plane with an angle of Cu(4)C(l)Cu(5) $\approx 80°$) yields type IV cluster (Figure 2.23d). In addition to the four main types, the structures of Cu(I) σ,π complexes can exhibit combined cluster types such as $[Cu_8]$ cluster in compound **37** (Table 2.9), which consists of type I and IV clusters with a bridging Cu atom between them. Cluster $[Cu_9]$ in compound **39** represents a combination of two type IV clusters. Clusters of types I–IV were originally discovered in Cu(I) chloride ethynyl complexes. The problems of a strong mutual influence of the σ and π interactions in these clusters and the coordination geometry of Cu(I) atoms in σ,π clusters are discussed in [23].

The structure of complex compounds isolated from solution cannot be directly related to the mechanism of their formation, since transformations of the initial intermediates of any type proceed toward the formation of structures with increased stability, while less soluble compounds tend to precipitate from solution. For example, an isolated ethynyl compound with a composition of $Cu_m(C_2H)$ can appear as a result of the interaction of $Cu_P(C_2H)$ with Cu_q $(p + q = m)$ in solution. At the same time, it would be of interest to analyze the possible "genetic" relationships between the structures of initial Cu(I) chloride anions and the products formed as a result of their interaction with RC≡CH.

Let us consider a group of crystalline compounds formed in the $(C_6H_5NH_3)Cl{-}CuCl{-}H_2O{-}HC{\equiv}CCH_2OH$ system. Colorless crystals of $(C_6H_5NH_3)_3Cu_4Cl_7$ (**11**, Table 2.9), which precipitate from an almost neutral concentrated solution of 4 moles CuCl and 3 moles $(C_6H_5NH_3)Cl$ in H_2O (20 moles) are rapidly dissolved when propargyl alcohol is added to the system. Instead, yellow crystals of ethynyl complex **38** are formed via reaction

$$2(C_6H_5NH_3)_3Cu_4Cl_{7\,\mathrm{solid}} + HC{\equiv}CCH_2OH + 2H_2O$$

$$\longrightarrow \underset{\mathbf{38}}{(C_6H_5NH_3)_3Cu_8Cl_{10}(C{\equiv}CCH_2OH)\cdot 2H_2O_{\mathrm{solid}}} + 3C_6H_5NH_3Cl + HCl.$$

which is accompanied by a decrease pH of the medium to ~1.5.

Let us analyze a relationship between the stereochemistry of compounds **11** and **38** and the possible mechanism of formation of the latter compound in solid-state process (Figure 2.24). Figure 2.24a shows a subunit of the $(Cu_4Cl_7{}^{3-})_n$ anion of structure **11**, which has the form of united coordination polyhedra of Cu atoms. A molecule of propargyl alcohol can substitute for a μ-Cl atom (shaded in Figure 2.24a) and incorporate into the initial macropolyhedron with the formation of a π complex (η^2-RC≡CH). As a result, the cavity in the macropolyhedron transforms from octahedral into quadratic-pyramidal. The subsequent deprotonation of the π ligand by H_2O molecule yields the proposed π complex with a $^-$C≡CR ligand, which results in the formation of a μ-η^2-$C{\equiv}CCH_2OH$ bridging ligand (Figure 2.24b). The formation of a π complex leads to the weakening of some Cu–Cl bonds (dashed in Figure 2.24b) followed by their rupture with the formation of two new η^1-Cu-C≡ bonds (Figure 2.24c) in the $[Cu_4(C{\equiv}CCH_2OH)]$ cluster, which is a structural element of complex **38**. In this complex structure, $[Cu_8Cl_{10}(C{\equiv}CCH_2OH)^{3-}]_n$ anions are formed as layers, in which σ,π-coordinated C≡CR groups of each propargilium anion is confined from four sides by Cu atoms inside a quadratic-pyramidal cavity. The sterical obstacles to penetration of the H_3O^+ ion inside this cavity (to reach an α-C atom of the $^-C{\equiv}CCH_2OH$ ligand) and a high strength of the σ,π-coordinated anion (in complex **38**, the length of one Cu–C bond of the acetylenide type is 1.89(2) Å and the Cu–C* distance is 1.97(1) Å) make structure **38** highly stable with respect to decomposition or transformation into π complex under the action of HCl.

In the $[Cu_4Cl_5(C{\equiv}CCH_2OH)^{2-}]_n$ anion of compound **34** (Table 2.9), the α-C atom of $^-C{\equiv}CCH_2OH$ fragment is accessible for the attack by proton, the more so that σ-coordinated Cu atoms in structure **34** are not as strongly bound to α-C atom [for Cu–C = 1.95(2)–2.00(2) Å] as in structure **38**. Apparently, for this reason the precipitated crystals of complex **34** are converted into π complex $[Cu_2Cl_3(HC{\equiv}CCH_2OH)^-]_n$ (**29**) on standing in mother liquor containing HCl (pH $\leq$ 1) [23].]

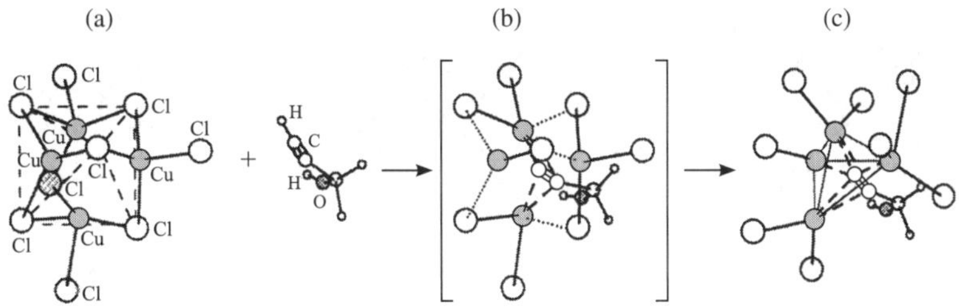

Figure 2.24 Transformation of complex $(C_6H_5NH_3)_3Cu_4Cl_7$ (11) into complex $(C_6H_5NH_3)_3Cu_8Cl_{10}$-$(C{\equiv}CCH_2OH)\cdot 2H_2O$ (38) in the presence of $HC{\equiv}CCH_2OH$ (stereochemical aspect) (reproduced with permission from The Russian Academy of Sciences and Turpion Ltd © 2000 [23])

2.3.5 Mechanisms of acetylene dimerization and hydrocyanation reactions. Crystallochemical aspects

Below we consider in more detail the possible mechanisms of two catalytic reactions, acetylene dimerization (in the Nieuwland catalytic system) and hydrocyanation (Section 2.3.2), based on the available information concerning the structure of polynuclear $(Cu_mCl_n)^{(n-m)-}$ anions, π complexes of acetylene and its derivatives, and σ ethynyl complexes with polynuclear Cu(I) anions.

The initial complex is modeled by a fragment of the crystal structure of complex **9a** ($Cu_3Cl_5^{2-}$ anion), and it is assumed that the short-range order in concentrated solution of dissolved complexes (that is close to the short-range order in the melt) corresponds to the complex structure in the crystalline state (T_m for compound **9a** is close to 80 °C). The process starts with complex **A** (Figure 2.25). The activation of acetylene, which leads to its deprotonation (step **1**, Section 2.3.2), most probably involves several reactions, the first of which results in the formation of π complexes **B**:

$$\underset{\mathbf{A}}{(Cu_mCl_n)^{(n-m)-}} + C_2H_{2\,(aq)} \rightleftharpoons \underset{\mathbf{B}}{[Cu_mCl_n(C_2H_2)]^{(n-m)-}} + H_2O.$$

A kinetic analysis shows that the displacement of Cl^- at this stage contradicts Eq. (2.349). The formation of a π complex can proceed with either the substitution of a coordinated water molecule or the rupture of the μ-Cl bond without replacing Cl^-. It is also possible that a bridging chlorine atom is replaced and moved to other coordination vacancies. In the course of water molecule substitution (**A** → **B**), the formation of π complex **B** may be accompanied by the transfer of the bridging chlorine atom in complex **A** to a coordination site of the leaving molecule of water (shaded in Figure 2.25). Intermediate **B** (namely, a coordination π node) has a structure that is analogous to one of π complexes (**20**, **21a**, **21b**, **22**, **23**).

The interaction of π complex **B** and water molecule leads to the formation of an acetylenide (ethynyl) intermediate **C**, while Cl^- ion either can leave the system simultaneously with H_3O^+ or can be liberated in a subsequent quasi-equilibrium step (without a change in the form of the kinetic equation):

$$\underset{\mathbf{B}}{[Cu_mCl_nC_2H_2]^{(n-m)-}} + H_2O \rightleftharpoons \underset{\mathbf{C}}{[Cu_mCl_{n-1}(C_2H)]^{(n-m)-}} + H_3O + Cl^-.$$

Among a great variety of the possible structures of ethynyl compounds, complexes **30** and **31** (Figure 2.21) correspond to intermediate C (which was originally suggested to exist, albeit on insufficient grounds, by Vestin [261]).

As noted above, by varying the partial pressure of acetylene $P_{C_2H_2}$ and $[H_3O^+]$ while maintaining the constant activities a_{Cu^+} and a_{Cl^-} (i.e., constant concentration of the active copper chloride complex), it is possible to establish which one of the σ organometallic compounds existing in solution is the intermediate in the catalytic reaction. In the VA synthesis, this is an asymmetric ethynyl complex. Depending

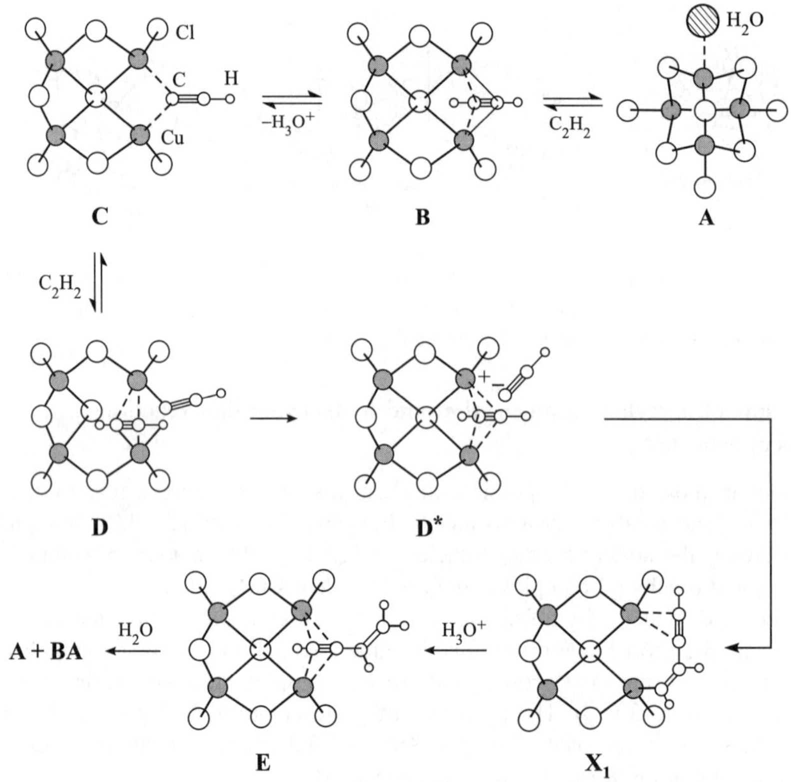

Figure 2.25 Structural transformations in the course of catalytic oligomerization of acetylene (reproduced with permission from The Russian Academy of Sciences and Turpion Ltd © 2000 [23])

on the pH of the solution, other "yellow" complexes can also exist, for example, symmetric acetylenide $[(Cu_2C_2)Cu_pCl_q)^{(q-p)-}$ complexes that are in equilibrium with $[Cu_mCl_{n-1}(C_2H)]^{(n-m)-}$ anions. This inactive complex enters into the material balance with respect to $[CuCl]_\Sigma$. The rate of VA synthesis passes through a maximum when the HCl concentration is varied and the $(Cu_mCl_n)^{(n-m)-}$ concentration is not maintained constant [211]. The maximum rate is observed at pH 1.5–0.9 and [HCl] = 0.05–0.15 $\tilde{m}$; an increase in the acidity (pH < 1.5) leads to the decomposition of a symmetric and the accumulation of an asymmetric $[Cu]C_2H$ acetylenide. As the acidity is increased further, this intermediate disappears and the reaction rate drops to zero. The formation of a complex of the type of compound **32** with the Cu_2C_2 group at pH 4 and low temperature is fully consistent with these considerations and experimental observations.

The next step in the Nieuwland process (step **2**) that leads to intermediate $\mathbf{X_1}$ apparently consists of the incorporation of acetylene at the Cu–C_2H bond with the formation of a CuCH=CHC≡CH fragment, the acidolysis of which yields C_4H_4. The steps of conversion of intermediate **C** into intermediate $\mathbf{X_1}$ (see Figure 2.25) are as follows:

$$\underset{\mathbf{C}}{[Cu_mCl_{n-1}C_2H]^{(n-m)-}} + C_2H_2 \rightleftharpoons \underset{\mathbf{D}}{[Cu_mCl_{n-1}(C_2H_2)(C_2H)]^{(n-m)-}} + H_2O.$$

$$\underset{\mathbf{D}}{[Cu_mCl_{n-1}(C_2H_2)(C_2H)]^{(n-m)-}} + H_2O \longrightarrow \underset{\mathbf{X_1}}{[Cu_mCl_{n-1}(C_4H_3)]^{(n-m)-}}.$$

Here, the step of acidolysis of intermediate $\mathbf{X}_1$ (electrophilic substitution of H^+ ion for Cu atom in the Cu–CH= fragment) undoubtedly takes place (as confirmed by the kinetic equation (2.351)). This reaction readily proceeds even with water [283, 284]. Steps of this type are also encountered in reactions of acetylene hydration, hydrochlorination, and hydrocyanation [8]:

$$\underset{\mathbf{X}_1}{[(Cu_mCl_{n-1}(C_4H_3)]^{(n-m)-}} + H_3O^+ \longrightarrow \underset{\mathbf{E}}{[Cu_mCl_{n-1}(CH_2{=}CHC{\equiv}CH)]^{(n-m+1)}}.$$

Intermediate **E** exhibits decomposition with the formation of VA (which can be displaced by acetylene) or participates in the trimerization of acetylene to hexa-1,5-dien-3-yne [divinylacetylene (DVA)]. The most important and least understood step of the process is the incorporation of acetylene with the formation of intermediate $\mathbf{X}_1$. The fact of acetylene incorporation seems to be reliably established, since the reactions of $R_2Cu \cdot MgX$ and $RCu \cdot MgX_2$ cuprates with alkynes are well known [283, 284] and proceed stereoselectively (100 % *syn*-addition) under mild conditions. A question can arise as to what is the mechanism of acetylene incorporation. In a mononuclear variant, the incorporation process

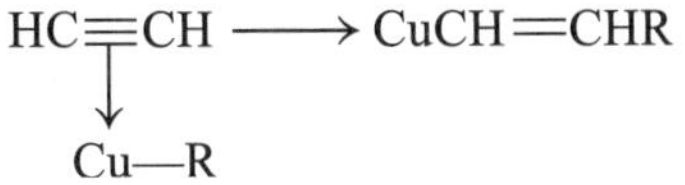

is not an elementary consistent reaction, since the d^{10} configuration and sp^3 hybridization of atomic orbitals in Cu(I) make this process symmetry-forbidden [285]. Formally, there are three possible variants:

1. Formation of metal carbenium ion **F** at the first irreversible step, followed by the fast transfer of the ethynyl group to carbenium-carbon and the formation of $\mathbf{X}_1$:

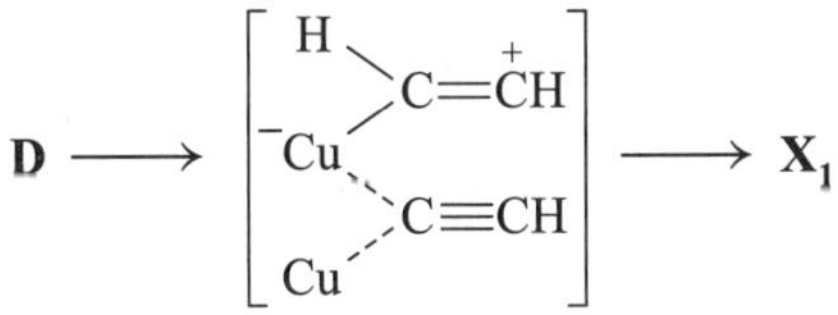

2. Dissociation of [Cu](μ-η^1–C≡CH) bond with the formation of a contact ion pair and the subsequent addition of carbanion to η^2-coordinated acetylene:

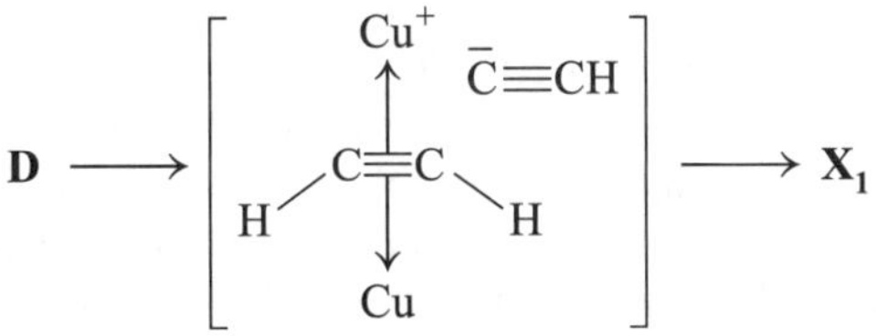

3. Dissociation of [Cu](μ-η^1–C≡CH) bridging bond with the transition of the ethynyl group to the η^2 state and the subsequent addition of η^2-HC≡C$^-$ carbanion to η^2-coordinated alkyne:

F

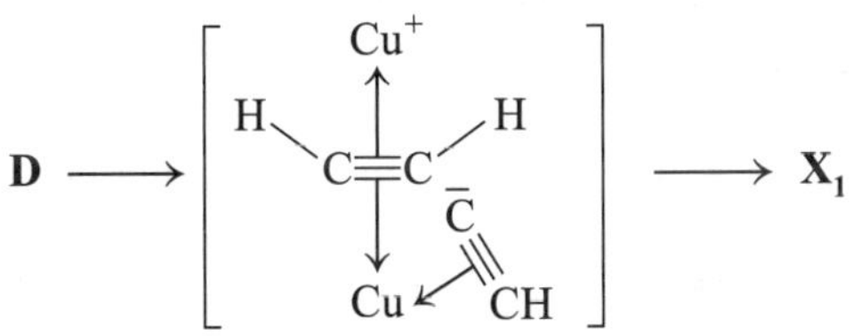

Variant 1 was considered for the interactions of Pt(II) cationic complexes with alkynes and suggested [8] to be operative in reactions involving Hg^{2+}. However, for Cu(I) halides this mechanism is less probable. Variant 3 is more attractive, since the reaction involves a softer nucleophile and a weak base (η^2-HC≡C$^-$). If complex **D** is formed via replacement of μ_4-Cl ligand in intermediate complex **C** by acetylene (μ-η^2,η^2-C_2H_2), then two ligands ([Cu_2]-η^1,η^1-C≡CH and η^2,η^2-C_2H_2) appear on the matrix. The conversion of intermediate **D** into $\mathbf{X}_1$ by variant 2 is preferred from the standpoint of stereochemistry, since it proceeds by rupture of the η^1-coordinated Cu–C bond with the formation of a contact ion pair **D*** (see Figure 2.25). Then, ^-C_2H ion rapidly attacks one of the $C^{\delta+}$ atoms of a π-bound C_2H_2 molecule. In the case of variant 3, the η^1,η^1-C≡CH ligand must first transform into η^1,η^2-C≡CH ligand, which is followed by rupture of the η^1-coordinated Cu–C bond and the attack of η^2-acetylene by η^2-coordinated anion. The protolysis of intermediate $\mathbf{X}_1$ leads to the formation of π-coordinated VA, which serves as intermediate **E**.

Intermediate **E**, representing $[Cu_mCl_n(C_4H_4)]^{(n-m)-}$ complex, can be deprotonated with the formation of intermediate **G** (of the type of compound **40**, Table 2.9). According to kinetic data [8], this intermediate converts into DVA:

$$\underset{\mathbf{G}}{[(Cu_mCl_n(C{\equiv}CCH{=}CH_2)]^{(n-m)-}} \xrightarrow{C_2H_2} \underset{\mathbf{H}}{[Cu_mCl_n(C_4H_3)(C_2H_2)]^{(n-m)-}} \longrightarrow \mathbf{X}_3 \xrightarrow{H_3O^+}$$

$$\longrightarrow \underset{\mathbf{I}}{Cu_2Cl_2(CH_2{=}CHC{\equiv}CCH{=}CH_2)} + (Cu_{m-2}Cl_{n-2})^{(n-m)-}.$$

The protolysis of intermediate $\mathbf{X}_3$ (analog of $\mathbf{X}_1$) leads to intermediate **I** (complex **34**, Table 2.9). Intermediate $\mathbf{X}_1$ is capable of interacting (via π complex) with another acetylene molecule that can be inserted into Cu–CH= bond, since the incorporation of alkynes at the Cu(I)–alkenyl bond is quite easy [283]. This results in the formation of hexa-1,3-dien-5-yne (DVA isomer). Under certain assumptions, the proposed scheme of formation of this intermediate agrees with the kinetic data [8].

The formation of polynuclear Cu(I) chloride–cyanide complexes in the reaction of acetylene hydrocyanation also seems to be reliably established, although these complexes have never been isolated from concentrated solutions. At the same time, an increase in the concentration of H_3O^+ during the introduction of HCN into solution, a high solubility of CuCN in the CuCl–NH_4Cl–H_2O ([CuCN] reaches 15 mass % at 85 °C), and the extremal dependence of the hydrocyanation reaction rate on P_{HCN} (see Fig. 2.13)–all these facts show evidence that mono- ($[Cu_mCl_{n-1}(CN)]^{(n-m)-}$) and polycyanide ($[Cu_mCl_{n-x}(CN)_x]^{(n-m)-}$) complexes can exist in solution, depending on [HCl].

The increased reactivity of polynuclear Cu(I) chloride–cyanide complexes is also related to their bridging structure, which is analogous to that in $NH_4Cu_2(CN)_3{\cdot}H_2O$ [286] and $KCu_2(CN)_3{\cdot}H_2O$ [287]. The bridging structure favors keeping CN^- as anion in the coordination sphere of the Cu^+C^-≡NCu complex during the insertion of acetylene into the Cu–CN bond. Acrylonitrile (formed upon acidolysis of the CuCH=CHC≡NCu intermediate species) can be coordinated with the formation of CuCl·CH_2=CHCN (type II) complexes [288], in which the acrylonitrile molecule plays the role of η^2 (via the C=C bond) and η^1 (via N atom) bridging ligand. It should be also noted that CN^- ions can be replaced in the structure of $KCu_2(CN)_3{\cdot}H_2O$ cyanide complex by C_2^{2-} dianions with the formation of $[K(H_3O^+)][Cu_2(CN)_2(C{\equiv}C)]$ complexes [188].

In concluding this section, let us summarize the important specific features of the coordination chemistry of Cu(I) halide complexes, which are essential in order to understand factors that are responsible for the high catalytic activity of these complexes in some conversions of acetylene in highly concentrated solutions.

(i) The formation of coordination polyhedra due to the bridging chloride (bromide) ligands, in which Cl^- ion plays the role of μ_2-, μ_3-, μ_4-ligand, leads to the formation of low-charge polynuclear $(Cu_mCl_n)^{(n-m)-}$ associates that are capable of effectively interacting with acetylene and alkynes, thus making possible the formation of a large variety of π and σ organocopper compounds. It is natural to suggest that low-charge polynuclear associates act as strong electron acceptors with respect to alkynes.

(ii) Polynuclear Cu(I) complexes feature the formation of bridging and nonbridging π complexes of alkynes with various degrees of activation (extension and polarization) of C≡C bonds.

(iii) High efficiency of the π-type interaction in Cu(I) ethynyl complexes (σ, π complexes), the formation of which in the MCl–CuCl–H_2O systems (according to the kinetic data) plays a determining role in many transformations of C_2H_2 and alkynes. This efficiency is related to the mutual enhancement of several types of interactions, in particular, Cu(I)–(η^1-C≡CR) and Cu(I)–(η^2-C≡CR). The phenomenon of mutual (synergistic) enhancement of the acetylenide- and π-type interactions, which was revealed in Cu(I) cluster acetylenides and ethynyl complexes, is frequently accompanied by unusual structural transformations and the appearance of qualitatively new properties of the π-activated C≡C group at the active metal center. In particular, C_2^{2-} dianions in structure **32** (Table 2.9, Figure 2.22) are capable of simultaneous π-coordination with three Cu(I) atoms.

(iv) Multiple binding of acetylene in the π complexes and ethynyl ligands in σ-organocopper compounds (as well as the multiple binding of cyanide groups) is among the most important factors that determine the catalytic activity of polynuclear Cu(I) complexes. The bridging character of coordinated ethynyl and cyanide groups increases the polarity of Cu–C bonds and the reactivity of $^-$C≡CH and $^-$CN groups. The bridging capacity of these σ ligands provides wide stereochemical possibilities for the incorporation of carbanion (RC≡C$^-$) and cyanide (NC$^-$) nucleophiles into the structure of complexes.

(v) Highly concentrated Cu(I) chloride solutions admit a high solubility of molecules such as CuC≡CH, CuC≡CR, and CuCN due to their interaction with polynuclear $(Cu_mCl_n)^{(n-m)-}$ anions and, hence, make possible a high concentration of intermediates in solution (e.g., the concentration of "yellow" acetylenide complexes in the Nieuwland reaction reaches 0.1 M).

(vi) The reactions of alkyne hydration and hydrochlorination, in which σ-organometallic intermediates are formed when Cu(I) π complexes are attacked by nucleophiles (Cl^-, H_2O) present in solution, can also proceed without participation of polynuclear complexes and, hence, must not be necessarily catalyzed by these complexes in concentrated solutions

Information on the catalysis using various forms of metal complexes, which was presented in Chapter 2, shows convincing evidence for the possible and highly significant nonuniformity of active centers in homogeneous catalysis (see also Chapters 3 and 4). Thus, the homogeneous catalysis with metal complexes, as well as the heterogeneous catalysis with metals and oxides, can encounter both uniform and nonuniform centers (with both inherited (biographical) and reactant-induced nonuniformity). Therefore, the uniformity or nonuniformity of the active centers cannot serve a basis for definitions of the homogeneous and heterogeneous catalysis (as was proposed in [288]).

References

1. J. Westley, *Enzymic Catalysis*, Harper & Row, New York, 1969.
2. L. Michaelis and M. L. Menten, *Biochem. Z.*, **49**, 333 (1913).
3. G. E. Briggs and J. B. S. Haldane, *Biochem. J.*, **19**, 338 (1925).
4. E. I. Schpitalsky, *Z. Phys. Chem.*, **122**, 257 (1926).
5. R. M. Flid, I. I. Moiseev and E. M. Kalmykova, *Zh. Fiz. Khim.*, **31**, 904 (1957).
6. I. I. Moiseev, *Issledovanie v oblasti mekhanizma zhidkofaznoi gidratatsii atsetilena* (Investigation of the Mechanism of Liquid-Phase Hydration of Acetylene), Candidate Dissertation in Chemistry, MITKhT (Moscow Institute of Fine Chemical Technology), Moscow, 1958.

7. O. N. Temkin and R. M. Flid, *Kataliticheskie prevrashcheniya atsetilenovykh soedinenii v rastvorakh kompleksov metallov* (Catalytic Transformations of Acetylene Compounds in Solutions of Metal Complexes), Nauka, Moscow, 1968.
8. O. N. Temkin, G. K. Shestakov and Yu. A. Treger, *Atsetilen: Khimiya, mekhanizmy reaktsii, tekhnologiya* (Acetylene: Chemistry, Reaction Mechanisms, and Technology), Khimiya, Moscow, 1991.
9. J. Bjerrum, *Metal Amine Formation in Aqueous Solution. Theory of the Reversible Step Reactions* (2nd ed.), P. Haase & Son, Copenhagen, 1957.
10. K. B. Yatsimirskii and V. P. Vasil'ev, *Konstanty nestoikosti kompleksnykh soedinenii* (Instability Constants of Complex Compounds), Izdat. Akad, Nauk SSSR, Moscow, 1959.
11. H. L. Schläfer, *Komplexbildung in Losung*, Springer-Verlag, Berlin, 1961.
12. F. J. S. Rossotti and H. Rossotti, *The Determination of Stability Constants and Other Equilibrium Constants in Solution*, McGraw-Hill, New York, 1961.
13. M. T. Beck, *Chemistry of Complex Equilibria*, Van Nostrand-Reinhold, London, 1970.
14. V. I. Belevantsev and B. I. Peshchevitskii, *Issledovanie slozhnykh ravnovesii v rastvore* (Investigation of Complex Equilibria in Solutions), Nauka, Novosibirsk, 1978.
15. I. I. Moiseev, *π-Kompleksy v zhidkofaznom okislenii olefinov* (π Complexes in Liquid-Phase Oxidation of Olefins), Nauka, Moscow, 1970.
16. A. N. Pryakhin, *Vestn. Mosk. Gos. Univ., Ser. 2: Khim.*, **24**, 337, 555 (1983); **27**, 243 (1986); **28**, 127 (1987).
17. A. N. Pryakhin, *Zhurn. Fiz. Khim.*, **58**, 20 (1984); **61**, 1879 (1987).
18. A. N. Pryakhin, *Kinetiko-termodinamicheskii analiz uravnenii statsionarnykh skorostei fermentativnykh reaktsii* (Kinetic-Thermodynamic Analysis of Equations of Steady-State Rates of Enzymatic Reactions), Izdat. MGU, Moscow, 1988.
19. K. H. Hsu (Xu) and L. M. Li, *Proceedings of the 21st Intern. Conf. on Coordination Chemistry (Toulouse, France, 1980)*, p. 284; L. M. Li and K. H. Hsu, *Kexue Tongbao*, **17**(1), 24 1966 (in English).
20. M. P. Rao, B. Sethuram, and T. N. Rao, *Proc. Indian Acad. Sci*., **55**(6), 858 (1989).
21. P. N. Char, S. Sondu, B. Sethuram and T. N. Rao, *J. Indian Chem. Sol*., **44**, 209 (1987).
22. O. N. Temkin, R. M. Flid and G. K. Shestakov, Features of Kinetics of Complex Catalytic Reactions in Solutions of Metal Complexes, in *Gomogennyi kataliz* (Homogeneous Catalysis), Ilim, Frunze, 1970, pp. 90–112.
23. B. M. Mykhalychko, O. N. Temkin and M. G. Mys'kiv, *Usp. Khim*. **69**(11), 1042 (2000).
24. F. Ungvary and L. Marko, *J. Organomet. Chem*., **20**, 205 (1969).
25. T. G. Sukhova, S. E. Zhukova, O. N. Temkin and R. M. Flid, *Dokl. Akad. Nauk SSSR*, **214**(6), 1365 (1974).
26. I. S. Akhrem, N. M. Chistovalova, H. N. Sofoyan et al., *J. Organomet. Chem*., **178**, 23 (1979).
27. J. P. Birk, J. Halpern and A. L. Pickard, *J. Am. Chem. Soc*., **90**, 449 (1968).
28. C. A. Tolman, W. C. Seidel and D. H. Gerlach, *J. Am. Chem. Soc*., **94**, 2669 (1972).
29. C. A. Tolman and J. N. Faller, Mechanistic Studies of Catalytic Reaction Using Spectroscopic and Kinetic Techniques, in *Homogeneous Catalysis with Metal Phosphine Complexes*, L. H. Pignolet (Ed), New York, 1983, pp. 13–110.
30. J. F. Harrod and C. A. Smith, *Can. J. Chem*, **48**, 870 (1970).
31. J. Halpern and C. S. Wong, *J. Chem. Soc. Chem. Commun*., 629 (1973).
32. K. I. Goldberg, J. Y. Yan and E. M. Breitung, *J. Am. Chem. Soc*., **117**, 6889 (1995).
33. V. R. Kartashov, T. N. Sokolova, T. V. Salikova and A. V. Shishulina, *Kinet. Katal*., **38**(4), 494 (1998).
34. E. S. Rudakov, A. P. Yaroshenko and V. V. Zamaschchikov, *React. Kinet. Catal. Lett*., **11**(2), 183 (1979).
35. L. J. Elding, *Inorg. Chim. Acta*, **6**, 647 (1972).
36. T. Ruhl, *Acta Chem. Scand*., **26**, 2961 (1972).
37. N. F. Gol'dshleger, M. B. Tyabin, A. E. Shilov and A. A. Shteinman, *Zh. Fiz. Khim*., **43**, 2174 (1969).
38. A. E. Shilov, *Pure Appl. Chem*., **50**, 725 (1978).
39. A. E. Shilov and G. B. Shul'pin, *Aktivatsiya i kataliticheskie reaktsii uglevodorodov* (Activation and Catalytic Reactions of Hydrocarbons), Nauka, Moscow, 1995.
40. J. L. Garnett, *Catal. Rev*., **5**, 229 (1971).
41. N. F. Gol'dshleger, V. V. Es'kova, A. E. Shilov and A. A. Shteinman, *Zh. Fiz. Khim*., **46**, 353 (1972).
42. V. V. Es'kova, A. E. Shilov and A. A. Shteinman, *Kinet. Katal*., **13**, 534 (1972).

43. M. B. Tyabin, A. E. Shilov and A. A. Shteinman, *Dokl. Akad. Nauk SSSR*, **198**, 380 (1971).
44. N. F. Gol'dshleger, A. I. Mustafin and A. A. Shteinman, *Kinet. Katal.*, **19**, 888 (1978).
45. N. F. Gol'dshleger, *Kataliticheskii deiteroobmen i okislitel'nye reaktsii alkanov v prisutstvii kompleksov dvukhvalentnoi platiny* (Catalytic Deuteroexchange and Oxidative Reactions of Alkanes in the Presence of Divalent Platinum Complexes), Candidate Dissertation in Chemistry, Inst. Chem. Phys., Akad. Nauk SSSR, Chernogolovka, 1979.
46. V. P. Tret'yakov, E. S. Rudakov, A. A. Galenin and R. N. Rudakova, *Dokl. Akad. Nauk SSSR*, **225**, 583 (1975).
47. E. S. Rudakov, V. P. Tret'yakov, A. A. Galenin and G. P. Zimtseva, *Dokl. Akad. Nauk UkrSSR, Ser. V*, 148 (1977).
48. V. P. Tret'yakov, E. S. Rudakov, A. V. Bogdanov et al., *Dokl. Akad. Nauk SSSR*, **249**, 878 (1979).
49. V. P. Tret'yakov, E. S. Rudakov, S. A. Mitchenko et al. *React. Kinet. Catal. Lett.*, **17**, 197 (1981).
50. V. V. Lavrushko, A. E. Shilov and A. A. Shteinman, *Kinet. Katal.*, **16**, 1479 (1975).
51. S. L. Litvinenko, V. V. Zamashchikov, V. I. Shologon and Z. Z. Rozhkova, *Metalloorg. Khim.*, **2**(3), 553 (1989).
52. V. I. Shologon, *Kinetika i mekhanizm reaktsii okislitel'nogo prisoedineniya dimetilsul'fata i etilenoksida k atsidokompleksam platiny(II)* (Kinetics and Mechanism of Reaction of Oxidative Addition of Dimethylsulfate and Ethylene Oxide to Platinum(II) Acidocomplexes), Candidate Dissertation in Chemistry, Inst. Phys. Chem. Carbochem., Donetsk, 2000.
53. S. L. Litvinenko, V. I. Shologon, V. V. Zamashchikov and S. V. Bilobrov, *Metalloorg. Khim.*, **4**(5), 1130 (1991).
54. V. I. Shologon, S. L. Litvinenko and V. V. Zamashchikov, *Izv. Akad. Nauk SSSR, Ser. Khim.*, **6**, 1177 (1995).
55. N. I. Bundina, O. L. Kaliya and O. L. Lebedev, *Zh. Org. Khim.*, **14**(10), 2447 (1978).
56. Yu. M. Golub, S. V. Barkanova, O. L. Kaliya and O. L. Lebedev, *Zh. Org. Khim.*, **15**(8), 1723 (1979).
57. O. L Kaliya, *Sostoyanie, reaktsionnaya sposobnost' i kataliticheskaya aktivnost' azaporfinov v rastvorakh* (State, Reactivity, and Catalytic Activity of Azaporphines in Solutions), Doctoral Dissertation in Chemistry, MNPO "NIOPIK", Moscow, 1991.
58. A. D. Ryabov, I. K. Sakodinskaya and A. K. Yatsimirsky, *J. Chem. Soc. Dalton Trans.*, **12**, 2629 (1985).
59. A. D. Ryabov, *Metalloorg. Khim.*, **1**(5), 994 (1988).
60. I. V. Zakharov, Yu. V. Geletii and V. A. Adamyan, *Kinet. Katal.*, **32**(1), 39 (1991).
61. K. K. S. Gupta, S. Maiti and S. K. Mandal, *Transit. Met. Chem.*, **9**, 416 (1984).
62. I. P. Skibida, *Usp. Khim.*, **44**(10), 1729 (1975).
63. G. A. Tolstikov, *Reaktsii gidroperoksidnogo okisleniya* (Hydroperoxide Oxidation Reactions), Nauka, Moscow, 1976.
64. V. N. Sapunov, *J. Mol. Catal.*, **7**, 149 (1987).
65. B. G. Bal'kov, I. P. Skibida and Z. K. Maizus, *Izv. Akad. Nauk SSSR, Ser. Khim.*, 1780 (1970).
66. V. A. Gavrilenko, E. I. Evzirikhin, V. A. Kolosov et al., *Izv. Akad. Nauk SSSR, Ser. Khim.*, No. 9, 1954 (1974).
67. V. A. Gavrilenko, E. I. Evzirikhin, V. A. Kolosov and I. I. Moiseev, *Izv. Akad. Nauk SSSR, Ser. Khim.*, No. 1, 29 (1977).
68. R. A. Akhmed'yanova, I. Yu. Litvintsev, A. Yu. Monaenkov and V. N. Sapunov, *Neftekhimiya*, **34**(6), 532 (1994).
69. N. N Protasov, *Gomogennye kobal'tsoderzhashchie katalizatory v reaktsiyakh radikal'nogo okisleniya uglevodorodov* (Homogeneous Cobalt-Containing Catalysts in Reactions of Radical Oxidation of Hydrocarbons), Candidate Dissertation in Chemistry, Mendeleev. Univ. Chem. Technol., Moscow, 1998.
70. R. A. Sheldon, J. A. Van Doorn, C. W. A. Schram and A. J. De Jong, *J. Catal.*, **31**, 427, 438 (1973).
71. V. A. Gavrilenko, E. I. Evzirikhin and I. I. Moiseev, *Izv. Akad. Nauk SSSR, Ser. Khim.*, No. 1, 34 (1977).
72. V. A. Gavrilenko, E. I. Evzirikhin and I. I. Moiseev, *Izv. Akad. Nauk SSSR, Ser. Khim.*, No. 10, 2172 (1975).
73. V. A. Gavrilenko, E. I. Evzirikhin and I. I. Moiseev, *Izv. Akad. Nauk SSSR, Ser. Khim.*, No. 6, 1266 (1977).
74. V. A. Gavrilenko, E. I. Evzirikhin, I. I. Moiseev and I. Sh. Fish, *Izv. Akad. Nauk SSSR, Ser. Khim.*, No. 8, 1746 (1977).
75. D. I. Kaminski, O. N. Temkin and D. G. Bonchev, *Appl. Catal. A*, **88**, 1 (1992).
76. P. M. Henry, *Palladium Catalyzed Oxidation of Hydrocarbons*, Reidel, Dordrecht/Boston, 1980.

77. M. N. Vargaftik, I. I. Moiseev and Ya. K. Syrkin, *Dokl. Akad. Nauk SSSR*, **147**, 399 (1962).
78. I. I. Moiseev, M. N. Vargaftik and Ya. K. Syrkin, *Dokl. Akad. Nauk SSSR*, **152**, 147 (1963).
79. I. I. Moiseev, M. N. Vargaftik and Ya. K. Syrkin, *Dokl. Akad. Nauk SSSR*, **153**, 140 (1963).
80. I. I. Moiseev, M. N. Vargaftik, S. V. Pestrikov et al., *Dokl. Akad. Nauk SSSR*, **171**, 1365 (1966).
81. O. G. Levanda and I. I. Moiseev, *Kinet. Katal.*, **12**(3), 567 (1971).
82. I. I. Moiseev, O. G. Levanda and M. N. Vargaftik, *J. Am. Chem. Soc.*, **96**(4), 1003 (1974).
83. P. M. Henry, *J. Am. Chem. Soc.*, **86**, 3246 (1964).
84. P. M. Henry, *J. Am. Chem. Soc.*, **88**, 1595 (1966); **94**, 4437 (1972).
85. J. Smidt, W. Hafner, R. Jira, R. Siber, S. Sedlmeier and A. Sabel, *Angew. Chem. Int. Ed.*, **1**, 80 (1962).
86. J. E. Bą̆ckvall, B. Ǎkermark and S. O. Ljunggren, *J. Am. Chem. Soc.*, **101**, 2411 (1979).
87. B. Ǎkermark, B. Soederberg and S. S. Hall, *Organometallics*, **6**(12), 2608 (1987).
88. S. M. Brailovsky, O. N. Temkin, V. S. Shestakova and A. F. Kuperman, *Kinet. Katal.*, **22**(6), 1446 (1981).
89. L. M. Zakharova, M. N. Vargaftik, I. I. Moiseev and E. A. Katsman, *Izv. Akad. Nauk SSSR, Ser. Khim.*, No. 3, 700 (1970).
90. W. Hafner, R. Jira, J. Sedlmeier and J. Smidt, *Chem. Ber.*, **95**, 1575 (1962).
91. V. M. Ignatenko and E. S. Rudakov, *Dokl. Akad. Nauk UkrSSR, Ser. B*, No. 3, 242 (1976).
92. E. S. Rudakov and V. M. Ignatenko, *Dokl. Akad. Nauk SSSR*, **241**(1), 148 (1978).
93. E. S. Rudakov and V. M. Ignatenko, *Kinet. Katal.*, **28**(3), 589 (1987).
94. V. M. Ignatenko and E. S. Rudakov, *Kinet. Katal.*, **29**(6), 1317 (1988).
95. V. M. Ignatenko and E. S. Rudakov, *Kinet. Katal.*, **30**(1), 44 (1989).
96. E. S. Rudakov and V. L. Lobachev, *React. Kinet. Catal. Lett.*, **22**(1–2), 81 (1983).
97. A. Zanardo, F. Pinna, R. A. Michelin and G. Strukul, *Inorg. Chem.*, **27**, 1966 (1988).
98. S. A. Mitchenko, V. V. Zamashchikov, A. A. Shubin and K. V. Malakhov, *Zh. Obshch. Khim.*, **61**(8), 1905 (1991).
99. S. A. Mitchenko, V. V. Zamashchikov, A. A. Shubin and O. N. Pryadko, *Kinet. Katal.*, **34**(2), 253 (1993).
100. V. P. Ananikov, S. A. Mitchenko, I. P. Beletskaja et al., *Inorg. Chem. Commun.*, **1**, 411 (1998).
101. S. A. Mitchenko, V. P. Ananikov, I. P. Beletskaja and Y. A. Ustynyuk, *Mendeleev Commun.*, **5**, 130 (1997).
102. J. A. Osborn, F. H. Jardine, J. F. Young and G. J. Wilkinson, *J. Chem. Soc. A*, 1711 (1966).
103. J. A. Osborn, G. Wilkinson et al., *J. Chem. Soc. A*, 1574 (1967); 1054 (1968).
104. G. O'Connor and G. Wilkinson, *Tetrahedron Lett.*, 1375 (1969).
105. R. S. Coffey, in *Aspects of Homogeneous Catalysis*, R. Ugo (Ed), Carlo Manfredi, Milan, 1970, Vol. 1, pp. 7–92.
106. B. R. James, *Homogeneous Hydrogenation*, John Wiley & sons, New York, 1973.
107. B. Hui and B. R. James, *Proceedings of the 4th Intern. Conf. on Organometallic Chemistry* (Bristol, 1969), L6.
108. M. G. Burnett and R. J. Morrison, *J. Chem. Soc. A*, 2325 (1971).
109. W. L. M. Van Gaal and F. L. A. Van Den Bekerom, *J. Organomet. Chem.*, **134**, 237 (1977).
110. S. Siegel and D. Ohrt, *Inorg. Nucl. Chem. Lett.*, **8**, 15 (1972).
111. N. G. Mekhryakova, L. G. Bruk, O. L. Kaliya et al., *Kinet. Katal.*, **20**(3), 629 (1979).
112. N. G. Mekhryakova, A. N. Nyrkova, N. F. Alekseeva et al., *Kinet. Katal.*, **30**, 98 (1989).
113. A. N. Nyrkova, G. A. Kukina, O. N. Temkin and M. A. Porai-Koshits, *Koord. Khim.*, **11**, 954 (1985).
114. A. P. Belov, N. G. Satsko and I. I. Moiseev, *Kinet. Katal.*, **13**, 892 1972; *Dokl. Akad. Nauk SSSR*, **202**, 81 (1972).
115. J. M. Aubry, *J. Am. Chem. Soc.*, **107**, 5844 (1985).
116. J. M. Aubry and B. Cazin, *Inorg. Chem.*, **27**(12), 2013 (1988).
117. L. N. Arzamaskova and A. P. Purmal', *Zh. Fiz. Khim.*, **45**, 186 (1971).
118. V. B. Zabrodin, O. V. Nesterov and S. G. Entelis, *Kinet. Katal.*, **11**, 114 (1970).
119. A. S. Goldman and J. Halpern, *J. Am. Chem. Soc.*, **109**, 7537 (1987).
120. M. Szwarc, *Carbanions, Living Polymers and Electron Transfer Processes*, John Wiley & Sons/Interscience, New York, 1968.

121. A. R. Gantmakher and R. V. Basova, in *Metody elementoorganicheskoi khimii: Li, K, Na, Rb i Cs* (Methods of Organoelemental Chemistry: Li, K, Na, Rb and Cs), T. V. Talalaeva and K. A. Kocheshkov (Eds), Nauka, Moscow, 1971, Vol. 2, p. 915.
122. V. V. Shamanin, E. Yu. Melenevskaya and V. N. Sgonik, *Acta Polym*., **33**(3), 175 (1982).
123. P. M. Henry, in *Advances in Organometallic Chemistry*, F. G. A. Stone and R. West (Eds), Academic Press, New York, 1975, Vol. 13, p. 363.
124. T. A. Stephenson, S. M. Morehouse, A. R. Powell et al., *J. Chem. Soc*., 3632 (1965).
125. A. C. Shapski and M. L. Smart, *J. Chem. Soc. Chem. Commum*., 658 (1970).
126. R. N. Pandey and P. M. Henry, *Can. J. Chem*., **52**, 1241 (1974).
127. A. K. Yatsimirsky and A. D. Ryabov, *Zh. Neorg. Khim*., **22**(7), 1863 (1977).
128. D. P. Bancroft, F. A. Cotton, L. R. Falvello and W. Schwotzer, *Polyhedron*, **7**(8), 615 (1988); A. S. Batsanov, G. A. Timko, Yu. T. Struchkov, et al., *Koord. Khim*., **15**(5), 688 (1989).
129. A. K. Yatsimirsky, S. A. Deiko and A. D. Ryabov, *Tetrahedron*, **39**, 1 (1983).
130. H. Kempter and R. Mecke, *Z. Phys. Chem. B*, **46**, 229 (1940).
131. N. D. Coggeshall and E. L. Saier, *J. Am. Chem. Soc*., **73**, 5414 (1951).
132. R. A. Akhmed'yanova, I. Yu. Litvintsev, A. Yu. Monaenkov and V. N. Sapunov, *Neftekhimiya*, **34**(5), 427 (1994).
133. R. A. Akhmed'yanova, A. Yu. Monaenkov, I. Yu. Litvintsev and V. N. Sapunov, *Neftekhimiya*, **36**(1), 76 (1996).
134. R. A. Akhmed'yanova, A. G. Liakumovich, I. Yu. Litvintsev et al., *Neftekhimiya*, **36**(1), 80 (1996).
135. V. V. Smirnov, E. N. Golubeva and T. N. Rostovshchikova, *Kinet. Katal*., **33**, 74 (1992).
136. T. N. Rostovshchikova, *Assotsiirovannye kataliticheskie sistemy v sinteze i prevrashcheniyakh galogenuglevodorodov* (Associated Catalytic Systems in Synthesis and Transformations of Halogenated Hydrocarbons), Doctoral Dissertation in Chemistry, MGU (Moscow State University), Moscow, 1998.
137. A. Yu. Bogomolov, V. V. Smirnov, T. N. Rostovshchikova and P. V. Gerasimov, *Kinet. Katal*., **36**, 253 (1995).
138. A. Yu. Bogomolov, P. V. Gerasimov, V. V. Smirnov and T. N. Rostovshchikova, *Kinet. Katal*., **36**, 665 (1995).
139. A. Yu. Bogomolov, V. V. Smirnov and T. N. Rostovshchikova, *Kinet. Katal*., **36**, 831 (1995).
140. F. Basolo and R. Pearson, *Mechanisms of Inorganic Reactions: A Study of Metal Complexes in Solutions* (2nd ed.), John Wiley & Sons, 1967.
141. N. N. Semenov, A. E. Shilov and G. N. Likhtenshtein, *Dokl. Acad. Nauk SSSR*, **221**, 1374 (1975).
142. G. N. Likhtenshtein, *Mnogoyadernye okislitel'no-vosstanovitel'nye metallofermenty* (Polynuclear Redox Metalloenzymes), Nauka, Moscow, 1979.
143. A. E. Shilov, *Metal Complexes in Biomimetic Chemical Reactions*, CRC Press, Boca Raton/New York, 1997.
144. J. Halpern, E. R. MacGregor and E. Peters, *J. Phys. Chem*., **60**, 1455 (1956).
145. T. S. Dzhabiev, *Izv. Akad. Nauk SSSR, Ser. Khim*., No. 10, 1750 (2001).
146. T. S. Dzhabiev, V. Ya. Shafirovich and A. E. Shilov, *React. Kinet. Catal. Lett*., **4**(1), 11 (1976).
147. T. S. Dzhabiev, *Kinet. Katal*., **30**(5), 1219 (1989).
148. T. S. Dzhabiev, *Kinet. Katal*., **39**(4), 487 (1998).
149. T. S. Dzhabiev, *Kinet. Katal*., **42**(4), 512 (2001).
150. T. S. Dzhabiev, S. A. Mironova and A. E. Shilov, *Kinet. Katal*., **40**, 844 (1999).
151. C. E. H. Bawn and A. G. White, *J. Chem. Soc*., 331 (1951).
152. J. H. Baxendale and C. F. Wells, *Trans. Faraday Soc*., **53**, 800 (1957).
153. P. A. Berlin, S. P. Bondarenko, R. P. Tiger and S. G. Entelis, *Khim. Fiz*., **5**, 722 (1984).
154. P. A. Berlin, R. P. Tiger, Yu. N. Chirkov and S. G. Entelis, *Khim. Fiz*., **3**(10), 1448 (1984).
155. P. A. Berlin, R. P. Tiger and S. G. Entelis, *Kinet. Katal*., **28**(6), 1347 (1987).
156. P. A. Berlin, R. P. Tiger, Yu. N. Chirkov and S. G. Entelis, *Kinet. Katal*., **28**(6), 1354 (1987).
157. P. A. Berlin, M. A. Levina, R. P. Tiger and S. G. Entelis, *J. Mol. Catal*., **64**, 15 (1991).
158. Y. R. Rao, K. C. Rajanna and P. K. Saiprakash, *Oxid. Commun*., **7**(3–4), 295 (1984).
159. F. K. Schmidt, N. M. Ryntina, V. V. Saraev et al., *React. Kinet. Catal. Lett*., **5**(2), 101 (1976).

160. F. K. Shmidt, *Kataliz kompleksami metallov pervogo perekhodnogo ryada reaktsii gidrirovaniya i dimerizatsii* (Catalysis of Hydrogenation and Dimerization Reactions by Complexes of Metals of the First Transition Series), Izdat. Irkutsk. Gos. Univ., Irkutsk, 1986, pp. 49–50.
161. B. Hui and B. R. James, *J. Chem. Soc. Chem. Commun.*, 198 (1969).
162. B. R. James, *Homogeneous Hydrogenation*, John Wiley & Sons, New York, 1973; *Gomogennoe gidrirovanie*, Mir, Moscow, 1976, p. 114 (Russian translation).
163. A. K. Yatsimirsky, A. D. Ryabov and I. V. Berezin, *Dokl. Acad. Nauk SSSR*, **236**(4), 924 (1977).
164. A. K. Yatsimirsky, A. D. Rybov and I. V. Berezin, *J. Mol. Catal.*, **5**, 399 (1979).
165. I. V. Kozhevnikov and K. I. Matveev, *React. Kinet. Catal. Lett.*, **4**, 451 (1976).
166. A. K. Yatsimirsky, A. D. Rybov and I. V. Berezin, *J. Mol. Catal.*, **4**, 151 (1978).
167. R. N. Pandey and P. M. Henry, *Can. J. Chem.*, **53**, 2223 (1975).
168. S. A. Deiko, A. S. Erokhin and A. K. Yatsimirsky, *Kinet. Katal.*, **22**(3), 613 (1981).
169. A. P. Belov, *Issledovanie vzaimodeistviya solei palladiya s olefinami v uksusnoi kislote* (Investigation of Interaction of Palladium Salts with Olefins in Acetic Acid), Candidate Dissertation in Chemistry, MITKhT (Moscow Institute of Fine Chemical Technology), Moscow, 1996.
170. A. P. Belov, I. I. Moiseev and N. G. Uvarova, *Izv. Akad. Nauk SSSR, Ser. Khim.*, 2224 (1966); 1642 (1965).
171. I. I. Moiseev, A. P. Belov, V. A. Igoshin and Ya. K. Syrkin, *Dokl. Acad. Nauk SSSR*, **173**, 863 (1967).
172. I. I. Moiseev, *π-Kompleksy v zhidkofaznom okislenii olefinov* (π Complexes in Liquid-Phase Oxidation of Olefins), Nauka, Moscow, 1970, pp. 123–144.
173. I. I. Moiseev, M. N. Vargaftik and Ya. K. Syrkin, *Dokl. Acad. Nauk SSSR*, **133**, 377 (1960).
174. V. I. Sokolov, V. V. Bashilov and O. A. Reutov, *Dokl. Acad. Nauk SSSR*, **188**, 127 (1969).
175. V. I. Sokolov, V. V. Bashilov, P. V. Petrovskii and O. A. Reutov, *Dokl. Acad. Nauk SSSR*, **213**, 1103 (1973).
176. N. Yu. Kozitsyna, M. V. Martens, I. P. Stolyarov et al., *Izv. Akad. Nauk SSSR, Ser. Khim.*, No. 9, 1965 (1999).
177. N. Yu. Kozitsyna, A. A. Bukharkina, M. V. Martens et al., *J. Organometal. Chem.*, **636**, 69 (2001).
178. P. E. Behnken, J. A. Belmont, D. C. Busby et al., *J. Am. Chem. Soc.*, **106**, 3011 (1984).
179. P. E. Behnken, D. C. Busby, M. S. Delaney et al., *J. Am. Chem. Soc.*, **106**, 7444 (1984).
180. I. I. Moiseev and M. N. Vargaftik, *Usp. Khim.*, **39**(12), 1931 (1990).
181. I. P. Stolyarov, M. N. Vargaftik and I. I. Moiseev, *Kinet. Katal.*, **28**(6), 1359 (1987).
182. V. P. Zagorodnikov, M. N. Vargaftik and I I. Moiseev, *Kinet. Katal.*, **27**, 851 (1986).
183. J. A. Nieuwland and R. R. Vogt, *The Chemistry of Acetylene*, Am. Chem. Soc., New York, 1945; J. Nieuwland and R. Vogt, in *Khimiya Atsetilena*, A. D. Petrov (Ed), GIIL, Moscow, 1947, Additional Chapters VI and VII (in Russian).
184. O. A. Chaltykyan, *Kuprokataliz* (Cuprocatalysis), Aipetrat, Erevan, 1963.
185. O. N. Temkin, G. K. Shestakov and N. Yu. Kozlova, *Kinet. Katal.*, **31**, 850 (1990).
186. M. G. Mys'kiv, *Sintez i kristallokhimiya olefinovykh i atsetilenovykh π-kompleksov galogenidov medi(I)* (Synthesis and Crystal Chemistry of Olefin and Acetylene π Complexes of Copper(I) Halides), Doctoral Dissertation in Chemistry, Inst. General. Inorg. Chem., Moscow, 1991.
187. S. I. Osechkin, *Sintez i rentgenostrukturnyi analiz π-kompleksov medi(1) s atsetilenom i ego proizvodnymi* (Synthesis and X-Ray Diffraction Analysis of Copper(I) π Complexes with Acetylene and Its Derivatives), Candidate Dissertation in Chemistry, L'viv State University, L'viv, 1991.
188. B. M. Mikhalichko, *π-Kompleksy galogenidov medi(1) s alkinami* (π Complexes of Copper(I) Halides with Alkynes), Doctoral Dissertation in Chemistry, L'viv State University, L'viv, 1999.
189. M. G. Mys'kiv and B. M. Mykhalichko, *Zh. Strukt. Khim.*, **35**(5), 120 (1994).
190. G. W. Parshal and S. D. Ittel, *Homogeneous Catalysis*, John Wiley & Sons, New York, 1992.
191. W. Keim (Ed), *Catalysis in C1 Chemistry*, D. Reidel Publ., 1983; W. Keim (Ed), *Kataliz v C1 khimii*, Khimiya, Leningrad, 1987 (I. I. Moiseev, Additional Comments to Russian Translation).
192. O. A. Chaltykyan, *Zh. Obshch. Khim.*, **18**, 1626 (1948).
193. S. I. Osechkin, M. G. Mys'kiv, P. Yu. Zavalii and A. N. Sobolev, *Metalloorg. Khim.*, **4**, 997 (1991).
194. I. S. Morozov and G. V. Ustavshchikova, *Izv. Akad. Nauk SSSR, Ser. Khim.*, 451 (1944).
195. T. G. Sukhova, *Issledovanie kompleksoobrazovaniya v zhidkofaznykh kataliticheskikh sistemakh na osnove kompleksov medi* (Investigation of Complex Formation in Liquid-Phase Catalytic Systems Based on Copper Complexes), Candidate Dissertation in Chemistry, MITKhT (Moscow Institute of Fine Chemical Technology), Moscow, 1969.

196. O. N. Temkin, R. M. Flid, T. G. Sukhova et al., *Zh. Prikl. Khim.*, **41**, 633 (1968).
197. T. G. Sukhova, O. N. Temkin and R. M. Flid, *Zh. Neorg. Khim.*, **15**, 1849 (1970).
198. S. Ahrland and J. Rawsthorne, *Acta Chem. Scand.*, **24**, 157 (1970).
199. A. A. Noyes and M. Chow, *J. Am. Chem. Soc.*, **40**, 739 (1918).
200. W. M. Latimer, *The Oxidation States of Elements and their Potentials in Aqueous Solutions* (2nd ed.), Prentice-Hall, Englewood Cliffs, 1952.
201. T. Hurlen, *Acta Chem. Scand.*, **15**, 1231 (1961).
202. T. G. Sukhova, O. N. Temkin, R. M. Flid and T. K. Kaliya, *Zh. Neorg. Khim.*, **13**, 2073 (1968).
203. T. G. Sukhova, N. Ya. Borshch, O. N. Temkin and R. M. Flid, *Zh. Neorg. Khim.*, **14**, 694 (1969).
204. L. T. Andrews and R. M. Keefer, *J. Am. Chem. Soc.*, **71**, 2379 (1949).
205. T. G. Sukhova, O. N. Temkin and R. M. Flid, *Zh. Neorg. Khim.*, **14**, 928 (1969).
206. H. A. C. McKay, *Trans. Faraday Soc.*, **49**, 237 (1953).
207. B. Hedström, *Acta Chem. Scand.*, **9**, 613 (1955).
208. R. M. Flid and I. I. Moiseev, *Zh. Prikl. Khim.*, **27**, 1145 (1954).
209. M. A. Paul and F. A. Long, *Chem. Rev.*, **57**, 1 (1957).
210. G. F. Tikhonov, O. N. Temkin and R. M. Flid, *Zh. Fiz. Khim.*, **40**, 3075 (1966).
211. G. F. Tikhonov, O. N. Temkin and R. M. Flid, *Kinet. Katal.*, **8**, 520 (1967).
212. H. Hartl and J. Fuchs, *Angew. Chem.*, **98**, 550 (1986).
213. L. P. Shevardina and V. M. Pinchuk, *Zh. Strukt. Khim.*, **30**(4), 17 (1989).
214. O. N. Temkin, *Kinetika i mekhanizm kataliticheskikh prevrashchenii atsetilena v rastvorakh kompleksov metallov* (Kinetics and Mechanism of Catalytic Transformations of Acetylene in Solutions of Metal Complexes), Doctoral Dissertation in Chemistry, MITKhT (Moscow Institute of Fine Chemical Technology), Moscow, 1972.
215. G. K. Shestakov, A. Ermakova, V. G. Mikhal'chenko et al., *Zh. Fiz. Khim.*, **44**, 406 (1970).
216. G. K., Shestakov, A. V. Massal'skaya, S. M. Airyan and O. N. Temkin, *Kinet. Katal.*, **18**, 399 (1977).
217. G. K. Shestakov, F. I. Bel'skii, S. M. Airyan and O. N. Temkin. *Kinet. Katal.*, **19**, 334 (1978).
218. O. N. Temkin, G. K. Shestakov and R. M. Flid, *Dokl. Acad. Nauk SSSR*, **184**, 362 (1969).
219. N. Yu. Vsesvyatskaya, G. K. Shestakov and O. N. Temkin, *Kinet. Katal.*, **27**, 1330 (1986).
220. O. N. Temkin, G. F. Tikhonov, R. M. Flid and Yu. R. Galiev, *Kinet. Katal.*, **8**, 123 (1967).
221. O. N. Temkin, R. M. Flid, G. K. Shestakov et al., *Kinet. Katal.*, **10**, 1230 (1969).
222. G. K. Shestakov, G. F. Tikhonov, O. N. Temkin and R. M. Flid, *Kinet. Katal.*, **11**, 575 (1970).
223. G. K. Shestakov, G. F. Tikhonov, O. N. Temkin et al., *Kinet. Katal.*, **11**, 875 (1970).
224. V. S. Vartanyan, G. K. Shestakov and O. N. Temkin, *Arm. Khim. Zh.*, **32**, 259, 264 (1979).
225. G. K. Shestakov, *Fiziko-khimicheskie osnovy protsessov sinteza monomerov i poluproduktov na osnove atsetilena i ego gomologov* (Physicochemical Principles of Synthesis Monomers and Semiproducts Based on Acetylene and Its Homologs), Doctoral Dissertation in Chemistry, MITKhT (Moscow Institute of Fine Chemical Technology), Moscow, 1989.
226. A. A. Khorkin, A. Ermakova, O. N. Temkin et al., *Kinet. Katal.*, **14**, 336 (1973).
227. G. F. Tikhonov, G. K. Shestakov, O. N. Temkin and R. M. Flid, *Kinet. Katal.*, **7**, 914 (1966).
228. S. M. Brailovsky, O. N. Temkin, R. M. Flid and N. G. Belova, *Kinet. Katal.*, **11**, 107, 1346 (1970).
229. S. M. Brailovsky, O. N. Temkin and R. M. Flid, *Kinet. Katal.*, **12**, 1152 (1971).
230. S. M. Brailovsky, O. N. Temkin, A. S. Kostyushin and K. Yu. Odintsov, *Kinet. Katal.*, **31**, 1371 (1990).
231. S. M. Brailovsky, Kh. M. Khoan and O. N. Temkin, *Kinet. Katal.*, **35**, 734 (1994).
232. R. M. Flid, O. N. Temkin, E. A. Afanas'eva et al., *Khim. Prom.*, **7**, 486 (1967).
233. A. I. Mogilyansky, O. N. Temkin, R. M. Flid et al., *Uchen. Zap. MITKhT im. M.V. Lomonosova*, **1**(2), 10 (1970).
234. O. N. Temkin, T. G. Sukhova, G. K. Shestakov et al., *Kinet. Katal.*, **10**, 1004 (1969).
235. S. I. Osechkin, M. G. Mys'kiv, P. Yu. Zavalii and A. N. Sobolev, *Metalloorg. Khim.*, **4**, 997 (1991).
236. B. M. Mykhalichko, M. G. Mys'kiv and L. G. Aksel'rud, *Koord. Khim.*, **19**, 722 (1993).
237. B. M. Mykhalichko, *Ukr. Khim. Zh.*, **64**, 16 (1998).
238. B. M. Mykhalichko, M. G. Mys'kiv and V. N. Davydov, *Zh. Neorg. Khim.*, **44**, 411 (1999).
239. K. Brantin, M. Häkansson and S. Jagner, *J. Organomet. Chem.*, **474**, 229 (1994).
240. B. M. Mykhalichko, M. G. Mys'kiv and E. A. Goreshnik, *Koord. Khim.* **25**, 65 (1999).
241. B. M. Mykhalichko and M. G. Mys'kiv, *Koord. Khim.*, **24**, 939 (1998).

242. M. G. Mys'kiv, P. Yu. Zavalii, B. M. Mykhalichko and V. S. Fundamenskii, *Koord. Khim.*, **14**, 1619 (1988).
243. B. M. Mykhalichko, *Zh. Neorg. Khim.*, **43**, 1840 (1998).
244. B. M. Mykhalichko and M. G. Mys'kiv, *Koord. Khim.*, **25**, 533 (1999).
245. B. M. Mykhalichko, M. G. Mys'kiv and V. N. Davydov, *Zh. Neorg. Khim.*, **44**, 46 (1999).
246. B. M. Mykhalichko, *Zh. Neorg. Khim.*, **43**, 1639 (1998).
247. B. M. Mykhalichko, *Koord. Khim.*, **25**, 359 (1999).
248. B. M. Mykhalichko and M. G. Mys'kiv, *Zh. Neorg. Khim.*, **44**, 1127 (1999).
249. B. M. Mykhalichko and M. G. Mys'kiv, *Koord. Khim.*, **24**, 596 (1998).
250. B. M. Mykhalichko and M. G. Mys'kiv, *Zh. Strukt. Khim.*, **40**, 1160 (1999).
251. B. M. Mykhalichko and M. G. Mys'kiv, *Koord. Khim.*, **25**, 461 (1999).
252. M. J. S. Dewar, *Bull. Soc. Chem. Fr.*, **18**, C71 (1951).
253. A. K. Barnard, *Theoretical Basis of Inorganic Chemistry*, McGraw-Hill, New York, 1965.
254. P. H. Kasai, D. McLeod, Jr. and T. Watanabe, *J. Am. Chem. Soc.*, **102**, 179 (1980).
255. M. Sodupe and C. W. Bauschlicher Jr., *J. Phys. Chem.*, **95**, 8640 (1991).
256. M. Bohme, T. Wagener and G. Frenking, *J. Organomet. Chem.*, **520**, 31 (1996).
257. M. G. Mys'kiv and V. V. Oliinik, *Koord. Khim.*, **21**, 290 (1995).
258. M. G. Mys'kiv, V. V. Oliinik, B. M. Mikhalichko and E. A. Goreshnik, *Prats. NTSh*, **1**, 201 (1997).
259. G. R. Desiraju, *Acc. Chem. Res.*, **24**, 290 (1991).
260. Z. Berkovitch-Yellin and L. Leizerovitz, *Acta Crystallogr. B*, **40**, 159 (1984).
261. R. Vestin, Svensk. *Kem. Tidskr.*, **66**, 65 (1954).
262. K. Osakada, T. Takizawa and T. Yamamoto, *Organometallics*, **14**, 3531 (1995).
263. M. P. Gamasa, J. Gimeno, E. Lastra et al., *J. Organomet. Chem.*, **378**, 11 (1989).
264. J. Diez, M. P. Gamasa, J. Gimeno et al., *Organometallics*, **12**, 2213 (1993).
265. J. Diez, M. P. Gamasa, J. Gimeno et al., *Organometallics*, **10**, 380 (1991).
266. V. W.-W. Yam, W.-K. Lee, K.-K. Cheung et al., *J. Chem. Soc. Dalton Trans.*, 3283 (1996).
267. D. M. Knotter, A. L. Spek and G. Van Koten. *J. Chem. Soc. Chem. Commun.*, 1738 (1989).
268. D. M. Knotter, A. L. Spek, D. M. Grove and G. Van Koten, *Organometallics*, **11**, 4083 (1992).
269. M. D. Janssen, J. G. D. Donkervoort, S. B. Van Berlekom et al., *Inorg. Chem.*, **35**, 4752 (1996).
270. R. W. M. Hoedt, J. G. Noltes, G. Van Koten and A. L. Spek, *J. Chem. Soc. Dalton Trans.*, 1800 (1978).
271. M. G. B. Drew, F. S. Esho and S. M. Nelson, *J. Chem. Soc. Chem. Commun.*, 1347 (1982).
272. V. W.-W. Yam, S. W.-K. Choi, C.-L. Chan and K.-K. Cheung, *J. Chem. Soc. Chem. Commun.*, 2067 (1996).
273. D. L. Reger, J. E. Collins, M. E. Huff et al., *Organometallics*, **14**, 5475 (1995).
274. V. W.-W. Yam, W.-K. Lee, K. K. Cheung et al., *J. Chem. Soc. Dalton Trans.*, 2889 (1996).
275. A. J. Edwards, M. A. Paver, P. R. Raithby et al, *Organometallics*, **13**, 4967 (1994).
276. V. W.-W. Yam, W.-K. Lee and T.-F. Lai, *Organometallics*, **12**, 2383 (1993).
277. L. Naldini, F. Demartin, M. Manassero et al., *J. Organomet. Chem.*, **279**, 42 (1985).
278. M. P. Gamasa, J. Gimeno, E. Lastra and X. Solaus, *J. Organomet. Chem.*, **346**, 277 (1988).
279. V. W.-W. Yam, W.-K. Lee and K.-K. Cheung, *J. Chem. Soc. Dalton Trans.*, 2335 (1996).
280. F. Olbrich, J. Kopf, and E. Weiss, *Angew. Chem. Int. Ed.*, **32**, 1077 (1993).
281. P. W. R. Corfield and H. M. M. Shearer, *Acta Crystallogr.*, **21**, 957 (1966).
282. V. W.-W. Yam, W. K.-M. Fung and K.-K. Cheung, *Angew. Chem. Int. Ed.*, **35**, 1100 (1996).
283. J. F. Normant, *Usp. Khim.*, **42**, 645 (1973).
284. J. F. Normant and A. Alexakis, *Synthesis*, 841 (1981).
285. Ya. A. Dorfman, *Zhidkofaznyi kataliz: orbital'noe modelirovanie* (Liquid-Phase Catalysis: Orbital Modelling), Nauka, Alma-Ata, 1981.
286. B. M. Mykhalichko, M. G. Mys'kiv and L. G. Aksel'rud, *Zh. Neorg. Khim.*, **41**, 75 (1996).
287. D. T. Cromer and A. C. Larson, *Struct. Rep.*, **27**, 416 (1962).
288. M. Massaux, M.-T. Le Bihan and R. Chevalier, *Acta Crystallogr. B*, **33**, 2084 (1977).

3

Multi-Route Mechanisms in Reactions Involving Metal Complexes

Multi-route mechanisms are frequently encountered in the catalytic chemistry in cases of both stoichiometrically many-valued ($Q_r \geq 2$) and single-valued ($Q_r = 1$) reactions, where Q_r is the number of linearly independent overall route equations. For some processes, the multi-route character reflects essential features of the nature of a process, such as asymmetric catalysis, chain reaction (including polymerization processes and acid-catalyzed conversions of hydrocarbons), and isotope exchange reactions. In other cases, the multi-route character relates only to the variety of possible pathways in the transformation of intermediates.

The principles of the theory of routes and the mathematical apparatus that can be employed to describe the kinetics of multi-route reactions were presented in Chapter 1. Some examples of multi-route reactions, in which several complexes, occurring in equilibrium, were involved in the same reaction (the case of "trivial" multi-route character), were considered in Chapter 2.

Taking into account that multi-route reactions are frequently encountered in the catalysis with metal complexes and that the study of reaction mechanisms for constructing theoretically justified kinetic models in the case of multi-route reactions is a quite difficult task, it is expedient to analyze the origins of multi-route reactions, consider the kinetic features of multi-route mechanisms, outline the possible approaches to the identification of their structure in reactions of various natures, and elucidate the essence of the phenomenon of kinetic and thermodynamic conjugation in the cases of chemical and enzymatic reactions.

3.1 Factors accounting for the appearance and kinetic features of multi-route mechanisms

The multi-route character of a process may be caused by various factors, so that it is possible to distinguish between several types of multi-route mechanisms. These types can be conveniently classified with respect to the way in which routes are linked in the scheme of a particular mechanism.

Homogeneous Catalysis with Metal Complexes: Kinetic Aspects and Mechanisms, First Edition. O. N. Temkin.

I. **Non-Linked Routes**
In the most frequently encountered case, both catalytic and noncatalytic reactions proceed in parallel, so that

$$r_{\Sigma} = r_{NC} + r_{C}.$$

In this case, the routes are interrelated only via the material balance with respect to reactants.

II. **Routes Linked via Common Equilibrium Step (Class A) or Common Intermediate (Class B)**
Typical kinetic graphs (KGs) of such mechanisms are depicted in Figure 3.1. This type can be further subdivided into the following groups.

(II.1) ***Different metal complexes, occurring in equilibrium, catalyze the same reaction.*** This case, corresponding to a "trivial" multi-route character, is represented by the following scheme:

$$\begin{array}{ccccc} ML & \overset{L}{\rightleftharpoons} & ML_2 & \overset{L}{\rightleftharpoons} & ML_3 \\ \downarrow & & \downarrow & & \downarrow \\ P & & P & & P \end{array}$$

(see Examples 2.10, 2.12, and 2.14–2.17). This group of reactions includes, for example, the noncatalytic hydrocarboxylation of acetylene in $Ni(CO)_4$ solutions, which is approximately represented by the following overall equation:

$$Ni(CO)_4 + 5C_2H_2 + 4H_2O + 2HCl \longrightarrow NiCl_2 + 4CH_2{=}CHCOOH + C_2H_4.$$

This reaction involves the conversion of various Ni(0) complexes:

$$Ni(CO)_4 \rightleftharpoons Ni(CO)_3L \rightleftharpoons Ni(CO)_2L_2 \rightleftharpoons Ni(CO)L_3,$$

where L is either solvent or acetylene [1, 2].
Schemes of this type are frequently encountered in the processes of hydrogenation, hydroformylation, and hydrocarboxylation of olefins (see examples below). Various sets of $Cu_mCl_n^{(n-m)-}$ complexes catalyze the reactions of dimerization ($Cu_4Cl_5^-, Cu_4Cl_6^{2-}, Cu_5Cl_6^-$) and hydrochlorination ($CuCl_2^-, Cu_2Cl_3^-$) of acetylene (see Section 2.3). Linear mechanisms of trivial multi-route processes (e.g., for $P = 3$) can be represented by kinetic graph KG1 (Figure 3.1), which belongs to the topological class A (subclass $\mathbf{A}^2$) according to the classification proposed in [3, 4].

(II.2) ***One metal complex simultaneously catalyzes several reactions.*** For example, the same $Cu_4Cl_5^-$ complex is active in the reactions of acetylene dimerization and hydrocyanation (see Section 2.3). One initial complex also serves as an origin for the routes of synthesis of two enantiomers in reactions of asymmetric catalysis (hydrogenation, epoxidation, carbonylation, etc.). The structure of such mechanisms is represented by kinetic graph KG3 (Figure 3.1) of the topological class B for two coupled routes.

(II.3) ***Different metal complexes, occurring in equilibrium, catalyze different reactions.*** The situation where predominant formation of various products in reactions catalyzed by different complexes is also frequently encountered in the catalysis with metal complexes. Examples are offered by the aforementioned hydrochlorination ($CuCl_2^-, Cu_2Cl_3^-$) and dimerization ($Cu_4Cl_p^{(p-4)-}, Cu_5Cl_q^{(q-5)-}$) of acetylene. For the same initial species of $Rh(CO)_2Cl_2^-$ in butanol, reactions in the presence of C_2H_2 and CO lead to the formation of acetaldehyde dibutylacetal and butylacrylate on different complexes of rhodium at the rates proportional to $[Rh]_{\Sigma}^{0.4}$ and $[Rh]_{\Sigma}^{2}$, respectively [5]. The structures of these mechanisms belong to the topological class A (Figure 3.1, graphs KG1 and KG2).
The routes in mechanisms of groups II.1–II.3 classes are interrelated via the material balance with respect to the catalyst.

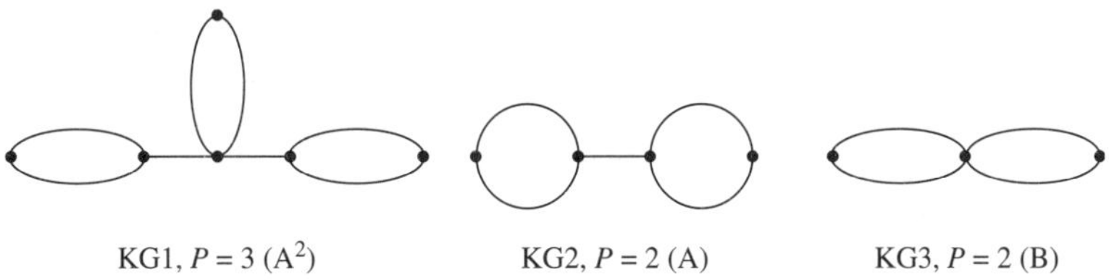

Figure 3.1 Kinetic graphs (KGs) for two- and three-route mechanisms of topological classes A, A^2, and B

(II.4) ***Combination of noncatalytic and catalytic routes within a common mechanism.*** This group of multi-route mechanisms is characteristic of chain reactions, which include the routes involving the steps of formation and termination of active centers and the steps of catalytic cycles (chain propagation or chain growth). These mechanisms are represented by graph KG3 (Figure 3.1) and are considered in more detail in Section 3.4 and Chapter 5.

III. **Routes linked via common steps and intermediates**
The appearance of branching in the steps involving conversion of intermediates, which implies the existence of parallel conversion pathways of intermediates that are common for various routes (processes), is the most general factor responsible for the multi-route character of catalytic processes involving organometallic intermediates. This type of relation between routes is one of the possible mechanisms of the kinetic conjugation of steps and routes (see Section 3.4).

The variety of conversion pathways and the high reactivity of σ organometallic compounds of transition metals, which account for the fact that these compounds could not be isolated from the reaction mixture for almost a century (from the middle of the 19th to the middle of the 20th centuries), are also the main factors that determine the multi-route character of processes in the catalytic chemistry of transition metals. Let us consider several examples of well-known reaction mechanisms.

According to the first scheme of conversions of metal alkyl intermediates in the presence of ethylene and ethane [6],

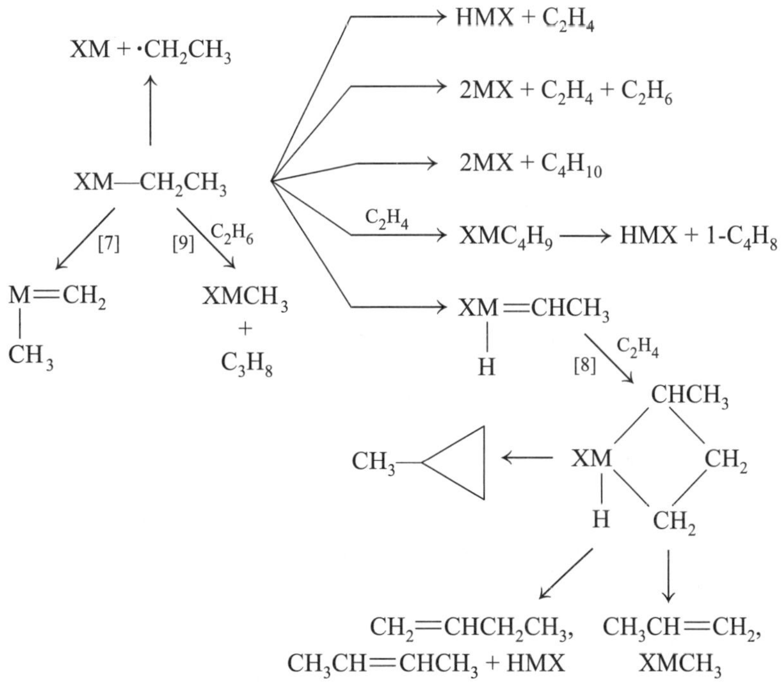

(which can predominantly develop in various directions depending on a particular metal M, its degree of oxidation, and the type of ligand X), there are eight possible organic products that can form via these reactions: C_2H_4, C_2H_6, C_4H_{10}, C_3H_8, C_3H_6, 1-C_4H_8, 2-C_4H_8, and methylcyclopropane.

Owing to diverse reactivity of organopalladium(II) σ complexes, the following conversions were established in aqueous solutions of $PdCl_2$ in the presence of C_2H_2, CO, and oxidant ($FeCl_3$) [10–12]:

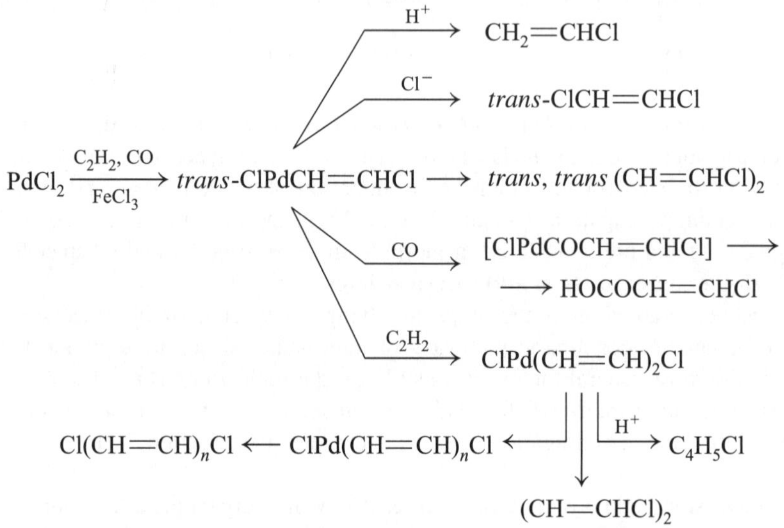

Another example is offered by C_2H_2 and CO conversions in Pd(II) bromide solutions in alcohols (both in the presence and absence of oxidizers) [13–15]:

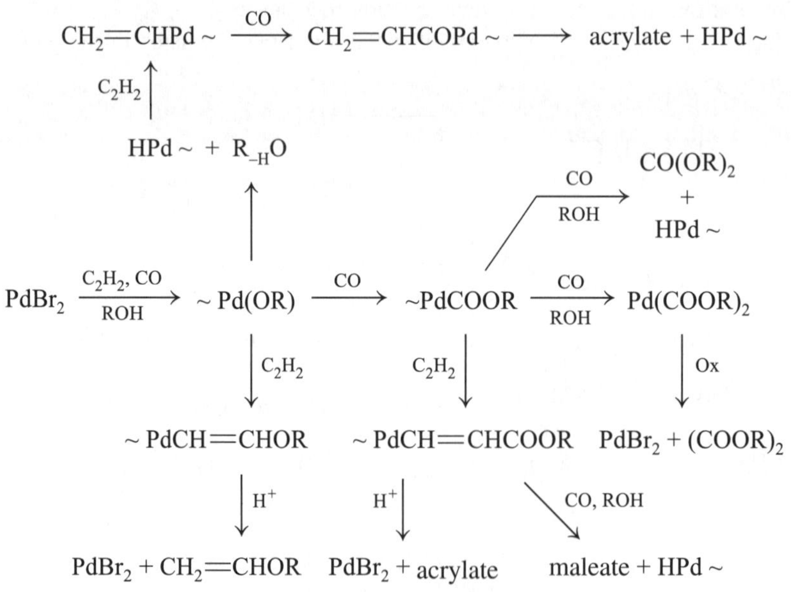

In the course of acetylene carbonylation in the $PdBr_2$–H_2O–CH_3CN system, succinic anhydride and acrylic acid are formed with the participation of different palladium complexes, namely, $Pd_2(\mu\text{-}CO)_2Br_4^{2-}$ and $HPdBr_n^{(1-n)}$, respectively [16].

Thus, in complicated processes under consideration, metal complexes of different natures, not occurring in equilibrium and possessing steady-state, quasi-steady-state, or nonsteady-state concentrations can participate in different reactions. As a result, both oxidation and reduction reactions can be conjugated in one catalytic system [17, 18] and different mechanisms of one reaction can be conjugated in the same catalytic system [18] (see Section 3.4).

The main effort of researchers working on increasing the chemoselectivity of catalytic reactions involving metal complexes is devoted to the attempts at influencing multi-route reactions of type III. The structure of a mechanism of this type is represented by a simple kinetic graph of the topological class C for a two-route mechanism (KG4, Figure 3.2) [4].

An example of a mechanism with a more complicated structure is offered by the reactions of oxidation of an allyl alcohol (S) in aqueous solutions of $PdCl_4^{2-}$ [19], the scheme of which is given in Figure. 3.3 (where π- and σ-type intermediates represent the palladium π and σ complexes, respectively). Here, the oxidation and reduction reactions involve the palladium complexes of three types: $PdCl_4^{2-}$, Pd(0), and HPdCl. Since the number of oxidation products (P_1, P_2, P_3) is significantly greater than the number of reduction products (P_4), $PdCl_4^{2-}$ species only partly perform the function of a catalyst.

This mechanism includes five routes that proceed under quasi-steady-state conditions (if $PdCl_4^{2-}$ is an intermediate species). In the case of a deuterated allyl alcohol $CD_2{=}CHCH_2OH$, an additional route of allyl isomerization appears [19]

The mechanisms of reactions in solutions of polyfunctional catalytic systems can include combinations of the multi-route reactions of various types (see Chapter 4). The main features of the kinetics of multi-route reactions are considered in Section 3.2. The principal feature of the kinetic models of these processes is the interrelation of routes, which increases in the series of topological classes A $\rightarrow$ B $\rightarrow$ C. The stronger the branching in the mechanistic scheme (i.e., the number of vertices with the in-degrees above 2 on a nonoriented graph), the greater the connectivity of this graph and the more pronounced the interrelation of routes.

In order to assess the influence of the topological structure of a given mechanism on the kinetics of involved reactions, it is convenient to use the indices of complexity of the mechanism. Linear mechanisms are characterized by the complexity index E that takes into account the number of weights (or rate constants) that enter into the rational fraction equation of the rate over a route (in the general case, over all routes) [3, 4, 20]. For an arbitrary mechanism, it is possible to use the information complexity index that characterized the information content of the stoichiometric matrix of the mechanism and its submechanisms [4, 21].

Let us consider in more detail the linear mechanisms of classes B and C. The case of a two-route ($P = 2$) catalytic reaction with four steps ($S = 4$) and three intermediates ($I = 3$) is represented by kinetic

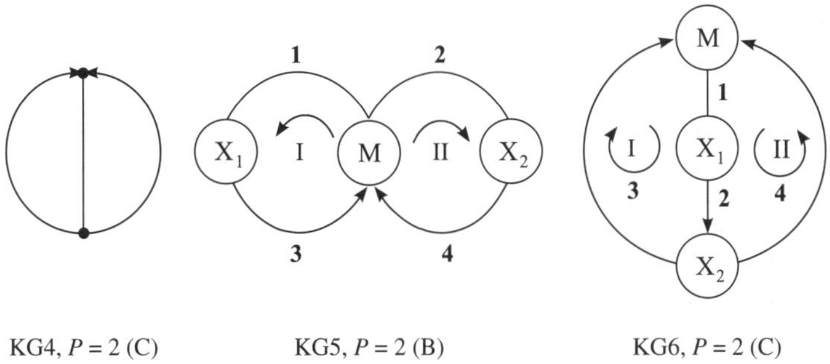

Figure 3.2 Kinetic graphs (KGs) for various mechanisms of classes C and B

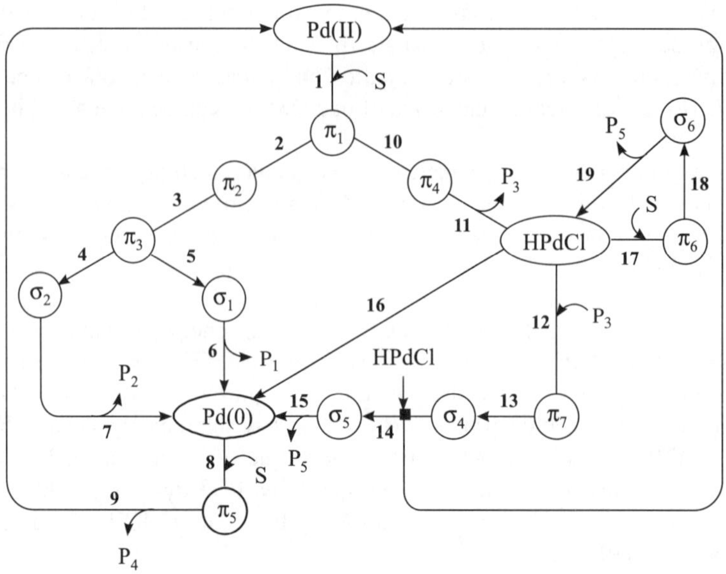

$S = CH_2{=}CHCH_2OH$ $P_3 = CH_2{=}CHCHO$

$P_1 = HOCH_2CH_2CHO$ $P_4 = CH_2{=}CHCH_3$

$P_2 = HOCH_2C(O)CH_3$ $P_5 = CH_3CH_2CHO$

Figure 3.3 Scheme of allyl alcohol oxidation in $PdCl_4^{2-}$ solutions ($2S = P_2 + P_4$, $2S = P_1 + P_4$, $2S = P_3 + P_4 + H_2O$, $S = P_5$, $3S = P_3 + P_4 + P_5 + H_2O$)

graphs KG5 and KG6 in Figure 3.2. For KG5, the rate over route I is described by the following equation (without allowance for the material balance with respect to a catalyst):

$$r_I = \frac{[M]\omega_1^+\omega_2^+}{\omega_1^- + \omega_2^+}. \tag{3.1}$$

Here, the number of weights of the steps in the numerator and denominator of Eq. (3.1), that is, the complexity index is $E^* = 4$. If this mechanism is described in terms of the concentration [M] of active centers, the rate r_I over route I is independent of the steps of route II. With allowance for the material balance ($[M]_\Sigma = [M] + [X_1] + [X_2]$) with respect to a catalyst, we obtain the following relation:

$$r_I = \frac{[M]\omega_1^+\omega_2^+(\omega_3^- + \omega_4{}^+)}{\omega_1^+(\omega_3^- + \omega_4{}^+) + \omega_3{}^+(\omega_1^- + \omega_2^+) + (\omega_1^- + \omega_2^+)(\omega_3^- + \omega_4{}^+)}. \tag{3.2}$$

In this case, $E = 22$ and the expression for rate r_I involves all steps of route II.

A different pattern is observed for a mechanism of class C with two common steps and three common intermediates (KG6, Figure 3.2). Here, the rate r_I without allowance for the material balance with respect to a catalyst is described by the following equation:

$$r_I = \frac{[M]\omega_1^+\omega_2^+\omega_3{}^+}{(\omega_1^- + \omega_2^+)(\omega_3{}^+ + \omega_4{}^+)}, \tag{3.3}$$

which involves all steps of route II, so that $E^* = 11$. In the case of using $[M]_\Sigma$, we obtain the following relation:

$$r_I = \frac{[M]\omega_1^+\omega_2^+\omega_3^+}{(\omega_1^- + \omega_2^+)(\omega_3^+ + \omega_4^+) + \omega_1^+(\omega_3^+ + \omega_4^+) + \omega_1^+\omega_2^+}, \tag{3.4}$$

according to which the complexity index is $E = 17$. In this case, the complexity index dos not exhibit as large increase on the passage from E^* to E.

In the case of a three-route mechanism ($P = 3, S = 6$, and $I = 4$) represented by graphs KG7 (class B^3) and KG8 (class C^3) in Figure 3.4, the kinetic equation becomes even more complicated. Indeed, the kinetic equation for one route (without allowance for the material balance with respect to a catalyst) will involve the weights of six steps ($E^* = 6$) for KG7 and 112 weights of steps ($E^* = 112$) for KG8 [22] (see also Section 1.3).

The presence of common steps in different routes leads to the appearance of high-order kinetic polynomials for the reaction participants and medium components in the numerator and denominator of a kinetic equation. This is the main circumstance that accounts for the absence of an algorithm for the passage from the mathematical description of a kinetic experiment to a scheme of the corresponding mechanism. The complexity of the kinetic equations of multi-route reactions in the absence of such single-valued algorithms for the passage from experimental data to mathematical models complicates the choice of an adequate model for a real process and in many cases even makes this impossible.

For example, investigation of the kinetics of a five-route reaction of acetylene carbonylation in alcohol solutions of palladium iodides [23] led to the following empirical dependence of the rate of the steady-state synthesis of acrylate (A) on $[H^+]$:

$$r_A = \frac{k'_{10}[H^+]^2}{1 + k'_1[H^+] + k'_2[H^+]^2}, \tag{3.5}$$

which significantly differs from the theoretical relation of $r_A = f([H^+])$ that follows from the simple scheme of a five-route mechanism [18]:

$$r_A = \frac{k_{10} + k_{11}[H^+] + k_{12}[H^+]^2 + k_{13}[H^+]^3 + k_{14}[H^+]^4}{k_1 + k_2[H^+] + k_3[H^+]^2 + k_4[H^+]^3 + k_5[H^+]^4 + k_6[H^+]^5}. \tag{3.6}$$

Evidently, this formula cannot be reconstructed from experimental data, but it can be checked in the presence of a mathematically formulated mechanistic hypothesis. These issues are considered in more detail in Chapter 7. Now we will consider examples of investigation of the kinetics of multi-route reactions and approaches to the establishment of a structure of the reaction mechanism.

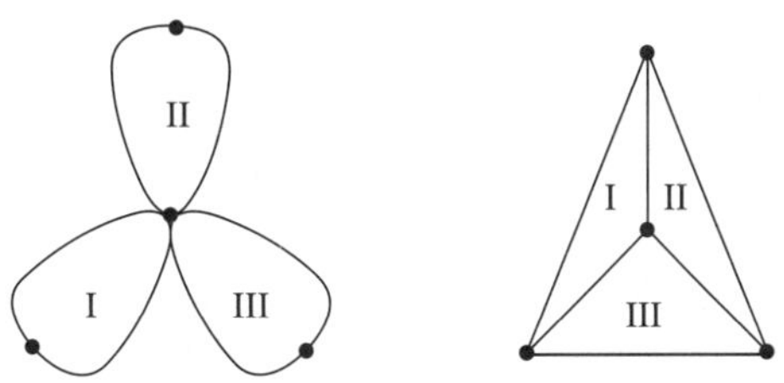

KG7 (B^3) KG8 (C^3)

Figure 3.4 Kinetic graphs (KGs) for mechanisms of classes B^3 and C^3

3.2 Analysis of multi-route reaction kinetics

Catalytic reactions with metal complexes in acid–base media frequently feature parallel routes of catalytic (with respect to the metal complex) and "noncatalytic" processes (acid or nucleophilic catalysis). For example, the process of methylacetylene hydration in $HCl–CuCl–H_2O$, $H_2SO_4–HgSO_4$, and $RuCl_3–HCl–H_2O$ solutions involves acid-catalyzed reaction in addition to those catalyzed by metal complexes. Another interesting example is offered by the process of allene–diene rearrangement in hydrochloric acid solutions containing CuCl [10, 24], which corresponds to the following equation,

$$CH_2{=}C{=}CHCH_2Cl \longrightarrow CH_2{=}CHC(Cl){=}CH_2 \tag{3.7}$$

and proceeds via two routes. Let us consider the kinetics of this reaction in more detail.

Example 3.1

► The kinetics of reaction (3.7) was studied at 50 °C in a continuous flow reactor (with respect to both reactant and product) of ideal mixing in the two systems: (A) $NH_4Cl–NH_4ClO_4–HClO_4–H_2O$ and (B) $NH_4Cl–HCl–H_2O$ with a constant total electrolyte concentrations of 4 M and 12 M, respectively. It was established that the reaction rate in system B (at $[CuCl]_\Sigma = 1.09$ M) is described by a first-order equation with respect to 4-chloro-1,2-butadiene (4-CBD) pressure (P_{4-CBD}). As the concentration of copper chloride is varied within rather narrow limits ([CuCl] ~ 0 – 0.6 M), the activity of Cu^+ ions (a_{Cu^+}) remained directly proportional to $[CuCl]_\Sigma$, which was indicative of the formation of a mononuclear complexes of copper (the method of measurements was described in Section 2.3). The rate of reaction (3.7) obeyed the following relation:

$$r = k_1 P_{4\text{-CBD}} + k_2 P_{4\text{-CBD}}[CuCl]_\Sigma, \tag{3.8}$$

or

$$r = k_1 P_{4\text{-CBD}} + k_2' P_{4\text{-CBD}} a_{Cu^+}. \tag{3.9}$$

where the Henry law constant for 4-CBD enters into the quantities k_1, k_2 and k_2'. Since the first terms in Eqs (3.8) and (3.9), which correspond to the first route, are also retained in the absence of copper, this route is catalyzed by H_3O^+ or Cl^- ions. The dependences of the reaction rate on these species were studied in system A, where it was possible to vary $[H_3O^+]$ and $[Cl^-]$ in a broader range at a constant total concentration of the electrolyte. It was found that the rate over the first route was a linear function of $[NH_4Cl]$ ($k_1 = k_3[NH_4Cl]$), while being independent of $[H_3O^+]$. Taking into account the entire set of kinetic data and the results of experiments with Cl^- replacement by Br^- and CNS^- anions via the reaction

$$CH_2{=}C{=}CHCH_2Cl + X^- \longrightarrow CH_2{=}\underset{\displaystyle X}{\underset{|}{C}}{-}CH{=}CH_2 + Cl^- \qquad (X = Br,\ CNS) \tag{3.10}$$

the mechanism of the allene–diene rearrangement can be represented by a scheme of the S_N2 type.

The dependence of the rate constant k_2 over the second route on $[H_3O^+]$ and $[Cl^-]$ was also studied in system A at [CuCl] = 0.05 M. As $[HClO_4]$ was varied from 0 to 2.5 M, the k_2 value remained unchanged. The activity a_{Cu^+} of Cu^+ ions (calculated with allowance for a change in the diffusion potentials) also remains constant. The dependence of k_2 on $[NH_4Cl]$ is described by a quadratic relation as $k_2 = k_4[NH_4Cl]^2$, so that eventually we obtain

$$r = k_3 P_{4\text{-CBD}}[NH_4Cl] + k_4 P_{4\text{-CBD}}\, a_{Cu^+}[NH_4Cl]^2, \tag{3.11}$$

where $[NH_4Cl]^2$ can be replaced by $a^2_{Cl^-}$, since these values are proportional under the conditions studied.

Thus, the behavior of the "cuprocatalytic" route is indicative of the participation of a 4-CBD complex with $CuCl_2^-$ in the limiting, irreversible step of the process that can be described by the following scheme:

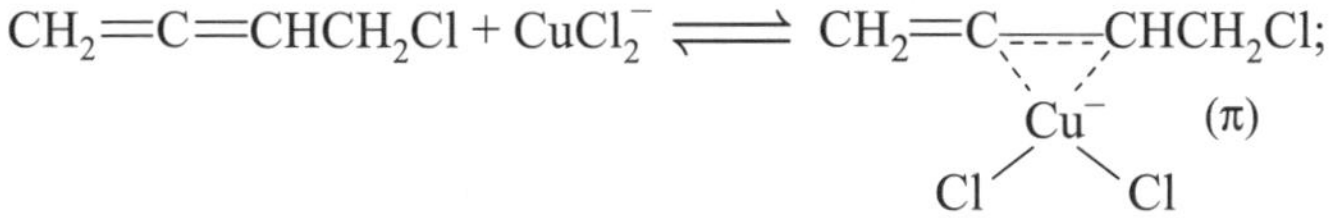
(3.12)

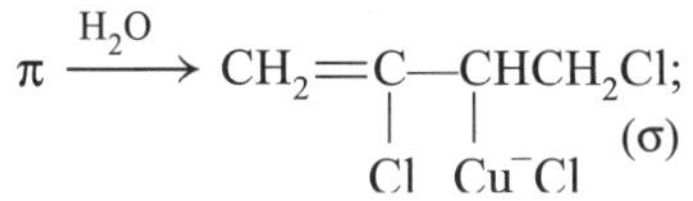
(3.13)

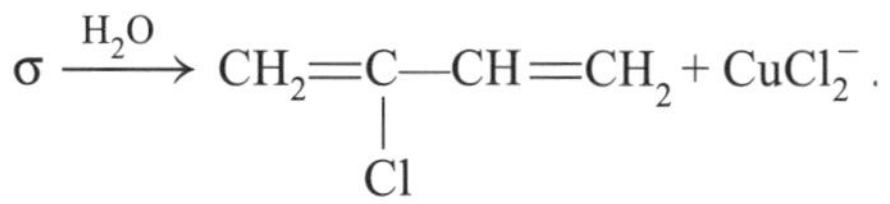
(3.14)

In more concentrated CuCl solutions, the reaction could also involve polynuclear Cu(I) complexes, but a high rate of the reaction (about 10 M/h) at [CuCl] > 1.5 M hindered studying this process in the kinetic region. Thus, the reaction mechanism can be classified as type I (see Section 3.1). ■

If experimental data lead to a two-term kinetic equation of the type of relation (3.8), it can turn out that the first term of this equation in a certain interval of the catalyst concentrations would also correspond to a catalytic route, provided that the catalyst is liberated in one of the quasi-equilibrium steps that precede the formation of one of the products. Therefore, it is necessary to thoroughly study the process in the absence of a catalyst. This situation was observed, for example, in studying the kinetics and mechanism of vinyl acetylene hydrochlorination (one possible method used for the commercial synthesis of chloroprene) in copper chloride solutions.

Example 3.2

► The kinetics of 2-chloro-1,3-butadiene (2-CBD or chloroprene) synthesis via the reaction

$$C_4H_4 + HCl \xrightarrow{Cu\,(I)} \underset{(2\text{-}CBD)}{CH_2{=}C(Cl)CH{=}CH_2} \qquad (3.15)$$

was studied [24, 25] in order to verify the hypotheses of Carothers and Klebansky concerning the process chemistry. According to these assumptions, the reaction is based on the noncatalytic (with respect to Cu(I)) formation of an intermediate 4-chloro-1,3-butadiene (referred to below as allene) followed by its Cu(I)-catalyzed isomerization to 2-CBD (Carothers) and on the direct catalytic 1,2-addition of HCl to vinylacetylene with the formation of 2-CBD (Klebansky).

The kinetics of reaction (3.15) was studied in the NH_4Cl–HCl system at 50 °C and a total electrolyte molality of $12\,\tilde{m}$, $[NH_4Cl] \geq 8\,\tilde{m}$, and $[CuCl] = 0 - 8\,\tilde{m}$ using the method described in Section 2.3. It was established that, at constant values of activities $a_{Cu^+} = (9.9 \pm 1.2) \cdot 10^{-8}, a_{Cl^-} = 4.0 \pm 0.5$, and $a_{Cl^-} = 4.0 \pm 0.5$ in a solution with $[NH_4Cl] = 10\,\tilde{m}$, $[HCl] = 2\,\tilde{m}$, and $[CuCl] = 4\,\tilde{m}$, the rate of 2-CBD formation obeys a first-order equation with respect to P_{VA}. In addition to 2-CBD, there also appeared 1-chloro-1,3-butadiene (1-CBD) with the first-order reaction rate constant in P_{VA} about 20 times as small as the corresponding rate constant for 2-CBD. For the same constant values of activities a_{Cu^+} and a_{Cl^-}, the formation of 2-CBD is also a first-order process with respect to [HCl]:

$$r_{2\text{-CBD}} = k_1' P_{VA}[HCl]. \qquad (3.16)$$

The study of dependences $k_1' = f([\mathrm{CuCl}])$ or $k_1' = f(a_{\mathrm{Cu^+}}, a_{\mathrm{Cl^-}})$ showed that the rate of 2-CBD formation in CuCl-free solutions was almost vanishing ($r_{2-\mathrm{CBD}} < 1 \cdot 10^{-5}$ M/h) but the dependence of k' on $a_{\mathrm{Cu^+}}$ and $a_{\mathrm{Cl^-}}$ is best described by the following two-term equation with the first term dependent on $a^2_{\mathrm{Cl^-}}$ and the second dependent on $a_{\mathrm{Cu^+}}a^3_{\mathrm{Cl^-}}$:

$$r_{2\text{-CBD}} = k_2' P_{\mathrm{VA}}[\mathrm{HCl}]a^2_{\mathrm{Cl^-}} + k_3' P_{\mathrm{VA}}[\mathrm{HCl}]a_{\mathrm{Cu^+}}a^3_{\mathrm{Cl^-}}, \tag{3.17}$$

which is valid in the interval of $[\mathrm{CuCl}] = 0.5 - 8\,\tilde{m}$. The obtained kinetic model agrees with the Carothers hypothesis, since rate of allene isomerization is about two orders of magnitude greater than $r_{2-\mathrm{CBD}}$, and the form of Eq. (3.17) is similar to that of Eq. (3.11). On the other hand, we should accept that 4-CBD can also be formed with the participation of Cu(I) complexes. There are two possible hypotheses that are consistent with Eq. (3.17).

Variant 1: According to this, the mechanistic scheme is as follows:

$$(1)\ \mathrm{C_4H_4}_{\ \mathrm{aq}} + \mathrm{Cu^I_{aq}} \rightleftharpoons \underset{(\pi_1)}{(\mathrm{C_4H_4})\mathrm{Cu^I}} + \mathrm{H_2O};$$

$$(2)\ \pi_1 + \mathrm{Cl^-} \rightleftharpoons \underset{(\sigma_1)}{\mathrm{Cu^ICH{=}C{=}CHCH_2Cl}};$$

$$(3)\ \sigma_1 + \mathrm{H_3O^+} \rightleftharpoons \text{4-CBD} + \mathrm{Cu^I_{aq}}$$

For quasi-equilibrium steps (1)–(3) with allowance for the Henry law constant of C_4H_4, the allene concentration varies as

$$[\text{4-CBD}] = P_{\mathrm{C_4H_4}}[\mathrm{H_3O^+}][\mathrm{Cl^-}], \tag{3.18}$$

i.e., does not depend on $[\mathrm{Cu^I_{aq}}]$. In this case, the active complex composition cannot be determined and

$$r_{2\text{-CBD}} = k_1[\text{4-CBD}][\mathrm{Cl^-}] + k_2[\text{4-CBD}][\mathrm{CuCl_2^-}]. \tag{3.19}$$

Relations (3.18) and (3.19) yield Eq. (3.17). The reversibility of step (3) is confirmed by a change in the solution acidity from pH 4.3 to 2.4 when 4-CBD is passed through the $CuCl–NH_4Cl–H_2O$ solution. In the case of 2-CBD passage, the level of pH remains unchanged.

Variant 2: In this case, the formation of chloroprene via allene is one possible route in the synthesis of 2-CBD via the following scheme:

$$(4)\ \mathrm{C_4H_4}_{\ \mathrm{aq}} + \mathrm{CuCl_2^-} \rightleftharpoons \underset{(\pi_1')}{(\mathrm{C_4H_4})\mathrm{CuCl_2^-}} + \mathrm{H_2O};$$

$$(5)\ \pi_1' + \mathrm{Cl^-} \xrightleftharpoons{\text{1,4-addition}} \underset{(\sigma_1')}{[\mathrm{Cl_2CuCH{=}C{=}CHCH_2Cl}]^{2-}};$$

$$(6)\ \pi_1' + \mathrm{Cl^-} \xrightleftharpoons{\text{1,2-addition}} \underset{(\sigma_2')}{[\mathrm{Cl_2CuCH{=}C(Cl){-}CH{=}CH_2}]^{2-}};$$

$$(7)\ \sigma_1' + \mathrm{H_3O^+} \rightleftharpoons \text{4-CBD} + \mathrm{CuCl_2^-};$$

$$(8)\ \sigma_2' + \mathrm{H_3O^+} \longrightarrow \text{2-CBD} + \mathrm{CuCl_2^-};$$

$$(9)\ \text{4-CBD} + \mathrm{Cl^-} \longrightarrow \text{2-CBD} + \mathrm{Cl^-},$$

according to which

$$r_{2\text{-CBD}} = k_1'[\text{4-CBD}][\text{Cl}^-] + k_2'[\sigma_2'][\text{H}_3\text{O}^+]. \tag{3.20}$$

Relation (3.20) also yields Eq. (3.17).

It is impossible to discriminate between the two hypothetical variants using the experimental kinetic data. It would be helpful to perform experiments with ^{13}C-labeled 4-CBD for establishing the real irreversibility of steps (1)–(3) or (4)–(6) and (7) through determination of the amount of ^{13}C-labeled vinylacetylene formed via both mechanistic schemes. Additional investigations are also necessary for determining the compositions of active complexes involved in the hydrochlorination reaction. ■

The mechanisms with a common catalyst (class B) are typical of the synthesis of various regioisomers using the reactions of addition to multiple bonds, asymmetric catalysis, kinetic catalytic splitting of racemates, and some other reactions. The scheme of a simple mechanism of the two-route reaction

$$\text{M} + \text{S}_1 \begin{cases} \underset{k_{-11}}{\overset{k_{11}}{\rightleftharpoons}} \text{X}_1 \xrightarrow[\text{S}_2]{k_{21}} \text{P}_1 + \text{M} \\ \underset{k_{-12}}{\overset{k_{12}}{\rightleftharpoons}} \text{X}_2 \xrightarrow[\text{S}_2]{k_{22}} \text{P}_2 + \text{M}. \end{cases} \tag{3.21}$$

is conveniently represented by kinetic graph KG9 in Figure 3.5, where the steps are given a new notation with the first subscript (1 or 2) denoting the step number along route (I or II, respectively) and the second subscript indicating the route number.

The main characteristic of this process (determining its selectivity with respect to a sum of products P_1 and P_2) is the ratio of the corresponding rates of formation (r_{P_1} and r_{P_2}). Under steady-state or

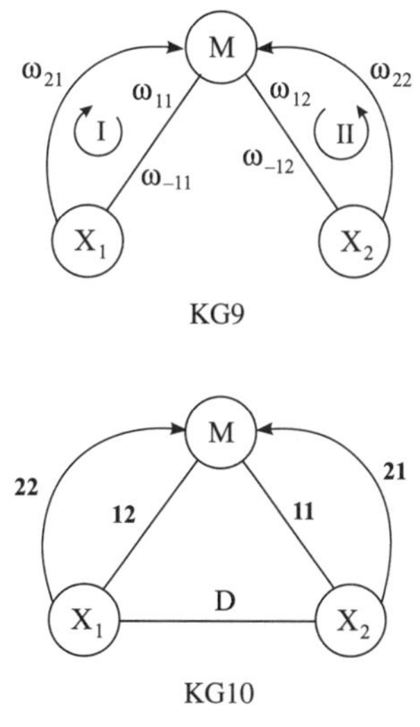

Figure 3.5 Kinetic graphs (KGs) for mechanisms of classes B and BC^2

quasi-steady-state conditions, we have

$$\frac{r_{P_1}}{r_{P_2}} = \frac{\omega_{21}[X_1]}{\omega_{22}[X_2]} = \frac{\omega_{11}\omega_{21}(\omega_{22}+\omega_{-12})}{(\omega_{21}+\omega_{-11})\omega_{12}\omega_{22}}, \tag{3.22}$$

where

$$[X_1] = \frac{\omega_{11}[M]}{\omega_{21}+\omega_{-11}}; \quad [X_2] = \frac{\omega_{12}[M]}{\omega_{22}+\omega_{-12}}.$$

For the reaction scheme (3.21), the kinetic equation (3.22) yields the following relation:

$$\frac{r_{P_1}}{r_{P_2}} = \frac{k_{11}k_{21}(k_{22}[S_2]+k_{-12})}{k_{12}k_{22}(k_{21}[S_2]+k_{-11})}. \tag{3.23}$$

If the $[S_2]$ values remain constant in the time, this relation simplifies to

$$\frac{r_{P_1}}{r_{P_2}} = \frac{[P_1]}{[P_2]}. \tag{3.24}$$

There are cases where steps (11) and (12) (see scheme (3.21)) are quasi-equilibrium ($k_{-11} \gg k_{21}[S_2]$ and $k_{-12} \gg k_{22}[S_2]$), which yields

$$\frac{[P_1]}{[P_2]} = \frac{k_{21}K_{11}}{k_{22}K_{12}} \tag{3.25}$$

and implies that steps (21) and (22) are "rate-limiting" for each route. If there is only one slow step (e.g., (21)), the rates over steps (12) and (22) are comparable. In any case, the rate of substrate S_2 consumption is determined by both constants k_{21} and k_{22}.

Two-route reactions of type (3.21) and more complicated multi-route schemes have been studied in much detail by Heller et al. [26–28]. For scheme (3.21), the rate of substrate S_1 consumption can be described using an equation of the Michaelis–Menten type (see Section 1.2.1):

$$-\frac{d[S_1]}{dt} = \frac{k_{obs}[M]_\Sigma[S_1][S_2]}{K_m+[S_1]}, \tag{3.26}$$

where

$$k_{obs} = \frac{k_{21}\dfrac{k_{11}}{k_{21}[S_2]+k_{-11}} + k_{22}\dfrac{k_{12}}{k_{22}[S_2]+k_{-12}}}{\dfrac{k_{11}}{k_{21}[S_2]+k_{-11}} + \dfrac{k_{12}}{k_{22}[S_2]+k_{-12}}};$$

$$K_m = \frac{1}{\dfrac{k_{11}}{k_{21}[S_2]+k_{-11}} + \dfrac{k_{12}}{k_{22}[S_2]+k_{-12}}}.$$

Multiplying by [M] the numerator and denominator of the expression for k_{obs}, we obtain the following formula that elucidates the meaning of this parameter:

$$k_{obs} = \frac{k_{21}[X_1]+k_{22}[X_2]}{[X_1]+[X_2]}. \tag{3.27}$$

At a constant substrate concentration $[S_2]$ (e.g., for $[S_2]=[H_2]=\text{const}$), the values of k_{obs} and K_m characterize the temporal course of the process.

In some cases, reaction (3.21) involves an additional step of intramolecular isomerization,

$$X_2 \underset{k_{-D}}{\overset{k_D}{\rightleftharpoons}} X_1,$$

and the scheme changes to

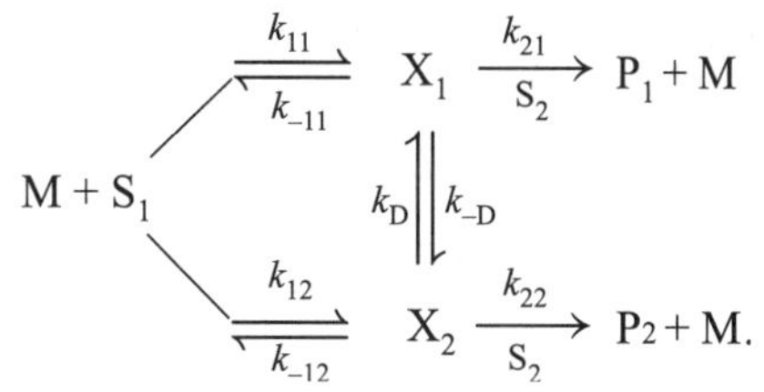

(3.28)

In a steady-state or quasi-steady-state regime (with respect to $[X_1]$ and $[X_2]$), the ratio of concentrations $[X_1]/[X_2]$ is described by the following formula:

$$\frac{[X_1]}{[X_2]} = \frac{k_{11}(k_{22}[S_2] + k_D) + k_{11}k_{-12} + k_{12}k_D}{k_{12}(k_{21}[S_2] + k_{-D}) + k_{12}k_{-11} + k_{11}k_{-D}}, \tag{3.29}$$

which coincides with the ratio of root determinants of the vertices X_1 and X_2 of kinetic graph KG10 (Figure 3.5). Under conditions where the rate of product formation is lower than the rate of isomerization (which is a frequently encountered situation) or $k_D \gg k_{22}[S_2]$ and $k_{-D} \gg k_{21}[S_2]$, it is possible to obtain the following expression:

$$\frac{[X_1]}{[X_2]} = \frac{k_D}{k_{-D}} \cdot \frac{\left(k_{11} + k_{12} + \dfrac{k_{11}k_{-12}}{k_D}\right)}{\left(k_{11} + k_{12} + \dfrac{k_{12}k_{-11}}{k_{-D}}\right)}. \tag{3.30}$$

Thermodynamic relations for the cyclic process ($K_{11} = K_{12}K_D$) and the principle of detailed balancing (see Section 1.4) lead to the following formula:

$$\frac{k_{11}k_{-12}}{k_D} = \frac{k_{-11}k_{12}}{k_{-D}}. \tag{3.31}$$

With allowance for this formula, Eq. (3.30) can be rewritten as follows:

$$\frac{[X_1]_{eq}}{[X_2]_{eq}} = \frac{k_D}{k_{-D}} = K_D, \tag{3.32}$$

which implies that the temperature dependence of the relative content of intermediates X_1 and X_2 is determined entirely by the temperature dependence of K_D. Thus, for the given scheme (3.28) and adopted assumptions, the plot of $\ln([P_1]/[P_2])$ versus inverse temperature ($1/T$) must be linear:

$$\frac{[P_1]}{[P_2]} = \frac{k_{21}[X_1]}{k_{22}[X_2]} = \frac{k_{21}}{k_{22}}K_D. \tag{3.33}$$

More complicated variants of the temperature dependence of the product ratio are considered below.

As was noted above, scheme (3.21) is typical of various reactions in the asymmetric catalysis [28]. In particular, it also describes the asymmetric hydrogenation (AH) of prochiral olefins [29–31]. Based on the results of investigations of the AH kinetics of α-acetamidocinnamic acid methyl ether in solutions of

Rh(I) complexes with achiral 1,2-bis(diphenylphosphino)ethane (dppe) [31, 32], nuclear magnetic resonance of intermediates in solutions [29–36], and X-ray diffraction of isolated intermediates [29, 35], as well as the kinetics of AH with chiral ligands [29–31], a mechanistic scheme was proposed and a general major/minor concept (common for the field of asymmetric catalysis) was developed. This approach explained the enantioselectivity of the asymmetric synthesis (with predominant formation of one enantiomer) and the mechanism of influence of a chiral ligand on the stereoselectivity of the process. The selectivity of the asymmetric synthesis is characterized by the ratio of concentrations of the two (R, S) enantiomers ($q = [P_R]/[P_S]$) and by the excess percentage enantiomer (*ee*, %) defined as

$$ee = \frac{[P_S] - [P_R]}{[P_S] + [P_R]} \cdot 100\,\%,$$

which is related to q by the equation

$$q = \frac{100 + ee}{100 - ee}.$$

Example 3.3

► The kinetics of the AH reaction

$$\underset{\text{(EAC)}}{PhCH{=}C(NHCOCH_3)COOC_2H_5} + H_2 \xrightarrow[\text{MeOH}]{\text{Rh(I)}} PhCH_2\overset{\bullet}{C}H(NHCOCH_3)(COOC_2H_5) \qquad (3.34)$$

was studied in methanol in a closed, ideally stirred, system at 25 °C and P_{H2} = const in the presence of rhodium complexes $RhL(MeOH)_2{}^+$, where L is S,S-chimphos or $(2S,3S)$-bis(diphenylphosphino)butane [28, 29]. The reaction product has the form of an R-enantiomer and is obtained with $ee > 95\,\%$. The rate of hydrogen consumption obeys the following empirical equation:

$$-\frac{d[H_2]}{dt} = k[H_2][RhL(EAC)^+]. \qquad (3.35)$$

Almost all Rh(I) occurs in the form of X_1 complex that is predominantly formed in solution according to the following scheme (S≡EAC):

$$ML + S \begin{cases} \underset{k_{-11}}{\overset{k_{11}}{\rightleftharpoons}} X_1 \xrightarrow[k_{21}]{H_2} X_1H_2 \xrightarrow{k_{31}} ML(SH)H \xrightarrow{k_{41}} \underset{(P_1)}{SH_2} + ML \\ \underset{k_{-12}}{\overset{k_{12}}{\rightleftharpoons}} X_2 \xrightarrow[k_{22}]{H_2} X_2H_2 \xrightarrow{k_{32}} ML(SH')H \xrightarrow{k_{42}} \underset{(P_2)}{S'H_2} + ML, \end{cases} \qquad (3.36)$$

which in fact reduces to (3.21), since steps 31, 41, 32 and 42 are rapid and irreversible. Obviously, the ratio r_{P_1}/r_{P_2} (or $[P_1]/[P_2]$), i.e., the process of enantioselectivity, is determined by the ratio of concentrations of the corresponding substrate diastereomer complexes $[X_1]/[X_2]$ and the ratio of rate constants k_{21}/k_{22} (at low concentrations of X_iH_2 and $ML(S_iH)(H)$).

For the phenomenon of asymmetric induction (i.e., the ligand-induced stereoselectivity), the essence of the major/minor concept consists of the experimentally confirmed fact that a predominantly formed complex (e.g., X_1) with a large equilibrium constant $[\sim(3-5) \cdot 10^3\ M^{-1}]$ reacts with H_2 at a much lower rate than does the other (weak) complex (X_2) that has a low concentration. In the case of X_2, the addition of H_2 to Rh(I) is sterically more favorable and yields a more stable product. Evidently, in a quasi-equilibrium regime of steps 11 and 12, the ratio of rates is determined by the value of $K_{11}k_{21}/K_{12}k_{22}$ (see Eq. (3.25)).

If $K_{12}k_{22} \gg K_{11}k_{21}$, then isomer P_2 (in the given case, *R*-isomer) is predominantly formed. The structure of complex X_1 (involving EAC) shows that it must be a precursor of *S*-isomer. With increasing P_{H_2}, the rates of steps 12 and 22 become comparable so that $k_{22}[H_2] \sim k_{-12}$ and, as a result

$$\frac{r_{P_1}}{r_{P_2}} = \frac{k_{21}K_{11}(k_{-12} + k_{22}[H_2])}{k_{22}K_{12}}. \tag{3.37}$$

According to this equation, a growth in P_{H_2} leads to an increase in the rate of product P_1 formation, while the *ee* value (i.e. excess of product P_2) will drop until inversion of the isomer ratio. In the region of high hydrogen pressures, steps 11 and 12 become slow and irreversible, so that $r_{P_1}/r_{P_2} = k_{11}/k_{12}$.

In the asymmetric catalysis, including hydrogenation reactions, the plots of $\ln([P_1]/[P_2])$ versus inverse temperature (1/*T*) in some cases are nonlinear [27, 28]. This behavior indicates that the condition of quasi-equilibrium is not attained at one or both steps (11 and/or 12) involving the substrate–catalyst complex formation and that these steps are not slow and $[P_1]/[P_2] \neq k_{11}/k_{12}$. The observed nonlinear dependence of $\ln([P_1]/[P_2])$ (or the enantiomer excess *ee*) on 1/T was attributed to a change in the nature of the "rate-limiting" (slow) stage with increasing temperature (see Eq. (3.23)), e.g., as a result of the passage from slow steps 21 and 22 to slow steps 11 and 12 in schemes (3.21) and (3.26) [28]. ■

Processes with a mechanistic structure represented by graph KG9 in Figure 3.5 also include reactions of the catalytic resolution of racemates. In this process, which is widely used to obtain pure enantiomer products, a chiral catalyst (Cat.*) selectively accelerates one of the reactions involving a certain enantiomer contained in the racemate $(R + S)$:

$$S_R + A \xrightarrow[\text{fast}]{\text{Cat.*}} P_R \quad (k_f);$$

$$S_S + A \xrightarrow[\text{slow}]{\text{Cat.*}} P_S \quad (k_s).$$

The mechanism of this process is usually described by the following scheme:

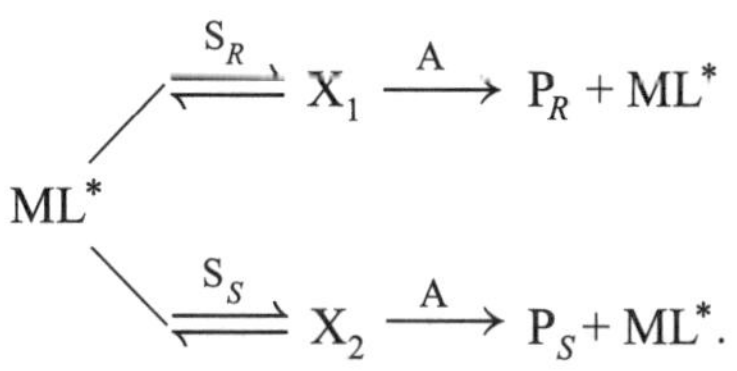

(3.38)

The selectivity of the process is determined by the ratio of the observed rate constants, $k_f/k_s = k_{rel}$, and the degree of substrate (S) conversion. Under conditions of the kinetic resolution of a racemate, it is possible to ensure high enrichment (*ee*) with one of the two isomers in the product (P_R versus P_S) and/or in the substrate (S_S versus S_R).

In the case of excess reactant (A) in the first-order process with respect to [S], the value of k_{rel} is related to the total substrate conversion *x* and the *ee* value in the substrate (enantiomer enrichment of the substrate) by the following equation:

$$k_{rel} = \frac{\ln[(1 - x)(1 + ee)]}{\ln[(1 - x)(1 - ee)]}. \tag{3.39}$$

The degree of conversion *X* that is necessary in order to obtain certain enrichment (*ee*) depends on the k_{rel} value. As k_{rel} increases, the value of *x* required to ensure high *ee* in the substrate decreases. Accordingly, the *ee* value grows with increasing *x*. In order to obtain high *ee* values in the product at conversions not exceeding 50 %, very high values of k_{rel} (above ∼450) are necessary because in this case the *ee* value drops with increasing x [37]:

$$k_{rel} = \frac{\ln[1 - x(1 + \mathrm{ee})]}{\ln[1 - x(1 - \mathrm{ee})]}. \tag{3.40}$$

Many examples of interesting catalytic reactions that are used for the kinetic resolution of racemates are given in review by Keith et al. [37]. In particular, these include the carbonylation of aziridines

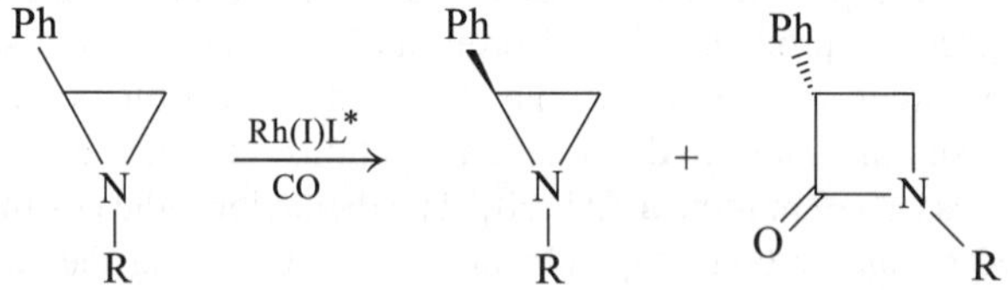

(3.41)

and the hydration of epoxides

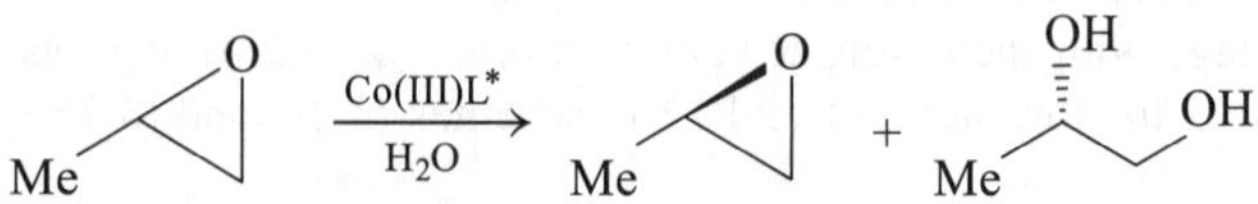

(3.42)

At the same time, the catalytic kinetic resolution of racemates can proceed via mechanisms that are represented by more complicated graphs than KG9 and belong to topological classes other than those considered above. Mechanisms with a common catalyst can include three, four, and greater number of routes. An interesting four-route mechanism of the asymmetric hydrogenation of prochiral olefins is operative in cases where the ligand possesses a single chiral center and two donor atoms, e.g., O and N (symmetry C_1) or two donor centers of different natures (e.g.,–OPR_2 and –NPR_2).

Example 3.4 [28]

► In the case of planar square Rh(I) complexes, it is possible that two pairs of diastereomers are formed with olefins coordinated in different planes (R_e and S_i). Figure 3.6 shows a kinetic graph for this mechanism, according to which the rates of formation of the *S*- and *R*-enantiomers are determined as the sums of rates over two routes. The presence of isomerization steps $\mathrm{I}_{R_e} \rightleftharpoons \mathrm{I}_{S_i}$ and $\mathrm{II}_{S_i} \rightleftharpoons \mathrm{II}_{R_e}$ somewhat complicates the topological structure of the mechanism, but does not influence the kinetics provided that these steps are quasi-equilibrium. ■

The structure of mechanisms belonging to the topological class C (with common steps between routes, see graph KG6 in Figure 3.2) has been encountered in the example of vinylacetylene and 2-chlorovinylacetylene synthesis in solutions of Cu(I)–Cu(II) complexes (Section 2.3). Additional examples of the mechanism of this class are given below.

Example 3.5 [28]

► A simplified mechanism of acetylene cocyclization with nitriles in solutions of Co(I) complexes [38, 39] is described by the following scheme (KG11, Figure 3.7):

$$\begin{array}{l} \qquad\qquad\qquad\qquad\qquad\nearrow^{\mathrm{RCN}}_{k_2}\ X_2 \longrightarrow \mathrm{ML} + P_1 \\ \mathrm{ML} + \mathrm{A} \underset{1}{\overset{K_0}{\rightleftharpoons}} X_0 \underset{2}{\xrightarrow[k_1]{\mathrm{A}}} X_1 \\ \qquad\qquad\qquad\qquad\qquad\searrow^{\mathrm{A}}_{k_3}\ X_3 \longrightarrow \mathrm{ML} + P_2, \end{array} \tag{3.43}$$

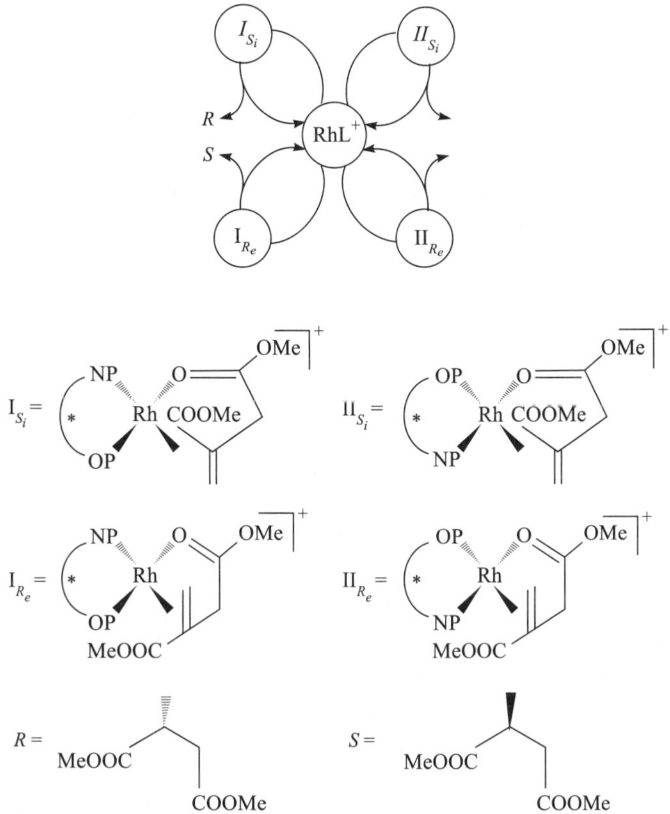

Figure 3.6 Kinetic graph for the mechanism of asymmetric hydrogenation of itaconic acid ester

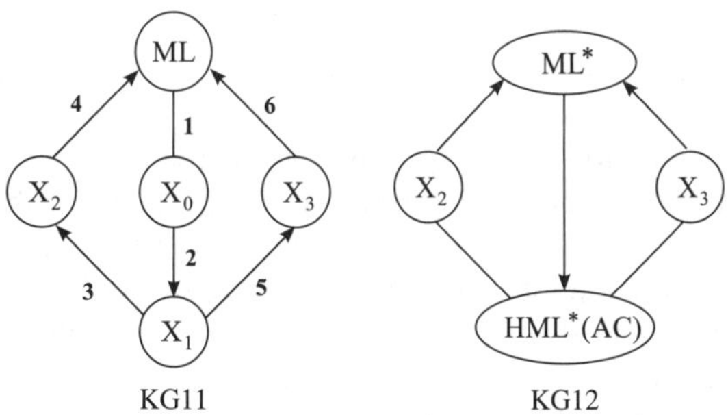

Figure 3.7 Kinetic graphs for mechanistic schemes (3.43) and (3.46)

where A $= C_2H_2$, ML $= CpCo(CO)_2$, $X_0 = CpCo(A)$, and

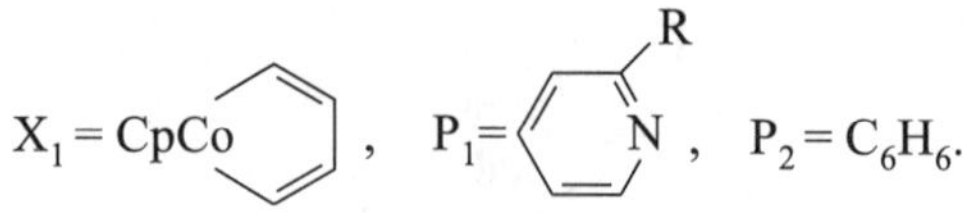

Under steady-state conditions at [RCN] $\gg$ $[Co]_\Sigma$ and $P_{C_2H_2}$ = const, the rate of alkylpyridine (P_1) formation can be expressed (assuming that $[Co]_\Sigma = [ML] + [X_0] + [X_1]$) as follows:

$$r_{P_1} = \frac{k_1 k_2 K_0 [A]^2 [RCN][Co]_\Sigma}{(k_2[RCN] + k_3[A])\left[1 + K_0[A] + \dfrac{k_1 K_0 [A]^2}{k_2[RCN] + k_3[A]}\right]}, \tag{3.44}$$

where K_0 is the equilibrium constant of formation of π complex X_0. According to this formula, the reaction order with respect to [RCN] varies from 1 to 0 and that with respect to [A] varies from 2 to 0. Accordingly, the ratio of the rates of product formation via the two-route reaction is

$$\frac{r_{P_1}}{r_{P_2}} = \frac{k_2[RCN]}{k_3[A]}.$$

If all the metal catalyst is in the form of ML and $[Co]_\Sigma \cong [ML]$, then the order of the reaction of P_1 formation with respect to [RCN] varies from 1 to 0 and that with respect to [A] is greater than 1 (in the presence of benzene) and amounts to 2 if the reaction via the second route almost does not proceed. Alternatively, if the catalyst is entirely in the form of X_0 and $[Co]_\Sigma \cong [X_0]$, then the reaction order with respect to [RCN] also varies from 1 to 0, while that with respect to [A] varies from 0 to 1. Finally, in the case of $[Co]_\Sigma \cong [X_1]$, the reaction order is 1 with respect to [RCN] and 0 with respect to [A]. Under the conditions of a process selective with respect to alkylpyridines, the experimentally established reaction order with respect to [A] was 2 and that with respect to nitrile was 0. These data are consistent with a single-route process (with no significant contribution from the route involving benzene formation) under the condition that $[Co]_\Sigma \cong [ML]$, since only in this case step (2) in scheme (3.43)

$$X_0 \xrightarrow{A} X_1$$

becomes the rate-limiting step.

The synthesis of pyridine derivatives in solutions of Co(I) complexes can also be carried out as a photocatalytic process, whereby the reaction proceeds at high rate and selectivity in the solution irradiated by light in a wavelength range of 350–500 nm [40]. It is interesting to note that the initial rate of pyridine formation under irradiation is proportional to the first power of $[RCN]_0$. Within the framework of scheme (3.43) and only under the conditions of high selectivity with respect to P_1, this fact can imply that the step (2) of X_1 formation becomes quasi-equilibrium and the limiting step is X_2 formation. The first order with respect to [RCN] can also be obtained for $[Co]_\Sigma \cong X_1$, which is possible in the photocatalytic process, but in this case one should expect a zero order with respect to [A]. In order to check for the hypothetical mechanism (3.43) and discriminate other hypotheses, it is necessary to study the reaction orders with respect to [A] and [RCN] in a broad range of r_{P_1}/r_{P_2} (from 1:10 to 10:1). ■

The reactions of asymmetric catalysis can also represent the mechanisms of class C, provided that the initial catalyst species ML* form intermediates that interact with a substrate possessing a prochiral center and lead to the predominant formation of one of the enantiomers.

Example 3.6 [41, 42, 43]

▶ The mechanism of ketone reduction by hydrogen transfer from an alcohol, e.g., from isopropyl alcohol (IPA) via the reaction

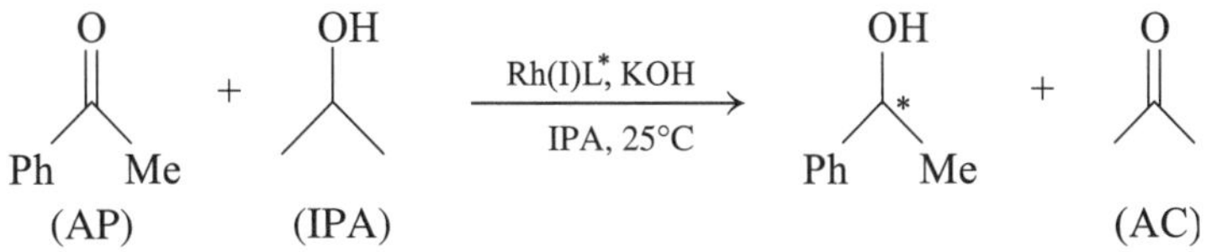

(3.45)

is described by the following scheme (KG12, Figure 3.7):

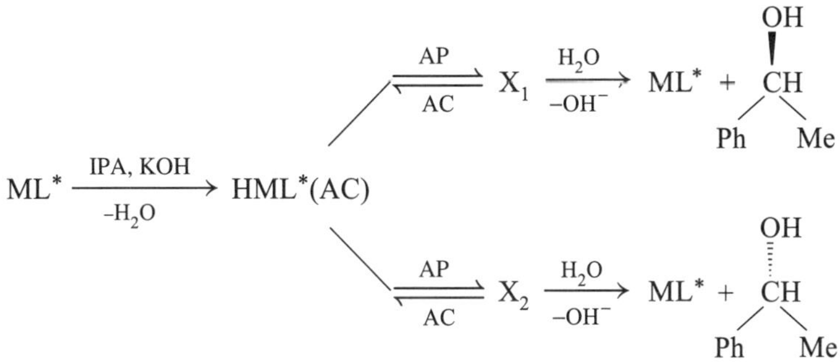

(3.46)

It was found [43] that an increase in the degree of AP conversion and the corresponding growth in the concentration of acetone in solution, as well as the initial introduction of acetone (product of reaction (3.45)) into the reaction medium leads to an increase in the *ee* value. This effect was explained [43–45] by a difference in the rate constants of the direct and reverse reactions in both routes of scheme (3.46) or analogous schemes. De Bellefon et al. [44] also considered the influence of the reactor type and the gas–liquid mass exchange rate on the *ee* value. ■

A two-route mechanism belonging to the topological class C, with the routes having three common steps, is operative in the reaction of olefin hydrogenation in solutions of Rh(I) complexes.

Example 3.7 [46]

▶ Investigations of the mechanism of olefin hydrogenation in solutions of Wilkinson's complexes ($RhClL_3$) (with the originally proposed scheme (2.199) and kinetic equation (2.200)) led to numerous variants of kinetic equations. In particular, a two-route scheme suggested by Tolman and Faller [46] was justified by the results of kinetic and isotope investigations. In the experiments with added ligands ($L = PAr_3$), a mechanistic scheme represented by kinetic graph KG13 (Figure 3.8) was proposed to explain the kinetic equations describing the hydrogenation of cyclohexene with the participation of $RhClL_3$ and $RhClL_2$:

$$r = (k_4K_1P_{H_2}/[L] + k_2P_{H_2})[RhClL_3], \tag{3.47}$$

$$r = \frac{(k_2[L] + k_4K_1)k_6'K_3K_5P_{H_2}[S][Rh]_\Sigma}{[L][(k_2[L] + k_4K_1)P_{H_2} + k_6'K_3K_5[S]]}, \tag{3.48}$$

where $k_6' = k_6k_7/(k_7 + k_{-6})$.

Equation (3.48) shows that the reaction rate cannot be considered as a product of the partial functions $r = f(x)$ (where $x = P_{H_2}$, [S], [L]) [47]. The extremal (nonmonotonic) dependence of $r = f([S])$ [47]

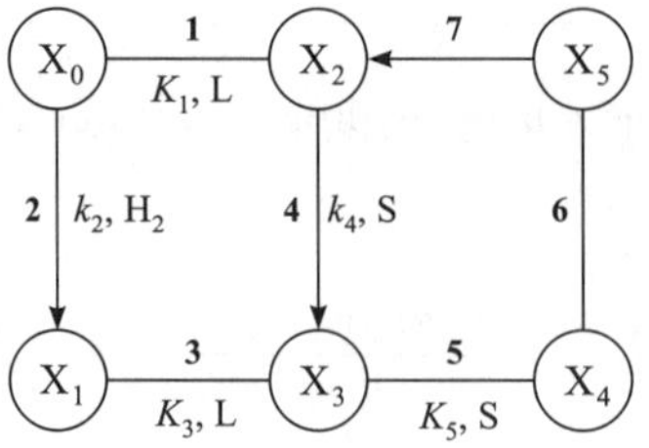

KG13

Figure 3.8 Kinetic graph for the mechanism of cyclohexene (S) hydrogenation in $RhClL_3$ solutions with additives of various ligands L: (X_0) $RhClL_3$; (X_1) H_2RhClL_3; (X_2) $RhClL_2$; (X_3) H_2RhClL_2(S); (X_4) $RhClL_2$; (X_5) $RRhH(Cl)L_2$

observed in the cyclohexene–benzene system was related to a very large variation of the composition (with a cyclohexene concentration of up to 2.8 M) and, hence, properties of the reaction medium. In the case of all quasi-equilibrium steps except 1, 4, and 7 (see KG13 in Figure 3.8) the mechanism separates into two routes that pass via different complexes X_0 and X_1 (thus, representing a "trivial" multi-route character). ■

The mechanisms of asymmetric catalysis considered above involve monoligand (with respect to the chiral ligand L*) metal complexes. Naturally, more complicated schemes are also encountered in practice, for example, in the following cases:

(i) Equilibrium between complexes of M and ML*, which catalyze the process and participate as reactants in a noncatalytic asymmetric synthesis [28].
(ii) Equilibrium between two (or more) initial complexes with chiral ligands (e.g., ML* and ML_2^*). In this case, each complex initiates a process that belongs to class B, but both the initial complexes and intermediates are related via additional steps that modify the topological structure of the mechanism. For example, [see graph KG14, (Figure 3.9)]:

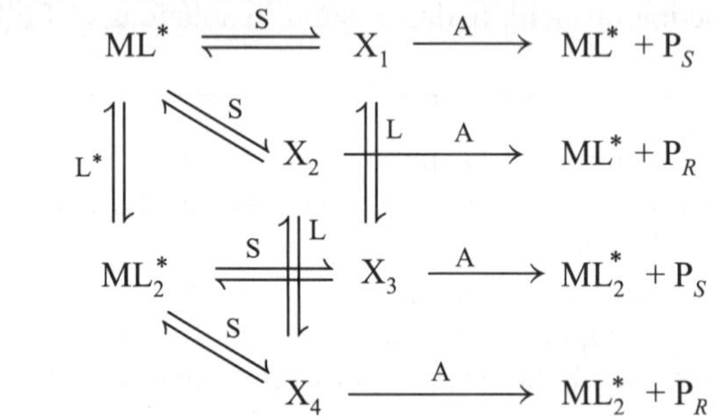

(3.49)

(iii) Equilibrium between complexes and ligands that are not pure enantiomers or diastereomers (*ee* < 100 % for the ligand) [48]. For example, in the case of stable complexes $M(L_R)_2$ and $M(L_S)_2$, a solution features the following equilibrium:

$$M(L_S)_2 + M(L_R)_2 \rightleftharpoons 2ML_SL_R. \tag{3.50}$$

Each of these complexes can catalyze the synthesis of enantiomers via the corresponding complexes with the substrate. A dependence of the *ee* value for the product on that of the ligand (enantiomer purity of the ligand) or on *ee* of the whole catalyst is determined by the character of equilibria in the solution

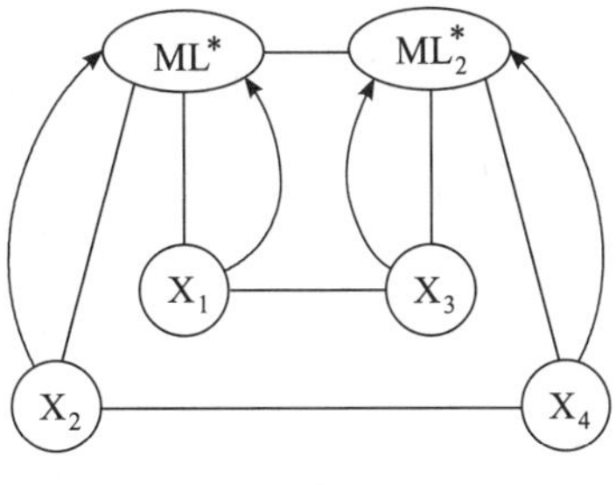

KG14

Figure 3.9 Kinetic graph for mechanistic scheme (3.49)

(type of the model) and the activity of a mixed complex ML_SL_R [48, 49]. Cases (i) and (ii) can be illustrated by the following examples.

Example 3.8 [50]

► Let us consider the following mechanism of the noncatalytic oxidation of olefins by osmium tetraoxide according to Sharpless in the presence of chiral ligand L [50]:

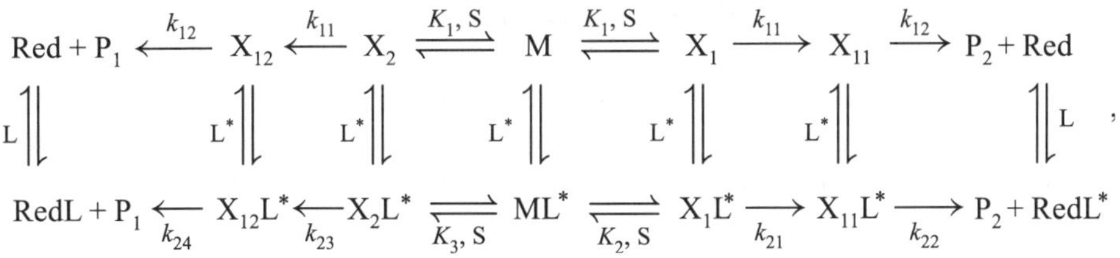

(3.51)

where

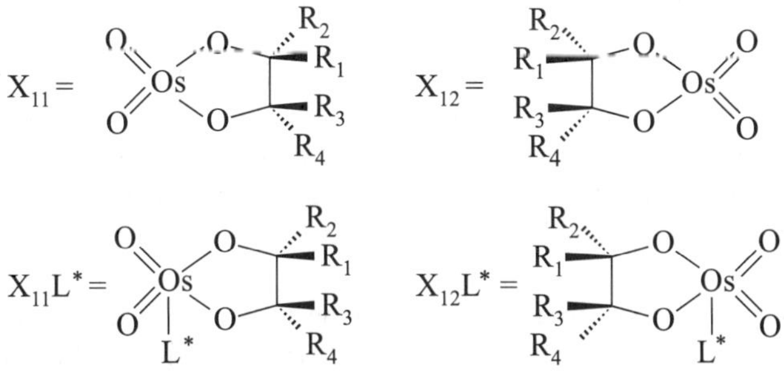

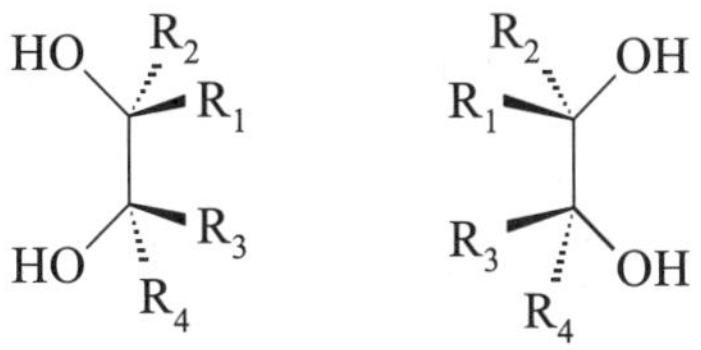

(P$_1$) (P$_2$)

Naturally, the rates of product (P_1 and P_2) formation via the first ligand-free sequence of steps are the same. Scheme (3.51) involves 17 steps, nine independent intermediates ($M = OsO_4$ is reactant), and eight independent routes. Assuming that all steps related to the passage from ligand-free to ligand-containing products and the steps of formation of X_1, X_2, X_2L, and X_1L are quasi-equilibrium, we obtain

$$\frac{r_{P_2}}{r_{P_1}} = \frac{[P_2]}{[P_1]} = \frac{k_{11}K_1 + k_{21}K_0K_2[L]}{k_{11}K_1 + k_{23}K_0K_3[L]}. \tag{3.52}$$

For $k_{21}K_2 \gg k_{23}K_3$ and $k_{11}K_1 \gg k_{23}K_0K_3[L]$, the ratio of $[P_2]/[P_1]$ will linearly increase with [L]. ■

Now let us consider an example of the reaction that proceeds on various initial complexes, which occur in equilibrium under the conditions where each complex is involved in a reaction mechanism of class C.

Example 3.9

► The process of olefin hydroformylation (oxosynthesis) in solutions of $HRh(CO)_xL_{4-x}$ complexes possesses a multi-route character that is related to the following factors:

- There exists a set of initial 18-electron Rh(I) complexes with compositions varying from $x = 4$ to $x = 1$, which are catalysts of the hydroformylation reaction

$$HRh(CO)_4 \underset{-CO}{\overset{L}{\rightleftharpoons}} HRh(CO)_3L \underset{-CO}{\overset{L}{\rightleftharpoons}} HRh(CO)_2L_2 \underset{-CO}{\overset{L}{\rightleftharpoons}} HRh(CO)L_3.$$

  The mechanistic steps involve 16-electron coordination-unsaturated compounds.
- The process leads to the formation of two regioisomers representing normal and iso-aldehydes.
- The process is accompanied by side reactions of olefin hydrogenation and isomerization [46, 51–53].

Although the dependences of the reaction rate on the concentrations of olefin, metal ($[Rh]_\Sigma$), ligand, P_{CO}, and P_{H_2} were determined for various olefins and phosphins, no general kinetic equation for this process has been suggested so far. For the most thorough analysis of this problem, see [46]. There is no kinetic equation even for a relatively simple classical cobalt $[Co_2(CO)_8]$ system, although the composition of intermediates and separate reaction steps have been studied in sufficient detail [52].

Using the information summarized in [46] and restricting the consideration to a two-route oxosynthesis (without hydrogenation and isomerization of olefins), it is possible to obtain a kinetic equation for the rate of the process on complexes containing RhL_2 fragments (assuming no significant contribution due to dimer complexes).

A mechanistic scheme for the synthesis of an aldehyde (e.g., with the normal structure) is generally analogous to the well-known Sternberg–Wender [54] and Breslow–Heck [55] schemes,

$$\underset{(X_1)}{HRh(CO)L_2} + S \overset{K_1}{\rightleftharpoons} \underset{(X_2)}{HRh(CO)L_2S}; \qquad X_4 \overset{K_4}{\rightleftharpoons} \underset{(X_5)}{RCORh(CO)L_2};$$

$$X_2 \overset{K_2}{\rightleftharpoons} \underset{(X_3)}{RRh(CO)L_2}; \qquad X_5 + H_2 \overset{k_5}{\longrightarrow} \underset{(X_6)}{RCORh(H)_2(CO)L_2}; \tag{3.53}$$

$$X_3 + CO \overset{K_3}{\rightleftharpoons} \underset{(X_4)}{RRh(CO)_2L_2}; \qquad X_6 \overset{k_6}{\longrightarrow} RCHO + X_1.$$

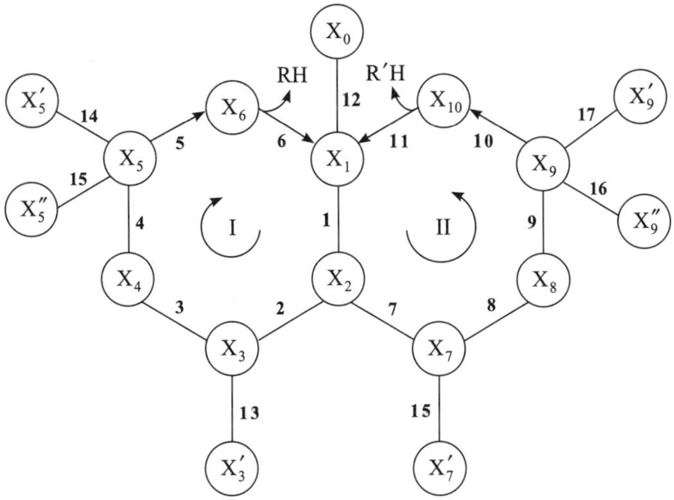

Figure 3.10 Kinetic graph for two-route mechanism of hydroformylation of α-olefins

to which it is expedient to add several reactions that describe the formation of coordination-saturated complexes:

$$X_1 + CO \underset{}{\overset{K_{12}}{\rightleftharpoons}} \underset{(X_0)}{HRh(CO)_2L_2};$$

$$X_3 + L \underset{}{\overset{K_{13}}{\rightleftharpoons}} \underset{(X'_3)}{RRh(CO)L_3};$$

$$X_5 + CO \underset{}{\overset{K_{14}}{\rightleftharpoons}} \underset{(X'_5)}{RCORh(CO)_2L_2};$$

$$X_5 + L \underset{}{\overset{K_{15}}{\rightleftharpoons}} \underset{(X''_5)}{RCORh(CO)L_3}.$$

Figure 3.10 shows graph KG15 that includes all intermediates and pendant vertices for two analogous routes of the synthesis of regioisomers. Here, route I corresponds to the formation of a normal aldehyde, and route II represents the formation of an iso-aldehyde. Assuming that edges **5** and **10** represent the only slow steps, we obtain the following equation for route I:

$$r_I = k_5[X_5]P_{H_2} = k_5K_1K_2K_3K_4K_{12}[S][X_0]P_{H_2}. \tag{3.54}$$

Assuming also that the main contributions to the material balance with respect to Rh are due to complexes X_0, X_1, X_5, X_9, X'_5, and X'_9 (in accordance with experiments for S = α-olefin), we obtain

$$[X_0] = [Rh]_\Sigma/F_{Rh};$$
$$F_{Rh} = 1 + K_{12}/P_{CO} + A[S] + B[S]P_{CO}, \tag{3.55}$$

where $A = K_1K_{12}(K_2K_3K_4 + K_7K_8K_9)$ and $B = K_1K_{12}(K_2K_3K_4K_{14} + K_7K_8K_9K_{17})$. Using Eqs (3.54) and (3.55), we arrive at the following formula:

$$r_\mathrm{I} = \frac{k_5K_1K_2K_3K_4K_{12}[\mathrm{S}]P_{\mathrm{H_2}}P_{\mathrm{CO}}[\mathrm{Rh}]_\Sigma}{P_{\mathrm{CO}} + K_{12} + A[\mathrm{S}]P_{\mathrm{CO}} + B[\mathrm{S}]P^2_{\mathrm{CO}}}. \tag{3.56}$$

which describes all the single-factor dependences observed in excess of [L]. With allowance for the complexes X''_5 and X''_9 in the complexity function F_{Rh}, we arrive at the following expression:

$$r_\mathrm{I} = \frac{k_{\mathrm{eff}}\ [\mathrm{S}]P_{\mathrm{H_2}}P_{\mathrm{CO}}[\mathrm{Rh}]_\Sigma}{P_{\mathrm{CO}} + K_{12} + A[\mathrm{S}]P_{\mathrm{CO}} + B[\mathrm{S}]P^2_{\mathrm{CO}} + C[\mathrm{S}][\mathrm{L}]}, \tag{3.57}$$

where $C = K_1K_{12}(K_2K_3K_4K_{15} + K_{17}K_8K_9K_{16})$. Equation (3.57) describes retardation of the process by added ligands.

Evidently, any scheme that takes into account only complexes of the $\mathrm{RhL_2}$ type (even with two $\mathrm{RhL_3}$ complexes (X''_5 and X''_9)) cannot describe the influence of $[\mathrm{L}]/P_{\mathrm{CO}}$ on the regioselectivity of the process, which is defined as

$$\frac{r_\mathrm{I}}{r_\mathrm{II}} = f\left(\frac{[\mathrm{L}]}{P_{\mathrm{CO}}}\right),$$

since both RhL and $\mathrm{RhL_3}$ species also catalyze the process of hydroformylation (KG15), although with a different ratio of the rates of regioisomer formation. Therefore, it is necessary to include the mechanism represented by graph KG15 for consideration for all complexes RhL, $\mathrm{RhL_2}$, and $\mathrm{RhL_3}$. In this multi-route scheme [46], the ratio $r_\mathrm{I}/r_\mathrm{II}$ (r_n/r_{iso}) will increase on the passage from one level to another in the series $\mathrm{RhL} < \mathrm{RhL_2} < \mathrm{RhL_3}$ due to sterical hindrances for the formation of iso-products in the presence of a large number of $\mathrm{PR_3}$ ligands in the complex. The total rate of olefin formation will decrease in the same order. Since the situation is too complicated for a quantitative kinetic analysis, it is necessary to obtain additional information in order to establish which complexes give the main contribution to the material balance for various $[\mathrm{L}]/P_{\mathrm{CO}}$. It is also necessary to study the possible appearance of nonlinear steps of CO ligand transfer between complexes. The appearance of these steps, which were considered as probable in the case of cobalt carbonyl complexes [52], can lead to the development of chain schemes, in particular, steps of the following type:

$$\mathrm{RRh(CO)L_2 + HRh(CO)_2L_2 \rightleftharpoons RRh(CO)_2L_2 + HRh(CO)L_2}. \tag{3.58}$$

■

An attempt to construct an exhaustive kinetic model with allowance for an almost complete set of initial Pd complexes and various transitions between routes (classes A and C) was undertaken by Petrov with co-workers [56–65] in an application to the reactions of hydrocarboxylation and hydrocarbalkoxylation (hydroalkoxycarbonylation) of olefins and alkynes. A commonly accepted scheme of the mechanism of these reactions [51, 66] includes the formation of a hydride complex HPdX and an alkyl intermediate RPdX that reacts with CO so as to form RCOPdX. The latter intermediate decomposes to RCOOH or RCOOR′ with the liberation of HPdX or Pd(0) (see postulate no. 1 in [16]):

$$\mathrm{Pd(0)} \xrightarrow{\mathrm{HX}} \mathrm{HPdX} \xrightarrow{\mathrm{S(olefin)}} \mathrm{RPdX} \xrightarrow{\mathrm{CO}} \mathrm{RCOPdX} \xrightarrow{\mathrm{H_2O}} \underset{(\mathrm{Pd(0)+HX})}{\mathrm{HPdX}} + \mathrm{RCOOH}. \tag{3.59}$$

The acyl complexes of Pd(II) with a formula of $\mathrm{RCOPdXL_2}$ were obtained according to scheme (3.59) from olefins and CO in alcohols [56, 67–69]. Investigations of the conversion of these complexes into products of alkene hydrocarboxylation and hydroalkoxycarbonylation reactions [56–65] elucidated the elementary steps, compositions, and structures of complex compounds involved in the hydride mechanism of said reactions, provided important information on the factors that determine the regioselectivity of processes obeying this mechanism, and ensured kinetic justification of the proposed scheme [64]. In particular, it was demonstrated that the experimental data on the hydrogen isotope exchange in the initial olefin in the $PdCl_2$–PPh_3–H_2O system (and that with additives of $SnCl_2$) are consistent only with the hydride mechanism [57].

It was established that an acid that is formed during palladium reduction plays an important role in the hydrocarboxylation of styrene. The addition of HCl lead to an increase in the rate and a decrease in regioselectivity of this process, while the addition of an alkali and $Pd(PPh_3)_4$ produce the reverse effect. It was suggested [58] that the acid influences the distribution of palladium complexes in the catalytic system by means of the participation of HCl in the step of oxidative addition to various Pd(0) complexes and, probably, in the step of hydrolysis of acyl compounds.

Information on the compositions and structures of palladium complexes was obtained predominantly by the method of IR spectroscopy *in situ* under conditions of ethylene and styrene hydrocarboxylation (100 – 110 °C, 0.8 – 2.4 MPa) [58, 59]. The main fraction of carbonyl complexes is represented by monocarbonyl $Pd(CO)(PPh_3)_2$ (absorbing at 1980 cm^{-1}) and dicarbonyl $Pd(CO)_2(PPh_3)_2$ (2024, 1986 cm^{-1}) compounds. These compounds apparently occur in equilibrium with the complexes such as $Pd(PPh_3)_3$ and $Pd(PPh_3)_2$ that contain no carbonyl groups.

The distribution of Pd(0), $Pd(CO)_xL_y$ complexes, and hydride species $HPdClL_2$, HPdCl(CO)L, and $HPdCl(CO)_2$ depends on P_{CO}, [L], and [HCl] and determines the regioselectivity of alkene hydrocarboxylation and hydroalkoxycarbonylation reactions described by the kinetic models of multi-route reactions [56, 60–64]. Complex species $Pd(PPh_3)_2$ catalyze the formation of products with the normal structure, while carbonyl complexes (except for $Pd(CO)(PPh_3)_3$) produce large contributions to the formation of branched products.

In $Pd(dba)_2$–PPh_3 – toluenesulfionic acid, a key intermediate in the hydride mechanism is the cationic palladium hydride $HPd(PPh_3)_3^+$, which is formed as a result of $Pd(PPh_3)_3$ protonation, while the carbonyl complexes of Pd(0) present in the system almost do not participate in the hydroalkoxycarbonylation of styrene (only retardation of the formation of both linear and branched products with increasing partial pressure of carbon oxide is observed) [65]. All intermediate compounds were identified by NMR in the reaction of ethylene hydromethoxycarbonylation in the presence of $C_6H_4(CH_2PBu^t_2)_2$ bidentate ligands. Relatively stable cationic hydride complexes of Pd(II) were obtained via the interaction of Pd(L-L)(dba) complex with HBF_4 or CF_3SO_3H in methanol [70].

Example 3.10 [60, 61]

► The kinetics of styrene hydrocarboxylation

$$\underset{\text{(St)}}{PhCH{=}CH_2} + CO + H_2O \longrightarrow \begin{cases} \underset{\textbf{(I)}}{PhCH_2CH_2COOH} \\ \underset{\textbf{(II)}}{PhCH(CH_3)COOH} \end{cases} \tag{3.60}$$

was studied in a closed ideally stirred system at 110 °C by measuring the rates of accumulation of products **I** and **II** in the H_2O – dioxane system (with water content within $[H_2O] = 1 - 10$ M) in the presence of $PdCl_2L_2$ complexes ($L = PPh_3$) and ligand L additions. The ratio of $\alpha = [L]_0/[PdCl_2L_2]_0$ in most experiments was varied from 6 to 32 (although experimental series with $\alpha = 1 - 6$ were also carried out).

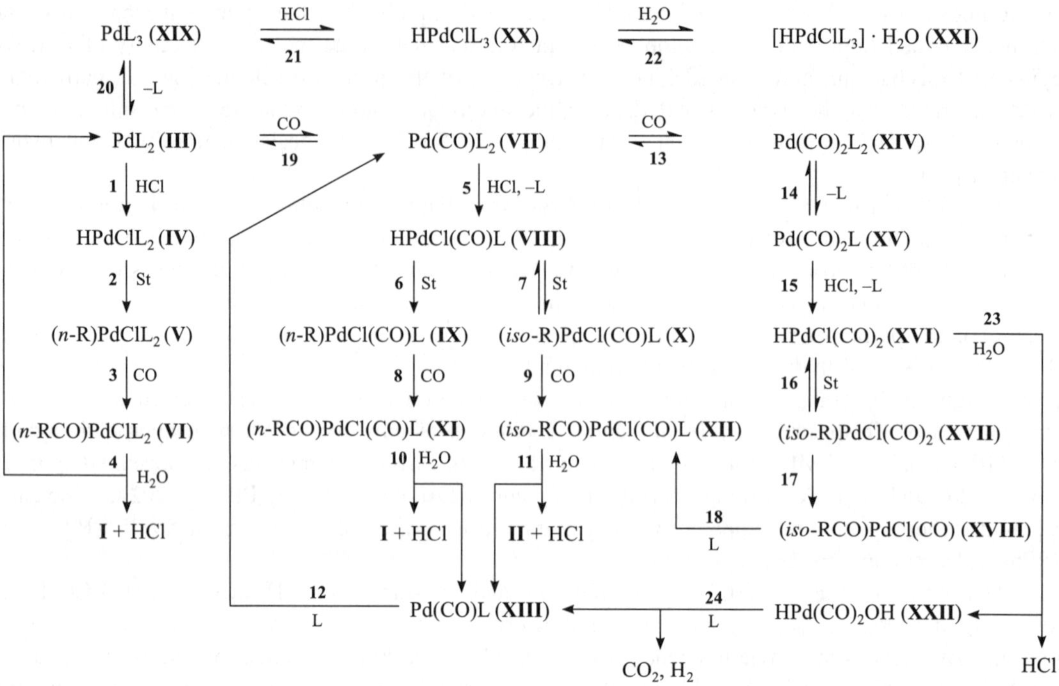

Figure 3.11 Mechanistic scheme of styrene hydrocarboxylation: n-R = CH_2CH_2Ph; iso-R = $CH(CH_3)Ph$; L = PPh_3; St = styrene (reproduced with permission from editorial board of *Kinet. Katal.* journal © 1994 [61])

Based on the obtained partial dependences of $r_i = f(P_{CO}, [St], [H_2O], [L]_0, [Pd]_0)$ and physicochemical data (see above) a mechanistic scheme (Figure 3.11) of the quasi-steady-state synthesis of **I, II**, and CO_2 (CO conversion) was proposed [60, 61].

The proposed mechanism includes two routes of the formation of product **I** ($\boldsymbol{\nu}^{(1)}$: steps **1, 2, 3, 4**; $\boldsymbol{\nu}^{(2)}$: steps **5, 6, 8, 10, 12**), two routes of the formation of product **II** ($\boldsymbol{\nu}^{(3)}$: steps **5, 7, 9, 11, 12**; $\boldsymbol{\nu}^{(4)}$: steps **14, 15, 16, 17, 18, 11, 12**), and one route of the formation of CO_2 and H_2 ($\boldsymbol{\nu}^{(5)}$: steps **12, 13, 14, 15, 23, 24**). The irreversibility of steps **2** and **6** and the reversibility of steps **7** and **16** were adopted based on the results of investigation of the isotope exchange in the initial styrene [57], while the irreversibility of the step of formation of hydride intermediates (**1, 5, 15**) was accepted based on the kinetic analysis. A special feature of the proposed scheme is the adopted formation of nonhydride complexes PdL_2 and Pd(CO)L at the steps of decomposition of acyl intermediates (**VI, XI, XII**). The mechanism also involves nonlinear steps **1, 5, 15, 21**, since HCl was an intermediate, the concentration of which was not determined.

The concentration of the phosphine ligand in most cases significantly exceeded that of $[Pd]_0$, which allowed the approximation of $[L]_0 \cong [L]$ to be used. The equation of material balance with respect to Pd included three complexes (**VI, VII**, and **XXI**):

$$[Pd]_0 = [(n\text{-RCO})PdClL_2] + [Pd(CO)L_2] + [HPdL_3Cl \cdot H_2O]. \quad (3.61)$$

In this case, $[HCl] = [Pd]_0 + [VII]$, since the total acid concentration was $[HCl]_0 = 2[Pd]_0 = [HCl] + [VI] + [XXI]$ (HCl is formed during the reduction of $Pd^{(II)}$). It was assumed that the active form of HCl that interacted with Pd(0) complexes was represented by hydrogen chloride hydrate (contact ion pair) species, the concentration of which was expressed through equilibrium hydration constant in the following equation:

$$HCl + H_2O \overset{K}{\rightleftharpoons} HCl \cdot H_2O, \tag{3.62}$$

The acid concentration [HCl] was calculated from the material balance as $[HCl] = [Pd]_0 + [VII]$. It was assumed that, under the adopted process conditions, this value was equal to the concentration of hydrogen chloride hydrate: $[HCl] \cong [HCl \cdot H_2O]$. The hydration constant implicitly enters into the rate and equilibrium constants.

Under all the adopted assumptions, a kinetic model was obtained that quite well described the process, including the effects of $[H_2O]$, $[L]_0$, and P_{CO} on the regioselectivity:

$$r_I = \frac{[H_2O][Pd]_0}{1 + A_1[H_2O]^2P_{CO}/([HCl] + A_2[L]_0[H_2O]^2)} \cdot \left[A_3 + \frac{A_4P_{CO}}{1 + A_5P_{CO}}\right];$$

$$r_{II} = \frac{[H_2O][Pd]_0P_{CO}^2}{1 + A_1[H_2O]^2P_{CO}/([HCl] + A_2[L]_0[H_2O]^2)}$$

$$\times \left[\frac{A_4A_5}{1 + A_5P_{CO}} + \frac{A_6[H_2O][St]}{([HCl] + A_7[L]_0[H_2O])([St] + A_8[H_2O])}\right]; \tag{3.63}$$

$$[HCl] = [Pd]_0 \cdot \frac{[HCl](1 + A_2[L]_0[H_2O]^2) + 2A_1[H_2O]P_{CO}^2}{[HCl](1 + A_2[L]_0[H_2O]^2) + A_1[H_2O]^2P_{CO}},$$

where

$$A_1 = k_4K_{19}/k_1;\ A_2 = k_4K_{20}K_{21}K_{22}/k_1;\ A_3 = k_4;$$
$$A_4 = k_4k_5K_{19}/k_1;\ A_5 = k_9K_7/k_6;\ A_6 = k_4K_{13}K_{14}K_{19}/k_1;$$
$$A_7 = k_{-14}/k_{15};\ A_8 = k_{23}/k_{17}K_{16}.$$

As can be seen from the mechanistic scheme (Figure 3.11) in the linear mechanism approximation with the corresponding kinetic graph ([HCl] enters into the weights of steps), the routes $\boldsymbol{\nu}^{(2)}$ and $\boldsymbol{\nu}^{(3)}$, as well as $\boldsymbol{\nu}^{(4)}$ and $\boldsymbol{\nu}^{(5)}$, are topologically interrelated as mechanisms of class C; route $\boldsymbol{\nu}^{(1)}$ is related to these pairs as the mechanism of class A. ■

Investigations of the kinetics of 1-heptene hydrocarboxylation [63] in the same system showed that the mechanistic scheme of this reaction is generally analogous to that for the above reaction with styrene, except for (i) the irreversible formation of alkyl complexes with both normal and iso-structure and (ii) the presence of an additional route for the formation of products with linear structures involving $HPdCl(CO)_2$ complex. The general scheme and kinetic models for various olefins are discussed in [64].

Example 3.11 [56]

► The kinetics of styrene hydroalkoxycarbonylation via reaction

$$PhCH{=}CH_2 + CO + ROH \longrightarrow \underset{(\mathbf{I})}{PhCH_2CH_2COOR} + \underset{(\mathbf{II})}{PhCH(CH_3)COOR} \tag{3.64}$$

was studied in the PdL_2Cl_2–n-BuOH–dioxane system at 110°C. In contrast to the case of aqueous media, the curves of the product concentration versus time for this system exhibited an induction period, which almost disappeared upon adding 0.1 M H_2O to the reaction medium. Figure 3.12 shows a mechanistic

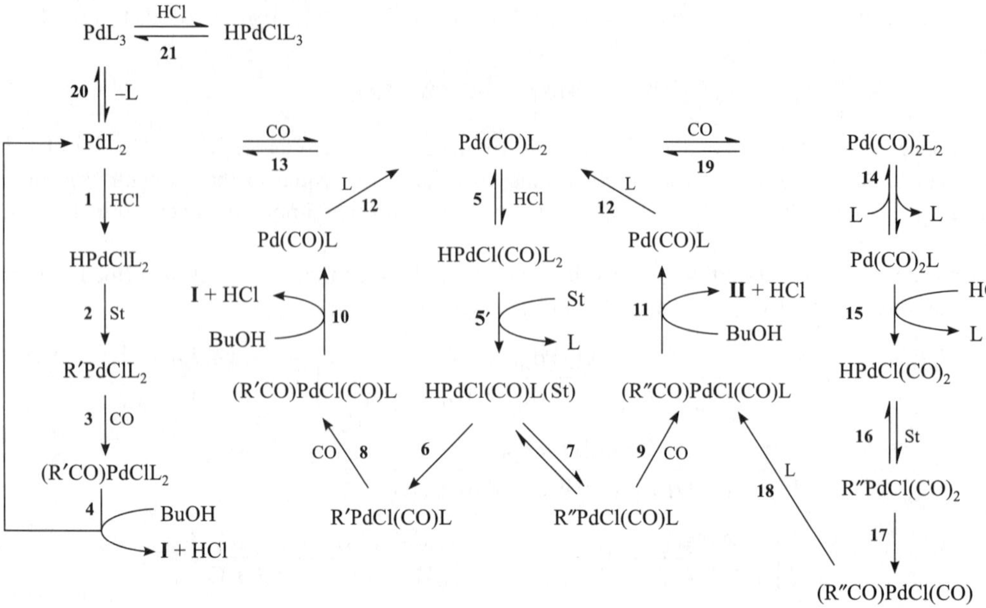

Figure 3.12 Mechanistic scheme of reaction (3.64) in $PdCl_2L_2$–PPh_3 system: $R' = PhCH_2CH_2$–; $R'' = Ph(CH_3)CH$–; steps 5, 7, 13, 16, and 19–21 are equilibrium (reproduced with permission from editorial board of *Kinet. Katal.* journal © 2000 [56])

scheme proposed for this reaction (3.64) based on the results obtained from an analysis of the partial kinetic dependences, isolation and identification of the intermediate compound $(PhCH_2CH_2CO)PdCl(CO)L$ significantly contributing to the material balance with respect to Pd, and comparison to the kinetics of hydrocarboxylation in an aqueous system. The scheme in Figure 3.12 differs somewhat from the mechanism considered above (Figure 3.11) in that the step of HCl addition to $Pd(CO)L_2$ (step **5**) is assumed to be reversible and an additional step **5′** is introduced. As a result, the concentration of styrene enters into the kinetic equation via routes $\boldsymbol{\nu}^{(2)}$ and $\boldsymbol{\nu}^{(3)}$; and the route of CO conversion via route $\boldsymbol{\nu}^{(5)}$ is naturally excluded.

Adopting that three complexes contribute to the material balance with respect to the catalyst,

$$[Pd]_0 = [(PhCH_2CH_2CO)PdClL_2] + [Pd(CO)L_2] + [HPdClL_3], \tag{3.65}$$

and assuming that $[HCl] \cong [Pd]_0$, the kinetic model of the process under consideration was as follows:

$$r_I = \frac{[BuOH][Pd]_0}{1 + A_1P_{CO}[BuOH]/[HCl] + A_2[L]_0[BuOH]} \cdot \left[A_3 + \frac{A_4P_{CO}[St]}{1 + A_5P_{CO}}\right];$$

$$r_{II} = \frac{[BuOH][Pd]_0P_{CO}^2}{1 + A_1P_{CO}[BuOH]/[HCl] + A_2[L]_0[BuOH]} \times \left[\frac{A_4A_5[St]}{1 + A_5P_{CO}} + \frac{A_6}{[HCl] + A_7[L]_0}\right], \tag{3.66}$$

where

$$[\mathrm{HCl}] \cong [\mathrm{Pd}]_0,\ A_1 = k_4K_{19}/k_1;\ A_2 = k_4K_{20}K_{21}/k_1;\ A_3 = k_4;$$
$$A_4 = k_5'K_5k_4K_{19}/k_1;\ A_5 = k_9K_7/k_6;\ A_6 = k_4k_{14}K_{13}K_{19}/k_1;\ A_7 = k_{-14}/k_{15}.$$

The proposed model adequately describes experimental results in the region of [BuOH] < 2.2 M. As can be seen from Eqs (3.66), the $r_{\mathrm{I}}/r_{\mathrm{II}}$ ratio depends, in a rather complicated manner, on P_{CO}, [St], $[\mathrm{Pd}]_0$, and $[\mathrm{L}]_0$. ■

A simpler kinetic model was obtained under the conditions of existence of an active cationic complex of $\mathrm{HPdL_3}^+$ [65, 71].

Example 3.12 [65]

▶ The kinetics of reaction (3.64) was also studied in the $\mathrm{Pd(dba)_2{-}TsOH{-}PPh_3}$ system using a closed reactor. This system featured the predominant formation of regioisomer **I** (with a selectivity of ≥65 %), the content of which was independent of $[\mathrm{L}]_0$. In contrast to the case of catalyst $\mathrm{PdCl_2L_2}$, the rates of formation of products **I** and **II** exhibited maxima with respect to P_{CO}, such that $r_{\mathrm{I}}/r_{\mathrm{II}} = 0.29 + 0.071P_{\mathrm{CO}}$. These observations give grounds to suggest that, in the reaction scheme depicted in Figure 3.13, step **2** is reversible and step **3** is quasi-equilibrium. The material balance with respect to Pd includes three complexes,

$$[\mathrm{Pd}]_0 = [\mathrm{HPdL_3}^+] + [\mathrm{Pd(CO)L_2}] + [\mathrm{Pd(CO)_2L_2}]; \tag{3.67}$$

$$\frac{[\mathrm{Pd}]_0}{[\mathrm{HPdL_3}^+]} = 1 + \frac{K_8K_9P_{\mathrm{CO}}}{[\mathrm{TsOH}][\mathrm{L}]} + \frac{K_8K_9K_{10}P_{\mathrm{CO}}^2}{[\mathrm{TsOH}][\mathrm{L}]}, \tag{3.68}$$

from which it follows that

$$[\mathrm{HPdL_3}^+] = \frac{[\mathrm{Pd}]_0[\mathrm{TsOH}]}{[\mathrm{TsOH}] + K_8K_9P_{\mathrm{CO}}/[\mathrm{L}] + K_8K_9K_{10}P_{\mathrm{CO}}^2/[\mathrm{L}]}. \tag{3.69}$$

For $\omega_{-3} \gg \omega_5$, the rates r_{I} and r_{II} are expressed as follows:

$$r_{\mathrm{I}} = \frac{\omega_1\,\omega_2\,\omega_4[\mathrm{HPdL_3}^+]}{(\omega_{-2}+\omega_4)\left(\omega_{-1} + \dfrac{\omega_3\,\omega_5}{\omega_{-3}}\right) + \omega_2\,\omega_4}; \tag{3.70}$$

$$r_{\mathrm{II}} = \frac{\omega_1\,\omega_3\,\omega_5(\omega_4+\omega_{-2})[\mathrm{HPdL_3}^+]}{(\omega_{-2}+\omega_4)\left(\omega_{-1} + \dfrac{\omega_3\,\omega_5}{\omega_{-3}}\right) + \omega_2\,\omega_4} \tag{3.71}$$

or in the original notation [65]:

$$r_{\mathrm{I}} = \frac{A_4[\mathrm{St}]P_{\mathrm{CO}}[\mathrm{HPdL_3}^+]}{([\mathrm{L}] + A_1P_{\mathrm{CO}})(1 + A_2P_{\mathrm{CO}}) + A_3P_{\mathrm{CO}}}; \tag{3.72}$$

$$r_{\mathrm{II}} = \frac{A_1A_4[\mathrm{St}]P_{\mathrm{CO}}(1 + A_2P_{\mathrm{CO}})[\mathrm{HPdL_3}^+]}{A_3[([\mathrm{L}] + A_1P_{\mathrm{CO}})(1 + A_2P_{\mathrm{CO}}) + A_3P_{\mathrm{CO}}]}; \tag{3.73}$$

$$[\mathrm{HPdL_3}^+] = \frac{[\mathrm{Pd}]_0[\mathrm{TsOH}]}{[\mathrm{TsOH}] + A_5P_{\mathrm{CO}}/[\mathrm{L}] + A_6P_{\mathrm{CO}}^2/[\mathrm{L}]}; \tag{3.74}$$

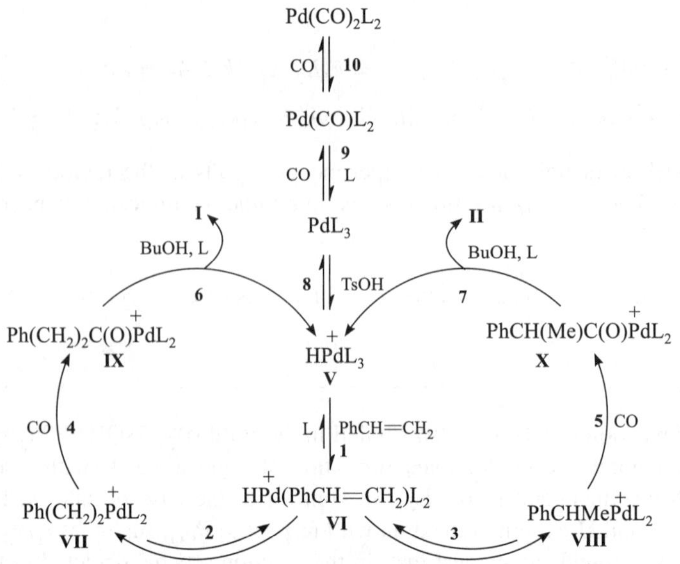

Figure 3.13 Mechanistic scheme of reaction (3.64) in Pd(dba)$_2$–TsOH–PPh$_3$ system (reproduced with permission from editorial board and staff of *Izv. Ross. Akad. Nauk, Ser. Khim.* journal © 2001 [65])

$$[\mathrm{TsOH}]_0 = [\mathrm{TsOH}] + [\mathrm{HPdL_3}^+]; \quad (3.75)$$

$$[\mathrm{L}]_0 = [\mathrm{L}] + 2[\mathrm{Pd}]_0 + [\mathrm{HPdL_3}^+], \quad (3.76)$$

where $A_1 = k_5K_3/k_{-1}$; $A_2 = k_4/k_{-2}$; $A_3 = k_2k_4/k_{-1}k_{-2}$; $A_4 = k_1A_3$; $A_5 = k_8K_9$; $A_6 = K_8K_9K_{10}$.

It is assumed that the active cationic complex $HPdL_3{}^{++}$ exists in the form of a solvation-separated ion pair $HPdL_3{}^+ \| TsO^-$. The butanolysis of cationic acyl intermediates is a fairly rapid process that cannot be considered as a slow step. The ratio $r_{II}/r_I = A_1(1 + A_2P_{CO})/A_3$ linearly increases with P_{CO}. ■

In the context of the examples considered above, it is also expedient to mention the results of Kron et al. [72, 73] on the kinetics of phenylacetylene hydroalkoxycarbonylation in a similar system of $Pd(dba)_2$–CF_3COOH–Ph_3P in n-butanol. The high regioselectivity of this process is manifested by the predominant formation of branched butyl-2-phenylpropyonate (95–97 %). According to [74], it was assumed that hydride complexes $HPdL_2X$ and $HPdL_3{}^+X^-$ did not participate in the steps of the catalytic cycle, while the vinyl derivative is formed via the protonation of the Pd^0L_2(alkyne) π complex. In this system, a single-route mechanism with more complicated complexity function was considered, according to which

$$[\mathrm{Pd}]_0 = [\mathrm{CH_2 = C(Ph)HPdL_2}^+\mathrm{X}^-] + [\mathrm{Pd(CO)L_2}] + [\mathrm{Pd(CO)_2L_2}] + [\mathrm{HPdL_3}^+\mathrm{X}^-], \quad (3.77)$$

and the material balance for the acid was expressed as follows:

$$[\mathrm{CF_3COOH}]_0 = [\mathrm{CF_3COOH}] + [\mathrm{HPdL_3}^+\mathrm{X}^-] + [\mathrm{CH_2 = C(Ph)HPdL_2}^+\mathrm{X}^-].$$

For the correct kinetic analysis, it would also be important to check whether there can be additional channels for the conversion of PPh_3 and CF_3COOH, which is quite probable for the system under consideration.

As an example of the multi-route reaction with a common step between all routes, let us consider the cyclodimerization of norbornadiene (C_7H_8).

Example 3.13 [75, 76]

▶ The kinetics of reaction

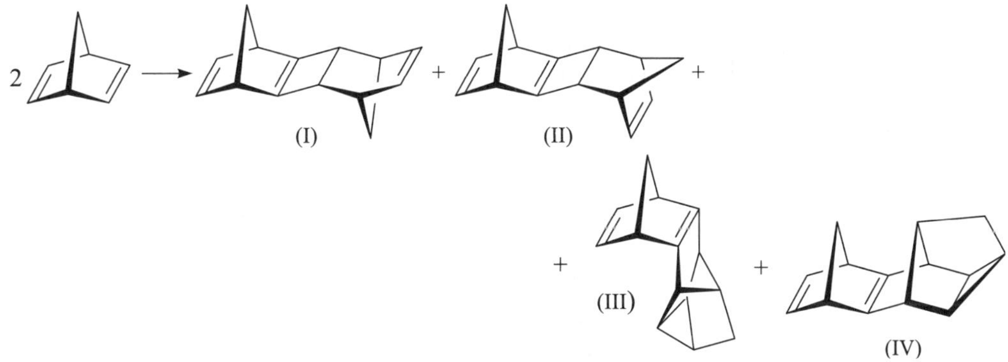

(3.78)

was studied in nonane and THF at temperatures within 10 – 110 °C and the concentrations of $Ni(C_3H_5)_2$ catalyst within 0.005–0.2 M and 2,5-norbornadiene (NBD) within 0.1–9.8 M [75]. It was established that reaction (3.78) is catalyzed by the complex $Ni(NBD)_2$ (V), which is formed from bis-π-allylnickel. The rates of formation of the main products I, II, and III are described by the following equations:

$$r_{\mathrm{I}} = k_1[\mathrm{Ni}]_0 C_{\mathrm{NBD}}^2; \tag{3.79}$$

$$r_{\mathrm{II}} = k_2[\mathrm{Ni}]_0 C_{\mathrm{NBD}}; \tag{3.80}$$

$$r_{\mathrm{III}} = k_3[\mathrm{Ni}]_0 C_{\mathrm{NBD}}. \tag{3.81}$$

A mechanistic scheme was developed based on the results of investigations of the stereochemistry of products; the equilibrium complex formation

$$\mathrm{Ni(NBD)_2} \overset{\mathrm{NBD}}{\rightleftharpoons} \mathrm{Ni(NBD)_3} \overset{\mathrm{NBD}}{\rightleftharpoons} \mathrm{Ni(NBD)_4}$$

and the kinetic data correspond to a four-route reaction (Figure 3.14) that is represented by kinetic graph KG16 (Figure 3.15). Despite the complexity of this scheme, quite simple kinetic equations are obtained under the assumption that all reversible steps are quasi-equilibrium and $[\mathrm{Ni}]_0 \cong [\mathrm{Ni(NBD)_2}]$.

A more complicated mechanistic scheme was proposed for the same reaction of NBD cyclodimerization in the $Ni(C_3H_5)_2$–PR_3 system [76]. In this case, Ni(0) complexes in solution exist predominantly in the form of two $\mathrm{NiL(NBD)_2}$ isomer complexes, from which four $\mathrm{NiL(NBD)_3}$ isomer complexes are formed ($[\mathrm{L}]_0/[\mathrm{Ni}]_0 = 1$). The conversions of these six complexes with the formation of five isomeric cyclodimers are described by the scheme of a nine-route reaction mechanism with kinetic equations:

$$r_i = k_i[\mathrm{Ni}]_0 C_{\mathrm{NBD}}, \tag{3.82}$$

which are simply due to the quasi-equilibrium character of all reversible steps of complex formation and a simple material balance condition of $[\mathrm{Ni}]_0 \cong [\mathrm{NiL(NBD)_2}]$. ■

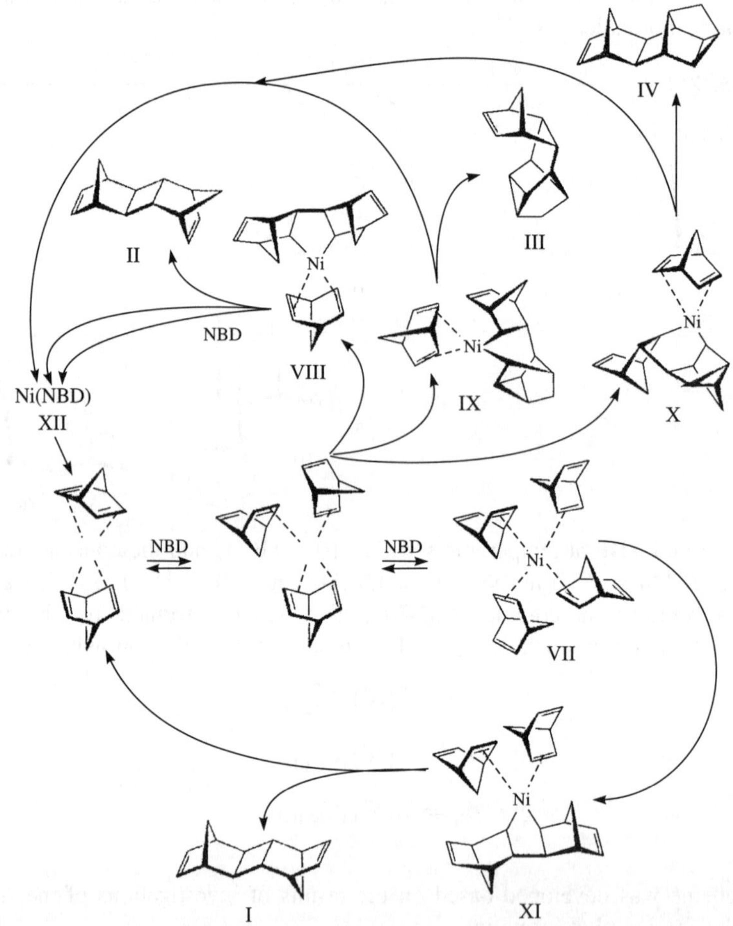

Figure 3.14 Mechanistic scheme of 2,5-norbornadiene (NBD) cyclodimerization in the presence of $Ni(NBD)_2$ complexes (reproduced with permission from editorial board of *Kinet. Katal.* journal © 2000 [75])

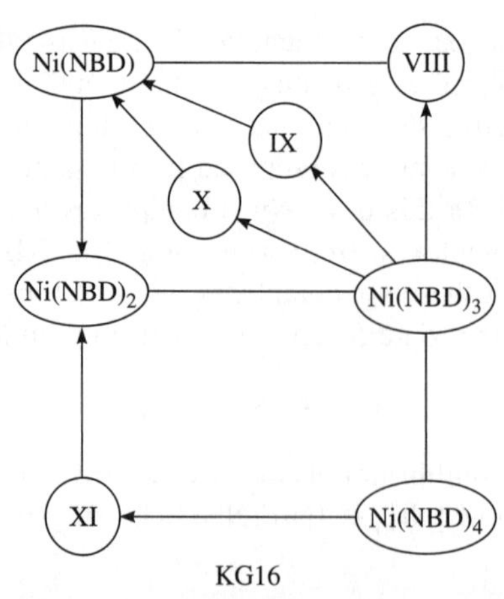

Figure 3.15 Kinetic graph for mechanism of NBD cyclodimerization

3.3 Conjugation nodes and artificial multi-route character

Multi-route reaction mechanisms with common steps contain special intermediates (i.e., vertices) at which the process exhibits branching. The fragments of a mechanism (i.e., of the corresponding topological structure) involving three (or more) steps related to the same intermediate X_i are called *conjugation nodes* [4, 18, 22, 77] (Figure 3.16). Under the conditions where stoichiometrically many-valued reactions are far from equilibrium and the routes (by which products P_k are formed) include irreversible steps, the ratios of the product formation rates (e.g., r_{P_1}/r_{P_2}) in a conjugation node are expressed by significantly simpler functions of the reactant concentrations as compared to expressions for the rates proper (r_{P_i}). Examples of the analysis of conjugation nodes were given above (see Section 3.2) for schemes (3.21) and (3.36) representing (d)-type nodes and for schemes (3.43) and (3.46) representing (a)-type nodes.

The r_{P_1}/r_{P_2} ratios for the nodes of various types (a, b, c, d) in Figure 3.16 are described by the following simple equations:

$$\text{(a)} \quad \frac{r_{P_1}}{r_{P_2}} = \frac{\omega_2}{\omega_3}; \tag{3.83}$$

$$\text{(b, c)} \quad \frac{r_{P_1}}{r_{P_2}} = \frac{\omega_2\,\omega_3}{\omega_4(\omega_{-2}+\omega_3)}; \tag{3.84}$$

$$\text{(d)} \quad \frac{r_{P_1}}{r_{P_2}} = \frac{\omega_2\,\omega_3(\omega_{-4}+\omega_5)}{\omega_4\,\omega_5(\omega_{-2}+\omega_3)}. \tag{3.85}$$

Clearly, these expressions do not contain catalyst concentrations and, hence, do not require taking into account the material balance with respect to the catalyst. In cases where the products are formed past the branching point (X_i), an analysis of the conjugation nodes is more useful for understanding the structure of the reaction mechanism (mechanistic scheme) and discriminating the hypotheses than the analysis of complete kinetic equations $r_{P_i} = f([C_i])$ [4, 22, 77]. For example, in a comparatively simple conjugation

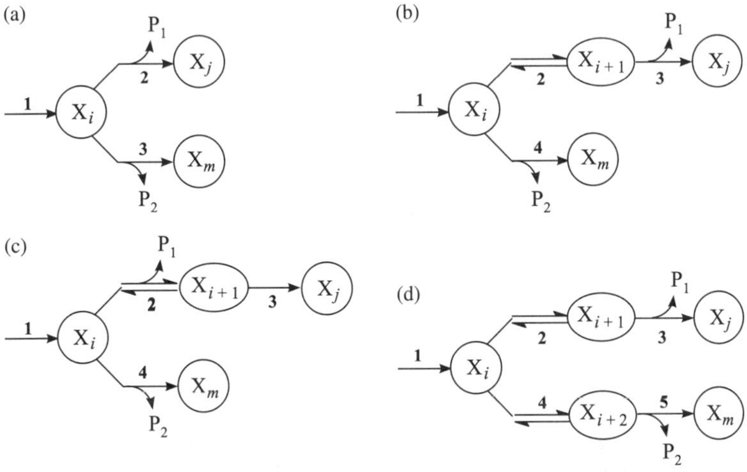

Figure 3.16 Conjugation nodes of linear mechanisms

node of a multi-route (including at least five routes) process of acetylene carbonylation [18]

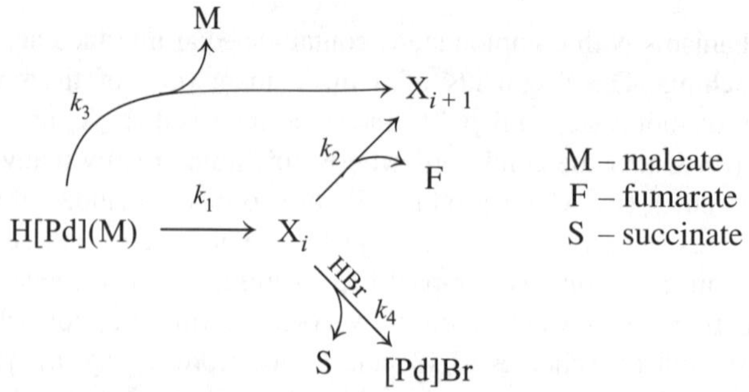

the ratio of the rates of succinate and fumarate formation is $r_S/r_F = k_4/k_2 \cdot [\mathrm{HBr}]$, whereas the dependences of r_S, r_F and r_M on [HBr] are described by expressions such as Eq. (3.6).

In the case of nonlinear mechanisms with elementary steps of molecularity $\mu \leq 2$, the ratio of rates is equal to the ratio of differential selectivities $\varphi_{P_1}/\varphi_{P_2}$ (i.e., to the relative selectivity $\varphi_{1,2}$).

In the case of nonlinear mechanisms

$$\begin{aligned}
&\text{(a)} \quad \xrightarrow{\mathbf{1}} X_i \quad \begin{matrix} X_i + X_j \xrightarrow{\mathbf{2}} X_k + P_1 \\ X_i + B \xrightarrow{\mathbf{3}} X_m + P_2 \end{matrix}; \\
&\text{(b)} \quad \xrightarrow{\mathbf{1}} X_i \quad \begin{matrix} X_i + X_j \overset{\mathbf{2}}{\rightleftharpoons} X_k \xrightarrow{\mathbf{3}} X_l + P_1 \\ X_i + B \xrightarrow{\mathbf{4}} X_m + P_2 \end{matrix}; \\
&\text{(c)} \quad \xrightarrow{\mathbf{1}} X_i \quad \begin{matrix} 2X_i \xrightarrow{\mathbf{2}} X_k + P_1 \\ X_i + B \xrightarrow{\mathbf{3}} X_m + P_2 \end{matrix}; \\
&\text{(d)} \quad \xrightarrow{\mathbf{1}} M \quad \begin{matrix} M \xrightarrow{\mathbf{2}} X_x + P_1 \\ 2M \overset{K_D}{\rightleftharpoons} M_2 \xrightarrow{\mathbf{3}} X_m + P_2 \end{matrix};
\end{aligned} \tag{3.86}$$

an analysis of the conjugation nodes (a)–(d) is not as unambiguous, but is still very informative for elucidating nonlinear steps:

$$\text{(a)} \quad \frac{r_{P_1}}{r_{P_2}} = \frac{\omega_2}{\omega_3}[X_j]; \tag{3.87}$$

$$\text{(b)} \quad \frac{r_{P_1}}{r_{P_2}} = \frac{\omega_2\,\omega_3}{\omega_4(\omega_{-2} + \omega_3)}[X_j]; \tag{3.88}$$

$$\text{(c)} \quad \frac{r_{P_1}}{r_{P_2}} = \frac{\omega_2}{\omega_3}[X_i]; \tag{3.89}$$

$$\text{(d)} \quad \frac{r_{P_1}}{r_{P_2}} = \frac{\omega_3}{\omega_2}\frac{[M_2]}{[M]} = \frac{\omega_3}{\omega_2}K_D[M]. \tag{3.90}$$

Node (d) in scheme (3.86) represents two routes coupled via the common intermediate [M]. In the case of equilibrium step **1**, we obtain Eq. (3.90). In all variants of nodes in scheme (3.86), the ratios r_{P_1}/r_{P_2}

must increase with the total catalyst $[M]_\Sigma$, which may be indicative of the presence of nonlinear steps in the proposed scheme.

In addition to the dimensionless ratio of rates (r_{P_i}/r_{P_j}), an analysis of the conjugation nodes in linear mechanisms can also provide other useful dependences of rates r_{P_i} on the total rates r_Σ over all routes or on the sums of ratios of the rates of formation of other products [4, 22]. For example, consider the following conjugation node with a reversible step of formation of intermediate X:

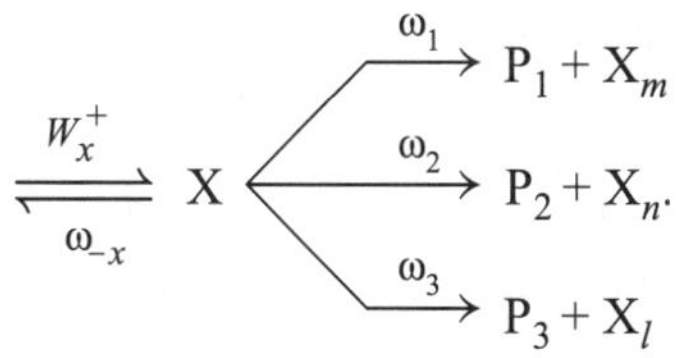

(3.91)

Under steady-state conditions, we have

$$W_x^+ - \omega_{-x}[X] = r_{P_1} + r_{P_2} + r_{P_3}$$

or

$$W_x^+ = \omega_{-x}[X] + \sum r_{P_i} = \omega_{-x}[X] + r_\Sigma. \tag{3.92}$$

Taking into account that $r_{P_1} = \omega_1[X]$, we can write

$$W_x^+ = \omega_{-x}\frac{r_{P_1}}{\omega_1} + r_\Sigma, \tag{3.93}$$

which yields

$$r_{P_1} = \frac{W_x^+\omega_1}{\omega_{-x}} - \frac{\omega_1}{\omega_{-x}}r_\Sigma \text{ or}$$

$$r_{P_1} = A - Br_\Sigma. \tag{3.94}$$

If reactants involved in the formation of products P_2 and P_3 at the second and third steps do not contribute to W_x^+, ω_1, and ω_{-x}, then coefficients A and B remain constant during any variation of the reactant concentrations entering into ω_2 and ω_3, so that relationship (3.94) is linear.

A relationship between r_{P_1} and r_{P_i}/r_{P_1} can also be derived in a different way. Upon dividing both parts of Eq. (3.92) by r_{P_1}, we obtain

$$\frac{W_x^+}{r_{P_1}} = \frac{\omega_{-x}[X]}{r_{P_1}} + \left(1 + \sum_{i \neq 1}\frac{r_{P_i}}{r_{P_1}}\right), \tag{3.95}$$

which yields

$$r_{P_1} = \frac{W_x^+}{\dfrac{\omega_{-x}}{\omega_1} + \left(1 + \sum_{i \neq 1}\dfrac{r_{P_i}}{r_{P_1}}\right)}. \tag{3.96}$$

If the concentrations of reactants entering into W_x^+, ω_1, and ω_{-x} remain constant when ω_2 and ω_3 are varied, then we can obtain the following dependences [78]:

$$r_{P_1} = f\left[\sum_{i \neq 1} \frac{r_{P_i}}{r_{P_1}}\right] = f\left(\frac{r_{P_2}}{r_{P_1}} + \frac{r_{P_3}}{r_{P_1}}\right). \tag{3.97}$$

By experimentally determining the form of dependences $r_{P_1}/r_{P_2} = f(C_k)$ and $r_{P_3}/r_{P_1} = f_1(C_k)$ (where C_k is the concentration of the kth reactant) jointly with equations of the types (3.94) or (3.96), it is possible to find the kinetic equations for each product P_i. Examples of using Eqs (3.94) and (3.97) for the analysis of conjugation nodes in mechanisms with nonlinear steps will be given below.

Thus, the appearance of a conjugation node in a mechanistic scheme allows information to be obtained about reactions following the first irreversible step (irrespective of the material balance structure) and facilitates determination of the form of dependences $r_{P_i} = f(C_k)$. Accordingly, for constructing a theoretically justified kinetic model, it is expedient to perform the following steps: (i) to study the reaction kinetics under conditions where the selectivity of product P_i formation is below 100 % and the main and side products are expected to form in the same reaction network (within one multi-route mechanism); and (ii) to create artificial branching points (conjugation nodes), that is, introduce an *artificial multi-route character*.

An artificial multi-route character arises either upon the introduction of several reactants interacting with one of the intermediates or upon the addition of several substrates interacting (competitive reactions) with an intermediate formed in a step that is common to all substrates. Various "inhibitors" capable of binding to an active center or intermediate product are widely used in the enzymatic catalysis. Competitive reactions are also frequently used in physical organic chemistry for determining the relative reactivity of various substrates or different reactive centers in the same substrate [79, 80]. A necessary condition for obtaining reliable information concerning relative rate constants is the existence of a conjugation node of type (a) (Figure 3.16) with irreversible steps of conversion for X_i. Then, for reactions of the first order with respect to a substrate under steady-state or quasi-steady-state conditions, we have

$$\frac{r_{P_1}}{r_{P_2}} = \frac{k_2\,[S_1]}{k_3\,[S_2]},$$

where

$$\frac{k_2}{k_3} = \frac{\lg(1 - x_1)}{\lg(1 - x_2)}, \tag{3.98}$$

for a closed system where x_i is the degree of conversion of the ith substrate.

The introduction of isotope-labeled molecules also creates an artificial multi-route character that can be used for studying reaction mechanisms. Competitive reactions with a substrate and its isotopomer make possible the measurement of kinetic isotope effects (KIEs). The measurements of KIEs with respect to the rates of formation of the labeled and unlabeled products are undoubtedly useful, but the interpretation of obtained values, such as r_H/r_D or $k_{\text{obs}}^{\text{H}}/k_{\text{obs}}^{\text{D}}$, is frequently rather ambiguous, since the label can be transferred via several steps in different routes. The use of a 100 %-labeled reactant even in an initially single-route reaction, e.g., of hydrogenation ($D_2 + C_2H_4$), can lead under steady-state conditions to the formation of a mixed reactant (HD) due to the reversible steps of β-elimination and, hence, to the appearance of routes involving the mixed reactant.

Application of the theory of steady-state reactions and the theory of routes to reactions with isotope-labeled participants was reviewed in [79, 80]. For a single-route reaction with a linear mechanism ($\nu_S = 1$), a given number of steps (S), and a single route of label transfer via $S^* \le S$ steps (the rate constants of which are assumed to be independent of the presence of the label), the rate ρ of label transfer can be expressed in terms of the free concentration of active centers:

$$\rho = \mu \frac{r_{S_1} r_{S_2} \cdot \ldots \cdot r_{S_n^*} u_{\text{A}} - r_{-S_1} r_{-S_2} \cdot \ldots \cdot r_{-S_n^*} u_{\text{B}}}{r_{S_2} r_{S_3} \cdot \ldots \cdot r_{S_n^*} + r_{-S_1} r_{S_3} \cdot \ldots \cdot r_{S_n^*} + \cdots + r_{-S_1} r_{-S_2} \cdot \ldots \cdot r_{S_{n-1}^*}}, \tag{3.99}$$

where μ is the number of labeled atoms transferred for one run of the overall reaction, and u_A and u_B are the atomic fractions of labeled atoms in the initial (A) and final (B) substances. For $u_A = u_B = 1$,

$$\rho = \mu r, \tag{3.100}$$

where r is the reaction rate in the absence of the label.

It can also be shown that

$$\frac{r_{-S_1} r_{-S_2} \cdot \ldots \cdot r_{-S_n^*}}{r_{S_1} r_{S_2} \cdot \ldots \cdot r_{S_n^*}} = \frac{\rho - \mu r u_{\text{A}}}{\rho - \mu r u_{\text{B}}}, \tag{3.101}$$

so that, for $S = S^*$,

$$\frac{r^-}{r^+} = \frac{\rho - \mu \cdot r \cdot u_{\text{A}}}{\rho - \mu \cdot r \cdot u_{\text{B}}}, \tag{3.102}$$

where $r^+ - r^- = r$.

This approach was successfully used in investigations of the kinetics of single-route heterogeneous reactions [80]; however, to the best of the author's knowledge, no cases of applying this method to studying the mechanisms of reactions in solutions of metal complexes have been reported so far.

It should be emphasized that the efficiency of using KIEs in order to determine the average stoichiometric number $\overline{\nu}$ or the stoichiometric number ν_L of the limiting step [80] depends on the validity of the assumption that the presence of a label does not influence the rate constants of related steps. Evidently, this assumption cannot be adopted with respect to labels such as H/D and O^{16}/O^{18}.

In the case of multi-route reactions, the study of label transfer rates is not very informative. For example, in a two-route reaction of ethylene hydrogenation via a linear mechanism (KG17, Figure 3.17), the use of an H_2/D_2 mixture even without allowance for the subsequent conversions of C_2H_3D would lead to the appearance of HZC_2H_4D, DZC_2H_5, DZC_2H_4D, ZD_2, and ZHD intermediate species (a total of seven linearly independent intermediates and 27 steps) and 20 linearly independent routes. The description of

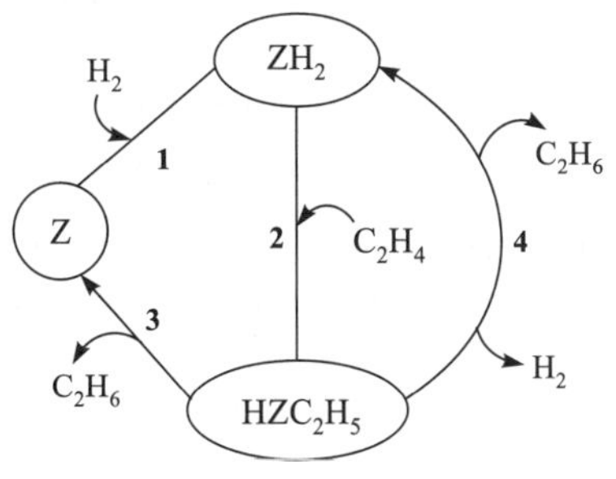

KG17

Figure 3.17 Kinetic graph for mechanism of ethylene hydrogenation such a reaction (label transfer rate) or the observed KIEs is evidently an unsolvable task. The use of isotope labels is more effective in the case of processes with limiting steps or competitive reactions with simple conjugation nodes (see below).

Let us consider several examples of reactions with the natural and artificial multi-route character that leads to the appearance of conjugation nodes. The natural multi-route character is manifested, for example, in processes including a route of catalyst destruction. The study of process kinetics under these conditions is useful for understanding the mechanism of the main catalytic reaction.

Example 3.14 [81]

▶ The kinetics of reaction

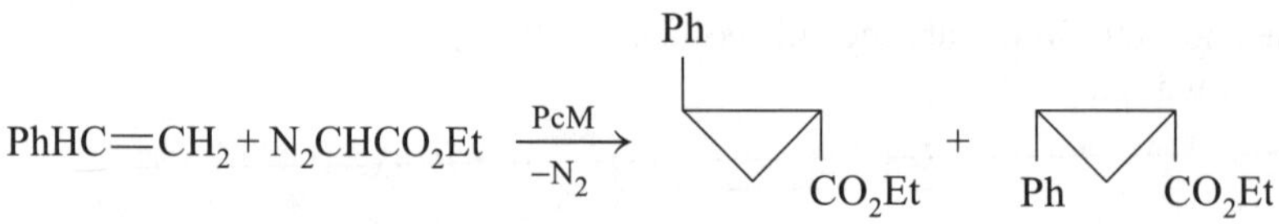

(I) (II) (3.103)

was studied by measuring the rates of product accumulation and N_2 evolution. The catalyst, Co(II) tetra-4-*tert*-butylphthalocyaninate (PcCo), ensures the formation of the sum of products I and II with high selectivity with respect to diaza ester at 25 °C. The rate of PcCo destruction was monitored by a spectrophotometric technique. It was found that the rates of nitrogen evolution and catalyst destruction are described by the following equations:

$$r_{N_{2,0}} = k_{obs}[PcCo]_\Sigma[N_2CHCO_2Et]_0; \tag{3.104}$$

$$r_{PcCo,0} = -\left.\frac{d[PcCo]_\Sigma}{dt}\right|_{t=0} = \frac{k'[PcCo]_\Sigma[N_2CHCO_2Et]_0}{b + c[PhCH{=}CH_2]}, \tag{3.105}$$

provided that $[N_2CHCO_2Et]_0$, $[PhCH = CH_2]_0 \gg [PcCo]_\Sigma$. The established dependences were explained using the following scheme:

$$\begin{aligned} &PcCo + N_2CHCO_2Et \underset{k_{-1}}{\overset{k_1}{\rightleftharpoons}} X_1 \xrightarrow{k_2} N_2 + X_2; \\ &X_2 + PhCH{=}CH_2 \xrightarrow{k_3} P + PcCo; \\ &X_2 \xrightarrow{k_4} \text{Products of destruction.} \end{aligned} \tag{3.106}$$

where $X_2 = (PcCo)(:CHCO_2Et)]$. This scheme yields the following equation that is analogous to the experimental relation:

$$r_{PcCo} = -\frac{d[PcCo]}{dt} = \frac{k_1k_2k_4[PcCo][N_2CHCO_2Et]}{(k_{-1} + k_2)(k_3[PhCH{=}CH_2] + k_4)}, \tag{3.107}$$

provided that the process is quasi-steady-state with respect to intermediates and $[PcCo]_\Sigma \cong [PcCo]$. This equation "reveals" olefin in step **3** due to the appearance of a conjugation node due to step **4**:

$$X_1 \xrightarrow[2]{k_2} X_2 \begin{cases} \xrightarrow[4]{k_4} P_i \\ \xrightarrow[3]{\omega_3} (I + II) + PcCo. \end{cases} \tag{3.108}$$

According to scheme (3.106) and the structure of conjugation node (3.108), the rate of formation of the product P (I + II) is described by the following equation,

$$r_P = \frac{dP}{dt} = \frac{k_1 k_2 k_3 [PcCo][N_2CHCO_2Et][PhCH{=}CH_2]}{(k_{-1} + k_2)(k_3[PhCH{=}CH_2] + k_4)}, \tag{3.109}$$

the ratio $r_{P,0}/r_{PcCo,0}$ can be expressed as

$$\frac{r_{P,0}}{r_{PcCo,0}} = \frac{k_3[PhCH{=}CH_2]_0}{k_4}. \tag{3.110}$$

and the sum of rates r_P and r_{PcCo} is equal to the rate of nitrogen evolution in step **2:**

$$r_{P,0} + r_{PcCo,0} = \left(\frac{k_1 k_2}{k_{-1} + k_2}\right)[PcCo]_\Sigma[N_2CHCOOEt]_0,$$

where

$$\frac{k_1 k_2}{k_{-1} + k_2} = k_{obs}$$

(see Eq. (3.104)). ■

As was noted above, an analysis of the natural and artificial conjugation nodes can help to establish the main features of the structure of a complex reaction mechanism and even find kinetic equations for the rates over separate routes.

In studying the mechanism of acetylene hydrochlorination in solutions of Cu(I) and Hg(II) complexes [82], it was suggested that intermediate compounds in these processes are *trans*-β-chlorovinyl chloride derivatives, namely, *trans*-Cu(I)CH=CHCl and *trans*-ClHgCH=CHCl. The oxidation of these intermediates by Cu(II) chloride on adding $CuCl_2$ to the CuCl–NH_4Cl–HCl–H_2O and $HgCl_2$–HCl–H_2O systems that catalyze the synthesis of vinyl chloride (VC) leads to a decrease in r_{VC} and the appearance of *trans*-1,2-dichloroethylene (*trans*-DCE), which disappears upon the complete consumption of $CuCl_2$ [82, 83]. It was also established that *trans*-ClHgCH=CHCl interacts with $CuCl_2$ to form *trans*-DCE, while a mixture of *cis* and *trans* isomers leads to a mixture of cis- and trans-DCE. Thus, the introduction of $CuCl_2$ into the $HgCl_2$–HCl–H_2O system creates an artificial conjugation node, which confirms the hypothesis concerning the mechanism of VC synthesis [84].

Example 3.15 [84]

► The kinetics of reactions

$$\underset{}{C_2H_2 + HCl \longrightarrow \underset{(VC)}{CH_2{=}CHCl}} \tag{3.111}$$

and

$$C_2H_2 + 2FeCl_3 \longrightarrow \underset{(DCE)}{trans\text{-}ClCH{=}CHCl} + 2FeCl_2 \tag{3.112}$$

were studied under the steady-state conditions with respect to $CuCl_2$ in the $HgCl_2$–$CuCl_2$–NH_4Cl–HCl–$FeCl_3$ system at $[NH_4Cl] = 2$ M, [HCl] = 1 M, $[FeCl_3] = 3$ M, and 80 °C in a continuous flow (with respect to reactants and products) ideally stirred reactor. It was established that, as $[HgCl_2]$ is varied from 0 to 0.1 M and $[CuCl_2]$ is within 0–0.5 M, the concentration of free chloride ions remains virtually constant. Therefore, the concentrations of free Cu^{2+} and Hg^{2+} ions and/or their active chloride complexes are proportional to $[CuCl_2]_\Sigma$ and $[HgCl_2]_\Sigma$ and the corresponding constant complexity functions F_{Cu} and F_{Hg} enter into the constants of kinetic equations.

The experimental kinetic equations (at [HCl] = const) are as follows:

$$r_{DCE} = k_4[CuCl_2]_\Sigma P_{C_2H_2} + \frac{k_5[CuCl_2]_\Sigma[HgCl_2]_\Sigma P_{C_2H_2}}{[CuCl_2]_\Sigma + k_6}; \tag{3.113}$$

$$r_{VC} = \frac{k_7[HgCl_2]_\Sigma P_{C_2H_2}}{[CuCl_2]_\Sigma + k_6}. \tag{3.114}$$

where the first term in Eq. (3.113) corresponds to the formation of DCE as catalyzed only by $CuCl_2$. Thus, DCE is formed via two independent routes, for which

$$r_{DCE} = r^{(I)}_{DCE} + r^{(II)}_{DCE},$$
$$\frac{r^{(II)}_{DCE}}{r_{VC}} = k_9[CuCl_2]_\Sigma. \tag{3.115}$$

A simplified mechanistic scheme of the second route (without description of the formation of metal chloride complexes) with a single conjugation node is as follows:

$$HgCl_2 + C_2H_2 \underset{}{\overset{1}{\rightleftharpoons}} Cl_2Hg(C_2H_2);$$

$$Cl_2Hg(C_2H_2) \overset{2}{\rightleftharpoons} ClHgCH{=}CHCl \tag{3.116}$$

$$ClHgCH{=}CHCl \xrightarrow[3]{H_3O^+,\ Cl^-} Cl_2Hg + CH_2{=}CHCl$$

$$ClHgCH{=}CHCl \xrightarrow[4]{CuCl_2} X \xrightarrow[5]{CuCl_2} ClCH{=}CHCl + 2CuCl;$$

$$2CuCl + 2FeCl_3 \longrightarrow 2CuCl_2 + 2FeCl_2.$$

The participation of two $CuCl_2$ molecules in steps **4** and **5** with the proposed formation of intermediate X representing a product of electrophilic replacement of Hg(II) chloride by Cu(II) chloride $[ClCu^{II}CH{=}CHCl]$ was confirmed by the results of investigation of the kinetics of step **4** [85].

The kinetics of reaction

$$\underset{(A)}{trans\text{-}ClHgCH{=}CHCl} + 2FeCl_3 \xrightarrow{CuCl_2} trans\text{-}ClCH{=}CHCl + 2FeCl_2 + HgCl_2 \tag{3.117}$$

was studied in 90% aqueous ethylene glycol at 25 °C in a flow of argon in the $LiClO_4$ (3 M)–LiCl (2 M) system. The concentration of reactant A was varied from $6 \cdot 10^{-5}$ to $4 \cdot 10^{-4}$ M and $[CuCl_2]_\Sigma$ was varied from 0.02 to 0.1 M. The reaction rate was monitored by measuring $[FeCl_3]$ using a potentiometric technique. The results showed evidence for a direct proportionality of $[Cu^{2+}] = a[CuCl_2]_\Sigma$.

Under steady-state conditions of [A] = const (with A in the solid phase), the rate of reaction (3.117) was described by a first-order equation in $[CuCl_2]_\Sigma$, while the initial reaction rates at various $[A]_0$ obeyed the following equation:

$$r_{DCE,0} = k[A]_0[CuCl_2]_\Sigma. \tag{3.118}$$

The results of experiments with a variable concentration of A in the time (at $[CuCl_2]_\Sigma = \text{const}$) showed that, at $[CuCl_2]_\Sigma < 0.1\,M$, the kinetic curves exhibit bending. The S-like shape of these curves was indicative of the presence of an intermediate and at least two steps in reaction (3.117):

$$A + CuCl_2 \xrightarrow{k_1} X \xrightarrow{k_2'} DCE + HgCl_2 + P. \tag{3.119}$$

An analysis of the kinetic curves showed that the behavior of the temporal derivative of r_{DCE} at $t \to 0$

$$\left.\frac{dr_{DCE}}{dt}\right|_{t\to 0} = k'[A]_0[CuCl_2]_\Sigma^2 \tag{3.120}$$

is indicative of the participation of a second $CuCl_2$ molecule in step **2** and $k' = k_1k_2$, so that

$$A + CuCl_2 \xrightarrow[\mathbf{1}]{k_1} \underset{(X)}{ClCuCH{=}CHCl} \xrightarrow[\mathbf{2}]{k_2,\ CuCl_2} DCE + HgCl_2 + P.$$

Measurements performed under steady-state conditions (Eq. (3.118)), yield the rate of the first step involving one $CuCl_2$ molecule. The irreversibility of step **1** is related to the high rate of demetallation of β-chlorovinylcopper chloride and other RCuCl compounds [85]. ■

Investigations of the kinetics of reactions in the $CuCl_2$–$HgCl_2$ system and the kinetics of conversion of the proposed intermediate (A) confirmed the hypothesis concerning the mechanism of hydrochlorination in Hg(II) complex solutions. This was analogous to how the study of the $CuCl_2$–CuCl system confirmed the proposed mechanism of VC formation in the conjugate process of synthesis of VC, DCE, and vinylidene chloride (VDC).

Example 3.16 [86]

► The kinetics of acetylene hydrochlorination to VC was studied in the NH_4Cl ($12\,\tilde{m}$)–CuCl ($3-9\,\tilde{m}$)–HCl (0.1–0.5 M)–$CuCl_2$ ($5\cdot10^{-3}$–$1.7\cdot10^{-1}$ M) system under the conditions of *trans*-DCE and VDC formation at 80 °C in a flow gradient-free reactor under the steady-state conditions with respect to $[CuCl_2]$, whereby CuCl is electrochemically oxidized to $CuCl_2$ via the following reaction:

$$2CuCl + 2HCl \longrightarrow 2CuCl_2 + H_2.$$

The overall reactions were described by the following equations:

$$C_2H_2 + HCl \xrightarrow{Cu(I)} \underset{(VC)}{CH_2{=}CHCl}; \tag{3.121}$$

$$C_2H_2 + 2HCl \xrightarrow{Cu(I)\text{-}Cu(II)} \underset{(DCE)}{trans\text{-}ClCH{=}CHCl} + H_2; \tag{3.122}$$

$$C_2H_2 + 2HCl \xrightarrow{Cu(I)\text{-}Cu(II)} \underset{(VDC)}{CH_2{=}CCl_2} + H_2. \tag{3.123}$$

The constancy of $[H_3O^+]$ and a_{Cu^+} in each series of experiments with variable $[CuCl_2]_\Sigma$ was checked using a pH-meter and a potentiometer, respectively. Under the indicated conditions, we have $a_{Cu^{2+}} \propto [CuCl_2]_\Sigma$ and $a_{Cu^{2+}} \propto [Cu^{2+}]$. The activities of Cu^{2+} and Cu^+ ions were determined from *in situ* measurements of the potentials of platinum and copper electrodes. Measurements of the dependence of

r_{DCE}/r_{VC} on the concentration of $CuCl_2$ ($a_{Cu^{2+}}$, $[Cu^{2+}]$) and $[H_3O^+]$ yielded the following equation,

$$\frac{r_{DCE}}{r_{VC}} = k_1' \frac{[CuCl_2]_\Sigma^2}{[H_3O^+]}, \tag{3.124}$$

which was indicative of a very simple structure of the conjugation node for the steps of formation of these products:

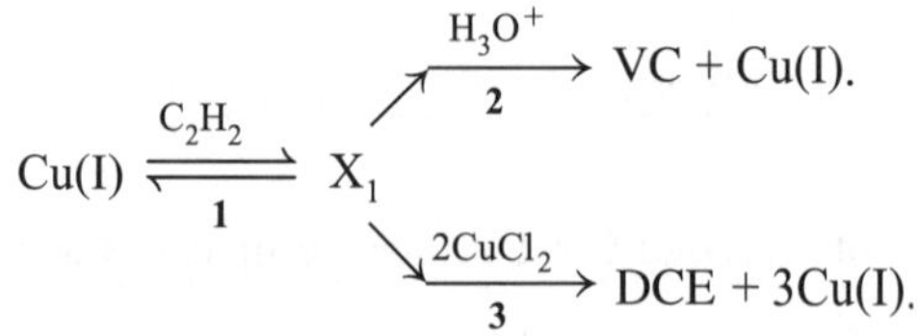

(3.125)

The dependence of r_{VDC}/r_{VC} on $[CuCl_2]_\Sigma$ came under a more complicated relation,

$$\frac{r_{VDC}}{r_{VC}} = k_2' \frac{[CuCl_2]_\Sigma^2}{[H_3O^+](k_3' + [CuCl_2]_\Sigma)}, \tag{3.126}$$

which showed evidence that the intermediates converted into VC and VDC are connected via at least one slow (not quasi-equilibrium) step **4**:

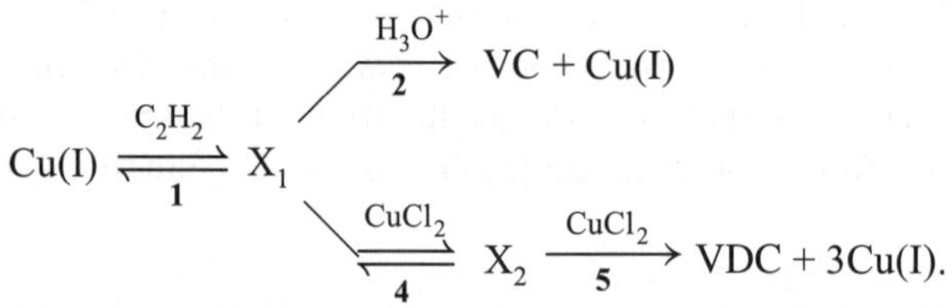

(3.127)

Combining the conjugation nodes (3.125) and (3.127) yields the following structure of a mechanism of conjugate processes with a single node:

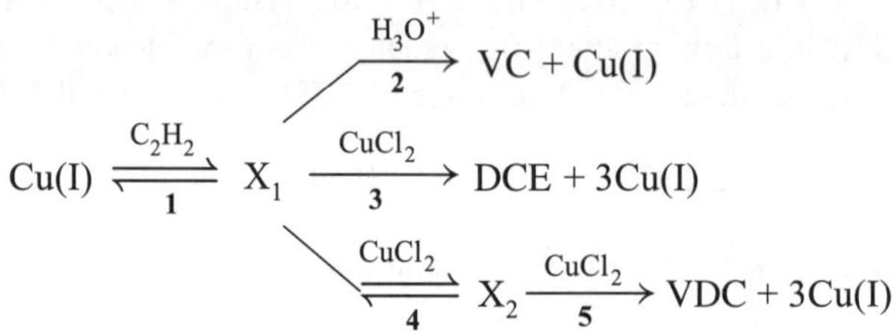

(3.128)

Since the reactions under indicated conditions yield no dimeric Cu(II) complexes (the Nernst equation is valid), step **3** is apparently a result of two steps similar to steps **4** and **5**, but with a quasi-equilibrium step **3**^{a}:

$$X_1 \underset{3^a}{\overset{CuCl_2}{\rightleftharpoons}} X_2' \xrightarrow[3^b]{CuCl_2} DCE + 3Cu(I).$$

All the three overall reactions are of the first order with respect to $P_{C_2H_2}$. In order to verify the proposed conjugation node structure and find the kinetic equations for each product (route), it is possible to use Eq. (3.94). Indeed, the dependence of r_{VC} on r_Σ ($r_\Sigma = r_{VC} + r_{VDC} + r_{DCE}$) is linear and can be described as

$$r_{VC} = k_4' - k_5' r_\Sigma \tag{3.129}$$

(provided that $[H_3O^+] = \text{const}$, $P_{C_2H_2} = \text{const}$, $a_{Cu^+} = \text{const}$, and $a_{Cl^-} = \text{const}$), where

$$k_4' = \frac{W_1{}^+\omega_2}{\omega_{-1}}, k_5' = \frac{\omega_2}{\omega_{-1}}.$$

Thus, the entire body of data for [CuCl] = const is described by the following equation:

$$\frac{r_{VC}}{P_{C_2H_2}[H_3O^+]} = k_6' - k_7'\frac{r_\Sigma}{P_{C_2H_2}}. \tag{3.130}$$

Solving the system of three equations ([(3.124), (3.126) and (3.130)) with three unknowns (r_{VC}, r_{VDC}, and r_{DCE}) leads to the corresponding kinetic equations. For example, an expression for r_{VC} is as follows:

$$r_{VC} = \frac{P_{C_2H_2}[H_3O]^+(k_3' + [CuCl_2]_\Sigma)}{\frac{k_3'}{k_6'} + \frac{k_7'k_3'}{k_6}[H_3O^+] + \frac{[CuCl_2]_\Sigma}{k_6'}(1 + k_7'[H_3O^+]) + \frac{k_7'}{k_6'}[CuCl_2]_\Sigma^2(k_2' + k_1'k_3') + \frac{\bar{k}_1k_7'}{k_6'}[CuCl_2]_\Sigma^3}.$$

By the same approach, for $P_{C_2H_2} = \text{const}$ and $[H_3O^+] = \text{const}$, we have

$$r_{VC} = \frac{\alpha + [CuCl_2]_\Sigma}{D}; \tag{3.131}$$

$$r_{DCE} = \frac{\varphi[CuCl_2]_\Sigma^2(\alpha + [CuCl_2]_\Sigma)}{D}; \tag{3.132}$$

$$r_{VDC} = \frac{\chi[CuCl_2]_\Sigma^2}{D}; \tag{3.133}$$

$$D = \varepsilon + \delta[CuCl_2]_\Sigma + \gamma[CuCl_2]_\Sigma^2 + \beta[CuCl_2]_\Sigma^3. \tag{3.134}$$

Obviously, it would be difficult to describe how the rate of product formation (VC and VDC) passes through a maximum with respect to $[CuCl_2]_\Sigma$ (according to the experimental data) and find equations (3.131)–(3.134) from dependences of the type $r_{P_i} = f([CuCl_2]_\Sigma)$.

Additional information on the mechanism of a conjugate process and the influence of activities a_{Cu^+} and a_{Cl^-} was obtained [87] from an analysis of the constants of a dimensionless "criterial" equation

$$\frac{r_{DCE}}{r_{VDC}} = k_1 + k_2[H_3O^+]^{1/2}\left(\frac{r_{DCE}}{r_{VC}}\right)^{1/2}, \tag{3.135}$$

which follows from Eqs (3.124) ad (3.126). ■

Example 3.17 [78, 88]

► The kinetics of chloroacetylene (CA) hydrochlorination was studied in a closed ideally stirred reactor in solutions of NH_4Cl (8 M), HCl, and CuCl with $CuCl_2$ additives under the conditions of conjugated reactions of *cis*-DCE formation (anti-Markovnikov HCl addition (3.136)) and oxidative CA chlorination to trichloroetylene (TCE):

$$\underset{(CA)}{ClC{\equiv}CH} + HCl \xrightarrow{Cu(I)} \underset{(DCE)}{cis\text{-}ClCH{=}CHCl}; \tag{3.136}$$

$$ClC{\equiv}CH + 2CuCl_2 \xrightarrow{Cu(I)} \underset{(TCE)}{Cl_2C{=}CHCl} + 2CuCl. \tag{3.137}$$

The initial reaction rates were determined from the kinetic curves plotted as product concentration in the gas phase versus time at a constant ratio of $V_{gas}/V_{liq} = \lambda$. Under the selected conditions, the concentrations of active chloride complexes of Cu(I) (or a_{Cu^+}) and Cu(II) (or $a_{Cu^{2+}}$) were proportional to $[CuCl]_\Sigma$ and $[CuCl_2]_\Sigma$, respectively, in a range of 0–1 M. In this system, TCE is formed via two routes, in which the first is independent of the CuCl concentration. It was also established that the $r_{TCE}^{(II)}/r_{DCE}$ ratio for the second route obeys the following simple equation:

$$\frac{r_{TCE}^{(II)}}{r_{DCE}} = k_1 \frac{[CuCl_2]_\Sigma}{[H_3O^+]}, \tag{3.138}$$

which is indicative of a simple structure of the conjugation node that appears in the presence of $CuCl_2$. The sum of rates $r_{TCE}^{(II)} + r_{DCE}$ depends on both $[CuCl_2]_\Sigma$ and $[H_3O^+]$, which points to the existence of a reversible step featuring the formation of a common intermediate:

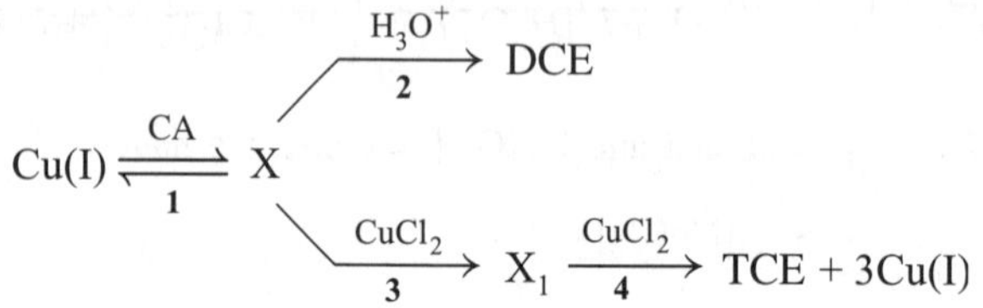

(3.139)

In contrast to the case considered in Example 3.16, the rate of the oxidative chlorination of CA to TCE is proportional to the first power of $[CuCl_2]_\Sigma$, which is probably due to the irreversibility of step **3** for CA. The kinetic equations for DCE formation and Cu(II)-catalyzed route of TCE formation were obtained in the form, of Eq. (3.96) as follows:

$$r_{DCE} = \frac{W_1^+}{\dfrac{\omega_{-1}}{\omega_2} + \left(1 + \dfrac{r_{TCE}^{(II)}}{r_{DCE}}\right)}. \tag{3.140}$$

For $W_1^+ = k_2' P_{CA}[CuCl]_\Sigma, \omega_{-1} = k_3'$, and $\omega_2 = k_4'[H_3O^+]$, the entire body of experimental data was well described by the linear relation (3.140) in the form of $y = a + bx$, where

$$y = \frac{P_{CA}[CuCl]_\Sigma[H_3O^+]}{r_{DCE}}, \qquad x = \left(1 + \frac{r_{TCE}^{(II)}}{r_{DCE}}\right)[H_3O^+],$$

which confirmed the correct choice of the structure of the conjugation node.

By jointly solving the system of Eqs (3.138) and (3.140), it is possible to obtain an expression for $r_{TCE}^{(II)}$ and r_{DCE}, where

$$r_{TCE}^{(II)} = \frac{k_1'k_2'k_4'[CuCl]_\Sigma[CuCl_2]_\Sigma P_{CA}}{k_3' + k_4'[H_3O^+] + k_1'k_4'[CuCl_2]_\Sigma}; \tag{3.141}$$

$$r_{DCE} = \frac{k_2'k_4'[CuCl]_\Sigma[H_3O^+]P_{CA}}{k_3' + k_4'[H_3O^+] + k_1'k_4'[CuCl_2]_\Sigma}. \tag{3.142}$$

The same approach was successfully applied to determining the form of kinetic equations for the process of hydrochlorination and oxidative chlorination of dicloroacetylene to TCE and perchloroethylene, respectively [88]. ■

As is known, a *cis*-dimethyl complex of Pt(IV) decomposes in $HCl{-}NaCl{-}H_2O$ chloride solutions (system I, $[Cl^-]_\Sigma = 3$ M) at 95 °C with the parallel formation of CH_3Cl, CH_4, and C_2H_6 [89]:

$$\underset{\textbf{(I)}}{Pt(CH_3)_2Cl_4^{2-}} \longrightarrow PtCl_4^{2-} + (CH_4, CH_3Cl, C_2H_6).$$

In a neutral system (system II, buffer solution, pH ≅ 7), the reaction products additionally contain ethylene, the formation of which in the presence of acid is already inhibited at [HCl] = 0.01 M. The initial rates of the formation of ethane (system I) and ethylene (system II) are described by equations of the first order with respect to the initial complex of Pt(IV). Let us consider the pathways of C_2-product formation in more detail.

Example 3.18 [89]

▶ The formation of CH_3Cl and C_2H_6 was considered as resulting from the outer-sphere (for CH_3Cl) and inner-sphere (for C_2H_6) reductive elimination involved in the conversion of the initial Pt(IV) complex (**I**). However, the investigation of this reaction in the $DCl{-}D_2O$ system under identical conditions showed that a multiple exchange takes places with the formation of $C_2H_{6-i}D_i$ in a ratio of 1 : 0.82 : 2 : 2 : 0.2 : 0.3 : 0.3. At the same time, there is no exchange between complex **I** and the medium (H/D in CH_3 groups) and between C_2H_6 and the medium in the presence of $PtCl_4^{2-}$. In order to explain these facts, an unusual scheme was proposed that included the formation of ethyl complexes of Pt(IV) and Pt(II):

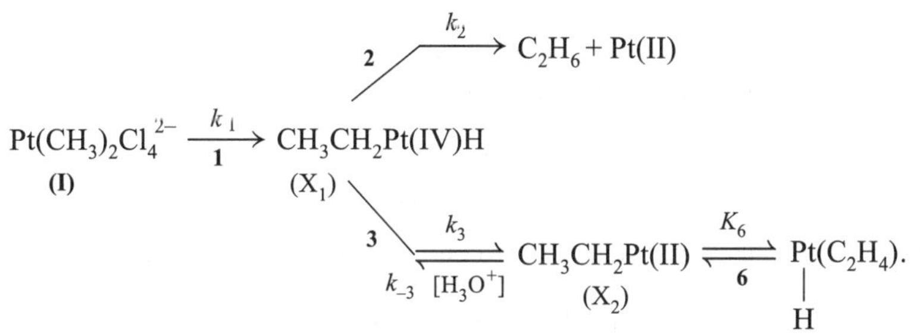

(3.143)

The appearance of X_1 is possible as a result of the following fast irreversible process:

$$Pt^{IV}(CH_3)_2 \longrightarrow \underset{\displaystyle H}{\overset{}{Pt^{IV}}}(CH_3)(:CH_2) \longrightarrow \underset{\displaystyle H}{Pt^{IV}}CH_2CH_3.$$

Since ethylene is virtually never observed in acid media (under deuterium exchange conditions), while step **6** can be considered as a quasi-equilibrium reaction involved in the exchange (β-elimination of Pt–H in X_2), it was decided to add a compound of Pt(IV) (e.g., $PtCl_6^{2-}$) to the system so as to oxidize X_2 with

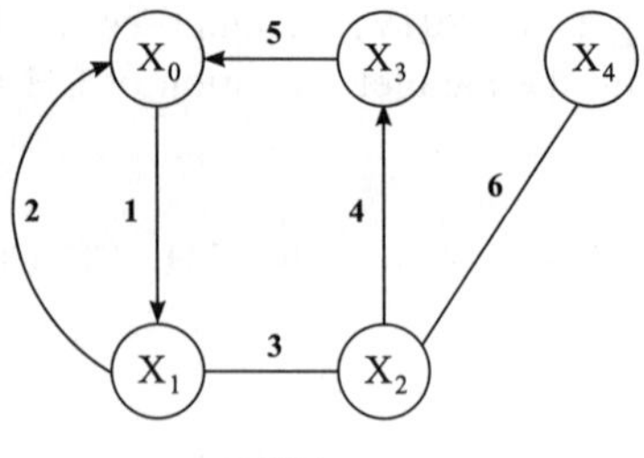

KG18

Figure 3.18 Kinetic graph for mechanism of demetallation of Pt(IV) dimethyl complex

the formation of CH_3CH_2Cl, thus introducing artificial branching into the system studied. The appearance of the new conjugation node,

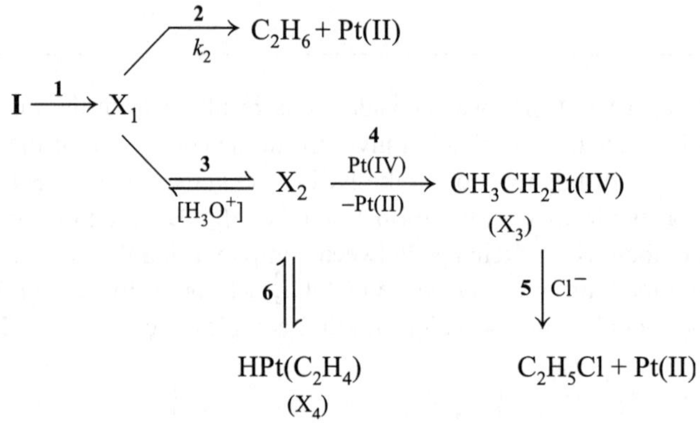

(3.144)

allowed the proposed mechanistic scheme (KG18, Figure 3.18) to be confirmed. The ratio of r_{RH}/r_{RCl} in this conjugation node (see Figure 3.16b)) can be expressed as follows:

$$\frac{r_{RH}}{r_{RCl}} = \frac{\omega_2(\omega_{-3} + \omega_4)}{\omega_3\,\omega_4} = \frac{k_2\,k_{-3}[H_3O^+]}{k_3\,k_4[Pt^{IV}]_\Sigma} + \frac{k_2}{k_3}, \tag{3.145}$$

and the sum of these rates under quasi-steady-state conditions can be written as

$$r_{RH} + r_{RCl} = W_1 = k_1[\mathbf{I}]. \tag{3.146}$$

Equations (3.145) and (3.146) under the condition that $[\mathbf{I}]_0 \cong [\mathbf{I}]$ yield

$$r_{RH} = \frac{k_1k_2[\mathbf{I}]_0}{k_2 + \dfrac{k_3k_4[Pt^{IV}]_\Sigma}{k_{-3}[H_3O^+] + k_4[Pt^{IV}]_\Sigma}} = k_{RH}[\mathbf{I}]_0 \tag{3.147}$$

with

$$k_{RH} = \frac{1 + \dfrac{k_4}{k_{-3}}\dfrac{[Pt^{IV}]_\Sigma}{[H_3O^+]}}{\dfrac{1}{k_1} + \dfrac{k_4}{k_1k_{-3}}\left(1 + \dfrac{k_3}{k_2}\right)\dfrac{[Pt^{IV}]_\Sigma}{[H_3O^+]}},$$

which agrees with the experiment. The second term in the denominator of Eq. (3.147) reflects the degree of influence of the second route (C_2H_5Cl formation) on the RH synthesis. It is interesting to note that the ratio of rate constants k_2/k_3 for the monomolecular conversion of X_1 is equal to that for its deuterated analog ($Cs_2Pt(CD_3)_2Cl_4$):

$$\frac{k_2}{k_3} \cong 0.10 \pm 0.04 \quad \text{and} \quad \frac{k_2^D}{k_3^D} \cong 0.12.$$

It is suggested that the deuterium exchange can involve, in addition to step **6** (exchange reaction $HPt(C_2H_4) + D^+ \leftrightarrow H^+ + DPt(C_2H_4)$), the step of intramolecular H/D transfer

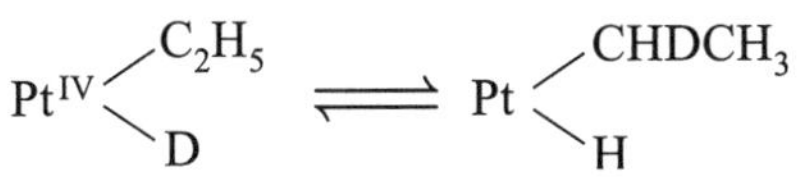

via a transition state of the agostic type, that is, intramolecular metathesis of σ bonds Pt–D and C–H:

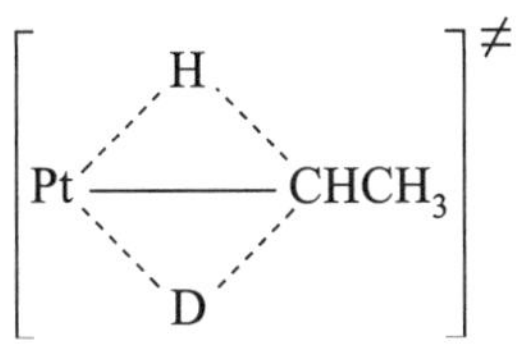

Since the activation of alkanes (CH_4, C_2H_6) by Pt(II) complexes also involves multiple exchange and leads to the formation of CH_3Cl and C_2H_5Cl in the presence of Pt(IV), it was suggested [89], that this process includes the stages of alkane activation via the same intermediates:

$$C_2H_6 \xrightarrow{Rt(II)} X_1 \rightleftharpoons X_2 \xrightarrow{Pt(IV)} X_3 \xrightarrow{Cl^-} C_2H_5Cl.$$
$$X_2 \rightleftharpoons X_4$$

■

Example 3.19 [89]

► An analogous relationship between the demetallation of *cis*-dimethyl complex of Pt^{IV} and the process of CH_4 activation was revealed by an analysis of the kinetics of CH_4 and CH_3Cl formation [89] and the kinetics of CH_4–D_2O exchange in K_2PtCl_4 solutions [90]. The former process is described by the following scheme:

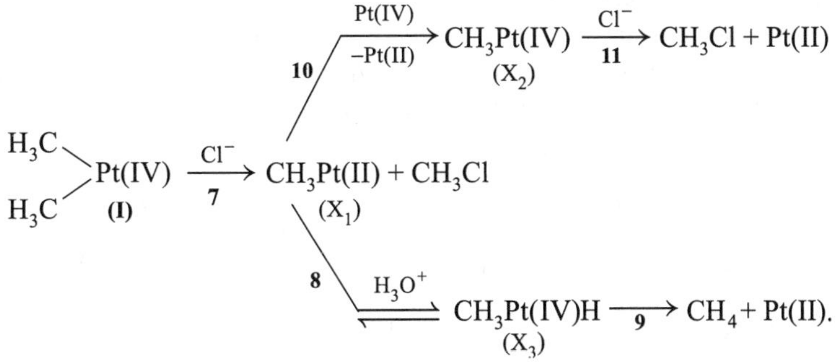

(3.148)

Within the framework of this scheme, the ratio of $r^{\Sigma}_{CH_3Cl} = W_7 + W_{10}$ and r_{CH_4} or the yields of these products under the condition of $[Pt^{IV}]_0 \gg [\mathbf{I}]$ is described by the following equation:

$$\frac{r^{\Sigma}_{CH_3Cl}}{r_{CH_4}} = \frac{2\,W_{10} + W_9}{W_9} = 1 + 2\left(\frac{k_{-8}}{k_9} + 1\right)\frac{k_{10}}{k_8}\frac{[Pt^{IV}]_{\Sigma}}{[H_3O^+]}, \tag{3.149}$$

where $W_7 = W_{10} + W_9$.

Assuming that the route of methane activation involves the intermediates (X_1 and X_3) of scheme (3.148), the obtained kinetics of multiple CH_4–D_2O exchange [89] was in very good agreement with experiment [90]. The proposed scheme

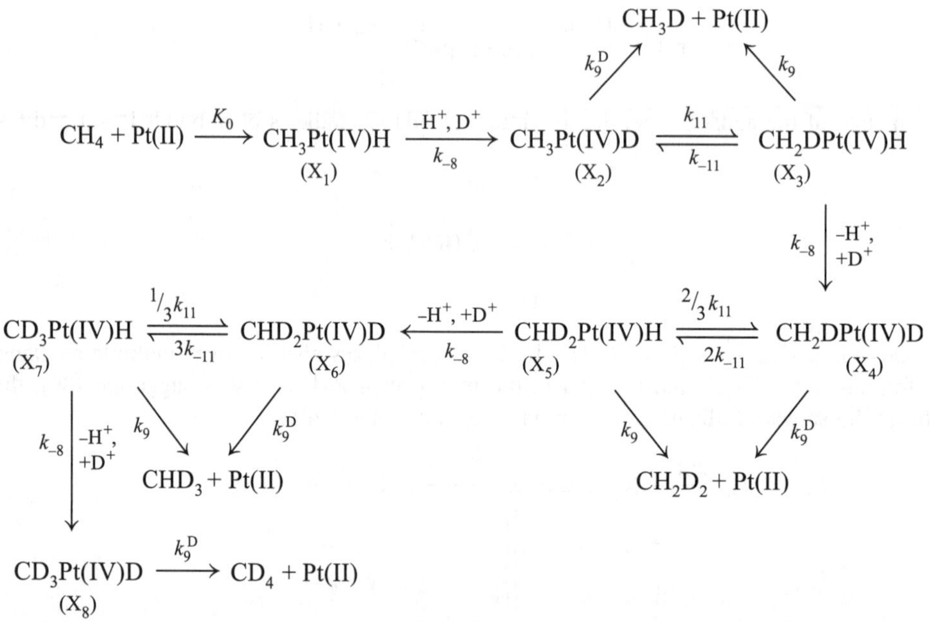

(3.150)

was written in a simplified form by omitting intermediates X_i ($CH_{3-i}D_iPt^{II}$) and considering the process (and analogous steps)

$$CH_3Pt(IV)H \xrightarrow[k_{-8}]{-H^+,\,+D^+} CH_3Pt(IV)D \tag{3.151}$$

as a fast and irreversible reaction with a rate constant of k_{-8}. The proposed scheme (3.150) can be represented by kinetic graph KG19 for a seven-route exchange reaction (Figure 3.19). For the steps of isomerization (intramolecular exchange $X_2 \rightleftharpoons X_3$, $X_4 \rightleftharpoons X_5$, and $X_6 \rightleftharpoons X_7$), the rate constants are given with allowance for the statistical factors relative to the rate of step $X_2 \rightleftharpoons X_3$.

Using the relative yields of deuterated methanes (d_2/d_1, d_3/d_2, and d_4/d_3), it is possible to find three ratios (a, b, and c) of the rate constants of steps in scheme (3.150), for example,

$$\frac{d_4}{d_3} = \frac{k_9^D[X_8]}{k_9[X_7] + k_9^D[X_6]} = \frac{1/3\ ca}{1 + c + 1/3\ a + 3b}, \tag{3.152}$$

where $a = k_{11}k_9^D, c = k_{-8}/k_9$, and $b = 3k_{-11}/k_9$ (note that writing $c = k_9/k_{-8}$ in [89] is probably a misprint).

It should be noted that the values of $k_9/k_{-8} = 0.12$ in scheme (3.148) and $k_9/k_{-8} = 0.07$ for various isotopomers according to the multiple exchange data in scheme (3.150) are rather close and differ but little from the ratios of k_2/k_3 and $k_2^Dk_3^D$ for ethane.

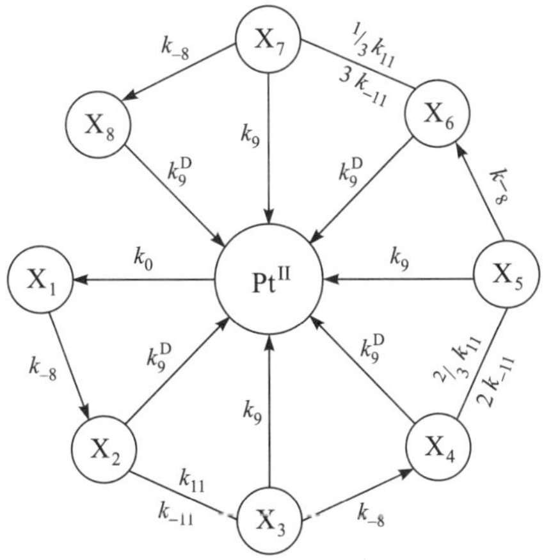

KG19

Figure 3.19 Kinetic graph for mechanism of H/D exchange in CH_4–D_2O–DCl system according to scheme (3.150)

Thus, the use of an isotope label increases the number of routes (two routes are added for each of CH_3D, CH_2D_2, and CHD_3) and thus provides conditions for obtaining additional information concerning the relative reactivity and confirming the proposed scheme. ■

In Example 3.14, a conjugation node useful for the understanding of the reaction mechanism appeared as a result of the route of catalyst destruction. A reaction route not leading to the formation of products and having an overall equation of the $0 = 0$ type (empty route) can also be involved in the appearance of an informative conjugation node.

Example 3.20 [91, 92]

► The kinetics of acetylene hydroiodination

$$C_2H_2 + HI \longrightarrow CH_2{=}CHI \tag{3.153}$$

was studied in the PtI_6^{2-}–NaI–H^+–H_2O system in a closed ideally stirred reactor at a constant electrolyte concentration of 5 M ([NaI] = 2.5 M, $HClO_4$–$NaClO_4$ = 2.5 M) and a temperature of 353 K, whereby the C_2H_2 and vinyl iodide (RI) concentrations in the gaseous phase were measured under conditions of the gas–solution equilibrium by the distribution method as described in [93].

The following reaction mechanism was proposed to explain the experimental dependences of the reaction rate (and RI product yield) on the concentrations of Pt(IV), H^+, and I_2 at $[I^-] = \text{const}$:

$$\begin{aligned}
&Pt^{IV} + C_2H_2 + I^- \underset{1}{\overset{K_1}{\rightleftharpoons}} \underset{(X_1)}{Pt^{IV}CH{=}CHI} \underset{2}{\overset{2I^-}{\rightleftharpoons}} \underset{(X_2)}{Pt^{II}CH{=}CHI} + I_2;\\
&X_2 \xrightarrow[3]{H^+} Pt^{II} + CH_2{=}CHI\\
&X_2 \xrightarrow[4]{} Pt^{II} + C_2H_2 + I^- \;;\\
&Pt(II) + I_2 \xrightarrow[5]{} Pt(IV) + 2I^-.
\end{aligned} \tag{3.154}$$

This scheme comprises two linearly independent routes, $\nu^{(1)} = [1\ 1\ 1\ 0\ 1]$ and $\nu^{(2)} = [1\ 1\ 0\ 1\ 1]$, with the following overall equations:

$$(1)\quad C_2H_2 + HI = CH_2{=}CHI;$$
$$(2)\quad 0 = 0.$$

The first-order reaction rate constant for route (1) is given by the following formula:

$$k_{eff} = \frac{K_1 k_2 k_3 [H^+][Pt^{IV}]}{k_{-2}[I_2] + k_3[H^+] + k_4}, \tag{3.155}$$

which includes the rate constant k_4 for the empty route. This expression describes the entire set of data for $[Pt^{IV}]_\Sigma \cong [Pt^{IV}]$. In the experiments performed without additives of I_2, where $k_{-2}[I_2] \ll k_3[H^+]$, the dependence of k_{eff} on $[H^+]$ appears only due to the presence of step **4** corresponding to the empty route. The possibility of intermediate X_1 formation was confirmed in the studies of reactions with acetylene derivatives [94, 95] and the reaction of 1,4-diiodobuta-1,3-diene formation [96, 97]. ■

Now let us consider some examples of using the method of competitive reactions, including those with isotopomers, for the elucidation of reaction mechanisms and discrimination of hypotheses in the catalysis with metal complexes. This method was successfully applied, in particular, to the discrimination between mechanisms proposed for the metathesis of olefins [98, 99], where an analysis of the product distribution at $t \to 0$ provided convincing arguments for the carbene mechanism of metathesis.

Example 3.21 [98, 99]

▶ The reaction of double cross-metathesis

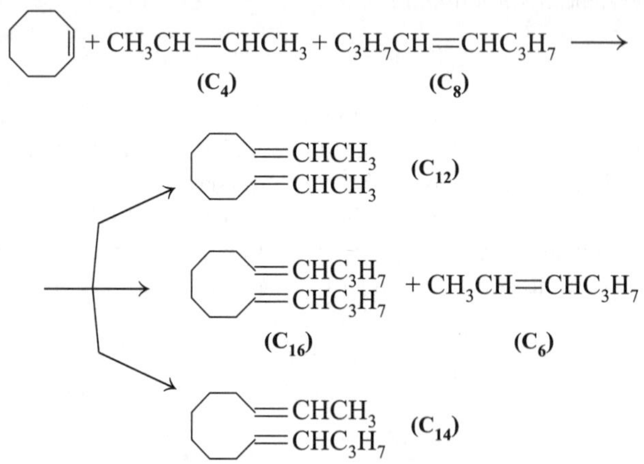

(3.156)

was studied in a chlorobenzene solution of $Mo(PPh_3)_2Cl_2(NO)_2$ and $MeAlCl_2$ at 0 °C and 25 °C by measuring the temporal variation of the concentrations of all reactants and products in a closed reactor. The values of $[C_{14}]/[C_{12}]$ and $[C_{14}]/[C_{16}]$ were extrapolated to $t \to 0$ [i.e., to the conditions of cyclooctene (Cy) conversion $\to 0$]. Since the indicated relative selectivities were significantly different from zero, it was concluded that product $\mathbf{C}_{14}$ (that appeared in parallel to $\mathbf{C}_{12}$ and $\mathbf{C}_{16}$) could not be formed via consecutive reactions

$$
\begin{aligned}
&\text{(I)} \quad \mathbf{C}_4 + \mathbf{C}_8 \rightarrow 2\mathbf{C}_6, \mathrm{Cy} + \mathbf{C}_6 \rightarrow \mathbf{C}_{14};\\
&\text{(II)} \quad \mathbf{C}_4 + \mathrm{Cy} \rightarrow \mathbf{C}_{12}, \mathbf{C}_{12} + \mathbf{C}_8 = \mathbf{C}_{14} + \mathbf{C}_6;\\
&\text{(III)} \quad \mathbf{C}_4 + \mathrm{Cy} \rightarrow \mathbf{C}_{12}, \mathbf{C}_8 + \mathrm{Cy} \rightarrow \mathbf{C}_{16}, \mathbf{C}_{12} + \mathbf{C}_{16} \rightarrow 2\mathbf{C}_{14}
\end{aligned}
\tag{3.157}
$$

within the framework of a "pairwise" (cyclobutane) mechanism, since the initial concentrations of intermediate products $\mathbf{C}_6$, $\mathbf{C}_{12}$, and $\mathbf{C}_{16}$ are very small compared to the concentrations of reactants.

The obtained results agree well with the following kinetic scheme, in which products $\mathbf{C}_{12}$, $\mathbf{C}_{14}$, $\mathbf{C}_{16}$, and $\mathbf{C}_6$ are formed via competitive reactions of the active carbene complexes

$$
\begin{aligned}
&CH_3CH{=}M(C),\ CH_3CH{=}[CH(CH_2)_6CH]_{n+1} = M(A_n),\\
&C_3H_7CH{=}M(D),\ C_3H_7CH{=}[CH(CH_2)_6CH]_{n+1} = M(B_n)
\end{aligned}
$$

with three olefins including Cy, 2-butene (Bu), and 4-octene (Oc):

$$
\begin{aligned}
&\mathrm{C} + \mathrm{Cy} \xrightarrow{k_3} \mathrm{A}_0 \\
&\mathrm{D} + \mathrm{Cy} \xrightarrow{k_4} \mathrm{B}_0 \\
&\mathrm{A}_0 + \mathrm{Cy} \xrightarrow{k_7} \mathrm{A}_1 \\
&\mathrm{B}_0 + \mathrm{Cy} \xrightarrow{k_7} \mathrm{B}_1 \\
\\
&\mathrm{A}_0 + \mathrm{Bu} \xrightarrow{k_1} \mathbf{C}_{12} + \mathrm{C} \\
&\mathrm{A}_0 + \mathrm{Oc} \xrightarrow{k_2} \mathbf{C}_{14} + \mathrm{D} \\
&\mathrm{B}_0 + \mathrm{Bu} \xrightarrow{k_1} \mathbf{C}_{14} + \mathrm{C} \\
&\mathrm{B}_0 + \mathrm{Oc} \xrightarrow{k_2} \mathbf{C}_{16} + \mathrm{D} \\
\\
&\mathrm{A}_1 + \mathrm{Bu} \xrightarrow{k_1} \mathbf{C}_{12+8} + \mathrm{C} \\
&\mathrm{A}_1 + \mathrm{Oc} \xrightarrow{k_2} \mathbf{C}_{14+8} + \mathrm{D} \\
&\mathrm{B}_1 + \mathrm{Bu} \xrightarrow{k_1} \mathbf{C}_{14+8} + \mathrm{C} \\
&\mathrm{B}_1 + \mathrm{Oc} \xrightarrow{k_2} \mathbf{C}_{16+8} + \mathrm{D} \\
\\
&\mathrm{C} + \mathrm{Oc} \xrightarrow{k_5} \mathbf{C}_6 + \mathrm{D} \\
&\mathrm{D} + \mathrm{Bu} \xrightarrow{k_6} \mathbf{C}_6 + \mathrm{C}
\end{aligned}
\qquad
\begin{aligned}
&\text{General case}\\
&\left\{\begin{aligned}
&\mathrm{A}_n + \mathrm{Cy} \xrightarrow{k_7} \mathrm{A}_{n+1}\\
&\mathrm{B}_n + \mathrm{Cy} \xrightarrow{k_7} \mathrm{B}_{n+1}
\end{aligned}\right\}\\
&\left\{\begin{aligned}
&\mathrm{A}_n + \mathrm{Bu} \xrightarrow{k_1} \mathbf{C}_{12+8n} + \mathrm{C}\\
&\mathrm{A}_n + \mathrm{Oc} \xrightarrow{k_2} \mathbf{C}_{14+8n} + \mathrm{D}\\
&\mathrm{B}_n + \mathrm{Bu} \xrightarrow{k_1} \mathbf{C}_{14+8n} + \mathrm{C}\\
&\mathrm{B}_n + \mathrm{Oc} \xrightarrow{k_2} \mathbf{C}_{16+8n} + \mathrm{D}
\end{aligned}\right\}
\end{aligned}
\tag{3.158}
$$

Since the obtained olefins contain several Cy molecules ($\mathbf{C}_{12+8n}$, $\mathbf{C}_{14+8n}$, and $\mathbf{C}_{16+8n}$) scheme (3.158) was restricted to the first terms with $n = 0$ and $n = 1$. This scheme with indicated steps of conversion for A_n and B_n corresponds to graph KG20 (Figure 3.20). The laws of kinetics and relative selectivities in the

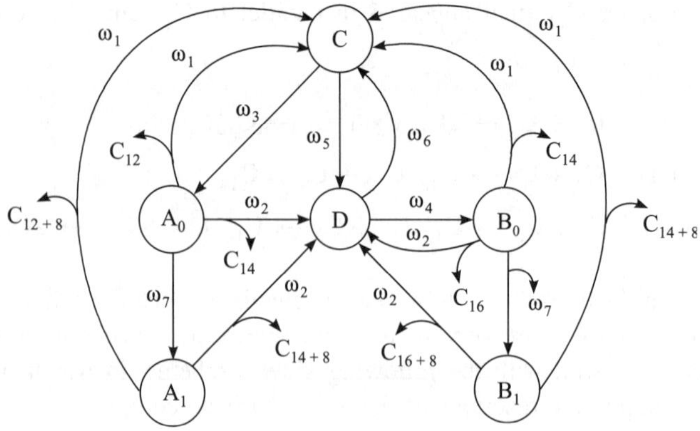

KG20

Figure 3.20 Kinetic graph for C_{m+8n} ($m=12$, 14, 16; $n=0$, 1) formation by means of double cross metathesis according to scheme (3.158)

reaction of double cross-metathesis are determined by the following multiply-repeated conjugation nodes (3.159) and (3.160) at A_n and B_n (at vertices A_0, A_1, and B_0, B_1 in KG20):

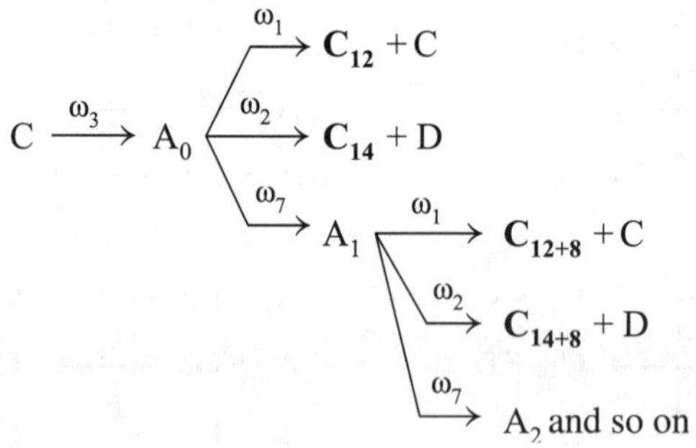

(3.159)

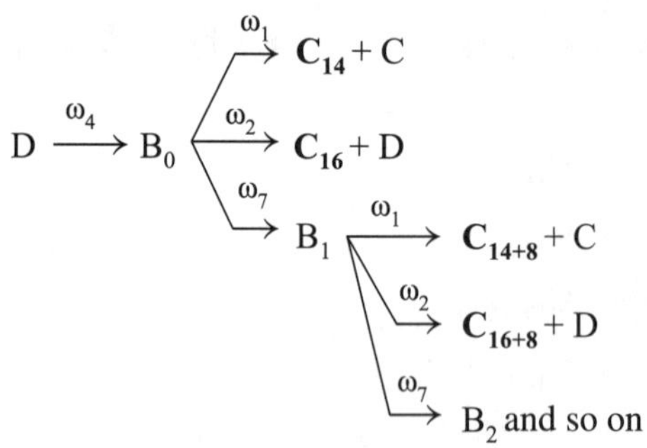

(3.160)

In an analysis of scheme (3.158), it was assumed that the rates of conversion for A_n and B_n are independent of n, but depend on the particular olefin (Bu vs. Oc and $\mathbf{C}_4$ vs. $\mathbf{C}_8$). For the conjugation node at A_1 in (3.159), it follows that

$$[A_1] = \frac{\omega_7[A_0]}{\omega_1 + \omega_2 + \omega_7}; \quad \frac{A_1}{A_0} = \frac{\omega_7}{\omega_1 + \omega_2 + \omega_7} = k_R,$$

so that

$$\frac{d[\mathbf{C_{12}}]}{d[\mathbf{C_{12+8}}]} = \frac{[A_0]}{A_1} = 1 + \frac{k_1[\text{Bu}]}{k_7[\text{Cy}]} + \frac{k_2[\text{Oc}]}{k_7[\text{Cy}]}. \tag{3.161}$$

Taking into account the relations

$$\frac{[A_1]}{[A_0]} = \frac{[A_n]}{[A_{n-1}]} = k_R, \quad \frac{[B_1]}{[B_0]} = k_R,$$

and the quasi-steady-state condition, we have $[A_n] = k_R^n[A_0]$ and $[B_n] = k_R^n[B_0]$, which implies that

$$\sum_{n=0}^{\infty}([A_n] + [B_n]) = ([A_0] + [B_0])\frac{1}{1 - k_R}.$$

Another assumption consists in that the ratio of the reaction rates for one carbene with different olefins (W_2/W_1) multiplied by the ratio of the reaction rates for one carbene with different olefins (W_3/W_4) is equal to the ratio of the reaction rates for different carbenes with different olefins:

$$\frac{W_2}{W_1} \cdot \frac{W_3}{W_4} = \frac{W_5}{W_6} \quad \text{or} \quad \frac{k_2}{k_1} \cdot \frac{k_3}{k_4} = \frac{k_5}{k_6},$$

which provides the following expressions for the $[A_0]/[B_0]$ ratio:

$$\frac{[A_0]}{[B_0]} = \frac{k_1[\text{Bu}]}{k_2[\text{Oc}]} = \frac{\omega_1}{\omega_2}. \tag{3.162}$$

Then, the product yields at $t \to 0$ are as follows:

$$\alpha_1 = \left.\frac{[\mathbf{C_{14}}]}{[\mathbf{C_{12}}]}\right|_{t \to 0} = \frac{\omega_2[A_0] + \omega_1[B_0]}{\omega_1[A_0]} = \frac{\omega_2}{\omega_1} + \frac{[B_0]}{[A_0]} = 2\frac{\omega_2}{\omega_1}; \tag{3.163}$$

$$\alpha_2 = \left.\frac{[\mathbf{C_{14}}]}{[\mathbf{C_{16}}]}\right|_{t \to 0} = \frac{\omega_2[A_0] + \omega_1[B_0]}{\omega_2[B_0]} = \frac{\omega_1}{\omega_2} + \frac{[A_0]}{[B_0]} = 2\frac{\omega_1}{\omega_2}. \tag{3.164}$$

Analogous expressions are also obtained for all homologs ($\mathbf{C}_{12}$, $\mathbf{C}_{14}$, and $\mathbf{C}_{16}$):

$$\alpha_1 = \frac{[\mathbf{C_{14+8n}}]}{[\mathbf{C_{12+8n}}]}; \alpha_2 = \frac{[\mathbf{C_{14+8n}}]}{[\mathbf{C_{16+8n}}]}.$$

Using relations (3.163) and (3.164) and taking into account the above assumptions, we conclude that

$$\alpha_1 \cdot \alpha_2 = 4.$$

The average experimental value of the $\alpha_1 \cdot \alpha_2$, product obtained by extrapolating α_1, α_2, or $\alpha_1 \cdot \alpha_2$ to $t \to 0$ amounted to 4.05 ± 0.05 and, in agreement with the theory, was independent of the time. This result confirmed the carbene hypothesis. Processing of the experimental data within the framework of various modifications of the pairwise mechanism allowed the other hypotheses to be fully rejected. Using the equations for α_1 and α_2, it is possible to determine the ratio of k_2/k_1, while Eq. (3.161) yields the relative rate constants k_1/k_7 and k_2/k_7. The values of k_1/k_2 for *trans* and *cis* olefins (Bu, Oc) amount to 3.08 ± 0.16 and 6.46 ± 0.87, respectively.

The simple cross-metathesis of cycloolefins with terminal olefins can lead, in view of the different stabilities of carbene species, to the predominant formation of one rather than three products according to the carbene mechanism:

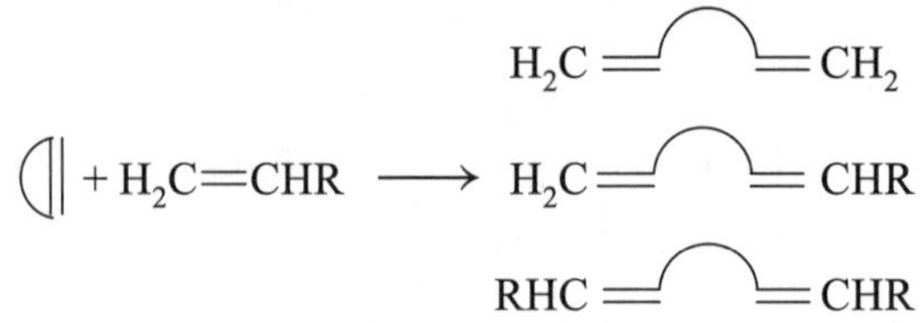

For example, the ratio of k/k' in the reaction

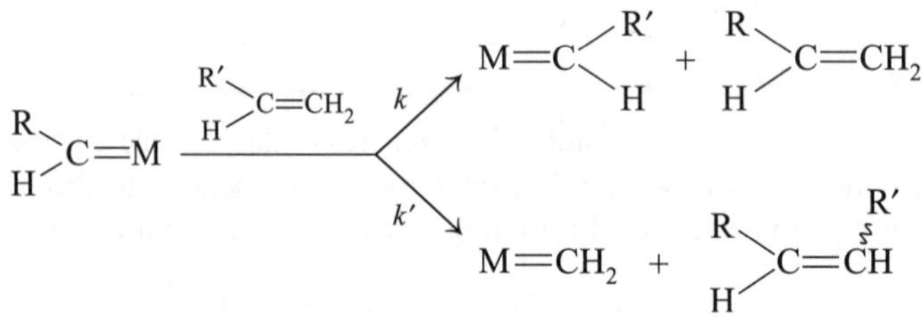

can reach up to 100 for tungsten catalysts [99], which is related to a higher stability of the carbene complex with $=$ CHR′ alkylidene group. ■

The kinetic features of conjugation nodes of the (d) type (Figure 3.16) with all irreversible steps (and the corresponding kinetic graphs KG11 (scheme (3.43), Figure 3.7) and KG12 (scheme (3.46), Figure 3.7) were also effectively used to study the mechanism of nonoxidative arylation of olefins by arylhalides (Heck reaction),

$$\mathrm{ArI + RCH{=}CH_2 + NR_3 \longrightarrow \underset{(P_1)}{ArCH{=}CHR} + \underset{(P_2)}{ArC(R){=}CH_2} + HNR_3{}^{+}I^{-}} \qquad (3.165)$$

by the method of competitive reactions [100, 101]. The commonly accepted mechanistic scheme of reactions (3.165) in solutions of palladium phosphine complexes includes two routes for the formation of regioisomers (P_1 and P_2) with five irreversible steps (with neglect of the steps of formation and loss of the active catalyst species):

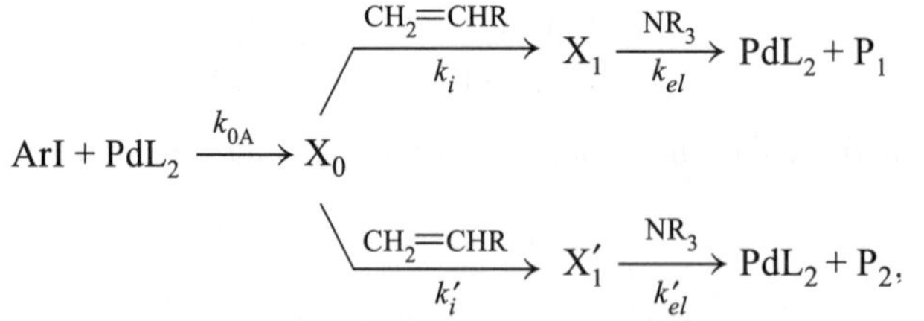

(3.166)

where k_i and k_i' are the rate constants of olefin insertion into the Pd–Ar bond and k_{el} and k_{el}' are the rate constants of the Pd–H bond β-elimination or deprotonation with the product formation.

The ratio and sum of the rates of formation of regioisomers in scheme (3.166) are, respectively

$$\frac{r_{P_1}}{r_{P_2}} = \frac{k_i}{k_i'}; \quad r_\Sigma = r_{P_1} + r_{P_2} = k_{0A}[\mathrm{ArI}][\mathrm{Pd}]_\Sigma / F_{\mathrm{Pd}}.$$

For $[Pd]_\Sigma \cong [PdL_2] + [X_1] + [X_1']$, the complexity function can be expressed as

$$F_{Pd} = 1 + \frac{k_i k_{0A}[ArI]}{k_{el}(k_i + k_i')} + \frac{k_i' k_{0A}[ArI]}{k_{el}'(k_i + k_i')}, \tag{3.167}$$

or as $F_{Pd} = 1 + (K_1 + K_{2)}[ArI]$, where

$$K_1 = \frac{k_i k_{0A}}{k_{el}(k_i + k_i')}, \quad K_2 = \frac{k_i' k_{0A}}{k_{el}'(k_i + k_i')}.$$

For $[Pd]_\Sigma \cong [X_1] + [X_1']$, we have

$$r_\Sigma = r_{P_1} + r_{P_2} = \frac{[Pd]_\Sigma k_{el} k_{el}'(k_i + k_i')}{k_i k_{el}' + k_i' k_{el}} = \frac{[Pd]_\Sigma k_{el} k_{el}'(L + 1)}{L k_{el}' + k_{el}}, \tag{3.168}$$

where $L = k_i/k_i' = r_{P_1}/r_{P_2}$.

Example 3.22 [100, 101]

► In order to obtain information concerning slow steps of the arylation reaction (3.165) and the structure of material balance for this process, the kinetics of the Heck reaction was studied for C_2H_4, C_2D_4, and a styrene/styrene-d_8 (1:1) mixture. For the scheme

$$ArI + PdL_2 \xrightarrow[k_{0A}]{} X_0 \xrightarrow[k_i^H]{C_2H_4} X_1 \xrightarrow[k_{el}^H]{NR_3} PdL_2 + P_{1H} \tag{3.169}$$

with quasi-steady-state steps for X_0 and X_1, the product formation kinetics is as follows:

$$r_{P_{1H}} = \frac{k_{0A}[ArI][Pd]_\Sigma}{1 + \dfrac{k_{0A}[ArI]}{k_i^H[C_2H_4]} + \dfrac{k_{0A}[ArI]}{k_{el}^H[C_2H_4]}}. \tag{3.170}$$

Accordingly,

$$r_{P_{1H}} = k_{0A}[ArI][Pd]_\Sigma$$

for $[Pd]_\Sigma \cong [PdL_2]$,

$$r_{P_{1H}} = k_i^H[C_2H_4][Pd]_\Sigma$$

for $[Pd]_\Sigma \cong [X_0]$, and

$$r_{P_{1H}} = k_{el}^H[Pd]_\Sigma$$

for $[Pd]_\Sigma \cong [X_1]$.

The kinetic isotope effect (KIE) of $r_{P_{1H}}/r_{P_{1D}} = 3.2$ observed on the passage from C_2H_4 to deuteroethylene is indicative of a significant contribution of the β-elimination (or deprotonation) step to the reaction rate, that is, of the proximity of $[Pd]_\Sigma$ and $[X_1]$. Therefore, the slow stage in the process under consideration must be the last step, for which

$$k_i^H[C_2H_4][X_0] \gg k_{el}^H[X_1]; \quad k_i^D[C_2D_4][X_0'] \gg k_{el}^D[X_0'].$$

These assumptions were confirmed in the experiments with an equimolar mixture of styrene (OI^H) and styrene-d_8 (OI^D). Under conditions of high regioselectivity, the mechanistic scheme for two isotopomers appears, by analogy with (3.166), as follows:

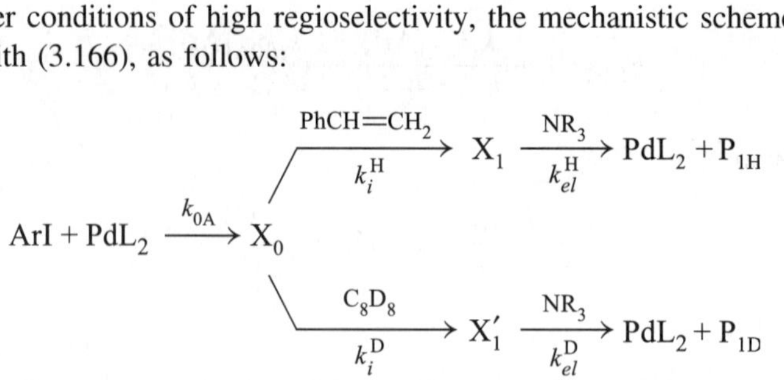

(3.171)

It was experimentally established that the product formation rates $r_{P_{1H}}$ and $r_{P_{1D}}$ are approximately the same, while the total rate r_Σ is about half that observed in the absence of deuterostyrene. The equality of $r_{P_{1H}}$ and $r_{P_{1D}}$ values under quasi-steady-state conditions is evidence for the equality of $k_{el}^H[X_1] = k_{el}^D[X_1']$, in which the distribution of $[Pd]_\Sigma$ between X_1 and X_1' is determined by the KIE magnitude:

$$KIE = \frac{k_{el}^H}{k_{el}^D} = \frac{[X_1']}{[X_1]}.$$

For equal concentrations of $[C_8H_8] = [C_8D_8]$, by analogy with Eq. (3.168) we have

$$r_\Sigma = r_{P_{1H}} + r_{P_{1D}} = \frac{[Pd]_\Sigma k_{el}^H k_{el}^D (L+1)}{L k_{el}^D + k_{el}^H}, \tag{3.172}$$

where $L = k_i^H / k_i^D$.

For the reaction with nondeuterated olefin at $[Pd]_\Sigma \cong [X_1]$, the rate is

$$r'_{P_{1H}} = k_{el}^H [Pd]_\Sigma.$$

Then, for the same rate of olefin insertion ($k_i^H = k_i^D, L = 1$), we obtain

$$\frac{r'_{P_{1H}}}{r_\Sigma} = \frac{k_{el}^D + k_{el}^H}{2k_{el}^D} = 1/2\left(\frac{k_{el}^H}{k_{el}^D} + 1\right). \tag{3.173}$$

For $r_{P_{1H}}/r_\Sigma \cong 2$ (experiment), we obtain $KIE = k_{el}^H/k_{el}^D = 3$, which is close to the value of 3.2 known for the reactions with C_2H_4 and C_2D_4.

In the general case of competitive reactions with a mixture of olefins (S_1 and S_2) at $L \neq 1$, we obtain

$$N = \frac{r'_{P_1}}{r_{P_1} + r_{P_2}} = \frac{L + k_{el}^{S_1}/k_{el}^{S_2}}{L+1}; \tag{3.174}$$

$$\frac{k_{el}^{S_1}}{k_{el}^{S_2}} = N(L+1) - L, \tag{3.175}$$

where L is determined (for the given scheme) from the r_{P_1}/r_{P_2} ratio.

Thus, the method of competitive reactions according to the mechanistic scheme (3.171) with the indicated ratios of step rates yields $r_{P_{1H}}/r_{P_{1D}} = 1$ for the last step with a *KIE* value close to 3. In this case, only a comparison of the sum of product formation rates in a competitive reaction to the rate of conversion of one isotopomer can be used for determining the *KIE* value and other relative rate constants.

For the adopted approximations, schemes (3.169) and (3.171) lead to the zero-order reaction rate with respect to olefins. At the same time, the observed reaction orders are ~0.3 for ethylene and ~0.6 for styrene [101]. This circumstance is indicative of a more complicated character of the real scheme (with the formation and termination steps) and the expediency of the KIE analysis with allowance for the participation of olefins in the β-elimination or deprotonation steps [100, 101]. In this case, at least two conjugation nodes appear in the scheme if each olefin forms a single isotopomer:

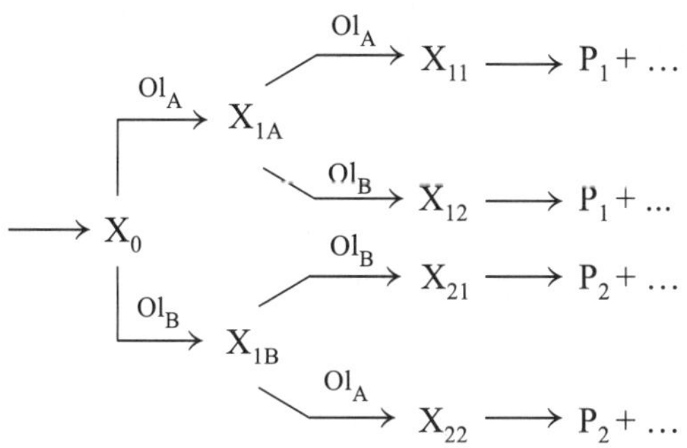

The *KIE* values and data on the nature of a slow step can also be obtained in experiments where the competition is due to the presence of H and D atoms in the same molecule (e.g., (*cis*- versus *trans*-$C_2H_2D_2$) rather than different isotopomers or olefins. In this case, the conjugation node for the formation of styrene isotopomers would be related to intermediate X_1 in scheme (3.169),

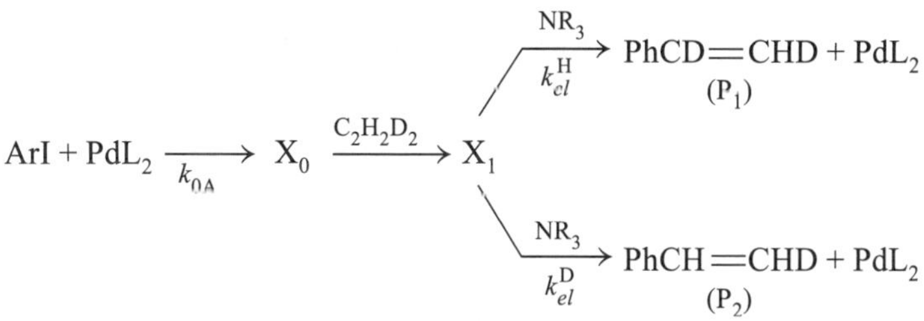

so that $r_{P_1}/r_{P_2} = k_{el}^H/k_{el}^D$. This approach was used [102] to clarify details of the mechanism of ethylene oxidation to acetaldehyde (see Example 2.22). The value of *KIE* ~1.9 does not contradict the assumption that the hydroxypalladation of ethylene is a slow step and the mechanism of σ-palladoethanol decomposition involves the hydride ion transfer from β to α position, probably, with the participation of a Pd atom:

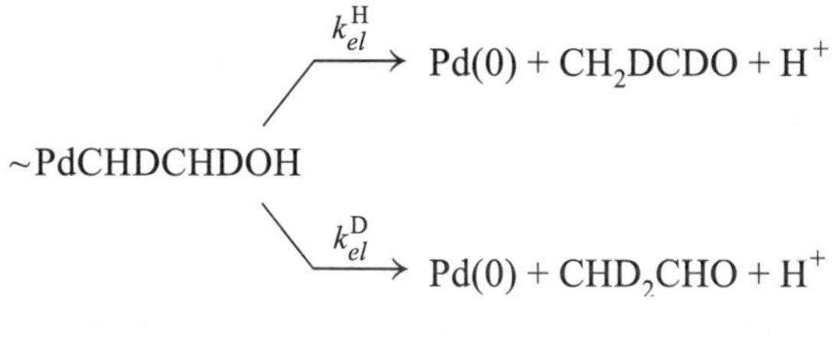

■

Competitive reactions can also be useful for studying mechanisms that contain nonlinear steps of intermediate interactions, that is, conjugation nodes with nonlinear steps. Two intermediates, X_i and X_j, which interact in steps

$$2X_i \longrightarrow P_1 + \ldots;$$
$$2X_j \longrightarrow P_2 + \ldots;$$
$$X_i + X_j \longrightarrow P_3 + \ldots;$$

can be formed from the same substrate during the parallel synthesis of regioisomers. For example, the oxidative dimerization of thiophene (T) in a reaction with c $Pd(OAc)_2$ in acetic acid leads to the formation of three dithiophene regioisomers [103]:

$$T + Pd(OAc)_2 \rightleftharpoons \underset{(X_1)}{\pi\text{-}(T)Pd(OAc)_2};$$

$$X_1 \xrightarrow[-AcOH]{k_1} \underset{(X_2)}{R^{(2)}PdOAc}; \qquad X_1 \xrightarrow[-AcOH]{k_2} \underset{(X_3)}{R^{(3)}PdOAc}$$

$$X_2 \xrightarrow[k_3]{X_2} \underset{(P_1)}{R^{(2)}\text{—}R^{(2)}} + Pd(0) + Pd(II); \qquad X_2 \xrightarrow[X_3]{k_5} \underset{(P_3)}{R^{(2)}\text{—}R^{(3)}} + Pd(0) + Pd(II) \tag{3.176}$$

$$X_3 \xrightarrow[k_4]{X_3} \underset{(P_2)}{R^{(3)}\text{—}R^{(3)}} + Pd(0) + Pd(II), \qquad X_3 \xrightarrow[X_2]{k_5} \underset{(P_3)}{R^{(2)}\text{—}R^{(3)}} + Pd(0) + Pd(II)$$

where $R^{(2)}$ and $R^{(3)}$ are 2- and 3-thienyl, respectively:

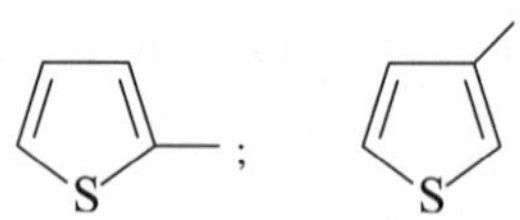

With the conjugation nodes at X_2 and X_3, the initial or current rates of formation of products P_1, P_2, and P_3 are interrelated as follows:

$$r_{P_1} r_{P_2} = \frac{k_3 k_4}{k_5^2} r_{P_3}^2. \tag{3.177}$$

In the case of two substrates with close properties (homologs), intermediates X_2 and X_3 can be formed from different substrates.

Example 3.23 [104, 105]

▶ The mechanism of the oxidative dimerization of terminal alkynes (Glyazer–Zalkind reaction [10, 106]) was studied in a closed ideally stirred reactor in the LiCl (5 M)–CuCl (0–0.5 M)–$CuCl_2$ (0–0.5 M)–H_2O system at 30–60 °C:

$$2RC{\equiv}CH + 2CuCl_2 \xrightarrow{Cu(I)} RC{\equiv}C{-}C{\equiv}CR + 2CuCl. \tag{3.178}$$

The initial rates of formation of diacetylene (DA) and dimethyldiacetylene (DMDA) were studied (for R = H, Me) as functions of the partial pressures of acetylene (P_A) and methylacetylene (P_{MA}), $[CuCl_2]_\Sigma$, $[CuCl]_\Sigma$, $[H_3O^+]$, concentration of Cl^- ($[LiClO_4]$ + [LiCl] = 5 M), and the nature of cation in MCl [104]. The concentrations of free copper ions (Cu^+ and Cu^{2+}) and active chloride complexes under the chosen conditions are proportional to $[CuCl]_\Sigma$ and $[CuCl_2]_\Sigma$, respectively. The rates of DA and DMDA formation for $[LiCl]_\Sigma$ = const are described by the same equation:

$$r_i = \frac{k_i [CuCl]_\Sigma^2 [CuCl_2]_\Sigma P_i}{[H_3O^+]}. \tag{3.179}$$

Here, the ratio of rate constants at 40 °C is $k_A/k_{MA} = 4.4$. A mechanistic reaction scheme corresponding to Eq. (3.179) was proposed, which can be written (for the sake of simplicity, in terms of CuCl and $CuCl_2$) as follows:

$$\begin{aligned} RC{\equiv}CH + CuCl &\underset{1}{\rightleftharpoons} \underset{(X_1)}{RC{\equiv}CCu} + H^+ + Cl^-;\\ X_1 + CuCl_2 &\underset{2}{\rightleftharpoons} \underset{(X_2)}{RC{\equiv}CCu_2Cl_2};\\ X_2 + CuCl &\xrightarrow[3]{} X_3;\\ 2X_3 &\xrightarrow[4]{} RC{\equiv}CC{\equiv}CR + 6CuCl. \end{aligned} \tag{3.180}$$

From the above kinetic equation (3.179), it follows that (i) the process has a limiting step and (ii) a transition state in this limiting step involves RC $\equiv C^-$, 2Cu(I), and Cu(II): $[(RC{\equiv}C^-)(Cu^I)_2Cu^{II}]^{\neq}$. The structure of intermediate X_3 and the mechanism of its subsequent transformation are unknown. For example, step **3** in scheme (3.180) can be replaced by the following reaction:

$$X_2 + CuCl \xrightarrow{3'} X_3' + 2CuCl,$$

where $X_3' = RC{\equiv}CCu^{II}Cl$, while the coupling of two X_3 (or X_3') can be replaced by the alkyne insertion into the C–Cu bond that can be described by the reaction

$$\underset{(X_3)}{RC{\equiv}CCu_3Cl_3} + RC{\equiv}CH \longrightarrow \underset{(X_4)}{RC{\equiv}C{-}CH{=}C(R)Cu_3Cl_3}. \tag{3.181}$$

The formation of free radicals RC≡C• from X_3 or X_3' via reaction

$$RC{\equiv}CH + Cu\,(II) + B \longrightarrow RC{\equiv}C\bullet + Cu\,(I) + BH^+.$$

and their subsequent dimerization are rejected because of a very high endothermicity ($\Delta H_{298} \cong$ 250 kJ/mole) of the process under consideration [10, 107]. Thus, the pathways of $X_3(X_3')$ conversion behind the limiting step can be traced by creating conjugation nodes with competitive reactions that involve a mixture of two substrates, acetylene (A) and methylacethylene (MA) [105]. In this case, both mechanism (3.180) and reaction (3.181) must yield DA, DMDA, and methyldiacethylene (MDA). In the former case, we have

$$\begin{array}{ll} \xrightarrow[1]{r_A} X_{3(A)}; & 2X_{3(A)} \xrightarrow{k_3} \underset{(P_1)}{DA} + \ldots; \\ \xrightarrow[2]{r_{MA}} X_{3(MA)}; & 2X_{3(MA)} \xrightarrow{k_4} \underset{(P_2)}{DMDA} + \ldots; \\ & X_{3(A)} + X_{3(MA)} \xrightarrow{k_5} \underset{(P_3)}{MDA} + \ldots. \end{array} \qquad (3.182)$$

There are two kinetic criteria following from scheme (3.182), which turn out to be consistent with experiment. The first is the relation

$$r_{MDA}^2 = k'(r_{DA} \cdot r_{DMDA}), \qquad (3.183)$$

where $k' = k_5^2/(k_3k_4)$. The second is the system of equations that follows from the quasi-steady-state condition with respect to $X_{3(A)}$ and $X_{3(MA)}$,

$$r_A = 2r_{DA} + r_{MDA}; \qquad (3.184)$$

$$r_{MA} = 2r_{DMDA} + r_{MDA}, \qquad (3.185)$$

where r_A and r_{MA} are the rates of formation of $X_{3(A)}$ and $X_{3(MA)}$, respectively, as described by Eq. (3.178) for r_{DA} and r_{DMDA} in reactions with the corresponding individual substrates. Equations (3.184), (3.185) and (3.179) lead to the following formula:

$$\varphi = \frac{(2r_i + r_{MDA})[H_3O^+]}{[CuCl_2]_\Sigma[CuCl]_\Sigma^2} = k_{0i}P_k, \qquad (3.186)$$

where $i = DA, k = A$; or $i = MA$, $k = MA$. The linear dependences of φ on P_A and P_{MA} describe the entire body of experimental data.

In the case of mechanism (3.181) with reactions

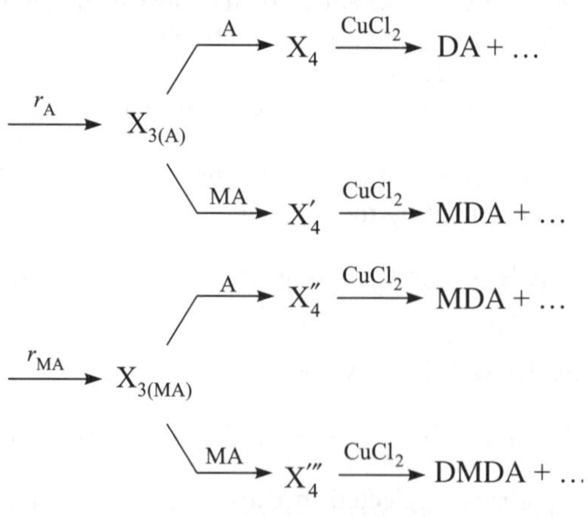

(3.187)

we obtain more complicated dependences that significantly differ from formulas (3.183) and (3.186). In particular, the quantity φ becomes a complex function of both P_A and P_{MA}.

The selectivity of the process with respect to MDA in the case of relation (3.183) is expressed as follows:

$$S_{\text{MDA}} = \frac{\sqrt{k'}}{\sqrt{k'} + \left(\dfrac{r_{\text{DMDA}}}{r_{\text{DA}}}\right)^{1/2} + \left(\dfrac{r_{\text{DA}}}{r_{\text{DMDA}}}\right)^{1/2}}, \tag{3.188}$$

where $\sqrt{k'} = k_5/\sqrt{k_3 k_4}$. In a temperature interval of $25 - 40\,°\text{C}$, this parameter is $\sqrt{k}' = 1.92 \pm 0.02$, and the selectivity with respect to MDA amounts to $\approx 30\,\%$ for a maximum value of $50\,\%$ that is reached under conditions where the selectivity with respect to the sum of products reaches $95\,\%$ [105]. ■

As was noted above, the use of a simple method of competitive reactions for determining the KIE and evaluating the relative reactivities of various centers (bonds) in a molecule or various pairs of substrates is inexpedient when several forms of a catalytically active center or a reactant occurring in solution are capable of interacting with one of the substrates (trivial multi-route character). In this situation, it is necessary to provide a detailed description of all intermediate states in solution and determine the rate constants for the interaction of a substrate with reactive species. It is even more difficult to study systems where the equilibrium between various forms of species is absent or various active species are irreversibly converted from one form to another in the course of reaction.

Example 3.24 [108–110]

► As is known, $KMnO_4$ in an acid medium oxidizes alkylbenzenes at the alkyl groups (with the formation of aromatic acids) and at the aromatic ring (with destruction of the ring). The observed second-order rate constants k_2 in the equation

$$-\frac{d\ln[\text{ArH}]}{dt} = k_2[\text{KMnO}_4] \tag{3.189}$$

were used to evaluate the KIE in the $KMnO_4$–$HClO_4$–H_2O system. It was established that the ratio $k_2^{\text{H}}/k_2^{\text{D}}$ for the attack on the aromatic ring is independent of the temperature and $[HClO_4]$ and amounts to unity (C_6H_6/C_6D_6, $C_6H_5CH_3/C_6D_5CH_3$). For the attack on the alkyl group, the ratio $k_2^{\text{H}}/k_2^{\text{D}}$ differs from unity and significantly decreases with increasing $[HClO_4]$:

(i) In the $C_6H_5CH_3/C_6H_5CD_3$ system at 70 °C, the *KIE* value drops from 13.6 to 1.45 when $[HClO_4]$ grows from 0 to 1.5 M; at 25 °C, the *KIE* value decreases from 8.4 to 1.3 when $[HClO_4]$ increases from 0.1 to 3 M;
(ii) At 70 °C and pH 5–7, the *KIE* value is 11.3 for *m*-xylene/*m*-xylene-d_{10} and 16.8 for *p*-xylene/*p*-xylene-d_{10}; in 0.5 M $HClO_4$, the *KIE* decreases to 1.18 and 1.23, respectively.

Investigations of the state of MnO_4^- in acid solutions and the dependence of k_2 on the acidity of the medium led to the following scheme of equilibria and conversions of ArH:

$$\begin{array}{ccccccc}
\text{MnO}_4^- & \underset{K_a}{\overset{\text{H}^+}{\rightleftharpoons}} & \text{HMnO}_4 & \underset{K_2}{\overset{\text{H}^+}{\rightleftharpoons}} & \text{H}_2\text{MnO}_4^+ & \underset{K_3}{\overset{-\text{H}_2\text{O}}{\rightleftharpoons}} & \text{MnO}_3^+ \\
k_-\downarrow \text{ArH} & & k_0\downarrow \text{ArH} & & & & k_+\downarrow \text{ArH} \\
\text{P}_1 & & \text{P}_2 & & & & \text{P}_3
\end{array} \tag{3.190}$$

where

$$K_a = \frac{a_{H^+}a_{MnO_4^-}}{a_{HMnO_4}} = \frac{C_{MnO_4^-}}{C_{HMnO_4}}h_-; \quad (3.191)$$

$$K_2 = \frac{a_{H^+}a_{HMnO_4}}{a_{H_2MnO_4^+}} = \frac{C_{HMnO_4}}{C_{H_2MnO_4^+}}h_0; \quad (3.192)$$

$$K_3 = \frac{a_{H_2MnO_4^+}}{a_{MnO_3^+}a_{H_2O}}; \quad (3.193)$$

$$K_2K_3a_{H_2O} = \frac{C_{HMnO_4}}{C_{MnO_3^+}}h_0, \quad (3.194)$$

h_- and h_0 are the values of acidity for the indicator equilibria $BH \rightleftharpoons B^- + H^+$ and $BH^+ \rightleftharpoons B + H^+$, respectively. Assuming that $[KMnO_4]_\Sigma \cong [MnO_4^-] + [HMnO_4]$, we obtain

$$[KMnO_4]_\Sigma = [HMnO_4](1 + K_a/h_-),$$

and then

$$k_2 = \frac{k_-K_a}{K_a + h_-} + \frac{k_0h_-}{K_a + h_-} + \frac{k_+h_0h_-/a_{H_2O}}{K_2K_3(K_a + h_-)}. \quad (3.195)$$

An analysis of the behavior of relation (3.195) in various regions of acidity gave the values of k_-, k_0, k_+, and *KIE* and provided information on the substrate selectivity and the relative reactivity of alkyl groups (k^{CH}) and aromatic ring (k^{Ar}) with respect to various forms of Mn(VII). For example, the relative parameters at 25 °C are as follows:

MnO_4^-	$k_-^H/k_-^D = 21.7,$	$k^{CH}/k^{Ar} \geq 200;$
$HMnO_4$	$k_0^H/k_0^D = 1.6 \pm 0.1,$	$k^{CH}/k^{Ar} = 0.6;$
MnO_3^+	$k_+^H/k_+^D = 1.1 \pm 0.1,$	$k^{CH}/k^{Ar} = 0.1.$

Thus, a decrease in the KIE with increasing $[HClO_4]$ is related both to changes in the contributions due to routes involving various forms of Mn(VII) and to variations in the contributions as related to the oxidizer attack on the alkyl group and aromatic ring. ■

Let us consider a situation with nonequilibrium forms of catalytically active species, which, in addition, are characterized by significantly different reactivity (specificity) with respect to various substrates.

Example 3.25 [111, 112]

► The oxidation of cyclohexane (CH), arenes, perfluoroarenes, terminal perfluoroolefins (1-PFO), internal perfluoroolefins (2-PFO), and perfluorocyclohexene (PFCH) by dihydroperoxide was studied in the V^V–H_2O_2–CF_3COOH system. The products of these reactions involving the rupture of C=C bonds represented the corresponding acids, diacids, and their fluoranhydrides. For the competitive oxidation reactions of benzene (S_1) in mixtures with other arenes (S_2), it was established that all these substrates are oxidized on the same catalytic center and the ratio of the corresponding rate constants $k_i/k_{C_6H_6}$ ([defined by formula (3.98)) is described by the Hammett equation with a reaction constant of $\rho \cong -1$. Hexafluorobenzene (HFB) is oxidized quite rapidly, although somewhat slower than benzene ($k_{C_6F_6}/k_{C_6H_6} = 0.63$). It was also found that the oxidation of all substrates in the system studied does not involve O_2, singlet $O_2(^1\Delta_g)$, O_3

(formed from H_2O_2 on vanadium complexes), OH^+ cation, and free radical species. The active species represent various peroxy complexes of vanadium(V).

A comparative study of the reactivity of indicated compounds showed that, at a constant ratio of $[H_2O_2]/[S_i]_0$, the degrees of conversion (x_i) of various substrates for the same time varied but not very significantly (to within a factor of 6.5) and decreased in the following order:

Substrate	1-PFO	CH	HFB	2-PFO	PFCH
x, %	92.0	78.0	57.0	48.0	15.0

At the same time, in the experiments with pairs of substrates (competitive reactions), the presence of CH virtually completely prevented the oxidation of C_6H_6 and HFB, HFB blocked the oxidation of 1-PFO and 2-PFO, and 1-PFO blocked the oxidation of 2-PFO and PFCH. Based on the results obtained for various pairs, the substrates can be arranged with respect to their blocking ability in the following order:

$$\text{CH} \gg \text{HFB} \gg \text{1-PFO} \gg \text{2-PFO} \gg \text{PFCH}.$$

It should be noted that, while in a mixture we have, $k_{CH}/k_{PFO-2} \sim 10^5$, the ratio of rate constants for the separate oxidation of the same substrates is about two. It was also established that the k_i/k_j values depended on the initial concentrations of S_1 and S_2.

An analysis of the aforementioned data and the results of numerous additional investigations led to a conclusion that four intermediate complexes X_n ($n = 1, 2, 3, 4$), which are sequentially formed from the initial vanadium(V) complex, participate in the oxidation reactions of various substrates: (X_1) for CH; (X_2) for arenes; (X_3) for 1-PFO; and (X_4) for 2-PFO and PFCH.

These intermediates are formed with the participation of perfluoroperacetic acid (CF_3COOOH) and H_2O_2. The following simplified scheme takes into account the steps of X_n formation (k_{n-1}), substrate oxidation (k''_n), irreversible conversion of X_n to X_{n+1} (k_n), and decomposition of vanadium peroxy complexes (k'_n):

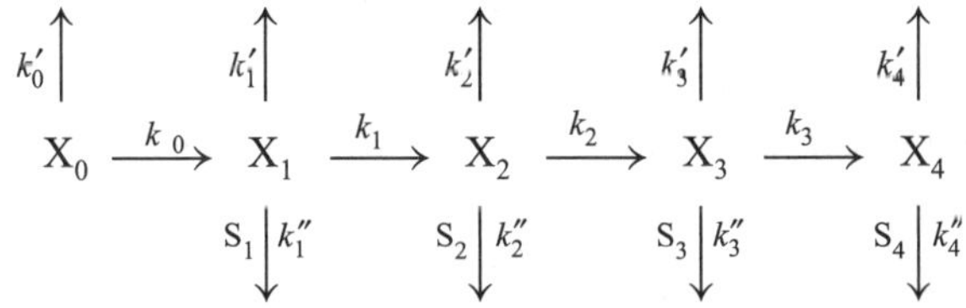

(3.196)

Figure 3.21 presents a more detailed scheme of the process under consideration.

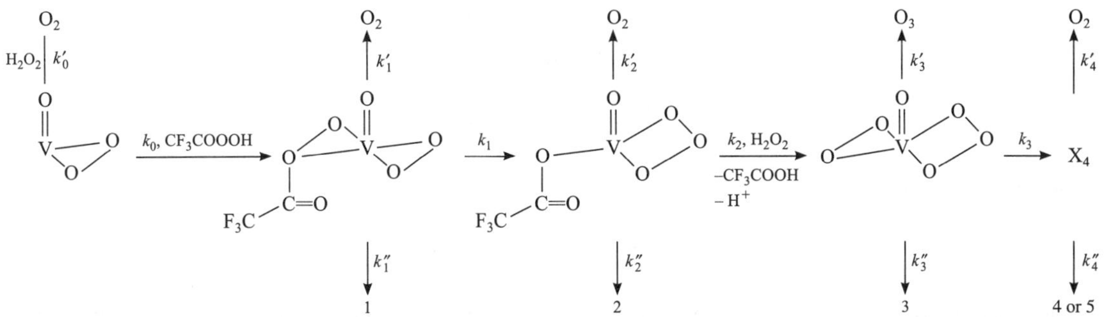

Figure 3.21 Mechanistic scheme of oxidation in mixtures of hydrocarbons and perfluorinated hydrocarbons: (1) cyclohexane; (2) arenes; (3) perfluorooctene-1; (4) perfluorooctene-2; (5) perfluorocyclohexene (reproduced with permission from editorial board and staff of *Izv. Ross. Akad. Nauk, Ser. Khim.* journal © 1986 [111])

According to scheme (3.196), there are four conjugation nodes of the following type:

$$\begin{array}{ccc} & \uparrow k_n' & \\ \xrightarrow[W_{n-1}]{k_{n-1}} & X_n & \xrightarrow[W_n]{k_n} \\ & W_n'' \downarrow k_n'' & \end{array} \tag{3.197}$$

(cf. conjugation nodes in Example 3.21). Assuming that the process is quasi-steady-state with respect to X_n, the concentration $[X_n]$ can be expressed as follows:

$$[X_n] = \frac{W_{n-1}}{k_n' + k_n + k_n''[S_n]}. \tag{3.198}$$

In the case of one substrate S_n reacting with one center of X_n, we have

$$[X_n] = \frac{k_0[X_0]}{k_n' + k_n + k_n''[S_n]} A_{n-1}'', \tag{3.199}$$

where $A_{n-1}' = \prod_{i=1}^{n-1} \frac{k_i'}{k_i + k_i'}$.

If two substrates are oxidized so that one of them (S_n) blocks the oxidation of another (S_{n+1}) on the adjacent center (i.e., $W_n'' \gg W_{n+1}''$), then

$$k_n''[S_n][X_n] \gg k_{n+1}''[S_{n+1}][X_{n+1}],$$

and, for close values of k_n'' and k_{n+1}'' and $[S_n] \approx [S_{n+1}]$, this implies that

$$[X_n] \gg [X_{n+1}].$$

This condition is valid for $W_n'' \gg W_n$ or $k_n''[S_n] \gg k_n$.

In the case of two substrates S_1 and S_2 oxidized on different centers X_1 and X_2, respectively, the oxidation rates are

$$r_1 = W_1'' = \frac{k_0 k_1''[X_0][S_1]}{k_1' + k_1 + k_1''[S_1]}; \tag{3.200}$$

$$r_2 = W_2'' = \frac{k_0 k_1 k_2''[X_0][S_2]}{(k_1' + k_1 + k_1''[S_1])(k_2' + k_2 + k_2''[S_2])}, \tag{3.201}$$

and the ratio of these rates can be expressed as follows:

$$\frac{r_1}{r_2} = \frac{k_1''[S_1]}{k_1 k_2''[S_2]}(k_2' + k_2 + k_2''[S_2]). \tag{3.202}$$

For $k_1''[S_1] \gg k_1$ and $k_2''[S_2] \gg k_1, k_2'$, this yields

$$\frac{r_1}{r_2} = \frac{k_1''[S_1]}{k_1} \gg 1.$$

Under certain assumptions, this model describes experimental dependences of the degrees of conversion on the initial reactant concentrations:

$$x_n = f\left(\frac{[S_n]_0}{[H_2O_2]_0}\right).$$

Thus, in the case of scheme (3.196), competitive reactions can provide distorted information about the relative reactivities for a pair of substrates converted on different active centers. ■

Now we will consider another example of mechanisms with a large number of conjugation nodes related by reversible and irreversible steps. Example 3.20 above was devoted to a reaction of acetylene hydroiodination in Pt(IV) solutions, which was described by a scheme that contained a single conjugation node between two routes. In solutions of Pt(II) complexes, this reaction proceeds according to a more complicated mechanism with three conjugation nodes and a route of acetylene reduction [113].

Example 3.26 [113]

▶ The conversion of acetylene into vinyl iodide (RI) and ethylene was studied in a closed ideally stirred reactor in the K_2PtCl_4–NaI–$HClO_4$–H_2O system at 80 °C for $[NaI]_0 = 2.5$ M and $[HClO_4] + [NaClO_4] = 2.5$ M. It was established that this system features the RI formation via two mechanisms that involve the intermediate compounds of different types: $Pt^{II}CH{=}CHI$ and $Pt^{IV}CH{=}CH_2$.

A mechanistic scheme that is consistent with the obtained experimental results and with published data on the mechanisms of C_2H_4 formation is as follows:

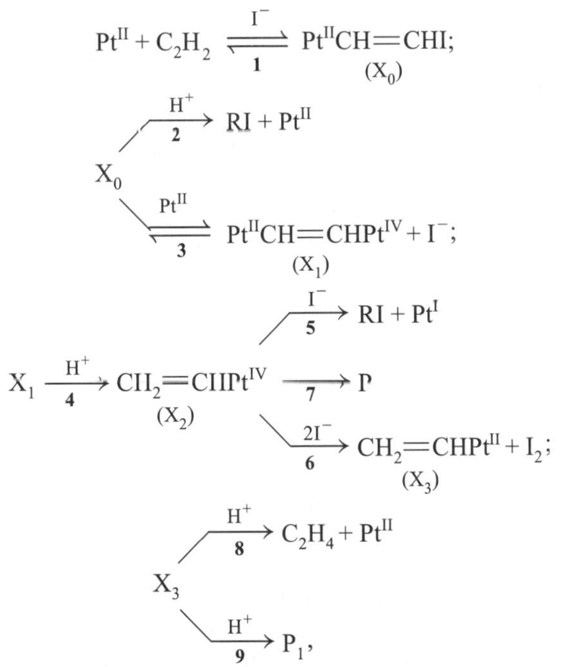

(3.203)

According to this scheme, there are four linearly independent intermediates and five routes (where P and P_1 are nonidentified products). Because of the presence of three conjugation nodes, an expression for the ratio of reaction rates $r_{RI}/r_{C_2H_4}$ for this scheme has a rather complicated form:

$$\frac{r_{RI}}{r_{C_2H_4}} = \frac{\omega_5[X_2] + \omega_2[X_0]}{\omega_8[X_3]}; \tag{3.204}$$

$$\frac{r_{RI}}{r_{C_2H_4}} = \frac{k_8 + k_9}{k_6 k_8}\left[k_5 + \frac{k_2(k_5 + k_6 + k_7)}{k_4 k_3 [Pt^{II}]_\Sigma}\right]. \tag{3.205}$$

In the case of

$$k_5 \ll \frac{k_2(k_5 + k_6 + k_7)}{k_4 k_3 [\mathrm{Pt^{II}}]_\Sigma},$$

the obtained equation is consistent with the experiment:

$$\frac{r_{\mathrm{RI}}}{r_{\mathrm{C_2H_4}}} = \frac{b}{[\mathrm{Pt^{II}}]_\Sigma}. \tag{3.206}$$

The effective first-order rate constant $k_{\mathrm{eff}}^{\mathrm{RI}}$ under these conditions is described by the following formula:

$$k_{\mathrm{eff}}^{\mathrm{RI}} = \frac{k_1 k_2 [\mathrm{Pt^{II}}]_\Sigma [\mathrm{H^+}]}{k_{-1} + k_2[\mathrm{H^+}] + k_4 K_3 [\mathrm{Pt^{II}}]_\Sigma [\mathrm{H^+}]}, \tag{3.207}$$

which is valid for the same assumptions ($k_2[\mathrm{X_0}][\mathrm{H^+}] > k_5[\mathrm{X_2}]$) and provided that $[\mathrm{Pt^{II}}]_\Sigma \cong [\mathrm{Pt^{II}}]$. The concentration $[\mathrm{I^-}]$ implicitly enters into the corresponding rate constants in Eqs (3.205) and (3.207). ■

In concluding this section, it should be emphasized that the study of relative selectivities, analysis of conjugation nodes, and use of various competitive reactions (artificial multi-route design) are effective methods for establishing the structure of reaction mechanisms and discriminating between hypotheses.

3.4 Conjugate processes

Descriptions of the phenomenon of conjugate chemical reactions and definitions of the related concepts appeared in the theory of reaction mechanisms [114] before the development of the theory of chain processes and long before the creation of the theory of routes that showed with confidence that the classical type of conjugate reactions is only a partial case of various manifestations of the *kinetic conjugation of steps and routes* related to the existence of common intermediates.[1]

The conjugation of steps within the framework of consequent single-route noncatalytic and catalytic reactions

$$\mathrm{A} \rightleftharpoons \mathrm{X} \longrightarrow \mathrm{P};$$

$$\mathrm{M} + \mathrm{A} \rightleftharpoons \mathrm{X_1} \overset{\mathrm{B}}{\rightleftharpoons} \mathrm{X_2} \longrightarrow \mathrm{P} + \mathrm{M} \tag{3.208}$$

as well as the conjugation of routes with one and more common steps, solves the problem of a thermodynamic conjugation of reactions [115, 116] (see Section 1.4.2). In cyclic reactions of type (3.208), every step is conjugated with at least one preceding and one following step. In single-route noncyclic reactions with n steps, there are $n-2$ steps that are conjugated from two sides.

Conjugate reactions occupy a special position among the processes that occur in a living cell [117, 118]. In biochemistry, it was established long ago [117] that the majority of enzymatic processes in the cell are conjugate reactions with numerous variants of the kinetic conjugation. In chemistry, the investigations of conjugate processes after Shilov were mostly concentrated on the reactions of one type and, according to this classical approach [114], the conjugate reactions type are still treated as those of a special kind (along with catalytic, chain, etc.) [119, 120]. In this context, it is expedient to analyze the nature of the kinetic conjugation so as to assess its position in the chemistry, kinetics, and the theory of mechanisms of complex reactions.

[1]The more common term "conjugate reactions" was introduced by N.A. Shilov [114] instead of the term "coupled reactions" (gekoppelte vorgänge) that was previously used by W. Ostwald.

3.4.1 Classical approach

Let us consider two reactions,

$$A + B \longrightarrow P_1; \tag{3.209}$$

$$A + C \longrightarrow P_2 \tag{3.210}$$

of which the former (exoergonic) process (primary reaction) proceeds spontaneously, while the latter process (secondary reaction) does not pass for thermodynamic or kinetic reasons. Then, the combination of reactions (3.209) and (3.210) in some cases makes the formation of product P_2 possible, that is, accelerates the secondary reaction, thus producing a kind of "chemical induction." These combinations are called *conjugate reactions*. The main features of conjugate reactions according to Shilov [114] are as follows:

- the presence of at least one common intermediate compound;
- mutual dependence of reaction rates;
- compensation of the endoergonic character ($\Delta G^\circ > 0$) of the secondary reaction by the exoergonic character of the primary reaction;
- only reactions with complex mechanisms are involved in conjugation.

Let us analyze the phenomenon of kinetic conjugation in commonly accepted terms, whereby reactant A is called *actor* (A), reactant B is referred to as *inductor* (Ind), and reactant C is *acceptor* (Ac). The simplest mechanistic scheme of a conjugate process

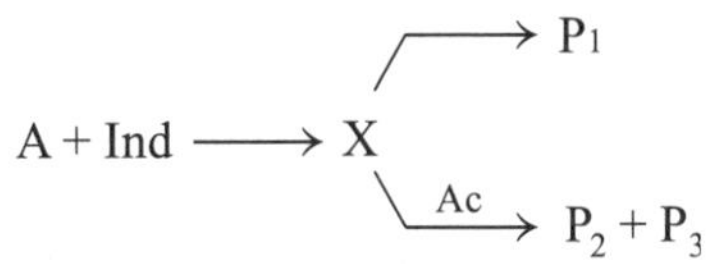

(3.211)

corresponds to a two-route reaction with the following overall route equations:

$$A + \text{Ind} \longrightarrow P_1; \tag{3.212}$$

$$A + \text{Ind} + \text{Ac} \longrightarrow P_2 + P_3, \tag{3.213}$$

According to this, a new reactant (Ind) and new product |(P_3) appear in the stoichiometry of secondary reaction (3.210). Thus, scheme (3.211) corresponds to the appearance of a new overall equation with modified stoichiometry of the secondary reaction (thus explaining the phenomenon of "chemical induction"), rather than to a trivial acceleration of the secondary reaction. The new overall equation can represent an exoergonic process, but the primary reaction does not perform useful work for the secondary reaction, since product P_1 in this case does not enter into the new overall equation. The useful change in ΔG° takes place prior to conjugation node X and results from the addition of process

$$\text{Ind} \longrightarrow P_3, \tag{3.214}$$

to reaction (3.210), which makes the combined process spontaneous. Thus, scheme (3.211) describes the conjugation of two routes and two overall equations, (3.212) and (3.213). Apparently, if the mechanism of a conjugate reaction differs from scheme (3.211), for example, as

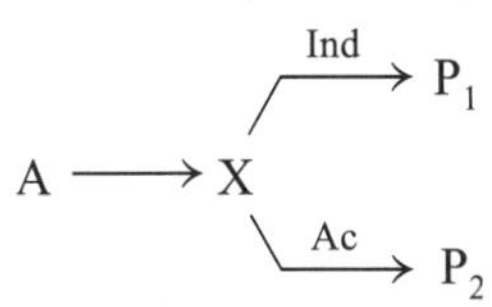

(3.215)

then no chemical induction and no thermodynamic conjugation of reactions (3.209) and (3.210) will take place. Indeed, this variant also provides common intermediate X (i.e., the kinetic conjugation of routes), but it does not modify the stoichiometry of the secondary reaction. The "synchronicity" of the reactions of P_1 and P_2 formation in scheme (3.215) has nothing in common with the phenomenon of chemical induction. A comparison of schemes (3.211) and (3.215) shows that the kinetic conjugation via a common intermediate is a more general case of the conjugate reactions than the conjugation with chemical induction in the classical variant.

We have considered noncatalytic reactions with a topological structure of type **C** (| Figure 3.22a), but the same considerations apply to catalytic processes with overall equations (3.212) and (3.213), where the thermodynamic conjugation can also be achieved by changing the stoichiometry of one route (Figure 3.22b) on the passage to a two-route reaction.

Examples of the classical variant of conjugation with chemical induction (Figure 3.22a) are offered by the reactions of Fe^{2+} oxidation by hydrogen peroxide conjugated with benzene oxidation [120, 121] and alkyne hydration [122, 123]. In particular, under the conditions of Fe^{2+} oxidation according to the Fenton scheme

$$Fe^{2+} + H_2O_2 \longrightarrow Fe(OH)^{2+} + \dot{O}H; \tag{3.216}$$

$$Fe^{2+} + \dot{O}H \longrightarrow Fe(OH)^{2+} \tag{3.217}$$

with the overall equation

$$2Fe^{2+} + H_2O_2 \longrightarrow 2Fe(OH)^{2+} \tag{3.218}$$

the intermediate $\dot{O}H$ radical is added to alkynes, e.g., methylacetylene,

$$\dot{O}H + CH_3C \equiv CH \longrightarrow CH_3\dot{C}HCHO \tag{3.219}$$

with the formation of $CH_3\dot{C}HCHO$ radical that interacts with Fe^{2+} (or even more rapidly with Cu^+, provided that Cu^+ is introduced into the Fenton reagent) and converts into propionaldehyde:

$$Cu^+ + CH_3\dot{C}HCHO + H_2O \longrightarrow Cu(OH)^+ + CH_3CH_2CHO. \tag{3.220}$$

The overall equation for the second route includes steps (3.216), (3.219) and (3.220),

$$Fe^{2+} + Cu^+ + H_2O_2 + C_3H_4 + H_2O \longrightarrow Fe(OH)^{2+} + Cu(OH)^+ + CH_3CH_2CHO, \tag{3.221}$$

i.e., step (3.216) that is shared by the two routes with the participation of H_2O_2 (inductor). It is this kinetic conjugation of the two routes, (3.218) and (3.221), that makes possible the otherwise kinetically hindered anti-Markovnikov addition of water molecule to alkyne.

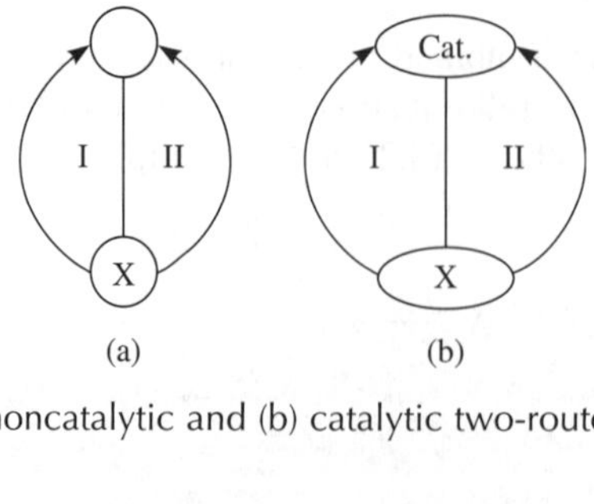

Figure 3.22 Kinetic graphs for (a) noncatalytic and (b) catalytic two-route reaction mechanisms of topological class C

Similarly, the interaction of $\dot{O}H$ radical with benzene leads to the consecutive steps

$$C_6H_6 + \dot{O}H \longrightarrow C_6\dot{H}_5 + H_2O; \quad (3.222)$$

$$C_6\dot{H}_5 + \dot{O}H \longrightarrow C_6H_5OH. \quad (3.223)$$

or the recombination of $C_6\dot{H}_5$ radicals with the formation of diphenyl. The overall equation of the oxidation reaction conjugated with process (3.218) is as follows:

$$2Fe^{2+} + 2H_2O_2 + C_6H_6 \longrightarrow 2Fe(OH)^{2+} + C_6H_5OH + H_2O. \quad (3.224)$$

Under certain conditions, the conjugated hydroxylation of benzene can also proceed catalytically, whereby the product of $\dot{O}H$ radical addition to benzene is oxidized by Fe^{3+} ion and then Fe^{2+} is regenerated.

The efficiency of a conjugated process, that is, the degree of the chemical induction of a secondary reaction, is characterized by the *induction factor* I, which in fact measures the selectivity of acceptor consumption with respect to the spent inductor. For the general overall equations (3.212) and (3.213), this value is defined as follows:

$$I = \Delta n_{Ac}/\Delta n_{Ind}. \quad (3.225)$$

For equations (3.218) and (3.224), the induction factor is as defined

$$I = 2\Delta n_{C_6H_6}/\Delta n_{H_2O_2}, \quad (3.226)$$

and the differential induction factor is determined as

$$I = 2r_2/(r_1 + 2r_2), \quad (3.227)$$

where r_1 and r_2 are the rates of reactions (3.218) and (3.224), respectively.

The classical variant of conjugated reactions can also be encountered in more complicated mechanisms [120]. For example, the reaction

$$2SO_3{}^{2-} + O_2 \longrightarrow 2SO_4{}^{2-} \quad (3.228)$$

proceeds as a chain process with Cu^{2+} catalyzed route of $SO_3^{\bullet-}$ active center generation (Figure 3.23), and the overall equations of the processes of chain initiation

$$2SO_3{}^{2-} + \tfrac{1}{2}O_2 + 2H^+ \longrightarrow 2SO_3{}^{\bullet-} + H_2O \quad (3.229)$$

and termination

$$2SO_3^{\bullet-} \longrightarrow S_2O_6{}^{2-} \quad (3.230)$$

(not depicted on the kinetic graph in Figure 3.23). The secondary reaction in this conjugated process is

$$2AsO_3{}^{3-} + O_2 \longrightarrow 2AsO_4{}^{3-}, \quad (3.231)$$

which does not proceed without primary reaction (3.228) for kinetic reasons. The overall equation of the new cyclic process

$$SO_3{}^{2-} + AsO_3{}^{3-} + O_2 \longrightarrow SO_4{}^{2-} + AsO_4{}^{3-} \quad (3.232)$$

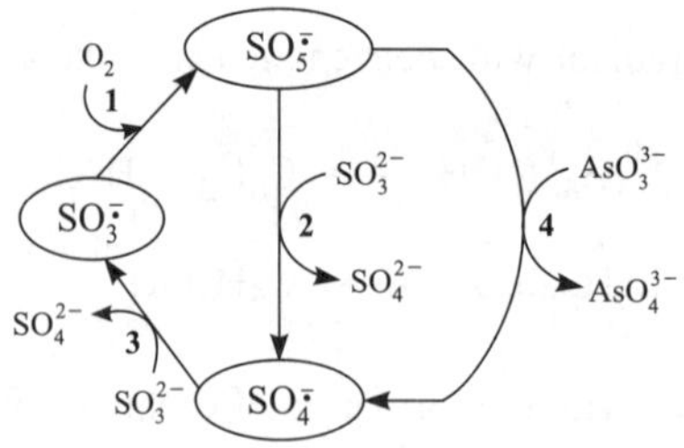

Figure 3.23 Kinetic graph for cyclic conjugated oxidation of $SO_3{}^{2-}$ and $AsO_3{}^{3-}$

is a half-sum of Eqs (3.228) and (3.231). As can be seen from Figure 3.23, this conjugation scheme significantly differs from the simple classical variants considered above. The process can be considered as catalytic with respect to $SO_3^{\bar{\bullet}}$ with two common steps (**1** and **3**) in two routes and three common intermediates in consecutive cyclic steps. In this case, both reactants and the product of primary reaction (3.228) enter into the overall equation of the new route (3.232).

In the case of catalytic multi-route processes with common steps (see Figure 3.22b), some specific features of conjugate reactions with chemical induction are manifested fairly infrequently. For example, in interaction of the catalase enzyme (E) with hydrogen peroxide and ethanol, the enzyme performs two functions in dilute H_2O_2 solution simultaneously, acting as catalase to decompose peroxide via reaction

$$2H_2O_2 \longrightarrow 2H_2O + O_2, \tag{3.233}$$

and as peroxidase to oxidize ethanol in reaction

$$H_2O_2 + C_2H_5OH \longrightarrow 2H_2O + CH_3CHO. \tag{3.234}$$

A simplified mechanism of this process is represented by a kinetic graph in Figure 3.24 [119]. The reaction is considered as a typical manifestation of the chemical induction phenomenon, with H_2O_2 as inductor in step **1**, H_2O_2 as actor in step **2**, and C_2H_5OH as acceptor in step **3** [119]. At the same time, both overall equations (3.233) and (3.234) are kinetically conjugated but the stoichiometry of reaction (3.234) is not changed in conjugation with reaction (3.233). The mutual relation of processes naturally takes place, as in any process with this structure of the mechanism. Numerous reactions of the oxidation and oxidative dehydrogenation of alkanes and alkenes in gas and liquid phases in the presence of H_2O_2, which are described in [119], belong to chain processes according to their mechanisms. Their classification as induced conjugate reactions are discussed in Section 3.4.3.

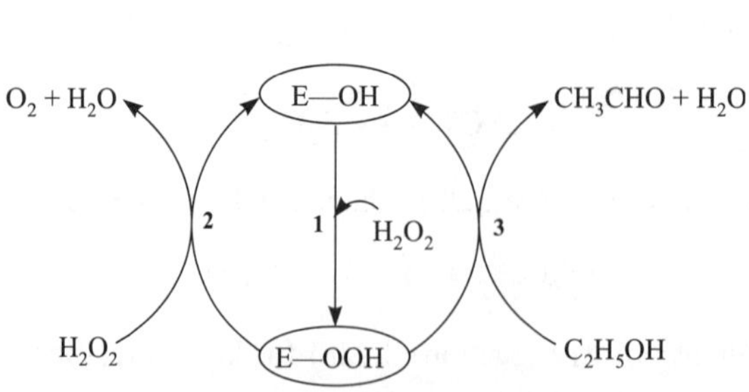

Figure 3.24 Kinetic graphs for two-route reaction with participation of enzyme *catalase*

The phenomenon of kinetic conjugation is not restricted to the variants with conjugated routes sharing common steps of the type represented by the kinetic graph in Figures 3.22a and 3.22b. Frequently encountered cases of kinetic conjugation also include:

- consequent single-route reactions (3.208) (frequently called coupled reactions);
- multi-route chain processes.

These variants do not fall within the framework of the classical definition of conjugate reactions, but both still represent conjugate processes. In the case of consecutive reactions (conjugated steps), there is thermodynamic conjugation with modification of the stoichiometry of overall equations. Although this type of conjugation does not apply to multi-route reactions considered in Chapter 3, any multi-route process involves series of consequent steps. This type of kinetic conjugation is analyzed below.

3.4.2 Kinetic and thermodynamic conjugation in consecutive reactions

Let us consider the consecutive reaction

$$\mathrm{A} \overset{K_1}{\rightleftharpoons} \mathrm{B}; \qquad \mathrm{B} \overset{K_2}{\rightleftharpoons} \mathrm{C}.$$

The direction of any chemical reaction is determined by the sign of the chemical affinity defined as

$$A_j = RT \ln \frac{K_j}{\prod C_i^{\beta_i}} \tag{3.235}$$

(see Section 1.4.2). In the case of equilibrium, $A_j = 0$ and $K_j = \prod [C_i]_{\text{eq}}^{\beta_i}$.[2]

According to the basic equation originally obtained by de Donde (see [124] and Section 6.2), the chemical affinity relates to the reaction kinetics (rate) as follows:

$$A_j = RT \ln(W_j^+/W_j^-), \tag{3.236}$$

where W_j^+ and W_j^- are the rates of the jth reaction in the forward and reverse directions, respectively.

Let the process $\mathrm{A} \rightleftharpoons \mathrm{B}$ be equilibrium ($A_j = 0$) at $K_1 \ll 1$. Then, $[\mathrm{A}]_{\text{eq}}/[\mathrm{B}]_{\text{eq}} = K_1 = 1$ and the equilibrium degree of conversion is also close to zero. In order to shift the reaction on the right, toward the formation of product B, it is necessary to violate the equilibrium by decreasing [B] or increasing [A], so as to ensure that $[\mathrm{B}]/[\mathrm{A}] < K_1$, $A_1 > 0$, and $W_j^+ > W_j^-$. A decrease in [B] can take place as a result of the kinetic conjugation of reaction $\mathrm{A} \rightleftharpoons \mathrm{B}$ with reaction $\mathrm{B} \rightleftharpoons \mathrm{C}$, for which $A_2 > 0$. By the same token, for $A_1 > 0$ and $A_2 \leq 0$ the conjugation will increase [B] and change the sign of A_2 to positive. In both cases, only the sign and magnitude of the chemical affinity are changed and the system deviates from equilibrium.

Under the conditions of a steady-state or quasi-steady-state many-step process, the rate of any step of a single-route reaction is related (see Section 1.1) to the rate for the overall equation (rate over the route) by the following relation:

$$W_j = \nu_j r, \tag{3.237}$$

where r is the rate for overall equation $\mathrm{A} \rightleftharpoons \mathrm{C}$ (with the mechanism considered above) and ν_j is the stoichiometric number of the jth step. For a consecutive reaction, all $\nu_j > 0$; then, for $r > 0$,

---

[2] In the absence of equilibrium, the quantity $\prod C_i^{\beta_i}$ is conventionally called the reaction product.

all $W_j = W_j^+ - W_j^- > 0$ and, according to Eq. (3.236), all $A_j > 0$. Thus, if the kinetic conjugation ensures the passage of process $A \rightleftharpoons B \rightleftharpoons C$ in the forward direction, then all $W_j > 0$, $A_j > 0$, and $\Sigma \nu_j A_j = A > 0$ [115].

If the consecutive process

$$A \overset{1}{\rightleftharpoons} B; \quad B \overset{2}{\rightleftharpoons} C$$

in a closed reactor begins with reactant B (parallel reactions $B \leftrightarrow A$ with a rate W_1 and $B \leftrightarrow C$ with a rate W_2), a situation is possible in which $W_2 > W_1$ and $A_2 \gg 0$. Then, at the initial stage of reaction, B will be rapidly converted into C (until $A_2 \approx 0$) and then B will be slowly converted into A, so that the concentration of B will decrease ($\prod C_i^{\beta_i} = [C]/[B] > K_2$) and A_2 will become negative. Then, the reaction will change its direction to reverse and proceed into A via B as the intermediate product:

$$C \rightleftharpoons B \rightleftharpoons A$$

Thus, the concentration of C at the initial stage is greater than its equilibrium value $[C]_{eq}$ at the end of the process. Let us consider an example of step conjugation in a consecutive reaction.

Example 3.27 [115]

► The simplest mechanism of the radical chain reduction of Br_2 to HBr can be represented by a sequence of three steps,

$$Br_2 \overset{K}{\rightleftharpoons} 2Br; \tag{3.238}$$

$$Br + H_2 \underset{1}{\rightleftharpoons} HBr + H; \tag{3.239}$$

$$H + Br_2 \underset{2}{\longrightarrow} HBr + Br \tag{3.240}$$

with an overall cyclic route equation of

$$H_2 + Br_2 \longrightarrow 2HBr. \tag{3.241}$$

Let us consider the variation of the chemical affinity of the endoergonic ($A^\circ < 0$) step (3.239) as a result of its conjugation with step (3.240). Since step (3.238) is equilibrium, the quasi-steady-state concentration ($[H]_{SS}$) of hydrogen atoms is

$$[H]_{SS} = \frac{k_1[Br]_{eq}[H_2]}{k_{-1}[HBr] + k_2[Br_2]}, \tag{3.242}$$

and for $[H_2] \cong [HBr] \cong [Br_2]$

$$[H]_{SS} = \frac{K_1[Br]_{eq}}{1 + k_2/k_{-1}}. \tag{3.243}$$

Writing the affinity of step (3.239) as

$$A_1 = RT \ln \frac{K_1[Br]_{eq}}{[H]_{SS}},$$

and using expression (3.243), we obtain

$$A_1 = RT \ln(1 + k_2/k_{-1}). \tag{3.244}$$

According to the experimental data, $k_2/k_{-1} \cong 10$ at $T = 500 - 600\,k$ (see [115]). Therefore, $A_1 > 0$ and reaction (3.239) under these conditions becomes spontaneous. Under the same assumptions, Eq. (3.243) and relation $[H]_{eq} = K_1[Br]_{eq}$ yield the following equation for the equilibrium concentration $[H]_{eq}$:

$$\frac{[H]_{eq}}{[H]_{SS}} = 1 + k_2/k_{-1}, \tag{3.245}$$

where $[H]_{eq} = K_1[Br]_{eq} = K'[Br_2]^{1/2}$, and K' is equilibrium constant of reaction

$$\tfrac{1}{2}Br_2 + H_2 \rightleftharpoons HBr + H \tag{3.246}$$

Formula (3.245) shows that the quasi-steady-state concentration $[H]_{SS}$ is about eleven times smaller than $[H]_{eq}$ for reaction (3.246), because H is scavenged in the fast step (3.240) about ten times faster than it returns to Br atom in reverse step (3.239). At the same time, $[H]_{SS}$ is still significantly greater than $[H]_{eq}$ for the equilibrium

$$H_2 \rightleftharpoons 2H.$$

In the limiting case, the kinetic conjugation makes the first (otherwise reversible) step (3.239) irreversible and ensures the maximum possible rate ($W_1{}^+$) of process (3.239). ■

Now let us consider the issue of thermodynamic conjugation in a consecutive reaction. The principle of energy conjugation was originally formulated in 1900 by Ostwald [125]. According to this, a reaction accompanied by energy evolution can initiate another process (even if it requires less energy than that liberated), only provided that there exists a common intermediate substance (gekoppelte vorgǎnge) within the framework of one new stoichiometric equation. Let us analyze some consecutive reactions in this context.

The oxidative dehydrogenation of alkanes, alkenes, and alcohols is frequently considered as the example of thermodynamic conjugation for exoergonic [e.g., (3.248)] and endoergonic [e.g., (3.247)] reactions (at temperatures below $200 - 300\,°C$),

$$RH \longrightarrow R_{-H} + H_2 \tag{3.247}$$

$$2H_2 + O_2 \longrightarrow 2H_2O \tag{3.248}$$

$$2RH + O_2 \longrightarrow 2R_{-H} + 2H_2O, \tag{3.249}$$

where $\Delta G°$ for the net reaction (3.249) is negative ($A^{\circ}_{(3.249)} > 0$):

$$\Delta G° = 2\Delta G^{\circ}_{(3.247)} + \Delta G^{\circ}_{(3.248)} < 0.$$

The value of the chemical affinity under equilibrium conditions is naturally independent of the pathways (and the nature) of reactions constituting the overall equations. In contrast, the affinity of reaction (3.249) on the passage to equilibrium depends on the pathway, that is, on the mechanism of a kinetic conjugation, since the values of A_j depend on the rates of particular steps (3.236). Therefore, the total chemical affinity $A = \sum \nu_j A_j$ must refer to a particular mechanism.

In order to involve reaction (3.247) into conjugation, it is necessary to ensure the process of dehydrogenation (3.247) and oxidation (3.248) of intermediary hydrogen with the aid of an appropriate catalyst.

According to thermodynamic estimations, the condition of $A_{(3.247)} > 0$ can be ensured provided that the partial pressure of hydrogen does not exceed $10^{-6} - 10^{-7}$ bar. A low partial pressure of hydrogen is probably reached in the course of cyclooctane dehydration at 200 °C in a boiling solution containing Ir(III) complexes [126]. However, a real mechanism of reaction (3.248) has nothing in common with Eqs (3.246) and (3.247) considered as probably contributing to the thermodynamic conjugation. A two-step mechanism of the heterogeneous catalytic reaction (3.249) can be written in a substantially simplified form, in which both first and second steps are oxidation reactions with a common intermediate representing a reduced form of the surface oxide on the metal catalyst:

$$\begin{aligned} RH + MO_x &\longrightarrow R_{-H} + H_2O + MO_{x-1}; \\ 2MO_{x-1} + O_2 &\longrightarrow 2MO_x \end{aligned} \tag{3.250}$$

or in a more detailed variant for M = Mo [115]:

$$\begin{aligned} RH + Mo^{6+} + 2O^{2-} &\longrightarrow Mo(OH)_2{}^{2+} + R_{-H}; \\ 2Mo(OH)_2{}^{2+} + O_2 &\longrightarrow 2Mo^{6+} + 2H_2O + 4O^{2-}. \end{aligned} \tag{3.251}$$

The proposed mechanisms include a kinetic conjugation of steps with $A_j > 0$, but no conjugation of the virtual (not actually taking place) overall reactions (3.247) and (3.248), no conjugation of routes, and no primary and secondary reactions. There is a single overall equation and no "synchronous interactions" of chemical reactions [119b].

The oxidative dehydrogenation of methanol (and other alcohols) on a silver catalyst [127] at 600 – 700 °C is accompanied by the dehydrogenation of alcohol with the formation of H_2. However, it was established that the rate of H_2 oxidation by oxygen is lower than the rate of formaldehyde formation from methanol, which implies that the dehydrogenation process (with $A^\circ > 0$) is a parallel route that is virtually not involved into the thermodynamic conjugation of type (3.247)–(3.248).

In an analysis of the thermodynamic conjugation in consecutive reactions, it is a common practice to use de Donde's criterion, according to which the sum of $A_j W_j$ for a consecutive spontaneous reaction must be positive [115, 119, 124]:

$$\sum_j A_j W_j > 0. \tag{3.252}$$

Let us consider the meaning of this inequality in more detail. Condition (2.252) is related to the production of entropy as a result of irreversible processes in the system and is valid for any set of reactions, including parallel and consecutive-parallel (see Section 6.2):

$$\frac{d_i S}{dt} = \nu \sum_j \frac{A_j W_j}{T} > 0. \tag{3.253}$$

It should be noted that, since the product

$$A_j W_j = R \ln \frac{W_j^+}{W_j^-} (W_j^+ - W_j^-) \tag{3.254}$$

is always positive for any step ($A_j > 0 \Rightarrow W_j > 0$, $A_j < 0 \Rightarrow W_j < 0$), the sum of $A_j W_j$ is also always positive. Positive and negative terms can only appear for the total standard chemical affinity A°_Σ. If a consecutive reaction proceeds in a preset direction, it can be concluded that all $A_j > 0$, all $W_j > 0$, and all

$A_j W_j > 0$, since only this ensures the thermodynamic conjugation for a particular mechanism of kinetic conjugation [115].

Numerous reactions that involve the transfer of hydrogen atoms [41], beginning with the Zelinsky reaction [128]

$$3C_6H_{10} \rightleftharpoons C_6H_6 + 2C_6H_{12}, \tag{3.255}$$

in fact do not include a conjugation of dehydrogenation and hydrogenation steps. They provide only a kinetic conjugation of the steps of hydrogen transfer from a substrate to the active centers of a metal catalyst with the formation of surface (or volume hydrides in the homogeneous case) and the steps of their interaction with an acceptor of hydrogen atoms. For example, the process with metal complexes as catalysts in the reaction:

$$PhCH_2OH + (CH_3)_2CO \longrightarrow PhCHO + (CH_3)_2CHOH; \tag{3.256}$$

is described by the following scheme:

$$\begin{aligned} PhCH_2OH + ML_n &\longrightarrow PhCHO + ML_n(H)_2; \\ ML_n(H)_2 + (CH_3)_2CO &\longrightarrow ML_n + (CH_3)_2CHOH. \end{aligned} \tag{3.257}$$

Analogous processes take place for alkanes and naphthenes. Complexes of Ir(I) take two H atoms from an alkane with the formation of an olefin and an Ir(III) dihydride complex. The latter complexes react with various acceptors of hydrogen atoms [129]. The process of membrane catalysis (e.g., on Pd membrane) [130]

$$C_6H_{12} + 3PhMe \longrightarrow 4C_6H_6 + 3CH_4. \tag{3.258}$$

also does not involve thermodynamic conjugation as a combination of the reactions of cyclohexane dehydrogenation and toluene hydrogenolysis.

Consecutive single-route reactions include some cases where the overall equation separates into two (or more) overall equations, which are kinetically conjugated within the framework of a common mechanism.

Example 3.28 [131]

▶ Using the kinetic conjugation of steps, it is possible to carry out the following exoergonic and endoergonic reactions:

$$CO + {}^1\!/\!_2O_2 \longrightarrow CO_2 \ (A_1^\circ > 0); \tag{3.259}$$

$$2AcOH \underset{<100\,^\circ C}{\longrightarrow} (Ac)_2O + H_2O \ (A_1^\circ < 0). \tag{3.260}$$

The mechanism of this conjugate process includes the following steps, in which $Pd(OAc)_2$ is considered as a catalyst:

$$\begin{aligned} &Pd(OAc)_2 \xrightarrow{CO} Pd(OAc)(\underset{\underset{O}{\|}}{C}OAc) \xrightarrow[-CO_2]{} Pd(OAc)(Ac) \xrightarrow[-(Ac)_2O]{} Pd^0; \\ &Pd^0 + 2AcOH + {}^1\!/\!_2O_2 \longrightarrow Pd(OAc)_2 + H_2O. \end{aligned} \tag{3.261}$$

The overall equation

$$CO + {}^1\!/\!_2O_2 + 2AcOH \longrightarrow CO_2 + (Ac)_2O + H_2O \tag{3.262}$$

is a sum of Eqs (3.259) and (3.260). The use of *para*-benzoquinone (Q) instead of O_2 leads to the following equation:

$$CO + Q + 2AcOH \longrightarrow CO_2 + (Ac)_2O + QH_2. \quad (3.263)$$

The ratio of the molar fractions of CO_2 and $(Ac)_2O$ according to Eqs (3.262) and (3.263) in the absence of side reactions must be unity. Investigation [132] of the mechanism of reaction (3.262) led to the conclusion that this process involves the formation of a Pd(I) carbonyl acetate cluster $Pd_4(CO)_4(OAc)_4$ followed by its reductive decomposition under the action of acetate ion (NaOAc) with the formation of Pd^0, CO_2, and Ac_2O.

It should be emphasized that the implementation of process (3.262) is ensured by the appearance of a new route with the overall equation (3.262) possessing a positive chemical affinity ($A > 0$), rather than by the summation of standard Gibbs potentials and/or standard chemical affinities of the overall reactions. Evidently, the equilibrium composition of the system including six substances and two linearly independent reactions must not depend on the choice of the basis set for the overall equations. For the equilibrium of reactions (3.259) and (3.260), the yield of acetic anhydride is close to zero for any linear combination of the overall equations. If the system has only one route with a single overall equation (3.262), the equilibrium yield of anhydride is determined by the standard chemical affinity for reaction (3.262). ■

A conjugation of the oxidative carbonylation of alkanes with the dehydration of trifluoroacetic acid also takes place in solutions of Pd(II) complexes according to the following scheme [133]:

$$CH_4 + CO + 2CF_3COOH + H_2S_2O_8 \longrightarrow CH_3COOH + (CF_3CO)_2O + 2KHSO_4; \quad (3.264)$$

$$CH_4 + CO + 2CF_3COOH + \tfrac{1}{2}O_2 \longrightarrow CH_3COOH + (CF_3CO)_2O + H_2O. \quad (3.265)$$

Example 3.29

► The synthesis of primary alcohols from α-olefins (anti-Markovnikov hydration) based on the hydroboration of olefins followed by the oxidation of alkylboranes with dihydroperoxide is an important synthetic method [134, 135]. In the presence of Rh(III) porphyrinates (e.g., (OEP)RhCl) it is possible to carry out the catalytic anti-Markovnikov hydration conjugated with a process of $NaBH_4$ oxidation according to the overall equation:

$$1\text{-}C_5H_{10} + H_2O + NaBH_4 + 2O_2 \longrightarrow 1\text{-}C_5H_{11}OH + NaB(OH)_4 \quad (3.266)$$

with a regioselectivity of 91 % and a catalyst turnover of TON = 210 [136]. During this process, (OEP)RhH and (OEP)RhOOH are formed from (OEP)RhCl. Then, alkylborane obtained from $BH_3(B_2H_6)$ and olefin is oxidized by Rh(III) hydroperoxide (as peroxide anion HO_2^- in the case of H_2O_2). ■

Example 3.30

► A conjugation of the steps of styrene hydration and the oxidation of ethanol and PPh_3 was discovered in the catalytic system of Co(salen)–PPh_3 – C_2H_5OH at 40°C [137]. Based on the analysis of kinetics and the absence of H_2O_2 activity in this process, it was suggested that the $[Co]O_2$ complex interacted with ethanol and styrene so as to convert them into acetaldehyde and ethylbenzene hydroperoxide, respectively, and the reaction

$$\underset{\displaystyle \text{OOH}}{\text{PhCHCH}_3} + \text{PPh}_3 \longrightarrow \underset{\displaystyle \text{OH}}{\text{PhCHCH}_3} + \text{O=PPh}_3 \qquad (3.267)$$

led to the formation of methylphenyl carbinol. The overall equation of the process includes four substrates:

$$PhCH{=}CH_2 + CH_3CH_2OH + PPh_3 + O_2 \longrightarrow PhCH(OH)CH_3 + CH_3CHO + O{=}PPh_3. \qquad (3.268)$$

■

Example 3.31

▶ Various olefins exhibit hydration in the course of a process conjugated with the oxidation of isopropanol. This reaction proceeds at 75 °C in a solution of Co(II) β-diketonates (or better, trifluoroacetyl acetonate) according to the following equation:

$$RCH{=}CH_2 + 2(CH_3)_2CHOH + O_2 \longrightarrow RCH(OH)CH_3 + 2(CH_3)_2CO + H_2O. \qquad (3.269)$$

with an alcohol yield of up to 90 %, a complete conversion of olefin, and small yields of ketone (from olefin), and an alkane [138].

It was also found that a selective oxidation of an olefin to a ketone, conjugated with the oxidation of a primary or secondary alcohol (monooxygenase mechanism) takes place in solutions of Rh(I) phosphine chelate complexes [139]. ■

Example 3.32

▶ The exoergonic process

$$2CO + O_2 \longrightarrow 2CO_2 \qquad (3.270)$$

($\Delta G_T^\circ = -135\,407 + 41.4\,T$ cal/mole) can be used for carrying out strongly endoergonic processes in a single-route conjugate reaction, e.g., for the oxidation of water as

$$2H_2O + O_2 \longrightarrow 2H_2O_2$$
$$(\Delta G_T^\circ = 47\,377 + 29.9\,T \text{ cal/mole}). \qquad (3.271)$$

This conjugate process was originally implemented [140] in the $PdX_2(PPh_3)_2$–PPh_3–H_2O system with an overall equation of

$$CO + H_2O + O_2 \longrightarrow H_2O_2 + CO_2. \qquad (3.272)$$

and it was suggested that CO_2 was formed according to the reaction

$$PdX_2L_2 + CO + H_2O + 2L \longrightarrow PdL_4 + CO_2 + 2HX, \qquad (3.273)$$

and hydrogen peroxide was formed according to the reaction

$$PdL_4 + O_2 + 2HX \longrightarrow PdX_2L_2 + H_2O_2 + 2L. \qquad (3.274)$$

Subsequent investigations showed that reaction (3.272) is catalyzed in two-phase aqueous organic systems containing Pd(II) complexes with substituted phenanthroline ligands (with the formation of a 7 % aqueous H_2O_2 solution) [141] and in homogeneous aqueous organic media containing only $PdBr_2$ [142–145]. In a homogeneous system, H_2O_2 (or O_2) is spent to a considerable extent for the oxidation of an organic solvent. The maximum rate of CO oxidation to CO_2 [0.7–1.5 M/h at 30 °C] without Pd black precipitation was achieved in aqueous solutions of THF (with the formation of γ-butyrolactone and 1,4-dioxane) and 1,4-dioxane (with the formation of dioxene, hydroxy-1,4-dioxane, and ethylene glycol mono- and diformate). In addition, the reaction products contained HCOOH. Based on the results of investigation of reaction (3.272) [144], it was suggested that the reduction $PdBr_2$ by carbon monoxide does not lead to the formation of Pd(0) complexes, although there are grounds to believe that complexes $Pd^0–Pd_2(\mu–CO)_2Br_2(CO)_2{}^{2-}$ exist [146]. Most probably, the reduction process involves the formation of a hydride complex of the $Pd^{II}–HPdBr_x(CO)_y^{(y-1)-}$ type, which is a key intermediate in the conjugation of CO and H_2O oxidation:

$$PdBr_2 + CO + H_2O \longrightarrow HPdBr + CO_2 + HBr; \tag{3.275}$$

$$HPdBr \xrightarrow{O_2} HOOPdBr \xrightarrow{HBr} PdBr_2 + H_2O_2. \tag{3.276}$$

The formation of $HOOPdXL_2$ was demonstrated in [147, 148]. The steps of solvent oxidation can involve palladium hydroperoxide (HOOPd∼) and H_2O_2. Since dioxygen does not oxidize solvents in the absence of CO, it is possible that the oxidation of THF and dioxane is related to the oxidation of CO and H_2O. Thus, reaction (3.145) is no longer a single-route process, which is due to the partial decomposition of hydrogen dihydroperoxide and its participation in side reactions of solvent oxidation.

The catalytic oxidation of THF to γ-butyrolactone conjugated with CO oxidation was also observed in the $PdCl_2–CuCl_2–CO–O_2–THF$ system at 25 °C and 24 bar ($CO : O_2 = 2 : 1$) [149], where ClPdOOH also probably plays the role of oxidizer species. Under the same conditions, diethyl ether was oxidized to ethyl acetate. ■

Example 3.33 [150]

► Highly effective epoxidation of trisubstituted internal olefins by O_2 at atmospheric pressure, conjugated with aldehyde oxidation and catalyzed by Ni(II) diketonates in dichloroethane at 25 °C, proceeds as a single-route process according to the reaction

$$RCH{=}CR_1R_2 + O_2 + RCHO \longrightarrow \underset{\diagdown O \diagup}{RCH{-}CR_1R_2} + RCOOH. \tag{3.277}$$

with a ∼100 % selectivity with respect to olefin at 100 % conversion (TON ∼ 1000). The absence of side products typically formed in radical chain oxidation suggests that Ni(II) complexes in this system catalyze the oxidation of aldehyde to peroxy acid, with the subsequent catalytic epoxidation of olefin by this acid. ■

The conjugate process of carbon monoxide oxidation and reduction in aqueous solutions proceeds almost like a single-route reaction [151]. The process of CO reduction by molecular hydrogen, even with the formation of hydrated formaldehyde,

$$(1)\ CO_g + H_{2g} \longrightarrow CH_2O_{aq},$$

is a thermodynamically unfavorable reaction ($\Delta G_1^\circ = 7.5\,\text{kJ/mole}$). However, the conjugation of this reaction with the exoergonic process of CO conversion by liquid water ($\Delta G_2^\circ = -11.5\,\text{kJ/mole}$)

$$(2)\ CO_g + H_2O_{liq} \longrightarrow CO_{2g} + H_{2g}$$

leads to a weakly exoergonic process with the overall equation

$$(3)\ 2CO_g + H_2O_{liq} \longrightarrow CH_2O_{aq} + CO_{2g}$$

and $\Delta G_\Sigma^\circ = -4\,\text{kJ/mole}$. This conjugate reaction proceeds at 80 °C (with TOF $\sim 1\,h^{-1}$) in aqueous solutions of dimeric cationic Mo(IV) oxocomplexes with Mo–Mo bonds [151]. It was suggested to be a two-route process with the formation of intermediate Mo hydride complexes at an insignificant contribution of reaction (2).

In concluding this subsection, it should be noted that the results of investigations of biochemical and chemical processes over seven decades after Ostwald's work [125] allowed Leninger to confirm the principle of energy conjugation, according to which *there is only one way of energy transfer from one reaction to another (besides fluorescence) under isothermal conditions, whereby two reactions (steps) are conjugated by sharing a common intermediate product* [117]. This statement is also confidently affirmed by the aforementioned results of investigations into the reactions with metal complexes, obtained in the four decades that have passed since the publication of Leninger's monograph [117].

It was shown in Section 3.4.1 that, in the classical variant of conjugate reactions with chemical induction, it is the energy of the first step ($A + Ind \rightarrow X$) that is eventually transferred via a sequence of steps of the induced route due to the appearance of an overall equation with a modified stoichiometry.

3.4.3 Conjugation in chain reactions

The theory of chain reactions belongs to most developed parts of the field of chemical kinetics and the theory of chemical reaction mechanisms [120, 152]. Chain processes represent multi-route reactions with the numbers of routes $P \geq 2$. One of these routes represents a non-closed (or open, according to Christiansen [153]) sequence of elementary steps of chain nucleation (initiation) and termination, i.e., the loss of an active center involved in one or several closed cycles of chain propagation. In rare cases, where the chain nucleation and termination steps represent the forward and reverse reactions (see, e.g., reaction (3.240)), the chain process becomes single-route. This radical-chain process (e.g., $RH + Cl_2$) is formally identical with a catalytic reaction, for example, with the acid-initiated alkylation of alkenes by alkenes according to the scheme

$$\begin{aligned} Cl_2 &\rightleftharpoons 2Cl\bullet; & C_4H_8 + HA &\rightleftharpoons C_4H_9{}^+A^-;\\ Cl\bullet + RH &\longrightarrow R\bullet + HCl; & C_4H_9{}^+A^- + C_4H_8 &\longrightarrow C_8H_{17}{}^+A^-;\\ R\bullet + Cl_2 &\longrightarrow RCl + \bullet Cl. & C_8H_{17}{}^+A^- + C_4H_{10} &\longrightarrow C_8H_{18} + C_4H_9{}^+A^-. \end{aligned}$$

in which $Cl^\bullet$ atoms and $C_4H_9{}^+A^-$ ion pairs are the catalysts, although this formalism is at variance with commonly accepted notions [154]. The presence of a catalytic cycle with the step of carbocation termination allows these reactions to be classified as nonbranched ion-chain processes [154].

Thus, in the case of a nonbranched chain reaction, a noncyclic route I is conjugated via one or two shared common intermediates with a cyclic route (or several routes) II. This situation is clearly illustrated by kinetic graphs KG21 and KG22 with null-vertices (Figure 3.25).

The reaction rate over a cyclic route II depends on the rates of steps in route I. The active centers X_i in chain mechanisms can represent atoms, radical species (R•), carbenium and carbonium ions, metal ions

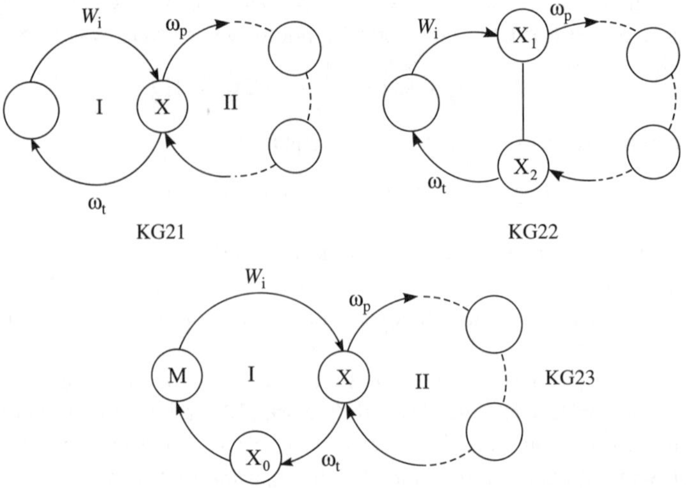

Figure 3.25 Kinetic graphs for nonbranched chain processes (W_i is the rate of initiation step; ω_p and ω_t are the weights of the chain propagation and termination steps, respectively)

and complexes [155–159], and various adsorbed species on the surface of metals and metal oxides [116] (see Chapter 5). A kinetic conjugation of two (or more) routes by the type of KG21 and KG22 is the main feature of chain processes. It is the presence of the first (noncyclic) route I that makes it possible to carry out the kinetically hindered route II, thus essentially providing a catalyst for the second route ($R\cdot, R^+, SO_3^{\bar{\bullet}}$, M–R, M–H, M=CHR, etc.). The overall equation of route II can include common reactants with the first route, but route I does not change the stoichiometry of the overall equation of chain propagation steps and, hence, there is no thermodynamic conjugation of the two routes. If the step of the active center loss in route I is followed by the step of regeneration of the chain initiator, the process will also become catalytic with respect to initiator M (KG23, Figure 3.25).

Chain processes of olefin polymerization in solutions of transition metal complexes and on the surface of heterogenized or solid-phase-dispersed metal complexes possess mechanisms that are more complicated as compared to the chain processes represented by KG21 and KG22. The reaction product – an oligomer or polymer molecule free of the initiator components (transition metals, organoaluminum compounds, etc.) is formed as a result of the quadratic termination of the active center (similar to the case of radical polymerization), rather than as the step of chain propagation. However, the main portion of the product appears as a result of the chain transfer to monomer (R_{-H} formation), for example, in reaction

$$\mathrm{MR} + \mathrm{CH_2{=}CH_2} \longrightarrow \underset{\text{active center}}{\mathrm{MC_2H_5}} + \mathrm{R_{-H}}; \tag{3.278}$$

or as a result of the kinetic and molecular chain termination (R_{-H} formation) with the subsequent re-initiation [150, 159]

$$\begin{aligned} \mathrm{MR} &\longrightarrow \mathrm{M{-}H} + \mathrm{R_{-H}}; \\ \mathrm{M{-}H} + \mathrm{C_2H_4} &\longrightarrow \mathrm{MC_2H_5}. \end{aligned} \tag{3.279}$$

In addition, the molecular chain terminates upon the chain transfer to AlR_xCl_{3-x} (in the case of Zigler–Natta systems) or as a result of the hydrogenolysis of M–R bonds in the reaction with H_2. It is the large contribution from steps (3.278) and (3.279) that accounts for the appearance of catalytic cycles and allows the processes of polymerization initiated by metal complexes to be considered as catalytic processes. The chain mechanism of olefin dimerization in such systems with metal complexes is virtually identical with the chain mechanism of catalytic oligomerization or polymerization, and up to 10,000 product molecules can form per atom of transition metal [160–163].

Many examples of conjugation between the reactions of oxidation of organic and inorganic molecules and the oxidative dehydration of alkanes and alkenes with participation of H_2O_2 in the gaseous and liquid phases are considered by Nagiev [119]. However, most of the proposed mechanisms illustrate typical chain processes rather than the classical variant of conjugated reactions with chemical induction. For example, the oxidative dehydrogenation of paraffins by hydrogen peroxide in the gas phase presents a kinetic conjugation of the route of H_2O_2 decomposition

$$2H_2O_2 \longrightarrow 2H_2O + O_2 \tag{3.280}$$

and oxidative dehydrogenation [119a]

$$H_2O_2 + RH \longrightarrow R_{-H} + 2H_2O. \tag{3.281}$$

within the framework of the following chain mechanism:

$$\begin{array}{lll} (1)\ H_2O_2 \longrightarrow 2\dot{O}H & \text{chain nucleation;} & \\ (2)\ H_2O_2 + \dot{O}H \longrightarrow H\dot{O}_2 + H_2O & \rbrace\ \text{chain} & \\ (3)\ H\dot{O}_2 + RH \longrightarrow R_{-H} + H_2O + \dot{O}H & \rbrace\ \text{propagation} & \\ (4)\ H\dot{O}_2 + \text{wall} \longrightarrow H\dot{O}_{2\,\text{wall}} & & \\ (5)\ \dot{O}H + \text{wall} \longrightarrow \dot{O}H_{\text{wall}} & \rbrace\ \text{chain termination} & \\ (6)\ H\dot{O}_{2\,\text{wall}} + \dot{O}H_{\text{wall}} \longrightarrow O_2 + H_2O & & \end{array} \tag{3.282}$$

Here, route I of the chain initiation and termination on the wall includes steps (1), (2), (4), (5), and (6), while route II of the chain propagation includes steps (2) and (3). Thus, the scheme presents a kinetic conjugation of reactions (3.280) and (3.281), rather than of reactions (3.280) ($\Delta G^{\circ}_{298\,^{\circ}\text{C}} = -56.3$ kcal/mole) and RH dehydrogenation (as was suggested by Nagiev). If the oxidative dehydrogenation were not a chain process, e.g.,

$$\begin{aligned} &2H_2O_2 \xrightarrow[1]{} [H\dot{O}_2 + \dot{O}H + H_2O] \xrightarrow[2]{} 2H_2O + O_2; \\ &RH + H\dot{O}_2 \xrightarrow[3]{} \dot{R} + H_2O_2; \\ &\dot{R} + \dot{O}H \xrightarrow[4]{} R_{-H} + H_2O. \end{aligned} \tag{3.283}$$

then the summation of steps **1, 3**, and **4** would give an equation of reaction (3.281) in the classical variant of conjugate reactions. However, since H_2O_2 in the case under consideration acts simultaneously as both inductor and actor, and the inductor molecule is isolated in step **3**, no new overall equation different from (3.281) is obtained.

In most reactions with H_2O_2 participation described in the literature and investigated by Nagiev [119], the consumption of H_2O_2 significantly (by a factor of 2–10) exceeds that of the substrate. For this reason, the kinetic chain length γ is significantly smaller than unity. For example, at a 10% selectivity of the oxidative dehydrogenation process, nine O_2 molecules are evolved and nineteen H_2O_2 molecules are spent per mole of obtained olefin. In an analysis of this situation, it was suggested [119a] to consider reaction with $\gamma \gg 1$ as the initiated chain processes and those with $\gamma < 1$ as the induced chain processes. However, this suggestion can hardly be accepted because the large consumption of inductor in the chain termination steps gives no contribution to the thermodynamics of chain propagating steps.

The summation of overall equations of the routes of chain reactions with allowance for the kinetic chain length of initiator consumption per mole target product naturally leads to the net overall equation. For example, this summation for reactions (3.280) and (3.281) yields

$$19H_2O_2 + RH \longrightarrow R_{-H} + 20H_2O + 9O_2.$$

However, this formal summation of the chemical affinity for separate routes does not change the affinity (A) of the overall equation (3.281). This procedure is equivalent (for chain processes) to the summation of overall equations of kinetically nonconjugate reactions.

The oxidation of organic compounds by hydrogen peroxide in the liquid phase can exhibit a chain character, but it can also be a nonchain catalytic process, as in the case of conjugated *catalase* and *peroxidase* mechanisms of action of the *catalase* enzyme (Figure 3.24).

Within the framework of the chain process, it is also possible to implement the classical variant of conjugation with a change in the overall equation of the conjugated reaction. For example, in the radical chain oxidation of aldehydes to peroxy acids [164–166],

$$RH + O_2 \longrightarrow ROOH\ (R = R'CO) \tag{3.284}$$

the route of chain propagation steps

$$\dot{R} + O_2 \xrightarrow[1]{} R\dot{O}_2;$$
$$R\dot{O}_2 + RH \xrightarrow[2]{} \dot{R} + ROOH \tag{3.285}$$

can be conjugated with the kinetically hindered reaction of olefin epoxidation

$$R\dot{O}_2 + {>}C{=}C{<} \xrightarrow[3]{} R\dot{O} + {>}\underset{\diagdown O \diagup}{C{-}C}{<}\ ;$$
$$R\dot{O} + RH \xrightarrow[4]{} ROH + \dot{R}. \tag{3.286}$$

Figure 3.26 shows a kinetic graph (KG24) representing the cyclic part of this chain reaction mechanism (with neglect of the possible steps of ROOH decay to RO· and ·OH). The overall equation of the new

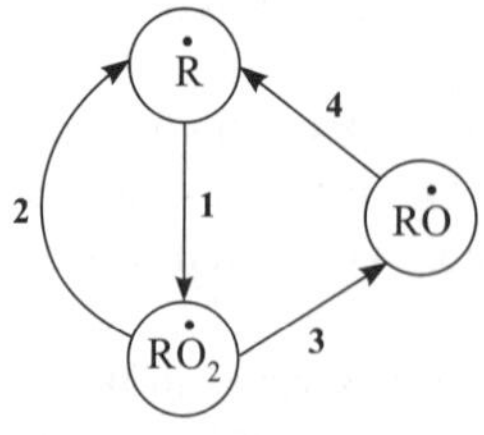

KG24

Figure 3.26 Kinetic graph for conjugated routes of RH oxidation and olefin epoxidation

route represents a new stoichiometry

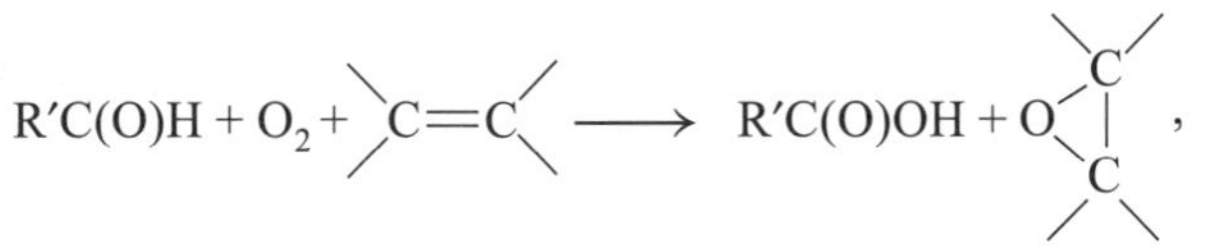
(3.287)

which combines two reactions – the oxidation of aldehyde to acid and the oxidation of olefin to epoxide. An analogous type of conjugation was considered above (Figure 3.23) for the reaction of SO_3^{2-} and AsO_3^{3-} oxidation, where the route of active center generation $SO_3^{\bullet -}$ included the Cu^{2+} catalyzed process of chain nucleation (3.229) and termination (3.230).

A chain variant of the kinetic conjugation of routes can also be implemented in a catalytic process (see KG25 in Figure 3.25 and Section 5.4). For example, the process of olefin (cyclohexene, CH) hydrocarboxylation to cyclohexanecarboxylic acid (CHCA) can be carried out under mild conditions in a solution of $PdBr_2$ in THF using conjugate consequent reaction (3.272) [144]:

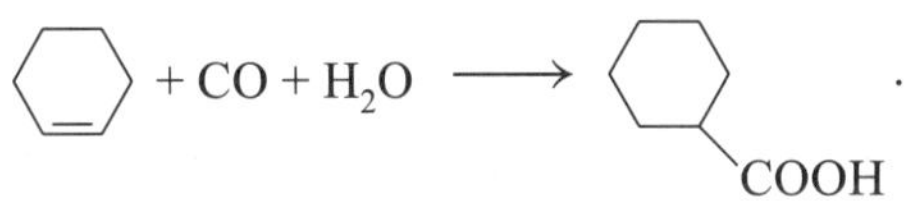
(3.288)

This reaction does not take place in the absence of oxygen (Pd(II) is reduced) and proceeds at a rate of 0.3–0.4 M/h only under the conditions of effective CO oxidation to CO_2. It was suggested that the intermediate hydride complex (HPdBr) involved in CO oxidation, which would be highly active in the absence of phosphine ligands, catalyzes reaction (3.288) at 30 °C and 1 bar. As was noted above, the absence of CO hinders the oxidation of THF to γ-butyrolactone (GBL). The scheme of possible conjugate processes for the formation of H_2O_2, CO_2, GBL, and CHCA is presented in Figure 3.27. The mechanism of THF oxidation by Pd(II) hydroperoxide is still not completely clear, while GBL is formed with probable participation of ~PdOOH via the following overall equation:

$$CO + THF + 1.5O_2 \longrightarrow CO_2 + GBL + H_2O. \quad (3.289)$$

It is also interesting to mention some oxidative systems proposed [167–169] for the nonoxidative hydrocarboxylation of olefins and alkynes. For example, the reaction

$$RCH{=}CH_2 + CO + H_2O \longrightarrow RCH(COOH)CH_3 \quad (3.290)$$

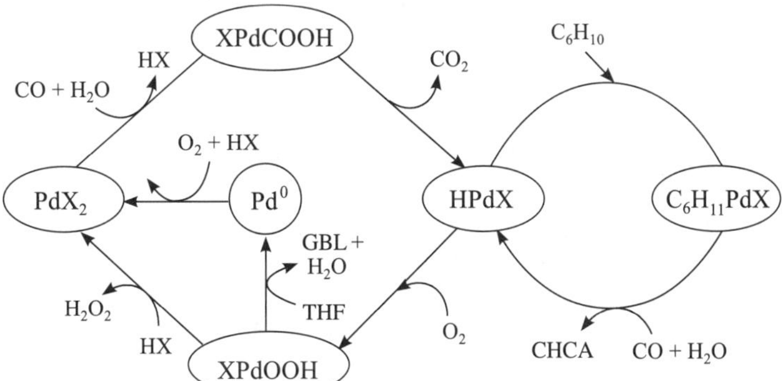

Figure 3.27 Kinetic graph for conjugated routes in the $CO–H_2O–THF–Cyclohexene–O_2–PdBr_2$ system

proceeds regioselectively under mild conditions in the $PdCl_2-CuCl_2-HCl-O_2$ system [167]. This process is suppressed by a large concentration of $CuCl_2$. The reaction mechanism was not studied, although the absence of the products of oxidative carbonylation of olefins, as well as the corresponding intermediates $RCH(COOH)CH_2Pd$ (α-substituted acrylates), suggests that this system also features a conjugation of the process of CO oxidation and reaction (3.290) via a common intermediate palladium hydride complex according to the chain mechanism.

An interesting case of the catalytic chain reaction with a KG23 type kinetic graph structure (Figure 3.25) is encountered in enzymatic processes with the participation of allosteric regulatory enzymes containing several linked active centers. A substrate (S) of this enzyme (E) can play the role of a positive modulator that increases the activity of other active centers. In this case, the substrate–enzyme interaction with the formation of the ES complex makes the enzyme more active with respect to the substrate conversion in the same reaction. These phenomena are called the *cooperative effects* [117].

A simple model of the substrate-induced cooperative effect is a four-step two-route scheme represented by graph KG25 in Figure 3.28. According to this, reaction

$$\mathrm{S} + \mathrm{A} = \mathrm{P} \tag{3.291}$$

can proceed on both centers E and X_1 of the enzyme. The latter center is capable of coordinating n molecules of the substrate and ensuring a higher rate of conversion than that on center E.

If steps **1** and **3** are quasi-equilibrium, then graph KG25 transforms into KG26, in which $X_1 = \mathrm{ES}$ and $X_2 = \mathrm{ES}_{n+1} = \mathrm{ES}_h$, so that $[\mathrm{E}]_0 = [\mathrm{E}](1 + K_1[\mathrm{S}] + K_h[\mathrm{S}]^h)$. Then, the rate of product P formation is described by the equation of a two-route reaction as

$$r_\mathrm{P} = \frac{(k_1K_1[\mathrm{S}][\mathrm{A}] + k_2K_h[\mathrm{S}]^h[\mathrm{A}])[\mathrm{E}]_0}{1 + K_1[\mathrm{S}] + K_h[\mathrm{S}]^h} = \frac{r_1^{\max}K_1[\mathrm{S}] + r_2^{\max}K_h[\mathrm{S}]^h}{1 + K_1[\mathrm{S}] + K_h[\mathrm{S}]^h}. \tag{3.292}$$

In the region of high substrate concentrations, the function $r_\mathrm{P} = f([\mathrm{S}])$ is described by an S-shaped curve that is plotted in the coordinates close to those of the Hill equation [120, 170]:

$$f = \left(\frac{r_\mathrm{P} - r_1^{\max}}{r_2^{\max} - r_\mathrm{P}}\right) = \frac{K_h}{K_1}[\mathrm{S}]^{h-1}. \tag{3.293}$$

In addition to the S-shaped curve, the initial reaction rates are sometimes described by a curve with two plateaus ($r_1^{\max}$ and $r_2^{\max}$), which corresponds to Eq. (2.292) with large values of exponent h. For example, in the reaction of acetylene reduction to ethylene and ethane by europium amalgam on an isolated nitrogenase cofactor FeMoCo [171, 172], the experimental values of the initial reaction rate (TOF, min^{-1}) were described by Eq. (3.292) with $h \cong 6$ for both ethane and ethylene (Figure 3.29). Thus, the cooperative effect can be described by the scheme of a chain process in which route I induces the activity

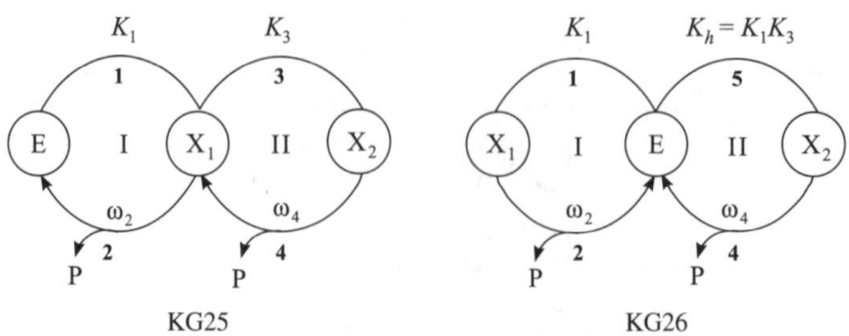

Figure 3.28 Kinetic graphs for processes with cooperative effects (see text for explanations)

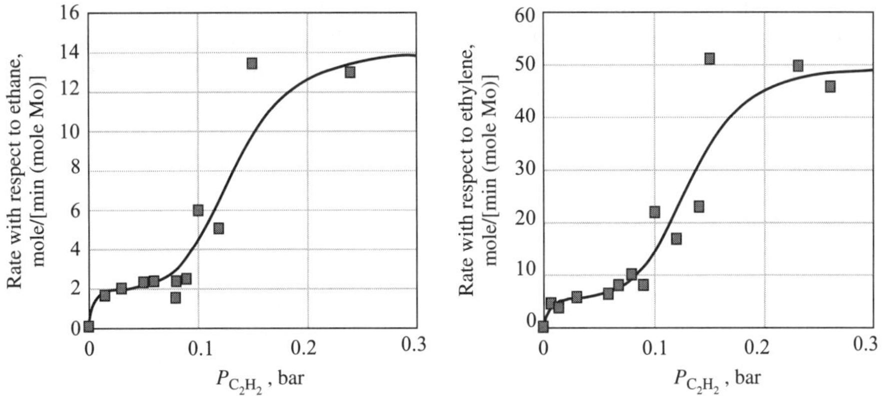

Figure 3.29 Plots of the initial acetylene reduction rates (with respect to ethane and ethylene) on a FeMo cofactor of nitrogenase. Points represent experimental data [173], curves show the results of model calculations performed by K.Yu. Odintsov

of other potentially active centers. There can be several types of these centers. If X_1 induces the activation of another one substrate molecule (X_2), while X_2 coordinates a total of h substrate molecules, then the kinetic equation will change and the terms proportional to $[S]^2$ will appear in both the numerator and denominator of Eq. (3.292). The cooperative models in enzymatic catalysis were considered in detail by Lescovac [173].

3.4.4 Conclusions

The aforementioned facts and their analysis convincingly confirm a conclusion that the kinetic conjugation of steps and routes via common intermediates is a more general concept than the classical variant of conjugated reactions usually considered in handbooks and monographs. This circumstance was originally pointed out by N.A. Shilov who wrote [14] that, "if the conjugation of reactions is treated as the dependence of the course of one process on the other chemical interactions in the same medium, then the possible variants of conjugate reactions are by no means restricted to the phenomenon of chemical induction."

One variant of conjugated reactions is called the chemical induction. This phenomenon is based essentially on the kinetic conjugation of steps in a consecutive reaction, which can either exhibit branching or not at one of the intermediate products. The main feature of the chemical induction is the participation of a new reactant (one or several substances that are foreign for the induced secondary reaction and are called inductors in the classical variant of conjugation) in the stoichiometry of the new overall equation. This modification is the essence of the thermodynamic conjugation. If we accept that any consumption of reactants for the formation of an active intermediate not leading to the appearance of a new overall equation (as in the classical variant of conjugated reactions) is a manifestation of the chemical induction, then any chain process with an arbitrary length of kinetic chain can be treated as a chemically induced process.

The kinetic conjugation in chain reaction mechanisms does not provide a thermodynamic conjugation, but makes it possible to carry out a kinetically hindered process provided that the primary reaction (route of chain initiation and termination) yields an intermediate (active center) for the secondary (cyclic) route As was demonstrated above, a kinetic conjugation with the participation of side reactants that ensure a new stoichiometry is the only chemical pathway to carry out endoergonic processes without using elevated temperatures and applying photochemical, radiative, ultrasonic, mechanical, microwave, etc. external actions upon substances. These conjugate reactions are still insufficiently used in practical chemistry, but the data presented above give convincing evidence for good prospects of this approach in designing novel catalytic systems and developing new types of chemical processes (see, e.g., reactions (3.221), (3.262), (3.266), (3.268), (3.269), (3.272), (3.277)).

Investigation into the mechanisms of conjugated reactions and the creation of new conjugation nodes is a powerful approach to discriminating hypotheses and obtaining new data about the nature of unstable intermediates.

Information gained during a century of research in the field of reactions involving metal complexes, organometallic compounds, and mechanisms of catalytic reactions opens the way to targeted development of various conjugate reactions.

The kinetic conjugation of steps and routes is manifested, depending on the method of conjugation, in the equations of reaction rates written according to the theory of complex reaction kinetics. There are no individual manifestations of the kinetics specific of the classical conjugation variant with chemical induction, which could separate this type of conjugation from the general group of multi-route mechanisms with common steps.

Complex multi-route catalytic processes can simultaneously feature various types of kinetic conjugation and different types of conjugate reactions. The conjugation of oxidative and reductive carbonylation of reactions with additive carbonylation processes was observed in solutions of Pd(I) iodide complexes [17, 18]. Figure 3.30 shows a kinetic graph for the acetylene carbonylation process, in which it is possible to reveal five independent conjugated routes and five corresponding overall equations:

$$\text{(I)}\quad \underset{\text{(steps } \mathbf{1,2,3}\text{)}}{CO + C_2H_2 + ROH} \longrightarrow \underset{\text{(A)}}{CH_2{=}CHCOOR};$$

$$\text{(II)}\quad \underset{\text{(steps } \mathbf{1,2,4,5}\text{)}}{2CO + C_2H_2 + 2ROH} \longrightarrow \underset{\text{(S)}}{ROCOCH_2CH_2COOR};$$

$$\text{(III)}\quad \underset{\text{(steps } \mathbf{1,2,4,6,9}\text{)}}{2CO + 2C_2H_2 + 2ROH} \longrightarrow \underset{\text{(M)}}{ROCOCH{=}CHCOOR} + \underset{\text{(E)}}{C_2H_4};$$

$$\text{(IV)}\quad \underset{\text{(steps } \mathbf{1,2,4,6,8,10}\text{)}}{3CO + 2C_2H_2 + 3ROH} \longrightarrow \underset{\text{(M)}}{ROCOCH{=}CHCOOR} + \underset{\text{(P)}}{CH_3CH_2COOR};$$

$$\text{(V)}\quad \underset{\text{(steps } \mathbf{7,8}\text{)}}{CO + C_2H_2 + ROH} \longrightarrow \underset{\text{(A)}}{CH_2{=}CHCOOR}.$$

Here, route V is conjugated according to the chain type with the steps of intermediate X_4 nucleation (**1, 2, 4, 6**) maleate (M) formation, and the steps (**9, 10**) of chain termination on X_4 (route III) and on X_5 (routes IV and V). The chain character of conjugation of the two routes of chain nucleation and termination (III, IV) with route V is illustrated by a simplified kinetic graph in Figure 3.31. The conjugation of the catalyst formation and conversion processes and the catalytic cycle has been studied for numerous reactions (see Chapter 5).

Various mechanisms of the kinetic and thermodynamic conjugation are widely spread in biochemical enzymatic processes [117, 118, 174, 175], where the oxidative reactions are conjugated with nonoxidative, and the nonoxidative exoergonic processes are conjugated with endoergonic. The Krebs cycle for citric acid oxidation is conjugated with the oxidation of one of the most important NADH coenzymes (the reduced form of nicotinamide adenine dinucleotide, NAD^+) and with the phosphorylation of adenosine diphosphate (ADP) to adenosine triphosphate (ATP). This conjugated process is among the most important elements of the cellular energy circuit [117, 118]. NAD^+ or nicotinamide adenine dinucleotide phosphate

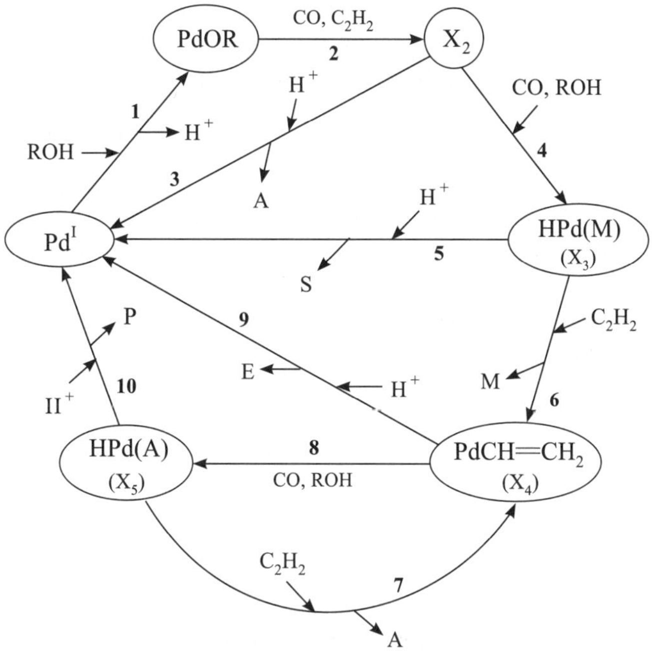

Figure 3.30 Kinetic graph for mechanism of acetylene carbonylation: (A) acrylate; (M) maleate; (S) succinate; (E) ethylene; (P) propionate

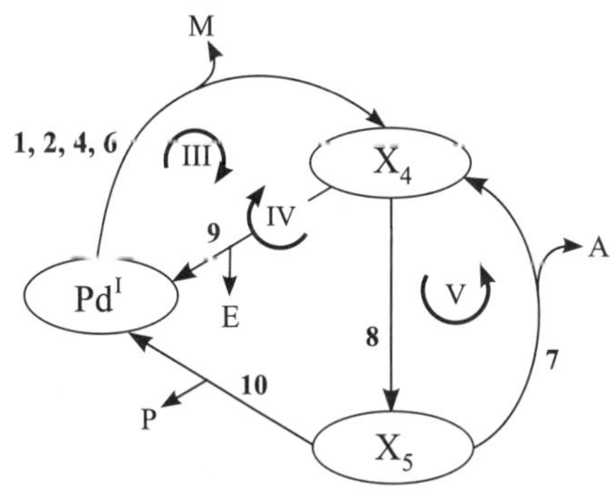

Figure 3.31 Kinetic graph for conjugated routes III, IV, and V in the scheme of catalytic chain process (notations are the same as in Figure 3.30)

(NADP) in combination with enzymes is involved in more than 150 biochemical processes. The essence of the process of NADH oxidation is the transfer of two electrons and proton (H^-) to an oxidant:

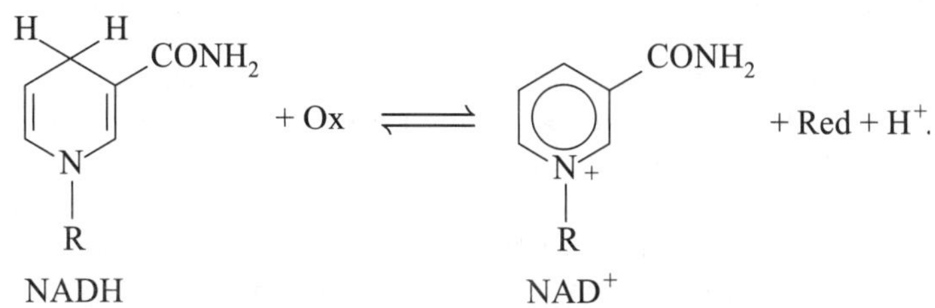

The synthesis of the energy-rich ATP molecule is conjugated with the processes of carbohydrate glycolysis to lactic acid and glucose oxidation to CO_2 in the course of oxidative (respiratory) phosphorylation. For example, the glycolysis of glucose proceeds according to the equation

$$\text{Glucose} + O_2 + 2\text{ADP} + 2P_n \longrightarrow 2\text{lactate} + 2\text{ATP} + 2H_2O. \tag{3.294}$$

where P_n is the free phosphate anion. The oxidation proceeds via the intermediate phosphoenolpyruvate anion, which transfers the phosphate group to ADP with the aid of the *pyruvate kinase* enzyme [117]. In most cases, the conjugate biochemical reactions were found to pass via special intermediates that ensure the conjugation of various processes. For example, the following mechanism ensures conjugation of the oxidation of glyceraldehyde-3-phosphate (RCHO) with the phosphoryalation of 3-phosphoglycerate ($RCOO^-$) to 1,3-diphosphoglycerate ($RCOOPO_3^{2-}$) [117]:

$$\text{RCHO} + \text{NAD}^+ + H_2O \longrightarrow \text{RCOO}^- + \text{NADH} + 2H^+;$$
$$\Delta G_1^\circ = -10.3 \text{ kcal/mole}$$

$$\text{RCOO}^- + H_2PO_4^- \longrightarrow \text{RC}\begin{matrix}\nearrow O \\ \searrow OPO_3^{2-}\end{matrix} + H_2O.$$
$$\Delta G_2^\circ = +1.5 \text{ kcal/mole}$$

The conjugation of these reactions with the participation of the enzyme *glyceralphosphate dehydrogenase* is provided by the following mechanism:

$$\text{E—SH} \xrightarrow[-H^+]{\text{NAD}^+} \text{E—S(NAD)} \xrightarrow{\text{RCHO}} \text{E(S—}\underset{\displaystyle O^-}{\underset{|}{\text{CHR}}}\text{)(NAD}^+\text{)} \longrightarrow \text{E(NADH)(S}\underset{\displaystyle O}{\underset{\|}{\text{C}}}\text{R)} \xrightarrow[\text{NADH}]{\text{NAD}^+}$$

$$\longrightarrow \text{E(NAD)(S}\underset{\displaystyle O}{\underset{\|}{\text{C}}}\text{R)} \xrightarrow{H_2PO_4^{2-}} \text{E—S(NAD)} + \text{RCOOPO}_3^{2-} + 2H^+.$$

in which E–S(NAD) is the active center. One NAD^+ molecule enters into the catalyst and another plays the role of the oxidant.

In the course of oxidative phosphorylation in the respiratory chain (with the formation of ATP in mitochondria) featuring the cycle of oxidation of carboxylic acids (a catalytic hypercycle, in which every step is a catalytic cycle and all steps also constitute a catalytic cycle), electrons and protons are generated that reduce NAD^+ to NADH (with the ATP formation). In turn, NADH is oxidized by oxygen in the process that is conjugated with the phosphorylation of ADP:

$$\text{NADH} + H^+ + 3\text{ADP} + 3P_n + \tfrac{1}{2}O_2 \longrightarrow \text{NAD}^+ + 4H_2O + 3\text{ATP}. \tag{3.295}$$

In these processes, which occur on the membranes inside mitochondria, another mechanism is responsible for the conjugation with the main participant (H^+ ion). The oxidation of NADH gives rise to a gradient of the H^+ ion concentration between the inside and outside of mitochondria, whereby excess OH^- ions appear inside and excess H^+ ions appear outside (in cytoplasm). The detailed mechanism of operation of the protein complex of ATP synthase is still unknown. It has been suggested that a conjugation of the phosphorylation process with the chemoosmotic effect responsible for the chemical potential ($\Delta\mu_{H^+}$) development and charge separation between the inner (−) and outer (+) surfaces of mitochondrial membranes is ensured by the following hypothetical scheme [118, 176]:

(i) synthesis of ATP ions (bound in complex with Mg^{2+}) without energy consumption via the reaction

$$\text{ADP Mg}^- + H_2PO_4^- \longrightarrow \text{ATP Mg}^{2-} + H_2O$$

from ADP and P_n ions bound by enzyme protein β-subunits;

(ii) synthesis of ATP causes a change in conformation of the enzyme protein β-subunit and opens a channel for the transport of protons inside mitochondria;

(iii) separation of bound ATP from protein and passage into solution, which is ensured by the energy of the neutralization reaction

$$H^+ + OH^- \rightleftharpoons H_2O \text{ (capacitor discharge)};$$

(iv) Transport of ATP^{4-} from mitochondria into cytoplasm and ADP^{3-} and phosphate anions from cytoplasm into mitochondria, which is ensured by the potential difference on the surface of the mitochondrial membrane.

NADH is frequently involved in *monooxygenase* mediated reactions of substrate oxidation:

$$S + O_2 + DH_2 \longrightarrow SO + D + H_2O; \quad (3.296)$$

$$\text{NADH} + O_2 + SH + H^+ \longrightarrow \text{NAD}^+ + H_2O + SOH. \quad (3.297)$$

Processes of this type were considered in Section 3.4.2.

In recent years, new data have been reported on the conjugate oxidation processes, in which the role of both proton (H^+) and electron donors is played by the molecular hydrogen [177–179]. In particular, using two Pt/SiO_2 catalysts for the synthesis of H_2O_2 from H_2 and O_2 and heteropolyacids for the hydroxylation of ArH by hydroperoxide, it is possible to carry out the following process [179]:

$$H_2 + O_2 + C_6H_6 \longrightarrow C_6H_5OH + H_2O. \quad (3.298)$$

An interesting example is the conjugation of nonoxidative reactions, namely, an exoergonic process of ATP hydrolysis and an endoergonic process of addition of tyrosine ($HOC_6H_4CH_2CH(NH_3{}^+)CO_2^-$, tyr) to transport ribonucleic acid (t-RNA) [180]:

$$E + \text{Tyr} + \text{ATP} \rightleftharpoons E\cdot\text{Tyr(ATP)};$$

$$E\cdot\text{Tyr(ATP)} \longrightarrow E(\text{Tyr}\cdot\text{AMP}) + PP_i;$$

$$PP_i = H_2P_2O_7^{2-}$$

$$E(\text{Tyr}\cdot\text{AMP}) + t\text{-PHK} \longrightarrow E + \text{AMP} + \text{Tyr}\cdot\text{PHK}.$$

The reaction of ATP hydrolysis is also used in reductive processes, in particular in the endoergonic reaction of N_2 reduction to NH_3 with the participation of *nitrogenase* enzymes [181]:

$$N_2 + 8e^- + 8H^+ + 16\ \text{ATP Mg}^{2-} \longrightarrow 2NH_3 + H_2 + 16\ \text{ADP Mg}^- + 16P_n. \quad (3.299)$$

Reactions such as the synthesis of formaldehyde, formic acid, and acetic acid

$$CO + H_2 \longrightarrow CH_2O; \quad (3.300)$$

$$CO + H_2O \longrightarrow HCOOH; \quad (3.301)$$

$$CH_4 + CO_2 \longrightarrow CH_3COOH. \quad (3.302)$$

and some other thermodynamically constrained processes can be carried out due to the conjugation with reactions of CO and H_2 oxidation. As noted above, formic acid was found in the conjugate process of CO and H_2O oxidation [144], formaldehyde was formed in the conjugate process of CO oxidation and reduction, and acetic acid was obtained via strongly endoergonic reactions (3.302) (in the gas phase and in the case of liquid acetic acid formation) in the $Pd(OAc)_2$–$Cu(OAc)_2$–$K_2S_2O_8$ в CF_3COOH system [133], most probably due to the conjugation of reaction (3.302) with methane oxidation:

$$2CH_4 + CO_2 + CF_3COOH + K_2S_2O_8 \longrightarrow CH_3COOH + CH_3OCOCF_3 + 2KHSO_4. \qquad (3.303)$$

The above examples of conjugate biochemical reactions show evidence of a considerable potential, still not utilized by chemists, which is offered by the idea of kinetic and thermodynamic conjugation, especially for carrying out thermodynamically constrained processes.

References

1. W. Reppe, *Lieb. Ann.*, **582**, 1 (1953); **596**, 1 (1955).
2. O. N. Temkin, S. M. Brailovskii and L. G. Bruk, *Tezisy dokladov Vsesoyuznoi konferentsii po mekhanizmam kataliticheskikh reaktsii* (Abstracts of Papers. All-Union Conf. on Mechnisms of Catalytic Reactions) (Moscow, 1978), Vol. 1, pp. 74–81.
3. E. Gordeeva, D. Bonchev, D. Kamenski, O. N. Temkin, *J. Chem. Inf. Comput. Sci*., **34**, 436 (1994).
4. O. N. Temkin, A. V. Zeigarnik and D. G. Bonchev, *Chemical Reaction Networks: A Graph-Theoretical Approach*, CRC Press, Boca Raton, FL, 1996.
5. L. G. Bruk and O. N. Temkin, unpublished results.
6. J. Collman, L. Hegedus J. Norton and R. Finke, *Principles and Applications of Organotransition Metal Chemistry* (Graduate Textbook), University Science Books, Mill Valley, CA, 1987.
7. R. H. Grubbs and A. Miyashita, in: *Fundamental Research in Homogeneous Catalysis*, M. Tsutsui (Ed.), Plenum Press, New York, 1979, Vol. 3, p. 151.
8. J. K. Stille, in: *The Chemistry of the Metal-Carbon Bond*, F. R. Hartley and S. Patai (Eds), John Wiley & Sons, 1985, Vol. 2, pp. 626–790.
9. J.-M. Basset, F. Lefebvre and C. Santini, *Coord. Chem. Rev*., **178–180**, 1703 (1998); F. Lefebvre and J.-M. Basset, *J. Mol. Catal. A: Chem*., **146**, 3 (1999).
10. O. N. Temkin, G. K. Shestakov and Yu. A. Treger, *Atsetilen: Khimiya, mekhanizmy reaktsii, tekhnologiya* (Acetylene: Chemistry, Reaction Mechanisms, and Technology), Khimiya, Moscow, 1991.
11. O. L. Kaliya, R. M. Flid, L. G. Volkova and O. N. Temkin, *Zh. Neorg. Khim*., **15**, 2561 (1970).
12. S. M. Brailovskii, V. S. Shestakova, T. N. Omel'chenko and O. N. Temkin, *Zh. Org. Khim*., **14**, 2346 (1978).
13. O. N. Temkin, O. L. Kaliya, R. M. Flid and N. G. Mekhryakova, *Dokl. Akad. Nauk SSSR*, **199**, 1321 (1971).
14. L. N. Zhir-Lebed', N. G. Mekhryakova, R. M. Flid, V. A. Golodov and O. N. Temkin, *Kinet. Katal*., **15**, 537 (1974).
15. L. N. Zhir-Lebed' and O. N. Temkin, *Kinet. Katal*., **25**, 316 (1984).
16. O. N. Temkin and L. G. Bruk, *Kinet. Katal*., **44**, 661 (2003).
17. N. Yu. Vsesvyatskaya, L. G. Bruk, T. A. Stromnova, N. F. Alekseeva and O. N. Temkin, *Zh. Org. Khim*., **14**, 473 (1978).
18. L. G. Bruk, *Metody vydvizheniya i diskriminatsii gipoteticheskikh mekhanizmov kataliticheskikh reaktsii kak osnova razrabotki tekhnologii karbonilirovaniya alkinov* (Method of Formulation and Discrimination of Hypothetical Mechanisms of Catalytic Reactions: Basis for Developing Alkyne Carbonylation Technlogy), Candidate Dissertation in Chemistry, MITKhT (Moscow Institute of Fine Chemical Technology), Moscow, 1996.
19. (a) K. Zaw, M. Lautens and P. M. Henry, *Organometallics*, **4**, 1286 (1985); (b) W. K. Wan, K. Zaw and P. M. Henry, *Organometallics*, **7**, 1667 (1988).
20. D. Bonchev, D. Kamenski and O. N. Temkin, *J. Math. Chem*., **1**, 345 (1987).
21. A. V. Zeigarnik and O. N. Temkin, *Kinet. Katal*., **37**, 372 (1996).

22. O. N. Temkin, L. G. Bruk and D. Bonchev, *Teor. Eksp. Khim.*, No. 3, 282 (1988).
23. L. G. Bruk, V. R. Flid, T. B. Belova and O. N. Temkin, in *Khimicheskaya hinetika v katalize: Kineticheskie modeli zhidkofaznykh reaktsii* (Chemical Kinetics in Catalysis: Kinetic Models of Liquid-Phase Reactions), Akad. Nauk SSSR, Chernogolovka, 1985, p. 120.
24. G. K. Shestakov, S. M. Airyan, F. I. Bel'skii and O. N. Temkin, *Zh. Org. Khim.*, **12**, 2053 (1976).
25. G. K. Shestakov, F. N. Bel'skii, S. M. Airyan and O. N. Temkin, *Kinet. Katal.*, **19**, 334 (1978).
26. D. Heller, R. Thede and D. Haberland, *J. Mol. Catal. A: Chem.*, **115**, 273 (1997).
27. D. Heller, H. Buschmann and H. Neuman, *J. Mol. Catal. A: Chem.*, **125**, 9 (1997).
28. D. Heller and H. Buschmann, *Top. Catal.*, **5**, 159 (1998).
29. A. S. C. Chan, J. J. Pluth and J. Halpern, *J. Am. Chem. Soc.*, **102**, 5952 (1980).
30. C. R. Landis and J. Halpern, *J. Am. Chem. Soc.*, **109**, 1746 (1987).
31. J. M. Brown and P. M. Chaloner, *J. Chem. Soc. Chem. Commun.*, 344 (1980).
32. J. Halpern, D. P. Riley, A. S. C. Chan and J. J. Pluth, *J. Am. Chem. Soc.*, **99**, 8055 (1977).
33. A. S. C. Chan, J. J. Pluth and J. Halpern, *Inorg. Chim. Acta*, **37**, L477 (1979).
34. A. S. C. Chan and J. Halpern, *J. Am. Chem. Soc.*, **102**, 838 (1980).
35. B. McCulloch, J. Halpern, M. R. Thompson and C. R. Landis, *Organometallics*, **9**, 1392 (1990).
36. J. A. Ramsden, T. D. Claridge and J. M. Brown, *J. Chem Soc. Chem. Commun.*, 2469 (1995).
37. J. M. Keith, J. F. Larrow and E. N. Jacobsen, *Adv. Synth. Catal.*, **343**, 5 (2001).
38. Y. Wakatsuki and H. Yamazaki, *Synthesis*, **26** (1996).
39. H. Bonnemann, *Angew. Chem. Int. Edit.*, **24**, 248 (1985).
40. B. Heller, D. Heller and G. Oehme, *J. Mol. Cat. A: Chem.*, **110**, 211 (1996).
41. G. Zassinovich, G. Mestroni and S. Gladiali, *Chem. Rev.*, **92**, 1051 (1992).
42. P. Gamez, F. Fache and M. Lemaire, *Tetrahedron-Asymmetr.*, **6**, 705 (1995).
43. C. de Bellefon and N. Tanchoux, *Tetrahedron-Asymmetr.*, **9**, 3677 (1998).
44. C. de Bellefon, N. Tanchoux and S. Caravielhes, *J. Organometal. Chem.*, **567**, 143 (1998).
45. C. de Bellefon, N. Tanchoux and D. Schweich, *Proceedings of the 2nd Eur. Congr. on Chemical Engineering (ECCE-2)* (Montpellier, France, 1999), p. 1.
46. C. A. Tolman and J. W. Faller, in *Homogeneous Catalysis with Metal Phosphine Complexes*, L. H. Pignolet (Ed.), Plenum Press, New York, 1983, pp. 13–110.
47. M. H. J. M. de Croon, P. F. M. T. Van Niscelroij, H. J. A. M. Kuipers and J. W. E. Coenen, *J. Mol. Catal.*, **4**, 325 (1978).
48. D. Guillaneux, S.-H. Zhao, O. Samuel and D. Rainford, *J. Am. Chem. Soc.*, **116**, 9430 (1994).
49. D. G. Blackmond, *Proceedings of the 11th Int. Symp. on Homogeneous Catatalysis*, (Scotland, UK, 1998), Book of Abstracts, p. 2.
50. D. J. Berrisford, C, Bolm and K. B. Sharpless, *Angew. Chem. Int. Edit.*, **34**, 1059 (1995).
51. D. Förster, A. Hershman and D. E. Morris, *Catal. Rev.*, **23**, 89 (1981).
52. V. Yu. Gankin and G. S. Gurevich, *Tekhnologiya oksosinteza* (Oxosynthesis Technology), Khimiya, Leningrad, 1981.
53. G. A. Korneeva, *Gidroformilirovanie nepredel'nykh uglevodorodov i kislorodsoderzhashchikh soedinenii v prisutstvii kompleksov kobal'ta i rodiya* (Hydroformylation of Unsaturated Hydrocarbons and Oxygen-Containing Compounds in the Presence of Cobalt and Rhodium Complexes), Candidate Dissertation in Chemistry, Topchiev Institute of Petrochemical Synthesis, Russian Academy of Sciences, Moscow, 1994.
54. H. Sternberg and I. Wender, *Proceedings of Int. Conf. on Coordination Chemistry* (London, 1959), *Chem. Soc. Spec. Publ.*, No. 13, 25 (1959).
55. R. F. Heck and D. S. Breslow, *J. Am. Chem. Soc.*, **53**, 4023 (1961).
56. Yu. G. Noskov, A. P. Simonov and E. S. Petrov, *Kinet. Katal.*, **41**, 564 (2000).
57. Yu. G. Noskov, N. A. Novikov, M. I. Terekhova and E. S. Petrov, *Kinet. Katal.*, **32**, 331 (1991).
58. N. E. Petrova, Yu. G. Noskov, M. I. Terekhova and E. S. Petrov, *Zh. Obshch. Khim.*, **63**, 678 (1993).
59. Yu. G. Noskov, M. I. Terekhova and E. S. Petrov, *Kinet. Katal.*, **34**, 1001 (1993).
60. Yu. G. Noskov and E. S. Petrov, *Kinet. Katal.*, **34**, 1005 (1993).
61. Yu. G. Noskov and E. S. Petrov, *Kinet. Katal.*, **35**, 728 (1994).

62. T. E. Kron, Yu. G. Noskov, M. I Terekhova and E. S. Petrov, *Zh. Fiz. Khim.*, **70**, 82 (1996).
63. Yu. G. Noskov and E. S. Petrov, *Kinet. Katal.*, **38**, 568 (1997).
64. E. S. Petrov and Yu. G. Noskov, *Zh. Ross. Khim. O-va im. D. I. Mendeleeva*, **42**, 149 (1998).
65. Yu. G. Noskov and E. S. Petrov, *Izv. Ross. Akad. Nauk, Ser. Khim.*, No. 10, 1756 (2001).
66. J. Tsiji, *Acc. Chem. Res.*, **2**, 144 (1969).
67. R. Bardi, A. Del Pra, A. M. Piazzesi and L. Toniolo, *Inorg. Chim. Acta*, **35**, L345 (1979).
68. A. V. Protchenko, R. R. Shifrina and E. S. Petrov, *Neftekhimiya*, **25**, 200 (1998).
69. G. Cavinato and L. Toniolo, *J. Organomet. Chem.*, **398**, 187 (1990).
70. G. R. Estham, B. T. Heaton, J. A. Iggo et al., *J. Chem. Soc. Chem. Commun.*, 609 (2000).
71. Yu. G. Noskov and E. S. Petrov, *Zh. Obshch. Khim.*, **70**, 1685 (2000).
72. T. E. Kron, N. E. Petrova, N. I. Terekhova and E. S. Petrov, *Zh. Obshch. Khim.*, **69**, 1171 (1999).
73. T. E. Kron, M. I. Terekhova, Yu. G. Noskov and E. S. Petrov, *Kinet. Katal.*, **42**, 204 (2001).
74. A. Scrivanti, V. Beghetto, E. Campagna et al., *Organometallics*, **14**, 630 (1998).
75. V. R. Flid, O. S. Manulik, A. A. Grigor'ev and A. P. Belov, *Kinet. Katal.*, **41**, 658 (2000).
76. V. R. Flid, V. B. Kuznetsov, A. A. Grigor'ev and A. P. Belov, *Kinet. Katal.*, **41**, 666 (2000).
77. O. N. Temkin, L. G. Bruk and A. V. Zeigarnik, *Kinet. Katal.*, **34**, 445 (1993).
78. L. V. Shchel'tsyn, S. M. Brailovskii and O. N. Temkin, *Kinet. Katal.*, **31**, 1361 (1990).
79. M. I. Temkin, in *Mekhanizmy slozhnykh reaktsii. Trudy IV Mezhdunarodnogo kongressa po katalizu* (Mechanisms of Complex Reactions. Proceedings of the 4th International Congress on Catalysis), Nauka, Moscow, 1970, p. 57.
80. S. L. Kiperman, *Osnovy khimicheskoi kinetiki v geterogennom katalize* (Principles of Chemical Kinetics in Heterogeneous Catalysis), Khimiya, Moscow, 1979, p. 197.
81. G. V. Pazynina, O. L. Kaliya, E. A. Luk'yanets and I. G. Bolesov, *Zh. Org. Khim.*, 23, 813 (1987).
82. O. N. Temkin, N. F. Alekseeva, R. M. Flid, O. L. Kaliya, G. K. Shestakov and T. I. Dolgina, *Dokl. Akad. Nauk SSSR*, **196**, 836 (1971).
83. S. M. Brailovskii, E. V. Boiko, O. N. Temkin, A. S. Kostyushin and R. M. Flid, *Kinet. Katal.*, **14**, 513 (1973).
84. S. M. Brailovskii, L. G. Bruk, A. S. Kostyushin, O. N. Temkin and R. M. Flid, *Kinet. Katal.*, **14**, 1222 (1973).
85. S. M. Brailovskii, O. N. Temkin, S. I. Khodakova and R. M. Flid, *Zh. Org. Khim.*, **11**, 672 (1975).
86. S. M. Brailovskii, O. N. Temkin, A. S. Kostyushin and K. Yu. Odintsov, *Kinet. Katal.*, **31**, 1371 (1990).
87. S. M. Brailovskii, M.-H. Huin and O. N. Temkin, *Kinet. Katal.*, **35**, 734 (1994).
88. L. V. Shchel'tsyn, *Fiziko-khimicheskie osnovy tekhnologii kataliticheskogo sinteza khloralkinov i polikhlorolefinov iz atsetilena i ego proizvodnykh* (Physicochemical Principles of Technology of Catalytic Synthesis of Chloroalkynes and Polychlorolefins from Acetylene and Its Derivatives), Candidate Dissertation in Chemistry, MITKhT (Moscow Institute of Fine Chemical Technology), Moscow, 1989.
89. V. V. Zamashchikov, V. G. Popov and E. S. Rudakov, *Kinet. Katal.*, **35**, 372 (1994).
90. A. I. Mustafin and A. A. Shteiman, *React. Kinet. Catal. Lett.*, **6**, 153 (1977).
91. S. A. Mitchenko, V. V. Zamashchikov, A. A. Shubin and O. N. Pryadko, *Kinet. Katal.*, **34**, 253 (1993).
92. S. A. Mitchenko, V. V. Zamashchikov and A. A. Shubin, *Kinet. Katal.*, **34**, 479 (1993).
93. S. A. Mitchenko, V. V. Zamashchikov and A. A. Shubin, *Kinet. Katal.*, **32**, 343 (1991).
94. S. A. Mitchenko, V. P. Ananikov, V. V. Zamashchikov and Yu. A. Ustynyuk, *Izv. Ross. Akad. Nauk, Ser. Khim.*, No. 2, 500 (1996).
95. S. A. Mitchenko, V. P. Ananikov and V. V. Zamashchikov, *Teor. Eksp. Khim.*, **32**, 234 (1996).
96. S. A. Mitchenko, V. P. Ananikov, I. P. Beletskaya and Yu. A. Ustynyuk, *Mendeleev Comm.*, No. 4, 130 (1997).
97. S. A. Mitchenko, *Kinet. Katal.*, **40**, 278 (1999).
98. T. J. Katz and J. McGinnis, *J. Am. Chem. Soc.*, **97**, 1592 (1975).
99. T. J. Katz and J. McGinnis, *J. Am. Chem. Soc.*, **99**, 1903 (1977).
100. A. F. Shmidt and V. V. Smirnov, *Kinet. Katal.*, **42**, 800 (2001).
101. A. F. Shmidt, *Sopryazhenie protsessov prevrashcheniya katalizatora i osnovnogo kataliticheskogo tsikla na primere reaktsii Kheka* (Conjugation of Catalyst Conversion and Main Catalytic Cycle in Heck Reaction), Doctoral Dissertation in Chemistry, Irkutsk State University, Irkutsk, 2003.
102. P. Henry, *J. Org. Chem.*, **38**, 2415 (1973); *Adv. Organometal. Chem.*, **13**, 381 (1975).

103. I. V. Kozhevnikov, Yu. V. Burov and K. I. Matveev, *React. Kinet. Catal. Lett*., **6**, 35 (1977).
104. M.-H. Huin, S. M. Brailovskii and O. N. Temkin, *Kinet. Katal*., **35**, 266 (1994).
105. M.-H. Huin, S. M. Brailovskii and O. N. Temkin, *Kinet. Katal*., **35**, 367 (1994).
106. I. L. Kotlyarevskii, M. S. Shvartsberg and L. B. Fisher, *Reaktsii atsetilenovykh soedinenii* (Reactions of Acetylene Compounds), Nauka, Novosibirsk, 1967.
107. L. G. Fedenok, V. M. Berdnikov and M. S. Shvartsberg, *Zh. Org. Khim*., **10**, 922 (1974).
108. E. S. Rudakov and V. L. Lobachev, *Kinet. Katal*., **35**, 198, 203; 1995, **36**, 538 (1994).
109. E. S. Rudakov, V. L. Lobachev and E. V. Zaichuk, *Kinet. Katal*., **37**, 180, 534 (1996); **38**, 825 (1997).
110. V. L. Lobachev, *Selektivnost' i mekhanizmy elektrofil'nykh i okislitel'nykh reaktsii alkilbenzolov* (Selectivity and Mechanisms of Electrophilic and Oxidative Reactions of Alkylbenzenes), Doctoral Dissertation in Chemistry, Institute of Physicoorganic Chemistry and Carbochemistry, National Academy of Sciences of Ukraine, Donetsk, 2002.
111. N. I. Moiseeva, A. E. Gekhman, S. G. Sakharov, I. P. Stolyarov and I. I. Moiseev, *Izv. Ross. Akad. Nauk, Ser. Khim*., No. 10, 2396 (1986).
112. A. E. Gekhman, I. P. Stoljarov, N. I. Moiseeva and I. I. Moiseev, *Comt. Rend. Acad. Sci., Ser. II: Chim*., **7**, 833 (2004).
113. S. A. Mitchenko, V. V. Zamashchikov and O. N. Pryadko, *Kinet. Katal*., **34**, 484 (1993).
114. N. A. Shilov, *O sopryazhennykh reaktsiyakh okisleniya* (On Conjugate Oxidation Reactions), Moscow, 1905.
115. M. J. Boudart, *Phys. Chem*., **87**, 2786 (1983).
116. O. V. Krylov, *Kinet. Katal*., **26**, 263 (1985).
117. A. L. Lehninger, *Principles of Biochemisrtry*, Worth Publishers, New York, 1976.
118. V. P. Skulachev, *Energetika biologicheskikh membranes* (Energetics of Biological Membranes), Nauka, Moscow, 1989.
119. (a) T. M. Nagiev, *Khimicheskoe sopryazhenie. Sopryazhennye reaktsii okisleniya perekis'yu vodoroda* (Chemical Conjugation. Conjugate Reactions of Oxidation with Hydrogen Peroxide). Nauka, Moscow, 1989; (b) T. M. Nagiev, *Vzaimodeistie sinkhronnykh reaktsii v khimii i biologii* (Interaction of Synchronous Reactions in Chemistry and Biology), ELM, Baku, 2001.
120. E. T. Denisov, O. P. Sarkisov and G. I. Likhtenshtein, *Khimicheskaya kinetika* (Chemical Kinetics), Khimiya, Moscow, 2000.
121. V. I. Metelitsa, *Modelirovanie okislitel'no-vosstanovitel'nykh fermentov* (Modeling Redox Enzymes), Nauka i Tekhnika, Minsk, 1984, p. 158.
122. Ch. Walling and G. El-Taliawi, *J. Am. Chem. Soc*., **95**, 848 (1973).
123. N. Yu. Vsesvyatskaya, *Gidratatsiya metilatsetilena i allena v polifunktsional'nykh kataliticheskikh sistemakh* (Hydration of Methylacetylene and Allene in Polyfunctional Catalytic Systems), Candidate Dissertation in Chemistry, MITKhT (Moscow Institute of Fine Chemical Technology), Moscow, 1983.
124. D. Kondepudi and I. Prigogine, *Modern Thermodynamics. From Heat Engines to Dissipative Structures*, John Wiley & Sons, Chichester, 1998.
125. W. Ostwald, *Z. Phys. Chem*., **34**, 248 (1900).
126. C. Jensen, *Chem. Commun*., 2443 (1999).
127. S. M. Brailovskii, I. V. Trofimova and O. N. Temkin, in *Problemy kinetiki i kataliza*, Vol. 19, *Partsial'noe okislenie organicheskikh soedinenii* (Problems in Kinetics and Catalysis, Vol. 19: Partial Oxidation of Organic Compounds), Nauka, Moscow, 1985, pp. 146–175.
128. N. D. Zelinski and G. S. Pavlov, *Chem. Ber*., **66**, 1420 (1933).
129. I. Gottker-Schnetmann and M. Brookhart, *J. Am. Chem. Soc*., **126**, 9330 (2004).
130. V. M. Gryaznov, V. S. Smirnov and L. K. Ivanova, *Vestn. Akad. Nauk SSSR*, No. 7, 142 (1971).
131. I. I. Moiseev, M. N. Vargaftik, O. N. Gentosh et al., *Dokl. Akad. Nauk SSSR*, **237**, 645 (1977).
132. T. A. Stromnova, M. N. Vargaftik and I. I. Moiseev, *Izv. Akad. Nauk SSSR, Ser. Khim*., No. 1, 31 (1983).
133. Y. Fujiwara, K. Takaki and Y. Takiguchi, *Synlett*, 591 (1996).
134. H. C. Brown, *Organic Synthesis via Boranes*, John Wiley & Sons, New York, 1975.
135. S. Narasimhan and R. Balakumar, *Aldrichimica Acta*, **31**, No. **1**, 19 (1998).
136. Y. Aoyama, Y. Tanaka, T. Fujisawa et al., *J. Org. Chem*., **52**, 2555 (1987).

137. M. A. Beckett, *Inorg. Chim. Acta*, **157**, 171 (1989).
138. S. Inoki, K. Kato, T. Takai, et al., *Chem. Lett.*, 515 (1989).
139. A. Morvillo and M. Bressan *J. Mol. Catal.*, **37**, 63 (1986).
140. V. N. Zudin, V. N. Likholobov and Yu. I. Ermakov, *Kinet. Katal.*, **18**, 921 (1977); 20, 1559 (1979).
141. D. Bianchi, R. Bortolo, R. D'Aloisio and M. Ricci, *J. Mol. Cat. A: Chem.*, **150**, 87 (1999).
142. A. S. Abdullaeva, E. A. Griva, I. V. Oshanina, L. G. Bruk and O. N. Temkin, *Tezisy dokladov VI Ross. konferentsii "Mekhanizmy kataliticheskikh reaktsii"* (Abstracts of Papers. The 6th All-Russia Conf. on Mechanisms of Catalytic Reactions), Moscow, 2002, Vol. 2, p. 44.
143. L. G. Bruk, A. S. Abdullaeva, A. V. Vyrodov, L. G. Bruk and O. N. Temkin, *Proceedings of the 14th Int. Symp. on Homogeneous Catalysis* (Munich, Germany, 2004), Book of Abstracts, p. 141.
144. A. S. Abdullaeva, *Printsip kineticheskogo sopryazheniya i ego ispol'zovanie dlya podbora kataliticheskikh sistem dlya okislitel'nykh protsessov* (Principle of Kinetic Conjugation and its Application to Selecting Catalytic Systems for Oxidative Processes), Dissertation in Chemistry, MITKhT (Moscow Institute of Fine Chemical Technology), Moscow, 2005.
145. L. G. Bruk, I. V. Oshanina, S. N. Gorodskii and O. N. Temkin, *Zh. Ross. Khim. O-va im. D.I. Mendeleeva*, **50** (11), 103 (2006).
146. A. V. Kulik, O. N. Temkin, L. G. Bruk et al., *Izv. Ross. Akad. Nauk., Ser. Khim.*, No. 6, 1349 (2005).
147. T. Hosokawa and S.-I. Murahashi, *Acc. Chem. Res.*, **23**, 49 (1990).
148. T. Nishimura, T. Onoue, K. Ohe and S. Uemura, *J. Org. Chem.*, **64**, 6750 (1999).
149. A. R. El'man, E. V. Slivinskii and S. M. Loktev, *Izv. Akad. Nauk SSSR, Ser. Khim.*, 2188 (1989).
150. T. Yamado, T. Takai, O. Rhode and T. Mukaiyama, *Chem. Lett.*, 1 (1991).
151. D. V. Korol'kov and A. V. Shchukarev, *Vestn. Sankt-Peterburg. Gos. Univ., Ser. 4, Fiz. Khim.*, No. 4, 34 (1992).
152. N. N. Semenov, *Izbrannye Trudy*, Vol. 1: *Tsepnye reaktsii* (Selected Scientific Works, Vol. 1: Chain Reactions), Nauka, Moscow, 2004.
153. E. Christiansen, in *Advances in Catalysis and Related Subjects*, W. Frankenburg, V. Komarewsky and E. K. Rideal (Eds), Academic Press, New York, 1953, Vol. 5, pp 311–353.
154. E. S. Rudakov and L. K. Volkova, *Khim. Fiz.*, **15** (11), 35 (1996).
155. I. I. Moiseev and S. V. Pestrikov, *Dokl. Akad. Nauk SSSR*, **171**, 151 (1966); *Kinet. Katal.*, **10**, 74 (1969).
156. O. N. Temkin, S. M. Brailovskii and L. G. Bruk, *Tezisy dokladov Vsesoyuznoi konferentsii po mekhanizmam kataliticheskikh reaktsii* (Abstracts of Papers. All-Union Conf. on Mechanisms of Catalytic Reactions), 1973, Vol. 1, p. 74.
157. A. P. Purmal', in *Problemy khimicheskoi kinetiki* (Problems in Chemical Kinetics), Nauka, Moscow, 1979, p. 193.
158. K. J. Ivin and J. C. Mol, *Olefin Metathesis and Metathesis Polymerization*, Academic Press, London, 1997.
159. *Handbook of Metathesis*, R. H. Grubbs (Ed.), Wiley-VCH, 2003, Vols. 1–3.
160. V. V. Mazurek, *Polimerizatsiya pod deistviem soedinenii perekhodnykh metallov* (Polymerization under the Action of Transition Metal compounds), Nauka, Leningrad, 1974.
161. P. E. Matkovskii, *Radikal'nye stadii v reaktsiyakh kompleksnykh metalloorganicheskikh i metallotsenovykh katalizatorov i ikh rol' v polimerizatsii* (Radical Steps in Reactions of Organometallic and Metallocene Complex Catalysts and Their Role in Polymerization Processes), Inst. Probl. Khim. Fiz., Chernogolovka, 2003.
162. J. Skupinska, *Chem. Rev.*, **91**, 613 (1991).
163. V. Kaim, *Vysokomol. Soedin., Ser. A*, **36**, 1644 (1994).
164. A. V. Bobolev, E. A. Blyumberg, Yu. D. Norikov and N. M. Emanuel', *Zh. Fiz. Khim.*, **44**, 1028 (1970).
165. S. A. Maslov, S. A. Vasil'chenko, E. A. Blyumberg and V. G. Dryuk, *Izv. Akad. Nauk SSSR, Ser. Khim.*, No. 11, 2467 (1972).
166. I. P. Skibida, *Usp. Khim.*, **44**, 1729 (1975).
167. H. Alper, J. B. Woell, B. Despeyroux and D. J. H. Smith, *Chem. Comm.*, No. 21, 1270 (1983).
168. T. E. Kron, L. F. Starosel'skaya, M. I. Terekhova and E. S. Petrov, *Zh. Obshch. Khim.*, **60**, 213 (1990).
169. H. Alper, B. Despeyroux and J. B. Woell, *Tetrahedron Lett.*, **24**, 5691 (1983).
170. J. D. Murray, *Lectures on Nonlinear Differential Equations: Models in Biology*, Clarendon Press, Oxford, 1981.

171. M. A. Bazhenova, G. N. Petrova and A. E. Shilov, *Proceedings of the 4th European Nitrogen Fixation Conference* (Seville, Spain, 2000), Book of Abstracts, p. 53.
172. M. A. Bazhenova, *Kataliticheskoe povedenie zhelezo-molibdenovogo kofaktora nitrogenazy vne belka* (Catalytic Ex-Situ Behavior of Iron–Molybdenum Cofactor of Nitrogenase), Candidate Dissertation in Chemistry, Institute for Problems in Chemical Physics, Chernogolovka–Moscow State University, Moscow, 2001.
173. V. Leskovac, *Comprehensive Enzyme Kinetics*, Kluwer Academic & Plenum Publishers, New York, 2003.
174. L. Stryer, *Biochemistry*, Freemen & Co., New York, 1995.
175. D. Voet, J. Voet and S. W. Pratt, *Fundamentals of Biochemistry*, John Wiley & Sons, New York, 1999.
176. A. D. Vinogradov, in *Entsiklopediya "Sovremennoe estestvoznanie"* (Encyclopedia of Modern Natural Sciences), Magistr-Press, Moscow, Vol. 8, p. 236.
177. E. D. Park, Y.-S. Hwang, C. W. Lee and J. S. Lee, *Appl. Catal.*, **247**, 269 (2003).
178. J. K. Remias, T. A. Pavlovsky and A. Sen, *J. Mol. Catal. A: Chem.*, **203**, 179 (2003).
179. N. I. Kuznetsova, N. V. Kirilova, L. I. Kuznetsova and V. A. Likholobov, *J. Mol. Catal. A: Chem.*, **204–205**, 591 (2003).
180. N. K. Nagradova, in *Entsiklopediya "Sovremennoe estestvoznanie"* (Encyclopedia of Modern Natural Sciences), Magistr-Press, Moscow, Vol. 8, p. 163.
181. J. Kim and D. C. Rees, *Biochemistry*, **33**, 389 (1994).

4

Polyfunctional Catalytic Systems

All metal complex catalysts in solutions can be treated as multicomponent catalytic systems (MCSs), since they always include, besides a metal (atom or ion), some ligands (neutral molecules and/or ions) and a solvent whose molecules are involved in the formation of the coordination shell. In addition, a catalytic system can also contain acids and bases, oxidants, and reductants that influence the state of a metal catalyst in the complex, as well as "inert" electrolytes. Salts and complexes of other metals capable of influencing the catalytic properties of a given MCS can also be present

Over a century ago, researchers began to investigate systems involving two (or more) different metal atoms in the context of the catalytic decomposition of hydrogen peroxide (H_2O_2) [1, 2]. The first generalization of the possible mechanisms of action of two catalysts in this reaction and an analysis of the probable kinetic models that explain acceleration of the process were made in 1926 by Shpitalskii [3] and later developed by Bogdanov et al [4–8]. During the past half a century, a large number of homogeneous MCSs involving two or three metals and several ligands, as well as MCSs including catalysts of different nature were studied and several variants of MCS classification were proposed [9–19].

The components of an MCS perform various chemical and kinetic functions [15, 17-19]. The *chemical polyfunctionality* of an MCS is manifested (i) in the organization of an active complex center, the components of which perform various chemical functions (sometimes synchronously as in the active centers of enzymes), and (ii) in that some MCS components perform the functions of acids, bases, oxidants, and reductants in the course of the process. The *kinetic polyfunctionality* is manifested in that the MCS components perform various kinetic functions (KFs) as catalysts of separate steps and blocks of steps – that is, in the catalysis of reactions according to various overall equations that correspond to new catalytic cycles (or, formally speaking, linearly independent circuits on a bipartite graph of the corresponding subnetwork of intermediates [18]). The overall equations for these circuits (for KFs) might, in contrast to the overall equations of routes, include some intermediates. Thus, the concept of a KF differs from the concept of a reaction route. Accordingly, the number (C) of KFs in a reaction mechanism can differ from the number (P) of linearly independent routes and from the number of circuits (oriented cycles) on a bipartite graph. For example, a two-route catalytic reaction ($P = 2$) with identical overall equations (S = P) corresponds to two catalytic cycles, but it has a single KF ($C = 1$) (kinetic graph KG1, Figure 4.1), whereas a single-route reaction ($P = 1$) with different overall equations in various catalytic cycles (circuits) and, hence, different

Homogeneous Catalysis with Metal Complexes: Kinetic Aspects and Mechanisms, First Edition. O. N. Temkin.

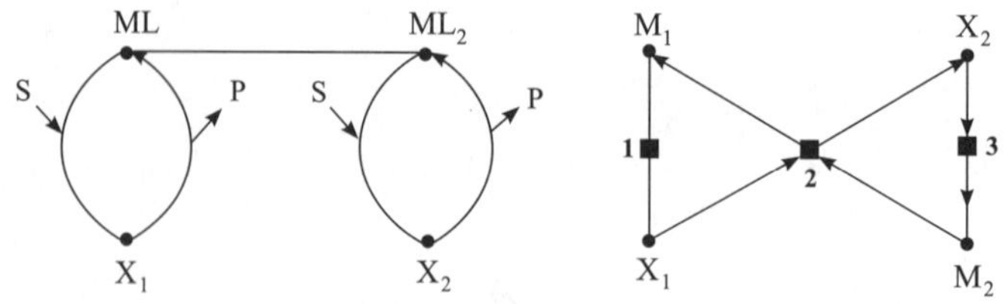

Figure 4.1 Kinetic graph (KG1) and bipartite graph (BG1) of the subnetwork of intermediates for reaction mechanism (4.1) (reaction **1** in this scheme is reversible)

KFs ($C = 2$) is characterized by $C > P$ (bipartite graph BG1, Figure 4.1):

$$\begin{aligned} M_1 + S &\rightleftharpoons X_1; \\ X_1 + M_2 &\longrightarrow X_2 + M_1; \\ X_2 &\longrightarrow M_2 + P. \end{aligned} \tag{4.1}$$

The overall equations for two KFs (two circuits of BG1) according to mechanism (4.1) can be written as follows:

$$S \xrightarrow{M_1} X_2; \tag{4.2}$$

$$X_1 \xrightarrow{M_2} P. \tag{4.3}$$

It was suggested [18] to use circuits on a bipartite graph of the corresponding subnetwork of intermediates in order to reveal KFs of the MCS components. Essential specific features of this approach are as follows:

(i) Any intermediate occurring in a vertex of the bipartite graph, as well as any intermediate in vertices of a kinetic graph of a linear mechanism, can, formally speaking, be considered as a catalyst of the corresponding reaction that proceeds according to an overall equation. On the other hand, it is conventional chemical practice to define as catalysts one or several initial components in the MCS or compounds formed from such initial precursors. For example, in the case of an initial compound PdL_2X_2, the catalyst in some reactions is considered to be Pd^0L_2. This conventional approach is also used in what follows.

(ii) The same overall equation can be obtained for two different circuits on a bipartite graph, which pass via different intermediates but include the same steps.

(iii) An overall equation can be unbalanced in cases where an intermediate compound is simultaneously a reaction product. In this case, one circuit leads to different overall equations with different sets of the stoichiometric numbers of steps.

(iv) The use of various circuits on a bipartite graph ensures the enlisting of all KFs (i.e., all cyclic sequences of intermediates), including those for which the overall equations have no conventional chemical sense, and thus do not clarify the role of components in the MCS.

With respect to the number of different KFs, the entire set of MCSs can be subdivided into two groups (A and B) as follows:

Group A	**Group B**
All components perform the same KF but different chemical functions	There are two (or more) components that perform different KFs and different or same chemical functions

Group A MCSs exhibit chemical polyfunctionality, while group B MCSs show kinetic polyfunctionality. The polyfunctional catalytic systems (PFCSs) characterized by the kinetic polyfunctionality represent systems of catalysts. Group A can, in turn, be subdivided into two subgroups (A1 and A2). According to this, the introduction of a new (ith) component into catalyst $\mathbb{K}_1$ retains the reaction mechanism in systems of the first subgroup (A1) and leads to a change in the mechanism in systems of the second subgroup (A2).

Let us analyze the MCSs of group A in more detail. Consider a catalytic cycle involving complex M_1L that is defined to be catalyst $\mathbb{K}_1$ (graph KG2, Figure 4.2). Let a new component (ligand, electrolyte salt, organic compound, complex of another metal M_2, etc.) be added to the system. Additives capable of changing the reaction rate are called promoters, modifiers, or activators [11]. If the new component performs the same KF as $\mathbb{K}_1$, that is, catalyzes the reaction with an overall equation $S = P$, and does not modify the structure of the mechanism (i.e., the number and sequence of intermediates X_i), while only changing the concentrations of X_i and their reactivities (rate constants of steps), the modified catalytic system should be classified into group A1. Alternatively, if the introduced component (e.g., acid or ligands) forms a new catalyst that still performs the same KF as $\mathbb{K}_1$, but changes the mechanism (i.e., the number and nature of intermediates X_i), then the system is classified into group A2. In the latter case, there appears a new active center ($\mathbb{K}_2$) with new chemical functions but with still a single KF (graph $\mathbb{K}$G3, Figure 4.2). For example, the $PdBr_2$–HBr–$P(OPh)_3$–n-BuOH system implements the mechanism of acrylate synthesis by alkyne hydroalkoxycarbonylation, which involves Pd(II) and ~PdCOOR as the active center and key intermediate, respectively [20]. In the modified system Pd(0)$(dba)_2$–HX–PPh_3–n-BuOH, the mechanism is changed so that Pd^0L_n complex becomes the new active center and $CH_2{=}C(R)PdL_2^+$ is the new key intermediate [21].

Previously (Section 2.3), we considered the influence of metal salts on the rate of acetylene dimerization into vinylacetylene (VA) in solutions of $Cu_mCl_n^{(n-m)-}$ complexes (Figure 2.12). For example, Zn(II) and Bi(III) salts increased the catalyst activity by binding chloride ions and increasing the concentration of active $Cu_4Cl_5^-$ complexes (MCS of A1 subgroup). At the same time, $CuCl_2$ and $HgCl_2$ decreased the rate of dimerization because they interacted with initial intermediates and formed new (intermediate) compounds.

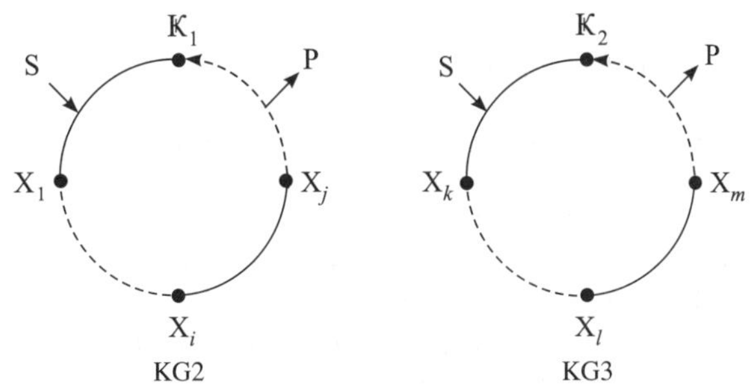

Figure 4.2 Kinetic graphs of catalytic cycles in a linear mechanism involving different catalysts ($\mathbb{K}_1$ and $\mathbb{K}_2$) for the same overall equation S = P

In this case, the intermediate X_2 formed in the reaction

$$\underset{}{Cu(I)} \underset{-H_3O^+}{\overset{C_2H_2}{\rightleftharpoons}} \underset{(X_1)}{CuC\equiv CH} \xrightarrow{C_2H_2} \underset{(X_2)}{CuCH{=}CHC\equiv CH} \xrightarrow{H_3O^+} \underset{(VA)}{C_4H_4} + Cu(I) \tag{4.4}$$

is bound by mercury chloride with the formation of a stable (with respect to H_3O^+) intermediate X_3 via reaction

$$\underset{(X_2)}{CuCH{=}CHC\equiv CH} + HgCl_2 \longrightarrow \underset{(X_3)}{ClHgCH{=}CHC\equiv CH} + CuCl, \tag{4.5}$$

which leads to a decrease in the concentration of X_2 and an increase in the concentration of hydrogen ions,

$$2C_2H_2 + HgCl_2 \xrightarrow{Cu(I)} \underset{(X_3)}{ClHgCH{=}CHC\equiv CH} + HCl, \tag{4.6}$$

thus hindering the formation of X_1 (MCS of A1 subgroup).

The reaction of $CuCl_2$ with X_2 leads to the formation of X_4 and a new product (2-chlorovinylacetylene, 2-CVA) upon the oxidative decomposition of X_4 in the reaction with the second $CuCl_2$ molecule:

$$\begin{gathered} \underset{(X_2)}{CuCH{=}CHC\equiv CH} + CuCl_2 \longrightarrow \underset{(X_4)}{(CuCl_2)CuCH{=}CHC\equiv CH}; \\ X_4 + CuCl_2 \underset{-CuCl}{\longrightarrow} X_5 \longrightarrow \underset{(CVA)}{CH_2{=}C(Cl)C\equiv CH} + 2CuCl. \end{gathered} \tag{4.7}$$

Thus, the introduction of a new component into an initial catalytic system leads to a change in the structure of the reaction mechanism and the appearance of a new intermediate compound and new kinetic functions. Under steady-state conditions (with the oxidation of CuCl to $CuCl_2$),

$$2CuCl + {}^1\!/\!_2\,O_2 + 2HCl \longrightarrow 2CuCl_2 + H_2O \tag{4.8}$$

the second route of 2-CVA formation involves two catalysts, Cu(I) and Cu(II). One of the KFs of CuCl is to catalyze the formation of X_5 via the following reaction:

$$2C_2H_2 + CuCl_2 \xrightarrow{Cu(I)} X_5 + HCl, \tag{4.9}$$

while the KF of $CuCl_2$ is to catalyze the conversion of X_2 into 2-CVA according to the overall equation

$$\underset{(X_2)}{CuCH{=}CHC\equiv CH} + {}^1\!/\!_2\,O_2 + 2HCl \longrightarrow CVA + CuCl + H_2O \tag{4.10}$$

so that both PFCS components catalyze the overall reaction

$$2C_2H_2 + {}^1\!/\!_2\,O_2 + HCl \xrightarrow{Cu(I)-Cu(II)} CVA + H_2O. \tag{4.11}$$

In the two-route reaction of VA and 2-CVA synthesis under consideration, it is possible to define six KFs of the two components, Cu(I) and Cu(II) (see Section 4.2). PFCSs of this type belong to the MCSs of group B. Naturally, the chemical functions of components in this system are different.

A classical MCS belonging to group B is the $PdCl_2$–$CuCl_2$ system used in the reaction of olefin oxidation to carbonyl compounds [22], where $PdCl_2$ catalyzes the step of olefin oxidation by Cu(II) chloride,

$$C_2H_4 + H_2O + 2CuCl_2 \xrightarrow{Pd(II)} CH_3CHO + 2CuCl + 2HCl, \tag{4.12}$$

and $CuCl_2$ catalyzes the step of Pd(0) oxidation by oxygen

$$Pd(0) + {}^{1}\!/_{2}\, O_2 + 2HCl \xrightarrow{Cu(II)} PdCl_2 + H_2O. \tag{4.13}$$

Note that, strictly speaking, there is the third catalyst, H_2O molecule, which catalyzes the overall brutto equation

$$C_2H_4 + {}^{1}\!/_{2}\, O_2 \xrightarrow{PdCl_2\text{-}CuCl_2\text{-}H_2O} CH_3CHO. \tag{4.14}$$

It has been suggested to classify the MCSs of group B based on the type of connections of the involved catalytic cycles (bipartite graph circuits) [15]. According to this, it is possible to distinguish between three classes of PFCSs. Below, the specific features of these classes are illustrated in the case of two catalysts ($\mathbb{K}_1$ and $\mathbb{K}_2$) with two KFs.

Class I. In this case, the catalysts ($\mathbb{K}_1$ and $\mathbb{K}_2$) performing different KFs cannot simultaneously enter either in the transition state or in the intermediates. The two catalytic cycles (or bipartite graph circuits) share an intermediate that involves neither $\mathbb{K}_1$ nor $\mathbb{K}_2$. Thus, the overall equations for the KFs also do not contain either the catalysts ($\mathbb{K}_1$ and $\mathbb{K}_2$) or intermediates involving $\mathbb{K}_1$ or $\mathbb{K}_2$. For example, the sequence of steps in the following mechanism belonging to class I,

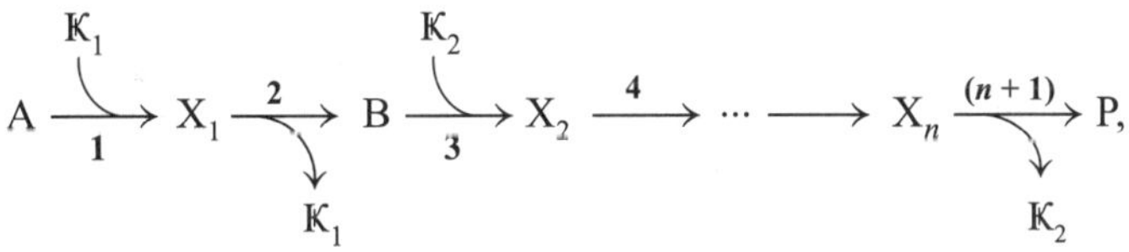

(4.15)

corresponds to bipartite graph BG2 depicted in Figure 4.3. The overall equations for the KFs of catalysts $\mathbb{K}_1$ and $\mathbb{K}_2$ are written as A = B and B = P, respectively, where B is an intermediate between the two circuits in BG2. Naturally, both catalysts ($\mathbb{K}_1$ and $\mathbb{K}_2$) are necessary for the formation of product P.

Class II. All intermediates of a given mechanism contain only one of the two catalysts $\mathbb{K}_i$, and both coexist only in the transition state of the catalyst exchange step **2**

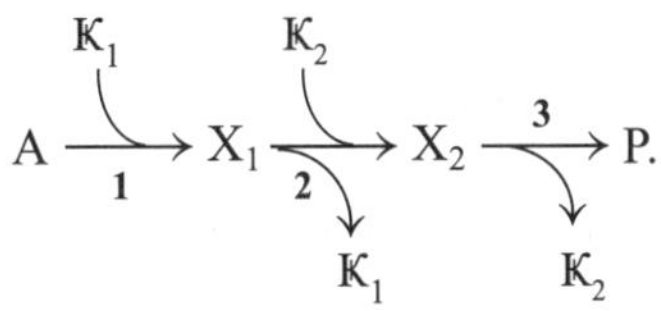

(4.16)

(see Figure 4.1, graph BG1 with all irreversible steps (4.1)–(4.3), where step **2** enters into both circuits of BG1). After the classical works by Moiseev [22, 24] and Heck [25], which showed the possibility of the transfer of organic groups from RHgX (organometallic compounds of nontransition metals) to PdX_2 (transition metal complexes), for example, in the system [23]

$$ClHgCH_2CH_2OH + PdCl_2 \xrightarrow[-HgCl_2]{} ClPdCH_2CH_2OH \longrightarrow Pd^0 + CH_3CHO + HCl, \tag{4.17}$$

the reactions of electrophilic substitution (well known in the chemistry of organometallic compounds)

$$RM_1X + M_2X_2 \longrightarrow RM_2X + M_1X_2 \tag{4.18}$$

and redox conversions of organometallic compounds [4, 27] have drawn the attention of researchers as possible ways of designing PFCSs [9]. Different variants of the reaction schemes of class II with step **2** were presented in [9].

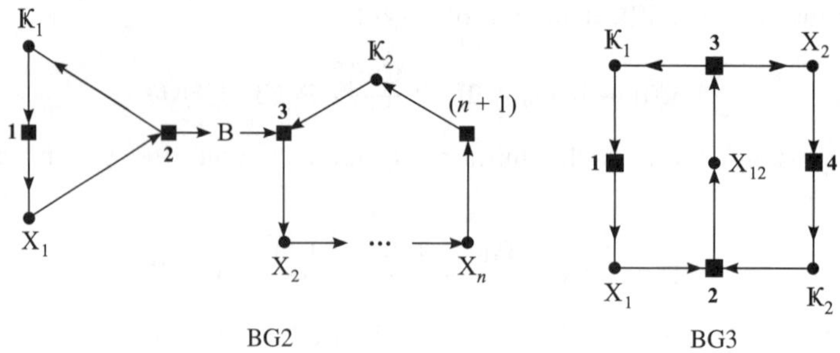

Figure 4.3 Bipartite graphs BG2 and BG3 of subnetworks of intermediates for reaction mechanisms (4.15) and (4.19), respectively

Class III. There is at least one intermediate that involves both catalysts $\mathbb{K}_1$ and $\mathbb{K}_2$ (BG3, Figure 4.3):

$$\mathrm{A} \xrightarrow[1]{\mathbb{K}_1} X_1 \xrightarrow[2]{\mathbb{K}_2} X_{12} \xrightarrow[\downarrow \mathbb{K}_1]{3} X_2 \xrightarrow[\downarrow \mathbb{K}_2]{4} \mathrm{P}. \tag{4.19}$$

The overall equations of catalytic cycles for components of MCS (KFs) are obtained by summing steps **1, 2** and **3** (for the circuit with $\mathbb{K}_1$) and steps **2, 3** and **4** (for the circuit with $\mathbb{K}_2$):

$$\mathrm{A} + \mathbb{K}_2 \xrightarrow{\mathbb{K}_1} X_2; \tag{4.20}$$

$$X_1 \xrightarrow{\mathbb{K}_2} \mathbb{K}_1 + \mathrm{P}. \tag{4.21}$$

Thus, the KF of catalyst $\mathbb{K}_1$ is to catalyze the formation of X_2 from A and $\mathbb{K}_2$, while the KF of catalyst $\mathbb{K}_2$ is to catalyze the conversion of X_1 into reaction product P and catalyst $\mathbb{K}_1$. The circuits of BG3 share the common intermediate X_{12}.

Naturally, various combinations of classes I–III in the mechanisms of complex PFCSs are also possible, examples of which are presented below.

If two catalysts are involved in the formation of an active center and isolated at the last step,

$$\mathbb{K}_1 + \mathbb{K}_2 \longrightarrow X_1 \xrightarrow{\mathrm{A}} X_2 \xrightarrow{\mathrm{B}} X_3 \longrightarrow X_1 \xrightarrow[\downarrow \mathrm{P}]{} \mathbb{K}_1 + \mathbb{K}_2, \tag{4.22}$$

the MCS will belong to group A, since both active centers perform the same KF of catalyzing the reaction $\mathrm{A} + \mathrm{B} \rightarrow \mathrm{P}$.

A close classification of PCFs was proposed by Golodov [16]. According to this approach, a concept of the level of catalysis is used instead of the KF, and the MCSs are classified in terms of these levels (i.e., the number of KFs). For example, catalysts $\mathbb{K}_2$ in reaction schemes (4.15), (4.16), and (4.19), as well as $CuCl_2$ in reaction (4.14), are referred to as catalysts of the second level.

For catalytic systems where the main species ($\mathbb{K}_1$) bearing the catalytic activity (first-level catalyst according to Golodov [16]), in the absence of which the reaction products are not formed, as well as additional active components (catalysts $\mathbb{K}_2$, $\mathbb{K}_3$, etc.) can be determined, still another classification has been proposed in [19]. According to this approach, which is based on differences in the nature of additional catalysts and is especially convenient for the analysis of oxidation processes, the PFCSs can be subdivided into the following groups:

(i) Additional components are not involved in the steps of product formation and only favor the regeneration of an active form of the main catalyst. An example is the Wacker process and analogous reactions [22], where the main components are palladium compounds, while the KF of an additional oxidant ($CuCl_2$, *para*-benzoquinone) is to catalyze oxidation of the reduced form of Pd^0 by oxygen or a third intermediate oxidant.

(ii) Additional components participate in the steps of product formation by interacting with intermediate (e.g., σ- or π-type) organometallic compounds. In this case, the nature of an additional component influences the direction and rate of decomposition of an intermediate. For example, in the reaction of 1,3-diene oxidation, *para*-benzoquinone influences the decay of intermediate η^3-allyl complexes [28].

(iii) Additional components participate in the steps of both the formation of key intermediates and their conversion into reaction products by interacting with the main catalytically active metal complex. These components (NO_2^-, $CuCl_2$) can modify the properties of the reactive center and, hence, the direction and rate of the reaction [10, 13, 29, 30].

Mechanisms of the third group are in fact related to the chemical polyfunctionality of a complex active center (whose components can perform different chemical functions at various steps of the process) rather than to the kinetic polyfunctionality. The KF of both components ($\mathbb{K}_1$ and $\mathbb{K}_2$) in this case is to catalyze the overall route equation. Therefore, this variant of the effect of additional components refers to the MCSs of group A. Examples of multicomponent PFCSs in both oxidative and nonoxidative reactions can be found in [10–15].

The next section is devoted to an analysis of the mechanisms and kinetics of catalytic reactions with participation of MCSs belonging to both groups A and B.

4.1 Oxidation reactions of organic and inorganic compounds

A very large number of two- and three-component PFCSs have been studied in oxidation reactions of various types. The second and third components in these systems are frequently oxidants, but in the presence of other chemical (O_2, H_2O_2) and/or electrochemical oxidation pathways, additional oxidants become catalysts of the process capable of performing specific chemical and kinetic functions.

4.1.1 Oxidation of alkenes

The process of ethylene oxidation to acetaldehyde (AA) in aqueous solutions has already been mentioned above (Section 2.1, Example 2.22). In the $PdCl_2$–$CuCl_2$ system, palladium chloride is a catalyst and $CuCl_2$ is a reactant-oxidant, while in the $PdCl_2$–$CuCl_2$–O_2 system both metal salts are catalysts. At the same time, the chemical functions of components remain unchanged on the passage from one system to another and only the kinetic functions (KFs) of $CuCl_2$ are changed. In the case of using other oxidants (Q, H_2O_2), the rate of ethylene oxidation is independent of [Ox], since the effect of these oxidants is believed to be manifested past the limiting step of the process (oxidation of the reduced form of palladium) [22].

The kinetic equations of AA synthesis by ethylene oxidation were determined in the $PdCl_2$–$CuCl_2$–H_2O system [31]. For component concentrations of $[CuCl_2] < 0.8\,M$ and $[LiCl]_\Sigma = 0.6$–$1.1\,M$ and an ionic strength within $I = 3\text{–}6\,M$ ($LiCl\text{–}HClO_4\text{–}LiClO_4$), the empirical equation was as follows:

$$r_{AA} = \frac{k[PdCl_2]_{\Sigma}^{0.84}\,(1 + 1.15[CuCl_2]_{\Sigma}^{2})\,P_{C_2H_4}}{[Cl^-]^{1.25}(0.253 + [H_3O^+])}. \tag{4.23}$$

The form of this expression indicates that the chloride ion concentration $[Cl^-]$ and complexity functions F_{Cu} and F_{Pd} do not remain constant when $[CuCl_2]_{\Sigma}$ and $[LiCl]_{\Sigma}$ are varied within the intervals of comparable concentrations. Thus, there are no grounds to believe that the oxidant ($CuCl_2$) participates in the process before and at the limiting step.

It was found that, in the presence of $[NaCl]_{\Sigma} = 0.25$ and $1.0\,M$, Cu(II) chloride even produces a weak inhibition of the reaction [32, 33]. Under these conditions, the empirical kinetic equation has the following form:

$$r = \frac{k_1[PdCl_2]_{\Sigma}}{[H^+]^{1.1}[Cl^-]_{\Sigma}^{1.86}[CuCl_2]_{\Sigma}^{0.25}} \tag{4.24}$$

at $P_{C_2H_4} = \text{const}$ and $[CuCl_2]_{\Sigma} = 0.037\text{–}0.15\,M$. This effect is probably also related to changes in the F_{Pd} and $[Cl^-]$ values during the variation of $[CuCl_2]_{\Sigma}$.

For the synthesis of acetaldehyde dimethylacetal (DMA) in the $PdCl_2\text{–}CuCl_2\text{–}NaCl\text{–}MeOH$ system, two kinetic equations were obtained [33] in different intervals of $[CuCl_2]_{\Sigma}$. In the interval of $[CuCl_2]_{\Sigma} = 5.8 \cdot 10^{-3}\text{–}4.9 \cdot 10^{-2}\,M$ at $P_{C_2H_4} = \text{const}$, the empirical equation was

$$r_{DMA} = \frac{k_2[PdCl_2]_{\Sigma}\,[CuCl_2]_{\Sigma}^{n}}{[H^+]^{1.36}[Cl^-]_{\Sigma}^{0.90}}, \tag{4.25}$$

where $n = 0.14$ (~ 0). At greater concentrations of $CuCl_2$ (within $4.9 \cdot 10^{-2}\text{–}0.123\,M$), the rate of DMA synthesis rapidly grows with increasing $[CuCl_2]_{\Sigma}$:

$$r_{DMA} = \frac{k_3[PdCl_2]_{\Sigma}\,[CuCl_2]_{\Sigma}^{1.56}}{[H^+]^{1.36}[Cl^-]_{\Sigma}^{0.90}}. \tag{4.26}$$

The observed orders of these kinetic equations with respect to $[Cl^-]$, which are close to -1, are explained by the influence of $CuCl_2$ on the concentration of free Cl^- ions ($[NaCl]_{\Sigma} = 0.35\,M$).

In the $PdCl_2\text{–}LiCl\text{–}Q\text{–}MeOH$ system containing *para*-benzoquinone (Q) [34], the kinetic equation of DMA synthesis has the same form as in aqueous solutions with Q [22] (see Chapter 2):

$$r_{DMA} = \frac{k'[PdCl_4^{2-}]_{\Sigma}\,[C_2H_4]}{[H^+][Cl^-]^2}. \tag{4.27}$$

If the kinetic equation of type (4.27) describes the process of σ-palladoethanol formation from the corresponding π complex,

$$\begin{gathered} C_2H_4 + PdCl_4^{2-} + H_2O \rightleftharpoons \underset{(\pi\ \text{complex})}{(HO)Cl_2Pd(C_2H_4)^-} + 2Cl^- + H^+; \\ (HO)Cl_2Pd(C_2H_4)^- \xrightarrow{lim} \underset{(\sigma\ \text{complex})}{Cl_2PdCH_2CH_2OH^-}, \end{gathered} \tag{4.28}$$

then the question naturally arises concerning the role of oxidant (Ox) in subsequent steps of the σ complex transformation. According to one of the first hypotheses [22], the σ complex decays into species containing Pd(0) such as $Pd(0)_{aq}$ or $PdCl_{n\ aq}^{n-}$, which are oxidized to Pd(II) during the subsequent interaction with Ox (Q, $CuCl_2$, $FeCl_3$). It should be recalled that the issue of $Pd(0)_{aq}$ or $PdCl_2^{2-}$ formation in elementary acts has already been discussed in Section 1.1. The ability of water molecules and Cl^- ions to coordinate $Pd(0)_{gas}$ particles with compensation for a very large endothermicity ($\Delta H^\circ_{298} = 93$ kcal/mole) of $Pd(0)_{gas}$ formation is unlikely. For example, reaction

$$PdCl_4{}^{2-} + C_2H_{4gas} + H_2O_{liq} \longrightarrow Pd(0)_{gas} + CH_3CHO_{gas} + 2H^+ + 4Cl^- \quad (4.29)$$

is characterized by $\Delta H^\circ_{298} = 78.8$ kcal/mole. Therefore, from the standpoint of thermodynamics, it is more probable that a σ complex decomposes with the formation of $HPdCl_2^-$. The proposed mechanism [34, 35] is consistent with the entire body of data available for the Wacker process.

After the limiting step, the oxidant can also participate in the σ complex decomposition, for, example, via the following reactions:

$$\sigma \text{ complex} \xrightarrow{PdCl_4^{2-}} Pd_2Cl_4^{2-} + CH_3CHO + HCl; \quad (4.30)$$

$$\sigma \text{ complex} \xrightarrow{CuCl_2} X \xrightarrow{CuCl_2} PdCl_4^{2-} + CH_3CHO + 2CuCl. \quad (4.31)$$

However, the high reactivity of σ-palladoethanol [23, 24] and, hence, its low steady-state concentrations can hardly provide for any significant rates of the σ complex interaction with an oxidant (Ox). The above considerations are also valid in cases with the other limiting step, whereby the σ complex is formed under quasi-equilibrium conditions with slow dissociation at the Pd–Cl bond [36]:

$$Cl_2PdCH_2CH_2OH^- \xrightarrow{lim} ClPdCH_2CH_2OH + Cl^-.$$

Thus, it is possible to recognize two KFs of an oxidant (Ox) in the process under consideration:

(i) the oxidation of species containing Pd(0) or HPdX (catalysis of this oxidation process);
(ii) the promotion of σ complex decay without the formation of Pd(0) or HPdX (catalysis of σ complex transformations).

When an iron phthalocyaninate (FePc) complex was used as a catalyst (electron transporter to O_2) of the oxidation of higher α olefins in the $Pd(OAc)_2$–PcFe–$HClO_4$–DMF–H_2O system [37], it was found that the addition of quinone (as QH_2) to the system leads to a significant increase in the oxidation rate. In this system, the oxidant Q is probably involved in the limiting step and/or influences the process via the material balance with respect to Pd(II) at significant concentration of σ complexes. It should be also noted that, in the $Pd(OAc)_2$–$HClO_4$–DMF–H_2O system [37], a catalyst precursor in the reaction of olefin oxidation most probably has the form of Pd^{2+} cation complexes. Previously, it was shown that the laws of olefin oxidation in the presence of *para*-quinone in systems of this type ($Pd(OAc)_2$–$HClO_4$, $PdSO_4$–H_2SO_4, $Pd(NO_3)_2$–$HClO_4$) [38–41] are principally different from the kinetics of analogous reactions in the $PdCl_2$–HCl – *para*-quinone system [22]. In particular, the rate of ethylene oxidation grows when $[HClO_4]$ is increased to 1 M [38], and the rates of cyclohexene and 1-decene oxidation grow with $[HClO_4]$ increasing up to 0.4–0.5 M [40]. These observations were confirmed in the case of cyclohexene oxidation in a chloride-free system at a constant ionic strength in acetonitrile–H_2O medium [42]. Measurements of the kinetics of cyclohexene oxidation by *para*-benzoquinone and other quinones in the $Pd(OAc)_2$–$HClO_4$–CH_3CN–H_2O system [42] showed that the initial reaction rate exhibits weak maxima with respect to the acid and quinone concentrations and is described by an S-shaped dependence on the palladium concentration, while the kinetics of substrate consumption has fast and slow stages. Based on these results and proposed kinetic model analysis, it was suggested that the complexes of Pd^{2+}, PdQ^0, and Pd_2Q^{2+} participate as active centers in the multi-route catalytic oxidation of cyclohexene. Reaction mechanisms involving solvated metal carbenium ions and the π-allyl and σ-vinyl intermediates were considered, which were consistent with the fact that the rate of propylene oxidation in solutions of palladium cation complexes is about 40 times greater than the rate of ethylene oxidation [39]. In these systems, *para*-benzoquinone species play the roles of ligands, oxidants [42], and catalysts [37].

Investigations into the oxidation of olefins by high concentrations of $CuCl_2$ and other types of oxidants in water [43], alcohols [28, 29], and acetic acid [44–50] showed that at least one new route appears in these systems for the formation of 1,2-disubstituted oxidation products (1,2-disubstituted ethanes, propanes, etc.), for example, XCH_2CH_2Y, where X = OH, Y = OAc; X = Cl, Y = OR, OAc; X = OAc, Y = OAc; and the X and Y groups of other types (amino, nitro, etc). In aqueous solutions of $PdCl_2$ with high concentrations of $CuCl_2$ (up to 4 M) and LiCl (up to 10 M), the main product of ethylene oxidation is ethylene chlorohydrin (ECH) [43],

$$C_2H_4 + 2CuCl_2 + H_2O \xrightarrow{PdCl_2} \underset{(ECH)}{HOCH_2CH_2Cl} + 2CuCl + HCl, \tag{4.32}$$

but this reaction proceeds under very hard conditions (50 bar, 100 °C). Reaction (4.32) can also be carried out with oxygen under conditions of catalysis by $CuCl_2$.

The synthesis of 1,2-disubstituted products of olefin oxidation in alcohols, water, and acetic acid was studied predominantly in PdX_2–Ox systems [13, 44–50] [Ox = $CuCl_2$, $Cu(OAc)_2$, $FeCl_3$, $Fe(NO_3)_3$, $Cu(NO_3)_2$, $Pb(OAc)_4$, $LiNO_3$, $NaNO_2$, $HAuCl_4$, $Tl(OAc)_3$, $LiEO_3$ (E = Cl, Br, I), EO_m^{n-} (E = Mn, Re, Cr, Mo, As)], but reactions in some other systems such as $PtCl_2$–$FeCl_3$, $RhCl_3$–$CuCl_2$, $IrCl_3$–$CuCl_2$ [44], and $CuCl_2$–I_2 [51] were also reported.

According to data on the reactions of formation of esters with a general formula of $AcOCH(R_1)CH(R_2)X$ (X = OAc, Cl, OH) in acetic acid media containing acetate and chloride ions [52–54], the reaction with ethylene in the $PdCl_2$–$CuCl_2$ system is described by the following equation:

$$r_{1,2} = k[PdCl_2]_\Sigma\,[CuCl_2]_\Sigma\,P_{C_2H_4}, \tag{4.33}$$

while in the case of $Cu(OAc)_2$ the kinetic equation appears as follows:

$$r_{1,2} = \frac{k[Pd]_\Sigma\,[Cu(OAc)_2]_\Sigma\,P_{C_2H_4}}{1 + K[Cu(OAc)_2]_\Sigma}. \tag{4.34}$$

The rate of formation of 1,2-disubstituted products of olefin oxidation passes through a maximum with respect to $[LiCl]_\Sigma$ [53]. Acetate ions inhibit the formation of saturated esters and accelerate the synthesis of vinyl acetate [54].

Investigations of the synthesis of ethylene glycol monoacetate in the $Pd(OAc)_2$–LiOAc–$LiNO_3$–AcOH system showed that the reaction rate

$$r_{1,2} = k[Pd(OAc)_2]_\Sigma\,P_{C_2H_4} \tag{4.35}$$

with respect to $LiNO_3$ exhibits a variable order that changes from unity to zero [55]. In $PdCl_2$–$Fe(NO_3)_3$–$LiNO_3$–LiCl solutions, the kinetic equation is

$$r_{1,2} = k\frac{[PdCl_2]_\Sigma\,[C_2H_4]}{[Cl^-]_\Sigma} \cdot (2.4 \cdot 10^{-3} + [H^+]) \cdot (1.5 \cdot 10^{-2} + [Fe^{III}]_\Sigma) \tag{4.36}$$

(E_{act} = 16 280 cal/mole at $[LiNO_3]_\Sigma > 5 \cdot 10^{-2}$ M) [56].

In $Pd(OAc)_2$–$LiXO_3$ systems (X = Cl, Br, I), the reaction rate can be expressed as

$$r_{1,2} = k[Pd^{II}]_\Sigma\,P_{C_2H_4}[XO_3^-]^n, (n > 1) \tag{4.37}$$

and exhibits a maximum with respect to $[LiCl]_\Sigma$ [49].

Unfortunately, the reported investigations (except for [54]) do not include measurements of the kinetics of vinyl acetate formation (Moiseev reaction), which hinders an analysis of the structure and mechanism of PFCS action in the process under consideration. According to the Henry hypothesis [57], both vinyl acetate and 1,2-disubstituted products of olefin oxidation are formed from a common intermediate that appears at the reversible step of olefin acetopalladation

$$C_2H_4 + Pd(II) + OAc^- \rightleftharpoons Pd(II)CH_2CH_2OAc \xrightarrow{-HPd\sim} CH_2{=}CHOAc.$$

$$Pd(II)CH_2CH_2OAc \xrightarrow{CuCl_2,\ X^-} AcOCH_2CH_2X \qquad (4.38)$$

However, the parallel formation of products within the framework of this scheme contradicts the observations [54]. For this reason, an alternative mechanism of the formation of VA and other oxidation products via different intermediates was considered [54]; in particular, it was suggested that glycol esters are formed with the participation of Pd(I) complexes.

In the other Pd(II)–Ox systems, where $Ox = EO_m^{n-}$, data obtained using the NMR on ^1H, ^{13}C, and ^{17}O nuclei showed evidence for the formation of intermediate complexes from two components, Pd(II) and EO_m^{n-}, which participate in the process from the first to last step [47–50], so that the given MCS belongs to the A2 subgroup. For example, in the case of $Ox = IO_4^-$, the proposed mechanistic scheme is as follows:

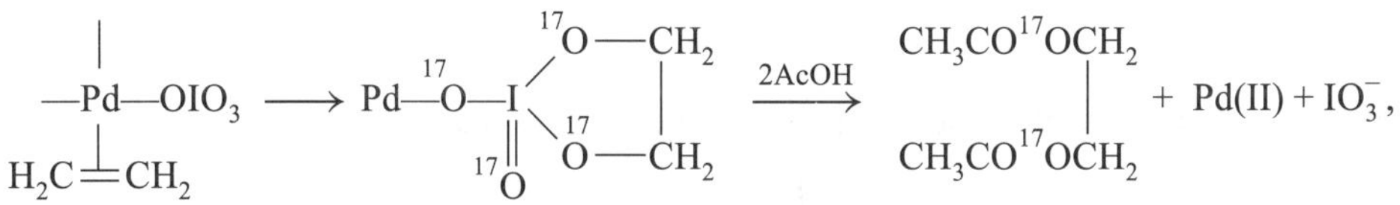

(4.39)

where IO_4^- is an oxidant rather than a catalyst. In the case of $Ox = NO_3^-$, there appear NO_2^- ions that act as catalysts of the process in which NO_3^- (or O_2) is the oxidant:

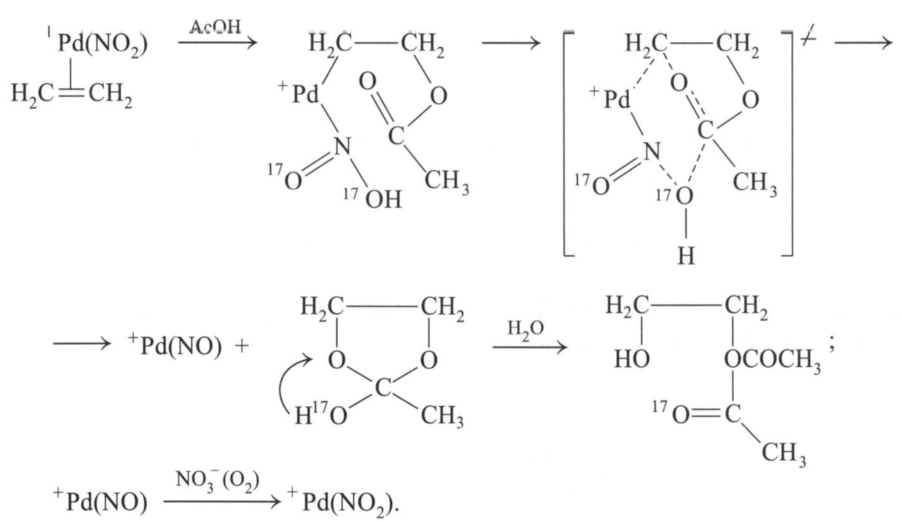

(4.40)

An attempt to elucidate the structure of the mechanism of action of the $PdCl_2$–$CuCl_2$ system was undertaken in the case of formation of β-halogenethyl alkyl ethers in alcohol media.

Example 4.1 [29, 30, 58, 59]

► Ethylene conversion in alcohol solutions of CuX_2 and PdX_2 leads to the formation of dialkylacetals $CH_3CH(OR)_2$ and β-halogenethyl alkyl ethers XCH_2CH_2OR. The replacement of CuX_2 by a quinone (Q) or $FeCl_3$ leaves dialkylacetals only. In contrast, in the case of $CuBr_2$, a system of 0.8 M $CuBr_2$–0.5 M LiBr–0.1 M $PdBr_2$–MeOH yields almost no acetal whereas the yield of bromoethyl methyl ether reaches ~97 %. As the concentration of LiBr is increased to 1.9 M, there appears 1,2-dibromoethane at a 74 % yield.

The kinetics of formation of acetals and halogenated ethers in these systems was studied at an ethylene pressure of 2 bar in a closed ideally stirred reactor, by monitoring the rates of ethylene consumption and product accumulation (including CuCl). The initial reaction rates were used. The concentration of acid in solution was varied so as to maintain the sum of [LiX] and [HX] constant. The following three systems were studied in methanol:

(I) $PdCl_2$–$CuCl_2$–LiCl–HCl (25 °C, 0.1–2 M $CuCl_2$, 0–1 M LiCl); main product: β-chloroethyl methyl ether (CME);
(II) $PdCl_2$–$CuCl_2$–LiCl–$LiClO_4$–HCl (50 °C); products: CME and acetaldehyde dimethyl acetal (DMA);
(III) $PdBr_2$–$CuBr_2$–LiBr–HBr (25 °C); main product: β-bromoethyl methyl ether (BME).

The rates of CME, DMA, and BME formation in all systems at $[CuX_2]_\Sigma = \text{const}$ and $[LiX]_\Sigma +$ $[HX] = \text{const}$ are described by the following general equation:

$$r_i = \frac{k_i [PdX_2]_\Sigma \, P_{C_2H_4}}{[HX]}. \tag{4.41}$$

In system I, the reaction order with respect to $[CuCl_2]_\Sigma$ depends on $[LiCl]_\Sigma$ as follows:[1]

$$r_{\mathbf{CME}} = \frac{k_{5.2}[PdCl_2]_\Sigma \, [CuCl_2]_\Sigma^2 P_{C_2H_4}}{[HCl]} \tag{4.42}$$

for $[LiCl]_\Sigma = 0$ and

$$r_{\mathbf{CME}} = \frac{[PdCl_2]_\Sigma \, [CuCl_2]_\Sigma \, P_{C_2H_4}}{[HCl](k_{5.3} + k_{5.4}[LiCl]_\Sigma^{1.88})} \tag{4.43}$$

for $[LiCl]_\Sigma = 0.1–1$ M.

In system III, the first order with respect to $[CuBr_2]_\Sigma$ is retained in the entire interval of $[LiBr]_\Sigma =$ 0–1.2 M:

$$r_{\mathbf{CME}} = \frac{[PdCl_2]_\Sigma \, [CuBr_2]_\Sigma \, P_{C_2H_4}}{[HBr](k_{5.23} + k_{5.24}[LiBr]_\Sigma + k_{5.25}[LiBr]_\Sigma^2)}. \tag{4.44}$$

In system II with $[CuCl_2]_\Sigma = 0.5$ M, $[LiCl]_\Sigma = 0.5$ M, and $[LiClO_4]_\Sigma = 2.25$ M, the rates of CME and DMA formation obey the following equations:

$$r_{\mathbf{CME}} = k_{2obs} \frac{[PdCl_2]_\Sigma \, P_{C_2H_4}}{[HCl]}; \tag{4.45}$$

$$r_{\mathbf{DMA}} = k_{3obs} \frac{[PdCl_2]_\Sigma \, P_{C_2H_4}}{[HCl]}; \tag{4.46}$$

---

[1] Notations of rate constants in Eqs 15. (4.42), (4.43) and throughout Example 4.1 are given according to [59].

$$k_{2obs} = \frac{(k_{5.16} + k_{5.17}[\text{LiCl}]_\Sigma)[\text{CuCl}_2]^2_\Sigma}{k_{5.18} + k_{5.19}[\text{LiCl}]_\Sigma}; \qquad (4.47)$$

$$k_{3obs} = \frac{k_{5.20} + k_{5.21}[\text{CuCl}_2]_\Sigma}{k_{5.22} + [\text{LiCl}]^2_\Sigma}. \qquad (4.48)$$

As can be seen from these expressions, the reaction of CME formation is second-order with respect to $CuCl_2]_\Sigma$. The weak dependence of k_{3obs} on $[CuCl_2]_\Sigma$ can be related to the influence of $[CuCl_2]$ on the concentration of "free" LiCl which, in turn, strongly affects r_{DMA} (see Eq. (4.25)). In order to clarify this extremely important issue, the reactions of equilibrium complex formation in system II were studied by spectrophotometric techniques [52]. The total concentration of LiCl and $LiClO_4$ in these experiments was maintained constant (2.75 M). The observed influence of $[PdCl_2]$ on the optical density of $CuCl_2$ solutions at $\lambda = 714$ nm (where the $PdCl_2$–LiCl–$LiClO_4$ solution does not absorb) showed evidence for the formation of mixed $PdCl_4 \cdot (CuCl_2)_n^{2-}$ complexes with $n = 1, 2$. Data were interpreted using 17 models of equilibrium in the $CuCl_2$–LiCl–$LiClO_4$–MeOH system. For ion equilibria, the dissociation constant for LiCl in methanol was assumed to be $K_{LiCl} = 0.316$ [60]. The best agreement between model calculations and the experimental data was obtained for the following model (No. 17):

$$\begin{array}{llll} CuCl_2 + LiCl & \rightleftharpoons Li[CuCl_3], & & K_1 = 0.83 \pm 0.02; \\ CuCl_2 + 2LiCl & \rightleftharpoons Li_2[CuCl_4], & & K_2 = 30.1 \pm 0.78; \\ 2CuCl_2 + 2LiCl & \rightleftharpoons Li_2[Cu_2Cl_6], & & K_3 = 2.43 \pm 1.62; \\ 2CuCl_2 & \rightleftharpoons Cu_2Cl_4, & & K_4 = (2.9 \pm 0.06) \cdot 10^{-2}. \end{array}$$

The calculated dependence of the concentration of free LiCl on $[LiCl]_\Sigma$ and $[CuCl_2]_\Sigma$ is as follows:

$$[\text{LiCl}] = 0.54[\text{CuCl}_2]_\Sigma^{-0.3}[\text{LiCl}]_\Sigma. \qquad (4.49)$$

The adequacy of a model of equilibria (No. 17) not involving ion species is probably related to a high concentration of the cationic background ($[LiClO_4]_\Sigma + [LiCl]_\Sigma = 2.75$ M). Taking into account low concentrations of $PdCl_2$ (not exceeding 0.01 M) in the kinetic experiments, it can be suggested that the equilibrium concentrations of LiCl and Cu(II) chloride complexes calculated for the $CuCl_2$–LiCl–$LiClO_4$ system are virtually uninfluenced by the introduction of $PdCl_2$.

The model expression (4.49) for [LiCl] and the known constant of equilibrium in methanol [34]

$$PdCl_3^- + Cl^- \overset{K_1}{\rightleftharpoons} PdCl_4^{2-} (K_1 = 1.0 \cdot 10^{-2} \text{ at } 25\,^\circ\text{C}) \qquad (4.50)$$

were used to evaluate the concentrations of $LiPdCl_3$ and Li_2PdCl_4 complexes in $CuCl_2$ and LiCl solutions. For example, in a solution of 0.01 M $PdCl_2$, 0.5 M $CuCl_2$, and 0.005 M $[LiCl]_\Sigma$, the concentration of Pd(II) in the form of $Li[PdCl_3]$ amounts to 97 %; at $[LiCl]_\Sigma = 0.1$ M, the concentration of this form decreases to 13 % Pd(II) and at $[LiCl]_\Sigma = 0.5$ M, to 3 % Pd(II). The concentration of a mixed copper–palladium complex was estimated [59] at 15–20 % of $[PdCl_2]_\Sigma$.

An analysis of the empirical expression (4.48) with allowance for the data on equilibrium constants showed that, if $[CuCl_2]_\Sigma$ is varied (from 0 to 1 M) and the concentration of free LiCl calculated using Eq. (4.49) is maintained constant (0.25 M) by changing $[LiCl]_\Sigma$ within 0.25–0.46 M, then the rate constant k_{3obs} and the rate r_{DMA} cease to depend on $[CuCl_2]_\Sigma$, while the quadratic dependence of k_{2obs} on $[CuCl_2]_\Sigma$ is retained. A change in [LiCl] probably accounts for the dependence of r_{DMA} on $[CuCl_2]$ at large $[CuCl_2]_\Sigma$, which was observed in [33].

The kinetic equations (4.45)–(4.48) are also indicative of the absence of common intermediates in the reactions under consideration and of a difference of the reaction mechanism from that according to the

Henry scheme (4.38) in the acetic acid. Accordingly, DMA and CME are formed in parallel via a two-route reaction on Pd(II) and Pd(II)–Cu(II) complexes, respectively. The presence of an oxidant in the σ complex of Pd(II)–CH_2CH_2OR formed from ethylene favors the transition of Pd(II) to a group that readily leaves the system upon the nucleophilic replacement of Pd(II)Ox by Cl^- or Br^- ions at a carbon atom.

Investigations into the stereochemistry of halogenated ethers formed upon the decomposition of organopalladium intermediates and determination of the reaction rate orders with respect to ROH provided a fairly complete pattern of the mechanism of 1,2-disubstituted product formation in the process under consideration. In particular, the oxidation of trans-1,2-dideuteroethylene in Pd(II) and Cu(II) solutions was studied under the conditions where Eqs (4.43) and (4.44) were valid, namely, for $[PdX_2]_\Sigma = 0.1$ M, $[CuX_2]_\Sigma = 2.0$ M, $[LiX]_\Sigma = 1.0$ M, and $[HX] = 0.02$ M at 25 °C (X=Cl, Br). The conformer structures were analyzed using their ^{1}H NMR spectra (measured using Eu(III) compounds as paramagnetic shift reactants for ClCHDCHDOMe). Both d_2–β-halogenethyl methyl ethers were obtained in a *threo* conformation from *trans*-1,2-dideuteroethylene, which is evidence for the total *cis*-addition of halogen and OCH_3 group at the double bond. Analogous stereochemical data were obtained in [36, 61], where d_2-ethylene chlorohydrin and deuterated halogenacetates were synthesized from *trans*-1,2- and *cis*-1,2-dideuteroethylenes (in the latter case, in the form of *erythro* isomers):

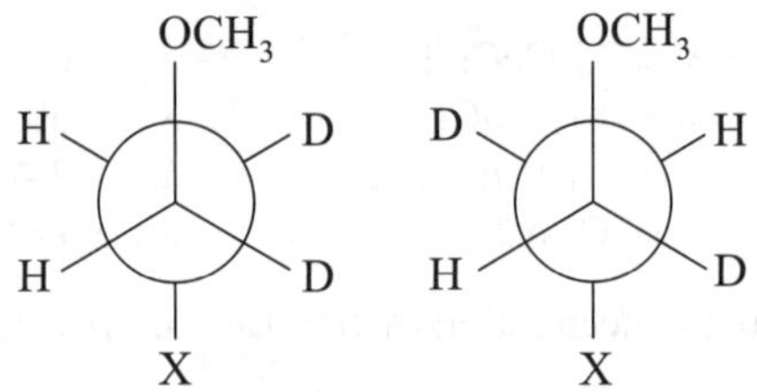

threo isomer *erythro* isomer

Since it was demonstrated that the nucleophilic replacement of palladium by halide ions at a carbon atom is accompanied by inversion of the configuration of the latter atom [36, 61], the overall stereochemical result of the two steps–alkoxypalladation of ethylene and nucleophilic substitution of Pd(II) stimulated by the oxidant – is indicative of the *trans*-addition of alcohol to the π complex of Pd(II) from solution, rather than of the *cis*-addition of ethylene at the Pd–OCH_3 bond. Thus, the two probable schemes of quasi-equilibrium (pre-equilibrium, PE) σ complex formation according to the kinetic equations (4.41) for CME and BME are as follows:

$$(1)\ \mathrm{Cu(II)\cdot Pd(II)(C_2H_4) + ROH \overset{PE}{\rightleftharpoons} \underset{(\sigma\ complex)}{Cu(II)\cdot Pd(II)CH_2CH_2OR^-} + H_3O^+;}$$

$$\sigma\ \text{complex} \xrightarrow[X^-]{lim} \mathrm{XCH_2CH_2OR} + \ldots$$

$$(2)\ \mathrm{\underset{(\pi\ complex\ I)}{Cu(II)\cdot Pd(II)(C_2H_4)} + ROH \overset{PE}{\rightleftharpoons} \underset{(\pi\ complex\ II)}{Cu(II)\cdot Pd(II)(OR)(C_2H_4)^-} + H_3O^+;}$$

$$\pi\ \text{complex II} + \mathrm{ROH} \xrightarrow{lim} \mathrm{Cu(II)\cdot Pd(II)(OR)(CH_2CH_2OR)^{2-} + H_3O^+}.$$

According to the second variant, the alkoxy group is *necessarily* formed on Pd(II), because ethylene is present as a strong *trans* ligand, but this group does not participate in the intra-sphere addition to ethylene. This variant also requires the participation of two alcohol molecules in the formation of halogenated ether. For this reason, the reaction order was studied in a bromide system under conditions of bromoethyl methyl ether (BME) and bromoethyl ethyl ether (BEE) formation in system III in a methanol–ethanol mixture.

The use of a MeOH–EtOH mixture and the r_{BME}/r_{BEE} ratio for determining the reaction order with respect to alcohol is justified by the following considerations:

(i) Most parameters that characterize the solvent properties (both specific and nonspecific) change by no more than 11 % on the passage from methanol to ethanol, except for the dielectric permittivity ε that decreases by ~25%. A significant change (>20 %) in the ε value of the MeOH–EtOH mixture is observed when the [MeOH]/[EtOH] ratio is reduced below 0.17 [62, 63]. The two alcohols have close donor numbers (19 and 20 for methanol and ethanol, respectively), and their acceptor numbers are also quite close (41.3 and 37.0, respectively [63], see also Section 8.5). The association equilibrium constants and ΔH°_{298} values for the formation of O–H ··· O hydrogen bonds in the MeOH–MeOH and EtOH–EtOH systems are 46.2 M^{-1}, 22.3 kJ/mole and 31.4 M^{-1}, 22.4 kJ/mole, respectively.

(ii) The use of the ratio of two product formation rates in the same solution (i.e., using competitive reactions) eliminates the problems related to differences between the free LiCl concentrations and the F_{Pd} and F_{Cu} values related to the variation of [MeOH], which would be encountered in any other (e.g., methanol–dioxane) mixture.

Thus, in the proposed series of experiments with variable concentrations of each of the two alcohols at constant values of $[PdBr_2]_\Sigma$, $[CuBr_2]_\Sigma$, $[LiBr]_\Sigma$, and $[HBr]_\Sigma$ in the solution, the reaction orders with respect to $[ROH]^n$ ($n = 1, 2$) must not significantly vary due to a change in the nature of the solvent.

The experimental dependences of r_{BME}/r_{BEE} on [MeOH]/[EtOH] proved to be nonlinear [59] and showed clear evidence for a quadratic contribution due to $([MeOH]/[EtOH])^2$. Writing the rates of formation of the two products as

$$r_{BME} = k_1[MeOH]^2 + k_2[MeOH][EtOH], \tag{4.51}$$

$$r_{BEE} = k_3[EtOH]^2 + k_4[MeOH][EtOH], \tag{4.52}$$

yields the ratio

$$\varphi = \frac{r_{BME}}{r_{BEE}} = \frac{k_1\left[\dfrac{MeOH}{EtOH}\right]^2 + k_2\left[\dfrac{MeOH}{EtOH}\right]}{k_3 + k_4\left[\dfrac{MeOH}{EtOH}\right]}. \tag{4.53}$$

The empirical second-order rate constants $k_i \cdot 10^4$ $[M^{-1}\ h^{-1}]$ in Eqs. (4.51) and (4.52) are as follows:

k_1	k_2	k_3	k_4
1.34	5.50	6.60	0.2

According to these data, k_4[MeOH/EtOH] < k_3 and, hence, the ratio φ is satisfactorily described by the following equation:

$$\varphi = k_5\left[\frac{MeOH}{EtOH}\right]^2 + k_6\left[\frac{MeOH}{EtOH}\right]. \tag{4.54}$$

Assuming that the obtained values of k_1 and k_3 correspond to the observed rate constants in pure alcohols and using the relations $r_{BME} = k_1^0[MeOH]^2$ and $r_{BEE} = k_3^0[EtOH]^2$ for pure alcohols at the same conditions, we obtain the following estimations for these alcohols under the same conditions: $k_1^0 = 1.34 \cdot 10^{-4}\ M^{-1}\ h^{-1}$

and $k_3^0 = 8.36 \cdot 10^{-4}\,M^{-1}\,h^{-1}$, which are in good agreement with the k_1 and k_3 values determined for the MeOH–EtOH mixture.

The data presented above on the kinetics of formation of DMA and halogenethyl alkyl ethers show that the mechanistic scheme of 1,2-disubstituted product formation according to Henry [52] and Bäckvall et al. [36] for ethylene chlorohydrin and its acetic acid ethers, which stipulates a conjugation node at the common σ complex,

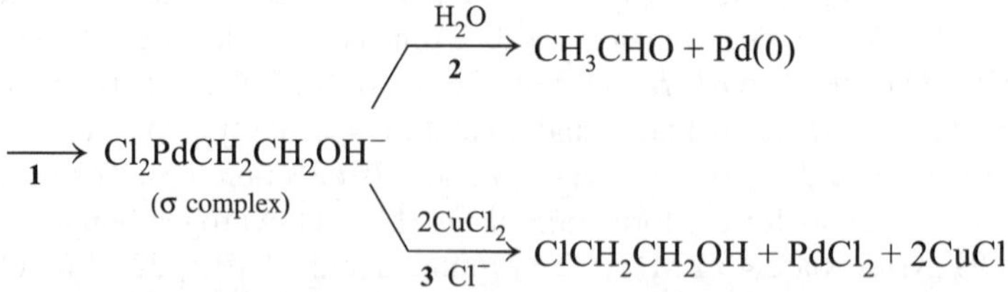

(4.55)

as well as the scheme with the dissociation of σ complexes at the limiting step [36],

$$\underset{1}{\overset{\text{PE}}{\rightleftharpoons}} \underset{(\sigma)}{Cl_2PdCH_2CH_2OH^-} \underset{-Cl}{\overset{lim}{\longrightarrow}} \underset{(\sigma_1)}{ClPdCH_2CH_2OH} \qquad (4.56)$$

do not agree with experimental data. The presence of a quasi-equilibrium (pre-equilibrium, PE) step involving the formation of an intermediate compound between $CuCl_2$ and σ complex,

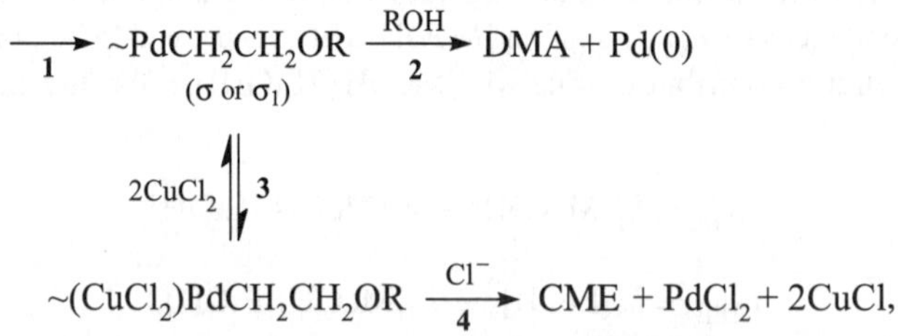

(4.57)

can also be excluded from consideration.

There are two other hypotheses that cannot be discriminated using formal kinetic criteria. Hypothesis I assumes quasi-equilibrium step **1** in scheme (4.57), according to which r_{DMA} ceases to depend on $[CuCl_2]$. Hypothesis II considers the formation of DMA and CME on different active centers from the very beginning of the process:

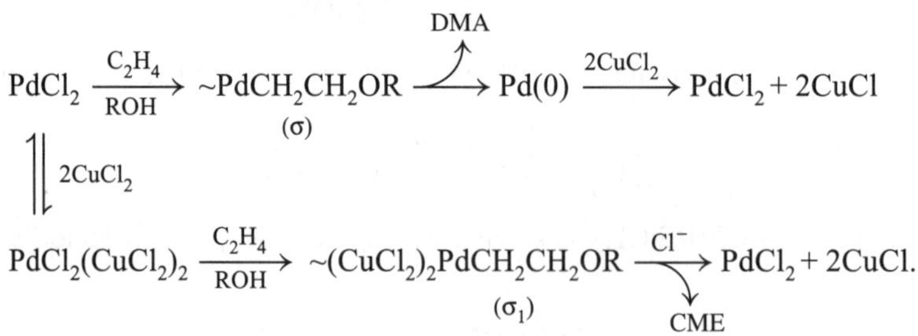

(4.58)

Hypothesis I cannot be used to explain the formation of ethylene chlorohydrin in water, since the kinetic isotope effect value (*KIE*) for the process of acetaldehyde formation is 1.07. Although this *KIE* (C_2H_4/C_2D_4) value was not determined at large concentrations of $CuCl_2$, step **2** of σ complex decomposition in schemes

(4.55) and (4.57) can hardly be considered slow compared to step **1** that becomes a fast quasi-equilibrium process in this scheme. Unfortunately, the KIE was also not measured under the conditions of DMA and CME formation in alcohols. Nevertheless, scheme (4.58) according to hypothesis II seems to be more likely. A scheme obtained by combining hypotheses I and II as

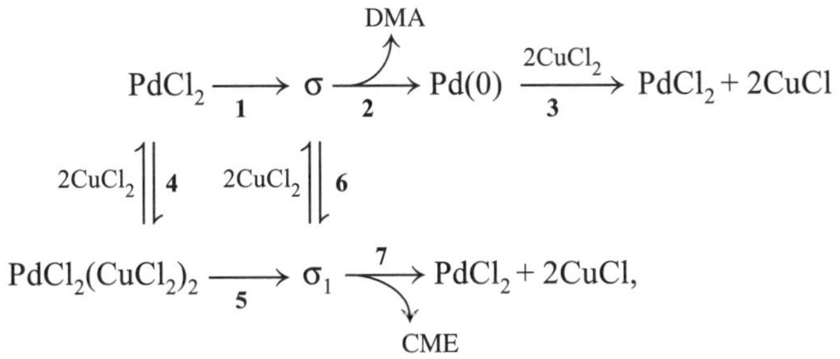

(4.59)

does not contradict the experiment, but the quasi-equilibrium of steps **1**, **5**, and **6** in this scheme also seems unlikely.

Taking into account observed kinetic laws, stereochemistry of halogenated ether formation, and the second reaction orders with respect to alcohols, let us write a kinetic equation for the synthesis of halogenated ethers within the framework of hypothesis II as represented by scheme (4.58). Note that the second order with respect to [ROH] in the halogenated ether synthesis according to this scheme by no means implies that the DMA synthesis is also second-order with respect to [ROH]. Within hypothesis II, data on the stereochemistry of 1,2-disubstituted product formation also do not suggest that the DMA formation involves the *trans*-addition of ROH to the π complex of Pd(II). Therefore, the conclusions of Bäckvall et al. [36] and Stille et al. [64] concerning the mechanism of olefin oxidation in aqueous solution (Wacker process) still cannot be treated as justified. Let us consider a more detailed mechanism of halogenated ether synthesis as represented by the following sequence of steps:

$$Li_2PdCl_4 + 2CuCl_2 \overset{1}{\rightleftharpoons} Li_2PdCu_2Cl_8; \qquad (4.60)$$

$$Li_2PdCu_2Cl_8 + C_2H_4 + ROH \overset{2}{\rightleftharpoons} \underset{(\pi)}{Li_2PdCu_2Cl_8(C_2H_4)(OR)^-} + H_s^+; \qquad (4.61)$$

$$\pi + ROH \xrightarrow{3} \underset{(\sigma)}{LiPdCu_2Cl_7(OR)(CH_2CH_2OR)^-} + H_s^+ + LiCl; \qquad (4.62)$$

$$\sigma + HCl \xrightarrow{4} Li_2PdCl_4 + 2CuCl_2^- + ClCH_2CH_2OR + ROH. \qquad (4.63)$$

The removal of H_s^+ in steps (4.61) and (4.62) during the formation of π and σ complexes can also take place without Cl^- evolution, since the coordination vacancies in this case appear due to the cleavage of chloride bridges in $PdCu_2Cl_6{}^{2-}$ anion. For the limiting step **3**, the reaction rate is

$$r_{CME} = \frac{k_3K_1K_2[Li_2PdCl_4][CuCl_2]^2[ROH]^2[C_2H_4]}{[H_s^+]}. \qquad (4.64)$$

Writing the material balances for Pd(II) and Cu(II) with allowance for only mononuclear metal complexes as

$$[PdCl_2]_\Sigma = [LiPdCl_3] + [Li_2PdCl_4] = [LiPdCl_3](1 + K_5[LiCl]); \tag{4.65}$$

$$\begin{aligned}[CuCl_2]_\Sigma &= [CuCl_2] + [LiCuCl_3] + [Li_2CuCl_4] \\ &= [CuCl_2](1 + K_6[LiCl] + K_7[LiCl]^2),\end{aligned} \tag{4.66}$$

where K_5, K_6, and K_7 are the equilibrium constants of reactions

$$LiPdCl_3 + LiCl \overset{K_5}{\rightleftharpoons} Li_2PdCl_4; \tag{4.67}$$

$$CuCl_2 + LiCl \overset{K_6}{\rightleftharpoons} LiCuCl_3; \tag{4.68}$$

$$CuCl_2 + 2LiCl \overset{K_7}{\rightleftharpoons} Li_2CuCl_4, \tag{4.69}$$

we obtain

$$r_{CME} = \frac{k_3K_1K_2'K_5[PdCl_2]_\Sigma\,[CuCl_2]_\Sigma^2\,[LiCl][ROH]^2P_{C_2H_4}}{[H_s^+](1 + K_5[LiCl])(1 + K_6[LiCl] + K_7[LiCl]^2)^2}, \tag{4.70}$$

where the Henry constant for ethylene is included into K_2'. A comparison of Eqs (4.45), (4.47), and (4.70) shows that function φ which is related to the concentration of free LiCl in (4.70) and defined as

$$\varphi = \frac{k_3K_1K_2'K_5[LiCl][ROH]^2}{(1 + K_5[LiCl])(1 + K_6[LiCl] + K_7[LiCl]^2)^2}, \tag{4.71}$$

can be approximated by function φ_1 related to the total LiCl concentration as:

$$\varphi_1 = \frac{k_{5.16} + k_{5.17}[LiCl]_\Sigma}{k_{5.18} + k_{5.19}[LiCl]_\Sigma}. \tag{4.72}$$

Assuming that the DMA synthesis is second-order with respect to ROH and adopting the traditional mechanistic scheme (ethylene insertion into Pd–OR bond) [22, 24, 34] with the steps

$$Li_2PdCl_4 + C_2H_4 + ROH \overset{K_8}{\rightleftharpoons} \underset{(\pi_1)}{Cl_2Pd(C_2H_4)(OR)^-} + H_s^+ + 2LiCl; \tag{4.73}$$

$$\pi_1 + ROH \xrightarrow{k_9} \underset{(\sigma_1)}{Cl_2Pd(CH_2CH_2OR)(ROH)^-}; \tag{4.74}$$

$$\sigma_1 \xrightarrow{k_{10}} HPdCl_2^- + CH_3CH(OR)_2; \tag{4.75}$$

we obtain the following equation:

$$r_{\mathrm{DMA}} = \frac{k_9 K_8 [\mathrm{Li_2PdCl_4}][\mathrm{C_2H_4}][\mathrm{ROH}]^2}{[\mathrm{H_s^+}][\mathrm{LiCl}]^2} = \frac{k_9 K_5 K_8' [\mathrm{PdCl_2}]_\Sigma \, P_{\mathrm{C_2H_4}} [\mathrm{ROH}]^2}{[\mathrm{H_s^+}][\mathrm{LiCl}](1 + K_5[\mathrm{LiCl}])}. \tag{4.76}$$

A comparison of Eqs (4.46), (4.48), and (4.76) and the above calculations of the concentration of free LiCl show that the function of [LiCl] in Eq. (4.76)

$$\varphi_3 = \frac{k_9 K_5 K_8' [\mathrm{ROH}]^2}{[\mathrm{LiCl}](1 + K_5[\mathrm{LiCl}])} \tag{4.77}$$

can be approximated by the function

$$\varphi_4 = \frac{k_{5.20} + k_{5.21}[\mathrm{CuCl_2}]_\Sigma}{k_{5.22} + [\mathrm{LiCl}]_\Sigma^2}. \tag{4.78}$$

The observed change in the reaction order with respect to $[\mathrm{CuCl_2}]_\Sigma$ as dependent on $[\mathrm{LiCl}]_\Sigma$ at 25 °C is probably related to the presence of two routes involving $\mathrm{CuCl}_n^{(n-2)-}$ and $\mathrm{Cu_2Cl}_m^{(m-4)-}$ complexes, the total contributions of which varies with increasing $[\mathrm{LiCl}]_\Sigma$ [58].

Within the framework of hypothesis II as represented by scheme (4.58) with the participation of dioxygen species, Cu(II) and Pd(II) complexes perform various kinetic functions. On the first route of DMA formation, $PdCl_2$ catalyzes the reaction

$$\mathrm{C_2H_4 + 2CuCl_2 + 2ROH \xrightarrow{PdCl_2} CH_3CH(OR)_2 + 2CuCl + 2HCl}, \tag{4.79}$$

while $CuCl_2$ catalyzes the reaction

$$\mathrm{Pd(0) + 1/2\,O_2 + 2HCl \xrightarrow{CuCl_2} PdCl_2 + H_2O}, \tag{4.80}$$

and both $PdCl_2$ and $CuCl_2$ catalyzes the process

$$\mathrm{C_2H_4 + 1/2\,O_2 + 2ROH \xrightarrow{PdCl_2\text{–}CuCl_2} CH_3CH(OR)_2 + H_2O}. \tag{4.81}$$

On the route of chloroethyl alkyl synthesis according to scheme (4.58), $CuCl_2$ jointly with $PdCl_2$ catalyzes the process

$$\mathrm{C_2H_4 + 1/2\,O_2 + ROH + HCl \xrightarrow{PdCl_2\text{–}CuCl_2} ClCH_2CH_2OR + H_2O}, \tag{4.82}$$

and $PdCl_2$ additionally catalyzes the reaction

$$\mathrm{C_2H_4 + 2CuCl_2 + ROH \xrightarrow{PdCl_2} ClCH_2CH_2OR + 2CuCl + HCl}. \tag{4.83}$$

It should be noted that, in the framework of scheme (4.60)–(4.63), it is also necessary to consider the question concerning the higher reactivity of a π complex with *trans*-methoxy ligand in step (4.62) as compared to the π complex containing only chloride ion and ethylene. ■

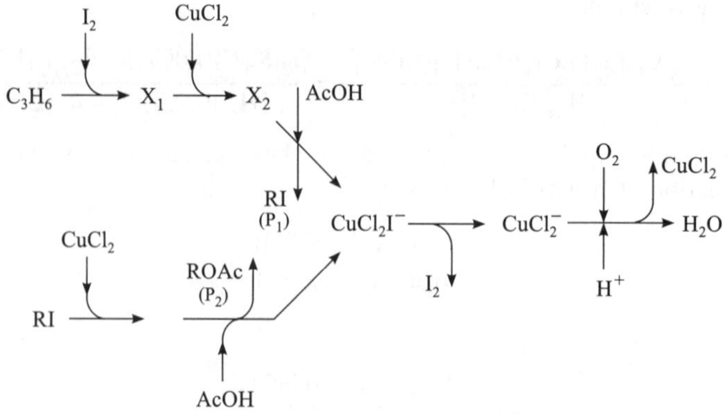

Figure 4.4 Mechanistic scheme of reaction (4.84)

Example 4.2 [51]

► Interesting kinetic functions (KFs) of the components of a catalytic system have been found in the reaction of 1,2-diacetoxypropane synthesis in the $CuCl_2$–I_2–AcOH system [51]:

$$C_3H_6 + {}^1\!/_2\,O_2 + 2AcOH \longrightarrow AcOCH(CH_3)CHOAc + H_2O. \tag{4.84}$$

The polyfunctional catalytic system $CuCl_2$–I_2 belongs to MCSs of group B, and the mechanism of reaction (4.84) contains MCSs of classes I–III (with respect to the conjugation of circuits in the corresponding bipartite graph). The following mechanistic scheme of reaction (4.84) was proposed (Figure 4.4) [51]:

$$\begin{aligned}
C_3H_6 + I_2 &\overset{1}{\rightleftharpoons} \underset{(X_1)}{C_3H_6I^+ \ldots I^-};\\
X_1 + CuCl_2 &\overset{2}{\rightleftharpoons} \underset{(X_2)}{C_3H_6I^+ \ldots CuCl_2I^-};\\
X_2 + AcOH &\overset{3}{\longrightarrow} \underset{(P_1)}{ICH_2CH(OAc)CH_3} + H^+ + CuCl_2I^-;\\
2CuCl_2I^- &\overset{4}{\longrightarrow} 2CuCl_2^- + I_2;\\
2CuCl_2^- + 2H^+ + {}^1\!/_2\,O_2 &\overset{5}{\longrightarrow} 2CuCl_2 + H_2O.
\end{aligned} \tag{4.85}$$

For a route with the vector of stoichiometric numbers of the steps $\nu^{(p)} = (2, 2, 2, 1, 1)$, the overall equation of the process catalyzed by $CuCl_2$ is as follows:

$$2C_3H_6 + I_2 + 2AcOH + {}^1\!/_2\,O_2 \longrightarrow 2P_1 + H_2O. \tag{4.86}$$

The kinetic function of Cu(II) chloride is to catalyze the conversion of X_1 into P_1 and I_2:

$$2X_1 + 2AcOH + {}^1\!/_2\,O_2 \longrightarrow 2P_1 + I_2. \tag{4.87}$$

At the second step of process (4.84), I^- in P_1 is replaced by OAc^- (or OH^-) with the formation of a propylene glycol diester. This step also involves $CuCl_2$ and the replacement proceeds as the reaction of oxidative substitution:

$$\begin{aligned}
&ICH_2CH_2(OAc)CH_3 + CuCl_2 \overset{6}{\rightleftharpoons} X_3;\\
&X_3 + AcOH \xrightarrow{7} \underset{(P_2)}{AcOCH_2CH_2(OAc)CH_3} - CuCl_2I^- + H^+;\\
&2CuCl_2I^- \xrightarrow{4} 2CuCl_2^- + I_2;\\
&2CuCl_2^- + 2H^+ + {}^1\!/_2\,O_2 \xrightarrow{5} 2CuCl_2 + H_2O.
\end{aligned} \tag{4.88}$$

For this stage, the vector of stoichiometric numbers is $\boldsymbol{\nu}^{(p)} = (2, 2, 1, 1)$ and the overall equation appears as

$$2P_1 + 2AcOH + {}^1\!/_2\,O_2 \xrightarrow{CuCl_2} 2P_2 + I_2. \tag{4.89}$$

Summation of Eqs (4.86) and (4.89) yields the process

$$2C_3H_6 + 4AcOH + O_2 \longrightarrow 2P_2 + 2H_2O, \tag{4.90}$$

which is catalyzed by $CuCl_2$ and I_2 (Figure 4.4).

The rate of the second step ($P_1 \rightarrow P_2$) with $CuCl_2$ is higher by an order of magnitude than that with $FeCl_3$ as the oxidant. It can be suggested that CuCl also participates in the steps of iodine substitution in RI (P_1). The steps of C–Hal bond cleavage with the participation of Cu(I) will be considered below. The bipartite graph of the mechanism under consideration (BG4, Figure 4.5) contains seven KFs, which correspond to the overall equations (4.86), (4.87), (4.89), and (4.90). ■

In Chapter 3, Example 3.8 presented the mechanism of noncatalytic asymmetric dihydroxylation (oxidation) of olefins according to Sharpless using OsO_4 as an oxidant. This process becomes catalytic when OsO_4 is oxidized by $K_3Fe(CN)_6$ or N-methylmorpholine oxide. It was shown [65] that this system can be rendered a polyfunctional system with three catalysts by introducing an additional catalyst (flavine hydroperoxide) of the N-methylmorpholine oxide regeneration and hydrogen peroxide as an additional

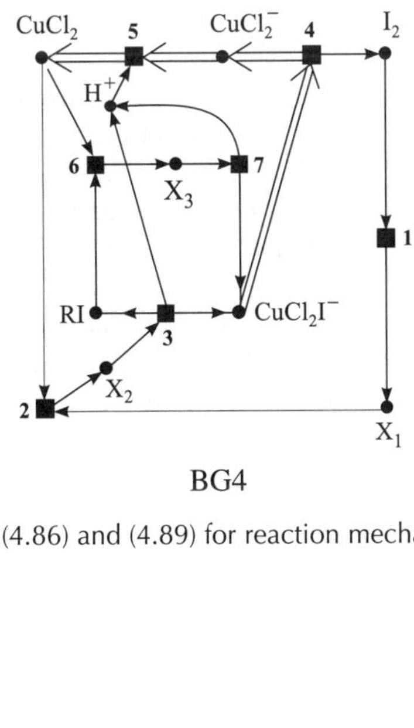

BG4

Figure 4.5 Bipartite graph of stages (4.86) and (4.89) for reaction mechanism (4.84) simplified by assuming all steps to be irreversible oxidant. The KFs of components in this PFCS are clearly illustrated in the following scheme, where F_{OH} and F_{OOH} are the derivatives of flavine (F) [65, 66]:

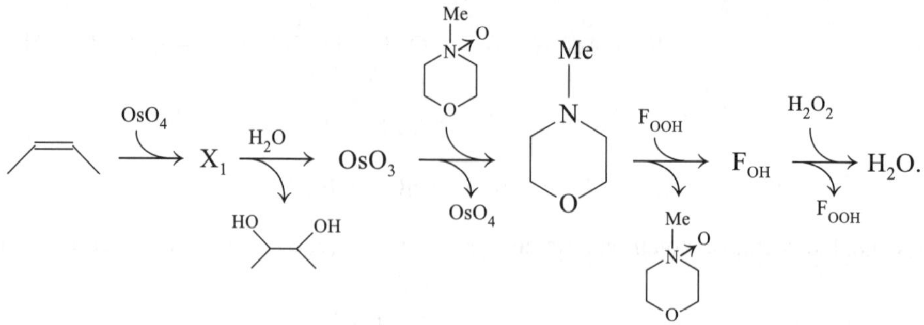

(4.91)

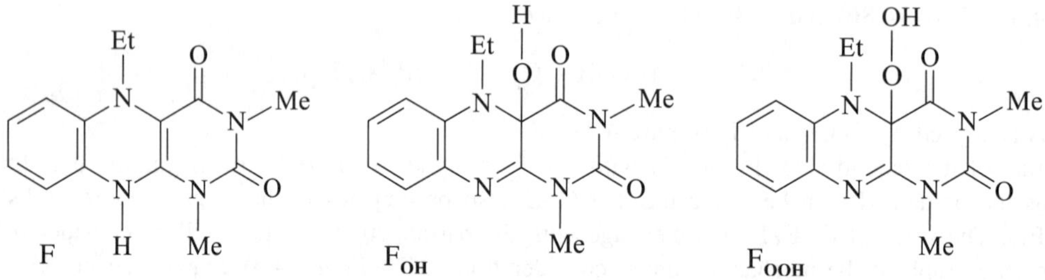

4.1.2 Oxidation of 1,3-dienes

Investigations of the oxidation of 1,3-dienes in solutions of MCSs revealed different variants of polyfunctionality in these reactions. In the PdX_2–CuX_2–Ox, CuX_2–I_2, and PdX_2–Q–ML–Ox systems (where ML denotes phthalocyaninates and porphyrinates of transition metals and Q are quinones), dienes are oxidized with the formation of various products [67–85].

Example 4.3 [67–70]

► In aqueous solutions of Pd(II) salts, 1,3-butadiene is oxidized to crotonaldehyde and furan, while 1,3-pentadiene is oxidized to α-methylfuran, halogenhydrins, and 2-pentene-1,4-diol. A scheme of the oxidation reaction in aqueous solutions includes the formation of a hydroxyl-containing η^3-allyl complex of Pd(II) and its redox decay with the formation of oxygen-containing products, which are consecutively oxidized to furan and α-methylfuran [67–70]. Two possible schemes were established for the conversion of η^3-allyl complexes of Pd(II).

In the $PdCl_2$–$CuCl_2$ system, the oxidation of 1,3-butadiene proceeds nonselectively [67, 68]. The yield of furan in a temperature interval of 80–105 °C reaches 27 %. Byproducts include crotonaldehyde, aldol (β-hydroxybutyric aldehyde), and their conversion products. The η^3-allyl complex (**I**) formed from 1,3-butadiene was isolated and characterized by ^{1}H NMR spectroscopy [68]. As the concentration of complex **I** in solution increases, it converts into complex **II**:

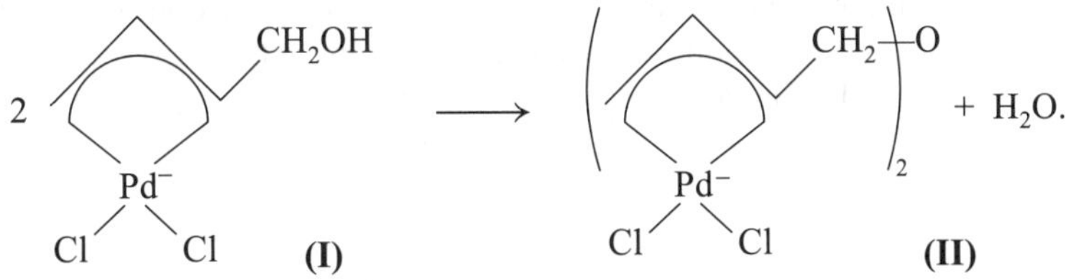

(4.92)

Upon a prolonged heating of complex **I** (at [HCl] = 0.1–0.3 M), it exhibits an intermolecular decay with the formation of crotonaldehyde (yield, ~60%) and Pd(0). The transfer of hydride ions from hydroxymethyl groups to Pd(II) favors an increase in the pH level and the ionization of hydroxymethyl groups [68, 70]. It was suggested that the formation of complexes between **I** and $CuCl_2$ also favors the formation of crotonaldehyde and chlorine-containing product. Based on the results of investigation of the oxidation of crotonaldehyde in $CuCl_2$ and $PdCl_2$, the following preliminary scheme of conversion of 1,3-butadiene and crotonaldehyde (CA) into furan and byproducts P_1, P_2, and P_3 at 80 ÷ 100 °C was proposed [67, 68]:

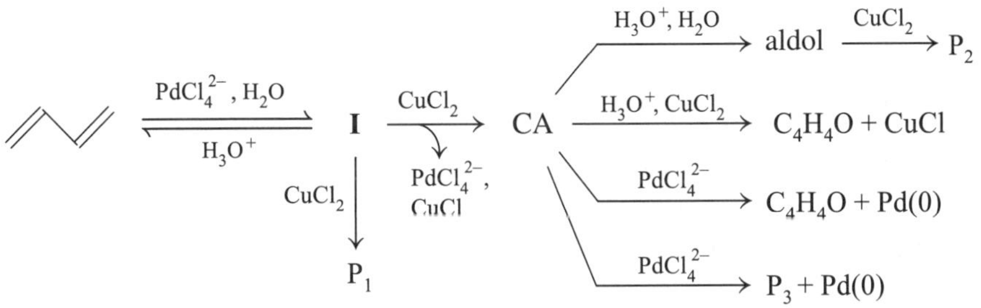

; (4.93)

$$Pd(0) + 2CuCl_2 \longrightarrow PdCl_2 + 2CuCl.$$

Under steady-state process conditions in the presence of *para*-benzoquinone or chloranil, CuCl is oxidized by these quinones (Q) to $CuCl_2$. Note that the kinetic functions of Pd(II) in this system are as follows:

- catalysis of diene oxidation to CA;
- catalysis of CA oxidation to furan;
- catalysis of CA oxidation to P_3 (β-ketobutyric aldehyde).

The kinetic functions of Cu(II) in this system are as follows:

- catalysis of **I** oxidation to CA and product P_1;
- catalysis of aldol oxidation to P_2;
- catalysis of CA oxidation to furan;
- catalysis of Pd(0) oxidation to $PdCl_4^{2-}$ by Q.

An alternative mechanism of pentadiene oxidation to methylfuran was proposed for the $PdCl_2$–$CuBr_2$–Ox system at 10–50 °C [69, 71]. Under these conditions, the oxidation of 1,3-pentadiene also proceeds via type **I** η^3-allyl complex of Pd(II), but the decay of this complex under the action of oxidant and nucleophile (Br^-) leads to the formation of bromohydrin (P_1). The substitution of OH for allyl bromine in P_1 and the oxidation of the corresponding 1,4-pentadiol (P_2) by Pd(II) chloride leads to the formation of hydroxycarbonyl compounds P_3 (and/or P_4), which in the presence of HCl covert into α-methylfuran:

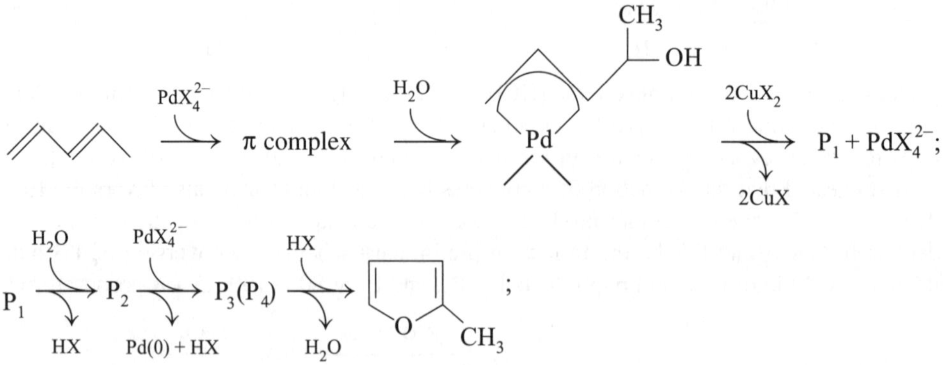

(4.94)

$$Pd(0) + 2CuX_2 \longrightarrow PdX_2 + 2CuX;$$

$$2CuX + {}^1/_2O_2 + 2HX \longrightarrow 2CuX_2 + H_2O;$$

$P_1 = BrCH_2CH{=}CHCH(CH_3)OH$ (drawn as $BrCH_2CH{=}CHCHOH$ with CH_3 on the carbinol carbon)

$P_2 = HOCH_2CH{=}CHCH(CH_3)OH$ (drawn as $HOCH_2CH{=}CHCHOH$ with CH_3 on the carbinol carbon)

$P_3 = O{=}CHCH{=}CHCH(CH_3)OH$ (drawn as $O{=}CHCH{=}CHCHOH$ with CH_3 on the carbinol carbon)

$P_4 = HOCH_2CH{=}CHC(CH_3){=}O$ (drawn as $HOCH_2CH{=}CHC{=}O$ with CH_3 on the carbonyl carbon)

The kinetic functions of PFCS components in this process are also rather diverse:

- Pd(II) catalyzes oxidation of 1,3-pentadiene to P_1 by copper(II) bromide:

$$C_5H_8 + 2CuBr_2 + H_2O \xrightarrow{Pd(II)} P_1 + 2CuBr + HBr; \qquad (4.95)$$

- Pd(II) catalyzes oxidation of glycol P_2 to P_3 or P_4 by copper(II) bromide:

$$C_5H_8(OH)_2 + 2CuBr_2 \xrightarrow{Pd(II)} P_3(P_4) + 2CuBr + 2HBr; \qquad (4.96)$$

- Cu(II) catalyzes oxidation of η^3-allyl complex to P_1 by oxygen:

$$BrPd(C_5H_8OH) + {}^1\!/_2\,O_2 + 2HBr \xrightarrow{Cu(II)} PdBr_2 + H_2O + P_1; \qquad (4.97)$$

- Cu(II) catalyzes oxidation of Pd(0) to $PdBr_2$:

$$Pd(0) + {}^1\!/_2\,O_2 + 2HBr \xrightarrow{Cu(II)} PdBr_2 + H_2O. \qquad (4.98)$$

Another important component of the PFCS under consideration is the acid (HBr or HCl), which accelerates the cyclization of P_3 (P_4) into methylfuran. Thus, the process according to the overall equation

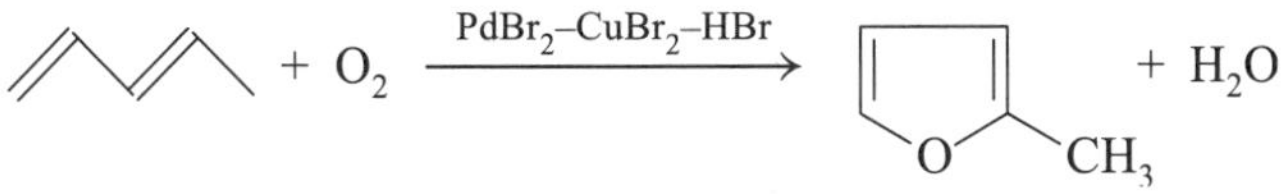

(4.99)

is catalyzed by a system of three catalysts. The synthesis of furans from various dienes and the oxidation of 4-hydroxy-η^3-allyl complexes of palladium by *para*-benzoquinone were considered in [72]. ■

In nonaqueous media (ROH, AcOH), 1,3-dienes are oxidized to ethers and esters of glycols (such as $ROCH_2CH{=}CHCH_2OR$, $ROCH_2CH(OR)CH{=}CH_2$) [66, 73, 75], 1,4-vinyldioxane and its derivatives in ethylene glycol [74, 76, 77], diacetates of butenediols and other unsaturated diols [80–85], and cyclopentene diesters [78, 79]. In the $PdCl_2$–$CuCl_2$–ROH [73, 74] and $PdCl_2$–Q–Ox systems [80–86], the reactions of diene oxidation include the steps of formation and conversion of η^3-allyl complexes Pd(II). A very effective system for these reactions is $CuCl_2$–I_2–Ox [87].

Example 4.4 [73, 74, 76–87]

► A general feature of the CuX_2–containing catalytic systems in alcohol and glycol solutions is the formation of intermediate dichloro- and dibromobutenes (from 1,3-butadiene) or the corresponding halogenated ethers.

The kinetics of 1,3-butadiene oxidation in ethanol was studied in the $CuCl_2$–I_2–LiCl (2 M) system at 303 K. The initial reaction rates were determined by the accumulation of Cu(I): $r_\Sigma = \frac{1}{2}\frac{d[Cu^I]}{dt}$. In the interval of $[CuCl_2]_\Sigma \le 1.6$ M, the behavior of $r_\Sigma|_{t\to 0}$ is described by the following equation:

$$r_\Sigma^0 = (k_1 + k_2[HCl])[CuCl_2]_\Sigma^3\,[I_2]^2 P_{C_4H_6}, \tag{4.100}$$

which corresponds to the conversion of the intermediate $mCuCl_2 \cdot nI_2 \cdot C_4H_6$ complex into isomeric iodoethers (IC_4H_6OR) and iodochlorobutenes (IC_4H_6Cl). The sequence of transformations can be represented by the following scheme:

$$\begin{array}{l} mCuCl_2 + nI_2 \underset{1}{\rightleftharpoons} B \overset{C_4H_6}{\underset{2}{\rightleftharpoons}} C \\ C \xrightarrow{3} IC_4H_6OR \xrightarrow{5} DE_1 + DE_2 \\ C \xrightarrow{4} IC_4H_6Cl \xrightarrow{6} DE_1 + DE_2 \\ IC_4H_6Cl \xrightarrow{7} ClC_4H_6Cl \xrightarrow{8} DE_1 + DE_2 \end{array} \tag{4.101}$$

Here, the steps of substitution of allylic halides (I, Cl) are catalyzed by CuCl and I_2 (in the presence of CuCl), while I^- liberated in steps **3–7** is oxidized by Cu(II) chloride to I_2 by analogy with the oxidation of olefins [51]. It is still unclear whether $CuCl_2$ participates in the steps of I^- replacement by OR^- and Cl^- (**5–7**). Taking into account the high mobility of allylic iodine, it can be suggested that the catalysis by Cu(I) chloride (as the aprotic acid) is sufficient for these substitution reactions. The synthesis of butenediol diethers and vinyl-1,4-dioxane from dichlorobutenes in the CuCl–$CuCl_2$–I_2 system has been studied in great detail [76, 77, 87]. Stage **8** in scheme (4.101) turns out to be a complicated multistep process, which includes the reactions of dichlorobutene isomerization, formation and isomerization of chlorinated ethers, and their conversion into diethers:

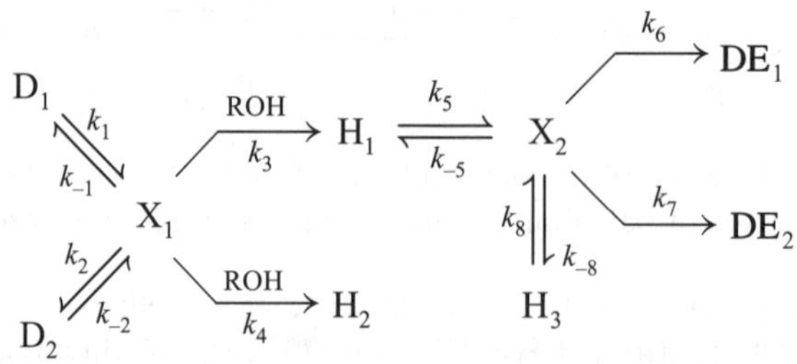

(4.102)

where D_1 is 3,4-dichlorobut-1-ene; D_2 is 1,4-dichlorobut-3-ene; H_1 is $ClCH_2CH{=}CHCH_2OR$; H_2 is $CH_2{=}CHCH(Cl)CH_2OR$; H_3 is $CH_2{=}CHCH(OR)CH_2Cl$; DE_1 is 3,4-dialkoxybut-1-ene; and DE_2 is dialkoxybut-2-ene. Steps **1, 2, 5**, and **8** are catalyzed by Cu(I) and I_2 (in the presence of Cu(I)).

Thus, the role of $CuCl_2$ in schemes (4.101) and (4.102) in the presence of an oxidant (Ox), which oxidizes CuCl to $CuCl_2$, consists in the catalysis of diene oxidation to IC_4H_6OR and IC_4H_6Cl. The kinetic functions of I_2 are as follows:

- catalysis of C_4H_6 oxidation to DE_1 and DE_2;
- catalysis of C_4H_6 oxidative chlorination to dichlorobutenes D_1 and D_2;
- catalysis (together with CuCl) of allylic halide substitution.

The kinetic function of CuCl is to catalyze the allylic substitution of halide ions (I, Cl) in halogenated esters, iodochlorobutenes, and dichlorobutenes. The $CuCl_2$–CuCl–I_2–Ox system features a three-component PFCS.

The kinetics of 1,3-butadiene oxidation in the $PdCl_2$–$CuCl_2$–LiCl (2 M) system was studied in ethylene glycol [74] and ethanol [73]. In the absence of oxidant (Ox), the reaction in ethylene glycol at 80 °C leads to the formation of 1-vinyl-1,4-dioxane (VD) with a 92 % yield:

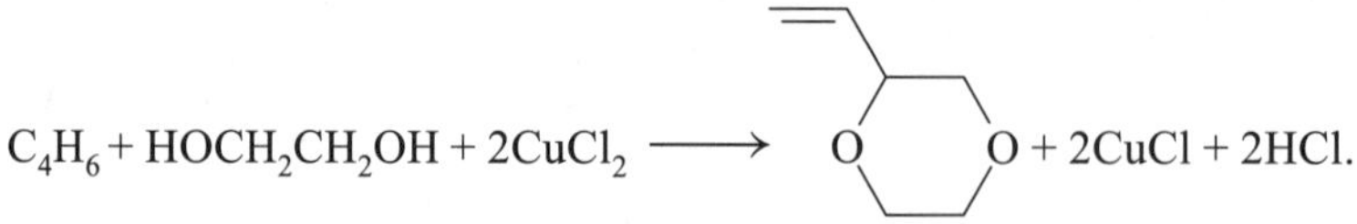

(4.103)

The reaction rate r_Σ is described by the following equation:

$$r_\Sigma = kP_{C_4H_6}[PdCl_2]_\Sigma\,[CuCl_2]_\Sigma^{4.3}. \tag{4.104}$$

Note that the high reaction order with respect to $[CuCl_2]_\Sigma$ was also observed in the reaction of diether synthesis in ethanol in the $PdCl_2$–$CuCl_2$–LiCl system at 30 °C [73, 87]:

$$r_\Sigma = P_{C_4H_6}[PdCl_2]_\Sigma\left(k'_{4.1}[CuCl_2]_\Sigma + k''_{4.1}[CuCl_2]_\Sigma^4\right). \tag{4.105}$$

The proposed mechanistic scheme of C_4H_6 oxidation in ethylene glycol implies the participation of polynuclear Cu(II) complexes:

$$PdCl_4^{2-} + nCuCl_2 \rightleftharpoons Pd(CuCl_2)_nCl_4^{2-} \underset{-2Cl^-}{\overset{C_4H_6}{\rightleftharpoons}} Pd(CuCl_2)_nCl_2(C_4H_6);$$

(π complex X)

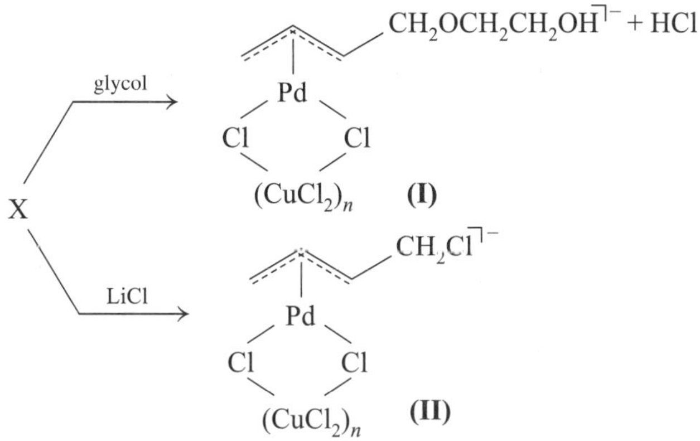

(4.106)

where η^3-allyl complexes **I** and **II** contain $CuCl_2$ and the fact of complex formation between Ox and η^3-allyl intermediate is essential for the reaction of C_4H_6 oxidation to VD (or diether formation). Note that Fe(III) chloride, which is also capable of oxidizing Pd(0) to Pd(II) but is less susceptible to forming polynuclear chloride complexes, does not lead to diene oxidation products in the $PdCl_2$–$FeCl_3$ system. The redox decomposition of complex **I** leads to the formation of chlorinated ethers A and B (the decomposition of complex **II** yields dichlorobutenes D_1 and D_2 that also convert into A and B), which yield VD:

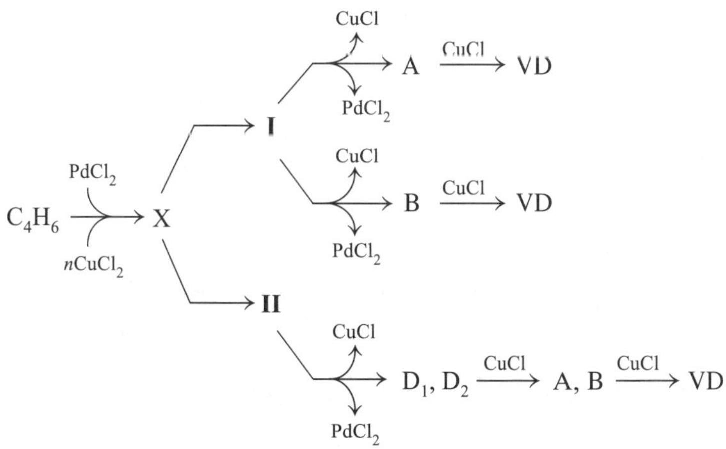

(4.107)

The oxidation of CuCl by oxygen to $CuCl_2$ makes the process catalytic with respect to three components – $PdCl_2$, $CuCl_2$, and CuCl – where the oxidation of CuCl must be incomplete. It was shown [77] that the steps of chlorine substitution in reactions of A, B, D_1, and D_2 are also catalyzed by $PdCl_2$, but CuCl has a significantly greater first-order reaction rate constant with respect to $MeCl_n$: $k_{CuCl} = 407\,M^{-1}\,min^{-1}$ against $k_{PdCl_2} = 67\,M^{-1}min^{-1}$.

The kinetic functions of $PdCl_2$ and I_2 in this system consist in the catalysis of C_4H_6 oxidation to products A, B, D_1, and D_2 by Cu(II) chloride. The role of $CuCl_2$ is to catalyze the oxidation of C_4H_6 to the same products by oxygen. The role of CuCl (and partly $PdCl_2$) is to catalyze the reactions of allyl substitution with the formation of A and B from D_1 and D_2 and the formation of VD from A and B. And all three components of the PFCS together catalyze the reaction

$$C_4H_6 + HOCH_2CH_2OH + {}^1\!/\!{}_2\,O_2 \xrightarrow{PdCl_2 - CuCl_2 - CuCl} VD. \tag{4.108}$$

■

The syntheses of diol diacetates by the oxidative acetoxylation (oxidative etherification) of conjugated dienes in the $Pd(OAc)_2$–Q–Ox system were described by Bäckvall et al. [28, 80–85]. It was established that the rate of formation of *cis*- and *trans*-diacetates of cyclohexenediol (and other cyclic diols) depends on the concentration of quinone Q under the conditions of a high regio- and stereoselectivity of processes in the $Pd(OAc)_2$–Q–AcOH–MnO_2 system [81]:

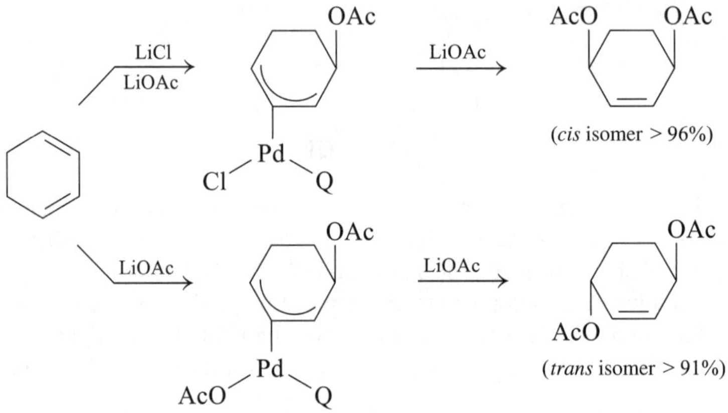

(4.109)

It was suggested [28, 80–83] that the main function of *para*-benzoquinone is to stabilize the leaving group in the form of a Pd(0)Q complex during nucleophilic substitution at the carbon atom. Therefore, quinone Q acts as a ligand that oxidizes Pd(0) at the subsequent steps of the process:

$$C_4H_6 \xrightarrow[LiOAc]{PdX_2,\ 1} \pi_X \overset{Q,\ 2}{\rightleftharpoons} \pi_X Q \xrightarrow[-P]{LiOAc,\ 3} Pd(0)Q \xrightarrow[-PdX_2]{2HX,\ 4} QH_2; \tag{4.110}$$

$$QH_2 + MnO_2(Ox) \xrightarrow[-Q]{5} MnO(Red) + H_2O.$$

For the fast and quantitative conversion of PdX_2 into η^3-allyl complex (π_X) with the limiting step **3**, the reaction rate must depend on [Q]. The kinetic function of PdX_2 in system (4.110) is to catalyze the oxidation of diene by the quinone

$$C_4H_6 + Q + 2AcOH \xrightarrow{PdX} \underset{(P)}{C_4H_6(OAc)_2} + QH_2,$$

while the kinetic function of quinone is to catalyze the oxidation of π_X by manganese dioxide (or by some other Ox):

$$\pi_x + MnO_2 + 2AcOH \xrightarrow{Q} P + Pd(OAc)_2 + MnO + H_2O.$$

The ability of quinone π complexes with Pd(0) to convert into Pd(II) and QH_2 was demonstrated in [86, 88]. The possibility of η^3-allyl complex transformation during the reductive elimination of the π complex of Pd(0) with quinone as a result of the weakly endothermal (2.5 kcal/mole) reaction

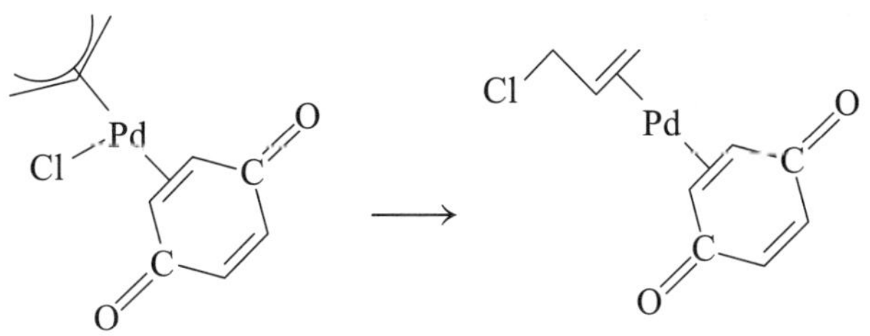

was theoretically justified by Szabo [89] on the B3PW91 density functional level (activation energy, ~15 kcal/mole). At the same time, the fact that the rate of η^3-allyl oxidation by quinone depends on $[H^+]$ [69, 90] cannot be consistent with this scheme and (4.110), even under the assumption that the material balance with respect to palladium includes the π_X, π_XQ, and Pd(0)Q complexes (see Example 2.30 illustrating features of the kinetics of η^3-$C_3H_5PdCl_2^-$ (π_y) complex oxidation by quinone). The empirical kinetic equation was as follows [90]:

$$r = \frac{[\pi_y][Q][H_3O^+][Cl^-]}{B[H_3O^+][Cl^-] + A}. \tag{4.111}$$

The kinetics of oxidation of 1-hydroxymethyl-η^3-allylpalladium chloride (π_x) complex by *para*-benzoquinone in the LiCl–HCl–$LiClO_4$ system at an ionic strength of 0.4 and 4.0 M was studied by Kalabin [69]. Based on the results, an equation of type (4.111) with different constants A and B was obtained and a reaction mechanism different from (4.110) was proposed, which implied a different chemical function of the quinone:

$$\begin{aligned}
\pi_X + Q &\underset{k_{-1}}{\overset{k_1}{\rightleftharpoons}} \pi_X Q;\\
\pi_X Q + H_3O^+ &\underset{k_{-2}}{\overset{k_2}{\rightleftharpoons}} \pi_X QH^+ + H_2O;\\
\pi_X QH^+ + Cl^- &\xrightarrow{k_3} ClCH_2CH{=}CHCH_2OH + HQPdCl_2^-;\\
HQPdCl_2^- + 2Cl^- + H_3O^+ &\xrightarrow[-H_2O]{} PdCl_4^{2-} + QH_2.
\end{aligned} \tag{4.112}$$

Assuming that step **2** in scheme (4.112) is quasi-equilibrium and $[\pi_X] \cong [\pi_X]_\Sigma = [\pi_X]_0 - [PdCl_4^{2-}]$, one can obtain Eq. (4.111), where $A = 1/(k_3K_1K_2)$ and $B = 1/k_1$. Within the framework of this scheme, an electron pair (rather than Pd(0) atom or $PdCl_4^{2-}$ group) is transferred to Q in step **3** and the degree of Pd(II) oxidation in hydroquinolate is retained:

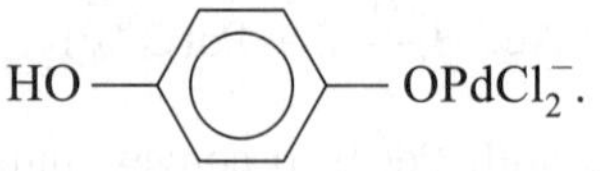

In other words, the electron pair in the course of the nucleophilic substitution is transferred to QH^+ with a leaving group of $HQPdCl_2^-$ rather than to Pd. Similar to the other cases considered above, the additional oxidant (catalyst) Q removes the thermodynamic constraints with respect to the formation of various forms of Pd(0), although quinone is probably also a sufficiently strong ligand that is capable (together with other ligands) of stabilizing Pd(0). The corresponding examples are presented in Section 4.6.

It has also been suggested to use complexes of transition metals (e.g., Co(II), Fe(II)) as additional oxidants of QH_2 to Q in reactions where oxygen plays the role of oxidant, which is equivalent to using three-component PFCSs [28, 83, 84].

Example 4.5

► It was shown [28] that the effective aerobic oxidation (diacetoxylation) of 1,3-dienes takes place at 25 °C in the ternary polyfunctional catalytic system $Pd(OAc)_2$–QH_2–ML–LiOAc–AcOH, where QH_2 is hydroquinone (in catalytic amounts) that is used instead of quinone Q, M = Co(II), Fe(II), and L represents various macrocyclic ligands such as

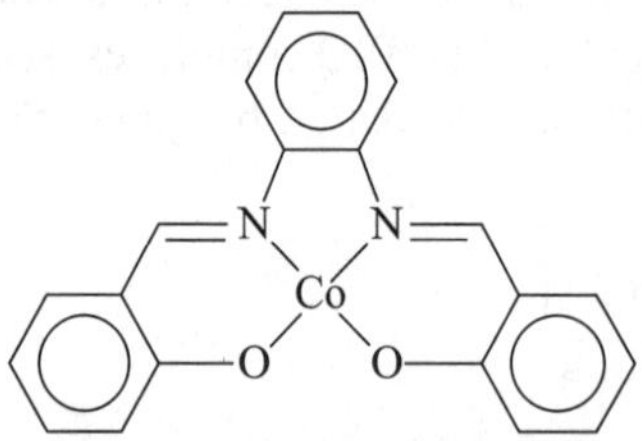

Co(Salophen)

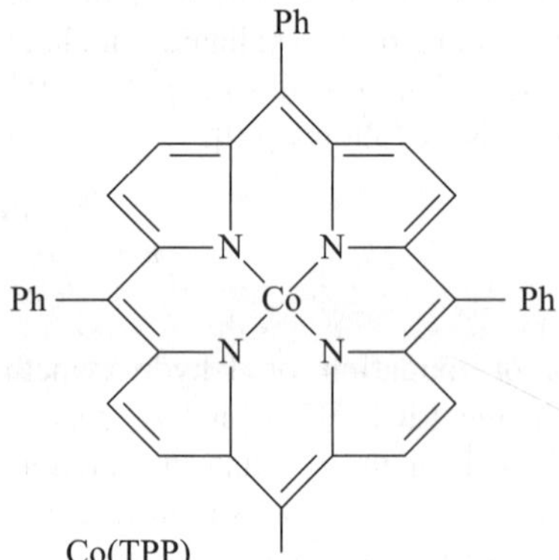

Co(TPP)

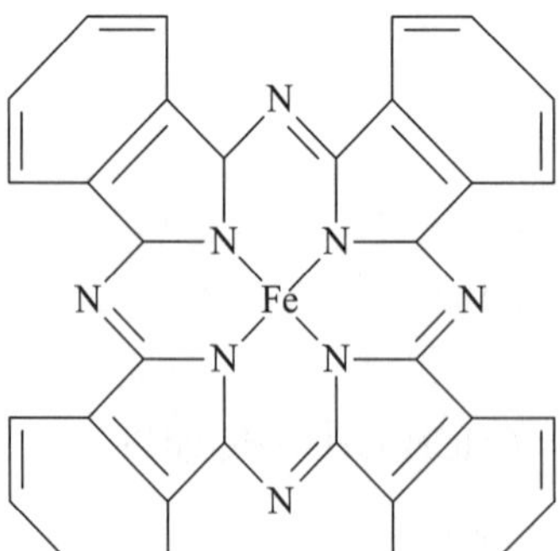

PcFe

The maximum activity in the reaction of 1,3-cyclohexadiene oxidation to *trans*-1,4-diacetoxycyclohexene was exhibited by a phthalocyaninate complex of iron (PcFe). In this system, complexes ML catalyze the oxidation of QH_2 to Q, e.g., by the following scheme:

$$ML + O_2 \longrightarrow MLO_2 \xrightarrow{QH_2} Q + ML + [H_2O_2].$$

However, it was pointed out [28] that the rate of QH_2 oxidation by oxygen in the absence of Pd(II) and dienes is higher than in the ternary system and, which is even more important, the orders of activities of the ML complexes for the oxidation of QH_2 and dienes in the ternary system (i.e., under the same conditions) are different. This circumstance can be related to the fact that some intermediate (e.g., XPdQH) rather than free QH_2 is oxidized by ML in the ternary system.

Two other observations [84, 85] showed evidence that the ternary system could also feature the formation of even more complicated intermediates, including Pd(II)QH–MLO_2. First, upon the introduction of two hydroxy groups into phenyl rings of TPP (in *para* position, Co(TQP)), the rate of O_2 uptake under otherwise identical conditions without QH_2 increases four times as compared to that in the Co(TPP) + QH_2 and Co(TDMPP) + QH_2 systems (TDMPP is a TQP analog with two OMe groups instead of two OH groups) [84]. In order to make possible that Q bound to a porphyrinate anion would participate in the decomposition of η^3-allyl complex (π_X), it is necessary for π_X to form a complex with TQP and, probably, with O_2. The same is evidenced by a difference in the stereoselectivity of the formation of the *trans* product of oxidation, which exceeds 90 % for Co(TPP) + QH_2 and is only 63 % for Co(TQP) [84].

The second experimental fact [85] was related to the attempts to discard Q as an intermediate ligand-oxidant. It was found that the activity of Co(TDMPP) complex in the oxidation of cyclohexadiene in the absence of QH_2 was greater than the activity of Co(TQP) complex and, in the presence of LiCl, a *cis* isomer of diacetate was predominantly formed. This suggested a direct electron transfer from η^3-allyl complex (for the *trans* attack of AcO^-) to oxygen coordinated by the CoL complex, with the assistance of the methoxy group of the phenyl ring of TDMPP macroligand:

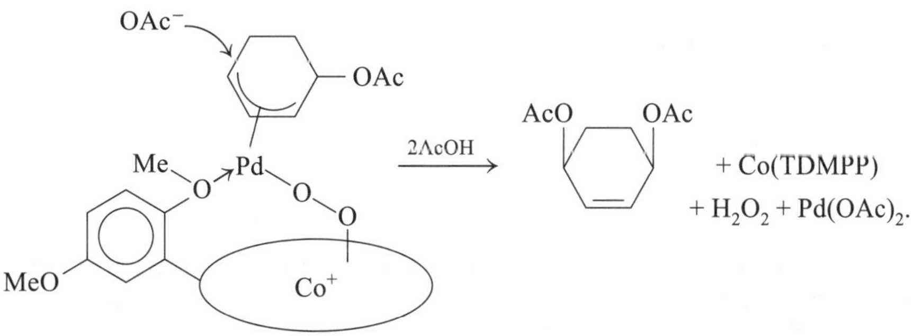

(4.113)

In this system, a substrate (diene) interacts with both catalysts (PdX_2 and CoL) and the oxidant, with the formation of [π_xO_2CoL] intermediate species that are converted with the yield of two catalysts:

$$\pi_XO_2CoL + 2AcOH \xrightarrow[\text{CoL}]{\text{PdX}_2} P + H_2O_2. \qquad (4.114)$$

Thus, each component of this catalytic system performs its own kinetic function:

- $Pd(OAc)_2$ catalyzes oxidation of substrate (S) by $CoLO_2$ component:

$$S + 2AcOH + CoLO_2 \xrightarrow{Pd(II)} S(OAc)_2 + CoL + H_2O_2;$$

- CoL catalyzes oxidation of η^3-allyl complex (π_X) by oxygen:

$$\pi_X + O_2 + 2AcOH \xrightarrow{CoL} S(OAc)_2 + Pd(OAc)_2 + H_2O_2.$$

Hydrogen peroxide can also participate in the oxidation of η^3-allyl intermediate (π_x). ■

4.1.3 Oxidation of alkynes and arenes

MCSs are employed in the reactions of alkyne oxidation at the C≡C– and ≡C–H bonds. [91]. Hydrogen peroxide oxidizes alkyne in Na_2MoO_4–$Hg(OAc)_2$ solutions to carboxylic acids (with rupture of the triple bond) and CO_2 [92]. Nitric acid oxidizes acetylene to glyoxal, and methylacetylene to methylglyoxal, under mild conditions in two-component MCl_n–$NaNO_2$ catalytic systems [93, 94]. The most active catalytic metal complexes are (in the order of decreasing activity) $PdCl_2 > AuCl_4^- > Hg(NO_3)_2$ [94].

Example 4.6 [93, 94]

► The reaction of acetylene oxidation

$$2C_2H_2 + 2HNO_3 \longrightarrow 2CHOCHO + N_2O + H_2O \quad (4.115)$$

in 3.5–8 M HNO_3 solutions was studied in the $PdCl_2$–$NaNO_2$ system in a temperature interval of 5–45 °C. The selectivity of this process with respect to acetylene is ~50%, and the byproducts are CO_2, CH_3CHO, CH_3COOH, $(COOH)_2$, and nitrogen-containing products of C_2H_2 oxidation by nitric acid. In the absence of $NaNO_2$ (1.5–5 M HNO_3, 1 M $PdCl_2$) glyoxal is not formed. It was established that the reaction rate is proportional to [HNO_3] in an 0.5–7 M interval and exhibits a narrow maximum with respect to [NO_2^-]. Based on these data, a mechanistic scheme that explained the chemical and kinetic functions of the catalyst components in this interesting process was developed. It should be noted that a scheme employing the joint action of $PdCl_2$ and NO_2^- such as

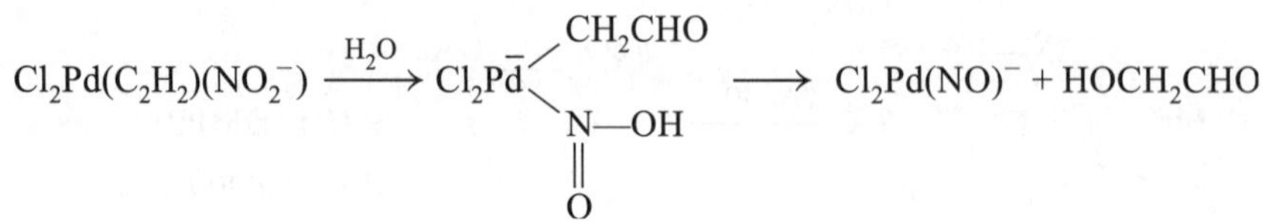

(4.116)

(by analogy with the mechanism (4.40) of ethylene oxidation to ethylene glycol acetate) is inapplicable in the case of acetylene oxidation because glycolaldehyde is not an intermediate in the formation of glyoxal (ethylene glycol and acetaldehyde also do not appear as intermediates in the system under consideration). The steps of consecutive participation of $PdCl_2$ (M^{n+}) and NO_2^- are stipulated by the following scheme:

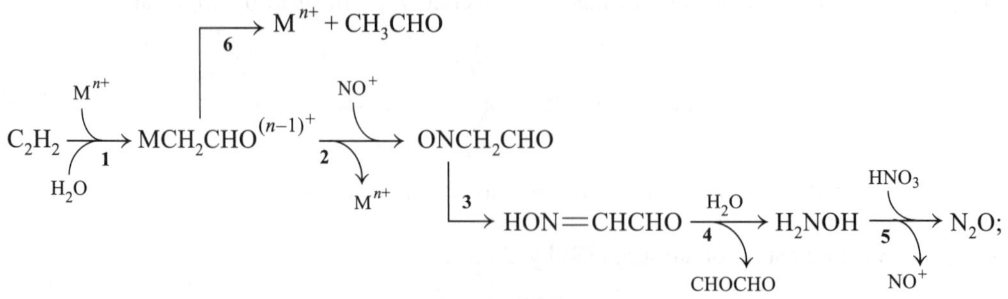

(4.117)

$$NO_2^- + 2HNO_3 \rightleftharpoons NO^+ + 2NO_3^- + H_2O,$$

where steps **1** and **6** are well known in the chemistry of acetylene.

Conversions of a metal-substituted acetaldehyde in the HNO_3–$NaNO_2$ system were studied for the reactions involving $RCOOHgCH_2CHO$ ($R = C_2H_5$) [94]. The reaction products contained glyoxal (with a selectivity of ~50 % at $[RCOOHgCH_2CHO] = 0.04$ M) and acetic acid. The formation of glyoxal from its monooxime (step **4**) can proceed either via oxime hydrolysis followed by the oxidation of hydroxylamine to N_2O,

$$2NH_2OH + 2HNO_3 \rightleftharpoons 2HONO + N_2O + 3H_2O \tag{4.118}$$

or via monooxime oxidation by nitric acid to N_2O and glyoxal [95, 96]. It is the oxidation of monooxime to nitrogen-containing products, which most probably accounts for a relatively low selectivity of the process.

The kinetic function of a metal [Pd(II), Hg(II), or Au(III)] complex is to catalyze the formation of nitrosoacetaldehyde from C_2H_2 and NO_2^- (NO^+), while the kinetic function of NO_2^- is to catalyze the oxidation of metalloacetaldehyde to glyoxal and N_2O by nitric acid. ■

The reaction of oxidative dehydrocondensation of alkynes (Glyazer–Zalkind reaction)

$$2RC{\equiv}CH + {}^1\!/\!_2\,O_2 \longrightarrow RC{\equiv}CC{\equiv}CR + H_2O \tag{4.119}$$

proceeds in aqueous chloride solutions of copper complexes and requires the simultaneous presence of Cu(I) and Cu(II). The mechanism of this reaction was discussed above (Section 3.3) in the context of an analysis of the method of competitive reactions. Let us return to this process for elucidating the kinetic functions of two components of the catalytic system [97, 98].

Example 4.7 [97, 98]

▶ Reaction (4.119) for methylacetylene (MA) in the $CuCl$–$CuCl_2$–$LiCl$ (5 M)–H_2O system is described by the following kinetic equation:

$$r_{DMDA} = \frac{kP_{MA}\,[CuCl]_\Sigma^2\,[CuCl_2]_\Sigma}{[H_3O^+]} \tag{4.120}$$

which was obtained in a closed ideally stirred reactor (in the absence of oxygen) by the method of initial rates at a constant concentration of free Cl^- and, hence, constant values of the complexity function for Cu(I) and Cu(II). This equation is also valid under steady-state conditions with the electrochemical regeneration of $CuCl_2$. Upon the discrimination of various hypotheses, the following sequence of steps (in a simplified variant) was proposed:

$$\underset{}{MeC{\equiv}CH} \underset{-H_3O^+}{\overset{CuCl\ \ 1}{\rightleftharpoons}} \underset{(X_1)}{MeC{\equiv}CCu} \overset{CuCl\ \ 2}{\rightleftharpoons} \underset{(X_2)}{MeC{\equiv}CCu_2Cl} \underset{2CuCl}{\overset{CuCl_2\ \ 3}{\longrightarrow}} \underset{(X_3)}{MeC{\equiv}CCuCl;}$$

$$2MeC{\equiv}CCuCl \overset{4}{\longrightarrow} \underset{(DMDA)}{MeC{\equiv}CC{\equiv}CMe} + 2CuCl. \tag{4.121}$$

In the presence of an oxidant ($Ox = O_2$), this is accompanied by the reaction

$$2CuCl + 2HCl + {}^1\!/\!_2\,O_2 \longrightarrow 2CuCl_2 + H_2O.$$

According to scheme (4.121), the kinetic function of CuCl is to catalyze the formation of Cu(II) ethynyl complex (X_3), which is not formed from $CuCl_2$ in the chloride system. The kinetic function of $CuCl_2$ (in the presence of Ox) is to catalyze the oxidation of Cu(I) ethynyl complex (X_2) by the oxidant with the formation of dimethyldiacetylene (DMDA). Solid acetylenides of Cu(I), Ag(I), and Au(I) ($RC{\equiv}CM$) are oxidized by copper(II) chloride with an almost quantitative yield of diacetylene derivatives [99, 100].

The formation of X_3 with the transfer of one electron from Cu(I) to X_2 also takes place under the action of other Ox on Cu(I) methylacetylenide. In particular, DMDA is formed in the reactions of $MeC{\equiv}CCu_{solid}$ in 5 M LiCl solution at 25 °C with various oxidants such as $FeCl_3$, $K_2Cr_2O_7$, $KMnO_4$, *para*-benzoquinone [101], and $K_3Fe(CN)_6$ (according to Glyazer):

$$MeC{\equiv}CCu_{solid} + Ox \xrightarrow[-Ox^{\bullet}]{} MeC{\equiv}CCu_{solid}{}^{+} \longrightarrow DMDA.$$

The formation of MeC≡CCu(II) is also possible without the participation of Cu(I), but only in the presence of strong bases – for example, in the $Cu(OAc)_2$ – pyridine system [102].

Solid methylacetylenide of Ag(I) is also oxidized to DMDA under the action of $FeCl_3$, $Na_2Cr_2O_7$ [101], and (as was noted above) $CuCl_2$. It is interesting to note that, in contrast to the ethynyl complexes of Cu(I) and Ag(I), a symmetric methylacetylenide of mercury(II) is not oxidized to DMDA by $FeCl_3$, $CuSO_4$, $Na_2Cr_2O_7$, and $KMnO_4$. At the same time, this methylacetylenide is oxidized by $CuCl_2$ with the formation of chloroalkyne,

$$Hg(C{\equiv}CMe)_2 + 4CuCl_2 \longrightarrow 2MeC{\equiv}CCl + 4CuCl + HgCl_2, \qquad (4.122)$$

probably, as a result of the electrophilic substitution of mercury ion at carbon atom of the ethynyl group:

$$Hg(C{\equiv}CMe)_2 + CuCl_2 \longrightarrow MeC{\equiv}CHgCl + MeC{\equiv}CCuCl.$$

The formation of chloroalkyne from MeC≡CCuCl under the action of $CuCl_2$ suggests that Cu(II) methylacetylenide (X_3), capable of converting into DMDA, contains at least one Cu(I) complex that stabilizes X_3 molecule and favors the coupling reaction (step **4** in scheme (4.121)). Representing the mechanism of step **4** (by analogy with addition of the ethynyl complexes of Ti(III) and Zr(III) [91, 103]) by the scheme as

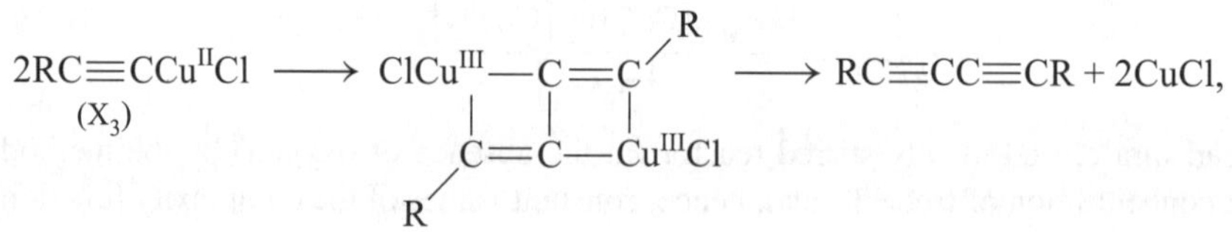

(4.123)

one can suggest that the presence of CuCl in X_3 would favor the oxidative addition with formation of metallacycle without the formation of Cu(III):

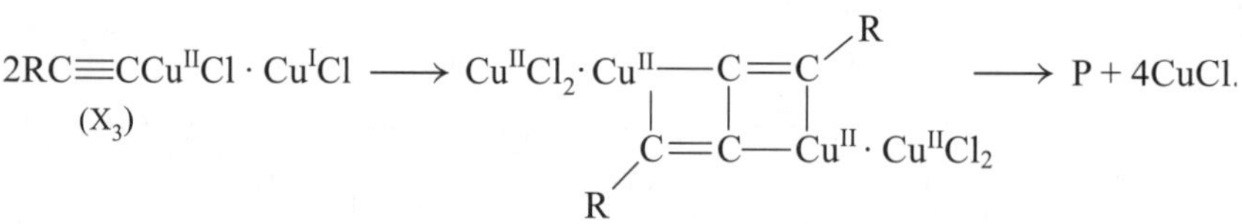

(4.124)

Bis(methylethynyl)mercury can also be quantitatively converted into DMDA, but this requires using other catalytic systems such as $PdCl_2$–$FeCl_3$ and $RhCl_3$–$FeCl_3$ [101].

Ethynyl complexes of Cu(II) formed without participation of Cu(I) in the presence of strong bases in non-aqueous media [91] and stabilized by ligands (such as L = NH_3, Py, Pipy), for example, $RC{\equiv}CCuOAc{\cdot}L_n$, are capable of converting into diacetylene derivatives also in the absence of Cu(I). In these systems, the kinetic equation has the following form [102]:

$$r = \frac{-d\left[Cu^{II}\right]_{\Sigma}}{dt} = \frac{k\,[RC{\equiv}CH]^2\left[Cu^{II}\right]_{\Sigma}^2}{\left[H^{+}\right]^2}, \qquad (4.125)$$

which is indicative of the participation of two ethynyl carbanions and two Cu(II) atoms in the limiting step of the condensation process. ■

MCSs are also used for the oxidative dehydrocondensation of ArH to Ar–Ar [9], as well as for the oxidative dehydrocondensation of ArH with olefins (Moritani–Fujiwara reaction) [104, 105]:

$$\mathrm{ArH} + \mathrm{CH_2{=}CHR} + \mathrm{Ox} \longrightarrow \mathrm{ArCH{=}CHR} + \mathrm{Red}. \quad (4.126)$$

Catalytic systems with $\mathrm{Ox} = K_2S_2O_8$ contain two components, for example, $Pd(OAc)_2$–AgOAc and $Pd(OAc)_2$–$Cu(OAc)_2$ [105]. A three-component system $Hg(OAc)_2$–$PdCl_2$–$CuCl_2$ with O_2 as the oxidant was proposed by Heck [25] for the reaction

$$C_6H_6 + CH_2{=}CHCH_3 + {}^1\!/_2\,O_2 \longrightarrow C_6H_5CH{=}CHCH_3 + H_2O. \quad (4.127)$$

The proposed mechanism of this reaction is as follows [9, 25]:

$$C_6H_6 \xrightarrow[-\mathrm{HOAc}]{Hg(OAc)_2} C_6H_5HgOAc \xrightarrow[\downarrow\,Hg(OAc)Cl]{PdCl_2} C_6H_5PdCl \xrightarrow{C_3H_6} C_6H_5CH_2CH(CH_3)PdCl \xrightarrow[\downarrow\,C_6H_5CH{=}CHCH_3]{}$$

$$\longrightarrow \mathrm{HPDCl} \xrightarrow[\downarrow\,PdCl_2]{2CuCl_2} 2CuCl + HCl; \quad (4.128)$$

$$2CuCl + 2HCl + {}^1\!/_2O_2 \longrightarrow 2CuCl_2 + H_2O.$$

The kinetic functions of three components of this PFCS in the framework of scheme (4.128) are as follows:

- $Hg(OAc)_2$ catalyzes the formation of C_6H_5PdCl;
- $PdCl_2$ catalyzes the oxidative dehydrocondensation of C_6H_5HgOAc with C_3H_6 with the participation of $CuCl_2$(Ox);
- $CuCl_2$ catalyzes the oxidation of HPdCl (or Pd(0)) by oxygen.

4.1.4 Oxidation of inorganic compounds

Various multicomponent PFCSs were also proposed for the oxidation of inorganic substrates such as PH_3, H_2S, SO_2, and CO [12, 106, 107]. Apparently, the $PdCl_2$–$CuCl_2$ multicomponent catalytic system was used for the first time to catalyze the oxidation of CO by oxygen in aqueous solutions [108]. The mechanism of CO oxidation in the $PdCl_2$–Ox system (where $CuCl_2$ can be replaced by another additional oxidant representing the second PFCS component) turns out to be more complicated than might be supposed for a simple sequence of reactions written by analogy with the Wacker process as:

$$PdX_2 + CO + H_2O \longrightarrow Pd(0) + CO_2 + 2HX; \quad (4.129)$$

$$Pd(0) + Ox + 2HX \longrightarrow PdX_2 + Red; \quad (4.130)$$

$$Red + {}^1\!/_2\,O_2 \longrightarrow Ox + H_2O. \quad (4.131)$$

Although the rate of CO_2 formation in this process can be independent of [Ox] up to large degrees of Ox conversion (for Ox = $CuCl_2$, $FeCl_3$, $Cr_2O_7^{2-}$, quinone) [107, 109], this rate still depends on the nature of the oxidant. In addition, it was shown that the catalytic CO oxidation rate with respect to PdX_2 (reactions (4.129) and (4.130) in the presence of Cu(II) and quinone) is approximately two orders of magnitude higher than the rate of reaction (4.129) in the absence of oxidant [107]. It was also established that the oxidant accelerates the decomposition of $Pd(CO)X_2$ complexes, the effect increasing with the oxidation potential ($E°$) [107]. The autocatalytic process observed in the course of Pd^{2+} reduction by carbon monoxide in acidic media is related to the formation and decomposition of Pd(I) carbonyl complexes [107, 109]. The oxidants with a standard oxidation potential above 1 V produce a decrease in the reaction rate and it is believed that this is caused by the oxidation of Pd(I) carbonyl complexes that are active in this process. The complexes of platinum-group metals in the oxidation state (0) can also participate in oxidation processes in the presence of an oxidant. In particular, the $[Pt(CO)_2]_n$ complex decomposes in the presence of Fe(III) with the formation of CO_2 and Pt(IV) complex $[Pt(H)_2(CO)Cl_3^-]$ [110], while the Pd(0) complex with quinones catalyzes the oxidative carbonylation of alkynes [111, 112].

Thus, the oxidation of CO in the PdX_2–Ox system can involve various forms of palladium complexes that appear in this system:

$$\mathrm{Pd(II)(CO)} \longrightarrow \mathrm{Pd_2(I)(CO)_2} \longrightarrow \mathrm{Pd(0)(CO)}_n. \qquad (4.132)$$

It was suggested that the chemical function of a quinone (Q) in the Pd^{2+}–Q–H_2SO_4 system consists of the stabilization of Pd(0) and $Pd_2^{2+}(CO)$ complexes with the participation of protonated intermediates $Pd(CO)QH^+$ and $Pd_2(CO)QH^{3+}$ in the catalytic cycles of CO_2 and QH_2 formation [109]. Investigations of the $PdCl_2$–$CuCl_2$ system [113–116] led to a conclusion that there are two additional routes of CO_2 formation besides the pathway analogous to the Wacker process.

- Route I:

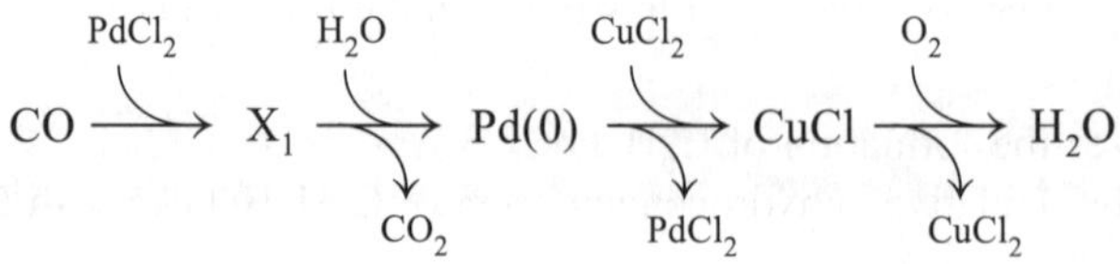

  where the kinetic functions of $PdCl_2$ and $CuCl_2$ are the same as in the reaction of ethylene oxidation.
- Route II:

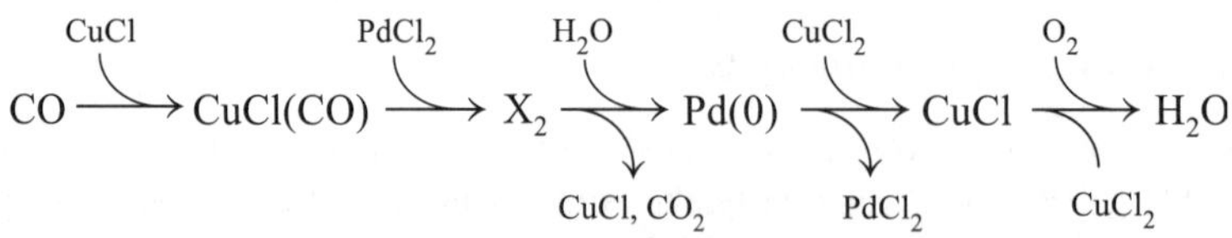

  where the kinetic function of CuCl is to catalyze the reduction of $PdCl_2$ by CO to Pd(0).
- Route III:

$$X_1 + PdCl_2 \xrightarrow{CO} Pd_2(CO)_2Cl_4^{2-} \longrightarrow \cdots \xrightarrow{CuCl_2} \cdots \longrightarrow CO_2 + CuCl.$$

  where it is still unclear whether the decomposition of Pd(I) carbonyl complexes takes place with or without the participation of Ox = $CuCl_2$; that is, whether Pd(I) is reduced to Pd(0) or the oxidation state of Pd(I) is retained. Accordingly, the kinetic functions of $PdCl_2$ and $CuCl_2$ on this route are also not completely clear.

The process rate under non-steady-state conditions is described by the following two-term equation [116]:

$$d[CO_2]/dt = k_1[Pd^{II}]_\Sigma\, P_{CO} + k_2[Pd^{II}]_\Sigma\, P_{CO}[Cu^I]_\Sigma, \tag{4.133}$$

while the steady-state oxidation of CO by oxygen at $P_{CO} > 20\,\%$ and $P_{O_2} < 20\,\%$ is described by the equation

$$r_{SS} = k_3\left[Cu^I\right]_\Sigma\, P_{CO}P_{O_2}, \tag{4.134}$$

which can be obtained under the conditions that Cu(II) is completely converted into Cu(I) and that the oxidation is limited by the block of steps related to the oxidation of Cu(I).

The oxidation of phosphine (PH_3) by copper(II) halides is significantly accelerated when Cu(I) halide, Hg(II) halide, or especially I^- (or I_2) species are added to the solution [10, 12]. It was suggested that Cu(I) and Hg(II) complexes activate PH_3 and that I_2 in the coordination sphere oxidizes phosphine to PH (or PH_2) species, which are subsequently oxidized by I_2 and CuX_2 to H_3PO_4. It was also assumed that mixed complexes $CuBr(PH_3)CuBr_2$ could form, with their subsequent decomposition to PH_2, HBr, and 2CuBr. In the chloride system with I^- (or I_2), a simplified mechanism can be described by the scheme

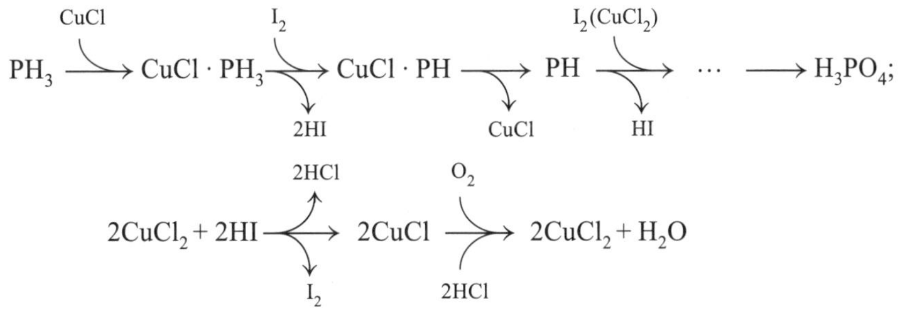

(4.135)

with the following kinetic functions of components:

- CuCl catalyzes the oxidation of PH_3 by iodine to PH;
- $CuCl_2$ catalyzes the oxidation of I^- by oxygen to I_2;
- $CuCl_2$ catalyzes the oxidation of PH by oxygen to H_3PO_4;
- I_2 catalyzes the oxidation of $CuCl{\cdot}PH_3$ complex by Cu(II) chloride to H_3PO_4 and CuCl;
- I_2 catalyzes the oxidation of PH by oxygen to H_3PO_4.

Thus, three components of the $CuCl_2$–CuCl–I_2 system catalyze the process

$$PH_3 \xrightarrow[O_2]{CuCl_2\,-\,CuCl\,-\,I_2} PH \xrightarrow[O_2]{I_2\,-\,CuCl_2} H_3PO_4. \tag{4.136}$$

An important function of iodine (I^-) was also revealed in the catalytic system of SO_2 oxidation to SO_3 [93]:

$$SO_2 + {}^1\!/\!{}_2\,O_2 + H_2O_2 \longrightarrow H_2SO_4. \tag{4.137}$$

It was suggested that I^0 present in the coordination sphere of Cu(I) participated in the oxidation of SO_2 as follows:

$$CuCl_2(I^-)(SO_2) \longrightarrow CuCl_2^-(I^0)(SO_2) \xrightarrow{H_2O} CuCl_2^- + H\dot{S}O_3 + HI;$$

$$CuCl_2 + H\dot{S}O_3 + H_2O \longrightarrow H_2SO_4 + CuCl_2^- + H^+.$$

Since I^- is very rapidly oxidized by Cu(II) to I_2, it is hardly possible to distinguish between the inner- and outer-sphere oxidation of SO_2 in this system.

Reaction (4.137) is also catalyzed by Cu(II) and Fe(III) chlorides [116–118]. In the $CuCl_2$–$FeCl_3$ two-component system, the oxidation rate significantly increases due to the participation of a 1:1 mixed complex in the process, according to which [118]:

$$FeCl_3 + CuCl_2 \rightleftharpoons FeCuCl_5 \xrightarrow[2H_2O]{SO_2} H_2SO_4 + CuCl + FeCl_2 + 2HCl;$$

$$CuCl + FeCl_2 + 2HCl + {}^1/_2O_2 \longrightarrow FeCuCl_5 + H_2O.$$

Here, the kinetic function of the system components ($CuCl_2$, $FeCl_3$) or the $FeCuCl_5$ complex is to catalyze reaction (4.137).

4.2 Reactions of chlorination and oxidative chlorination of organic compounds

In contrast to the case of halogenation, the functions of the oxidant and source of halogen in the reactions of oxidative halogenation are separated. In particular, the role of oxidant (Ox) in the reactions of oxidative chlorination can be played by any compound with the oxidation potential sufficient to provide for the reaction

$$S + Ox + Cl^- \longrightarrow P + Red. \qquad (4.138)$$

The oxidation of reduced species makes this process catalytic with respect to the oxidant. The functions of a catalyst and chlorinating agent in reaction (4.138) can be performed by salts and complexes of transition metals, such as $CuCl_2$, $PdCl_2$, $AuCl_3$, $CeCl_4$, $PtCl_4$, and CrO_2Cl_2 [119]. Reaction (4.138) can also be accelerated by metal (Cu(I), Hg(II), Pt(II) etc.) complexes not possessing chlorination properties. In this case, the catalytic systems become PFCSs.

Examples of reactions with the formation of oxidative chlorination products were considered above (see Chapters 2 and 3 and Section 4.1). These were the syntheses of 2-chlorovinylacetylene (Section 2.3) and 1.2-dichloroethylene, 1,1- dichloroethylene, and trichloroethylene (Section 3.3) from acetylene in the $CuCl_2$–CuCl system; chlorinated esters from olefins, and dienes and dichlorobutenes from 1.3-butadiene in various systems (Section 4.1); and methyl chloride and ethyl chloride from alkanes in Na_2PtCl_6–Na_2PtCl_4 solutions (Section 3.3). Below, some of these reactions are considered in more detail in order to reveal the kinetic functions of the corresponding PFCS components, together with some new examples not mentioned above.

4.2.1 Oxidative chlorination of alkynes

The most exhaustively studied catalytic system for the reactions of substitutive and additive oxidative chlorination of alkynes in both aqueous and nonaqueous media is the CuCl–$CuCl_2$–HCl–NH_4Cl system [91, 119]. Depending on the intervals of concentrations of Cu(I), Cu(II), and HCl, this system catalyzes the formation of various products from alkynes at 60–90 °C (Table 4.1).

The proposed scheme of alkyne conversion was proved by kinetic and chemical methods [91, 120–124], in particular by measurements of the relative selectivity (defined as the ratio of the rates of product formation in a closed system as a function of the reaction time τ) and by analysis of the temporal

Table 4.1 Conditions and products of oxidative chlorination of acetylene

[CuCl], M	$[CuCl_2]$, M	[HCl], M	Products of oxidative chlorination	Other products
>2	$<10^{-2}$	<0.1	$CH_2{=}C(Cl)C{\equiv}CH$	C_4H_4
>2	$<10^{-2}$	$0.1 \div 0.4$	*trans*-ClCH=CHCl, $CH_2{=}CCl_2$	$CH_2{=}CHCl$
<0.6	>1	pH= $1 \div 4$	HC≡CCl, ClC≡CCl, $Cl_2C{=}CHCl$, $Cl_2C{=}CCl_2$	C_4H_2

dependence of the product concentration in the gas phase, which was proposed for the investigation of monomolecular reactions [125]. The latter method can be used to determine the number of organic intermediates preceding the formation of a given (ith) product, the concentrations of which are comparable with that of this product. According to this method, the power index n in the relation

$$C_i = k_i \tau^n \tag{4.139}$$

is determined at constant values of [CuCl], $[CuCl_2]$, and $[H_3O^+]$ under the conditions of pseudo-first-order reaction with respect to organic substances. In particular, it was established that *trans*-1,2-dichloroethylene (DCE) and monochloroacetylene (MCA) are formed directly from acetylene ($n = 1$). A quadratic ($n = 2$) dependence of the concentrations of dichloroacetylene (DCA) and trichloroethylene (TCE) on the time τ is indicative of the existence of an intermediate whose concentration is comparable with that of the products. In the given example, this intermediate is MCA. The cubic ($n = 3$) dependence of the concentration of tetrachloroethylene (PE) on τ is consistent with a hypothesis about the consecutive PE formation from acetylene via MCA and DCA. The general scheme of alkyne conversions is presented in Figure 4.6. Below we will consider in more detail separate groups of reactions according to this scheme.

Example 4.8 [122, 126, 127]

▶ The kinetics of reversible reactions **1** and **2** (Figure 4.6) with stoichiometric equations

$$HC{\equiv}CH + 2CuCl_2 \rightleftharpoons HC{\equiv}CCl + 2CuCl + HCl; \tag{4.140}$$

$$HC{\equiv}CCl + 2CuCl_2 \rightleftharpoons ClC{\equiv}CCl + 2CuCl + HCl \tag{4.141}$$

was studied in closed ideally stirred reactors by the method of initial rates. Reactions in the forward direction were studied in at 60–90 °C in the NH_4Cl (8 M)–$CuCl_2$ (0.25–1.3 M)–CuCl (10^{-3}–10^{-2} M)–HCl

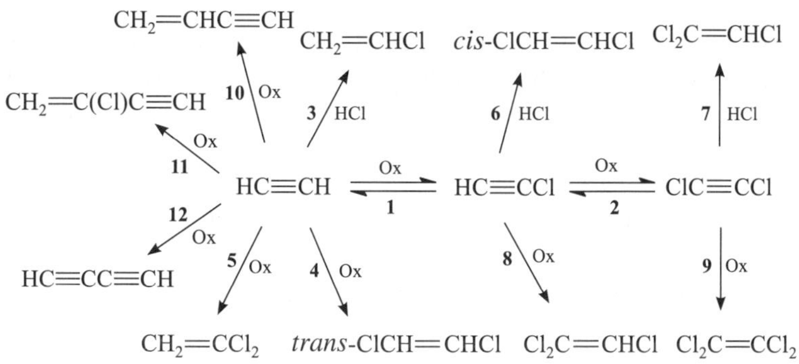

Figure 4.6 Scheme of acetylene and chloroalkyne conversions in CuCl–$CuCl_2$ system

(10^{-2}–$4 \cdot 10^{-1}$ M)–H_2O system under the conditions of a constant concentration of free Cl^- species, i.e., for the proportionality of

$$[CuCl_2] \propto [CuCl_2]_\Sigma \ (a_{Cu^{2+}} \propto [CuCl_2]_\Sigma)$$

and

$$[CuCl] \propto [CuCl]_\Sigma \ (a_{Cu^{+}} \propto [CuCl]_\Sigma).$$

In these investigations, the following kinetic equations were obtained for the rates of MCA and DCA formation:

$$r_{MCA} = k_1' \frac{[CuCl]_\Sigma \, [CuCl_2]_\Sigma^2 \, P_{C_2H_2}}{[H_3O^+]} \tag{4.142}$$

$$r_{DCA} = k_2' \frac{[CuCl]_\Sigma \, [CuCl_2]_\Sigma^2 \, P_{MCA}}{[H_3O^+]}. \tag{4.143}$$

Analogous equations were obtained for the rates of MetCA synthesis from methylacetylene [97, 122] and 4-chlorovinylacetylene synthesis from vinylacetylene [122].

An increase in [CuCl] makes significant the reverse processes of reductive dechlorination of chloroalkynes. The kinetics of reactions (4.140) and (4.141) that proceed in the reverse direction is described by the following equations:

$$r_{C_2H_2} = k_1' \frac{[CuCl]_\Sigma^3 \, P_{MCA}}{k_2 + [CuCl]_\Sigma}; \tag{4.144}$$

$$r_{MCA} = k_2' \frac{[CuCl]_\Sigma^3 \, P_{DCA}}{k_3 + [CuCl]_\Sigma}. \tag{4.145}$$

Equations (4.142) and (4.143) show that the process of substitutive chlorination has a limiting step of redox conversion of the intermediate complex **A** = ($RC{\equiv}C^-$) (Cu^+) $(CuCl_2)_2$.

The kinetics of reverse reactions (4.144) and (4.145) with allowance for the principle of microscopic reversibility leads to the conclusion that the decomposition of complex **A** is at least a two-step process. At the first step, a CuCl molecule is eliminated as

$$\underset{\mathbf{A}}{RC{\equiv}CCu(CuCl_2)_2} \rightleftharpoons \underset{\mathbf{B}}{RC{\equiv}CCu^{II}Cl(CuCl_2)} + CuCl,$$

and at the second step, complex **B** converts into $RC{\equiv}CCl$ and 2CuCl. A possible valent isomer of complex **B** is a copper(III) ethynyl complex **C** = $RC{\equiv}CCu^{III}Cl_2(CuCl)$, which as a result of the reductive elimination converts into $RC{\equiv}CCl$ and 2CuCl. A simplified mechanistic scheme of the reversible process consists of at least three steps:

$$RC{\equiv}CH + CuCl \overset{K_1}{\rightleftharpoons} RC{\equiv}CCu + HCl; \tag{4.146}$$

$$RC{\equiv}CCu + 2CuCl_2 \underset{k_{-2}}{\overset{k_2}{\rightleftharpoons}} RC{\equiv}CCu_2Cl_3 + CuCl; \tag{4.147}$$

$$RC{\equiv}CCu_2Cl_3 \underset{k_{-3}}{\overset{k_3}{\rightleftharpoons}} RC{\equiv}CCl + 2CuCl, \tag{4.148}$$

which correspond to the kinetic equation

$$r = \frac{k_2k_3K_1[\mathrm{CuCl}][\mathrm{CuCl_2}]^2P_{\mathrm{RC_2H}} - k_{-2}k_{-3}[\mathrm{CuCl}]^3[\mathrm{H_3O^+}]\,P_{\mathrm{RC_2Cl}}}{[\mathrm{H_3O^+}](k_3 + k_{-2}[\mathrm{CuCl}])}. \quad (4.149)$$

The experimental equations (4.142) and (4.143) are identical to (4.149) at $k_3 \gg k_{-2}$[CuCl] (far from equilibrium) with the limiting step (4.147). At large concentrations of CuCl, the reverse reaction proceeds with the irreversible reverse step (4.147) under the conditions of comparable rates of steps (4.147) and (4.148) in the reverse direction ($k_3 \approx k_{-2}$[CuCl]). Within the framework of mechanism (4.146)–(4.148), reaction (4.148) in the reverse direction represents the oxidative addition at the ≡C–Cl bond with the transfer of two electrons from two Cu(I) atoms. It is possible that the presence of a second CuCl molecule favors the stabilization of intermediate **C** due to a redistribution of electrons from Cu(I) to Cu(III) and the transition to complex **B**. Thus, the chemical function of the second CuCl molecule in the reverse process is to remove the thermodynamic constraints for the formation of $RC{\equiv}CCu^{III}Cl_2$.

The kinetic functions of CuCl and $CuCl_2$ (in the case of CuCl oxidation to $CuCl_2$ by an external oxidant–oxygen) in this process can be interpreted as follows:

- CuCl catalyzes the formation of complex B (C) from alkyne and $CuCl_2$ via reaction

$$\mathrm{RC{\equiv}CH + 2CuCl_2 \xrightarrow{CuCl} RC{\equiv}CCu_2Cl_3 + HCl};$$

- CuCl catalyzes the oxidation of alkyne by Cu(II) chloride via autocatalytic reaction

$$\mathrm{RC{\equiv}CH + 2CuCl_2 \xrightarrow{CuCl} RC{\equiv}CCl + 2CuCl + HCl}$$

- $CuCl_2$ catalyzes the oxidation of RC≡CCu by oxygen via reaction

$$\mathrm{RC{\equiv}CCu + {}^1\!/_2\,O_2 + 2HCl \xrightarrow{CuCl_2} RC{\equiv}CCl + CuCl + H_2O}$$

  (in the circuit with intermediate CuCl, which is unbalanced since CuCl enters into the overall equation);
- CuCl–$CuCl_2$ catalyzes the overall reaction

$$\mathrm{RC{\equiv}CH + HCl + {}^1\!/_2\,O_2 \xrightarrow{CuCl\,-\,CuCl_2} RC{\equiv}CCl + H_2O}.$$

■

The additive oxidative chlorination of acetylene (see Figure 4.6) in the CuCl–$CuCl_2$ system under the conditions of vinyl chloride (VC) formation (at 0.1–0.5 M HCl) corresponds to reactions

$$\mathrm{C_2H_2 + 2CuCl_2 \longrightarrow \mathit{trans}\text{-}\underset{(DCE)}{ClCH{=}CHCl} + 2CuCl}; \quad (4.150)$$

$$\mathrm{C_2H_2 + 2CuCl_2 \longrightarrow \underset{(VDC)}{CH_2{=}CCl_2} + 2CuCl}; \quad (4.151)$$

$$\mathrm{C_2H_2 + HCl \longrightarrow \underset{(VC)}{CH_2{=}CHCl}} \quad (4.152)$$

which proceed via a common intermediate CuCH=CHCl (see Section 3.3). Let us analyze the kinetic functions of components of this catalytic system using a mechanism proposed by Brailovsklii et al.

[128–130]:

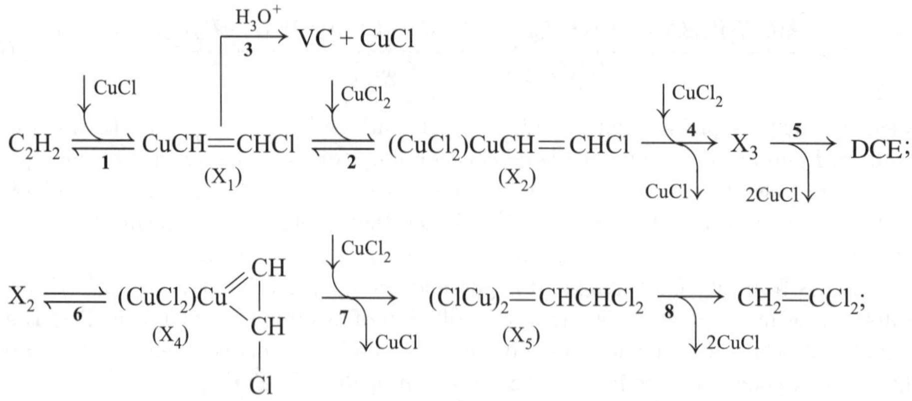

(4.153)

$$2CuCl + {}^{1}/_{2}O_2 + 2HCl \xrightarrow[9]{} 2CuCl_2 + H_2O.$$

A kinetic model for this system was obtained at constant values of a_{Cu^+} and a_{Cl^-} under the conditions of existence of polynuclear Cu^+ complexes (3–9 $\tilde{m}$ CuCl). Therefore, symbols CuCl and Cu in scheme (4.153) only indicate the participation of this molecule or atom in the process. The composition of intermediates with respect to $CuCl_2$ corresponds to the kinetic model. The kinetic functions of CuCl depend on the step in which the given species are involved. The elimination of CuCl in steps **4** and **7** of scheme (4.153) does not follow from the kinetics and is written by analogy with the mechanism of oxidative chlorination (4.146)–(4.148), which was proposed on the basis of investigation of the kinetics of a reversible reaction. The kinetic functions of components of the system under consideration (under the conditions of CuCl oxidation to $CuCl_2$ by atmospheric oxygen) within the framework of (4.153) are listed in Table 4.2.

An interesting feature of the route of vinylidene chloride (VDC) formation is the participation of Cu(I) in the steps of formation of intermediates X_4 and X_5 (probably Cu(I) carbene complexes). A chlorovinyl derivative of Cu(II), which is probably formed from ClHgCH=CHCl via the exchange reaction (see Section 3.3)

$$ClHgCH{=}CHCl + CuCl_2 \longrightarrow ClCuCH{=}CHCl + HgCl_2,$$

Table 4.2 Kinetic functions of PFCSs according to scheme (4.153)

Catalyst component	Steps	Overall equation for kinetic function
CuCl	**1, 2, 3**	$C_2H_2 + HCl \rightarrow VC$;
	1, 2, 4	$C_2H_2 + 2CuCl_2 \rightarrow X_3$;
	1, 2, 6, 7	$C_2H_2 + 2CuCl_2 \rightarrow X_5$;
	1, 2, 4, 5	$C_2H_2 + 2CuCl_2 \rightarrow DCE + 2CuCl$;
	1, 2, 6, 7, 8	$C_2H_2 + 2CuCl_2 \rightarrow VDC + 2CuCl$
$CuCl_2$	**2, 4, 5, 9**	$X_1 + {}^{1}/_{2}O_2 + 2HCl \rightarrow DCE + CuCl + H_2O$
	2, 6, 7, 8, 9	$X_1 + {}^{1}/_{2}O_2 + 2HCl \rightarrow VDC + CuCl + H_2O$
CuCl, $CuCl_2$	**1,2, 4, 5, 9**	$C_2H_2 + {}^{1}/_{2}O_2 + 2HCl \rightarrow DCE + H_2O$;
	1, 2, 6, 7, 8, 9	$C_2H_2 + {}^{1}/_{2}O_2 + 2HCl \rightarrow VDC + H_2O$

does not form intermediate X_4 in the presence of CuCl and is much faster demetallated to 1,2-dichloroethylene (DCE) via interaction with $CuCl_2$ [131].

It is also not excluded that complex X_4 can exhibit isomerization (1,2-shift of hydrogen atom) with the formation of α-chlorovinyl intermediate via free carbene bound to copper atom [91]:

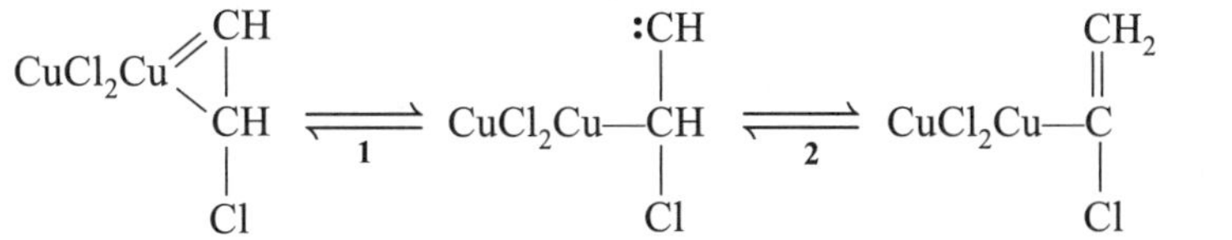

X_4, η^2-β-chlorovinyl complex η^1-α-chlorovinyl complex

The formation of VDC was demonstrated in a model reaction of α-chlorovinylmercury chloride with $CuCl_2$:[2]

$$ClHgC(Cl){=}CH_2 + 2CuCl_2 \longrightarrow CH_2{=}CCl_2 + HgCl_2 + 2CuCl,$$

which gives grounds to suggest the possible formation of α-chlorovinyl derivative of Cu(II) as

$$ClHgC(Cl){=}CH_2 \xrightarrow[-HgCl_2]{CuCl_2} ClCuC(Cl){=}CH_2 \xrightarrow{CuCl_2} VDC + 2CuCl.$$

The additive oxidative chlorination of acetylene to *trans*-1,2-dichloroethylene (DCE) at sufficiently large concentrations of $CuCl_2$ (0.1–1 M) and small concentrations of CuCl (0.05–0.5 M) and HCl (below 0.05 M) under the conditions of monochloroacetylene synthesis [126] is described by a two-route kinetic equation of the first-order with respect to $[CuCl_2]_\Sigma$:

$$r_{DCE} = (k_I + k_{II}[CuCl]_\Sigma^2)[CuCl_2]_\Sigma\, P_{C_2H_2}. \quad (4.154)$$

It can be suggested that the formation of an intermediate of the ClCuCH=CHCl type is also possible without the participation of Cu(I) (first route) and that the step of formation of this intermediate via both routes is irreversible:

$$C_2H_2 + CuCl_2 \xrightarrow[1]{} \underset{(X_1)}{ClCuCH{=}CHCl} \xrightarrow[2]{CuCl_2} ClCH{=}CHCl + 2CuCl;$$

$$C_2H_2 + 2CuCl + Cl^- \underset{3}{\rightleftharpoons} \underset{(X_2)}{Cl_2Cu_2(CH{=}CHCl)^-} \xrightarrow[4,\ -CuCl]{+CuCl_2} \underset{(X_3)}{Cl_3Cu_2CH{=}CHCl^-} \xrightarrow[5,\ -CuCl]{+CuCl_2} \quad (4.155)$$

$$\longrightarrow \underset{(X_4)}{Cl_4Cu_2CH{=}CHCl^-} \xrightarrow[6]{} DCE + 2CuCl + Cl^-.$$

It can also be suggested that, at large $CuCl_2$ concentrations, intermediate X_3 does not isomerize into η^2-vinyl complex X_4 (4.153) and only converts into DCE.

In contrast to C_2H_2, chloroalkynes under the same conditions participate in additive oxidative chlorination and hydrochlorination with the formation of common intermediates.

[2] α-Chlorovinyl derivative of mercury(II) was synthesized by M. A. Kazankova.

Example 4.9 [122, 127]

► The conversions of MCA were studied under the conditions described above for the synthesis of *trans*-DCE (4.154), for the following three directions:

$$\begin{aligned} & HC{\equiv}CCl \xrightarrow{CuCl,\ HCl} HC{\equiv}CH && \text{(a)} \\ & HC{\equiv}CCl \xrightarrow{HCl} cis\text{-}ClCH{=}CHCl \ (cis\text{-DCE}) && \text{(b)} \\ & HC{\equiv}CCl \xrightarrow{CuCl_2} ClCH{=}CCl_2 \ \text{(TCE)} && \text{(c)} \end{aligned} \tag{4.156}$$

The kinetics of the reductive dechlorination reaction (a) was considered in Example 4.8 above. It was also established that reactions (b) and (c) are conjugated via the common intermediate and the rate of TCE formation, like the rate of DCE formation (4.154), is described by the equation of a two-route process as ($r_{TCE} = r^{I}_{TCE} + r^{II}_{TCE}$), where

$$r^{I}_{TCE} = k'_{I}[CuCl_2]_{\Sigma}\, P_{MCA}. \tag{4.157}$$

An analysis of the conjugation node of processes (b) and (c) with a common intermediate [127] (see Section 3.3) showed that

$$\frac{r^{II}_{TCE}}{r_{cis\text{-}DCE}} = k'\frac{[CuCl_2]_{\Sigma}}{[H_3O^+]}; \tag{4.158}$$

$$r^{II}_{TCE} = \frac{\alpha[CuCl]_{\Sigma}\,[CuCl_2]_{\Sigma}\,P_{MCE}}{1+\beta[H_3O^+]+\gamma[CuCl_2]_{\Sigma}}; \tag{4.159}$$

$$r_{cis\text{-}DCE} = \frac{\chi[CuCl]_{\Sigma}\,[H_3O^+]\,P_{MCA}}{1+\beta[H_3O^+]+\gamma[CuCl_2]_{\Sigma}}, \tag{4.160}$$

where $k' = \alpha / \chi$.

A simple mechanistic scheme

$$\begin{aligned} & HC{\equiv}CCl + CuCl_2 \xrightarrow{k_1} ClCuC(Cl){=}CHCl; \ (X_0) \\ & X_0 + CuCl_2 \xrightarrow{k_2} Cl_2C{=}CHCl + 2CuCl; \\ & HC{\equiv}CCl + CuCl \underset{k_{-3}}{\overset{k_3,\ Cl^-}{\rightleftharpoons}} CuC(Cl){=}CHCl; \ (X_1) \\ & X_1 + H_3O^+ \xrightarrow{k_4,\ Cl^-} CuCl + ClCH{=}CHCl; \\ & X_1 + CuCl_2 \xrightarrow{k_5} X_2; \\ & X_2 + CuCl_2 \xrightarrow[-CuCl]{k_6} X_3 \xrightarrow{k_7} Cl_2C{=}CHCl + 2CuCl \end{aligned} \tag{4.161}$$

leads to Eqs (4.157), (4.159), and (4.160). For example,

$$r_{cis\text{-DCE}} = \frac{k_3' k_4 [\mathrm{CuCl}]_\Sigma [\mathrm{H_3O^+}] P_{\mathrm{MCA}}}{k_{-3} + k_4[\mathrm{H_3O^+}] + k_5[\mathrm{CuCl_2}]_\Sigma}, \qquad (4.162)$$

where the constant k_3' includes both $[Cl^-]$ and the Henry constant for HC≡CCl.

It is interesting to note that the HCl addition in the case of HC≡CCl and Cu(I) complexes proceeds against the Markovnikov rule and probably involves *trans*-addition of Cl^- at CH group of chloroalkyne in π complex CuCl(CH≡CCl). In contrast, in $HgCl_2$ solutions, the addition of HCl to MCA obeys the Markovnikov rule and involves (similar to the reaction in solutions of typical acid catalysts such as $HFeCl_4$ [122]), the formation of $CH_2{=}CCl_2$ [131]. The kinetic functions of CuCl and $CuCl_2$ in the mechanism (4.161) under conditions of CuCl oxidation by oxygen are quite evident.

Analogous kinetic equations were also obtained for the additive oxidative chlorination of DCA and MetCA according to reactions **7** and **9** in Figure 4.6,

$$\mathrm{ClC{\equiv}CCl} \begin{cases} \xrightarrow{\mathrm{HCl}} \mathrm{ClCH{=}CCl_2} \ \text{(TCE)} \\ \xrightarrow{\mathrm{CuCl_2}} \mathrm{Cl_2C{=}CCl_2} \ \text{(PE)} \end{cases} ; \quad \mathrm{CH_3C{\equiv}CCl} \begin{cases} \xrightarrow{\mathrm{HCl}} \mathrm{CH_3C(Cl){=}CHCl} \\ \xrightarrow{\mathrm{CuCl_2}} \mathrm{CH_3C(Cl){=}CCl_2}, \end{cases}$$

where the reactivity (k_I) of alkynes in the first route of additive oxidative chlorination according to scheme (4.155) decreases in the following order [122]:

$$\mathrm{ClC{\equiv}CH} \approx \mathrm{ClC{\equiv}CCl} > \mathrm{HC{\equiv}CH} > \mathrm{CH_2{=}CHC{\equiv}CH} > \mathrm{CH_3C{\equiv}CCl} > \mathrm{CH_3C{\equiv}CH}.$$

■

The synthesis of 2-chlorovinylacetylene (CVA), as well as that of vinylidene chloride, occupies a special place in the class of oxidative chlorination processes and in the whole catalytic chemistry of acetylene. This is related to the fact that the reaction

$$\mathrm{C_2H_2} + \mathrm{X^-} + \mathrm{Y^-} + \mathrm{Ox} \longrightarrow \mathrm{CH_2{=}C}\begin{matrix}\diagup \mathrm{X} \\ \diagdown \mathrm{Y}\end{matrix} + \mathrm{Red}$$

involves the 1,2-shift of hydrogen atom, whereby $X^- = Y^- = Cl^-$ (see scheme (4.153)) and $X^- = Cl^-$, $Y^- = C{\equiv}CH^-$.

This interesting reaction was mentioned above for the first time in substantiating the existence of intermediate CuCH=CHC≡CH during the synthesis of vinylacetylene in solutions of polynuclear Cu(I) complexes (Section 2.3), and for the second time, in the context of discussion of the role of components and classification of multicomponent catalytic systems (Section 4.1). Now let us turn again to CVA synthesis for a more detailed discussion of the kinetic functions of components and the mechanism of the 1,2-shift of hydrogen atom.

Example 4.10 [120, 132, 134]

► In the case of CVA synthesis, it was demonstrated that neither vinylacetylene and diacetylene nor chloroacetylene, for which the processes of formation are suppressed by H_3O^+ ions, are involved as intermediates in the CVA formation [120]. For the reactions in concentrated CuCl and NH_4Cl solutions, the character of products strongly depends on $[CuCl_2]$. In the interval of $[CuCl_2] = 10^{-4}$–10^{-2} M, the main products are CVA and VA (Figure 4.7). Let us consider in more detail the possible mechanism of CVA formation and the kinetic functions of components of the $CuCl-CuCl_2$ system in this process.

The kinetic equations of CVA and VA formation are consistent with the existence of a conjugation node of the reactions of dimerization and oxidative chlorination of C_2H_2 [132–134]:

$$r_{2-\mathrm{CVA}} = \frac{k_{\mathrm{CVA}}[\mathrm{CuCl_2}]_{\Sigma}\, P^2_{\mathrm{C_2H_2}}}{([\mathrm{H_3O^+}] + K_{\mathrm{A}} P_{\mathrm{C_2H_2}})(k_1[\mathrm{H_3O^+}] + [\mathrm{CuCl_2}]_{\Sigma})} \tag{4.163}$$

$$r_{\mathrm{VA}} = \frac{k_{\mathrm{VA}}[\mathrm{H_3O^+}]\, P^2_{\mathrm{C_2H_2}}}{([\mathrm{H_3O^+}] + K_{\mathrm{A}} P_{\mathrm{C_2H_2}})(k_1[\mathrm{H_3O^+}] + [\mathrm{CuCl_2}]_{\Sigma})}. \tag{4.164}$$

These equations involve the total concentration of $CuCl_2$, which is proportional under conditions studied to the free concentration and activity of Cu^{2+} ($a_{Cu^{2+}}$). The activities of Cu^+ and Cl^- ions were maintained constant (see Section 2.3). The anodic oxidation of Cu(I) can be represented by the following equation:

$$2\mathrm{CuCl} + 2\mathrm{HCl} \xrightarrow{\text{electric current}} 2\mathrm{CuCl_2} + \mathrm{H_2}. \tag{4.165}$$

In the mechanism of CVA synthesis considered above (Eqs (4.4) and (4.7)–(4.10)), CuCl was oxidized by oxygen. Denoting Cu(I) by the general symbol CuCl allows the process mechanism to be written as follows:

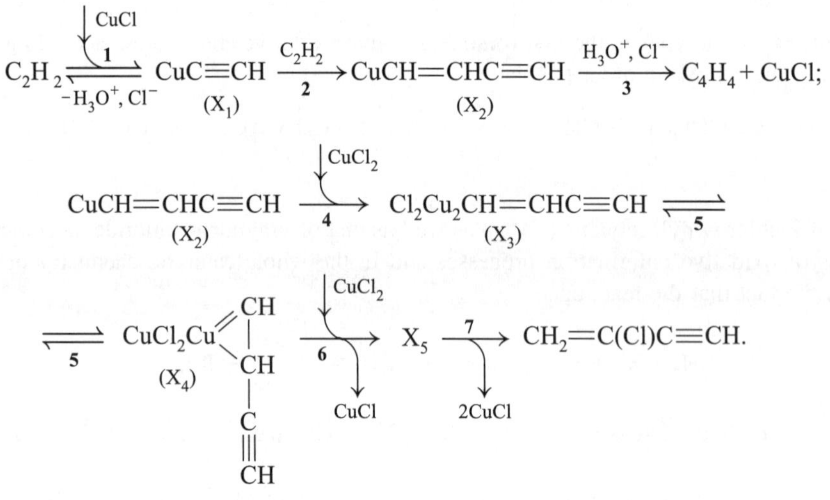

(4.166)

Here, intermediate X_5 can be considered (by analogy with VDC synthesis) as a carbene complex with two CuCl molecules:

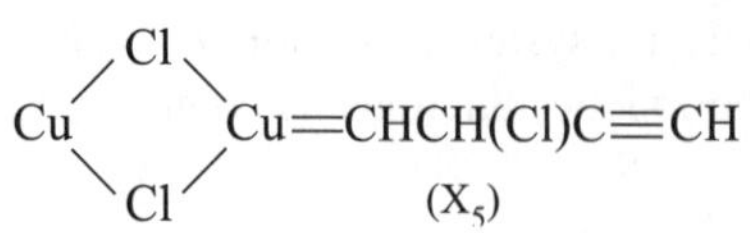

The subsequent decomposition of this complex into free carbene and the isomerization of carbene species lead to CVA. It also cannot be excluded that two σ bonds Cu(II)–C are formed in the X'_5 intermediate as

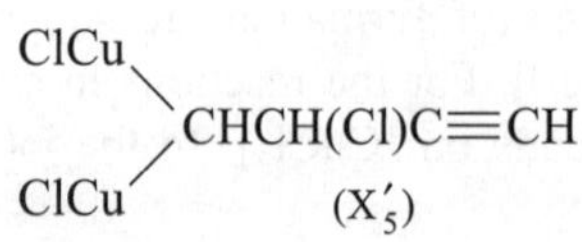

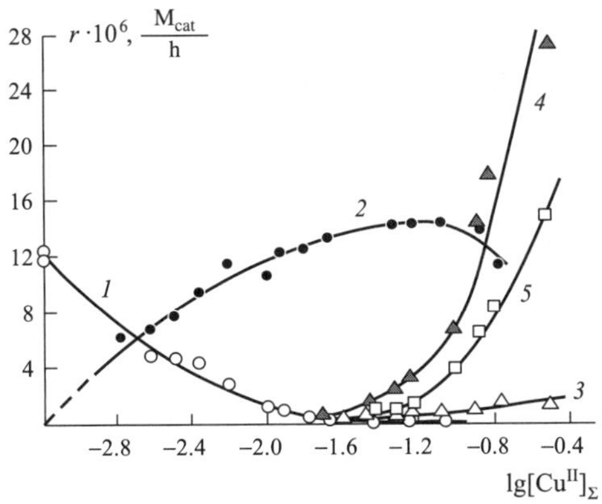

Figure 4.7 Rates of acetylene conversion product formation in 12 $\tilde{m}$ CuCl–12 $\tilde{m}$ NH_4Cl solution versus $[Cu^{II}]_\Sigma$: (1) monovinylacetylene; (2) 2-chlorovinylacetylene; (3) diacetylene; (4) trans-1, 2-dichloroethylene; (5) vinylidene chloride (reproduced with permission from editorial board of *Kinet. Katal.* journal © 1970 [120])

and then the reductive elimination of free carbene from X_5' with the formation of 2CuCl leads to the same result.

The kinetic functions are as follows:

- Cu(I) catalyzes the synthesis of VA and the formation of X_5 and CVA:

$$2C_2H_2 + 2CuCl_2 \longrightarrow X_5,$$
$$2C_2H_2 + 2CuCl_2 \longrightarrow CVA + 2CuCl + HCl;$$

- $CuCl_2$ catalyzes the conversion of X_2 into X_5 and CVA:

$$CuCH{=}CHC{\equiv}CH + CuCl + 2HCl \longrightarrow X_5 + H_2,$$
$$CuCH{=}CHC{\equiv}CH + 2HCl \longrightarrow CVA + CuCl + H_2.$$

Both components catalyze the reaction according to the overall equation as

$$2C_2H_2 + HCl \xrightarrow{\text{electric current}} CVA + H_2. \quad (4.167)$$

■

Throughout this chapter, as well as in other chapters, the compounds considered as intermediates in the reaction mechanisms include organometallic σ complexes of Cu(II), in particular RC≡CCuX, ClCuCH=CHCl, and ClCuCH=CHC≡CH (both in complexes with and without Cu(I)). Although RCuX type compounds have never been isolated because of their extremely high reactivity in various processes (especially in those accompanied by one-electron transfer), the formation of RCu^+ species from $\dot{R}$ and Cu^+ was confirmed by the electronic absorption spectroscopy. The kinetics, products, and mechanisms of conversions involving RCu^+ species were thoroughly studied (see, e.g., [135] and references therein). Here are some reactions that characterize the properties of Cu(II)–C bonds:

$$Cu^{II}{-}R \rightleftharpoons Cu^I + \dot{R}\text{(homolysis)};$$
$$CuCH_2COOH^+ + H_3O^+/H_2O \longrightarrow Cu^{2+} + CH_3COOH + H_2O/OH^-$$

[k_{298} (pH = 7) = 7000 s^{-1} (heterolysis)];

$$Cu^{II}{-}CR^1R^2CR^3R^4X \longrightarrow Cu^{II} + R^1R^2C{=}CR^3R^4 + X^-$$

[X = OH, OR, NR_2 (β-elimination)];

$$CuCH_3^+ + CO \longrightarrow CuC(=O){-}CH_3^+ \xrightarrow{H_2O} Cu^{2+} + CH_3CHO + OH^-$$

$$2CuCH_3^+ \xrightarrow{k_c} 2Cu^+ + C_2H_6.$$

[$k_c = 1.5 \cdot 10^8\ M^{-1} \cdot s^{-1}$(298 K, pH 2)] (the rate of the later bimolecular reaction is independent of [Cu^+]). The reactions of organo-Cu(II) and -Cu(III) intermediates were considered in [136–139].

Thus, there are no grounds for excluding the possible formation of reactive RCuCl intermediates in the processes of oxidation and oxidative chlorination of alkynes. The presence of π electrons in the organic groups of ethynyl and chlorovinyl intermediates makes these compounds even more stable. Another factor that stabilizes RCuCl compounds is the formation of mixed heterovalent complexes containing Cu(I) and Cu(II). The very high rate of electron transfer via the $Cu^I{-}Cl{-}Cu^{II}$ chain [140, 141] ($k = 0.5 \cdot 10^8\ M^{-1} \cdot s^{-1}$) in these compounds and, hence, the possible appearance of valent isomers such as

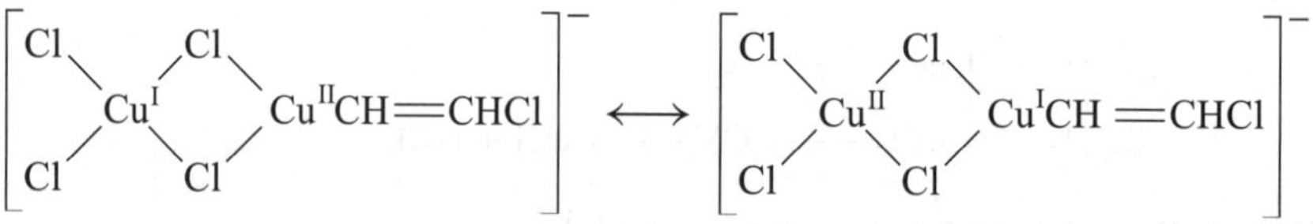

undoubtedly leads to an increase in the stability of $Cu^{II}{-}C$ bonds. Thus, the electron transfer from RCu(I) to $CuCl_2$ is both the pathway of $RCu^{II}Cl$ formation and the factor that stabilizes RCu(II) compounds. For this reason, the order of kinetic equations with respect to $[CuCl_2]_\Sigma$ for oxidative chlorination reactions at small concentrations of $CuCl_2$ in the CuCl–$CuCl_2$ system frequently amounts to 2, while that in the $HgCl_2$–$CuCl_2$ system is 1. The first order with respect to $[CuCl_2]_\Sigma$ always takes place in the routes of the additive oxidative chlorination of C_2H_2, HC≡CCl, and ClC≡CCl in the absence of Cu(I), since the rate of conversion of RCuCl intermediates not stabilized by coordination with Cu(I) in reactions with c $CuCl_2$ is very high.

The reasons for two $CuCl_2$ molecules to be involved in the oxidative chlorination processes were considered in the context of investigations of the kinetics of reactions [91, 130, 142, 143] (see also Section. 3.3):

$$C_2H_2 + 2FeCl_3 \xrightarrow{CuCl_2-HgCl_2} trans\text{-DCE} + 2FeCl_2; \qquad (4.168)$$

$$trans\text{-}ClHgCH{=}CHCl + 2FeCl_3 \xrightarrow{CuCl_2} HgCl_2 + trans\text{-DCE} + 2FeCl_2. \qquad (4.169)$$

Let us analyze the kinetic functions of PFCS components in reaction (4.168).

Example 4.11 [142, 143]

► For the oxidative chlorination of acetylene under steady-state conditions in the absence of CuCl in the $HgCl_2$ (0–0.5 M)–$CuCl_2$ (0–0.09 M)–NH_4Cl (2 M)–$FeCl_3$ (3 M)–HCl (1 M) system at 70 °C, the following equations were obtained for the rate of *trans*-DCE formation

$$r_{DCE} = k_1[CuCl_2]_\Sigma\, P_{C_2H_2} + \frac{k_2[CuCl_2]_\Sigma\,[HgCl_2]_\Sigma\, P_{C_2H_2}}{k_3 + [CuCl_2]_\Sigma} \tag{4.170}$$

(which describes the routes both with and without the participation of $HgCl_2$) and for the rate of VC synthesis (as conjugated with the DCE formation)

$$r_{VC} = \frac{k_4[HgCl_2]_\Sigma\, P_{C_2H_2}}{k_3 + [CuCl_2]_\Sigma}. \tag{4.171}$$

The following scheme of a three-route process was proposed [142]:

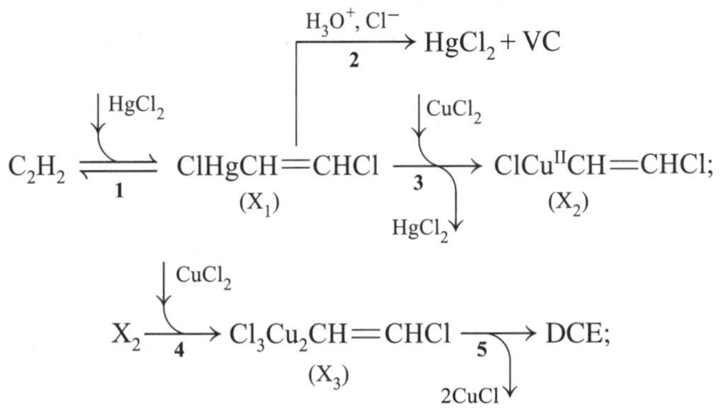

(4.172)

$$2CuCl + 2FeCl_3 \xrightarrow[6]{} 2CuCl_2 + 2FeCl_2.$$

The subsequent kinetic investigation [143] of the key electrophilic substitution reaction **3**, in which the geometric structure of the chlorovinyl group is retained (so that *cis*-β-chlorovinylmercury chloride converts into *cis*-DCE) [9], confirmed scheme (4.172) with the consecutive participation of two $CuCl_2$ molecules. Judging from the results of thermodynamic calculations, the step

$$ClCuCH{=}CHCl \xrightarrow{Cl^-} Cu^o_{aq} + ClCH{=}CHCl \tag{4.173}$$

can hardly proceed as a two-electron process of reductive elimination. Indeed, the ΔH°_{298} value for reaction

$$C_2H_{2gas} + Cu^{2+} + 2Cl^- \longrightarrow Cu^\circ_{gas} + ClCH{=}CHCl_{gas} \tag{4.174}$$

(i.e., for the overall equation of product formation in step (4.173)), amounts to 393 kJ/mole. It is unlikely that the heats of hydration of Cu°_{gas} (to Cu°_{aq}) and DCE, as well as the heat of complex formation of Cu°_{aq} with Cl^- ($CuCl^{n-}_{n\ aq}$) would significantly reduce the endothermicity of process (4.174). At the same time, the endothermicity of the decay of RCuCl with the participation of a second $CuCl_2$ molecule is significantly lower, which makes the process of demetallation kinetically probable. The ΔH°_{298} value for reaction

$$C_2H_{2gas} + 2Cu^{2+} + 6Cl^- \longrightarrow 2CuCl^{2-}_{aq} + ClCH{=}CHCl_{gas} \qquad (4.175)$$

is about 100 kJ/mole, while the ΔH°_{298} value for the electrode process

$$CuCl_2 + e^- \longrightarrow CuCl_2^-$$

in aqueous solutions is small [–(12–29) kJ/mole]. Therefore, $CuCl_2$ in water is a weak one-electron oxidant that can hardly lead to the formation of free radicals via the demetallation of RHgCl (in contrast to the reaction of $CuBr_2$ with RHgCl in DMF [144]).

In the reactions of RMX_n with $CuCl_2$, the main processes leading to the RCl(RX) products are frequently interpreted as homolytic steps with the formation of free radicals $\dot{R}$ from RCuCl [27]. In reactions with the participation of organic substrates or RHgCl considered above, the kinetics was indicative of the participation of a second $CuCl_2$ molecule immediately after the step of RCuCl formation, which is consistent with the kinetics observed and the mechanism proposed by Tamura and Kochi [145] (in THF solutions):

$$RMgBr + CuCl_4^{2-} \longrightarrow RCuCl_3{}^{2-} + MgBrCl;$$

$$RCuCl_3{}^{2-} + CuCl_4^{2-} \longrightarrow RCl + 2CuCl_2^- + 2Cl^-;$$

$$2RCuCl_3{}^{2-} \longrightarrow R{-}R + 2CuCl_2^- + 2Cl^-.$$

Note that the second order with respect to $CuCl_2$ also takes place in equations (4.142) and (4.143). A scheme with the participation of radical intermediates in the synthesis of RC≡CCl is inconsistent with the observed kinetics. Indeed, the endothermicity of process

$$RC{\equiv}CH + Cu^{2+} \xrightarrow{B} RC{\equiv}\dot{C} + BH^+ + Cu^+ \qquad (4.176)$$

is too large ($\Delta H^\circ_{298} \cong 250$ kJ/mole [102b]) for radical $RC{\equiv}\dot{C}$ to be involved as an intermediate in the oxidative dehydrocondensation and the substitutive oxidative chlorination of alkynes.

Returning to scheme (4.172), the kinetic functions of components of the $HgCl_2$–$CuCl_2$ system can be quite evidently assigned as follows:

- $HgCl_2$ catalyzes the synthesis of VC (steps **1** and **2**) and the synthesis of X_2 (steps **1**and **3**):

$$C_2H_2 + CuCl_2 \xrightarrow{HgCl_2} \underset{(X_2)}{ClCuCH{=}CHCl};$$

- $CuCl_2$ catalyzes the oxidation of X_1 by $FeCl_3$ (steps **3–6**):

$$X_1 + 2FeCl_3 \rightarrow DCE + HgCl_2 + 2FeCl_2;$$

- $HgCl_2$ and $CuCl_2$ catalyze reactions (4.168). ■

4.2.2 Oxidative chlorination of 1,3-dienes

Dichlorobutenes (DCBs) have been considered above (see Section 4.1) as intermediates in the reactions of diene oxidation to diethers and diesters and vinyl dioxane. However, DCBs are also of independent interest as initial materials for the synthesis of chloroprene (3, 4-dichloro-1-butene), butenediols, dinitriles of hexenedicarboxylic acids, etc. The 100 % selectivity with respect to the sum of DCBs is achieved in the $CuCl_2{-}I_2{-}CH_3CN$ (adiponitrile) and $CuCl_2{-}PdCl_2{-}CH_3CN$ systems at $T = 60\,°C$ [146]. Moreover, even when using the $CuCl_2{-}PdCl_2{-}LiCl$ system with ethanol under certain conditions (T, [HCl]), it is possible to obtain DCBs with a selectivity of 80 % (for C_4H_6). The latter system was used to study the kinetics of DCB synthesis in the presence of O_2 and without oxygen. The initial rate of reaction (in the absence of O_2)

$$C_4H_6 + 2CuCl_2 \xrightarrow{PdCl_2} \underset{\text{(isomer sum)}}{DCB} + 2CuCl \tag{4.177}$$

is described by the empirical equation

$$r_{DCB,0} = \frac{k_1 P_{C_4H_6}[PdCl_2]_\Sigma\,[CuCl_2]^3_\Sigma}{[LiCl]^2_\Sigma(k_2 + [CuCl_2]^2_\Sigma\,[LiCl]^4_\Sigma)}. \tag{4.178}$$

Here, the complicated dependences on $[CuCl_2]_\Sigma$ and $[LiCl]_\Sigma$ are related to the fact that the F_{Pd} and F_{CuCl_2} complexity functions do not remain constant when $[CuCl_2]_\Sigma$ is increased up to 2 M while the $[LiCl]_\Sigma +$ $[HCl]_\Sigma$ sum does not exceed 4 M. At the same time, by analogy with the kinetics of β-halogenated ethyl alkyl ethers (4.71), butenediol diesters (4.100), and vinyl-2,4-dioxane (4.104), it is possible to consider the participation of polynuclear Cu(II) complexes (at least, Cu_2Cl_4) in the decomposition steps of the corresponding *syn*- and *anti*-η^3-allyl complexes of palladium(II) to *trans*-1,4-DCB, *cis*-1,4-DCB, and 3,4-DCB (racemate, R,S-forms). The fact that the ratio of 1,4-DCB concentrations does not vary with the time can be indicative of the equilibrium isomerization of *syn*- and *anti*-η^3-allyl complexes:

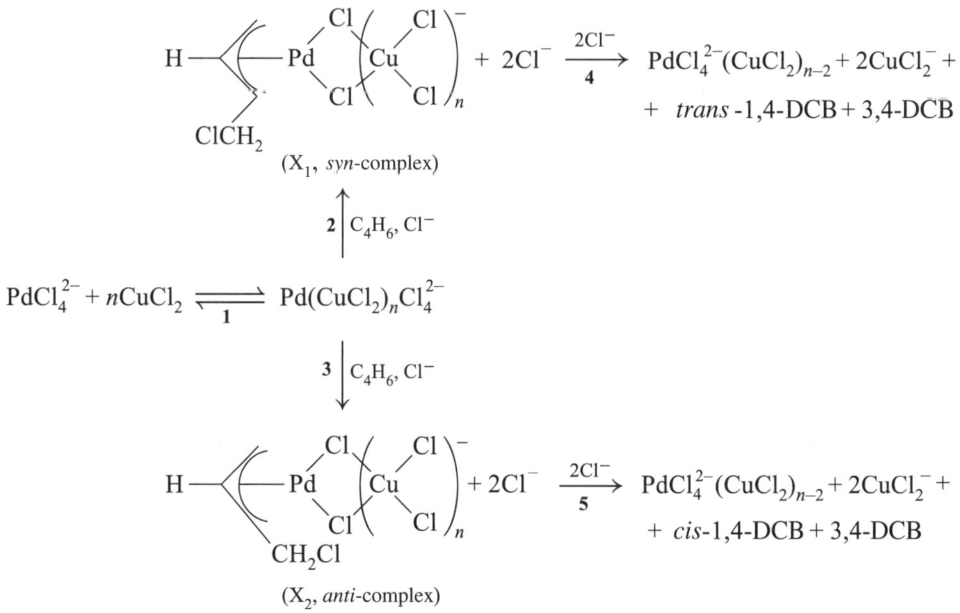

(4.179)

In the presence of O_2, the rapid oxidation of CuCl to $CuCl_2$ by oxygen allows the process to be carried out under steady-state conditions and the reaction selectivity to be increased. Indeed, as indicated above, one of the kinetic functions of CuCl in alcohol solutions is to catalyze the replacement of Cl^- by alcoholate ions with the formation of diethers. In addition, CuCl catalyzes the isomerization of DCB, while $PdCl_2$ and

$CuCl_2$ as components of the catalytic system with oxygen as the oxidant perform the following specific kinetic functions:

- $PdCl_2$ catalyzes the oxidation of C_4H_6 by Cu(II) chloride to the sum of DCB isomers;
- $CuCl_2$ catalyzes the oxidation of C_4H_6 to DCB by oxygen via the reaction

$$C_4H_6 + {}^1\!/\!_2\,O_2 + 2HCl \longrightarrow DCB + H_2O. \tag{4.180}$$

Investigation of this reaction showed [146] that, in the presence of oxygen, the initial rate of Cu(II) reduction to Cu(I) in the course of attaining steady-state conditions is higher than that during a non-steady-state process with respect to Cu(II). This phenomenon is worthy of separate investigation.

4.2.3 Polyfunctional catalytic systems in chlorination reactions

Among the catalysts of the homogeneous chlorination of aromatic compounds, a high activity is inherent in sulfur-containing compounds. In particular, it was established that the most effective catalyst for the chlorination of 2,6-dicarbomethoxynaphthalene (ArH) in tetrachloroethane is a mixture of Sb_2S_5 and elemental sulfur, the amount of the latter significantly influencing the reaction rate [147]. It was suggested that Sb(V) sulfide and S are chlorinated to a mixture of $SbCl_5$ and S_2Cl_2, and S_2Cl_2 can be further chlorinated to SCl_2 and SCl_4. The rate of ArH chlorination in the $SbCl_5$ ($8 \cdot 10^{-4}$ M)–S_2Cl_2 ($2 \cdot 10^{-3}$ M) system is significantly higher than the rate of chlorination by separate components of this system, which implies a pronounced synergistic effect of the two chlorides. The contribution of sulfur chloride to the total reaction rate was evaluated as 2.5 % and that of $SbCl_5$ as 12.7 %. It is known that $SbCl_5$ forms 1:1 complexes with S_2Cl_2 and that Sb(V) can also form complexes with other sulfur chlorides [148].

The mechanism of functioning of this complex catalytic system was elucidated, in particular, by studying the kinetics of an aromatic monochloride (ArCl) formation by the initial rate method in $SbCl_5$, S_2Cl_2, and $SbCl_5$–S_2Cl_2 solutions [148, 149]. Under the conditions of $[ArH]_0 \gg [S_2Cl_2]_0$, the reaction kinetics in S_2Cl_2 solutions was first-order with respect to $[S_2Cl_2]_0$ and variable-order with respect to $[ArH]_0$. Then, assuming the quasi-equilibrium formation of ArH complex with S_2Cl_2, the reaction rate was described by the equation

$$r_0 = \frac{k_2K_1[S_2Cl_2]_\Sigma\,[ArH]_0}{1 + K_1[ArH]_0}, \tag{4.181}$$

and the following reaction scheme was proposed:

$$\begin{aligned} &ArH + S_2Cl_2 \overset{K_1}{\rightleftharpoons} ArH(S_2Cl_2);\\ &ArH(S_2Cl_2) + Cl_2 \xrightarrow{k_2'} ArCl + S_2Cl_2 + HCl. \end{aligned} \tag{4.182}$$

In Eq. (1.81), the concentration of Cl_2 in solution (which is constant in the time for the process in a flow reactor) enters into the k_2 constant. The equilibrium constant K_1 at 55 °C is 5.3 M^{-1}. An analogous scheme was proposed for the reaction in $SbCl_5$ solutions:

$$\begin{aligned} &ArH + SbCl_5 \overset{K_3}{\rightleftharpoons} ArH \cdot SbCl_5;\\ &ArH \cdot SbCl_5 + Cl_2 \xrightarrow{k_4'} ArCl + SbCl_5 + HCl \end{aligned} \tag{4.183}$$

with an analogous equation of

$$r_0 = \frac{k_4K_3[SbCl_5]_\Sigma\,[ArH]_0}{1 + K_3[ArH]_0}, \tag{4.184}$$

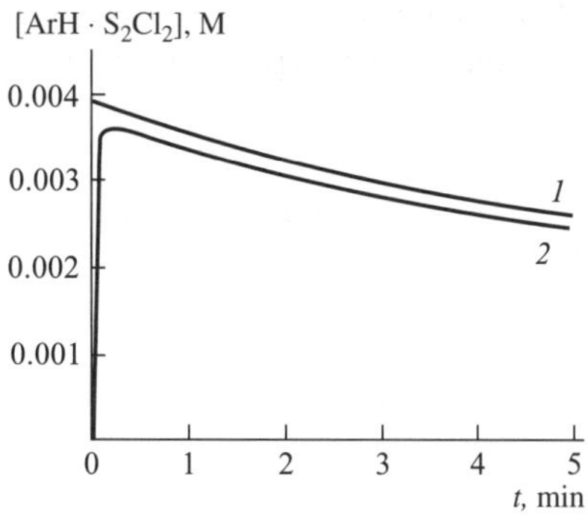

Figure 4.8 Temporal dependence of $ArH \cdot S_2Cl_2$ complex concentration calculated using (1) quasi-equilibrium and (2) non-steady-state models (reproduced with permission from editorial board of *Kinet. Katal.* journal © 1983 [149])

where $K_3 = 110\ M^{-1}$. It was found that [HCl] did not influence the rate of chlorination in both individual component solutions. For both these systems, the rate constants for the steps of mechanisms (4.182) and (4.183) were determined within the framework of a non-steady-state system using k_2, K_1, K_4, and K_3 as initial estimations, and the adequacy of models obtained in the quasi-equilibrium approximation was proved. A comparison of the results of integration of the non-steady-state and quasi-equilibrium models with refined values of the constants obtained upon solving the inverse problem of chemical kinetics ($K_1 = 12.8\ M^{-1}$), showed that the temporal variation, e.g., of the concentration of $ArH{\cdot}S_2Cl_2$ complexes in both cases is approximately the same (Figure 4.8) and the equilibrium complex formation regime is attained virtually instantaneously.

The kinetics of chlorination in the $SbCl_5{-}S_2Cl_2$ two-component system was studied under the conditions of $[S_2Cl_2]_\Sigma \gg [SbCl_5]_\Sigma$ and $[ArH]_0 \gg [S_2Cl_2]_\Sigma$, that is, where $[ArH] \approx [ArH]_0$ [149]. It was established that the reaction orders with respect to $[SbCl_5]_\Sigma$ and $[S_2Cl_2]_\Sigma$ amount to 2 and 1.5, respectively. With allowance for the formation of $SbCl_5{\cdot}S_2Cl_2$ (**A**) and $ArH{\cdot}S_2Cl_2$ (**B**) complexes and the possible appearance of ternary complexes, one of the possible mechanistic schemes is as follows:

$$
\begin{aligned}
&(1)\ ArH + S_2Cl_2 \overset{K_1}{\rightleftharpoons} \underset{(\mathbf{B})}{ArH \cdot S_2Cl_2};\\
&(2)\ SbCl_5 + S_2Cl_2 \overset{K_3}{\rightleftharpoons} \underset{(\mathbf{A})}{(S_2Cl^+)(SbCl_6^-)};\\
&(3)\ \mathbf{B} + \mathbf{A} \overset{K_4}{\rightleftharpoons} \underset{(\mathbf{D})}{(ArH \cdot S_4Cl_3^+)(SbCl_6^-)};\\
&(4)\ \mathbf{D} + SbCl_5 \xrightarrow{k_4} \underset{(\mathbf{E})}{(ArH \cdot S_4Cl_2^{2+})(SbCl_6^-)_2};\\
&(5)\ \mathbf{E} \xrightarrow{k_5} ArCl + HCl + SbCl_5 + \underset{(\mathbf{F})}{S_4Cl_2 \cdot SbCl_5};\\
&(6)\ \mathbf{F} + Cl_2 \xrightarrow{k_6} 2S_2Cl_2 + SbCl_5.
\end{aligned}
\tag{4.185}
$$

Here, the conversion of complex **D** into complex **E** can proceed via two mechanisms. The first is the addition of strongly electrophilic Cl^+ from $S_4Cl_2^{2+}$ species to the ring as

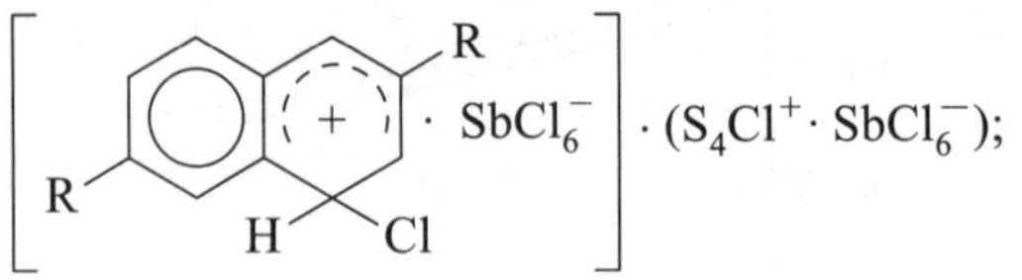

and the second is the addition of $S_4Cl_2^{2+}$ cation to the ring followed by the chlorinolysis of the C–S bonds as

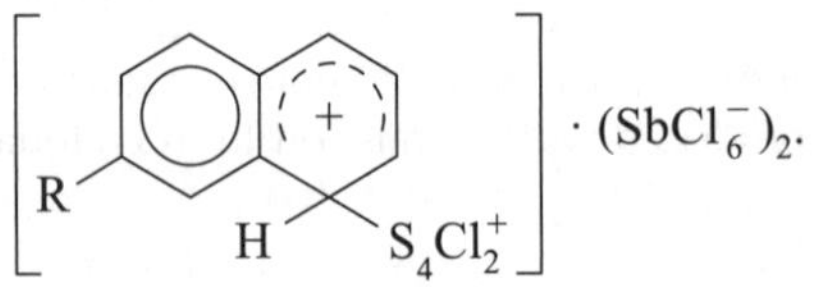

The possible limiting steps are both the elimination of Cl^- under the action of a second $SbCl_5$ molecule [step (4)] and the conversion of complex **E**, although the case of reaction of (4) as the limiting step seems to be more probable. Ignoring the possible formation of ternary complexes **D** and **E** and taking into account the material balance equations

$$[S_2Cl_2]_\Sigma \cong [S_2Cl_2] + [ArH \cdot S_2Cl_2];$$

$$[SbCl_5]_\Sigma = [SbCl_5] + [SbCl_5 \cdot S_2Cl_2],$$

we obtain the following kinetic equation:

$$r_0 = \frac{k_4 K_1 K_3 K_4 [ArH]_0 [SbCl_5]^2_\Sigma [S_2Cl_2]^2_\Sigma}{(1 + K_1[ArH]_0 + K_3[S_2Cl_2]_\Sigma)^2}, \qquad (4.186)$$

which quite adequately describes the experimental results. In deriving this equation, it was taken into account that the formation of complexes between two catalysis under the condition $[S_2Cl_2]_\Sigma \gg [SbCl_5]_\Sigma$ leads to a situation where the complexity term $F_{S_2Cl_2}$ is canceled and the complexity function F_{SbCl_5} involves both $F_{S_2Cl_2}$ and the total concentration of S_2Cl_2:

$$[S_2Cl_2][SbCl_5] = \frac{[S_2Cl_2]_\Sigma\, [SbCl_5]_\Sigma}{F_{S_2Cl_2}\left(1 + K_3 \dfrac{[S_2Cl_2]_\Sigma}{F_{S_2Cl_2}}\right)}.$$

Other hypotheses adopting the participation of $SbCl_5$ and SCl_4 complexes within the framework of a three-route scheme were also considered [149], but the estimations showed that, in the interval of $[S_2Cl_2]_0 = (2\text{–}20) \cdot 10^{-3}$ M, a contribution of the route under consideration (scheme (4.185)) amounts to 87–99.5 %. Equation (4.186) also well describes the kinetics of chlorination of 1, 4, 7, 8-tetrachloro-2, 6-dicarbomethoxynaphthalene.

With allowance for the equilibrium of the first three steps in scheme (4.185), let us write a simplified variant of the mechanism under consideration so as to reveal the kinetic functions of the catalyst components

($S_2Cl_2 = \mathbb{K}_1$, $SbCl_5 = \mathbb{K}_2$) as follows:

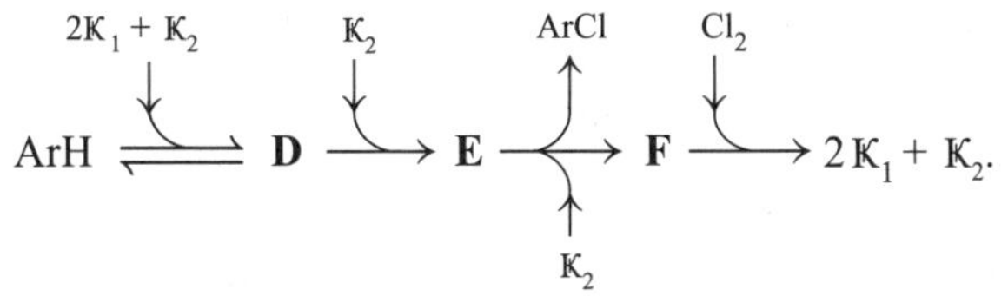

(4.187)

As can be seen from this scheme, the kinetic function of $\mathbb{K}_2$ ($SbCl_5$) is to catalyze the conversion of intermediate **D** into **F** with the formation of a product (ArCl), and the kinetic function of both $\mathbb{K}_1$ and $\mathbb{K}_2$ is to catalyze the overall process with the overall equation

$$ArH + Cl_2 \longrightarrow ArCl + HCl.$$

The synergistic effect in this PFCS is related to the formation of species that interact with ArH (with the participation of both components $S_2Cl^+ \cdot SbCl_6^-$, $S_4Cl_3{}^+ \cdot SbCl_6^-$, and $S_4Cl_2{}^{2+} \cdot (SbCl_6^-)_2$), which are more electrophilic than the active reactants $Cl_2 \cdot SbCl_5$ and $S_2Cl_2 \cdot$ (SCl_2, SCl_4) formed in the case of single-component catalysts.

4.3 Oxidative carbonylation of organic compounds

The class of oxidative carbonylation processes includes reactions with the participation of carbon monoxide (CO), which are accompanied by an increase in the total degree of oxidation of the elements transferred from reactants to carbonylation products. If the oxidant does not enter into the reaction products, the oxidative carbonylation process can be represented by two semireactions (by analogy with the electrode reactions in electrochemistry). For example, the process

$$2CH_3OH + CO + {}^1\!/_2\,O_2 \longrightarrow (CH_3O)_2CO + H_2O \qquad (4.188)$$

is described by semireactions

$$2CH_3OH + CO \longrightarrow (CH_3O)_2CO + 2H^+ + 2e^-; \qquad (4.189)$$

$${}^1\!/_2\,O_2 + 2H^+ + 2e^- \longrightarrow H_2O. \qquad (4.190)$$

For the reaction

$$2RONO + 2CO \longrightarrow (COOR)_2 + 2NO, \qquad (4.191)$$

in which the oxidation number of carbon changes from +2 to +3 upon the passage of every CO molecule into oxalate, the semireactions are written as

$$2ROH + 2CO \longrightarrow (COOR)_2 + 2H^+ + 2e^-;$$

$$2RONO + 2H^+ + 2e^- \longrightarrow 2ROH + 2NO.$$

Typical oxidants in the oxidative carbonylation reactions are O_2, H_2O_2, R_2O_2, RONO, *para*-benzoquinone (Q), NaOCl, $K_2S_2O_8$, $FeCl_3$, and $CuCl_2$, and typical catalysts are the salts of Pd(II), Rh(I, III), Cu(II) and various multicomponent catalytic systems [91, 112, 150–154]. Let us analyze some reactions of oxidative carbonylation that involve multicomponent catalytic systems and consider the functions of components in these systems.

4.3.1 Oxidative carbonylation of HY molecules (Y = OR, OPh, NR_2, Ar, Alk)

4.3.1.1 Alcohols, phenols, and amines

These reactants mostly participate in the carbonylation reactions of two types:

$$HX + HY + CO \longrightarrow XCOY + 2H^+ + 2e^-; \quad (4.192)$$

$$HX + HY + 2CO \longrightarrow XCOCOY + 2H^+ + 2e^-, \quad (4.193)$$

where X = Y or X ≠ Y and X, Y = OR, OPh, NHR. The salts of Pd(II), Cu(II), and Pb(IV) are known to be the simplest oxidants. The oxidation of the reduced forms of palladium and copper renders them catalysts for the process of oxidative carbonylation. PFCSs for the synthesis of dialkyl carbonates include the $PdCl_2$–$CuCl_2$ [155] and CuCl–$CuCl_2$ systems (with O_2 as the oxidant) [151, 154, 156].

In a catalytic variant of reaction (4.191) for the NO oxidation to alkyl nitrite by oxygen, the PFCS includes $PdCl_2$ and RONO [157]. The two-component system of CuCl(2,6-dimethylpyridine)$_n$–Pd(Acac)$_2$ in the presence of $Bu^t_2O_2$ catalyzes the synthesis of oxalate (oxalate/carbonate ~30) with a total yield of ~90% [158]. During the carbonylation of Bu^tOH alcohol in the reaction

$$Bu^t_2O_2 + 2CO \longrightarrow (COOBu^t)_2, \quad (4.194)$$

the oxidation numbers of two carbon atoms in CO molecules increase by two units, while the oxidation numbers of two oxygen atoms in the peroxide decrease by two units ($O_2^{2-} \rightarrow 2O^{2-}$).

The two- and three-component systems of the $PdBr_2$–Mn(Acac)$_2$ [159, 160] and $PdCl_2$–Q–MX_n types, where M = Ce(III), Cu(II), Co(III), Mn(III) [161, 162], have been also used for the synthesis of diaryl carbonates via reaction

$$2PhOH + CO + {}^1\!/_2\,O_2 \longrightarrow (PhO)_2CO + H_2O. \quad (4.195)$$

The main products of the oxidative carbonylation of amines are ureas, carbamates, oxamides, and oxamates. Some reactions of primary amines yield RNCO type isocyanates, which are sometimes also considered as intermediates in the synthesis of ureas and carbamates [154a]. Similar to the case of alcohols and phenols, the reactions of amines are also catalyzed by PFCSs such as $PdCl_2$–$CuCl_2$ [163–165] and CuCl–$CuCl_2$ [166]. Both these systems exhibit an autocatalytic character [163, 165]. The $PdCl_2$–CuI–CH_3CN–HC(OCH_3)$_3$ features the formation of oxamates as [167]:

$$R^1R^2NH + R^3OH + 2CO + {}^1\!/_2\,O_2 \longrightarrow R^1R^2NCOCOOR^3 + H_2O. \quad (4.196)$$

Since no strict kinetic models for the synthesis of dialkyl carbonates, dialkyl oxalates, ureas, and carbamates have been formulated so far, the kinetic functions of the components of catalytic systems can only be judged from the results of investigations of the corresponding stoichiometric reactions, conversions of the expected intermediates, and some known kinetic laws. In the systems involving Pd(II), the key intermediates are conventionally assumed to be alkoxy- and phenyloxycarbonyl ClPdCOOR complexes in the syntheses of carbonates and oxalates [112, 155, 168–173] and carbamoyl ClPdCONHR complexes in the syntheses of ureas, carbamates, and isocyanates [163, 165, 174–176].

According to the available data, the role of the second component ($CuCl_2$) does not reduce entirely to the oxidation of Pd(0) and HPdCl formed upon the decomposition of ClPdCOOR and ClPdCONHR, respectively. The kinetic function of $CuCl_2$ apparently consists of the catalysis of oxidation of the organopalladium intermediates XPdCOY (Y = OR, NHR) by oxygen with the formation of reaction products [163, 170] or the intermediate chloroformates or chloroformamides [165, 175, 176] (or the corresponding iodinated derivatives if I_2 is used instead of $CuCl_2$ [165]). These hypotheses are consistent with the following stoichiometric reactions:

$$L_2Pd(COOCH_3)_2 + 4CuCl_2 \longrightarrow 2ClCOOCH_3 + 4CuCl + LPdCl_2 \ (L = Phen, bipy)\ [175];$$

$$L_2PdCl(COOCH_3) + 2CuCl_2(RNH_2)_2$$
$$\longrightarrow RNHCOOR + 2CuCl + [RNH_3]Cl + 2RNH_2 + L_2PdCl_2 \ [165];$$

$$L_2PdCl(CONRR') + 2CuCl_2 \longrightarrow 2CuCl + ClCONRR' + L_2PdCl_2 \ [176];$$

$$L_2PdCl(CONHR) + 2CuCl_2 \longrightarrow 2CuCl + RNCO + HCl \ (L = PPh_3), \ [176].$$

Let us consider two variants of action of the second component in the systems under consideration.

Example 4.12 [163]

► The oxidative carbonylation of piperidine (and other amines) was studied in the $PdCl_2$–$CuCl_2$–Py system at 40 °C and atmospheric pressure. A simplified mechanistic scheme of this process includes a step of the oxidative demetallation of ClPdCONHR with the participation of one $CuCl_2$ molecule and the formation of dialkylurea, Cu(I), and Pd(I):

$$RNH_2 \xrightarrow[1,\ -HCl]{PdCl_2} ClPdNHR \xrightarrow[2]{CO} ClPdCONHR \xrightarrow[3]{Cu(II),\ RNH_2} X_4 \xrightarrow[4,\ \rightarrow Cu(I)]{\uparrow P} Pd(I) \xrightarrow[5,\ \rightarrow PdCl_2]{Cu(I)} Cu(I); \qquad (4.197)$$

$$2Cu(I) + 2H^+ + {}^1/_2O_2 \xrightarrow[6]{} 2Cu(II) + H_2O.$$

where symbol H^+ refers to RN^+H_3 species. The participation of a single $CuCl_2$ molecule in this step is only justified by the presence of a maximum in the initial reaction rate at Pd(II)/Cu(II) close to unity.

The kinetic functions of the two components of PFCS under consideration are as follows:

- $PdCl_2$ catalyzes the oxidative carbonylation of amine via reaction

$$2RNH_2 + CO + 2Cu(II) \xrightarrow{PdCl_2} P + 2Cu(I) + 2H^+ \qquad (4.198)$$

  and catalyzes [together with Cu(II)] the overall reaction

$$2RNH_2 + CO + {}^1\!/\!_2\,O_2 \longrightarrow P + H_2O; \qquad (4.199)$$

- $CuCl_2$ catalyzes the reaction

$$2Pd(I) + {}^1\!/\!_2\,O_2 + 2H^+ \longrightarrow 2Pd(II) + H_2O \qquad (4.200)$$

  (steps **5** and **6**) and two reactions of ClPdCONHR decomposition with unbalanced overall equations

$$2ClPdCONHR + 2RNH_2 + {}^1\!/\!_2\,O_2 \longrightarrow 2P + 2Pd(I) + H_2O \qquad (4.201)$$

  (steps **3, 4, 6**) or

$$2ClPdCONHR + 2RNH_2 + {}^1\!/\!_2\,O_2 + H^+ + Cl^- \longrightarrow P + PdCl_2 + H_2O \qquad (4.202)$$

  (steps **3–6**).

The autocatalytic character of this process is related to the kinetic functions of Cu(I), which consist in the catalysis of the formation of ClCuCONHR intermediate

$$RNH_2 + CO + CuCl_2 \xrightarrow{Cu(I)} ClCuCONHR + HCl \tag{4.203}$$

and the catalysis of the reaction with unbalanced overall equation (autocatalysis)

$$2RNH_2 + CO + 2Cu(II) \xrightarrow{Cu(I)} P + 2Cu(I) + 2H^+. \tag{4.204}$$

■

Example 4.13 [165, 175, 176]

► The oxidative carbonylation of amines effectively proceeds in the $PdCl_2$–L–$CuCl_2$–R_4N^+ I^-. It was suggested that the role of additional oxidants ($CuCl_2$, I_2) is to catalyze the oxidation of intermediate ClPdCONHR (or, in the case of alcohols, ClPdCOOR) by oxygen to intermediate ClCONHR (or ClCOOR), which rapidly interacts with amines (or alcohols) to form ureas, carbamates, or carbonates (for ClCOOR) according to the following scheme:

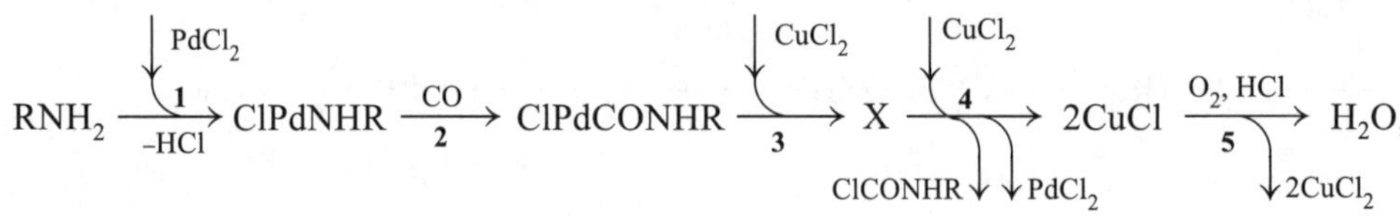

(4.205)

$$ClCONHR + RNH_2 \xrightarrow[6]{} (RNH)_2CO + HCl.$$

In the framework of this scheme, the kinetic function of $CuCl_2$ is to catalyze the overall process

$$ClPdCONHR + \tfrac{1}{2}O_2 + 2HCl \xrightarrow{Cu(II)} ClCONHR + PdCl_2 + H_2O. \tag{4.206}$$

It is also possible that $PdCl_2$ is eliminated in step **3** with the formation of ClCuCONHR [176]. The action of I_2 on ClPdCONHR leads to the formation of ICONHR, Pd (II), and I^-. The oxidation of I^- to I_2 by oxygen makes iodine the second catalyst in the $PdCl_2$–I_2 system. ■

The kinetic functions of components in the CuCl–$CuCl_2$ system can be considered by example of the synthesis of dialkyl carbonate. [151].

Example 4.14

► Investigations of the kinetics of catalytic reactions in the $CuCl_2$–CuCl–MeOH and CuCl–MeOH systems in the presence of O_2 [177], as well as the stoichiometric carbonylation reactions of Cu(II) salts and alkoxides [178, 179] and $XCuOCH_3$ (X = Cl, Br, OAc) according to the equation

$$2XCuOCH_3 + CO \longrightarrow (CH_3O)_2CO + 2CuX \tag{4.207}$$

allowed a mechanistic scheme to be formulated [180], which resembled that for the oxidative chlorination of alkynes in analogous systems. For a system simultaneously containing $CuCl_2$ and CuCl, this scheme can be written as follows [180]:

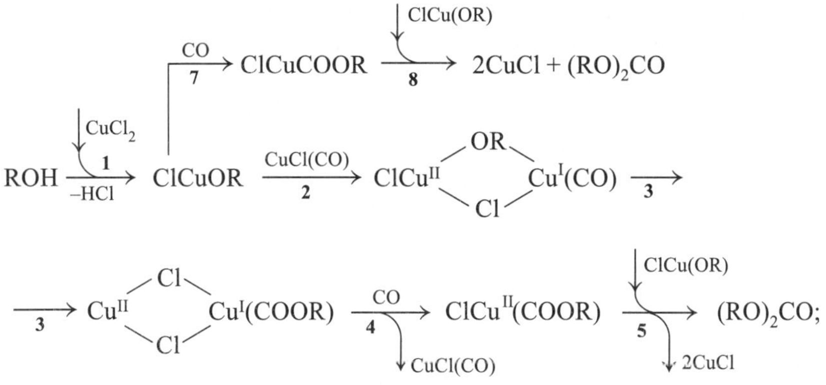

(4.208)

$$2CuCl + 2HCl + {}^1/_2O_2 \xrightarrow[6]{} 2CuCl_2 + H_2O,$$

If CuCl (present in excess) is oxidized via the reaction,

$$2CuCl + 2ROH + {}^1\!/_2\,O_2 \longrightarrow 2ClCuOR + H_2O, \tag{4.209}$$

then Cu(II) alkoxy chloride becomes one of the catalysts. The route **1** → **7** → **8** → **6** makes a small contribution to the total rate and characterizes the latent (induction) period of the process (in the absence of CuCl), the rate of which increases when CuCl appears in step **8**. The kinetic function of Cu(I) in scheme (4.208) is to catalyze the conversion of ClCuOR into $ClCu^{II}(COOR)$ (steps **2**–**4**) according to the overall equation

$$ClCuOR + CO \xrightarrow{Cu(I)} ClCu^{II}(COOR), \tag{4.210}$$

and the kinetic function of $CuCl_2$ is to catalyze the reaction

$$2ROH + CO + {}^1\!/_2\,O_2 \xrightarrow{Cu(II)} (RO)_2CO + H_2O. \tag{4.211}$$

A bipartite graph of mechanism (4.208) contains two other circuits with the corresponding unbalanced overall equations (steps **1, 5, 6** and **1–4, 6**). It also cannot be excluded that other pathways can also exist for the conversion of ClCu(COOR). For example, with allowance for the susceptibility of ClCuR compounds to the reductive elimination of RCl during the interaction with $CuCl_2$, step **5** can be written as

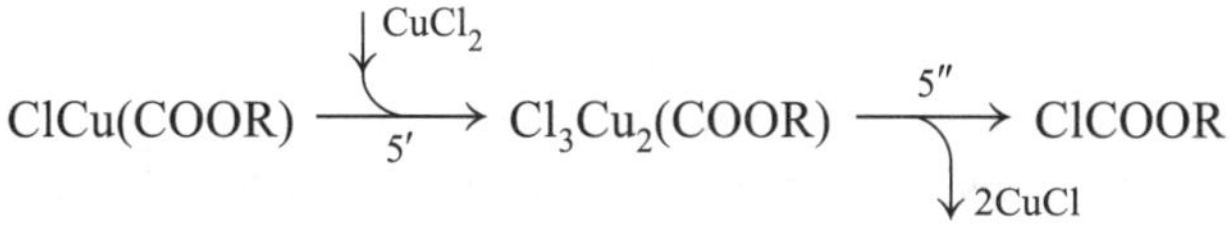

(4.212)

followed by the rapid conversion of chloroformate into carbonate. The possibility of ClCu(COOR) formation is consistent with the results of investigation of the reaction between AcOHgCOOMe and $CuCl_2$ in methanol [181], which leads to the formation of dimethyl carbonate. ■

Quinones catalyze the oxidative carbonylation as well as other oxidation processes (see Section 4.1) such as reaction (4.195). The presence of a quinone as the oxidant in $PdCl_2$–*para*-benzoquinone (Q) and $[Pd(CO)Cl]_n$–Q systems influences the selectivity of the oxidative carbonylation of alcohols [112, 170, 171].

Example 4.15 [171]

► It was established that an increase in the quinone concentration [Q] in the $PdCl_2$–LiCl (0.7 M)–MeOH system at 30 °C in the presence of Et_3N, MeONa, or benzoate buffer ([PhCOOH] = 0.5 M, [PhCOONa] = 0.5 M, pH 4.0–4.5) leads to a decrease in the rate of dimethyl carbonate (DMC) formation and an increase in the rate of dimethyl oxalate (DMO) formation. At $[Q]/[Pd]_\Sigma > 20$, DMO becomes virtually the only product in this process (with the selectivity reaching 96–98 %). Thus, the oxidant interacting with intermediates changes the direction of their decomposition. The sequence of intermediate conversions in the presence of quinone can be described by the following scheme:

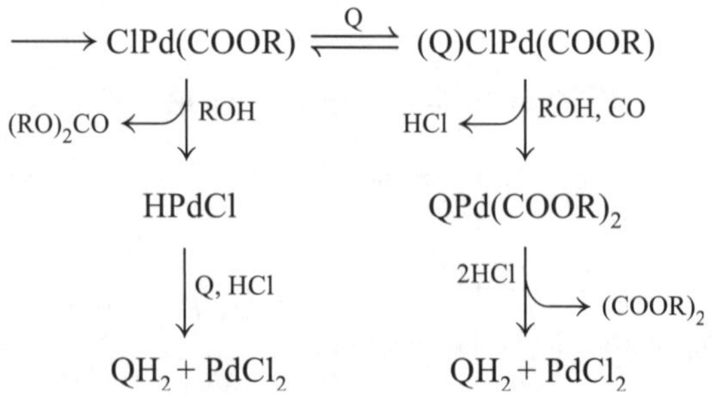

(4.213)

The steps of this scheme were modeled based on the results of studying the kinetics of the reaction of methoxycarbonylmercuric acetate and $PdCl_2$ in the absence and presence of quinone. For the equimolar amounts of reactants, the reaction without Q yields methyl chloroformate as

$$PdCl_2 + AcOHgCOOMe \longrightarrow ClCOOMe + Pd_{solid} + AcOHgCl. \tag{4.214}$$

with a 87 % yield. In the presence of a three-fold excess of AcOHgCOOMe, this reaction yields dimethyl carbonate with a yield of 81 %. The introduction of PPh_3 in several minutes after the mixing of reactants allowed the $ClPd(COOCH_3)(PPh_3)_2$ complex to be isolated with a good yield. This product was crystallized and its structure was studied by X-ray diffraction [173]. The introduction of Q into the solution leads, similarly to the case of the oxidative carbonylation of methanol by carbon monoxide, to the predominant yield of DMO and a drop in the rate of DMC formation.

The rates of DMO and DMC formation as functions of [Q] in the $PdCl_2$–Q–AcOHgCOOMe–LiCl system in benzoate buffer (Figure 4.9) are described by the following equations:

$$r_{DMO} = \frac{k_1[Q]}{1 + K_1[Q]}; \tag{4.215}$$

$$r_{DMC} = \frac{k_2[Q]}{1 + K_1}, \tag{4.216}$$

which are consistent with scheme (4.213). Thus, quinone in this system plays the role of a ligand that changes the direction of conversions of the organopalladium intermediate ClPdCOOR and, being an oxidant, participates in the step of demetallation of the probable bis-alkoxycarbonyl $Pd(COOR)_2$ complex, thus favoring the removal of thermodynamic constraints in the step of Pd^0_{aq} formation. This behavior is analogous to the participation of Q in the reactions of demetallation of η^3-allyl Pd(II) complexes (see Section 4.1). On heating in the absence of the oxidant, $(PPh_3)_2Pd(COOR)_2$ complex decomposes into DMO and $Pd(PPh_3)_2$ [172]. ■

Example 4.16

► The mechanism of the RONO action as an oxidant in the reactions of formation of oxalate [151], polyoxalate [182], and an oxalate–succinate mixture in the presence of an olefin [183] or as the component

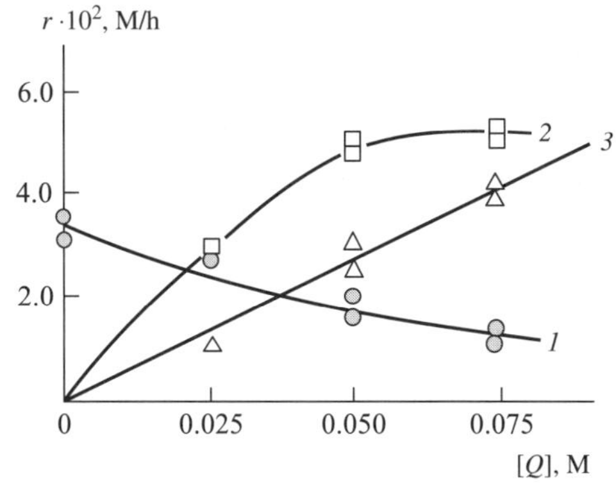

Figure 4.9 Product formation rate *r* versus *para*-benzoquinone concentration [BQ] in $4{\cdot}10^{-2}$ M $PdCl_2$–0.125 M AcOHgCOOMe solution: (1) r_{DMC}; (2) r_{DMO}; (3) r_{DMO}/r_{DMC} (reproduced with permission from editorial board of *Kinet. Katal.* journal © 1984 [171])

of a catalytic system (when the NO oxidation to RONO by oxygen is conjugated with the oxidative carbonylation) includes the steps of the oxidative addition of RONO to the palladium complex

$$[Pd] + RONO \longrightarrow [Pd]\langle^{NO}_{OR} \qquad (4.217)$$

with the formation of a nitrosyl-containing palladium alcoholate and the incorporation of CO at the [Pd]–OR bond. In the reactions of polycondensation of CO and ONOR′ONO (R=R′ONO) [182, 184], the $Pd(PhCN)_2Cl_2$ complex was used as the initial catalyst and the $L_2Pd(0)$ complex was treated as featuring the active form of palladium.

The oxidative carbonylation of ethylene at 80 °C with the formation of dibutyl succinate and dibutyl oxalate was studied [183] in butyl nitrite with the same benzonitrile complex in the absence and presence of PPh_3 (2 mole per mole Pd). The absence of the activity of $PdCl_2(PPh_3)_2$ and $ClPd(COOR)(PPh_3)_2$ complexes, close results obtained for the systems with $PdCl_2(PhCN)_2$ and $PdCl_2(PhCN)_2 + 2PPh_3$, and the composition and structure of the Pd(IV) complex (complex **I**) isolated in the latter case allowed the active form of [Pd] to be identified as initial complex **II:**

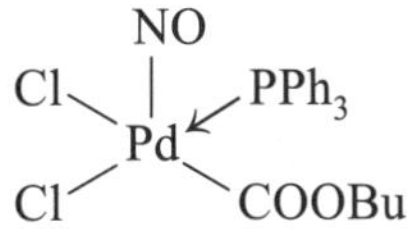

I

(ν_{NO} = 1690 см$^{-1}$)

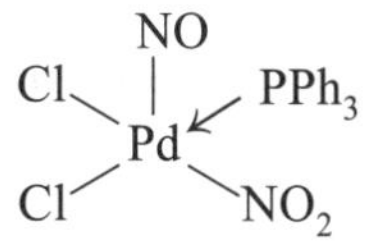

II

(ν_{NO} = 1550 см$^{-1}$)

Complex I reacts with ethylene at room temperature and atmospheric pressure in the absence of CO as follows:

$$\mathbf{I} + C_2H_4 \xrightarrow{BuONO} \underset{(25\,mol\,\%)}{CH_2{=}CHCOOBu} + \underset{(25\,mol\,\%)}{CH_3CH_2COOBu} + \underset{(50\,mol\,\%)}{(BuO)_2CO} + \mathbf{II}. \qquad (4.218)$$

The formation of acrylate and propionate is evidence of the disproportionation of an intermediate with the $\sim PdCH_2CH_2COOBu$ fragment, while the formation of dibutyl carbonate is indicative of repeated oxidative addition (or some other form of nitrite RO group participation) via the reaction

$$\sim\text{Pd–COOBu} + \text{RONO} \longrightarrow \sim\text{Pd(NO)} + (\text{BuO})_2\text{CO}. \qquad (4.219)$$

As the ethylene pressure was increased to 3.0 MPa, the reaction of complex **I** with ethylene yielded only acrylate (75 %) and propionate (25 %). Therefore, an increase in the ethylene concentration hindered reaction (4.219). The formation of oxalate under these conditions was indicative of a probable existence of the following intermediates:

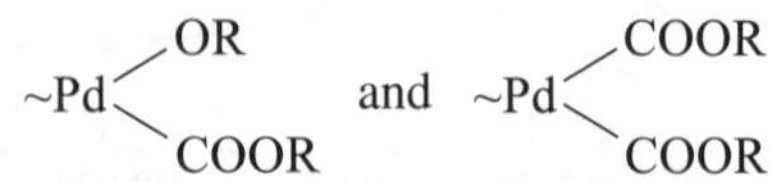

The data presented above are still insufficient, but the known schemes of oxalate formation in other systems nevertheless allow a possible sequence of steps with the participation of Pd(II) and Pd(IV) complexes to be proposed, the more so that the Pd(IV) nitrosyl complexes (complex **III**) containing alkyl and aryl groups were also obtained with other sets of ligands [185]:

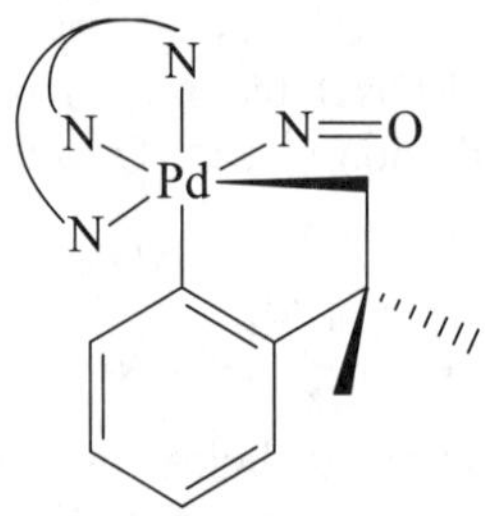

III

($\nu_{NO} = 1650\ \text{cm}^{-1}$)

where N~N~N moiety belongs to a hydro-tris(pirazolyl)borate anion. The probable mechanism of oxalate synthesis (without details of ligand environment) can be described by the following scheme:

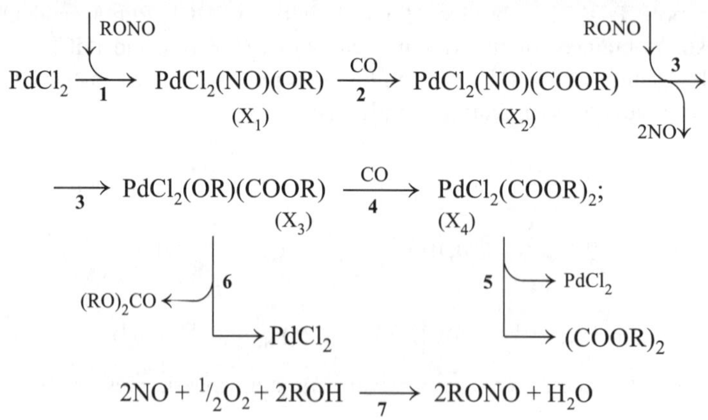

(4.220)

(for NO oxidation *in situ*).

In a catalytic variant with respect to RONO, the kinetic functions of components can be represented by five balanced overall equations as follows:

- $PdCl_2$ catalyzes the oxidative carbonylation of alkyl nitrite to dialkyl carbonate and dialkyl oxalate:

$$2RONO + CO \xrightarrow{PdCl_2} (RO)_2CO + 2NO,$$

$$2RONO + 2CO \xrightarrow{PdCl_2} (ROCO)_2 + 2NO;$$

- RONO catalyzes the oxidation of $PdCl_2$ to X_3 (steps **1–3, 7**):

$$PdCl_2 + CO + {}^1\!/_2\,O_2 + 2ROH \xrightarrow{RONO} PdCl_2(OR)(COOR) + H_2O;$$

- $PdCl_2$ and RONO together catalyze the oxidative carbonylation of alcohol according to overall equations

$$CO + 2ROH + {}^1\!/_2\,O_2 \longrightarrow (RO)_2CO + H_2O,$$

$$2CO + 2ROH + {}^1\!/_2\,O_2 \longrightarrow (ROCO)_2 + H_2O.$$

The mechanism of the NO oxidation in alcohols is not completely clear. The oxidation of both free NO and palladium nitrosyl complexes is possible [185]. Indeed, the formation of PdNOCl nitrosyl complexes during the reduction of nitrite complexes by olefins [47, 186] and the oxidation of these complexes [47] were observed. In addition, gaseous products of reactions with RONO were reported [183] to contain CO_2, N_2O, N_2, and even O_2, which is indicative of a complicated character of processes in the system under consideration that is not adequately described by scheme (4.220). ■

4.3.1.2 Arenes

The oxidative carbonylation of arenes was carried out using single- and multicomponent systems containing Pd(II) or Rh(III) [154a]. In particular, Henry in 1967 obtained a mixture of toluic acids from toluene and CO in the $PdCl_2$–$Hg(OAc)_2$–$CuCl_2$–98 % H_2SO_4 system at 120 °C and 14 MPa, while the reaction in a PdX_2–HgX_2(TlX_3)–$NaNO_3$–CF_3COOH system yielded a mixture of these acids at 20–50 °C and atmospheric pressure for 4 h [187]. Various systems containing $Pd(OAc)_2$ and a strong oxidant (^tBuOOH, $K_2S_2O_8$) in CF_3COOH were studied [188], in which the oxidative carbonylation process can be described by two semireactions as follows:

$$ArH + CO + 2RCOOH \longrightarrow ArCOOH + (RCO)_2O + 2H^+ + 2e^-;$$

$$K_2S_2O_8 + 2H^+ + 2e^- \longrightarrow 2KHSO_4.$$

More detailed investigations were performed for the $Pd(OAc)_2$–$Cu(OAc)_2$–CF_3COOH–$(CF_3CO)_2O$ system in the presence of O_2 [188, 189]. At $P_{CO} = 0.2$ MPa, $P_{O_2} = 1.0$ MPa, and $T = 140$ °C, the catalyst turnover number was TON = 362 at a 19.3 % yield of toluic acids (with respect to toluene). This system showed evidence for the formation of a 1:1 Pd(II)·Cu(II) complex [188], in which Pd(II) was suggested to exhibit increased electrophilic properties. Recently, a $Pd_2Cu(OAc)_6$ complex was isolated and its structure was determined [190, 191]. The properties of Rh(III) were studied in the $RhCl_3$–$CuCl_2$–R_3N–CH_3OH–CH_2Cl_2 systems, where salicylic acid phenyl ester was formed from phenol at a selectivity of up to 100 % [154a]. In the RhX_3–CuX_2–CF_3COOH (X = Cl, Br) and $[Rh(CO)_2Cl]_2$–$Cu(OCOCF_3)_2$–CF_3COOH–$(CF_3CO)_2O$ systems in the presence of oxygen [154a, 192], *para*-toluic acid was obtained at the selectivity on a level of 90 % at $T = 138$ °C, $P_{CO} = 0.8$ MPa, $P_{O_2} = 0.3$ MPa with TON = 679 and an 18 % yield. Nonaqueous systems containing Rh(I) exhibited the evolution of CO_2 parallel to the formation of ArCOOH and at a close rate, which was interpreted [192] to be the result of the conjugation of the oxidative carbonylation of ArH and the oxidation of CO:

$$\mathrm{ArH + 2CO + O_2 \longrightarrow ArCOOH + CO_2.} \tag{4.221}$$

The mechanism of the oxidative carbonylation of arenes in multicomponent systems has not been much studied to date. There is no doubt that the reaction of electrophilic substitution in the aromatic nucleus with the participation of Pd(II) or Rh(III) complexes is possible. The role of a second component in these reactions can be different, including both the formation of mixed ($Pd^{II}{\sim}Cu^{II}$) complexes increasing the electrophilic character of Pd(II) and the catalysis of the substitution reaction (HgX_2, TlX_3),

$$\mathrm{ArH + HgX_2 \longrightarrow ArHgX + HX;}$$

$$\mathrm{ArHgX + PdX_2 \longrightarrow ArPdX + HgX_2,}$$

shown to be possible [9, 25]. At the subsequent steps of the oxidative carbonylation process, the demetallation of acyl derivatives $ArCOMX_n$ can also proceed with the participation of CuX_2 as the oxidant. It was suggested [188] that the demetallation step involves an RCOOH acid molecule as the nucleophile with the formation of mixed aromatic acid anhydrides, for example:

$$\mathrm{ArCOMX}_n \xrightarrow{\mathrm{RCOOH}} \mathrm{ArCOOCOR + HMX}_n\,(\mathrm{MX}_{n-1} + \mathrm{HX});$$

$$\mathrm{ArCOOCOR + RCOOH \longrightarrow ArCOOH + (RCO)_2O.}$$

4.3.1.3 Alkanes

The original investigation [193] of the oxidative carbonylation of both cyclic and acyclic alkanes in the $Pd(OCOCH_3)_2–CF_3COOH–K_2S_2O_8$ system showed the possibility of alkanecarboxylic acid synthesis under relatively mild conditions ($T = 80\,°C$, $P_{CO} = 3.0\,MPa$). Methane, ethane, and propane are oxidatively carbonylated into acids in the $Pd(OAc)_2–Cu(OAc)_2–K_2S_2O_8–CF_3COOH$ and $Pd(OAc)_2–Cu(OAc)_2–CF_3COOH–O_2$ [188, 194]. In the latter system, the oxidative carbonylation of methane takes place at $P_{CH_4} = 2.0\,MPa$, $P_{CO} = 1.5\,MPa$, $P_{O_2} = 1.5\,MPa$, and $T = 80\,°C$.

Promising catalytic systems were also found among those involving Rh(III) salts as the main components [195–197]. For example, the catalyst turnover number in the $RhCl_3–HI–HCl–CH_3COOH–H_2O–O_2$ system at 95 °C and $P_{O_2} + P_{CO} = 8\,MPa$ reached TON = 10 for 24 h [197]. It was shown that HI and HCl are necessary components in this system and that CH_3OH and CH_3I are not intermediates of the process. The catalytic system capable of operating under most mild conditions contains an ytterbium salt, manganese acetate, and NaOCl as the oxidant [198]. For the reaction

$$\mathrm{CH_4 + CO + NaOCl \longrightarrow CH_3COOH + NaCl} \tag{4.222}$$

in the system with $Yb(OCOCH_3)_3/Mn(OCOCH_3)_2 = 10$, the catalyst turnover number at $P_{CH_4} = 2.0\,MPa$ and $P_{CO} = 0.5\,MPa$ at 40 °C for 2 h was TON = 5.5.

Based on the kinetic isotope effect ($k_H/k_D = 3 \div 3.2$) for the reactions of cyclohexane in Pd(II)- and Pd(II)–Cu(II)-containing systems [188] and the formation of 1:1 $Pd(OCOCF_3)_2–Cu(OCOCF_3)_2$ complexes in trifluoroacetic acid [194], it was suggested [188] that these reactions involve the electrophilic replacement of H^+ in alkane by Pd(II) occurring in a Pd(II) · Cu(II) complex. The subsequent steps of CO insertion and redox decomposition under the action of CH_3COOH or CH_3COO^- are similar to the analogous steps in the oxidative carbonylation of arenes.

All catalytic systems under consideration are characterized by the use of strong oxidants such as $K_2S_2O_8$ and O_2 for Pd(II), $HI{-}O_2$ for Rh(III), and NaOCl for Yb(III) [together with Mn(II)]. Taking into account a low redox potential of the semireaction

$$CH_4 + CO + H_2O_{liq} \longrightarrow CH_3COOH_{liq} + 2H^+ + 2e^- (E^\circ = -0.17\,V), \tag{4.223}$$

the necessary use of strong oxidants with high E° values in the oxidative carbonylation of alkanes suggests that overcoming the barrier at the most difficult step of electrophilic substitution in methane requires electrophiles that are stronger than Pd(II) or Rh(III). These can be Pd(IV), palladium and rhodium oxo complexes, and oxygenyl rhodium complexes participating in the elimination of hydrogen ion. In the case of Yb(III)–Mn(II)–NaOH solutions, it was suggested that $[Yb{=}O]^{2+}$ oxo complexes participate in the elimination of H from CH_4 with the formation of $CH_3\cdot$ radicals [198]. The possible mechanism of action of the PFCS components in this system with $Yb{=}O^{2+}$ active species can be described by the following scheme:

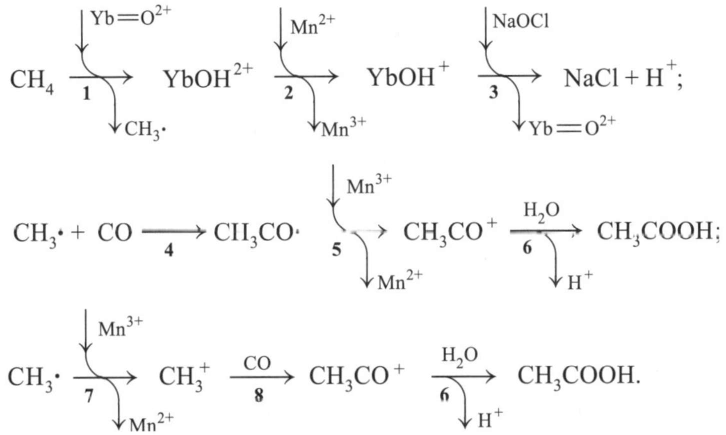

(4.224)

The kinetic functions of components in this catalytic system are as follows:

- $Yb{=}O^{2+}$ species catalyze the oxidation of CH_4 and Mn^{2+}:

$$CH_4 + NaOCl + Mn^{2+} \longrightarrow CH_3\bullet + Mn^{3+} + NaCl + OH^-;$$

- Mn(II) catalyzes the oxidation of $CH_3CO\bullet$ radical by Yb(III) (steps **2** and **5**),

$$YbOH^{2+} + CH_3CO\bullet \xrightarrow{Mn^{2+}} Yb^{2+} + CH_3CO^+$$

or the oxidation of $CH_3\bullet$ radical (steps **2** and **7**):

$$YbOH^{2+} + CH_3\bullet \xrightarrow{Mn^{2+}} Yb^{2+} + CH_3{}^+;$$

- $Yb{=}O^{2+}$ and Mn(II) together catalyze the reaction according to overall equation (4.222).

Investigations of reactions in the rhodium-iodide system in the presence of O_2 can provide ground for useful hypotheses about the participation of oxygenyl Rh(III) complexes and strong oxidants of the IO_x^- type [199]. Interesting ideas concerning the possible roles of components of the Rh–Cu(II)–Cl^- system in the reactions of alkanes with O_2 and CO were recently proposed in [200].

4.3.2 Oxidative carbonylation of alkenes, dienes, and alkynes

In the Pd(II)–Cu(II) two-component catalytic system $Pd(OAc)_2-CuCl_2-LiCl-Ac_2O-AcOH$, the target product, β-acetoxycarboxylic acid, is formed according to the overall reaction

$$C_2H_4 + CO + {}^1\!/_2\,O_2 + AcOH \longrightarrow AcOCH_2CH_2COOH \qquad (4.225)$$

which proceeds at 140–150 °C and 6.0 MPa with a selectivity of ~85% and a rate of ~2 M/h [201]. This reaction was modified by the Union Oil Company so as to obtain acrylic acid as

$$C_2H_4 + CO + {}^1\!/_2\,O_2 \longrightarrow CH_2{=}CHCOOH, \qquad (4.226)$$

and the selectivity of this process amounted to ~85% at $P_{CO} = 7$ MPa and 135 °C [157]. In alcohols, this process yields β-alkoxy esters and succinates [155]. Reactions of 1,3-dienes in $PdCl_2-CuCl_2$ solutions yield Δ^3-dihydromuconate. The kinetic function of $CuCl_2$ in all these reactions (by analogy with the Wacker process) is the same and consists of the catalysis of the Pd(0) or HPdX oxidation to Pd(II).

The $PdCl_2-RONO-O_2$ system is also active in the synthesis of succinates from olefins, CO, and ROH [157, 183] and the synthesis of polysuccinates from olefins, CO, and ONOR′ONO [184]. The role of RONO in processes of this type was considered above (see Example 4.16).

The known reactions of the oxidative carbonylation of alkynes [154a] include interesting PFCSs that have been used for the synthesis of derivatives of β-chloroacrylic acid, maleic anhydride, and β-substituted esters of propiolic acid. Let us consider some of these processes in more detail.

The synthesis of *trans*-β-chloroacrylic acid and its esters from acetylene and CO via the reaction

$$C_2H_2 + CO + HCl + {}^1\!/_2\,O_2 \longrightarrow ClCH{=}CHCOOH \qquad (4.227)$$

offers a good example of the directional construction of a PFCS from separate non-catalytic reactions (blocks), whereby the kinetic functions of components revealed by the analysis of a kinetic model are quite adequate to the kinetic functions used in "assembling" the PFCS.

Example 4.17 [202–204]

► It is known that *trans*-β-chlorovinylmercury chloride (CVM) is readily formed from C_2H_2 and $HgCl_2$ in aqueous and alcohol hydrochloric solutions [205] and appears as was shown above (see, e.g., Example 3.15) as an intermediate in the synthesis of vinyl chloride and *trans*-dichloroethylene. The acidolysis of CVM was studied [206], and it was established that the interaction of CVM with $PdCl_2(PhCN)_2$ in benzene [207] and polar solvents (methanol, acetone, acetonitrile) [208] is accompanied by the formation of *trans*-ClPdCH=CHCl [207] and its decomposition products $Cl(CH{=}CH)_nCl$ ($n = 1$–6) [208]. The yield of 1, 4-dichloro-1,3-butadiene on $Pd(II)_\Sigma$ in all experiments amounted to 60–75 %. The oxidative addition of various substituted vinylmercury derivatives in reactions with $PdCl_2$ was studied in [209, 210]. The formation of $C_4H_4Cl_2$ dimers was also observed in reactions of CVM with Au(III), Rh(III), and Pt(IV) complexes, albeit at lower yields [208]. In the presence of an oxidant (Ox = $FeCl_3$, Q), the process becomes catalytic with respect to $PdCl_2$:

$$n\,ClHgCH{=}CHCl + Ox \xrightarrow{PdCl_2} Cl(CH{=}CH)_nCl + n\,HgCl_2 + Red, \qquad (4.228)$$

and the catalyst turnover number in the $C_4H_4Cl_2$ synthesis reaches 1960. The fact of formation of the products of the coupling reaction with $n > 2$ in the absence of acetylene is probably related to the demetallation of palladium vinyl chloride derivative and the appearance of an acetylene π complex, followed by the insertion of acetylene into the Pd–C bond, e.g., as

$$\mathrm{ClPdCH{=}CHCl} \xrightarrow{\mathrm{ClHgCH{=}CHCl}} \mathrm{Pd}\begin{matrix}\diagup \mathrm{CH{=}CHCl} \\ \diagdown \mathrm{CH{=}CHCl}\end{matrix} \xrightarrow[-\mathrm{Cl}^-]{} \tag{4.229}$$

$$\longrightarrow \overset{\displaystyle \mathrm{HC{\equiv}CH}}{\downarrow} \atop {}^{+}\mathrm{Pd{-}CH{=}CHCl} \longrightarrow \mathrm{PdCH{=}CHCH{=}CHCl}.$$

In the presence of CO, reaction (4.228) is suppressed, and the main products are *trans*-β-chloroacrylic acid esters in alcohols,

$$\mathrm{ClHgCH{=}CHCl + CO + ROH + 2FeCl_3 \longrightarrow ClCH{=}CHCOOR + HgCl_2 + 2FeCl_2 + HCl}, \tag{4.230}$$

and *trans*-β-chloroacrylic acid in water [211].

The rate of *trans*-β-chloroacrylic acid esthers (E) formation in CVM-saturated solutions (solid phase) at $P_{\mathrm{CO}} = 2$–2.2 bar, $[\mathrm{FeCl_3}] = 1$ M, and $T = 25\,°\mathrm{C}$ is described by the first-order equation with respect to $[\mathrm{PdCl_2}]_\Sigma$ [211]:

$$r_{\mathrm{E}} = k_1'[\mathrm{PdCl_2}]_\Sigma. \tag{4.231}$$

The synthesis of *trans*-β-chloroacrylic acid (CAA) in aqueous solutions is accompanied by the side reactions of CO_2 and C_2H_2 formation (from CVM) and obeys the equation

$$r_{\mathrm{CAA}} = \frac{k_2'[\mathrm{PdCl_2}]_\Sigma}{1 + k_3'[\mathrm{PdCl_2}]_\Sigma}. \tag{4.232}$$

Based on the obtained results, a catalytic synthesis of *trans*-β-chloroacrylic acid from acetylene and CO was performed in a three-component catalytic system according to the following reactions [202–204]:

$$\mathrm{C_2H_2 + CO + H_2O + 2FeCl_3 \xrightarrow{HgCl_2-PdCl_2} CAA + HCl + 2FeCl_2}; \tag{4.233}$$

$$\mathrm{C_2H_2 + CO + H_2O + HCl \xrightarrow[\text{electric current}]{HgCl_2-PdCl_2-FeCl_3} CAA + H_2}. \tag{4.234}$$

The kinetics of reaction (4.233) was studied in the $\mathrm{HgCl_2}$ (2 M)–$\mathrm{PdCl_2}$ [(0.5–3)$\cdot 10^{-3}$ M]–$\mathrm{FeCl_3}$ (1 M)–LiCl (4 M)–HCl (1 M) system at 70 °C in a continuous gas-flow ideally stirred reactor [203]. It was found that the rate of *trans*-β-chloroacrylic acid formation was independent of [HCl] in the 1–3 M interval and of $[\mathrm{HgCl_2}]_\Sigma$ in the 0.5–2 M interval. Investigation of the side reaction kinetics in the presence of C_2H_2 (vinyl chloride, acetaldehyde, CO_2) under conditions of *trans*-β-chloroacrylic acid synthesis gave the following kinetic model of reaction (4.233),

$$r_{\mathrm{CAA}} = \frac{[\mathrm{PdCl_2}]_\Sigma\, P_{\mathrm{CO}}^2}{k_4' + k_5'[\mathrm{PdCl_2}]_\Sigma + k_6' P_{\mathrm{CO}}}, \tag{4.235}$$

and allowed the following mechanistic scheme to be proposed:

$$
\begin{aligned}
&C_2H_2 + HgCl_2 \underset{}{\overset{K_1}{\rightleftharpoons}} \underset{(X_1)}{ClHgCH{=}CHCl}; \\
&X_1 + HCl \xrightarrow{k_2} CH_2{=}CHCl + HgCl_2; \\
&X_1 + PdCl_2 \overset{K_3}{\rightleftharpoons} \underset{(X_2)}{ClPdCH{=}CHCl} + HgCl_2; \\
&X_2 + CO \underset{k_{-4}}{\overset{k_4}{\rightleftharpoons}} \underset{(X_3)}{(CO)ClPdCH{=}CHCl}; \\
&X_3 + CO \xrightarrow{k_5} \underset{(X_4)}{(CO)ClPdCOCH{=}CHCl}; \\
&X_3 + PdCl_2 \xrightarrow[H_2O]{k_6} C_2H_2 + CO_2 + 2HCl + Pd_2Cl_2; \\
&X_4 + H_2O \xrightarrow{k_7} ClCH{=}CHCOOH + HCl + Pd(0); \\
&\underset{(\text{or } Pd_2Cl_2)}{Pd(0)} + Ox + 2HCl \longrightarrow PdCl_2 + Red.
\end{aligned}
\tag{4.236}
$$

In the interval of concentrations studied, the steps of X_1 and X_2 formation in scheme (4.236) of the *trans*-β-chloroacrylic acid synthesis were assumed to be quasi-equilibrium, since r_{VC} is independent of $[PdCl_2]_\Sigma$, and r_{CAA} is independent of $[HgCl_2]_\Sigma$. Scheme (4.236) takes into account the appearance of a new route for the CO_2 formation in the presence of acetylene (such that the order with respect to $[PdCl_2]_\Sigma$ in the CO_2 synthesis becomes ~1.5).

In deriving an equation for r_{CAA} according to scheme (4.236), it was necessary to take into account the contributions of $PdCl_3^-$, $PdCl_4^{2-}$, and σ complexes X_2 (formed from $PdCl_3^-$ and $PdCl_4^{2-}$) to the material balance with respect to Pd(II). This yields the following relation:

$$F_{Pd} = \frac{[PdCl_2]_\Sigma}{[PdCl_2]} = A + BP_{C_2H_2}, \tag{4.237}$$

where term A depends on $[Cl^-]$ and the equilibrium constants of steps involving the formation of $PdCl_3^-$ (K') and $PdCl_4^{2-}$ (K'') complexes ($A = K'[Cl^-] + K''[Cl^-]^2$) and factor B depends on the $[Cl^-]$ and equilibrium constants of the steps of X_1 and X_2 formation for various complexes $PdCl_n^{(n-2)-}$. The quantities A and B remain unchanged at a constant and high value of $[Cl^-]_\Sigma$. The expression for F_{Pd} can also reflect the presence of the π complexes of Pd(II) with acetylene, but this would only change the meaning of constant factor B. Under the above assumptions, the kinetic equation is as follows:

$$r_{CAA} = \frac{k_4 k_5 K_1 K_3 P_{C_2H_2} P_{CO}^2 [PdCl_2]_\Sigma}{(A + BP_{C_2H_2})(k_{-4} + k_5 P_{CO}) + k_6 [PdCl_2]_\Sigma}. \tag{4.238}$$

The presence of the side conversions of acetylene does not allow high selectivity to be reached in the synthesis of *trans*-β-chloroacrylic acid, which amounts only to ~60 % at $r_{CAA} = 0.26$ M/h [203].

The role of $HgCl_2$ is kinetically not manifested because of the quasi-equilibrium character of the first and third steps in mechanism (4.236), but the exclusion of mercuric chloride from the system at 70 °C significantly decreases r_{CAA}. The synthesis under these conditions is accompanied by the rapid loss of catalytic activity as a result of the binding of Pd(II) in nonreactive organopalladium compounds. As the temperature is increased further (>90 °C), the process still remains steady-state and proceeds in the presence of a single component ($PdCl_2$). A large concentration of X_1 at 70 °C favors the appearance of X_2 in large amounts and a decrease in the concentration of free $PdCl_2$. The conclusion concerning the quasi-equilibrium character of the step of X_2 formation is also confirmed by the lack of influence of P_{CO} on the rate of one more process, the reaction of 1-chloro-1,3-butadiene formation with the participation of X_2 [204],

$$X_2 + C_2H_2 \xrightarrow{HCl} ClPdCH{=}CHCH{=}CHCl \xrightarrow{HCl} CH_2{=}CH{-}CH{=}CHCl + PdCl_2,$$

the mechanism of which will be considered below.

The kinetic functions of the PFCS under consideration, which belongs to the multicomponent systems of group B, within the framework of scheme (4.236) are as follows:

- $HgCl_2$ catalyzes the synthesis of vinyl chloride and catalyzes the formation of X_2 from C_2H_2 and $PdCl_2$;
- $PdCl_2$ catalyzes the oxidation of X_1 to CAA according to equation

  $$X_1 + CO + H_2O + 2FeCl_3 \longrightarrow CAA + HgCl_2 + 2FeCl_2 + HCl,$$

  catalyzes the oxidation of CO to CO_2 according to equation

  $$X_1 + CO + H_2O + 2FeCl_3 \longrightarrow CO_2 + HgCl_2 + C_2H_2 + 2FeCl_2 + 2HCl,$$

  and catalyzes the oxidation of X_3 to CO_2 according to unbalanced equation

  $$(CO)ClPdCH{=}CHCl + 2FeCl_3 + H_2O \longrightarrow CO_2 + PdCl_2 + C_2H_2 + 2FeCl_2 + 2HCl$$

  (i.e., $PdCl_2$ is the catalyst and the product simultaneously);
- $HgCl_2$ and $PdCl_2$ catalyze the overall reaction according to equations (4.233) and

$$CO + 2FeCl_3 + H_2O \longrightarrow CO_2 + 2FeCl_2 + 2HCl. \qquad (4.239)$$

Judging from the experimental data, a contribution from the route of CO oxidation to CO_2 on a single $PdCl_2$ in the presence of acetylene is not large. The rate of CO_2 formation according to scheme (4.236) increases with $P_{C_2H_2}$ and passes through a maximum at large values of $P_{C_2H_2}$. In the presence of O_2 (or in the anode chamber of a reactor-electrolyzer), $FeCl_3$ becomes the third PFCS component, the kinetic function of which is to catalyze the oxidation of reduced palladium species to $PdCl_2$ by oxygen (or electric current). ■

An important position among the oxidative carbonylation reactions of alkynes belongs to the syntheses of α,β-unsaturated diacids, their ethers, and ahydrides [150, 153, 154]. The possibility of obtaining maleic anhydride from acetylene via the reaction

$$C_2H_2 + 2CO + {}^1/_2O_2 + H_2O \longrightarrow$$

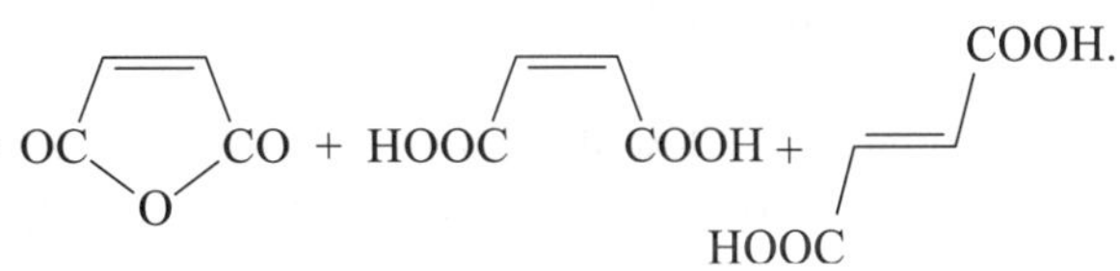

(4.240)

was originally patented by L.V. Levchenko and D.V. Sokol'skii in the PdX_2–Ox–HY [X = Cl, Br; HY = H_2SO_4, HCl; Ox = $FeCl_3$, $Fe_2(SO_4)_3$] systems (see, e.g., [150]).

Investigation of the mechanism of nonoxidative (additive) carbonylation of acetylene to succinic anhydride in the $PdBr_2$–LiBr–acetone (acetonitrile) system according to the reaction [212]

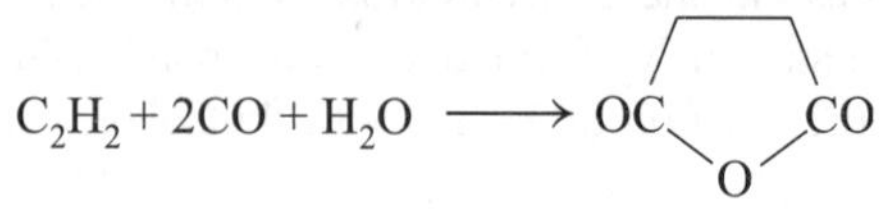

(4.241)

led to conclusions that the acting catalyst species in this system are Pd(I) carbonyl bromide complexes and an intermediate product in the succinic anhydride (SA) formation is maleic anhydride (MA) coordinated to a palladium hydride complex:

$$\longrightarrow \mathrm{H[Pd](MA)} \begin{array}{l} \rightleftharpoons \mathrm{MA + H{-}[Pd]} \\ \xrightarrow{\mathrm{H-[Pd]}} \mathrm{SA + 2Pd(I)} \\ \longrightarrow \mathrm{MA + Pd(I)} \end{array} \quad (4.242)$$

The introduction of oxygen into the initial gaseous mixture significantly increases r_{MA} and decreases r_{SA} according to this scheme [19, 213]. The same effects are observed under the action of other oxidants such as V_2O_5, $CuCl_2$, and $CuBr_2$ (at small concentrations) [213].

The maximum rate of MA synthesis is achieved in the presence of O_2, but only at a sufficiently high pressure of this oxidant. Under certain conditions, even a small increase in the O_2 concentration in the gaseous phase leads to a sharp growth in r_{MA}, which is a manifestation of the phenomenon of bistability (see Section 6.4.3) [214]. In order to reduce the P_{O_2} level down to a lower critical limit in a mixture with acetylene and CO, it is possible to introduce additional catalysts capable of accelerating the oxidation process

$$\mathrm{H{-}[Pd]} + O_2 \xrightarrow{\text{catalyst}} \mathrm{Pd(I,II)}. \quad (4.243)$$

It was found that catalysts capable of oxidizing palladium hydride complexes include the phthalocyaninate complexes of cobalt ($PcCo_{solid}$) and iron ($Pc^*Fe_{solution}$), where Pc^*H_2 is bis(dibutylsulfamoyl)phthalocyanine [19]. It was established that PcM species not only accelerate the regeneration of palladium in the active form, but also modify characteristic features of the process. The rate of MA formation depends on the concentrations of $PdBr_2$ and PcM (Figure 4.10) and increases with P_{O_2} [19], while r_{SA} in the presence of PcM significantly decreases (until almost completely vanishing) at $P_{O_2} = 0.1$ bar.

Based on these results, it was suggested [19] that oxygen activated by PcM is present in the Pd-containing reactive complex. The introduction of strong oxidants ($CuCl_2$, $CuBr_2$) into the $PdBr_2$–PcM–nitrile (acetonitrile, adiponitrile) system showed that the use of $CuBr_2$ instead of O_2 in the presence of $PcCo_{solid}$ leads to a sharp decrease in both r_{SA} and r_{MA} until the onset of $CuBr_2$ action. Accordingly, it was concluded that Pd(I) is active in this system. In the case of Pc* Fe, the introduction of strong oxidants into the system did not lead to a decrease in the rates of SA and MA formation. It was suggested that this system is catalyzed by Pd(II).

The same investigation [19] allowed preliminary hypotheses to be formulated about the functioning of multicomponent systems in the MA synthesis. Both systems under consideration feature the formation of complexes of the catalytic system components. The given two-component systems belong to the MCS of group A and to PFCSs of class III, in which additional components (PcCo, Pc* Fe) participate in the steps of formation of the key intermediates.

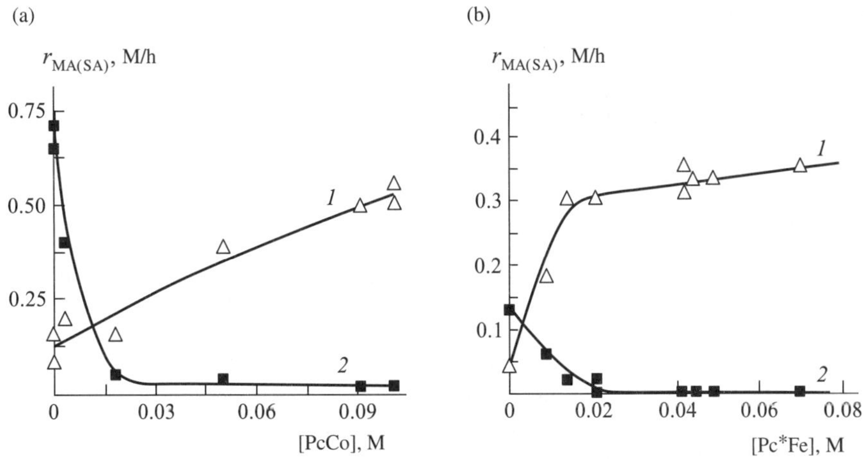

Figure 4.10 Rates of (1) maleic anhydride and (2) succinic anhydride formation versus (a) [PcCo] in $PdBr_2$–LiBr–PcCo–AN and (b) [Pc*Fe] in $PdBr_2$–LiBr–Pc*Fe–AN systems at $PO_2 \cong 0.1$ bar (reproduced with permission from editorial board and staff of *Izv. Akad. Nauk, Ser. Khim.* journal © 1999 [19])

For the $PdBr_2$–$PcCo_{solid}$ system, the mechanistic scheme can be represented by the following sequence of steps:

$$PcCo_{solid} + Pd(I) \rightleftharpoons PcCo \cdot Pd(I)_{solid} \xrightarrow[H_2O,O_2]{C_2H_2,CO} \longrightarrow (O_2)PcCo(HPdBr)_2(MA)_{solid} \longrightarrow PcCo \cdot Pd(I)_{solid} + MA + H_2O. \quad (4.244)$$

According to this, Pd(I) complexes adsorbed on the surface of $PcCo_{solid}$ favor an increase in the rate of Pd hydride oxidation and r_{MA}, and a decrease in the rate of MA interaction with H–[Pd].

In contrast, the sequence of steps in the $PdBr_2$–Pc* Fe system involves Pd(II):

$$Pc^{*}Fe + Pd(II) \rightleftharpoons Pc^{*}Fe \cdot Pd(II) \xrightarrow[H_2O,\,O_2]{C_2H_2,CO} \longrightarrow (O_2)Pc^{*}Fe(HPdBr)_2(MA) \longrightarrow Pc^{*}Fe \cdot Pd(II) + MA + H_2O. \quad (4.245)$$

Here, Pc*Fe hinders the formation of Pd(I) and fully inhibits the formation of SA.

The kinetic functions of PcM and Pd(I, II) in both systems consist in the catalysis of the overall oxidative carbonylation reaction,

$$C_2H_2 + 2CO + {}^1/_2O_2 \longrightarrow \text{OC–CH=CH–CO (cyclic, –O–)}, \quad (4.246)$$

the selectivity of which with respect to acetylene reaches 95 %. The chemical functions of PcM in this process are probably not restricted to the activation of O_2 in the reaction of hydride ligand oxidation and include an increase in the catalytic activity of Pd(I) and Pd(II) complexes [19].

In 1980, Tsuji et al. [215] established that terminal alkynes can exhibit the oxidative carbonylation of $\equiv$C–H bonds in the $PdCl_2$–$CuCl_2$ system according to the reaction

$$\begin{aligned}&RC{\equiv}CH + CO + MeOH + 2NaOAc + 2CuCl_2\\ &\longrightarrow RC{\equiv}CCOOMe + 2AcOH + 2NaCl + 2CuCl.\end{aligned} \tag{4.247}$$

The mechanism and kinetics of this reaction are well illustrated by the following example.

Example 4.18 [216–220]

► An analysis of the laws of reaction (4.247) showed [216] that the side products in this process are chloroalkynes RC≡CCl, dialkyl(diaryl)diacetylenes, and CO_2. The kinetic curves revealed an induction period. In contrast to the Wacker process, the introduction of oxygen as an oxidant for CuCl led to an increase in the induction period and a decrease in the yield of substituted propiolic acid esters. In contrast, the additives of CuCl to the initial solution decreased the induction period until complete vanishing and significantly increased the main reaction rate [216, 217]. As a result of these investigations, a new method was developed for the synthesis of alkenylcarboxylic acids in the CuCl–$PdCl_2$–$CuCl_2$ three-component catalytic system [217] and the mechanism of this process was studied according to the overall reaction [218–220]

$$RC_2H + CO + MeOH + {}^1\!/_2\,O_2 \longrightarrow RC{\equiv}CCOOMe + H_2O. \tag{4.248}$$

The rate of reaction (4.248) was studied in the absence of O_2, and the constant concentration of $[H^+]$ was maintained using an AcOH–Et_3N buffer system (pH 5.5–5.8). The rates of formation of an ester (r_E) and phenylchloroalkyne (r_{PCA}) from PhC≡CH was studied in LiCl (3 M) solution in methanol at 20 °C and $P_{CO} = 1$ bar. It was found that both r_E and r_{PCA} grow with $[CuCl]_\Sigma$; r_E increases and r_{PCA} decreases with increasing $[PdCl_2]_\Sigma$, whereas the inverse effects are produced by increasing $[CuCl_2]_\Sigma$. At the same time, the ratio of these rates

$$\frac{r_E}{r_{PCA}} = k\frac{[PdCl_2]_\Sigma}{[CuCl_2]_\Sigma^2} \tag{4.249}$$

is independent of $[CuCl]_\Sigma$. Upon the discrimination of various hypotheses [193, 194], the following mechanistic scheme was proposed:

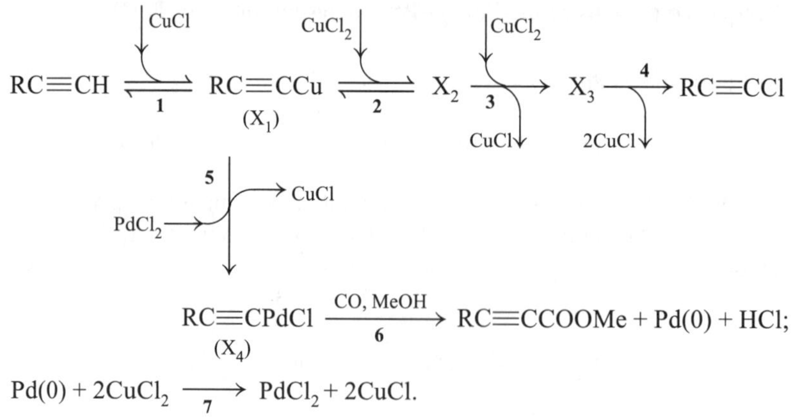

(4.250)

In the presence of O_2, CuCl under steady-state conditions is partly oxidized by oxygen as

$$2CuCl + 2HCl + {}^1\!/_2\,O_2 \longrightarrow 2CuCl_2 + H_2O. \tag{4.251}$$

According to the experimental data, the initial rates, $r_{0\,\mathrm{E}}$ and $r_{0\,\mathrm{PCA}}$, can be described by the following equations:

$$r_{0\,\mathrm{E}} = \frac{k_{\mathrm{I}}[\mathrm{RC{\equiv}CH}][\mathrm{CuCl}]_{\Sigma}\,[\mathrm{PdCl_2}]_{\Sigma}}{[\mathrm{H^+}] + k_{\mathrm{II}}[\mathrm{CuCl_2}]^2_{\Sigma} + k_{\mathrm{III}}[\mathrm{PdCl_2}]_{\Sigma}}; \tag{4.252}$$

$$r_{0\,\mathrm{PCA}} = \frac{k_{\mathrm{IV}}[\mathrm{RC{\equiv}CH}][\mathrm{CuCl}]_{\Sigma}\,[\mathrm{CuCl_2}]^2_{\Sigma}}{[\mathrm{H^+}] + k_{\mathrm{II}}[\mathrm{CuCl_2}]^2_{\Sigma} + k_{\mathrm{III}}[\mathrm{PdCl_2}]_{\Sigma}}, \tag{4.253}$$

which are written under the assumption of quasi-equilibrium step **2**, quasi-steady-state reactions with respect to $[\mathrm{X_1}]$ and $[\mathrm{X_4}]$, and the following assumptions concerning the material balance with respect to $[\mathrm{CuCl_2}]$, $[\mathrm{PdCl_2}]$, and [CuCl]:

$$[\mathrm{X_1}] \ll [\mathrm{CuCl}]_{\Sigma}; \quad [\mathrm{X_2}], [\mathrm{X_3}] \ll [\mathrm{CuCl_2}]_{\Sigma}; \quad [\mathrm{X_4}] \ll [\mathrm{PdCl_2}]_{\Sigma}.$$

The kinetic equations (4.252) and (4.253) are consistent with scheme (4.250) provided that $k_{\mathrm{I}} = k_5 K_1, k_{\mathrm{II}} = k_3 K_2/k_{-1}, k_{\mathrm{III}} = k_5/k_{-1}$, and $k_{\mathrm{IV}} = k_3 K_1 K_2$.

The kinetic functions of the three components of PFCSs under consideration can be represented by the following overall equations of the catalyzed reactions:

- CuCl catalyzes the formation of $\mathrm{X_3}$ from alkyne

  $$\mathrm{RC{\equiv}CH + 2CuCl_2 \longrightarrow X_3 + HCl},$$

  catalyzes the oxidation of alkyne to RC≡CCl with the unbalanced overall equation (autocatalytic process)

  $$\mathrm{RC{\equiv}CH + 2CuCl_2 \longrightarrow RC{\equiv}CCl + 2CuCl + HCl},$$

  and catalyzes the formation of $\mathrm{X_4}$ as

  $$\mathrm{RC{\equiv}CH + PdCl_2 \longrightarrow X_4 + HCl};$$

- $\mathrm{PdCl_2}$ catalyzes the oxidative carbonylation of $\mathrm{X_1}$ as

  $$\mathrm{X_1 + CO + MeOH + 2CuCl_2 \longrightarrow RC{\equiv}CCOOMe + 3CuCl + HCl};$$

- $\mathrm{CuCl_2}$ (in a steady-state process) catalyzes the oxidative chlorination of $\mathrm{X_1}$

  $$\mathrm{X_1 + 2HCl + {}^1\!/_2\,O_2 \longrightarrow RC{\equiv}CCl + CuCl + H_2O},$$

  and catalyzes the oxidation of Pd(0) as

  $$\mathrm{Pd(0) + 2HCl + {}^1\!/_2\,O_2 \longrightarrow PdCl_2 + H_2O};$$

- CuCl together with $\mathrm{PdCl_2}$ catalyze the reaction (unbalanced with respect to CuCl)

  $$\mathrm{RC{\equiv}CH + 2CuCl_2 + CO + MeOH \longrightarrow RC{\equiv}CCOOMe + 2CuCl + 2HCl};$$

- CuCl together with $\mathrm{CuCl_2}$ catalyze the reaction according to the overall equation of PCA synthesis:

  $$\mathrm{PhC{\equiv}CH + HCl + {}^1\!/_2\,O_2 \longrightarrow RC{\equiv}CCl + H_2O};$$

- CuCl–$\mathrm{PdCl_2}$–$\mathrm{CuCl_2}$ system catalyzes the overall reaction (4.248). ■

4.4 Additive carbonylation of alkynes, alkenes, dienes, and alcohols

The reactions of hydrocarboxylation and hydroalkoxycarbonylation of olefins, dienes, and alkynes are in some cases performed using multicomponent catalysts, but the functions of components in these systems are by no means always convincingly justified.

The catalytic synthesis of acrylic acid from acetylene is carried out under homogeneous conditions in the $NiBr_2$–CuI–THF system (BASF process) [221], but the kinetic function of CuI in this process is yet not established. In the case of olefin hydrocarboxylation in the $PdCl_2(PPh_3)$–PPh_3–dioxane–water system [222–224] and their hydroalkoxycarbonylation in the $PdCl_2(PPh_3)$–PPh_3–ROH system [225], it was shown that $CoCl_2$ additives produce a two- to threefold increase in the reaction rate and (which is more important) in the regioselectivity with respect to the products with normal structures. The regioselectivity of the process can also be increased using $Co_2(CO)_8$ additives instead of $CoCl_2$, but the mechanisms of their action are probably different [223]. While $CoCl_2$ significantly increases the rate of formation of RCH_2CH_2COOH (product I), $Co_2(CO)_8$ decreases the yield of its isomer $RCH(CH_3)COOH$ (product II). Investigation of the process kinetics in a two-component (with respect to metals) system of $PdCl_2$–$CoCl_2$ [224] and a comparison to the kinetic model in the system without $CoCl_2$ (see Section 3.2) led to a conclusion that the introduction of $CoCl_2$ into the given catalytic system does not principally modify the reaction mechanism and only changes the reactivity of intermediates and the relative reaction rates over different routes. In particular, there is an increase in the fraction of $Pd(CO)L_2$ monocarbonyl species that lead to product I. According to the IR spectroscopy data obtained *in situ* upon adding $CoCl_2$ to the $PdCl_2$–PPh_3 system, there is a significant growth in intensity of the ν_{CO} absorption band in the region of $1980\,cm^{-1}$, which corresponds to the Pd(0) monocarbonyl intermediates. Thus, in the presence of $CoCl_2$ in addition to $Pd(CO)L_2$ complexes, there appears a new species containing a monocarbonyl derivative of cobalt and palladium, which converts into the acyl derivative faster than does the $Pd(CO)L_2$ complex alone.

Analogous features were observed in the solutions of alcohols [225]. Upon adding $CoCl_2$ to a palladium-containing system, the rate of product II formation remains virtually unchanged, whereas the rate of isomer I formation in the reaction of styrene hydroalkoxycarbonylation in a dioxane–methanol solution increases by a factor of 8–9. The regioselectivity also exhibits inversion, the ratio [I]/[II] being about 0.2 without $CoCl_2$ and about 1.9 in the presence of $CoCl_2$. The systems under consideration can be classified to the first approximation as the multicomponent catalytic systems of group A.

As for the well-studied reaction of olefin hydroalkoxycarbonylation in solutions of $HCo(CO)_4$ cobalt carbonyl complexes [226], it was established that a slow step in this process is the conversion of acylcobalt carbonyl into the product of reaction

$$RCOCo(CO)_x + R'OH \longrightarrow RCOOR' + HCo(CO)_x.$$

The process can be catalyzed by pyridine additives playing the role of a second (nucleophilic) catalyst [227–230]. A simplified mechanism is presented by the following scheme (HA = HOH, ROH):

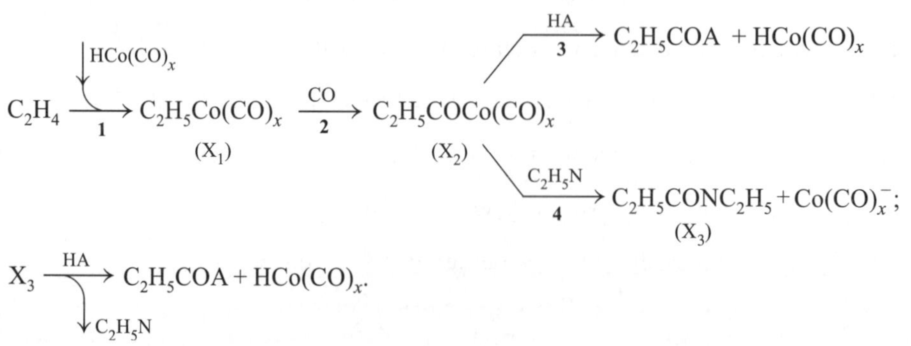

(4.254)

which reflects the main features of the process, although the real mechanism is probably more complicated, since the excess C_5H_5N leads to the formation of a free pyridine and $C_5H_5N^+$ H $Co(CO)_x^-$ salt. The kinetic function of pyridine in this system is to catalyze the conversion of X_2 into the main reaction product. The rate of reaction **4** is 15–25 times that of step **3** [230]. The kinetic function of $HCo(CO)_x$ is to catalyze the conversion of ethylene into the reaction product according to the overall equation of carbonylation:

$$C_2H_4 + CO + AH \longrightarrow C_2H_5COA. \qquad (4.255)$$

It should be noted that the use of a two-component catalytic system [C_5H_5N–$HCo(CO)_4$] makes possible the hydroalkoxycarbonylation of 1,3-dienes (e.g., 1,3-butadiene) to an adipic acid ester [227, 231, 232].

It is interesting to note that the additives of $CuCl_2$ to $PdCl_2$ or the use of a $CuCl_2$–$PdCl_2$–O_2 system in nonoxidative reactions involving CO significantly influence both regio- and stereoselectivity. For example, the $PdCl_2$–$CuCl_2$–HCl–MeOH–O_2 system features room-temperature regioselective (~100%) reactions of nonoxidative hydromethoxycarbonylation and hydrocarboxylation of olefins [233, 234]:

$$RCH{=}CH_2 + CO + MeOH \longrightarrow R\text{–}CH(Me)CO_2Me. \qquad (4.256)$$

The $PdCl_2$–$CuCl_2$–L–MeOH system (where L is a chiral phosphine-containing ligand) features an analogous reaction of 2-vinyl-6-methoxynaphthalene with a 85 % optical yield [235], while the presence of acids (TsOH) makes possible the regioselective carbonylation of 1-(6′-methoxy–2′-naphthyl) ethanol [235]:

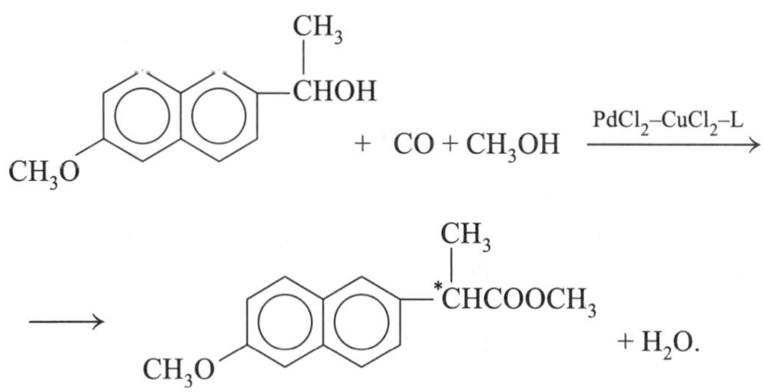

(4.257)

The introduction of $CuCl_2$ ([Cu]/[Pd] = 4) leads to an increase in the regioselectivity (yield of a branched product grows from 85 to 97 %) and stereoselectivity (*ee* increases from 42 to 77 %). The mechanism of $CuCl_2$ action in these systems has not been studied so far. In the case of naphthylethanol, the third component of catalytic system (4.257) is an acid (TsOH).

Classical examples of PFCSs with two components ML_n and HX are offered by methanol carbonylation in the Rh(I)–HI system (Monsanto process) [236–240] and Ir(I)–HI system (BP Chemicals, Cativa process) [241–248]. The kinetic functions of components are illustrated by the following example.

Example 4.19 [236–248]

► The mechanism of processes in the $Rh(CO)_2I_2^-$–HI – MeOH–H_2O system (without the steps of formation, inactivation, and side reactions) is presented in Figure 4.11a (bipartite graph BG5) and the

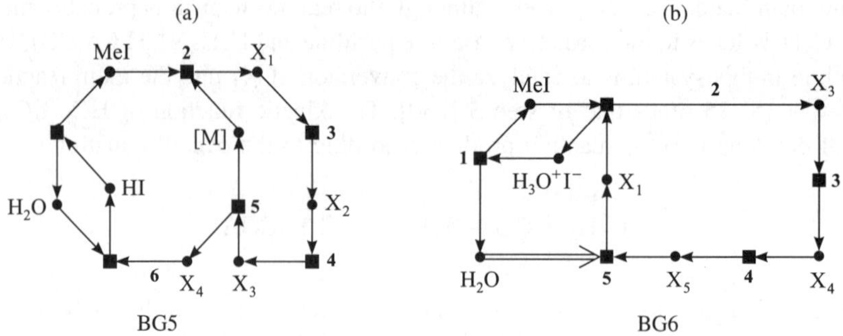

Figure 4.11 Bipartite graphs of subnetworks of intermediates for reaction mechanisms (a) (4.258) and (b) (4.260) simplified by assuming all steps to be irreversible

following scheme:

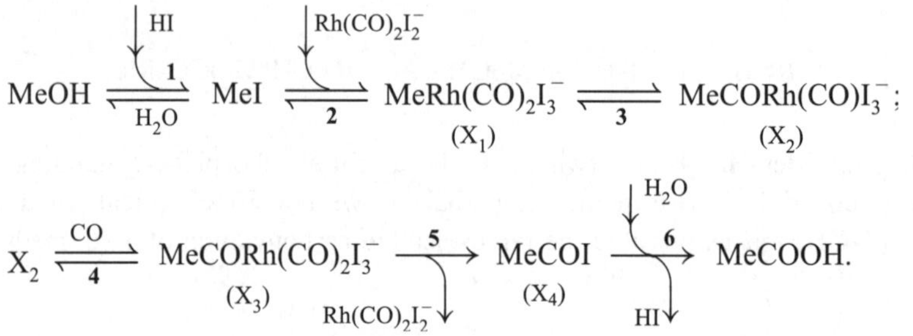

(4.258)

For $[H_2O] > 8$ wt % in methanol, the limiting stage in this process is step **2** as the irreversible step. In this case, the reaction rate is

$$r = k'[Rh]_\Sigma\,[MeI],$$

provided that $[Rh]_\Sigma \cong [Rh(CO)_2I_2^-]$. For $[H_2O] < 8$ wt %, the limiting stage is step **5** (reductive elimination of MeCOI), so that the reaction rate begins to depend on P_{CO}. It also follows from scheme (4.258) that the components are featuring the following main kinetic functions:

- Rh(I) catalyzes the carbonylation of MeI to MeCOI (steps **2–5**)

$$MeI + CO \longrightarrow MeCOI;$$

- HI catalyzes the overall process (steps **1–6**)

$$MeOH + CO \longrightarrow MeCOOH;$$

Another, non-evident overall equation follows from bipartite graph BG5 (Figure 4.11) that includes a third circuit (steps **1** and **6**) involving HI and H_2O:

$$MeOH + MeCOI \overset{HI}{\longrightarrow} MeCOOH + MeI.$$

The mechanism of the reaction of methanol carbonylation in IrI_3 solutions (Cativa method) is essentially analogous to that in scheme (4.258), but there are several important distinctions. According to Foerster [242] the reaction proceeds via two routes with the participation of both neutral and charged acyl complexes [$AcIr(CO)_2I_2$ and $AcIr(CO)_2I_3^-$], and the main iridium-containing intermediate is $MeIr(CO)_2I_3^- \cdot H_3O^+$ (X_2). The formation of this compound as a result of the oxidative addition of MeI to $Ir(CO)_2I_2^-$ is about 150 times as fast as in the case of Rh(I), and the step of X_2 formation ceases to be limiting at very low H_2O concentrations (~0.5 wt %). The formation of an acyl intermediate is preceded by the dissociation of X_2 with the evolution of I^- (which is just what inhibits the carbonylation process), while the step of the acyl complex formation is a slow stage. The investigations of Maitlis, Haynes et al. [241, 248], refined the Foerster scheme [2.242] and showed that the X_2 complex very slowly converts into an acyl derivative and becomes a balance compound (pendant vertex of a KG or BG) rather than an intermediate.

Let us consider the modified Foerster scheme based on the above results with allowance for the following equilibrium:

$$\mathrm{MeOH + AcOH \rightleftharpoons AcOMe + H_2O.} \tag{4.259}$$

At low concentrations of water and methanol, AcOMe (rather than MeOH) was considered to be the initial reactant according to the scheme

$$\mathrm{AcOMe + H_3O^+ \cdot I^- \underset{1}{\rightleftharpoons} MeI + AcOH + H_2O;}$$

$$\mathrm{MeI \xrightarrow[2]{+\,Ir(CO)_2I_2^- \cdot H_3O^+\,(X_1);\ -\,H_3O^+ \cdot I^-} \underset{(X_3)}{MeIr(CO)_2I_2} \underset{3}{\overset{CO}{\rightleftharpoons}} \underset{(X_4)}{MeIr(CO)_3I_2} \xrightarrow[4]{} \underset{(X_5)}{MeCO(CO)_2I_2} \xrightarrow[5]{2H_2O;\ -X_1} AcOH,} \tag{4.260}$$

the five steps of which yield the following overall equation:

$$\mathrm{AcOMe + CO + H_2O \longrightarrow 2AcOH.} \tag{4.261}$$

The retardation of process involving I^- is explained by the transition of complex X_3 into inactive X_2 as

$$\mathrm{X_3 + H_3O^+ \cdot I^- \rightleftharpoons \underset{(X_2)}{MeIr(CO)_2I_3^- \cdot H_3O^+}.}$$

In addition to the step of X_2 complex formation, scheme (4.260) should be supplemented by some other equilibria influencing the complexity of iridium:

$$\mathrm{\underset{(X_5)}{AcIr(CO)_2I_2} \overset{CO}{\rightleftharpoons} \underset{(X_6)}{AcIr(CO)_3I_2};}$$

$$\mathrm{X_5 + H_3O^+ \cdot I^- \rightleftharpoons \underset{(X_7)}{AcIr(CO)_2I_3^- \cdot H_3O^+};}$$

$$\mathrm{\underset{(X_1)}{Ir(CO)_2I_2^- \cdot H_3O^+} \rightleftharpoons \underset{(X_8)}{Ir(CO)_2I} + H_3O^+ \cdot I^-;}$$

$$\mathrm{X_8 + CO \longrightarrow \underset{(X_9)}{Ir(CO)_3I}.}$$

The content of X_2 complexes in the system reaches 90–95 % of $[\mathrm{Ir}]_\Sigma$.

Using bipartite graph BG6 (Figure 4.11b), it is possible to assign the following kinetic functions for the PFCS Ir(I)–HI:

- $H_3O^+ \cdot I^-$ species catalyze reaction

$$AcOMe + X_1 \longrightarrow AcOH + X_3 + H_2O$$

  (steps **1** and **2**);
- $Ir(CO)_2I_2^-$ (X_1) complex catalyzes reaction

$$MeI + CO + 2H_2O \longrightarrow AcOH + H_3O^+ \cdot I^-$$

  (steps **2–5**);
- X_1 and $H_3O^+{\cdot}I^-$ complexes catalyze reaction (4.261) (steps **1–5**).

Note that BG6 has an unbalanced circuit of steps **1**, **2**, and **5**, since H_2O is both reactant and intermediate in the process according to scheme (4.260).

The influence of metal salts and complexes on the activity of the Ir(I)–HI system [248] in the reaction of methanol or methyl acetate carbonylation has been studied. At 22 bar CO and 190 °C, the rate of methyl acetate carbonylation in the presence of promoters increases from 8 to 21.6 M/h (for $Ru(CO)_4I_2$)), 18.6 M/h (for $Os(CO)_4I_2$), and 19.4 M/h ($InI_3/Ru(CO)_4I_2$). It is suggested that these promoters favor the conversion of X_2 into X_3 and X_4, thus significantly increasing the concentration of X_4. ■

4.5 Substitution and addition reactions in alkyne chemistry

Various multicomponent PFCSs have been studied based on the reactions of hydrogen ion substitution and the addition of molecules with acid hydrogen atom (HX) in the catalytic chemistry of alkynes [91]. In particular, the condensation of acetylene

$$RC{\equiv}CH + R'X + R_3N \longrightarrow RC{\equiv}CR' + Et_3NH^+X^-$$

very effectively proceeds in the Sonagashira $Pd(PPh_3)_4$–CuI system [249]. In the case of phenylacetylene and iodobenzene, the reaction was carried out for 3 h at 25 °C with a tolane yield on a level of 90 %. The reaction kinetics was not studied, but it is believed [250] that the role of CuI reduces to the catalysis of Pd(II) ethynyl complex formation by the analogy with the kinetic function of Cu(I) in the reactions of oxidative carbonylation of alkynes at the ≡C–H bond (see, e.g., [18]), for example:

$$R'I + PdL_4 \longrightarrow R'PdIL_2 + 2L;$$

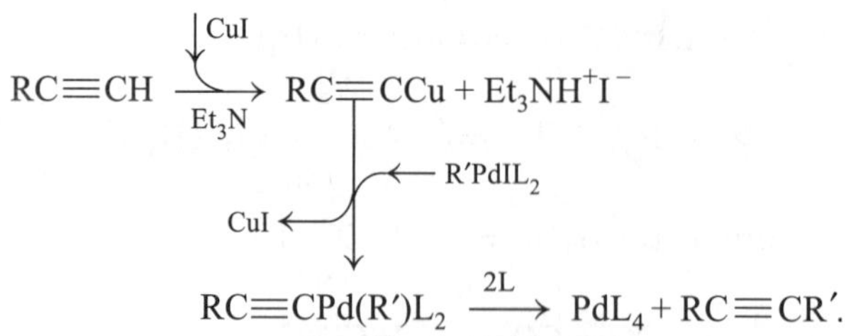

(4.262)

Multicomponent PFCSs containing metal-complex (CuCl), nucleophilic (RSH), and acid (HCl) catalysts catalyze the reactions of alkyne hydration with a change in the regioselectivity toward the anti-Markovnikov addition of H_2O molecules [91]. In 1957–1959, it was originally reported that the reaction of acetylene hydration in the $CuCl–NH_4Cl–HCl–H_2O$ system is significantly accelerated when small amounts of H_2S, Na_2S, or Cu_2S are added to the system [251, 252]. An analogous effect is produced by the additives of NaSCN and CuSCN [251]. It was suggested that copper sulfide Cu_2S (or S^{2-} ion) in CuCl solutions acts as a ligand (modifier) that changes the properties of Cu(I) species and influences the kinetics of acetylene hydration [253]. Since the solubility product of Cu_2S in water is very small ($\sim 10^{-50}$), the dissolution of Cu_2S in the $CuCl–NH_4Cl$ system at [CuCl] $= 12\tilde{m}$ and [NH_4Cl] $= 12\tilde{m}$ was explained by the formation of complexes with c $Cu_mCl_n^{(n-m)-}$ anions:

$$qCu_2S_{solid} + Cu_mCl_n^{(n-m)^-} \rightleftharpoons Cu_{m+2q}Cl_nS_q^{(n-m)^-}. \tag{4.263}$$

The total concentration of Cu_2S in the solution studied was on the order of $1 \cdot 10^{-3}$ mole per 1000 g H_2O. Under the conditions where the activities a_{Cu^+} and a_{Cl^-} are constant, the rate of hydration on the sulfide complexes was expressed as [253]

$$r_S = r_\Sigma - r, \tag{4.264}$$

where r_Σ is the total rate of hydration on copper chloride and sulfide complexes and r is the rate of hydration in the absence of Cu_2S. The dependence of r_S on $P_{C_2H_2}$ and $h_0 a_{H_2O}$ (where h_0 is the acidity of solution) can be described by the following equation:

$$r_S = k_I P_{C_2H_2} + k_{II} P_{C_2H_2} h_0 a_{H_2O}. \tag{4.265}$$

The reaction orders with respect to a_{Cu^+} and a_{Cl^-} amount to $m = 2.6$ and $n = 3.6$, respectively, and are higher than the corresponding values in the absence of Cu_2S ($m = 1.44, n = 2.58$) [254].

Subsequent investigations showed that, in addition to increasing the catalyst activity, Cu_2S additives change the regioselectivity of the reaction of methylacetylene hydration [255]. For example, the rate of this reaction in CuCl ($12\tilde{m}$) NH_4Cl ($12\tilde{m}$)–HCl ($2.5\tilde{m}$) solution upon adding Cu_2S_{solid} increases from $1 \cdot 10^{-3}$ to $8.3 \cdot 10^{-3}$ M/h (for acetone, AC) and from $0.16 \cdot 10^{-3}$ to $3.4 \cdot 10^{-3}$ M/h (for propionaldehyde, PA). An increase in the CuCl concentration in the system leads to a decrease in r_{AC}/r_{PA} from 10.5 to 1.6. By varying the concentrations of CuCl and HCl and by adding other inorganic and organosulfur compounds, conditions were found under which r_{PA}/r_{AC} reached 1.5 (CuS, mercaptobenzothiazole) [255] and it was shown that the activating and regioselectivity-modifying effects were produced by compounds of the RSH or RSSR (readily converting into RSH) types, whereas R_2S and $R_2C{=}S$ groups exhibited no activity in the hydration reaction. These observations, together with the absence of a "modification" effect in the reaction of HCl addition in the presence of any sulfur-containing compounds, led to the assumption that a different mechanism may account for the influence of the RSH type compounds [255]. This alternative mechanism can be a Cu(I)-catalyzed addition of RSH (as a soft bulky nucleophile) to alkyne with the formation of intermediate alkenyl thioethers, the subsequent hydrolysis of which yields acetone and propionaldehyde. As is known, mercaptans can sometimes add to alkynes against the Markovnikov rule in the presence of RONa [256]. It was shown that vinyl and propenyl sulfides introduced into copper chloride solutions were hydrolyzed at rates that exceeded the rates of alkyne hydration [257].

The assumption about the formation of intermediate thioethers

$$RC{\equiv}CH + R'SH \xrightarrow{Cu(I)} RCH{=}CHSR + RC(SR){=}CH_2$$

and their subsequent hydrolysis was confirmed by the results of investigation of the influence of steric factors (Charton's steric parameters) on the regioselectivity of methylacetylene hydration. In the series of RSH [R = $(C_2H_5)_2NC_2H_4$, C_3H_7, C_5H_{11}, *iso*-C_3H_7, *tert*-C_4H_9], the r_{PA}/r_{AC} ratio obeys the following relation [258]:

$$\lg \frac{r_{PA}}{r_{AC}} = A + \varphi\vartheta_R, \tag{4.266}$$

where $\varphi = 1.93 \pm 0.14$, $A = -0.90 \pm 0.06$, and ϑ_R is the Charton constant. At the same time, r_{PA}/r_{AC} is independent of P_{MA}, [HCl], $[CuCl]_\Sigma$, and $[RSH]_\Sigma$.

Additional information about the mechanism of alkyne hydration was obtained from the results of investigations of the reaction kinetics and the composition of active copper chloride complexes [259, 260].

Example 4.20 [259, 260]

► The kinetics of methylacetylene hydration was studied in a gradient-free flow reactor at 80 °C for $[NH_4Cl] + [HCl] = 12\tilde{m}$. The concentration of H_3O^+ and the a_{Cu^+} value were determined by the potentiometry with glass and copper electrodes. The conditions were selected such that $a_{Cu^+} \propto [Cu^+]$ and $a_{H_3O^+} \propto [H_3O^+]$. In order to maintain a_{Cu^+} = const during the variation of $[RSH]_\Sigma$, $[HCl]_\Sigma$, and P_{MA}, an additional amount of CuCl (up to 0.2 % of $[CuCl]_\Sigma$) was introduced into the catalytic system. The first series of experiments was performed in a solution of $10\tilde{m}$ CuCl. Special preliminary tests in anhydrous media (CuCl–DMF–HCl) showed that methylacetylene in the presence of RSH (3-mercaptopentane) converted into a mixture of propenyl sulfides, which were isolated using the extraction with n-decane and identified by the signals of vinyl protons in the ^{1}H NMR spectra. Since neither the variation of [CuCl] within 3 – $10\tilde{m}$ nor a change in P_{MA}, [RSH], and $[H_3O^+]$ influenced r_{PA}/r_{AC}, it was concluded that the same Cu(I) complexes are involved in both reactions. Then, an analysis of the dependences of r_i on $[RSH]_\Sigma$ and $[H_3O^+]$ (Figure 4.12) at constant a_{Cu^+} and a_{Cl^-} allowed one of the two hypotheses concerning the mechanism of RSH action to be discriminated.

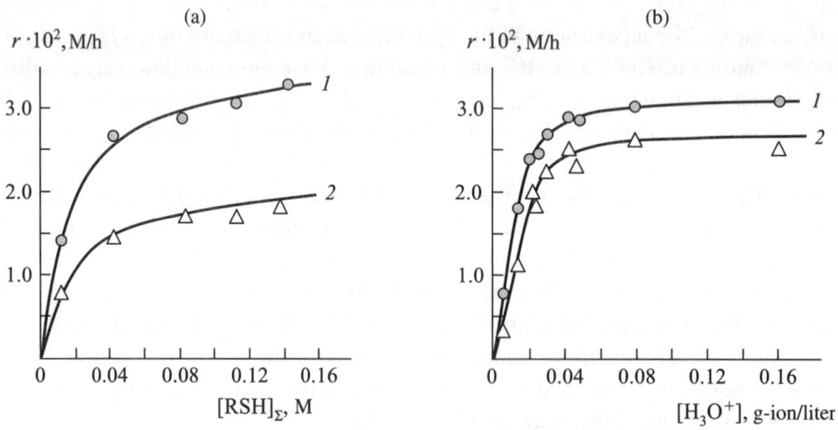

Figure 4.12 Rates of (1) propionaldehyde and (2) acetone formation versus (a) total concentration of 3-mercaptopentane at $[H_3O^+]$ = 1.9 g-ion/liter and (b) steady-state concentration of H_3O^+ at $[RSH]_\Sigma$ = 0.10 M in CuCl–DMF–HCl system with $a_{Cu+} = 33.0 \cdot 10^{-8}$ ([CuCl] = 10 $\tilde{m}$) (reproduced with permission from editorial board of *Kinet. Katal.* journal © 1986 [259])

Hypothesis I. The reaction proceeds via the *trans*-addition of mercaptan group to a Cu(I) π complex by analogy with the addition of H_2O, Cl^- [254], or RS^- in alkaline media:

$$\begin{array}{c} CH_3C(Cu){=}CHSR + H_3O^+ \\ (\sigma_1) \\ \uparrow k_2' \;\; RSH, H_2O \\ Cu(I) + CH_3C{\equiv}CH \underset{k_{-1}}{\overset{k_1}{\rightleftharpoons}} Cu(CH_3C{\equiv}CH)\ (\pi\ \text{complex}) \\ \downarrow k_2'' \;\; RSH, H_2O \\ CH_3C(SR){=}CHCu + H_3O^+ \\ (\sigma_2) \end{array}$$

Then, the protolysis of σ_1 and σ_2 followed by the acid-catalyzed hydrolysis lead to the formation of PA (from σ_1) and AC (from σ_2).

Hypothesis II. The reaction proceeds via the insertion of MA into Cu–SR bond in a Cu(I) π complex:

$$Cu(I) + RSH \underset{K_1}{\overset{H_2O}{\rightleftharpoons}} CuSR + H_3O^+;$$

$$\begin{array}{c} \sigma_1 \\ \uparrow k_3' \\ CuSR + CH_3C{\equiv}CH \underset{k_{-2}}{\overset{k_2}{\rightleftharpoons}} Cu(SR)(CH_3C{\equiv}CH). \\ (\pi\ \text{complex}) \\ \downarrow k_3'' \\ \sigma_2 \end{array}$$

The initial increase in r_i with $[RSH]_\Sigma$ and $[H_3O^+]$ followed by the plateaus (Figure 2.14) could only be related to the participation of RSH in the second step of the process according to hypothesis I, with a quasi-steady-state character of the process with respect to the Cu(I) π complexes (a_{Cu^+} and a_{Cl^-} are constant and, hence, $[RSH] \propto [RSH]_\Sigma$). If RSH were involved in the second step, but with dissociation and CuSR formation as

$$Cu(I) + CH_3C{\equiv}CH \rightleftharpoons Cu(CH_3C{\equiv}CH) \overset{RSH, H_2O}{\rightleftharpoons} Cu(SR)(CH_3C{\equiv}CH) + H_3O^+,$$

then the observed dependence on H_3O^+ also would not take place. Thus, the possible formation of CuSR can only enter into the material balance with respect to RSH:

$$[RSH]_\Sigma = [RSH] + [CuSR];$$
$$[RSH] = \frac{[RSH]_\Sigma\,[H_3O^+]}{[H_3O^+] + K_1[Cu^I]}. \tag{4.267}$$

Then, according to hypothesis I and experimental data, the reaction rates can be expressed as

$$r_i = \frac{k_1 k_2^i P_{MA}[Cu^I][RSH]}{k_{-1} + (k_2' + k_2'')[RSH]}, \tag{4.268}$$

or as

$$r_i = \frac{k_1 k_2^i P_{MA}[Cu^I][RSH]_\Sigma [H_3O^+]}{k_{-1}[H_3O^+] + K_1[Cu^I] + k_2[RSH]_\Sigma [H_3O^+]}, \tag{4.269}$$

where $k_2 = k_2' + k_2''$ and K_1 is the equilibrium constant of the first reaction of RSH dissociation in hypothesis II.

According to Eq. (4.269), the ratio $r_{PA}/r_{AC} = k_2'/k_2''$ must be independent of all parameters of the process except for the nature of RSH (see also Eq. (4.266)). Similar to other processes with multicomponent catalysts capable of forming complexes containing both components (see, e.g., Section 4.2.3), the material balance with respect to one component (RSH) also involves the second component (Cu^I). Under conditions of zero-order reactions with respect to $[RSH]_\Sigma$ and $[H_3O^+]$, the reaction rates with respect to PA and AC can be expressed as follows:

$$r_{PA} = k'_{PA} P_{MA} = k_{eff}^{PA}[Cu_m Cl_n^{(n-m)-}]\, P_{MA};$$
$$r_{AC} = k'_{AC} P_{MA} = k_{eff}^{AC}[Cu_m Cl_n^{(n-m)-}]\, P_{MA}.$$

Based on the established reaction orders with respect to a_{Cu^+} and a_{Cl^-} (i.e., the average composition indices m and n of the complexes involved in the catalysis with 3-mercaptopentane (A) and N,N-diethylaminoethylmercaptan (B)) [260], it was concluded that:

(i) In the case of mercaptan A, equiprobable contributions to the process are due to complexes $Cu_2Cl_3^-, Cu_3Cl_4^-$ $(n - m = 1)$ and $Cu_3Cl_5^{2-}, Cu_4Cl_6^{2-}$ $(n - m = 2)$, whereas in the case of mercaptan B, the active complexes include $Cu_4Cl_5^-, Cu_5Cl_6^-$ $(n - m = 1)$ and $Cu_5Cl_7^{2-}, Cu_6Cl_8^{2-}$ $(n - m = 2)$;

(ii) In the system with RSH, the reaction orders with respect to a_{Cu^+} are much greater than those in the $CuCl{-}NH_4Cl{-}H_2O$ system [254], i.e., the main contribution to the observed hydration rates are due to more polynuclear Cu(I) complexes.

Figure 4.13 shows a subgraph of intermediates for the following nonlinear mechanism of hydration (simplified scheme with irreversible steps, alkyne = acetylene, and Cu(I) = CuCl):

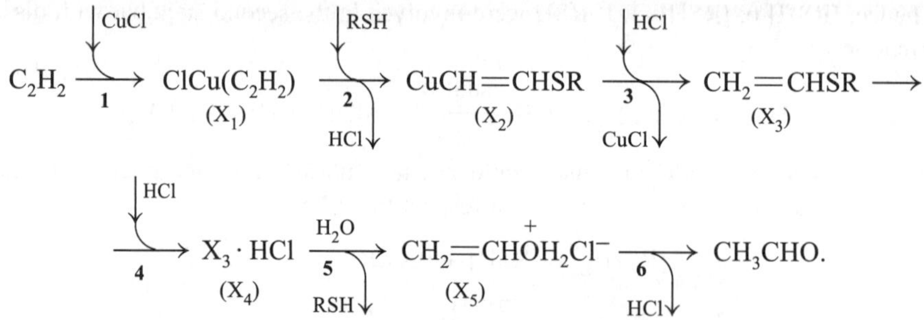

(4.270)

This subgraph contains seven circuits that characterize the kinetic functions of components, some of which are as follows:

- CuCl catalyzes the formation of X_3 (steps **1–3**, participants CuCl, X_1, and X_2)

$$C_2H_2 + RSH \longrightarrow X_3;$$

- HCl catalyzes the hydrolysis of X_3 (steps **4–6**)

$$X_3 + H_2O \longrightarrow CH_3CHO + RSH;$$

- RSH catalyzes the conversion of X_1 into X_5 (steps **2**–**5**)

$$X_1 + HCl + H_2O \longrightarrow X_5 + CuCl.$$

The same steps also form an unbalanced circuit passing through the vertex of HCl (reactant and intermediate). Another unbalanced circuit includes steps **3**–**6**. RSH also catalyzes the process consisting of steps **2, 4**, and **5**, but the overall equation of this circuit is deprived of clear chemical sense.

The results of investigations of the activity of RSH species as nucleophilic catalysts also suggest an alternative point of view on the schemes with Cu_2S. Assuming that sulfide ion in a complex containing one Cu_2S group is involved in the equilibrium,

$$Cu_mCl_n(Cu_2S) + H_3O^+ \rightleftharpoons Cu_{m+2}Cl_n(\overset{+}{S}H) + H_2O, \qquad (4.271)$$

and that both disubstituted Cu(I) sulfide ($\sim Cu_2S$) and monosubstituted Cu(I) hydrosulfide ($\sim$Cu–SH) are capable of acting as nucleophilic catalysts, the scheme of conversions in this multicomponent system can be presented in the following form:

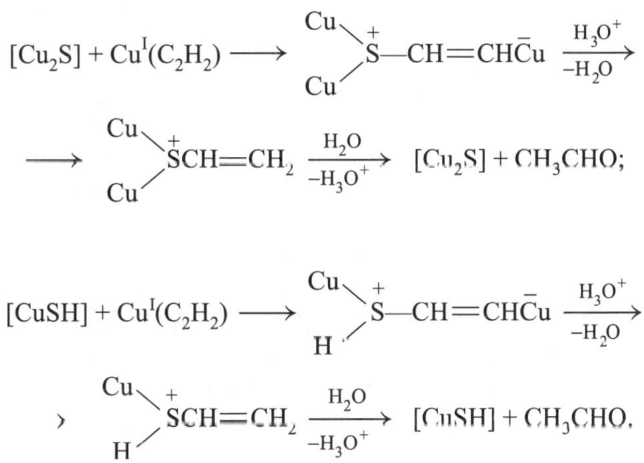

(4.272)

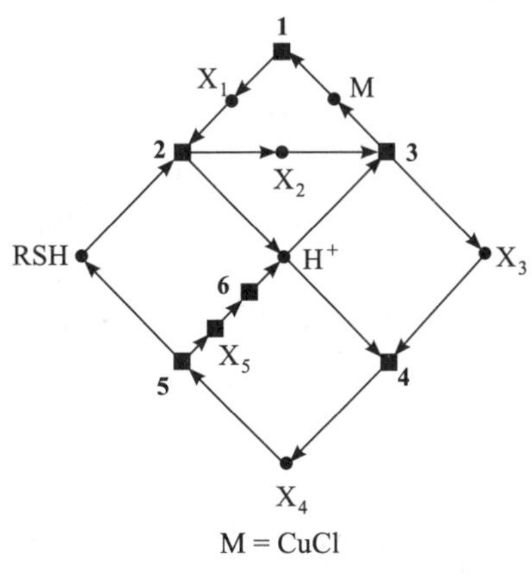

Figure 4.13 Bipartite graph of subnetwork of intermediates for reaction mechanism (4.270)

The interactions of Cu_2S and CuSH with $Cu(C_2H_2)$ can also take place within the same polynuclear Cu(I) complex. At a constant activity of Cu^+ and Cl^- and, hence, almost constant (on the background of variable $[H_3O^+]$) concentrations of dissolved $(Cu_2S)Cu_mCl_n^{(n-m)-}$, the concentration of CuS^+H would grow in proportion to $[H_3O^+]$. This circumstance probably explains the meaning of two-route equation (4.265), the second term of which can be related to an active form of [CuSH]. ■

It is known that the activity of the $CuCl–ZnCl_2–H_2O$ two-component catalyst system in the process of acetylene hydration at 172 °C is comparable with the activity of a mercury catalyst in the $HgSO_4–Fe_2(SO_4)_3–H_2SO_4–H_2O$ system, but the former system is characterized by a much higher stability [91]. A high concentration of $ZnCl_2$ (up to 75 wt %) in the melt of $ZnCl_2 \cdot 2H_2O$ crystal hydrate makes possible the formation of mixed complexes of the (H_3O^+) $[Cu(Zn_mCl_{2m+1})(OH)^-]$ type (with a $ZnCl_2/CuCl$ ratio of ~7.5 in solution). Taking into account the high acidity of these solutions and a low activity of water ($a_{H_2O} = 0.11$) at 172 °C, it can be suggested that a slow step in this system is the addition of a bridging OH group to the π complex of acetylene with Cu(I):

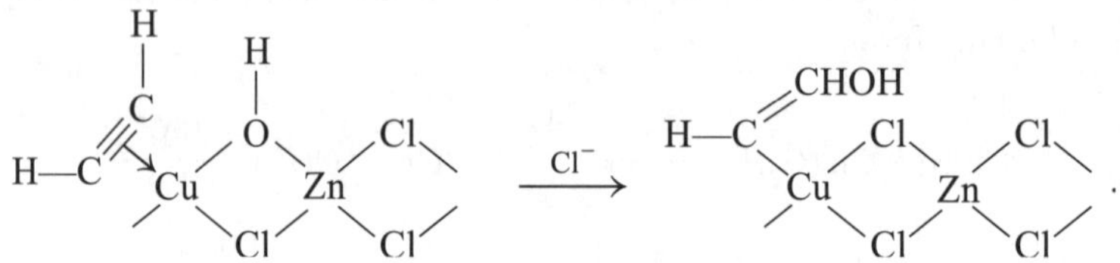

(4.273)

Accordingly, the chemical functions of Zn(II) and Cu(I) consist on the activation of water and acetylene molecules, respectively. The system with this mechanism should be classified as a multicomponent catalyst system belonging to group A.

Among the reactions of HX molecule addition to alkynes, the process of hydrochlorination has been studied most thoroughly [91]. The hydrochlorination of acetylene and alkynes was catalyzed by the complexes of Hg(II) [142, 261, 262], Cu(I) [254, 263], Ru(III) [264], Au(III) 4. [91, 265], Pt(II) [91, 266–270], Rh(III) [271, 272], and Pd(II) [273] and by the two-component $CuCl–HgCl_2$ system [274].

Just as expected, the reactions in $PdCl_2$ solutions yielded, in addition to vinyl chloride (VC), 1-chloro-1,3-butadiene (CB) [273] because the anticipated intermediate of *trans*-ClPdCH=CHCl readily reacts with acetylene to form organopalladium σ complexes [275, 276]. As was demonstrated above (see Example 4.17), the formation of polymeric $ClPd(CH{=}CH)_nCl$ is also possible and, at $n = 2$, the acidolysis of ClPdCH=CHCH=CHCl yields CB since *cis*-, *trans*-δ-chlorobutadienylpalladium chloride converts into *trans*-CB. The stereochemistry of these conversions is presented by the following scheme [277]:

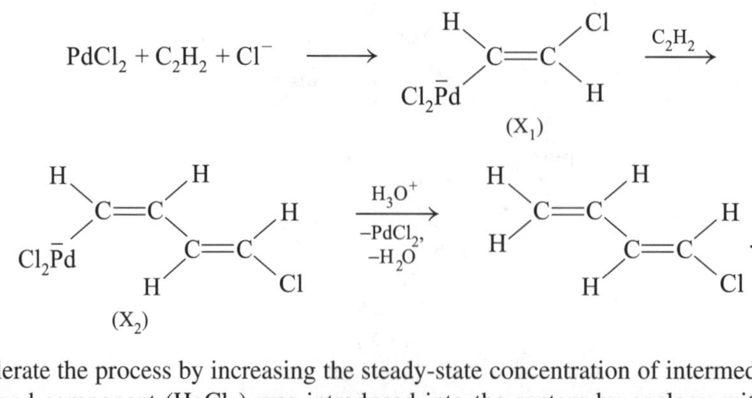

(4.274)

In order to accelerate the process by increasing the steady-state concentration of intermediate X_1 in scheme (4.274), the second component ($HgCl_2$) was introduced into the system by analogy with the synthesis of

β-chloroacrylic acid (see Example 4.17). However, the study of the process kinetics led to a conclusion [278] that both kinetic and chemical functions of $HgCl_2$ in the $HgCl_2$–$PdCl_2$ system during the reaction of CB formation are principally different from those in the synthesis of chloroacrylic acid.

Example 4.21 [204, 277, 278]

► Taking into account the possibility of a reductive decomposition of X_1 to 1,2-dichloroethylene (DCE) (see scheme (4.274)) and X_2 to 1,4-dichloro-1,3-butadiene and $Cl(CH{=}CH)_nCl$, the process of CB synthesis was studied in the presence of $FeCl_3$ in the $HgCl_2$(2 M)–$FeCl_3$ (2 M)–NH_4Cl (1.5 M)–HCl (1 M)–$PdCl_2$ system. In this system, the CB synthesis can be carried out at 60–90 °C in a steady-state regime for 7 h [277]. The selectivity with respect to CB reaches a maximum level of 40 % at $r_{CB} = 0.13$ M/h. The introduction of $PdCl_2$ into the $HgCl_2$–$FeCl_3$–NH_4Cl–HCl system under the conditions of VC synthesis, leads to a sharp drop in the VC formation rate and the appearance of CB among the products. The drop in the rate of VC formation is apparently related to a decrease in the steady-state concentration of *trans*-β-chlorovinylmercury chloride as a result of its interaction with $PdCl_2$ via the reaction

$$ClHgCH{=}CHCl + PdCl_2 \longrightarrow ClPdCH{=}CHCl + HgCl_2. \qquad (4.275)$$

An analogous pattern was observed when $CuCl_2$ was introduced into the reaction solution instead of $PdCl_2$ [262].

If the introduced amount of $PdCl_2$ is sufficiently large, the passage from one steady state to another can proceed via the so-called "jump" analogous to that observed previously for the oxidation processes [279]. The "jump" over a steady-state rate of the VC formation (Figure 4.14) indicates that the steady-state concentration of free Pd(II) chloride is significantly lower than that existing at the moment of its introduction into the reaction solution.

Immediately after the introduction of Pd(II) chloride into the solution, *trans*-DCE is detected in the reactive gases. The rate of its formation passes through a maximum, decreases, and vanishes in about 20–30 min after the introduction of $PdCl_2$ (Figure 4.14, curve 4). In a steady-state regime of CB synthesis, the formation of *trans*-DCE is not observed. The absence of *trans*-DCE in the reaction

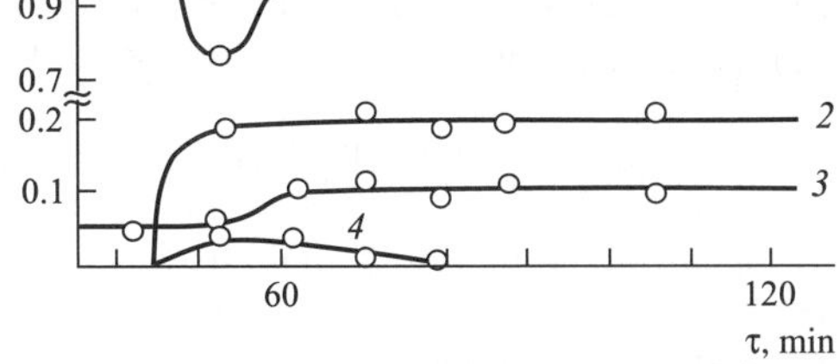

Figure 4.14 Temporal variation of product formation rates in $HgCl_2$–$PdCl_2$–$FeCl_3$–NH_4Cl–HCl system (Example 4.21): (1) vinyl chloride; (2) 1-chloro-1, 3-butadiene; (3) acetaldehyde; (4) trans-1, 2-dichloroethylene (reproduced with permission from editorial board of *Kinet. Katal.* journal © 1978 [277])

products can be treated as indicative of the fact that the rate of demetallation of the intermediate *trans*-β-chlorovinylpalladium chloride is significantly smaller than the rate of its conversion into X_2 (see scheme (4.274)). The formation of *trans*-DCE at the initial period of time can probably be observed because the concentration of *trans*-β-chlorovinylpalladium chloride for this time is increased compared to its steady-state level (the rate of trans-metallation reaction (4.275) is high), while the concentration of dissolved acetylene is below the steady-state level as a result of its interaction with divalent mercury liberated in this reaction.

After the introduction of $PdCl_2$, the concentration of trivalent iron in the solution begins to decrease, which is indicative of the fact that palladium chloride also catalyzes the oxidation of acetylene. In addition to *trans*-DCE, the oxidation products contain dichlorobutadiene and chlorinated oligomers with a general formula of $C_nH_nCl_2$ ($n = 10, 12$).

An interesting feature of this catalytic system for the synthesis of CB is the existence of a certain critical ratio $[PdCl_2]/[HgCl_2]$, above which the catalytic system fails to operate. An increase in this ratio is accompanied by the formation of a significant amount of polymeric products $ClPd(CH{=}CH)_nCl$, which produce the irreversible binding of $PdCl_2$, thus leading to a drop in the catalytic activity.

Based on the above results, the following simple scheme was proposed to explain the operation of the two-component system under consideration [277]:

$$\begin{aligned}
C_2H_2 + HgCl_2 &\rightleftharpoons \underset{(X_0)}{ClHgCH{=}CHCl};\\
X_0 + PdCl_2 &\rightleftharpoons \underset{(X_1)}{ClPdCH{=}CHCl} + HgCl_2;\\
X_1 + C_2H_2 &\longrightarrow X_2;\\
X_2 + HCl &\longrightarrow PdCl_2 + CB.
\end{aligned} \tag{4.276}$$

Detailed investigation of the kinetics of CB synthesis and the side reactions of VC and acetaldehyde formation [278] led to a change in the interpretation of the mechanism. The kinetics was studied in the $HgCl_2$ (2 M)–$FeCl_3$ (1 M)–[LiCl + HCl] (5 M)–$PdCl_2$ [$(0.5–6.7)\cdot 10^{-3}$ M] system at 70 °C. The observed kinetics admitted the following simplest description (for $P_{CH_2} = \text{const}$ or $[PdCl_2]_\Sigma = \text{const}$, $[H_3O^+] = \text{const}$):

$$r_{CB} = \frac{[H_3O^+][PdCl_2]_\Sigma}{(k_3[H_3O^+] + k_4) + (k_6 + k_5[H_3O^+])[PdCl_2]_\Sigma}; \tag{4.277}$$

$$r_{CB} = \frac{P^2_{C_2H_2}}{k_7 + k_8P_{C_2H_2} + k_9P^2_{C_2H_2} + k_{10}P^3_{C_2H_2}}. \tag{4.278}$$

The concentration dependence of $r_{CB} = f[HgCl_2]_\Sigma$ was studied at 95 °C in a range $[HgCl_2]_\Sigma = 0.1–0.5$ M in a solution with $FeCl_3$ (2 M), LiCl (5 M), HCl (3 M), and $PdCl_2$ ($2 \cdot 10^{-3}$ M), so as to ensure a steady-state character of the process and maintain the constant complexity of $HgCl_2$. The obtained relation

$$r_{CB} = k_{11} + k_{12}[HgCl_2]_\Sigma, \tag{4.279}$$

was consistent with the results of other observations and could be treated as evidence for the existence of two routes for the CB formation.

The fact that the rate of VC synthesis under these conditions (quasi-equilibrium step of X_0 formation according to scheme (4.276)) was independent of $[PdCl_2]_\Sigma$ and a large ratio of the first-order rate constants with respect to the metal salt in the acetylene hydrochlorination process for $PdCl_2$ and $HgCl_2$ ($k_{PdCl_2}/k_{HgCl_2} = 137$ [273]) led to a conclusion that the main role of $HgCl_2$ in this system is to increase the rate of acidolysis for X_2 (see scheme (4.276)) rather than to accelerate the formation of X_1. This hypothesis allowed the entire set of experimental facts to be rendered consistent with the formal kinetic relations. The assumption concerning the possibility of reactions

$$\mathrm{ClPdCH{=}CHCH{=}CHCl + HgCl_2 \rightleftharpoons ClHgCH{=}CHCH{=}CHCl + PdCl_2}; \tag{4.280}$$

$$\mathrm{ClHgCH{=}CHCH{=}CHCl + HCl \longrightarrow HgCl_2 + CB} \tag{4.281}$$

explained both the catalytic effect of $HgCl_2$ and the observed reaction order (below unity) with respect to $[PdCl_2]_\Sigma$ (4.277). According to (4.279), the CB is formed at 95 °C even in the absence of $HgCl_2$. It can be suggested that RHgCl compounds are protolyzed faster than their palladium analogs. In particular, under the conditions of a higher rate of ClPdCH=CHCl formation at 25 °C, VC is found only upon the introduction of $HgCl_2$, while at temperatures above 60 °C it is formed in the absence of $HgCl_2$. The proposed mechanistic scheme [278] takes into account the ability of $ClPd(CH{=}CH)_nCl$ and $ClHg(CH{=}CH)_2Cl$ to participate in the reductive demetallation with the formation of $Cl(CH{=}CH)_nCl$ ($P_2, n = 2$; $P_1, n = 3, 4$):

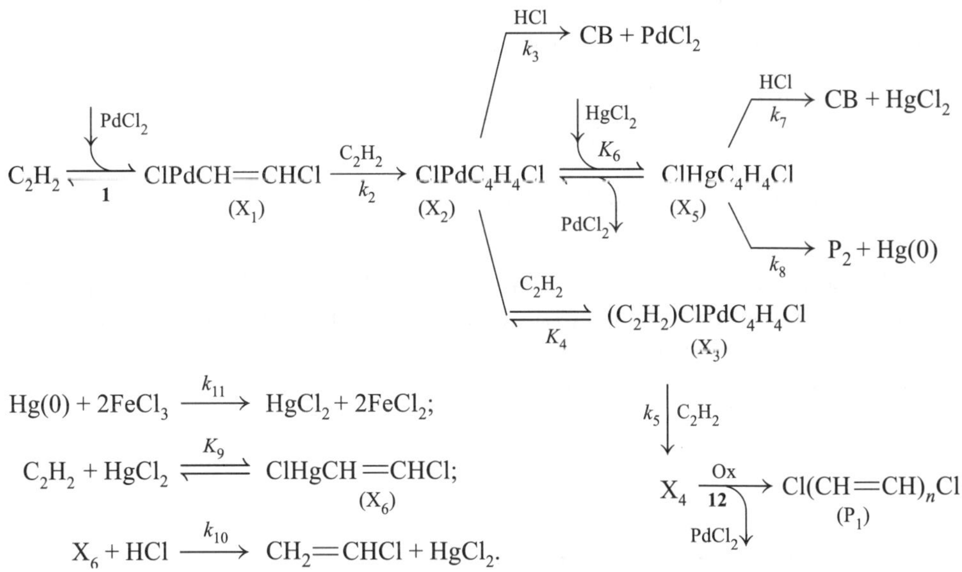

(4.282)

The rate of CB synthesis at 70 °C via the "mercuric" route according to this scheme is described by the following equation:

$$r_{CB}^{II} = \frac{k_2 k_7 K_1 K_6 [HgCl_2][PdCl_2][H_3O^+]\, P_{C_2H_2}^2}{(k_3[H_3O^+] + k_5 K_4 P_{C_2H_2}^2)[PdCl_2] + (k_8 + k_7[H_3O^+])K_6[HgCl_2]}. \tag{4.283}$$

Assuming that the complexity function F_{Pd} of palladium depends on the chloride complexes of Pd(II) and X_1, while the complexity function F_{Hg} of mercury depends on the chloride complexes of Hg(II) and X_6, these functions can be expressed as follows:

$$F_{Pd} = \frac{[PdCl_2]_\Sigma}{[PdCl_2]} = A + K_1 P_{C_2H_2}; \qquad F_{Hg} = \frac{[HgCl_2]_\Sigma}{[HgCl_2]} = B + K_9 P_{C_2H_2},$$

where

$$A = 1 + \sum_{n=3}^{4} \beta_n [\text{Cl}^-]^n; \quad K_1' = \Sigma K_m [\text{Cl}^-]^m \ (m = 1, 2);$$
$$B = 1 + \sum_{m=3}^{4} \beta_m [\text{Cl}^-]^m; \quad K_9' = \Sigma K_l [\text{Cl}^-]^l \ (l = 1, 2);$$

β_m are the equilibrium constants of formation of the corresponding $PdCl_n^{(n-2)-}$ and $HgCl_m^{(m-2)-}$ complexes; and K_m and K_l are the constants involving β_n, K_1 and β_m, K_9, respectively (quantities A, K_1', B, and K_9' are constant at $[Cl^-] = \text{const}$).

Using expressions for the complexity functions F_{Pd} and F_{Hg}, Eq. (4.283) can be represented as follows:

$$r_{CB}^{(II)} = \frac{k_2 k_7 K_1 K_6 [\text{HgCl}_2]_\Sigma [\text{PdCl}_2]_\Sigma [\text{H}_3\text{O}^+] P_{\text{C}_2\text{H}_2}^2}{D} +$$
$$D = (B + K_9' P_{\text{C}_2\text{H}_2})(k_3[\text{H}_3\text{O}^+] + k_5 K_4 P_{\text{C}_2\text{H}_2}^2)[\text{PdCl}_2]_\Sigma + \quad (4.284)$$
$$(A + K_1' P_{\text{C}_2\text{H}_2})(k_8 + k_7[\text{H}_3\text{O}^+])K_6[\text{HgCl}_2]_\Sigma.$$

The contribution due to the first (non-mercury) route of CB formation can be expressed as

$$r_{CB}^{(I)} = \frac{k_2 k_3 K_1 [\text{PdCl}_2]_\Sigma^2 [\text{H}_3\text{O}^+] P_{\text{C}_2\text{H}_2}^2 (B + K_9' P_{\text{C}_2\text{H}_2})}{D}$$
$$D = (A + K_1' P_{\text{C}_2\text{H}_2})(B + K_9' P_{\text{C}_2\text{H}_2})(k_3[\text{H}_3\text{O}^+] + k_5 K_4 P_{\text{C}_2\text{H}_2}^2)[\text{PdCl}_2]_\Sigma + \quad (4.285)$$
$$(A + K_1' P_{\text{C}_2\text{H}_2})^2 (k_8 + k_7[\text{H}_3\text{O}^+])K_6[\text{HgCl}_2]_\Sigma$$

and becomes significant only at 95 °C. A comparison of Eqs (4.279), (4.284), and (4.285) shows that, at 95 °C, a contribution due to the term with $[HgCl_2]_\Sigma$ in the denominator of kinetic equations is not very large in the interval of 0.1–0.5 M $HgCl_2$.

Figure 4.15 shows a bipartite subgraph of intermediates corresponding to scheme (4.282), which contains six linearly independent circuits that characterize the kinetic functions of components of the multicomponent catalytic system under consideration:

- $PdCl_2$ catalyzes the following reactions:

$$2\text{C}_2\text{H}_2 + \text{HCl} \longrightarrow \text{CB (step } \mathbf{1\text{–}3}),$$
$$2\text{C}_2\text{H}_2 + \text{HgCl}_2 \longrightarrow \text{X}_5 \text{ (step } \mathbf{1, 2, 6}),$$
$$n\text{C}_2\text{H}_2 + 2\text{FeCl}_3 \longrightarrow \text{Cl}_2(\text{C}_2\text{H}_2)_n + 2\text{FeCl}_2 \text{ (step } \mathbf{1, 2, 4, 5, 12});$$

- $HgCl_2$ catalyzes the following reactions:

$$\text{C}_2\text{H}_2 + \text{HCl} \longrightarrow \text{CH}_2\text{=CHCl (step } \mathbf{9, 10}),$$
$$\text{X}_2 + \text{HCl} \longrightarrow \text{CB} + \text{PdCl}_2 \text{ (step } \mathbf{6, 7}),$$
$$\text{X}_2 + 2\text{FeCl}_3 \longrightarrow \text{P}_2 + 2\text{FeCl}_3 + \text{PdCl}_2 \text{ (step } \mathbf{6, 8, 11}).$$

- A complex circuit involving steps **1, 2, 6**, and **7** corresponds to the reaction

$$2\text{C}_2\text{H}_2 + \text{HCl} \longrightarrow \text{CB},$$

  which is catalyzed by both components ($PdCl_2$ and $HgCl_2$).

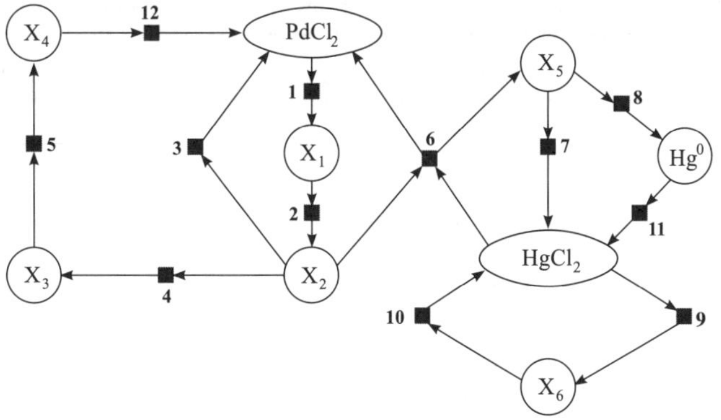

Figure 4.15 Bipartite graph of subnetworks of intermediates for 1-chlorobuta-1, 3-diene synthesis according to scheme (4.282) simplified by assuming all steps to be irreversible

The chemical function of $HgCl_2$ in this process, in contrast to the synthesis of β-chloroacrylic acid, reduces to the formation of intermediate X_5 that is more reactive than X_3 in the step of electrophilic substitution of a metal. It can be suggested that a strongly pronounced conjugation in ClPdCH=CHCH=CHCl leads to the appearance of a multiple Pd=C bond and an effective positive charge on Pd and the adjacent carbon atom of the carbene-like organic ligand

$$Cl\overset{+}{Pd}=CH-CH=CH-\overset{-}{C}HCl.$$

4.6 General problems in PFCS theory and practice

Upon an analysis of the catalytic reactions of various types and the kinetic and chemical functions of the components of multicomponent catalytic systems (MCSs) in these reactions (Sections 4.1–4.5), it is expedient to compare the properties of catalytic system employed in both the redox and nonoxidative processes, consider their functions, and outline the possible ways of developing effective PFCSs.

4.6.1 PFCSs and principles of their functioning

Table 4.3 presents catalytic systems with two and three components that are capable of catalyzing reactions of various types, from which it is seen that these catalysts represent compounds of different classes including complexes of transition, posttransition, and nontransition metals, metal oxides (OsO_4), inorganic anions (NO_2^-, NO_3^-, etc. [13, 39]), acids and nucleophiles (HI, HCl, I^-, RSH), inorganic compounds (S_2Cl_2, I_2), and organic molecules (C_5H_5N, quinones, N-methylmorpholine, flavine).

The maximum number of MCSs is employed in oxidation processes. In these reactions, the second component of an MCS performs the function of either an oxidation catalyst or an oxidant reactant (in the absence of additional oxidants such as O_2, H_2O_2, NaOCl, etc.). In organic oxidative reactions, the kinetic functions of the second component are as follows:

- catalysis of the oxidation of a reduced form of the first component, i.e., of the electron transfer to an oxidant from M^0L_n or HMX_n;
- catalysis of the oxidation of organometallic intermediates, i.e., of the redox-demetallation of organometallic σ compounds.

Table 4.3 Two- and three-component polyfunctional catalytic systems

Component I[a]	Component II[b]	Component III[c]
$PdCl_2$	$CuCl_2$	
–″–	$FeCl_3$	
–″–	$Cu(OAc)_2$	
–″–	$HAuCl_4$	
–″–	$HgCl_2$	
–″–	$Pb(OAc)_4$	
–″–	$NaNO_2$	
–″–	$LiNO_3$	
–″–	PcM[d]	
–″–	TPPM[e]	
–″–	Q	
$Pd(OAc)_2$	$Cu(OAc)_2$	
–″–	Q	
$PdBr_2$	$CuBr_2$	
–″–	PcM *	
–″–	$Mn(OAc)_2$	
$PdCl_2$	CuCl	$CuCl_2$
–″–	$HgCl_2$	$CuCl_2$
–″–	$HgCl_2$	$FeCl_3$
–″–	TlX_3	$LiNO_3$
–″–	$Hg(OAc)_2$	$CuCl_2$
–″–	Q	Ce(III),Co(III),Mn(III)
–″–	Q	PcM[d]
$Pd(OAc)_2$	Q	PcM[d], TPPM[e]
PdL_2X_2	$CoCl_2$	
–″–	CuI	
$AuCl_3$	$NaNO_2$	
$RhCl_3$	$CuCl_2$	
–″–	HI	HCl
Rh(I)	CuX_2	
–″–	HI	
Ir(I)	HI	
YbX_3	$Mn(OAc)_2$	
Mo(VI)	$Hg(OAc)_2$	
–″–	FeX_3	
OsO_4	N-Methylmorpholine	Flavine[f]
$HCo(CO)_4$	C_5H_5N	
–″–	$Ru_3(CO)_{12}$	
$Co(OAc)_2$	$Mn(OAc)_2$	
$HgCl_2$	$CuCl_2$	
$SbCl_5$	S_2Cl_2	
$CuCl_2$	CuCl	I_2
–″–	$FeCl_3$	
CuCl	$CuCl_2$	
–″–	$HgCl_2$	

Table 4.3 (continued)

Component I[a]	Component II[b]	Component III[c]
–″–	$BiCl_3$	
–″–	$ZnCl_2$	
–″–	RSH	HCl
–″–	H_2S	HCl

Notes: [a]Frequently, the main component according to classification [4.19] or the first-level catalyst according to [16]; [b,c]In the absence of an additional oxidant (O_2, H_2O_2, ROOH, NaOCl, MnO_2, $K_2S_2O_8$, etc.), second and third components of MCS can perform the oxidant function; [d]Phthalocyaninate ligands with various substituents; [e]Tetraphenylporphyrinate ligand (Q=benzoquinone and other quinones); [f]Flavine in the OH form.

Examples are offered by fragments of the aforementioned processes:

(1) 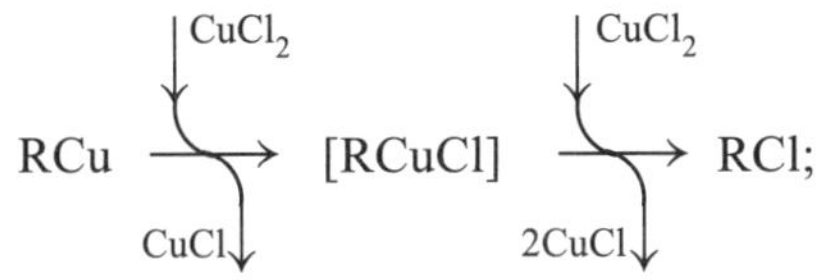

(2) 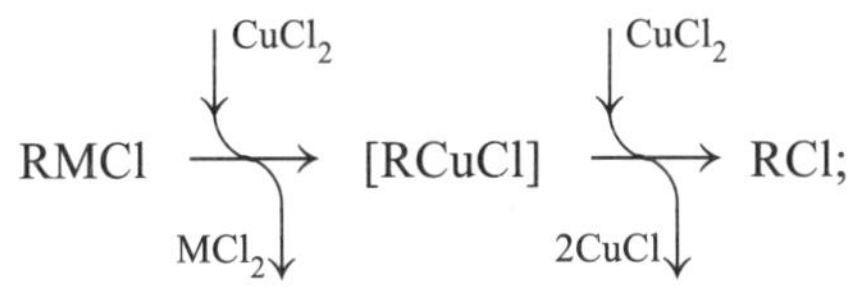

(3) 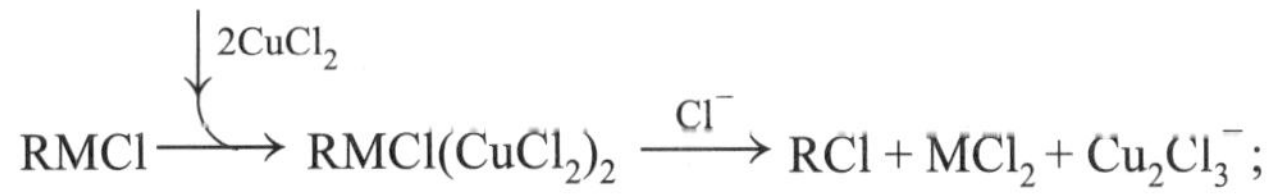

$$\text{(1)–(3)}\ CuCl + Ox \longrightarrow CuCl_2 + Red;$$

(4) 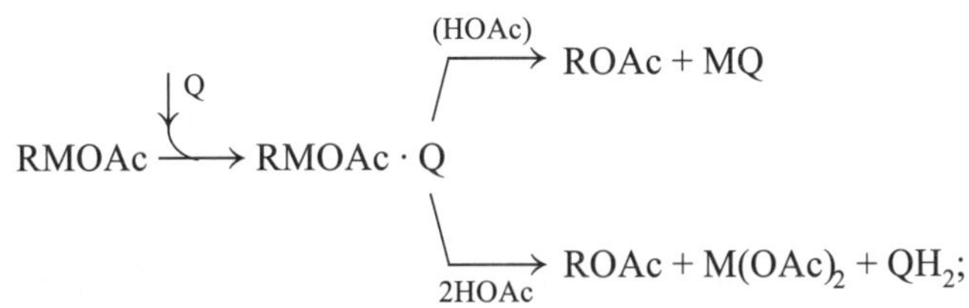

$$MQ + 2HOAc \longrightarrow M(OAc)_2 + QH_2;$$
$$QH_2 + Ox \longrightarrow Q + Red;$$

(5) 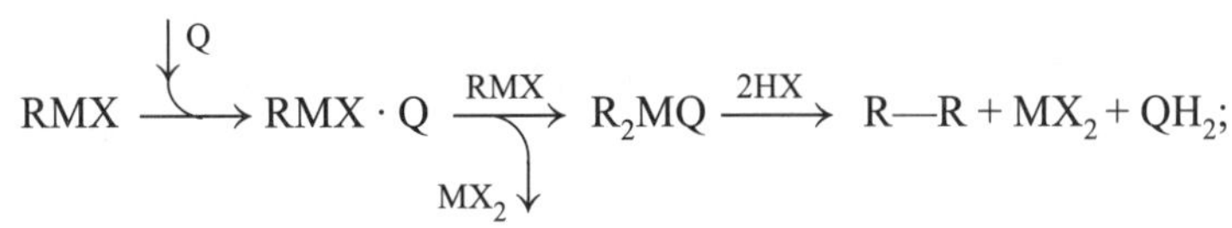

$$QH_2 + Ox \longrightarrow Q + Red$$

where Q is a quinone.

The thermodynamic characteristics of the second components acting as oxidants must ensure the possibility of the necessary reaction [13, 22, 46]. At the same time, their kinetic characteristics are no less important in ensuring the interaction with intermediates and the formation of intermediate complexes or the outer-sphere electron transfer [13, 46]. By oxidizing intermediates, second components frequently act as a factor that controls the selectivity of processes.

In non-oxidative reactions, the kinetic functions of the second component are as follows:

- catalysis of electrophilic and nucleophilic substitution;
- catalysis of nucleophilic addition.

  Here are several characteristic examples:

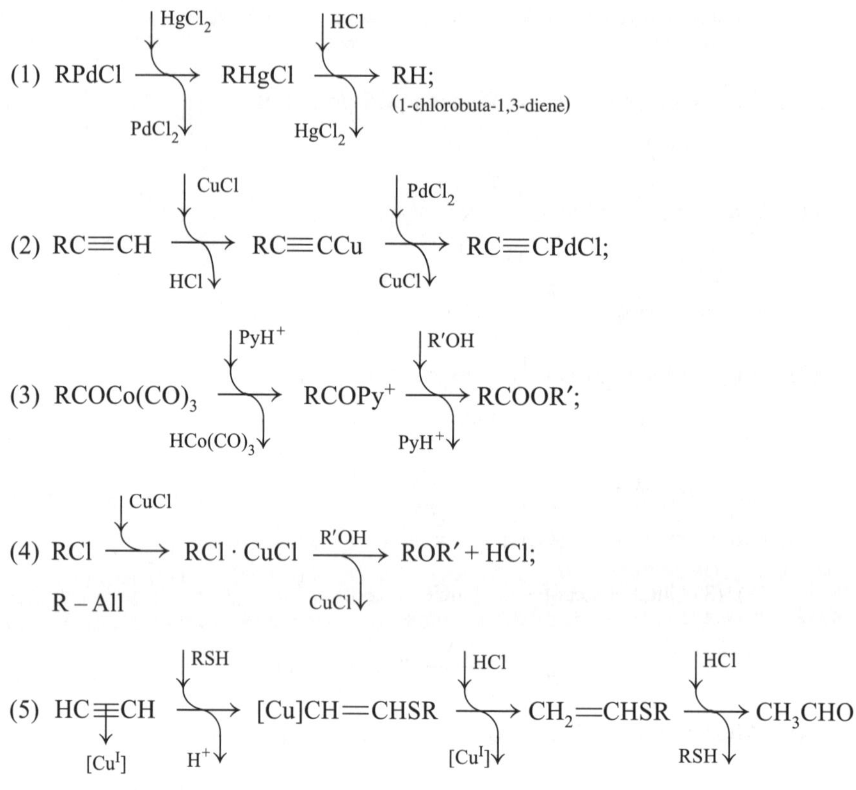

Information gained so far on the catalytic processes of nonoxidative types in MCSs is yet insufficient for making generalizations and predicting catalytic properties of these systems. Various properties of oxidants and ligands in both oxidative and nonoxidative processes have been observed for molecules such as para-quinones. These properties are considered below in some detail.

4.6.2 Kinetic and chemical functions of *p*-benzoquinone and other quinones in PFCSs

The main chemical function of quinones in numerous chemical and biological processes is the electron transfer. For example, ubiquinones

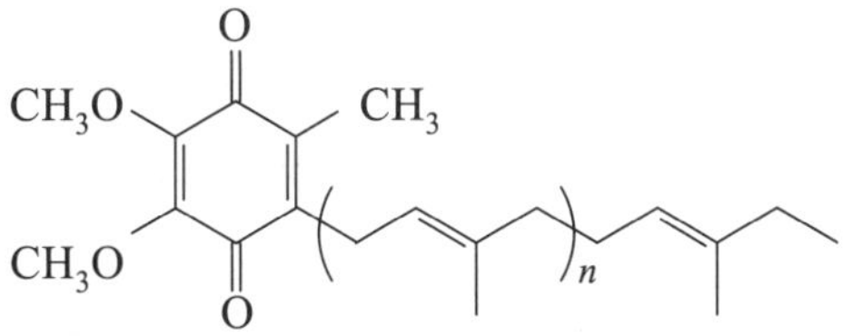

ubiquinones (n = 5–9)

with $n = 5-9$ are the only nonprotein electron transporters involved in the human metabolism, being necessary components in the chains of electron and proton transfer across mitochondrial membranes [280]. Quinones (Q) are classified as one- and two-electron oxidants of medium strength. Accordingly, hydroquinones (QH_2) are relatively easily oxidizable molecules. For this reason, quinones in catalytic chemistry can serve as both oxidants and catalysts. In addition, in the catalysis with metal complexes, quinones sometimes also play the role of ligands in oxidative and nonoxidative processes. In considering the chemical and kinetic functions of quinones in catalytic processes [112], it is convenient to subdivide material with respect to the chemical functions of these compounds.

4.6.2.1 Oxidation of reduced forms of catalytically active metal complexes

Reduced complexes such as $Pd(II)-Pd^0_{sol}, PdCl^-_{sol}, PdCl^{2-}_{2sol}, Pd(CO)_2Br_2^{2-}$ or hydride species such as HPdX are oxidized by quinones to the initial active complexes of Pd (II) in olefin oxidation reactions [122, 34, 112] (see Section 4.1.1). The kinetic function of Q in these processes is to oxidize M(0) or HMX, or catalyze their oxidation by oxygen or other oxidants (e.g., MnO_2). The rates of these processes are independent of [Q], since Q is involved in reactions that follow the limiting (irreversible) step of the process, while the contributions due to reduced particles to the material balance with respect to $[Pd]_\Sigma$ can hardly be significant.

Hydride complexes of metal catalysts are evidently formed in a conjugated process of CO and H_2O oxidation:

$$CO + H_2O + O_2 \xrightarrow{Pd(II)} CO_2 + H_2O_2, \tag{4.286}$$

but the anticipated palladium hydrides are oxidized in this reaction directly by oxygen, not requiring Q species as catalysts-transporters [281, 282]:

$$H-[Pd] \xrightarrow{O_2} HOO[Pd] \xrightarrow{HX} H_2O_2 + [Pd]-X. \tag{4.287}$$

In the reaction of oxidative dehydrogenation of alcohols, hydride complexes of ruthenium are effectively oxidized by quinones, which act as second components of the MCS [283]. The mechanism of oxidation of hydride ligands by quinones has still not been sufficiently studied, but the possible schemes (in the case of HPdX) can be as follows:

$$HPdX \xrightarrow{Q} HPd(Q)X \longrightarrow PdX^+(QH)^- \xrightarrow{HX} PdX_2 + QH_2; \tag{4.288}$$

$$\mathrm{HPdX} \underset{-\mathrm{HX}}{\overset{\mathrm{Q}}{\rightarrow}} \mathrm{Pd^0(Q)} \overset{\mathrm{HX}}{\longrightarrow} \mathrm{PdX^+(QH)^-} \overset{\mathrm{HX}}{\longrightarrow} \mathrm{PdX_2 + QH_2}; \tag{4.289}$$

$$\mathrm{HPdX} \overset{\mathrm{Q}}{\rightarrow} \bullet\mathrm{QH + PdX} \longrightarrow \mathrm{PdX^+(QH)^-} \overset{\mathrm{HX}}{\longrightarrow} \mathrm{PdX_2 + QH_2}. \tag{4.290}$$

Here, the PdX^+ $(QH)^-$ complex can be palladium(II) hydroquinolate or the η^3-allyl complex of QH^- in the semiquinone form. A complex of this type was also obtained for Pt(II) [284]:

$\mathrm{Pt(DQ)(Cod) + HBF_4} \longrightarrow$ 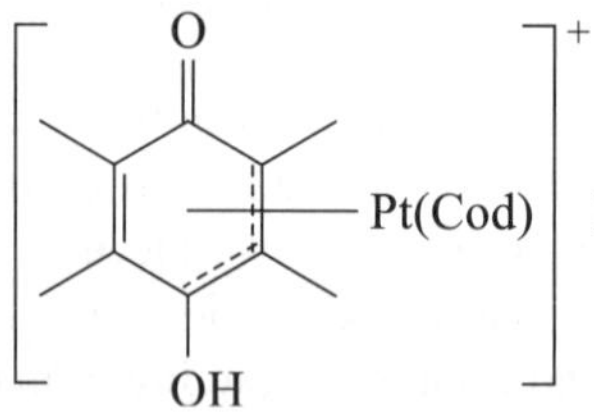 BF_4^-, (4.291)

where DQ is duroquinone and Cod is cycloocta-1,5-diene.

For example, let us consider the kinetic functions of MCS components in the reaction of catalytic oxidation of alcohols in $RuCl_2L_3$ solutions.

Example 4.22 [283]

► The oxidation of primary benzyl alcohols was studied at 20° in $RuCl_2L_3$ or $RuCl(OAc)L_3$ ($L = PPh_3$) solutions with *p*-benzoquinone (BQ). Molecular oxygen slowly oxidizes the intermediate hydride complexes of $HRuClL_3$, while BQ makes this much faster according to the reaction

$$\mathrm{HRuClL_3 + BQ + H^+ + Cl^-} \longrightarrow \mathrm{QH_2 + RuCl_2L_3}.$$

the rate of which is independent of [BQ]. The kinetic isotope effect in the case of $PhCD_2OH$ amounts to 3.3. The oxidation of QH_2 by oxygen is catalyzed with macrocyclic complexes of Co(II)–Co(salophen)PPh_3. Thus, the ternary $RuCl(OAc)L_3$–QH_2–Co(salophen)PPh_3 system in CH_2Cl_2 at 20 °C catalyzes the formation of PhCHO according to the following scheme:

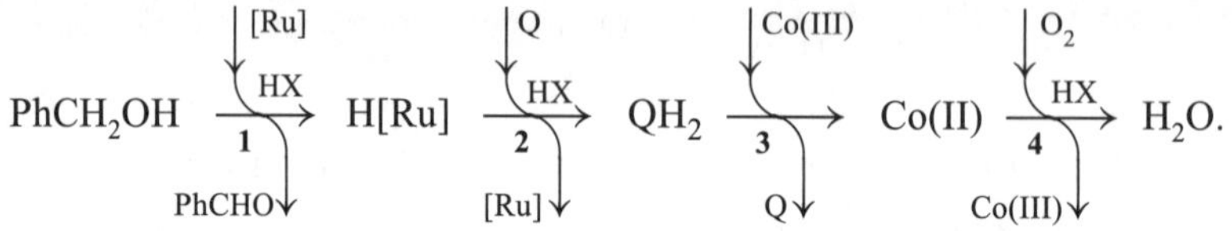

(4.292)

Within the framework of this scheme, the kinetic functions of components are as follows:

- Ru(II) catalyzes the oxidation of ROH by quinone (steps **1** and **2**);
- Q catalyzes the oxidation of H–[Ru] by the oxidized form of CoL(PPh_3) complex (steps **2** and **3**);
- CoL(PPh_3) (oxidized) catalyzes the oxidation of QH_2 by oxygen.

The oxidation of secondary benzyl alcohols was carried out in the presence of bases (K_2CO_3). For this reason, an additional oxidant in these reactions was 2,6-di-*tert*-butyl-1,4-benzoquinone (Q) known to be stable in the presence of K_2CO_3. The best results were obtained with $[RuCl_2(p\text{-cymene})]_2$ complex in THF. In the [Ru]–Q–MnO_2 system, the yield of ketone from PhCH(Me)OH amounted to 90 %. In the presence of O_2, the reaction was also studied with other complexes of Ru and Co. An increase in the rate of alcohol oxidation on the passage from O_2 to Q in the case of scheme (4.292) is indicative of a significant contribution from H[Ru] to the material balance with respect to Ru, while the independence of

[Q] is evidence for a high rate of step **2**, i.e., for a low concentration of H[Ru] as compared to that of $[Ru]_\Sigma$. This discrepancy is removed by assuming the formation of H[Ru]·Q complex as

$$\mathrm{H[Ru]} \underset{K_2}{\overset{\mathrm{Q}}{\rightleftharpoons}} \mathrm{H[Ru]Q} \underset{\mathbf{2'}}{\overset{\mathrm{HX}}{\longrightarrow}} \mathrm{QH_2} + [\mathrm{RuX}]. \tag{4.293}$$

For $[Ru]_\Sigma \cong \{H[Ru]Q\}$, the rate of aldehyde (or ketone) formation would depend on the nature of Q, while being independent on its concentration:

$$r_{\mathrm{RCHO}} = \omega_1[\mathrm{Ru}] \cong \omega_2'[\mathrm{Ru}]_\Sigma.$$

■

Investigations of the mechanism of intermediate hydride formation in the reaction of hydrogen transfer to ketone from chiral α-deutero-α-phenylethanol showed that the initial $RuCl_2L_3$ complex in the presence of bases in the reaction with $PhCD(CH_3)(OD)$ forms RuD_2L_3 [285], while in the reaction with $PhCD(CH_3)(OH)$ the same initial complex forms $RuH(D)L_3$ [286], which implies the oxidative addition of alcohol at the O–H bond to RuL_3. The use of Ru(0) cyclopentadienone complex in reaction with $PhCD(CH_3)(OH)$ leads to the formation of an intermediate monodeuterated Ru(II) cyclopentadienyl complex [286],

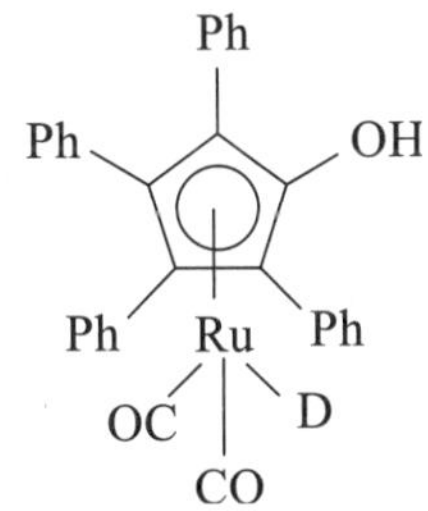

which participates in the processes of hydrogen transfer or alcohol oxidation.

The three-component polyfunctional Ru(0)–Q–Co(II)L system can be considered as a biomimetic model in the aerobic biooxidation of alcohols, where the electron transfer involves the NAD^+/NADH–ubiquinone–cytochrome C system [287]:

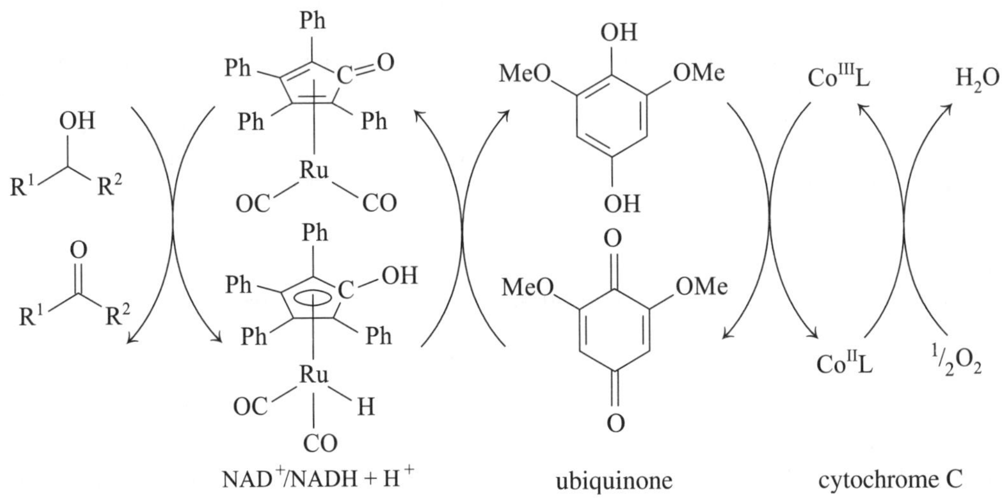

4.6.2.2 *Transfer of one electron to substrate during oxidation of hydrogen donors*

The reduction of some substrates by hydrogen (H_2) or hydride ion donors (BH_4^-, dihydropyridines) involves single-electron transporters such as flavines, quinones, and heterocycles with conjugated electron systems (methylene blue dye, etc.) [288–290]. Typical one-electron substrates in these reactions are O_2, RNO_2, RNO, R•, and $M^{n+}/M^{(n-1)+}$. In fact, catalysts-transporters catalyze noncomplementary reactions by converting the reduction mechanism from two- to one-electron [289, 290]. For example, reaction

$$H_2 + RNO_2 \longrightarrow RNO + H_2O$$

in the presence of metal catalysts or complexes of transition metals (Pt, Rh, Ir) is accelerated by quinones according to the following scheme:

$$\begin{aligned} ML_n + H_2 &\longrightarrow ML_n(H)_2; \\ ML_n(H)_2 + 2Q &\longrightarrow ML_n + 2\dot{Q}H; \\ \dot{Q}H + RNO_2 &\longrightarrow RNO_2^{\bar{\cdot}} + H^+ + Q; \\ RNO_2^{\bar{\cdot}} + \dot{Q}H &\xrightarrow[Q]{} RNO^-(OH) \xrightarrow{H^+} RNO + H_2O. \end{aligned} \tag{4.294}$$

The chemical function of Q in this process is to transfer one electron and proton to a substrate with the formation of an intermediate semiquinone. The role of an electron donor can be played by BH_4^-, dihydropyridine, and other two-electron reductants. The kinetic function of quinone in scheme (4.294) is to catalyze the reduction of RNO_2 by metal hydride. These processes reveal the extremal dependence of the activity of Q on the oxidation potential $E°$. It was suggested that the two-component system under consideration models the enzymatic functions of hydrogenase [290].

4.6.2.3 *Oxidation of catalytically active metal complex species*

By oxidizing active species of a catalytic complex, quinones perform the kinetic function of an inhibitor of the given catalytic reaction and suppress the reaction. This function of BQ was originally discovered during investigations of the positional isomerization of olefins in solutions of Pd (II) complexes under conditions of the side process of olefin oxidation to ketone [291, 292]. It was found that Pd (I) complexes formed as intermediates in the course of olefin oxidation catalyze the isomerization process, and it was suggested that the oxidation of $Pd_2Cl_4^{2-}$ complexes by benzoquinone terminates the isomerization process. This hypothesis was confirmed by kinetic [292] and chemical [293] experiments with Pd (I) complexes. Analogous observations were reported for the reactions of alkyne carbonylation in the PdI_2–NaI–HCl–BuOH system [170, 294]. The introduction of oxidants (Ox = *p*-benzoquinone, I_2) leads to the appearance of an induction period that is proportional to the initial oxidant concentration.

4.6.2.4 *Participation of quinones in oxidative demetallation of intermediates*

The participation of *para*-quinones in the oxidative decomposition of organometallic compounds was kinetically substantiated for the first time by Moiseev with co-workers [90] in the investigation of the mechanism of nucleophilic substitution during the reaction of Pd (II) η^3-allyl complex with *p*-benzoquinone (see Eq. (2.210) and Example (2.31)). Figure 4.16 shows a plot of the first-order rate constant k_I versus [BQ] for this process [90]. Analogous dependences were observed for the oxidation of other η^3-allyl complexes (4.111) and dienes [69]. The main features of this mechanism are the formation of a protonated

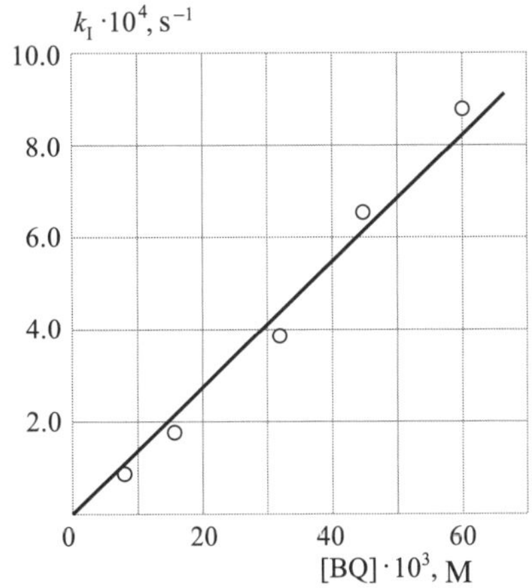

Figure 4.16 Plot of k_I for equation $r = k_I[\eta^3\text{-}C_3H_5PdCl]$ versus para-benzoquinone concentration [BQ] (reproduced with permission from editorial board of *Kinet. Katal.* journal © 1972 [90])

$(\eta^3\text{-}RC_3H_4)PdCl$ complex with BQ and the electron transfer from a nucleophile (Cl^-) via Pd(II) to BQH^+ with the formation of a BQH^- hydroquinolate anion:

$$Cl^- + (\eta^3\text{-}RC_3H_4)PdCl(BQH^+) \longrightarrow RC_3H_4Cl + PdCl(BQH) \quad (4.295)$$

where R=H, $HOCH_2$, $CH_3C(OH)$.

An alternative hypothesis concerning the role of BQ in oxidative reactions was suggested in the aforementioned numerous investigations by Bäckvall with co-workers [28, 80–85]. According to this, the kinetically proved formation of π complexes between BQ and intermediates (see Figures 4.17 and 4.18) is treated as evidence of the participation of BQ in the stabilization of Pd(0) formed during the electron transfer from nucleophile to Pd(II),

$$Nu^- + (\eta^3\text{- All})PdX(BQ) \longrightarrow Nu\text{-}All + Pd(BQ) + X^- \quad (4.296)$$

with the subsequent protonation and reduction of BQ by Pd(0) to BQH_2 [86, 88]. Note that BQ in this scheme performs the chemical functions of both ligand and oxidant. In addition to the aforementioned acceleration of the oxidation of α-olefins by p-benzoquinone [37] (see Section 4.1.1), some other facts are also indicative of the ability of quinones to interact with organopalladium(II) compounds of the RPdX and R_2Pd types with the formation of intermediate complexes. In particular, it was shown that BQ influences the selectivity of conversion of the intermediate alkoxycarbonyl compounds ClPd(COOR) in oxidative carbonylation of alcohols (see Example 4.15 [170, 171, 181]). An increase in [BQ] accelerates the formation of dialkyloxalate via the hypothetical $(BQ)Pd(COOR)_2$ complex. In the $PdCl_2$–LiCl–PhCOOH–PhCOONa–MeOH–BQ system, the rate of dimethyloxalate formation at 20 °C linearly depends on [BQ] (Figure 4.19). An increase in the rate of diphenylcarbonate formation with increasing BQ concentration was also observed in the $Pd(OAc)_2L_2$–$Co(Acac)_3$–PhOH–Bu_4NBr system [296].

Interesting conversions of acryloyl derivatives of Pd were reported in [297, 298]. For example, the reaction in acetonitrile at 50 °C in the presence of 2 M aqueous HBF_4 yields ~85% of dihydromuconate

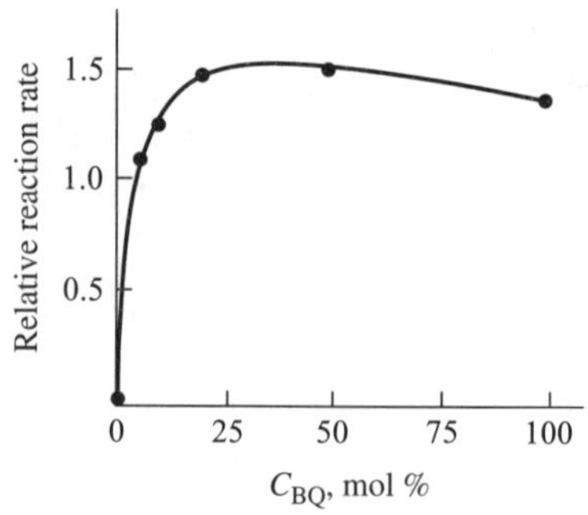

Figure 4.17 Relative rate of 2-cyclohexene trans-1, 4-diacetate formation versus *para*-benzoquinone concentration in $Pd(OAc)_2$–AcOH system (reproduced with permission from American Chemical Society © 1984)

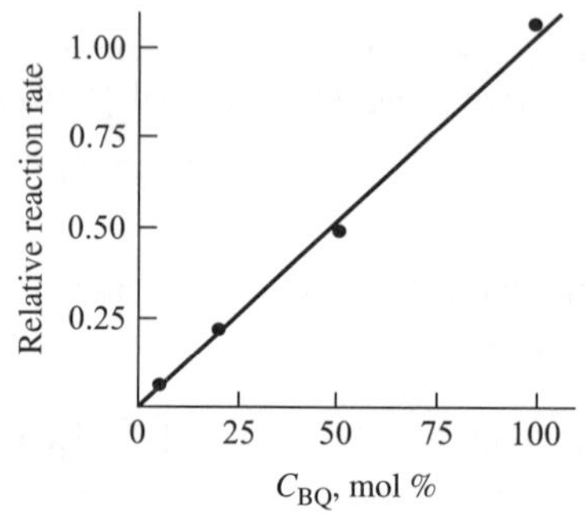

Figure 4.18 Relative rate of 2-cyclohexene *cis*-1, 4-diacetate formation versus *para*-benzoquinone concentration in $Pd(OAc)_2$–LiCl–AcOH system (reproduced with permission from American Chemical Society © 1984)

(Δ^3-DHM) with symmetric arrangement of substituents with respect to the double bond

$$2(\text{P–P})\text{PdCl–CH=CHCOOMe} + 2\text{HBF}_4 \longrightarrow \text{MeOOCCH}_2\text{CH}\underset{E}{=}\text{CHCH}_2\text{COOMe} + 2(\text{P–P})\text{PdCl(BF}_4), \quad (4.297)$$

where (P–P) = 1, 2-bis(diphenylphosphino)ethane. The reaction of $\sim$PdCH=CDCOOCH$_3$ fragment in the DCl–D_2O system yields Δ^3-DHM with a high regioselectivity,

$$\text{MeOOCCD}_2\text{CH}\underset{E}{=}\text{CHCD}_2\text{COOMe},$$

whereas the addition of 1 M BQ per mole Pd(II) complex prior to the introduction of HBF_4 leads to the formation of dimethylmuconate with almost quantitative yield:

$$2(\text{P–P})\text{PdCl–CH}\underset{E}{=}\text{CHCOOMe} + \text{BQ} + 2\text{HBF}_4 \longrightarrow \text{MeOOCCH=CHCH=CHCOOMe} + \text{BQH}_2 + 2(\text{P–P})\text{PdCl(BF}_4). \quad (4.298)$$

Similar to the case of ClPd(COOR), BQ also probably favors symmetrization with the formation of (BQ)Pd(CH=CHCOOMe)$_2$ complexes, which exhibit demetallation in the reaction with c HBF_4. In this case, BQ also plays the roles of both a ligand, which changes selectivity of the process, and an oxidant.

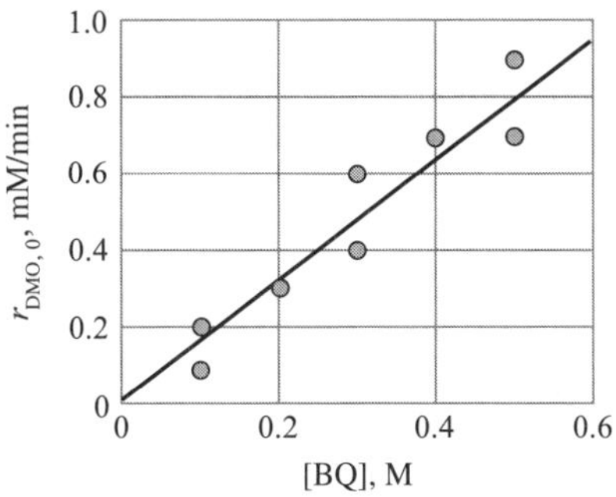

Figure 4.19 Rate of dimethyloxalate formation versus *para*-benzoquinone concentration in $PdCl_2$–LiCl–Ph COOH–PhCOONa–MeOH–BQ system at 20 °C [295]

The participation of Q in the conversions of intermediates in the reaction of CO oxidation to CO_2 in solutions of Pd(II) compounds is also confirmed by an increase in the rate of this reaction in the PdX_2–Q system as compared to the initial rate of a stoichiometric process in the absence of Q [107]. At the same time, it was suggested that BQ plays a stabilizing role in the formation of Pd(0) and Pd(I) complexes such as $Pd(CO)_x(BQH)^+$ and $Pd_2(CO)_y(BQH)^+X_2$ [109].

Investigations [299, 300] of the kinetics of reaction

$$CO + H_2O + Q \longrightarrow CO_2 + QH_2 \tag{4.299}$$

in THF solutions containing 0.03 M $PdCl_2$, 0.8 M LiCl, and 1 M H_2O at 25 °C showed that the initial reaction rate increases with the concentration of quinone and depends on its nature. In the region where the reaction rate becomes independent of $[Q]/[Pd]_\Sigma$, this rate in the system with BQ is about three times greater than in the system with chloranil (CQ) (Figure 4.20). In addition, it was established that the process is inhibited by *para*-toluenesulfonic acid ($r_0 \propto [TsOH]^{-n}, n \cong 1$). It was suggested that quinone is involved in fast reactions preceding a limiting step of the decomposition of (Q)ClPd(COOH).

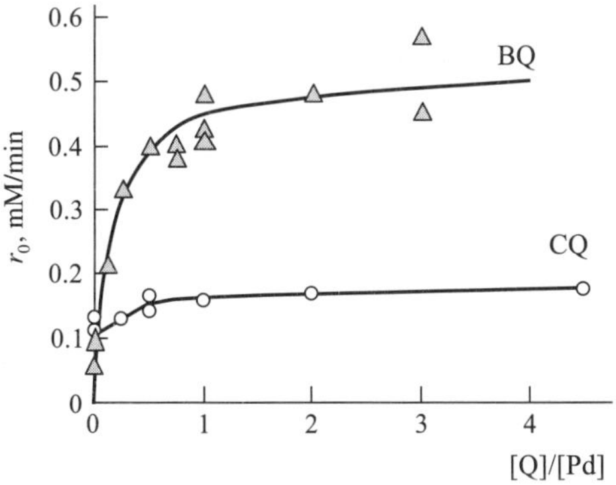

Figure 4.20 Initial rate r_0 of reaction (4.299) versus quinone concentration [Q] in THF solution with [LiCl] = 0.8 M, $[PdCl_2]$ = 0.03 M, and $[H_2O]_0$ = 1 for Q = *para*-benzoquinone (BQ) and chloranil (CQ)

4.6.2.5 Activation of substrates by quinone acting as ligand and oxidant

Transition metal complexes with *para*-quinones (Q), such as (DQ)Fe$(CO)_3$ (where DQ is duroquinone), have been known since 1958 [301]. Up to now, numerous complexes have been synthesized and characterized, including those of Co(I) [302–305], Ni(0) [306–308], Ru(0) [309], Rh(I) [302–304, 310] and Rh(II) [311, 312], Pd(0) [88, 111, 313–319], Pt(0) [284, 320, 321], Ir(I) [310], Ir(III) [310, 322] etc. However, no complexes of Pd(II) or RPdX with Q have been reported so far. Shitova et al. [323] synthesized a complex of Pd(I) with the composition (BQ)$Pd_2SO_4 \cdot 3H_2O$, but its structure remains unstudied.

X-ray diffraction was used to study the structures of quinone complexes with Rh(II) [311, 312], Ru(0) [309], Pt(0) [284, 321], Ni(0) [308], and Pd(0) [88, 111, 317, 324]. In typical complexes with transition metals, the molecules of Q are bound according to η^4 type (**A**), although π complexes of the η^3 type and even complexes where the metal is simultaneously bound to π bond of one quinone and carbonyl group of another quinone are also known. Homoligand complexes of a different type (**B**) were also obtained, but their structures are yet not established:

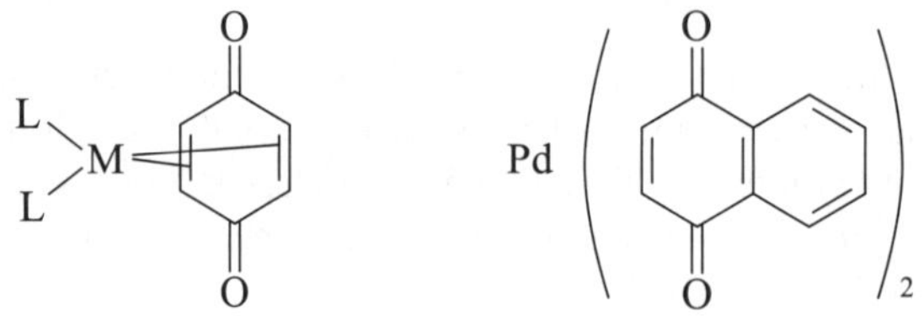

A **B**

Complexes of Pd(0) with BQ proved to be active in some catalytic reactions [111, 315]. For example, $(PPh_3)_2$Pd(BQ) complex participates in the allyl oxidation, and there are grounds to suggest that BQ is involved in the elimination of allyl hydrogen from olefin [315]:

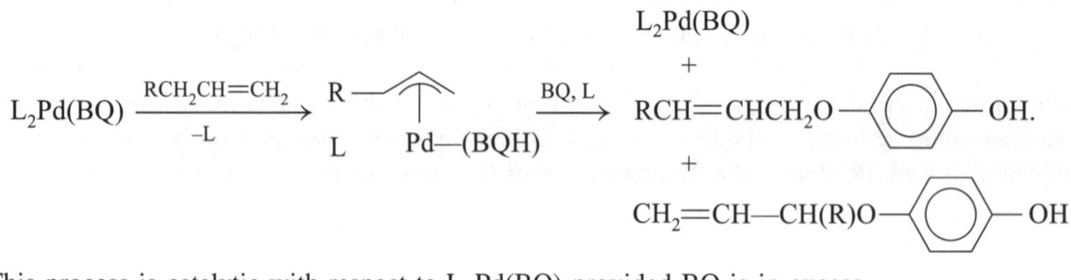

(4.300)

This process is catalytic with respect to L_2Pd(BQ) provided BQ is in excess.

In the presence of water under the same conditions, olefin exhibits oxidative dehydrogenation to 1, 3-diene with the subsequent formation of anthraquinone from 1, 3-diene and BQ. The following scheme was proposed for this reaction:

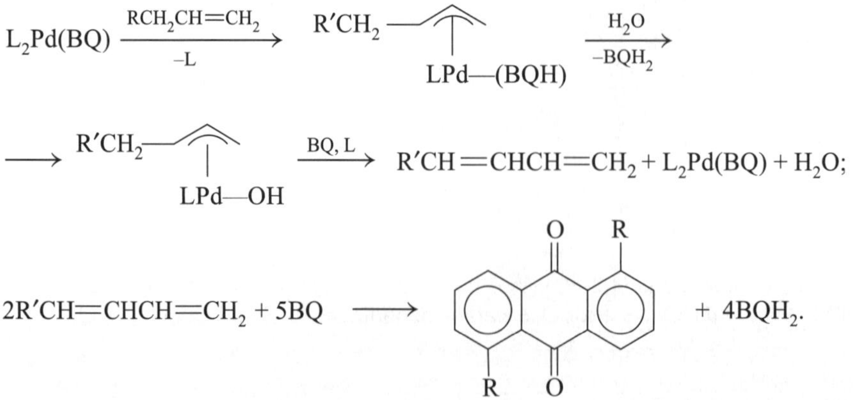

(4.301)

It was suggested that quinones can play an analogous role in the reaction of oxidative carbonylation of alkynes at the C–H bonds in the $Pd(OAc)_2–PPh_3–Q–MeOH$ system [111, 215, 325]. Investigation of the mechanism of active complex formation [325] led to a conclusion that the reduction of palladium acetate leads by the formation of Pd(0) complexes with quinone,

$$L_2Pd(OAc)_2 \xrightarrow[\text{CO+MeOH}]{\text{CO};\ \text{PPh}_3} L_2Pd \xrightarrow{Q} L_2PdQ, \tag{4.302}$$

which are the active catalytic species in this process. Two new complexes of Pd(0) with BQ were synthesized [111, 300, 325], which proved to be active in the synthesis of unsaturated acid esters from alkynes (for more detail, see Section 5.3.6). Investigation of the effect of H^+ showed the absence of protonic acids in the reaction medium, which allowed the following mechanistic scheme to be considered with the steps of proton transfer (probably via a palladium hydride complex) from alkyne to coordinated Q molecules (*p*-benzoquinone or chloranil):

$$L_2PdQ + RC{\equiv}CH \longrightarrow L_2Pd\begin{matrix} C{\equiv}CR \\ QH \end{matrix} \xrightarrow[\downarrow\ RC{\equiv}CCOOMe]{\text{CO, MeOH}} L_2PdH(QH) \xrightarrow{Q} L_2PdQ + QH_2. \tag{4.303}$$

Within the framework of this scheme, quinone molecules perform the functions of ligands, hydrogen acceptors (activation of C–H bonds), and oxidants.

4.6.2.6 *Participation of quinones as ligands in nonoxidative reactions*

The Heck reaction with aryl chlorides

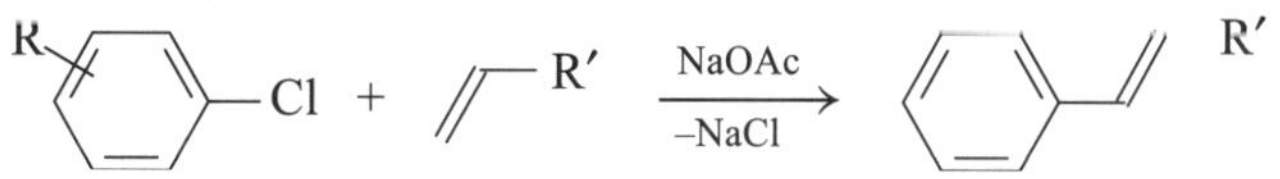

(4.304)

is carried out in an ionic liquid medium (n-Bu_4NBr) at 140 °C in the presence of catalysts such as Pd(0) complexes (LPdQ) with carbene (L) and quinone (Q) ligands (Figure 4.21) [318]. This reaction involves nonactivated aryl chlorides and quinones (BQ or naphthoquinone, NQ), and it is a quinone ligand that makes complexes stable under the reaction conditions. The quinone π complexes of palladium (with Q = NQ) showed activity in the Heck reaction of aryldiazonium salts with olefins [319] and in the Suzuki-Miyaura reactions of the synthesis of asymmetric bisaryls [319]:

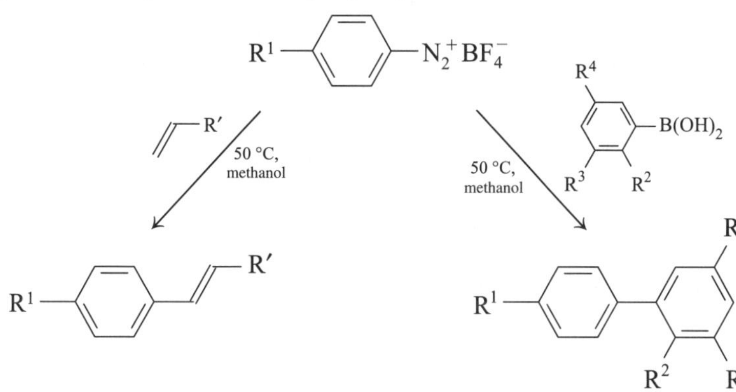

(4.305)

Thus, quinone molecules in reactions (4.304), (4.305) play the role of ligands that stabilize the active catalyst species and prevent the formation of metallic palladium.

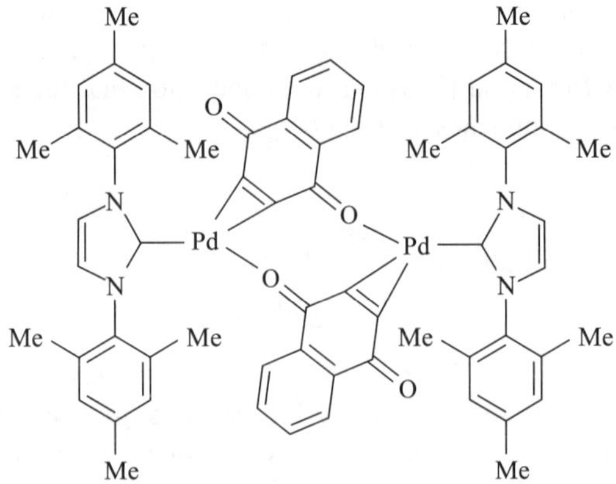

Figure 4.21 Structure of palladium complex with carbene ligand and η^2-naphthoquinone (reproduced with permission from Wiley-VCH Verlag GmbH & Co. KGaA © 2002)

4.6.3 Variants of association of catalytic reactions and catalytic systems

The association of reactions catalyzed by the same or different catalysts in a common reaction system (one-pot reactions) or, in other words, the creation of a PFCS that acts according a class I mechanism as

$$A \xrightarrow{K_1} B \xrightarrow{K_2} P, \tag{4.306}$$

allows in some cases interesting practical results to be obtained. This method of PFCS design is not unambiguously effective, since only the subdivision of the process into separate technological steps ($A \rightarrow B$, $B \rightarrow P$) would allow each reaction to be carried out under optimum conditions. The association of processes is sometimes principally impossible (for incompatible conditions), while in other cases the association is established spontaneously (irrespective of the researcher will), but directional control is yet more and more frequently achieved. Let us consider some examples of the association of catalytic reactions and catalytic systems.

In the oxidation of 1, 3-dienes to diethers in alcohol solutions of $CuCl_2$, $CuCl_2$ and I_2 (see Example 4.4), Cu(I) chloride formed during the oxidation of dienes to iodochlorobutenes and dichlorobutenes catalyzes the replacement of halide ion by alcohol. By the same token, an intermediate Cu(I) chloride in the $PdCl_2$–$CuCl_2$ system catalyzes the conversion of chloroethers and dichlorides into vinyl-1,4-dioxane in ethylene glycol solutions, e.g., via the reaction

$$C_4H_6 \xrightarrow[-HCl]{CuCl_2—PdCl_2} ClC_4H_6OCH_2CH_2OH \xrightarrow[-HCl]{CuCl}$$ 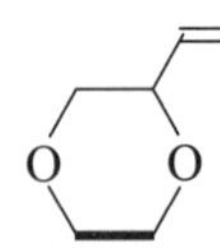 . (4.307)

In the $CuCl_2$–I_2 system, Cu (II) catalyzes both the formation of iodoacetate from propylene and the substitution of I^- by acetate ion in the organoiodine compound (see, e.g., Example 4.2).

Using a $HCo(CO)_3PBu_3$ catalyst with KOH (K_2CO_3, $Mg(OEt)_2$, or other bases) in the reaction of propylene hydroformylation, it is possible to combine hydroformylation, aldehyde condensation, and hydrogenation processes in the synthesis of 2-ethylhexanol from propylene according to the following scheme

[14, 326–328]:

$$
\begin{aligned}
&(1)\ C_3H_6 + CO + H_2 \xrightarrow{[Co]} \underset{(P_1)}{CH_3CH_2CH_2CHO};\\
&(2)\ 2P_1 \xrightarrow{KOH} \underset{(P_2)}{CH_3CH_2CH_2CH(OH)CH(C_2H_5)CHO};\\
&(3)\ P_2 \xrightarrow[-H_2O]{t,\,^\circ C} P_3;\\
&(4)\ P_3 + H_2 \xrightarrow{[Co]} \underset{(P_4)}{C_4H_9CH(C_2H_5)CHO};\\
&(5)\ P_4 + H_2 \xrightarrow{[Co]} C_4H_9CH(C_2H_5)CH_2OH.
\end{aligned}
\quad (4.308)
$$

It was pointed out [14] that, in the $HCo(CO)_3L$–KOH two-component system, it would be difficult to ensure simultaneously high activity in the hydrogenation and selectivity in the formation of 2-ethylhexanol. The hydrogenation process [steps (4) and (5) in scheme (4.308)] would also require a higher temperature [326].

As is known, nickel cyanide complexes are capable of regioselectively catalyzing the carboxylation of alkynes [329]. The use of a two-component system including Co(II) cyanide complexes and Ni(II) cyanide in a two-phase aqueous organic solution (5 N KOH–H_2O–toluene/poly(ethylene glycol) (PEG-400)) allowed the reactions of alkyne carboxylation and unsaturated acid hydrogenation to be combined [330]. As a result, the addition of $Ni(CN)_2$ to Co(II) cyanide solution increased the yield of 2-phenylpropionic acid from phenylacetylene in the reaction

$$RC{\equiv}CH + 2CO + 3KOH \longrightarrow RCH(COOK)CH_3 + K_2CO_3. \quad (4.309)$$

from 21 % to 65 % (90 °C, 1 bar).

It was demonstrated [331], that the reaction of methanol homologization to ethanol in the $Co_2(CO)_8$–CH_3I system is in fact a combination of two consequent reactions:

$$CH_3OH + CO + H_2 \longrightarrow CH_3CHO + H_2O; \quad (4.310)$$

$$CH_3CHO + H_2 \longrightarrow CH_3CH_2OH. \quad (4.311)$$

The introduction of $RuCl_3$ as an additional catalyst of reaction (4.311) led to a significant (approximately fourfold) increase both in the ethanol yield and in the selectivity of the process. The kinetics of this process with 1,4-dioxane in a two-component catalytic system including cobalt and ruthenium complexes was studied in a broad range of H_2 pressures (20–200 bar), CO pressures (70–130 bar) and temperatures (160–200 °C) by the initial rate method under quasi-steady-state conditions with respect to intermediates and CH_3CHO. The observed kinetic laws and results of IR spectroscopic measurements *in situ* allowed Watanabe et al. [331] to propose a mechanism in which ruthenium iodocarbonyl complexes [$RuI_2(CO)_4$ and

$RuI_3(CO)_3^-$] catalyze the hydrogenation of acetaldehyde. It was shown that both homo- and heterometallic clusters of Ru and Co decompose in the course of this process into mononuclear complexes of Ru(II) and Co(I). This mechanistic scheme consists of the following steps:

$$
\begin{aligned}
&(1)\ \underset{(X_1)}{Co_2(CO)_8} + H_2 \rightleftharpoons \underset{(X_2)}{2HCo(CO)_4};\\
&(2)\ X_2 \rightleftharpoons \underset{(X_3)}{HCo(CO)_3} + CO;\\
&(3)\ HCo(CO)_4 + MeOH \rightleftharpoons \underset{(X_4)}{CH_3\overset{+}{O}H_2Co(CO)_4^-};\\
&(4)\ CH_3I + X_4 \rightleftharpoons \underset{(X_5)}{CH_3Co(CO)_4} + HI + MeOH;\\
&(5)\ HI + MeOH \rightleftharpoons CH_3I + H_2O;\\
&(6)\ X_5 + CO \rightleftharpoons \underset{(X_6)}{CH_3COCo(CO)_4};\\
&(7)\ X_6 + H_2 \longrightarrow HCo(CO)_4 + CH_3CHO;\\
&(8)\ RuI_x(CO)_y + H_2 + MeOH \rightleftharpoons \underset{(X_7)}{HRuI_x(CO)_y^-(Me\overset{+}{O}H_2)};\\
&(9)\ X_7 + CH_3CHO \rightleftharpoons \underset{(X_8)}{CH_3CH_2ORuI_x(CO)_y^-(Me\overset{+}{O}H_2)};\\
&(10)\ X_8 + H_2 \longrightarrow CH_3CH_2OH + X_7.
\end{aligned}
\qquad (4.312)
$$

The corresponding kinetic model was derived with allowance for the two material balances, for HI and Co:

$$
\begin{aligned}
[HI]_\Sigma &= [CH_3I]_0 - x[RuCl_3]_0;\\
[Co]_\Sigma &= 2[X_1] + [X_2] + [X_4] + [X_5] + [X_6].
\end{aligned}
$$

The optimum ratio of $[Ru]_\Sigma/[Co]_\Sigma$ is related to the binding of HI by the Ru(III) complex and the resulting decrease in $[CH_3I]$.

The hydrogenation of unsaturated acids in the presence of ruthenium complexes was used for the directional association of a monosubstrate reaction of alkyne carbonylation to unsaturated hydroxylactones [$Co_2(CO)_8$] and the hydrogenation of unsaturated intermediate to γ-ketoacid [332], although this PFCS does not belong to class I reaction (4.306) with respect to the mechanism of action. In alkaline solutions of $Co_2(CO)_8$ the reaction proceeds by the following scheme:

$$RC{\equiv}CH + CH_3I + 2CO \xrightarrow{[Co]} [CH_3COCH{=}C(R)CO[Co]I] \longrightarrow$$

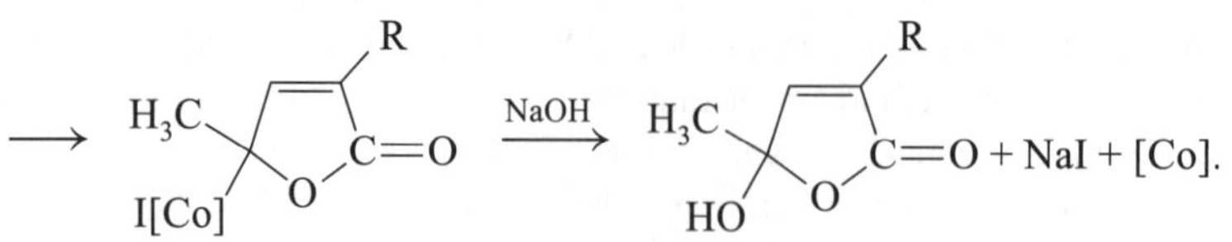

(4.313)

The addition of $Ru_3(CO)_{12}$ (which forms $HRu_3(CO)_{11}^-$ complexes in the alkaline medium) leads to the reduction of the first intermediate X_1 to γ-ketoacid $CH_3COCH_2CH(R)COONa$ according to the following overall equation:

$$RC{\equiv}CH + CH_3I + 2CO + 2NaOH \xrightarrow{Co_2(CO)_8-Ru_3(CO)_{12}} \longrightarrow CH_3COCH_2CH(R)COONa + NaI + CO_2. \quad (4.314)$$

This process formally consists of the formation of hydroxylactone, its reduction, and hydrolysis.

As is known, $Co_2(CO)_8$ complexes also catalyze the following reaction of aldehyde amidocarbonylation to amino acids [333, 334]:

$$RCHO + CH_3CONH_2 + CO \xrightarrow{[Co]} RCH(NHCOCH_3)COOH. \quad (4.315)$$

This reaction was combined with the synthesis of aldehydes, isomerization of allyl alcohol and oxiranes, or hydroformylation of olefins [335]. For example, the two-component $Ti(i\text{-}OPr)_4-Co_2(CO)_8$ system catalyzes the synthesis of amino acids as

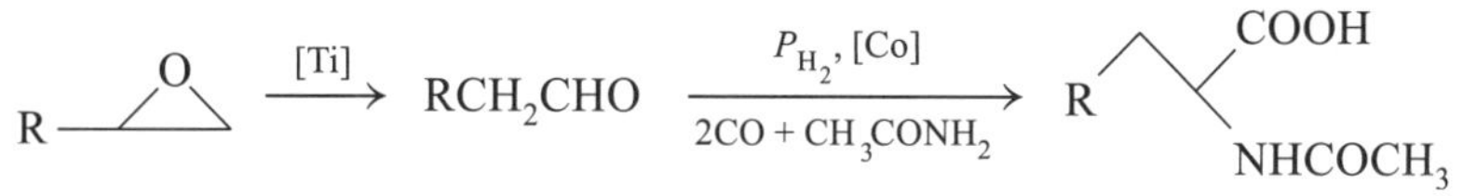

(4.316)

in CO–H_2 mixture at 100 bar and 110 °C.

Let us consider the results of investigation of the synthesis of amino acids from fluorinated olefins.

Example 4.23 [335, 336]

► The hydroformylation of fluorinated olefins $CF_3CH{=}CH_2$ and $C_6F_5CH{=}CH_2$ in $Co_2(CO)_8$ solutions yields predominantly the products with normal structures. This trend is also manifested under the conditions of amidocarbonylation, e.g., via the following reaction:

$$CF_3CH{-}CH_2 + 2CO + H_2 + H_2NCOCH_3 \longrightarrow$$

$$\longrightarrow \underset{\mathbf{I}\,(n)}{CF_3CH_2CH_2CH(COOH)(NHCOCH_3)} + \underset{\mathbf{II}\,(iso)}{CF_3CH(CH_3)CH(COOH)(NHCOCH_3)} \quad (4.317)$$

which yields 96 % $n-$ and 4 % *iso*-products. It was shown that the regioselectivity is determined entirely by the step of hydroformylation, although the presence of amide can somewhat increase the yield of n-products.

In the case of $C_6F_5CH{=}CH_2$, the n-isomer is formed with a selectivity of 90–92 % (in dioxane). When $Rh_4(CO)_{12}$ is used as a catalyst, the regioselectivity exhibits inversion and the *iso*-product (**II**) is obtained with a 98 % selectivity (in dioxane) at 90 °C. The study of $Co_2(CO)_8-Rh_4(CO)_{12}$ mixed systems showed for both substrates that an increase in the $[Co_2]_\Sigma/[Rh_4]_\Sigma$ leads to a decrease in the proportion of the *iso*-product, but even at $[Co_2]_\Sigma/[Rh_4]_\Sigma = 50$ the selectivity with respect to product **II** reaches 94 %. The selectivity for the *iso*-product obtained from perfluorophenylstyrene reaches 80 %, and the further increase in the $[Co]_\Sigma/[Rh]_\Sigma$ ratio almost does not change the ratio of n- and *iso*-products. These results indicate that a new catalyst (CoRh cluster) is formed in the two-component system, which is significantly more active than $Co_2(CO)_8$ in the hydroformylation process:

$$Rh_4(CO)_{12} + 2Co_2(CO)_8 \overset{K_1}{\rightleftharpoons} 4CoRh(CO)_7, \quad (4.318)$$

where $K_1 = 3.5 \cdot 10^{-3}$ in hexane at 84 °C. A complex of this type has been described by Horvath et al. [337, 338]. It was suggested that it may involve bridging CO groups with $\nu_{CO} = 1940\,cm^{-1}$. The ratio of three species in solution is [CoRh] : $[Rh_4]$: $[Co_2]$ = 1: $2.9 \cdot 10^{-3}$: 19.7 at $[Co]_\Sigma/[Rh]_\Sigma = 40$ and 1 : $4.6 \cdot 10^{-4}$: 50 at $[Co]_\Sigma/[Rh]_\Sigma = 100$. The relative activities of the three species were calculated using their relative concentrations of $[X_c]$, $[X_a]$, and $[X_b]$ in the scheme

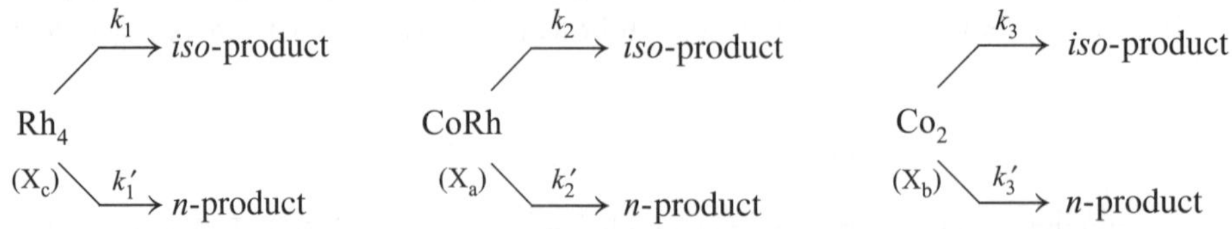

and the ratios of N_{iso}/N_n observed for pentafluorophenylpropanal. It was experimentally established that, under certain conditions, $k_1/k_1' = 76, k_2/k_2' = 7.5 \quad k_2/k_2' = 7.5$, and $k_3/k_3' = 0.47$ (where k_2/k_2' was determined under the conditions where all Rh_4 is converted into $CoRh(CO)_7$, the activity of which is about 2200 times that of $Co_2(CO)_8$). Using the calculated values of $[X_c]$, $[X_a]$ and $[X_b]$ for various $[Co]_\Sigma/[Rh]_\Sigma$ ratios and the experimentally determined k_{iso}/k_n values for all species under consideration, the relative amounts of isomers for arbitrary [Co]/[Rh] ratios (assuming that the reactions are first-order with respect to active species and the regioselectivity remains constant during the reaction) can be described by the following equations:

$$\frac{N_{iso}}{N_n} = \frac{k_1[X_c] + k_2[X_a] + k_3[X_b]}{k_1'[X_c] + k_2'[X_a] + k_3[X_b]} = \frac{76[X_c] + 7.5[X_a]x + 0.47[X_b]y}{[X_c] + [X_a]x + [X_b]y}, \tag{4.319}$$

where $x = k_2'/k_1'$ and $y = k_3'/k_1'$.

Then, by setting various values of x at preset $y = \text{const}$ for each set of X_c, X_a and X_b with increasing $[Co_2]$, i.e., for various values of the relative catalytic activity of $Rh_4(CO)_{12}$ and $CoRh(CO)_7$ complexes defined as

$$\alpha = \frac{k_1 + k_1'}{k_2 + k_2'} = \frac{(k_1/k_1' + 1)k_1'}{(k_2/k_2' + 1)k_2'} = \frac{77}{8.5}\frac{1}{x}, \tag{4.320}$$

it is possible to calculate N_{iso}/N_n and compare these values to the experimental data. It was found [336] that the best fit of calculations and experiment with the variation of $[Co]_\Sigma/[Rh]_\Sigma$ is achieved for $x = 0.23, y = 1.2 \cdot 10^{-4}$ at $\alpha = 40$ for $[Rh_4]$/[CoRh] and $\alpha = 10$ (with recalculation per atom Rh). The relative catalytic activity β (per metal atom) for $Rh_4(CO)_{12}/Co_2(CO)_8$ complexes amounts to ~220 000. Thus, the ratio of the activities of components (per metal atom) in the given system is as follows:

$$Rh_4(CO)_{12} : CoRh(CO)_7 : Co_2(CO)_8 = 10 : 1 : 4.54 \cdot 10^{-5}.$$

Estimations of the relative catalytic activity determined by this method are close to the estimations obtained directly from the measured first-order rate constants with respect to separate catalytically active species [336]. The kinetic function of the $CoRh(CO)_7$ catalyst in the reaction of aldehyde amidocarbonylation combined (associated) with olefin hydroformylation is to catalyze the first step of aldehyde synthesis with a high regioselectivity (~88 % for the *iso*-product), while the kinetic function of $Co_2(CO)_8$ is to catalyze the amidocarbonylation process. As a result of the association of two processes and two catalysts, the activity of the catalytic system decreases by a factor of approximately ten as compared to the activity of the rhodium complex alone, but an active and regioselective cluster is formed. ■

An interesting association of the reactions of alkylbenzene oxidation and propylene epoxidation to propylene oxide (single-step chalcone process) in the FeX_3–Mo (VI) system was reported [339]. Other olefins were also employed, and the selectivity with respect to olefin oxide in the reaction

$$RH \xrightarrow[O_2]{Fe(III)} ROOH \xrightarrow[R'CH=CH_2]{Mo(VI)} R'\underbrace{CH—CH_2}_{O} + ROH. \qquad (4.321)$$

in the best cases reached 88 %.

A two-route process of the kinetic resolution of substrates (racemates) considered in Section 3.2 has a significant disadvantage, since the yield of product (e.g., P_R) obtained using a chiral catalyst (metal complex or enzyme) from one enantiomer of the pair (e.g., R-isomer) in the most favorable case (very slow conversion of S-isomer into P_S product) can only approach 50 % per substrate. Only after five cycles with the isolation of products and racemization of the residual S-isomer, the theoretical yield of P_R reaches 95 % [340]. The problem can be solved by combining *in situ* the process of enantioselective synthesis of P_R from R-enantiomer and the catalytic racemization (i.e., the conversion of S-enantiomer into racemate):

$$\begin{array}{ccc} R & \xrightarrow[\text{fast}]{k_R} & P_R \\ k_{rac} \upharpoonleft\downharpoonright & & \\ S & \xrightarrow[\text{slow}]{k_S} & P_s, \end{array} \qquad (4.322)$$

where R and S are the initial enantiomers and k_{rac} is the racemization rate constant ($S \rightarrow R$ or $R \rightarrow S$) that characterizes the speed of attaining the equilibrium ($[R]/[S] = 1$).

This associated process is called dynamic kinetic resolution (DKR). The catalytic conversion $R \rightleftharpoons S$ must include the formation of a chiral intermediate, which can be represented by an achiral enolate

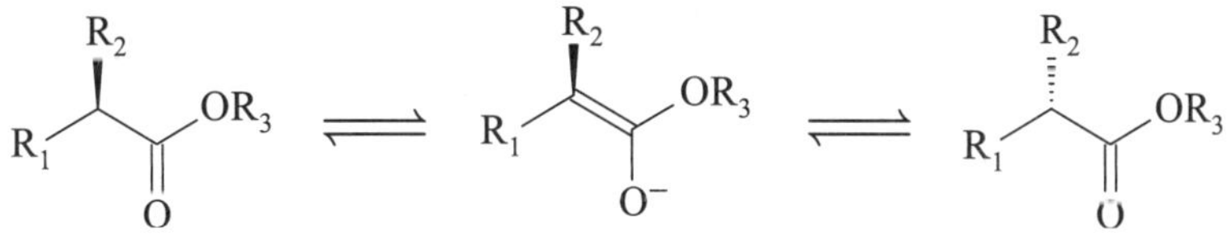

or a ketone (aldehyde)

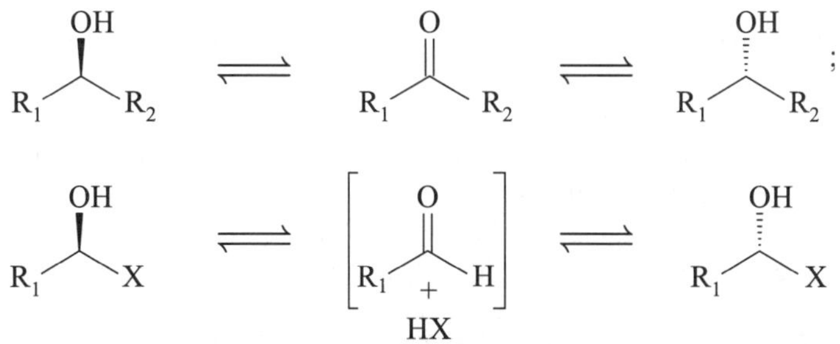

X = OR, SR, NHR, CN

where X = OR, SR, NHR, CN. Thus the DKR process involves a two-component catalytic system with different kinetic functions of the components. An advantageous combination in the case of secondary alcohols and secondary amines was achieved using enzymatic and metal-complex catalysts [341, 342]. In particular, enzyme lipase B (CALB) from *Candida Antarctica* supported on acrylic polymer (known under the commercial name of Novozym 435), catalyzes the transfer of acyl groups from various acylating agents (*para*-chlorophenyl acetate, etc.) to the R-enantiomer of methyl phenylcarbinol, while a Ru(II) complex catalyzes fast conversion of the $S-$enantiomer alcohol into the R-enantiomer (i.e., racemization of the substrate). Their combined action ensures a $\sim$100% yield of the R-enantiomer of the target acetate (92 % yield of isolated product) with a high optical purity (*ee* $> 99\,\%$) [341]. The mechanism of hydrogen transfer from alcohol to Ru(II) complex and from the hydride complex to ketone was considered in Example 4.22.

Conditions of successful operation of the DKR system are as follows:

- kinetic irreversibility of the reactions of formation of P_R and P_S;
- existence of a single chiral center;
- sufficiently large enantiomer ratio: $k_R/k_S > 20$ for obtaining P_R (or $k_S/k_R > 20$, for P_S);
- $k_{\mathrm{rac}} \geq k_R$ (for acceptable enantioselectivity, $k_{\mathrm{rac}}/k_R \geq 10$);
- absence of interaction between the catalyst of formation of P_R (P_S) and the catalyst of racemization;
- absence of any side conversions of substrates and products.

The use of chemoenzymatic catalytic systems was also useful in enantioselective synthesis of amino acids and some other chiral compounds [343, 344]. Another interesting but still insufficiently studied direction is the association of reactions involving metal complexes (and/or metal-containing anions) and electrochemical processes. For example, it was demonstrated that the association of the propylene oxidation by Hg(II, I) complexes with electrochemical Hg(0, I) oxidation allows the development of an effective electrocatalytic synthesis of acrolein [345]. By the same token, the reaction of linalool oxidation to citral in sodium bichromate solutions can also be developed as an electrocatalytic process [346].

The associated processes considered in this section (involving PFCSs of class I according to reaction (4.306)) can be treated as conjugate processes (see Section 3.4), since product B ensures a kinetic conjugation between two catalytic reactions. On the other hand, there is certain difference between the kinetic conjugation of reaction steps and routes in conjugate and associated processes. In contrast to the conjugate processes, associated reactions can be also carried independently of each other as

$$\mathrm{A} \rightarrow \mathrm{B}; \quad \mathrm{B} \rightarrow \mathrm{P},$$

while in conjugate processes with chain mechanisms, as well as in most catalytic and noncatalytic conjugate reactions, this is impossible.

Tandem, cascade, and domino reactions [347, 348], multicomponent reactions [349], and competitive tandem catalysis [350] in the organic chemistry are closer to associated processes, but can also involve unstable intermediates (in the presence of catalysis) as in conjugate processes with different types of polyfunctionality. A classification of the types of tandem catalysis that was recently proposed [351, 352] is very restricted in view of the large variety of action of the multicomponent catalytic systems (see above). Considering the tandem catalysis, it would be expedient to distinguish between the tandem of catalysts operating in (i) the same overall reaction (that is, kinetic polyfunctionality, see Section 4.4) and (ii) different overall reactions in the same reaction system (class I according to the adopted classification).

The results of investigations of PFCSs considered in Chapter 4, as well as the other results reported for the oxidation of alkanes [188, 200, 353] and alcohols [353–357] (except for [287]) in these systems provide convincing evidence for the abilities of PFCSs as a powerful tool in commercial and synthetic chemistry.

References

1. T. S. Price, *Z. Phys. Chem.*, **27**, 474 (1898).
2. J. Brode, *Phys. Chem.*, **37**, 257 (1901).

3. E. Spitalsky, *Z. Phys. Chem.*, **122**, 257 (1926).
4. G. A. Bogdanov, *Zh. Fiz. Khim.*, **25**, 323, 332 (1951).
5. G. A. Bogdanov and N. N. Petin, *Zh. Obshch. Khim.*, **12**, 369, 598 (1942).
6. G. A. Bogdanov and S. I. Pastukhova, *Zh. Fiz. Khim.*, **27**, 1556 (1953).
7. V. A. Shcherbinin and G. A. Bogdanov, *Zh. Fiz. Khim.*, **32**, 1942 (1958).
8. G. A. Petrova and G. A. Bogdanov, *Zh. Fiz. Khim.*, **32**, 2334 (1958).
9. O. N. Temkin, O. L. Kaliya, G. K. Shestakov and R. M. Flid, *Dokl. Akad. Nauk. SSSR*, **190**, 398 (1970).
10. D. V. Sokol'skii and Ya. A. Dorfman, *Kataliz ligandami v vodnykh rastvorakh* (Catalysis by Ligands in Aqueous Solutions), Nauka, Alma-Ata, 1972.
11. P. Bonchev, *Kompleksoobrazovanie i kataliticheskaya aktivnost'* (Complex Formation and Catalytic Activity), Nauka I Izkustvo, Sofia, 1972; Mir, Moscow, 1975 [Translated from Bulgarian].
12. V. A. Golodov, *J. Res. Inst. Catal. Hokkaido Univ.*, **29** (1), 49 (1981).
13. V. A. Likholobov and Yu. I. Ermakov, *Kinet. Katal.*, **21** (4), 904 (1980).
14. N. S. Imyanitov, *Koord. Khim.*, No. 11, 1443 (1984).
15. O. N. Temkin and S. M. Braylovskii, in: *Fundamental Research in Homogeneous Catalysis*, A. E. Shilov (Ed.), Gordon & Breach, Philadelphia, 1986, Vol. 2, p. 621.
16. V. A. Golodov, in: *Fundamental Research in Homogeneous Catalysis*, A. E. Shilov (Ed.), Gordon & Breach, Philadelphia, 1986, Vol. 2, p. 1131.
17. O. N. Temkin, A. V. Zeigarnik and D. G. Bonchev, *Chemical Reaction Networks: A Graph-Theoretical Approach*, CRC Press, Boca Raton, FL, 1996.
18. A. V. Zeigarnik and O. N. Temkin, *Kinet. Katal.*, **40** (4), 550 (1999).
19. L. G. Bruk, A. P. Kozlova, O. V. Marshakha, I. V. Oshanina, O. N. Temkin and O. L. Kaliya, *Izv. Akad. Nauk., Ser. Khim.*, No. 10, 1899 (1999).
20. N. G. Mekhryakova, L. G. Bruk, O. L. Kaliya, A. Yu. Prudnikov, and O. N. Temkin, *Kinet. Katal.*, **20**, 629 (1979).
21. T. E. Kron, M. I. Terekhova, Yu. G. Noskov and E. S. Petrov, *Kinet. Katal.*, **42** (2), 204 (2001).
22. I. I. Moiseev, *Kompleksy olefinov v zhidkofaznom okislenii* (Olefin π Complexes in Liquid-Phase Oxidation), Nauka, Moscow, 1970.
23. I. I. Moiseev and M. N. Vargaftik, *Dokl. Akad. Nauk SSSR*, **166**, 370 (1966).
24. I. I. Moiseev, in: *Kompleksoobrazovanie v katalize* (Complex Formation in Catalysis), Nauka, Moscow, 1968; *Probl. Kinet. Katal.*, **13**, 36 (1968).
25. R. F. Heck, *J. Am. Chem. Soc.*, **91**, 5518 (1968).
26. H. Zeiss (Ed.), *Organometallic Chemistry*, Reinhold Publ. Corp., Chapman & Hall, New York-London, 1960.
27. O. A. Reutov, I. P. Beletskaya, T. A. Artamkina and A. N. Kashin, *Reaktsii metalloorganicheskikh soedinenii kak redoks-protsessy* (Reactions of Organometallic Compounds as Redox Processes), Nauka, Moscow, 1971.
28. J.-E. Bäckvall, R. B. Hopkins, H. Grennberg et al., *J. Am. Chem. Soc.*, **112**, 5160 (1990).
29. A. B. Svetlova, S. M. Brailovskii and O. N. Temkin, *Kinet. Katal.*, **19**, 1588 (1978).
30. A. B. Pshenichnikova, S. M. Brailovskii and O. N. Temkin, *Kinet. Katal.*, **30**, 92 (1989).
31. K. I. Matveev, I. F. Bukhtoyarov, N. N. Shul'ts and O. A. Emel'yanova, *Kinet. Katal.*, **5**, 572 (1964).
32. P. Francois and Y. Trambouze, *Bull. Soc. Chim. Fr.*, **51** (1969).
33. P. Francois, *Ann. Chim (Paris)*, **14** (4), 370 (1969).
34. H.-B. Lee and P. M. Henry, *Can. J. Chem.*, **54**, 1726 (1976).
35. R. Jira, J. Sedlmeier and J. Smidt, *Lieb. Ann. Chem.*, **693**, 99 (1966).
36. J.-E. Bäckvall, B. äkermark and S. O. Ljunggren *J. Am. Chem. Soc.*, **101**, 2411 (1979).
37. J.-E. Bäckvall and B. R. Hopkins, *Tetrahedron. Lett.*, **29**, 2885 (1988).
38. K. I. Matveev, N. B. Shitova and E. G. Zhizhina, *Kinet. Katal.*, **17** (4), 893 (1976).
39. E. G. Zhizhina, N. B. Shitova and K. I. Matveev, *Kinet. Katal.*, **22** (6), 1451 (1981).
40. D. G. Miller and D. D. M. Wayner, *J. Org. Chem.*, **55**, 2924 (1990).
41. M. Skumov and E. Balbolov, *Catal. Lett.*, **69**, 103 (2000).
42. O. N. Temkin, L. G. Bruk and E. Sudachenko, unpublished results.
43. L. H. Stang and R. Jira, *Tetrahedron Lett.*, **41**, 3589 (1970).

44. P. M. Henry, *J. Org. Chem.*, **38**, 1681 (1973).
45. P. M. Henry, *J. Org. Chem.*, **39**, 3871 (1974).
46. V. A. Likholobov, *Proceedings of the 4th Int. Symp. on Homogeneous Catalysis (Leningrad, USSR, 1984)*, Book of Abstracts, p. 20.
47. N. I. Kuznetsova, V. A. Likholobov, M. A. Fedotov and Yu. I. Yermakov, *J. Chem. Soc. Chem. Comm.*, 973 (1982).
48. V. A. Likholobov, N. I. Kuznetsova and Yu. I. Yermakov, *React. Kinet. Catal. Lett.*, **12** (3), 247 (1979).
49. N. I. Kuznetsova, V. A. Likholobov and Yu. I. Yermakov *React. Kinet. Catal. Lett.*, **22** (2), 139 (1983).
50. N. I. Kuznetsova, M. A. Fedotov, V. A. Likholobov and Yu. I. Yermakov, *J. Mol. Cata* l., **38**, 263 (1986).
51. E. V. Gusevskaya, V. A. Likholobov and Yu. I. Ermakov, *Kinet. Katal.*, **21** (3), 639 (1980).
52. P. M. Henry, *J. Org. Chem.*, **32**, 2575 (1967).
53. D. Clark, P. Hayden and R. D. Smith, *Disc. Faraday Soc.*, **46**, 98 (1968).
54. A. I. Cheboksarov, *Kinetika i mekhanizm atsetoksilirovaniya etilena pri katalize kompleksami palladiya* (Kinetics and Mechanism of Ethylene Acetoxylation Catalyzed by Palladium Complexes), Candidate Dissertation in Chemistry, Mendeleev. Univ. Chem. Technol., Moscow, 1982.
55. M. Jamura and J. Yasui, *J. Chem. Soc., Chem. Commun.*, 1209 (1968).
56. M. H. Volkonskii, V. A. Likholobov, L. I. Shtokalo and Yu. I. Yermakov, *React. Kinet. Catal. Lett.*, **21** (3), 225 (1982).
57. P. M. Henry, *J. Am. Chem. Soc.*, **94** (21), 7305 (1972).
58. A. B. Pshenichnikova, S. M. Brailovskii and O. N. Temkin, in *Khimicheskaya kinetika v katalize: Kineticheskie modeli zhidkofaznykh reaktsii* (Chemical Kinetics in Catalysis: Kinetic Models of Liquid-Phase Reactions), Akad. Nauk SSSR, Chernogolovka, 1985, pp. 67–74.
59. A. B. Pshenichnikova, *Kataliticheskii sintez β-galoidalkil'nykh efirov v rastvorakh kompleksov Pd(II) i Cu(II)* (Catalytic Synthesis of β-Alkyl Halide Ethers in Solutions of Pd(II) and Cu(II) Complexes), Candidate Dissertation in Chemistry, MITKhT (Moscow Institute of Fine Chemical Technology), Moscow, 1985.
60. S. E. Manahan and R. T. Iwamoto *J. Electroanalyt.Chem.*, **13** (4), 411 (1967).
61. J.-E. Bäckvall, *Tetrahedron Lett.*, No. 3, 467 (1977).
62. G. A. Al'per and M. Yu. Nikiforov, in *Problemy khimii rastvorov* (Problems in Chemistry of Solutions), Nauka, Moscow, 1998, pp. 21–62.
63. R. Schmid and V. N. Sapunov, *Non-Formal Kinetics*, Verlag-Chemie, Weinheim, 1982.
64. J. K. Stille and R. Dwakaruni, *J. Am. Chem. Soc.*, **100**, 1303 (1978).
65. K. Bergstad, S. Y. Jonsson and J.-E. Bäckvall, *J. Am. Chem. Soc.*, **121**, 10424 (1999).
66. S. Y. Jonsson, K. Färnegardh and J.-E. Bäckvall, *J. Am. Chem. Soc.*, **123**, 1365 (2001).
67. (a) L. Elefteriu, *Izuchenie okislitel'nykh prevrashchenii butadiena-1,3 v rastvorakh kompleksov perekhodnykh metallov* (Studying Oxidative Transformations of 1,3-Butadiene in Solutions of Transition Metal Complexes), Candidate Dissertation in Chemistry, MITKhT (Moscow Institute of Fine Chemical Technology), Moscow, 1979; (b) L. Elefteriu, S. M. Brailovskii, O. N. Temkin and I. V. Moskaleva, *Zh. Org. Khim.*, **14**, 2490 (1978).
68. L. Elefteriu, S. M. Kalabin, M. K. Andari, S. M. Brailovskii, A. P. Belov and O. N. Temkin, *Koord. Khim.*, **8**, 1230 (1982).
69. S. M. Kalabin, *Kinetika i mekhanizm reaktsii okisleniya sopryazhennykh dienovykh uglevodorodov pod deistviem kompleksov palladiya(II) v vodnykh rastvorakh* (Kinetics and Mechanism of Oxidation of Conjugated Diene Hydrocarbons under the Action of Palladium (II) Complexes in Aqueous Solutions), Candidate Dissertation in Chemistry, MITKhT (Moscow Institute of Fine Chemical Technology), Moscow, 1984.
70. S. M. Kalabin and A. P. Belov, *Kinet. Katal.*, **25** (2), 287 (1984).
71. E. P. Nazarenko, S. M. Kalabin, A. V. Krylov et al., *Kinet. Katal.*, **29** (5), 1050 (1988).
72. J.-E. Nystrom, B. Soderberg, B. Akermark and J.-E. Bäckvall, *J. Organometal. Chem.*, **334**, 169 (1987).
73. S. M. Brailovskii, L. Elefteriu, O. N. Chernysheva, A. P. Belov and O. N. Temkin, *Kinet. Katal.*, **23** (1), 54 (1982).
74. M. S. Shlapak, S. M. Brailovskii and O. N. Temkin, *Kinet. Katal.*, **24** (6), 1380 (1983).
75. S. M. Brailovskii, M. S. Shlapak and O. N. Temkin, *Kinet. Katal.*, **26** (1), 69 (1985).
76. M. S. Shlapak, S. M. Brailovskii, O. N. Temkin and V. B. Frolov, *Kinet. Katal.*, **24** (6), 1375 (1983).

77. M. S. Shlapak, S. M. Brailovskii and O. N. Temkin, *Kinet. Katal.*, **27**, 1481 (1986).
78. M. Ya. Skumov, S. M. Brailovskii and O. N. Temkin, *Izv. Akad. Nauk, Ser. Khim.*, No. 4, 572 (2002); *Zh. Org. Khim.*, **39** (2), 199 (2003).
79. M. Skumov, S. Brailovskii and O. Temkin, *Oxid. Comm.*, **30** (4), 814 (2007).
80. J.-E. Bäckvall, *Accounts Chem. Res.*, **16**, 335 (1983).
81. J.-E. Bäckvall, S. E. Byström and P. E. Nördberg, *J. Org. Chem.*, **49**, 4619 (1984).
82. J.-E. Bäckvall, *Bull. Soc. Chim. Fr.*, No. 4, 665 (1987).
83. J.-E. Bäckvall, A. K. Awasthi and Z. D. Renko, *J. Am. Chem. Soc.*, **109**, 4752 (1987).
84. H. Grennberg, S. Faizon and J.-E. Bäckvall, *Angew. Chem. Int. Ed.*, **32** (2), 263 (1993).
85. H. Grennberg and J.-E. Bäckvall, *J. Chem. Soc., Chem. Commun.*, 1331 (1993).
86. H. Grennberg, A. Gogoll and J.-E. Bäckvall, *Organometallics*, **12**, 1790 (1993).
87. M. S. Shlapak, *Kataliticheskii sintez prostykh nenasyshchennykh atsiklicheskikh i tsiklicheskikh diefirov na osnove 1,3-butadiena* (Catalytic Synthesis of Unsaturated Acyclic Diethers Based om 1,3-Butadiene), Candidate Dissertation in Chemistry, MITKhT (Moscow Institute of Fine Chemical Technology), Moscow, 1983.
88. B. Milani, A. Anzilutti, L. Vicentini et al., *Organometallics*, **16**, 5064 (1997).
89. K. Szabo, *Organometallics*, **17**, 1677 (1998).
90. A. P. Belov, N. G. Satsko and I. I. Moiseev, *Kinet. Katal.*, **13**, 892 (1972).
91. O. N. Temkin, G. K. Shestakov and Yu. A. Treger, *Atsetilen: khimiya, mekhanizmy reaktsii, tekhnologiya* (Acetylene: Chemistry, Reaction Mechanisms, Technology), Khimiya, Moscow, 1991.
92. F. P. Ballistreri, S. Failla and G. A. Tomaselli, in: *The Role of Oxygen in Chemistry and Biochemistry, Proceedings of Int. Symp. (Tsukuba, Japan, 1987)*, Series: Studies in Organic Chemistry, W. Ando and Y. Morooka (Eds.), Elsevier, 1988, Vol. 33, p. 341.
93. A. I. Kozlov, S. M. Brailovskii and O. N. Temkin, *Kinet. Katal.*, **35** (4), 551 (1994).
94. A. I. Kozlov, S. M. Brailovskii and O. N. Temkin, *Kinet. Katal.*, **36** (2), 225 (1995).
95. M. M. Frojmovic and G. Just, *Can. J. Chem.*, **46**, 3719 (1968).
96. H. C. Aaraujo, G. A. L. Ferreira and J. R. Mahajan, *J. Chem. Soc. Perkin Trans. I*, 2257 (1974).
97. M.-H. Huin, S. M. Brailovskii and O. N. Temkin, *Kinet. Katal.*, **35** (2), 266 (1994).
98. M.-H. Huin, S. M. Brailovskii and O. N. Temkin, *Kinet. Katal.*, **35** (3), 367 (1994).
99. A. M. Sladkov and I. R. Gol'ding, *Dokl. Akad. Nauk SSSR*, **200**, 132 (1971).
100. A. M. Sladkov and L. Yu. Ukhin, *Usp. Khim.*, **37**, 1750 (1968).
101. M.-H. Huin, S. M. Brailovskii and O. N. Temkin, *Kinet. Katal.*, **35** (6), 889 (1994).
102. L. G. Fedenok, V. M. Berdnikov and M. S. Shvartsberg, *Zh. Org. Khim.*, (a) **9**, 1781 (1973); (b) **10**, 922 (1974); (c) **12**, 1395 (1976); (d) **14**, 1423 (1978); (e) **14**, 1429 (1978).
103. D. G. Sekutowcki and G. D. Stucky, *J. Am. Chem. Soc.*, **98**, 1376 1976).
104. I. Moritani and Y. Fujiwara, *Tetrahedron Lett.*, 1119 (1967).
105. Y. Fujiwara, T. Jintoku and K. Takaki, *Chemtech*, 636 (1990).
106. Ya. A. Dorfman, T. L. Rakitskaya, T. V. Petrova et al., in: *Gomogennoe okislenie (Trudy Instituta organicheskogo kataliza i elektrokhimii AN Kaz. SSR)* [Homogeneous Oxidation (Scientific Works of Institute of Organic Catalysis and Electrochemistry, Acad. Sci. Kazakh SSR)], Nauka, Alma-Ata, 1978, Vol. 17, pp. 34–89.
107. L. N. Kuznetsova, K. I. Matveev and E. G. Zhizhina, *Kinet. Katal.*, **26** (5), 1029 (1985).
108. H. Drägerwerk and B. Dräger, FRG Patent. No. 713791 (1934).
109. (a) E. G. Zhizhina, K. I. Matveev and L. N. Kuznetsova, *Kinet. Katal.*, **26** (2), 461 (1985); (b) E. G. Zhizhina, L. N. Kuznetsova and K. I. Matveev, *Kinet. Katal.*, **26** (6), 1349 (1985).
110. K. I. Matveev, N. K. Eremenko and N. K. Rachkovskaya, *Izv. Sib. Otd. Akad. Nauk SSSR, Ser. Khim.*, No. 7, 113 (1968).
111. A. V. Kulik, L. G. Bruk, O. N. Temkin, V. R. Khabibulin, V. K. Belsky and V. E. Zavodnik, *Mendeleev Commun.*, 47 (2002).
112. O. N. Temkin and L. G. Bruk, *Kinet. Katal.*, **44** (5), 661 (2003).
113. V. A. Golodov, E. L. Kuksenko and G. V. Taneeva, *Kinet. Katal.*, **23** (1), 248 (1982).
114. V. A. Golodov, E. L. Kuksenko and D. V. Sokol'skii, *Dokl. Akad. Nauk SSSR*, **272** (3), 628 (1983).
115. V. A. Golodov, E. L. Kuksenko, G. V. Taneeva et al., *Kinet. Katal.*, **26** (2), 330 (1984).

116. N. G. Gruzinskaya, B. S. Dzhumakaeva and V. A. Golodov, *Kinet. Katal.*, **36** (2), 211 (1995).
117. W. C. Higginson, *J. Chem. Soc.*, 447 (1957).
118. V. A. Golodov and L. V. Kushnikova, *J. Mol. Catal.*, **19**, 113 (1983).
119. S. M. Brailovskii and O. N. Temkin, *Zh. Vses. Khim. o-va im. D.I. Mendeleeva*, **30** (3), 331 (1985).
120. S. M. Brailovskii, O. N. Temkin, R. M. Flid and N. G. Belova, *Kinet. Katal.*, **11** (5), 1346 (1970).
121. L. G. Bruk, A. S. Kostyushin, S. M. Brailovskii, O. N. Temkin and R. M. Flid, *Zh. Org. Khim.*, **10** (11), 2262 (1974).
122. L. V. Shchel'tsyn, *Fiziko-khimicheskie osnovy tekhnologii kataliticheskogo sinteza khloralkinov i polikhlorolefinov iz atsetilena i ego proizvodnykh* (Physico-chemical Principles of Technology of Catalytic Synthesis of Chloroalkynes and Polychlorolefins from Acetylene and its Derivatives), Candidate Dissertation in Chemistry, MITKhT (Moscow Institute of Fine Chemical Technology), Moscow, 1989.
123. L. V. Shchel'tsyn, S. M. Brailovskii, Mai Suan Tin' and O. N. Temkin, *Kinet. Katal.*, **29** (3), 581 (1988).
124. L. V. Shchel'tsyn, S. M. Brailovskii and O. N. Temkin, *Kinet. Katal.*, **31** (2), 469 (1990).
125. N. M. Rodigin and E. N. Rodigina, *Posledovatel'nye khimicheskie reaktsii* (Consequent Chemical Reactions), Izd-vo AN SSSR, Moscow, 1960, p. 26.
126. L. V. Shchel'tsyn, S. M. Brailovskii, E. Yu. Murugova and O. N. Temkin, *Kinet. Katal.*, **29** (5), 1044 (1988).
127. L. V. Shchel'tsyn, S. M. Brailovskii and O. N. Temkin, *Kinet. Katal.*, **31** (6), 1361 (1990).
128. S. M. Brailovskii, O. N. Temkin, A. S. Kostyushin and K. Yu. Odintsov, *Kinet. Katal.*, **31** (6), 1371 (1990).
129. S. M. Brailovskii, M.-H. Huin and O. N. Temkin, *Kinet. Katal.*, **35** (5), 734 (1994).
130. S. M. Brailovskii, E. V. Boiko, A. S. Kostyushin and R. M. Flid, *Kinet. Katal.*, **14** (2), 513 (1973).
131. L. G. Bruk, S. M. Brailovskii, O. N. Temkin and R. M. Flid, *Kinet. Katal.*, **15** (2), 516 (1974).
132. S. M. Brailovskii, O. N. Temkin, R. M. Flid and N. G. Belova, *Zh. Fiz. Khim.*, **44** (4), 1112 (1970).
133. S. M. Brailovskii, O. N. Temkin, R. M. Flid and N. G. Belova, *Kinet. Katal.*, **11** (1), 107 (1970).
134. S. M. Brailovskii, O. N. Temkin and R. M. Flid, *Kinet. Katal.*, **12** (5), 1154 (1971).
135. N. Navon, G. Golub, H. Cohen and D. Meyerstein, *Organometallics*, **14**, 5670 (1995).
136. M. Masarawa, H. Cohen, J. Saar and D. Meyerstein, *Isr. J. Chem.*, **30**, 361 (1990).
137. M. Masarawa, H. Cohen and D. Meyerstein, *Inorg. Chem.*, **30**, 1849 (1991).
138. S. Goldstein, G. Czapski, H. Cohen and D. Meyerstein, *Inorg. Chem.*, **31**, 2439 (1992).
139. S. Goldstein and D. Meyerstein, *Accounts Chem. Res.*, **32** (7), 547 (1999).
140. H. M. McConnell and H. E. Weaver, *J. Chem. Phys.*, **25**, 304 (1956).
141. H. M. McConnell and H. M. Davidson, *J. Am. Chem. Soc.*, **72**, 3168 (1950).
142. S. M. Brailovskii, L. G. Bruk, A. S. Kostyushin, O. N. Temkin and R. M. Flid, *Kinet. Katal.*, **14** (5), 1222 (1973).
143. S. M. Brailovskii, O. N. Temkin, S. I. Khodakova and R. M. Flid, *Zh. Org. Khim.*, **11** (4), 672 (1975).
144. G. A. Artamkina, I. P. Beletskaya and O. N. Reutov, *Zh. Org. Khim.*, **9**, 1769 (1973).
145. M. Tamura and J. K. Kochi, *Bull. Chem. Soc. Jpn.*, **44**, 3063 (1971).
146. L. A. Zakharova, *Okislitel'noe khlorirovanie etilena i 1,3-butadiena i reaktsii izomerizatsii khlorolefinov* (Oxidative Chlorination of Ethylene and 1,3-Butadiene and Reaction of Isomerization of Chlorinated Olefins), Candidate Disser tation in Chemistry, MITKhT (Moscow Institute of Fine Chemical Technology), Moscow, 1985.
147. L. Ya. Mesh, O. N. Temkin and Yu. A. Shapirovskii, *Zh. Prikl. Khim.*, **51** (4), 883 (1978).
148. L. Ya. Mesh, O. N. Temkin and S. S. Gluzman, *Kinet. Katal.*, **24** (1), 98 (1983).
149. L. Ya. Mesh, O. N. Temkin and S. S. Gluzman, *Kinet. Katal.*, **24** (2), 336 (1983).
150. L. G. Bruk and O. N. Temkin, *Khim. Promst.*, No. 5, 57 (1993).
151. R. Ugo, in: *Catalysis in C1 Chemistry*, W. Keim (Ed.), Reidel, Dordrecht, 1983, pp. 135–167.
152. A. Klausener and J.-D. Jentsch, in: *Applied Homogeneous Catalysis with Organometallic Compounds*, B. Cornils and W. A. Herrmann (Eds), VCH, Weinheim, 1996, Vol. I, pp. 169–187.
153. J. Tsuji, *Palladium Reagents and Catalysis: Innovation in Organic Synthesis*, J. Wiley & Sons, New York, 1996.
154. (a) O. Temkin and L. Bruk, in: *Encyclopedia of Catalysis*, I. T. Horvath (Ed.), J. Wiley & Sons Hoboken, 2003, Vol. 5, pp. 394–424; (b) B. Gabriele, G. Salerno and M. Costa, *Topics Organomet. Chem.*, **18**, 239 (2006).
155. D. M. Fenton and P. J. Steinward, *J. Org. Chem.*, **39**, 701 (1974).
156. A. Chauvel, B. Delmon and W. F. Hölderich, *Appl. Catal.*, **115**, 173 (1994).

157. J. Tsuji, *Synthesis*, 739 (1990).
158. G. E. Morris, D. Oakley, D. A. Pippard and D. J. H. Smith, *J. Chem. Soc., Chem. Commun.*, 410 (1987).
159. J. E. Hallgren, G. M. Lukas and R. O. Matthews, *J. Organomet. Chem.*, **204**, 135 (1981).
160. J. E. Hallgren, G. M. Lukas and R. O. Matthews, *J. Organomet. Chem.*, **212**, 135 (1981).
161. M. Goyal, R. Nagahata, J. Sigiyama et al., *Catal. Lett.*, **54**, 29 (1998).
162. M. Goyal, R. Nagahata, J. Sigiyama et al., *J. Mol. Cat. A: Chem.*, **137**, 147 (1999).
163. Yu. L. Sheludyakov, V. A. Golodov and D. V. Sokol'skii, *Dokl. Akad. Nauk SSSR*, **249**, 658 (1979).
164. P. J. Giannoccaro, *Organomet. Chem.*, **336**, 271 (1987).
165. P. J. Giannoccaro, *Organomet. Chem.*, **470**, 249 (1994).
166. W. Brackmann, *Disc. Faraday Soc.*, **46**, 122 (1968).
167. S. I. Murahashi, Y. Mitsue and K. Ike, *J. Chem. Soc., Chem. Commun.*, 125 (1987).
168. O. L. Kaliya, O. N. Temkin, N. G. Mekhryakova and R. M. Flid, *Dokl. Akad. Nauk SSSR*, **119**, 132 (1971).
169. M. Graziani, P. Uguagliati and G. J. Carturan, *Organomet. Chem.*, **27**, 275 (1971).
170. O. N. Temkin, O. L. Kaliya, L. N. Zhir-Lebed', L. G. Bruk, V. A. Golodov and N. G. Mekhryakova, in: *Gomogennoe okislenie (Trudy Instituta organicheskogo kataliza i elektrokhimii AN Kaz. SSR)* [Homogeneous Oxidation (Scientific Works of Institute of Organic Catalysis and Electrochemistry, Acad. Sci. Kazakh SSR)], Nauka, Alma-Ata, 1978, Vol. 17, pp. 1–33.
171. L. N. Zhir-Lebed' and O. N. Temkin, *Kinet. Katal.*, **25**, 316, 325 (1984).
172. F. Rivetti and U. Romano, *J. Organomet. Chem.*, **154**, 323 (1978).
173. L. N. Zhir-Lebed', L. G. Kuz'mina, Yu. T. Struchkov, V. A. Golodov and O. N. Temkin, *Koord. Khim.*, **4**, 1046 (1978).
174. R. L. Angelici, *Accounts Chem. Res.*, **5**, 335 (1972).
175. P. Giannoccaro, N. Ravasio and M. Aresta, *J. Organomet. Chem.*, **451**, 243 (1993).
176. P. Giannoccaro, I. Tommasi and M. Aresta, *J. Organomet. Chem.*, **476**, 13 (1994).
177. U. Romano, R. Tesei, M. Massi Mauri and P. Rebora, *Ind. Eng. Chem. Prod. Res. Dev.*, **19**, 396 (1980).
178. T. Saegusa, T. Tsuda, K. Isayama and K. Nishijima, *Tetrahedron Lett.*, No. 7, 831 (1968).
179. T. Saegusa, T. Tsuda and K. Isayama, *J. Org. Chem.*, **35** (9), 2976 (1970).
180. P. Koch, L. G. Cipriani and E Perrotti, *Gazz. Chim. Ital.*, **104**, 599 (1974).
181. L. N. Zhir-Lebed', N. G. Mekhryakova, V. A. Golodov and O. N. Temkin, *Zh. Org. Khim.*, **11** (11), 2297 (1975).
182. J. H. Pawlow, A. D. Sadow and A. Sen, *Organometallics*, **16**, 1339 (1997).
183. P. Brechot, Y. Chavin, D. Commereuc and L. Saussine, *Organometallics*, **9**, 26 (1990).
184. J. H. Pawlow, A. D. Sadow and A. Sen, *Organometallics*, **16**, 5659 (1997).
185. J. Campora, P. Palma, D. Del Rio et al., *Organometallics*, **22**, 3345 (2003).
186. J. Smidt and R. Jira, *Chem. Ber.*, **93**, 162 (1960).
187. R. Ugo and A. Chiesa, *J. Chem. Soc., Perkin Trans.*, 2625 (1987).
188. Y. Fujiwara, K. Takaki and Y. Taniguchi, *Synlett*, 591 (1996).
189. I. O. Kalinovskii, A. A. Leshcheva, M. M. Kuteinikova and A. I. Gel'bshtein, *Zh. Obshch. Khim.*, **60**, 123 (1990).
190. N. Yu. Kozitsyna, M. N. Vargaftik and I. I. Moiseev, *J. Organomet. Chem.*, **274**, 593 (2000).
191. N. Yu. Kozitsyna, S. E. Nefedov, F. M. Dolgushin, M. N. Vargaftik and I. I. Moiseev, *Inorg. Chim. Acta*, **359**, 2072 (2006).
192. A. A. Leshcheva, I. O. Kalinovskii, V. V. Pogorelov et al., *Kinet. Katal.*, **39**, 381 (1998).
193. Y. Fujiwara, T. Jintoku and Y. Uchida, *New J. Chem.*, **13**, 649 (1989).
194. K. Nakata, T. Miyata, Y. Taniguchi et al., *J. Organomet. Chem.*, **489**, 71 (1995).
195. M. Lin and A. Sen, *Nature*, **368**, 613 (1994).
196. M. Lin, T. E. Hogan and A. Sen, *J. Am. Chem. Soc.*, **118**, 4574 (1996).
197. E. G. Chepaikin, A. P. Boyko, A. P. Bezruchenko, A. A. Lescheva and E. H. Grigoryan, *J. Mol. Catal. A: Chem.*, **129**, 15 (1998).
198. M. Asadullah, Y. Taniguchi, T. Kitamura and Y. Fujiwara, *Appl. Organomet. Chem.*, **12**, 277 (1998).

199. E. G. Chepaikin, A. P. Bezruchenko, A. A. Lescheva, G. N. Boiko, I. V. Kuzmenkov, E. H. Grigoryan and A. E. Shilov, *J. Mol. Catal. A: Chem*., **169**, 89 (2001).
200. (a) E. G. Chepaikin, A. P. Bezruchenko, G. N. Boiko, A. E. Gekhman, I. I. Moiseev, *Kinet. Katal*., **47** (1), 16 (2006); (b) E. G. Chepaikin, *Usp. Khim*., **80**(4), 384 (2011).
201. D. Medema, R. Van Helden and F. Kohel, *Inorg. Chim. Acta*, **3**, 255 (1969).
202. V. S. Shestakova, S. M. Brailovskii and O. N. Temkin, *Kinet. Katal*., **18** (6), 1585 (1978).
203. S. M. Brailovskii, O. N. Temkin and V. S. Shestakova, *Kinet. Katal*., **24** (2), 326 (1983).
204. V. S. Shestakova, *Razrabotka novykh sintezov na osnove atsetilena i oksida ugleroda s primeneniem gomogennoi polifunktsional'noi kataliticheskoi sistemy* (Development of New Syntheses Based on Acetylene and Carbon Oxide Using Homogeneous Polyfunctional Catalytic System), Candidate Dissertation in Chemistry, MITKhT (Moscow Institute of Fine Chemical Technology), Moscow, 1980.
205. L. G. Makarova and A. N. Nesmeyanov, *Metody elementoorganicheskoi khimii: Rtut'* (Methods of Organoelemental Chemistry: Mercury), Nauka, Moscow, 1965), p. 166.
206. L. P. Kur'yaninova, O. L. Kaliya, O. N. Temkin and R. M. Flid, *Zh. Obshch. Khim*., **39**, 2695 (1969).
207. O. L. Kaliya, O. N. Temkin, R. M. Fli, and L. G. Volkova, *Zh. Neorg. Khim*., **15**, 2561 (1970).
208. V. S. Shestakova, T. N. Omel'chenko, S. M. Brailovskii and O. N. Temkin, *Zh. Org. Khim*., **14**, 2346 (1978).
209. C. Döbler, H. Kreuzfeld and H. Krause, *Z. Chem*., **11**, 147 (1971).
210. R. Larock, *J. Org. Chem*., **41**, 2241 (1976).
211. S. M. Brailovskii, G. A. Kabalina, V. S. Shestakova and O. N. Temkin, *Zh. Org. Khim*., **13**, 1158 (1977).
212. L. G. Bruk, I. V. Oshanina, A. P. Kozlova, E. V. Vorontsov and O. N. Temkin, *J. Mol. Catal. A: Chem*., **104**, 9 (1995).
213. L. G. Bruk, I. V. Oshanina, A. P. Kozlova, K. Yu. Odintsov and O. N. Temkin, *Izv. Akad. Nauk, Ser. Khim*., **47**, 1104 (1998).
214. L. G. Bruk, I. V. Oshanina, A. S. Zakieva and O. N. Temkin, *Kinet. Katal*., **39**, 183 (1998).
215. J. Tsuji, M. Takahashi and T. Takahashi, *Tetrahedron Lett*., **21**, 849 (1980).
216. T. T. Zung, L. G. Bruk and O. N. Temkin, *Izv. Akad. Nauk, Ser. Khim*., No. 10, 1806 (1993).
217. T. T. Zung, L. G. Bruk and O. N. Temkin, *Mendeleev Commun*., 2 (1994).
218. L. G. Bruk, S. N. Gorodskii, A. V. Zeigarnik, R. E. Valdes-Perez and O. N. Temkin, *J. Mol. Catal. A: Chem*., **130**, 29 (1998).
219. L. G. Bruk, S. N. Gorodskii, A. V. Zeigarnik, R. E. Valdes-Perez and O. N. Temkin, *Izv. Akad. Nauk, Ser. Khim*., No. 5, 882 (1999).
220. L. G. Bruk and O. N. Temkin, *Inorg. Chim. Acta*, **280**, 202 (1998).
221. M. Beller, B. Cornils, C. D. Frohning and C. W. Kohlpaintner, *J. Mol. Catal. A: Chem*., **104**, 17 (1995).
222. Yu. G. Noskov, M. I. Terekhova and E. S. Petrov, *Zh. Obshch. Khim*., **60** (6), 1336 (1990).
223. Yu. V. Tel'naya, Yu. G. Noskov and E. S. Petrov, *Zh. Obshch. Khim*., **65** (8), 1391 (1995).
224. T. E. Kron, M. I. Terekhova, Yu. G. Noskov and E. S. Petrov, *Zh. Fiz. Khim*., **72** (10), 1834 (1998).
225. T. E. Kron, M. I. Terekhova, Yu. G. Noskov and E. S. Petrov, *Zh. Obshch. Khim*., **67** (1), 108 (1997).
226. V. Yu. Gankin and G. S. Gurevich, *Tekhnologiya oksosinteza* (Oxo Synthesis Technology), Khimiya, Leningrad, 1981.
227. N. S. Imyanitov and D. M. Rudkovskii, *Zh. Org. Khim*., **2**, 23 (1966).
228. N. S. Imyanitov and D. M. Rudkovskii, *Zh. Prikl. Khim*., **41**, 172 (1968).
229. N. S. Imyanitov, N. M. Bogoradovskaya and T. A. Semenova, *Kinet. Katal*., **19** (3), 573 (1978).
230. V. Yu. Gankin, M. G. Katsnel'son and D. M. Rudkovskii, in: *Karbonilirovanie nenasyshchennykh uglevodorodov* (Carbonylation of Unsaturated Hydrocarbons), D. M. Rudkovskii (Ed.), Khimiya, Leningrad, 1968, p. 178.
231. N. S. Imyanitov and D. M. Rudkovskii, *Zh. Prikl. Khim*., **39**, 2335 (1966).
232. A. Matsuda, *Bull. Chem. Soc. Jpn*, **46**, 524 (1973).
233. B. Despeyroux and H. Alper, *Ann. New York Acad. Sci*., **415**, 148 (1984).
234. H. Alper, J. B. Woell, B. Despeyroux and D. J. H. Smith, *J. Chem. Soc. Chem. Commun*., 1270 (1983).
235. B.-H. Xie, C.-G. Xia S.-J. Lu, et al., *Tetrahedron Lett*., **39**, 7365 (1998).
236. J.-F. Roth, J. H. Craddock, A. Hershman and F. E. Paulik, *Chem. Technol*., 600 (1971).
237. D. Förster, *J. Am. Chem. Soc*., **98**, 846 (1976).

238. D. Förster, *Adv. Organometal. Chem.*, **17**, 255 (1979).
239. D. Förster and T. C. Singleton, *J. Mol. Catal.*, **17**, 299 (1982).
240. T. W. Dekleva and D. Förster, *Adv. Catal.*, **34**, 81 (1986).
241. P. M. Maitlis, A. Haynes, G. J. Sanley and M. J. Howard, *J. Chem. Soc. Dalton Trans.*, 2187 (1996).
242. D. Förster, *J. Chem. Soc. Dalton Trans.*, 1639 (1979).
243. P. R. Ellis, J. M. Pearson, A. Haynes and P. M. Maitlis, *Organometallics*, **13**, 3215 (1994).
244. M. J. Howard, G. J. Sunley, A. D. Poople et al., *Stud. Surf. Sci. Catal.*, **121**, 61 (1999).
245. G. J. Sunley and D. J. Watson, *Catal. Today*, **58**, 293 (2000).
246. J. H. Jones, Plat. *Met. Rev.*, **44**, 94 (2000).
247. A. Haynes, *Educ. Chem.*, **38**, 99 (2001).
248. A. Haynes, P. M. Maitlis, G. E. Morris et al., *J. Am. Chem. Soc.*, **126**, 2847 (2004).
249. K. Sonogashira, Y. Tohda and N. Hagihara, *Tetrahedron. Lett.*, 4467 (1975).
250. K. Sonogashira, *J. Organometal. Chem.*, **145**, 101 (1978).
251. M. L. Spector, G. L. M. Cristopher and E. L. Carpenter, US Patent no. 2784236 (1957); *Ref. Zh. Khim.*, No. 24, 8558OP (1958).
252. S. A. Vartanyan, S. K. Pirenyan and N. G. Manasyan, *Izv. Akad. Nauk Arm. SSR, Ser. Khim.*, **12**, 345 (1959).
253. O. N. Temkin, G. K. Shestakov and R. M. Flid, *Dokl. Akad. Nauk SSSR*, **184** (2), 362 (1969).
254. G. K. Shestakov, A. Ermakova, V. G. Mikhal'chenko, O. N. Temkin and R. M. Flid, *Zh. Fiz. Khim.*, **44**, 406 (1970).
255. G. K. Shestakov, O. N. Temkin, N. Yu. Vsesvyatskaya and A. M. Stepanov, *Zh. Org. Khim.*, **15**, 248 (1979).
256. V. I. Laba, A. V. Sviridova, A. A. Kron and E. N. Prilezhaeva, *Izv. Akad. Nauk SSSR, Ser. Khim.*, No. 7, 1566 (1977).
257. G. K. Shestakov, N. Yu. Vsesvyatskaya, T. Yu. Gulina and O. N. Temkin, *Zh. Org. Khim.*, **16** (2), 324 (1980).
258. N. Yu. Vsesvyatskaya, D. V. Tsyganov, A. M. Bykov, G. K. Shestakov and O. N. Temkin, *Zh. Org. Khim.*, **21** (10), 2041 (1985).
259. N. Yu. Vsesvyatskaya, G. K. Shestakov and O. N. Temkin, *Kinet. Katal.*, **27** (7), 1330 (1986).
260. N. Yu. Vsesvyatskaya, G. K. Shestakov and O. N. Temkin, *Kinet. Katal.*, **29** (4), 965 (1988).
261. N. F. Alekseeva, O. N. Temkin and R. M. Flid, *Kinet. Katal.*, **11** (4), 1071 (1970).
262. O. N. Temkin, N. F. Alekseeva, R. M. Flid, O. L. Kaliya, G. K. Shestakov and T. I. Dolgina, *Dokl. Akad. Nauk SSSR*, **196** (4), 836 (1971).
263. S. M. Airyan, A. M. Masal'skaya, G. K. Shestakov and O. N. Temkin, *Kinet. Katal.*, **18**, 399 (1977).
264. O. N. Temkin, O. Yu. Antonova, S. A. Kezin, N. G. Akhmedov and Yu. A. Ustynyuk, *Proceedings of the 12th Int. Symp. on Homogeneous Catalysis (Stockholm, 2000)*, Book of Abstracts, p. 224.
265. G. J. Hutchings and R. Joffe, *Appl. Catal.*, **20**, 215 (1986).
266. L. A. Sil'chenko, S. A. Panova, G. K. Shestakov and O. N. Temkin, *Kinet. Katal.*, **38** (6), 861 (1997).
267. L. A. Sil'chenko, S. A. Panova, G. K. Shestakov and O. N. Temkin, *Kinet. Katal.*, **39**, 28 (1998).
268. L. A. Sil'chenko, *Razrabotka fiziko-khimicheskikh osnov tekhnologii sinteza vinilkhlorida iz atsetilena na nertutnykh katalizatorakh* (Development of Physico-chemical Principles of Vinyl Chloride Synthesis from Acetylene on Nonmercuric Catalysts), Candidate Dissertation in Chemistry, MITKhT (Moscow Institute of Fine Chemical Technology), Moscow, 1987.
269. N. Dan and S. D. Khue, in: *Fundamental Research in Homogeneous Catalysis*, A. E. Shilov (Ed.), Gordon & Breach, Philadelphia, 1986, Vol. 2, p. 675.
270. D. I. Kamenski, S. D., Dimitrov, L. A. Sil'chenko, G. K. Shestakov and O. N. Temkin, *Appl. Catal.*, **67**, 159 (1990).
271. S. A. Panova, *Fiziko-khimicheskie osnovy gidrokhlorirovaniya atsetilena na rodievykh katalizatorakh* (Physico-chemical Principles of Acetylene Hydrochlorination on Rhodium Catalysts), Candidate Dissertation in Chemistry, MITKhT (Moscow Institute of Fine Chemical Technology), Moscow, 1990.
272. S. A. Panova, G. K. Shestakov and O. N. Temkin, *J. Chem. Soc., Chem. Commun.*, 977 (1994).
273. S. M. Brailovskii, O. N. Temkin and E. S. Klimova, *Kinet. Katal.*, **24** (5), 1091 (1983).
274. G. K. Oparina, R. N. Gurskii, I. L. Vaisman and R. M. Flid, *Khim. Promst*, No. 9, 15 (1967).
275. S. M. Brailovskii, O. L. Kaliya, O. N. Temkin and R. M. Flid, *Kinet. Katal.*, **9** (1), 177 (1968).

276. P. M. Maitlis, *Accounts Chem. Res.*, **9**, 93 (1976).
277. V. S. Shestakova, S. M. Brailovskii, O. N. Temkin and G. I. Tarasova, *Kinet. Katal.*, **19** (1), 157 (1978).
278. S. M. Brailovskii, O. N. Temkin and V. S. Shestakova, *Kinet. Katal.*, **24** (4), 848 (1983).
279. I. P. Skibida, Z. K. Maizus and N. M. Emanuel', *Usp. Khim.*, **38**, 3 (1969).
280. W. H. Elliott and D. C. Elliott, *Biochemistry and Molecular Biology*, Oxford University Press, Oxford, UK, 2001.
281. A. S. Abdulaeva, K. Yu. Philushova, A. N. Vyrodov et al., *Proceedings of the Int. Symp. Modern Trends in Organometallic Catalytic Chemistry (Moscow, 2003)*, Book of Abstracts, no. 17.
282. L. G. Bruk, I. V. Oshanina, S. N. Gorodskii and O. N. Temkin, *Zh. Khim. o-va im. D.I. Mendeleeva*, **50** (11), 103 (2006).
283. J.-E. Bäckvall, R. L. Chowdhury, U. Karlsson and G.-Z. Wang, in: *Perspectives in Coordination Chemistry*, A. E. Williams, C. Floriani and A. E. Merbac (Eds), VCH, Basel, 1992, pp. 463–486.
284. M. J. Chetcuti, J. A. K. Howard, M. Pteffer et al., *J. Chem. Trans.*, **276** (1981).
285. A. Aranyos, G. Csjernyik, K. J. Szabo and J.-E. Bäckvall, *J. Chem. Soc. Chem. Commun.*, 351 (1999).
286. Y. R. Santosh Laxmi and J.-E. Bäckvall, *J. Chem. Soc. Chem. Commun.*, 611 (2000).
287. G. Csjernyik, A. H. Ell, L. Fadini et al., *J. Org. Chem.*, **67**, 1657 (2002).
288. M. L. Khidekel', in: *Kompleksoobrazovanie v katalize* (Complex Formation in Catalysis), Nauka, Moscow, 1968, pp. 126–140.
289. S. N. Zelenin and M. L. Khidekel', *Usp. Khim.*, **39**, 209 (1970).
290. E. G. Chepaikin and M. L. Khidekel', *Koord. Khim.*, **4** (5), 643 (1978).
291. I. I. Moiseev and S. V. Pestrikov, *Dokl. Akad. Nauk SSSR*, **171**, 151 (1966).
292. I. I. Moiseev, S. V. Pestrikov and S. V. Sverzh, *Kinet. Katal.*, **10**, 74 (1969).
293. L. G. Bruk, O. N. Temkin, Z. V. Goncharova, I. S. Ustenko and V. R. Flid, *React. Kinet. Catal. Lett.*, **9**, 303 (1978).
294. L. G. Bruk, *Metody vydvizheniya i diskriminatsii gipoteticheskikh mekhanizmov kataliticheskikh reaktsii kak osnova razrabotki tekhnologii karbonilirovaniya alkinov* (Methods of Generation and Discrimination of Hypothetical Mechanisms of Catalytic Reactions: Basis for Developing Alkyne Carbonylation Technology), Doctoral Dissertation in Chemistry, MITKhT (Moscow Institute of Fine Chemical Technology), Moscow, 1996.
295. S. N. Gorodskii, L. G. Bruk and O. N. Temkin, unpublished results.
296. A. Vavasori and L. Toniolo, *J. Mol. Cat. A: Chem.*, **151**, 37 (2000).
297. G. Oehme and S. Pracejus, *J. Prakt. Chemie*, **322** (5), 798 (1980).
298. G. Oehme, *Uber σ-organo-Palladium-Verbindungen von Acetonitril, Acrylnitril und Methylacrylat und deren Beziehungen zu C-C Knüpfungsreactionen*, Dissertation des Grades Doktor der Wissenschaften, W. Pieck Universitat, Rostock, 1980.
299. A. V. Kulik, A. A. Goncharov, L. G. Bruk and O. N. Temkin, *Proceedings of the 14th Int. Symp. on Homogeneous Catalysis (Munich, Germany, 2004)*, Book of Abstracts, p. 109.
300. A. V. Kulik, *Funktsii para-khinonov v kataliticheskikh protsessakh okislitel'nogo karbonilirovaniya* (Functions of Para-Quinones in Catalytic Processes of Oxidative Carbonylation), Candidate Dissertation in Chemistry, MITKhT (Moscow Institute of Fine Chemical Technology), Moscow, 2004.
301. H. W. Sternberg, R. Markby and I. Wender, *J. Am. Chem. Soc.*, **80**, 1009 (1958).
302. R. S. Dickson and H. P. Kirsch, *Aust. J. Chem.*, **25**, 61 (1972).
303. R. S. Dickson and H. P. Kirsch, *Aust. J. Chem.*, **27**, 61 (1974).
304. R. S. Dickson and S. H. Johnson, *Aust. J. Chem.*, **29**, 2189 (1976).
305. C. I. Jewell, L. S. Liebeskind and M. Williamson, *J. Am. Chem. Soc.*, **107**, 6715 (1985).
306. G. N. Schrauzer and H. Tsyret, *J. Am. Chem. Soc.*, **82**, 6420 (1960).
307. M. D. Glick and L. F. Dahl, *J. Organomet. Chem.*, **3**, 200 (1965).
308. H.-F. Klein, E. Auer, A. Dal et al., *Inorg. Chim. Acta*, **287**, 167 (1999).
309. Y. Ura, Y. Sato, M. Shiotsuki et al., *Organometallics*, **22**, 77 (2003).

310. G. N. Schrauser and K. C. Dewhirst, *J. Am. Chem. Soc.*, **86**, 3265 (1964).
311. M. Handa, A. Takata, T. Nakao et al., *Chem. Lett.*, 2085 (1992).
312. M. Oh, G. B. Carpenter and D. A. Sweigart, *J. Chem. Soc., Chem. Commun.*, 2168 (2002).
313. S. Takahashi and N. Hagihara, *J. Chem. Soc. Jpn. (Pure Chem. Sect.)*, **88**, 1306 (1967).
314. T. Ukai, H. Kawazura, Y. Ishii et al., *J. Organomet. Chem.*, **65**, 203 (1974).
315. P. Roffia, G. Gregorio, F. Conti et al., *J. Mol. Catal.*, **2**, 191 (1977).
316. M. Hiramatsu, K. Shiozaki, T. Fujinami and S. Sakai, *J. Organomet. Chem.*, **246**, 203 (1983).
317. M. Tschoerner, G. Trabensinger, A. A. Albinaty and P. S. Pregosin, *Organometallics*, **16**, 3447 (1997).
318. K. Selvakumar, A. Zapf and M. Beller, *Org. Lett.*, **4** (18), 3031 (2002).
319. K. Selvakumar, A. Zapf, A. Spannenberg and M. Beller, *Chem. Eur. J.*, **8** (17), 3901 (2002).
320. S. Centini, R. Ugo and G. La Monica, *J. Chem. Soc. A*, 416 (1971).
321. M. J. Chetcuti, J. A. Herbert, J. A. K. Howard et al., *J. Chem. Soc., Dalton Trans.*, 284 (1981).
322. J. La Braas, H. Amouri and J. Vaissermann, *Organometallics*, **17**, 1116 (1998).
323. N. B. Shitova, L. I. Kuznetsova, E. N. Yurchenko, I. A. Ovsyannikova and K. I. Matveev, *Izv. Akad. Nauk SSSR, Ser. Khim.*, No. 7, 1453 (1973).
324. R. A. Klein, P. Witte, R. van Belzen et al., *Eur. J. Inorg. Chem.*, 319 (1998).
325. V. R. Khabibullin, A. V. Kulik, V. M. Nosova, Yu. A. Ustynyuk, I. V. Oshanina, L. G. Bruk and O. N. Temkin, *Izv. Akad. Nauk, Ser. Khim.*, No. 2, 244 (2007).
326. Yu. Fal'be, *Sintezy na osnove oksida ugleroda* (Carbon Oxide Based Syntheses), Khimiya, Leningrad, 1971, p. 63.
327. K. A. Alekseeva, M. P. Vysotskii, N. S. Imyanitov and V. A. Rybakov, *Zh. Vses. Khim. o-va im. D.I. Mendeleeva*, **22**, 45 (1977).
328. N. S. Imyanitov, *Neftekhimiya*, **21** (6), 872 (1981).
329. I. Amer and H. Alper, *J. Organomet. Chem.*, **383**, 573 (1990).
330. J.-Tae Lee and H. Alper, *Tetrahedron Lett.*, **32** (15), 1769 (1991).
331. K. Watanabe, K. Kudo and N. Sugita, *Bull. Chem. Soc. Jpn.*, **58**, 2029 (1985).
332. H. Alper and J.-F. Petrignani, *J. Chem. Soc., Chem. Commun.*, 1154 (1983).
333. H. Wakamatsu, J. Uda and N. Yamakami, *J. Chem. Soc., Chem. Commun.*, 1540 (1971).
334. J. J. Paraud, G. Camperi and P. Pino, *J. Mol. Catal.*, **6**, 341 (1979).
335. I. Ojima, K. Hirai, M. Fujita and T. Fuchikami, *J. Organomet. Chem.*, **279**, 203 (1985).
336. I. Ojima, M. Okabe, K. Kato, et al., *J. Am. Chem. Soc.*, **110**, 150 (1988).
337. I. T. Horväth, G. Bor, M. Garland and P. Pino, *Organometallics*, **5**, 1441 (1986).
338. I. T. Horväth, G. Bor, M. Garland and P. Pino, *Proceedings of the 193rd Natl. Meeting of ACS (Denver CO, 1987)*, Abstracts of Papers.
339. H. Arzoumanian, H. Bitar and J. Metzger, *J. Mol. Catal.*, **7**, 373 (1980).
340. U. T. Strauss, U. Felfer and K. Faber, *Tetrahedron: Asymm.*, **10**, 107 (1999).
341. B. A. Persson, A. L. E. Larsson, M. Le Ray and J.-E. Bäckvall, *J. Am. Chem. Soc.*, **121**, 1645 (1999).
342. R. P. G. Gatti, A. L. E. Larsson and J.-E. Bäckvall, *J. Chem. Soc. Perkin Trans. 1*, **4**, 577 (1997).
343. C. Simons, U. Hanefeld, I. W. C. E. Arends and Th. Maschmeyer, *Adv. Synth. Catal.*, **348** (4-5), 471 (2006).
344. C. Simons, U. Hanefeld, I. W. C. E. Arends et al., *Topics Catal.*, **40** (1-4), 35 (2006).
345. I. A. Esikova, *Issledovanie mekhanizma nekotorykh prevrashchenii olefinov v rastvorakh solei rtuti* (Studying Mechanisms of Some Olefin Transformations in Solutions of Mercury Salts), Candidate Dissertation in Chemistry, MITKhT (Moscow Institute of Fine Chemical Technology), Moscow, 1970.
346. L. A. Il'ina, *Fiziko-khimicheskie osnovy malootkhodnoi tekhnologii sinteza parfyumernogo tsitralya* (Physicochemical Principles of Low-Waste Technology of Cosmetic Citral Synthesis), Candidate Dissertation in Chemistry, MITKhT (Moscow Institute of Fine Chemical Technology), Moscow, 1992.
347. A. Ajamian and J. L. Gleason, *Angew. Chem.*, **116**, 3842 (2004).

348. L. F. Tietze, G. Brasche and K. M. Gericke, *Domino Reactions in Organic Chemistry*, Wiley-VCH, Weinheim, 2007.
349. J. Zhu and H. Bienayme (Eds.), *Multicomponent Reactions*, Wiley-VCH, Weinheim, 2005.
350. J.-C. Wasilke, S. J. Obrey, R. T. Baler and G. C. Bazan, *Chem. Rev*., **105**, 1001 (2005).
351. D. E. Fogg and E. N. dos Santos, *Coord. Chem. Rev*, **248**, 2365 (2004).
352. N. Shindoh, Y. Takemoto and K. Takasu, *Chem. Eur. J*., **15**, 12168 (2009).
353. Z. An, X. Pan, X. Liu, X. Han and X. Bao, *J. Am. Chem. Soc*., **128**, 16028 (2006).
354. A. Dijksman, I. W. C. E. Arends and R. A. Sheldon, *J. Chem. Soc., Chem. Commun*., 1591 (1999).
355. A. Dijksman., A. Marino-Gonzales, A. Maizatai Payeras et al., *J. Am. Chem. Soc*., **123**, 6826 (2001).
356. R. A. Sheldon, I. W. C. E. Arends, G.-J. Ten Brick and A. Dijksman, *Accounts Chem. Res*., **35**, 774 (2002).
357. F. Minisci, F. Recupero, A. Cecchetto et al., *Eur. J. Org. Chem*., 109 (2004).

5

Mechanisms of Formation of Catalytically Active Metal Complexes

One of the most important problems in constructing a theoretically justified kinetic model, planning a kinetic experiment, and implementing this plan is related to establishing the nature of catalytically active species in solution or on the surface of a catalyst. In the case of catalysis with metal complexes, this problem reduces to determining the degree of oxidation of the metal, the number of metal atoms in the active species, and the composition of a coordination sphere of the complex that enters into a reaction with substrates of the catalytic process.

In heterogeneous catalysis with metals and their oxides, the formation of a catalyst proceeds with the direct participation of the reaction medium [1, 2]. This process involves the recrystallization of a catalyst (or "catalytic corrosion" in Roginsky's expression), reduction or oxidation of the surface atoms, and their irreversible interaction with reactants and reaction products. The nonequilibrium surface states tend toward equilibrium or steady states. It is this interaction of the medium with the catalyst which leads to the appearance of the proper "active centers" that catalyze the process. It was convincingly established that a reconstruction of the catalyst surface can be reversible and vary depending on the occupation of the surface with reactants [3–5], sometimes even in an oscillatory regime [6, 7]. In 2007, G. Ertl was awarded the Nobel Prize in chemistry for his studies of chemical processes on solid surfaces.

The phenomenon of catalyst formation is especially important in homogeneous catalysis, where two other factors, solvent and ligands, appear that can lead to changes in the composition of the coordination sphere and the degree of metal oxidation. In cases where the steps of the formation of a catalytic system lead to the appearance of metal complexes in a stable form which is incapable of converting in reactions other than steps of the catalytic process, it is possible to exclude the formation steps from the consideration of reaction kinetics and study the process as occurring under the conditions of steady-state functioning of the catalyst. In such cases, the effect of the reaction medium on the reaction rate is determined by the kinetics of a catalytic route (routes).

However, practice shows that *real catalysts* possess limited working lifetimes. An active center (catalytic complex) which appears upon the aforementioned formation process only performs a finite number of catalytic acts (cycles), which is frequently rather small. As a result of the loss of catalytic centers and/or the side conversions of intermediates, the catalyst passes to an inactive state, frequently with a change in the degree of metal oxidation. Naturally, the more complicated the catalyst, the greater the number of

Homogeneous Catalysis with Metal Complexes: Kinetic Aspects and Mechanisms, First Edition. O. N. Temkin.

pathways for its passage to the inactive state. Accordingly, the problem of formation and conversion of catalytically active centers is especially important for the catalysis with metal complexes and the enzymatic catalysis. The enzymes – biological catalysts with very complicated organization – are characterized by very restricted lifetimes and exhibit continuous transformation during the catalytic process. The loss of active enzymes in the organism is compensated by their synthesis [8, 9].

In catalytic reactions with the participation of low-stability (or highly reactive) complexes, the steps of catalyst formation and the loss of active centers significantly influence the kinetics of the main catalytic reaction. The process may acquire a chain character and exhibit all features of the chain reaction kinetics. The possible variants of the kinetics of heterogeneous catalytic reactions, with allowance for the formation and conversion of catalytically active centers, were originally considered by Rozovskii [10–12]. This chapter is devoted to an analysis of the chemical and kinetic aspects of this problem in the catalysis with metal complexes. First, let us consider the main stages constituting a real catalytic process in solution of metal complexes.

5.1 Main stages of catalytic process

Figure 5.1 shows six possible stages of the working life of a real catalyst in the case of mononuclear metal complexes, as represented by six blocks of possible elementary steps. In the first stage (block of steps **1**), an active center (complex) ML_n is formed from a precursor (precatalyst) ML_mX_n. In the simplest cases, the initial compound can either be the active center (as, e.g., Na_2PdCl_4 in the reaction of olefin oxidation) or convert into an active complex as a result of dissociation (i.e., ML_n is a precursor):

$$ML_n \rightleftharpoons ML_{n-1} + L.$$

However, even in these simple cases, various geometric isomers or allotropic modifications of the metal complex can possess different activities. For example, red and orange modifications of the Wilkinson's complex $RhClL_3$ [13] exhibit different activities in the hydroformylation of allyl alcohol [14].

In stage **1**, the typical elementary steps are as follows:

- reactions of ligands (reactant) addition and ligand substitution and/or exchange with components of the catalytic system;

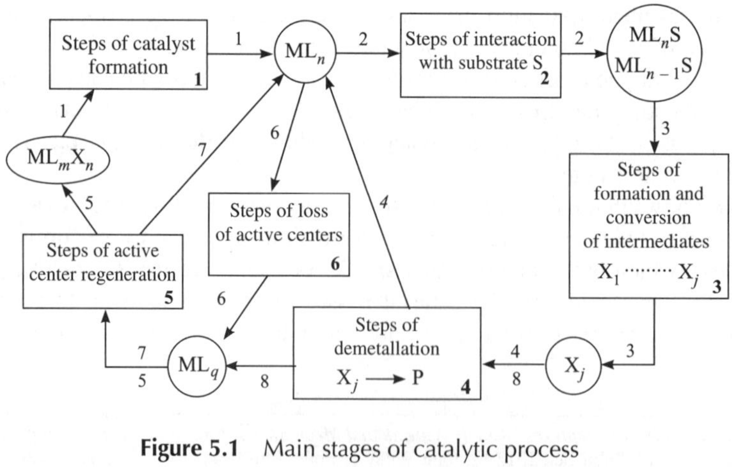

Figure 5.1 Main stages of catalytic process

- oxidation–reduction reactions with reactants, ligands, solvents, and other components of the system, including the steps of oxidative addition, for example,

$$Co_2(CO)_8 + H_2 \rightleftharpoons 2HCo(CO)_4.$$

In stage **2**, the active complexes (ML_n) react with molecules of a substrate or substrates (H_2, CO, CO_2, N_2, O_2, HX, olefin, alkyne, arene, diene, etc.) with the formation of primary complexes (ML_nS or $ML_{n-1}S$), in which the so-called activation of substrate takes place. Ligands (L) in the primary complex can be represented by the molecules and groups such as H_2O, PR_3, NR_3, H^-, R^-, CN^-, Cl^-, OH^-, NR_2^-, $R_2C{=}$, $RC{\equiv}$, etc. The most typical reactions in this stage are as follows:

- addition of substrate;
- substitution of ligand;
- oxidative addition;
- electron transfer.

In stage **3**, the primary complexes ML_nS transform into other intermediates, in which the substrate or substrates are gradually converted into compounds X_j which yield the target reaction product P. This stage can involve almost any type of elementary steps characteristic of reactions with coordinated ligands, including:

- electrophilic substitution

$$M(SH) \rightleftharpoons MS^- + H^+,$$

$$M_1R + MX \longrightarrow M_1X + MR;$$

- oxidative addition

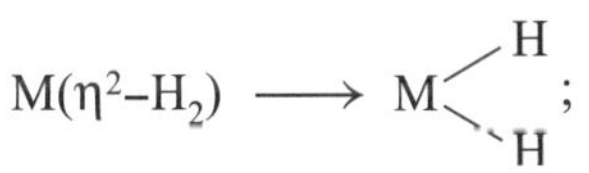

- α, β-elimination (and reverse insertion reactions);
- 2 + 2-cycloaddition (and reverse reaction of metallacycle decomposition);
- nucleophilic addition to π complexes;
- electrophilic (oxidative) addition to π complexes

$$M(\eta^2\text{-}C_2H_4) + HX \longrightarrow M(CH_2CH_3)^+X^-;$$

- isomerization of complexes ($\eta^1 \rightarrow \eta^2$, $\eta^3 \rightarrow \eta^1$, etc.)

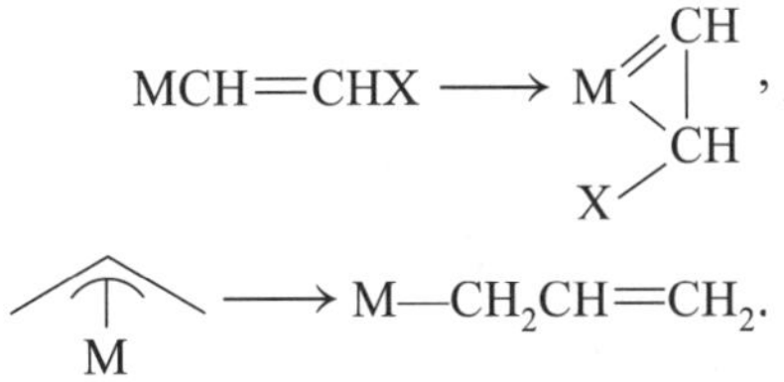

It should be noted that if there are several reaction products, some of them can be eliminated in the steps preceding the formation of intermediate X_j that enters into stage **4**.

In stage **4**, demetallation of intermediate X_j proceeds with the formation of reaction product P and one of the metal-containing catalyst species. If complex ML_n is formed, then a catalytic cycle arises that consists of stages **2**, **3**, and **4**. If stage **4** leads to the formation of inactive species ML_q (or these species appear as a result of a side conversion of intermediates at the early steps of stage **3**), it becomes necessary to regenerate active centers ML_n in stage **5** either immediately from ML_q species or by means of their return to the initial state (i.e., via stages **5** + **1**). Note also the possible appearance of an additional block of steps (stage **6**) involving the loss of active centers ML_p. The composition of a coordination sphere and the degree of metal oxidation in ML_n and ML_q complexes can be different. The demetallation stage **4** can include the following steps:

- electrophilic substitution

$$L_2M(R)X + E^+ \longrightarrow L_2MX^+ + RE;$$

- reductive elimination

$$L_2M(R)X \longrightarrow L_2M + R\text{–}X;$$

- reductive elimination with participation of oxidant (Ox)

$$RMX(Ox) \xrightarrow{2HX} RX + Red + 2H^+ + MX_2;$$

- nucleophilic substitution at carbon atom

$$Y^- + RMX_2 \longrightarrow YR + MX_2^-;$$

- nucleophilic substitution at carbon atom with participation of oxidant (Ox)

$$Y^- + RMX + Ox \xrightarrow{2HX} YR + MX_2 + Red + 2H^+;$$

- β-elimination;
- homolytic rupture of M–R bonds.

Taking one reaction (step) from each block (stage), it is possible to construct a kinetic graph (Figure 5.2) that reflects the topology of a real catalytic process with a chain mechanistic scheme and consists of four independent routes, including two routes for the formation and loss of active centers (steps 1–6–5 and 7–6) and two cyclic routes for the chain propagation (steps 2–3–4 and 2–3–8–7). In the absence of

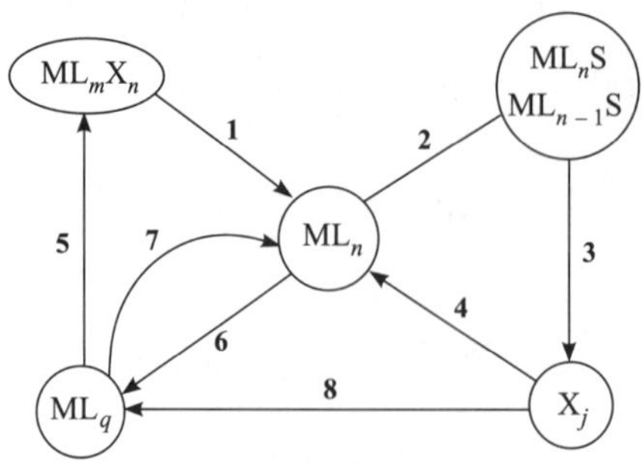

Figure 5.2 Kinetic graph corresponding to main stages of catalytic process

steps 5 and 7 (stage **5**), the kinetic graph transforms into that of a typical nonbranched chain process. Examples of chain mechanisms in the catalysis with metal complexes will be considered below.

Numerous variants of elementary steps in organic catalysis with metal complexes were considered in Chapters 2–4 and can be found in [15–18]. Some special cases for particular processes are analyzed below. In the analysis of reaction mechanisms, often insufficient attention is devoted to the steps of active complex formation (stage **1**) and loss (stage **6**), the interrelation of these steps with the main catalytic routes, the pathways of kinetic conjugation of these processes (see Section 3.4), and the chain mechanisms of real processes. In the author's opinion, a most detailed analysis of these issues was given in [19–23].

5.2 Chemical reactions involved in the formation of active centers

Let us consider possible variants of the interaction of precursors with all components of the reaction medium in the homogeneous catalysis with metal complexes.

Reactants (substrates and products) of catalytic reactions. The replacement of a volatile ligand L (e.g., N_2) in the precursor by substrate S (e.g., cyclohexene or styrene) leads to the formation of a substrate-containing metal (e.g., Co) complex that is active in the hydrogenation reaction [24–26]:

$$HCoN_2L_3 + S \longrightarrow HCoSL_3 + N_2\uparrow \,. \tag{5.1}$$

Since the rate of cyclohexene substitution for N_2 is significantly higher than the rate of hydrogenation, almost all cobalt will occur in $HCoSL_3$ species. In the case of initial complex $Co(N_2)L_3$ styrene (S) forms 15-electron tricoordinated complexes CoS_2L and $CoSL_2$, which are active in the styrene hydrogenation reaction [25, 26].

The most characteristic reactions of the initial metal complexes are redox interactions involving reactants. Several examples of such catalytic systems were considered above. Four systems, in which active centers are formed with the participation of carbon monoxide and olefin, are presented by reactions (5.2)–(5.5) below. The initial rhodium iodide (RhI_3) is reduced by the substrate (CO) with the formation of Rh(I) complex that is active in the CH_3COOH synthesis according to the process developed at the Monsanto company:

$$RhI_3 + 3CO + H_2O \longrightarrow Rh(CO)_2I_2^- + CO_2 + HI + H^+ \,. \tag{5.2}$$

The initial $PdCl_4^{2-}$ complex is reduced by butenes to Pd(I) chloride complex $Pd_2Cl_4^{2-}$, which catalyzes the process of positional isomerization of linear butenes [27]:

$$2PdCl_4^{2-} + C_4H_8 + H_2O \longrightarrow Pd_2Cl_4^{2-} + C_4H_8O + 2HCl + 2Cl^- \,. \tag{5.3}$$

The catalyst of acetylene carbonylation to acrylates [18] is formed via the following reaction:

$$NiBr_2L_2 + 3CO + H_2O \longrightarrow Ni(CO)_2L_2 + CO_2 + 2HBr. \tag{5.4}$$

The catalyst of ethylene dimerization [28] is formed from a precursor as

$$RhCl_3 \cdot 3H_2O + 3C_2H_4 \xrightarrow{H_2O\,-\,\text{ethanol}} [Rh(C_2H_4)_2Cl_2]^- + CH_3CHO + HCl + H^+ \,. \tag{5.5}$$

In the reactions of hydroalkoxycarbonylation of various substrates in alcohol solutions of $Co_2(CO)_8$, the molecules of substrate (alcohol) participate in the process of Co(0) complex (precursor) disproportionation [29] as

$$3Co_2(CO)_8 + 12ROH \longrightarrow 2[Co(ROH)_6{}^{2+}][Co(CO)_4{}^-]_2 + 8CO$$

with the formation of catalytically active $Co(CO)_4^-$ nucleophiles.

The methods and mechanisms of reduction of Pd(II) and Pt(II) complexes by hydrogen, CO, and other reducing agents were exhaustively reviewed in [30]. Here are several examples of the formation of low-valence complexes in the systems where CO and H_2 are substrates in catalytic processes. Tetranuclear Pd(I) cluster is formed in glacial acetic acid from $Pd(OAc)_2$ and CO as follows:

$$4Pd(OAc)_2 + 6CO \longrightarrow Pd_4(CO)_4(OAc)_4 + 2Ac_2O + 2CO_2. \tag{5.6}$$

The reduction of dimeric Pd(II) complexes by hydrogen yields Pd(I) complexes

$$[(PPh_3)Pd(OAc)_2]_2 + H_2 \longrightarrow [PPh_3Pd(OAc)]_2 + 2AcOH, \tag{5.7}$$

which participate in the subsequent steps of catalyst formation for the hydrogenation of olefins, alkynes, and dienes (see Section 5.3).

In the reactions of hydrosilylation of olefins and alkynes in the Speyer system (H_2PtCl_6–isopropanol), the catalytically active species are Pt(IV) and Pt(II) complexes, the latter being formed upon the reduction of Pt(IV) complexes by alcohols (or $HSiR_3$). In the case of alkyne hydrosilylation, it was shown [31] that the substrate ($^tBuC{\equiv}CH$) participates in the formation of catalytically active Pt(II) carbene complexes

$$Cl_2Pt \longleftarrow C\begin{cases} OCH(CH_3)_2 \\ CH_2Bu^t \end{cases}.$$

A Pt(II) carbene complex with phosphine ligands was also obtained from cationic Pt(II) complexes [32] via the following reaction:

$$MePtClL_2 + RC{\equiv}CH + MeOH \xrightarrow[-AgCl]{AgPF_6} MePtL_2\left(C\begin{cases} OMe \\ CH_2R \end{cases}\right)^+ PF_6^-. \tag{5.8}$$

Probably, both types of active species are formed in the course of ROH addition to Pt(II) vinylidene complexes (Pt=C=CHR).

Solvents. In homogeneous catalysis with metal complexes, solvents perform various functions, ensuring both nonspecific and specific solvation of participants and transition states at each step of the process, acting as ligands that enter into the coordination sphere of metal atoms, and being reactants that either convert the metal into a new oxidative state or supply new types of ligands.

Alcohols, glycols, and glycol monoethers. These solvents are capable of reducing transition metal complexes in the presence of ligands L (L = PR_3, AsR_3, etc.) with the formation of hydride, hydride–carbonyl, and carbonyl complexes as follows [15, 16, 33]:

$$RhCl_3 \cdot 3H_2O + ROH \xrightarrow{PPh_3} RhCl(CO)L_2; \tag{5.9}$$

$$Na_2IrCl_6 + MeOCH_2CH_2OH \xrightarrow{L} HIr(CO)L_3 \longrightarrow IrCl(CO)L_3; \tag{5.10}$$

$$Na_2OsCl_6 + MeOCH_2CH_2OH \xrightarrow{L,\ KOH} (H)_2Os(CO)L_3; \tag{5.11}$$

$$RuCl_3 + EtOH \xrightarrow{L,\ KOH} HRuCl(CO)L_3. \tag{5.12}$$

The mechanism of carbonylation of ruthenium and rhodium complexes by alcohols and aldehydes was originally described by James [34]. The well-known Wilkinson's complex (red modification) is obtained in boiling ethanol as [35]

$$RhCl_3 \cdot 3H_2O + 3PPh_3 + EtOH \longrightarrow RhClL_3 + CH_3CHO.$$

The reduction of $Pd(OAc)_2$ to Pd(0) complexes by ethanol is employed in some catalytic systems for the Heck reaction [36].

Alcohols are used as sources of not only hydride ligands and CO, but also olefins. The first π complex in the history of chemistry was $K[Pt(C_2H_4)Cl_3]$ (Zeise's salt) synthesized from $PtCl_4$ and ethanol as a result of Pt(IV) reduction and ethanol dehydration to olefins. Reactions of $PtCl_4$ with various alcohols at 25 °C also yielded π complexes of Pt(II) with ethylene, propylene, and 1-butene [37]:

$$PtCl_4 + ROH \xrightarrow{Bu_4NCl} Bu_4N[PtLCl_3], \tag{5.13}$$

where L is olefin.

The interaction of alcohols with both σ and π ligands leads to the synthesis of new ligands. For example, this takes place in the formation of Pt(II) carbene complexes with the participation of alkynes. Alcohols also readily interact with isonitrile ligands [38] to yield the corresponding carbene complexes as

$$PtX_2L(RNC) + EtOH \longrightarrow cis\text{-}X_2(L)Pt \leftarrow C(OEt)(NHR), \tag{5.14}$$

where $L = PEt_3$.

The process of Pd(II) reduction by alcohol and Ph_2C_2 leads to the synthesis of Pd(I) complexes with η^5-C_5Ph_5 ligands via the following remarkable process:

$$3Pd(OAc)_2 + 6MeOH + 7Ph_2C_2 \longrightarrow Pd_2(\eta^5\text{-}C_5Ph_5)_2(\mu - \eta^2\text{-}Ph_2C_2) + 2PhC(OMe)_3 + 6AcOH + Pd^0, \tag{5.15}$$

the mechanism of which was proposed by Ban et al. [39].

Dimethylformamide (DMF), dimethylacetamide, and other amides. These solvents are capable of reducing and carbonylating transition metal complexes on heating (below 100 °C) as follows:

$$RhCl_3 \cdot 3H_2O + DMF \xrightarrow{L} RhCl(CO)L_2; \tag{5.16}$$

$$PdCl_2 + DMF \xrightarrow{L} PdL_4. \tag{5.17}$$

In contrast to $PdCl_2$, $RhCl_3$ heated in DMF yields a set of carbonyl compounds not containing phosphine ligands, including $Rh^{III}(CO)$, $Rh^{III}(CO)_2$, and $Rh(CO)_2Cl_2^-$ complexes [40].

Dimethylacetamide exhibits decarbonylation under conditions of hydrogenation in the presence of Rh(I) complexes with the formation of CH_4 and, probably, dimethylamine [41]. It should also be noted that amides in acid media are readily hydrolyzed to ammonium salts of hydroxylic acids.

By analogy with reaction (5.16), iridium compounds $IrCl_3$ and $(NH_4)_2IrCl_6$ are reduced by DMF with the formation of the Vaska complex [42]:

$$IrCl_3 \cdot 3H_2O + DMF \xrightarrow{PPh_3} IrCl(CO)L_2.$$

Similar reactions are possible with ethylene glycol and 2-ethoxyethanol.

Cyclic and acyclic ethers. Ethers are also subjected to various transformations, including decarbonylation (tetrahydrofuran, dioxane) in the presence of Rh(III) and dehydrogenation (dioxane → dioxene) in Rh(I) and Pd(II) solutions. In solutions of cationic Pt (II) complexes, diethyl ether and tetrahydrofuran (THF) exhibit intramolecular rupture of C–H bonds according to the following scheme:

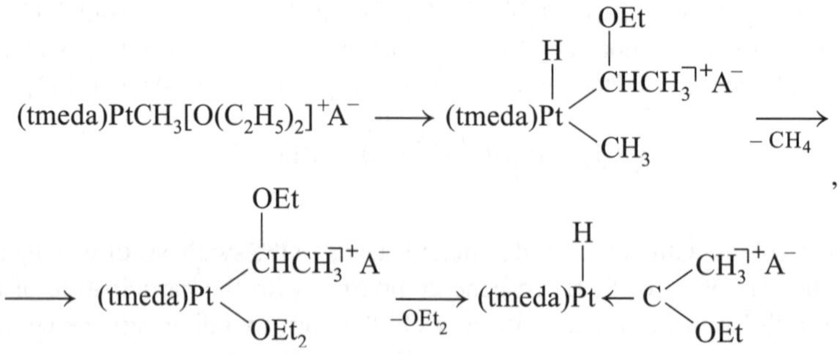

(5.18)

(where tmeda = tetramethylenediamine) with the formation of Pt(II) carbene complexes [43].

Acetone, other ketones, and various nitriles. These solvents usually play the role of ligands that form coordination spheres of active complexes. However, in a few cases, these solvents can exhibit significant transformations. For example, the interaction of acetone with W_2L_6 clusters (where L is an anion) leads to the formation of W (V) oxo complexes [44]:

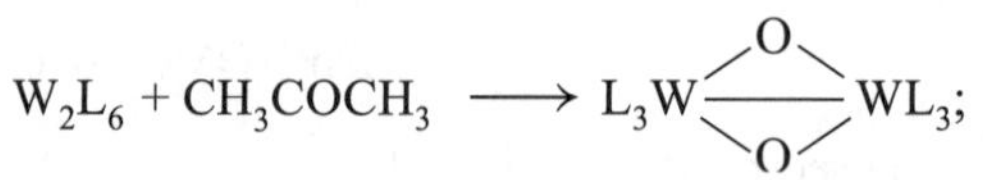

(5.19)

The products of acetone conversion in this system were not determined, but the formation of tetramethylethylene was quite probable.

Nitriles are readily converted in acid media, especially in the presence of water and alcohols, with the formation of amides, acids, imines, and even compounds with C–Hal bonds. For example, a CH_3CN solution containing HBr, which was used as a solvent for $PdBr_2$ in the process of acetylene carbonylation, featured the formation of a product with a composition of $2CH_3CN \cdot 3HBr$ and a structure of $[H_2N{=}C(Me)NHC(Me)Br_2]^+Br^-$ as determined by X-ray diffraction [45]. Compounds of this type could influence the catalytic process.

A catalyst for acetylene hydrocyanation was prepared by dissolving CuCl in adiponitrile (ADN) at 130 °C [46], which yielded a CuCl · ADN complex that precipitated from solution upon cooling. This complex was obtained from CuCl solutions with concentrations within 5–30 wt%. At the same time, crystals precipitated from solutions with [CuCl] < 1 wt% upon cooling and 1-day standing represented CuCN · 2ADN complex with a structure depicted in Figure 5.3 [47] (by the results of primary analyses, the complex composition was misinterpreted as Cu · 2ADN [46]). The formation of copper cyanide in this system is indicative of the possible elimination of HCN from ADN catalyzed by CuCl.

Nitriles usually form the coordination spheres of mononuclear metal complexes, but these solvents are also capable of stabilizing binuclear Pd (I) clusters. Quite recently, Murakashi et al. [48, 49] obtained

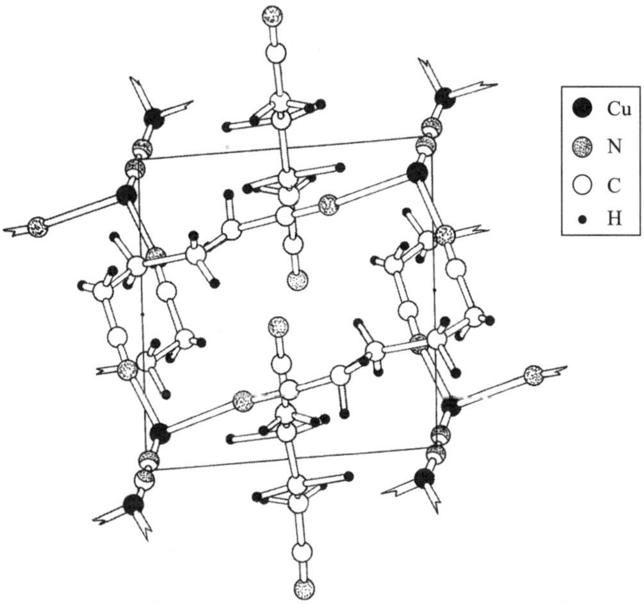

Figure 5.3 Elementary cell of the CuCN · 2ADN complex

$[Pd_2(CH_3CN)_6{}^{2+}][BF_4{}^-]_2$ complexes via the following reaction:

$$PdL_4^{2+} + {}^1/_2Pd_2(dba)_3 \xrightarrow[L\text{—}CH_3CN]{CH_3CN\ \ CH_2Cl_2} \left[L_3Pd\text{—}PdL_3\right]^{2+}, \tag{5.20}$$

where dba is dibenzylideneacetone.

Benzene and its homologs (toluene, xylene, etc.). These solvents are frequently used as inert nonpolar hydrocarbon solvents for weakly polar metal complexes. However, ArH solvents in some cases play the role of η^6-ligands capable of forming a coordination sphere of these complexes. For example, $(C_6H_6)_2Fe[PF_6]_2$ complex was isolated upon boiling a solution of $FeCl_3$ and $AlCl_3$ in benzene by treating the mixture with aqueous methanol and NH_4PF_6 [50]. Analogous complexes of Pd(II), Ni(II), and Co(II) were obtained in the MCl_2–$AlCl_3$–C_6H_6 system [51]. Upon a prolonged boiling of $PdCl_2$ and $AlCl_3$ in benzene, Pd(II) reduces to Pd(I) and forms a complex of $Pd_2(\mu\text{-}C_6H_6)_2[AlCl_4]_2$ [52].

Chloroform, methylene chloride and some other chloro-organic solvents. These solvents are capable in some cases of exhibiting the oxidative addition to metal in low-valence metal complexes with the resulting rupture of C–Cl bonds. For example, chloroform adds to PdL_4 ($L = PPh_3$) in xylene as [53]

$$PdL_4 + CHCl_3 \xrightarrow[-2L]{} ClPd(CHCl_2)L_2, \tag{5.21}$$

and the resulting complex converts into $PdCl_2L_2$ (and, probably, $HPdClL_2$) with the formation of an active center in the synthesis of acrylates from acetylene and CO under the conditions where PdL_4 and

$PdCl_2L_2$ fail to catalyze this reaction in xylene-ROH [54]. A complex of Pd(0) with dba in the presence of $L = P(^tBu)_3$ produces rupture of three C–Cl in chloroform with the formation of a tetranuclear cluster of palladium with μ_3-CH group [55]:

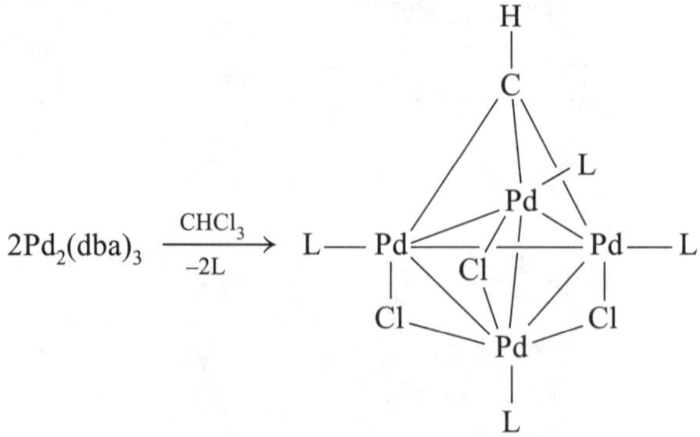

(5.22)

Analogous behavior was observed for CH_2Cl_2 and $Pd_2(dppm)_3$ complex [56]:

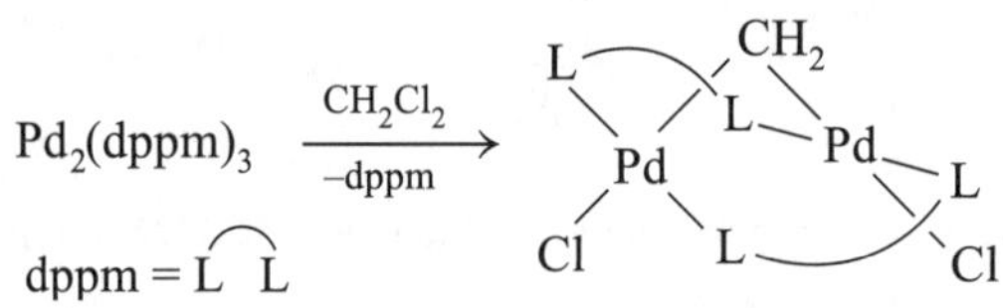

(where dppm = diphenylphosphinometane).

The well-known reaction of CCl_4 and $CHCl_3$ with metal hydride complexes leads to the exchange of H and Cl [33]:

$$HML_n + CCl_4 \longrightarrow ClML_n + HCCl_3. \qquad (5.23)$$

Exchange of Cl and R groups is also possible in reactions of chloroform with R–M compounds [33, 57]. For example, a Cu (I) ethynyl complex reacts with chloroform at 25 °C with the formation of C–C bonds [57]:

$$L_3CuC{\equiv}CPh + CHCl_3 \longrightarrow L_3CuCl + Cl_2CHC{\equiv}CPh. \qquad (5.24)$$

Acetic acid. Acetic acid as a solvent favors the formation of Pd (II) hydride complexes that are active in alkyne reactions [58]:

$$PdL_4 + AcOH \rightleftharpoons HPdL_4{}^{+}OAc^{-}. \qquad (5.25)$$

Ligands. Ligands not only form the coordination sphere of an active catalyst and form various inactive species, but also participate (as a result of chemical transformations) in redox processes, thus radically changing their nature and forming cluster complexes. As was noted above, allotropic modifications of the Wilkinson's complex with fairly insignificant differences in their geometry [13, 14] are capable of forming various isomers in reactions with solvents and CO (as a reactant). In the orange modification (<70 °C), the phenyl ring of *trans*-phosphine ligand (with respect to Cl) is linked via H atom to Rh, whereas in the red modification, it is the *cis* ligand that is linked to Rh in the same manner. Accordingly, the linkage between phenyl ring and the metal disappears upon the dissociation of the *trans* ligand:

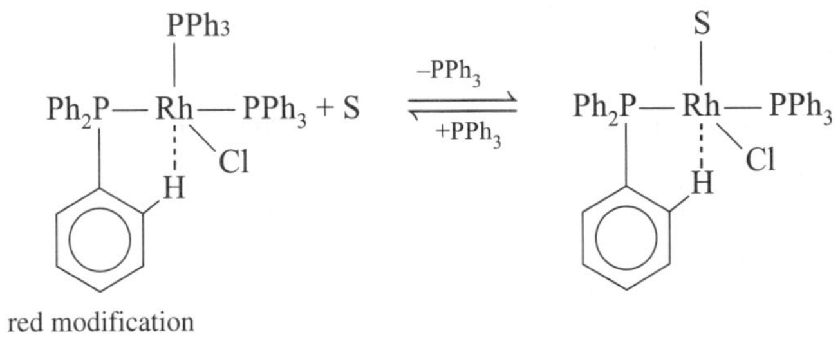

(5.26)

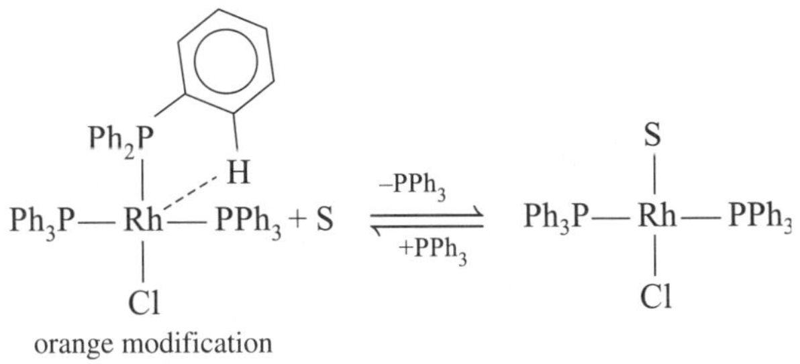

(5.27)

Carbon monoxide acts similarly to solvents, by replacing ligand L and forming two modifications of *trans*-$RhCl(CO)(PPh_3)_2$. The rate of allyl alcohol hydroformylation at $T < 70\,°C$ in the case of the red modification is about five times higher than that in solutions of the orange modification. Even the treatment of a carbonyl complex solution with a synthesis gas at 65 °C and 1 MPa leads to the formation of compounds with the NMR spectra different for the two initial modifications. At temperatures above 78 °C, only the red modification exists. Thus, the geometry of these metal complexes, which differ even by weak interactions of the agostic type, influences their catalytic properties.

It was found that PPh_3 and $Pd(Alk)_3$ ligands can perform the functions of labile bridging groups [59, 60]. In particular, a unique coordination of μ-PPh_3 ligands was demonstrated in the case of Pd(I) clusters, which can include the agostic donation of P–C bonds and the η^1-phenyl coordination with a Pd–C_{ipso} bond length of 2.336 Å. The Pd–Pd bond length in these clusters is 2.5833(7) Å. The equilibrium in solution is as follows:

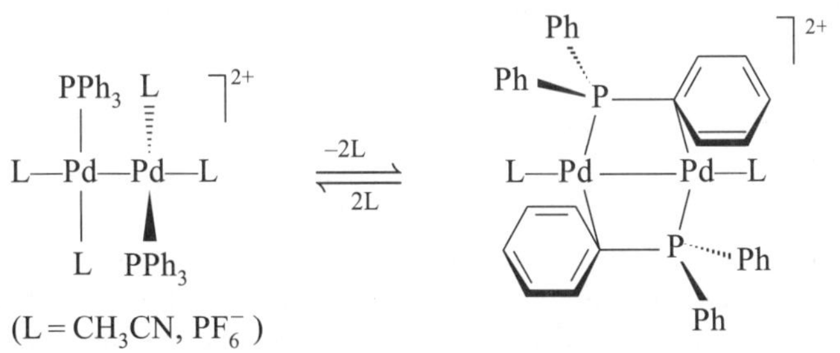

(5.28)

The bridging ligand (η^1-Ph) is readily replaced by various molecules, in particular, alkynes [35].

A more significant difference between the structures of *cis* and *trans* isomer planar square Pd(II) complexes is manifested by a much stronger difference in their activities. For example, *cis*-$PdBr_2[P(OPh)_3]_2$ is active in the reaction of acetylene carboalkoxylation (hydroalkoxycarbonylation) to butyl acrylate in butanol, while the *trans* complex of the same composition exhibits no activity [61]. The mechanism of the formation of the active *cis* complexes will be considered in the next section. It was also established that a *cis*-bipyridyl complex of $Cu^{2+}[Cu(bipy)_2(H_2O)_2{}^{2+}]$ exhibited catalase activity (favoring H_2O_2 decomposition), while its *trans* isomer is inactive [62].

The interest in ligands containing P(III) (such as PR_3, $P(OR)_3$, etc.) in the homogeneous catalysis was inspired by the investigations of W. Reppe and related primarily to the possibility of controlling the activity and selectivity of catalytic processes. The research was stimulated by the successful use of asymmetric catalysis, mostly with bidentate phosphine and phosphite ligands. It was found later that the reactivity of PR_3 and $P(OR)_3$ compounds increases in the coordination sphere of a metal so that they become susceptible to various transformations, including (i) transfer of alkyl and aryl groups R to the metal with the formation of phosphide (PR_2) and phosphinidene (PR) ligands capable of stabilizing polynuclear and cluster complexes; (ii) reduction of the metal with the formation of $O{=}PR_3$; and (iii) metallation of R groups in PR_3 and $P(OR)_3$. Various transformations of these ligands can lead to both the appearance of highly catalytically active complexes and, on the contrary, to inactivation of a catalyst.

The ability of triarylphosphines to transfer aryl groups to a metal was originally established in 1968 by Coulson [63] during the investigation of reactions of $PdCl_2$ with PdL_4 ($L = PPh_3$), where the *trans*-$PhPdCl(PPh_3)_2$ complex was isolated in addition to diphenylphosphide clusters such as $[Pd_3(PPh_3)_3(\mu\text{-}PPh_2)_2Cl^+]Cl^-$. A platinum cluster $Pt_3(PPh_3)_2(\mu\text{-}PPh_2)_3Ph$ with close structure was synthesized in 1966 (see, e.g., review [30]). The dephenylation of PPh_3 is observed in various reductive media upon the replacement of Cl^- and Br^- by acetate groups under the conditions of oxidation of a fraction of PPh_3 groups in PdL_4 complexes. Here are some examples:

(i) The reaction of L_2PdCl_2 ($L = PPh_3$) with NaOEt is accompanied by the formation of benzene [64]. It was established that, for the reaction with NaOEt in DMF at 20 °C under argon atmosphere, the yields of benzene and diphenyl (on Pd) amount to 40 % and 55 %, respectively [65].

(ii) The heating of $(PAr_3)_2Pd(OAc)_2$ complexes leads to the formation of bis-aryls [64] and is accompanied by the arylation of olefins, which is analogous to the Heck reaction. The kinetics and mechanism of olefin arylation by $(PPh_3)_2Pd(OAc)_2$ complexes in glacial AcOH were studied by Ryabov and Yatsimirskii [66a]. Recently it was established [66b] that Pd–Ar/P–Ar exchange takes place in the reaction of ArBr with PAr'_3, which is catalyzed by palladium acetate or Pd(0) complexes. The reaction of $(PPh_3)_2PdBr_2$ with $Cu(OAc)_2$ (105 °C, DMF) gives a 100 % yield of diphenyl (on Pd) in addition to $(CH_3CO)_2O$ [65]. The same products are formed in the reaction of PdL_4 with $Cu(OAc)_2$ [65]. Acetic anhydride is formed during the oxidation of PR_3 by metal acetates according to the following reactions:

$$Hg(OAc)_2 + PPh_3 \longrightarrow Hg^0 + O{=}PPh_3 + (CH_3CO)_2O;$$

$$Cu(OAc)_2 + PPh_3 \longrightarrow Cu^0 + O{=}PPh_3 + (CH_3CO)_2O.$$

The reductive properties of phosphines are also used for the synthesis of Pd(0) and Pt(0) complexes [30] and for the formation of low-valence metal complexes in the catalysis. For example, in the phosphine-containing catalytic systems for the Heck reaction, the reduction of Pd(II) to Pd(0) by phosphine is the most probable pathway. The reduction is also favored by hard nucleophiles (OH^-, RO^-, OAc^-, H_2O) as a result of the probable nucleophilic substitution at phosphorus [36]:

$$Cl_2Pd^{II}{-}PR_3 \xrightarrow{Nu^-} Cl_2Pd^{0]2-} + R_3P^+{-}Nu;$$

$$R_3P^+{-}OH \longrightarrow R_3P{=}O + H^+;$$

$$R_3P^+{-}OAc \xrightarrow{AcOH} R_3P{=}O + (AcO)_2O + H^+.$$

(iii) The interaction of $[Ph_3PPd(OAc)_2]_2$ with HCOOH is accompanied by the appearance and oxidation of phosphide ligands, which leads to the formation of $[PhPd(\mu\text{-}O_2PPh_2)(PPh_3)]_2$ [67].

The conversions of ligands during the formation of clusters and metal-containing species in the reactions with H_2 will be considered in the next section.

Reactions characteristic of PAr_3, P $(OAr)_3$, and some nitrogen-containing aromatic ligands also include the following processes of the aromatic ring metallation (via electrophilic substitution or oxidative addition at the C–H bond) [36, 68, 69]:

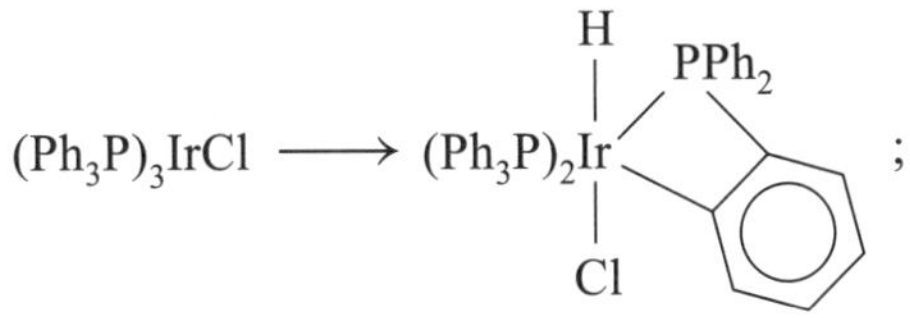

(5.29)

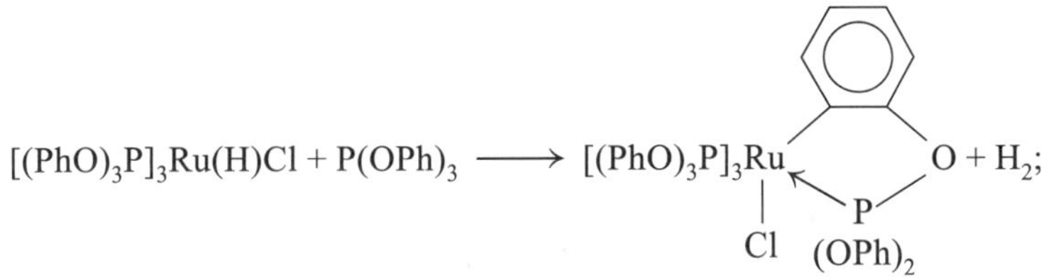

(5.30)

$$cis\text{-}PtCl_2[P(OPh)_3]_2 \xrightarrow[-HCl]{} ClPt[P(OC_6H_4)(OPh)_2][P(OPh)_3] \quad (5.31)$$

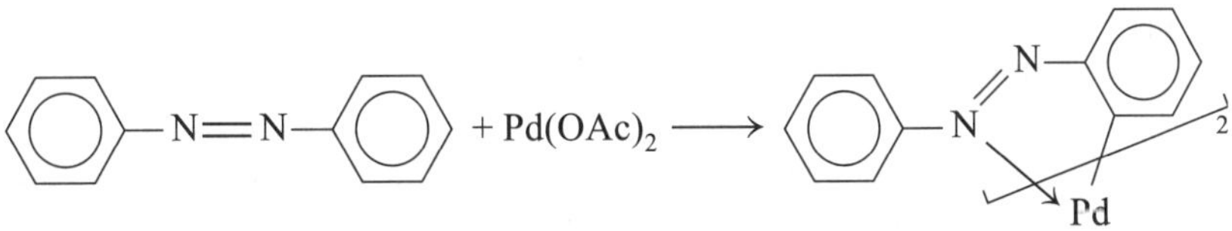

(5.32)

It was established that a catalyst precursor for the Heck reaction in the $Pd(OAc)_2$–$P(o\text{-}Tol)_3$ system has the form of a dimeric metallacycle,

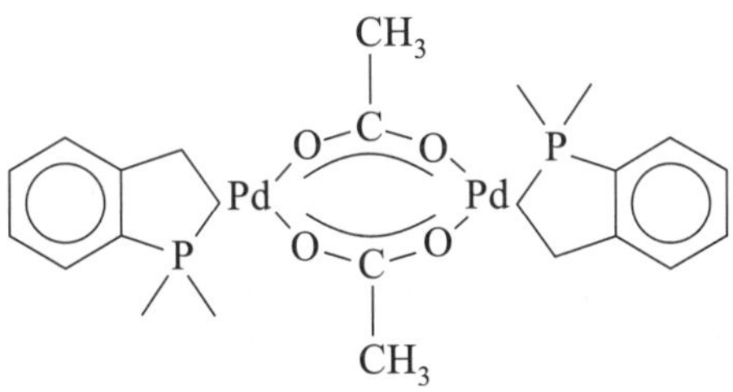

with Pd : P = 1 : 1 [70]. Numerous N-, O-, and S-containing palladacycles were shown to be excellent precursors for the Heck reaction catalysis [36] by analogy with the so-called pincer complexes,

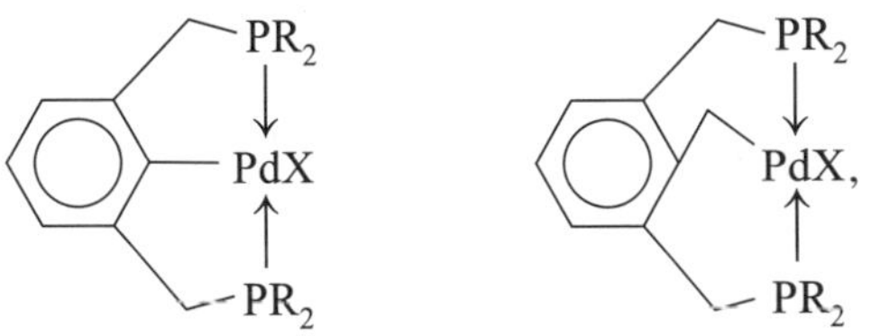

which are known to be active both in the Heck reaction [36] and in many other catalytic processes [71].

Phosphines are susceptible to the oxidation by oxygen, addition to polar olefins, alkynes, and quinones, and participation in the reactions of alkylation, dealkylation [$P(OR)_3$], and various rearrangements, as a result of which the phosphor oxidation number increases to five. These properties make phosphines generally fairly unsuitable for commercial processes, although some of these (e.g., oxo synthesis) were originally based on the phosphine complexes of transition metals. However, the general tendency of passing to more stable nitrogen-containing ligands in catalytic reactions of various types, which is pronounced in the past two decades, is quite understandable.

Even stable ligands such as η^5-C_5H_5 and η^5-C_5R_5, which are effectively employed in the processes of polymerization (as well as in other reactions), in some cases enter into intra-sphere reactions with M–H and M–R groups, thus determining the pathways of loss of the active catalysts. For example, cyclopentadienyl ligands enter into the diene synthesis from alkynes activated by electron-acceptor groups [72] such as

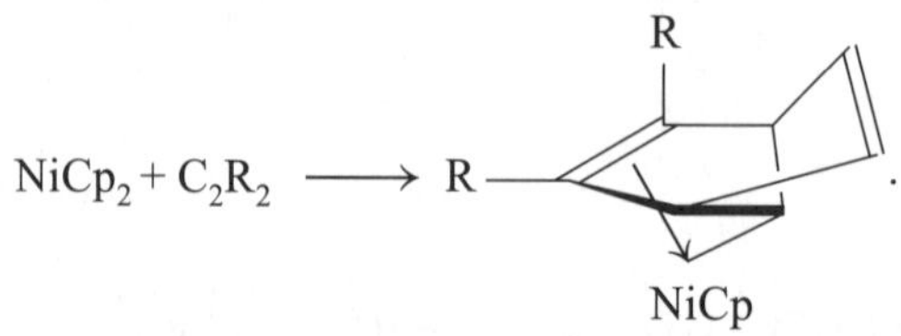

(5.33)

Components of catalytic system. In addition to reactants, solvents, and ligands, the formation of active catalysts can also involve some other components of a catalytic system that either accidentally occur in (O_2, H_2O) or are intentionally added to the system ($NaBH_4$, ${}^tBu_2AlH$, Li[AlH(o-${}^tBu$)$_3$], AlR_3, AlR_2Cl, R_4NX, imidazolium salts, R_4PX, NaOAc, R_3N, etc.). These components can perform various functions such as the reduction or oxidation of metal catalysts, ligands, and bases. For example, acetate ions can act as a base, nucleophilic agents (e.g., in the oxidation of phosphines considered above), and ligands favoring the conversion of trimeric $Pd_3(OAc)_6$ complex (low-active in the reactions of olefin and arene oxidation and substitution at the sp^2-carbon atom) into active $Na_2Pd_2(OAc)_6$ dimers or $Na_2Pd(OAc)_4$ monomers.

It was established that L_2PdCl_2 complexes (not activating H_2 under mild conditions) does not exhibit activity in the reactions of olefin hydrogenation even upon the reduction by $NaBH_4$ to Pd(0) complexes, unless the system would feature the contact of Pd(0) with oxygen [30, 73–75]. Accordingly, it was assumed that the heterolytic activation of H_2 requires the presence of M–O bonds (M–OH, M–OAc, M–O–O, etc.) [76]. Thus, the following chain of conversions can be proposed:

$$L_2PdCl_2 \xrightarrow[L]{BH_4^-} L_nPd \xrightarrow{O_2} L_2PdO_2 \xrightarrow{H_2} L_2PdH(OOH) \xrightarrow{H_2} L_2Pd(H_2), \qquad (5.34)$$

which may be also accompanied by the processes of dephenylation and cluster formation (see below).

The aforementioned components such as imidazolium or phosphonium salts, which usually perform the functions of phase transfer catalysts and/or ionic liquids (as media in which the catalytic process takes place), proved also to be capable of participating in the formation of active catalysts. For example, the electrophilic replacement of H^+ by palladium acetate in imidazolium salts leads to the formation of a new carbene ligand in the coordination sphere of Pd(II) [36, 77]:

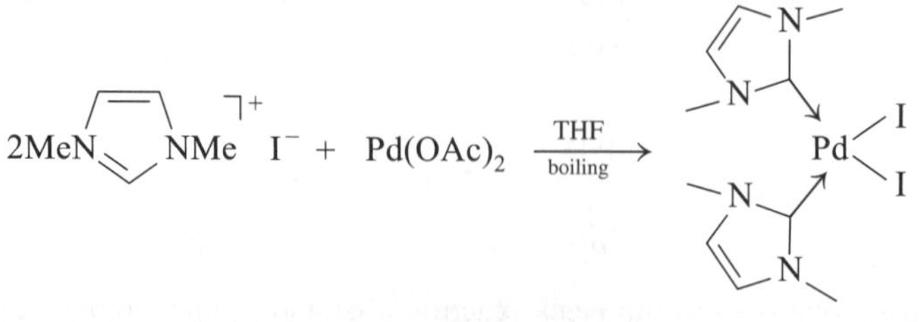

(5.35)

Imidazolidene carbene complexes of Pd(II) and Pd(0) showed activity in the Heck reaction [36] and some other processes.

An interesting pathway for the formation of a monophosphine complex of Pd(II) was discovered in the reaction of $Pd(OAc)_2$ with Ph_4PCl [78], where the transfer of a phenyl group to the reduced Pd(0) atom yields a phenyl-containing derivative of Pd(II):

$$Pd(OAc)_2 + Ph_4PCl \xrightarrow[25\text{–}60\ ^\circ C]{EtOH\text{–}DMF}$$ 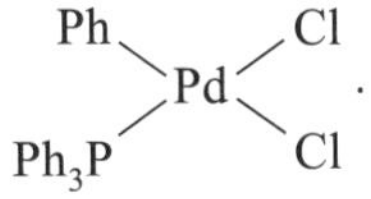 . (5.36)

The reactions of organometallic compounds of nontransition metals (such as LiR, RMgX, AlR_3, AlR_2Cl, SnR_4, which are components of Ziegler's catalytic systems) with transition metal complexes lead to the transfer of alkyl groups to the transition metal that results in the formation of M–R bonds, reduction of the transition metal, formation of hydride and carbene complexes, and the appearance of free radicals [26, 79, 80]. The main stages of decomposition of the alkylated complexes of transition metals are as follows:

$$2RMX_{n-1} \longrightarrow 2MX_{n-1} + R\text{—}R;$$

$$2RMX_{n-1} \longrightarrow 2MX_{n-1} + RH + R_{-H};$$

$$RMX_{n-1} \longrightarrow MX_{n-2} + RX;$$

$$RMX_{n-1} \longrightarrow HMX_{n-1} + R_{-H};$$

$$RMX_{n-1} \longrightarrow R'CH{=}\underset{\substack{|\\ H}}{M}X_{n-1} \longrightarrow R'CH{=}MX_{n-2} + HX;$$

$$RMX_{n-1} \longrightarrow R\bullet + MX_{n-1}.$$

These reactions may be accompanied by the formation of bridging complexes of transition metals or heterometallic bridging complexes such as

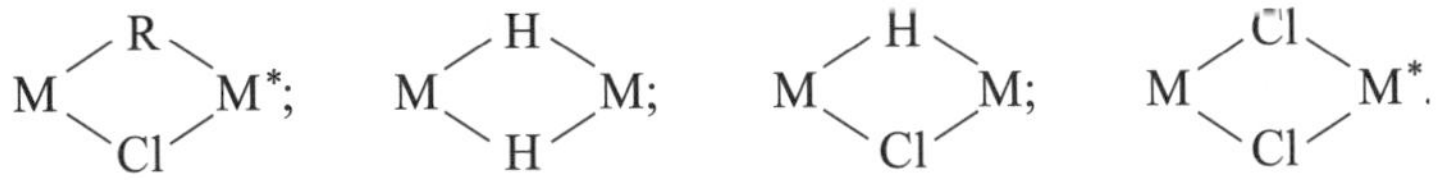

During investigations of the fourth-generation metallocene catalysts (Cp_2TiCl_2, Cp_2ZrCl_2 and their derivatives) for polymerization processes, it was established that an important role in these systems (in addition to the organoaluminum component) is played by small additives of water. Accordingly, the products of the partial hydrolysis of AlR_3 (alumoxanes, in particular, methylalumoxane) were introduced as second components. It was suggested at the first stage of investigations that the functions of methylalumoxane (MAO) include the alkylation of metallocene [$Cp_2Zr(Me)Cl$] and the binding of Cl^- with the formation of ion pair Cp_2ZrMe^+–$(MAO)Cl^-$ containing a large-size anion. Later, it was also found that important activators of the M–R bonds are strong protonic acids formed during the hydrolysis of AlR_3 and the formation of methylalumoxane [80].

It was established [26] that the Ziegler type catalytic system NiL_2Cl_2–$AlEtCl_2$ of olefin dimerization features, in addition to the reduction of Ni(II) to Ni(I) and Ni(0) complexes and the formation of a catalytically active hydride complex $LNi(H)Cl \cdot AlEtCl_2$, the reactions of disproportionation of Ni(I) complexes as

$$2NiLCl(C{=}C) \xrightarrow{L} LNi(C{=}C)_2 + L_2NiCl_2 \qquad (5.37)$$

and the oxidation of Ni(0) complexes by ethylaluminum dichloride. This results in an almost quantitative formation of Ni(I) complex (NiL_2Cl) in the NiL_4–$AlEtCl_2$ and $NiL_2(C_2H_4)$–$AlEtCl_2$ systems [26]. Analogous properties of $AlEtCl_2$ with respect to $Ti(C_6H_6)_2$ complexes were reported in [81].

In the case of processes catalyzed by two-component (or multi-component) bimetallic systems, heterometallic complexes can be formed between the components that are involved in the processes but not in the consequent steps (kinetic polyfunctionality, see Chapter 4). For example, the participation of an active $Pd(OAc)_2 \cdot Cu(OAc)_2$ complex in the reaction of oxidative carbonylation of alkanes [82] was established spectrophotometrically and confirmed by the kinetic data. Recently, Kozitsyna et al. [83] synthesized a complex with the composition of $Cu(OAc)_2[Pd(OAc)_2]_2$. Analogous bimetallic complexes were also obtained with some other pairs of metals. A complex with the composition of $2PdCl_2 \cdot CuCl\ (DMF)_4$ was isolated from the $PdCl_2$–$CuCl_2$–DMF–H_2O system catalyzing the oxidation of olefins [84].

The above analysis of the mechanisms of formation of the catalytically active metal complexes or potentially active complexes shows that, in practice, there are no types of element–element bonds (C–Hal, C–H, H–H, C≡O, O_2, N_2, C=O, etc.) that cannot be ruptured by low-valence transition metal complexes in the steps of their interaction with reactants, ligands, and solvents. Even bonds such as C–C in cyclopropane and cyclobutane, ≡C–C≡ in dialkynes, ≡C–C≡N in cyanoalkynes, and C–C in benzene $[C_6(CF_3)_6]$ can be broken by these complexes. The high reactivity of transition metal complexes is the main factor that accounts for the difficulties encountered in the attempts to elucidate the nature of an active center and mechanism of its formation.

5.3 Mechanisms of active center formation in particular processes

Data presented in the preceding section lead with confidence to a conclusion that the composition of a catalytic system (metal, ligands, solvent, additional components) and the nature of reactants are factors that determine the mechanism of formation of the catalytically active complexes in this system. In cases where several catalytic reactions take place in the same system, the active centers for these reactions can be different and, hence, there are no general laws for the formation (and loss) of active centers in the catalysis with metal complexes. At the same time, an analysis of particular catalytic systems in respect of the formation and loss of catalytically active complexes and the interrelations between the routes of active complex formation and the routes of catalytic processes makes possible some generalizations of the character of kinetics and provides extensive data on the chemistry of the main stages of a real catalytic process.

Below we will consider representative catalytic processes with the participation of various complexes of Pd, Rh, Ni, Co, Mo, W, Re, and Ru. However, most of these examples are related to the catalytic chemistry of palladium, which is among the most extensively developed fields in catalysis with metal complexes.

5.3.1 Mechanisms of active metal complex formation in $PdBr_2$–LiBr–$P(OPh)_3$–HBr–n-C_4H_9OH catalytic system for acrylate synthesis

The synthesis of butyl acrylate (BA) according to reaction

$$C_2H_2 + CO + BuOH \longrightarrow CH_2{=}CHCOOBu, \tag{5.38}$$

and partly of its dimerization product, dibutyldihydromuconate (BM), proceeds under mild conditions (40–70 °C, 1 atm) with a total selectivity with respect to BA and BM about 95 % [85–87]. It was established that this reaction is catalyzed by Pd(II) phosphite bromide complexes formed in the $PdBr_2$–LiBr–$P(OPh)_3$–HBr–n-C_4H_9OH system as a result of rather complicated reaction pathways, whereas $Pd[P(OPh)_3]_4$ complexes are inactive and can only be activated provided that an oxidant

($FeBr_3$) is introduced. Investigations of the mechanism of formation of active centers for the catalysis of reaction (5.38), which were performed by the methods of UV and IR spectrophotometry, ^{31}P NMR spectroscopy, and X-ray diffraction [88–92], revealed three stages in the formation of these active centers.

At the **first stage**, the dissolution of $PdBr_2$ in the LiBr–HBr–C_4H_9OH (ROH) system leads to the predominant formation of two species, $PdBr_3^-$ (~35 %) and $PdBr_4^{2-}$ (~65 %), in solution, while $Pd_2Br_6^{2-}$ dimers are not present in significant amounts. The concentration equilibrium constant K_c of reaction

$$PdBr_3^- + Br^- \rightleftharpoons PdBr_4^{2-} \tag{5.39}$$

weakly depends on the electrolyte concentration: $K_c = 5\,M^{-1}$ at [LiBr] + [HBr] = 0.3 M, $K_c = 5.2\,M^{-1}$ at [LiBr] + [HBr] = 0.5 M, and $K_c = 6.2\,M^{-1}$ at a total electrolyte concentration of 1 M.

At small values of the parameter $\alpha = [L]/[Pd]$ ($\alpha < 1$), the addition of $P(OPh)_3$ to the solution leads to the formation of $LPdBr_3^-$ and $Pd_2LBr_5^-$ complexes [61, 89, 92]. Figure 5.4 shows a fragment of the interesting supramolecular structure of a chain of Pd(II) monophosphite complex, which was first isolated from solution for the complex of $LPdBr_3^-$ anion linked by hydrogen bonds to $MeNH_2^+$ cation [90].

At **stage II**, $P(OPh)_3$ is added at an amount of $\alpha = 2.6$ to the system containing $PdBr_3^-$ and $PdBr_4^{2-}$. This leads to a rapid precipitation of solid complexes with the composition PdL_2Br_2. It was found that, in order to obtain an active catalyst at **stage III** (where the solid precipitate is dissolved in mother liquor on heating to 70–100 °C), it is necessary that water be present in solution in stage II (prior to adding L) at an amount of no less than 0.95 M. A reference system was prepared with 1.3 M H_2O. Investigations showed that a solid *cis*-PdL_2Br_2 isomer was formed in this "hydrated" aquatic system, whereas the "anhydrous" system (with $[H_2O] < 0.1$ M) yielded *trans*-PdL_2Br_2. In butanol, the *trans* → *cis* isomerization process

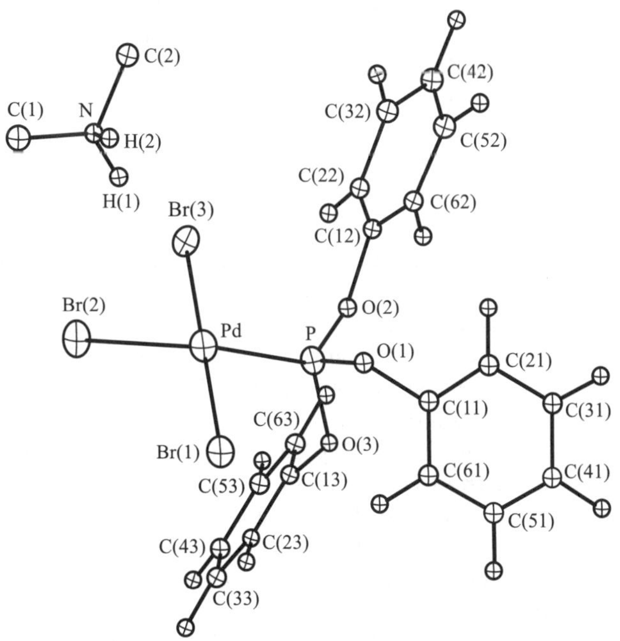

Figure 5.4 Structure of the $(CH_3)_2NH_2^+Pd[P(OPh)_3]Br_3^-$ complex (reproduced with permission from AIPC Nauka © 1982)

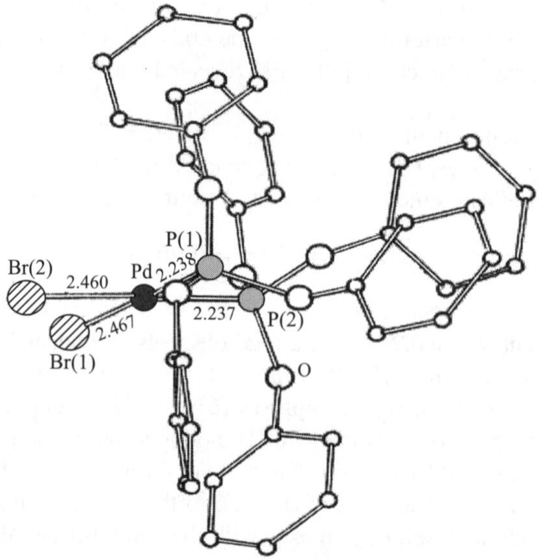

Figure 5.5 Structure of *cis*-PdL_2Br_2 complex (reproduced with permission from AIPC Nauka © 1984 [91])

is extremely slow and almost does not take place during the dissolution of complexes at stage III. Crystals of the *cis* isomer (Figure 5.5), which is thermodynamically more stable and catalytically active, were obtained using the isomerization of the *trans* complex (formed in an anhydrous system) in benzene [91].

In addition, it was established that the dissolution of the *cis* isomer at stage III was accompanied by the partial trans-etherification of $P(OPh)_3$ ligands by butanol with the formation of $P(OPh)_n(OBu)_{3-n}$ ligands and, hence, a mixture of *cis*-$Pd[P(OPh)_n(OBu)_{3-n}]_2Br_2$ complexes in solution [61]. The degree of transetherification of the active complexes amounts to 20–60 % (depending on the temperature and time of heating). In the case of *trans* complexes, the degree of transetherification reaches 70–95 %. It is interesting to note that $P(OBu)_3$ ligands in butanol solutions form active, but very unstable, complexes that are rapidly reduced by carbon monoxide. In the phenol – *p*-xylene solvent system, P $(OPh)_3$ ligands probably form *cis* complexes that are active in the synthesis of phenyl acrylate. Thus, the transetherification is not a necessary step in the formation of active centers, but takes place by virtue of the ability of free ligands (occurring in a small excess in solution at $\alpha = 2.2–2.6$) to the transetherification in acid media. The resulting alkyl phosphites in the free state are susceptible to the Arbuzov rearrangement with the formation of $BuP(O)(OBu)_2$ and subsequent hydrolysis, which leads to the loss of L from the coordination sphere and, thus, stimulates the reduction of Pd(II).

As for the role of water at the stages of *cis*-PdL_2Br_2 complex formation, it is also possible to make some assumptions. The more labile *trans* coordinate in $LPdBr_3^-$ complexes is naturally occupied by the molecules of alcohol and water that are present in solution at large excess relative to the molecules of ligand. It can be suggested that, as a result of the formation of a trigonal bipyramid, the fifth ligand (L) would replace a softer ROH ligand and form a *trans* complex (Figure 5.6a) or it would replace Br^- in the aquacomplex so as to form a *cis* complex (Figure 5.6b).

The formation of a *cis* complex possessing two *trans* coordinates L–Pd–Br evidently favors the formation of intermediates [$L_2PdBr(OR)$, $L_2Pd(CO)(OR)^+$, $L_2PdBr(COOR)$, $L_2Pd(C_2H_2)(COOR)^+$, etc.] within the framework of the kinetically and chemically established "alcoholate" mechanism of processes in the

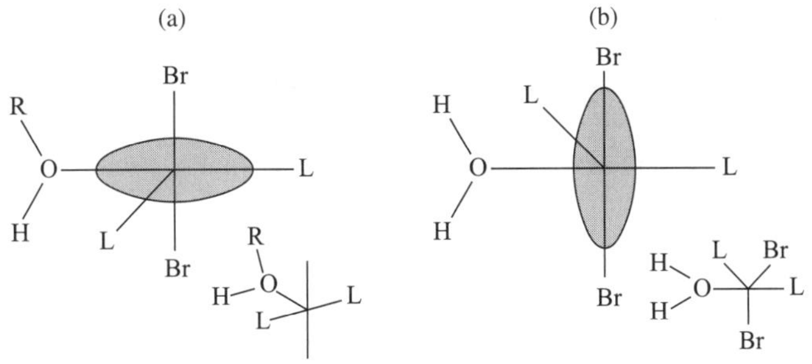

Figure 5.6 Scheme of ligand substitution in aqueous and nonaqueous systems

system under consideration [85–87, 93, 94]. In addition, as was shown above, the *cis* complexes are less susceptible to the transetherification that leads to unstable and less selective catalysts.

5.3.2 Carbene metal complexes in metathesis of olefins and analogous processes

Reactions initiated (or catalyzed) by metal complexes with carbene ligands include the following processes:

- metathesis of olefins (homometathesis or cross-metathesis, CM):

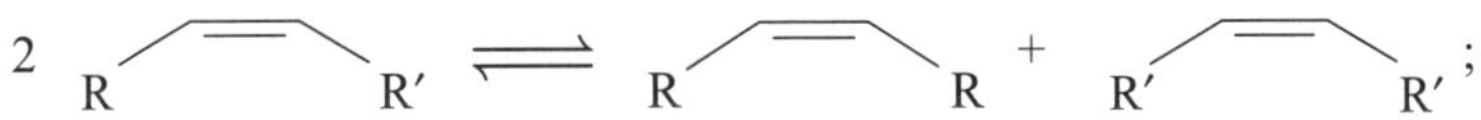

(5.40)

- formation of cycles from cyclic dienes (ring-closing metathesis, RCM):

(5.41)

- cross-metathesis of cyclic and acyclic olefins (ring opening metathesis, ROM):

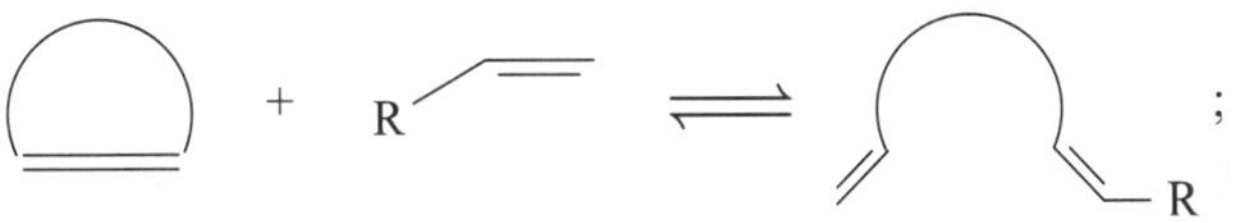

(5.42)

- polymerization of cyclic olefins with ring opening (ring-opening metathesis polymerization, ROMP)

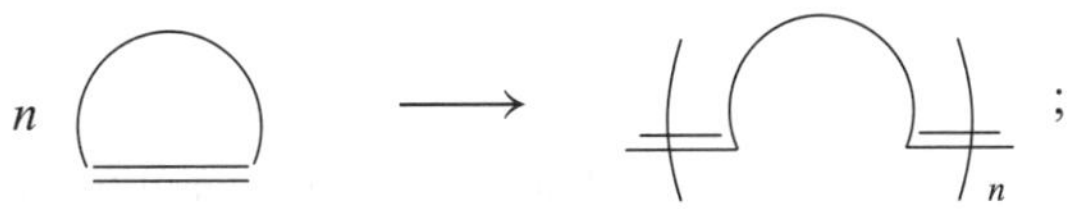

(5.43)

- polymerization of acyclic dienes:

(5.44)

- synthesis of dienes from olefins and alkynes (ene-yne metathesis):

$$CH_2{=}CH_2 + RC{\equiv}CR \longrightarrow CH_2{=}C(R){-}C(R){=}CH_2; \quad (5.45)$$

- polymerization of alkynes:

$$n\mathrm{RC{\equiv}CH} \longrightarrow \left(\mathrm{C(R){=}CH}\right)_n. \quad (5.46)$$

After the synthesis of the first carbene complex by E. Fischer in 1964 with a structure of

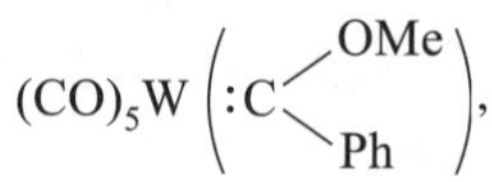

the analogous complexes were synthesized for almost all transition metals. It was demonstrated that they can participate in various catalytic and noncatalytic reactions [95–98], including cyclopropanation of olefins [98] and metathesis of olefins [21, 98–102]. At present, it is commonly adopted to distinguish between two types of carbene ligands and, accordingly, two types of carbene complexes.

(i) *Fischer carbene complexes (Fischer carbenes)*, in which the carbene ligand contains one or two heteroatoms (O, N, S) linked to the carbene carbon atom:

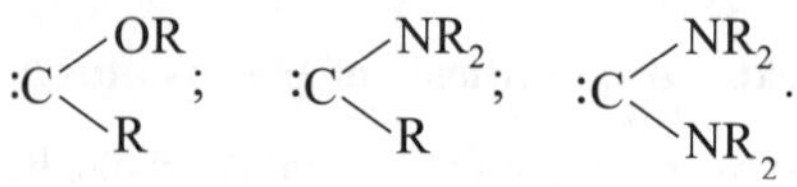

For example, Fischer carbenes obtained from imiodazolium salts,

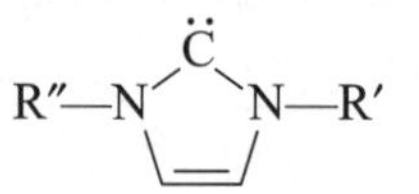

are very popular in recent years in the catalysis with metal complexes. Carbene ligands of this type are σ-donor ligands with electrophilic carbene C atom. Fischer complexes admit rotation about the $M \leftarrow L$ bond. The activation barrier for this rotation does not exceed 25–30 kJ/mole [98].

(ii) *Schrock carbene complexes (Schrock carbenes)*, in which the carbene ligand contains the following groups,

with alkyl, aryl, vinyl, and other substituents of any types. Complexes with these ligands contain explicit multiple bonds,

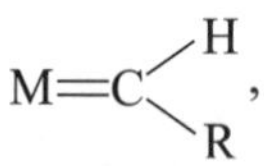

and the free rotation about the metal – carbene bond is hindered (rotation barrier, >85 kJ/mole [98]). Formally, the appearance of Schrock carbenes can be treated either as a result of the oxidative addition of a triplet carbine (:CHR) to a metal with the formation of two covalent M–C bonds and an increase in the metal oxidation number by two units,

$$M^0 + :\mathrm{CHR} \longrightarrow M^{II}{=}\mathrm{CHR},$$

or as a result of the oxidative addition of two heme-dihalide derivatives

$$M^0 + \mathrm{Cl_2CHR} \longrightarrow \mathrm{Cl_2}M^{IV}{=}\mathrm{CHR}.$$

A convenient model for the formation of $M{=}CR_2$ multiple bonds is based on the notion of a strong donating π-donor bond supplementary to a σ-donor bond:

$$M \rightleftharpoons CR_2.$$

The Schrock carbene ligands usually possess nucleophilic properties. It was suggested that the metathesis process requires carbene ligands with not too strongly pronounced electrophilic or nucleophilic properties [103], which also depend to a significant extent on the ligand environment of the metal.

The "carbene" mechanism of reactions (5.40)–(5.44) was proposed by Herrison and Chauvin [104] and refined by Dolgoplosk, Makovetskii, and Tinyakova [105]. According to Herrison and Chauvin, a carbene complex is the active center of metathesis, an intermediate metallacyclobutane is formed from the carbene complex and olefin, and a new olefin and new carbene complex (new ligand) are formed upon the decomposition of the metallacycle. The same mechanism was used to explain the polymerization of cycloolefins (ROMP):

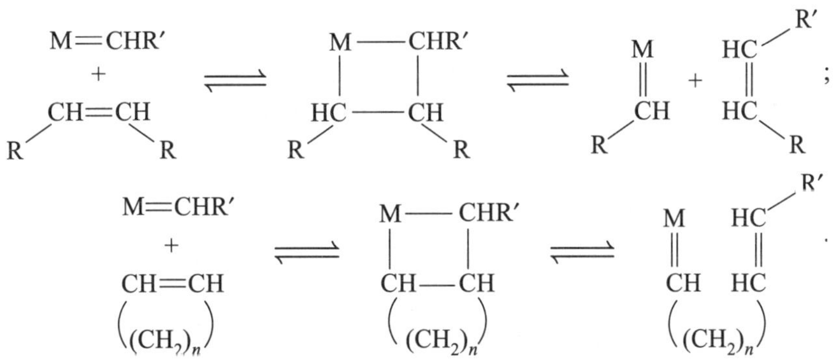

Dolgoplosk et al. [105] were the first who clearly pointed out that the ROMP and olefin metathesis are processes with chain mechanisms, which can be explained by adopting the presence of active carbene centers. Shortly after (in 1974), Casey and Burkhardt [106] showed that the reaction of the $Ph_2C{=}W(CO)_5$ carbene complex with isobutene involves the exchange of alkylidene groups and leads to the formation of a new olefin (1,1-diphenylethylene) and new carbene ligand:

$$(CO)_5W{=}CPh_2 + Me_2C{=}CH_2 \longrightarrow Ph_2C{=}CH_2 + (CO)_5W{=}CMe_2, \qquad (5.47)$$

where the ratio of $Ph_2C{=}CH_2$ and $Ph_2C{=}CMe_2$ exceeds 1200:1. The investigation of co-metathesis kinetics in a system of three olefins performed in 1975 by Katz [107] finally confirmed the validity of the "carbene" chain mechanism in the metathesis of olefins according to reactions (5.40)–(5.44)] (see also Section 3.33).

Reaction (5.46) passes via the formation and decomposition of metallacyclobutene,

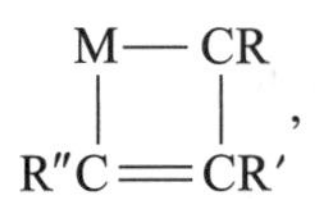

whereas reaction (5.45) involves the formation and conversion of both metallacyclobutene and metallacyclobutane:

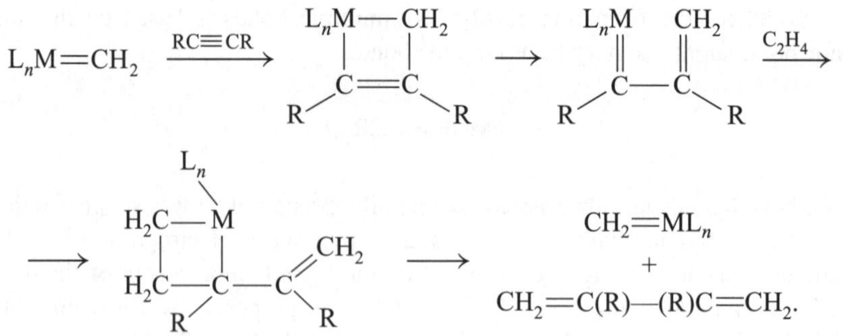

(5.48)

There are two known types of precursors in processes of the metathesis type [98, 100–102]:

(i) Catalytic systems of the Ziegler type such as WCl_6–$EtAlCl_2$–EtOH, $W(OMe)_6$–$EtAlCl_2$, $MoCl_2(NO)_2L_2$–$Al_2Me_3Cl_2$, WCl_6–SnR_4, and

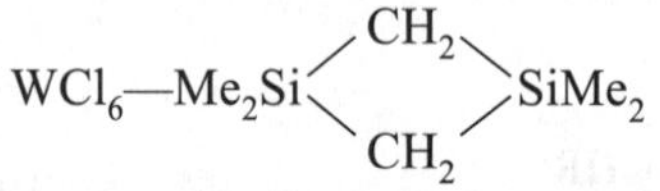

[108], WCl_6—Ph_2SiH_2 [109].

(ii) Stable carbene complexes $L_nM{=}CR_1R_2$ of transition metals, among which the maximum activity and stability are observed for the complexes of Mo and W (Schrock carbene) and Ru (Hoveyda–Grubbs carbenes), for example:

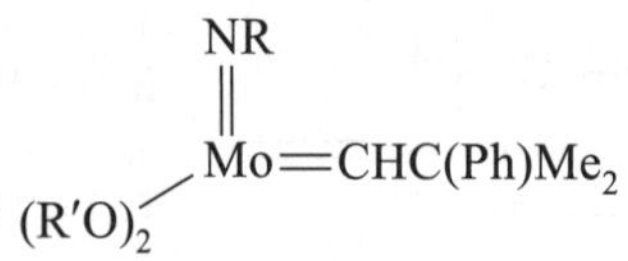

Schrock catalyst;

R= 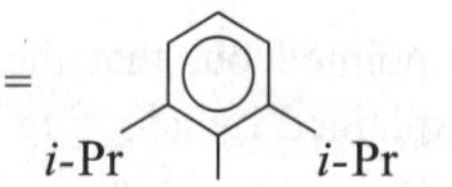

R′ = $(CF_3)_2(CH_3)C$

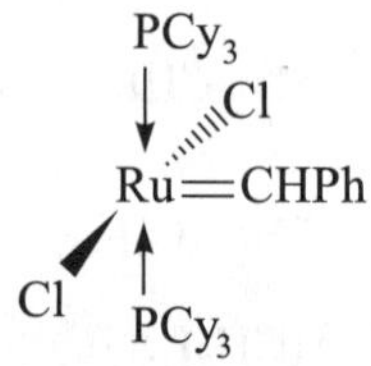

Grubbs' catalyst of first generation

Cy = cyclohexyl;

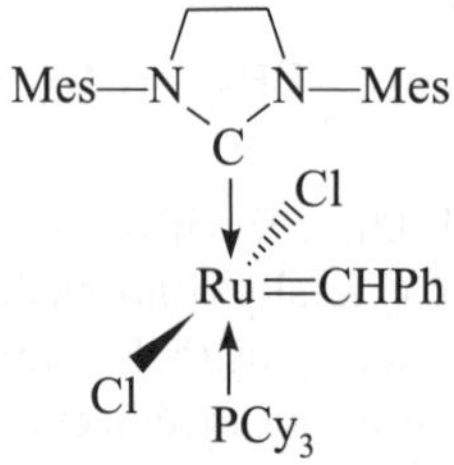

Grubbs' catalyst of second generation

Mes = trimethylphenyl

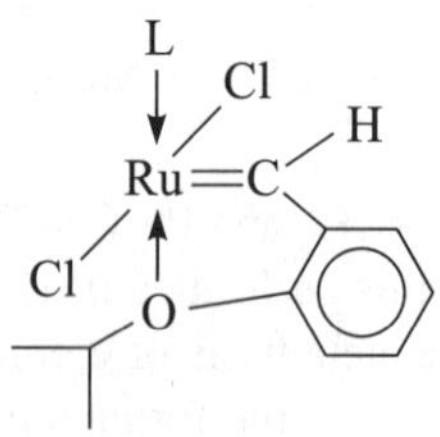

Hoveida catalyst
L = PCy3 (first generation)
L = imidazolin-2-ylidene (second generation)

Commercially available Grubbs' ruthenium catalysts are much more tolerant with respect to the functional groups (including α-substituted olefins) [110], but less active than Schrock's molybdenum and tungsten complexes. In order to create more active catalysts with longer working lives, extensive investigations are performed for establishing the mechanisms of active center formation in the Ziegler type systems and the loss mechanisms of active carbene complexes (and metallacyclobutane intermediates) for all types of precursors and initiators.

There are several known pathways for the formation of primary carbene complexes in the Ziegler type systems. The result of metal alkylation in these cases, as well as in those considered above (Section 5.2), is the formation of M–R and M–H bonds and the reduction of metal M to lower degrees of oxidation upon their rupture.

(i) The main pathway for carbene complex formation from M–R groups is via the α-elimination reaction [98, 100]:

$$L_nM{-}CH_2R \rightleftharpoons \underset{\displaystyle H}{\underset{|}{L_nM}}{=}CHR; \qquad (5.49)$$

$$L_nM{-}(CH_2R)_2 \longrightarrow L_nM{=}CHR + CH_3R; \qquad (5.50)$$

$$2L_nM{-}CH_2R \longrightarrow L_nM{=}CHR + CH_3R + L_nM. \qquad (5.51)$$

(ii) In the presence of an olefin, the M–H group converts into an M–R group with the subsequent α-elimination and the formation of carbene ligands:

$$L_nM{-}H + C_2H_4 \longrightarrow L_nMCH_2CH_3 \longrightarrow \underset{\displaystyle H}{\underset{|}{L_{n-1}M}}{=}CHCH_3 + L. \qquad (5.52)$$

This mechanism is probably operative in the WCl_6–$LiAlH_4$ system.

(iii) There are model reactions indicative of the possibility of the intramolecular 1,2-shift of hydrogen atom in the π complex of olefin [97, 98]:

$$L_nM(C_2H_4) \rightleftharpoons L_nM{=}CHCH_3. \qquad (5.53)$$

(iv) In the case of strongly electrophilic (cationic) π complexes of metals or strongly nucleophilic complexes of low-valence metals, vinyl (alkenyl) fragments can be formed from olefins. In the latter case, the olefin is oxidatively added to a metal via the =C–H bond [111]:

$$L_nM(CH_2{=}CHR) \longrightarrow \underset{\displaystyle H}{\underset{|}{L_nM}}{-}CH{=}CHR \longrightarrow L_nM{=}CHCH_2R. \qquad (5.54)$$

This reaction proceeds quite readily in the case of cluster complexes such as $Os_3(CO)_{12}$.

(v) Low-valence metals in complexes can participate in the oxidative addition of two olefin molecules with the formation of metallacyclopentanes [98, 101] and the subsequent conversion via metallacyclobutane into carbene complexes:

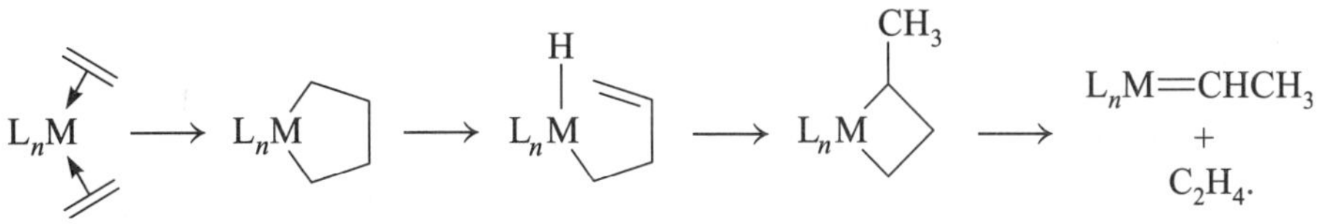

(5.55)

There is also an alternative pathway for the conversion of saturated metallacycles with the formation of carbene ligands and smaller metallacycles. Indeed, the decomposition of metallacyclobutane via the reaction

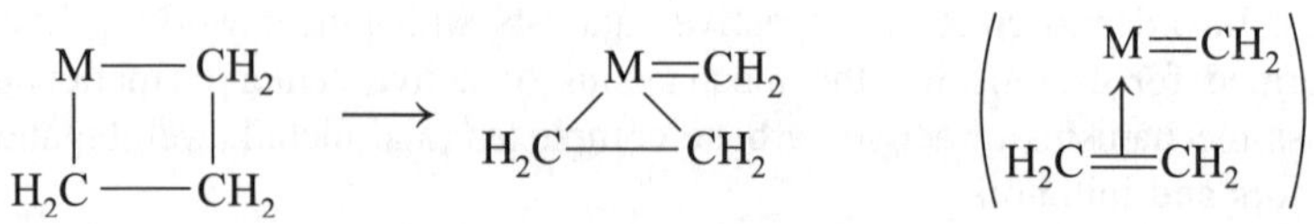

(5.56)

is in fact the reaction of carbene α-elimination with the C–C bond rupture. The possibilities of these reactions were convincingly demonstrated by Grubbs in the case of metallacyclohexane [112]:

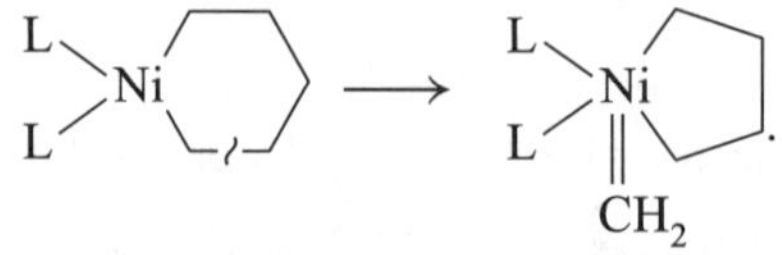

(5.57)

It can be suggested that the intermediate metallacyclopentane is also susceptible to this reaction (by analogy with reactions (5.56) and (5.57)). Then, the reaction

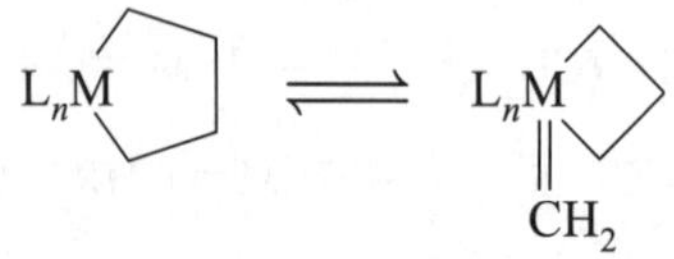

(5.58)

opens the way to carbene and metallacyclobutane (and, accordingly, to carbene as well).

In order to explain the isotope composition of the products of diene metathesis (on a $MoO_3/CoO/Al_2O_3$ heterogeneous catalyst)

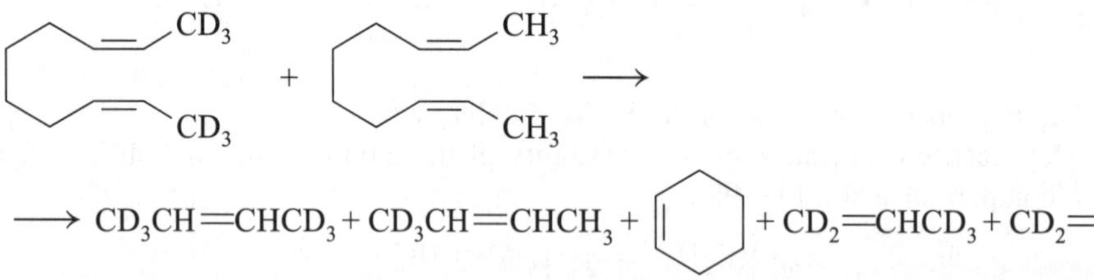

the following mechanism was proposed for the formation of metallacyclobutane intermediates via η^3-allyl complexes [113]:

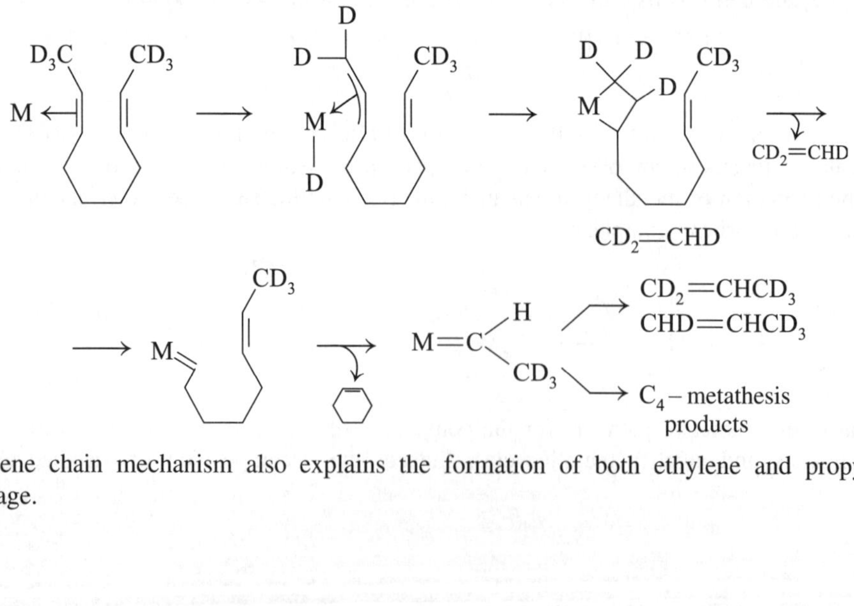

The carbene chain mechanism also explains the formation of both ethylene and propylene at the initiation stage.

A more complicated mechanism for the formation of a primary (initiating) carbene was revealed in a system containing Re(I)–$Re(CO)_5Cl$–$EtAlCl_2$ complexes [114], where the absence of gas evolution (C_2H_6, CO), which could be expected to accompany the formation of a primary ethylidene complex [Re]=$CHCH_3$, and the results of analysis of the primary products of 1,7-octadiene (RCM) or *trans*-4-octene metathesis and the products of primary carbene termination (i.e., products of demetallation of metallacyclobutane intermediates) led to a conclusion concerning the formation of propylidene carbene as a result of alkylation of a carbonyl ligand. The initial stage of reaction

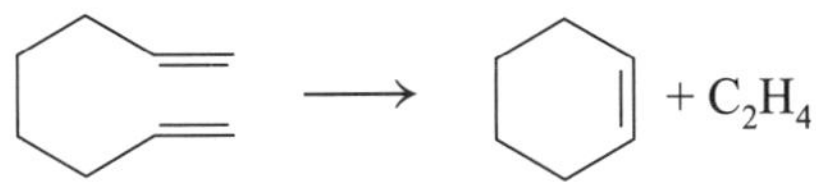

(5.59)

yields 1-butene and 1,7-decadiene in an amount close to the initial concentration of [Re]. Both these products are formed as a result of the reaction of [Re]=$CHCH_2CH_3$ with 1,7-octadiene. The minor products included three cyclopropane derivatives:

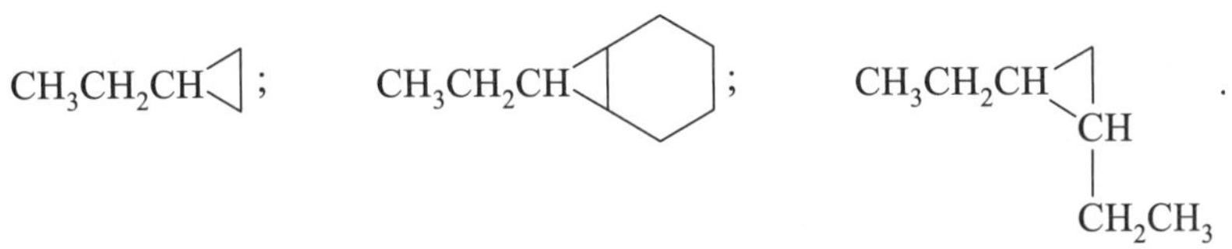

Although ethylcyclopropane can also appear as a result of the addition of methylene to 1-butene, the combination of these products is convincing evidence for the formation of [Re]=$CHCH_2CH_3$. It was suggested that rhenium carbonyl chloride is attacked by ethylaluminum dichloride at the carbonyl group with the formation of a carbene ligand of the Fischer type:

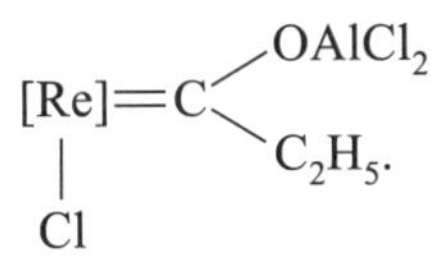

It should be noted that acyl ligands also partly possess a carbene character that depends on the donor properties of a complex-forming metal in the L_nMCOR complex:

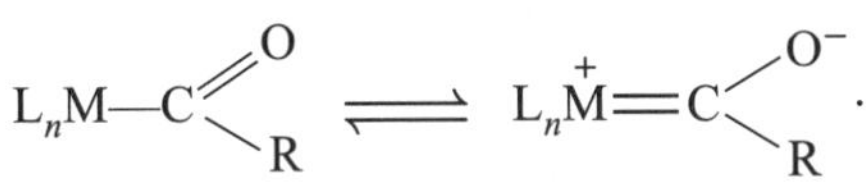

(5.60)

The reduction of $=\overset{|}{C}$—$OAlCl_2$ groups to C–H could take place either as a result of the formation of Re–H (from Re–C_2H_5) or due to the transfer of hydrogen from 1,7-octadiene. The latter variant was adopted in the following mechanism [114]:

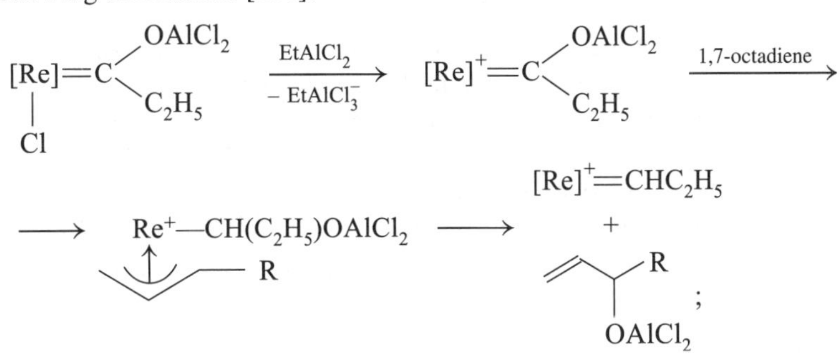

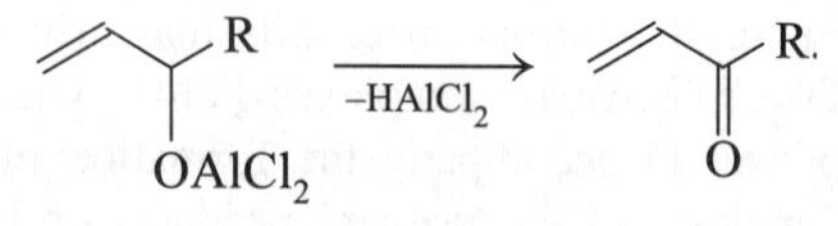

The product of 1,7-octadiene oxidation (1,7-octaden-3-one) was also identified in the reaction products.

An interesting mechanism for the formation of active centers was proposed for the systems containing WCl_6 and 1, 1, 3, 3-tetramethyl-1, 3-disilacyclobutane (DSCB) [115] or $W(CO)_6$ and Ph_2SiH_2 [102]. The opening of DSCB ring upon preliminary partial reduction of WCl_6 leads to the formation of σ-organometallic ligand

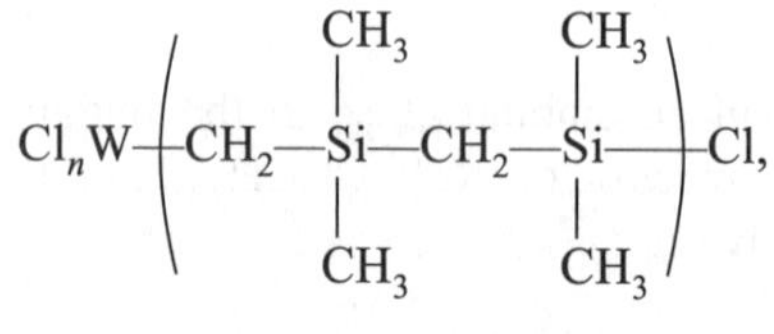

which is converted as a result of α-elimination or γ-elimination into carbene or metallasilacyclobutane:

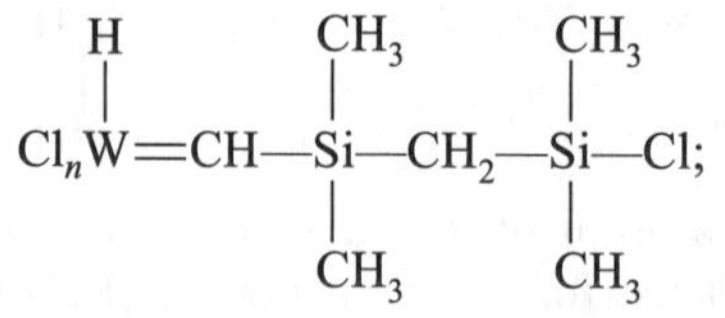

α-H

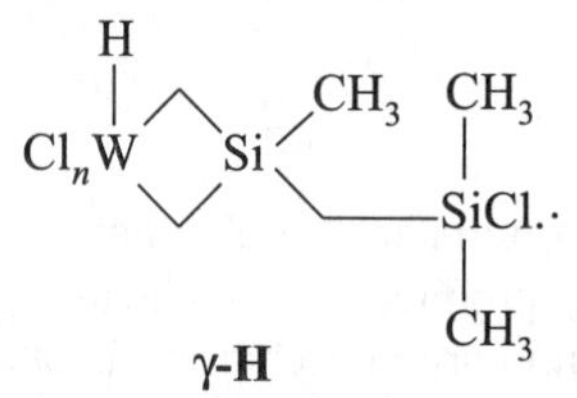

γ-H

The decomposition of a metallacycle leads to $Cl_qW{=}CH_2$ and $ClSi(CH_3)_2CH_2Si(CH_3)_2Cl$. In the $W(CO)_6$–Ph_2SiH_2 system, the process of carbene formation probably proceeds via the silylene complex of tungsten:

$$W(CO)_6 + Ph_2SiH_2 \xrightarrow[\text{-H}_2\text{,CO}]{} (CO)_5W{=}SiPh_2. \qquad (5.61)$$

The reaction of this complex with olefin is just what leads to the formation of an active carbene complex that is involved in the steps of kinetic chain propagation. The cyclodimerization of intermediate silylene complexes yields dimers

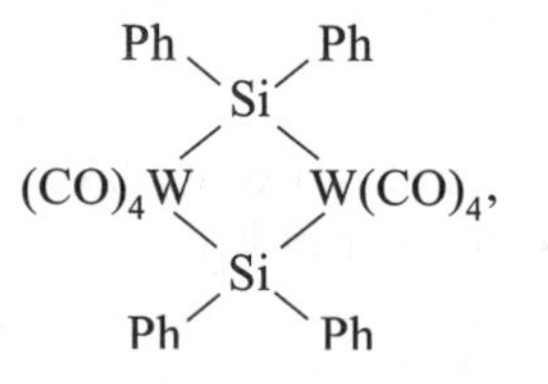

which are inactive at 45 °C, and, hence, leads to inactivation of the catalyst. These dimers are active in the metathesis at temperatures above 100 °C or under UV irradiation.

Using diazo derivatives for the synthesis of carbenes *in situ*, it is possible to form active centers for the metathesis of olefins in the WCl_6–$PhCHN_2$ and WCl_6–$N_2CHCOOEt$ systems [116]. The interaction of diazo compounds with metal chlorides in high degrees of oxidation may be accompanied by their preliminary reduction analogous to the reactions of $CuCl_2$ with $N_2CHCOOEt$ [117]:

$$2CuCl_2 + N_2CHCOOEt \xrightarrow[-N_2]{} Cl_2CHCO_2Et + 2CuCl;$$

$$2CuCl_2 + Ph_2CN_2 \xrightarrow[-N_2]{} Ph_2CCl_2 + 2CuCl;$$

$$4CuCl_2 + 2N_2CHCOOEt \xrightarrow[-N_2]{} EtOCOC(Cl){=}C(Cl)COOEt + 2CuCl + 2HCl.$$

Various reactions and reactants can be used for the synthesis of comparatively stable carbene complexes of Ru, Mo, and W, which can be introduced as initiators into reactions of the metathesis type. Naturally, the reactions of diazo derivatives are convenient for the formation of carbene ligands, in particular, for the synthesis of 1st generation Grubbs' catalysts [101]:

$$L_3RuCl_2 + PhCHN_2 \xrightarrow[-L]{} L_2RuCl_2(CHPh) + N_2. \quad (5.62)$$

It is also possible to use reactions of the oxidative addition of dihalogenalkanes to Ru (0):

$$L_3Ru + PhCHCl_2 \xrightarrow[-L]{} L_2RuCl_2(CHPh). \quad (5.63)$$

In particular, this reaction was performed using Ru (COD) (cyclooctatriene) as Ru (0) complexes and phosphine ligands. Convenient precursors of Ru (0) are represented by Ru (II) dihydrogen dihydride complexes in the presence of olefins [118]:

$$(H)_2Ru(\eta^2\text{-}H_2)_2L_2 + PhCHCl_2 \xrightarrow{\text{cyclohexene}} Cl_2Ru{=}CHPh(L)_2 + 2H_2 + \text{cyclohexane}.$$

Olefins (ethylene, cyclohexene) were introduced for the removal of two hydride ligands (and, probably, H_2). Complex Ru $(C_6H_{10})_2(H)_2L_2$ was isolated from CH_2Cl_2.

Schrock complexes of Mo and other metals are obtained by deprotonating their alkyl derivatives with strong bases (B) or by eliminating H^- with the aid of carbenium ions:

$$[Mo]CH_2R + B \longrightarrow [Mo]{=}CHR + BH^+, \quad (5.64)$$

$$[W]CH_2R + Ph_3C^+BF_4^- \longrightarrow [W]{=}CHR^+BF_4^- + Ph_3CH. \quad (5.65)$$

The addition of strong nucleophiles (H^-, R^-) to η^3-allyl complexes leads to the formation of metallacyclobutane, which exhibits subsequent decomposition into carbene complex and olefin [119]:

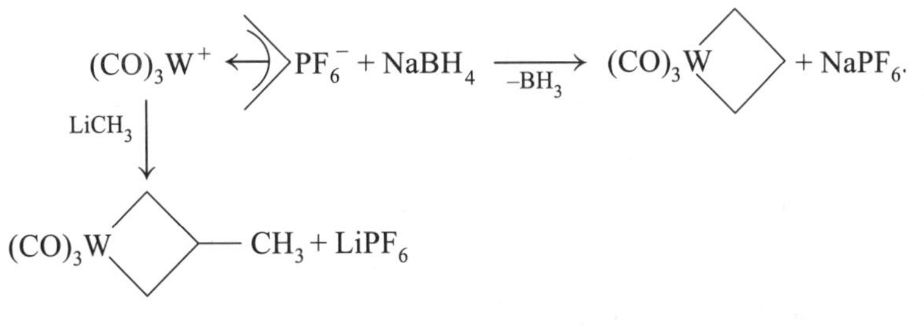

(5.66)

Another source of carbene ligands is provided by the conversions of alkynes in the coordination sphere of a complex-forming metal. The isomerization of alkyne to the vinylidene group is readily achieved in the case of transition metal complexes [18, 102, 120–123]. For example, π complexes of Ru(II) in methanol or acetonitrile are converted into vinylidene complexes via the following first-order reaction at

15–40 °C [124]:

$$Cp(PMe_3)_2Ru(HC{\equiv}CMe)^+PF_6^- \longrightarrow Cp(PMe_3)_2Ru{=}C{=}CHMe^+PF_6^-. \qquad (5.67)$$

The 1,2-shift in alkyne also takes place in disubstituted alkynes, provided that one of the substituents is a $SiMe_3$ group [125]:

$$[RhClL_2]_2 + Me_3Si{-}C{\equiv}C{-}SiMe_3 \xrightarrow[-20\,°C]{pentane} ClRhL_2\left({=}C{=}C\begin{matrix} SiMe_3 \\ SiMe_3 \end{matrix}\right). \qquad (5.68)$$

A vinylidene complex of Ru(II) is also formed upon the protonation of its ethynyl complex [102, 126]:

$$(\eta^6\text{-}C_6Me_6)RuCl(PMe_3)(C{\equiv}CPh) \xrightarrow{H^+} (\eta^6\text{-}C_6Me_6)RuCl(PMe_3)({=}C{=}CHPh)^+. \qquad (5.69)$$

Propargyl derivatives $HC{\equiv}C{-}C(X)R_2$ usually yield allenylidene ligands, which can also serve as initiators in metathesis reactions [102, 126]:

$$(\eta^6\text{-}C_6Me_6)RuCl_2(PMe_3) + HC{\equiv}CC(OH)Ph_2$$

$$\xrightarrow[\text{methanol, } 25\,°C]{NaPF_6} (\eta^6\text{-}C_6Me_6)RuCl(PMe_3)({=}C{=}C{=}CPh_2)^+PF_6^-. \qquad (5.70)$$

The formation of vinylidene complexes was reported for mononuclear complexes of Mo, W, Mn, Re, Fe, Ru, Co, Rh, Ir and some clusters [18].

Alkynes also participate in the formation of other types of carbene ligands with a general formula of $\eta^n{-}C_nR_{n+1}$, where $n = 2$, 3, and 4. In 1979, Davidson et al. [127] proposed a new mechanism for the step of alkyne insertion into the M–Nu bond, which included the formation of intermediate η^2-vinyl derivative (i.e., carbene with $n = 2$) containing metallacyclopropene:

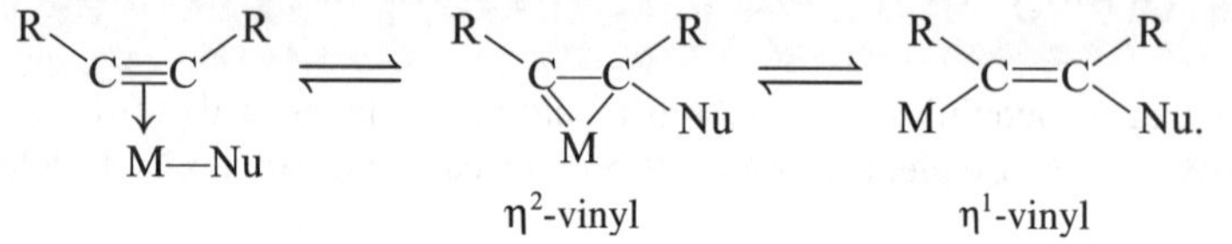

(5.71)

Stable η^2-vinyl complexes were synthesized for Mo, W, and Re [18]. The η^2-vinyl ligands can also be obtained via the protonation of π complexes [128],

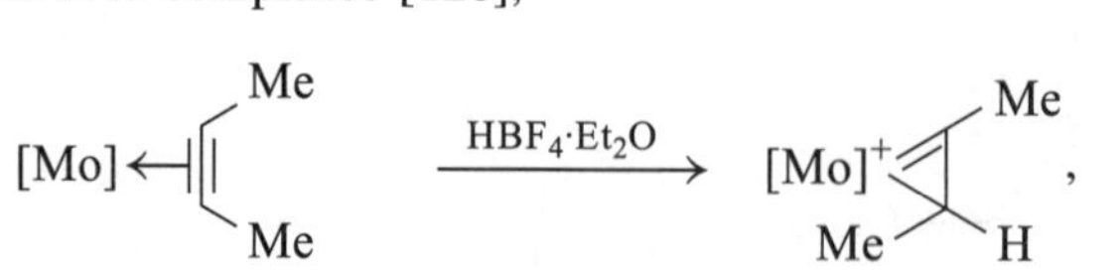

(5.72)

and the deprotonation of substituents in alkynes [129]:

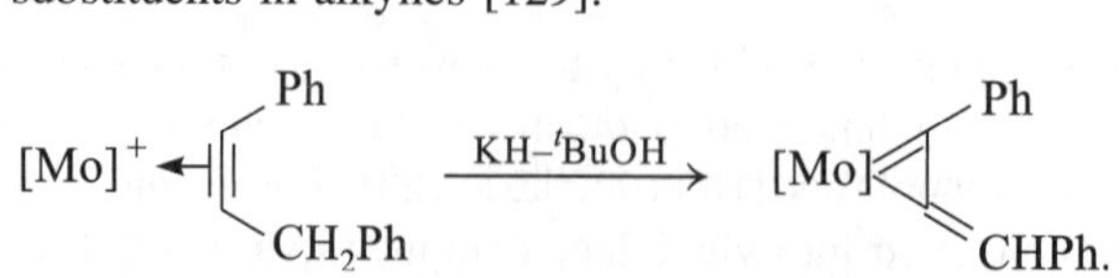

(5.73)

Ligands of the η^3-C_3R_4 type (vinyl carbenes) were formed via the reaction of carbenes ($n = 1$) with alkynes (see mechanism (5.48)), for example as [130]

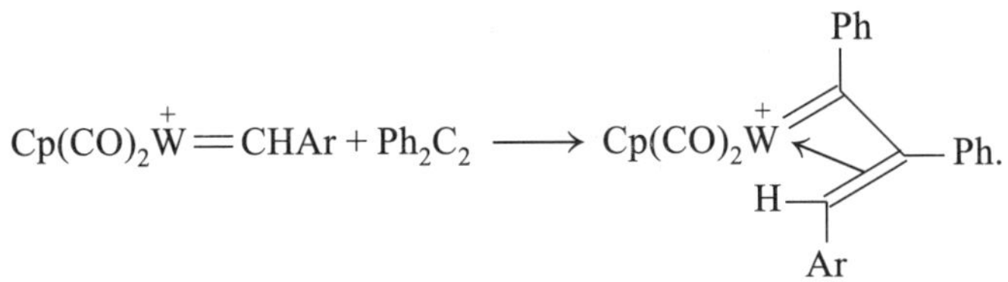

(5.74)

Vinyl carbene complexes were also obtained from Ru(II) hydride η^2-dihydrogen complexes and propargyl chlorides [118]:

$$HRuCl(\eta^2\text{-}H_2)L_2 + Me_2C(Cl)C{\equiv}CH \xrightarrow[-H_2]{} RuCl_2L_2({=}CH{-}CH{=}CMe_2). \quad (5.75)$$

The insertion of alkyne into M–C σ-bond in metallacyclopropene (η^2-vinyl complex) yields η^4-C_4R_5 (η^4-butadienyl) ligand [128]:

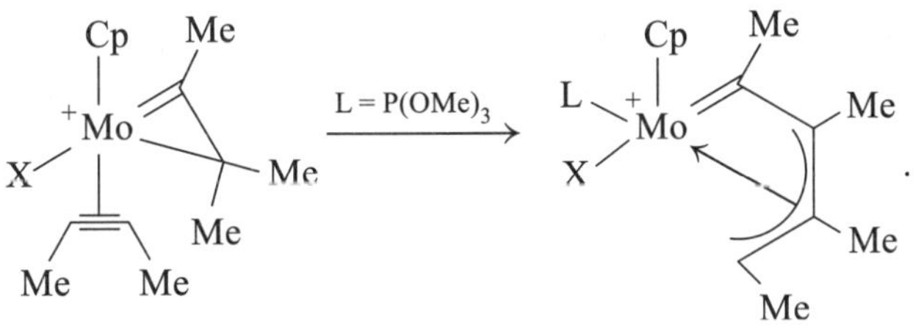

(5.76)

The obtained complex can also be represented as metallacyclopentadiene:

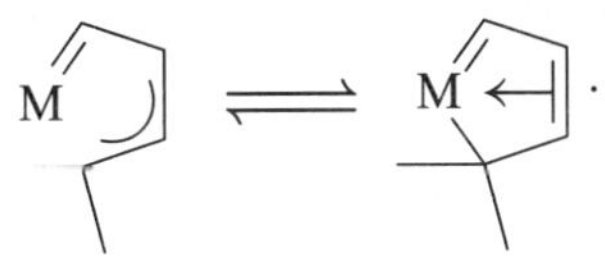

The same ligand is obtained upon the protonation of bis-acetylene π complex [130]:

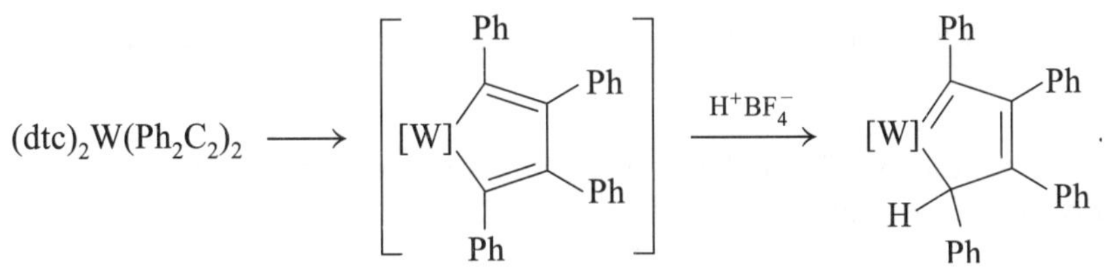

(5.77)

where (dtc) is diethylthiocarbamate.

Various conversions of η^4-C_4R_5 ligands (in particular, to vinyl carbene η^3-C_3R_4) were described in [130]. It can be suggested that all types of carbenoids with $n = 2$, 3, and 4 can be initiators of metathesis processes.

Romero et al. [103] proposed an unusual pathway for the synthesis of Ru complexes with heterosubstituted electrophilic carbenes, which are very active in the metathesis (RCM) of diethyl diallyl malonate at 273 K. Using this method, ligands containing phosphorus atoms (Ru=C(H)P^+R_3) were obtained from the first- and second-generation Grubbs' complexes:

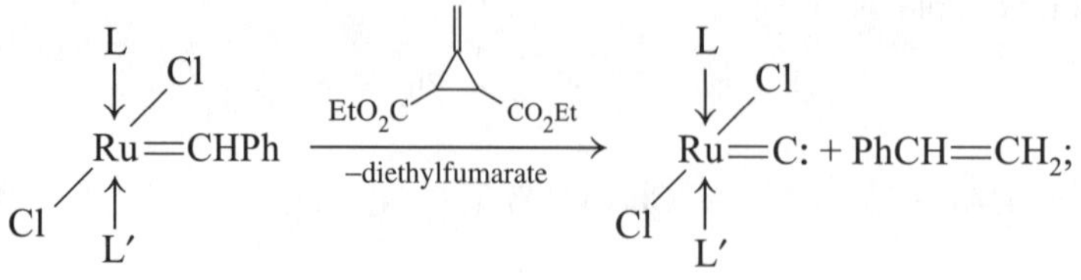

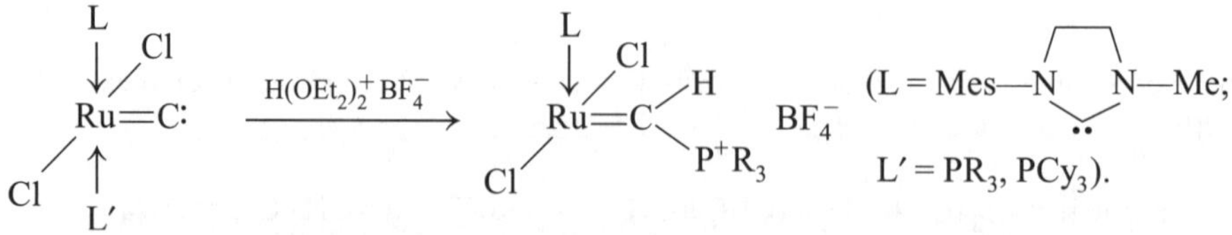

For the indicated ligands (L and L′), the obtained 14-electron complexes exhibited higher activity as compared to the Schrock complexes and second-generation Grubbs' catalysts. Some other active metathesis catalysts with heteroatoms at carbene carbon (Fischer's carbenes) were also proposed, for example, $RuCl_2(PCy_3)_2$(=CHSPh) (Cy = cyclohexyl) [131].

The appearance of active initiators based on ruthenium complexes *in situ* upon adding alkynes to the catalytic system can be related to the formation of species such as [Ru]=C=CHR or (in alcohol systems)

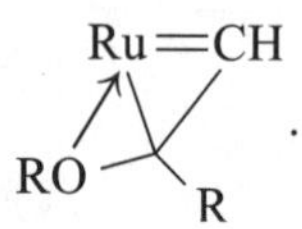

In particular, the $RuCl_3$ ($RuBr_3$)–PR_3–RC_2H–ROH system in H_2 atmosphere catalyzes the metathesis of linear olefins [132, 133], while the [(*p*-cumene)RuCl]$_2$ – methylimidazolium chloride (IMeCl)–^{t}BuONa–^{t}BuC≡CH–hexane system features the cross metathesis of olefins [134] and the RCM process [134, 135].

In the course of catalytic reactions, both the initiators formed *in situ* and carbenoids synthesized as complexes of known compositions and structures are subjected to various transformations determined by the high and diverse reactivity of these species. These transformation of carbene complexes lead to the loss of active centers and produce (like the side reactions of metallacyclobutanes) inactivation of the catalytic system.

The main mechanisms of the kinetic chain termination in catalytic processes of the metathesis type are as follows [98]:

(i) In comparison to reaction (5.53), the reverse process of carbene isomerization to olefin is thermodynamically more favorable:

$$Cp_2W{=}C(H)(CH_3) \longrightarrow Cp_2W(C_2H_4). \tag{5.78}$$

(ii) The appearance of additional hydrides or alkyls on the metal in a carbene complex (excess alkylation in Ziegler systems) leads to reactions that are reverse to α-elimination:

$$M(R'){=}CHR \longrightarrow MCHRR'; \tag{5.79}$$

$$\underset{\displaystyle \mathrm{H}}{\overset{|}{\mathrm{M}}}\mathrm{{=}CHR} \longrightarrow \mathrm{MCH_2R}. \tag{5.80}$$

(iii) The appearance of electrophilic species in the case of nucleophilic carbenes also leads to the formation of alkyls,

$$\mathrm{M{=}CHR + H^+ \longrightarrow M^+{-}CH_2R}, \tag{5.81}$$

where as in the case of vinylidene initiators, the protonation of the vinylidene ligand at the β position even by weak acids converts carbene into carbyne [136]:

$$\mathrm{ReCl(dppe)_2({=}C{=}CHPh) + Et_3NH^+ \longrightarrow ReCl(dppe)_2({\equiv}CCH_2Ph)^+ + Et_3N} \tag{5.82}$$

(dppe •)

(dppe = diphenylphosphiniethane).

(iv) Conversions of metallacyclobutane (demetallation) with the formation of cyclopropane or olefin are also possible pathways to inactivation, since metal M in this case is reduced with a decrease in the oxidation number by two units:

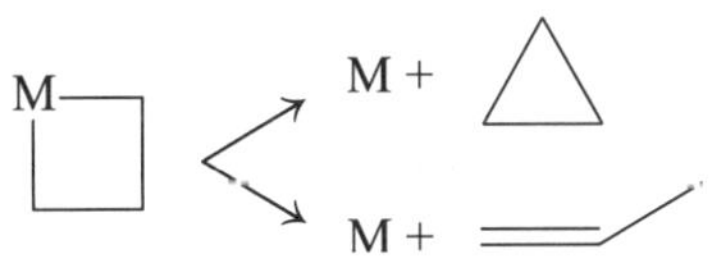

(5.83)–(5.84)

(v) In the case of unstable carbenes of the Fischer type, carbene can be displaced from the coordination sphere by phosphine with the formation of ylide species:

$$\mathrm{M \longleftarrow C(OR)R' + PR_3 \longrightarrow M + R_3P{=}C\langle^{OR}_{R'}}. \tag{5.85}$$

The addition of nucleophiles (ROH, H_2O, OAc^-) to vinylidene carbenoids leads to the formation of Fischer's carbenes or the yield of products of HX addition to alkynes (X–OH, OR, OAc). In view of the impressive synthetic possibilities offered by the first- and second-generation Grubbs' catalysts (G1 and G2, respectively) and the Hoveyda catalysts in the reactions of functionally substituted olefins [101, 110, 137], the pathways of the loss of active centers in these systems have been especially thoroughly studied. It was established that a G1 catalyst with PhCH = carbene in benzene at 55 °C was highly stable, with a half-decomposition period of about $t_{1/2} = 190$ h [138, 139]. In contrast, a G1 (=CH_2) catalyst exhibited half-decomposition within 0.6 h, while a G2 (=CH_2) catalyst was significantly more stable ($t_{1/2} = 5.6$ h) but exhibited a 70 times lower initiation rate [139]. The carbene complexes with PhCH= and RCH= decomposed according to a second-order kinetics and the decomposition rate decreased upon adding PR_3. It was suggested that the decomposition proceeded according to the following scheme,

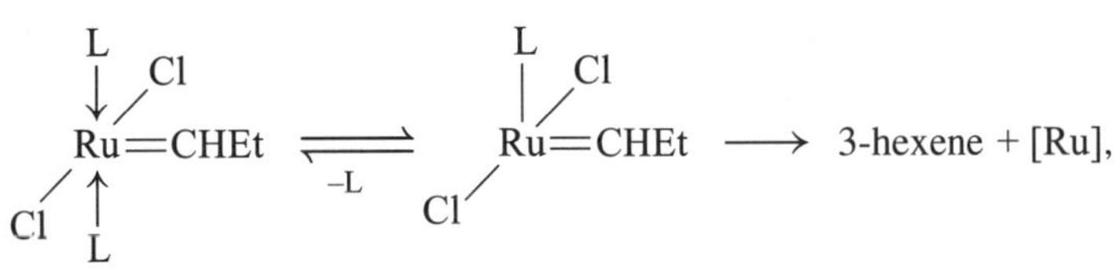

(5.86)

with the possible formation of dimeric Ru–Ru and/or Ru=Ru complexes.

The step of dissociation of Ru–L bonds is necessary for the creation of a vacant site and the coordination of olefin during the initiation, but the same step probably also provides conditions for the dimerization of complexes and the elimination of olefin.

In the case of methylidene carbene, other decomposition pathways also probably exist, since both G1(=CH_2) and G2(=CH_2) decompose according to the first-order kinetics and their decomposition is not inhibited by phosphines. Grubbs with co-workers [140] analyzed the products of decomposition of G2 (=CH_2) complexes in benzene at 55 °C and isolated a ruthenium-containing complex with carbyne ligand:

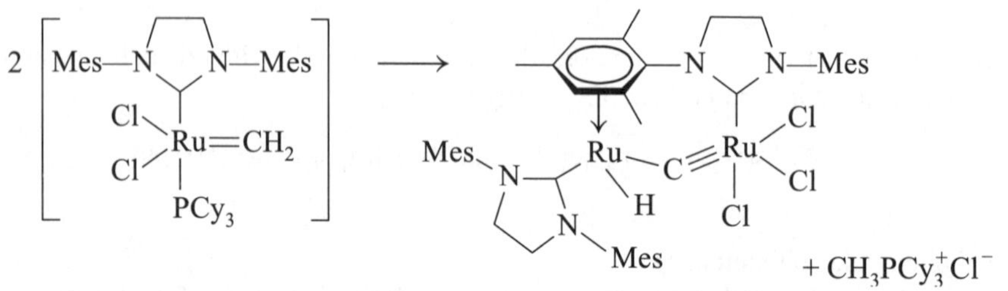

(5.87)

The process of methylidene transformation during this reaction can readily be explained by admitting the dimerization of carbenoids by analogy with the inactivation of silylene complexes via reaction (5.61) [115]:

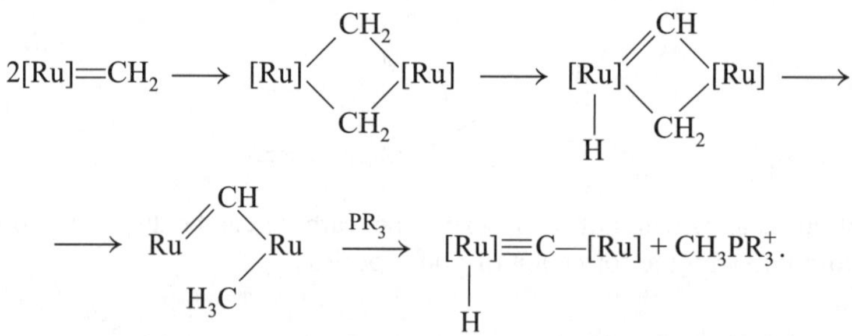

(5.88)

The influence of the admixtures of alcohols, water, and oxygen on the transformations of Grubbs' catalysts was studied for both G1 [141] and G2 [142, 143]. It was established that a G1(=CHPh) catalyst reacted with MeOH in toluene at 70 °C with the formation of a $RuCl_2L_2(H)(CO)$ carbonyl hydride complex, and this process was accelerated by the additives of bases. The obtained complex catalyzed the isomerization of olefins. In the case of benzyl alcohol, the reaction led to the formation of the $RuCl_2L_2(Ph)(CO)$ phenyl carbonyl complex, which was also formed under the action of O_2 on G1(=CHPh). Analogous complexes were obtained in reactions of G2(=CHPh) [142, 143].

The reactions of G1(=CH_2) with ethylene were used to study the conversions of an intermediate metallacyclobutane [144]. For the reaction in benzene at 40 °C, the two main products were C_3H_6 and 1-C_4H_8, while the minor products contained cyclopropane (see reactions (5.83) and (5.84)). It was suggested that 1-butene was a secondary product of the interaction between methylidene and propene, and the following mechanism of propene elimination via the formation of η^3-allyl hydride complex **2**, its isomerization to η^1-allyl **3**, and conversion into propene by means of the reductive elimination:

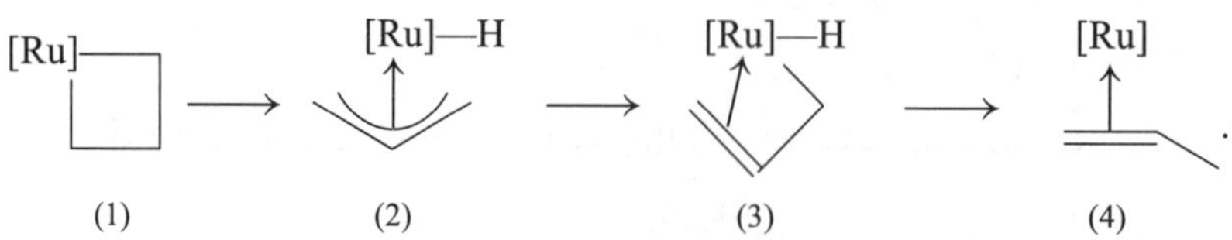

(1) (2) (3) (4) (5.89)

Investigations into the ethenolysis of a second-generation Hoveyda complex in benzene at 55 °C [145] revealed, in addition to various derivatives of 2-isopropoxybenzylidene ligand, the following hydrocarbon products: ethane (~7 %), propylene (~79 %), *trans*-2-butene (~7 %), *cis*-2-butene (~2.5 %), 1-butene (~1 %), isobutene (~0.4 %), and C_5 olefins (~2.5 %):

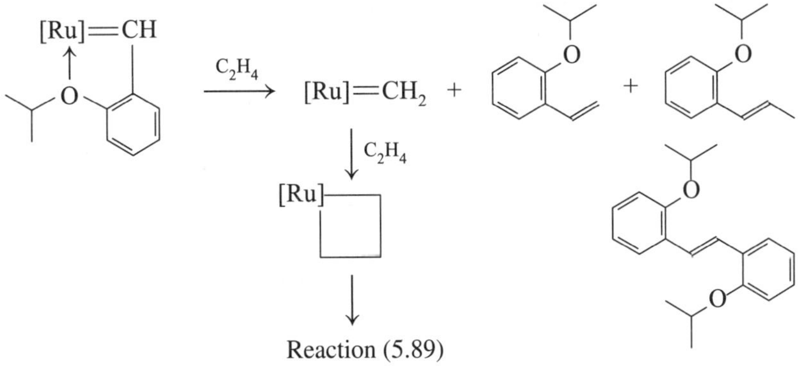

(5.90)

The mechanism of ethane formation was probably related to the appearance of hydride ligands.

It was established that G1 (=CHPh) complexes are readily subject to hydrogenolysis, which converts them from metathesis catalysts to olefin hydrogenation catalysts [146]. This circumstance makes it possible to combine metathesis (CM, ROMP) with hydrogenation of the metathesis products. Propargyl chloride additives convert ruthenium hydride complexes into vinyl carbene complexes (C_3R_4) (see reaction (5.75)), which are initiators of olefin metathesis [146],

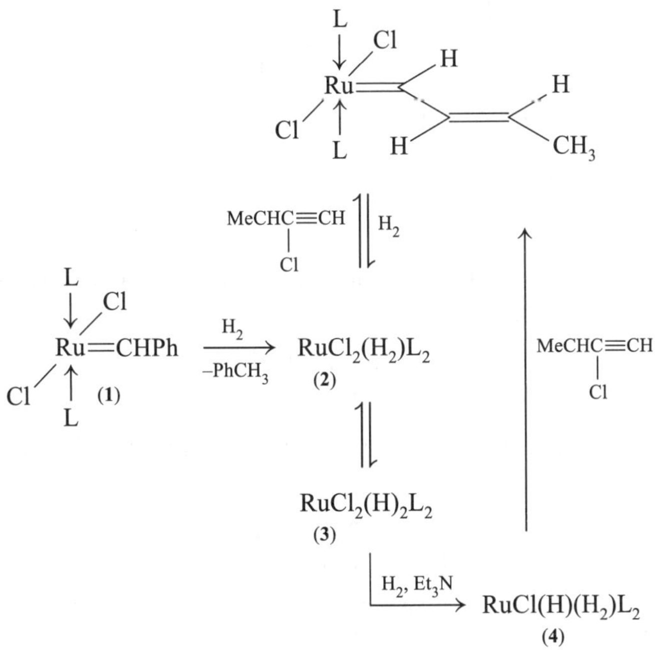

(5.91)

and complex **4** is a catalyst of olefin hydrogenation.

Grubbs with co-workers [139] thoroughly studied the kinetics and mechanism of the process of initiation, i.e., the formation of active centers for the chain propagation steps from the initial carbene species (precursors) via the following reaction for 14 complexes representing G1 and G2 with various anions, phosphines, and carbenes:

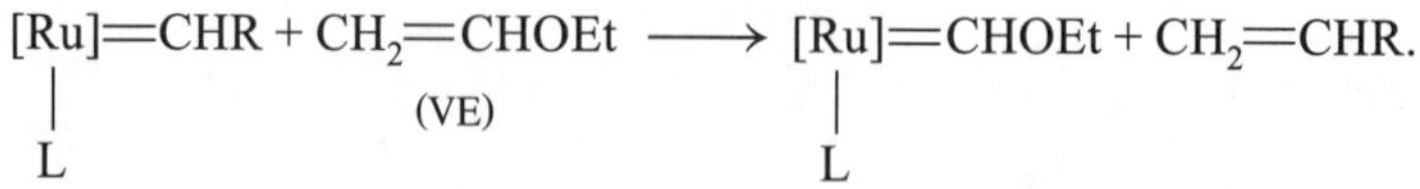

(5.92)

This reaction rapidly proceeds with quantitative yield and high regioselectivity, and results in the formation of stable Fischer's carbenes. The reaction rate is described by an equation of the first order with respect to [Ru]=CHR and a variable order with respect to olefin. It was established that the reaction is inhibited by phosphine, and the exchange rate of free and coordinated phosphine is determined by the dissociation step,

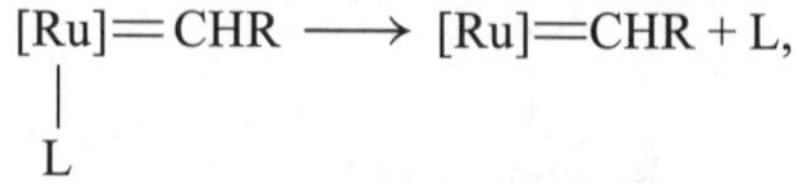

i.e., by the transformation of a 16-electron complex into a 14-electron complex. This step was also adopted for the mechanism of initiation in reaction (5.92). The process of carbene formation from vinyl ether (VE) is described by the following multistep scheme:

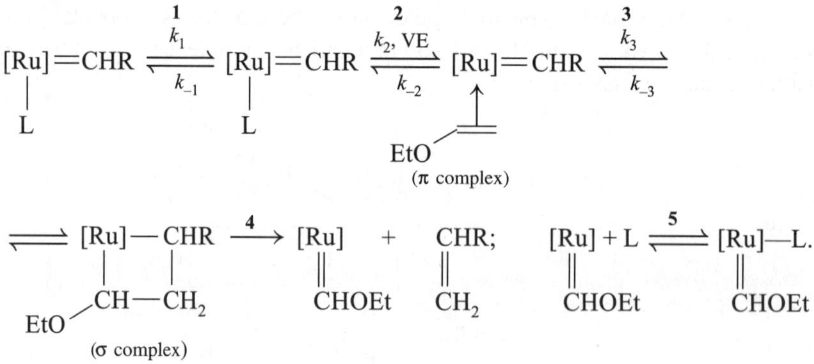

(5.93)

Step **2** in this scheme was assumed to be irreversible [139], and the process kinetics in the first two steps was described by the following equation:

$$r = \frac{k_1 k_2 [\mathrm{Ru}][\mathrm{VE}]}{k_{-1}[\mathrm{L}] + k_2[\mathrm{VE}]}. \tag{5.94}$$

In the case of adding excess L and a large excess of VE, this relation is linearized in the following coordinates:

$$\frac{1}{k_{\mathrm{obs}}} = \frac{k_{-1}[\mathrm{L}]}{k_1 k_2 [\mathrm{VE}]} + \frac{1}{k_1}, \tag{5.95}$$

where k_{obs} is the observed first-order rate constant with respect to the loss of [Ru]=CHR. The obtained values of k_1 in Eq. (5.95), k_1 in Eq. (5.94) under the conditions where r is independent of [VE] (without adding L), and rate constant k_1 of L exchange rate (without olefin) are close to within experimental error.

It is also interesting to compare the standard Grubbs' complexes of G1 and G2 types. The rate constant of the [Ru]–PR_3 dissociation step (k_1) decreases on the passage from G1 to G2 by about two orders of magnitude, while the ratio of k_{-1}/k_2 is $1.3 \cdot 10^4$ for G1 and 1.25 for G2 (at 50 °C). Thus, the G1 complex rapidly dissociates, but the reverse reaction (k_{-1}) prevails over the reaction with olefin (k_2), whereas the G2 complex possesses higher selectivity with respect to olefin as compared to PR_3. Therefore, the kinetic chain length (i.e., the number of cycles of chain propagation steps per act of active center termination by phosphine) in the case of G2 is greater. It should be noted that, if step **3** (rather than step **2**) were assumed to be the irreversible slow step in scheme (5.93), which is more probable, the interpretation of the meaning of k_{-1}/k_2 ratio would be changed.

The results of investigations of the stage of active center nucleation with allowance for all the possible mechanisms of decomposition of active centers and intermediates lead to a relatively complete scheme of the chain mechanism of metathesis in the case of ethenolysis of higher olefins (CM process) via the reaction

$$C_2H_4 + RCH{=}CHR' \rightleftharpoons RCH{=}CH_2 + R'CH{=}CH_2. \qquad (5.96)$$

In order to somewhat simplify the very complicated scheme, let us set R = R′ and consider only the most probable variants of the decomposition of active centers and intermediates (Figure 5.7). In the presence of excess olefin RCH=CHR, the initiation process should begin with this olefin. The initiation mechanism (5.93) can be simplified according to Grubbs by adopting the irreversible step **2**, which determines the initiation rate W_{in}. Note that the steps of formation of K1C and K2C complexes ("sleeping" active centers) are equilibrium (pendant vertices on the corresponding kinetic graph), and the steps of formation of metallacycles σ_3 and σ_4 represent the steps of degenerate metathesis, which can also lead (as well as the steps of decomposition of σ_1 and σ_2) to chain termination (W_{term1} and W_{term2}, respectively).

In the case of Hoveyda complexes with chelate benzylidene carbene, the initiation takes place in the step of dechelation (i.e., chelate cycle opening) with dissociation of the Ru ← O–bond,

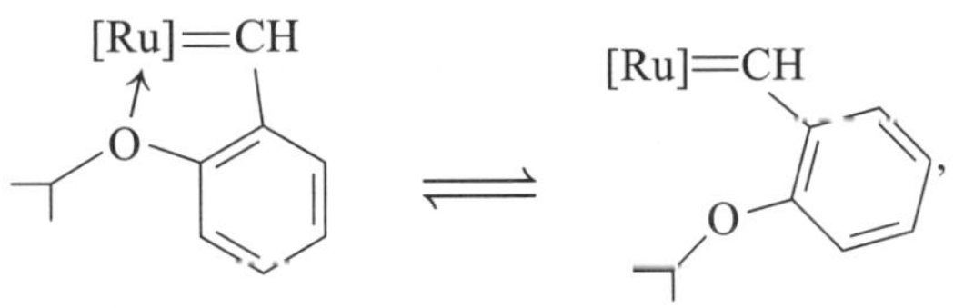

(5.97)

and the appearance of a 14-electron complex that coordinates olefin. This ligand provides better protection of the initiator against loss (as compared to the protection by phosphine in solution, since a bimolecular process is one to two orders of magnitude less effective than monomolecular). However, after the very first initiation event (e.g., with $CH_2{=}CH_2$), a styrene derivative passes to solution and becomes as external ligand (for –O– group) as PR_3. The protection of the active center in metathesis catalysts in this case still presents an unsolved problem. Simultaneous increase in the chain propagation rate, i.e., concentrations of K1 and K2 and the rates of their interaction with the substrate, as well as in the stability of active centers is in a certain sense a contradictory task. Indeed, to solve this problem, it is necessary to increase both the initiation rate (i.e., concentrations of K1 and K2) and the chain propagation rate at a decrease in the rates of decomposition of the active centers and metallacycles. It should be noted that metallacycles (σ_1, σ_2, etc.) with high probability belong to intermediates rather than transition states, although the results obtained in [139] do not give an unambiguous answer to this question. With allowance for the heuristic selection rule of elementary steps (based on the topology of the transition state or the structure of the topological identifier) of the given reaction (see Section 1.4), it is evident that the redistribution of two double bonds cannot take place synchronously at one elementary step via a single transition state.

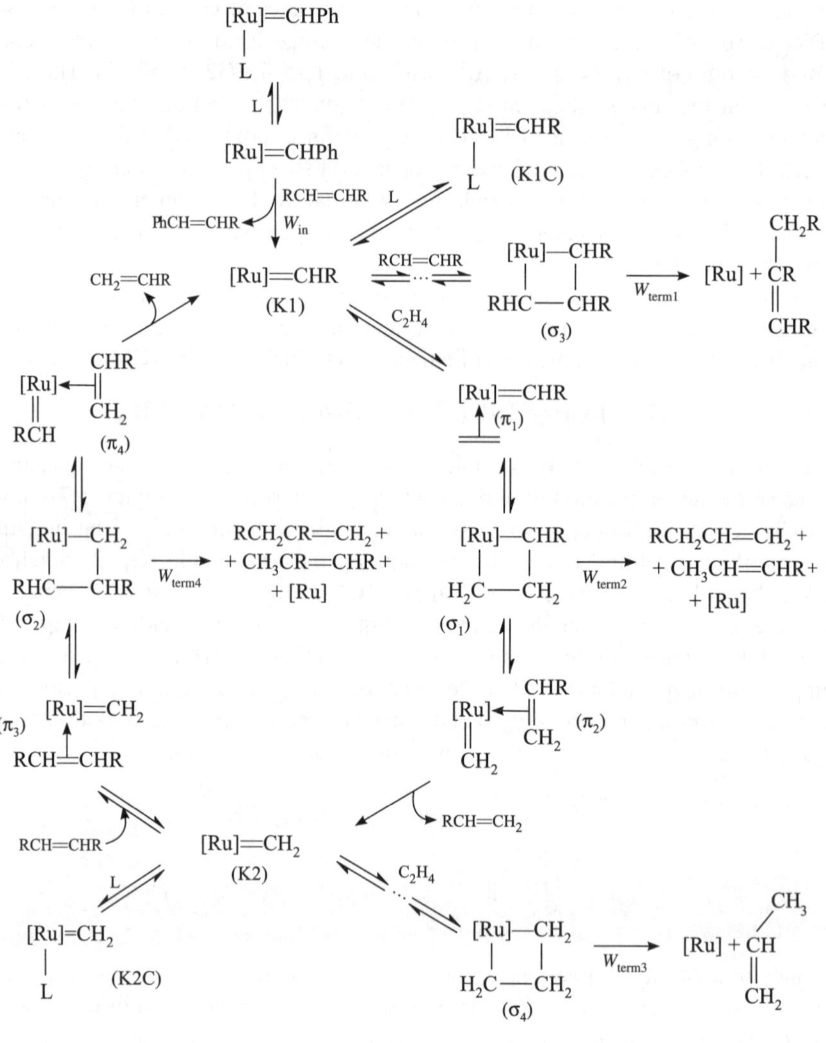

Figure 5.7 Mechanistic scheme of cross-metathesis on Grubbs' catalyst

5.3.3 Mechanisms of 1-butene isomerization in $Ni[P(OEt)_3]_4$–H_2SO_4–MeOH system

The mechanism of butene isomerization in this system was studied in detail by Tolman [147, 148]. The process of active center formation was studied by kinetic methods and by NMR and electronic absorption spectroscopy techniques. The isomerization of butenes into an equilibrium mixture of isomers at 25 °C requires several minutes. In contrast, the dissociation of NiL_4 ($L = P(OEt)_3$) complexes is a rather slow process with a conversion half-period of several hours and an equilibrium dissociation constant of the first L below 10^{-10} M^{-1} even at 70 °C. For this reason, the first step in the formation of a catalytically active center in the system under consideration is believed to be the reaction of protonation:

$$NiL_4 + H^+ \underset{\mathbf{1}}{\overset{K_1}{\rightleftharpoons}} HNiL_4{}^+. \qquad (5.98)$$

The kinetics of this reaction was studied at 25 °C by the method of stopped flow with $HClO_4$, which gave $k_1 = 1550\,M^{-1}\cdot s^{-1}$ and $k_{-1} = 45\,s^{-1}$. The calculated constant $K_1 = 35\,M^{-1}$ was consistent with the value obtained in the investigation of equilibrium ($K_1 = 48 \pm 14\,M^{-1}$). The protonation of NiL_4 complexes leads to approximately a 100-fold acceleration of the dissociation of Ni–L bonds via reaction

$$HNiL_4^+ \overset{K_2}{\rightleftharpoons} HNiL_3^+ + L, \tag{5.99}$$

where the upper limit of K_2 is about $4 \cdot 10^{-5}$ M. The resulting coordination-unsaturated complex $HNiL_3^+$ is the active center for 1-butene isomerization according to a mechanism with the scheme presented in Figure 5.8. The process kinetics at the initial stage (not involving the steps of decomposition of the intermediate products and active centers) is described by the first-order equation with respect to butene (S):

$$-\frac{d[S]}{dt} = k_{obs}[S], \tag{5.100}$$

where the observed rate constant k_{obs} depends on the concentration of ligand L and $[H^+]$:

$$k_{obs} = \frac{kK[Ni]_0[H^+]}{[L](1 + K[H^+])}. \tag{5.101}$$

The equation of material balance for $[Ni]_0$ in accordance with formula (5.101) includes NiL_4 and $HNiL_4^+$ complexes and does not account for $HNiL_3^+$, X_1, X_2, X_3, and X_4 species. The equilibrium constant K in Eq. (5.101) amounts to $50\,M^{-1}$, which is close to K_1 values determined by independent methods (see above). The slow steps are assumed to be the insertion of olefin into Ni–H bond (steps **4** and **7**), while steps **1** and **2** are considered quasi-equilibrium. The rate constant k in formula (5.101) amounts to $0.7\,s^{-1}$.

At high activity, the catalyst's turnover number is not large (about 300 at 25 °C), which is related to the decomposition of intermediates with the formation of butane and Ni(II) (steps **8** and **9**) and/or the loss of active $HNiL_3^+$ species in the absence (or low concentration) of olefin in step **10**:

$$HNiL_3^+ + H^+ \xrightarrow[\mathbf{10}]{} H_2 + Ni(II). \tag{5.102}$$

The catalyst turnover number increases to 3000 at 0 °C, which indicates that the activation energy of chain termination steps is greater than that of olefin isomerization. It is interesting to note that the investigation of this process in the $CH_3OD–D_2SO_4$ system at 25 °C under conditions where 39 % of 1-butene was isomerized within 15 s showed that over 99 % of 2-butenes did not contain deuterium and the content

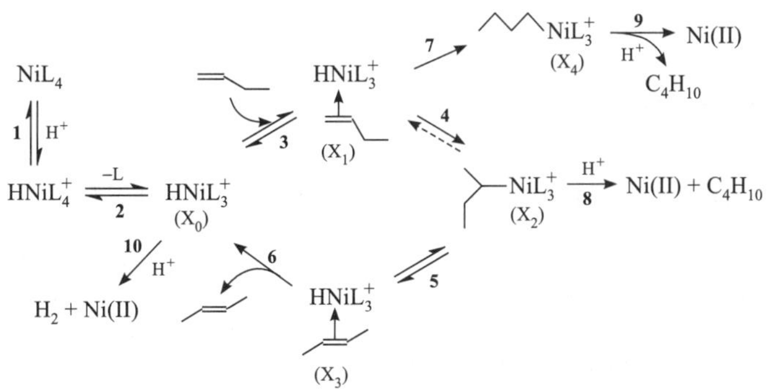

Figure 5.8 Mechanistic scheme of 1-butene isomerization in the $NiL_4–H_2SO_4–MeOH$ system

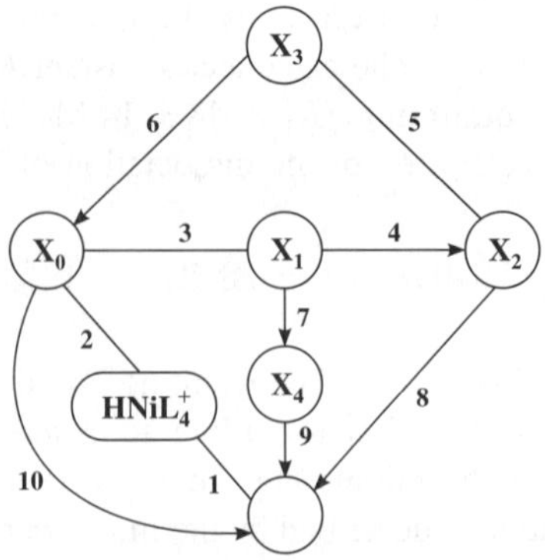

Figure 5.9 Kinetic graph corresponding to 1-butene isomerization mechanism

of d_1-1-butene was as small as 0.5 %. The ratio of the amount of isomerization products to that of the deuterated products was close to 170. Apparently, the regeneration of $HNiL_3{}^+$ as a result of the first cycle with $DNiL_3{}^+$ allows about 170 catalytic cycles to take place prior to the HNi exchange via steps **−2** and **−1** with D_2SO_4. Prolonged keeping of 1-butene under the same conditions leads to the formation of an equilibrium mixture of 1-butene (6 %), *cis*-2-butene (25 %), and *trans*-2-butene (69 %) with a statistical distribution of D over all carbon atoms in C_4H_8 [132]. For a long-term isomerization process, an expression for the reaction rate must include the rates of the steps of formation and the weights of steps of the decomposition of intermediates (C_4H_{10}) and active centers (H_2), which makes it difficult to explain the observed low value of the ratio $k_H/k_D < 1$, which reflects the contribution of steps **1**, **4**, **7**, and **8**.

Since the initial NiL_4 species (reactant, initiator) under real conditions are consumed and Ni(II) complexes are formed (reaction products), the corresponding mechanistic scheme can be clearly represented by a kinetic graph with the null-vertex (Figure 5.9). This graph reveals three routes for the formation and loss of active centers (**1–2–10**; **1–2–3–7–9**; **1–2–3–4–8**) and one route for the kinetic chain propagation (**3–4–5–6**). Thus, in respect of the mechanism, a real catalytic process has a chain character.

5.3.4 Features of the formation and decay of active centers in acrylic derivatives synthesis by the Reppe Method

The synthesis of acrylic and methacrylic acids (and their esters) in catalytic systems with $Ni(CO)_4$–HX–H_2O (ROH) has many features in common with the process considered above for the NiL_4–HX system. There are three well-known variants of the participation of $Ni(CO)_4$ in these reactions [18, 149]:

- Noncatalytic synthesis takes place at 25–40 °C and atmospheric pressure and is described by the following brutto equation:

$$4C_2H_2 + 4H_2O + Ni(CO)_4 + 2HCl \longrightarrow 4CH_2{=}CHCOOH + NiCl_2 + 2[H]. \qquad (5.103)$$

  The yield of acrylic acid (AA) in this process reaches 95 % for the consumed acetylene, which implies that almost all of the four CO groups in $Ni(CO)_4$ are involved in the reaction. Hydrogen in the free state was not found, but ethylene, ethane, and propionic acid were detected. The yield of methylacrylate (MA) in methanol reaches 80 %.
- "Semicatalytic" synthesis takes place in the presence of CO, when the pressures of CO and C_2H_2 are within 0.5–1 MPa, in which case $Ni(CO)_4$ delivers about 15–20 % CO for the AA and MA synthesis.

The main side products in this case are propionic and allylacetic acids (or their esters). The yields of AA and MA reach 85–90% [150].

Catalytic synthesis of AA proceeds in the $Ni(CO)_4$–$CuBr_2$–HBr–THF–H_2O (160–200 °C, 4.5–5.0 MPa) system with an AA yield of 87–90% [151]. The rate of the catalytic synthesis of methyl methacrylate (MMA) from methyl acetylene and CO in methanol in the $Ni(CO)_4$–HX (130 °C, 2 MPa) system is described by the following equation [152]:

$$r = k\left[\mathrm{Ni(CO)_4}\right]_0 [\mathrm{HX}]\frac{[\mathrm{C_3H_4}]}{P^n_{\mathrm{CO}}}, \quad (5.104)$$

which shows evidence for the mechanism of active Ni(II) hydride carbonyl complex formation by analogy with the NiL_4–H_2SO_4 system (see Section 5.3.3), but with a lower equilibrium constant. The presence of steps of the formation and decomposition of active centers and the formation of reductive carbonylation products allows these processes in $Ni(CO)_4$ solutions to be considered as chain reactions. Unfortunately, no data are available on the state of nickel in the $Ni(CO)_4$–$CuBr_2$–HBr system. However, judging from the known behavior of the $PdBr_2$–$CuBr_2$ system (see below) at small concentrations of $CuBr_2$ (comparable with $[Ni]_0$ in the system under consideration), it is possible to admit the existence of $HNi(CO)_4{}^+CuBr_3{}^-$ species.

The aforementioned "semicatalytic" synthesis is of special interest, since the partial consumption of CO from $Ni(CO)_4$ implies that the reaction takes place on several active centers with different content of CO in the coordination sphere of Ni(II), so that the rate of regeneration of the coordination sphere of the metal is lower than the rate of transition between various centers. Let us consider a possible mechanistic scheme of the "semicatalytic" process, which takes into account all of the known facts.

The process of active center formation by the analogy with NiL_4 (see Section 5.3.3) and in accordance with the kinetics of MMA synthesis can be described by the following reactions:

$$\mathrm{Ni(CO)_4} + \mathrm{H^+} \underset{1}{\rightleftharpoons} \mathrm{HNi(CO)_4}^+; \quad (5.105)$$

$$\mathrm{HNi(CO)_4}^+ \underset{2}{\rightleftharpoons} \mathrm{HNi(CO)_3}^+ + \mathrm{CO}. \quad (5.106)$$

Taking into account the possible formation of side products detected in the case of acetylene in the systems with $Ni(CO)_4$, the steps of decomposition of the active centers and intermediates can be represented by the equation

$$\mathrm{HNi(CO)_n}^+ + \mathrm{CH_2{=}CHCOOMe} \xrightarrow[-n\mathrm{CO}]{\mathrm{H^+}} \mathrm{Ni(II)} + \mathrm{CH_3CH_2COOMe}, \quad (5.107)$$

where the vinyl derivative of Ni(II) can be involved (in the framework of the hydride mechanism of acetylene hydroalkoxycarbonylation) in the following conversions:

$$\mathrm{CH_2{=}CHNi(CO)_n}^+ \xrightarrow[-n\mathrm{CO}]{\mathrm{H^+}} \mathrm{C_2H_4} + \mathrm{Ni(II)}; \quad (5.108)$$

$$\mathrm{CH_2{=}CHNi(CO)_n}^+ \xrightarrow{\mathrm{C_2H_2}} \mathrm{Ch_2{=}CH{-}CH{=}CHNi(CO)_n}^+ \xrightarrow{\mathrm{CO,\ MeOH}}$$
$$\longrightarrow \mathrm{CH_2{=}CH{-}CH{=}CH{-}COOMe} + \mathrm{HNi(CO)_n}^+; \quad (5.109)$$

$$\mathrm{HNi(CO)_n}^+ + \mathrm{C_4H_5COOMe} \xrightarrow{\mathrm{H^+}} \mathrm{Ni(II)} + \mathrm{CH_2{=}CH{-}CH_2CH_2COOMe}. \quad (5.110)$$

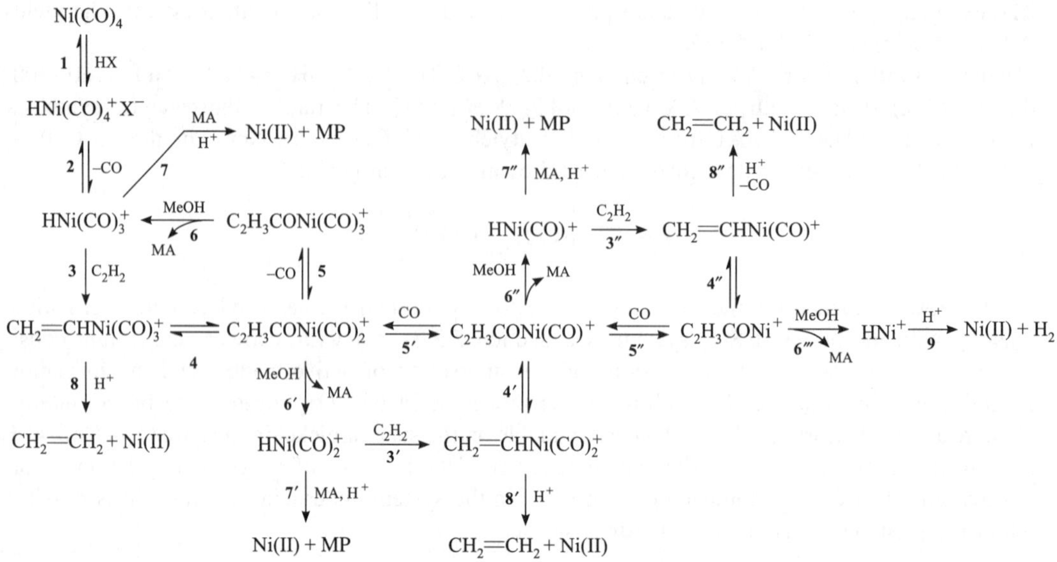

Figure 5.10 Mechanistic scheme of semicatalytic synthesis of methyl acrylate (MA) in $Ni(CO)_4$–HX–MeOH system (MP = methyl propionate)

During gradual decarbonylation, the primary active center ($HNi(CO)_3{}^+$) sequentially converts into $HNi(CO)_2{}^+$, $HNi(CO)^+$, and HNi^+. Since Ni(II) species in this series become (in the course of decarbonylation) increasingly hard, the appearing $HNi(CO)_n{}^+$ species with $n = 0, 1, 2$ have decreasingly low probability of interacting with soft CO ligands so as to increase the coordination number with respect to carbon monoxide. At the same time, the probability of chain termination on harder Ni(II) species via the reaction with H^+ increases. If the process of decarbonylation proceeds to the formation of HNiX, these species either no longer participate in the acrylate synthesis ($HNiX + HX \rightarrow H_2 + NiX_2$) or are rapidly converted into NiX_2 at the step of vinyl intermediate acidolysis.

Taking into account information concerning the steps of the "hydride" mechanism of alkyne carbonylation [18], a scheme of the semicatalytic chain acrylate synthesis can be represented as depicted in Figure 5.10. This scheme shows only three types of stages of the decomposition of active centers and intermediates (**7**, **8**; **7′**, **8′**; **7″**, **8″**, and **9**), steps of the active center formation (**1**, **2**), and three cycles of kinetic chain propagation – that is, catalytic cycles with the participation of $HNi(CO)_3^+$ (**3**, **4**, **5**, **6**), $HNi(CO)_2{}^+$ (**3′**, **4′**, **5′**, **6′**), and $HNi(CO)^+$ (**3″**, **4″**, **5″**, **6″**). Within this scheme, the rate of MA formation is determined by the sum of rates of the four steps:

$$r_{MA} = W_6 + W_{6'} + W_{6''} + W_{6'''}.$$

In constructing a kinetic model of this process, one should also take into account the possibility of chain termination via the formation of coordination saturated species, for example, as

$$CH_2{=}CHCONi(CO)_3{}^+ + CO \rightleftharpoons CH_2{=}CHCONi(CO)_4{}^+. \qquad (5.111)$$

5.3.5 Protecting active centers by catalytic process from destruction

An interesting phenomenon of self-regulation of the working life of a catalyst was discovered in the process of olefin hydroformylation in solutions of carbonyl complexes of rhodium not modified by phosphines [153,

154]. Investigations of the hydroformylation kinetics of propylene, 1-hexene, 2-butenes, and isobutene in $Rh_4(CO)_{12}$ solutions showed a drop in the catalyst activity upon treatment of the initial mixture with synthesis gas (syngas) (363 K, 6 MPa, 4 h). The drop was more pronounced for olefins that were less active in the hydroformylation process (i.e., for 2-butenes and isobutene). An analysis of the IR absorption spectra measured *in situ* showed that the initial $Rh_4(CO)_{12}$ in the presence of H_2 or synthesis gas was partly converted into rhodium hydride complexes, which irreversibly transformed the catalyst into hydride-containing clusters inactive in the hydroformylation process.

Under the action of alkenes and synthesis gas, the cluster complexes of rhodium exhibit fragmentation and thus convert into catalytically active mononuclear forms. In the case of more "active" olefins (propylene, 1-hexene), $Rh_4(CO)_{12}$ gradually transforms into an acyl derivative $R'CORh(CO)_3$. The conversion of this derivative into aldehydes is a limiting step of the process (Figure 5.11). Thus, a high rate of conversion of the active centers $[HRh(CO)_3]$ in the corresponding steps of the catalytic cycle leads to a decrease in their stationary concentration, which is just what prevents the clusterization process. The mononuclear rhodium complexes are stabilized in the form of an acyl derivative. Therefore, the catalytic process protects the active mononuclear species from the possible inactivation via the reactions

$$2HRh(CO)_4 \rightleftharpoons Rh_2(CO)_8 + H_2; \quad (5.112)$$

$$Rh_2(CO)_8 + 2HRh(CO)_4 \rightleftharpoons Rh_4(CO)_{12} + H_2 + 4CO; \quad (5.113)$$

$$Rh_4(CO)_{12} + 2HRh(CO)_4 \rightleftharpoons Rh_6(CO)_{16} + H_2 + 4CO; \quad (5.114)$$

$$nRh_4(CO)_{12} + mHRh(CO)_4 \rightleftharpoons H_xRh_{(4n+m)}(CO)_y + {}^1\!/_2(m - x)H_2 + (12n - 4m - y)CO. \quad (5.115)$$

An increase in the hydrogen pressure and the presence of olefin that binds $HRh(CO)_4$ (or $HRh(CO)_3$) drive reactions (5.112)–(5.115) to reverse their direction and leads to the regeneration of species that are active in the catalytic cycle. In the case of less active olefins (2-butenes, isobutene), step **1** becomes the limiting step (see Figure 5.11), therefore $HRh(CO)_3$ and $HRh(CO)_4$ are involved in the clusterization and the catalytic activity gradually decreases as a result of inactivation of the catalytic system. In the case of 2-butene, it was shown [155] that the concentration of binuclear rhodium complexes in solution is significantly higher than that of mononuclear complexes and the initial $Rh_4(CO)_{12}$ clusters, and the reaction rate is proportional to $\sqrt{[Rh]_\Sigma}$. At a low concentration of 2-butene or high CO pressure, the conversions of binuclear clusters lead to the irreversible inactivation of the catalyst.

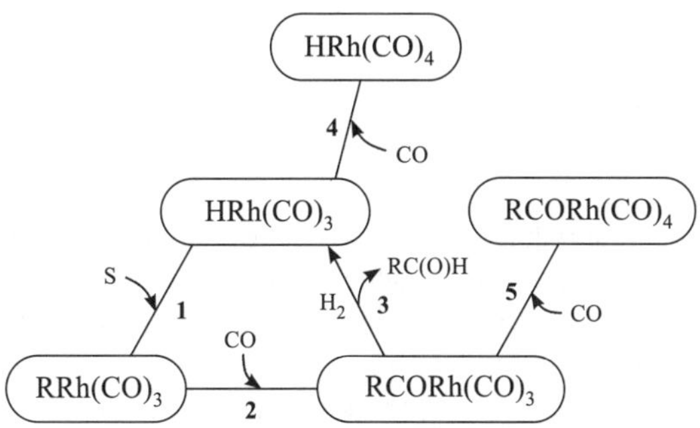

Figure 5.11 Kinetic graph corresponding to mechanism of olefin (S) hydroformylation with catalytic cycle and two pendant vertices

As demonstrated above (see Sections 1.1.2 and 3.2), the presence of a common step for the two routes (catalytic reaction and decomposition of active centers) is just what accounts for the fact that the step rates of one route influence the rate of the other route:

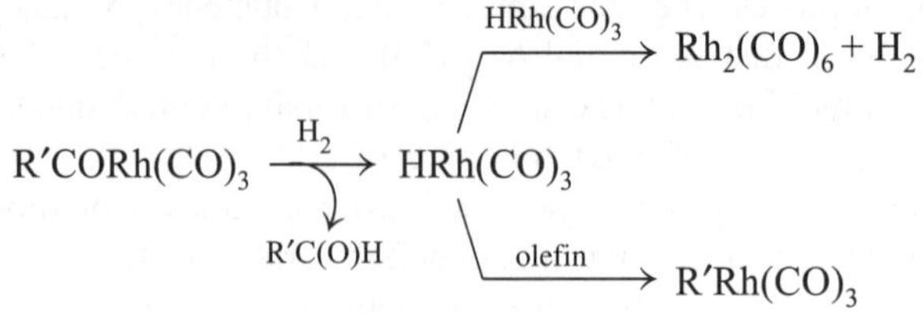

(5.116)

5.3.6 Mechanism of active center formation in $Pd(OAc)_2$–PPh_3–*p*-benzoquinone–MeOH catalytic system for alkyne oxidative carbonylation at ≡C–H bond

Alkynes (PhC≡CH, MeC≡CH) exhibit oxidative carbonylation at ≡C–H bonds under mild conditions ($T = 30$–$40\,°C$, $P_{CO} = 1$ bar) in the $Pd(OAc)_2$–Q–MeOH system (Q = *para*-quinone) [156] that can be improved by introducing PPh_3 [157]:

$$RC{\equiv}CH + CO + Q + MeOH \longrightarrow RC{\equiv}CCOOMe + QH_2. \quad (5.117)$$

The oxidative carbonylation of phenylacetylene proceeds with a selectivity of up to 85 %, depending on the catalytic system composition, added solvent, and the nature of quinone Q (e.g., *p*-benzoquinone (BQ), chloranil (CA)). It should be emphasized that the presence of *p*-quinones is a factor of key importance for this reaction. Other oxidants (e.g., Cu(II) and Fe(III)) capable of oxidizing Pd(0) do not ensure the formation of phenylpropiolic acid methyl ester (MEPA) in this system.

The particular nature of quinone (e.g., BQ or CA) significantly influences the process rate and selectivity. For example, the initial rate of the reaction with CA is about one order of magnitude higher than that with BQ (the initial turnover frequency for a catalyst with CA is in the order of $70\,h^{-1}$). These data suggest that the role of quinones in this process does not reduce to the oxidant function. Recent investigation [158] of the formation of catalytically active complexes from $Pd(OAc)_2$, which was performed by ^1H, ^{13}C, and ^{31}P NMR techniques *in situ* and in model systems, led to a conclusion that palladium acetate under strongly reductive conditions (see Section 5.3.6) in the presence of PPh_3, CO, and MeOH is reduced to Pd(0) complexes containing quinones. Judging from NMR data, the initial $Pd(OAc)_2$ in methanol in the presence of PPh_3 ligands (L) exists in the form of monomer and dimer complexes of $L_3Pd(OAc)^+$ cations:

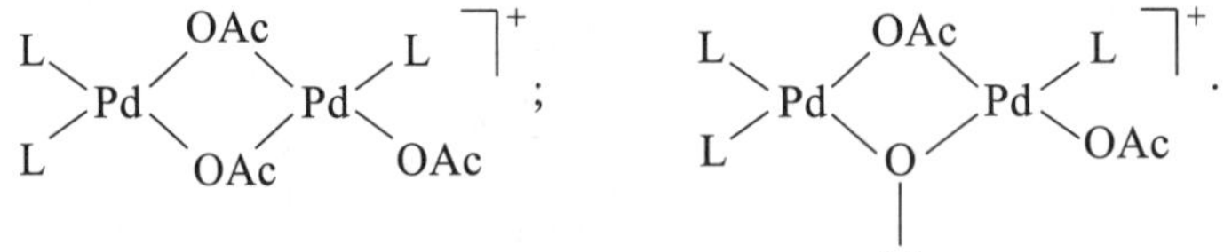

These complexes participate in the reduction of Pd(II) by triphenylphosphine (phosphine oxide was found in all systems), methanol, and carbon monoxide, which is accompanied by an increase in the concentration of acid (CH_3COOH) in solution. Figure 5.12 shows the structure (determined by X-ray diffraction) of a Pd(II) methoxycarbonyl acetate complex with a composition of $[L_2Pd(OAc)(COOMe)]_2 \cdot QH_2$ (**I**), which was isolated from the $Pd(OAc)_2$–PPh_3–Q–(methanol/acetone, 1:1) in the presence of CO [158]. This complex is an intermediate compound in the oxidative carbonylation of alcohol, namely, the intermediate

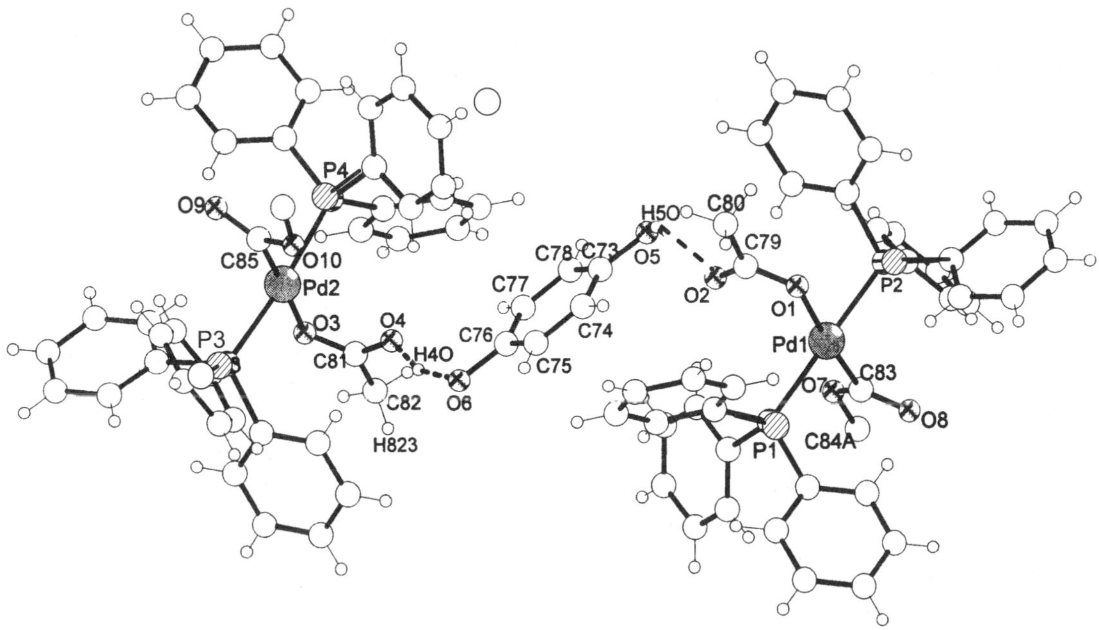

Figure 5.12 Molecular structure of $[Pd(COOCH_3)(OAc)(PPh_3)_2]_2H_2Q \cdot CH_3OH$ (I) complex (disordered molecule of methanol is not depicted) (reproduced with permission from AIPC Nauka and editorial board of *Kinet. Katal.* journal © 2007 [158])

in the reduction of Pd(II) to Pd(0) by methanol and CO:

$$\begin{aligned} Pd(OAc)_2 + CO + MeOH &\longrightarrow AcOPdCOOMe + AcOH; \\ AcOPdCOOMe + MeOH &\longrightarrow (MeO)_2C{=}O + AcOH + Pd(0). \end{aligned} \tag{5.118}$$

Previously, $L_2PdX(COOMe)$ monomer complexes were obtained with X = Cl [159] and X = OAc [160]. The partial oxidation of Pd(0) by BQ and the appearance of hydroquinone (QH_2) stabilizes complex **I** in the form of an interesting supramolecular structure. A methoxycarbonyl complex with $X = TsO{-}L_2Pd(OTs)(COOMe)$ was synthesized in the $MeOH{-}PdL_2(OTs)_2{-}CO$ system [161].

Thus, the process of active center formation with the participation of quinones can be represented by the following sequence of reactions:

$$Pd(OAc)_2 + PPh_3 \xrightarrow[Q]{CO,\ MeOH} L_nPd(0) \xrightarrow{Q} L_nPdQ. \tag{5.119}$$

Complexes of Pd(0) with BQ were known for a long time in systems with various ligands (bipy, Phen, Cod, PPh_3). However, until the recent systematic study [162], structural investigations were restricted to complexes with nitrogen-containing ligands (see references in [162]), while the structures of complexes with PPh_3 were only treated based on the results of elemental analyses and some indirect data [163–165]. The synthesis, isolation, and characterization of new complexes containing Pd(0), PPh_3, and Q are described in [158, 162]. In particular, new complexes were synthesized with the compositions $[PdL_2Q]_2 \cdot QH_2 \cdot MeOH$ (**II**) (Figure 5.13) and $(PdLQ)_2 \cdot 1.5$ MeOH

(**III**) (Figure 5.14). Complex **III** was obtained using the dissolution of complex **II** in acetone. A complex, in which the chemical shift of ^{31}P signal (31.13 ppm) in the ^{31}P NMR spectrum is close to that for complex **II** was also observed in the PdL_4–Q–$CDCl_3$ system at 25 °C. The NMR spectra of the $Pd(OAc)_2$–Q–PPh_3 system free of CO and alkyne displayed the signals characteristic

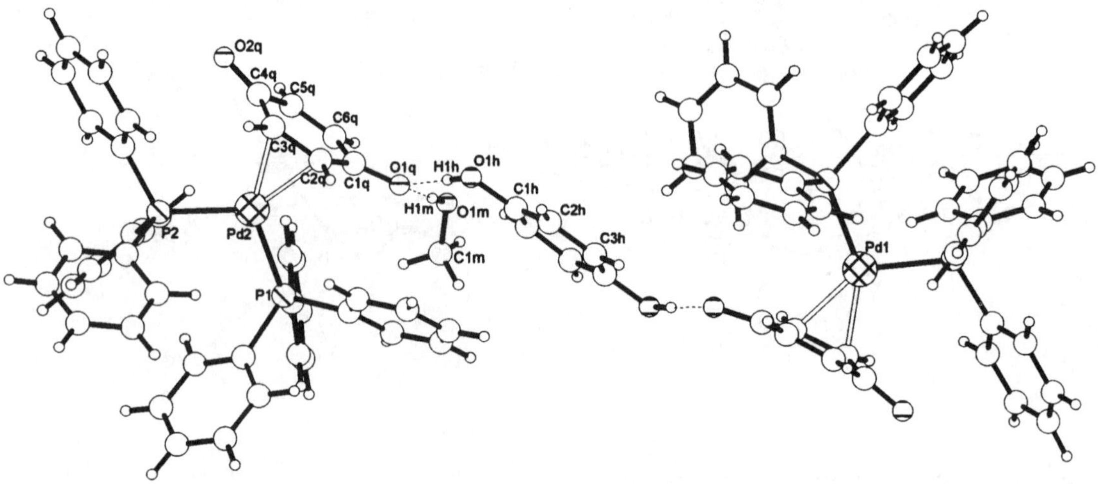

Figure 5.13 Molecular structure of the $[PdL_2Q]_2QH_2 \cdot MeOH$ (II) complex ($L = PPh_3$) (reproduced with permission from AIPC Nauka and editorial board of *Kinet. Katal*. journal © 2007 [158])

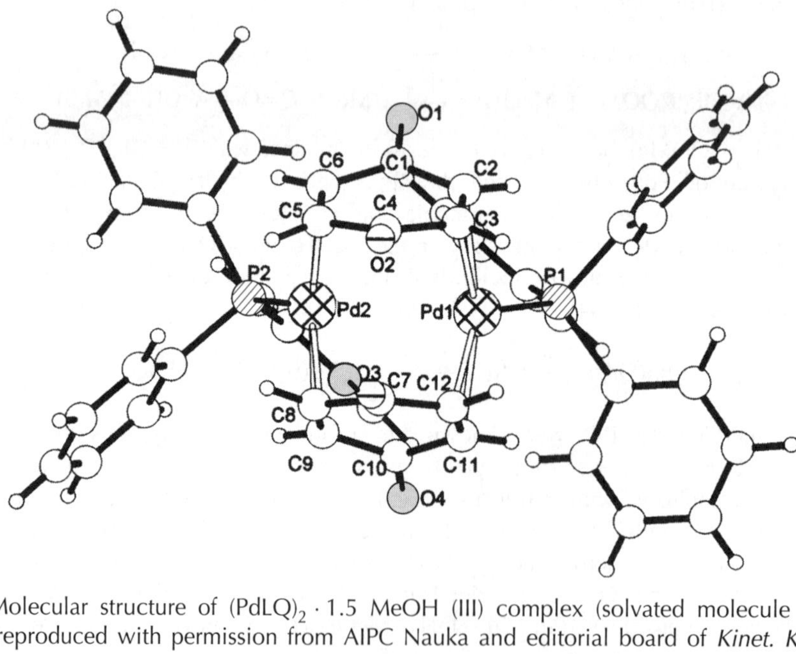

Figure 5.14 Molecular structure of $(PdLQ)_2 \cdot 1.5$ MeOH (III) complex (solvated molecule of methanol is not depicted) (reproduced with permission from AIPC Nauka and editorial board of *Kinet. Katal*. journal © 2007 [158])

of complexes **II** and **III**. The content of complex **II** in methanol solutions (CD_3OD, 25 °C) varied within 3 to 14 % (relative to $[Pd]_0$).

The investigations of systems with chloranil provided additional evidence for the formation of Pd(0) complexes with quinones. In particular, a product of the oxidative addition of chloranil to $L_2Pd(0)$ was isolated in the form of complex **IV** with a composition of $L_2PdCl[C_6Cl_2O_2(OMe)]$ (Figure 5.15), which most probably formed via the reaction

$$L_2Pd(C_6Cl_4O_2) \longrightarrow L_2PdCl(C_6Cl_3O_2), \tag{5.120}$$

with the subsequent replacement of one chlorine atom in the para position to Pd by the methoxy group:

$$L_2PdCl(C_6Cl_3O_2) + MeOH \underset{-HCl}{\longrightarrow} L_2PdCl[C_6Cl_2O_2(OMe)]. \tag{5.121}$$

A comparison of the catalytic activity of various precursors in reaction (5.117) showed that, irrespective of the particular system [$Pd(OAc)_2$–PPh_3–Q, $Pd(PPh_3)_4$–Q, $Pd(dba)_2$–PPh_3–Q, complex **II**, or complex **III**], the activity of a catalyst at [L]/[Pd] = 2 and [Q]/[Pd] = 17 was the same and amounted to 0.04–0.08 M/h (in the quasi-steady-state regime) [158, 162]. These observations are consistent with the notions about the formation of the same active center (L_nPdQ) in processes with various precursors.

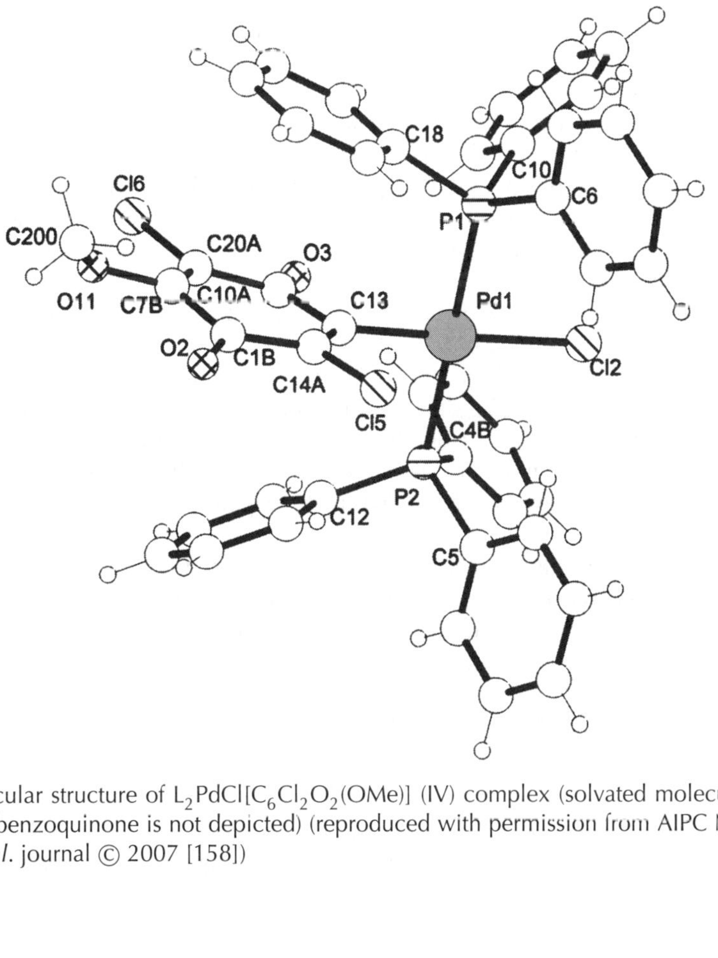

Figure 5.15 Molecular structure of $L_2PdCl[C_6Cl_2O_2(OMe)]$ (IV) complex (solvated molecule of 2,5-dichloro-3,6-dimethoxy-1,4-benzoquinone is not depicted) (reproduced with permission from AIPC Nauka and editorial board of *Kinet. Katal.* journal © 2007 [158])

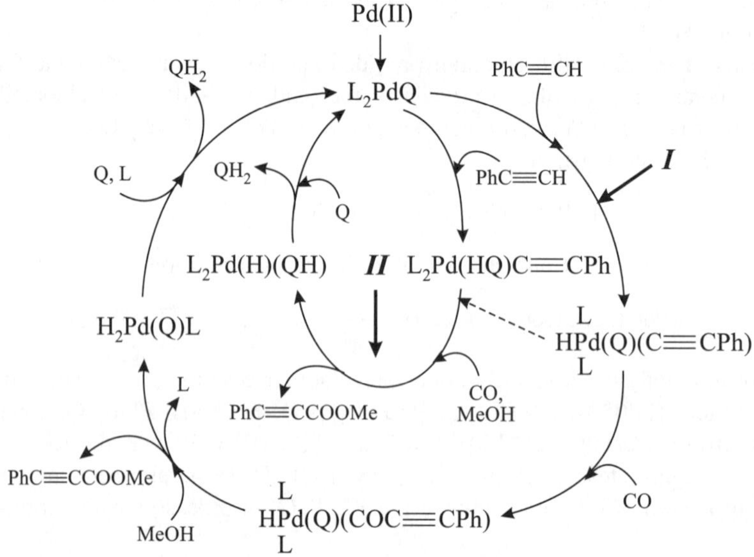

Figure 5.16 Mechanistic scheme of *phenylpropiolic methyl* esther synthesis [reaction (5.117)] (Q = *para*-benzoquinone; L = PPh_3) (reproduced with permission from AIPC Nauka and editorial board of *Kinet. Katal.* journal © 2007 [158])

The low content of protons in the system and the independence of the reaction rate of pH in a 5.9–7.7 interval suggest that Pd(0) complexes in the systems under consideration are not oxidized by quinones. In the case of a much stronger acid ($[CF_3COOH]/[Pd] = 2$) with pH 0.4, the reaction rate drops to zero [158].

The results concerning the formation of active quinone complexes, low concentrations of protons in the optimum acidity interval (pH 5.9–7.7), sharp inhibition of the reaction by strong acid, and the proximity of the kinetic isotope effects to unity in the $Pd(OAc)_2$–PPh_3–XA–CH_3OH/CH_3OD systems made it possible to discriminate between various hypotheses and propose a mechanistic scheme of reaction (5.117) [158, 162, 166], which includes the formation of L_2PdQ complexes and their reactions with PhC≡CH and CO via two possible routes (Figure 5.16). Route **I** includes the formation of an ethynyl hydride complex with the subsequent conversion into H_2PdQL_n dihydride intermediates, which yield QH_2 and active centers. Route **II** seems to be more probable and, in addition, explains the important role of quinone Q as a ligand involved in the proton transfer (from alkyne via π complex to Q) with the formation of a hydroquinolate intermediate and its subsequent conversion into Pd(II) hydridehydroquinolate and the reductive elimination of QH_2.

The mechanisms of decomposition of the active centers and intermediates were not studied, but it is quite evident that the formation of by-products involving both ethynyl and hydride ligands, which are responsible for a decrease in the selectivity below 70–80 % and a low turnover number ($\leq$13) of active centers, can also lead to the decomposition of the catalyst:

$$\sim\text{Pd–C}\equiv\text{CPh} \xrightarrow{\text{PhC}\equiv\text{CH}} \sim\text{Pd–CH=C(Ph)C}\equiv\text{CPh} \longrightarrow \text{polymer;}$$

$$\sim\text{Pd–H} \xrightarrow{\text{PhC}\equiv\text{CH}} \text{Pd–CH=CHPh} \xrightarrow{\text{PhC}_2\text{H}} \text{polymer.}$$

One also should not exclude the possible isomerization of π complexes with PhC_2H and/or the conversion of ethynyl ligand (as result of protonation) into vinylidene ligand, which can participate in various reactions including alkyne polymerization. If the following investigations will show the presence of termination steps, process (5.117) would acquire all the characteristic signs of a chain process.

5.3.7 Catalysis with small palladium(I) halide and carbonyl halide clusters

The first hypotheses concerning the catalytic activity of Pd(I) complexes appeared in 1966–1969 in the investigations by Moiseev with co-workers [27, 167, 168], Davidson [169], and Chiusoli [170]. The state of the coordination and catalytic chemistry of Pd(I) clusters is discussed in reviews [49, 166, 171–175]. Let us consider in more detail the mechanisms of formation and decomposition of halogen-containing Pd(I) clusters. By the metal cluster complexes we will imply M_nL_m complexes with M–M bonds beginning with $n = 2$.

Pd (I) halide complexes. Original investigations [167, 168] of the mechanism of geometric isomerization of butenes

$$\text{1-butene} \rightleftharpoons \text{2-butene} \tag{5.122}$$

in aqueous solutions of $PdCl_4^{2-}$ complexes showed that this process kinetics was first-order with respect to [1-C_4H_8]. The dependence of the reaction rate on $[PdCl_4^{2-}]_\Sigma$ at a constant ionic strength and almost constant concentration of free Cl^- ions was described by an equation of the second order. With allowance for the observed dependence of the third-order rate constant on $[Cl^-]$ and $[H_3O^+]$, the kinetic equation of process (5.122) in the forward direction at an almost constant concentration $[C_4H_8]_\Sigma$ was as follows:

$$r'_{\text{isom}} = k'\frac{[1-C_4H_8][PdCl_4^{2-}]_\Sigma^2}{[Cl^-]^3[H_3O^+]}. \tag{5.123}$$

In view of the facts that the sum of butenes slowly converted into methyl ethyl ketone and there appeared traces of metallic palladium, it was decided to maintain the concentration of Pd(II) constant using an oxidant (p-benzoquinone) by analogy with the reaction of ethylene [27]. As a result, it was rather unexpectedly established that p-benzoquinone suppressed the isomerization process, which was restored only after the complete reduction of the oxidant. Based on these results, it was suggested that the isomerization was catalyzed by a low-valence palladium complex (e.g., $Pd_2Cl_4^{2-}$) formed from Pd(II), rather than by $PdCl_4^{2-}$.

This hypothesis also explained both the second order with respect to [Pd(II)] for the route leading to $Pd_2Cl_4^{2-}$ and the suppression of isomerization by quinone. These considerations led to an important conclusion on the chain character of the isomerization process, in which the oxidation of a sum of olefins

$$C_4H_8 + 2PdCl_4^{2-} + H_2O \xrightarrow[\mathbf{1}]{W_1} C_4H_8O + Pd_2Cl_4^{2-} + 2HCl + 2Cl^- \tag{5.124}$$

is the step of chain initiation, $Pd_2Cl_4^{2-}$ is the active center of butenes isomerization (propagation steps)

$$\underset{(X_1)}{Pd_2Cl_4^{2-}} + 1\text{-}C_4H_8 \underset{\mathbf{2}}{\rightleftharpoons} X_2; \tag{5.125}$$

$$X_2 \underset{\mathbf{3}}{\rightleftharpoons} X_3; \tag{5.126}$$

$$X_3 \underset{\mathbf{4}}{\rightleftharpoons} 2\text{-}C_4H_8 + Pd_2Cl_4^{2-} \tag{5.127}$$

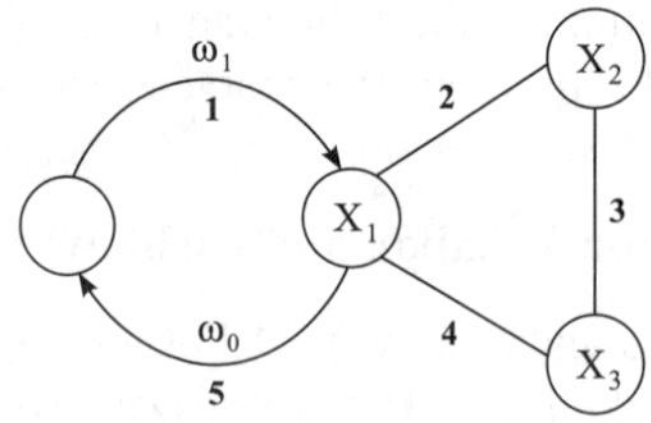

Figure 5.17 Kinetic graph corresponding to chain mechanism of butene isomerization in $PdCl_4^{2-}$ solutions

(where (5.126) is the limiting step), and

$$Pd_2Cl_4{}^{2-} \xrightarrow[5]{k_0} Pd^0 + PdCl_4{}^{2-} \tag{5.128}$$

is the step of chain termination.

Experiments confirmed that the rate of isomerization in the forward direction depends not only on $[1\text{-}C_4H_8]$, but on the $[C_4H_8]_\Sigma$ as well, with

$$k' = k[C_4H_8]_\Sigma. \tag{5.129}$$

Thus, the rate of the nonbranched chain process (with a kinetic graph presented in Figure 5.17) is described by the following equation:

$$r'_{isom} = \frac{W_1}{\omega_0}\omega_{isom}, \tag{5.130}$$

where

$$W_1 = k_1\frac{[C_4H_8]_\Sigma[PdCl_4{}^{2-}]^2}{[Cl^-]^3[H_3O^+]}, \tag{5.131}$$

is the rate of chain initiation, $\omega_0 = k_0$ is the weight of the chain termination step (5.128); and $\omega_{isom} = k_{isom}[1\text{-}C_4H_8]$ is the weight of the chain propagation step.

Equation (5.130) agrees with the experiment, provided that the complexity functions of Pd(II) and Pd(I) with respect to butenes under conditions of $[Pd]_\Sigma \cong [PdCl_4{}^{2-}]$ are close to unity. The kinetic chain length is close to 30. The catalysis of butene isomerization by Pd(I) clusters was confirmed in the case of $[Pd(CO)Cl]_n$ and $Cs_2Pd_2(CO)_2Cl_4$ complexes in the DMF–LiCl system [176], where the isomerization began without an induction period, was inhibited by quinone (BQ) already at [BQ]/[Pd] = 1, and continued after the consumption of quinone.

The results obtained in [167, 168] probably provided the first kinetic confirmation of the chain mechanism and the participation of the steps of active center formation in the kinetics of the main catalytic process in the catalysis with metal complexes.

Another case of the participation of Pd(I) halide clusters was found in the investigation of carbonylation of acetylene and other alkynes in the PdI_2–LiI (KI)–HCl–ROH system [87, 177–179]:

$$C_2H_2 + CO + n\text{-BuOH} \longrightarrow \text{A, S, P, M, F, E} \tag{5.132}$$

where A, S, P, M, F and E are butyl acrylate, dibutyl succinate, butyl propionate, dibutyl maleate, and dibutyl fumarate, and ethylene, respectively. The selectivity with respect to butyl acrylate amounts to

~50 % (70 °C, 1 bar). The results of investigations suggested that a catalytically active form of the catalyst participating in the conjugate process of formation of the products of additive (A, S), oxidative (M, F), and reductive (P) carbonylation is a complex of palladium in a low degree of oxidation (probably, $Pd_2I_4{}^{2-}$). This complex is formed during an induction period from $PdI_4{}^{2-}$ via the reaction

$$2PdI_4{}^{2-} + C_2H_2 + CO + BuOH \longrightarrow Pd_2I_4{}^{2+} + \text{Products.} \quad (5.133)$$

The introduction of I_2, *p*-benzoquinone, and O_2 (at high P_{O_2}) terminates the carbonylation process:

$$Pd_2I_4{}^{2+} + I_2 \xrightarrow{\text{LiI}} 2PdI_4{}^{2-}. \quad (5.134)$$

The synthesis of Pd(I) carbonyl iodide complexes by replacing Cl^- with I^- in $Pd_2(CO)_2Cl_4{}^{2-}$ in hydrochloric acid solutions under CO atmosphere leads to the substitution of CO in palladium carbonyl complex with partial water gas shift reaction (the gas phase contains CO_2 and H_2) and the formation of black diamagnetic complexes with compositions $LiPdI_2 \cdot 2H_2O$ and $CsPdI_2 \cdot H_2O$ [176]. These complexes were found to be active in reaction (5.132) [176, 179] without an induction period of the catalyst formation. Recently, analogous complexes of KPd_4I_5, $CsPd_3I_4$, and $Ph(Me)_3N[Pd_3I_4]$ were isolated under close conditions and identified using the data of elemental analyses and X-ray photoelectron spectroscopy [180], which also exhibited activity in the reactions of oxidative carbonylation. A mechanistic scheme of the multi-route conjugate process of carbonylation in the iodide system, which was obtained upon the discrimination of possible hypotheses, was considered in Chapter 3 (see Figures 3.30 and 3.31). This mechanism exhibits all features of a catalytic chain process, including steps of initiation (5.133) and the decomposition of intermediates leading to regeneration of the Pd(II) initiator (not depicted in Figure 3.30). The irreversible loss of intermediates can be related to the formation of Pd-containing polyacetylene products.

Gabriele et al. [181] proposed the oxidative carbonylation of alkynes (PhC≡CH, MeC≡CH, etc.) in the acid-free PdI_2–KI–MeOH system for the synthesis of *Z,E*-diesters of α,β-unsaturated acids (20 °C, 20 bar). It was found that, at certain concentrations of reactants, PdI_2, and KI, definite PH values, and atmospheric pressure of CO and O_2, this system is characterized by self-oscillations of the Pt electrode potential and the pH value and by a step-like gas ($CO + O_2$) uptake [182–184]. The catalysis of the auto-oscillating synthesis of phenylmaleic and phenylfumaric acid diesters from phenylacetylene (PA) by KPd_4I_5 complexes [183], the detection of H_2O_2 in the reaction products, and a small concentration of HI acid formed in the steps of Pd(II) reduction to Pd(I)

$$PdI_2 + CO + H_2O \longrightarrow HPdI + CO_2 + HI; \quad (5.135)$$

$$HPdI + PdI_2 \longrightarrow Pd_2I_2 + HI, \quad (5.136)$$

suggested that the oxidative carbonylation of PA in the system under consideration is autocatalytic with respect to HPdI. One of the proposed mechanisms included step (5.136), reaction

$$Pd_2I_2 + PA + 2CO + 2MeOH \longrightarrow \text{Products} + 2HPdI, \quad (5.137)$$

and the steps of decomposition of HPdI (O_2, I_2) and Pd_2I_2 (H_2O_2) [166, 183]. Figure 5.18 shows a bipartite graph representing a simplified mechanistic scheme of this process.

Kinetic models and specific features of this catalytic chain process in an oscillatory regime are considered in Section 6.4.3. It should be noted that all palladium species (Pd(II), HPdI, and Pd_2I_2) play more important roles in the appearance of oscillations as compared to the intermediate products of alkyne carbonylation, which is represented in this scheme by a single pseudo-elementary reaction (5.137).

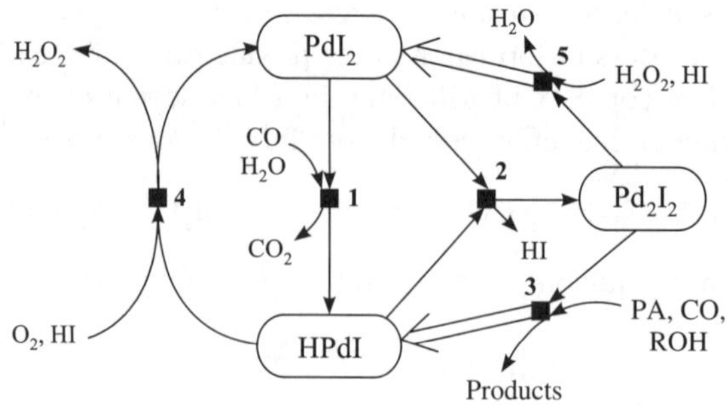

Figure 5.18 Bipartite graph corresponding to a simplified mechanism of oscillatory process including steps that are linear (1, 4) and nonlinear (2, 3, 5) with respect to palladium complexes (only Pd-containing intermediates are indicated at the vertices)

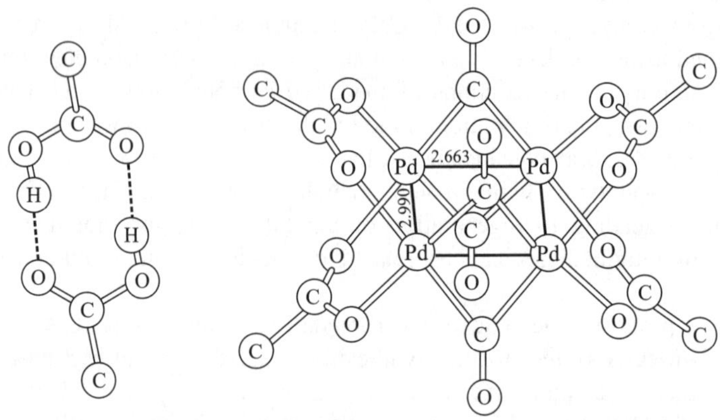

Figure 5.19 Structure of $Pd_4(CO)_4(OAc)_4 \cdot 2AcOH$ cluster (reproduced with permission from The Russian Academy of Sciences and Turpion Ltd © 2000 [23])

Pd(I) carbonyl halide complexes. Carbonyl complexes of Pd(I) with reliably established compositions were originally obtained by Manchot and König ($[Pd(CO)Cl]_n$, 1926) [185]. Other well-known complexes were synthesized by Gel'man and Meilakh ($(NH_4)_2Pd_2(CO)_2Cl_4$, 1942) [186] and Colton et al. ($Pd_2(\mu\text{-CO})(dppm)_2X_2$, 1977) [187]. Moiseev with co-workers [175, 188] synthesized and structurally characterized the first Pd(I) carbonyl acetate cluster with composition $Pd_4(CO)_4(OAc)_4$ (Figure 5.19). Goggin et al. [189] used X-ray diffraction to study the structure of the $[(Bu)_4N]_2Pd_2(\mu\text{-CO})_2Cl_4$ complex. Figure 5.20 shows the structure of the Gel'man complex (as described in [190]) that is frequently used in catalytic reactions. In this complex anion, each Pd (I) atom has a planar-ligand environment of two terminal chlorine atoms and two bridging CO molecules, and the Pd–Pd distance is 2.6892 Å. Each NH_4^+ group is bound by a system of hydrogen bonds and via oxygen atoms of the bridging CO molecules into a very interesting packing of dimer anion layers (Figure 5.21). This character of packing sharply distinguishes the structure of the Gel'man complex from that of the $[(Bu)_4N]_2Pd_2(\mu\text{-CO})_2Cl_4$ complex with like geometry of complex anion.

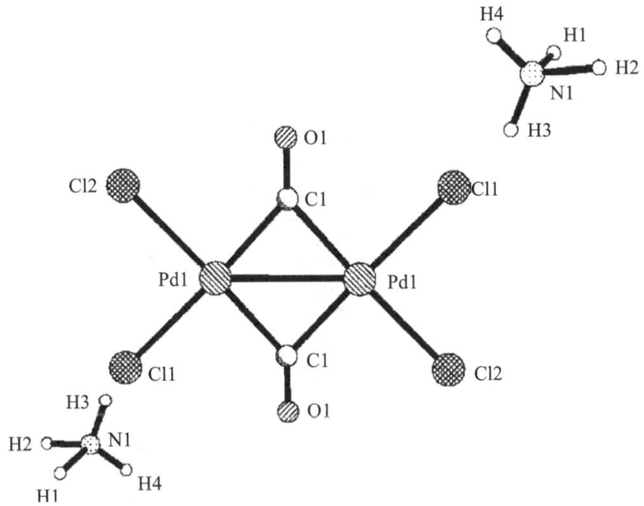

Figure 5.20 Structure of Gel'man complex anion (reproduced with permission from AIPC Nauka © 2002 [190])

Complex anion $Pd_2(CO)_2Cl_4{}^{2-}$ catalyzes the oxidative carbonylation of methanol to dimethyl oxalate [166, 191]:

$$2CO + 2MeOH + Q \longrightarrow (COOMe)_2 + QH_2. \tag{5.138}$$

The rate of reaction (5.138) in solutions of Pd(I) complexes is about four times higher than that in $PdCl_2$ solutions [192], which is indicative of the absence of oxidation of Pd(I) to Pd(II) by quinone. The reaction rate at 30 °C and a CO pressure of 1 bar in the presence of PhCOOH–PhCOONa buffer (pH ≈ 4) in $Li_2Pd_2(CO)_2Cl_4$ [191] and Li_2PdCl_4 [193] solutions is described by the first-order equation with respect to [BQ]. It was suggested that benzoquinone was involved in the formation of intermediates according to the following scheme [166, 191]:

$$Pd_2(CO)_2Cl_4{}^{2-} \underset{-Cl}{\overset{Q}{\rightleftharpoons}} QPd_2(CO)_2Cl_3{}^- \xrightarrow[-PhCOOH]{MeOH,\ PhCOONa} QPd_2(CO)(COOMe)Cl_2{}^- \longrightarrow$$
$$\xrightarrow[-PhCOOH]{MeOH,\ PhCOONa} QPd_2(COOMe)_2Cl^- \xrightarrow{2PhCOOH,\ 2CO,\ 3Cl^-}$$
$$\longrightarrow (COOMe)_2 + QH_2 + Pd_2(CO)_2Cl_4{}^{2-}. \tag{5.139}$$

The participation of Pd(I) carbonyl chloride complexes in other processes was reviewed in [171, 172, 174, 194].

The interest in Pd(I) carbonyl bromide complexes was caused by their high activity in the $PdBr_2$–LiBr–acetone (acetonitrile)–H_2O systems used in the syntheses succinic [166, 177, 195, 196] and maleic anhydrides [197, 198] from acetylene, in the syntheses of succinic acid and its esters in the $PdBr_2$–HBr–*n*-BuOH–$(CH_3)_2CO$ system [166], and in the oxidative carbonylaton of phenol to diphenyl carbonate [194].

Since the structures of carbonyl bromide anions and polymer complexes of Pd(I) had been previously only discussed based on the data of IR spectroscopy and elemental analyses, an attempt was

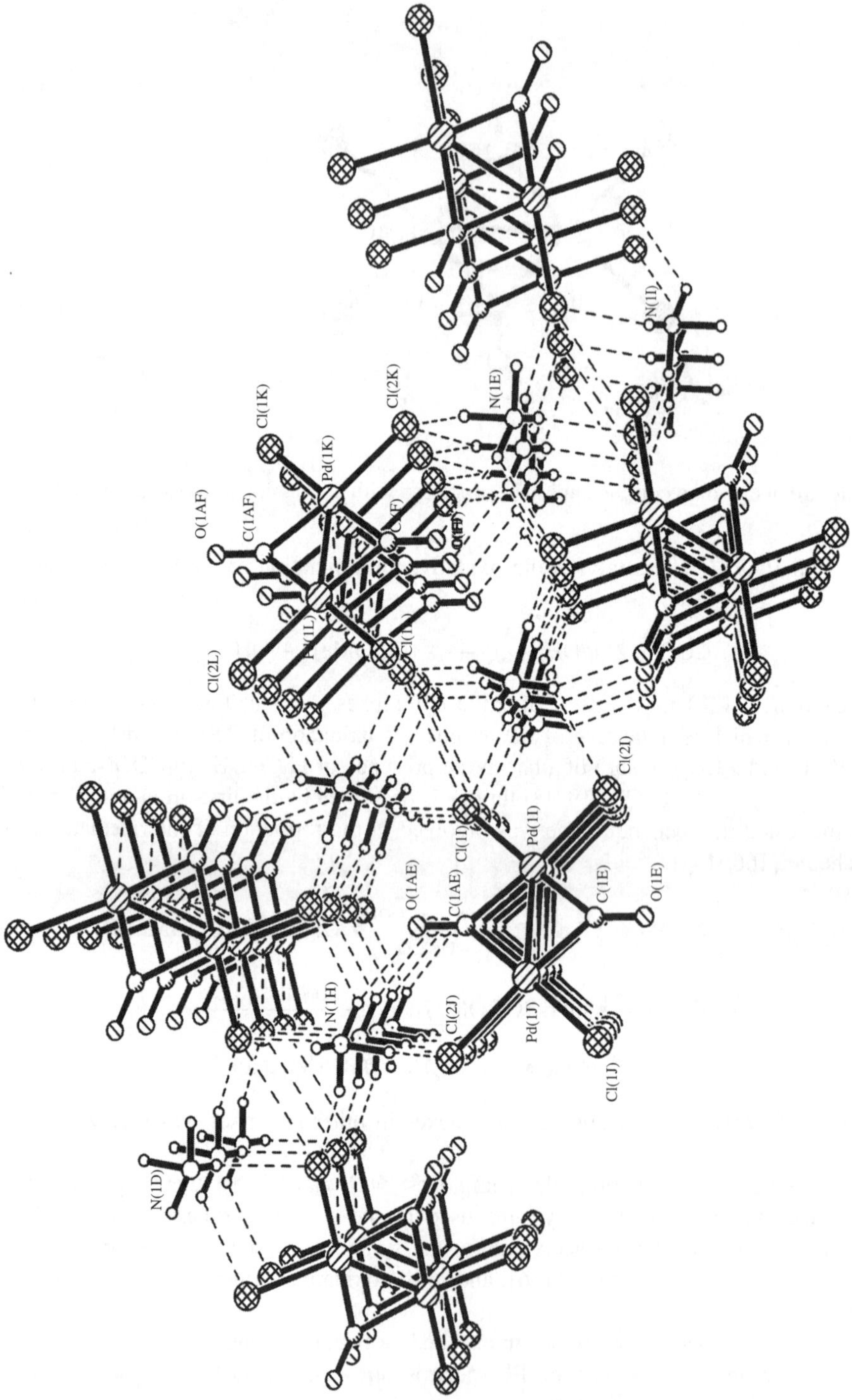

Figure 5.21 Supramolecular environment of $(NH_4)_2Pd_2(\mu\text{-}CO)_2Cl_4$ Gel'man complex anion (reproduced with permission from AIPC Nauka © 2002 [190])

Table 5.1 Component ratios in crystalline complexes I–III by X-ray diffraction data [199]

Crystal	Br/Pd	CO/Pd	Br/CO	$Br/CO_{(terminal)}$
I	1.3	1.7	0.76	1.86
II	1.63	1.38	1.18	4.29
III	1.5	1.50	1.00	3.00

undertaken [199] to study the mechanisms of bromide cluster formation in aqueous and nonaqueous solutions of $PdBr_4^{2-}$ anions in reactions with CO. The carbonylation of Li_2PdBr_4 solutions in THF containing small amounts of water is accompanied by the formation of CO_2, and the IR spectra of solutions exhibit absorption bands belonging to bridging CO groups in Pd(I) complexes (1905–1913 cm^{-1}) and another band due to terminal CO groups (2102–2106 cm^{-1}) [199]. The intensity of the latter band did not decrease for a long time in the case of a reaction in the CO atmosphere, and both bands were manifested under the conditions where carbonyl bromide complexes were synthesized by Br^- substituting for Cl^- in $Li_2Pd_2(CO)_2Cl_4$. With the aid of bis (triphenylphosphoranylidene) ammonium (PPN^+) as cations, three crystalline carbonyl bromide complexes (I, II, III) with different compositions were isolated from solutions depending on the particular conditions [199]. The complexes have different ratios of Br/Pd and Br/CO for the same ratio of $PPN^+/Pd \cong 1$ (Table 5.1). According to the X-ray diffraction data, all complexes (irrespective of the Br/CO ratio), represented sets of complex anions $Pd_2Br_n(CO)_m^{2-}$ ($m + n = 6$) comprising $Pd_2(\mu\text{-}CO)_2$ fragments and PPN^+ cations (two cations per Pd_2 dimer). Figure 5.22 shows the structure of such a complex anion with $Pd_2(\mu\text{-}CO)_2$ fragment and four partly disordered terminal ligand (Br, CO) (Figure 5.22). Two PPN^+ cations are bound to the anion by short contacts between the hydrogen atoms of phenyl rings and halogen atoms of terminal ligands (Figure 5.23).

Experimental data on the structures of complexes I–III were explained by assuming that a mixture of Pd(I) and Pd(0) complexes with different ratios of terminal ligands (Br/CO) exists in solution and

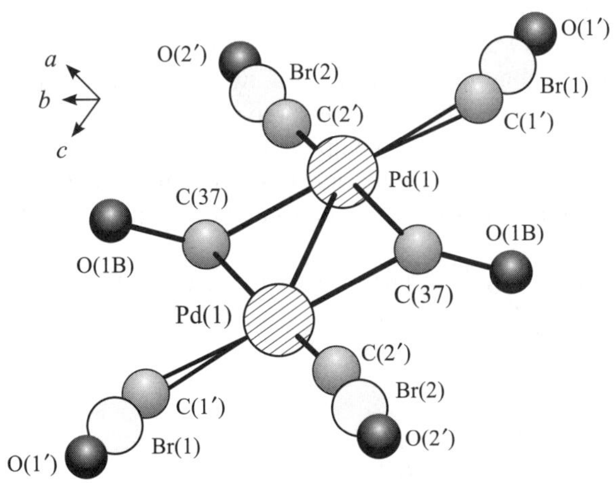

Figure 5.22 Structure of $Pd_2Br_n(CO)_m^{2-}$ complex anion with partly disordered terminal ligands (reproduced with permission from editorial board of *Izv. Akad. Nauk, Ser. Khim.* journal © 2005 [199])

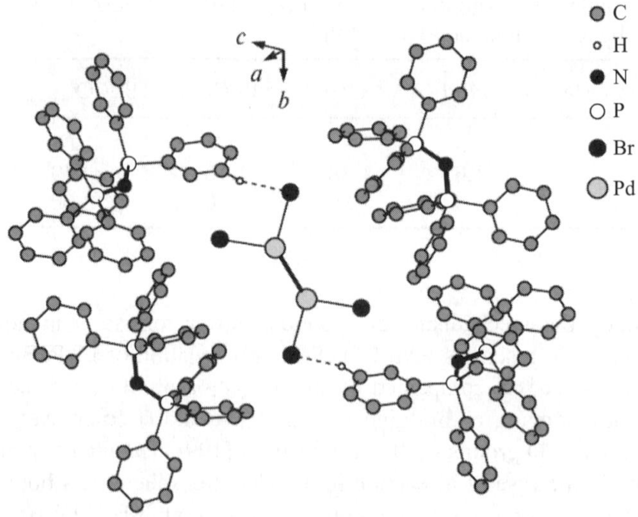

Figure 5.23 Structure of $Pd_2Br_n\,(CO)_m^{2-}$ complex anion surrounded by four cations, two of which are linked to the complex by hydrogen bonds (reproduced with permission from editorial board of *Izv. Akad. Nauk, Ser. Khim.* journal © 2005 [199])

exhibits cocrystallization in the same crystalline form. This assumption agrees with the data of X-ray photoelectron spectroscopy for complex III with a brutto composition of $Pd_2(CO)_3Br_3^{2-} \cdot 2PPN^+$. The X-ray photoelectron spectrum exhibits two Pd $3d_{5/2}$ lines with the binding energies of 337.3 and 335.6 eV, which are attributed to Pd(I) and Pd(0), respectively. A structure containing Pd(I) and Pd(0) simultaneously in one $Pd_2(CO)_2$ fragment was rejected because signals of such paramagnetic particles were observed neither at 25 °C nor at liquid nitrogen temperature (with allowance for a large distance of 9.5 Å between Pd_2 fragments and the absence of direct channels for exchange interaction).

Thus, the reduction of tetrabromopalladate(II) or the replacement of Cl^- by Br^- in Pd(I) carbonyl chloride anion leads to the formation of Pd(I) and Pd(0) bromide complexes in various ratios. For example, in the case of complex III, anions $Pd_2(\mu\text{-}CO)_2Br_4^{2-}$ and $Pd_2(\mu\text{-}CO)_2Br_2(CO)_2^{2-}$ are crystallized in a 1:1 ratio. This hypothesis also explains the appearance of two bands in the IR absorption spectrum. The formation of Pd(0) clusters can be a result of the disproportionation of Pd_2(I) complexes into Pd(II) and Pd(0) and/or the subsequent reduction of Pd(I) complexes by carbon monoxide. The appearance of terminal CO in the Pd(I) complex as a result of Cl^- replacement by Br^- makes the complex capable of interacting with H_2O, with the resulting Pd_2(I) reduction to Pd_2(0) and the formation of CO_2:

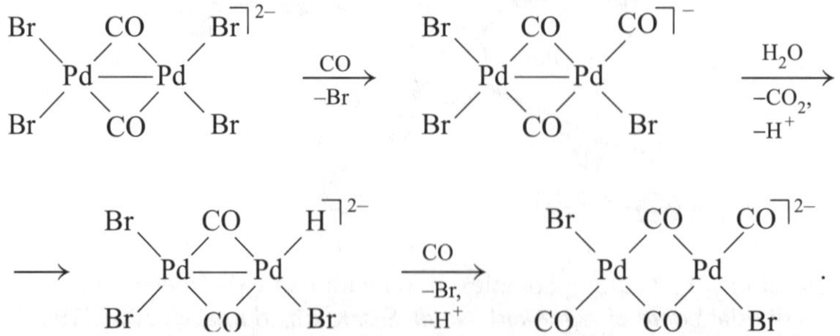

(5.140)

Trebbe et al. [200] synthesized a complex of Pd(0) with bridging CO and diphosphine ligands, in which the Pd–Pd distance was relatively short (2.714 Å) and could feature a weak "metallophilic" Pd–Pd interaction [199].[1] The possible formation of Pd(0) halide (PdX_n^{n-}) and carbonyl halide ($Pd(CO)_2X_n^{n-}$) complexes was frequently taken into account when considering the mechanisms of redox reactions with the participation of palladium compounds [27, 201, 202].

Carbonyl bromide complexes of Pd(I) obtained *in situ* during the reduction of Pd(II) bromide by carbon monoxide are capable of catalyzing various reactions (see previous page), including the oxidative and nonoxidative carbonylation of acetylene and alkynes. It was suggested that the syntheses of succinic and maleic anhydrides from acetylene involved an intermediate compound formed by the incorporation of 2CO and C_2H_2 at the Pd–Pd bond [166, 197, 198]:

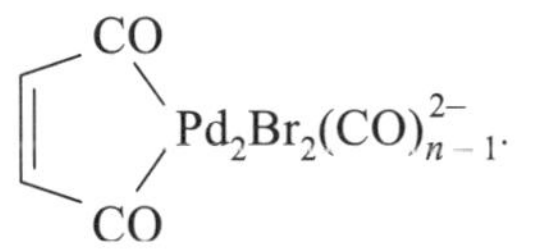

It can also be suggested that Pd(0) complexes with $Pd_2(CO)_4Br_2^{2-}$ is a possible form of Pd(0) stabilization that can participate in steps of Pd(II) or Pd(I) regeneration in the reactions with oxidants (O_2, Q, etc.) or serve as catalyst, e.g., in reactions of ArBr carbonylation in ligand-free systems [203]. At the same, the transition of Pd(I) clusters into binuclear Pd(0) complexes represents the steps of active center decomposition in processes catalyzed by Pd(I) complexes.

As was noted above, attempts at substituting I^- for Cl^- in solution did not allow new Pd(I) carbonyl complexes to be obtained. The reaction led to the appearance of an absorption band at 2086 cm^{-1}, which was typical of terminal CO, whereas the bands due to bridging CO were absent [180]. However, in the course of carbonylation of a suspension of $PdCl_2$ and Bu_4NI in THF, the introduction of PPN^+ cations led to the isolation of single crystals of a Pd(I) carbonyl chloride-iodide complex [180]. The composition of this complex (as determined from the X-ray diffraction data) suggested that it could be a product of the cocrystallization of $Pd_2(CO)_2Cl_4^{2-}$ anions (A) with either $Pd_2(CO)_2Cl_3I^{2-}$ (B) or $Pd_2(CO)_2Cl_2I_2^{2-}$ (C) anions in a ratio of A/B ≅ 2 or A/C ≅ 4.

5.3.8 Mechanisms of formation of large cluster complexes and microheterogeneous nanoparticles

The development of cluster chemistry as the independent branch of coordination chemistry [204], the discovery of medium-sized ($n > 12$), large ($13 < n < 150$) and, especially, giant metal clusters M_nL_m ($n > 150$) close to colloidal metals and nanodimensional particles (with dimensions within 20–300 Å) [204] stimulated the interest in the catalytic chemistry of large metal clusters and the catalysis with colloidal metals and heterogeneous nanoparticles. The most interesting of large clusters include $(Pt_3(CO)_6)_6^{2-}$ [205], $Pt_{38}(CO)_{44}^{2-}$ [206], $Rh_{15}(CO)_{27}^{3-}$, $Rh_{17}(CO)_{30}^{3-}$ [207], $Rh_{22}(CO)_{37}^{4-}$ [206], $Rh_{22}(CO)_{37}^{4-}$ [206], $Pd_{23}(CO)_{20}L_8$ and $Pd_{23}(CO)_{22}L_{10}$ [208], $Pd_{38}(CO)_{30}L_{12}$ [209], $Pd_{34}(CO)_{24}L_{12}$, $Pd_{38}(CO)_{28}L_{12}$ [210], $Ni_{38}Pt_6(CO)_{48}H_2^{4-}$ [211], $Au_{55}(PPh_3)_{12}Cl_6$ [212], $Pd_{147}(Phen)_{32}(OCOR)_{30}O_{60}$ [213], $Pd_{145}(CO)_x(PEt_3)_{30}$, ($x \cong 60$) [214], and $(\mu_{12}\text{-Pt})Pd_{164-x}Pt_x(CO)_{72}(PPh_3)_{20}$ ($x \cong 7$) [215]. Giant clusters include $Pd_{561}Phen_{60}(OAc)_{180}$ and $Pd_{561}Phen_{60}O_{60}(PF_6)_{60}]$ complexes containing approximately 560 metal atoms [30, 172–174, 216] and $Pt_{309}(Phen^*)_{36}O_{30\pm10}$ complexes containing 309 metal atoms (where Phen* = bis-sulfophenyl phenanthroline) [217].

[1]For d^{10}–d^{10} interactions see, e.g., S. Scalfort and P. Braunstein, Chem. Soc. Rev., **40**, 2741 (2011).

The formation of large clusters frequently proceeds via a sequential reductive addition of new vertices to the cluster [204a]:

$$[Rh_{13}(CO)_{24}H]^{4-} + [Rh(CO)_2(CH_3CN)_2]^+ \longrightarrow [Rh_{14}(CO)_{25}H]^{3-} \xrightarrow{Na_2CO_3}$$

$$\longrightarrow [Rh_{14}(CO)_{25}]^{4-} \xrightarrow{[Rh(CO)_2(CH_3CN)_2]^+} [Rh_{15}(CO)_{27}]^{3-}$$

Alternative approaches employ the oxidation of clusters of smaller dimensions [209],

$$Pd_{10}(CO)_{14}L_4 + Pd(OCOCF_3)_2 \longrightarrow Pd_{38}(CO)_{30}L_{12},$$

and some other methods [204].

It was convincingly demonstrated in various research centers that reductive conditions (H_2, ROH, R_3N, RR_3) favored the conversion of Pt, Pd and Co mononuclear complexes into large and giant clusters and nanodimensional particles, the surfaces of which contain the active centers of catalytic hydrogenation and oxidation [30, 172–174, 218], Heck reactions [36, 219–222], and other processes on monometallic [223–225] and bimetallic [226, 227] colloids.[2]

Although the complicated hypothetical mechanisms of the formation of these complexes and heterogeneous nanoparticles are not yet convincingly justified, there are significant achievements that make expedient consideration of the present-day status of research in this interesting field of coordination, colloid, and catalytic chemistry.

$Pd(OAc)_2$–PPh_3–H_2 – (organic solvent) systems. The aforementioned hypothesis concerning the necessity of Pd–O bonds for the activation of H_2 in the hydrogenation reactions (see Section 5.2) explains the activity of $Pd(OAc)_2$ complexes with PPh_3 in DMF in these processes [74–76]. However, subsequent investigations have showed that the situation is not that simple [228–230]. Monophosphine dimer complexes formed in DMF or CH_2Cl_2 media via reaction

$$2[Pd(OAc)_2]_3 + 6PPh_3 \longrightarrow 3[PPh_3Pd(OAc)_2]_2, \quad (5.141)$$

interact with H_2 with the formation of Pd_2L_2 ($L = PPh_3$) complexes via an intermediate Pd(I) compound:

$$\begin{aligned} &[LPd(OAc)_2]_2 + H_2 \longrightarrow [LPd(OAc)]_2 + 2AcOH; \\ &[LPd(OAc)]_2 + H_2 \longrightarrow Pd_2L_2 + 2AcOH. \end{aligned} \quad (5.142)$$

This step virtually does not involve the dephenylation of PPh_3. An analogous complex is formed from $[LPd(OAc)_2]_2$ and sodium amalgam in CH_2Cl_2 [30].

Investigations of the kinetics of reaction (5.142) showed that the formation of Pd_2L_2 is accompanied by the autocatalytic formation of $[LPd(OAc)]_2$ from Pd_2L_2 as

$$[LPd(OAc)]_2 + Pd_2L_2 \longrightarrow 2[LPd(OAc)]_2. \quad (5.143)$$

Thus, the reduction of Pd(II) to Pd(0) via reaction (5.142) is catalyzed by Pd(I) complexes. Figure 5.24 shows the kinetics of the autocatalytic formation of Pd_2L_2 [228]. However, in this step of catalyst formation, there are no active species capable of activating H_2 or the substrate. Indeed, Pd_2L_2 complexes absorb about

---

[2]Additional useful information on catalysis with heterogeneous nanoparticles, nanoclusters, and colloids can be found in *Nanocatalysis*, U. Heiz and U. Landman (Eds) (Springer, 2007) and *Nanoparticles and Catalysis*, Vol. 1, D. Astruc (Ed.) (Wiley-VCH, 2008).

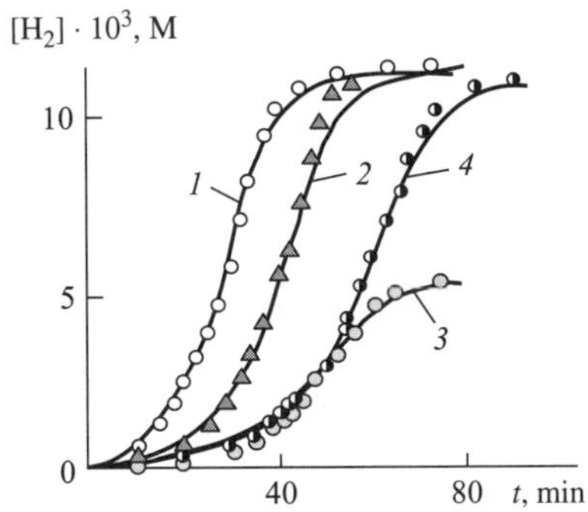

Figure 5.24 Kinetic curves of H_2 uptake by $[Ph_3PPd(OAc)_2]_2$ solution in CH_2Cl_2. Curves show results of model calculations, points present experimental data (20 °C, total pressure, 1 bar): *(1)* $[[Ph_3PPd(OAc)_2]_2]_0 = 4.6 \cdot 10^{-3}$ M, $[(Ph_3P)_2Pd_2]_0 = 2.6 \cdot 10^{-4}$ M; *(2)* $[[Ph_3PPd(OAc)_2]_2]_0 = 4.6 \cdot 10^{-3}$; *(3)* $[[Ph_3PPd(OAc)_2]_2]_0 = 2.2 \cdot 10^{-3}$ M; *(4)* $[[Ph_3PPd(OAc)_2]_2]_0 = 4.9 \cdot 10^{-3}$ M ($H_2/Ar = 1 : 1$, v/v). (reproduced with permission from editorial board of *Izv. Akad. Nauk, Ser. Khim.* journal © 1978 [228])

1 mole H_2, but the resulting adduct does not catalyze the hydrogenation of unsaturated compounds. The introduction of trace amounts of O_2 or H_2O_2 for a partial oxidation of Pd_2L_2 to $Pd(OAc)_2$ or the direct introduction of $Pd(OAc)_2$ into DMF in a stoichiometric ratio of $3Pd(OAc)_2$ per Pd_2L_2 molecule in a flow of H_2 leads to an intense dephenylation of PPh_3 via the reaction

$$[Pd(OAc)_2]_3 + Pd_2(PPh_3)_2 + 5H_2 \longrightarrow (PhP)_2Pd_5 + 6AcOH + 4C_6H_6 \tag{5.144}$$

and the formation of polynuclear $[(PPh)_2Pd_5]_n$ clusters, which are soluble in DMF and are the true catalysts of the hydrogenation of various organic substrates [231–233]. The molecular masses and sizes of these clusters were initially not determined, but the subsequent investigations of the process of active center formation in analogous systems showed that $[(PPh)_2Pd_5]$ clusters could belong to the family of colloidal metals or nanoparticles stabilized on the surface by the residual dephenylated PPh_3.

The activity of this catalyst in the hydrogenation of acetylene admixtures in the gas phase products of gasoline pyrolysis was retained for 400 h at 90 °C, a pyrogas volume rate of 12 000 h^{-1}, a pressure of 10 bar, and cluster concentration of up to 10^{-5} M. The catalyst turnover frequency in the process of penta-1, 3-diene hydrogenation to pentenes was in the order of 60 000 h^{-1} [232]. If $Pd(OAc)_2$ and PPh_3 are introduced with H_2 flow containing O_2 into DMF in the presence of a substrate, the catalyst is formed in the course of the catalytic process, but a more effective approach consists of preparing the active centers prior to catalysis. Figure 5.25 shows a bipartite graph that well illustrates a complicated structure of the proposed mechanism of formation of $[L'_2Pd_5]_n$ clusters.

A no less complicated pattern of the formation of clusters and nanodimensional particles was established for $Pd(Acac)_2$–PR_3–H_2 systems.

$Pd(Acac)_2$–PR_3–H_2–toluene. The initial stage in more than a quarter-century history of investigations of this system was described in sufficient detail in review [30]. More recent studies of the reduction of $Pd(Acac)_2 \cdot PPh_3$ complexes by hydrogen revealed the formation of Pd(I) complexes, $[(PPh_2)Pd(Acac)]_2$ dimer species with bridging phosphine ligands, and $[Pd_3(PPh_3)_3(\mu\text{-}PPh_2)_2Acac]^+$ $Acac^-$ trinuclear clusters [234, 235] similar to chloride analogs [52] (see Section 5.2). In addition, a solid substance with a particle

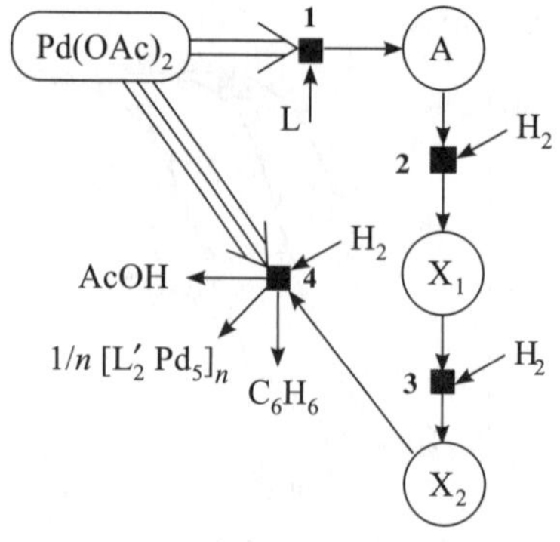

Figure 5.25 Bipartite graph corresponding to mechanism of active catalyst formation for hydrogenation of unsaturated compounds (A = $[LPd(OAc)_2]_2$; X_1 = $[LPd(OAc)]_2$; X_2 = Pd_2L_2; L = PPh_3; L′ = PPh)

size of about 3 nm and a high degree of PPh_3 dephenylation was isolated and assigned a formula of $[(Pd_3P)_8 \cdot PPh_3]_n$ [235–237]. These nanoparticles proved to be the active hydrogenation catalysts.

Subsequently, it was established that $Pd(Acac)_2PPh_3$ palladium phosphine complexes (A) with the structure

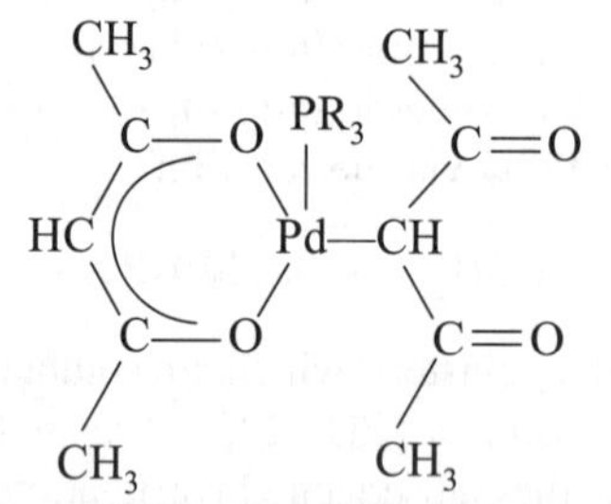

(A)

contained $Acac^-$ groups in different forms, O,O-chelate and C-bound [238]. The scheme of complex A reduction by hydrogen in some steps resembles the mechanism of $Pd(OAc)_2$ reduction. Indeed, the reduction of complex A leads to the formation of $[PdPPh_3]$ species, which are involved in both the reduction of A and the process of PPh_3 dephenylation to $[PdPPh_2]_n$ [237–239]. The formation of $[PdPPh_2]_n$ clusters via different routes is of key importance, since it is the interaction of these clusters with H_2 and Pd(0) complexes that yields an X-ray-amorphous substance with a composition of $Pd_5P_4Ph_6$. It was suggested [240–242] that this product consists of associated polynuclear clusters (with PPh_2 and PPh ligands), on which the ensembles (clusters) of $Pd(0)_x(PdPPh_2)_y(Pd_2PPh)_zPPh_3$ are immobilized. In addition, these associates contain phosphide fragments (Pd_3P, Pd_6P, etc.).

It was established that Pd(0)-containing nanoparticles with dimensions of 3–5 nm exhibited the maximum activity in the hydrogenation of olefins and alkynes in toluene. For example, in a system containing $[Pd]_\Sigma = 5 \cdot 10^{-4}$ M with [substrate]/[Pd] = 1800, $[PPh_3]/[Pd] = 0.75$ at 30 °C and $P_{H_2} = 1$ bar, the catalyst turnover frequency reached 4500 h^{-1}. At the same time, the Pd(I) complexes ($L_2Pd_2(Acac)_2$) and the aforementioned trinuclear clusters did not exhibit catalytic activity. Trioctylphosphine favors the formation of colloidal systems with increased solubility in toluene, where the catalyst activity in penta-1, 3-diene hydrogenation reached 36 000 h^{-1} (comparable with the activity of a system based on $[(PhP)_2Pd_5]_n$). Note that $Pd_5P_4Ph_6$ species formally differ in composition from the aforementioned $Pd_5(PPh)_2$ cluster by the presence of two extra diphenylphosphide groups, $Pd_5(PPh)_2(PPh_2)_2$. Systems based on $Pd(Acac)_2$ with some other types of phosphorus-containing ligands (Ph_2PH, $PhPH_2$) and reductant (NaH_2PO_2, $AlEt_3$) were also studied (see, e.g., [243]).

A principal difference of the process of $Pd(Acac)_2PPh_3$ reduction by hypophosphite from the reduction by hydrogen consists of a lower degree of PPh_3 dephenylation. As a result, about half of palladium is retained in the form of nanosized Pd(0) clusters stabilized on the surface by PPh_2 groups. An analysis of the $Pd(Acac)_2PPh_3{-}NaH_2PO_2$ system functioning in the hydrogenation reactions includes three routes: (i) formation and decomposition of active centers (dephenylation of PPh_3 groups by hydrogen); (ii) catalytic hydrogenation; and (iii) regeneration of inactivated species via reactions with the initial components. This is a typical variant of the chain mechanism of a catalytic process [243]. It was suggested that the regeneration of active $Pd(0)_xPPh_z$ centers takes place when an intermediate responsible for the chain termination interacts with the initial components:

$$Pd_xPPh_{z-1} \begin{cases} \xrightarrow{H_2} Pd_xP + C_6H_6\,(\text{termination}) \\ \xrightarrow[NaH_2PO_2]{Pd(Acac)_2PPh_2} Pd(0)_xPPh_z \end{cases} \qquad (5.145)$$

$Pd(Acac)_2{-}AlEt_3$–toluene. The reduction of Pd(II) by triethylaluminum proceeds with the formation of C_2H_6 and C_2H_4 (in nearly equimolar amounts) and heterogeneous nanoparticles with mean dimensions about 3–4 nm (maximum on a histogram of particle size distribution). These particles are stabilized by organometallic ligands of $AlEt_2(Acac)$ (for $[Al]/[Pd] < 3$) or $AlEt_2(Acac)\cdot AlEt_3$ (for $[Al]/[Pd] > 3$) [243, 244]. Systems of still higher dispersion are formed via a reaction of $Pd(Acac)_2PPh_3$ and $AlEt_3$, in which case the obtained nanoparticles have a mean size of 2–3 nm. The maximum activity of this system with PPh_3 in the styrene hydrogenation was observed at $[Al]/[Pd] = 4$ in the system, where the catalyst turnover frequency reached $TOF = 12\,000\ h^{-1}$.

$Co(Acac)_2{-}AlEt_3$–toluene (heptane). In the case of heptane, this system is characterized by a sharp maximum on the dependence of the rate of styrene hydrogenation on the total metal content ($[Co]_\Sigma$) [245]. As $[Co]_\Sigma$ is increased from $0.25\cdot10^{-3}$ M (corresponding to the maximum rate) to $10\cdot10^{-3}$ M, the TOF value drops from 850 to $5\ min^{-1}$. A colloidal state of the reduced metal particles appears at small Co concentrations. In the case of toluene, the system features the formation of Co(0) arene π complexes $(\eta^6\text{-ArH})CoEt\cdot Et_2Al(Acac)^-AlEt_2^+$ or hexene π complexes $(\eta^6\text{-ArH})Co(C_6H_{12})\cdot Et_2Al(Acac)$, which exhibit evolution into ferromagnetic nanoparticles in the flow of H_2. At $[Al]/[Co] = 4$, these particles have sizes within 2–5 nm and form agglomerates with dimensions up to 10–50 nm. The surface of Co(0) clusters bears adsorbed $AlEt_3$ and $AlEt_2(Acac)$ ligands and/or $AlEt_2(Acac)\cdot AlEt_3$ complexes, which stabilize the colloidal metal. The formation of weakly bound $AlEt_2(Acac)\cdot AlEt_3$ complexes on the surface leads to the formation of the most catalytically active nanoparticles.

The results of investigations of the formation of active species in another Ziegler system, $CoBr_2{-}Li[AlH(^tBuO)_3]$–THF, which catalyzes the hydrogenation of anthracene at 298 K with an activity of $TOF = 120{-}150\ h^{-1}$ [246, 247] showed that the formation of a catalyst in H_2 atmosphere in the absence of a substrate is accompanied by the evolution of H_2 from $CoBr_2{-}Li[AlH(^tBuO)_3]$ with the reduction of Co(II) to Co(I) and Co(0). It was suggested that the catalytically active species contain the reducing agent (H^-), which acts as a stabilizing ligand. A prolonged presence of H_2 in the system at the stage of catalyst formation (prior to substrate injection) leads to inactivation of the catalyst.

Apparently, this system also features the formation of colloidal metal nanoparticles, and the presence of H_2 favors their aggregation.

The rate of hydrogenation in the system under consideration is described (after the induction period, which disappears at [Al]/[Co] > 5) by a first-order equation with respect to a substrate. The kinetics of catalyst inactivation and H_2 evolution in the steps of catalyst formation is described by the following equation [246]:

$$n_{H_2} = a_0 + a_1(1 - e^{-k_6 t}) + a_2(1 - e^{-k_7 t}), \qquad (5.146)$$

where a_0 is the amount of hydrogen rapidly evolved during the contact of $CoBr_2$ with aluminum hydride (~0.5 mole H_2 per mole $CoBr_2$) and the two other terms correspond to a slow evolution of H_2 in two unidentified reactions of the first order with respect to the catalyst.

The aforementioned nanoparticles formed during the reduction of Pd(II) and Co(II) by triethylaluminum and trialkoxyaluminum hydride anions, as well as metal particles that appear during the reduction of Cr(III), Mn(II), Co(II), Ni(II), Ru(III), Os(III), Ir(III), Rh(III), Pd(II), and Pt(II) salts by complex boron hydride $R_4N^+[BEt_3H]^-$ species [225], are classified to colloidal metals. In large metal-containing clusters, in contrast to colloids, a multilayer metal core is surrounded by a certain composition of ligands adsorbed on the outermost metal layer [248]. In the case of colloids, the particles exhibit a broad distribution with respect to their size (or the number of metal atoms) [248, 249]. The synthesis of large [213] and especially giant clusters also leads to a mixture of various complexes, but this mixture admits the separation of compounds with more or less definite structures.

5.3.9 Synthesis and characterization of giant palladium clusters

The giant clusters of Pd and Pt mentioned in the preceding section should probably be classified [30] as compounds that occupy an intermediate position between large molecular clusters and colloidal metals (or other nanoparticles described in Section 5.3.8). The synthesis of giant clusters and the verification of their proposed structures have been considered in much detail in reviews [30, 172–174]. Here, let us only briefly consider the known pathways of the synthesis of these unusual compounds.

The reaction of palladium acetate reduction by hydrogen in acetic acid in the presence of 1,10-phenanthroline (Phen) yielded an X-ray amorphous complex with the composition $Pd_4Phen(OAc)_4H_4$ (B) via the following reaction:

$$4Pd_3(OAc)_6 + 3Phen + 15H_2 \longrightarrow {}^3/_n[Pd_4(Phen)(OAc)_2H_4]_n + 18AcOH.$$

According to the data of transmission electron microscopy (TEM), the particles of this complex contain a metal core with a diameter of 20 ± 5 Å. These core dimensions allowed the value of n in the formula $[Pd_4Phen(OAc)_4H_4]_n$, to be evaluated as close to $n = 100$ [250]. Upon the oxidation of compound B by oxygen, the major fraction of complexes converts into a cluster compound with the composition $Pd_9(Phen)(OAc)_3$ (C), which is soluble in water and polar organic solvents. Estimation of the molecular mass of complex C gave $(1 \pm 0.05) \cdot 10^5$ Da. These data and the results of investigations by a combination of methods including TEM, small-angle X-ray scattering (SAXS), and extended X-ray absorption fine structure (EXAFS) allowed the main structural features of the cluster compound to be established, according to which it represents regular spherical particles with a diameter of 26.0 ± 3.5 Å with a polyhedral metal core (Pd icosahedron). The molecule of complex C was assigned a composition of $Pd_{570\pm30}Phen_{63\pm3}(OAc)_{190\pm10}$. This formula well agrees with the ideal model of a five-layer icosahedron containing 561 atoms of Pd and with the estimated molecular mass. The idealized formula of cluster C was adopted as $Pd_{561}Phen_{60}(OAc)_{180}$. In this cluster, a positively charged core (+180) is surrounded by outer-sphere acetate ligands that are responsible for the affinity to the polar solvents. The precipitation of clusters C with KPF_6 salt from an aqueous solution yielded a new giant cluster (D) possessing a core charge of +60 and assigned the idealized formula $Pd_{561}Phen_{60}O_{60}(PF_6)_{60}$.

The use of nitrogen-containing ligands, which are not decomposed in the course of reduction and cover (as bidentate ligands) the surface of a metal core, hinders the further aggregation of metal particles and stabilizes the obtained molecular associates.

Giant clusters with dimensions greater than Pd_{561} were also reported. In particular, particles with cores containing seven and eight layers of palladium Pd_7 (31.5 Å) and Pd_8 (36 Å) were obtained and assigned the idealized formulas of $Pd_{1415}Phen_{60}O_{1100}$ and $Pd_{2057}Phen_{84}O_{1600}$, respectively [251]. The structures of Pd_{145} [214] and $Pd_{157}Pt_7$ [215] complexes were determined by X-ray diffraction.

Specific catalytic properties of these giant clusters in the oxidation [173] and some other reactions [172, 174] allow these systems to be considered as a new class of soluble metal complex catalysts.

5.3.10 Approaches to identification of the nature of catalytically active species in solutions of metal complexes

The number of questions concerning the nature of active species participating in the homogeneous catalysis with metal complexes and the pathways of transformation of various precursors (molecular complexes and/or heterogeneous particles) into active centers considerably increased due the fact of giant cluster formation and the catalysis with these clusters [173, 174] and colloidal metals [218, 248], which are jointly referred to below as metal-containing nanoclusters (MNCs). The problem of discriminating between various hypotheses pertaining to the catalysis with MNCs and homogeneous molecular complexes is actively discussed in the literature [36, 218, 222, 252, 253].

Catalytic systems with MNCs were called "soluble heterogeneous catalysts" [218], since it was suggested that centers with different activities could be present on the MNC surface. As was noted at the end of Section 2.2, the uniformity or non-uniformity of the active centers cannot serve as a criterion for the homogeneity or heterogeneity of a catalyst. A more significant basis is provided by the approach to the MNC surface as the nanoparticle–solution interface. The formation of MNCs with or without ligands adsorbed on the surface is favored by the presence of readily reducible metal precursors and a reducing medium (H_2, CO, olefins, alcohols, alkynes, PR_3, etc.). Although the formation of MNCs (and the more so, solid metal particles) is a thermodynamically highly favorable process for many transition metals, the presence of stabilizers that prevent the agglomeration of particles is also very important. The role of stabilizers can be played by anion species (Hal^-, $RCOO^-$), large organic molecules adsorbed on the MNC surface, R_4N^+ X^- (and other ionic liquids) [254], and polar solvents. For example, palladium-containing nanoclusters (colloids) are stable in propylene carbonate ($\varepsilon = 69$) at 140–155 °C for several days [218].

The issue of discriminating between homogeneous molecular catalysis and homogeneous catalysis with MNCs (microheterogeneous catalysis) is especially important for the reactions of hydrogenation, hydrosilylation, oxidation of alkanes, carbonylation of alcohols, and coupling reactions in solutions of palladium complexes (Heck reaction, Suzuki reaction, Stille reaction, etc.). In the case of heterogeneous precursors, a homogeneous character of catalysis was established for oxidation and some coupling reactions [218, 222].

Difficulties in the discrimination of hypotheses concerning the nature of active centers are already encountered at the first phenomenological stage of analysis, where it necessary to choose between the homogeneous molecular catalysis and the catalysis with nanoparticles (i.e., objects possessing a particle–solution interface with a large number of potential active centers) and, in the latter case, determine whether the catalytic activity is related to whole particles or whether they serve as a reservoir of active homogeneous molecular species, e.g., Pd(0) complexes:

$$Pd_n \rightleftharpoons Pd_{n-1} + Pd(0). \quad (5.147)$$

Presently, no unambiguous simple criteria are known that would make it possible to distinguish between the homogeneous molecular catalysts and "soluble microheterogeneous catalysts," so that a combination of criteria is employed that will be commented on below.

Approaches to the second stage of investigation, a quantitative kinetic description of these systems, are also not yet developed, but some simple kinetic laws have been used as criteria for discriminating hypotheses of the nature of active centers [218, 222]. In particular, it is possible to use the mean TOF values and TOF upon the induction period [222, 253] as characteristics of the catalyst recycling ability, effect of catalyst concentration on the TOF, etc.

Evidently, descriptions of the kinetics of reactions where MNCs act as sources of active homogeneous centers or as catalysts featuring the catalytic process on the MNC surface are considerably different. In the former case, which was described in detail in Section 2.2, we deal with a simple equilibrium (if it takes place) of the type of reaction (5.147) after the stage of MNC formation. At very small concentrations of molecular complexes (or ligand-free Pd(0) species) as compared to $[Pd]_\Sigma$ (i.e., at $[Pd]_\Sigma \cong n[Pd_n]$) we have

$$[Pd(0)] = \sqrt[n]{K_n[Pd]_\Sigma/n},$$

where K_n is the equilibrium constant of reaction

$$Pd_n \rightleftharpoons nPd(0).$$

In this case, the reaction rate and TOF are given by the following formulas:

$$r = \omega[Pd(0)] = \omega K_n^{1/n} n^{-1/n}[Pd]_\Sigma^{1/n}; \quad TOF = \frac{r}{[Pd]_\Sigma} = \omega'[Pd]_\Sigma^{(1/n)-1} \cong \omega'[Pd]_\Sigma^{-1} \ (\text{for } n \gg 1).$$

The TOF therefore decreases with increasing metal concentration in systems of soluble associates with low numbers of nuclei (for arbitrary $n > 1$). For example, during the anionic polymerization of olefins and dienes in solutions of organolithium initiators, the reaction rate is

$$R = k[M][C]_0^{1/n},$$

where M is monomer and C is initiator.

For a degree of association $n = 6$, TOF $\propto [C]_0^{-5/6}$ (Section 2.2) and, hence, TOF decreases with increasing $[C]_0$. The increased activity of very small concentrations of palladium complexes is well characterized by the term "homeopathic" catalysis [36, 256]. The homeopathic character of palladium catalysts in some cases was repeatedly confirmed [218]. Leadbeater with co-workers [257] showed that the Heck reaction, Suzuki reaction, and Sonogashira reaction can even be catalyzed by traces of palladium contained in sodium carbonate of commercial purity. As is known, the palladium concentration on a level of 50 ppb is sufficient to ensure the reliable catalysis of these reactions [222]. The above considerations suggest that a decrease in the TOF with increasing total metal content in the system (at a simultaneous growth in the reaction rate) can be related to the fact that only a part of the initial catalyst is active in the given reaction [258] and is not necessarily due to the presence of a homeopathic amount of catalytically active species.

A more complicated situation corresponds to cases where MNCs of various dimensions (e.g., associates, see Section 2.2) are formed and the reaction involves the steps of MNC formation and decomposition under the action of reactants and/or agglomeration with the formation of solid metal nanoparticles (M_{sol} [258]). In the case of active molecular complexes, the formation of MNCs (associates) can also lead under certain conditions to the independence of the reaction rate on the total metal concentration (see Example 2.34 in

Chapter 2) and even a decrease in the reaction rate with increasing concentration of the catalyst (in the region of its low concentrations) [282, 283, 259]. The mechanism of MNC nucleation upon the formation of uncharged and charged nanoparticles (e.g., $Pd(0)_n$ and Pd_n^{m+}) is still unclear, although some interesting models of MNC formation [253, 258] and even universal models of colloid aggregation [260, 261] were proposed.

The kinetics of processes on the surface of appearing MNCs must depend to a considerable extent on specific features of each particular catalytic system. In this respect, the following questions can arise.

(i) what is the distribution function of the number (N_n) of MNCs with various numbers n of atoms, that is, $\varphi(n) = dN_n/dn$;
(ii) whether the form of this function changes (and how, if it does) during the interaction of catalyst with the substrate (e.g., with phenyl iodide in the Heck reaction);
(iii) whether the activity of centers on the MNC surface depends (for uniform centers) on the MNC size (n);
(iv) whether the distribution of centers with respect to their activity or the constants of substrate binding for non-uniform centers (e.g., $\xi = \Delta G°/RT$, where $\Delta G°$ is the change in the Gibbs energy for the substrate–active center complex dissociation on the MNC surface) depend on the MNC size, in other words, how does the degree of non-uniformity of the surface of MNCs depends on their dimensions?.

If the surface of each MNC is uniform and the rate constants are independent of n, the rate of a catalytic reaction can be described by simple Langmuir relations [262], since the total number of accessible centers will be proportional to the amount of metal in the system. However, it is possible only provided that the distribution $\varphi(n)$ of particles in the system during this reaction remains unchanged, i.e., the specific "surface" of the MNC is constant.

If the surface of each MNC is uniform but the corresponding rate constants depend on their dimensions, the description requires setting the dependence of these rate constants (and the reaction rate) on n. Then, the reaction rate is described by the following equation:

$$R = \int_{n=2} r(n)\varphi(n)dn.$$

In the most general case, the reaction rate in this "soluble microheterogeneous" system must depend on the MNC distribution function $\varphi(n)$ and the activity distribution function $r(n)$ and on the form of the dependence of the distribution function with respect to physical properties (reactivity) on the MNC size. The approaches to a solution to these tasks are still unclear, but it is believed that the experience gained in the investigations of heterogeneous catalysis on non-uniform surfaces can be useful in this respect.

Let us consider the existing criteria for discrimination of the hypotheses of catalysis with molecular complexes and MNCs based on exhaustive reviews [36, 218, 222] with some comments.

1. **Kinetic criteria (tests).** The use of TOF values and the dependences of reaction rates on the catalyst concentration were considered above. Another important kinetic criterion for the existence of catalytically active MNCs is believed to be σ-shaped curves of the reactant consumption or S-shaped curves of the reaction product formation. The appearance of these curves is considered as the manifestation of autocatalytic phenomena (e.g., in hydrogenation reactions) [218]. Rapid buildup of concentration of active centers (e.g., MNC concentration) can take place where the process of metal salt (complex) reduction to M(0) is autocatalytic with respect to the reduced metal or the process of nucleation with the formation of active species is autocatalytic. This model (reduction autocatalysis) was analyzed by Wedegren, Finke, et al. [217, 263, 264]. At the same time, one

should take into account that homogeneous molecular systems also frequently exhibit S- and σ-shaped kinetic curves. In addition, the catalysis on MNCs can be slower than that in a solution of ML_n complexes. In this situation, the process would exhibit self-inhibition. In order to derive better justified conclusions from the results of primary kinetic measurements, it is expedient to compare the kinetics of precursor conversion and catalytic reaction (e.g., of hydrogenation), with special attention devoted to the kinetics of substrate hydrogenation after the complete conversion (reduction) of the initial molecular complex. An σ-shaped kinetics cannot be manifested if the MNCs are more rapidly inactivated than formed (e.g., as a result of conversion into M_{solid}).

2. **Transmission electron microscopy (TEM).** Using the TEM method, it is possible to evaluate the number and determine the size distribution of MNCs (with linear dimensions above 1 nm) after evaporation of the solvent in vacuum from an appropriate support material.
3. **Light scattering.** Using the classical light scattering technique, it is also possible to determine the presence of particles with linear dimensions above 1 nm. An advantage of this method is the possibility of performing measurements *in situ* (in solution), without any disturbing action of the probing beam (e.g., electron beam in TEM) and vacuum on the sample.
4. **Poisoning with catalytic poisons (ligands).** Complete poisoning of a soluble catalyst with strong ligands L (L = thiophene, CS_2, RSH, SCN^-, I^-) at $[L] < 1$ eq. for $[M]_\Sigma$ evidences with high probability for a catalytic activity of the MNC surface [218, 222]. For example, it was demonstrated for the propylene oxidation on giant palladium clusters that the rate of reaction in solution drops to vanish at a ratio of ~15 mole ligand per mole Pd cluster containing ~560 metal atoms for C_2H_5SH, I^-, and SCN^- ligands [265, 266]. This result indicates that specific centers on the cluster surface participate in the catalysis of oxidation reactions, the more so that bulky ligands such as triphenylphosphine and phenanthroline are coordinated by the cluster, while not influencing the oxidation rate [266]. In the process of cyclohexene hydrogenation in solutions of rhodium nanoclusters at 25 °C, the introduction of 0.05 mole CS_2 per mole rhodium also completely inhibited the reaction [218].
   Note that, in using poisons for the discrimination of hypotheses concerning the nature of the catalyst, it is necessary to know the concentration of metal-containing species in the solution. The stoichiometric factor ($[L]/[M]_\Sigma \ll 1$) alone cannot be considered as a criterion unless the concentration of active species is known (which can be present in concentration on a level of ppm and even ppb).
   Another test employing the poisoning effect is based on the use of monomer and polymer ligands capable of selectively binding the molecular complexes of a metal. The Crabtree test employs dibenzocyclooctatetraene as a ligand that binds $RhClL_3$ complexes but does not interact with $Rh(0)_n$ and $Pd(0)_n$ nanoclusters and with heterogeneous Pd/C catalysts [217]. Crosslinked polyvinylpyridine binds and removes molecular complexes of palladium from the solution, but does not react with palladium nanoclusters [222].
5. **Poisoning with metallic mercury.** It is commonly accepted that the introduction of Hg(0) into a catalytic system under the reaction conditions suppresses the catalysis with MNCs as a result of amalgamation (e.g., for Pt, Pd, Ni), while not inhibiting the process in the case of metals (such as Rh, Ir, and Ru) that are not forming amalgams. In all cases, it is also necessary to check for the possible reactions of mercury with precursor complexes and bear in mind that metal atoms and/or M(0) complexes can also participate in the catalytic cycle. In the latter case, the termination of the kinetic chain by mercury would also take place when MNCs do not participate in the catalytic process.
6. **Filtration.** According to the Maitlis test, powdered porous materials (cellulose, Celite, graphite) are added to a catalyst solution during a reaction or after its end. After filtration, the deposit on the filter is

washed by the solvent and returned back to the reactor. Fresh solvent and substrate are added and the system is examined for the possible reaction. The manifestation of any catalytic activity is considered to be evidence for a heterogeneity (or microheterogeneity) of the catalyst. It is recommended to use various sorbents and also check for the activity of a filtrate. Another variant of this test makes use of microporous membranes, but in all cases it is necessary to establish whether the MNCs are retained on the filter or pass to the filtrate.

7. **Centrifugation.** A safer method of separating MNCs from a solution consists of centrifuging it at a speed of above 20 000 min^{-1}, which excludes the possible decomposition of MNCs during treatment on filters and membranes. After deposition of the heavy fraction, the decanted solution and the suspension of deposited material are separately tested for catalytic activity.
8. **The Collman test.** According to this criterion, it is suggested that in most cases a homogeneous catalyst must be more active in the hydrogenation of a polymer-bound substrate in comparison to a heterogeneous or nanocluster catalyst. This assumption is probably also valid for other reactions. In the case of MNCs, it is expected that the activity would exhibit a sharp drop in reaction with the same amount of substrate grafted on a polymer matrix [218].
9. **Three-phase test.** This test is more labor-consuming than the Collman test. A polymer-bound substrate and the catalyst that is also bound to a different polymer or supported on a carrier are introduced into a solution containing a secondary substrate (a three-phase system). If the catalyst passes to the solution, then the dissolved substrate can react with a polymer-bound substrate.
10. **Sensitivity of process parameters to catalyst preparation conditions.** It is believed that, in the case of MNCs, the reproducibility of activity, selectivity, working life (TOF), and recycling capacity of a catalyst more significantly depend on the method of catalyst preparation and on the variation of process conditions. The selectivity (especially the stereo- and regioselectivity) can provide a good discrimination test, since this characteristic must for many reasons be significantly different in the cases of catalysis with molecular complexes and MNCs. Indeed, the like distribution of the products of Heck reactions with various catalyst precursors (palladacycles, $Pd(OAc)_2$, $Pd_2(dba)_3$, Pd/C, Pd/TiO_2) can be treated as evidence for the identical nature of active particles in solution, i.e., for a homogeneous character of the process [267] (see also Example 5.8). It is also very helpful to monitor the relative yields of competitive reactions during a process where the formation of MNCs can be expected.
11. **Reactivity of substrates.** Investigations of the hydrogenation processes showed that (i) benzene is converted in most cases on heterogeneous catalysts or nanoclusters, and (ii) these catalysts are more active in the hydrogenation of benzene and cumene than polystyrene [218]. For the identification of active species in the case of a new catalyst, it is recommended to use less active, nonactivated aryl bromides and aryl chlorides instead of aryl iodides, which would ensure more reliable discrimination of the nature of active species [222].

An analysis of the results of investigations, in which the aforementioned criteria (tests) were employed, shows that the only way of reliably identifying the nature of catalytically active species consists of the use of a combination of tests [218, 222]. It should be also borne in mind that, in the case of processes possessing the features of chain mechanisms with catalytic cycles conjugated with the routes of formation and decomposition of active centers (including MNCs and solid metal particles), the use of tests where one of the catalytic species is removed from the process (poisoning, polymer binding, filtration, etc.) would unavoidably disturb the catalytic system, probably with the complete loss of its activity.

5.4 Examples of chain mechanisms and chain carriers of various natures

Preceding sections (5.3.2, 5.3.3, 5.3.4, 5.3.7, and 5.3.8), of this chapter gave some examples of chain mechanisms encountered in real catalytic processes, where transition metal complexes or related metal-containing nanoclusters (MNCs) were the kinetic chain carriers or active centers of catalytic cycles. The development of chain mechanisms, in which ions and/or metal complexes rather than radicals played the role of chain carrier was originally recognized at the end of 1950s by Baxendale with co-workers (see, e.g., [268]) and then in the 1960s by Moiseev with co-workers [27, 167, 168] in investigations of the isomerization of olefins (see Section 5.3.7). The problem of implementing chain mechanisms, considered as a class of processes irrespective of the nature of a chain carrier, was posed [19, 20] at the end of the 1970s (see also review [269]). Let us consider some other examples of cyclic routes possessing all features of chain processes, including those from the works cited.

Example 5.1

► Purmal' [20] analyzed the following chain mechanism of the photocatalytic decomposition of H_2O_2 [268, 270], in which the steps of chain propagation involved $HO^•$ radicals and Cu^{2+} and Cu^+ ions:

$$\begin{aligned}
&(1)\quad H_2O_2 \xrightarrow{h\nu} 2\dot{O}H;\\
&(2)\quad \dot{O}H + H_2O_2 \longrightarrow H\dot{O}_2 + H_2O;\\
&(3)\quad H\dot{O}_2 + Cu^{2+} \longrightarrow Cu^+ + O_2 + H^+;\\
&(4)\quad Cu^+ + H_2O_2 \longrightarrow Cu^{2+} + \dot{O}H + OH^-;\\
&(5)\quad Cu^+ + H\dot{O}_2 \longrightarrow Cu^{2+} + HO_2^-.
\end{aligned}$$

In the cyclic route of steps (3) and (4), Cu^{2+} ions catalyze the process according to the overall equation

$$H\dot{O}_2 + H_2O_2 \xrightarrow{Cu^{2+}} \dot{O}H + O_2 + H_2O, \qquad (5.148)$$

while $HO^•$ radicals and Cu^{2+} ions in steps (2)–(4) catalyze the process according to the overall equation

$$2H_2O_2 \longrightarrow 2H_2O + O_2. \qquad (5.149)$$

Reactions (1) and (5) represent the steps of chain origination and termination, respectively. Thus, catalysts $HO^•$ and Cu^{2+} in this process perform different kinetic functions (see Chapter 4). ■

Example 5.2

► The oxidation of ascorbic acid (DH_2) catalyzed by Cu^+ ions [20, 271] is described by the following chain mechanism:

$$\begin{aligned}
&Cu^{2+} + DH^- \rightleftharpoons CuDH^+;\\
&(1)\quad CuDH^+ + Cu^{2+} \longrightarrow 2Cu^+ + D + H^+;\\
&(2)\quad Cu^+ + O_2 \longrightarrow CuO_2^+;\\
&(3)\quad CuO_2^+ + DH^- \longrightarrow HO_2^- + D + Cu^+;\\
&(4)\quad CuO_2^+ + Cu^+ \xrightarrow{H^+} 2Cu^{2+} + HO_2^-,
\end{aligned}$$

where two Cu^{2+} ions are involved in step (1) of chain origination and diamagnetic Cu^+ ions serve as active chain carriers. The overall equation of the cyclic route is

$$DH^- + O_2 \xrightarrow{Cu^+} HO_2^- + D, \tag{5.150}$$

and the overall equation of a route producing the active centers Cu^+ for the cyclic route (steps (1), (2), and (4)) is

$$CuDH^+ + O_2 \longrightarrow Cu^{2+} + D + HO_2^- \tag{5.151}$$

Reactions (5.150) and (5.151) correspond to two circuits on a bipartite graph (Figure 5.26) representing the mechanism of this chain process.

Investigations of the kinetics of ligand substitution in transition metal complexes showed that reactions of the metals complexes with 17-electron shells frequently proceed at a rate that is significantly (sometimes, by a factor of 10^9) higher than the same reactions in 18-electron complexes [272, 273]. In other words, the $17e^- \rightleftharpoons 19e^-$ transition (associative mechanism) is much more effective than $18e^- \rightleftharpoons 16e^-$ transitions. Accordingly, the isolated $19e^-$ complexes rapidly dissociate into $17e^-$ complexes. It was also demonstrated that one-electron oxidation or reduction of the $18e^-$ complexes significantly accelerates the ligand exchange, which becomes a chain process with the catalytic cycle of electron transfer. These processes were classified as representing the electron transfer chain (ETC) catalysis [274–276]. The catalyst (electron) turnover number in some cases reaches 1000 (moles of product per mole one-electron oxidant) [275].

Numerous examples of both inorganic and organic reactions involving ETC catalysis are considered in review [277]. The general scheme of ETC catalysis includes an initiation step of the oxidation (reduction) of ML_n complex, in which the replacement of ligand L by L′ must be performed according to reaction

$$ML_n + L' \rightleftharpoons ML_{n-1}L' + L. \tag{5.152}$$

The ferricenium cation $FeCp_2{}^+PF_6{}^-$ ($17e^-$) is frequently used as an oxidant and the Fe(I)–$CpFe(C_6Me_6)$ ($19e^-$) complex is used as a reductant. The redox reactions can also be implemented as electrode processes. The general mechanistic scheme is as follows:

$$\begin{aligned}
&(1)\quad ML_n \longrightarrow ML_n^{\pm};\\
&(2)\quad ML_n^{\pm} + L' \longrightarrow ML_{n-1}L'^{\pm} + L;\\
&(3)\quad ML_{n-1}L'^{\pm} + ML_n \longrightarrow ML_n^{\pm} + ML_{n-1}L';\\
&(4)\quad ML_n^{\pm} \longrightarrow ML_{n-1} + L^{\pm},
\end{aligned} \tag{5.153}$$

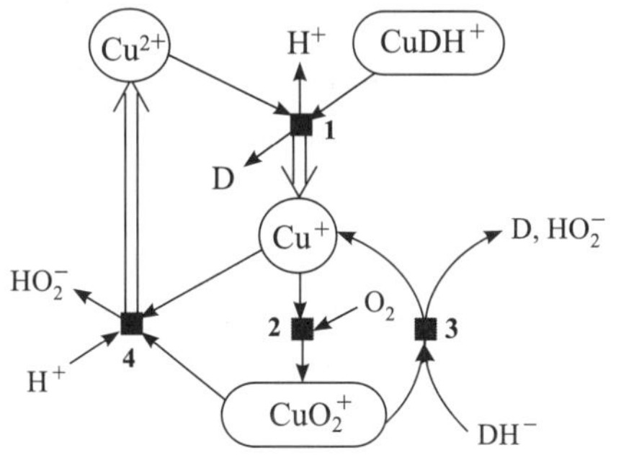

Figure 5.26 Bipartite graph corresponding to mechanism of ascorbic acid oxidation

where step (4) is the kinetic chain termination as a result of the decomposition of $ML_n^{\pm}$ species or their deeper oxidation or reduction. ■

Let us consider an example of the process of ligand substitution according to the chain scheme (5.153).

Example 5.3 [278]

► A difficult process of the arene π ligand removal from $CpFe(\eta^6\text{-toluene})^+$ ($18e^-$) complex during the replacement of toluene by methylphosphine can be accelerated and accomplished quantitatively within several minutes by introducing a small amount of $CpFe(\eta^6\text{-toluene})$ ($19e^-$) reductant into a solution of the complex in acetonitrile or THF. The reductant can also be obtained from the initial $CpFe(\eta^6\text{-toluene})^+$ reactant by donating it one electron as

$$\begin{aligned}
&(1)\quad \underset{18e^-}{CpFe(\eta^6\text{-toluene})^+} \xrightarrow{e^-} \underset{19e^-}{CpFe(\eta^6\text{-toluene})};\\
&(2)\quad \underset{19e^-}{CpFe(\eta^6\text{-toluene})} + 3PMe_3 \longrightarrow \underset{19e^-}{CpFe(PMe_3)_3} + \text{toluene};\\
&(3)\quad \underset{19e^-}{CpFe(PMe_3)_3} + \underset{18e^-}{CpFe(\eta^6\text{-toluene})^+} \longrightarrow \underset{18e^-}{CpFe(PMe_3)_3{}^+} + \underset{19e^-}{CpFe(\eta^6\text{-toluene})}.
\end{aligned} \qquad (5.154)$$

In order to carry out the ETC process, it is necessary to ensure that steps (3) in schemes (5.153) and (5.154) would be exergonic [273, 277]. In particular, step (3) in scheme (5.154) has $\Delta G^\circ = -76.9\,\text{kJ/mole}$. ■

The chain electron transfer catalysis is also employed in organic chemistry as illustrated by the following examples.

Example 5.4

► An effective chain substitution process, whereby Cl^- in ArCl is replacing by a nucleophile containing C-, P-, S-, As-, and Sb-centered anions or by other Hal^- ions, proceeds when one electron is transferred to PhCl with the formation of anion radical [278]:

$$PhCl + Nu \xrightarrow{e^-} PhNu + Cl^-. \qquad (5.155)$$

A possible mechanistic scheme of reaction (5.155) includes the steps of active center origination and the chain propagation and termination:

$$\begin{aligned}
&(1)\quad PhCl \xrightarrow{e^-} PhCl^{\bar{\bullet}};\\
&(2)\quad PhCl^{\bar{\bullet}} + Nu^- \longrightarrow PhNu^{\bar{\bullet}} + Cl^-;\\
&(3)\quad PhNu^{\bar{\bullet}} + PhCl \longrightarrow PhNu + PhCl^{\bar{\bullet}};\\
&(4)\quad PhCl^{\bar{\bullet}} \longrightarrow Ph^{\bullet} + Cl^-\\
&\qquad (Ph^{\bullet} = Ph_2, PhH \text{ and other products}).
\end{aligned} \qquad (5.156)$$

This variant of the ETC process proved to be useful for implementing a kinetically hindered reaction of olefin arylation by aryl chlorides (Heck reaction). Attempts to activate ArCl by nickel complexes showed that the $Pd_2(dba)_3$–$P(o-Tol)_3$–Et_3N–DMF system in the presence of $NiBr_2$ and NaI at 140 °C catalyzes the Heck reaction of $p\text{-}YC_6H_4Cl$ (Y = H, Me, OMe, COOMe, CN) and polar olefins

(5.157)

with very high yields [279]. Investigation of the effects of reductants, Ni(0) complexes, and the presence of $PhNiCl(PPh_3)_2$ led to a conclusion that the only scheme consistent with the entire set of facts is that combining the $NiBr_2$ initiated ETC process of Cl replacement by I in ArCl with a traditional mechanism of the Heck reaction on Pd(0) complexes:

$$\begin{aligned}
&(1)\ NiBr_2 + 4I^- \rightleftharpoons NiI_4^{2-} + 2Br^-;\\
&(2)\ NiI_4^{2-} + PhCl \longrightarrow PhCl^{\overline{\bullet}} + NiI_4^-;\\
&(3)\ PhCl^{\overline{\bullet}} + I^- \longrightarrow PhI^{\overline{\bullet}} + Cl^-;\\
&(4)\ PhI^{\overline{\bullet}} + PhCl \longrightarrow PhCl^{\overline{\bullet}} + PhI;
\end{aligned} \qquad (5.158)$$

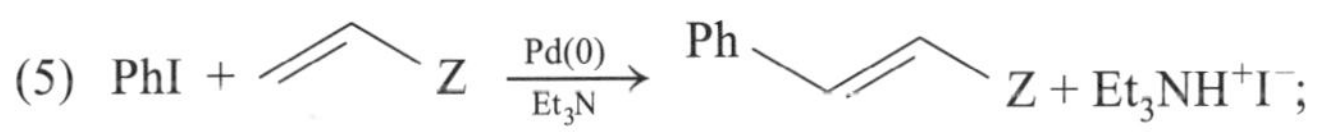

$$(6)\ PhCl^{\overline{\bullet}} \longrightarrow Ph\cdot + Cl^-$$

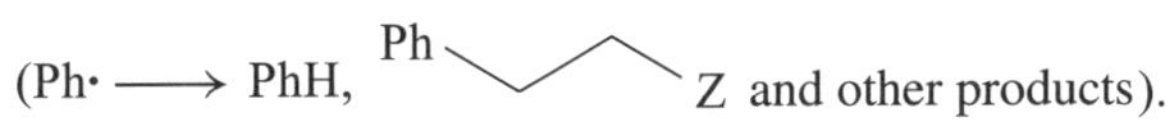

The steps of chain termination in this scheme were not studied, but the product of reduction ($PhCH_2CH_2Z$) was found. ■

Example 5.5

► An interesting application of the ETC process in the catalysis with metal complexes was described by Desbois and Astruc [273, 277]. W(0) complexes $W(CO)_3(CH_3CN)_3$ very slowly catalyze the polymerization of alkynes (A), which can be significantly accelerated by introducing $FeCp_2{}^+PF_6{}^-$ as an oxidant. At a molar ratio of $[Fe]/[W] = 0.2$, the rate of PhC_2H polymerization in THF at 25 °C exhibits approximately a 26-fold increase.

In accordance with the carbene mechanism of alkyne polymerization, it was shown that the role of oxidant consists in the initiation of the ETC process for replacing two CH_3CN molecules in the complex with the formation of tungsten vinylidene and π-complexes: $W(CO)_3(CH_3CN)_2(=C=CHPh)$ and $W(CO)_3(CH_3CN)(=C=CHPh)(PhC\equiv CH)$. Then, the polymer with a molecular mass of the order of 25 000 Da is formed within 10 min. Figure 5.27 presents a scheme of this ETC process.

It was established that $FeCp_2{}^+PF_6{}^-$ oxidant does not initiate the polymerization process, which is strongly retarded if the [Fe]/[W] ratio is increased up to unity. Thus, the ETC process of ligand substitution represents a step of origination of a carbene complex that is active in the polymerization reaction. The mechanism of $[W]A_2$ complex formation in fact involves two catalytic processes with the overall equations

$$A + [W]S_2 \xrightarrow{[WS_2^+]} [W]SA + S;$$

$$[W]SA + A \xrightarrow{[W]SA^+} [W]A_2 + S,$$

which correspond to two circuits on a bipartite graph representing the mechanism of active center formation. The reaction of alkyne polymerization on carbene complexes is also a chain process, since there are many well-known pathways of chain termination on the carbene complexes (see Section 5.3.2). ■

The role of an active kinetic chain carrier in chain processes can also be played by a ligand (in reactions with metal complexes) [20] or even by a vacant site in the coordination sphere of a metal.

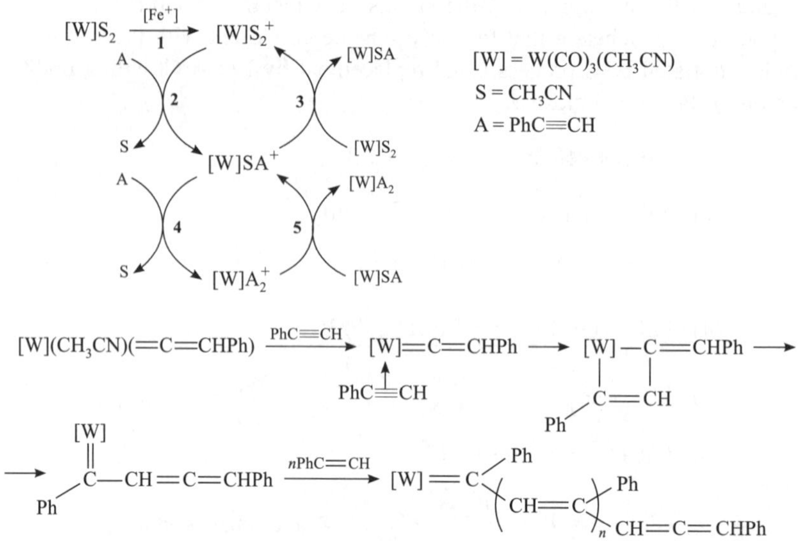

Figure 5.27 Carbene mechanism of alkyne polymerization conjugated with ETC initiation process

Example 5.6

► Olson and Margerum [280] observed a high rate of attaining equilibrium in the reaction of ligand substitution

$$\mathrm{Ni(II)L_1 + Cu(II)L_2 \rightleftharpoons Ni(II)L_2 + Cu(II)L_1}, \tag{5.159}$$

which was many times greater than the rate of dissociation of the reacting complexes (L_1 = triethylenetetramine, L_2 = ethylenediamine tetraacetate). The following chain mechanism of reaction (5.159) was proposed and justified:

$$\begin{aligned} &(1)\quad \mathrm{Ni(II)L_1 \rightleftharpoons Ni(II) + L_1};\\ &(2)\quad \mathrm{L_1 + Cu(II)L_2 \longrightarrow L_1Cu(II) + L_2};\\ &(3)\quad \mathrm{L_2 + Ni(II)L_1 \longrightarrow L_2Ni(II) + L_1};\\ &(4)\quad \mathrm{Ni(II) + L_2 \longrightarrow Ni(II)L_2}, \end{aligned} \tag{5.160}$$

where the chain propagation steps (2) and (3) involve a cyclic process of ligand substitution for which a high concentration of ligand L_1 is not necessary.

Example 5.7

► Gankin and Gurevich [281] proposed ligand exchange with the transfer of a vacancy in the coordination sphere through the steps of kinetic chain propagation for the process of olefin hydroformylation. This mechanism is believed to be quite probable in some reactions of ligand substitution. This type of chain mechanism was called coordination-unsaturated (conens) chain reaction (by analogy with "conens," a particle with vacant coordination site).

Investigations of the noncatalytic hydroformylation of olefins by the reaction

$$2RCH{=}CH_2 + 4HCo(CO)_4 \longrightarrow 2R'CHO + Co_2(CO)_8 + {}^1\!/\!_2[Co(CO)_3]_4 \qquad (5.161)$$

showed that the induction period in this process, which was usually assigned to the dissociation step

$$(1)\quad HCo(CO)_4 \rightleftharpoons HCo(CO)_3 + CO, \qquad (5.162)$$

disappeared upon adding only about ~5 vol % of solution taken from the system where the reaction continued after the induction period. It was suggested that the system featured processes that additionally led to the formation of coordination-unsaturated $HCo(CO)_3$ complex (X_1). In this case, reaction (5.162) is a step of origination of the active center X_1, which also appears in the subsequent steps of chain propagation:

$$\begin{aligned}
&(2)\quad X_1 + RCH{=}CH_2 \longrightarrow \underset{(X_2)}{R'Co(CO)_3};\\
&(3)\quad X_2 + CO \rightleftharpoons \underset{(X_3)}{R'COCo(CO)_3};\\
&(4)\quad X_3 + HCo(CO)_4 \longrightarrow R'CHO + \underset{(X_4)}{Co_2(CO)_7};\\
&(5)\quad X_4 + HCo(CO)_4 \longrightarrow X_1 + Co_2(CO)_8.
\end{aligned} \qquad (5.163)$$

Here, X_4 is "conens" particle that disappears in the following steps:

$$\begin{aligned}
&(6)\quad X_4 + CO \longrightarrow Co_2(CO)_8;\\
&(7)\quad X_4 \longrightarrow {}^1\!/\!_2[Co_4(CO)_{12}] + 2CO.
\end{aligned}$$

Figure 5.28 shows a simplified scheme of this chain mechanism as represented by a kinetic graph with null-vertex, since $HCo(CO)_4$ is reactant, while $Co_2(CO)_8$ and $Co_4(CO)_{12}$ are the reaction products. Scheme (5.163) includes three independent routes:

- Catalytic route (2)–(3)–(4)–(5) with the overall equation

$$RCH{=}CH_2 + 2HCo(CO)_4 + CO \longrightarrow R'CHO + Co_2(CO)_8 \qquad (5.164)$$

  and two routes including the formation of catalytic centers and the decomposition of intermediate compounds:
- Route (1)–(2)–(3)–(4)–(6) with the same overall equation (5.164);
- Route (1)–(2)–(3)–(4)–(7) with the overall equation

$$RCH{=}CH_2 + 2HCo(CO)_4 \longrightarrow R'CHO + {}^1\!/\!_2Co_4(CO)_{12} + CO. \qquad (5.165)$$

It should be noted that, according to the mechanistic scheme represented by the kinetic graph in Figure 5.28, the stationary concentration of X_1 is determined by the rates of steps (1), (6), and (7) and is independent of the "conens" transfer step (5). Therefore, introduction of the concept of "conens" as a kinetic chain carrier does not significantly modify the description of the chain process with catalytic cycle

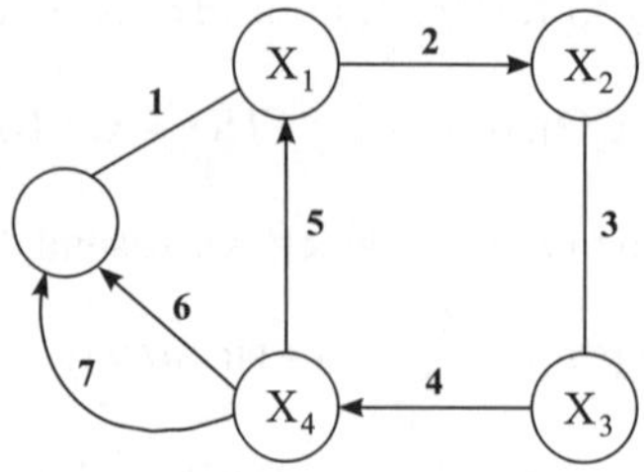

Figure 5.28 Kinetic graph corresponding to chain mechanism of hydroformylation (Example 5.7)

(2)–(3)–(4)–(5). A new detail is the step of $HCo(CO)_3$ complex regeneration by the ligand transfer to $Co_2(CO)_7$ "conens" particle. ■

The consideration of pathways of the kinetic conjugation of various routes in Section 3.4 was illustrated by several examples of catalytic reactions with chain mechanisms. Let us return to two cases which are also of interest in the context of problems considered in the present chapter.

The coordination polymerization of olefins in Ziegler catalyst solutions is correctly treated as a catalytic chain process, since the reaction mechanism includes, in addition to the steps of active center origination and the molecular chain propagation and termination, the steps of chain transfer to monomer and the steps of β-elimination of R_{-H} from M–R with the formation of M–H and the subsequent re-initiation (formation of M–R from M–H) that leads to the appearance of usual cyclic routes as shown in Figure 5.29. Therefore, this reaction offers an example of the catalytic process with a chain mechanism.

The catalytic hydrocarboxylation of olefins in $PdBr_2$–LiBr–THF solutions, where an active center (HPdBr) appears in the course of the second catalytic route of CO oxidation to CO_2 and water oxidation to H_2O_2 (see Section 3.4, Figure 3.27) according to the following reaction:

$$CO + O_2 + H_2O \longrightarrow CO_2 + H_2O_2. \quad (5.166)$$

In this case, a chain scheme of the active center formation operates under the conditions where both processes are catalytic.

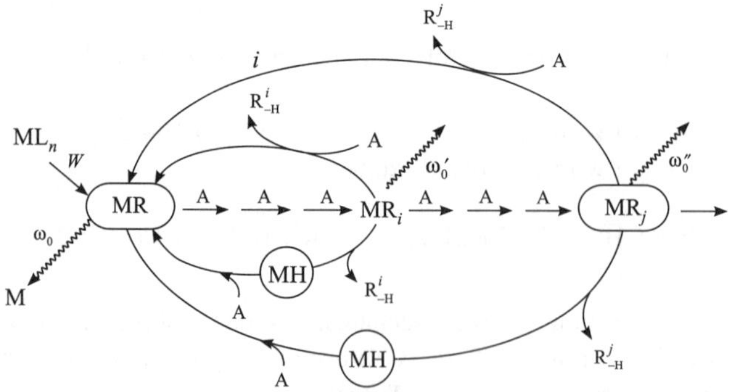

Figure 5.29 Kinetic graph corresponding to chain mechanism of catalytic polymerization of olefins: W is the initiation rate; w_0, w_0', w_0'' are the weights of steps of quadratic termination of active center MR and intermediates MR_i and MR_j, respectively

Let us consider a mechanism of the Heck reaction, in which the steps of formation, decomposition, and regeneration of active centers are taken into account [22, 23, 282, 283].

Example 5.8

► Example 3.22 described a mechanism of the Heck reaction of olefin arylation in solutions of palladium complexes. This mechanism was established based on the results of investigations of competitive reactions, the kinetics of which provided information on slow steps of the main catalytic cycle. Investigations of this reaction in a ligand-free system (without PR_3 ligands) performed by Schmidt with co-workers [22, 23] allowed the nature of the steps of formation, decomposition, and regeneration of active centers to be established as follows.

1. In the $Pd(Acac)_2{-}NBu_3{-}DMF$ system, $Pd(Acac)_2$ was reduced to Pd(0) complexes by tributylamine (and other trialkylamines). Since the reaction of ArI with olefins was accompanied by the evolution of HI,

$$ArI + RCH{=}CH_2 + Bu_3N \longrightarrow ArCH{=}CHR + Bu_3NH^+I^-, \qquad (5.167)$$

the process of active center formation was also studied with PdI_2 and $PdI_4{}^{2-}$. The reduction of Pd(II) by amine at 100–130 °C corresponds to the following equations:

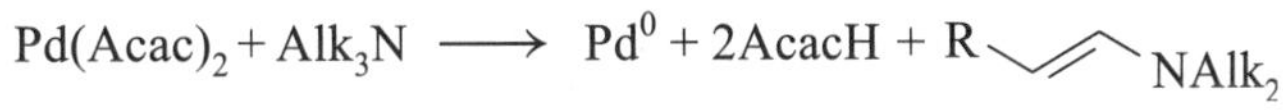
(5.168)

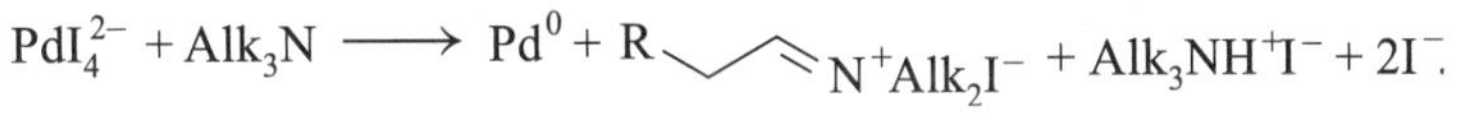
(5.169)

The rate of reduction of Pd(II) iodide compounds is much higher than the rate of reduction of $Pd(Acac)_2$ precursor species. The appearance of I^- during this reaction and the conversion of $Pd(Acac)_2$ into PdI_2 and $PdI_4{}^{2-}$ are among the main factors that account for the autoacceleration in the stage of formation of Pd(0) complexes (probably, in the form of $PdX_n{}^{n-}$) that are active in the reaction under consideration. It was suggested that, in the initial stage of the process, Pd(0) species also catalyze the process of Pd(II) reduction (i.e., autocatalysis takes place at the step of Pd(0) formation):

$$Pd^{II} + Pd^0 \xrightarrow{\text{reductant}} 2Pd^0.$$

It was demonstrated that, if about ~ 5 vol % of solution (A) was taken from the system upon attaining the maximum reaction rate (i.e., after the induction period), and transferred to solution (B) where the process was initiated, then the induction period in solution B almost disappears and the reaction rate rapidly attained a constant level. Unfortunately, these experiments were not performed in solutions with excess amounts of $Alk_3NH^+I^-$ (exceeding the concentration of ArI). It was also found that PdL_2Cl_2 phosphine complexes were reduced by Bu_3N.

2. It was established that side reactions of the conversion of intermediate ArPdI (or $ArPdI_3{}^{2-}$) species led to the formation of PdI_2 (i.e., these reactions contributed to the oxidation of Pd(0) to Pd(II)). The reductive coupling of ArI to Ar–Ar and the reduction of ArI to ArH are catalyzed by Pd(0) complexes in the presence of Bu_3N according to the following scheme:

$$\begin{gathered} Pd^0 + ArI \longrightarrow ArPdI \xrightarrow{ArI} Ar{-}Ar + PdI_2 \\ (\text{or } 2ArPdI \longrightarrow Ar{-}Ar + PrI_2 + Pd^0); \end{gathered} \qquad (5.170)$$

$$ArPdI + Bu_3NH^+I^- \longrightarrow ArH + PdI_2 + Bu_3N; \qquad (5.171)$$

$$PdI_2 + 2Bu_3N \longrightarrow Pd(0) + RCH_2CH{=}N^+Bu_2I^- + Bu_3NH^+I^-. \qquad (5.172)$$

If the reaction with ArI is performed in the absence of olefin, the rate of PdI_2 reduction by amine becomes so high that PdI_2 is detected in solution only after the complete consumption of Bu_3N. Thus, amine performs three functions in the process under consideration:

- reduction of Pd(II) precursor to Pd(0) complexes active in the process;
- neutralization of evolving HI to a weaker $R_3NH^+I^-$ acid species;
- regeneration of Pd(0) from PdI_2 formed during the decomposition of the main intermediate (i.e., as a result of termination of the kinetic chain).

3. One of the most important problems concerning the mechanism of the Heck reaction consists of establishing the nature of the active reduced Pd(0) species, i.e., determining whether the process is homogeneous with molecular complexes as participants or the active species represent nanodimensional colloidal clusters. The role of microheterogeneous forms, palladium black, and ordinary heterogeneous supported catalysts has been actively discussed in the literature [36, 219–221, 284, 285] (see also Section 5.3.10).

The results of investigations of the behavior of heterogeneous catalysts (Pd/C, Pd/SiO_2, Pd black) in the Heck reaction under e identical conditions [22] led to a conclusion that the active species in both $Pd(Acac)_2$–DMF–Bu_3N and Pd_{Het}–DMF–Bu_3N systems represent homogeneous Pd(0) complexes (see also [284]). Determination of the pseudo-first-order rate constant k_{eff} in the equation $r_{SS} = k_{eff}[Pd]_\Sigma$ (i.e., the catalyst turnover frequency) for the homogeneous systems and heterogeneous catalysts (with allowance for the amount of Pd passing from the surface of Pd/C, Pd/SiO_2, and Pd black to solution) showed that k_{eff} is almost constant (within 1–1.5 min^{-1}) for all studied systems. The removal of a heterogeneous Pd/C phase during the arylation process almost does not change the rates of conversion of the remaining reactants. Thus, the formation of active catalyst species from a heterogeneous precursor consists of the partial dissolution of Pd from the surface by aryl iodide and the conversion into ArPdI or $ArPdI_3^{2-}$, since I^- favors the partial passage of Pd_{solid} into solution. Even in the presence of a reductant (NaOCOH), the supported Pd/C acts as a homogeneous catalyst and the reductant converts PdI_4^{2-} into Pd(0). By the end of the process in the presence of a reductant, all Pd(II) is converted into Pd(0) adsorbed on the carbon carrier (activated charcoal). The in-phase variation of the rates of the Heck reaction with ArBr and ArCl and the amount of Pd passing from the carrier to solution was also observed in [284].

The ability of ArI to oxidize even supported Pd(0) and Pd black indicates that ArI has another function in addition to participating in the catalytic cycle and side reactions of the decomposition of intermediates. If the unstable PdI_n^{n-} complexes exhibit aggregation with the formation of colloidal nanoparticles (and probably blacks), ArI oxidizes these species to ArPdI, thus returning them to the catalytic cycle and participating in the regeneration of Pd(0) or intermediates of the process. It was suggested [22, 23] that the mechanistic scheme of this reaction represents a multi-route process in which the routes of formation, decomposition, and regeneration of active centers and intermediates are combined with the main catalytic cycle of the Heck reaction. Thus, the typical chain process with the routes of formation and decomposition of active centers and intermediates (Figure 5.30a) in fact transforms into a catalytic process with the steps of Pd(0) and PhPdI regeneration (Figure 5.30b). This process is described by a chain scheme with the overall equations of four routes corresponding to the scheme in Figure 5.30b:

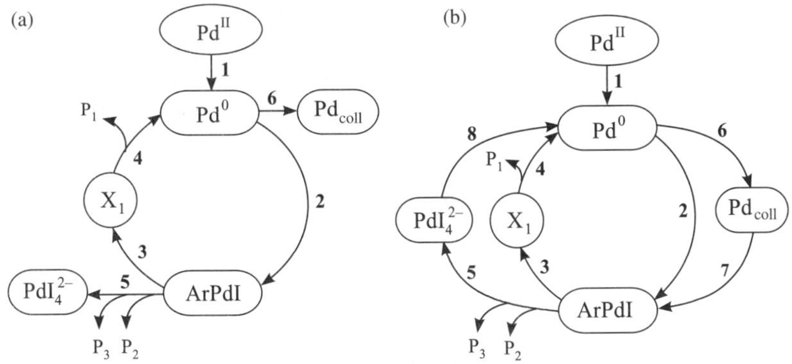

Figure 5.30 Schemes of real catalytic process of olefin arylation according to Heck: P_1 is the product of olefin arylation; P_2 and P_3 are the products of reductive reactions of formation of Ar–Ar and ArH, respectively

- route I of chain propagation (catalytic cycle of steps (2)–(3)–(4)):

$$ArI + RCH{=}CH_2 + Bu_3N \longrightarrow P_1 + Bu_3NH^+I^-; \qquad (5.173)$$

- route II of chain propagation, including the steps of formation of the new (microheterogeneous, colloidal) active catalyst species Pd_{coll} (which interact with the substrate in step (7)) and steps (6)–(3)–(4) with the same overall equation (5.175);
- routes III and IV of the formation of active Pd(0) species and the decomposition of ArPdI intermediate via (2)–(5)–(8) steps with the overall equations

$$2ArI + 2Bu_3N \longrightarrow Ar{-}Ar + R\diagdown\!\diagup{=}N^+Bu_2I^- + Bu_3N^+HI^- \qquad (5.174)$$

$$ArI + Bu_3N \longrightarrow ArH + R\diagdown\!\diagup{=}N^+Bu_2I^-. \qquad (5.175)$$

The fact of metal dissolution from the surface of heterogeneous particles is indicative of the reaction of oxidative addition of ArI to the metal with the formation of Ar–Pd bonds on the surface of palladium:

The principal possibility of olefin insertion into Ar–Pd bond on the surface seems to be evident, but the rate of this process on the surface is probably much lower than on a homogeneous $ArPdI_n^{(n-1)-}$ complex.

Strictly speaking, route II is a homogeneous–heterogeneous (or homogeneous–microheterogeneous) process, the contribution of which in the case of ArI is not large.

In the case of a less active aryl halide such as ArBr, the colloidal Pd is oxidized slowly and, hence, has time to convert back into palladium black that is inactive in the system under consideration. Unusual laws observed in the $PdCl_2$–NaOAc, $PdCl_2$–Bu_3N–NaOAc, and $PdCl_2$–$2PPh_3$–Bu_3N–NaOAc systems [282, 283] led to a conclusion that the reduced Pd(0) species catalyze both the reduction of $PdCl_2$ and the aggregation of Pd(0) with the formation of colloidal metal (Pd_{coll}), which plays an important role in these systems, being one of the catalysts in the Heck reaction. ■

In the past two decades, numerous investigations of the Heck (or Heck–Mizoroki) reaction (some of which were considered in this section) led to a consistent conclusion that this reaction, both in the presence of ligands and in ligand-free systems, is a homogeneous catalytic process with the participation of molecular Pd(0) complexes and ArPdX intermediates, at least in the case of ArI (see reviews [36, 222, 286–290]). These complexes can also be formed in solution via the routes involving microheterogeneous clusters (Pd_{coll}) and even Pd_{solid} particles as precursors. These routes in the scheme of Heck reaction were originally proposed by Schmidt and co-workers [22, 291] and more recently used in [285] and other investigations.

A different situation was encountered in the case of the Suzuki–Miyaura coupling reaction [292], where a significant contribution to the process rate is due to the catalytic cycles involving microheterogeneous or heterogeneous species or proceeding on their surface. Recent results obtained using a combination of methods including the Operando XAS, XPS and kinetic investigations provided the first proof for the fact that the Suzuki addition can proceed as a heterogeneous process on Pd nanoclusters stabilized by polyvinylpyrrolidone without the transition of intermediate into solution [293–295].

In various tests for the coexistence of active centers of various natures in the Heck and Suzuki reactions, an important role belongs to the investigations of competitive reactions with evaluation of the selectivities of different processes [267, 293, 296, 297].

It should be emphasized that the existence of routes (catalytic cycles) with the participation of microheterogeneous or heterogeneous sources of molecular intermediates leading to the formation of products of the Heck and Suzuki reactions is indicative of a homogeneous–heterogeneous character of these reactions with various contributions due to purely homogeneous (Heck), homogeneous–microheterogeneous (Heck, Suzuki), and fully heterogeneous (Suzuki) routes.

5.5 Classification of mechanisms of real catalytic processes

Evidently, any catalytic process is based on a cyclic (closed) sequence (chain) of steps, which begins and ends on the same intermediate. Any intermediate in this cycle has the "right" to be called a catalyst (or active center) [286]. Since real catalytic processes frequently involve the steps of formation and decomposition of active centers, it is very important for both ensuring the catalytic system operation and constructing a kinetic model of the process to establish the character and mode of the interaction of steps or routes of the formation and decomposition of active centers with the main catalytic cycles. As was repeatedly emphasized above, it is the structure and the topological features of a reaction mechanism (i.e., the interrelations between intermediates and routes in the space of intermediates) that determine the degree of mutual influence of the steps and routes. Based on a strict graph-theory model of the complex reaction mechanisms, a classification of the multi-route reaction mechanisms was developed [299, 300], including the classification of mechanisms for polyfunctional catalytic systems [298]. The topological features of mechanisms belonging to the main classes of catalytic processes can be demonstrated using the kinetic graphs of linear mechanisms, which clearly reflect the interrelations of the catalytic cycle with the other steps and routes of the formation and decomposition of active centers.

Let us assign index 0 (class 0) to the ideal catalytic cycle (with kinetic graph KG1, Figure 5.31). Kinetic graphs will be presented with a minimum number of intermediates which makes it possible to perform the topological and kinetic analysis of the mechanism. The simplest case of a real catalytic process is that with a step of active center formation from a precursor (KG2 and KG3, Figure 5.31).These mechanisms are assigned class I. In the mechanism represented by KG2, the step of ML_n complex formation from ML_m is irreversible. If this process is fast compared to the steps of the catalytic cycle, it will not influence

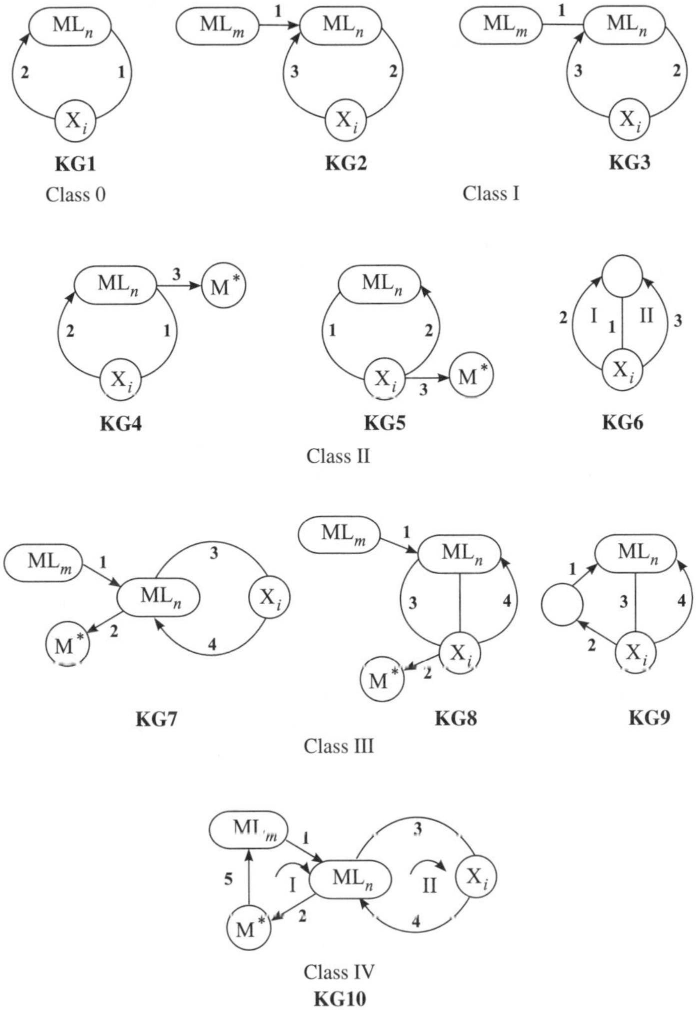

Figure 5.31 Kinetic graphs and classes of mechanisms of real catalytic processes

the kinetics of a steady-state or quasi-steady-state process. An example is offered by the formation of a π complex from $CoL_3(N_2)$ in the process of olefin hydrogenation (see Section 5.2). If the rate of the step of active center formation is comparable with the rate of the catalytic reaction, the process in a closed system would proceed with a significant self-acceleration due to a constant supply of active ML_n species to the system. For example, a slow reduction of $RhCl_3$ by olefin to $RhCl_2(\text{olefin})_2^-$ complexes that are active in the olefin dimerization [28] would influence the kinetics of the process of olefin dimerization. The reversible step I in the mechanism represented by KG3 (Figure 5.31) is always an equilibrium step. This step, as well as irreversible steps, does not enter into the overall equation of the catalytic process ("defect" of a mechanism, see Section 1.5 [300]) and only influences the material balance and the structure of the kinetic equation (Section 2.1). If the corresponding pendant vertex represents the step of ligand dissociation,

$$ML_m \rightleftharpoons ML_n + L,$$

the catalytic process is inhibited by ligand L with increasing concentration [L], while in the absence of added ligand, the reaction order with respect to $[ML_m]_\Sigma$ decreases below unity. In some cases, an increase (rather than decrease) in the content of L in the coordination sphere makes the catalytic centers more active. For example, the passage from neutral HPd(CO)Cl(L) complexes to cations via the reaction

$$\text{HPd(CO)Cl(L)} + \text{CO} \rightleftharpoons \text{HPd(CO)}_2\text{L}^+\text{Cl}^-$$

leads to an increase in the activity of cation complexes in the reaction of olefin carbonylation (see Section 3.2) In this case, if CO participates in the catalytic cycle in a reversible step preceding the limiting step and $[\text{Pd}]_\Sigma \cong [\text{HPd(CO)Cl(L)}] + [(\text{HPd(CO)}_2\text{L}^+)\ \text{Cl}^-]$, then the order of the catalytic reaction with respect to P_{CO} would be greater than unity, for example, as

$$r = \frac{k[\text{Pd}]_\Sigma P_{CO}^2[\text{olefin}]}{1 + K_1 P_{CO}}. \tag{5.176}$$

Mechanisms involving steps of the decomposition of active centers ML_n (KG4) and intermediates X_i (KG5) to the inactive state are attributed to class II of the mechanisms of catalytic processes. In these processes, the content of the active catalyst species decreases with time which is frequently encountered in real catalysis. In both cases (KG4 and KG5), the rate of inactivation (step **3**) depends on the rates of steps of the catalytic cycle. The rapid binding of ML_n (step **1**, KG4) or rapid conversion of X_i (step **2**, KG5) decreases the concentrations of ML_n and X_i and reduces the rate of catalyst decomposition (see Section 5.3). Since ML_n is the active center and M* is the product of the irreversible conversion of X_i, a mechanism represented by KG5 is conveniently described by a kinetic graph with a null-vertex (null-intermediate compound). In this case, the cyclic graph KG6 clearly reveals two cycles corresponding to the catalytic (**1–2**) and noncatalytic routes. Assuming a quasi-steady-state regime for X_i for a certain period of time in a closed system (and in a steady-state regime in an open system), we obtain

$$r_I = \frac{\omega_1\omega_2[ML_n]}{\omega_{-1} + \omega_2 + \omega_3}; \quad r_{II} = \frac{\omega_1\omega_3[ML_n]}{\omega_{-1} + \omega_2 + \omega_3}. \tag{5.177}$$

The formula for r_{II} shows that an increase in the weight of step 2 (ω_2) leads to a decrease in the catalyst inactivation rate. In Example 3.14, an increase in the olefin concentration decreases the rate of degradation of the cobalt phthalocyaninate catalyst of olefin cyclopropanation. For a process represented by graph KG5 with $[X_i] \ll [ML_n]$ and $[M]_\Sigma = [ML_n] + [M^*]$, the inactivation rate r_{II} obeys a simple equation

$$r_{II} = \frac{d[M^*]}{dt} = A([M]_\Sigma - [M^*]), \tag{5.178}$$

where

$$A = \frac{\omega_1\omega_3}{\omega_{-1} + \omega_2 + \omega_3},$$

$$[M^*] = [M]_\Sigma(1 - e^{-At}),$$

which yields

$$r_{II} = \frac{\omega_1\omega_3[M]_\Sigma e^{-At}}{\omega_{-1} + \omega_2 + \omega_3}; \tag{5.179}$$

$$r_I = \frac{\omega_1\omega_2[M]_\Sigma e^{-At}}{\omega_{-1} + \omega_2 + \omega_3}. \tag{5.180}$$

In some cases, the active ML_n species are decomposed step-by-step to inactive state M*, but the intermediate compounds still exhibit catalytic activity:

$$ML_n \xrightarrow{\omega_1} M_i \xrightarrow{\omega_2} M^*.$$

A simple catalytic reaction of pseudo-first order with respect to substrate S, which proceeds on active ML_n and M_i species, obeys the following equation:

$$-\frac{d \ln S}{dt} = k_{I}[ML_n] + k_{II}[M_i]. \tag{5.181}$$

Since $[ML_n] = [M]_{\Sigma} e^{-\omega_1 t}$, the quasi-steady-state regime with respect to $[M_i]$ implies that $[M_i] = \omega_1/\omega_2 [M]_{\Sigma} e^{-\omega_1 t}$, which yields

$$-\frac{d \ln S}{dt} = \left(k_{I} + k_{II}\frac{\omega_1}{\omega_2}\right)[M]_{\Sigma} e^{-\omega_1 t}. \tag{5.182}$$

Accordingly, the concentration of substrate S varies with the time as described by the formula

$$\ln \frac{S_0}{S} = \left(\frac{k_{I}}{\omega_1} + \frac{k_{II}}{\omega_2}\right)[M]_{\Sigma}(1 - e^{-\omega_1 t}). \tag{5.183}$$

A more complicated mechanism with chain termination on two centers was proposed for the process of 1-hexene oxidation to 2-hexanone and 2-hexanol by molecular oxygen in the (TPP)MnCl–$NaBH_4$–DMF system (TPP = tetra(*p*-methoxyphenyl) porphyrinate) [301]. The existence of a limiting conversion of 1-hexene (15–20 %) was explained by the steps of catalyst decomposition during the reaction. The following five-step mechanism was proposed to explain the observed process kinetics [301]:

$$\begin{aligned}
&P_1 + N \underset{k_{-1}}{\overset{k_i}{\rightleftharpoons}} P_2 + C;\\
&P_2 + M + O_2 \underset{k_{-2}}{\overset{k_2}{\rightleftharpoons}} P_3;\\
&P_3 \xrightarrow[(N)]{k_2} P_1 + B;\\
&P_2 \xrightarrow{k_4} Q_1;\\
&P_3 \xrightarrow{k_5} Q_2.
\end{aligned} \tag{5.184}$$

where P_1 = (TPP)Mn^{III}Cl; P_2 = (TPP)Mn(II); P_3 = intermediate cyclic complex; N = BH_4^-; M = olefin; B = olefin oxidation products; Q_1, Q_2 = catalyst decomposition products; and C = BH_4^- conversion products.

Figure 5.32 shows a kinetic graph for mechanism (5.184), which differs from graphs KG4 and KG5 (Figure 5.31) by the presence of two steps for the decomposition of intermediate compounds. In the case of $[M]_0 \gg [P_0] = p_0$ and $[N]_0 \gg [P_0]$, the amount of products $[B] = b(t)$ is described by the following equation:

$$b(t) = k_3 \int_0^t p_3(t)dt, \tag{5.185}$$

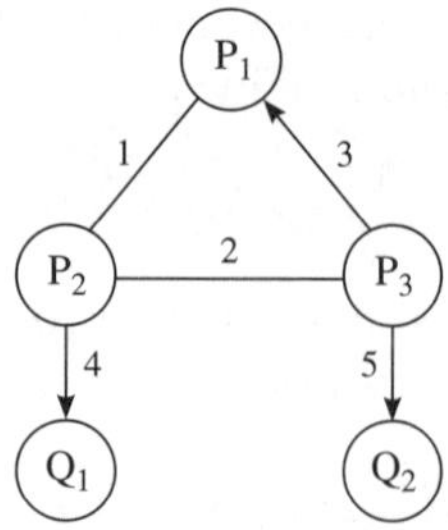

Figure 5.32 Kinetic graph corresponding to mechanism of 1-hexene oxidation (5.184)

where p_3 is the concentration of intermediate complexes P_3. The $p_3(t)$ function was determined by solving the equation

$$\frac{d^3p_3}{dt^3} + A_1\frac{d^2p_3}{dt^2} + A_2\frac{dp_3}{dt} + A_3p_3 = 0, \tag{5.186}$$

which was derived from a system of differential equations for the concentrations p_1, p_2, and p_3 of intermediate complexes P_1–P_3, respectively, according to mechanism (5.184). The obtained solutions of Eqs. (5.185) and (5.186) yielded expressions for $b(t)$, $p(t) = p_2(t) + p_3(t)$, and $p_1(t)$, which were compared to the experimental curves of $b(t)$, $p_1(t)$, and $p_2(t)$ constructed using the data of electronic absorption spectroscopy in the visible range. The estimations of parameters obtained in this way provided adequate description of the process under consideration [301].

An interesting variant of the loss of active centers was studied in enzymatic catalysis [302], but it can also be encountered in the catalysis with metal complexes. In systems with oligomeric enzymes, a possible mechanism of inactivation includes the step of dissociation (dimer $\rightleftharpoons$ monomer) that precedes the irreversible loss of a monomeric enzyme (e.g., via protein denaturation):

$$E_2 \underset{k_{-1}}{\overset{k_1}{\rightleftharpoons}} 2E_1; \qquad E_1 \xrightarrow{k_D} E^*. \tag{5.187}$$

Approximate methods for determining the constants of scheme (5.187) were developed in application to a model enzymatic reaction [303–305]. Using the dependences of $\ln(r/r_0)$ on the time, where r and r_0 are the rates of this reaction under the conditions of substrate saturation at the current time t and $t = 0$, respectively, the positions of bending points, and the slopes of $\ln\frac{r}{r_0}$ *versus* t curves, it is possible to calculate the values of k_1, k_{-1}, and k_D (deactivation), or k_1, k_D and $K_{DIS} = k_1/k_{-1}$ with high accuracy. Approximate formulas were proposed that can be used to directly determine k_1, K_{DIS}, and k_D from the experimental data [302, 303].

If the mechanism of a process includes, in addition to the catalytic cycle, the steps of formation and decomposition of active centers and/or intermediates (i.e., a new noncyclic route) corresponding to a noncatalytic reaction (KG7 and KG8 in Figure 5.31), this process should be classified as a typical chain process. Let a simple submechanism consisting of steps **1** and **2** (KG7),

$$ML_m + S_1 \xrightarrow{\mathbf{1}} ML_n + S_2; \qquad ML_n \xrightarrow[S_3]{\mathbf{2}} +M^* + S_4,$$

with the overall equation

$$ML_m + S_1 + S_2 \longrightarrow M^* + S_3 + S_4 \tag{5.188}$$

be a source of active ML_n species in the catalytic cycle, the concentration of which is usually maintained on a quasi-steady-state level (exceeding the equilibrium concentration for the steps of ML_n decomposition). In this case, steps **1** and **2** are no longer "defects" of the mechanism (as in systems of classes I and II) and enter into one of the routes. Real catalytic processes with mechanisms possessing all features of the chain reactions (see Sections 5.3, 5.4, and 3.4) will be referred to as belonging to class III. The chain mechanism with linear steps can also be represented by a bicyclic kinetic graph with null-vertex, e.g., such as KG9 (Figure 5.31), which clearly illustrates the two-route structure of a simple chain mechanism. Under quasi-steady-state conditions with respect to $[ML_n]$ for KG7 (Figure 5.31), we have $W_1 = W_{init} = \omega_o[ML_n]$, where W_{init} is the rate of kinetic chain initiation (step **1**) and ω_o is the weight of chain termination step (step **2**). Then, $[ML_n] = W_{init}/\omega_o$ and the rate of chain propagation steps (**3** and **4**) is given by the formula

$$r = \frac{\omega_3 \omega_4 W_{init}}{\omega_{-3} + \omega_4}, \tag{5.189}$$

and the kinetic chain length γ can be expressed as

$$\gamma = r/W_{init}. \tag{5.190}$$

A finite length of the kinetic chain is the distinguishing feature of a chain process. A theoretically unlimited (infinite) kinetic chain length ($\gamma = \infty$) for the ideal catalytic cycle (KG1, Figure 5.31) and mechanisms of class I (KG2 and KG3) is what differentiates these processes from the chain processes of class III. Indeed, even in the case of a slow formation of active centers (KG2), the catalyst turnover number in the absence of chain termination steps can be arbitrarily large and, after the consumption of ML_m precursor species, the process becomes the ideal catalytic process. For this reason, it is impossible to accept the proposal of Purmal' [20] to consider the mechanism

$$A \longrightarrow A^* \longrightarrow B^* \longrightarrow \cdots \longrightarrow A^* + P, \tag{5.191}$$

(corresponding to KG2, Figure 5.31) as corresponding to the general scheme of chain mechanisms (with A^* and B^* representing intermediate compounds and P denoting the reaction products). Thus, it is the *combination of at least two routes*, cyclic and noncyclic (noncatalytic, representing the active center formation route) of class III, rather than "multiple repetition of a cycle" [306] that is characteristic of a chain reaction.

Within the framework of the approach to chain processes as presented here, the concepts of *chain process* and *catalytic process* are not treated as opposing each other. It was demonstrated above that real catalytic processes possess all features of the typical chain processes. At the same time, a free-radical chain process of liquid-phase oxidation, e.g., of aldehydes, is catalyzed by metal (Cu^{II}, Mn^{II}, Co^{II}) complexes, which act as catalysts of the steps of chain origination, degenerate branching, termination, and (in some cases) propagation [306].

The proximity of the mechanisms of real heterogeneous catalytic processes and those of chain reactions (the presence of steps of the origination and loss of active centers, which maintain a steady-state concentration of these centers) was discussed by Boudart [307] and Krylov [308] (see Section 3.4). It should be noted that most of the examples of reaction mechanisms encountered in the catalysis with metal complexes and mechanisms of chain reactions with nonradical carriers of kinetic chains, which were considered in this chapter, showed that chain reactions and mechanisms represent a special type of kinetic conjugation, which is invariant to the chemical nature of active centers and chain carriers. Therefore, it would be expedient to speak of the realization of many real catalytic processes via chain mechanisms rather than of the "proximity" or "analogy" [308] of the mechanisms of catalytic and chain processes.

If a block of the steps of regeneration of active species (e.g., step **5** that converts KG7 into KG10 in Figure 5.31) is added to a catalytic process with a chain mechanism (class III), the process becomes a multi-route "ideal" catalytic process (class IV). At the same time, under certain conditions (e.g., of material balance with respect to the catalyst) this process retains all the kinetic features of a chain process. In particular, the formation of an active center for route II takes place in the course of route I. For example, the rate (r_{II}) of route II is expressed with allowance for the steps of reversible chain initiation (**1**), active center decomposition (**2**), and quasi-equilibrium (**3**) as follows:

$$r_{II} = \frac{\omega_4 K_3 \omega_1 [ML_m]}{\omega_{-1} + \omega_2}. \tag{5.192}$$

For $[M]_\Sigma \cong [ML_m]$ ($[X_i]$, $[ML_n] \ll [M]_\Sigma$), the value of r_{II} is independent of the rate of step **5**, i.e., the passage from KG7 to KG10 does not change the form of kinetic relations. For example, if it were possible to find a selective oxidant capable of oxidizing Pd(0) to $PdCl_4{}^{2-}$ while not affecting $Pd_2Cl_4{}^{2-}$ (X_1) and use it in the isomerization of butenes in $PdCl_4{}^{2-}$ solutions (see Figure 5.17), then this isomerization process would become fully catalytic, while retaining all features of the chain process (with the same assumptions concerning the structure of material balance with respect to Pd). Under the above assumptions, the rate r_I of the route of ML_n formation

$$r_I = \frac{\omega_1 \omega_2 [ML_m]}{\omega_{-1} + \omega_2} \tag{5.193}$$

is independent of the steps of the second route and the kinetic chain length formally is $\gamma = \omega_4 K_3/\omega_2$.

If [M*] enters into the material balance with respect to the catalyst with a significant weight, the rate of the second route is described by the equation

$$r_{II} = \frac{\omega_1 \omega_4 K_3 \omega_5 [ML_m]_\Sigma}{(\omega_{-1} + \omega_2)\omega_5 + \omega_1 \omega_5 + \omega_1 \omega_2}, \tag{5.194}$$

which includes all steps of the routes (as in the case of any multi-route catalytic process). Under the same assumptions, the rate over route I is given by the formula

$$r_I = \frac{\omega_1 \omega_2 \omega_5 [ML_m]_\Sigma}{(\omega_{-1} + \omega_2)\omega_5 + \omega_1 \omega_5 + \omega_1 \omega_2}, \tag{5.195}$$

while the kinetic chain length is still expressed as $\gamma = \omega_4 K_3/\omega_2$.

The classical mechanism of acetylene hydration according to Kucherov in the commercial $HgSO_4$–H_2SO_4–$Fe_2(SO_4)_3$ catalytic system [309] can also be considered as belonging to Class IV. The mechanism of acetylene hydration in the $HgSO_4$–H_2SO_4 system can be described by the following simplified scheme:

$$HgSO_4 + C_2H_2 \xrightarrow[\mathbf{1}]{H_2O} X_i \xrightarrow[\mathbf{2}]{H_3O^+} HgSO_4 + CH_3CHO;$$

$$X_i \xrightarrow[\mathbf{3}]{H_2O} Hg(0)_{liquid} + CH_3COOH. \tag{5.196}$$

The formation of acetic acid takes place upon the redox decomposition of intermediate compound X_i, which is probably $(SO_4)^-$ $HgCH_2CHO$ [309]. Scheme (5.196) is represented by the kinetic graph KG of class II in Figure 5.33a. The lifetime of $HgSO_4$ catalyst at 90 °C in 10 % H_2SO_4 solution does not exceed

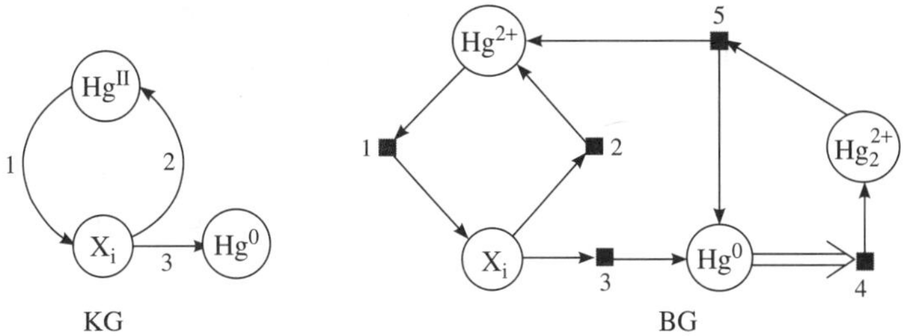

Figure 5.33 Kinetic (KG) and bipartite (BG) graphs corresponding to acetylene hydration: (a) KG of mechanism (5.196); (b) BG of subnetwork of intermediates of mechanism (5.195–5.197)

about 10 min. Since Fe (III) salt plays the role of oxidant for Hg(0), the $HgSO_4$–H_2SO_4–$Fe_2(SO_4)_3$ system acquires a block of steps featuring the regeneration of Hg(II) active species, which was shown by Flid and Moiseev [310, 311] to include the following reactions:

$$\begin{aligned} 2Hg(0)_{liquid} + 2Fe(III) &\xrightarrow[4]{} Hg_2SO_4 + 2Fe(II); \\ Hg_2SO_4 &\xrightarrow[5]{} HgSO_4 + Hg(0)_{liquid}. \end{aligned} \quad (5.197)$$

Since steps (5.197) are not linear, the mechanism of acetylene hydration via reactions (5.196) and (5.197) can be represented by a bipartite graph BG of subnetwork of intermediates (Figure 5.33b), which represents a mechanism of class IV. This graph contains three circuits, which correspond to three simple submechanisms and three overall equations for two routes (5.198) and (5.199) and one submechanism (5.200):

$$C_2H_2 + H_2O \longrightarrow CH_3CHO \quad (5.198)$$

[steps (1) (2)];

$$C_2H_2 + 2Fe(III) + 2H_2O \longrightarrow CH_3COOH + 2Fe(II) + H_2SO_4 \quad (5.199)$$

[steps (1)–(3)–(4)–(5)];

$$Hg(0)_{liquid} + 2Fe(III) \longrightarrow HgSO_4 + 2Fe(II) \quad (5.200)$$

[steps (4)–(5)].

If the rate of a hydration process according to BG (Figure 5.33b) is described as a function of [M*] (in this particular case, as a function of the surface area of liquid mercury ($Hg(0)_{liquid}$), the process on the whole can be considered as a chain process with a mechanism where the active $HgSO_4$ species are formed from Hg(0) precursor with a chain length of ~70–100 at 90 °C. However, if the hydration rate is described as a function of Hg(II) in the absence of a liquid mercury phase ($[HgSO_4]_\Sigma \gg [Hg(0)]$), then the kinetic model of the process will not differ from that of the usual two-route catalytic process. The consumption of Fe(III) decreases $E^o_{Fe(III)/Fe(II)}$ to such an extent that the oxidation of Hg(0) ceases and a liquid mercury phase ($Hg(0)_{liquid}$) appears. Since the lifetime of this catalyst [0.2 % $HgSO_4$, 20 % $Fe_2(SO_4)_3$, 10 % H_2SO_4]

at 90 °C amounts to approximately 3 h, it is necessary to perform continuous regeneration of the catalyst by oxidizing Fe(II) to Fe(III). Additives of β-thiodiglycol (TDG) ligands at an amount of ~0.1 M to this system lead to an additional 7- to 8-fold increase in the catalyst lifetime at about of 20 % decrease in the initial activity and about 5-fold growth in the total catalyst turnover number [309, 312]. The possible interaction of TDG ligands with the intermediate organomercuric compound (~$HgCH_2CHO$) probably stabilizes it with respect to the redox decomposition to CH_3COOH. The stability constant of $Hg(DTG)^{2+}$ complexes at 90 °C was estimated at $(11.2 \pm 2) \cdot 10^3$ [313].

It should be noted that another two processes, which were considered above, can also be assigned to the catalytic processes of class IV. These are the hydrocarboxylation of olefins in phosphine-free system in $PdBr_2$–THF solutions (see Figure 3.27) and the Heck reaction in phosphine-free system (Example 5.8, Figure 5.30b) with the steps of formation, decomposition, and regeneration of active centers.

References

1. G. K. Boreskov, *Zh. Fiz. Khim*., **32**, 2739 (1958).
2. G. K. Boreskov, *Kinet. Katal*., **13**, 543 (1972).
3. J. Falta, R. Imbihl and M. Henzler, *Phys. Rev. Lett*., **64**, 1409 (1990).
4. R. Krischer, M. Eiswirth and G. Ertl, *J. Chem., Phys*., **96** (12), 9161 (1992).
5. R. Koch, M. Sturmat and J. Schulz, *Surf. Sci*., **454**, 543 (2000).
6. M. I. Monine, L. M. Pismen and R. Imbihl, *J. Chem. Phys*., **121**, 11332 (2004).
7. M. I. Monine and L. M. Pismen, *Phys. Rev. E*, **66**, 051601 (2002); **69**, 021606 (2004).
8. L. A. Nikolaev, *Osnovy fizicheskoi khimii biologicheskikh protsessov* (Principles of Physical Chemistry in Biological Processes), Vysshaya Shkola, Moscow, 1971.
9. A. Lehninger, *Principles of Biochemistry*, Worth, New York, 1982.
10. A. Ya. Rozovskii, *Dokl. Akad. Nauk SSSR*, **151**, 1390 (1963).
11. A. Ya. Rozovskii, *Kinet. Katal*., **5**, 609 (1964).
12. A. Ya. Rozovskii, *Katalizator i reaktsionnaya sreda* (Catalyst and Reaction Medium), Nauka, Moscow, 1988.
13. M. J. Bennett and P. E. Donaldson, *Inorg. Chem*., **16** (3), 655 (1977).
14. G. A. Korneeva, M. P. Filatova, R. V. Chekhova et al., *Koord. Khim*., **18** (6), 636 (1992).
15. J. P. Collman, L. S. Hegedus, J. R. Norton and R. G. Finke, *Principles and Applications of Organotransition Metal Chemistry*, University Sci. Book, Mill Valley, CA, 1987.
16. G. B. Shul'pin, *Organicheskie reaktsii, kataliziruemye kompleksami metallov* (Organic Reactions Catalyzed by Metal Complexes), Nauka, Moscow, 1988.
17. O. A. Reutov, I. P. Beletskaya, G. A. Artamkina and A.I. Kashin, *Reaktsii metalloorganicheskikh soedinenii kak redoks-protsessy* (Reactions of Organometallic Compounds as Redox Processes), Nauka, Moscow, 1981.
18. O. N. Temkin, G. K. Shestakov and Yu. A. Treger, *Atsetilen: khimiya, mekhanizmy reaktsii, tekhnologiya* (Acetylene: Chemistry, Reaction Mechanisms, Technology), Khimiya, Moscow, 1991.
19. O. N. Temkin, S. M. Brailovskii and L. G. Bruk, *Tezisy dokladov II vsesoyuznoi konferentsii po mekhanizmam kataliticheskikh reaktsii* (Abstracts of Papers. All-Union Conf. on Mechanisms of Catalytic Reactions), 1978, Vol. 1, p. 74.
20. A. P. Purmal', in *Problemy khimicheskoi kinetiki* (Problems in Chemical Kinetics), Nauka, Moscow, 1979, p. 193.
21. K. J. Ivin and J. C. Mol, *Olefin Metathesis and Metathesis Polymerization*, Academic Press, London, 1997.
22. A. F. Shmidt, V. V. Smirnov, O. V. Starikova and A. V. Elaev, *Kinet. Katal*., **42** (2), 223 (2001); *Kinet. Catal*., **42**, 199 (2001).
23. A. F. Shmidt, *Sopryazhenie protsessov prevrashcheniya katalizatora i osnovnogo kataliticheskogo tsikla na primere reaktsii Kheka* (Conjugation of Catalyst Conversion and Main Catalytic Cycle in Heck Reaction), Doctoral Dissertation in Chemistry, Irkutsk State University, Irkutsk, 2003.
24. E. Balogh-Hergovich, G. Speier and L. Marko, *J. Organomet. Chem*., **66**, 303 (1974).

25. F. K. Shmidt, V. V. Saraev, L. O. Nindakova et al., *Kinet. Katal.*, **19**, 316 (1978).
26. F. K. Shmidt, *Kataliz kompleksami metallov pervogo perekhodnogo ryada reaktsii gidrirovaniya i dimerizatsii* (Catalysis with First Transition Series Metal Complexes in Hydrogenation and Dimerization Reactions), Irkutsk State University, Irkutsk, 1986.
27. I. I. Moiseev, *π-Kompleksy v zhidkofaznom okislenii olefinov* (π Complexes in Liquid-Phase Oxidation of Olefins), Nauka, Moscow, 1970.
28. R. Cramer, *Inorg. Synth.*, **15**, 14 (1974).
29. I. Wender, H. W. Sternberg and M. Orchin, *J. Am. Chem. Soc.*, **74** (5), 1216 (1952).
30. N. Yu. Kozitsyna and I. I. Moiseev, *Usp. Khim.*, **64** (1), 51 (1995).
31. V. B. Pukhnarevich, Yu. T. Struchkov, G. G. Aleksandrov et al., *Koord. Khim.*, **5** (10), 1537 (1979).
32. M. H. Chisholm and H. C. Clark, *J. Chem. Soc., Chem. Comm.*, 763 (1970).
33. J. P. Candlin, K. A. Taylor and D. T. Thompson, *Reactions of Transition-Metal Complexes*, Elsevier, New York, 1968.
34. B. R. James, *Inorg. Chim. Acta*, **4**, 73 (1970).
35. J. A. Osborn, F. H. Jardine, J. F. Young and G. Wilkinson, *J. Chem. Soc. A*, No. 12, 1711 (1966).
36. I. P. Beletskaya and A. V. Cheprakov, *Chem. Rev.*, **100**, 3009 (2000).
37. F. R. Hartley, *Inorg. Chim. Acta Rev.*, **5** (2), 197 (1971).
38. E. M. Badley, J. Chatt, R. L. Richards and G. A. Sim, *J. Chem. Soc., Chem. Comm.*, 1322 (1969).
39. E. Ban, P.-T. Cheno, T. Jack et al., J. *Chem. Soc., Chem. Comm.*, 368 (1973).
40. I. B. Bondarenko, N. A. Buzina, Yu. S. Varshavskii et al., *Zh. Neorg. Khim.*, **16** (11), 3071 (1961).
41. O. N Efimov and V. B. Panov, *Izv. Akad. Nauk SSSR, Ser. Khim.*, 491 (1970).
42. J. P. Collman, C. T. Sears and M. Kubota, *Inorg. Synth.*, **11**, 101 (1968).
43. M. W. Holtcamp, J. A. Labinger and J. E. Bercaw, *J. Am. Chem. Soc.*, **119** (4), 848 (1977).
44. T. P. Blatchford, M. H. Chisholm, K. Folting and J. C. Huffman, *J. Chem. Soc., Chem. Comm.*, 1295 (1984).
45. I. V. Oshanina, N. A. Kokoreva, L. G. Bruk, O. N. Temkin, I. L. Eremenko and E. L. Nefedov, *Izv. Akad. Nauk, Ser. Khim.*, 2274 (1998).
46. N. V. Lappo, K. I. Petrov, O. N. Temkin and R. M. Flid, *Zh. Neorg. Khim.*, **13** (6), 1710 (1968).
47. N. V. Lappo, O. N. Krasochka, L. O. Atovmyan and O. N. Temkin, unpublished results.
48. T. Murahashi, T. Nagai, T. Okuno, T. Matsutani and H. Kurosava, *J. Chem. Soc., Chem. Comm.*, 1689 (2000).
49. T. Murahashi and H. Kurosawa, *Coord. Chem. Rev.*, **231**, 207 (2002).
50. K. G. Powell and F. J. McQuillin, *J. Chem. Soc., Chem. Comm.*, 931 (1971).
51. G. A. Chukhadzhyan, Zh. I. Abramyan, V. P. Kukolev et al., *Arm. Khim. Zh.*, **31** (8), 607 (1978).
52. G. Allegra, A. Immirzi and L. Porri, *J. Am. Chem. Soc.*, **87**, 1394 (1965).
53. O. L. Kaliya, O. N. Temkin, R. M. Flid, Ya. M. Kimel'feld et al., *Izv. Akad. Nauk SSSR, Ser. Khim.*, 2854 (1969).
54. O. L. Kaliya, G. S. Kirchenkova, O. N. Temkin and R. M. Flid, *Kinet. Katal.*, **10**, 1186 (1969).
55. A. D. Burrows, D. M. P. Mingos, S. Menzer et al., *J. Chem. Soc., Dalton Trans.*, 2107 (1995).
56. S. J. Higgins and B. L. Shaw, *J. Chem. Soc., Dalton Trans.*, 457 (1988).
57. F. Klages and K. Mönkemeyer, *Chem. Ber.*, **85**, 109 (1952).
58. I. Kadota, L. M. Lutete, A. Shibuya and Y. Yamamoto, *Tetrahedron Lett.*, **42**, 6207 (2001).
59. T. Murahashi, T. Otani, T. Okuno et al., *Angew. Chem. Int. Ed.*, **39**, 537 (2000).
60. P. Braustein and N. M. Boag, *Angew. Chem. Int. Ed.*, **40**, 2427 (2001).
61. N. G. Mekhryakova, A. N. Nyrkova, N. F. Alekseev and O. N. Temkin, *Kinet. Katal.*, **30** (1), 98 (1989).
62. R. Griesser, B. Prijs and H. Sigel, *J. Am. Chem. Soc.*, **91** (27), 7758 (1969).
63. D. R. Coulson, *J. Chem. Soc., Chem. Comm.*, No. 23, 1530 (1968).
64. K. Kikukawa, T. Yamane, M. Takagi and T. Matsuda, *J. Chem. Soc., Chem. Comm.*, No. 11, 695 (1972).
65. T. G. Sukhova and O. N. Temkin, *Koord. Khim.*, **2** (11), 1540 (1976).
66. (a) A. D. Ryabov and A. K Yatsimirsky, *J. Mol. Catal.*, **4** (6), 449 (1978); (b) F. Y. Kwong, C. W. Lai, M. Yu and K. S. Chan, *Tetrahedron*, **60**, 5635 (2004).

67. N. Yu. Kozitsyna, A. M. Ellern, M. Yu. Antipin et al., *Mendeleev Comm.*, 92 (1991).
68. G. W. Parshall, *Accounts Chem. Res.*, **3**, 139 (1970).
69. A. D. Ryabov, *Metalloorg. Khim.* **1** (5), 994 (1988).
70. W. A. Herrmann, C. Brossmer, K. Ofele et al., *Angew. Chem. Int. Ed.*, **34**, 1844 (1995).
71. M. E. Van der Boom and D. Milstein, *Chem. Rev.*, **103**, 1759 (2003).
72. M. Dubeck, *J. Am. Chem. Soc.*, **82** (23), 6193 (1960).
73. E. W. Stern and P. K. Maples, *J. Catal.*, **27**, 120 (1972).
74. A. S. Berenblyum, L. I. Lakhman and M. L. Khidekel', *Izv. Akad. Nauk SSSR, Ser. Khim.*, 910 (1974).
75. A. S. Berenblyum, L. I. Lakhman, I. I. Moiseev and E. D. Radchenko, *Izv. Akad. Nauk SSSR, Ser. Khim.*, No. 11, 2417 (1975).
76. A. S. Berenblyum, L. I. Lakhman, I. I. Moiseev and E. D. Radchenko, *Koord. Khim.*, **2**, 841 (1976).
77. W. A. Herrmann, M. Elison, J. Fischer et al., *Angew. Chem. Int. Ed.*, **34**, 2371 (1995).
78. M. T. Reetz, G. Löhmer and R. Schwickardi, *Angew. Chem. Int. Ed.*, **37**, 481 (1998).
79. N. N. Korneev, A. F. Popov and B. A. Krentsel', *Kompleksnye metalloorganicheskie katalizatory* (Complex Organometallic Catalysts), Khimiya, Leningrad, 1969.
80. P. E. Matkovskii, *Radikal'nye stadii v reaktsiyakh kompleksnykh metalloorganicheskikh i metallotsenovykh katalizatorov i ikh rol' v polimerizatsii* (Radical Steps in Reactions of Organometallic and Metallocene Complex Catalysts and Their Role in Polymerization Processes), Inst. Probl. Khim. Fiz., Chernogolovka, 2003.
81. V. M. Akhmedov, M. T. Anthony, M. L. H. Green and D. Young, *J. Chem. Soc. Dalton Trans.*, 1412 (1975).
82. K. Nakata, T. Miyata and Y. Taniguchi, *J. Organomet. Chem.*, **489**, 71 (1995).
83. N. Yu. Kozitsyna, S. E. Nefedov, F. M. Dolgushin et al., *Inorg. Chim. Acta*, **359**, 2072 (2006).
84. T. Hosokawa, T. Nomura and S. I. Murahashi, *J. Organomet. Chem.*, **551**, 387 (1998).
85. O. L. Kaliya, *Issledovanie reaktsii oligomerizatsii i kataliticheskogo karbonilirovaniya atsetilena v rastvorakh kompleksov palladiya* (Investigation of Reactions of Acetylene Oligomerization and Catalytic Carbonylation in Solutions of Palladium Complexes), Candidate Dissertation in Chemistry, MITKhT (Moscow Institute of Fine Chemical Technology), Moscow, 1970.
86. N. G. Mekhryakova, *Issledovanie zakonomernostei i mekhanizma kataliticheskogo karbonilirovaniya atsetilena v rastvorakh kompleksov palladiya* (Investigation of the Laws and Mechanism of Catalytic Carbonylation of Acetylene in Solutions of Palladium Complexes), Candidate Dissertation in Chemistry, MITKhT (Moscow Institute of Fine Chemical Technology), Moscow, 1974.
87. N. G. Mekhryakova, L. G. Bruk, O. L. Kaliya, O. N. Temkin and A. Yu. Prudnikov, *Kinet. Katal.*, **20**, 629 (1979).
88. A. N. Nyrkova, O. L. Kaliya, O. N. Temkin, R. M. Flid, and N. K. Vorob'eva, *Zh. Neorg. Khim.*, **20**, 3017 (1975).
89. A.N. Nyrkova, O. L. Kaliya and O. N. Temkin, *Koord. Khim.*, **5** (11), 1683 (1979).
90. A. N. Nyrkova, L. G. Kuz'mina, Yu. T. Struchkov and O. N. Temkin, *Koord. Khim.*, **8** (1), 82 (1982).
91. A. N. Nyrkova, G. A. Kukina, O. N. Temkin and M. A. Porai-Koshits, *Koord. Khim.*, **11** (7), 954 (1985)
92. A. N. Nyrkova, *Issledovanie protsessov kompleksoobrazovaniya i mekhanizma formirovaniya katalizatora karbalkoksilirovaniya atsetilena v sisteme PdBr2–P(OPh)3–n-C4H9OH* (Investigation of Complex Formation Processes and Mechanism of Catalyst Formation for Acetylene Carbalkoxylation in PdBr2–P(OPh)3–*n*-C4H9OH System), Candidate Dissertation in Chemistry, MITKhT (Moscow Institute of Fine Chemical Technology), Moscow, 1982.
93. O. N. Temkin, O. L. Kaliya, R. M. Flid and N. G. Mekhryakova, *Dokl. Akad. Nauk SSSR*, **199**, 1321 (1971).
94. O. N. Temkin, O. L. Kaliya, L. N. Zhir-Lebed', L. G. Bruk and N. G. Mekhryakova, in *Gomogennoe okislenie (Trudy Instituta organicheskogo kataliza i elektrokhimii AN Kaz. SSR)* [Homogeneous Oxidation (Scientific Works of Institute of Organic Catalysis and Electrochemistry, Acad. Sci. Kazakh SSR)], Nauka, Alma-Ata, 1978, Vol. 17, pp. 3–33.
95. K. H. Dötz, H. Fischer, P. Hofmann, F. R. Kreissl, U. Schubert and K. Weiss, *Transition Metal Carbene Complexes*, Verlag Chemie, Deerfield Beach, FL, 1984.

96. M. P. Doyle, *Chem. Rev*., **86**, 919 (1986).
97. I. I. Moiseev, *Usp. Khim*., **58** (7), 1175 (1989).
98. O. M. Nefedov, A. I. Ioffe and L. G. Menchikov, *Khimiya karbenov* (Chemistry of Carbenes), Khimiya, Moscow, 1990.
99. K. J. Ivin, *Olefin Metathesis*, Academic Press, London, 1983.
100. *Metathesis Chemistry: From Nanostructure Design to Synthesis of Advanced Materials* (NATO Science Series II: Mathematics, Physics and Chemistry, Vol. 243), Springer, Berlin, 2007.
101. *Handbook of Metathesis*, Vols. 1–3, R. H. Grubbs (Ed.), Wiley–VCH, New York–Weinheim, 2003.
102. *Ruthenium Catalysts in Fine Chemistry*, C. Bruneau and P. H. Dixneuf (Eds) (Topics in Organometallic Chemistry, Vol. 11), Springer, Berlin, 2004.
103. P. E. Romero, W. E. Piers and R. McDonald, *Angew. Chem. Int. Ed*., **43**, 6161 (2004).
104. J. L. Herisson and I. Chauvin, *Makromol. Chem*., **141**, 161 (1971).
105. B. A. Dolgoplosk, K. L. Makovetskii and E. I. Tinyakova, *Dokl. Akad. Nauk SSSR*, **202**, 871 (1972).
106. C. P. Casey and T. J. Burkhardt, *J. Am. Chem. Soc*., **96**, 7808 (1974).
107. T. J. Katz and J. McGinnis, *J. Am. Chem. Soc*., **97**, 1592 (1975).
108. N. B. Bespalova, M. A. Bovina, A. V. Popov and J. C. Mol, *J. Mol. Catal. A: Chem*., **160** (1), 157 (2000).
109. N. S. Nametkin, V. M. Vdovin, E. D. Babich et al., *Dokl. Akad. Nauk SSSR*, **225**, 573 (1975).
110. A. K. Chatterjec, T.-L. Choi, D. P. Sanders and R. H. Grubbs, *J. Am. Chem. Soc*., **125**, 11360 (2003).
111. J. R. Sharpley, S. J. Richter, M. Tachikawa and J. B. Keister, *J. Organomet. Chem*., **94**, C43 (1975).
112. R. H. Grubbs and A. Miyashita, in *Fundamental Research in Homogeneous Catalysis*, M. Tsutsui (Ed.), Plenum Press, New York, 1979, Vol. 3, p. 151.
113. R. H. Grubbs and S. J. Swetnick, *J. Organomet. Chem*., **8**(1–3), 25 (1980).
114. W. S. Greenlee and M. F. Farona, *Inorg. Chem*., **15** (9), 2129 (1976).
115. N. B. Bespalova, *Metallokompleksnye kremniisoderzhashchie kataliticheskie sistemy dlya reaktsii metatezisa olefinov* (Metal-Complex Silicon-Containing Catalytic Systems of Olefin Metathesis Reactions), Candidate Dissertation in Chemistry, Topchiev Institute of Petrochemical Synthesis, Moscow, 2003.
116. K. L. Makovetskii, B. A. Dolgoplosk and Yu. V. Korshak, *Zh. Vses. Khim. o-va im. D.I. Mendeleeva*, **24**, 513 (1974).
117. T. Saegusa, Y. Ito, T. Shimizu and S. Kobayashi, *Bull. Chem. Soc. Jpn*., **42**, 3535 (1969).
118. R. Grubbs et al., US Patent No. 0022733 A1 (2002).
119. G. J. A. Adam, S. G. Davies, K. A. Ford et al., *J. Organomet. Chem*., **8**(1–3), 15 (1980).
120. M. I. Bruce and A. G. Swincer, *Adv. Organomet. Chem*., **59** (1983).
121. A. B. Antonova and A. A. Iogansen, *Usp. Khim*., **58**, 1197 (1989).
122. H. Werner, *Angew. Chem. Int. Ed*., **29**, 1077 (1990).
123. M. I. Bruce, *Chem. Rev*., **91**, 197 (1991).
124. R. M. Bullock, *J. Chem. Soc., Chem. Comm*., 165 (1989).
125. D. Schneider and H. Werner, *Angew. Chem*., **103** (6), 710 (1991).
126. H. Le Bozec, D. Devanne and P. H. Dixneuf, in *Advances in Metal Carbene Chemistry*, U. Schubert (Ed.), Kluwer Acad. Press, 1989, pp. 107–121.
127. J. L. Davidson, M. Shiralian, L. Manojlovic-Muir and K. Muir, *J. Chem. Soc., Chem. Comm*., 30 (1979).
128. G. C. Conole, M. Green, M. McPartlin et al., *J. Chem. Soc., Chem. Commum*., 1310 (1979).
129. F. J. Feher, M. Green and R. A. Rodriges, *J. Chem. Soc., Chem. Comm*., 1206 (1987).
130. S. G. Feng, A. S. Gamble and J. L. Templeton, *Organometallics*, **8**, 2024 (1989).
131. C. W. K. Gstöttmayr and P. A. Van der Schaaf, *Proceedings of the 14th Int. Symp. on Homogeneous Catalysis (Munich, Germany, 2004)*, Book of Abstracts, p. 475.
132. P. O. Nubel and C. L. Hunt, *J. Mol. Catal. A: Chem*., **145**, 323 (1999).
133. C. Van Schalkwyk, H. C. M. Vosloo and J. M. Botha, *J. Mol. Cat. A: Chem*., **190**, 185 (2002).
134. J. P. Morgan and R. H. Grubbs, *Org. Lett*., **2**, 3153 (2000).

135. J. Louie and R. H. Grubbs, *Angew. Chem. Int. Ed.*, **40**, 247 (2001).
136. M. F. Carvalho, R. A. Henderson, A. J. L. Pombeiro and R. L. Richards, *J. Chem. Soc., Chem. Comm.*, 1796 (1989).
137. A. H. Hoveida, D. G. Gillingham, J. J. Van Veldhuizen et al., *Org. Biomol. Chem.*, **2**, 8 (2004).
138. M. Ulman and R. H. Grubbs, *J. Org. Chem.*, **64**, 7202 (1999).
139. M. S. Sanford, J. A. Love and R. H. Grubbs, *J. Am. Chem. Soc.*, **123**, 6543 (2001).
140. S. H. Hong, M. Day and R. H. Grubbs, *J. Am. Chem. Soc.*, **126**, 7414 (2004).
141. M. B. Dinger and J. C. Mol, *Organometallics*, **22**, 1089 (2003).
142. M. B. Dinger and J. C. Mol, *Eur. J. Inorg. Chem.*, **2827** (2003).
143. D. Banti and J. C. Mol, *J. Organomet. Chem.*, **689**, 3113 (2004).
144. W. J. Van Rensburg, P. J. Steynberg, W. H. Meyer et al., *J. Am. Chem. Soc.*, **126**, 14332 (2004).
145. (a) A. V. Nizovtsev and N. B. Bespalova, *Proceedings of the 16th Int. Symp. on Olefin Metathesis and Related Chemistry* (Poznan, Poland, 2005), Book of Abstracts, p. 98; (b) A. V. Nizovtsev, V. V. Afanasiev, E. V. Shutko and N. B. Bespalova, in *Metathesis Chemistry: From Nanostructure Design to Synthesis of Advanced Materials* (NATO Science Series II: Mathematics, Physics and Chemistry, Vol. 243), Springer, 2007, pp. 125–135.
146. D. E. Fogg, D. Amoroso, S. D. Druin et al., *J. Mol. Catal. A: Chem.*, **190**, 177 (2002).
147. C. A. Tolman, *J. Am. Chem. Soc.*, **94**, 2994 (1972).
148. C. A. Tolman and J. W. Faller, in *Homogeneous Catalysis with Metal Phosphine Complexes*, L. H. Pignolet (Ed.), New York,Plenum Press, 1983, pp. 13–110.
149. W. Peppe, *Lieb. Ann.*, **582**, 1 (1953).
150. J. E. Ehrreich, R. G. Nickerson and C.E. Ziegler, *Ind. Eng. Chem. Proc. Des. Dev.*, **4**, 77 (1965)
151. H. C. Clark and R. K. Mittal, *Can. J. Chem.*, **51**, 1511 (1973).
152. S. Kinichika, Y. Sakakibara and T. Nakamura, *Bull. Chem. Soc. Jpn.*, **41**, 390 (1968).
153. Yu. B. Kagan, A. Ya. Rozovskii, E. V. Slivinskii et al., *Kinet. Katal.*, **28** (6), 1508 (1987).
154. G. A. Korneeva, *Gidroformilirovanie nepredel'nykh uglevodorodov i kislorodsoderzhashchikh soedinenii v prisutstvii kompleksov kobal'ta i rodiya* (Hydroformylation of Unsaturated Hydrocarbons and Oxygen-Containing Compounds in the Presence of Cobalt and Rhodium Complexes), Candidate Dissertation in Chemistry, Topchiev Institute of Petrochemical Synthesis, Moscow, 2003.
155. A. R. El'man, V. I. Kurkin, E. V. Slivinskii and S. M. Loktev, *Neftekhimiya*, **30** (1), 46 (1990).
156. J. Tsuji, M. Takahashi and T. Takahashi, *Tetrahedron Lett.*, **21**, 849 (1980).
157. L. G. Bruk and O. N Temkin, *Inorg. Chim. Acta*, **280**, 202 (1998).
158. V. R. Khabibulin, A. V. Kulik, I. V. Oshanina et al., *Kinet. Katal.*, **48** (2), 1 (2007).
159. L. N. Zhir-Lebed', L. G. Kuz'mina, Yu. T. Struchkov, O. N. Temkin and V. A. Golodov, *Koord. Khim.*, **4**, 1046 (1978).
160. F. Rivetti and U. Romano, *J. Organomet. Chem.*, **154**, 323 (1978).
161. G. Cavinato, L. Toniolo and A. Vavasori, *J. Mol. Catal. A: Chem.*, **219** (2), 233 (2004).
162. A. V. Kulik, L. G. Bruk, O. N. Temkin, V. R. Khabibulin, V. K. Bel'skii and V. E. Zavodnik, *Mendeleev Comm.*, 47 (2002).
163. S. Takahashi and N. Hagihara, *J. Chem. Soc. Jpn. (Pure Chem. Sect.)*, **88**, 1306 (1967).
164. P. Roffia, G. Gregorio, F. Conti, G. F. Pregaglia and R. Ugo, *J. Mol. Catal.*, **2**, 191 (1977).
165. B. Milani, A. Anzilutti, L. Vicentini et al., *Organometallics*, No. 16, 5064 (1997).
166. O. N. Temkin and L. G. Bruk, *Kinet. Katal.*, **44** (5), 661 (2003).
167. I. I. Moiseev and S.V. Pestrikov, *Dokl. Akad. Nauk SSSR*, **171**, 151 (1966).
168. I. I. Moiseev, S. V. Pestrikov and S. V. Sverzh, *Kinet. Katal.*, **10**, 74 (1969).
169. J. Davidson and C. Triggs, *J. Chem. Soc. A*, 1324 (1968).
170. G. P. Chiusoli, C. Venturello and S. Merzoni, *Chem.& Ind.*, 977 (1968).
171. O. N. Temkin and L. G. Bruk, *Usp. Khim.*, **52**, 206 (1983).
172. I. I. Moiseev, *Kataliz kompleksami so svyaz'yu Pd–Pd, in: Itogi nauki i tekhniki: Kinetika i kataliz* (Advances in Science and Technology: Kinetics and Catalysis), VINITI, Moscow, 1984, Issue 13, p. 147.
173. I. I. Moiseev and M. N. Vargaftik, *Usp. Khim.*, **39** (12), 1931 (1990).

174. I. I. Moiseev and M. N. Vargaftik, in *Catalysis by Di- and Polynuclear Metal Complexes*, F. A. Cotton and R. L. Adams (Eds), Wiley–VCH, New York, 1998, Issue 4, p. 395.
175. T. A. Stromnova and I. I. Moiseev, *Usp. Khim.*, **67**, 542 (1998).
176. L. G. Bruk, O. N. Temkin, Z. V. Goncharova, I. S. Ustenko and V. R. Flid, *React. Kinet. Catal. Lett.*, **9** (3), 303 (1978).
177. L. G. Bruk and O. N. Temkin, *Khim. Promst*, 57 (1993).
178. L. G. Bruk, N. Yu. Vsesvyatskaya, T. A. Stromnova, N. F. Alekseeva and O. N. Temkin, *Zh. Org. Khim.*, **14**, 473 (1978).
179. L. G. Bruk, *Metody vydvizheniya i diskriminatsii gipoteticheskikh mekhanizmov kataliticheskikh reaktsii kak osnova razrabotki tekhnologii karbonilirovaniya alkinov* (Methods of Generation and Discrimination of Hypothetical Mechanisms of Catalytic Reactions: Basis for Developing Alkyne Carbonylation Technology), Doctoral Dissertation in Chemistry, MITKhT (Moscow Institute of Fine Chemical Technology), Moscow, 1996.
180. O. N. Temkin, L. G. Bruk, and A.V. Kulik, unpublished results.
181. B. Gabriele, M. Costa, G. Salern and G. P. Chiusoli, *J. Chem. Soc., Perkin Trans.*, No. 1, 83 (1994).
182. A. V. Malashkevich, L. G. Bruk and O. N. Temkin, *J. Phys. Chem. A.*, **101**, 9825 (1997).
183. S. N. Gorodskii, A. N. Zakharov, A. V. Kulik, L. G. Bruk and O. N. Temkin, *Kinet. Katal.*, **42** (2), 280 (2001).
184. S. N. Gorodskii, E. S. Kalenova, L. G. Bruk and O. N. Temkin, *Izv. Akad. Nauk, Ser. Khim.*, No. 7, 1452 (2003).
185. W. Manchot and J. König, *Chem. Ber.*, **59**, 883 (1926).
186. A. D. Gel'man and E. Meilakh, *Dokl. Akad. Nauk SSSR*, **36**, 188 (1942).
187. R. Colton, M. J. McCormick and C. D. Pannan, *J. Chem. Soc., Chem. Comm.*, 823 (1977).
188. I. I. Moiseev, T. A. Stromnova, M. N. Vargaftik et al., *J. Chem. Soc., Chem. Comm.*, 27 (1978).
189. P. L. Goggin, R. J. Goodfellow, I. R. Herbert, A. G. Orpen, *J. Chem. Soc., Chem. Comm.*, 1077 (1981).
190. A. V. Kulik, L. G. Bruk, O. N. Temkin et al., *Dokl. Akad. Nauk*, **42**, 218 (2002).
191. L. N. Zhir-Lebed' and O. N. Temkin, *Kinet. Katal.*, **25** (2), 325 (1984).
192. L. N. Zhir-Lebed' and O. N. Temkin, *Kinet. Katal.*, **25** (2), 316 (1984).
193. S. N. Gorodskii, L. G. Bruk and O. N. Temkin, unpublished data (2005).
194. O. N. Temkin and L. G. Bruk, Oxidative Carbonylation: Homogeneous, in *Encyclopedia of Catalysis*, I. T. Horvath (Ed.), Wiley, Hoboken, 2003, Vol. 5, pp. 394–424.
195. L. G. Bruk and O. N. Temkin, *Proceedings of the 9th Soviet-Japan Seminar on Catalysis (Novosibirsk, 1990)*, p. 83.
196. L. G. Bruk, I. V. Oshanina, A. P. Kozlova, E. V. Vorontsov and O. N. Temkin, *J. Mol. Catal.*, **104**, 9 (1995).
197. L. G. Bruk, I. V. Oshanina, A. S. Zakieva and O. N. Temkin, *Kinet. Katal.*, **39** (2), 183 (1998).
198. L. G. Bruk, I. V. Oshanina, A. P. Kozlova, O. N. Temkin and K. Yu. Odintsov, *Izv. Akad. Nauk, Ser. Khim.*, No. 6, 1104 (1998).
199. A.V. Kulik, O. N. Temkin, L. G. Bruk et al., *Izv. Akad. Nauk, Ser. Khim.*, No. 6, 1349 (2005).
200. R. Trebbe, R. Goddard, A. Rufinska et al., *Organometallics*, **18**, 2466 (1999).
201. A. B. Fasman, G. G. Kutyukov and D. V. Sokol'skii, *Dokl. Akad. Nauk SSSR*, **158**, 1176 (1964).
202. C. Amatore, E. Carre, A. Jutland et al., *Organometallics*, **14**, 5605 (1995).
203. M. Beller and A. F. Indolese, *Chimia*, **55**, 684 (2001).
204. (a) S. P. Gubin, *Khimiya klasterov* (Chemistry of Clusters), Nauka, Moscow, 1987; (b) *Metal Clusters in Chemistry*, P. Braunstein, L. A. Oro and P. R. Raithby (Eds), Wiley–VCH, Weinheim, 1999, Vol. 1.
205. G. Longoni and P. Chini, *J. Am. Chem. Soc.*, **98**, 7225 (1976).
206. P. Chini, *J. Organomet. Chem.*, **200**, 37 (1980).
207. G. Ciani, A. Magni, A. Sironi and S. Martinengo, *J. Chem., Soc., Chem. Comm.*, 1280 (1981).
208. E. G. Mednikov, *Izv. Akad. Nauk SSSR, Ser. Khim.*, **42** (7), 1299 (1993).
209. E. G. Mednikov, N. K. Eremenko, Yu. L. Slovokhotov and Yu. T. Struchkov, *J. Chem. Soc., Chem. Comm.*, 218 (1987).
210. E. G. Mednikov and N. I. Kanteeva, *Izv. Akad. Nauk, Ser. Khim.*, **44** (1), 167 (1995).
211. A. Ceriotti, F. Demartin and G. Longoni, *Angew. Chem. Int. Ed.*, **24**, 697 (1985).

212. G. Schmid, R. Pfeil, R. Boese et al., *Chem. Ber.*, **114**, 3634 (1981).
213. I. R. Stolyarov, Yu. V. Gaugash, G. N. Kryukova et al., *Izv. Akad. Nauk, Ser. Khim.*, **53** (6), 1147 (2004).
214. N. T. Tran, D. R. Powell and L. F. Dahl, *Angew. Chem. Int. Ed.*, **39**, 4121 (2000).
215. E. G. Mednikov, M. C. Jewell and L. F. Dahl, *J. Am. Chem. Soc.*, **129**, 11619 (2007).
216. G. Schmid, *Polyhedron*, **7**, 2321 (1988).
217. J. A. Wedegren and R. G. Finke, *J. Mol. Catal. A: Chem.*, **198**, 317 (2003).
218. G. Schmid, B. Morun and J.-O. Malm, *Angew. Chem.*, **101**, 772 (1989).
219. M. T. Reetz and E. Westermann, *Angew. Chem. Int. Ed.*, **39**, 165 (2000).
220. M. T. Reetz and M. Maase, *Adv. Mater.*, **11**, 773 (1999).
221. S.-W. Kim, M. Kim, W. Y. Lee and T. Hyeon, *J. Am. Chem. Soc.*, **124**, 7643 (2002).
222. N. T. S. Phan, M. Van der Sluys and C. W. Jones, *Adv. Synth. Catal.*, **348**, 609 (2006).
223. L. N. Lewis, *Chem. Rev.*, **93**, 2693 (1993).
224. H. Bönnemann and W. Brijoux, *Nanostruct. Mater.*, **5**, 135 (1995).
225. H. Bönnemann, W. Brijoux, R. Brinkmann et al., *Angew. Chem. Int. Ed.*, **30**, 1312 (1991).
226. M. T. Reetz, R. Breinbauer and K. Wanniger, *Tetrahedron Lett.*, **37**, 4499 (1996).
227. M. B. Thathagar, J. Beckers and G. Rothenberg, *J. Am. Chem. Soc.*, **124**, 11858 (2002).
228. A. S. Berenblyum, A. G. Knizhnik, S. L. Mund and I. I. Moiseev, *Izv. Akad. Nauk SSSR*, *Ser. Khim.*, No. 12, 2711 (1978).
229. A. S. Berenblyum, A. G. Knizhnik, S. L. Mund and I. I. Moiseev, *Izv. Akad. Nauk SSSR*, *Ser. Khim.*, No. 8, 1687 (1979).
230. A. S. Berenblyum, A. G. Knizhnik, S. L. Mund and I. I. Moiseev, *Izv. Akad. Nauk SSSR*, *Ser. Khim.*, No. 9, 2157 (1979).
231. A. S. Berenblyum, S. L. Mund, L. G. Danilova and I. I. Moiseev, *Izv. Akad. Nauk SSSR, Ser. Khim.*, 906 (1981).
232. A. S. Berenblyum, S. L. Mund, T. P. Goranskaya and I. I. Moiseev, *Izv. Akad. Nauk SSSR, Ser. Khim.*, 2472 (1981).
233. A. G. Knizhnik, *Mekhanizm formirovaniya klasterov palladiya, kataliziruyushchikh gidrirovanie nepredel'nykh soedinenii* (Mechanism of Formation of Palladium Clusters that Catalyze Hydrogenation of Unsaturated Hydrocarbons), Candidate Dissertation in Chemistry, Kurnakov Institute of General and Inorganic Chemistry, Moscow, 1981.
234. V. A. Khutoryanskii, T. I. Bakunina, S. V. Zinchenko et al., *Koord. Khim.*, **13**, 809 (1987).
235. T. I. Bakunina, *Bis-atsegilatsetonatnye kompleksy palladiya v kataliticheskoi reaktsii gidrirovaniya* (Bis-Acetylacetonate Complexes of Palladium in Catalytic Hydrogenation Reaction), Candidate Dissertation in Chemistry, Irkutsk Institute of Organic Chemistry, Irkutsk, 1988.
236. T. I. Bakunina, S. V. Zinchenko, V. A. Khutoryanskii et al., in *Metallokompleksnyi kataliz (Sbornik trudov)* [Catalysis with Metal Complexes (Collection of Scientific Works)], Irkutsk,Izdat. IrGU, 1989, p. 132.
237. I. P. Stolyarov, M. N. Vargaftik, O. M. Nefedov and I. I. Moiseev, *Kinet. Katal.*, **23**, 376 (1982).
238. L. B. Belykh, F. K. Shmidt, O. V. Burlakova et al., *Koord. Khim.*, **16** (11), 1549 (1990).
239. L. B. Belykh, F. K. Shmidt, T. V. Dmitrieva and S. V. Zinchenko, *Koord. Khim.*, **18** (6), 642 (1992).
240. L. B. Belykh, T. V. Cherenkova and F. K. Shmidt, *Koord. Khim.*, **25** (6), 446 (1999) [*Russ. J. Coord. Chem.*, **25**, 418 (1999)].
241. F. K. Shmidt, L. B. Belykh and T. V. Cherenkova, *Kinet. Katal.*, **42** (2), 182 (2001) [*Kinet. Catal.*, **42**, 163 (2001)].
242. F. K. Shmidt, L. B. Belykh and T. V. Goremyka, *Kinet. Katal.*, **44** (5), 683 (2003) [*Kinet. Catal.*, **44**, 623 (2003)].
243. L. B. Belykh, *Formirovanie, priroda aktivnosti i svoistva nanorazmernykh katalizatorov gidrirovaniya na osnove kompleksov palladiya s fosfororganicheskimi ligandami* (Formation, Nature of Activity, and Properties of Nanodimensional Hydrogenation Catalysts Based Palladium Complexes with Organophosphor Ligands), Doctoral Dissertation in Chemistry, Irkutsk State University, Irkutsk, 2005.
244. L. B. Belykh, T. V. Goremyka, N. I. Skripov et al., *Kinet. Katal.*, **47** (3), 373 (2006) [*Kinet. Catal.*, **47**, 367 (2006)].

245. F. K. Schmidt, L. O. Nindakova, B. A. Shainyan et al., *J. Mol. Catal. A: Chem.*, **235**, 161 (2005).
246. B. Schuldt, D. Heller, F. Storz and K. Madeja, *J. Mol. Catal.*, **81**, 195 (1993).
247. F. Storz, D. Heller and K. Madeja, *J. Mol. Catal.*, **84**, 33 (1993).
248. L. N. Lewis, in *Catalysis by Di- and Polynuclear Metal Cluster Complexes*, R. D. Adams and F. A. Cotton (Eds), Wiley–VCh, New York, 1998, p. 373.
249. H. Bönnemann and R. Richards, in *Nanofabrication Toward Biomedical Applications*, C. S. S. R. Kumar, J. Hormes and C. Leuschner (Eds), Wiley–VCH, New York, 2005, pp. 1–32.
250. M. N. Vargaftik, V. P. Zagorodnikov, I. P. Stolyarov et al., *Izv. Akad. Nauk SSSR, Ser. Khim.*, 2381 (1985).
251. G. Schmid, M. Harms, J.-O. Malm et al., *J. Am. Chem. Soc.*, **115**, 2046 (1993).
252. Y. Lin and R. G. Finke, *Inorg. Chem.*, **33**, 4891 (1994).
253. J. D. Aikin and R. G. Finke, *J. Am. Chem. Soc.*, **120**, 9545 (1998).
254. V. I. Parvulesku and C. Hardacre, *Chem. Rev.*, **107**, 2615 (2007).
255. I. P. Beletskaya, A. N. Kashin, N. B. Karlstedt et al., *J. Organomet. Chem.*, **622**, 89 (2001).
256. A. H. M. De Vries, J. M. C. A. Mulders, J. H. M. Mommers et al., *Org. Lett.*, **5** (18), 3285 (2003).
257. R. K. Arvela, N. E. Leadbeater, M. S. Sangi et al., *J. Org. Chem.*, **70**, 161 (2005).
258. A. F. Schmidt, A. Al-Halaiqa and V. V. Smirnov, *J. Mol. Catal. A: Chem.*, **250** (5), 131 (2006).
259. A. F. Shmidt, A. Al-Halaiqa and V. V. Smirnov, *Kinet. Katal.*, **49** (3), 416 (2008) [*Kinet. Catal.*, **49**, 395 (2008)].
260. P. Meakin, *Adv. Colloid Interface Sci.*, **28**, 249 (1988).
261. M. Y. Lin, H. M. Lindsay, D. A. Weitz et al., *Nature*, **339**, 360 (1989).
262. G. S. Fonseca, J. B. Domingosa, F. Nobeb and J. Dupont, *J. Mol. Catal. A*, **248**, 10 (2006).
263. J. A. Widegren, J. D. Aiken, S. Ozkar and R. G. Finke, *Chem. Mater.*, **13**, 312 (2001).
264. J. A. Widegren, M. A. Bennett and R. G. Finke, *J. Am. Chem. Soc.*, **125**, 10301 (2003).
265. I. P. Stolyarov, M. N. Vargaftik and I. I. Moiseev, *Kinet. Katal.*, **28**, 1359 (1987).
266. M. N. Vargaftik, V. P. Zagorodnikov, I. P. Stolyarov et al., *J. Mol. Catal.*, **53**, 315 (1989).
267. L. Djakovitch, M. Wagner, C. G. Hartung, M. Beller and K. Koehler, *J. Mol. Catal. A: Chem.*, **219**, 121 (2004).
268. J. H. Baxendale and J. A. Wilson, *Trans. Soc.*, **53** (2), 344 (1957).
269. N. M. Emanuel', G. E. Zaikov and V. A. Kritsman, *Tsepnye reaktsii. Istoricheskii aspect* (Chain Reactions: A Historical Aspect), Nauka, Moscow, 1989, p. 230.
270. V. M. Berdnikov and Yu. N. Kozlov, *Zh. Strukt. Khim.*, **47**, 598 (1973).
271. E. V. Shtamm, A. P. Purmal' and Yu. I. Skurlatov, *Zh. Fiz. Khim.*, **48**, 1454, 2229 (1974).
272. D. Astruc, *Chem. Rev.*, **88**, 1189 (1988).
273. M.-H. Desbois and D. Astruc, *New J. Chem.*, **13**, 595 (1989).
274. M. Chanon, *Bull. Soc. Chim. Fr.*, 197 (1985); 209 (1982).
275. J. K. Kochi, *J. Organometal. Chem.*, **300**, 139 (1986).
276. L. Eberson, *Electron Transfer Reactions in Organic Chemistry*, Springer, Berlin, 1987, Ch. 10.
277. D. Astruc, *Accounts Chem. Res.*, **24**, 36 (1991).
278. J. F. Bunnett, *Accounts Chem. Res.*, **11**, 413 (1978).
279. J. J. Bozell and C. E. Vogt, *J. Am. Chem. Soc.*, **110**, 2655 (1988).
280. D. S. Olson and D. W. Margerum, *J. Am. Chem. Soc.*, **85**, 297 (1963).
281. V. Yu. Gankin and G. S. Gurevich, *Tekhnologiya oksosinteza* (Oxo Synthesis Technology), Khimiya, Leningrad, 1981.
282. A. F. Schmidt and V. V. Smirnov, *J. Mol. Cat.*, **203**, 75 (2003).
283. A. F. Schmidt and V. V. Smirnov, *Topics Catal.* **32**(1–2), 71 (2005).
284. S. S. Pröckl, W. Kleist and K. Kohler, *Tetrahedron*, **61**, 9855 (2005).
285. C. S. Consorti, F. R. Flores and J. Dupont, *J. Amer. Chem. Soc.*, **127**, 12054 (2005).
286. A. F. Schmidt, A. Al-Halaiqa and V. V. Smirnov, *Synlett*, No. 18, 2861 (2006).
287. K. Koehler, S. S. Proeckl and W. Kleist, *Current Org. Chem.*, **10** (13), 1585 (2006).
288. J. G. De Vries, *Dalton Trans.*, 421 (2006).
289. K. Koehler, W. Kleist and S. S. Proeckl, *Inorg. Chem.*, **46** (6), 1876 (2007).
290. D. Astruc, *Inorg Chem.*, **46** (6), 1884 (2007).

291. A. F. Shmidt and A. Al-Halaiqa, *Kinet. Katal.*, **39** (6), 875 (1998) [*Kinet. Catal.*, **39**, 803 (1998)].
292. N. Miyaura, T. Yanagi and A. Suzuki, *Synth. Comm.*, **11**, 513 (1981).
293. A. A. Kurokhtina and A. F. Schmidt, *Arkivoc*, **11**, 185 (2009).
294. P. J. Ellis, I. J. S. Fairlamb, S. F. J. Hackett, K. Wilson and A. F. Lee, *Angew. Chem. Int. Ed.*, **49** (10), 1820 (2010).
295. A. F. Lee, P. J. Ellis, I. J. S. Fairlamb and K. Wilson, *Dalton Trans.*, **39**, (43), 10473 (2010).
296. A. F. Shmidt, A. Al-Halaiqa and V. V. Smirnov, *Kinet. Katal.*, **48**, 716 (2007) [*Kinet. Catal.*, **48**, 716 (2007)].
297. A. F. Shmidt, A. Al-Halaiqa and V. V. Smirnov, *Kinet. Katal.*, **49**, 416 (2008) [*Kinet. Catal.*, **49**, 395 (2008)].
298. A. V. Zeigarnik and O. N. Temkin, *Kinet. Katal.*, **40** (4), 550 (1999).
299. A. V. Zeigarnik, O. N. Temkin and D. Bonchev, *J. Chem. Int. Comput. Sci.*, **36**, 973 (1996).
300. A. V. Zeigarnik and O. N. Temkin, *Kinet. Katal.*, **40** (4), 538 (1999).
301. A. B. Solov'eva, A. I. Samokhvalova, E. I. Karakozova et al., *Kinet. Katal.*, **25** (5), 1080 (1984).
302. O. M. Poltorak, E. S. Chukhrai and I. Yu. Torshin, *Biokhimiya*, **63** (3), 360 (1998).
303. O. M. Poltorak and E. S. Chukhrai, in *Itogi nauki i tekhniki. Biotekhnologiya* (Advances in Science and Technology: Biochemistry] VINITI, Moscow, 1987, Issue 5, pp. 50–86.
304. O. M. Poltorak, E. S. Chukhrai and A. N. Pryakhin, *Zh. Fiz. Khim.*, **59**, 1585 (1985).
305. O. M. Poltorak and E. S. Chukhrai, *Zh. Fiz. Khim.*, **69**, 330 (1995).
306. E. T. Denisov and V. V. Azatyan, *Ingibirovanie tsepnykh reaktsii* (Inhibition of Chain Reactions), Inst. Khim. Fiz., Ross. Akad. Nauk, Chernogolovka, 1997; E. T. Denisov and V. V. Azatyan, *Inhibition of Chain Reactions*, Gordon & Breach, London. 2000.
307. M. Boudart, *J. Phys. Chem.*, **87**(15), 2786 (1983).
308. O. V. Krylov, *Kinet. Katal.*, **26** (2), 263 (1985).
309. O. N. Temkin and R. M. Flid, *Kataliticheskie prevrashcheniya atsetilenovykh soedinenii v rastvorakh kompleksov metallov* (Catalytic Transformations of Alkynes in Solutions of Metal Complexes), Nauka, Moscow, 1968.
310. R. M. Flid and I. I. Moiseev, *Zh. Prikl. Khim*, **28**, 711 (1955).
311. I. I. Moiseev, *Issledovaniya v oblasti kinetiki i mekhanizma zhidkofaznoi gidratatsii atsetilena v prisutstvii solei rtuti* (Study of Kinetics and Mechanisms of Liquid-Phase Acetylene Hydration in the Presence of Mercury Salts), Candidate Dissertation in Chemistry, MITKhT (Moscow Institute of Fine Chemical Technology), Moscow, 1958.
312. O. N. Temkin, R. M. Flid and M. M. Strelets, USSR Inventor's Certificate No.175943, *Byull. Izobret.*, No.21, 1965.
313. A. I. Mogilyanskii, T. G. Sukhova, O. N. Temkin and R. M. Flid, *Zh. Neorg. Khim.*, **16** (7), 2017 (1971).

6

Nonlinear Effects (Critical Phenomena) in Reaction Dynamics in Homogeneous Catalysis with Metal Complexes

Investigations into the kinetics and mechanisms of homogeneous reactions in solutions of metal complexes (as well as in general catalytic chemistry) in the past 50–60 years were based almost entirely upon inveterate notions concerning linear mechanisms and linear kinetic models. This situation related primarily to the level of knowledge about the mechanisms of catalytic reactions achieved by the middle of the 20th century, but another, probably even more essential, reason for the wide spread of the linearity paradigm was the possibility of obtaining in explicit form a single stationary solution of a system of algebraic equations written in the quasi-steady-state approximation for all or a part of intermediates formed in a closed system or a continuous ideally stirred-tank reactor under steady-state conditions (see Section 1.1.2):

$$r_{\mathrm{X}} = B_{\mathrm{X}}^{T} W_j = 0. \tag{6.1}$$

Solving this system makes possible an analysis of kinetic models having clear physicochemical sense. The latter circumstance was of considerable importance, since the traditional strategy of investigations into the mechanisms of reactions was based upon the passage from experimental kinetic equation to reaction mechanism (see Section 7.1).

Still another popular model approach approximation was the quasi-equilibrium (pre-equilibrium) approximation that led to simple models, including cases where nonlinear steps were assumed to be limiting in single-route reactions. In open systems, the dynamic behavior of systems described by fractional-rational kinetic equations nonlinear with respect to reactant concentrations was mostly studied in enzymatic catalysis.

In contrast to heterogeneous catalysis, where nonlinear effects are inherent in the object of investigation (nonuniformity of the surface, mutual influence of adjacent adsorbed particles, nonisothermicity of the surface in exothermal reactions, significant role of transfer processes in near-surface layers and pores of a catalyst, etc.), the homogeneous systems readily attain isothermicity, and the processes of mass transfer in homogeneous heterophase reactions are eliminated. For this reason, nonlinear effects in homogeneous processes predominantly have a chemical nature and are related primarily to the presence of nonlinear elementary steps, although nonlinear dynamic effects may still appear due to mass-transfer processes in liquids and, in some cases, gas–liquid systems.

Homogeneous Catalysis with Metal Complexes: Kinetic Aspects and Mechanisms, First Edition. O. N. Temkin.

In simple cases, the nonlinearity of a mechanism can be eliminated and the corresponding kinetic model simplified. For example, if a mechanism includes steps that are nonlinear with respect to intermediates, such as

$$\text{(a)}\quad [M] + SH \rightleftharpoons [M]S^- + H^+;$$

$$\text{(b)}\quad [M]Cl + S \rightleftharpoons [M]S^+ + Cl^-;$$

$$\text{(B)}\quad M[Cl] + SH \rightleftharpoons [M]S + H^+ + Cl^-,$$

(where all species in the right-hand parts are intermediates), it is possible to provide conditions where $[H^+]$ and $[Cl^-]$ are either fixed or measured, and then consider these steps as linear with respect to $[M]S^-$, $[M]S^+$, and [M]S. In processes involving polynuclear complexes, it is possible to monitor the activities or concentrations of these complexes (or quantities that are proportional to these variables) and keep them constant while varying the concentrations of reactants and solution components (see Chapter 3). Even in studying polyfunctional catalytic systems, which in most cases feature nonlinear mechanisms (essential for the concept of kinetic polyfunctionality), it is possible to provide conditions where the concentration of active species would be proportional to their known (measured) total concentration.

This kind of "linearization" of the kinetic models provided a basis for the development of basic notions concerning the mechanisms of catalysis with metal complexes and favored the formulation of kinetic models for many commercial homogeneous catalytic processes. At the same time, this approach certainly limited a deeper insight into the essence of a complicated phenomenon such as catalysis with metal complexes.

An analysis of the dynamic behavior of nonlinear mechanisms and kinetic models, which was carried out in basic investigations of D.A. Frank-Kamenetskii [1], R. Aris [2], B.V. Vol'ter and I.E. Sal'nikov [3], A.M. Zhabotinskii [4, 5], P. Glansdorff, I. Prigogine, and G. Nicolis [6, 8], R.J. Field, R.M. Noyes, P. de Kepper, J. Boissonade, I.R. Epstein [9], J.J. Tayson [10], G.R. Ivanitskii, B.I. Krinskii, and E.E. Sel'kov [11], G.S. Yablonsky, V.I. Bykov, and V.I. Elokhin [12], V.I. Bykov [13, 14], M.G. Slin'ko and M.M. Slin'ko [15], M.M. Slin'ko and N.I. Jaeger [16], P. Gray and S.K. Scott [17], and many other researchers of homogeneous and heterogeneous processes led to a revolution in understanding the nature of complex reaction kinetics, stable operation of chemical reactors, nature of periodic biochemical processes, and the driving forces and mechanisms of development of complex temporal patterns and spatial structures. It became clear that using the nonlinear dynamics approach provides a powerful tool for the elucidation of detailed mechanisms of catalytic reactions [15–20].

The discovery of critical phenomena in branched chain reactions of ignition, combustion, and explosion, and especially the investigations in the field of self-oscillating redox reactions which have been extensively developed since the beginning of the 1960s [5, 9, 10], could bring chemists to the realization that nonlinear effects must be also frequently encountered in catalysis with metal complexes. Apparently, the presence of strong one-electron oxidants (Ce^{IV}/Ce^{III}, Mn^{III}/Mn^{II}, Co^{III}/Co^{II}, etc.) and the corresponding intermediate radicals and radical ions in these processes allowed researchers to suggest that this field of catalytic chemistry deals with processes featuring nonlinear dynamics that are characteristic specifically to radical chemistry. Non-radical periodic enzymatic processes were probably out of the scope of interest of the chemists engaged in catalysis.

Besides the aforementioned circumstances that retarded not only the development of investigations into the critical phenomena but even experimental observations of the nonlinear effects both in homogeneous catalysis (rather low level of knowledge in the mechanisms of homogeneous catalytic reactions, "transparency" of kinetic models with linear mechanisms, and evident "convenience" of using these models, expected radical character of intermediates) and (to a lower degree) in heterogeneous catalysis, there were some other factors that also favored the "linear thinking" of chemists engaged in catalysis [21]:

(a) a rather low level of mathematical and physical education, especially in the fields of the qualitative theory of differential equations and the theory of nonlinear oscillations;
(b) insufficient knowledge of the thermodynamics of irreversible processes, in particular, the results of the schools of T. de Donde and I. Prigogine.

It was not accidental that the problems of "chemical" oscillations discovered at the end of 1940s have drawn the attention of physicists, in particular, those belonging to the scientific school of L.I. Mandelshtam and A.A. Andronov (see [1, 3]), scientists engaged in biophysics (Shnol' [22], Zhabotinskii [4, 5], Walter [23]), and many others (see, e.g., [5] and [11]), including mathematicians working in cooperation with biophysicists. Investigations of the periodic biological processes gave especially strong impact to studies of the concentration oscillations in chemical systems [5]. It should be noted that the oscillatory reactions, threshold phenomena, and other critical effects were always treated by chemists as exotic cases or mistakes. Professor R.M. Noyes (University of Oregon), one of the most authoritative specialists in the mechanisms of oscillatory reactions, wrote in his review article of 1990: "Twenty-one years ago, most chemists, including myself, either had never heard of oscillatory reactions or else believed that the few alleged examples were artifactual". [24].

Specialists at the Laboratory of Kinetics and Catalysis of the M.V. Lomonosov Moscow State Institute of Fine Chemical Technology observed manifestations of critical phenomena for the first time at the beginning of the 1980s in investigations of the kinetics of heterogeneous oxidation of isoamyl alcohol (IAA) on supported copper catalysts (such as CuO/Al_2O_3(corundum)) [25] in a continuous-flow gradientless Carberry reactor. At 340 °C, the process exhibited a jumplike transition to a new steady state where, in a certain interval of variation of the oxygen pressure P_{O_2} (dependent on P_{IAA}), it was impossible to determine the dependence of the reaction rate on P_{O_2} and a hysteresis (bistability) was observed (Figure 6.1a) for the rates of formation of both the main product, isovaleric aldehyde (IVA) and all side-reaction products, e.g., r_{CO_2} (Figure 6.1b). The bistability was also clearly manifested on the dependence of P_{O_2} at the output on P_{O_2} at the input (Figure 6.1c).

From 1983 to1985, Shulyakovskii and Temkin [26, 27] discovered an oscillatory regime in a reaction of acetylene carbonylation to maleic and fumaric acid (a typical process of the catalysis with metal complexes) in the $PdBr_2$–KBr–HBr–DMSO–ROH system and in 1997 Malashkevich, Bruk, and Temkin described an oscillatory reaction of the synthesis of phenylmaleic acid diesters in the PdI_2–KI–MeOH system [28]. The phenomenon of bistability (the presence of two stable steady-state regimes) was also observed in the synthesis of maleic anhydride from acetylene in a continuous gas-flow (C_2H_2–CO–O_2) gradientless reactor [29], where the transition from one stable state to another was accompanied by a sharp change (increase) in the process selectivity.

Below we will consider the state-of-the art in the field of dynamics of nonlinear homogeneous catalytic processes with special attention paid to the factors responsible for the appearance of nonlinear effects and the methods of description and analysis of these phenomena. Although this field has still not reached as clear an understanding as that in the kinetics of linear systems, the achievements in the chemistry, kinetics, and mathematical description of these processes are very impressive.

It might seem that, after the appearance of excellent monographs by Zhabotinskii [5], Bykov [14], and Gray and Scott [17] it would be possible to restrict the presentation to descriptions of the kinetics and mechanisms of reactions involving nonlinear effects encountered in the catalysis with metal complexes. However, in that case, the material would be not completely clear to a reader who is unfamiliar with this field, and the passage from linear kinetics to nonlinear dynamics could remain vague and ambiguous. In the author's opinion, it is expedient to present the problem from the standpoint of a chemist that employs the kinetic method in studying reaction mechanisms so as to make the problem clear for other chemists.

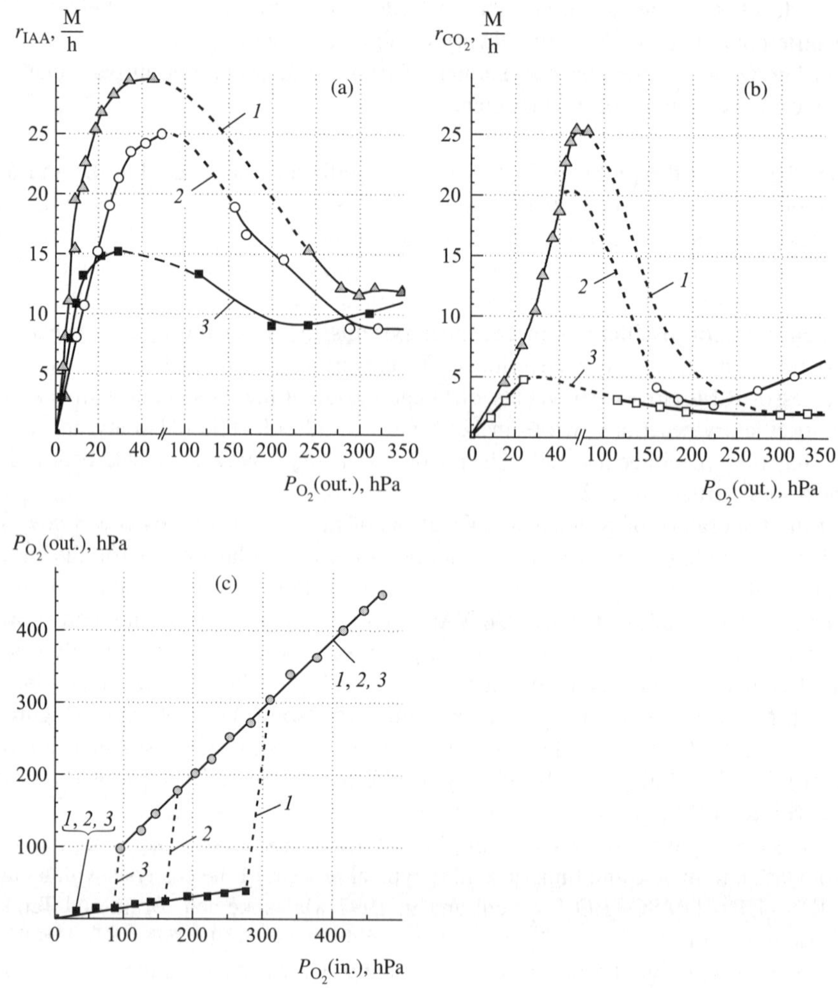

Figure 6.1 Kinetics of isoamyl alcohol (IAA) oxidation on 25% CuO/Al_2O_3(corundum) catalyst at 340 °C and various partial pressures P_{IAA} (hPa): (1) 200; (2) 100; (3) 60. Dashed portions of curves correspond to unstable regimes

6.1 Historical notes

Below, a brief history of investigations into periodic chemical reactions and the development of theoretical approaches to these phenomena is presented according to B.V. Vol'ter [30, 31] and some other researchers [1, 5, 9, 12, 17, 21], giving the year of a discovery or important scientific publication with minimum necessary comments.

1828 G.T. Fechner observed oscillations of current in an electrochemical cell.

1834 M.L. Rosenshield, followed by M.F. Joubert (1874), observed oscillations of emission from phosphorus during its oxidation.

1886 H.H. Landolt, critical phenomena in the reaction of IO_3^- reduction by sulfite ion (Landolt clock reaction) [32–35]. The Landolt reaction served as a basis for the development of minimal chemical oscillators [9].

1896 W. Ostwald, oscillations in the rate of chromium dissolution in acids.

1897 J. Liesegang, concentric circles formed during the crystallization of silver bichromate (wave character of silver salt formation) [36].

1916 J.C. Morgan, periodic evolution of CO during formic acid dehydration in concentrated H_2SO_4 (Morgan reaction) [37].

1910, 1920 A. Lotka, first serious attempts at modeling periodic phenomena in chemistry [38, 39].

1913 R. K. Kremann, review *Die periodischen Erscheinungen in der Chemie* (Periodic Phenomena in Chemistry), Ferdinand Enke, Stuttgart, 1913 (cited by [41]).

1921 W.C. Bray discovered the oscillatory reaction of H_2O_2 decomposition under the action of periodate ions and sulfuric acid [40a]. The first periodate oscillator was also studied for five decades by H.A. Liebhafsky [40b, 40c]. No less than 26 of his publications were devoted to this process (see [24]), which is presently known as the Bray–Liebhafsky reaction.

1926 E. S. Hedges and J.E. Meyers, *The Problem of Physico-Chemical Periodicity* (Arnold, London) devoted to heterogeneous periodic reactions.

1929 A.A. Andronov (at that time, a post-graduate student of L.I. Mandelstam) wrote his Candidate's Dissertation *Poincaré Limit Cycles and Theory of Oscillations*, where he showed that the problem of studying self-sustained oscillations (including those in chemistry) reduces to finding limit cycles in the phase space and determining their parameters [41–43].

1937 A.A. Andronov, A.A. Vitt and S.E. Khaikin, first edition of *Theory of Oscillations*, presenting a mathematical apparatus for the analysis of nonlinear phenomena [43] (see also a history of this monograph in [31]).

1938 F.M. Shemyakin and P.F. Mikhalev, monograph *Physico-Chemical Periodic Processes* [44].

1948 I.E. Sal'nikov wrote his Candidate's Dissertation *On the Theory of Periodic Homogeneous Chemical Reactions*. The work was suggested by A.A. Andronov, performed at the Institute of Chemical Physics (USSR Academy of Sciences, Moscow) under the supervision of D.A. Frank-Kamenetskii, and defended at the Gorki State University. A positive Referee's Report on this dissertation (not accepted for defense at the Institute of Chemical Physics) was written by academician N.M. Emanuel' [31].

1951 B.P. Belousov published results of his investigations of chemical oscillations in a closed system of bromate anion–Ce^{IV}–citric acid–sulfuric acid [45]. Repeated short communication was published in 1959 [46]. Extensive investigation of this process by A.M. Zhabotinskii over the period of 1960–1974 [4, 5] and his discovery of the family of bromate oscillators gave grounds to call this class of oscillatory processes the Belousov–Zhabotinskii (BZ) reactions. The BZ reaction has been known worldwide since its publication in *Nature* (1970) [47].

1952 A.M. Turing theoretically predicted the phenomena of spatial self-organization in nonlinear systems, i.e., the effects of concentration nonuniformity in a homogeneous medium (layers, waves, spirals, etc. [48], which were experimentally discovered by Belousov, Zhabotinskii [49, 5], and other researchers [9, 50].

Since the 1960s, a large number of old and new homogeneous oscillators have been studied [5, 9, 17, 51], including the popular Briggs–Rausher system [52] obtained by combining the Bray–Liebhafsky reaction components and the BZ reaction in the HIO_3–H_2O_2–Mn(II)–malonic acid system. Another important discovery was made somewhat later, when Schmitz, Graziani and Hudson [53] apparently observed chaotic oscillations in the BZ reaction.

Important events also took place in heterogeneous catalysis:

1970–1972 P. Hugo observed periodic concentration oscillations in the reactions of N_2O decomposition on CuO and CO oxidation on platinum grid under isothermal conditions [54, 55].
1972 H. Bensch, P. Fieguth and E. Wicke studied self-sustained oscillations in the reaction of CO oxidation on Pt/Al_2O_3 on platinum wire [56].
1973 V.D. Belyaev, M.M. Slin'ko, V.I. Timoshenko and M.G. Slin'ko reported on self-sustained oscillations in the oxidation of H_2 on Ni [57].

The above pioneering works stimulated an avalanche of investigations devoted to self-sustained oscillations and other critical phenomena in heterogeneous catalysis with metals and metal oxides [12–14, 58].

In homogeneous catalysis, critical phenomena were discovered in the catalytic processes of hydrocarbon oxidation:

1969 J. Ogata and T. Morimoto observed clearly pronounced decaying oscillations during the autoxidation of toluene catalyzed by $CoBr_2$ [59].
1983 J.H. Jensen studied oscillatory regimes in the reaction of benzaldehyde oxidation in the Co(II)–Br^- system [60, 61].

Some other systems containing Pd(II), discovered after 1980, have already been mentioned above [20, 26–28]. Oscillatory reactions in biological systems were described in reviews [5, 11, 88, 89]. By the time of writing this book, homogeneous oscillatory reactions catalyzed by metal complexes only included the oxidation of organic and inorganic substrates with the participation of strong oxidants and the oxidative carbonylation of alkynes, where nontrivial dynamic behavior was observed including the multiplicity of steady states in open systems, self-sustained oscillations of concentration in closed and open systems, chaotic oscillations, and formation of periodic spatial structures.

Table 6.1 gives a brief summary of data on the metals catalyzing oxidation reactions, the dynamics of which show manifestations of the critical phenomena. The substrates of these oxidation processes include organic acids, ketones, diketones (AcacH), phenols, aniline, NADH, PhCHO, RH, RC≡CH, carbohydrates, S^{2-}, SCN^-, $S_2O_3^{2-}$, AsO_3^{3-}, I^-, $Fe(CN)_6^{2-}$, and CO.

The above brief historical survey shows that, by the end of the 20th century, the scientific community eventually recognized the ideas of nonlinear dynamics (nonlinear chemical kinetics) and the "nonlinear" point of view about natural phenomena in chemistry and biology, which gradually have become commonly accepted.

The importance of nonlinear approaches in heterogeneous catalysis, where almost no room is now left for linear models, is undoubted [12–16, 21, 82, 83]. The necessity of using the methods of nonlinear dynamics in modeling reactors and analyzing their stability has been recognized for many years [1, 3, 84]. Evidently, homogeneous catalysis with metal complexes should also expect a revision of the kinetic models for many reactions with chain mechanisms, nonlinear steps, and the manifestations of autocatalysis and autoinhibition. The material presented in Chapters 3–5 shows serious prospects in this respect.

The mechanisms of some organic reactions involving elements of autocatalysis (e.g., the Butlerov reaction of carbohydrate synthesis from formaldehyde [85, 86], asymmetric synthesis [87], etc.) also give grounds to suggest the possible appearance of critical phenomena. Investigations of the oscillatory regimes in enzymatic glycolysis, oxidation, and many other processes are actively developing [5, 11, 88].

For many years at the N.G. Chernyshevsky Saratov State University, Professor D.I. Trubetskov read lectures on nonlinear dynamics and synergetics even to students on humanities courses [42]. This is a

Table 6.1 Catalysts for oxidation reactions

Oxidant	Metal catalyst, ligands	References
$HBrO_3$	Ce(IV/III)	[4]
	FeL_3(III/II)	[51, 62]
	Mn(III/II), acetylacetonate and other salts	[51, 63–66]
	Ag(II/I)L_2,	[67]
	Cr(III)L_3	[68]
	Cu(III/II)L'_3	[69]
	Cu(III/II)L''	[51, 70]
	Ni(III/II)L'_3	[69]
	Ni(III/II)L''	[51, 70]
	Co(III/II)L'_3	[51, 69]
	Co(III/II)L''	[51, 70]
	Os(III/II)L_3	[68]
	Ru(III/II)L_3	[71, 72]
HIO_3	Ce(IV)–Mn(II)–ferroin	[73]
H_2O_2	Fe(III/II)	[74]
	Cu(II/I)	[75, 76]
HIO_3–H_2O_2	Mn(III/II)	[9, 52, 77]
$K_2S_2O_8$	Cu(II)	[78]
	Ag(I)	[79]
O_2	Co(III/II)	[59–61, 80]
O_2	Cu(II/I)	[81]
O_2	Pd(II/I)	[19, 20, 28]

Notes: L = bipy, Phen; L′ = bipy, Phen; phenylenediamine; L″ = tetraaza macrocycles [51].

promising step toward increasing the general level of knowledge in the field of self-organization processes in chemistry, physics, biology, ecology, medicine, sociology, demography, and economy.

6.2 Physicochemical factors responsible for the critical phenomena in homogeneous reactions

The kinetic models of the aforementioned reactions with linear or nonlinear but artificially "linearized" mechanisms possess globally stable steady states (or stable solutions) representing the so-called *attractors*, to which all trajectories in the space of variables tend as the system approaches a steady (or quasi-steady) state. These systems (reactions) in close reactors exhibit no nontrivial kinetic critical phenomena [3].

In open systems, such as a continuous stirred-tank reactor (CSTR), even linear mechanisms in the case of nonlinear complexity functions with respect to the reagent concentrations is that $F_M = 1 + \Sigma K_i[A_i]^n$ at $n \geq 1$ or in nonisothermal reactors can lead under certain values of parameters (rate and equilibrium constants, contact times, temperature, etc.) to myriad nonlinear effects in the vicinity of an attractor [2, 9, 17, 50, 88]. The presence of nonlinear steps in the reaction network and positive and negative feedbacks in the CSTR under isothermal conditions lead to the appearance of stable self-sustained oscillations and chaotic oscillations, the generation of waves, and the formation of spatial structures in a homogeneous medium. Under certain conditions, a complicated dynamic behavior may be caused entirely by the action of kinetic factors also in closed systems.

The huge amount of material available on the complex dynamics of chemical systems cannot be exhaustively presented in one chapter, the more so as the mathematical apparatus used in this field is highly sophisticated. Below, we will briefly consider the thermodynamic, chemical, kinetic and the related mathematical reasons for the appearance of critical phenomena, analyze the most typical models, and consider the examples of some reactions with the aid of a minimal mathematical apparatus that is only necessary for correctly understanding the essence of the observed phenomena and the main ideas of nonlinear chemical dynamics.

6.2.1 Thermodynamic features of nonequilibrium processes near and far from equilibrium

From the standpoint of thermodynamics, chemical reactions belong to irreversible processes since their spontaneous course has only one direction, that is, leads to the equilibrium state. The time to attaining the equilibrium can be either very large or small, and is determined by the reaction rate.

Using classical thermodynamics of reversible equilibrium processes, it is possible to judge on the final equilibrium state of a system. The evolution to this state is determined by the signs of thermodynamic potentials such as the Gibbs ($dG_{T,P} \leq 0$) or Helmholtz potentials ($dF_{T,V} \leq 0$), which exhibit minima in a stable equilibrium state ($d^2G_{T,P} > 0$ and $d^2F_{T,V} > 0$). The important role of irreversible processes in the system evolution is already inherent in formulation of the Second Law of thermodynamics, according to which the entropy S of an isolated system always increases ($dS > 0$) until reaching $S_{\max}$ in the equilibrium state.

According to Clausius, an uncompensated heat $\delta Q'$ produced in an irreversible process is always positive and the $\delta Q'/T$ value characterizes the entropy increment (d_iS) due to internal irreversible processes [8]:

$$d_iS = \delta Q'/T > 0. \tag{6.2}$$

The expediency of separating the total entropy increment dS into two components, d_iS and d_eS (the latter representing the heat and matter exchange with the external medium), was originally pointed out in 1922 by T. de Donde, the founder of the Belgian school of thermodynamics [8], who introduced the concept of chemical affinity and related it to the kinetics of chemical reactions. Using this concept, de Donde also included the uncompensated heat $\delta Q' = Td_iS$ into the formalism of the second law of thermodynamics. He was also the first in considering the thermodynamic potentials as functions of the time, thus laying foundation of the thermodynamics of nonequilibrium irreversible processes (including those in open systems), which was subsequently developed by Onsager, de Groot and Masur [204], von Bertalanffy, Glansdorff, Prigogine, and others [6, 8, 50].

It was established that, according to the second law of thermodynamics, open systems (or closed systems occurring far from equilibrium) can feature stable steady (or quasi-steady) states with high degrees of organization and ordering and complicated spatiotemporal structures. The investigations performed by Schrödinger, von Bertalanffy, Glansdorff, Prigogine, Turing, and later by Zhabotinskii [5], Ivanitskii, Krinskii, and Sel'kov [11], Eigen and Schuster [89] and many others showed the important role of thermodynamics of nonequilibrium processes in open systems for the abiogenesis and self-organization phenomena in living nature.

Prior to formulating the main principles and conclusions of the thermodynamics of nonequilibrium states and the problems of thermodynamic stability of these states (for detail see, e.g., [6, 8, 90, 91]), let us briefly consider the interrelationship between the thermodynamic potentials used in classical equilibrium thermodynamics and the main concepts of the thermodynamics of irreversible processes.

6.2.1.1 Entropy production in irreversible processes

The variations of thermodynamic potentials related to a change in the amount (number of moles) of a substance in a system due to diffusion, chemical reactions, or exchange between various phases are described by the following system of equations [8]:

$$dU = TdS - PdV + \sum \mu_i dn_i; \tag{6.3}$$

$$dG = -SdT + VdP + \sum \mu_i dn_i; \tag{6.4}$$

$$dF = -SdT - PdV + \sum \mu_i dn_i, \tag{6.5}$$

where

$$\left(\frac{\partial U}{\partial n_i}\right)_{s,V,n_j} = \left(\frac{\partial G}{\partial n_i}\right)_{T,P,n_j} = \left(\frac{\partial F}{\partial n_i}\right)_{T,V,n_j} = \mu_i \tag{6.6}$$

is the chemical potential; U is the internal energy; and G and F are the Gibbs and Helmholtz potentials, respectively.

Using Eqs (6.3) and (6.4) and taking into account the relations $dS = d_eS + d_iS$ and $dU = Td_eS - PdV$ (for closed systems featuring energy exchange with an external medium) and the equation $dQ = Td_eS$, one can derive the following important relationship:

$$dG_{P,T} = -Td_iS. \tag{6.7}$$

For a chemical reaction that evolves toward equilibrium, we have $d_iS > 0$ and, hence, $dG_{P,T} < 0$.

On the other hand, a change in the number of moles dn_i can be expressed in terms of the chemical variable ξ (degree of completion), which characterizes a particular reaction:

$$dn_i - \beta_i\, d\xi. \tag{6.8}$$

For $P, T =$ const, Eq. (6.4) yields

$$dG_{P,T} = \sum \mu_i dn_i = \left(\sum \beta_i \mu_i\right) d\xi = \Delta\mu_{\mathrm{P}}\, d\xi, \tag{6.9}$$

where $\Delta\mu_{\mathrm{P}} = \sum \beta_i \mu_i$ is a change in the chemical potential of the reaction. de Donde suggested that the later quantity, taken with the minus sign, $A = -\Delta\mu_{\mathrm{P}}$, was called the chemical affinity of reaction. Using this quantity, we can rewrite the above relationships as

$$dG_{P,T} = -Ad\xi, \tag{6.10}$$

and, accordingly,

$$\left(\frac{\partial U}{\partial \xi}\right)_{S,V} = \left(\frac{\partial G}{\partial \xi}\right)_{P,T} = \left(\frac{\partial F}{\partial \xi}\right)_{T,V} = -A \tag{6.11}$$

Equations (6.7) and (6.10) lead to the following important conclusion:

$$d_iS = \frac{A}{T}\, d\xi. \tag{6.12}$$

Since it is commonly accepted in chemical kinetics to write the reactions (and steps of a mechanism) from left to right so as to follow the evolution of a system toward equilibrium (i.e., in the direction of decreasing chemical affinity), the $d\xi$ value is by definition positive (ξ varies from 0 to a certain finite positive value) and the chemical affinity is positive ($A \geq 0$), so that

$$d_i S = \frac{A}{T} d\xi \geq 0. \tag{6.13}$$

Thus, we have alternatively arrived at the inequality $d_i S \geq 0$.

The rate of entropy increase (entropy production) can be expressed as follows:

$$\mathsf{P} = \frac{d_i S}{dt} = \frac{A}{T}\frac{d\xi}{dt} = V\frac{A}{T}W > 0 \tag{6.14}$$

where W is the reaction rate and V is the volume of the system. In the case of a system involving several chemical reactions,

$$\mathsf{P} = \frac{d_i S}{dt} = \frac{V}{T}\sum A_j\, W_j > 0 \tag{6.15}$$

$$\sigma = \frac{d_i s}{dt} = \frac{1}{T}\sum_j A_j W_j > 0 \tag{6.16}$$

where s is the entropy "density" (entropy per unit volume) and σ is the local entropy production.

Thus, the conditions

$$d_i S > 0, \quad \frac{d_i S}{dt} > 0, \quad \frac{dG_{P,T}}{dt} = V\sum A_j W_j < 0$$

express the second law of thermodynamics for any irreversible process. In the equilibrium state, $d_i S = 0$ and $d_i S/dt = 0$.

Taking into account that the chemical affinity is related to the reaction rates in the forward and reverse directions (see Section 1.12) as

$$A_j = RT \ln \frac{W_j^+}{W_j^-}, \tag{6.17}$$

and $W_j = W_j^+ - W_j^-$, we obtain the following explicit relationship between entropy production as a result of irreversible processes (chemical reactions) inside the system and chemical process kinetics:

$$\mathsf{P} = \frac{d_i S}{dt} = RV\sum_j \ln \frac{W_j^+}{W_j^-}(W_j^+ - W_j^-) > 0. \tag{6.18}$$

This equation leads to the unambiguous conclusion that, for a spontaneous chemical reaction evolving toward equilibrium in the forward or reverse direction (for the same writing of stoichiometric equation), each term of the $\sum A_j W_j$ sum is always positive (for both $W_j^+ > W_j^-$ and $W_j^+ < W_j^-$) and (as was noted in Section 3.4) the $A_j W_j$ terms with different signs for kinetically conjugated processes cannot be summarized. Moreover, since the conditions $d_i S > 0$ are also valid for all subsystems (chemical reactions), the inequality $d_i S = \sum_k d_i S_k > 0$ is satisfied at $d_i S_k > 0$ for each subsystem [8].

The equation (introduced by de Donde)

$$dS = d_e S + d_i S \tag{6.19}$$

(together with Eq. (6.15)) expresses one of the most important principles in the thermodynamics of irreversible processes. Let us consider this principle in more detail.

In an *isolated* system, we have

$$d_e S = 0, dS = d_i S \geq 0$$

(if the process is fully reversible, which is an ideal situation, then $d_i S = 0$).

In a *closed* system:

$$d_e S = \frac{dU + PdV}{T} = \frac{dQ}{T}, \; d_i S \geq 0.$$

In an open system:

$$dn_k = d_e n_k + d_i n_k;$$

$$dU = d_e U + d_i U;$$

$$dS = d_e S + d_i S.$$

From the first law of thermodynamics, $d_i U = 0$ (law of energy conservation), it follows that the changes in $dU = d_e U, d_e n_k$, and $d_e S$ can be greater or smaller than zero and equal to zero, and $d_i S \geq 0$. Evidently, the dS value in an open system can be negative provided that $d_e S < 0$ and $|d_e S| > d_i S$. For $d_e S < 0$, the system gives the entropy to an external medium (in other words, it absorbs the negative entropy ("negentropy") from outside). *This circumstance explains the ability of nonequilibrium open systems to pass to the states that are self-organized in the time (oscillations) and/or space (running waves, Turing structures).*

In open systems, the process can involve the appearance of steady states ("current equilibrium"), in which

$$dU = 0 \text{ (and, hence, } d_e U = 0);$$

$$dn_k = 0 \text{ and } d_e n_k = -d_i n_k = -\sum \beta_{kj}\, d\xi_j;$$

$$dS = 0 \text{ and } d_e S = -d_i S < 0.$$

As can be seen, these steady states are characterized by finite increments and finite rate of entropy production:

$$d_i S > 0, \; P > 0.$$

Since any system featuring a chemical reaction is characterized by an increase in entropy, all processes in the system are accompanied by the dissipation of energy and a decrease in working capacity. The value of energy dissipation per unit volume can be characterized by the product of $T\sigma$. The dissipation of energy and the appearance of dissipative structures (i.e., states that are self-organized in time and/or space), which appear and pass to a thermodynamically equilibrium state only in a jumplike manner (Prigogine), is a characteristic feature of irreversible processes far from equilibrium.

In this context, an important part of the thermodynamics of irreversible processes is the analysis of stability of the equilibrium states and nonequilibrium steady states.

6.2.1.2 *Stability of equilibrium Gibbs states [8]*

The fact that the entropy of an isolated system reaches a maximum in the equilibrium state implies that this state is stable with respect to any perturbation ($\delta T, \delta V, \delta n_k$, etc.) decreasing S, since irreversible processes ($d_i S > 0$) return the system to equilibrium.

Expansion of the entropy into Taylor's series in the vicinity of the equilibrium state yields

$$S = S_{eq} + \delta S + {}^1\!/\!_2\delta^2 S + \ldots, \tag{6.20}$$

where the first-order term vanishes ($\delta S = 0$) at the point of maximum. Since the fluctuations can only decrease S, it follows that $\delta^2 S < 0$.

If the temperature fluctuations in one part of the system lead to a perturbation δU in the energy flux, the corresponding value of $\Delta S = S - S_{eq}$ can be expressed (for $V = \text{const}$) by the following equation:

$$\Delta S = \delta S + \frac{1}{2}\delta^2 S = \left(\frac{1}{T_1} - \frac{1}{T_2}\right)\delta U + \left[\frac{\partial}{\partial U_1}\frac{1}{T_1} + \frac{\partial}{\partial U_2}\frac{1}{T_2}\right]\frac{(\delta U)^2}{2}. \tag{6.21}$$

At the point of equilibrium, $T_1 = T_2$ and $\delta S = 0$. An analysis of this equation leads to the following expression that describes the thermal stability of the system:

$$\frac{1}{2}\delta^2 S = -\frac{C_V(\delta T)^2}{2T^2}. \tag{6.22}$$

Since C_V (heat capacity at constant volume) is positive, then $\delta^2 S < 0$ and, hence, the system is always stable with respect to thermal fluctuations.

During an equilibrium chemical reaction, the fluctuations may also be caused by variations in the molar concentrations of substances and described as fluctuations of the chemical variable ξ:

$$S - S_{eq} = \Delta S = \delta S + \frac{1}{2}\delta^2 S = \left(\frac{\partial S}{\partial \xi}\right)_{U,V}\delta\xi + \frac{1}{2}\left(\frac{\partial^2 S}{\partial \xi^2}\right)_{U,V}(\delta\xi)^2. \tag{6.23}$$

The relationship

$$\left(\frac{\partial S}{\partial \xi}\right)_{U,V} = \frac{A}{T}$$

(see Eqs (6.13)), implies that

$$\Delta S = \left(\frac{A}{T}\right)_{eq}\delta\xi + \frac{1}{2T}\left(\frac{\partial A}{\partial \xi}\right)_{eq}(\delta\xi)^2. \tag{6.24}$$

In the equilibrium state, $A/T = 0$ and $\delta S = 0$. Then, the second variance of the entropy is negative,

$$\frac{1}{2}\delta^2 S = \frac{1}{2T}\left(\frac{\partial A}{\partial \xi}\right)_{eq}(\delta\xi)^2 < 0, \tag{6.25}$$

provided that

$$\left(\frac{\partial A}{\partial \xi}\right)_{eq} < 0.$$

Thus, inequality (6.25) is the condition of stability of the equilibrium state. In the case of a system featuring several reactions, this condition appears as

$$\frac{1}{2}\delta^2 S = \sum_{i,j} \frac{1}{2T}\left(\frac{\partial A_i}{\partial \xi_j}\right)_{\text{eq}} \delta\xi_i \delta\xi_j < 0. \quad (6.26)$$

The existence of extrema in the other thermodynamic potentials ($G_{\min}, F_{\min}$) also makes the equilibrium system stable, since irreversible processes would return the system back to these extremum points ($G_{\min}, F_{\min}, S_{\max}$). Far from an equilibrium state, a system can either be unstable or pass from the unstable to stable state with a higher degree of organization.

6.2.1.3 Main principles of the thermodynamics of irreversible processes

Let us consider the main postulates and inferences that underlie any analysis of the thermodynamics of nonequilibrium (irreversible) processes, with special attention to those not mentioned above.

(i) Representation of the entropy increment as separated into two components ($dS = d_e S + d_i S$).
(ii) Establishment of the most general form of the second law of thermodynamics ($d_i S > 0$) for any system.
(iii) Introduction of the chemical affinity A of a reaction as the most universal thermodynamic function (see Eq. (6.11)).
(iv) Establishment of a relationship between $d_i S$ and the chemical reaction rate (see Eqs (6.14) and (6.15)). The appearance of time in the thermodynamic equations allows the concepts of flows and thermodynamic driving forces to be used and linear phenomenological laws to be formulated for the thermodynamics of nonequilibrium processes in the linear region (Onsager).
(v) Formulation of the second law of thermodynamics in terms of thermodynamic driving forces (F_k) and flows (J_k):

$$\frac{d_i S}{dt} = \sum_k F_k \frac{dx_k}{dt} \geq 0. \quad (6.27)$$

For chemical reactions, $F_k = A_k/T$ is the thermodynamic driving force that causes the flow of matter (conversion of reactants into products) defined as

$$\frac{dx_k}{dt} = \frac{d\xi_k}{dt} = VW_k = J_k,$$

so that

$$P = \frac{d_i S}{dt} = \sum_k F_k J_k > 0 \quad (6.28)$$

This condition implies that the production of a finite entropy increment $\Delta_i S$ due to fluctuations near the equilibrium is described by a second-order equation with respect to the force and flux perturbations [8],

$$\frac{d\Delta_i S}{dt} = \sum \delta F_k \delta J_k > 0,$$

or in terms of the excess entropy production

$$\frac{1}{2}\frac{d(\delta^2 S)}{dt} = \sum_k \delta F_k \delta J_k > 0. \tag{6.29}$$

Thus, if a nonequilibrium system is isotropic with respect to T and P, then the parameters of nonequilibrium states and thermodynamic potentials are represented by functions of the time [90]. However, the general approach to the thermodynamics of nonequilibrium is based on the concept of local equilibrium, which makes possible the analysis of spatially nonunifom processes.

(vi) The concept of local equilibrium, according to which intensive variables such as temperature and pressure are uniquely determined in any elementary volume, makes it possible to use concepts possessing a statistical sense. Extensive variables (such as number of moles, entropy, and internal energy) are replaced by their densities (per unit volume or mass) [8]. Accordingly, $s(x,t)$ is the entropy per unit volume, $u(x,t)$ is the internal energy per unit volume; $n'_k(x,t)$ is the number of moles per unit volume (local concentration), and x is the radius vector of the given volume element in the space.

The chemical potential μ is a function of the coordinates and time. In this case, the Gibbs relation (6.3) takes the form of the following equation:

$$Tds = du - \sum \mu_k \, dn'_k. \tag{6.30}$$

It was pointed out (see, e.g. [8]) that the first and second laws of thermodynamics have a local character.[1] If the local entropy production is defined as

$$\sigma(x,t) = \frac{d_i s}{dt} \geq 0, \tag{6.31}$$

then the total entropy production in the system is given by the following formula:

$$\frac{d_i S}{dt} = \int_V \sigma(x,t) dV. \tag{6.32}$$

The theory of nonequilibrium thermodynamics allows the local entropy production σ (x,t) to be represented explicitly as a function of all thermodynamic forces and flows and to be determined from experimental data via de Donde relations [8, 91]:

$$\sigma = \sum_\alpha F_\alpha J_\alpha. \tag{6.33}$$

For example, a thermodynamic force related to the temperature gradient ($F_1 = \nabla(1/T)$), induces heat flow (J_u), the force $F_2 = A_j/T$ causes the substance flow due to the chemical reaction (W_j), and force $F_3 = \nabla\mu_k/T$ induces the diffusion flow (J_k) (where ∇ is the Hamiltonian operator nabla). In the absence of external fields:

$$\sigma = \sum F_\alpha J_\alpha = J_u \nabla \frac{1}{T} - \sum_k J_k \nabla \frac{\mu_k}{T} + \sum_j \frac{A_j W_j}{T} \geq 0. \tag{6.34}$$

---

[1]This approach resembles that used to describe the kinetics of catalytic reactions on a biographically nonuniform catalyst surface, where the adsorption on centers of one type is described by the ideal Langmuir model with the subsequent integration over the entire surface [92].

6.2.1.4 Thermodynamics of irreversible processes near equilibrium (linear region)

Approaches to the description of steady-state processes occurring near the equilibrium were developed by Onsager, de Groot, and Prigogine (see [8] and references therein). The Onsager method of generalized driving forces was developed based on three main assumptions:

(i) It is postulated that the fluxes and forces in the state near equilibrium are linearly interrelated:

$$J_k = L_k F_k.$$

In the case of irreversible processes, there are many phenomenological laws of this kind (such as Fick's, Fourier's, and Ohm's laws). The ith flux is determined in the general case by the set of all driving forces:

$$J_i = \sum L_{ik} F_k. \tag{6.35}$$

Linear relations of this type are only valid provided a small deviation from equilibrium and correspond to the first terms of conventional expansions with respect to a small parameter. The diagonal elements L_{ik} $(i = k)$ in the $\boldsymbol{L}_{ik}$ matrix characterize the influence of gradients (forces) on the corresponding flows (e.g., of the temperature gradient on the heat flux). The off-diagonal elements L_{ik} $(i \neq k)$ represent the interactions (conjugation) between various forces and flows.
In some cases, the conjugation between various forces (irreversible processes) is absent. For example, the chemical affinity A of a reaction (representing a scalar thermodynamic force) cannot induce a directed heat flux (vector force). In these cases, $L_{ik} = 0$.

(ii) A theorem was proved that the $\boldsymbol{L}_{ik}$ matrix is symmetric, that is, $L_{ik} = L_{ki}$ (Onsager reciprocity relations).

(iii) Another theorem states that, if the thermodynamic driving forces are selected so that

$$T\sigma = \sum J_i F_i, \tag{6.36}$$

then the Onsager reciprocity relations are always satisfied. This theorem establishes a relationship between the local entropy production, which is defined as

$$\sigma = \frac{1}{V}\frac{d_i S}{dt},$$

and linear combinations of forces. In the region of validity of the reciprocity relations, the system evolves toward a *stable steady state* with minimum entropy production (Prigogine theorem):

$$d\sigma/dt \leq 0. \tag{6.37}$$

Therefore, critical phenomena leading to an increase in the entropy production cannot take place in the region of linearity.

Let us consider some features of kinetics in the linear region. The rate of the elementary step is defined as

$$W_j = W_j^+ - W_j^- = W_j^+\left(1 - \frac{W_j^-}{W_j^+}\right) = W_j^+(1 - e^{-A_j/RT}). \tag{6.38}$$

Near equilibrium (at $A_j \ll RT$), we have

$$W_j = W_j^+ \frac{A_j}{RT}. \tag{6.39}$$

Taking into account that (in this region) $W_j^+ \cong W_{j\,\text{eq}}^+$, equation (6.39) can be written in the following form:

$$W_j = W_{j\,\text{eq}}^+ \frac{A_j}{RT}, \tag{6.40}$$

which can also be obtained by expanding (6.38) with respect to small deviations ΔW_j from the equilibrium rate. Here, A_j/T, is the thermodynamic force and W_j is the flow of matter. Then, the relation $J_j = L_j F_j$ yields $W_j = L_j A_j/T$, where $L_j = W_j^+/R$. In the case of several forces and flows, the above relation is written as

$$W_k = \sum_j L_{kj} A_j/T. \tag{6.41}$$

For a steady-state single-route process involving $(m + 1)$ linear steps, we have $r = W_j$, where r is the rate over the route (overall equation) and $W_j = W_j^+ A_j/(RT)$. Then, taking the sum of A_j over all steps, we obtain

$$\sum_j \frac{r}{W_j^+} = \sum_j \frac{A_j}{RT} = \frac{A_\Sigma}{RT}, \tag{6.42}$$

which yields

$$r\left(\sum_j \frac{1}{W_j^+}\right) = \frac{A_\Sigma}{RT}. \tag{6.43}$$

This equation can be rewritten as

$$r = r_{\text{eff}} \frac{A_\Sigma}{RT}, \tag{6.44}$$

where

$$r_{\text{eff}} = \frac{\prod_{j=1}^{m+1} W_j^+}{D};$$
$$D = W_2^+ W_3^+ \cdot \ldots \cdot W_{m+1}^+ + W_1^+ W_3^+ \cdot \ldots \cdot W_{m+1}^+ + \ldots + W_1^+ W_2^+ + \ldots + W_1^+ W_2^+ \cdot \ldots \cdot W_m^+$$

(see also Section 1.2.1).

Note that, for a kinetically reversible steady-state process with a linear mechanism, the process rate far from the equilibrium can also be expressed using the equation proposed by M.I. Temkin:

$$r = \frac{\prod_j W_j^+ - \prod_j W_j^-}{D}, \tag{6.45}$$

where $D = W_2^+ W_3^+ \cdot \ldots \cdot W_{m+1}^+ + W_1^- W_3^+ \cdot \ldots \cdot W_{m+1}^+ + \ldots + W_1^- W_2^- \cdot \ldots \cdot W_m^-$. Representing r as the difference of r^+ and r^-, we obtain

$$r = r^+ - r^- = r^+\left(1 - \frac{r^-}{r^+}\right) = r^+\left(1 - \frac{\prod W_j^-}{\prod W_j^+}\right) = r^+(1 - e^{-A_\Sigma/(RT)}), \tag{6.46}$$

where

$$r^{+} = \prod W_j{}^{+}/D. \tag{6.47}$$

If the steady state is close to the equilibrium, $W_j{}^{-} \cong W_j{}^{+}$, then

$$1 - e^{-A_\Sigma/(RT)} \cong A_\Sigma/RT$$

and we eventually arrive at the expression (6.44) for r_{eff}.

In the case of a consecutive-parallel mechanism with a simple conjugation node (common step and common intermediate for two steps), the relationship between entropy production (or energy dissipation rate) $T\sigma$ and $A_j W_j$ will be different. In particular, for the mechanism

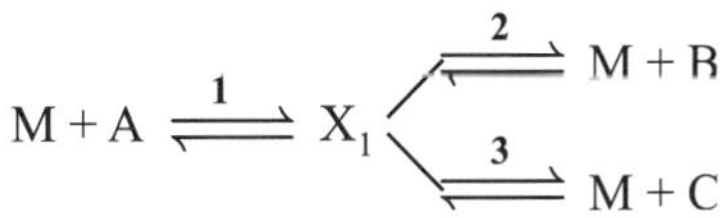

(6.48)

the rate of the first step enters into expressions for the rates of the first (B) and second (C) routes:

$$W_1 = r_{\text{B}} + r_{\text{C}}; \tag{6.49}$$

$$W_1 = W_2 + W_3. \tag{6.50}$$

The latter equation yields

$$W_1{}^{+}\left(1 - \frac{W_1{}^{-}}{W_1{}^{+}}\right) = W_2{}^{+}\left(1 - \frac{W_2{}^{-}}{W_2{}^{+}}\right) + W_3{}^{+}\left(1 - \frac{W_3{}^{-}}{W_3{}^{+}}\right). \tag{6.51}$$

Taking into account that, near the equilibrium,

$$1 - \frac{W_j{}^{-}}{W_j{}^{+}} \cong \frac{A_j}{RT},$$

we obtain the following relationship:

$$W_1{}^{+}A_1 = W_2{}^{+}A_2 + W_3{}^{+}A_3. \tag{6.52}$$

This equation leads to two interesting conclusions:

$$A_1 = \frac{W_2{}^{+}}{W_1{}^{+}}A_2 + \frac{W_3{}^{+}}{W_1{}^{+}}A_3, \tag{6.53}$$

which implies that the affinity at the first step is a sum of the fractions of affinities A_2 and A_3, and

$$W_1{}^{+} = \frac{A_2}{A_1}W_2{}^{+} + \frac{A_3}{A_1}W_3{}^{+}, \tag{6.54}$$

which implies that the rate of the first (common) step is a sum of the fractions of rates $W_2{}^{+}$ and $W_3{}^{+}$ determined by the ratios of A_2/A_1 and A_3/A_1, respectively.

Let us use the principle of minimum entropy production in the linear region for deriving the condition of the steady state of steps (6.49) and (6.50). In the steady-state process, the flows of compounds A, B, and C are maintained constant, which implies the constancy of the corresponding chemical potentials ($\mu_{\text{A}}, \mu_{\text{B}}$, and μ_{C}). In this case, the chemical affinity for the first and second routes can be expressed as

$A_{\rm I} = \mu_{\rm A} - \mu_{\rm B} = A_1 + A_2 = {\rm const}$ and $A_{\rm II} = \mu_{\rm A} - \mu_{\rm C} = A_1 + A_3 = {\rm const}$, respectively. Now let us write equation (6.16) for the local entropy production as $\sigma = f(A_1)$, using $A_{\rm I}$ and $A_{\rm II}$ and the linear relationships for W_j:

$$W_1 = L_{11}\frac{A_1}{T}, \quad W_2 = L_{22}\frac{A_2}{T}, \quad W_3 = L_{33}\frac{A_3}{T},$$

$$\sigma = L_{11}\frac{A_1^2}{T^2} + L_{22}\frac{(A_{\rm I} - A_1)^2}{T^2} + L_{33}\frac{(A_{\rm II} - A_1)^2}{T^2}. \quad (6.55)$$

Then, the condition $\partial\sigma/\partial A_1 = 0$ for the minimum of entropy production σ with respect to A_1 (in the steady-state process) yields

$$\frac{2L_{11}A_1}{T^2} - \frac{2L_{22}(A_{\rm I} - A_1)}{T^2} - \frac{2L_{33}(A_{\rm II} - A_1)}{T^2} = 0.$$

Returning to the affinities of steps $A_2 = A_{\rm I} - A_1$ and $A_3 = A_{\rm II} - A_1$, we obtain the following condition of the steady state of steps ($W_1 = W_2 + W_3$):

$$\frac{L_{11}A_1}{T} - \frac{L_{22}A_2}{T} - \frac{L_{33}A_3}{T} = 0, \quad (6.56)$$

which turns out to be valid both near and far from equilibrium.

Using the condition of the steady state of steps written as

$$\mathbf{W}_j = G\mathbf{r}_p, \quad (6.57)$$

it is possible to obtain expressions that relate the local entropy production (see Eq. (6.16)) to the rates over routes for any position of the steady-state process.

For a steady-state single-route process, matrix G is a column-vector of the stoichiometric numbers of steps (ν_j) and Eq. (6.16) yields

$$T\sigma = \sum A_j W_j = \sum A_j \nu_j r = r\sum A_j \nu_j = rA_\Sigma > 0. \quad (6.58)$$

Equation (6.58) for mechanism (6.48), appears as

$$T\sigma = \sum A_j W_j = A_1 W_1 + A_2 W_2 + A_3 W_3 > 0. \quad (6.59)$$

Using the values of affinity for the two routes

$$A_{\rm I} = A_1 + A_2; \quad A_{\rm II} = A_1 + A_3,$$

expression (6.50), relations $W_2 = r_{\rm B}$ and $W_3 = r_{\rm C}$, and Eq. (6.59), we obtain

$$T\sigma = A_{\rm I}r_{\rm B} + A_{\rm II}r_{\rm C} > 0, \quad (6.60)$$

since for a linear mechanism all $\nu_j = 1$. This equation can be generalized to the case of j steps and p routes:

$$T\sigma = \sum A_j W_j = \sum A_j G r_p = \sum A_p r_p > 0. \quad (6.61)$$

The product of row-vector $\mathbf{A}_j$ and matrix G of the stoichiometric numbers of steps yields column-vector $\mathbf{A}_p$ of the affinities of the routes (chemical affinity for the overall equations of routes).

6.2.1.5 Thermodynamics of irreversible processes far from equilibrium (nonlinear region)

In the universe of instability and the evolution of systems toward new organized structures, the destiny of a system can be determined by very small factors (e.g., fluctuations) [8].

As established above for a steady-state process in the linear region, the presence of two entropy functions obeying the conditions

$$P = \frac{d_i S}{dt} > 0 \text{ and } \frac{dP}{dt} < 0,$$

which imply a positive increment P of the entropy (or positive rate of energy dissipation, $Td_i S/dt > 0$) and a minimum entropy production, and so ensures the stability of any steady state.

In a nonlinear region far from the equilibrium, there are no functions that have extrema in the steady state. Hence, the entropy production P must not tend to a minimum and therefore its fluctuations due to the variation of driving forces can be both positive and negative.

Let us consider the application of Lyapunov functions to solving the problem of stability of steady states for an arbitrary dynamical system [8]. If the variable x_k has a steady state x_{ks} ($dx_k/dt = 0$), then a positive definite function of fluctuations δx can be found such that

$$L(\delta x) > 0,$$

where $\delta x = x - x_{ks}$. The steady state will be stable, provided that this deviation monotonically decreases with the time, i.e., $dL(\delta x)/dt < 0$. The state will also be stable for $L(\delta x) < 0$ and $dL(\delta x)/dt > 0$. Thus, for the stability of states, it is important that the function L with constant sign and its derivative would have opposite signs. It is convenient to select L in the form of quadratic functions, which always have a constant sign. For example, in the case of two variables, it is possible to use $L(\delta x, \delta y) = [(\delta x)^2 + (\delta y)^2] > 0$.

Since the second variance of the entropy for any thermodynamic system is negative ($\delta^2 S < 0$), this function can be used as the Lyapunov function for any dynamical system [6, 90]. In a nonequilibrium system, a deviation of the density of entropy (specific deviation Δs) from a steady state can be expressed as

$$\Delta s = \delta s + \frac{1}{2}\delta^2 s [\text{J}/(\text{liter K})],$$

where δs is nonzero (in contrast to the case of a linear region) and the second variance is always negative ($\delta^2 s < 0$). It was demonstrated [6, 90] that the Lyapunov function can be defined as

$$L = \frac{1}{2}\delta^2 s,$$

and then the condition $dL/dt > 0$ of the stability of a steady state appears as (see Eq. (6.29))

$$\frac{dL}{dt} = \frac{1}{2}\frac{d(\delta^2 s)}{dt} = \sum \delta F_k \delta J_k > 0. \tag{6.62}$$

Moreover, this criterion of stability is valid both near and far from equilibrium, but close to a steady state under consideration.

For $dL/dt < 0$, a system *can be* (but not necessarily) unstable, so that Eq. (6.62) is a sufficient but not necessary condition of stability.

In order to pass from specific entropy s to S, it is possible to use the integral of the former function,

$$\frac{\partial(\delta^2 S)}{dt} = \int_0^V \frac{\partial(\delta^2 s) dv}{\partial t} = \int_0^V \left(\sum \delta F_k \delta I_k\right) dV,$$

and call it *excess entropy production* [6, 90].

A system can go far away from the equilibrium but, as long as condition (6.62) is satisfied, it will be stable (i.e., the system occurs on a so-called thermodynamic branch [8]). In a certain region of parameters or concentrations, this condition is violated and the system can go to another branch. On the new branch, the system can exhibit oscillations in the concentrations and rates of steps but, since the affinities A_j and rates W_j have the same sign (irrespective of the direction of reaction), the derivative $d_iS/dt = P$ would oscillate while remaining positive. The values of critical parameters and concentrations and the character of a new steady state can be determined by special mathematical methods (developed by Poincaré, Lyapunov, and other researchers), which are considered below. Here we only consider examples of using condition (6.62) [8].

Consider a reaction

$$\mathrm{A + B \rightleftharpoons C + D},$$

which is carried out in a steady-state regime in a CSTR. The values of affinity (F_k) and flux (J_k) are as follows:

$$A = RT \ln W^+/W^-; \qquad W = W^+ - W^-,$$

so that

$$\frac{dL}{dt} = \frac{1}{2}\frac{d(\delta^2 s)}{dt} = \sum \delta F_k \delta I_k = \frac{\delta A}{T}\delta W. \tag{6.63}$$

For a perturbation (δB) in the concentrations of reactants in the steady states $\mathrm{A^{ss}}$, $\mathrm{B^{ss}}$, $\mathrm{C^{ss}}$, and $\mathrm{D^{ss}}$, we have:

$$A = RT \ln \frac{k_1 C_A^{ss} C_B}{k_{-1} C_C^{ss} C_D^{ss}}; \qquad A^{ss} = RT \ln \frac{k_1 C_A^{ss} C_B^{ss}}{k_{-1} C_C^{ss} C_D^{ss}}; \qquad \frac{\delta A}{T} = \frac{A - A^{ss}}{T}.$$

For $C_B = C_B^{ss} + \delta C_B$, this yields

$$\frac{\delta A}{T} = R \ln \frac{C_B^{ss} + \delta C_B}{C_B^{ss}} \cong R\frac{\delta C_B}{C_B^{ss}};$$

$$\delta W = W - W^{ss} = k_1 A^{ss}(C_B^{ss} + \delta C_B) - k_1 A^{ss} C_B^{ss} = k_1 A^{ss} \delta C_B.$$

Thus, we have

$$\frac{\delta A}{T}\delta W = Rk_1 \frac{A^{ss}}{C_B^{ss}}(\delta C_B)^2 > 0$$

and, hence, the steady state is stable.

For an autocatalytic reaction

$$\mathrm{2X + A \rightleftharpoons 3X}$$

we have

$$W^{ss} = k_1 (C_x^{ss})^2 A^{ss} - k_{-1}(C_x^{ss})^3;$$

$$W = k_1 (C_x^{ss} + \delta C_x)^2\, A^{ss} - k_{-1}(C_x^{ss} + \delta C_x)^3.$$

Rejecting the terms with $(\delta C_x)^2$ and $(\delta C_x)^3$, we obtain

$$\delta W = W - W^{ss} = \left(2k_1 A^{ss} C_x^{ss} - 3k_{-1}(C_x^{ss})^2\right)\delta C_x;$$

$$\frac{\delta A}{T} = R\ln\frac{A^{ss}}{C_x^{ss}+\delta C_x} - R\ln\frac{A^{ss}}{C_B^{ss}} = R\ln\frac{C_B^{ss}}{C_x^{ss}+\delta C_x}$$

$$= -R\ln\left(1+\frac{\delta C_x}{C_x^{ss}}\right) \cong -R\frac{\delta C_x}{C_x^{ss}}.$$

From this we obtain the following function:

$$\frac{\delta A}{T}\delta W = -R\left(2k_1 A^{ss} C_x^{ss} - 3k_{-1}(C_x^{ss})^2\right)\frac{(\delta C_x)^2}{C_x^{ss}}. \tag{6.64}$$

which can be both negative and positive. In particular, for $k_1/k_{-1} > \frac{3}{2}\frac{C_x^{ss}}{A^{ss}}$, the derivative is $dL/dt < 0$ and, hence, the steady state can be unstable.

Consider a simple reaction

$$n\text{X} \xrightarrow{W} \text{C} \tag{6.65}$$

in a CSTR under conditions of the fluctuating concentration X ([X]) and temperature. An analysis [91] shows that, even with neglect of the excess entropy production due to the energy and mass exchange with the medium, the excess entropy production due to a chemical reaction in the reactor volume V includes three terms:

$$\frac{d(\partial^2 S)}{dt} = V\left[Rn^2W\left(\frac{\delta C_x}{C_x}\right)^2 + (E_{\text{act}} - Q)nW\frac{\delta X}{X}\frac{\delta T}{T^2} - \frac{EQ}{RT^4}W(\partial T)^2\right]. \tag{6.66}$$

Evidently, (i) the higher the thermal effect Q of the reaction, the greater the third term and (ii) for $Q > E$, the second term can be negative. The energy exchange with the medium does not introduce negative terms, while the mass exchange introduces a negative contribution to the entropy production. The forward reaction is a factor that stabilizes the process, the role of which increases with W and n. If the total derivative is $\frac{d(\partial^2 S)}{dt} < 0$, then the process can be unstable.

6.2.2 Dynamic behavior of systems with linear mechanisms in open reactors with complete mixing

It should be emphasized that, by definition, linear mechanisms contain sequences of only linear steps that form reaction routes. A kinetic equation and, especially, the mathematical model of a process, can be nonlinear with respect to reactants due to the corresponding nonlinear material balance equations.

Specific features of the dynamical behavior of kinetic models for the processes in open reactors with complete mixing (continuous stirred-tank reactors, CSTRs) under isothermal conditions have been repeatedly studied [2, 3, 5, 17, 93]. For example, let us analyze the behavior of a kinetic model of catalytic reaction A → B that proceeds according to the following two-step mechanism (also considered in the preceding chapters and sections):

$$\text{A} + \text{M} \underset{}{\overset{1}{\rightleftharpoons}} \text{AM} \xrightarrow{2} \text{B} + \text{M}. \tag{6.67}$$

The overall reaction rate under quasi-steady-state conditions is described by the following equation:

$$r = \frac{k_2 C_A[M]_\Sigma}{\dfrac{k_{-1}+k_2}{k_{-1}} + C_A} = \frac{k_2 C_A[M]_\Sigma}{K_m + C_A}, \tag{6.68}$$

which is known as the Michaelis–Menten equation. Under the conditions of quasi-equilibrium at the first step, we have

$$r = \frac{k_2 K_0 C_A[M]_\Sigma}{1 + K_1 C_A}. \tag{6.69}$$

The material balance with respect to reactant A in a CSTR is described by the following equation:

$$VdC_A = W_0 C_{A0}\,dt - WC_A\,dt - Vdt|r_A| \tag{6.70}$$

or

$$\frac{dC_A}{dt} = \frac{W_0 C_{A0} - WC_A}{V} - |r_A|, \tag{6.71}$$

where dC_A is a change in the concentration of A in the reactor volume V; C_{A0} and C_A are the concentrations of A at the reactor input and output, respectively; W_0 and W are the volume flow rates (e.g., l/h) at the reactor input and output, respectively; and r_A is the rate of reactant A conversion (loss) [$r = |r_A|$ in reaction (6.67)]. In a steady-state regime, we have $dC_A/dt = 0$ and, for $W_0 = W$

$$r = \frac{W_0(C_{A0} - C_A)}{V} = \frac{C_{A0} - C_A}{\tau}, \tag{6.72}$$

where τ is the average reactant residence (contact) time in the reactor.

It is important to note that, in an open system, reactant A formally also becomes an intermediate due to the channel of its introduction at a rate of $k_0 C_{A0}$ and a channel of removal at a rate of $k_0 C_A$, besides the channel of its conversion at a rate of r ($k_0 = 1/\tau$). According to Eq. (6.71) a steady-state regime is attained when the reaction rate (see Eq. (6.69)) is equal to the difference r_τ of reactant flow rates (expressed in, e.g., mole/(l h)) at the reactor input and output:

$$r_\tau = (C_{A0} - C_A)/\tau = k_0(C_{A0} - C_A), \tag{6.73}$$

which characterizes the reactor ($\tau = V/W_0$) and depends on the degree of conversion (χ) of reactant A, the contact or residence time (τ) and the initial concentration (C_{A0}):

$$r_\tau = C_{A0}\chi/\tau. \tag{6.74}$$

A single solution of the second-power equation with respect to χ, which is given by the formula (6.75) where $r = r_\tau$ (in $0 \le \chi \le 1$ interval)

$$\frac{C_{A0}\chi}{\tau} = \frac{k' C_{A0}(1-\chi)}{1 + K_1(1-\chi)C_{A0}}. \tag{6.75}$$

is derived at the point of intersection of the straight line $r_\tau = C_{A0}\chi/\tau$ and a monotonic hyperbolic curve of the reaction rate $r = f(\chi)$ that decreases in the $\chi = 0 - 1$ interval from $r = k_1' C_{A0}/(1 + K_1 C_{A0})$ to $r = 0$. An analogous situation takes place in the second-order reaction with respect to A for $r = k' C_A^2 = k' C_{A0}^2(1-\chi)^2$, since $r(\chi)$ is a monotonic function of the degree of conversion.

A quite different dynamical behavior of the kinetic model is observed for the same linear mechanism with an additional reaction that binds the catalyst in an inactive complex (represented by a pendant vertex on the kinetic graph), for example, as

$$M + 2A \underset{}{\overset{K_1}{\rightleftharpoons}} MA_2. \tag{6.76}$$

Under the same assumptions concerning the quasi-equilibrium first step in mechanism (6.67), we obtain

$$r = \frac{k_2 K_0 [C_A][M]_\Sigma}{1 + K_1 C_A + K_2 C_A^2}, \tag{6.77}$$

where $[M]_\Sigma = [M](1 + K_1 C_A + K_2 C_A^2)$. Here, for the same linear mechanism, the process exhibits a nonlinear material balance with respect to C_A (with respect to the catalyst), and the function $r = f(C_A)$ becomes nonmonotonic. This kinetic model was originally analyzed by Seelig [94a]. Later, an analogous model of a heterogeneous catalytic reaction in the CSTR was considered by Kol'tsov et al. [94b] for the following scheme:

$$A + Z \rightleftharpoons AZ;$$
$$AZ \rightleftharpoons B + Z;$$
$$Z + 2A \rightleftharpoons ZA_2.$$

Let us simplify the kinetic model (6.77) by assuming that the concentration of intermediate AM is relatively small ($[AM] \ll [M]_\Sigma$ and $K_1 C_A \ll 1$). This leads to the following equation:

$$r = \frac{k' C_{A0}(1-\chi)}{1 + K_1 C_{A0}^2 (1-\chi)^2}, \tag{6.78}$$

according to which the reaction rate r exhibits a maximum with respect to χ (Figure 6.2). The temporal variation of χ is described by the following equation:

$$\frac{d\chi}{dt} = \frac{k' C_{A0}(1-\chi)}{1 + K_2 C_{A0}^2 (1-\chi)^2} - \frac{C_{A0}\chi}{\tau}. \tag{6.79}$$

At $d\chi/dt = 0$, this is a cubic equation for χ that possesses three roots and, accordingly, three steady-state solutions χ^{ss}. Three real roots at $C_{A0} = 1\,M$ exist in the time interval of $\tau = 17.8 - 26\,s$. For the straight lines $a(\tau = 26\,s)$ and $c(\tau = 17.8\,s)$ in Figure 6.2, two of these real roots coincide (at the point of touching the $r = f(\chi)$ curve).

The straight line b (reactor equation at $\tau = 22\,s$) gives three different real roots (and three steady-state χ^{ss} values). In the case of line d ($\tau = 30\,s$), we have one real and two imaginary complex conjugate roots. In the region of $17.8\,s < \tau < 26\,s$, where three steady-state points exist in the interval of $\chi = 0.2 - 0.9$, the system features a positive feedback (whereby r increases with χ) and, accordingly, an unstable state $\chi^{ss(2)}$ appears that is characterized by the relation $\frac{dr}{d\chi} > \frac{dr_\tau}{d\chi}$. The system cannot occur in this state and shifts toward either the first stable state $\chi^{ss(1)}$ (for $\chi < \chi^{ss(2)}$) or the second sable state $\chi^{ss(3)}$ (for $\chi > \chi^{ss(2)}$).

Thus, the condition of stability for steady states χ^{ss} requires that

$$\frac{dr}{d\chi} < \frac{dr_\tau}{d\chi}$$

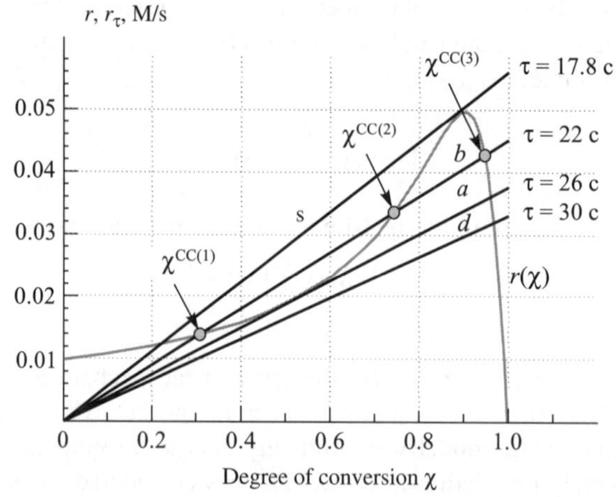

Figure 6.2 Plots of r and r_τ versus χ at $C_{A0} = 1.0\,M$, $k' = 1\,s^{-1}$, and $K_1 = 100\,M^{-2}$

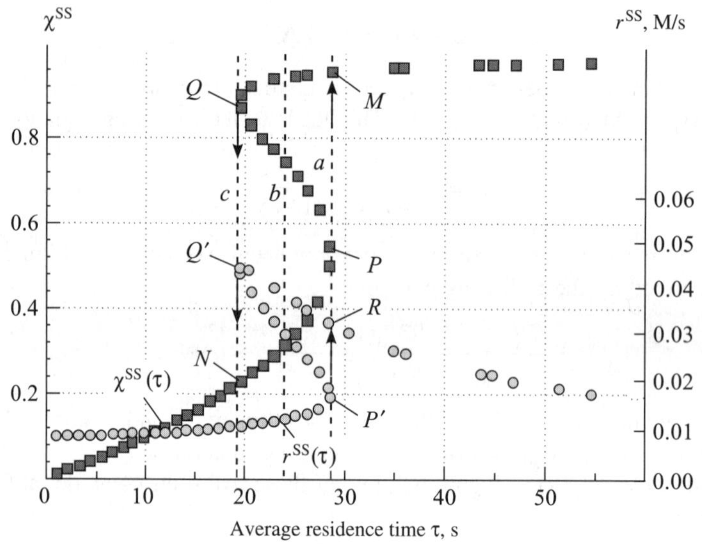

Figure 6.3 Hysteresis loops on $r = f(\tau)$ and $\chi^{SS} = f(\tau)$ curves, P and Q are turning points on $\chi^{SS} = f(\tau)$ curve; P', Q' are turning points on $r^{SS} = f(\tau)$ curve

at the intersection points of $r_\tau(\chi)$ and $r(\chi)$. The existence of a region of τ where the χ^{ss} values are unstable leads to the appearance of hysteresis on the $\chi^{ss} = f(\tau)$ and $r^{ss} = f(\tau)$ curves (Figure 6.3). As the contact time τ increases to ~20 s in the region of existence of a single steady state, both χ^{ss} and r^{ss} values slowly increase (Figure 6.3).

At point N that corresponds to the higher tangential state at $r^{ss} = r_\tau$ and $\tau = 17.8$ s, two new steady states appear but the system still occurs on lower branches of the hysteresis loops $\chi^{ss}(\tau)$ and $r^{ss}(\tau)$ up to the turning point P (or P'), although the other steady states are already available. However, the further, even very slight increase in τ (for the second tangential steady state) drives the system jump to the upper branch of the $\chi^{ss}(\tau)$ or $r^{ss}(\tau)$ hysteresis loop at point M (or R) (*ignition* point). The upper part of each loop is called thermodynamic, since the system on this branch approaches the equilibrium state (in the given case, full completion of the reaction).

Then, if variable τ is decreased so as to approach point M from the right, the system keeps on the upper branch until reaching the turning point Q, at which the upper steady state disappears and χ^{ss} falls at point N. When τ decreases in the $M-Q$ interval, the reaction rate increases to reach a maximum at point Q' (for $\chi^{ss} \cong 0.9$), at which it jumps down to the lower branch of the $r^{ss}(\tau)$ loop. Points Q and Q' are referred to as the reaction quenching points. Thus, the multiplicity of steady states (MSS) and the phenomena of reaction ignition and quenching are manifested at the tangential points, and the $Q-P$ branch bounded by these points contains the unstable steady states.

The jumps of the reaction rates and concentrations and the related system transition to a new state are referred to as bifurcation, and a parameter whose variation leads to these changes is called the bifurcation parameter. In the example under consideration, this parameter is represented by the residence time τ, whose variation in the interval of 17.8–26 s leads to the intersection of r and r_τ curves. In other words, their approach to $\tau = 17.8$ s and $\tau = 26$ s leads to the proximity of stable and unstable steady states, whereby the two roots coincide at the tangential points. At this very moment, the system goes away from a stable branch of the hysteresis loop.

The above example illustrates the phenomenon of bistability, that is, the transition of a system from one stable steady state to another in a certain interval of conditions. This is a particular case of the general phenomenon of polysteady states or multistability. A nonlinear (with respect to reactant) equation with a positive feedback (whereby an increase in the parameter χ decreases the degree of reaction inhibition by reactant A and thus increases the reaction rate) leads even in the case of a mechanism comprising a set of linear elementary steps with an overall equation A $\rightarrow$ B to a nonlinear kinetic model with respect to C_A, nonlinear effects, and interesting critical phenomena.

It should be noted that the kinetic equations of high orders with respect to [A], representing monotonic functions $r - f(C_A)$ such as

$$r = \frac{k' C_{A0}^2 (1 - \chi)^2}{1 + K_1 C_{A0} (1 - \chi)}, \tag{6.80}$$

give a single steady-state solution according to which the reaction rate is a decreasing hyperbolic function of τ, which tends to zero as $\tau \rightarrow \infty$.

Thus, even a simple example (considered in Section 6.2.2) reveals a number of issues that are important for the analysis of dynamical systems:

(i) existence, number, uniqueness, and multiplicity of steady states;
(ii) stability of steady states;
(iii) presence of positive feedback;
(iv) ignition and quenching of a process, which lead to bifurcations in the (r) rate and (χ) concentration of reactants;
(v) dependence of r and χ on τ, which determines the parametric portrait of a given dynamical system.

Evidently, the passage from a model with one unknown to a system of differential equations with a greater (two, three, etc.) number of unknowns and the passage from linear to nonlinear mechanism lead to more diverse dynamic behavior, including the oscillation of concentrations and other nonlinear effects.

In this context, it would be of interest to consider the chemical features of mechanisms and the physicochemical conditions of process organization that lead to nonlinear effects in the dynamics of chemical reactions.

6.2.3 Nonlinearity of kinetic models

There are several possible factors related to the nature of elementary steps, structure of the mechanism, and methods of conducting reactions, which lead to the nonlinearity of a homogeneous dynamical system, i.e., of the corresponding kinetic model [88].

Nonlinearity of steps. Examples of the mechanisms with bimolecular steps nonlinear with respect to intermediates ($X_1 + X_2 \rightarrow X_3$, $X_1 + X_2 \rightarrow X_3 + X_4$, $2X_1 \rightarrow X_2$, etc.) were considered in Chapter 1. As was noted there, the molecularity of an elementary step must not exceed $m = 2$. This condition was among several criteria of the elementarity of steps, according to which the participation of a reactant (e.g., substrate) in an elementary step is only possible provided that this step is linear with respect to the intermediate. A description of the step as

$$A + 2X_i \xrightarrow{k} 2X_j, \quad W = kC_A C_{X_i}^2 \tag{6.81}$$

is indicative of the use of a pseudo-elementary step, that is, of a block of elementary steps that can be obtained from two elementary steps with the first quasi-equilibrium reaction:

$$\begin{aligned} A + X_i &\rightleftharpoons AX_i; \\ AX_i + X_i &\longrightarrow 2X_j. \end{aligned} \tag{6.82}$$

The appearance of a product of variables (concentrations of intermediates) on the right-hand side of the system of differential equations of the model is among the most characteristic manifestations of nonlinearity.

Nonlinearity of the complexity functions of a catalyst or an intermediate. The presence of pendant vertices on a kinetic graph, that is, the formation of dead-end complexes from intermediates of the catalytic cycle, reactants, and reaction products leads to the appearance of additional terms in the material balance with respect to the catalyst. These terms correspond to fractional-rational expressions on the right-hand side of the system of differential equations that describes the given dynamical system. One example of the fractional-rational nonlinearity with respect to a reactant was considered in Section 6.2.2. In addition, fractional-rational terms on the right-hand side of the system of differential equations can appear in cases where some intermediates are described in the quasi-steady-state approximation, whereby an expression of the quasi-steady-state concentration of one or several intermediates enters into the corresponding differential equation.

The quasi-equilibrium approximation for the steps of formation of an intermediate can also be used for reducing the dimensionality of the system of equations and writing a fractional-rational term that describes the material balance:

$$\longrightarrow X_i \underset{1}{\overset{Y}{\rightleftharpoons}} X_j \underset{2}{\longrightarrow} X_k \longrightarrow;$$

$$W_2 = k_2 X_j = \frac{k_2 K_1 [X_i]_\Sigma [Y]}{1 + K_1 [Y]}, \quad ([X_i]_\Sigma = [X_i] + [X_j]). \tag{6.83}$$

The inhibition or self-inhibition reactions of the nth order are described in kinetic models [5, 88] by equations of the following type:

$$\frac{dC_i}{dt} = \frac{f(C_i, C_j)}{\alpha + C_j^n}; \tag{6.84}$$

$$\frac{dC_i}{dt} = \frac{f(C_i, C_j)}{\alpha + C_i^n}. \tag{6.85}$$

The inhibition (suppression) processes are frequently encountered in enzymatic processes. For example, the inhibition of the first step by the final product of a chain of enzymatic reactions

$$S_0 \xrightarrow{E_1} S_1 \xrightarrow{E_2} \cdots \longrightarrow S_n$$

leads to the appearance of terms such as (6.84) and self-sustained oscillations for $n = 3, 4$ [5] (see below). This type of nonlinearity is related to the material balance with respect to a catalyst described by relations of the type of Eqs (6.84) and (6.85).

Exponential nonlinearity. For processes carried out in a nonisothermal regime (e.g., in an adiabatic reactor), the equations of mass and heat balance contain nonlinear terms that are related to the exponential dependence of the rate constants and equilibrium constants on the temperature [$k_i = A\exp(-E/RT)$, $K_i = B\exp(-\Delta H/RT)$]. Kinetic models with a nonlinearity of this type are usually employed in analyzing the stability and modeling the behavior of industrial reactors [1–3, 17, 84]. This nonlinearity leads to the appearance of so-called thermokinetic oscillations in the temperature and concentrations. An exponential nonlinearity can also arise under isothermal conditions if the activation energies and/or enthalpies of steps depend on the concentrations of reactants and intermediates, which is frequently the case in the heterogeneous catalysis where the thermodynamic ($\Delta G^\circ, \Delta H^\circ$) or kinetic ($E_{act}$) characteristics of the steps involving adsorption depend on the degree of coverage of the surface by reactants.

It should be also noted that the exponential nonlinearity of various types leads to nonlinear models even if all steps are linear with respect to their intermediates.

One possible manifestation of the exponential nonlinearity is the appearance of positive and negative feedbacks in a kinetic model. For example, an increase in the temperature can lead to a growth in the rate constant that, in turn, increases the temperature of the catalyst and/or solution. In another example, an increase in the fraction of the surface area (Θ_A) occupied by reactant A on a nonuniform surface can decrease the heat of adsorption of this reactant, increase the adsorption activation energy (E_{act}), and reduce the rate constant.

Positive and negative feedback in homogeneous systems under isothermal conditions. The most frequently encountered type of nonlinearity in investigations of the dynamic behavior of kinetic models under homogeneous conditions is the autocatalysis, whereby a jth step is accelerated by a product of this step. A mathematical expression for this circumstance is as follows:

$$\frac{dC_i}{dt} = kC_i^n, \quad (n = 1, 2). \tag{6.86}$$

These steps represent one possible mechanism of the appearance of positive feedbacks.

Kinetic models frequently employ the so-called pseudo-elementary steps. For example, a reaction with the stoichiometry

$$A \longrightarrow B + C,$$

can also be written as follows:

$$\text{(a) } A \xrightarrow[k_a]{B} B + C$$

(where the autocatalyst is indicated above the arrow) or

$$\text{(b) } A + B \xrightarrow{k_a} 2B + C.$$

In both cases, the autocatalysis is described as

$$\frac{dC_{\mathrm{B}}}{dt} = k_a C_{\mathrm{A}} C_{\mathrm{B}}, \tag{6.87}$$

and both variants reflect the main fact that the rate of formation of product B is proportional to C_{B}. Variant (b) is referred to as the quadratic autocatalysis, while the reaction written as

$$\text{(c) A} + 2\text{B} \longrightarrow 3\text{B} + \text{C}$$

is called the cubic autocatalysis [17]. However, Zhabotinskii [5] believed that these types of autocatalysis should be more correctly referred to as linear (b) or quadratic (c) autocatalysis with respect to n in Eq. (6.86). Equations of the types (a), (b), and (c) can also be interpreted in terms of elementary steps by introducing additional variables (intermediates). For example, by introducing X_1 and using the limiting approximations (quasi-equilibrium or quasi-steady-state), variant (b) can be written as follows:

$$\text{A} + \text{B} \overset{K_1}{\rightleftharpoons} \text{X}_1; \qquad \text{X}_1 \xrightarrow{k_2} 2\text{B} + \text{C};$$

$$\frac{dC_{\mathrm{B}}}{dt} = k_2 K_1 C_{\mathrm{A}} C_{\mathrm{B}}.$$

It is interesting to note that autocatalyst U in the step [5]

$$\text{A} \xrightarrow[k]{\text{U}} \text{B} \tag{6.88}$$

can be formed from B in a quasi-equilibrium step as

$$n\text{B} \overset{K_1}{\rightleftharpoons} \text{U} \tag{6.89}$$

which leads to an equation that is analogous to a relation describing the autocatalytic step:

$$\frac{dC_{\mathrm{B}}}{dt} = kC_{\mathrm{A}}C_{\mathrm{U}} = kK_1 C_{\mathrm{A}} C_{\mathrm{B}}^n, \ (n = 1, 2). \tag{6.90}$$

Autocatalytic steps are used in the mechanisms of many oscillatory reactions and in models involving other nonlinear effects. For example, in the Landolt reaction, where the concentration of I_2 spontaneously increases after the consumption of a reductant (e.g., SO_3^{2-} or AsO_3^{3-}), the rate of periodate IO_3^- reduction by I^- is proportional to $[I^-]^2$. The iodine ions evolved as a result of the periodate reduction by SO_3^{2-} species, in turn, catalyze the reduction of periodate by I^-. In one of the steps, two intermediate molecules appear that lead to the formation of I_2 [33–35]:

$$\text{HIO}_2 + \text{I}^- + \text{H}^+ \longrightarrow 2\text{HOI};$$

$$\text{HOI} + \text{I}^- + \text{H}^+ \longrightarrow \text{I}_2 + \text{H}_2\text{O}. \tag{6.91}$$

Examples of autocatalysis were considered above (see Sections 4.1–4.3) in an analysis of oxidation processes in the CuCl–$CuCl_2$ and $PdCl_2$–$CuCl_2$. For examples, Cu (I) complexes in reactions

$$\text{RC}{\equiv}\text{CH} + 2\text{CuCl}_2 \xrightarrow{\text{Cu(I)}} \text{RC}{\equiv}\text{CCl} + 2\text{CuCl} + \text{HCl}; \tag{6.92}$$

$$2\text{RC}{\equiv}\text{CH} + 2\text{CuCl}_2 \xrightarrow{\text{Cu(I)}} \text{RC}{\equiv}\text{CC}{\equiv}\text{CR} + 2\text{CuCl} + 2\text{HCl}; \tag{6.93}$$

$$\text{RC}{\equiv}\text{CH} + \text{CO} + \text{MeOH} + 2\text{CuCl}_2 \xrightarrow{\text{Cu(I)}-\text{Pd(II)}} \text{RC}{\equiv}\text{CCOOMe} + 2\text{CuCl} + 2\text{HCl} \tag{6.94}$$

play the roles of both the catalysts and reaction products.

It should be emphasized that by no means all cases of self-acceleration are considered as autocatalysis. As was demonstrated above, an increase in the reaction rate could be related to a decrease in the degree of process inhibition by a reactant [78]. If a catalytic reaction is accompanied by the formation of active centers (as in catalytic processes of class I, see Section 5.5), the reaction rate would also increase. The self-acceleration observed in radical chain reactions in the course of consumption of an inhibitor or deceleration of the recombination of radicals at high degrees of conversion in polymerization processes should not be treated as autocatalytic phenomena.

The phenomenon of autoinhibition is manifested in cases where a product of some step decelerates this or some other step by reversibly (or irreversibly) binding a participant of this step into an inactive compound. This is an example of negative feedback. A typical case of autoinhibition is encountered in acid catalysis, where an acid catalyst is also a reactant. For example, the hydrochlorination or hydrobromination of olefins in nonaqueous media at low temperatures proceeds according to the following reaction:

$$\mathrm{RCH{=}CH_2 + HX \xrightarrow{HX} RCHXCH_3},$$

the rate of which is proportional to $[\mathrm{HX}]^2$.

In multi-route reactions of the catalysis with metal complexes involving the appearance of conjugation nodes between steps, the initial active complex (or an intermediate compound) can interact with one of the key intermediates of a given route I, making it inactive or converting it into an intermediate of another route II:

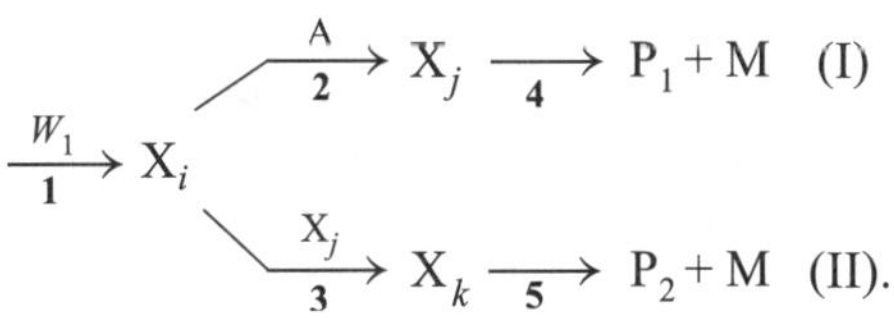

This block of steps has a single steady state for X_j, the position of which depends on the ratio of ω_2/ω_3 and ω_1/ω_4. For $X_j \to 0$ and $\omega_2 \gg \omega_3$ ($\omega_3 = k_3 X_j$), we have $r_{P_1} \to W_1$ and $r_{P_2} \to W_1$, whereas for $\omega_2 \ll \omega_3$ we have $r_{P_1} \to 0$ and $r_{P_2} \to W_1$. Thus, an increase in $[X_j]$ in step **2** can retard the formation of P_1, by decreasing $[X_i]$ in step **3** and thus reducing $[X_j]$. This is also a case of negative feedback.

Before proceeding to an analysis of nonlinear kinetic models, it is expedient to briefly recall the main concepts of the theory of dynamical systems and present the main approaches to the mathematical description and analyses of the stability of steady states and the structure of phase, dynamic, and parametric portraits of dynamical systems. The following brief presentation of the mathematical apparatus of the theory of dynamical systems is based on the monographs [1–3, 5, 9, 17, 91, 96–99] and handbooks [50, 100–102].

6.2.4 Main principles and methods of analysis of the dynamic behavior of nonlinear systems

As was demonstrated in Section 6.2.1, an analysis of the thermodynamic stability of nonequilibrium irreversible processes and, in particular, determination of the sign of the quadratic function $d(\delta^2 s)/dt$ for fluctuating thermodynamic forces and flows is an important tool for the investigation of dynamical systems [6, 8, 90, 91]. The thermodynamics of irreversible processes (e.g., chemical reactions) shows that various critical phenomena and nonlinear effects can take place far from the equilibrium. The nature of these effects, as related to specific features of the corresponding steady states, can be revealed using the mathematical methods developed by Poincaré. Andronov [41, 43] (see also [42]) was probably the first who understood and recognized this fact.

In application to homogeneous chemical processes, a dynamical system appears as a system of ordinary differential equations of the following type:

$$\boldsymbol{C} = \boldsymbol{B}_m^{\mathrm{T}}\boldsymbol{W}, \tag{6.95}$$

where $\boldsymbol{C}$ is the column-vector of the derivatives of concentrations C_i $\left(\frac{dC_i}{dt}\right)$, $\boldsymbol{W}_j$ is the column-vector of the reaction rates W_j of the jth step, and B_m^{T} is the transposed matrix of stoichiometric coefficients of the participants of the jth step. In the case of an open system, the right-hand side of this equation contains additional terms due to the rates of input and output flows of the reaction participants (see Section 6.2.2).

An important factor that determines the behavior of a dynamical system is the dimensionality. The dimensionality of a system is the minimum number of linearly independent variables (and, hence, linearly independent differential equations) necessary to describe it. Solutions obtained in the spaces of various dimensions are different because of the differences in their geometries. The space of variables is called the phase space. By monitoring the variables $x_i = f(t)$ as functions of the time t that is varied from $-\infty$ to $+\infty$, one obtains an oriented curve that is called the phase trajectory (e.g., $x_i = f(x_j)$). Quasi-steady states (solutions) in closed systems ($dx_i/dt = 0$) or steady states in open systems ($dC_i/dt = 0$), where C_i are the concentrations of all reactants and intermediates, are called singular points. The trajectories attain and leave these points only in the limit (i.e., as $t \to +\infty$ and $t \to -\infty$, respectively). The phase trajectories can be either open or closed. The closed trajectories correspond, e.g., to the motion of a harmonic oscillator, which is imaged by an ellipse on the (x_i, x_j) phase plane.

In the analysis of dynamical systems, a special important issue is the concept of closed trajectories as introduced by Poincaré. A limiting set of open trajectories occurring inside or outside a periodic trajectory for two-dimensional systems is called the limit-cycle. Simple steady states that attract trajectories in the phase space (and are called attractors) include singular points, periodic trajectories of the limit-cycle type, and stable toruses. To provide a mathematically strict description of a dynamical system means introducing a topological (qualitative) subdivision of the phase space into the regions of existence of the trajectories belonging to various topological types and to construct a phase portrait of the system [99]. The main problem encountered in the construction of phase portraits of a given system consists of determining the topological nature of singular points (steady states) and evaluating their stability, including the stability of limit-cycles.

In any limited region of the phase space, a dynamical system can have a finite number of isolated steady states (SS). Among the different possible steady states (if there are several of these), the system usually features the states that are stable with respect to unavoidable fluctuations in the concentrations or parameters (temperature, rate of reactant supply into a CSTR, etc.). If the ith reactant concentration at a steady state is x_i^{ss}, then the fluctuation δx_i brings a new value to this variable:

$$x_i(t) = x_i^{\mathrm{ss}} + \delta x_i. \tag{6.96}$$

In cases where a fluctuation that arises at a moment of time t_0 spontaneously decays within a period of time $t - t_0 \gg \tau$ (where τ is called the system relaxation time), then the given state is called asymptotically stable. If δx_i remains virtually unchanged rather than grows, then the state is classified as stable [50]. The condition of fluctuation vanishing with the time,

$$|\delta x_i(t)| = |x_i(t) - x_i^{\mathrm{ss}}| \longrightarrow 0 \quad \text{for} \quad t \longrightarrow \infty$$

is called the *Lyapunov criterion* for the asymptotic stability. An unstable system, once deviating from a steady state, would move toward another (close) stable steady state. Let us consider the scheme of an analysis for the stability according to Lyapunov in the vicinity of a steady state for a two-dimensional nonlinear system on the plane represented by a system of two differential equations:

$$\begin{cases} dx/dt = f_1(x, y) \\ dy/dt = f_2(x, y) \end{cases}, \tag{6.97}$$

where x and y are the variable concentrations. Singular points (steady states) of this system are determined by solving a system of algebraic equations for $dx/dt = 0$ and $dy/dt = 0$:

$$f_1(x^{ss}y^{ss}) = 0; \qquad f_2(x^{ss}y^{ss}) = 0. \tag{6.98}$$

A conclusion concerning the stability or instability of a steady state is based on the behavior of trajectories near this point. Upon excluding the time from Eqs (6.97), we arrive at the following equation:

$$\frac{dx}{dy} = \frac{f_1(x, y)}{f_2(x, y)}, \tag{6.99}$$

which determines the slope of the trajectory as dx/dy at each point of the phase plane (x, y), while the direction of the trajectory is determined from Eqs (6.97). The derivative (6.99) is indeterminate only at the singular point, where $f_1(x, y) = 0$ and $f_2(x, y) = 0$.

A simpler analysis of the nature of singular points and their stability can be performed using the method of linearization of system (6.97) with respect to the deviations from steady states. Defining the deviations from steady states as

$$\varepsilon = x - x^{ss}; \quad \eta = y - y^{ss}, \tag{6.100}$$

expanding (6.97) into Taylor's series with respect to η and ε, and retaining only the first derivatives of $f_1(x, y)$ with respect to ∂x and ∂y at point (x^{ss}, y^{ss}),

$$\begin{cases} \dfrac{d\varepsilon}{dt} = \dfrac{\partial f_1(x^{ss}, y^{ss})}{\partial x}\varepsilon + \dfrac{\partial f_1(x^{ss}, y^{ss})}{\partial y}\eta \\ \dfrac{d\eta}{dt} = \dfrac{\partial f_2(x^{ss}, y^{ss})}{\partial x}\varepsilon + \dfrac{\partial f_2(x^{ss}, y^{ss})}{\partial y}\eta \end{cases}, \tag{6.101}$$

we obtain the following linearized system relative to the new origin of coordinates:

$$\begin{cases} \dfrac{d\varepsilon}{dt} = a_{11}\varepsilon + a_{12}\eta \\ \dfrac{d\eta}{dt} = a_{21}\varepsilon + a_{22}\eta \end{cases}. \tag{6.102}$$

The second-order terms in the Taylor's series,

$$\frac{1}{2}\left\{\frac{\partial^2 f_i(x^{ss}, y^{ss})}{\partial x^2}\varepsilon^2 + 2\frac{\partial^2 f_i(x^{ss}, y^{ss})}{\partial x \partial y}\varepsilon \cdot \eta + \frac{\partial^2 f_i(x^{ss}, y^{ss})}{\partial y^2}\eta^2\right\}, \tag{6.103}$$

as well as higher-order terms in the local stability analysis are not taken into consideration. Equations (6.102) can also be obtained in a still simpler way by substituting Eqs (6.100) and (x^{ss}, y^{ss}) into system (6.97) and excluding quadratic terms with ε^2, η^2, and $\varepsilon\eta$.

Let us seek for a solution to the linear system (6.102) in the form of exponents:

$$\varepsilon = c_1 e^{\lambda t}; \quad \eta = c_2 e^{\lambda t}. \tag{6.104}$$

Substituting expressions (6.104) into Eq. (6.102), we obtain the following equations:

$$\begin{cases} (a_{11} - \lambda)c_1 + a_{12}c_2 = 0 \\ a_{21}c_1 + (a_{22} - \lambda)c_2 = 0 \end{cases}; \tag{6.105}$$

which can be rewritten in the matrix form as follows:

$$\begin{vmatrix} a_{11}-\lambda & a_{12} \\ a_{21} & a_{22}-\lambda \end{vmatrix} \times \begin{vmatrix} c_1 \\ c_2 \end{vmatrix} = 0; \qquad \boldsymbol{Mc} = 0. \tag{6.106}$$

For the coefficients c_1 and c_2 to be nonzero, the determinant (det) of the characteristic matrix $\boldsymbol{M}$ must obey the condition

$$\det|\boldsymbol{A} - \lambda \boldsymbol{E}| = 0, \tag{6.107}$$

where

$$\boldsymbol{A} = \begin{vmatrix} a_{11} & a_{12} \\ a_{21} & a_{22} \end{vmatrix}$$

is the Jacoby matrix (with Jacobian $\boldsymbol{J}$); λ are the eigenvalues of the characteristic matrix $\boldsymbol{M}$, and $\boldsymbol{E}$ is the unity matrix. Thus, the linearized system of equations (6.102) in the matrix form is as follows:

$$\frac{d\boldsymbol{U}}{dt} = \boldsymbol{AU}, \tag{6.108}$$

where $\boldsymbol{U}$ is the column-vector of deviations (ε, η) from the steady -state. The eigenvalues of matrix $\boldsymbol{M}$ are given by the roots of the characteristic equation. For the two-dimensional system (6.102), this equation can be written as

$$\lambda^2 - \sigma\lambda + \Delta = 0, \tag{6.109}$$

and its roots appear as

$$\lambda_{1,2} = \frac{\sigma}{2} \pm \frac{1}{2}\sqrt{\sigma^2 - 4\Delta}, \tag{6.110}$$

where $\sigma = \mathrm{tr}\boldsymbol{A} = a_{11} + a_{22}$ (trace of matrix $\boldsymbol{A}$) and $\Delta = \det\boldsymbol{A}$. These values determine both the type of singular point (steady state) and its stability.

In respect of their stability under the conditions of fluctuating parameters, dynamical systems can be subdivided into robust – the dynamic behavior of which is not changed upon small variations in the right-hand side of Eqs (6.97) – and non-robust. The singular points (i.e., steady states) are also classified as robust and non-robust. According to Poincaré and Lyapunov, the behavior of the initial nonlinear system (6.97) in the vicinity of robust steady states is the same as that of the linearized system (6.102). If all roots λ_i are different and their real parts $\mathrm{Re}(\lambda_i)$ are nonzero, the corresponding steady states are robust. In addition, if all $\mathrm{Re}(\lambda_i) < 0$, the steady states are asymptotically stable according to Lyapunov. If at least one $\mathrm{Re}(\lambda_i) > 0$, then the steady states are unstable. As can be seen from Eq. (6.110), the steady states corresponding to $\sigma \neq 0$ and $\Delta \neq 0$ are robust. If one or both of the two parameters (σ, Δ) equal zero, the steady state is non-robust. In this case, the phase portrait of a dynamical system depends on nonlinear terms of the expansion of system (6.97) into Taylor's series.

Let us consider the topology and stability of particular singular points (steady states) in more detail.

6.2.4.1 Robust singular points

1. $\Delta > 0,\ \sigma \neq 0,\ \sigma^2 - 4\Delta > 0$:
   - λ_i are real numbers of the same sign; singular point is a *node*.
   - $\sigma < 0$, stable node; trajectories monotonically approach singular point on the phase plane (Figure 6.4a).
   - $\sigma > 0$, unstable node; trajectories monotonically go away from singular point.

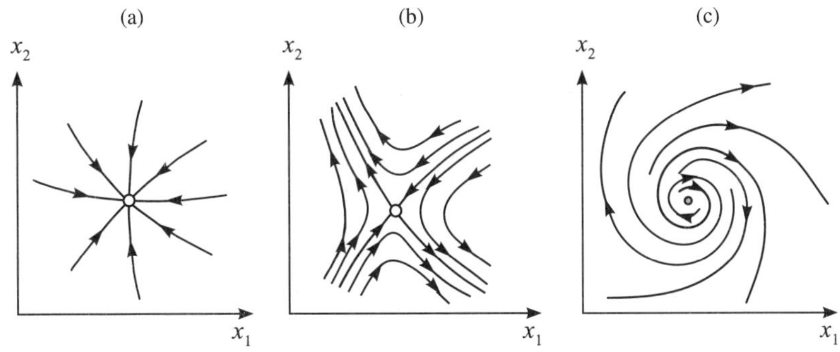

Figure 6.4 Robust singular points: (a) stable node; (b) saddle; (c) unstable focus (reproduced with permission from AIPC Nauka © 1974 [5])

2. $\Delta < 0,\ \sigma \neq 0,\ \sigma^2 - 4\Delta > 0$:
   - λ_i are real numbers of opposite signs; singular point of the *saddle* type. Two integral curves passing through this singular point are called *separatrices* (Figure 6.4b). Trajectories near the saddle point are nonmonotonic. The saddle is an unstable singular point.

3. $\Delta > 0,\ \sigma \neq 0,\ \sigma^2 - 4\Delta < 0$:
   - λ_i are complex conjugate numbers; singular point is a *focus*.
   - $\sigma < 0,\ Re(\lambda_i) < 0$, stable focus; the variable (concentration) exhibits decaying oscillations on approach to steady state.
   - $\sigma > 0,\ Re(\lambda_i) > 0$, unstable focus (Figure 6.4c); oscillations increase on going away from steady state.

6.2.4.2 Non-robust singular points

1. $\Delta > 0,\ \sigma = 0,\ \sigma^2 - 4\Delta < 0,\ Re(\lambda_i) = 0$:
   - Singular point of the *center* type (Figure 6.5). In the linear approximation, phase trajectories are closed curves. In a nonlinear system, phase trajectories appear as densely wound spirals (complex focus); the variable exhibits nondecaying oscillations, but the steady state is asymptotically unstable.

2. $\Delta = 0,\ \sigma \neq 0,\ \sigma^2 - 4\Delta > 0$:
   - Singular point of the *saddle–node* type (Figure 6.6).
   - $\lambda_1 = 0$; $\lambda_2 > 0$ and $\lambda_2 < 0$ correspond to exponential increase and decrease, respectively, in the variable (concentration); if the parameter is changed, this singular point separates into several robust points.

3. $\Delta = 0,\ \sigma = 0,\ \lambda_{1,2} = 0$:
   - Most complicated variant of singular point (Figure 6.7).

According to Poincaré and Lyapunov, phase portraits of a dynamical system and its linearized model in the vicinity of an immobile singular point are qualitatively equivalent, provided that this point of not of the center type.

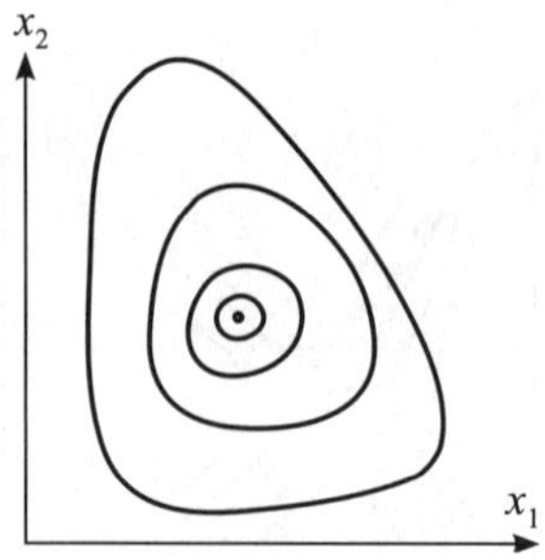

Figure 6.5 Center-type singular point (reproduced with permission from AIPC Nauka © 1974 [5])

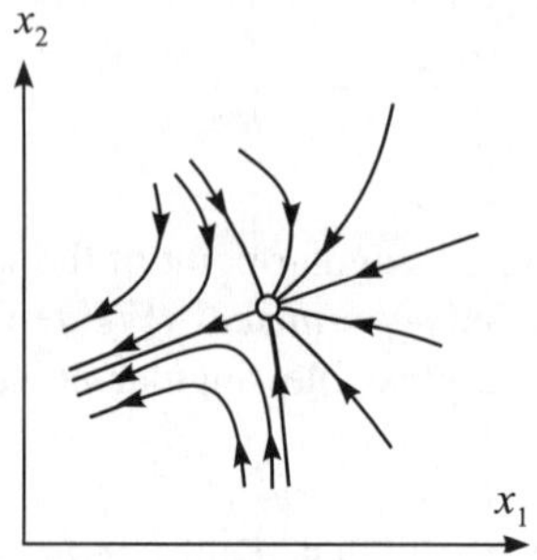

Figure 6.6 Saddle-node type singular point (reproduced with permission from AIPC Nauka © 1974 [5])

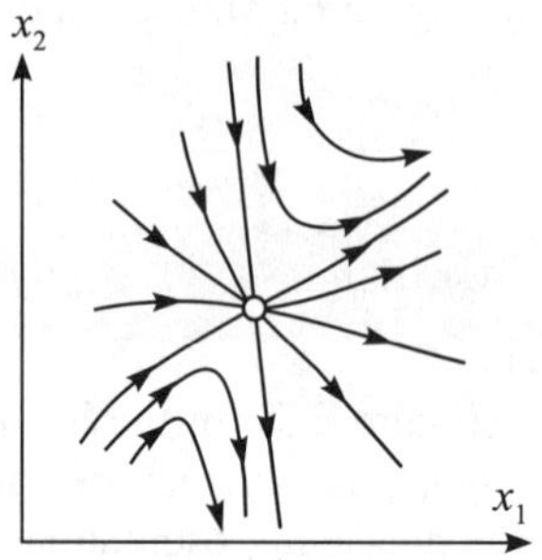

Figure 6.7 Complex singular point (reproduced with permission from AIPC Nauka © 1974 [5])

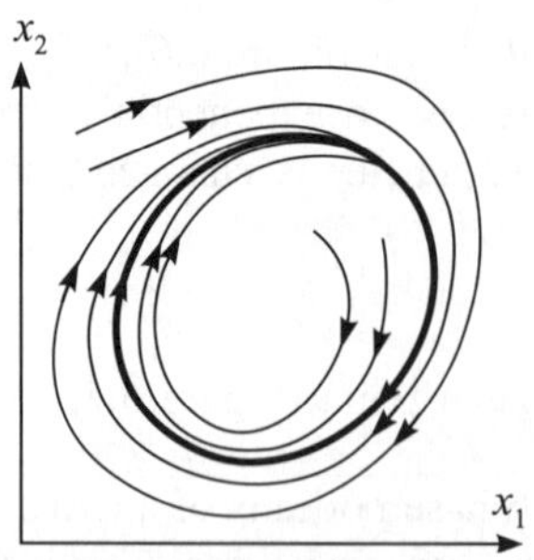

Figure 6.8 Stable limit cycle (reproduced with permission from AIPC Nauka © 1974 [5])

There are various types of system trajectories in the phase space, among which two special types are the separatrices of saddle points and the limit-cycles (LCs).

The first special case is the LC, which represents closed trajectories. A stable LC is the mathematical model (image) of self-sustained oscillations. If the trajectories appear as winding on an LC, this LC is stable (Figure 6.8); if the trajectories are unwinding, the LC is unstable. There are several theorems concerning the existence of LCs according to Poincaré and Andronov, which are reproduced here from monograph [5].

(i) A closed trajectory (in particular, LC) contains at least one singular point of the node or focus type (but not of the center type).
(ii) If there is more than one singular point inside a closed trajectory, the number of nodes (foci) is always greater by one than the number of saddle points.
(iii) If there is a single unstable focus (node) and the infinity is unstable (i.e., all phase trajectories cross a circle of radius R centered at $x = y = 0$ in the inward direction), then there exists at least one stable LC (Figure 6.9).

If a system has a center that transforms into an unstable focus when the parameters (σ, Δ) are varied, then there is a quite high probability of the appearance of an LC, whereby the system with $\sigma = 0$ and $\Delta > 0$ changes sign and passes to a state with $\sigma > 0$ (see below).

The second special kind of trajectories is the saddle-point separatrix, which can terminate in a node, focus, or in the same or different saddle (these cases represent non-robust singular points).

In an analysis of the phase portraits of nonlinear systems upon their linearization performed using the method described above (i.e., local phase portraits in the vicinity of a steady state), one should bear in mind that the possible existence of several singular points (polysteady states) can result in that global phase portraits would differ from local ones. In particular, the LCs can be correctly determined only by an analysis of the global phase portrait. In linear systems, a periodic motion by an elliptic trajectory in

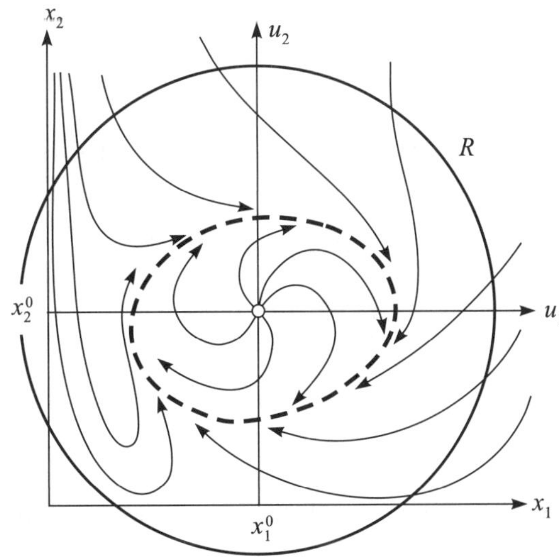

Figure 6.9 Illustration to theorems concerning the existence of limit-cycle (reproduced with permission from AIPC Nauka © 1974 [5])

the presence of a singular point of the center type is possible only for the dynamical systems describing a harmonic oscillator without energy losses (see below). As was noted above, these trajectories are unstable.

The values of parameters (constants of a given system, initial concentrations) at which the behavior of a dynamical system qualitatively changes are called bifurcation parameters. The values of $\sigma = 0$ and $\Delta = 0$ corresponding to these parameters divide the phase plane into regions with different numbers, characters, and stabilities of singular points. In particular, on the passage via $\sigma = 0$ ($\Delta > 0$), a stable focus becomes unstable and this may be accompanied by the separation of an LC. On the passage via $\sigma = 0$, a complex focus can separate into a simple focus and LC. All these bifurcations, which involve a change in the character of singular points, are called the Andronov–Hopf bifurcations. Bifurcations of some other kinds and the possible scenarios of LC decay will be considered below (see Section 6.3) in an analysis of particular dynamical systems.

For a two-dimensional problem with two parameters (α, β) is the following special lines can be constructed on the parametric portrait so as to separate it into regions with different properties of singular points [5].

- *Line of neutrality* is determined by solving the system of equations

$$\begin{cases} f_1(x^{ss}, y^{ss}, \alpha, \beta) = 0 \\ f_2(x^{ss}, y^{ss}, \alpha, \beta) = 0 \\ \sigma(x^{ss}, y^{ss}, \alpha, \beta) = 0 \end{cases}$$

  and shows the boundary between stable and unstable steady states.
- *Line of monostationarity*, which separates the regions with one and three steady states, is determined by the following system of equations:

$$\begin{cases} f_1(x^{ss}, y^{ss}, \alpha, \beta) = 0 \\ f_2(x^{ss}, y^{ss}, \alpha, \beta) = 0. \\ \Delta(x^{ss}, y^{ss}, \alpha, \beta) = 0 \end{cases}$$

- *Line of multiplicity* of roots of the characteristic equation separates the regions of foci and nodes and is determined by the following system of equations:

$$\begin{cases} f_1(x^{ss}, y^{ss}, \alpha, \beta) = 0 \\ f_2(x^{ss}, y^{ss}, \alpha, \beta) = 0, \\ D(x^{ss}, y^{ss}, \alpha, \beta) = 0 \end{cases}$$

$$D = \sigma^2 - 4\Delta.$$

Let us briefly consider three-dimensional problems. In this case, the characteristic equation has the following form:

$$\lambda^3 + b\lambda^2 + c\lambda + d = 0, \tag{6.111}$$

where the coefficients are determined using the Jacoby matrix of the system with three variables:

$$\mathbf{A} = \begin{vmatrix} a_{11} & a_{12} & a_{13} \\ a_{21} & a_{22} & a_{23} \\ a_{31} & a_{32} & a_{33} \end{vmatrix}; \tag{6.112}$$

The coefficients in Eq. (6.111) are as follows:

$$b = \text{-tr}\mathbf{A} = -(a_{11} + a_{22} + a_{33});\ \ d = -\det\mathbf{A}.$$

$$c = \begin{vmatrix} a_{11} & a_{12} \\ a_{21} & a_{22} \end{vmatrix} + \begin{vmatrix} a_{22} & a_{23} \\ a_{32} & a_{33} \end{vmatrix} + \begin{vmatrix} a_{11} & a_{13} \\ a_{31} & a_{33} \end{vmatrix}. \tag{6.113}$$

According to the Routh–Hurwitz criterion, a steady state is stable provided that $b > 0$ ($\text{tr}\mathbf{A} < 0$), $c > 0, d > 0$, and $bc - d > 0$. If at least one of the higher coefficients in the characteristic polynomial (here, coefficient b in Eq. (6.111)) is negative, then the corresponding steady state is unstable. When, in addition, the last term (d) is positive, then a single steady state is unstable, i.e. the system admits oscillations of the concentrations.

For three-dimensional systems, there are five types of robust singular points and a large number of possible combinations of the λ_i values [99]. Restricting the consideration to attractors (i.e., stable steady states), the general pattern for the systems of various dimensions (n) is as follows [91]:

- $n = 1$ (1) $\lambda_1 < 0$ stable node;
- $n = 2$ (1) $\lambda_1 < 0,\ \lambda_2 < 0$ node and focus;
  (2) $\lambda_1 = 0,\ \lambda_2 < 0$ LC;
- $n = 3$ (1) $\lambda_i < 0$ stable singular points;
  (2) $\lambda_1 = 0,\ \lambda_{2,3} < 0$ stable LC;
  (3) $\lambda_1 = 0,\ \lambda_2 = 0,\ \lambda_3 < 0$ stable two-dimensional torus;
  (4) $\lambda_1 > 0,\ \lambda_2 = 0,\ \lambda_3 < 0$ strange attractor, chaotic oscillations.

For systems of any dimensionality, the number of steady states and their stability are determined from an analysis of coefficients of the characteristic polynomial by solving the equation

$$P(\lambda) = \lambda^n + a_1\lambda^{n-1} + a_2\lambda^{n-2} + \cdots + a_{n-r}\lambda^r = 0,$$

where n is the number of unknown variables, and r is the number of conservation laws. For example, in the case of a single catalyst participating in the reaction mechanism, we have $r = 1$. In this case, $\lambda = 0$ is among the eigenvalues, since the polynomial appears as

$$\lambda(\lambda^{n-1} + a_1\lambda^{n-2} + a_2\lambda^{n-3} + \cdots + a_{n-1}) = 0.$$

For $n = 4$, the eigenvalues are determined from the equation

$$\lambda^3 + a_1\lambda^2 + a_2\lambda + a_3 = 0,$$

which is equivalent to Eq. (6.111) with $a_1 = b,\ a_2 = c$, and $a_3 = d$.

Special approaches have been developed for studying the behavior of a kinetic model without recourse to an analysis of the corresponding system of differential equations. Using these approaches, it is possible to predict the presence of one or several steady states and judge their stability based on an analysis of the topological structure of the proposed mechanism. Here, the topological structure implies (i) an interrelationship between intermediates with allowance for the stoichiometry of steps and (ii) an interrelationship between complexes of reactants (i.e., combinations of their symbols with nonnegative integer coefficients on both sides from the arrow in the equation of each step).

The first, most strict but very laborious solution of this problem was obtained by Clarke [103–110] with the aid of graph theory methodology. An important step in this direction was made in the investigations by Ivanova and co-workers [111–115], where bipartite graphs were used to develop a method [116] for determining the polynomial coefficients, general conditions were formulated for the existence of a single steady state, and criteria were found for separating the regions of parameters corresponding to poly-steady states and unstable single steady states. An algorithm for determining the polynomial coefficients was implemented as a personal computer program [113]. However, the most complicated step, involving the analysis of critical inequalities for determining the signs of coefficients a_i, is still not formalized. In continuation of these investigations, an algebraic method of determining a_i values was proposed [117, 118], which employed the matrix of stoichiometric coefficients of the mechanism and led to simpler inequalities.

An alternative approach to discriminating the classes of mechanisms with a single positive and (both locally and globally) stable steady state was developed by Horn, Jackson, and Feinberg [119–125]. According to the Feinberg theorem of complex balancing (deficiency zero) of a reaction network, if the deficiency δ defined as the difference between the number n of complexes and the numbers of connectivity classes of the mechanism (l) and linearly independent steps (S, or the rank of matrix $\boldsymbol{B}_m$), is zero

$$\delta = n - l - S = 0, \tag{6.114}$$

then the mechanism exhibits a "quasi-thermodynamic" behavior [116] and possesses a single stable steady state. This rule is also applicable to open systems [122]. Nonzero deficiency values, $\delta = 1$ and $\delta = 2$, are indicative of the possibility of nontrivial dynamic behavior (including oscillatory regimes). Feinberg also proposed a method for the qualitative analysis of the properties of special bipartite graphs consisting of the vertices of two kinds, substances (S) and connectivity classes (L). Edges connecting the vertices of types S and L correspond to complexes (C) involved in steps of the given mechanism. Based on the analysis of SCL graphs [122–125], it is possible to judge on the possible dynamic behavior of a kinetic model.

6.3 Analysis of simple nonlinear kinetic models

In 1987, Razon and Schmitz [126] reviewed a large number of investigations (more than 40) in which the necessary and sufficient (or both) conditions for attaining single or multiple steady states in reactors of various types and configurations were established for different kinetic models. The reviewed results were mostly obtained in heterogeneous reactions and under nonisothermal conditions. Heterogeneous catalytic reactions were also considered in monographs [12–14].

Many interesting homogeneous catalytic processes (including enzymatic reactions) were analyzed [5, 9, 11, 17, 50]. Below, these results and an analysis of some reaction mechanisms by Yablonsky et al. [12] are used to consider the stability of steady states and the dynamic behavior of nonlinear models and nonlinear mechanisms involving various types of nonlinearity. Prior to this analysis, let us analyze the stability of nonlinear kinetic models with linear mechanisms for some reactions in continuous stirred-tank reactors (CSTRs) considered above.

Example 6.1

► Consider catalytic reaction A → B in a CSTR with the two-step mechanism (6.67), where the concentration of reactant A obeys the following equation:

$$\frac{dC_A}{dt} = \frac{C_{A0} - C_A}{\tau} - \frac{kC_A}{1 + K_1 C_A} = P(C_A). \tag{6.115}$$

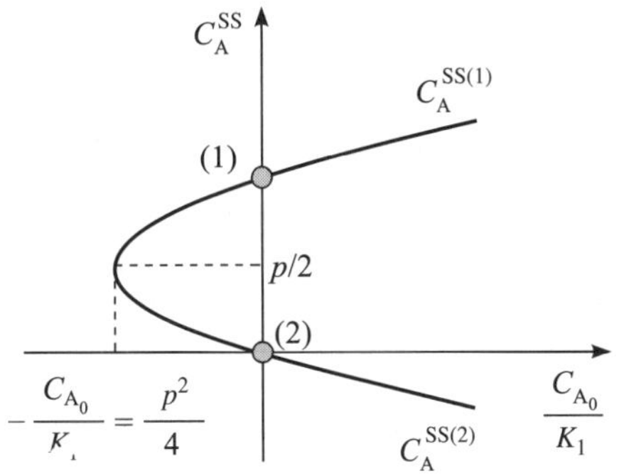

Figure 6.10 Plot of steady-state C_A^{SS} values versus C_{A0}/K_1 [see Eq. (6.117)]

Under steady-state conditions $[P(C_A) = 0]$, the steady-state concentrations $C_A^{ss(1,2)}$ are determined by solving the quadratic equation

$$C_A^2 + pC_A + q = 0, \tag{6.116}$$

where $p = -(K_1C_{A0} - k\tau - 1)/K_1$ and $q = -C_{A0}/K_1$, which yields

$$C_A^{ss(1,2)} = \frac{K_1C_{A0} - k\tau - 1}{2K_1} \pm \sqrt{\frac{(K_1C_{A0} - k\tau - 1)^2}{4K_1^2} + \frac{C_{A0}}{K_1}}. \tag{6.117}$$

Depending on parameter q (in the region of $q \leq 0$), the steady-state solutions $C_A^{ss(1,2)}$ occur on the upper and lower branches, respectively, of a parabolic curve (Figure 6.10). The physical meaningful solutions correspond to the upper branch on the right of point (1), at which $C_A^{ss(1)} = |p|$. At a point with $-C_{A0}/K_1 = p^2/4$, the system exhibits the bifurcation and passes into a state where Eq. (6.116) has no steady-state solutions.

Let us assess the stability of steady states $C_A^{ss(1)}$ and $C_A^{ss(2)}$ in a linear approximation described by the equation

$$d\varepsilon/dt = a\varepsilon, \tag{6.118}$$

where ε is the fluctuation of concentration C_A and coefficient a is the derivative $\partial P(C_A)/\partial C_A$ at points $C_A^{ss(1)}$ and $C_A^{ss(2)}$:

$$a = -\frac{1}{\tau} - \frac{k}{(1 + K_1C_A^{ss})^2}. \tag{6.119}$$

Since $a < 0$ and eigenvalue $\lambda < 0$ $(a = \lambda)$ for all C_A^{ss}, the steady states on $C_A^{ss(1)}$ and $C_A^{ss(2)}$ branches are stable, and the singular point corresponds to a stable node. The value of concentration C_A fluctuation ε exponentially approaches the singular point as $t \to \infty[\varepsilon = c_1 \exp(-at)]$. ■

Example 6.2

▶ Let us check for the stability of three steady states determined by Eq. (6.79) in the following form:

$$\frac{dC_A}{dt} = \frac{C_{A0} - C_A}{\tau} - \frac{k'C_A}{1 + K_1C_A^2} = P(C_A). \tag{6.120}$$

In this case, the value of coefficient a in linear equation (6.118) is

$$a = \left.\frac{\partial[P(C_A)]}{\partial C_A}\right|_{C_A^{ss}} = -\frac{1}{\tau} - \frac{k'\left(1 - K_1(C_A^{ss})^2\right)}{\left[1 + K_1(C_A^{ss})^2\right]^2}, \tag{6.121}$$

and it can be seen that the sign of this coefficient depends on the sign of term $[1 - K_1(C_A^{ss})^2]$ and the relationship between the terms $-\partial r_\tau/\partial C_A$ and $\partial r/\partial C_A$. For example, in the case of $\tau = 22$, $C_{A0} = 1$, $k' = 1$, and $K_1 = 100$, the values of coefficient a for three steady states C_A^{ss}, namely $C_A^{ss(1)} = 0.06, C_A^{ss(2)} = 0.25$, and $C_A^{ss(3)} = 0.68$ (see Figure 6.2), are as follows:

$$a_1 = -0.045 - 0.346 = -0.391 < 0, \text{ (stable node)};$$

$$a_2 = -0.045 + 0.0998 = 0.0549 > 0, \text{ (unstable node)};$$

$$a_3 = -0.045 + 0.0203 = -0.0247 < 0, \text{ (stable node)}. \quad ■$$

Example 6.3

► An analogous analysis of the stability of steady states for the nonlinear equation (6.80) written as

$$\frac{dC_A}{dt} = \frac{C_{A0} - C_A}{\tau} - \frac{k'C_A^2}{1 + K_1 C_A} = P(C_A) \tag{6.122}$$

showed that positive $C_A^{ss(1)}$ values occur on the upper branch of the curve in Figure 6.10 (similar to the situation in Example 6.1 with Eq. (6.117)), but the q value is expressed as

$$q = \frac{C_{A0}}{k'\tau + K_1}.$$

The derivative of $P(C_A)$ with respect to C_A yields

$$a = -\frac{1}{\tau} - \frac{(2k'C_A^{ss} + k'K_1[C_A^{ss}]^2)}{(1 + K_1 C_A^{ss})^2}, \tag{6.123}$$

so that $a < 0$ (stable node) for all $C_A^{ss(1)} > 0$. The steady states $C_A^{ss(2)}$ will be stable ($a < 0$) provided that $|C_A^{ss(2)}| > 2/K_1$. ■

The kinetic models with linear mechanisms, in which the catalysts exhibit complexation with the substrates, products, or inhibitors in solution, are widely used in enzymatic catalysis, where any kind of catalyst binding to the substrate is called the substrate-induced suppression [11]. As a result, the reactions in CSTRs have to be described by nonlinear models that are more complicated than those considered in Section 6.2.2 and Examples 6.1–6.3. Let us consider two processes of this kind [11].

Example 6.4

► A single-substrate reaction of the type

$$S_1 \overset{M}{\rightleftharpoons} S_2 \tag{6.124}$$

proceeds according to a linear mechanism with respect to the intermediates. This mechanism is represented by the kinetic graph depicted in Figure 6.11, where substrate S_1 forms inactive complexes with intermediates MS_1 and MS_2 and it can be assumed that $K_i = k_i^-/k_i^+$ are the same for both pendant vertices. In

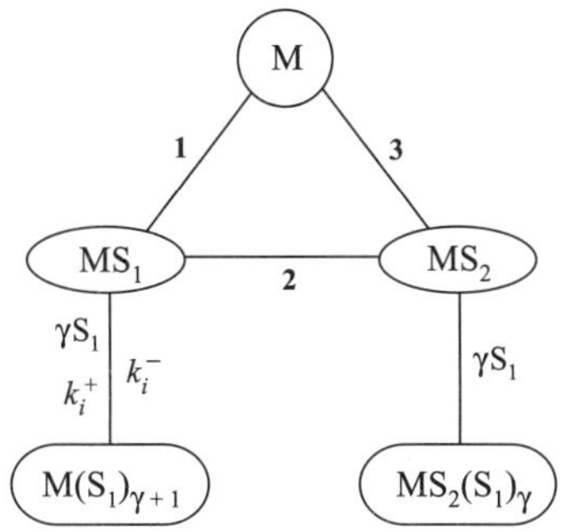

Figure 6.11 Mechanism of substrate-inhibited reaction (6.124)

the quasi-steady-state approximation, a dimensionless kinetic equation for the rate of reaction (6.124) is as follows:

$$\nu = \frac{\sigma_1 - \alpha\sigma_2}{\xi + (\sigma_1 + \sigma_2)(\xi + \sigma_1^{\gamma})}, \tag{6.125}$$

where $\xi = (K_i/K_1)^{\gamma}$ and $\alpha = \nu_-/\nu_+$ are the parameters (ν_- and ν_+ being the maximum rates of reaction (6.124) in the forward and reverse directions, respectively), $\sigma_1 = C_{S_1}/K_1$, $\sigma_2 = C_{S_2}/K_2$, and $K_i^{\gamma} = (k_i^-/k_i^+)^{\gamma}$ $(i = 1, 2)$, K_1 and K_2 are the complexes of rate constants for the corresponding elementary steps of mechanism in Figure 6.11.

A kinetic model of process (6.124) with a continuous supply of substrates S_1 and S_2 into the CSTR is as follows:

$$\begin{cases} \varepsilon \dfrac{d\sigma_1}{d\tau} = \nu_{1m} - \beta_1\sigma_1 - \nu \equiv \varepsilon P(\sigma_1,\ \sigma_2) \\ \dfrac{d\sigma_2}{d\tau} = \nu_{2m} - \beta_2\sigma_2 + \nu \equiv Q(\sigma_1,\ \sigma_2) \end{cases}, \tag{6.126}$$

where $\tau = \nu_+(K_1/K_i)^{-\gamma}/K_2$; $\varepsilon = K_1/K_2$, $\nu_{1m} = \eta_{1m}/(\nu_+\xi)$; $\nu_{2m} = \eta_{2m}/(\nu_+\xi)$; $\beta_1 = k_1K_1/(\nu_+\xi)$; $\beta_2 = k_2K_2/(\nu_+\xi)$; η_{1m} and η_{2m} are the rates of supply of S_1 and S_2, respectively; and k_1 and k_2 are the corresponding rate constants for the supply of S_1 and S_2.

Writing the steady-state condition as $\varepsilon(d\sigma_1/dt) = 0$, we obtain $\sigma_1^{ss} = (\nu_{1m} - \nu^{ss})/\beta_1$. Then, substituting σ_1^{ss} into Eq. (6.125) yields the function $\nu^{ss} = f(\sigma_2^{ss})$ that can be written as follows:

$$\sigma_2^{ss} = \frac{\sigma_1^{ss} - \nu^{ss}(\xi + \sigma_1^{ss}(\xi + (\sigma_1^{ss})^{\gamma}))}{\alpha + \nu^{ss}(\xi + (\sigma_1^{ss})^{\gamma})}. \tag{6.127}$$

Figure 6.12 shows one of the hysteresis curves of the ν^{ss} versus σ_2^{ss} plot for relation (6.127), where points O_1, O_2, and O_3 of its intersection with the line $\nu = \beta_2\sigma_2 - \nu_{2m}$ correspond to the steady states.

The quantitative and qualitative analysis of model (6.126) led to the following [11] conclusions:

1. Steady states O_1 and O_3 in various regions of the parametric portraits (β_1, ν_{1m}) and (β_2, ν_{2m}) can represent stable nodes (SNs), unstable nodes (UNs), stable foci (SFs), and unstable foci (UFs).
2. Stable state O_2 always represents a saddle.
3. Single steady state can represent UN, SF, and UF (on the $\Delta = 0$ line).

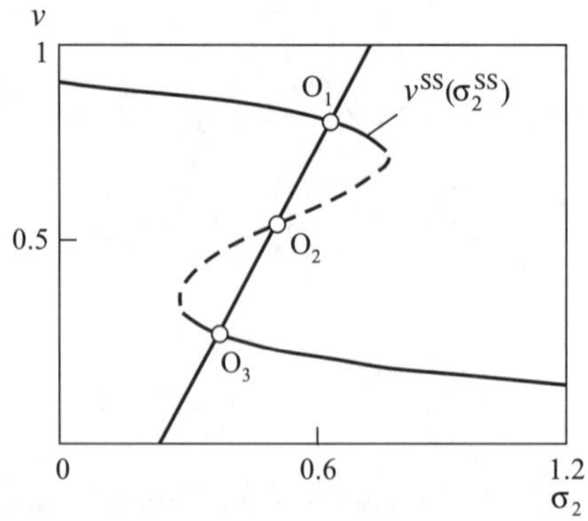

Figure 6.12 Graphical determination of steady states of reaction (6.124) for a kinetic model with the parameters $\alpha = 0.2,\ \gamma = 2,\ \beta_1 = 0.4, \beta_2 = 2.5, \nu_{1m} = 0.95, \nu_{2m} = 0.5, \xi = 0.1$ (reproduced with permission from AIPC Nauka and authors © 1978 [11])

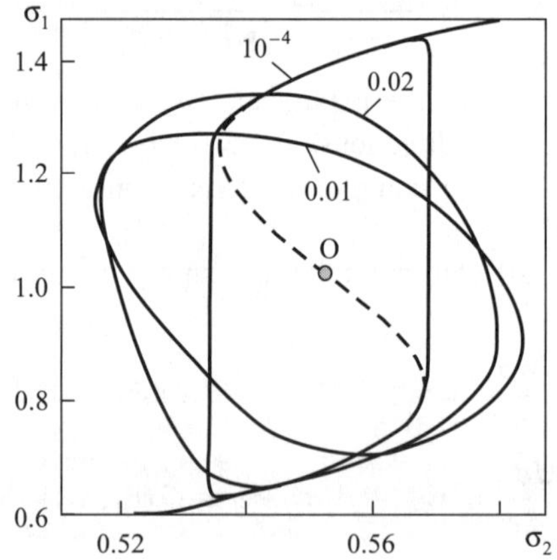

Figure 6.13 Limit-cycles on (σ_1, σ_2) phase plane of model (6.126) for reaction (6.124) with $\alpha = 0.5,\ \gamma = 2,\ \beta_1 = 0.32; \beta_2 = 0.9, \nu_{1m} = 0.72, \nu_{2m} = 0.1, \xi = 0.1$ and various values of parameter ε (indicated at the curves) (reproduced with permission from AIPC Nauka and authors © 1978 [11])

An analysis of the parametric portraits shows that simple reaction (6.124) can have one, two (on the $\Delta = 0$ line), or three steady states, which can vary in both stability and character. In some regions of parameters, all steady states are unstable and, hence, there are conditions for the appearance of limit-cycles. Figure 6.13 shows an example of the stable limit-cycle surrounding a single steady-state point in the system under consideration. ■

Example 6.5

► Various types of dynamic behavior can also be observed in the following irreversible two-substrate reaction with the substrate-induced suppression:

$$S_1 + S_2 \xrightarrow[E]{r} S_1' + S_2', \tag{6.128}$$

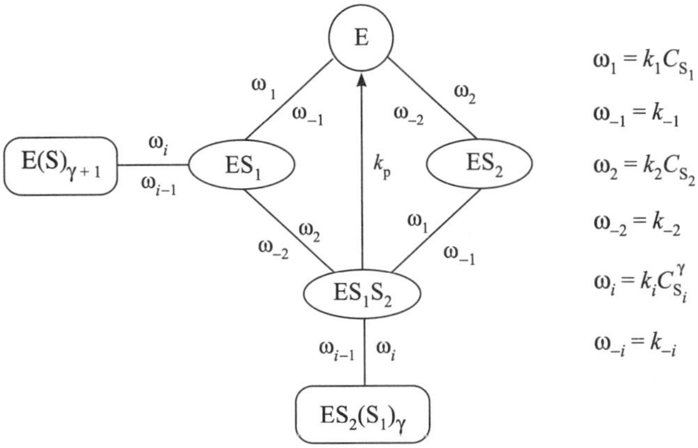

Figure 6.14 Kinetic graph of mechanism of substrate-inhibited reaction (6.128)

which proceeds via a two-route linear mechanism represented by the graph depicted in Figure 6.14. In the quasi-steady-state approximation, the rate of reaction (6.128) is as follows:

$$r = \frac{\nu C_{S_1} C_{S_2}}{\left(K_1 + C_{S_1} + \frac{C_{S_1}^{\gamma+1}}{K_i^{\gamma}}\right)\left(K_2 + C_{S_2}\right)}, \qquad (6.129)$$

where $\nu = k_p E_0$; $K_1 = k_{-1}/k_1$; $K_2 = k_{-2}/k_2$, and $K_i^{\gamma} = (k_{-i}/k_i)^{\gamma}$.

With an additional step of substrate S_2 regeneration

$$S_2' \xrightarrow{W_2} S_2, \qquad (6.130)$$

where $W_2 = k_2 C_{s_2'}$ and $C_{s_2'} + C_{s_2} = C_0$, the process in a CSTR is described by the following system of equations:

$$\begin{cases} \dfrac{dC_{s_1}}{dt} = \eta_{1m} - k_1 C_{S_1} - r \\ \dfrac{dC_{s_2}}{dt} = k_2(C_0 - C_{s_2}) - r \end{cases} \qquad (6.131)$$

that can be rewritten in a dimensionless form as

$$\begin{cases} \varepsilon \frac{d\sigma_1}{d\tau} = \nu_{1m} - \beta_1 \sigma_1 - \nu \\ \frac{d\sigma_2}{d\tau} = 1 - \sigma_2 - \nu \end{cases}, \qquad (6.132)$$

where

$$\nu = \frac{\alpha \sigma_1 \sigma_2}{(1 + \delta \sigma_2)\left[\xi + \sigma_1 (1 + \sigma_1^{\gamma})\right]}; \; \sigma_1 = C_{s_1}/K_i; \; \sigma_2 = C_{s_2}/C_0;$$

$$\nu = r/(k_2 C_0); \; \tau = k_2 t; \; \alpha = \nu/(k_2 K_2); \; \beta_1 = k_1 K_i/(k_2 C_0);$$

$$\delta = C_0/K_2; \; \varepsilon = K_i/C_0; \; \nu_{1m} = \eta_{1m}/(k_2 C_0); \; \xi = K_1/K_i;$$

and K_1 and K_2 are the corresponding complexes of constants.

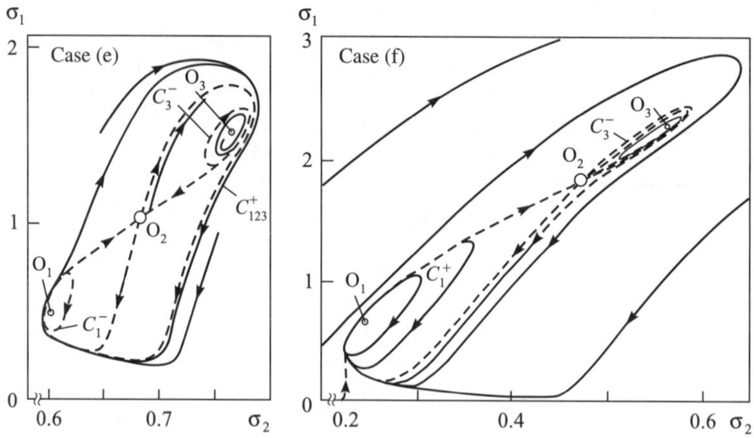

Figure 6.15 Singular points and limit-cycles on phase portraits $\sigma_1 = f(\sigma_2)$ of reaction (6.128) in cases (e) and (f) (reproduced with permission from AIPC Nauka and authors © 1978 [11])

It was found [11] that the quasi-steady-state characteristic $\nu^{ss} = f(\sigma_2^{ss})$ also has a hysteresis shape and system (6.132) has one, two, or three steady states and limit cycles on the phase portraits (Figure 6.15). An analysis of these phase portraits showed that there are several possible types of limit-cycles $C_i^{\pm}$, where subscript i indicates the number of a singular point (steady state), and symbols $\pm$ indicate stabile (+) and unstable (−) limit-cycles:

$$\text{(a) } C_{123}{}^{+}; \quad \text{(b) } C^{-} \text{ and } C^{+}; \quad \text{(c) } C_1{}^{-}; \quad \text{(d) } C_3{}^{-} \text{ and } C_{123}{}^{+};$$
$$\text{(e) } C_1{}^{-}, C_3{}^{-}, \text{ and } C_{123}{}^{+}; \quad \text{(f) } C_1{}^{+} \text{ and } C_3{}^{-}.$$

Figure 6.15 shows the phase portraits in cases (e) and (f). ■

As can be seen, each of the two models considered in Examples 6.4 and 6.5 exhibits variable behavior, even in the regions where they possess hysteresis quasi-steady-state characteristics and only single steady states.

The kinetic curves plotted as the concentration or reaction rate versus time exhibit a variety of shapes and their behavior can depend on the initial conditions. If there is a single stable steady state, then the same reaction rate is established in the course of reaction irrespective of the initial conditions, but this steady-state level can be attained in different regimes, e.g., monotonic (Figure 6.16, case *1*), relaxation (case *2*), or relaxation via damping oscillations (cases *3* and *4*). In the case of a single steady state surrounded by two limit-cycles, after a transient period the reaction attains a non-oscillatory of self-oscillatory regime (case *5*) dependent on the initial conditions. The shape of the self-oscillations can vary from almost harmonic (case *6*) to sharp relaxation (case *7*), depending on the features of the parametric portrait. In particular, the character of relaxation can vary depending on whether the reactant concentrations are small (case *7*) or large (case *8*).

In cases where a quasi-steady-state model exhibits a hysteresis and there are three steady states (Figure 6.17), one of the pair of states can be attained, depending on the initial conditions, either monotonically (case *1*) or nonmonotonically (cases *2* and *3*) or in a self-oscillatory regime (cases *4*

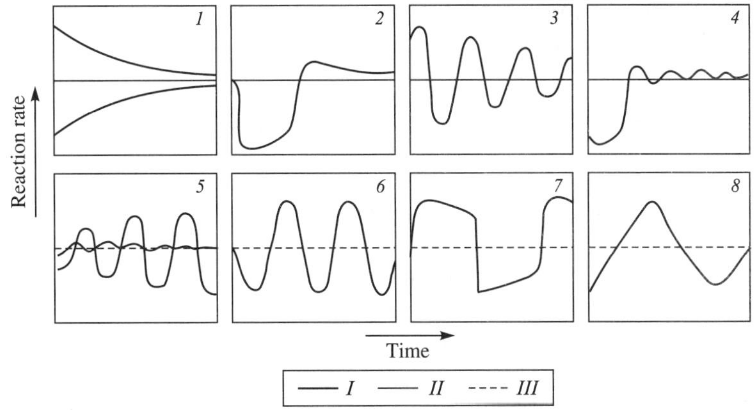

Figure 6.16 Forms of dynamic behavior of reaction in CSTR with a single steady-state and quasi-steady-state hysteresis characteristic for various initial conditions (1–8): (I) temporal variation of reaction rate; (II) stable steady-state rate; (III) unstable steady-state rate (reproduced with permission from AIPC Nauka and authors © 1978 [11])

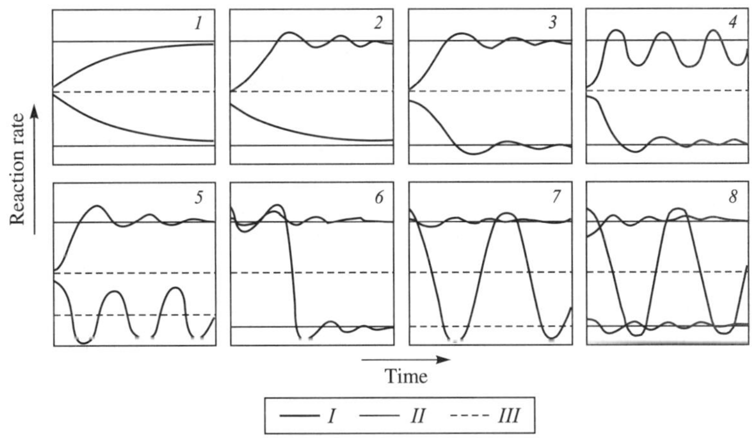

Figure 6.17 Possible types of dynamic behavior of reaction in CSTR with three alternative steady states: (I) temporal variation of reaction rate corresponding to three steady-state points (O_1, O_2, O_3 in Figure 6.13); (II) stable steady-state reaction rate; (III) unstable steady-state reaction rate (reproduced with permission from AIPC Nauka and authors © 1978 [11])

and *5)*. If one of the stable steady-state points is surrounded by an unstable limit-cycle, then either one of the two stable steady states (depending on the initial conditions) is established (case *6)* or a self-oscillatory regime is developed (case *7)*. If the phase portrait of a dynamical system with three singular points has two unstable limit cycles and one stable limit cycle, which surrounds the unstable ones (see Figure 6.15, case (e)), then the system can either attain one of the two non-oscillatory steady states with significantly different rates or occur in a self-oscillatory regime with large amplitude ranging between the two potential steady states (Figure 6.17, case *(8)*).

Sel'kov [127] and Zhabotinskii [5] showed that, in a chain of linear conversions of substrates S_i catalyzed by enzymes E_i according to the scheme

$$S_0 \underset{}{\overset{r_0, E_0}{\rightleftharpoons}} S_1 \overset{r_1, E_1}{\rightleftharpoons} S_2 \longrightarrow \cdots \overset{r_n, E_n}{\rightleftharpoons} S_n$$

$$\frac{dS_1}{dt} = \frac{r_0}{1+\alpha S_n^{\gamma}} - r_1, \quad r_{i(i\neq n)} = l_i S_i, \quad r_n = \frac{k_n S_{n-1}}{k_m + S_{n-1}}, \tag{6.133}$$

the inhibition of the first step by a product of the last step leads to the concentration of self-oscillations for $n \geq 3$ and $\gamma = 1$.

In a system of reactions catalyzed by two enzymes (E_1 and E_2) according to the scheme

$$\begin{aligned} \longrightarrow S_1\ S_1 + E_1 &\rightleftharpoons E_1S_1 \longrightarrow E_1 + P_1,\ P_1 \longrightarrow; \\ 2P_1 + E_2 &\rightleftharpoons I_2; \\ S_1 + E_2 &\rightleftharpoons E_2S_1 \longrightarrow E_2 + P_2,\ P_2 \longrightarrow; \\ 2P_2 + E_1 &\rightleftharpoons I_1, \end{aligned} \tag{6.134}$$

the cross-inhibition by products P_1 and P_2 (material balance effect) also leads to the development of self-oscillations [5].

The examples considered above show that the nonlinear models, in which the nonlinearity is related to the appearance of fractional-rational functions in the right-hand part of differential equations, exhibit a nontrivial dynamic behavior in CSTR (including self-oscillations of concentrations and reaction rates) even for the mechanisms that are linear with respect to intermediates. An analogous effect in the case of linear mechanisms is produced by the exponential nonlinearities.

It should be noted that a model of the catalytic reaction with a linear mechanism (with neglect of the material balances for a catalyst) naturally gives a single steady state. However, if the number of intermediates in the cycle is $n \geq 3$, then the concentrations and reaction rates exhibit decaying oscillations on approach to this steady state. In the case of $n = 4$ (with three linearly independent intermediates), the oscillations are slowly decaying [5]. This phenomenon was originally observed by Christiansen [128].

Example 6.6

▶ Let us analyze a simple chain of reactions in a closed system:

$$X_1 \xrightarrow{\alpha_1} X_2 \xrightarrow{\alpha_2} X_3 \xrightarrow{\alpha_3} X_1, \tag{6.135}$$

where $\alpha_j = k_j C_{s_j}$ are the weights of stages, $C_{s_j} \cong$ const, and $C_{X_0} = C_{X_1} + C_{X_2} + C_{X_3}$. For the sake of simplicity, let us assume that $\alpha_1 = \alpha_2 = \alpha_3 = \alpha$. Then the kinetic equations can be written as

$$\begin{cases} \dfrac{dC_{X_1}}{dt} = \alpha(C_{X_3} - C_{X_1}) = \alpha(C_{X_0} - 2C_{X_1} - C_{X_2}) = f_1 \\ \dfrac{dC_{X_2}}{dt} = \alpha(C_{X_1} - C_{X_2}) = f_2 \end{cases} \tag{6.136}$$

The corresponding elements of Jacobian $\boldsymbol{A}$ are

$$a_{11} = -2\alpha; \quad a_{12} = -\alpha; \quad a_{21} = \alpha; \quad a_{22} = -\alpha,$$

so that

$$\sigma = \mathrm{tr}\boldsymbol{A} = -3\alpha(\sigma < 0);\ \ \Delta = 3\alpha^2 > 0;\ \ \sigma^2 - 4\Delta = -3\alpha^2 < 0;$$

$$\lambda_{1,2} = -\frac{3}{2}\alpha \pm \frac{i\alpha\sqrt{3}}{2} \quad \text{and} \quad \mathrm{Re}(\lambda_{1,2}) < 0,$$

which implies that the singular point is a stable focus. ■

According to Bendixson's criterion, the limit cycle and self-oscillations cannot exist in a two-dimensional ($n = 2$) system, if the sum of derivatives

$$\frac{\partial f_1}{\partial x} + \frac{\partial f_2}{\partial y}$$

does not change sign in the region of $x > 0$ and $y > 0$. In the case of a linear mechanism, this sum is always negative, since $a_{11} + a_{22} < 0$ [98].

Now let us consider several examples of processes with nonlinear steps of different types.

Example 6.7

▶ For a two-step reaction with one nonlinear step in a closed system,

$$A \rightleftharpoons X;\quad 2X \longrightarrow B, \tag{6.137}$$

$$\frac{dC_X}{dt} = k_1 C_A - k_{-1} C_X - 2k_2 C_X^2 = P(C_X), \tag{6.138}$$

a quasi-steady-state solution is

$$C_X^{ss(1,2)} = -\frac{p}{2} \pm \sqrt{\frac{p^2}{4} + q}, \tag{6.139}$$

where $p = k_{-1}/2k_2$ and $q = k_1 C_A/2k_2$. For large C_{A0} and $C_X \ll C_{A0}$, the initial stage is characterized by $q \cong k_1 C_{A0}/(2k_2)$ and the C_X^{ss} value depends on the two parameters (p and q) as depicted in Figure 6.18a. The physically meaningful solutions correspond to the upper branch of the parabola, where $C_X^{ss} > 0$. In particular, $C_X^{ss(1)} \geq 0$ for $q \geq 0$.

For an analysis of the stability of steady-state solutions, coefficient a of the linear approximation is determined as

$$a = \left.\frac{\partial P(C_X)}{\partial C_X}\right|_{C_X^{ss}} = -k_{-1} - 4k_2 C_X^{ss}. \tag{6.140}$$

Substituting expressions for $C_X^{ss(1,2)}$ into Eq. (6.140), we obtain

$$a_1 = -k_{-1} - 4k_2 C_X^{ss(1)} = -\sqrt{(k_{-1})^2 + 8k_1 k_2 C_{A0}},$$

$$a_2 = -k_{-1} - 4k_2 C_X^{ss(2)} = \sqrt{(k_{-1})^2 + 8k_1 k_2 C_{A0}},$$

so that $a_1 < 0$ and $a_2 > 0$ and, hence, the steady state for $C_X^{ss(1)}$ is a stable node. The negative branch $\left(C_X^{ss(2)}\right)$ corresponds to unstable solutions. The concentration fluctuations $\varepsilon = c_2 e^{a_2 t}$ increase with the time. The phase portrait of this one-dimensional problem with the given parameters p and q can also be constructed on the C_X^{ss} axis (Figure 6.18b), where the trajectories exhibit attraction to point $C_X^{ss(1)}$ (attractor) and repulsion from point $C_X^{ss(2)}$ (repeller). In the given kinetic model, point $C_X^{ss(2)}$ has no physical meaning.

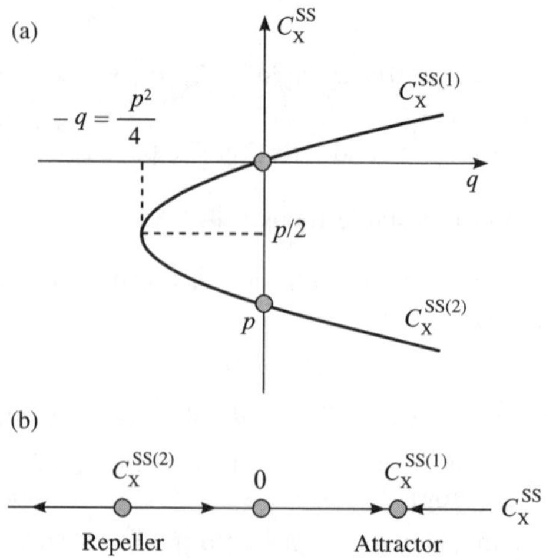

Figure 6.18 Dependence of C_X^{SS} on parameter $q = k_1 C_{A0}/(2k_2)$ entering in Eq. (6.139), plotted as (a) two-dimensional and (b) one-dimensional phase portraits

An analysis of the stability of steady states can also be based on the relationship between the derivatives of the rates of formation $[f(C_X) = k_1 C_A]$ and consumption $[g(C_X) = k_{-1}C_X + 2k_2 C_X^2]$ of X with respect to C_X (see, e.g., the relationship between the derivatives of r_τ and R in Section 6.2.2). For example, the inequality

$$\frac{\partial f(C_X)}{\partial C_X} < \frac{\partial g(C_X)}{\partial C_X} \tag{6.141}$$

at $f(C_X) = g(C_X)$ (i.e., at the steady-state point) is the criterion of the stability of steady states [12, 17]. The derivative of $f(C_X)$ vanishes at $C_A \cong C_{A0}[f'(C_X) = 0]$, while the derivative of $g(C_X)$ is given by the formula

$$g'(C_X) = K_{-1} + 4k_2 C_X^{SS}$$

and is positive at $C_X^{SS} > 0$. Thus, $g'(C_X) > f'(C_X)$ and the steady state is stable. ■

Now let us consider process (6.137) as carried out in a continuous stirred-tank reactor (CSTR).

Example 6.8

► First, let us use the condition of large concentration C_A that is constant in time ($C_A \cong C_{A0}$). In this case, an additional term enters into the kinetic equation (6.138), which is related to the outflow of X from the reactor:

$$\frac{dC_X}{dt} = k_1 C_{A0} - k_{-1}C_X - 2k_2 C_X^2 - k_0 C_X = P(C_X) \tag{6.142}$$

where $k_0 = 1/\tau$ and τ is the contact time. Since the appearance of this term only changes the coefficient at C_X in Eq. (6.138) from k_{-1} to $k_{-1} + k_0$, the analysis of solutions for their stability remains generally the same. In particular, the coefficient a of the linear approximation is determined as

$$a_1 = \lambda_1 = -\sqrt{(k_{-1} + k_0)^2 + 8k_1 k_2 C_{A0}} < 0,$$

but the position of stable steady state $C_X^{ss(1)}$ now depends on the contact time τ.

In cases where C_A significantly varies under the CSTR conditions, an analysis for the stability is based on a system of two differential equations, which involves variables C_X and C_A and has a dimensionality of $n = 2$:

$$\begin{cases} \dfrac{dC_X}{dt} = k_1C_A - k_{-1}C_X - k_0C_X - 2k_2C_X^2 = P(C_X) \\ \dfrac{dC_A}{dt} = k_0C_{A0} - k_1C_A - k_{-1}C_X - k_0C_A = Q(C_X) \end{cases} . \quad (6.143)$$

The steady-state concentrations $C_X^{ss(1,2)}$ and $C_A^{ss(1,2)}$ are determined from the following equations:

$$2k_2C_X^2 + (k_0k_{-1} + k_1k_0 + k_0^2)C_X - k_1k_0C_{A0} = 0; \quad (6.144)$$

$$C_A^{ss} - \frac{k_{-1}C_X^{ss} + k_0C_{A0}}{k_1 + k_0}. \quad (6.145)$$

Physically meaningful solutions $C_A^{ss(2)} > 0$ are obtained provided that $k_0C_{A0} > k_{-1}\left|C_X^{ss(2)}\right|$. The coefficients of linear approximation determined from the derivatives $P(C_X)$ and $Q(C_X)$ using a linearized system of equations for fluctuations ε and η are as follows:

$$a_{11} = -(k_{-1} + k_0) - 4k_2C_X^{ss}; \quad a_{12} = k_1, \quad a_{21} = k_{-1}; \quad a_{22} = -k_1 - k_0.$$

The eigenvalues of the corresponding characteristic equation

$$\lambda^2 - \sigma\lambda + \Delta = 0$$

are determined by its coefficients σ and Δ:

$$\sigma = \mathrm{tr}A = -\left[(k_{-1} + k_2) + 4k_2C_X^{ss} + k_1 + k_0\right] < 0;$$

$$\Delta = a_{11}a_{22} - a_{21}a_{12} > 0 \quad \text{for} \quad C_X^{ss} > 0;$$

$$\sigma^2 - 4\Delta > 0 \quad \text{for} \quad k_{-1} \geq k_1.$$

Thus, physically meaningful steady-state solutions ($C_X^{ss(1)} > 0$, $C_X^{ss(2)} > 0$) correspond to stable nodes ($\lambda_{1,2} < 0$). For $k_{-1} \geq k_1$, this system cannot exhibit cyclic phenomena of the hysteresis type or self-sustained oscillations. For $k_{-1} \leq k_1$, we have $\sigma^2 - 4\Delta < 0$, and the system passes to a singular point of the stable focus type with negative real parts of complex-conjugate solutions for $\lambda_{1,2}$ that correspond to decaying oscillations in the vicinity of the steady state. ■

Yablonsky et al. [12, 129–132] carried out a detailed analysis of three-step mechanisms for heterogeneous catalytic processes involving nonlinear steps. Mechanisms of the Langmuir–Hinshelwood type were also studied by other researchers (see, e.g., [88]). Since all the adsorption mechanisms were based on a model of the ideal surface, the results are also applicable to homogeneous metal complex systems. In this analysis, it is also convenient to employ dimensionless concentrations of the intermediates and $[M]_\Sigma$:

$$[M]_\Sigma = [M] + \Sigma[X_i],$$

$$1 = \frac{[M]}{[M]_\Sigma} + \Sigma\frac{[X_i]}{[M]_\Sigma} \quad \text{or} \quad 1 = \Theta_0 + \Sigma\Theta_i,$$

where Θ_0 and Θ_I are the free and occupied fractions of the surface.

Let us consider several variants of three-step mechanisms without passing from the "adsorption" terminology to the notions of coordination chemistry. Mechanisms of the type

$$\begin{aligned}
&(1)\quad A_n + nZ \rightleftharpoons nAZ;\\
&(2)\quad B_m + mZ \rightleftharpoons mBZ;\\
&(3)\quad pAZ + qBZ \longrightarrow (p+q)Z + A_pB_q
\end{aligned}\tag{6.146}$$

were analyzed [12] assuming that the molecularity of each step is $\mu \leq 3$. As was noted above, the probability of elementary steps with $\mu = 3$ is rather small, so that mechanisms with $\mu = 1$ and $\mu = 2$ are more likely. Note that the steps of type (1) with $n = 2$ require that two adjacent active centers would exist, but then the two sites can be considered as a center of the new type Z^* on the surface. In the catalysis with metal complexes, a collision of A_2 molecule with two complexes Z is of low probability. At the same time, each of these steps can be considered as two consecutive events, the first of which is quasi-equilibrium

$$\begin{aligned}
&A_2 + Z \rightleftharpoons A_2Z;\\
&A_2Z + Z \rightleftharpoons 2AZ,
\end{aligned}$$

and then a kinetic equation for the reaction rates in the forward and reverse directions would correspond to step (1). Thus, the steps of type (1) will be taken into consideration.

It should also be noted that the kinetic models of type (6.146) were analyzed in a closed system, but the concentrations of A_n and B_m were maintained constant by means of a controlled introduction of these reactants into the system (without changing its volume) or due to very small degrees of conversion. The constant concentrations (or partial pressures) of A_n and B_m can also be ensured in a CSTR at very high flow rates such that $C_{i0} \approx C_i$. Then, only the concentrations of intermediate compounds X_i will be variables, while the C_{i0} values will be parameters. However, for a homogeneous system with the solution of catalyst and reactants continuously supplied to and removed from the reactor, the differential equations for the intermediates would contain the terms with $[X_i]/\tau$.

With a view to gradually increasing the complexity of the problem, let us begin with a simple variant of mechanism (6.146) with $m = n = p = q = 1$ and a single nonlinear step of interaction between the intermediates [12].

Example 6.9

► In this case, system (6.146) appears as

$$\begin{aligned}
&(1)\quad A + Z \rightleftharpoons AZ;\\
&(2)\quad B + Z \rightleftharpoons BZ;\\
&(3)\quad AZ + BZ \longrightarrow AB + 2Z,
\end{aligned}\tag{6.147}$$

and the still simpler case of $k_{-1} = k_{-2} = 0$ yields the kinetic equations

$$\begin{cases}
\dfrac{dx}{dt} = k_1P_A(1-x-y) - k_3xy = P(x,\ y)\\[2ex]
\dfrac{dy}{dt} = k_2P_B(1-x-y) - k_3xy = Q(x,\ y),
\end{cases}\tag{6.148}$$

where P_A and P_B are the partial pressures of reactants (constant under the given conditions); x and y are the dimensionless concentrations of AZ and BZ, respectively; and $1 - x - y$ is the dimensionless concentration of active centers (metal complexes).

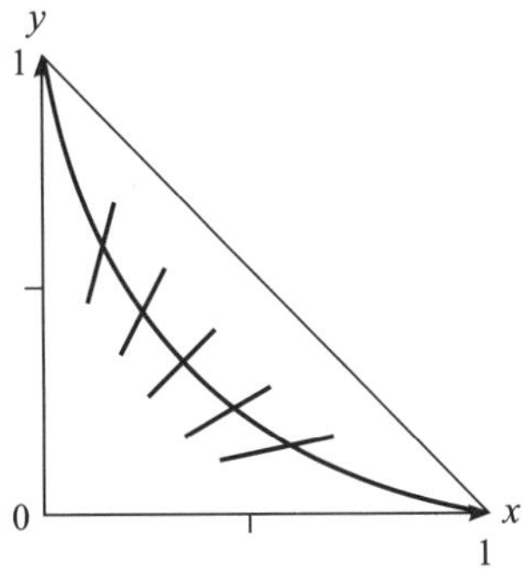

Figure 6.19 Line of steady states plotted in x–y concentration simplex for $k_1 P_A = k_2 P_B$

For $P(x, y) = 0$ and $Q(x, y) = 0$, there are two boundary steady states (points at the boundaries of a concentration simplex in the phase space) with ($x_1^{ss} = 1$, $y_1^{ss} = 0$) and ($x_2^{ss} = 0$, $y_2^{ss} = 1$), where $W_3 = k_3 x^{ss} y^{ss} = 0$. In addition, for a certain ratio of parameters $P_A/P_B = k_2/k_1$, there appear internal steady-state points. At this ratio of parameters, the two main isoclines, $P(x, y) = 0$ and $Q(x, y) = 0$, on the phase portrait $x = f(y)$ merge into a common curve representing a continuous set of steady-state points with indeterminate ($dx/dy = 0/0$) derivatives (Figure 6.19). It is this curve of steady states which the system attains proceeding from any initial state.

An analysis of the characteristic equation $\lambda^2 - \sigma\lambda + \Delta = 0$ for system (6.148) yields $\sigma < 0$, $\Delta = 0$, $\lambda_1 = 0$, and $\lambda_2 = \sigma < 0$. At this point, a stable node transforms into a saddle. For this saddle–node bifurcation, a bifurcation parameter is P_A/P_B ratio. In particular, at

$$\frac{P_A}{P_B} \neq \frac{k_2}{k_1},$$

the value of Δ ceases to be zero and the system becomes unstable.

Now let us consider mechanism (6.147) for $k_{-1} \neq k_{-2} \neq 0$. In this case, the system

$$\begin{cases} \dfrac{dx}{dt} = k_1 P_A (1 - x - y) - k_{-1} x - k_3 xy = P(x, y) \\ \dfrac{dy}{dt} = k_2 P_B (1 - x - y) - k_{-2} y - k_3 xy = Q(x, y) \end{cases} \quad (6.149)$$

no longer has boundary steady-state points. The equivalent equation

$$Ay^2 + By - C = 0,$$

where $A = k_3(k_2 P_B + k_{-2} - k_1 P_A)$, $B = k_1 k_3 P_A + k_1 k_{-2} P_A + k_{-1} k_{-2} - k_2 k_3 P_B$, and $C = k_{-1} k_2 P_B$, has a single positive solution with a steady-state point at $y_1^{ss} > 0$, $x_1^{ss} > 0$. An analysis for the stability of a linearized system yields

$$a_{11} + a_{22} < 0; \qquad \sigma < 0; \qquad \det \mathbf{A} = \Delta > 0,$$

so that the single steady state represents a stable node. ■

Thus, the presence of one nonlinear step in both two- and three-step mechanisms (Examples 6.7 and 6.9, respectively) is not a sufficient condition for the appearance of critical phenomena. The implementation of process (6.147) in a CSTR at constant P_A and P_B would only introduce an additional constant $k_0 = 1/\tau$ into Eqs. (6.149) [with the terms $(k_{-1} + k_0)x$ and $(k_{-2} + k_0)y$, respectively], while changing nothing in the dynamic behavior of the model under consideration.

Then, let us consider mechanism (6.146) with two nonlinear steps for $n = 2, m = 1$, and $p = q = 1$.

Example 6.10 [12, 129–131]

► The presence of two nonlinear steps in the mechanism

$$\begin{aligned} &(1) \quad A_2 + 2Z \rightleftharpoons 2AZ; \\ &(2) \quad B + Z \rightleftharpoons BZ; \\ &(3) \quad AZ + BZ \longrightarrow AB + 2Z \end{aligned} \tag{6.150}$$

leads to nontrivial dynamic behavior of the kinetic model. In the case of $k_{-1} = k_{-2} = 0$, the equations

$$\begin{cases} \dfrac{dx}{dt} = 2k_1P_{A2}(1 - x - y)^2 - k_3xy = P(x,\ y) \\ \dfrac{dy}{dt} = k_2P_B(1 - x - y) - k_{-2}y - k_3xy = Q(x,\ y) \end{cases} \tag{6.151}$$

determine two boundary steady states, $(x_1^{ss} = 1,\ y_1^{ss} = 0)$ and $(x_2^{ss} = 0,\ y_2^{ss} = 1)$, where $W_3 = k_3x^{ss}y^{ss} = 0$. Equations (6.151) can be reduced to a relationship

$$1 - x - y = \frac{k_2P_B}{2k_1P_{A_2}}, \tag{6.152}$$

which has a physical meaning at $2k_1P_{A2} > k_2P_B$, i.e., for $1 - x - y < 1$. Two positive solutions of the appropriate quadratic equation exist provided that

$$k_3(2k_1P_{A_2} - K_2P_B)^2 \geq 8k_1k_2^2P_{A_2}P_B^2. \tag{6.153}$$

In the region of $2k_1P_{A_2} > k_2P_B$, condition (6.153) takes the following form:

$$\frac{2k_1P_{A_2}}{1 + 2\sqrt{2k_1P_{A_2}/k_3}} \geq k_2P_B. \tag{6.154}$$

Thus, there are two inner steady states with $x^{ss} > 0$ and $y^{ss} > 0$. The multiplicity of steady states in this system leads to interesting critical phenomena. An analysis of the stability of boundary states $(x_1^{ss} = 0,\ y_1^{ss} = 1)$ and $(x_2^{ss} = 1,\ y_2^{ss} = 0)$ (Figure 6.20a) shows that the first singular point has $\sigma < 0$, $\Delta > 0$ (stable node), while the second point has $\sigma < 0$, $\Delta < 0$ and, hence, real λ values with opposite signs (unstable singularity of the saddle type). For the inner steady states, $\sigma < 0$ and the sign of Δ depends on the relationship between x^{ss} and y^{ss}. For $x^{ss} < y^{ss}$, we have $\Delta > 0$ and a stable node, while for $x^{ss} > y^{ss}$ we have $\Delta < 0$ and a saddle. When the criterion (6.154) is valid, the steady-state reaction rate for the initial conditions in the region of attraction of the stable inner steady state is as follows:

$$R = W_3 = k_3xy = \frac{k_2^2{P_B}^2}{2k_1P_{A_2}}. \tag{6.155}$$

In all other cases, $W_3 = 0$. Since the value of $P_{A_2} < k_2P_B/(2k_1)$ implies that $z = 1 - x - y > 1$, the reaction rate in this region $[P_{A_2} = 0 \div k_2P_B/(2k_1)]$ vanishes (zero branch for the inner steady states).

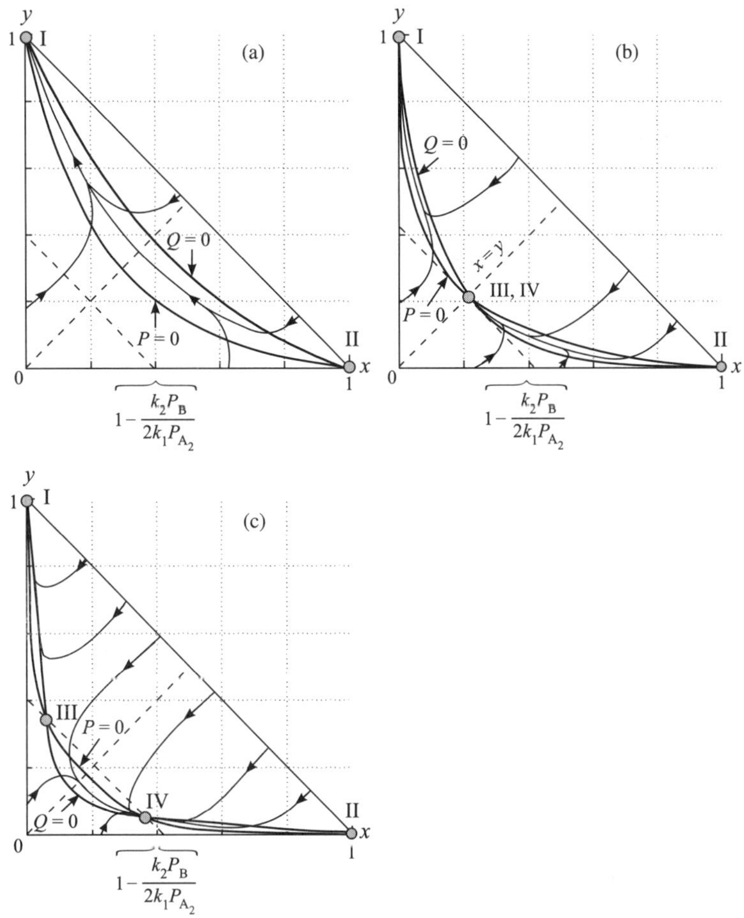

Figure 6.20 Possible variants of $y = f(x)$ isoclines for system (6.151) with $Q = 0$ and $P = 0$ (reproduced with permission from authors © 1984 [12])

At the point where $P_{A_2} = k_2P_B/(2k_1)$, the entire catalyst occurs in the initial state z ($x = 0,\ y = 0$). On the passage through this point at $P_{A_2} > k_2P_B/(2k_1)$, the rate exhibits a jumplike increase to W_{crit} and then gradually decreases with growing P_{A_2}. The system exhibits a jumplike transition to a stable inner steady state. Similarly, the increase in P_B leads to a quadratic growth of the rate from P_B up to a critical value of $P_B = (2k_1/k_2)P_{A_2}$, after which the rate exhibits a jumplike drop to zero and passes to the zero branch. The phase portraits of this system are considered in more detail in [12]. Here we only briefly consider the phase portraits with various possible arrangements of the main isoclines $x = f(y)$ for $P(x,\ y) = 0$ and $Q(x,\ y) = 0$ and the stable (I, IV) and unstable (II, III) singularities (Figure 6.20).

The first inner steady state (III, IV, Figure 6.20b) corresponds to the intersection of isoclines at the point with

$$x = y = \frac{1}{2}\left(1 - \frac{k_2P_B}{2k_1P_{A_2}}\right),$$

which yields

$$1 - 2x = z_{\max} = \frac{k_2 P_B}{2k_1 P_{A_2}},$$

where $z_{\max}$ is the maximum concentration of active centers for both isoclines. For certain values of parameters, the isoclines can exhibit two intersections (on the $\left(x + y = 1 - \frac{k_2 P_B}{2k_1 P_{A_2}}\right)$ curve), which implies that

$$1 - 2x = z_{\max} > \frac{k_2 P_B}{2k_1 P_{A_2}}$$

for each isocline at $x = y$ (Figure 6.20c). The latter inequality is a sufficient condition for the multiplicity of steady states [12]:

$$\max z|_{Q=0} > \frac{k_2 P_B}{2k_1 P_{A_2}}; \quad \max z|_{P=0} > \frac{k_2 P_B}{2k_1 P_{A_2}}.$$

The necessary and sufficient condition for the multiplicity of steady states consists in that x^{ss} at the point of intersection of the isocline with the line $x = y$ for $Q = 0$ is smaller than that for $P = 0$ or, in other words:

$$\max z|_{Q=0} > \max z|_{P=0}.$$

It would be also of interest to elucidate the influence of the reversibility of steps (1) and (2) in mechanism (6.150) on the behavior of a more complicated dynamical system

$$\begin{cases} \dfrac{dx}{dt} = 2k_1 P_{A_2}(1 - x - y)^2 - 2k_{-1}x^2 - k_3 xy = P(x, y) \\ \dfrac{dy}{dt} = k_2 P_B(1 - x - y) - k_{-2}y - k_3 xy = Q(x, y) \end{cases} . \tag{6.156}$$

Expressing y from the equation $Q(x, y) = 0$ and substituting it into the equation $P(x, y) = 0$, we obtain an equation of the fourth power

$$2k_1 P_{A_2}(k_{-2} + k_3 x)^2(1 - x)^2 = x(k_2 P_B + k_{-2} + k_3 x)$$
$$\times \{k_2 P_B k_3(1 - x) + 2k_{-1}x(k_2 P_B + k_{-2} + k_3 x)\},$$

where the left side $f(x)$ represents the rate of increase in x and the right side $g(x)$ is the rate of decrease in x. At $x = 0$, we have $f(0) > 0$ and $g(0) = 0$; at $x = 1$, we have $f(1) = 0$ and $g(1) > 0$.

Therefore, in the segment of $x \in [0, 1]$ these functions possess at least one steady state. The necessary and sufficient condition for the existence of one unstable (x_2^{ss}) and two stable ($x_1 < x_2 < x_3$) states is as follows:

$$\frac{\partial f(x)}{\partial x} > \frac{\partial g(x)}{\partial x} \tag{6.157}$$

at $f(x) = g(x)$.

An explicit criterion for the multiplicity of steady states can be obtained from the condition that inequality (6.157) would be valid at the bending point x^* of the $f(x)$ function. Equating the second derivative of this function to zero, $f(x)'' = 0$, we obtain the following equation for determining x^*,

$$x^*_{1,2} = \frac{(k_3 - k_{-2}) \pm (k_3 + k_{-2})\sqrt{1/3}}{2k_3}, \tag{6.158}$$

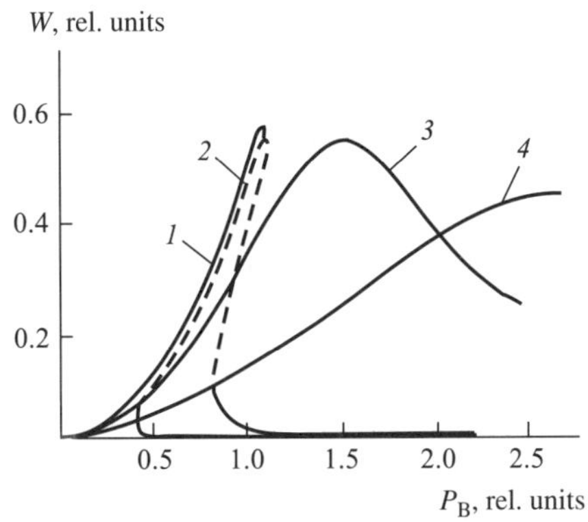

Figure 6.21 Plots of steady-state reaction rate W versus partial pressure P_B at constant values of $P_{A2} = 1, k_{-1} = 0.01$, and various k_{-2}: (*1*) 0.01; (*2*) 0.1; (*3*) 1; (*4*) 7; dashed curve shows the behavior of W in an unstable steady-state (reproduced with permission from editorial board of *Kinet. Katal.* journal © 1979 [130])

and the necessary condition for the multiplicity of steady states at $x_2^* > 0$:

$$k_{-2}/k_3 \leq 0.227, \tag{6.159}$$

which is equivalent to the requirement of a weak reversibility of step (2) in mechanism (6.150).

The presence of three steady states with a single unstable singularity leads to the appearance of hysteresis in the curves of steady-state reaction rate $W = k_3xy$ (Figure 6.21) [130]. With increasing k_{-2} (at $k_3 = 10$), the hysteresis gradually vanishes and, at $k_{-2} > k_3$, the dependence $W = f(P_B)$ becomes monotonic ($k_{-2}/k_3 = 0.7$). The hysteresis in $W = f(P_B)$ and $W = f(P_{A_2})$ has opposite directions. In the former case, it is counter-clockwise, whereby the maximum rate is achieved at small partial pressures P_{A_2}. Figure 6.22 shows a dynamic portrait of the system and its projection onto the plane of parameters.

In dynamical system (6.156), as well as in the other aforementioned systems containing no autocatalytic steps, all derivatives in the concentration simplex $x \in [0, 1]$ and $y \in [0, 1]$ were nonpositive: $\dot{P}_x, \dot{P}_y, \dot{Q}_x, \dot{Q}_y \leq 0$. Therefore, the sum of the diagonal elements is negative: $\dot{P}(x) + \dot{Q}(x) = a_{11} + a_{22} < 0$ (or $\sigma < 0$ in the characteristic equation). This inequality is called the *Bendixson's negative criterion* for the absence of limit-cycles. As for the sign of determinant Δ, it depends on the values of parameters. In the case of three steady states, two of these (x_1 and x_3) are stable nodes, while x_2 is a saddle (i.e., unstable steady state with $\lambda_1 < 0$ and $\lambda_2 > 0$). A numerical analysis of this system for $k_1 = k_2 = 1, k_{-1} = 0.01, k_{-2} = 0.1$, and $k_3 = 10$ at $P_{A_2} = 1$ bar [131] showed that one of the eigenvalues (λ_1) changes its sign from $\lambda_1 < 0$ to $\lambda_1 > 0$ within a very narrow interval of parameter P_B ($P_B = 0.77 - 1.1$ bar). For example, at $P_B = 0.7$ bar, there is a single steady state with $y^{ss} \cong 0.05, x^{ss} \cong 0.6$ ($\Delta > 0$), whereas at $P_B = 1.1$ bar there are three steady states (Figure 6.23):

	y	x	Δ
I	~0.25	~0.22	>0
II	~0.3	~0.18	< 0
III	~0.9	0.02	>0

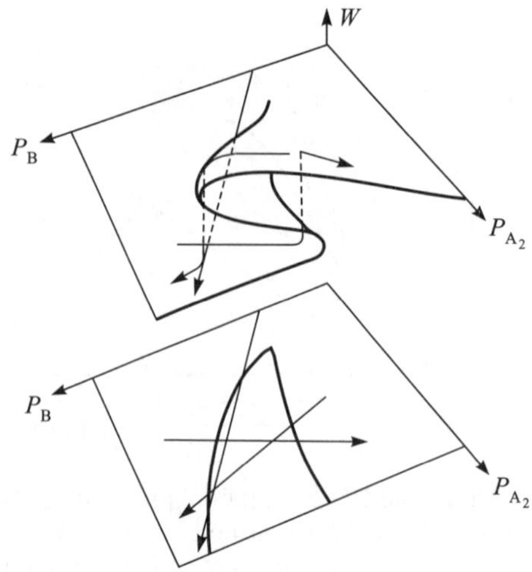

Figure 6.22 Surface of steady-state reaction rate $W(P_{A_2}, P_B)$ at T = const. (reproduced with permission from editorial board of *Kinet. Katal.* journal © 1979 [130])

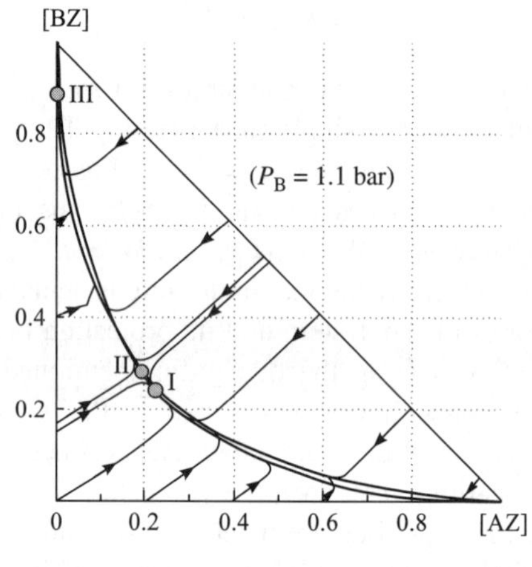

Figure 6.23 Phase portraits of mechanism (6.150) with three steady-states I–III (states I and III are stable; state II is unstable) [12] (reproduced with permission from authors © 1984 [12])

One separatrix of saddle point II separates the regions of steady states III–II and II–I and passes between the two main isoclines, $P(x, y) = 0$ and $Q(x, y) = 0$. Another separatrix in this saddle originates at the boundary of simplex $x + y = 1$. ■

An analysis of different variants of mechanism (6.146) [12] led to the following important conclusions:

(i) A necessary condition for the existence of multiple steady states (MSS) is the presence of a step with nonlinear interaction between intermediates:

$$X_i + X_j \rightarrow$$

(ii) A sufficient condition for the existence of a single inner steady state in scheme (6.146) is $n = m \geq p, q$. In the case of $n = m < p$, one or three inner steady states can exist.
(iii) A sufficient condition for the existence of MSS is $n > m$ for $m \leq p$.
(iv) The existence of MSS depends on restrictions with respect to rate constants and other parameters.
(v) If the aforementioned step $X_i + X_j \rightarrow$ is present but the deficiency zero theorem $\delta = 0$ (6.114) is satisfied, then a single and stable steady state does exist [116].
(vi) A sufficient condition for the uniqueness of a steady state in a catalytic reaction with arbitrary mechanism is the absence of interactions between intermediates ($X_i + X_j \rightarrow$) [116].

Using the kinetic polynomial technique (see Section 1.1.3), Yablonsky and Lazman [132] obtained interesting kinetic and thermodynamic characteristics of "critical" points on the hysteresis curves of reaction rate $W = f(P_B)$ for mechanism

$$\begin{array}{ll} (1) & A_2 + 2Z \longrightarrow 2ZA; \\ (2) & B + Z \rightleftharpoons ZB; \\ (3) & ZA + ZB \longrightarrow 2Z + AB. \end{array} \qquad (6.160)$$

with the following system of kinetic equations:

$$\begin{cases} \dfrac{dx}{dt} = k_1 P_{A_2}(1 - x - y)^2 - k_3 xy = P(x, y); \\ \dfrac{dy}{dt} = k_2 P_B (1 - x - y) - k_{-2} y - k_3 xy = Q(x, y). \end{cases} \qquad (6.161)$$

Figure 6.24 shows a dynamic portrait for this system with four critical points. Point A on the hysteresis curve represents the activation point, where the system jumps to point B on the upper branch of stable steady states. At point C, the system jumps down to the lower branch of stable steady states, which corresponds to the quenching of the process. At large values of k_3, the following expressions were obtained for the lower (W_I) and upper (W_{III}) branches of stable states and the middle (W_{II}) branch of unstable states:

$$W_I = k_{-2}\left(\sqrt{H} - \sqrt{H - 1}\right)^2; \qquad (6.162)$$

$$W_{II} = k_{-2}\left(\sqrt{H} + \sqrt{H - 1}\right)^2; \qquad (6.163)$$

$$W_{III} = k_2^2 P_B^2 / 2k_1 P_{A_2}, \qquad (6.164)$$

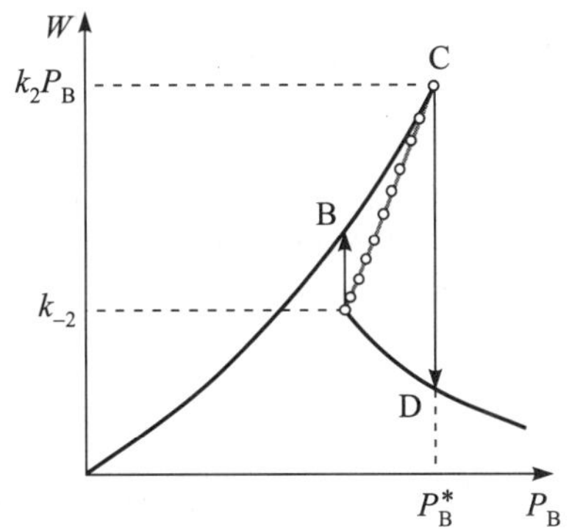

Figure 6.24 Dynamic portrait of system (6.161) for P_{A_2} = const and $P_{A_2} > P_B$ (reproduced with permission from authors © 1996)

where

$$H = \frac{(k_2 P_B + k_{-2})^2}{8k_{-2}k_1 P_{A_2}}.$$

Expressions (6.162–6.164) are valid provided that

$$k_2 P_B > k_{-2}, \qquad k_2 P_B/2 \le k_1 P_{A_2} \le \frac{(k_2 P_B + k_{-2})^2}{8k_{-2}},$$

i.e., for $H \ge 1$. Therefore, point A corresponds to the condition of $W_I = W_{II}$, which is satisfied for $H = 1$. Thus, at this point we have

$$W_A = W_I = W_{II} = k_{-2}, \tag{6.165}$$

which implies that the reaction rate at the ignition point is determined entirely by the k_{-2} value. The critical point C corresponds to $W_{II} = W_{III}$, which is valid for $H > 1$ (at large values of P_B), so that

$$W_C = W_{II} = W_{III} = \frac{k_2^2 P_B{}^2}{2k_1 P_{A_2}}. \tag{6.166}$$

In the case of $k_2 P_B^* = 2k_1 P_{A_2}$ (where P_B^* is the partial pressure of B at the quenching point C) we have

$$W_C = k_2 P_B^* = 2k_1 P_{A_2}. \tag{6.167}$$

The above special cases are called "critical" in order to distinguish them from other approximations (quasi-steady-state and quasi-equilibrium). The meaning of these simplifications is revealed by relations (6.165) and (6.167), from which it can be seen that the rates at the critical points are equal to the weights of the corresponding steps. The equation $W_A = k_{-2} = \omega_{-2}$ indicates that the surface is completely covered (i.e., the catalyst is fully bound) by compound B ($y = 1$). At point C ($W = W_C$), the entire surface is free ($z = 1$). The above equations lead to some interesting thermodynamic relations. For example,

$$\frac{W_C}{W_A} = \frac{k_2 P_B^*}{k_{-2}} = K_2 P_B^*, \tag{6.168}$$

where K_2 is equilibrium constant of step (2) in mechanism (6.160).

Thus, the reaction rates at the aforementioned critical points can be controlled by the thermodynamic characteristics of steps (K_2). Some other relations of this kind for kinetic models with MSS were presented in monograph [12].

If the aforementioned mechanism (6.150) includes an additional step, in which the active center interacts with a reactant or product related so as to form a complex that does not participate in the catalytic cycle (hence, represented by a pendant vertex on the corresponding graph), the dynamic behavior of the system becomes more complicated and there appear limit-cycles and self-oscillations of the concentrations of intermediate products [133].

Example 6.11 [133]

► Consider mechanism (6.150) with the following additional step (pendant vertex):

$$B + Z \underset{(q)}{\rightleftharpoons} BZ^* \ .$$

In this case, the system of differential kinetic equations acquires the following form:

$$\begin{cases} \dfrac{dx}{dt} = 2k_1z^2 - 2k_{-1}x^2 - k_3xy = P(x,\ y,\ q); \\ \dfrac{dy}{dt} = k_2z - k_{-2}y - k_3xy = Q(x,\ y,\ q); \\ \dfrac{dq}{dt} = k_4z - k_{-4}q = R(x,\ y,\ q), \end{cases} \tag{6.169}$$

where $z = 1 - x - y - q$; and the constant partial pressures P_{A_2} and P_B are introduced in the constants k_1 and k_2, respectively. Under steady-state conditions, equation $R(x,\ y,\ q) = 0$ implies $z = k_{-4}/k_4q = \alpha q$.

Upon solving the system of equations $P = Q = 0$ with respect to x and y, one can obtain the function $z = f(q)$, where variable q plays the role of a parameter ($0 \le q \le 1$). This S-shaped curve (Figure 6.25) has a region showing the existence of three steady states. The z versus q curve exhibits a hysteresis character, whereby the concentration of active centers (z) exhibits a jumplike change at $q = q^*$. An analysis of the stability of system $P = Q = 0$ indicates that a single steady state is stable, whereas among the three steady states existing at $q = q^*$ (inflection point) one (middle state z_2) is unstable. An analysis of the complete system of equations (6.169) showed that, under the conditions of existence of a single steady state, state z_2 is unstable and its projection onto the phase surface is surrounded by limit-cycles [133], which corresponds to oscillations in the component concentrations and the reaction rate W. The necessary and sufficient condition of instability of a steady state (point * in Figure 6.25) in the given system is $c < 0$ in the characteristic equation (6.111), where c is the sum of principal minor determinants of the matrix of a linearized system of equations.

The bifurcation values of parameters lead to $c = 0$, which implies a transition from the stable steady state (corresponding to $c > 0$) to an unstable state at $c < 0$. Oscillations are observed in the regions of coordinates (z, q and (x, q) that are close to the hysteresis curve $z = f(q)$ depicted in Figure 6.25. The smaller the values of k_4 and k_{-4} at a constant ratio of $\alpha = k_{-4}/k_4$, the more the limit-cycle shape is like a hysteresis. The main condition of the stability of self-oscillations is the slow equilibrium of step (4), such that $K = k_4/k_{-4} > 1$ and $k_4,\ k_{-4} \ll 1$.

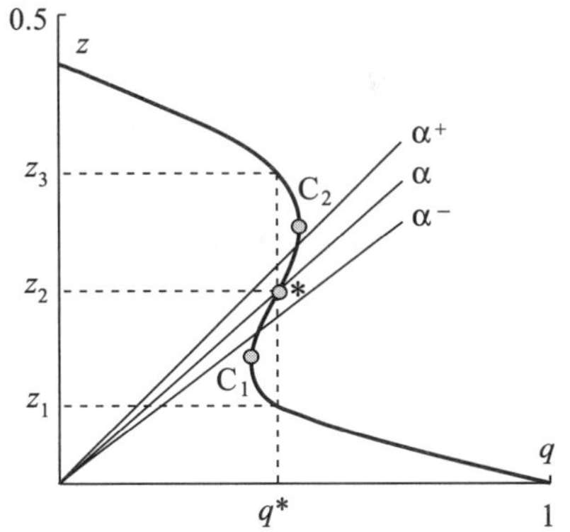

Figure 6.25 Steady-state points on z versus q curve (reproduced with permission from AIPC Nauka © 1978 [133])

The example under consideration illustrates the transition from bistability to self-oscillations as a result of the variation of parameters. The presence of slow reversible step (4) causes the relaxation self-oscillations upon a rapid (jumplike) change in the number of vacant sites ensured by three-step mechanism (6.150).

An analysis of the parametric portrait of system (6.169) [134] on the (k_2, α) plane depending on the variation of constant k_4 (control parameter) revealed 23 phase portraits containing the regions of parameters where the bistability and self-oscillations are observed simultaneously and the regions with stable and unstable limit-cycles. ■

The appearance of self-oscillations in simple four-step mechanisms was also established for the systems represented by graphs with pendant vertices of some other types, in particular [114]:

$$AZ + D \rightleftharpoons DZ + A;$$

$$2Z + D \rightleftharpoons DZ_2,$$

where D is a foreign component or a reaction product.

Based on an analysis of the coefficients of characteristic polynomials for simple four-step mechanisms of catalytic reactions, it was found that the condition of the existence of a single unstable steady state ($c < 0$, $d > 0$) was valid for eight mechanisms not involving autocatalytic steps, with a molecularity of $\mu \leq 2$ with respect to intermediates [117].

Now let us consider changes in the dynamic behavior of a responding system, which are introduced by the appearance of an autocatalytic step or an autocatalytic chain of such steps. First, let us analyze the case of so-called quadratic autocatalysis:

$$A + X \longrightarrow 2X + B. \tag{6.170}$$

This reaction is linear with respect to catalyst X, and its special kinetic feature is the appearance of a positive term in the differential equation of rate balance, which increases with [X].

Example 6.12

► For a closed system, the rate of step (6.170) can be expressed as follows:

$$\frac{dC_X}{dt} = kC_A C_X = f(C_X). \tag{6.171}$$

In the case of $C_A = \text{const}$, this yields

$$C_X = C_{X0} \exp[kC_A t]. \tag{6.172}$$

Evidently, for $C_{X0} = 0$, we have $C_X = 0$ and $dC_X/dt = 0$. Thus, it can formally be accepted that $f(C_X) = 0$ for $C_X = 0$, and this is a single steady state in the system under consideration. This steady state is unstable, since $\partial f(C_X)/\partial C_X = a = kC_A > 0$ [50]. However, without an arbitrary finite initial concentration [X] (C_{X0}), reaction (6.170) cannot proceed, since

$$\frac{dC_X}{dt} = kC_A(C_{X0} + C_X) = f(C_X) \tag{6.173}$$

and $f(C_X) = 0$ at $C_X = -C_{X0}$, which is deprived of physical sense. Equation (6.173) has no physically meaningful steady-state solutions and the concentration C_X exhibits infinite growth according to Eq. (6.172), since C_A is constant.

For a closed system with $C_0 = C_{A0} + C_{X0} = C_A + C_X$ and $C_A = C_0 - C_X$, one can obtain the equation [50]

$$\frac{dC_X}{dt} = k_1 C_X (C_0 - C_X) = f(C_X) \tag{6.174}$$

and the following integral expression for C_X:

$$C_X = \frac{(C_{A0} + C_{X0})C_{X0}}{C_{X0} + C_{A0}e^{-k_1(C_{A0}+C_{X0})t}}, \tag{6.175}$$

in which $C_X = C_{X0}$ at $t = 0$.

An analysis of the steady-state solutions of Eq. (6.174) [50] for $f(C_X) = 0$ shows that this quadratic equation yields two steady-state solutions, $C_X^{ss(1)} = 0$ and $C_X^{ss(2)} = C_0$. For the derivative $f'(C_X) = k_1 C_0 - 2k_1 C_X^{ss} - a$, we have $a_1 > 0$ (unstable steady state) for $C_X^{ss(1)}$ and $a_2 = -k_1 C_0 < 0$ (stable node) for $C_X^{ss(2)} = C_0$. However, the situation with $C_X^{ss(1)} = 0$ is again deprived of physical sense, since it can only take place for $C_X = -C_{X0}$. Alternatively writing Eq. (6.174) as

$$\frac{dC_X}{dt} = k_1 (C_{X0} + C_X)(C_{A0} - C_X) = f(C_X)$$

we obtain

$$(C_X)^2 = (C_{A0} - C_{X0})C_X - C_{A0}C_{X0} = 0,$$

where $C_X^{ss(1)} = C_{A0}$ and $C_X^{ss(2)} = -C_{X0}$. The second solution has no physical meaning, but is stable since $a_2 < 0$. The first solution $C_X^{ss(1)} = C_{A0}$ yields $a_1 > 0$ and is unstable. If we subdivide reaction (6.170) into two elementary steps, with the first quasi-equilibrium reaction

$$\mathrm{A + X \rightleftharpoons Y;}$$

$$\mathrm{Y \longrightarrow 2X + B,} \tag{6.176}$$

the behavior of the system would not change both for $C_A = \text{const}$ and $C_A = \text{var}$ (at $C_Y \ll C_A$), since the law of mass action (6.171) will have the same form (with constant k_1 replaced by $k_2 K_1$). The situation will also remain the same if we exclude the quasi-equilibrium so as to obtain a quasi-steady-state value of C_Y.

Let us analyze the system of two equations for $C_A = C_{A0} - C_X$:

$$\begin{cases} \dfrac{dC_Y}{dt} = k_1 C_A C_X - k_{-1} C_Y - k_2 C_Y = P(C_X, C_Y) \\ \dfrac{dC_X}{dt} = -k_1 C_A C_X + k_{-1} C_Y + 2k_2 C_Y = Q(C_X, C_Y) \end{cases} \tag{6.177}$$

$$\begin{cases} \dfrac{dC_Y}{dt} = k_1 (C_{A0} C_X (C_{X0} + C_X) - (k_{-1} + K_2) C_Y = P(C_X, C_Y) \\ \dfrac{dC_X}{dt} = k_1 C_{A0} - C_X)(C_{X0} + C_X) + (k_{-1} + k_2) C_Y = Q(C_X, C_Y) \end{cases} \tag{6.178}$$

with the steady states $C_X^{ss(1)} = C_{A0}$ and $C_Y^{ss(1,2)} = 0$.

For system (6.178), the determinant of the characteristic equation is $\Delta > 0$, while the signs of σ and $\sigma^2 - 4\Delta$ depend on the ratio of the rate constants and the sum of $C_{A0} + C_{X0}$. For example, $\sigma = k_1(C_{A0} + C_{X0}) - (k_{-1} + k_2)$. For this reason, both the nature of the steady state and its stability in this system can substantially vary, even including a singular point of the center type ($\sigma = 0, \Delta > 0, \sigma^2 - 4\Delta < 0$). ■

Now let us supplement reaction (6.170) by a step of conversion of X species, i.e., the step that terminates the autocatalyst formation.

Example 6.13

▶ The modified mechanism appears as follows:

$$A + X \longrightarrow 2X + B; \quad X \longrightarrow F. \tag{6.179}$$

In a closed system at $C_A = \text{const}$, the kinetic equation can be written as

$$\frac{dC_X}{dt} = k_1 C_A (C_{X0} + C_X) - k_2 (C_{X0} + C_X) = \\ = (k_1 C_A - k_2)(C_{X0} + C_X) = f(C_X), \tag{6.180}$$

where $f(C_X) = 0$ for $C_X = -C_{X0}$ or $C_A^{ss} = k_2/k_1 = C_A^{crit}$ and C_X in this case is the increment in the autocatalyst concentration, which can be both positive and negative.

The derivative dC_X/dt changes sign at $C_A = C_A^{crit}$, which leads to the corresponding change in the stability. For $C_A > C_A^{crit}$, we have $a = k_1 C_A - k_2 > 0$ and the system is unstable, whereas for $C_A < C_A^{crit}$ the derivative is negative ($a < 0$) and the steady state is stable. ■

Mechanism (6.179) with the first reversible step was analyzed in detail in [50, 135, 136].

Example 6.14

▶ The mechanism with a reversible step appears as

$$A + X \rightleftharpoons 2X; \quad X \longrightarrow F, \tag{6.181}$$

and the corresponding kinetic equation at $C_A = \text{const}$ is as follows:

$$\frac{dC_X}{dt} = k_1 C_A C_X - 2k_{-1} C_X^2 - k_2 C_X = f(C_X). \tag{6.182}$$

The steady-state values $C_X^{ss(1,2)}$ are determined from the condition

$$f(C_X) = C_X (k_1 C_A - 2k_{-1} C_X - k_2) = 0, \tag{6.183}$$

which yields $C_X^{ss(1)} = 0$ and $C_X^{ss(2)} = (k_1 C_A - k_2)/(2k_{-1})$.

It should be emphasized that $C_{X0} = 0$ implies that the process does not start at all, while $C_{X0} \neq 0$ with $C_X^{cc(1)} = 0$ implies that $C_X = -C_{X0}$, i.e., catalyst X completely disappears. The critical concentration $C_A^{crit} = k_2/k_1$ is the bifurcation point and, at $C_A = C_A^{crit}$, the reaction rate drops to zero due to the complete loss of autocatalyst X $\left(C_X^{cc(2)} = 0\right)$. In this case, even the increase to $C_A > C_A^{crit}$ does not make the passage to the branch with a positive reaction rate possible unless the autocatalyst is added [135]. This model (and more complicated variants) was used in discussions of the problem of selection and evolution of the so-called prebiotic autocatalysts [135, 136], since the value of C_A^{crit} (critical amount of "food") depends on k_2 and k_1 and, if several autocatalysts (X_i) are operative, those with higher C_A^{crit} must be consecutively lost. An analysis of the stability of steady-state solutions $C_X^{ss(1)}$ and $C_X^{ss(2)}$ shows that the derivative of $f(C_X)$ with respect to C_X is expressed as

$$f'(C_X) = k_1 C_A - 2k_{-1} C_X - k_2 = a.$$

For $C_A^{ss(1)} = 0$ and $C_A > C_A^{crit}$, we have $a_1 > 0$ and the solution is unstable, whereas for $C_X^{ss(2)} = (k_1 C_A - k_2)/(2k_1)$, and $C_A > C_A^{crit}$, we have $a_2 = 0$ and the steady state corresponds to a saddle–node singularity with $\lambda = 0$ [98]. ■

Now let us consider the case where there is a constant supply of catalyst X to the reaction medium.

Example 6.15 [50]

▶ A mechanism (6.184) with the constant supply of catalyst X to the reaction medium at $C_A = \text{const}$ (or $d_e C_X/dt = k_0$ [50]) is described by the following equations:

$$\xrightarrow{k_0} \mathrm{X};$$

$$\mathrm{X} + \mathrm{A} \xrightarrow{k_1} 2\mathrm{X} + \mathrm{B};$$

$$\mathrm{X} \xrightarrow{k_2} \mathrm{F}; \tag{6.184}$$

$$\frac{dC_X}{dt} = k_1 C_A C_X - k_2 C_X + k_0 = P(C_X), \tag{6.185}$$

for which $P(C_X) = 0$ at

$$C_X^{ss(1)} = \frac{k_0}{k_2 - k_1 C_A}.$$

Evidently, $C_X^{ss} \geq 0$ for $k_2/k_1 > C_A$, and this system also has a critical point at $C_A^{crit} = k_2/k_1$. For $C_A \geq C_A^{crit}$, the steady-state solution has no physical sense and $C_X^{ss} = 0$. For $C_A^{ss} < C_A^{crit}$, C_X^{ss} grows with increasing C_A and asymptotically approaches $C_X^{ss} = \infty$. The C_A^{crit} value is a bifurcation parameter. The derivative

$$P'(C_X) = k_1 C_A - k_2 = a_1 \tag{6.186}$$

for $C_A < C_A^{crit}$ is negative ($a_1 < 0$) and the solution is stable, whereas for $C_A > C_A^{crit}$ we have $a_1 > 0$ and the solution at $C_X^{ss} = 0$ is unstable.

If the autocatalytic step in mechanism (6.184) is reversible (at $C_B = \text{const}$), then we obtain the equation

$$\frac{dC_X}{dt} = (k_1 C_A - k_2) C_X - k_{-1} C_X^2 + k_0 = P(C_X), \tag{6.187}$$

for which the condition $P(C_X) = 0$ yields

$$C_X^{ss(1,2)} = -\frac{\alpha}{2k_{-1}} \pm \sqrt{\frac{\alpha^2}{4k_{-1}^2} + \frac{k_0}{k_{-1}}},$$

where $\alpha = k_2 - k_1 C_A$. Only one root is positive ($C_X^{ss} > 0$),

$$C_X^{ss(1)} = -\frac{\alpha}{2k_{-1}} + \sqrt{\frac{\alpha^2}{4k_{-1}^2} + \frac{k_0}{k_{-1}}} > 0.$$

The corresponding value of derivative

$$P'(C_X) = k_1 C_A - k_2 - 2k_1 C_X = a$$

is

$$a_1 = -2k_1 \sqrt{\frac{\alpha^2}{4k_{-1}^2} + \frac{k_0}{k_{-1}}} < 0,$$

so that the steady-state solution with $C_X^{ss(1)}$ is stable. The steady-state solution with $C_X^{ss(2)} < 0$ corresponds to $a_2 > 0$ and, hence, is unstable. ■

Mechanism (6.181) is a fragment of the Butlerov reaction of formaldehyde condensation [85, 86] with a somewhat more complicated dependence on C_A. At a constant concentration of formaldehyde (C_1), the steps involving one of the possible autocatalysts, glycol aldehyde (C_2), are as follows:

$$2C_1 + C_2 \rightleftharpoons 2C_2;$$
$$C_2 + C_1 \longrightarrow C_3, \tag{6.188}$$

where C_3 is the product of condensation of three CH_2O molecules. Denoting the concentrations of these substances by C_1, C_2, and C_3, we can write the kinetic equation as

$$\frac{dC_2}{dt} = k_1 C_1^2 C_2 - k_{-1} C_2^2 - k_2 C_1 C_2 = P(C_2). \tag{6.189}$$

For $P(C_2) = 0$, we obtain $C_2^{ss(1)} = 0$ and $C_2^{ss(2)} = [k_1 C_1/k_{-1}](C_1 - k_2/k_1)$. Thus, we have $C_2^{ss(2)} = 0$ for $C_1 = 0$ and for $C_1 = C_1^{\text{crit}} = k_2/k_1$ (a critical point).

Mechanism (6.188) can also be supplemented by adding the step of formation of glycol aldehyde (C_2):

$$2C_1 \xrightarrow{k_0} C_2;$$
$$2C_1 + C_2 \rightleftharpoons 2C_2;$$
$$C_1 + C_2 \longrightarrow C_3. \tag{6.190}$$

In this case, a physically meaningful steady-state solution with $C_2^{ss(1)} > 0$ exists for all relations between C_A and C_A^{crit}, but this steady state is unstable ($a_1 > 0$). Under the condition $(k_1 C_A - k_2)^2 \gg 4k_{-1}k_0$, a solution of the quadratic equation with respect to C_2

$$C_2^{ss(1)} = 0 \quad \text{and} \quad C_2^{ss(2)} = \frac{k_1 C_1}{k_{-1}} \left(C_1 - \frac{k_2}{k_1} \right),$$

transforms into a solution of Eq. (6.189) with the critical (threshold) concentration $C_1 = k_2/k_1$.

The above examples lead to a conclusion that the appearance of the quadratic autocatalysis step (6.170) in the mechanisms of reactions in closed and semi-closed systems provides conditions for the appearance of bifurcation parameters (k_2/k_1), threshold concentrations, and simple critical kinetic phenomena (reaction rate jumps).

If mechanism (6.184) is modified so as to allow the introduced reactant X to interact with autocatalyst Y, then we obtain the historically first two-dimensional ($n = 2$) dynamic Lotka model [38], which was constructed based on the law of mass action and led to the pattern of decaying oscillations.

Example 6.16

▶ Consider the first Lotka model with the following scheme:

$$\xrightarrow{k_0} X;$$

$$X + Y \xrightarrow{k_1} 2Y;$$

$$Y \xrightarrow{k_2} F, \tag{6.191}$$

where $k_0 = d_e C_X/dt$ is the rate of introduction of reactant X (k_0 can also be expressed as $k_0 = k_0' C_0$ for $C_0 = \text{const}$). The corresponding kinetic equations are as follows:

$$\begin{cases} \dfrac{dC_X}{dt} = k_0 - k_1 C_X C_Y = P(C_X, C_Y); & (6.192) \\ \dfrac{dC_Y}{dt} = k_1 C_X C_Y - k_2 C_Y = Q(C_X, C_Y). & (6.193) \end{cases}$$

The steady-state values of C_X and C_Y for this system are $C_X^{ss} = k_2/k_1$ and $C_Y^{ss} = k_0/k_2$. In the linear approximation, we have

$$\begin{cases} \dfrac{d\varepsilon}{dt} = -\dfrac{k_1 k_0}{k_2}\varepsilon - k_2\eta; \\ \dfrac{d\eta}{dt} = \dfrac{k_1 k_0}{k_2}\varepsilon, \end{cases} \tag{6.194}$$

which yields $\sigma = -k_1 k_0/k_2 < 0$ and $\Delta = k_1 k_0 > 0$. The discriminant is positive ($\sigma^2 - 4\Delta > 0$) for $k_1 k_0/k_2^2 > 4$ (stable node) and negative ($\sigma^2 - 4\Delta < 0$) for $k_1 k_0/k_2^2 < 4$. A singular point with $Re(\lambda) < 0$ represents a stable focus, to which the system approaches by decaying oscillations. The condition $k_1 k_0 = 4k_2^2$ corresponds to the bifurcation point, where the regime changes from decaying oscillations to the exponential motion toward a steady state.

The system of equations (6.194) can readily be reduced to a second-order differential equation with a single variable:

$$\frac{d^2\varepsilon}{dt^2} + \frac{k_1 k_0}{k_2}\frac{d\varepsilon}{dt} + k_1 k_0 \varepsilon = 0. \tag{6.195}$$

In the general case, this equation can be written as

$$\frac{d^2x}{dt^2} + \gamma\frac{dx}{dt} + \omega_0^2 x = 0, \tag{6.196}$$

which represents the mathematical model of a harmonic linear oscillator (e.g., pendulum) with the energy dissipation as a result of friction. This model describes decaying oscillations, but does not provide for self-sustained oscillations. The value of $\omega_0 = \sqrt{k_1 k_0}$ gives the oscillation frequency, while the coefficient $\gamma = k_1 k_0/k_2$ describes the contribution of energy dissipation processes. ■

Example 6.17

► The second Lotka model [39] includes two autocatalytic steps:

$$(1)\quad A + X \xrightarrow{k_1} 2X;$$

$$(2)\quad X + Y \xrightarrow{k_2} 2Y;$$

$$(3)\quad Y \xrightarrow{k_3} P. \tag{6.197}$$

and can be represented by the bipartite graph that is depicted in Figure 6.26. Sometimes, this scheme is supplemented by the fourth step

$$X + Y \xrightarrow{k_4} P_1,$$

but this only modifies the structure of constants, while not changing the signs of σ and Δ and, hence, the dynamic behavior of the system.

Let us consider system (6.197) assuming that $C_A = \text{const}$ (or A is present in a large excess) and the initial concentrations of X and Y are small compared to the current quasi-steady-state values of C_X and C_Y ($C_{X0} \ll C_X, C_{Y0} \ll C_Y$). Similarly, it can be assumed that the rates of supply of X and Y (d_eC_X/dt and d_eC_Y/dt, respectively) are constant and much smaller than the rates of steps (1) and (2):

$$\left\{\begin{aligned} \frac{dC_X}{dt} &= k_1C_AC_X - k_2C_XC_Y = P(C_X, C_Y); \\ \frac{dC_Y}{dt} &= k_2C_XC_Y - k_3C_Y = Q(C_X, C_Y) \end{aligned}\right. . \tag{6.198)–(6.199}$$

Conditions $P(C_X, C_Y) = 0$ and $Q(C_X, C_Y) = 0$ yield

$$C_X(k_1C_A - k_2C_Y) = 0; \quad C_Y(k_2C_X - k_3) = 0,$$

so that $C_X^{ss(1)} = 0, C_Y^{ss(1)} = 0$, and $C_X^{ss(2)} = k_3/k_2, C_Y^{ss(2)} = k_1C_A/k_2$ (more strictly, for the adopted assumptions we have $C_{X'}^{ss(2)} = k_3/k_2 - C_{X0} \cong k_3/k_2$ and $C_{Y'}^{ss(2)} = k_1C_A/k_2 - C_{Y0} \cong k_1C_A/k_2$).

The coefficients of the linearized system for $C_X^{ss(2)}$ and $C_X^{ss(2)}$ are as follows:

$$a_{11} = 0; \quad a_{22} = 0; \quad a_{21} = -k_3; \quad a_{21} = k_1C_A,$$

which yields $\sigma = 0$, $\Delta = k_1k_3C_A$, $\lambda_{1,2} = \sqrt{\sigma^2 - 4\Delta} = \pm 2i\sqrt{k_1k_3C_A}$, and $Re(\lambda_{1,2}) = 0$. Accordingly, the singular point is of the center type, and the closed trajectories surrounding the center have nearly elliptical shapes.

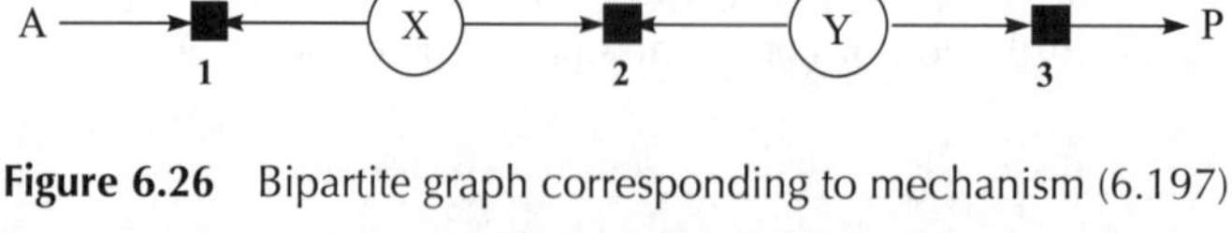

Figure 6.26 Bipartite graph corresponding to mechanism (6.197)

Let us construct the second-order equation for fluctuations ε. For $d\varepsilon/dt = -k_3\eta$ and $d\eta/dt = k_1 C_A \varepsilon$, one can obtain

$$\frac{d^2\varepsilon}{dt^2} = -k_1 k_3 C_A \varepsilon, \quad (6.200)$$

where the value of $k_1 k_3 C_A = \omega_0^2$ has a dimensionality of s^{-2}, i.e., squared oscillation frequency. Thus, small fluctuations in the vicinity of a steady state are periodic functions with a frequency of $\omega_0 = \pm\sqrt{k_1 k_3 C_A}$.

Writing Eq. (6.200) in the general form as

$$\frac{d^2x}{dt^2} + \omega_0^2 x = 0, \quad (6.201)$$

we obtain an equation that describes a harmonic oscillator, i.e., a mathematical pendulum without friction, which performs nondecaying oscillations. Denoting $dx/dt = y$, we reduce this equation to $dy/dt = -\omega_0^2 x$. Excluding the time and integrating the resulting equation

$$\frac{dx}{dy} = -\frac{y}{\omega_0^2 x}, \quad (6.202)$$

we obtain the equation of ellipse

$$\frac{x^2}{c^2} + \frac{y^2}{(\omega_0 c)^2} = 1, \quad (6.203)$$

where C is the integration constant. Thus, the derivative of the concentration with respect to time (i.e., y considered as the function of x, which is a dynamic portrait of the system) appears as an ellipse on the dx/dt versus x plane with the origin at $(dx/dt = 0, x = 0)$. Closed curves on the (C_X, C_Y) plane can also be obtained from equations (6.198) and (6.199), since this system has the following analytical integral:

$$\frac{dC_X}{dC_Y} = \frac{C_X(k_1 C_A - k_2 C_Y)}{C_Y(k_2 C_X - k_3)};$$

$$\left(C_X - \frac{k_3}{k_2}\ln C_X\right) + \left(C_Y - \frac{k_1 C_A}{k_2}\ln C_Y\right) = H(C_X, C_Y) = \text{const.} \quad (6.204)$$

These closed curves are surrounding the steady state $(C_X^{ss(1)} = 0, C_Y^{ss(1)} = 0)$, which represents a saddle point.

The closed curves on the phase portrait plotted as C_X versus C_Y reflect the existence of oscillations, but do not represent a limit-cycle that is a mathematical image of self-oscillations. This circumstance is related both to specific features of a singular point of the center type and to the features of Eqs (6.198), (6.199), and (6.201). The integral equation (6.204) shows that a change in the initial conditions (C_{A0}) and/or fluctuations of the concentrations [C_X, C_Y, and $C_A(H)$] generate a dense set of closed curves surrounding the steady state of the center type. In addition, this singular point is asymptotically unstable, since there are no mechanisms that would cause the decay of fluctuations and return the system to $(C_X^{ss(2)}, C_Y^{ss(2)})$.

The first Lotka model with mechanism (6.197) has a continuous spectrum of rotation frequencies over an infinite set of cycles, each one corresponding to certain initial conditions. Since each cycle represents a state at the boundary of stability, there is no "mean" orbit in the vicinity of which the dynamical system can occur [6].

System (6.198)–(6.199) is also structurally unstable, since small changes in the right-hand parts of equations transform closed trajectories into spirals [98]. According to a special theorem [98], if a singular point of system A is of the center type, then for any $\delta > 0$ there exists a modified system $\tilde{A}$ that is δ-close to A and has a steady state of the focus type. Thus, even small fluctuations convert the center into a focus and the ellipse into a damping spiral. ■

Dynamical systems of the type (6.198), (6.199) and (6.201) belong to so-called *conservative systems*, which do not exhibit self-oscillations. These systems are characterized by the following specific features [98]:

- there exists an analytical integral of the type $H(x, y) = const$;
- singular points can only be of the center and saddle-node types;
- closed trajectories are not isolated and fill the entire region of the phase space.

Conservative systems are essentially non-robust. The theory of nonlinear oscillations deals with non-conservative systems which, on the contrary, are essentially robust. In these systems, all trajectories from one region of the phase space tend to the same attractor as $t \to +\infty$, and they tend to the same repeller as $t \to -\infty$. Closed trajectories in nonconservative systems always represent isolated, structurally stable limit-cycles. Mathematically, a limit-cycle corresponds to the existence of at least one solution of the system not coinciding with the analytical integral $H = \text{const}$.

The introduction of additional terms into the Lotka equations (6.198–6.199), which take into account the supply and removal of reactants X and Y, makes the system nonconservative, with a rich set of steady states and diverse dynamic behavior, including self-oscillations. The Lotka model also becomes nonconservative if the autocatalytic steps (1) and (2) in scheme (6.197) are replaced by reactions that possess a certain chemical sense, for example:

$$\mathrm{A+X} \xrightarrow{k_1'} \mathrm{X_1}; \quad \mathrm{B+X_1} \xrightarrow{k_2'} \mathrm{2X};$$

$$\mathrm{X+Y} \xrightarrow{k_3'} \mathrm{X_2}; \quad \mathrm{C+X_2} \xrightarrow{k_4} \mathrm{2Y};$$

$$\mathrm{Y} \xrightarrow{k_5} \mathrm{D}.$$

The system of differential equations corresponding to this scheme admits both self-oscillations (for $C_A, C_B, C_C = \text{const}$) and decaying oscillations [137]. An increase in the number of intermediate steps or the implementation of process (6.197) in a continuous-flow stirred tank reactor (CSTR) increases the system dimension ($n \geq 3$). In this context, it should be noted that Tyson and Light [144] proved the impossibility of oscillatory behavior for any mechanisms with $n = 2$ and bimolecular steps (without allowance for nonlinear complexity functions of the catalyst or intermediates). Variants of some other model systems based on the Lotka models are described in [5].

Alfred Lotka used his second model to describe the evolution of populations in the predator–prey system. Somewhat later, analogous population models were proposed by the Italian mathematician Vito Volterra in his famous book *Leçons sur la théorie mathématique de la lutte pour la vie* (Gauthier-Villars, Paris, 1931; Reissued 1990, edited by J. Gabay) (see also [24]) to describe oscillations in the predator–prey type systems within the framework of the same conservative dynamical system. In excess of plant food ($C_A = \text{const}$), the number of prey (rabbits) N_1 exhibits autocatalytic growth according to the law

$$dN_1/dt = e_1 N_1, \tag{6.205}$$

while their loss takes place According to the second-order equation in proportion to the probability of meeting predators (wolves), i.e., to their number N_2:

$$dN_1/dt = -\gamma_2 N_1 N_2. \tag{6.206}$$

As the population of prey decreases, predators are short of food and their population also starves according to the equation

$$dN_2/dt = -e_2N_2, \tag{6.207}$$

(cf. third step in (6.197)), which favors growth in the prey number N_1 and, in turn, in the predator number:

$$dN_2/dt = \gamma_1 N_1 N_2. \tag{6.208}$$

This corresponds to a system of type (6.198)–(6.199) with the corresponding consequences, in particular, oscillations in the N_1 and N_2 values at a frequency of $\omega_0 = e_1e_2$. It should be noted that modern models of population development are more complicated and involve three parties (predator–prey–food) [11, 42, 96].

Harmonic oscillator (6.201) is called linear because the oscillations are sinusoidal. In more complicated dynamical systems, the oscillations acquire a pronounced non-sinusoidal shape, and a significant nonlinearity leads to relaxation oscillations (see Figure 6.16 (case *7*)). These oscillations appear when a dynamical system possesses at least two stable steady states and oscillates (switches) between them. Systems with relaxation oscillations exhibit a threshold character as manifested by the existence of critical values of variables (concentrations), the attaining of which drives the system to pass (switch) from one state to another.

A stronger impact on the behavior of a dynamical system is related to the appearance of *cubic autocatalysis*, i.e., a step of the type

$$\mathrm{A} + 2\mathrm{X} \xrightarrow{k} 3\mathrm{X} + \mathrm{C}. \tag{6.209}$$

In a closed system with a small initial concentration of catalyst X ($C_{X0} \ll C_{A0}$), the concentration of A varies very slowly until a certain moment of time, when C_X exhibits rapid growth, C_A drops, and a sharp maximum appears on the dC_A/dt versus t curve (Figure 6.27). For $k = 100\,\mathrm{M}^{-2}\cdot\mathrm{min}^{-1}$, $C_{A0} = 0.1$ M, and $C_{X0} = 10^{-3}$M, one can obtain a maximum in $-dC_A/dt$ at $t \approx 100$ min, where $C_A = (C_{A0} + C_{X0})/3$. This behavior of the system is much like the phenomenon of color variation in the Landolt reaction considered above (see Section 6.2.3).

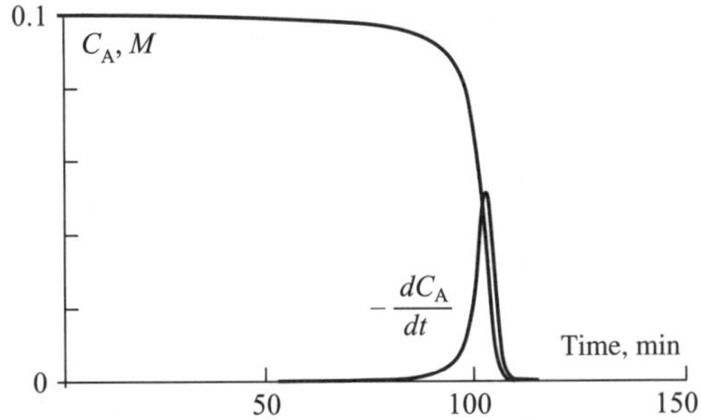

Figure 6.27 Plots of concentration C_A and reaction rate dC_A/dt versus time according to Eq. (6.210) for $C_{A0} = 0.1$ M, $C_{B0} = 10^{-3}$M, and $k = 100\,\mathrm{M}^{-2}\,\mathrm{min}^{-1}$ (reproduced with permission from Oxford University Press © 1994 [17])

Let us analyze the behavior of a system with the catalyst kinetics described by the equation

$$\frac{dC_X}{dt} = kC_A C_X^2 \tag{6.210}$$

for a process with step (6.209) carried out in a CSTR.

Example 6.18 [17]

► For the process in a CSTR, the equation of rate balance for C_A is

$$\frac{dC_A}{dt} = k_0(C_{A0} - C_A) - kC_A(C_{A0} + C_{X0} - C_A)^2 \tag{6.211}$$

and the linear conservation law appears as

$$C_{A0} + C_{X0} = C_A + C_X.$$

In order to determine C_A^{ss} and solve the cubic equation with respect to C_A, it is necessary to find the points of intersection for the curves representing the reactor equation

$$r_\tau = k_0(C_{A0} - C_A) \tag{6.212}$$

and the rate equation for reaction (6.210) written as

$$r = kC_A(C_{A0} + C_{X0} - C_A)^2. \tag{6.213}$$

This system admits the appearance of hysteresis and bistability with jumps in the values of rates and concentrations C_A at the tangential intersection points of Eqs (6.212) and (6.213), where

$$r = r_\tau \quad \text{and} \quad \frac{dr}{dC_A} = \frac{dr_\tau}{dC_A}. \tag{6.214}$$

Using conditions (6.214), it is possible to determine C_A^{ss} values at the tangential (critical) points. Indeed, the relations

$$kC_A(C_{A0} + C_{X0} - C_A)^2 = k_0(C_{A0} - C_A); \tag{6.215}$$

$$k(C_{A0} + C_{X0} - C_A)(C_{A0} + C_{X0} - 3C_A) = -k_0 \tag{6.216}$$

lead to the equation

$$\frac{k(C_{A0} + C_{X0} - C_A)}{(C_{A0} + C_{X0} - 3C_A)} = -(C_{A0} - C_A), \tag{6.217}$$

which yields

$$C_A^{ss(1,2)} = \frac{1}{4}\left(3C_{A0} \pm [C_{A0}(C_{A0} - 8C_{X0})]^{1/2}\right). \tag{6.218}$$

The greater value $C_A^{ss(1)}$ corresponds to a tangential point at low degrees of conversion of reactant A (extinction of the process). The lower value of $C_A^{ss(2)}$ corresponds to higher degrees of conversion, where a jump in the concentration C_A and the rate corresponds to activation (ignition) of the process. Here, multiple and tangential steady states are possible for $C_{A0} > 8\,C_{X0}$. In the case of $C_{A0}/8 < C_{X0}$, the curves described by Eqs (6.212) and (6.213) intersect at a single steady-state point. ■

Thus, the cubic autocatalysis by itself (not taking into account the other steps of a mechanism) in a CSTR leads to only two types of behavior, which are manifested on the dynamical portraits (bifurcation diagrams) by (i) monotonic dependence of the degree of conversion $\chi^{ss} = (C_{A0} - C_A)/C_{A0}$ on parameter k_0 and (ii) hysteresis curve $\chi^{ss} = f(k_0)$. It is interesting to note that the hysteresis curve of χ^{ss} versus τ in this case virtually coincides with the hysteresis curve in Figure 6.3.

Let us complicate the above model (called the "cubic autocatalator" [17]) by adding a step of the decomposition of autocatalyst X in the process carried out in a CSTR.

Example 6.19 [17, 138]

► Consider the process in a CSTR with the steps of cubic autocatalysis and catalyst decomposition

$$\begin{aligned}&(1)\quad A + 2X \xrightarrow{k_1} 3X + C;\\ &(2)\quad X \xrightarrow{k_2} D\end{aligned} \tag{6.219}$$

which is described by the following kinetic equations:

$$\begin{cases}\dfrac{dC_A}{dt} = k_0(C_{A0} - C_A) - k_1 C_A {C_X}^2 = P(C_A, C_X)\\[2ex] \dfrac{dC_X}{dt} = k_0(C_{X0} - C_X) + k_1 C_A {C_X}^2 - k_2 C_X = Q(C_A, C_X)\end{cases}. \tag{6.220}$$

Equations (6.220) can be written in a simpler dimensionless form as

$$\begin{cases}\dfrac{d\alpha}{d\tau} = k_0'(1 - \alpha) - \alpha\beta^2 = P(\alpha, \beta)\\[2ex] \dfrac{d\beta}{d\tau} = k_0'(\beta_0 - \beta) + \alpha\beta^2 - \beta/\gamma = Q(\alpha, \beta)\end{cases}, \tag{6.221}$$

where $\alpha = C_A/C_{A0}$; $\beta = C_X/C_{A0}$; $\beta_0 = C_{X0}/C_{A0}$; $k_0' = k_0/(k_1 C_{A0}^2)$; $\tau = k_1 C_{A0}^2 t$; and $\gamma = k_1 C_{A0}^2/k_2$. In the case of a single tangential point for β, the steady-state values are as follows:

$$\beta^{ss} = \frac{2k_0'\beta_0\gamma}{k_0'\gamma + 1}; \tag{6.222}$$

$$\alpha^{ss} = \frac{k_0'(k_0'\gamma + 1)^2}{k_0'(k_0'\gamma + 1)^2 + 4k_0'^2\beta_0^2\gamma^2}. \tag{6.223}$$

Investigation of the stability of system (6.221) at this point shows that the values of σ and Δ can change sign, so that the given steady state is unstable. This state admits the condition of $\sigma = 0$ at $\Delta > 0$, which is evidence for the possible appearance of a limit-cycle and periodic oscillations. An analysis of all steady states for variable $\tau = 1/k_0$ revealed five typical (boundary) cases [17] on the parametric portraits of $1 - \alpha^{ss} = f(\tau)$ (Figure 6.28). Note the appearance of isolated steady states (called isoles) and mushroom-shaped double hysteresis loops.

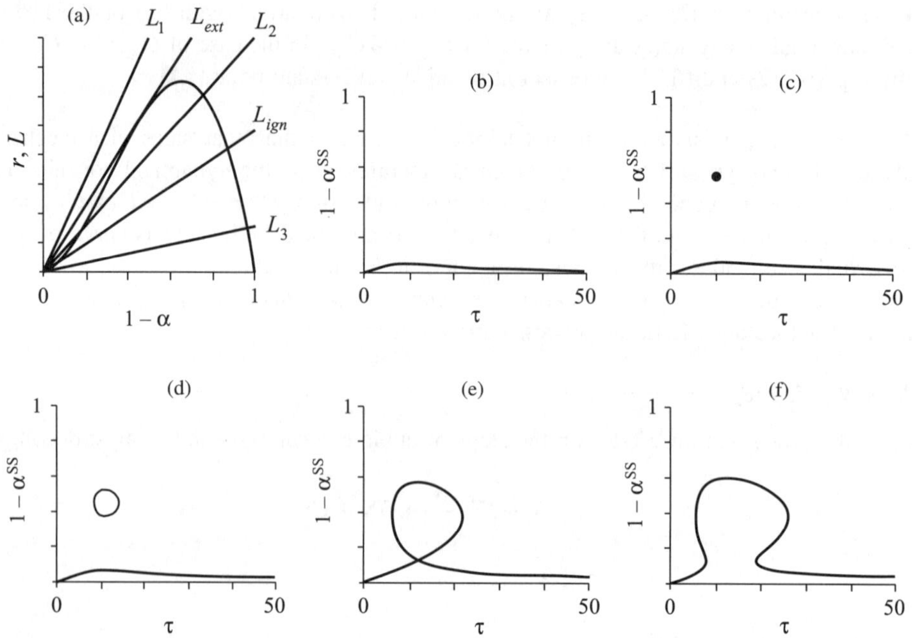

Figure 6.28 (a) Plots of flows $r_\tau = L$ and reaction rate r versus degree of conversion $(1-\alpha)$ and (b–f) plots of $1-\alpha^{ss}$ versus contact time τ for steady-states of various types (reproduced with permission from Oxford University Press © 1994 [17])

The point of a single real root of the cubic equation (L_1, Figure 6.28a) corresponds to a monotonic variation of $1-\alpha^{ss} = f(\tau)$ at large k_0 (small τ and small $1-\alpha^{ss}$) (Figure 6.28b). The first (higher) tangential point of extinction (L_{ext}) corresponds to the appearance of an isolated steady state representing a point (Figure 6.28c) that transforms into an isole (Figure 6.28d) with increasing τ (L_2). The subsequent transformation of the isole takes place at the lower tangential point of ignition (L_{ign}) (Figure 6.28e) with the formation of a typical mushroom-shaped portrait with the further increase in τ (Figure 6.28f). ■

Cook at al. [139] analyzed some model mechanisms, in which the "chemically pathological" trimolecular reaction (1) in scheme (6.219) was replaced by a sequence of bimolecular steps.

Example 6.20 [139]

► Let us consider two model schemes, representing process (6.219) with step (1) replaced by bimolecular steps (3) and (4) as follows:

Model I

$$(3)\ \ A + X \rightleftharpoons Y$$

$$(4)\ \ Y + X \longrightarrow 3X$$

$$A \longrightarrow C$$

Model II

$$
\begin{array}{ll}
(3) & 2\mathrm{X} \rightleftharpoons \mathrm{Y} \\
(4) & \mathrm{Y} + \mathrm{A} \longrightarrow 3\mathrm{X} \\
(2) & \mathrm{X} \longrightarrow \mathrm{C} \\
\hline
 & \mathrm{A} \longrightarrow \mathrm{C}
\end{array}
$$

In an analysis of the stability of the three-dimensional $(C_{\mathrm{A}}, C_{\mathrm{X}}, C_{\mathrm{Y}})$ model of reactions I and II in a CSTR, conditions were selected so as to obtain the cubic characteristic equation in the following form [43]:

$$(\lambda + \alpha)(\lambda^2 + \omega^2) = 0.$$

Then, one root is $\lambda_1 = -\alpha < 0$, while the two other roots have $Re(\lambda) = 0$ and are imaginary: $\lambda_{2,3} = \pm i\sqrt{\omega^2} = \pm i\omega$. The latter eigenvalues $(\lambda_{2,3})$ correspond to a point of the Andronov–Hopf bifurcation in the three-dimensional system under consideration. The regime of multiple steady states (MSS) appears at a large dimensionless contact time $\tau' = k_1 C_{\mathrm{A0}}^2/\tau$ within $299 \le \tau' \le 414$. The position of a single (unstable) steady state surrounded by a limit-cycle in both models is determined by two dimensionless parameters:

$$k' = k_3 C_{\mathrm{A0}}/k_{-3}; \qquad \mu = k_4 C_{\mathrm{A0}}/k_{-3}.$$

For example, at $k' = 3$ and $\mu = 1$, the Andronov–Hopf bifurcation arises at $\tau' = 290$. For $k' = 4$, the bifurcation appears at $\tau' = 295$ and evolves toward the MSS regime $(292 \le \tau' \le 405)$. Oscillations are still observed, albeit in a very restricted interval of parameters, prior to vanishing. Indeed, the bifurcation point is retained up to $k' = 15$, but at very large values of the equilibrium constant $k_3/k_{-3} = K_3$ it shifts to the branch of saddle points.

Thus, in a three-dimensional system $(n = 3)$, the replacement of trimolecular steps of type (6.209) retains the oscillatory behavior of a dynamical system. ■

Mechanism (6.219) enters as a block into more complicated mechanisms exhibiting nonlinear behavior. From 1969 to 1971, Lefevr and Nicolis ([142], see also in [7]) developed the well-known scheme that is now analyzed in many monographs and handbooks [5, 6, 9, 50, 96, 101] and called the "brusselator model" in honor of the famous Belgian school of thermodynamics (Prigogine et al.). A close model was proposed in 1968 by Sel'kov [141].

Example 6.21 [5, 6]

► Let us consider the brusselator model process

$$
\begin{array}{ll}
(1) & \mathrm{A} \xrightarrow{k_1} \mathrm{X}; \\[2ex]
(2) & \mathrm{B} + \mathrm{X} \xrightarrow{k_2} \mathrm{Y}; \\[2ex]
(3) & \mathrm{Y} + 2\mathrm{X} \xrightarrow{k_3} 3\mathrm{X} + \mathrm{C}; \\[2ex]
(4) & \mathrm{X} \xrightarrow{k_4} \mathrm{P}.
\end{array}
$$

in a semi-open system at $C_A = const$ and $C_B = const$:

$$\begin{cases} \dfrac{dC_X}{dt} = k_1C_A - k_2C_BC_X + k_3C_YC_X^2 - k_4C_X \\ \dfrac{dC_Y}{dt} = k_2C_BC_X - k_3C_YC_X^2 \end{cases} . \quad (6.224)$$

This system of equations contains four constants and two reactant concentrations (C_A and C_B), which form a set of six parameters. The system can be rendered dimensionless with two parameters (a and b) and two concentrations (x and y):

$$\begin{cases} \dfrac{dx}{d\tau} = a + yx^2 - bx - x = P(x, y) \\ \dfrac{dy}{d\tau} = bx - yx^2 = Q(x, y) \end{cases} , \quad (6.225)$$

where

$$\alpha C_X = x;\ \beta C_Y = y;\ \gamma t = \tau;\ \delta C_A = a;\ \Delta C_B = b.$$

Substituting the values of C_X, C_Y, t, C_A, and C_B from these expressions into system (6.224) and comparing this system to (6.225), we obtain six equations with five unknowns, including four equations,

$$\frac{k_1\alpha}{\delta\gamma} = 1;\ \frac{k_2}{\Delta\gamma} = 1;\ \frac{k_3}{\alpha\beta\gamma} = 1;\ \frac{k_4}{\gamma} = 1,$$

from the relation for dC_X/dt and two equations,

$$\frac{k_2\beta}{\alpha\Delta\gamma} = 1;\ \frac{k_3}{\alpha^2\gamma} = 1,$$

from the relation for dC_Y/dt. The set of these six equations yields

$$\gamma = k_4;\ \alpha = \sqrt{\frac{k_3}{k_4}};\ \Delta = \frac{k_2}{k_4};\ \beta = \sqrt{\frac{k_3}{k_4}};\ \delta = \frac{k_1}{k_4}\sqrt{\frac{k_3}{k_4}}.$$

Substituting the corresponding dimensionless values,

$$x = \sqrt{\frac{k_3}{k_4}}C_X;\ y = \sqrt{\frac{k_3}{k_4}}C_Y;\ \tau = k_4t;\ a = \frac{k_1}{k_4}\sqrt{\frac{k_3}{k_4}}C_A;\ b = \frac{k_2}{k_4}C_B,$$

into Eqs (6.225) must lead to system (6.224).

Let us find a steady-state solution, $x^{cc} = a$ and $y^{cc} = b/a$, of system (6.225) for $P(x, y) = 0$ and $Q(x, y) = 0$. An analysis of the stability of this system in a linear approximation yields

$$a_{11} = b - 1;\quad a_{12} = a^2;\quad a_{21} = -b;\quad a_{22} = -a^2.$$

The coefficients σ and Δ of the characteristic equation are as follows:

$$\sigma = b - 1 - a^2;\quad \Delta = a^2 > 0.$$

For $b < 1 + a^2$, we have $\sigma < 0$ and, hence, the system has stable nodes and foci. On the contrary, for $b > 1 + a^2$, we have $\sigma > 0$, and the system has unstable nodes and foci. For example, at $a = 1$ and

$b = 3, \sigma = 1 > 0$ and $2\lambda_{1,2} = 1 \pm i\sqrt{3}$ with $Re(\lambda) > 0$ (unstable focus). On the plane of parameters (b, a), the regions of foci are separated from that of nodes by the curves $b = (a+1)^2$ determined from the condition $\sigma^2 - 4\Delta = 0$.

For $b > b_{crit} = a^2 + 1(\sigma > 0)$, the system exhibits an Andronov–Hopf bifurcation, whereby a stable focus splits into unstable focus and stable limit-cycle at a sufficiently large distance from the steady state. Thus, the value of $b_{crit}(\sigma = 0)$ corresponds to a bifurcation parameter. As was noted above, a linear approximation in the general case is insufficient for determining limit-cycles in steady states of the center type.

On the passage through $\sigma = 0$, stable oscillations can arise far from a steady state of the center type, but closed trajectories can disappear with the passage to a stable focus as a result of negligibly small fluctuations. With diffusion terms, $D_x \frac{\partial^2 x}{\partial t^2}$ and $D_y \frac{\partial^2 y}{\partial t^2}$, introduced into Eqs (6.225), the system describes spatiotemporal nonuniformities (concentration waves) along the tube length l [6]. Detailed analysis of the brusselator model and a proof of the existence of a limit cycle are given in [142, 143]. ■

A very similar model system with a different nature of the autocatalyst (that is, Y instead of X), was proposed by Tyson and Light [143, 144] and analyzed both qualitatively and quantitatively by Gray and Scott [17].

Example 6.22 [17]

► Let us consider the following system:

$$\begin{aligned}
&(1)\quad \mathrm{A} \longrightarrow \mathrm{X};\\
&(2)\quad \mathrm{X} \longrightarrow \mathrm{Y};\\
&(3)\quad \mathrm{X} + 2\mathrm{Y} \longrightarrow 3\mathrm{Y};\\
&(4)\quad \mathrm{Y} \longrightarrow \mathrm{C}.
\end{aligned} \tag{6.226}$$

In the original notation, it appears as [17]

$$\begin{aligned}
&(1)\quad \mathrm{P}_0 \xrightarrow{k_0} \mathrm{A};\\
&(2)\quad \mathrm{A} \xrightarrow{k_u} \mathrm{B};\\
&(3)\quad \mathrm{A} + 2\mathrm{B} \xrightarrow{k_1} 3\mathrm{B};\\
&(4)\quad \mathrm{B} \xrightarrow{k_2} \mathrm{C}.
\end{aligned} \tag{6.227}$$

An analysis of the corresponding system of kinetic equations

$$\left\{\begin{aligned}
\frac{da}{dt} &= k_0 p - k_1 ab^2 - k_u a\\
\frac{db}{dt} &= k_1 ab^2 + k_u a - k_2 b
\end{aligned}\right., \tag{6.228}$$

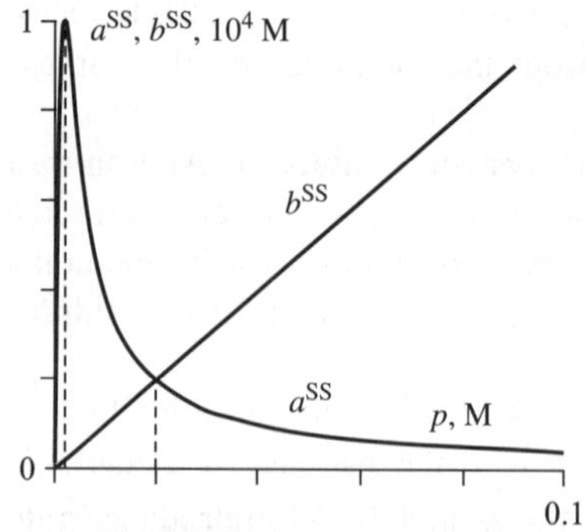

Figure 6.29 Plots of steady-state concentrations a^{ss} and b^{ss} versus reactant concentration p (exact position of maximum, $a^{cc}_{max} = 2 \cdot 10^{-4}$ M at $p = 2 \cdot 10^{-3}$M) (reproduced with permission from Oxford University Press © 1994 [17])

where a, b, and p are the concentrations of A, B, and P, respectively, yields the following steady-state values of a and b:

$$b^{ss} = \frac{k_0 p}{k_2}; \qquad a^{ss} = \frac{k_2^2 k_0 p}{k_1 k_0^2 p^2 + k_2^2 k_u}.$$

Figure 6.29 shows the curves of a^{ss} and b^{ss} versus p for $p_0 = 0.1\,\text{M}$, $k_0 = 10^{-3}\text{s}^{-1}$, $k_u = 10^{-2} \cdot \text{s}^{-1}$, $k_1 = 2.5 \cdot 10^9 \text{M}^{-2} \cdot \text{s}^{-1}$, and $k_2 = 1\text{s}^{-1}$. As can be seen, the quasi-steady-state concentrations a^{ss} and b^{ss} are three orders of magnitude lower than the initial reactant concentration ($p_0 = p + a + b + c$).

A numerical analysis of the system of equations (6.228) showed that, in the interval of concentrations p between the point of intersection $a^{ss} = b^{ss}$ (t_{cross}) and the point of maximum $a^{ss}(t_{max})$, the system exhibits concentration oscillations in the time interval of 1752–3940 s (Figure 6.30). At the point of intersection ($t = t_{cross}$):

$$a^{ss} = b^{ss} = [(k_2 - k_u)/k_1]^{1/2};$$

$$p_{cross} = (k_2/k_1)[(k_2 - k_u)/k_1]^{1/2},$$

while the point of maximum corresponds to

$$a^{ss}_{max} = (1/2)k_2/(k_1 k_u)^{1/2} \quad \text{at} \quad p = [k_2^2 k_u/(k_1 k_0^2)]^{1/2}.$$

An analysis of the sum of derivatives $\sigma = a_{11} + a_{22}$ in the linear approximation of system (6.228) for the obtained steady-state solutions shows that $a_{11} = -k_1(b^{ss})^2 - k_u$ is negative, while $a_{22} = 2k_1 a^{ss} b^{ss} - k_2$ changes sign. Then, the sum $\sigma = a_{11} + a_{22}$ can also change its sign and, hence, the system can be unstable. Thus, the stability changes where $a_{11} + a_{22} = \sigma = 0$. Using this circumstance, one can find the critical values $p^*_{1,2}$ by solving the following quadratic equation,

$$\frac{k_2(k_0^2 k_1 p^2 - k_u k_2^2)}{k_0^2 k_1 p^2 + k_u k_2^2} - \left[k_1 \left(\frac{k_0 p}{k_2}\right)^2 + k_u\right] = 0, \tag{6.229}$$

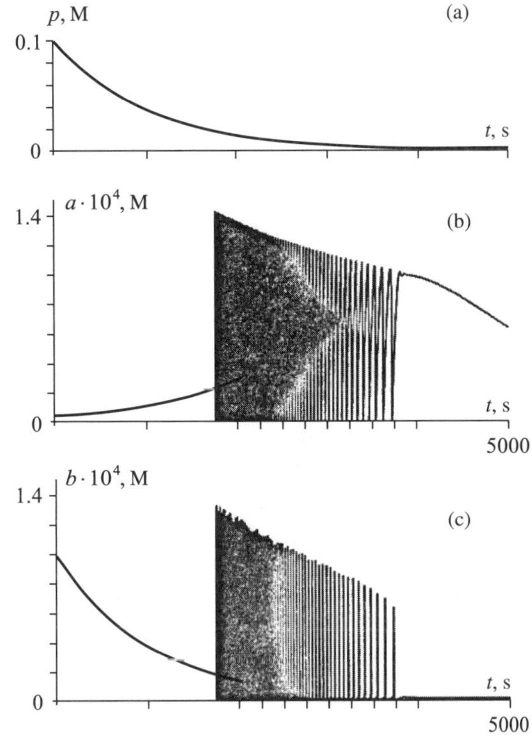

Figure 6.30 Plots of concentrations (a) p, (b) a, and (c) b versus time (reproduced with permission from Oxford University Press © 1994 [17])

which leads to the following approximate formula:

$$p_{1,2}^{*} = \left(\frac{k_2^2}{2k_1k_0^2}\right)^{1/2} \left\{(k_2 - 2k_u) \pm [k_2(k_2 - 8k_u)]^{1/2}\right\}. \tag{6.230}$$

For the given rate constants, the interval of p^* values for which the instability is possible ranges from $p_1^* = 1.97 \cdot 10^{-2}$M to $p_2^* = 2 \cdot 10^{-3}$M. However, this interval is only realistic provided that $k_2 > 8k_u$ (see Eq. (6.230)). If the reaction described by Eqs (6.228) proceeds in a closed system at $p = \text{const}$ (e.g., $p = 0.01$ M), the integration leads to the pattern of oscillations in a and b and a limit cycle in the phase portrait on the (a, b) plane (Figure 6.31).

As k_u increases to $k_2 = 8k_u$, the values of roots coincide ($p_1^* = p_2^*$), the region of instability disappears, and oscillations vanish. The reversibility of steps

$$A \rightleftharpoons B; \qquad A + 2B \rightleftharpoons 3B$$

also leads to a decrease of the region of unstable behavior of system (6.128), but retains its oscillatory behavior [17].

For a more detailed analysis of bifurcations in system (6.228), let us reduce it to a dimensionless form with a smaller number of control parameters. This is achieved by introducing dimensionless concentrations

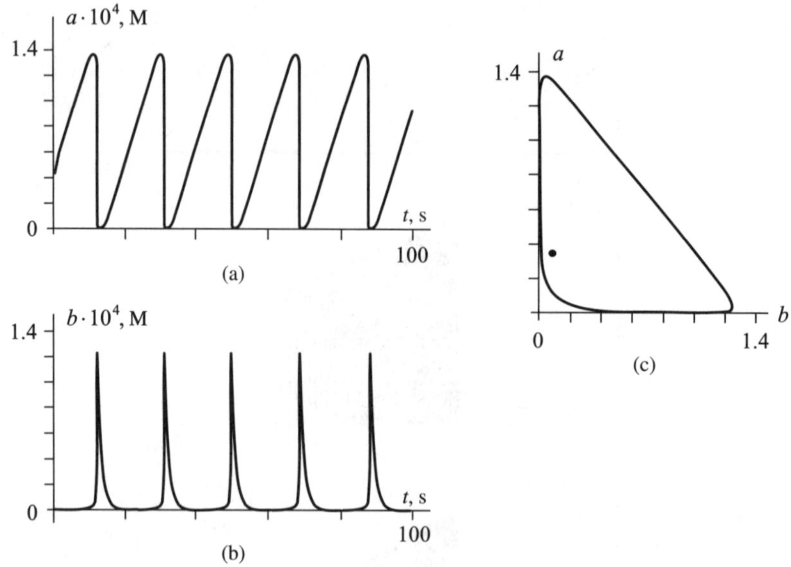

Figure 6.31 (a, b) Stable oscillations of concentrations a and b of intermediates A and B, respectively, at $p = 0.01$ M and (c) plot of limit-cycle (reproduced with permission from Oxford University Press © 1994 [17])

normalized to $a^{ss} = (k_2/k_1)^{1/2}$ (for $k_2 \gg k_u$) as

$$\pi = p\left(\frac{k_1}{k_2}\right)^{1/2}; \tau = k_2 t; \quad \alpha = a\left(\frac{k_1}{k_2}\right)^{1/2}; \beta = \left(\frac{k_1}{k_2}\right)^{1/2},$$

where the initial value of π at the given constants and initial conditions is $\pi_0 = 5000$. Defining the dimensionless constants as $\varepsilon = k_0/k_2$ and $k'_u = k_u/k_2$ and considering ε to be a small parameter, we can eventually rewrite Eqs (6.228) as the system of three equations

$$\begin{cases} \dfrac{d\pi}{d\tau} = -\varepsilon\pi; \\ \dfrac{d\alpha}{d\tau} = \varepsilon\pi - \alpha\beta^2 - k'_u\alpha; \\ \dfrac{d\beta}{d\tau} = \alpha\beta^2 + k'_u\alpha - \beta, \end{cases} \tag{6.231}$$

with three variables (π, α, β), two parameters (ε, k'_u), and the initial conditions $\pi = \pi_0$ and $\alpha = \beta = 0$ at $\tau = 0$. Note that π is a slow variable since ε is a small parameter. Taking this into account and introducing a new small parameter as

$$\varepsilon\pi = \varepsilon\pi_0 e^{-\varepsilon\tau} = \mu_0 e^{-\varepsilon\tau} = \mu; \; \varepsilon\pi_0 = \mu_0,$$

we arrive at a system with two variables and two parameters (μ and k'_u):

$$\begin{cases} \dfrac{d\alpha}{d\tau} = \mu - \alpha\beta^2 - k'_u\alpha = Q(\alpha, \beta) \\ \dfrac{d\beta}{d\tau} = \alpha\beta^2 + k'_u\alpha - \beta = P(\alpha, \beta) \end{cases} . \tag{6.232}$$

The steady-state values of α and β are as follows:

$$\beta^{ss} = \mu; \quad \alpha^{ss} = \frac{\mu}{\mu^2 + k_u'}; \quad \alpha_{max}^{ss} = \frac{1}{2}(k_u')^{-1/2}; \quad \mu_{max} = k_u'^{1/2}. \tag{6.233}$$

Evidently, the small parameter k_u' must always be nonzero ($k_u' \neq 0$), otherwise $\alpha^{ss} \to \infty$ as $\mu \to 0$.

For an analysis of the local stability with respect to small perturbations, let us assume μ to be constant (by analogy with the assumption that $p = const$). Then, we obtain the following expressions for the partial derivatives in Eqs (6.232),

$$a_{11} = -(\mu^2 + k_u'); \quad a_{12} = -\frac{2\mu^2}{\mu^2 + k_u'}; \quad a_{21} = \mu^2 + k_u'; \quad a_{22} = \frac{\mu^2 - k_u'}{\mu^2 + k_u'},$$

the value of $a_{11} + a_{22} = \sigma$,

$$\sigma = -\frac{\mu^4 - (1 - 2k_u')\mu^2 + k_u'(1 + k_u')}{\mu^2 + k_u'}, \tag{6.234}$$

and the Jacobian determinant

$$\Delta = \mu^2 + k_u' > 0.$$

For $k_u' = 10^{-2}$ and $\mu = 5$, we have $\sigma = -24 < 0$ and the system is stable. However, as p is decreased at $\mu = 0.5$, we have $\sigma = 0.66$, so that the system is locally unstable.

Let us preliminarily consider a system with $k_u' = 0$ (absence of the step of autocatalyst formation), which is physically nonrealistic but offers a simpler case for the analysis. In this case, we have

$$\sigma = 1 - \mu^2; \quad \Delta = \mu^2; \quad \sigma^2 - 4\Delta = \mu^4 - 6\mu^2 + 1.$$

First, let us determine roots μ for the boundary values of $\sigma^2 - 4\Delta = 0$, i.e., for the points where the discriminant changes sign. Denoting $\mu^2 = y$, we obtain the equation

$$y^2 - 6y + 1 = 0,$$

which yields $y_{1,2} = 3 \pm \sqrt{8} = 3 \pm 2\sqrt{2}$ and, hence, $\mu = \pm\sqrt{3 \pm 2\sqrt{2}}$. Thus, we have four roots

$$\mu_1 \cong 2.41; \quad \mu_2 = -2.41; \quad \mu_3 = 0.414; \quad \mu_4 = -0.414,$$

for which $\sigma^2 - 4\Delta = 0$, and there are the following possible situations (Figure 6.32):

(i) $\mu > 1 + \sqrt{2}(> 2.41)$, $\sigma < 0$, $\Delta > 0$, $\sigma^2 - 4\Delta > 0$ (stable node (SN));
(ii) $1 < \mu < 1 + \sqrt{2}$, $\sigma < 0$, $\Delta > 0$, $\sigma^2 - 4\Delta < 0$ (stable focus (SF); $Re(\lambda_{1,2}) < 0$);
(iii) $1 > \mu > \sqrt{2} - 1$, $\sigma > 0$, $\Delta > 0$, $\sigma^2 - 4\Delta < 0$ (unstable focus (UF));
(iv) $0 < \mu < \sqrt{2} - 1$, $\sigma > 0$, $\Delta > 0$, $\sigma^2 - 4\Delta > 0$ (unstable node (UN));
(v) $\mu = 1$, $\sigma = 0$, $\Delta = 1$, $\sigma^2 - 4\Delta = -4 < 0$ (center).

Since this system (with $k_u' = 0$) always has $\Delta > 0$, the saddle point never appears. The point of intersection of the $\alpha^{ss}(\tau)$ and $\beta^{ss}(\tau)$ curves at $\mu = 1$ (Figure 6.32) makes possible the appearance of the Andronov–Hopf bifurcation and limit-cycle, but the linear approximation is insufficient for unambiguously judging on the appearance of oscillations.

The behavior of a realistic system with $k_u' \neq 0$ has much in common with the simplified case considered above, but there some important distinctions also appear [17]. The equation for $\sigma^2 - 4\Delta$ becomes rather

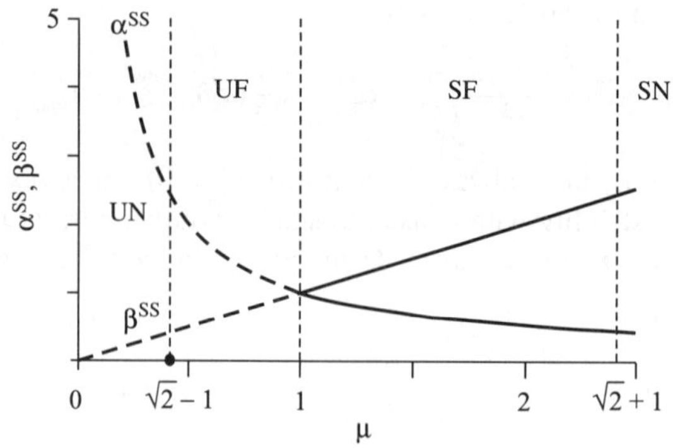

Figure 6.32 Regions of local stability for $\alpha^{ss}(\mu)$ and $\beta^{ss}(\mu)$ at $k'_u = 0$ (reproduced with permission from Oxford University Press © 1994 [17])

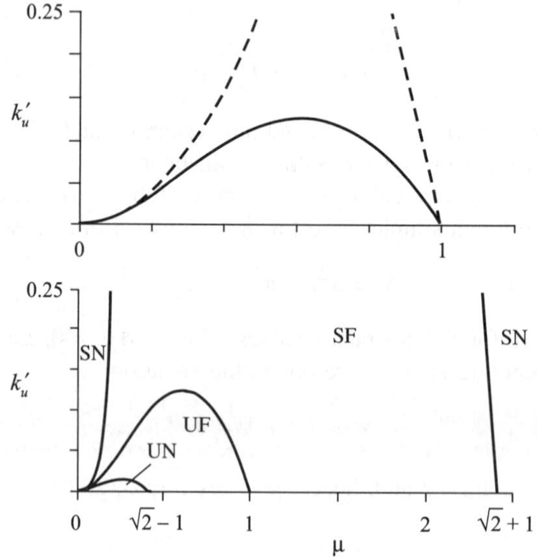

Figure 6.33 Regions of local stability on $k'_u - \mu$ parametric portrait. Dashed curves refer to maximum of $\alpha(\mu)$ and to point of crossing (t_{cross}) at $\alpha(\mu) = \beta(\mu)$. Region of instability is bounded by these points (reproduced with permission from Oxford University Press © 1994 [17])

complicated (8th power in μ), and a condition for the passage from nodes to foci can only be determined by numerical methods. The above analysis indicates that the roots would tend to $\sqrt{2} \pm 1$ as $k'_u \to 0$. Figure 6.33 shows a parametric portrait of this system with the regions of nodal and focal steady states. The condition for a change in the character of local stability is $\sigma = 0$ in Eq. (6.234). The roots of this equation for $\sigma = 0$ are given by the formula

$$(\mu^*_{1,2})^2 = \frac{1}{2}[(1 - k'_u) \pm (1 - 8k'_u)^{1/2}], \tag{6.235}$$

which yields real values for $k'_u < 1/8$. The boundary values, for which $\sigma^2 - 4\Delta = 0$, are, e.g., $\mu_1^* = 0.9847$ and $\mu_2^* = 0.102$ for $\pi = 5000$, $\mu = 0.1$, and $k'_u = 0.01$. As k'_u increases to 0.1, the $\mu_{1,2}^*$ values decrease and the region of possible oscillations exhibits narrowing. It is interesting to note that the point of intersection of the $\alpha^{ss}(\tau)$ and $\beta^{ss}(\tau)$ curves and the α_{max} point are close to $\kappa\mu_1^*$ and μ_2^*, respectively. ■

Now it is expedient to consider the ways bifurcations appear, that is, the transformation of steady states as a result of instability. Bifurcation analysis is a special field of the theory of differential equations and nonlinear dynamics, which is treated in monographs and handbooks (see, e.g., [43, 91, 97, 98, 101]). Here we will only briefly consider the main notions of this field, in particular, those concerning the Andronov–Hopf bifurcation and the other bifurcation pathways.

According to the Poincaré–Bendixson theorem [17], if there is a single unstable steady state (or multiple steady states, each one being unstable) and it is known that variables x and y (e.g., concentrations of intermediates) can neither grow to infinity nor become negative, then there must exist a non-steady-state attractor to which the trajectories would tend. According to this theorem, the concentrations cannot walk randomly within a finite domain and must stop at some point. For a two-dimensional problem, the only possible attractor of this kind is a limit cycle.

As repeatedly noted above, a singular point in a two-dimensional system at $\sigma > 0, \Delta > 0$, and $\sigma^2 - 4\Delta < 0$ represents an unstable focus with $\lambda_{1,2} = \text{Re} \pm i\sqrt{|\sigma^2 - 4\Delta|}$. The transition of eigenvalues λ across the imaginary axis for $\sigma = 0$ and $Re(\lambda) = 0$ is called the Andronov–Hopf bifurcation, which is accompanied by the appearance of a limit-cycle and is manifested by self-oscillations of the concentrations. The limit-cycle arises from an unstable focus, and this case represents the bifurcation of limit-cycle formation. Strictly speaking, a stability analysis in this case cannot be restricted to local analysis (linear approximation). With allowance for the high-order terms in the expansion into Taylor's series, a simple oscillatory behavior of the $\cos(\omega_0\tau)$ type can change or vanish [17]. At the same time, changes in the local stability related to $\sigma = 0$ (Andronov–Hopf bifurcation) or $\Delta = 0$ (saddle–node bifurcation) are characteristic of the probable appearance of oscillations. For the model system of equations (6.232) considered above, the conditions of Andronov–Hopf bifurcation were encountered at points μ_1^* and μ_2^*. Let us consider some additional factors related to the Andronov–Hopf bifurcation.

Degeneracy of Andronov–Hopf bifurcation. The transition of $Re(\lambda)$ via zero and the appearance of roots with $Re(\lambda) > 0$ in a certain interval of parameters may not take place, provided that $Re(\lambda)$ has a maximum at this zero point. In order to avoid this situation, it necessary that $d\sigma/d\mu$ at points μ_1^* and μ_2^* would not be zero at $\sigma = 0$ (see Example 6.22):

$$\frac{d\sigma}{d\mu} = 2\mu\frac{\mu^4 + 2k'_u\mu^2 - k'_u(2 - k'_u)}{(\mu^2 + k'_u)^2} \neq 0.$$

This inequality is satisfied at all bifurcation points μ_1^* and μ_2^*, provided that $k'_u < 1/8$. At $k'_u = 1/8$, the Andronov–Hopf bifurcation disappears and σ does not exceed zero at any μ. If $\sigma = 0$ and $d\sigma/d\mu = 0$ simultaneously, the Andronov–Hopf bifurcation is called degenerate. In this case, two bifurcation points exhibit a mutual approach and cancel each other.

Stability of oscillations. A limit cycle that appears at the point of bifurcation can, like steady-state points, be either stable or unstable (see, e.g., Figure 6.15). Accordingly, the oscillatory solutions can also be stable or unstable. Therefore, the bifurcation analysis must help in establishing whether a limit cycle grows with increasing parameter (μ), assessing the rate of growth, and determining how the oscillation period is changing. These tasks are solved by determining the sign of Floquet's special factor β_2 (Andronov–Hopf bifurcation analysis [17, Sections 6.4 and 6.5]).

In the model considered in Example 6.22, the size of a limit-cycle and the amplitude and period of oscillations vary depending on parameter μ. As this quantity decreases and passes through the upper Andronov–Hopf bifurcation point (μ_1^*), a stable limit-cycle appears around the singular point that became unstable. At the bifurcation point, the limit-cycle size (and the amplitude of oscillations) is zero, but the cycle appears with a finite period. As the μ value decreases further, the oscillations immediately behind point μ_1^* are virtually sine-shaped and their amplitude reaches maximum within the interval of $\mu_2^* < \mu < \mu_1^*$. On approach to μ_2^*, the oscillation amplitude decreases (Figure 6.34). This model revealed no unstable limit-cycles.

In three-dimensional ($n = 3$) systems under certain conditions ($\lambda_1 > 0, \lambda_2 = 0, \lambda_3 < 0$), a trajectory can be infinite line, but neither leaving a given region for $t \to \infty$ nor attracting to steady-state points or closed orbits. In particular, the trajectory walks between an unstable node and two unstable limit cycles. This behavior of a dynamical system is referred to as "strange attractor". This type of attractor was described in 1963 by Lorentz [145] (see also [146]. After the discovery of chaotic oscillations in the Belousov – Zhabotinskii (BZ) reaction [53] and the enzymatic (peroxidase) oxidation of NADH by oxygen in an open (with respect to oxygen) system [147], the interest in chaotic oscillations in the chemical and biochemical systems, as well as the general interest in the problems of chaotic behavior in nonlinear models, triggered an avalanche of investigations (see, e.g., [17, 147–149]). Strange attractors serve mathematical models of deterministic chaos. Figure 6.35 gives examples of various phase portraits of chaotic oscillations. According to one of possible scenario, the transition to chaos proceeds via quasi-periodic regimes. After the first two Andronov–Hopf bifurcations, the third bifurcation transforms the limit cycle in the three-dimensional space ($n = 3$), into a torus filled by non-closed and non-intersecting trajectories, which corresponds to the appearance of a strange attractor. Intermittency of regular and chaotic behavior can also take place, whereby the motion on a limit cycle switches to chaos. Another possibility of the appearance of chaos is related to period-doubling bifurcations [17, 91].

Among the types of bifurcations that are characteristic of two-dimensional systems, a special case is presented by the saddle–node bifurcation. If a system has two steady states, saddle and node, then the points of the two types can approach one another when the parameters vary and, should one of the eigenvalues turn zero, a bifurcation takes place and the two points merge together with the formation of a singularity of the saddle–node type.

By using an external action upon variables (e.g., reactant concentration) or changing a parameter (contact time, temperature) in a system with multiple steady states (hysteresis), it is possible to switch the system from one stable state to another, e.g., from one node via a saddle separatrix to another node. However, strictly speaking, this will not be a bifurcation, since the nature of the attractor remains unchanged.

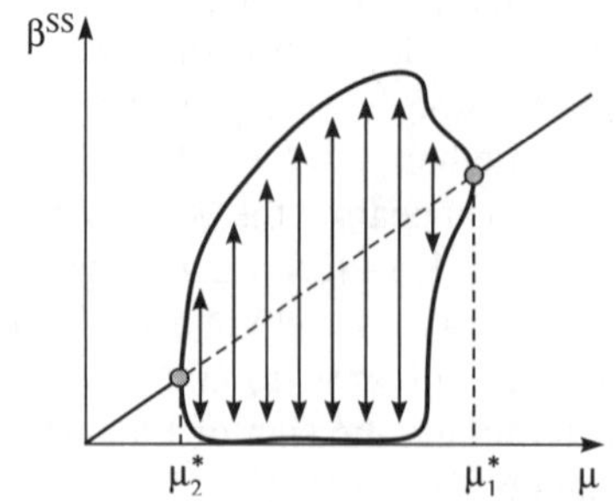

Figure 6.34 Buildup and disappearance of oscillations during variation of μ from μ_1^* to μ_2^* (reproduced with permission from Oxford University Press © 1994 [17])

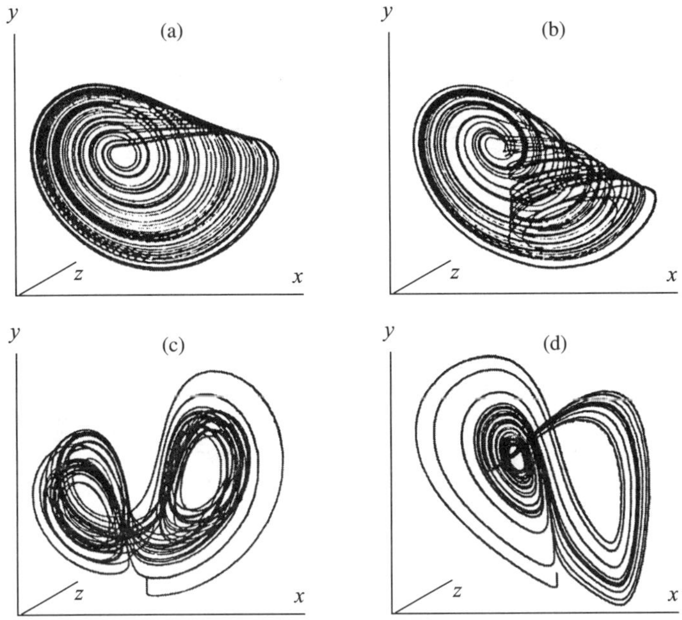

Figure 6.35 Phase portraits of chaotic behavior: (a) spiral; (b) screw; (c) inverted spiral + saddle (Lorentzian chaos); (d) noninverted spiral + saddle (reproduced with permission from Springer Sci.+ Business Media © 1983)

Section 6.3 presented both abstract schemes and several real kinetic models in order to reveal and illustrate the main laws in the dynamic behavior of kinetic models with various structures and degrees of complexity. Now it is possible to generalize the results. Table 6.2 gives a summary of the results of more than 20 examples considered in this section, which takes into account the following factors:

- dimensionality (n) of kinetic models;
- nonlinearity of kinetic models;
- number of nonlinear steps in mechanism;
- presence, number, and type of autocatalytic steps;
- other types of nonlinearity (material balance with respect to catalyst);
- reaction system (reactor) type: open, semi-open, closed.

An analysis of the data in Table 6.2 leads to the following important conclusions:

(i) The presence of nonlinearity in the kinetic model is a necessary condition for the appearance of critical effects (multiple steady states (MSS), limit cycles).
(ii) The presence of nonlinear steps involving intermediates in a given mechanism is not a sufficient condition for nonlinear behavior of the kinetic model.
(iii) The nonlinearity in the kinetic model can be ensured by allowance of the material balance with respect to a catalyst within the steady-state or quasi-equilibrium approximation, including all types of inhibition (pendant vertices on kinetic and bipartite graphs).
(iv) Self-oscillations appear in the case of mechanisms not involving autocatalytic steps or phenomena (Examples 6.4, 6.5, and 6.11 ($n = 3$), schemes 6.133 and 6.134).

Table 6.2 Analysis of examples

n	Non-linearity of kinetic model	Number of nonlinear steps	Autocatalytic steps (type/number)	Other types of non-linearity	Reaction system type	Nonlinear effects	Example no. (equation)
1	Yes	None	None	MB	O	None	6.1 (6.115)
	Yes	None	None	MB	O	MSS, HS	6.2 (6.120)
	Yes	None	None	MB	O	None	6.3 (6.122)
	Yes	1	None	None	S (C_{A0} = const)	None	6.7 (6.138)
	Yes	1	None	None	S ($C_A \cong C_{A0}$)	None	6.8 (6.142)
	Yes	1	Quadratic/1	None	S	CP, TC	6.13 (6.180) 14 (6.182)
	Yes	1	Cubic/1	None	O	MSS, HS	6.15 (6.185) 18 (6.211)
2	Yes	None	None	MB	O	MSS, HS, LC	6.4 (6.126)
	Yes	None	None	MB	O	MSS, HS, LC	6.5 (6.131)
	Yes	None	None	MB	C	LC	(6.134)
	Yes	1	None	None	O	DO	6.8 (6.143)
	None	None	None	None	C	DO	6.6 (6.136)
	Yes	1	None	None	S (C_A, C_B = const)	IS, saddle–node bifurcation	6.9 (6.148)
	Yes	1	None	None	–	None	6.9 (6.149)
	Yes	2	None	None	S (C_A, C_B = const)	MSS, DO	6.10 (6.151)
	Yes	2	None	None	S	MSS, HS	6.11 (6.169)
	Yes	1	Quadratic/1	None	S	DO	6.16 (6.192–6.193)
	Yes	2	Quadratic/2	None	S	Oscillations, fluctuations near center	6.17 (6.198–6.199)
	Yes	1	Cubic/1	None	O	MSS, DO, isols, mushrooms	6.19 (6.220)
	Yes	1	Cubic/1	None	S	LC	6.21 (6.224)
	Yes	1	Cubic/1	None	C	LC	6.22 (6.228)
3	Yes	2	None	None	S	MSS, LC	6.11 (6.169)
	Yes	None	None	MB	C	LC	(6.133)
	Yes	1, 2	two-step	None	O	MSS, LC	6.20

Notes: MB = material balance; O = open system; S = semi-open system; C = closed system; HS = hysteresis; CP = critical phenomena; TC = threshold concentration; IS = instability; DO = decaying oscillations; MSS = multiple steady states; LC = limit cycle

(v) Dimensionality of the kinetic model is an important factor that influences the appearance of critical phenomena. In particular, the models with $n = 1$ can exhibit threshold concentrations and MSS, while those with $n = 2$ and $n = 3$ exhibit MSS and limit cycles. Systems with $n \leq 2$ and all bimolecular steps cannot exhibit self-oscillations, while MSS can only appear in the case of processes in CSTR [143].

(vi) The use of open reaction systems (reactors) increases the degree of nonlinearity of the kinetic model, since the reactants and products formally become intermediates and increase the dimensionality of the model. The dynamic behavior of an irreversible open system for a certain period of time coincides with that of the completely reversible closed system far from the equilibrium [5]:

$$\mathrm{A} \longrightarrow \mathrm{X} \longrightarrow \mathrm{B} \quad \text{(open system);}$$

$$\mathrm{A} \rightleftharpoons \mathrm{X} \rightleftharpoons \mathrm{B} \quad \text{(closed system).}$$

Based on the material presented in Sections 6.2 and 6.3, it is possible to separately formulate the thermodynamic, chemical, and mathematical conditions for the appearance of self-oscillations in the concentrations of the participants of catalytic chemical reactions.

Thermodynamic conditions

- Reaction is carried out far from equilibrium (i.e., in open system for any mechanisms).
- Reaction is carried out in a closed system for mechanisms, in which at least one step is irreversible ($W_j^+ \gg W_j^-$).
- Thermodynamic instability of steady states under any reaction conditions:

$$\sum_k \delta F_k \delta I_k < 0,$$

where δF_k and δI_k are the fluctuations of thermodynamic forces and flows. Under isothermal conditions, in the absence of diffusion retardation in heterophase systems:

$$\sum_j \frac{\delta A_j}{T} \delta W_j < 0.$$

Chemical conditions

- A *necessary* chemical condition for the appearance of limit cycles and self-oscillations is the presence of nonlinear steps in the mechanistic scheme or significant values of nonlinear complexity functions of the catalyst ($F_\mathrm{M} \gg 1$) with respect to reactants, products, and/or foreign inhibitors.
- Presently, *sufficient* chemical conditions for the appearance of self-oscillations are difficult to formulate, since the critical phenomena are determined by combinations of a large number of factors (dimensionality of the system, number of nonlinear steps, type of nonlinear step, open or closed reactor, etc.). As was shown above, three types of nonlinearity are encountered under isothermal conditions:

(i) Nonlinear step of the $\mathrm{X}_i + \mathrm{Y}_i \rightarrow$ type [116]. In a system with $n = 3$, the presence of one step of this type among two nonlinear steps plus one pendant vertex are sufficient for the appearance of a limit cycle [Example 6.11, system (6.169)]. For $n = 4$, one nonlinear step of this type and one pendant vertex are also sufficient for the appearance of a limit cycle [150].

(ii) Autocatalytic step or a block of elementary steps that results in autocatalysis. For $n = 3$, a three-step mechanism with bimolecular steps (including one nonlinear step) and a common autocatalytic effect in CSTR leads to self-oscillations (Example 6.20).

(iii) Nonlinearity related entirely to a material balance. This factor leads in a system with $n = 2$ to limit cycles in CSTR (Examples 6.4 and 6.5, schemes (6.133) and (6.134)).

The latter two types of nonlinearity are responsible for the appearance of positive and negative feedbacks in kinetic models.

- **Necessary mathematical conditions**
  - System of differential equations must be nonconservative.
  - Dimensionality of system must be $n \geq 2$.
  - For the instability of steady states in a system with $n = 2$, the sum of derivatives $\partial P/\partial x|_{x^{ss},\ y^{ss}} + \partial Q/\partial y|_{x^{ss},\ y^{ss}}$ (where $P(x,\ y) = dx/dt$ and $Q(x,\ y) = dy/dt$) must change sign.
  - For system with $n \geq 3$, the negative value of one of the higher coefficients in characteristic polynomial for eigenvalues (e.g., $b < 0$ in polynomial (6.111) for $n = 3$) implies instability of the steady state.
  - Complicated dynamic behavior is inherent in systems of equations with three temporal scales [5, 151]:

$$\frac{dR_k}{dt} = \varepsilon\varphi(R, n, u);$$

$$\frac{dn_i}{dt} = f_i(R, n, u);$$

$$\frac{du_i}{dt} = 1/\varepsilon\Psi(R, n, u);$$

where R_k are slow variables (reservoir, reactants and products), n_i are the main variables, and u_i are fast variables ($[R_k] \gg [n_i] \gg [u_i]$).

Many of the aforementioned examples were considered under the conditions of $dR_k/dt \cong 0$ (e.g., $C_A = const$), using dependences of the type $n_i = f(t)$ and excluding u_i in the steady-state approximation as $\varepsilon du_i/dt \cong 0$, where $\varepsilon = [M]_0/[R_{k0}] \ll 1$ (M is a catalyst). In the study of system evolution for $t \to \infty$, the main variables are R_k and the fast variables are n_i. The latter condition follows from the Korzukhin theorem [151]. According to this theorem, a system of kinetic equations can always be constructed based on the law of mass action, the behavior of which coincides to within any preset accuracy with the behavior of any given system

$$\frac{dx}{dt} = A_i(x),$$

where $A_i(x)$ are any polynomials with integer positive powers.

6.4 Mechanisms of oscillatory catalytic reactions

This section provides more detailed information on the mechanisms and kinetic models of several reactions catalyzed by metal complexes, in which various critical phenomena have been observed as manifestations of nonlinear dynamics.

6.4.1 Belousov–Zhabotinskii reaction (BZ reaction)

The Belousov–Zhabotinskii (BZ) reaction is among the most famous of the chemical oscillators that have been studied since 1951. This reaction has been described in numerous research articles and analyzed in much detail for different variants of mechanisms in monographs [4, 9, 10, 17]. Experiments in continuous-flow stirred tank reactors (CSTRs) revealed regular self-oscillations, bistability [9, 152, 153], packets of oscillations separated by non-oscillatory steady states [152–154], quasi-periodic oscillations and chaotic regimes [53,154–157], as well as running waves and periodic spatial structures [5, 9, 17]. Chaotic oscillations in enzymatic reactions were observed for the first time in 1977 by Olsen et al. [158] (simultaneously with Hudson [53]) in the reaction of NADH oxidation with oxygen catalyzed by peroxidase.

The first mechanism and kinetic model describing the appearance of an induction period and self-oscillations in a closed system were proposed by Zhabotinskii with co-workers [4, 5, 9] for the $KBrO_3$–$Ce(SO_4)_2$–H_2SO_4–malonic acid (MA) system. It was demonstrated that the oxidation of Ce^{3+} by bromate in strong acid media corresponds to the reaction

$$HBrO_3 + 4Ce^{3+} + 4H^+ \longrightarrow HOBr + 4Ce^{4+} + 2H_2O, \qquad (6.236)$$

which has an autocatalytic character and is inhibited by Br^- ions that catalyze the decomposition of $HBrO_2$ autocatalyst. The products of bromate reduction ($HBrO_2$, HOBr, Br_2) brominate malonic acid, while the oxidation of bromomalonic acid by Ce^{4+} leads to the appearance of Br^- species. It was also established that the introduction of radical scavengers that bind radicals (R·) led to the termination of the oscillatory process, but the participation of radical species in the proposed reaction scheme was ignored. The kinetic model involved three variables $[Ce^{4+}]$ (x), $[HBrO_2]$ (y), and $[Br^-]$ (z), and consisted of five steps, including autocatalysis with respect to y (step (1)):

$$(1)\quad A \xrightarrow{k_1' y} Y; \qquad (2)\quad Y \xrightarrow{k_2 z} P_1;$$

$$(3)\quad Y \xrightarrow{k_3} X; \qquad (4)\quad X \xrightarrow{k_4} Z;$$

$$(5)\quad Z \xrightarrow{k_5} P_2.$$

It was also assumed that steps (1) and (3) involved Ce^{3+} ($[Ce^{3+}] = [Ce]_0 - x$). The resulting model

$$\frac{dx}{dt} = k_3(C_0 - x)y - k_4 x,$$

$$\frac{dy}{dt} = k_1(C_0 - x)y - k_2 yz,$$

$$\frac{dz}{dt} = k_4 x - k_5 z, \qquad (6.237)$$

(where ($k_1 = k_1' - k_3$) satisfactorily described the process in a certain region of reactant concentrations.

In 1972, Field, Körös and Noyes (FKN) [159] proposed a mechanism that included 11 steps and involved seven intermediates, which served a basis for a simplified kinetic model called Oregonator [160]. The most complete model proposed in 1990 by Györgyi, Turanyi and Field (GTF) [161] included 80 reactions and involved 26 intermediates. An analysis of this scheme led to a simpler mechanism that consisted of 42 steps and involved 22 reactants [162]. The latter model, which shows good quantitative agreement with experimental data, contains two negative feedbacks that control the autocatalysis by inhibition with Br^- ions and with radicals of malonic acid.

Let us consider in more detail the FKN mechanism and the Oregonator scheme (as described in [17]). Among numerous reactions involved in the BZ reaction system, the FKN scheme separates three overall processes of key importance:

Process A (oxidation of Br^- by bromate) includes the following reactions,

$$\text{(R1)}\quad BrO_3^- + Br^- + 2H^+ \longrightarrow HBrO_2 + HOBr;$$

$$\text{(R2)}\quad HBrO_2 + Br^- + H^+ \longrightarrow 2HOBr;$$

$$\text{(R3)}\quad HOBr + Br^- + H^+ \longrightarrow Br_2 + H_2O.$$

which contribute to the overall process as (R3) + (R2) +3(R1):

$$\textbf{(A)}\quad BrO_3^- + 5Br^- + 6H^+ \longrightarrow 3Br_2 + 3H_2O.$$

This process strongly decreases $[Br^-]$, slightly changes $[BrO_3^-]$, and virtually retains $[Ce^{4+}]$.

Process B (oxidation of Ce^{3+} by bromate) is switched on when $[Br^-]$ strongly decreases. Then, $HBrO_2$ is oxidized by bromate in step (R5) with the formation $Br\dot{O}_2$ radical:

$$\text{(R5)}\quad BrO_3^- + HBrO_2 + H^+ \longrightarrow 2Br\dot{O}_2 + H_2O;$$

$$\text{(R6)}\quad Br\dot{O}_2 + Ce^{3+} + H^+ \longrightarrow HBrO_2 + Ce^{4+}.$$

These steps constitute an autocatalytic overall process as (R5) + 2(R6):

$$BrO_3^- + 2Ce^{3+} + 3H^+ + HBrO_2 \longrightarrow 2HBrO_2 + 2Ce^{4+} + H_2O.$$

Evidently, reaction (R5) is competitive with (R2). The rates of these reactions are leveled when $[Br^-]$ attains a critical level of $[Br^-]_{crit}$ that is defined as follows:

$$k_{R5}[HBrO_2][H^+][BrO_3^-] = k_{R2}[HBrO_2][H^+][Br^-];$$

$$[Br^-]_{crit} = \frac{k_{R5}}{k_{R2}}[BrO_3^-] = 10^{-5}[BrO_3^-].$$

At $[Br^-] > [Br^-]_{crit}$, process **A** predominates and, hence, $[Br^-]$ decreases. At $[Br^-] < [Br^-]_{crit}$, the autocatalytic process B is dominating, which indirectly leads to an increase in $[Br^-]$ via reactions of block C.

Process C (oxidation of malonic acid (MA) and bromomalonic acid (Br-MA)) consists of the following steps:

$$\text{(C1)}\quad MK \longrightarrow \text{enol};$$

$$\text{(C2)}\quad \text{enol} + Br_2 \longrightarrow BrMK + H^+ + Br^-;$$

$$\text{(C3)}\quad BrMK + MK + 2Ce^{4+} \longrightarrow 2Ce^{3+} + fBr^-;$$

The chain branching in process B does not proceed infinitely and is restricted at high concentrations of $HBrO_2$ by the quadratic termination step as

$$\text{(R4)}\quad 2HBrO_2 \longrightarrow HOBr + BrO_3^- + H^+.$$

With allowance for reaction (R4), the total stoichiometry of process **B** [2(R5) + 4(R6) + (R4)] is expressed by the following equation:

$$\textbf{(B)}\quad BrO_3^- + 4Ce^{3+} + 5H^+ \longrightarrow HOBr + 4Ce^{4+} + 2H_2O$$

(cf. Eq. (6.236)).

Since the consumption of Ce^{4+} is determined by reactions with various organic products, the coefficient f in reaction (C3) is not constant. As will be shown below, the oscillations are possible within a certain interval ($f = 0.5 - 2.4$). In order to simplify the FKN mechanism, some intermediates (e.g., $BrO_2\bullet$) are assumed to be quasi-steady-state and quasi-equilibrium, while the concentrations of $[BrO_3^-]$, $[H^+]$, and $[RH]_\Sigma$ (organic substrate) are set constant. The Oregonator model includes five steps and is written as follows:

$$\text{(O1)}\quad A + Y \longrightarrow X + P \qquad (W_{O1} = k_3AY);$$

$$\text{(O2)}\quad X + Y \longrightarrow 2P \qquad (W_{O2} = k_2XY);$$

$$\text{(O3)}\quad A + X \longrightarrow 2X + 2Z \qquad (W_{O3} = k_5AX);$$

$$\text{(O4)}\quad 2X \longrightarrow A + P \qquad (W_{O4} = k_4X^2);$$

$$\text{(O5)}\quad B + Z \longrightarrow \tfrac{1}{2}fY \qquad (W_{O5} = k_0BZ);$$

where $[HBrO_2] = X, [Br^-] = Y, [Ce^{4+}] = Z, [HOBr] = P, [BrO_3^-] = A,$ and $[MA] = B$ and it is assumed that $[Ce^{4+}] \ll [Ce]_0$.

The corresponding system of kinetic equations appears as

$$\begin{cases} \dfrac{dX}{dt} = k_3AY - k_2XY + k_5AX - 2k_4X^2; \\ \dfrac{dY}{dt} = -k_3AY - k_2XY + \tfrac{1}{2}fk_0BZ; \\ \dfrac{dZ}{dt} = 2k_5AX - k_0BZ, \end{cases} \qquad (6.238)$$

and can be written on a dimensionless form according to Tayson [10]:

$$\begin{cases} \varepsilon\dfrac{dX}{d\tau} = qy - xy + x(1-x); \\ \varepsilon'\dfrac{dY}{d\tau} = -qy - xy + fz; \\ \dfrac{dZ}{d\tau} = x - z, \end{cases} \qquad (6.239)$$

where $x = X/X_0$; $y = Y/Y_0$; $z = Z/Z_0$; $\tau = t/T_0$; $X_0 = k_5A/(2k_4)$; $Y_0 = k_5A/k_2$; $Z_0 = (k_5A)^2/(k_4k_0B)$; $T_0 = 1/(k_0B)$; $\varepsilon = k_0B/(k_5A)$; $\varepsilon' = 2k_0k_4B/(k_2k_5A)$; $q = 2k_3k_4/(k_2k_5)$.

Model (6.239) was numerically simulated for the following parameters: $[H^+] = 0.8\,M$, $k_2 = 8\cdot 10^5\,M^{-1}s^{-1}$, $k_3 = 1.28\,M^{-1}s^{-1}$, $k_4 = 2\cdot 10^3\,M^{-1}s^{-1}$, $k_5 = 8\,M^{-1}s^{-1}$, $[A] = [BrO_3^-] = 0.06\,M$, $B = [RH]_0 = 0.02\,M$, $k_0 = 1\,M^{-1}s^{-1}$, $\varepsilon = 4\cdot 10^{-2}$, $\varepsilon' = 4\cdot 10^{-4}$, $q = 8\cdot 10^{-4}$.

Calculations showed the presence of relaxation oscillations in the system. An analysis of the linearized system in the vicinity of steady-state points showed validity of the first condition for the appearance of Andronov–Hopf bifurcations in a system with three variables, whereby it has a pair of eigenvalues $\lambda_{1,2} = \pm i\omega$, and λ_3 has a negative real part ($\mathrm{tr}|\boldsymbol{J}| < 0$, $\det|\boldsymbol{J}| < 0$).

Conditions can be found (as determined by parameters $\varepsilon, \varepsilon', q$, and f) for which system (6.239) exhibits oscillatory behavior. As can be seen from the expressions for ε and ε', the derivatives of x and y in Eqs (6.239) are multiplied by small factors. Therefore, the system has fast and slow variables and, hence, the appearance of relaxation oscillations can be expected. First, let us assume that $\varepsilon' \frac{dy}{dt} \cong 0$ i.e., that $[Br^-]$ is always close to steady-state values

$$y = \frac{fz}{q + x}. \tag{6.240}$$

As a result, the Oregonator model reduces to the following system with two variables:

$$\begin{cases} \varepsilon \dfrac{dX}{dt} = x(1 - x) + \dfrac{f(q - x)z}{q + x} = g(x, z) \\ \dfrac{dz}{dt} = x - z = h(x, z) \end{cases} \tag{6.241}$$

Figure 6.36 shows isoclines $g(x,\ y) = 0$ and $h(x, z) = 0$ for this system. The latter isocline represents the straight line $z = x$. The former equation $g(x, z) = 0$ describes a cubic curve, which drops from the initial $z = \infty$ at $x = q$, exhibits minimum and maximum, and crosses the abscissa axis at $x = 1$. In segment B–D, where a steady-state point (intersection of isoclines) occurs, the autocatalysis is observed for which variable z (concentration of Ce^{4+}) increases with x (concentration of $HBrO_2$). Loop ABCD can be expected to feature instability and oscillations with rapid variation of x (B → C, D → A) and slow relaxation of z (C → D, A → B). An analysis of the $g(x, z) = 0$ curve allows the coordinates of points A, B, C, and D to be determined and conditions for the region of oscillations to be established as

$$(1 + \sqrt{2})q = x_B < x^{ss} < x_D = \frac{1}{2} - q(z^{ss} = x^{ss}),$$

$$0.5 \approx \frac{1}{2} + q < f < (1 + \sqrt{2})[1 - (1 + \sqrt{2})q] \approx 2.41 \quad \text{(i.e., } 0.5 < f < 2.41).$$

System (6.241) has three steady-state solutions for x^{ss} (and, accordingly, for z^{ss}): $x_1^{ss} = 0, x_2^{ss} > 0$, and $x_3^{ss} < 0$. For the appearance of Andronov–Hopf bifurcations in the two-dimensional Oregonator ($n = 2$), it is necessary that $\mathrm{tr}|\boldsymbol{J}| = 0$ ($\sigma = 0$) and $\det|\boldsymbol{J}| > 0$. Using the condition $\sigma = a_{11} + a_{22} = 0$, we obtain

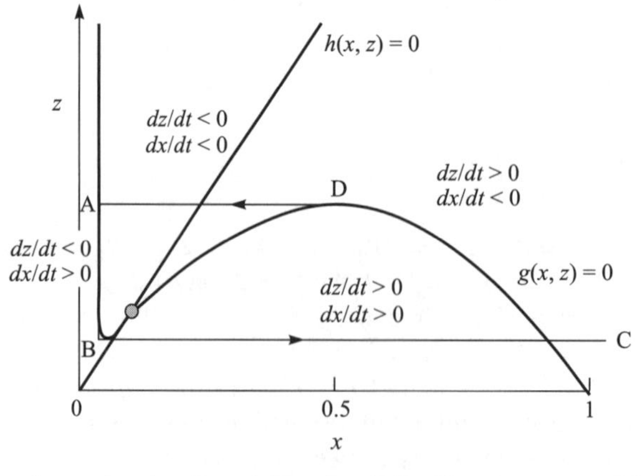

Figure 6.36 Schematic isoclines $g(x, y) = 0$ and $h(x, y) = 0$ for Oregonator model (6.241) (reproduced with permission from Oxford University Press © 1994 [17])

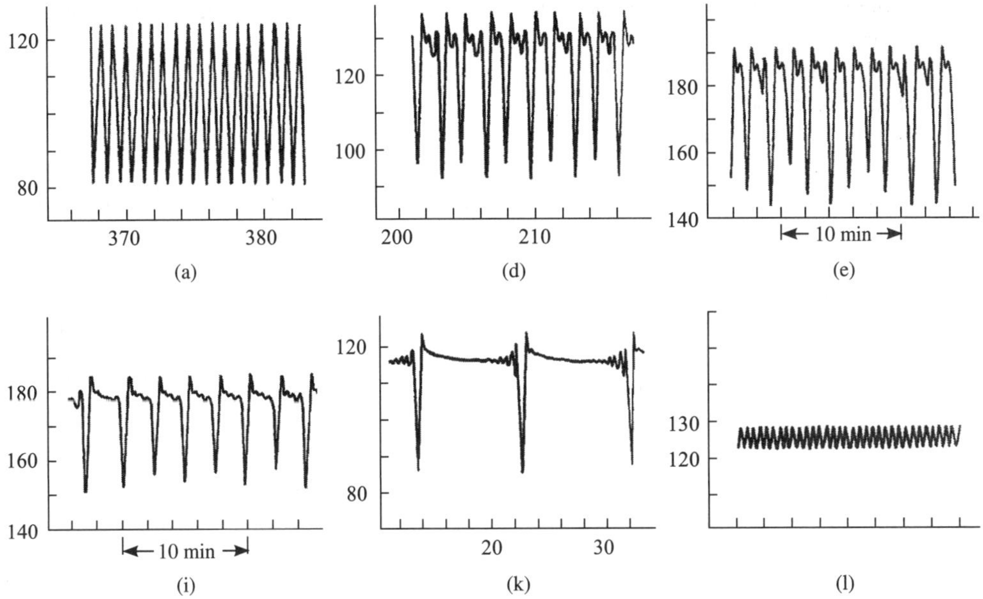

Figure 6.37 Complicated oscillations in BZ reaction in CSTR [155] (see text for explanations) (reproduced with permission from American Institute of Physics © 1979)

the dependence of $\varepsilon = \varphi\ (f, q, x^{ss})$. Then, by varying the values of f and q, we can determine various solutions x^{ss} from the equation for the positive root,

$$x^{ss} = \frac{1}{2}\left\{(1-f-q) + [(1-f-q)^2 + 4q(1+f)]^{1/2}\right\},$$

and the parametric dependence of $\varepsilon = \varphi(f)$ for Andronov–Hopf bifurcations in the entire interval of f from 0.5 to 2.4. Calculations showed that the oscillations are possible for all $\varepsilon < 0.88$. This result implies that $k_0[RH]_0 < 0.88k_5\ [BrO_3^-]$ and determines the possible values of constants and the initial concentrations of BrO_3^- and the organic substrate. Estimations of Y_{crit} (i.e., $[Br^-]_{crit}$), for which the mechanism shifts from process **A** to **B** and there begins the autocatalytic growth of X ($HBrO_2$), yield $[Br^-]_{crit} \cong 1.71(k_{R5}/k_{R2})[BrO_3^-] \cong 1.71 \cdot 10^5[BrO_3^-]$ (see above) in agreement with experiment ($[Br^-]_{crit} = 2 \cdot 10^{-5}[BrO_3^-]$) [159].

Investigations of the BZ reaction in a CSTR showed stable self-oscillations under the conditions of existence of true limit cycles, in which the frequency, amplitude, and shape of oscillations could be controlled by varying the contact time τ. At large times τ, the oscillations exhibit the same wave nature as that in a closed reactor [155]. As τ decreases, the amplitude of oscillations drops and their frequency grows. With a further decrease in τ the system passes through the bifurcation point, the limit cycle breaks, the amplitude of oscillations vanishes, and a stable steady state appears. The oscillation shapes pass via a sequence of complicated bifurcation patterns. Figure 6.37 shows examples of these patterns as observed in the experiment of Hudson, Hart and Marinko [155, 169] and described according to Gray and Scott [17]:

(a) $\tau = 8.73$ min; simple oscillations of large amplitude A; response type 1^0 (only peaks of large A, no small-amplitude variations).
(b) $\tau = 6.76$ min; alternation between 1^0 and 1^1; response type 2^1.

(c) $\tau = 6.26$ min; fully periodic oscillations; every total cycle contains one peak of large A and one small peak; response type 1^1.
(f) $\tau = 5.85$ min; additional small peaks appear; response type 1^2.
(g) $\tau = 5.50$ min; response type 1^3.
(h) $\tau = 5.28$ min; response type 1^4.

Intermediate stages between (b) and (f), (f) and (h), (h) and (j) correspond to more complicated patterns, in particular:

(a) $\tau = 6.18$ min; alternation of types 1^1 and 1^2.
(b) $\tau = 5.89$ min; response types 1^1 and 1^2, but without regular alternation and strict periodicity (aperiodic mixture of two shapes).
(c) $\tau = 5.63$ min; chaotic patterns resulting from superposition of types 1^2 and 1^3.
(d) $\tau = 5.34$ min; aperiodic superposition of types 1^3 and 1^4; this pattern was highly stable and existed for 28 h.
(e) $\tau = 4.73$ min; response type 1^n with $n \gg 1$; number of peaks with small A increases and that of large peaks decreases; response type tends to 0^1 (that is attained at $\tau = 4.69$ min).
(f) $\tau = 4.6$ min; no oscillations.

The first attempt to describe these complicated oscillations observed in the BZ reaction was undertaken in 1978 by Showalter et al. [163]. For this purpose, the Oregonator model was modified by setting all steps reversible except the last (O5), subdividing step (O3) into two substeps as

$$\text{(S3)}\quad \mathrm{A + X \rightleftharpoons 2W} \qquad (\mathrm{W = Br\dot{O}_2});$$

$$\text{(S4)}\quad \mathrm{C + W \rightleftharpoons X + Z} \qquad (\mathrm{C = Ce^{3+}}),$$

and adding Ce^{3+} into step (O5). The modified model with somewhat changed parameters exhibited complicated oscillations with responses of type 1^5, but did not show aperiodic (chaotic) behavior.

In 1980, Janz et al. [164] returned to the Oregonator model with irreversible structure, but took into account the temporal variation of the f factor by considering it to be a function of [HOBr] $= P$ as follows:

$$f = \frac{FP^2}{k + P^2}, \tag{6.242}$$

i.e., f was allowed to oscillate following variations of the HOBr concentration (P). This variant could describe oscillations of the M^N type with large M and N values. However, this model, as well as the later (1989) model of Showalter (see [17]) still demonstrated only strictly periodic oscillations with bifurcations between various patterns following a definite logical scenario. From these results, it was concluded that the aperiodicity was not entirely related to kinetic features of the BZ reaction mechanism. It was suggested that perturbations in the regime of stirring and/or rate of mixture supply to the reactor could lead to the experimentally observed aperiodic patterns.

An alternative approach to the observation and modeling of chaotic oscillations was developed by a Bordeaux–Texas group [165–168], which considered the transition to chaos in the BZ reaction as period-doubling bifurcations based on a discrete (Poincaré mapping) map model [17, 91, 101, 102] and employed the Fourier analysis of the experimental oscillation spectrum, $E_{\mathrm{Br}}(t_i + \Delta t) = \varphi E_{\mathrm{Br}}(t_i)$ delay maps, and $E_{n+1} = \varphi E_n$ cubic return map (texattractor). This algorithm is described in detail elsewhere [17]. As a result, a scheme based on the FKN mechanism was proposed that consisted of nine equations (including three reactions of organic substances) with seven variables. An analysis of the system of equations showed the possibility of chaotic oscillations intermittent with periodicity windows. This pattern corresponds to the ideas of Feigenbaum concerning the behavior of a cubic mapping [101, 102]:

$$X_{n+1} = AX_n(1 - X_n)^2. \tag{6.243}$$

In 1984, Hudson and Rössler [169] showed that the kinetic scheme

$$\begin{array}{lll}
(1) & \mathrm{P} \longrightarrow \mathrm{A} & (W_1 = k_1 p); \\
(2) & \mathrm{Q} \longrightarrow \mathrm{B} & (W_2 = k_2 q); \\
(3) & \mathrm{A} + \mathrm{B} \longrightarrow 2\mathrm{B} & (W_3 = k_3 ab); \\
(4) & \mathrm{B} \longrightarrow \mathrm{R} & (W_4 = k_4 b/(b + K_2)); \\
(5) & \mathrm{A} \rightleftharpoons \mathrm{C} & (W_5 = k_5 a - k_{-5} c)
\end{array} \tag{6.244}$$

(where a, b, and c are variables) leads to aperiodic oscillations provided that there are two types of nonlinearity – quadratic autocatalysis and nonlinear material balance with respect to the autocatalyst B in step (4) in the form $W_4 = k_4 b/(b + K)$.

The problem of determinacy of chaotic behavior of dynamical systems, which is caused entirely by kinetic factors, has been actively discussed [17]. In the case of a BZ reaction, chaos that develops as a result of period-doubling bifurcations is deterministic, which means that any initial state corresponds to a strictly predictable result. The approach to an attractor (limit cycle) is accompanied by a decrease in the spacing between trajectories and leads to chaotic behavior.

In the case of a strange attractor, chaos is also deterministic, which means that there is a system of equations determining the positions of trajectories in a finite phase volume, and every initial condition provides a reproducible result (upon the second and third integration, etc.), although the behavior for an adjacent starting point cannot be predicted.

In the analysis of mechanisms based on the FKN scheme and variants of the Oregonator model, much doubt relates to a poorly studied block of organic reactions, which at least involves brominating species (Br_2, HOBr) and numerous steps of the oxidation and bromination of organic species (up to CO_2 formation). Since the role of substrate RH reduces eventually to the regeneration of Br^- and reduction of the catalyst (Ce^{4+} to Ce^{3+}), it would be natural to replace the organic block by the direct supply of Br^- and Ce^{3+} into the flow reactor (CSTR), thus excluding the entire block of organic reactions from the mechanistic scheme. This system, demonstrating both bistability and oscillatory behavior, is called the minimal oscillator [9, 17].

Bistable behavior in a flow reactor was observed in a broad concentrations range (about two orders of magnitude) for both $[Br^-]_0$ and $[BrO_3^-]_0$ [170]. The replacement of Ce^{3+} by Mn(II) revealed a clear pattern of transition between the upper and lower branches ($[Br^-]$) of the hysteresis loop (Figure 6.38) [171].

In this context, very interesting results were obtained by Bar-Eli [172] in the quantitative analysis of a minimal bromate oscillator ($BrO_3^-, H^+, Br^-, Ce^{3+}$) in a continuous flow reactor, which was based on the early scheme of Noyes and Field (1971) that included seven reversible steps with 14 constants. By decreasing the number of steps and varying the reversibility of steps, a simpler model was found that consisted of three steps with four constants and involved three independent substances. The model exhibited both bistability and self-oscillations and qualitatively reproduced all the main features of the complete scheme. The model included the following steps [172]:

$$\begin{array}{ll}
(1) & \mathrm{BrO_3^-} + \mathrm{Br^-} + 2\mathrm{H^+} \longrightarrow \mathrm{HBrO_2} + \mathrm{HOBr}; \\
(2) & \mathrm{HBrO_2} + \mathrm{Br^-} + \mathrm{H^+} \longrightarrow 2\mathrm{HOBr}; \\
(4'') & \mathrm{Ce^{3+}} + \mathrm{HBrO_2} \rightleftharpoons 2\mathrm{HBrO_2} + \mathrm{Ce^{4+}}.
\end{array} \tag{6.245}$$

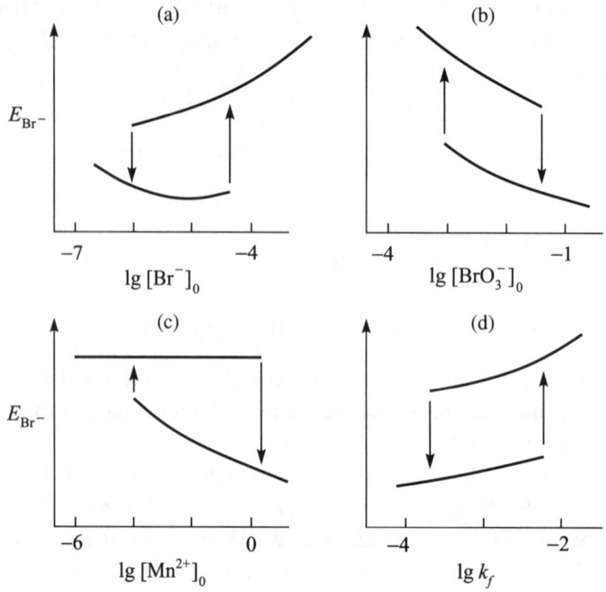

Figure 6.38 Hysteresis loops (bistability) observed in minimal bromate oscillator in CSTR during variation of (a) $[Br^-]_0$, (b) $[BrO_3^-]_0$, (c) $[Mn^{2+}]_0$, and (d) flow rate k_f (reproduced with permission from the American Chemical Society © 1982)

where reversible reaction (4″) (representing quadratic autocatalysis) replaced two reactions in the complete scheme,

$$(4)\quad BrO_3^- + HBrO_2 + H^+ \longrightarrow 2\dot{B}rO_2 + H_2O;$$
$$(5)\quad Ce^{3+} + \dot{B}rO_2 + H^+ \longrightarrow Ce^{4+} + HBrO_2$$

with an overall equation of (4) + 2(5):

$$BrO_3^- + 2Ce^{3+} + HBrO_2 + 3H^+ \longrightarrow 2Ce^{4+} + 2HBrO_2,$$

A minimal model was based upon the following scheme:

$$\begin{aligned}&(B1)\quad A + Y \longrightarrow X + P;\\&(B2)\quad X + Y \longrightarrow 2P;\\&(B3)\quad X + Z \rightleftharpoons 2X,\end{aligned} \tag{6.246}$$

where $X \equiv [HBrO_2]$, $Y \equiv Br^-$, and $Z \equiv Ce^{3+}$.[2] The corresponding system of equations was as follows:

$$\begin{cases}\dfrac{dx}{dt} = k_1y - k_2xy + k_{4''}xz - k_{-4''}x^2;\\[2mm] \dfrac{dy}{dt} = -k_1y - k_2xy + k_0(y_0 - y);\\[2mm] \dfrac{dz}{dt} = -k_{4''}xz + k_{-4''}x^2 + k_0(z_0 - z),\end{cases} \tag{6.247}$$

[2] Reasons for the absence of Ce^{4+} in step (B3) were not considered in [172].

where $x \equiv [HBrO_2]$, $y \equiv [Br^-]$, and $z \equiv [Ce^{3+}]$. The system was studied for the following parameters:

$$k_1 = 2.1 M^{-3}s^{-1} (\text{including } [BrO_3^-]_0);\ k_2 = 2 \cdot 10^9 M^{-2}s^{-1};$$

$$k_{4''} = k_5 = 9.75 \cdot 10^5 M^{-1}s^{-1} \text{ or } k_{4''} = 9.75 \cdot 10^{-5}[BrO_3^-]_0;$$

$$k_{4''} = 4 \cdot 10^7 M^{-1}s^{-1} \text{ or } k_{-4''} = 4 \cdot 10^9 M^{-1}s^{-1}.$$

It was also assumed that $[H^+]_0 = 1.5\,M$, specific flow rate $k_0 = 4 \cdot 10^{-3}s^{-1}$, $[Ce^{3+}]_0 = 1.5 \cdot 10^{-4}M$, $[BrO_3^-] \cong [BrO_3^-]_0$, and $[Ce]_\Sigma \cong [Ce^{4+}]$.

Figure 6.39 shows a parametric portrait for this scheme, where solid curves bound the region of bistability (hysteresis loop) and dashed curves show the region of oscillatory behavior (boundary of Andronov–Hopf bifurcations).

At the initial stage of investigations of the BZ reaction, special attention was devoted to proofs of the homogeneous character of the process. The fact that the oscillatory regime was independent of the intensity of stirring was interpreted as the absence of the influence of processes at the gas–liquid interface (dissolution of oxygen from air, desorption of CO_2) on the critical phenomena in the system [4, 5]. Then, organic substrates were found that exhibited conversion into CO_2, but in a dissolved form (up to saturation of the solution) without the formation of macroscopic bubbles (e.g., acetone–$H_2PO_2^-$, arabinose–acetone) [65]. In these systems, as well as in minimal oscillators (free of RH), the self-oscillations were retained.

However, subsequent investigations showed that the saturation of a solution with oxygen [173] and the intensification of stirring [174–180] in a certain interval of concentrations can influence the induction period, frequency, and amplitude of oscillations up to their complete vanishing. For example, HOBr additions to the initial solution led to a decrease in the induction period. However, the introduction of HOBr into an O_2-saturated solution induced a train of several oscillations followed by the passage into a reduced steady state within 10 min. In an identical experiment under argon atmosphere, the oscillations began immediately upon the introduction of HOBr and continued for ~10 h. It was suggested that the formation of radicals $\dot{R}O_2$ and $\dot{R}$ (from RH) favors the accelerated formation of Br^- from bromomalonic acid.

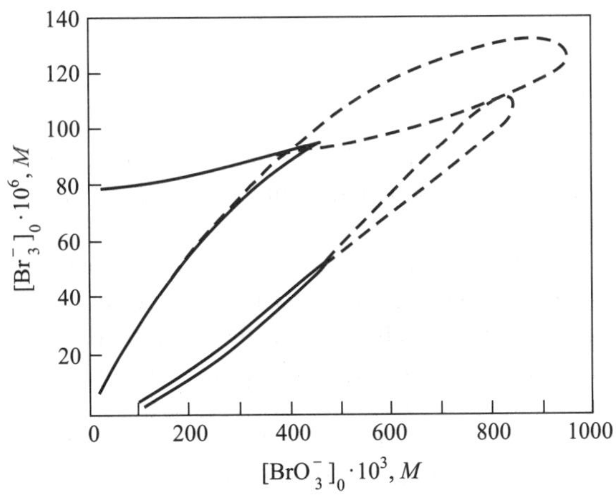

Figure 6.39 Regions of bistability and oscillations in model (6.247). Upper curves: $k_{4''} = 9.75 \cdot 10^5 M^{-1}s^{-1}$, $k_{-4''} = 4 \cdot 10^9 M^{-1}s^{-1}$; lower curves: $k_{4''} = 9.75 \cdot 10^5 [BrO_3^-]_0$, $k_{-4''} = 4 \cdot 10^9 M^{-1}s^{-1}$; $[Ce^{3+}]_0 = 1.5 \cdot 10^{-4}M$, $[H^+]_0 = 1.5M$, and $k_0 = 4 \cdot 10^{-3}s^{-1}$ (reproduced with permission from the American Chemical Society. © 1985)

It was also demonstrated [177–180], that the intensity of stirring influences the character of oscillations in the atmosphere of nitrogen and argon. It was suggested that a BZ reaction can lead to the formation of a bromine-oversaturated solution (up to the appearance of liquid bromine microbubbles) [18]. Therefore, an increase in the interface area on stirring favors the oscillatory desorption of bromine. Since the effect of stirring was also observed in some other systems (e.g., in the Briggs–Rauscher iodate system, [182]), alternative hypotheses have been formulated as well.

The effect of stirring on the induction period and character of oscillations is explained by the formation of nuclei (concentration nonuniformities) due to large-scale fluctuations that result from the interaction between microscopic volumes of solution featuring small-scale fluctuations [7, 180]. In the absence of stirring within thin layers, the nuclei give rise to wave processes and structure formation due to the molecular diffusion of active species. At weak stirring, coarse nuclei are formed and then gradually broken so that their fragments are spread over the entire volume. As the intensity of stirring increases, the nuclei become smaller until the point comes with their complete breakage, leveling of concentration, and vanishing of oscillations.

Within the framework of this hypothesis, the nonlinearity of the mechanism and the kinetic model is a necessary condition for the appearance of critical phenomena, which determines the number of steady states and their stability. The development of a temporal or spatial organization in a system far from thermodynamic equilibrium takes place as a result of fluctuations that cause bifurcations in the phase space with the transition to new steady states. The appearance of fluctuations and concentration nonuniformities, with their spreading over the entire volume on stirring, leads to cooperative interactions between microvolumes (elements), which render the system organized in time and make the concentration oscillations synchronized over the entire volume. In the absence of stirring, the fluctuations induce spatial structures and wave processes such as spiral waves during the BZ reaction [183]. The scale of fluctuations and the extent of the stirring effect probably depend on the type of reaction, structure of the mechanism, nature of solvent, and some other factors.

It was suggested (A. Winfree [9]) that microheterogeneous dust particles can play the role of centers for the excitation of wave processes in reaction media in the absence of stirring. These centers (pacemakers) can act as sources of large-scale fluctuations. Undoubtedly, elucidation of the role of diffusion and hydrodynamic factors as well as microheterogeneity in the dynamic behavior of nonlinear systems requires further investigation.

6.4.2 Liquid-phase oxidation of organic compounds by oxygen in $Co(OAc)_2$–Br–CH_3COOH system

Investigation [59] of the effect of organic bromides on the oxidation of toluene in the $Co(OAc)_2$–Br^-–CH_3COOH system showed that the oxidation rate exhibited significant growth when [PhBr] was increased in the interval from 0 to 2 M, and this was accompanied by the appearance of oscillations in the rate of oxygen uptake in a continuous-O_2-flow stirred tank reactor. An analogous effect was observed on adding 1,2-dibromoethane, *n*-butyl bromide, and cyclohexyl bromide. Note that the concentrations of all these bromides in the course of reaction remain unchanged. At the same time, the oscillations in the O_2 uptake rate were also observed in the case of adding Br_2 and benzyl bromide, which delivered Br^- into the solution. The process of toluene oxidation consists of two stages: (i) slow accumulation of benzaldehyde; and (ii) fast oxidation with rapid consumption of benzaldehyde. The concentration of benzaldehyde exhibits oscillations until toluene is completely consumed (Figure 6.40).

The mechanism of benzaldehyde oxidation in an self-oscillatory regime [60] was studied in much detail [24, 184]. In the $Co(OAc)_2$–Br^-–90 % acetic acid system at 70 °C, the reaction is accompanied by periodic oscillations of the solution color, potential of platinum electrode, and concentration of dissolved

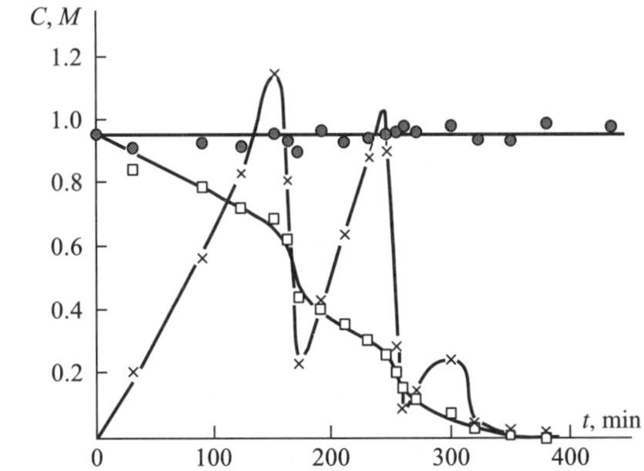

Figure 6.40 Variation of concentrations of toluene (□), bromobenzene (•), and benzaldehyde (×) ($C \cdot 10$, M) with time at 90 °C ($[PhMe]_0 = 0.951$ M, $[PhBr]_0 = 0.95$ M, $[Co]_0 = 0.120$ M) (reproduced with permission from the Royal Chemical Society © 1969)

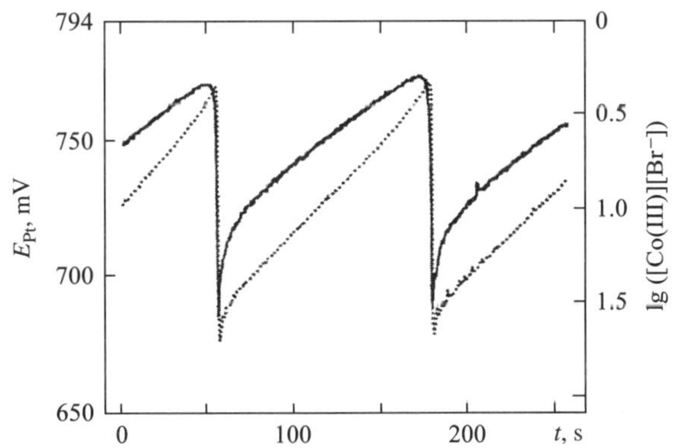

Figure 6.41 Oscillations of (solid curve) electrode potential E_{Pt} and (dotted curve) lg([Co(III)/[Br⁻]) value determined by colorimetry at $[Br^-]_0 = 15\ \mu$M. The measurements were performed 32 min after introduction of benzaldehyde; vertical axes are matched so that $E = 794 + 69$ lg([Co(III)/[Br⁻]) (reproduced with permission from the American Chemical Society © 1987)

oxygen (Figures 6.41 and 6.42). Roelofs et al. [61] proposed a kinetic model that included 21 steps and satisfactorily reproduced the observed pattern of oscillations (Figure 6.43). Let us briefly consider the proposed scheme as described in the review by Noyes [24].

All hypotheses concerning the mechanism of this chemical oscillator included two important steps (PhCO ≡ Bz):

$$\text{(R17)}\quad O_{2gas} \rightleftharpoons O_{2sol};$$

$$\text{(R18)}\quad O_{2sol} + Bz \longrightarrow Bz\dot{O}_2.$$

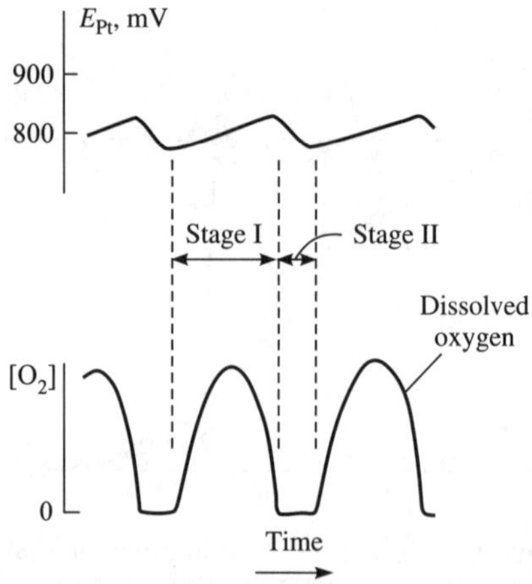

Figure 6.42 Comparison of E_{Pt} and dissolved oxygen concentration profiles at $[Br^-]_0 = 2\ \mu M$ (reproduced with permission from the American Chemical Society © 1987)

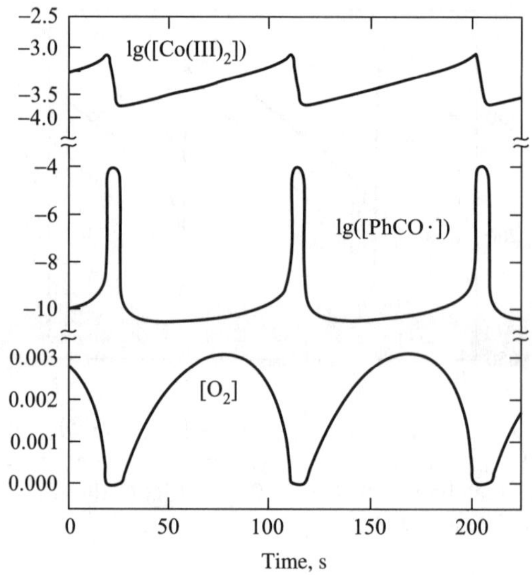

Figure 6.43 Temporal profiles of $Co(III)_2$, PhCO•, and O_2 concentrations in solution calculated according to the adopted kinetic mechanism (reproduced with permission from the American Chemical Society © 1987)

The other two steps convert peroxybenzoyl radical into peroxy acid (BzOOH):

$$\text{(R20)}\quad Bz\dot{O}_2 + Co(II) + H^+ \longrightarrow BzOOH + Co(III)$$

(cf. step (R6) in the FKN mechanism),

$$\text{(R19)}\quad Bz\dot{O}_2 + BzH \longrightarrow BzOOH + \dot{B}z.$$

and this peroxy acid oxidizes Co(II) via the reaction

$$\text{(L)} \quad \mathrm{BzOOH + Co(II) + BzH + H^+ \longrightarrow BzOH + Co(III) + H_2O + \dot{B}z}.$$

Reactions (R17), (R18), (R20), and (L) contribute to the overall process

$$\text{(M)} \quad \mathrm{O_{2gas} + BzH + 2Co(II) + 2H^+ \longrightarrow BzOH + 2Co(III) + H_2O},$$

while reactions (R17), (R18), (R19), and (L) contribute to the overall process

$$\text{(N)} \quad \mathrm{O_{2gas} + 2BzH + Co(II) + H^+ \longrightarrow BzOH + \dot{B}z + Co(III) + H_2O}.$$

Processes (M) and (N) lead to the exponential growth in [Co(III)] and the O_2 uptake rate as long as step (R17) of O_2 transport from the gas phase can provide for these reactions. Co(III) that is formed via reactions (M) and (N) oxidizes BzH

$$\text{(O)} \quad \mathrm{Co(III) + BzH + H_2O \longrightarrow BzOH + Co(II) + 2H^+};$$

$$\text{(P)} \quad \mathrm{Co(III) + \dot{B}z + H_2O \longrightarrow BzOH + Co(II) + H^+}.$$

The rate of process (O) in argon is much greater than the rate of O_2 transport (step (R17)), almost independent of [Co(III)]. Process (O) is inhibited by radical traps and catalyzed by Br^-. The stoichiometry of the overall process

$$\mathrm{2BzH + O_{2gas} \longrightarrow 2BzOH}$$

is determined by the sum of processes (M) + (O) or (N) + (P). It is believed that the alternation of two routes in this process is a necessary but insufficient condition for the appearance of oscillations.

At the process start, $[O_2]_{sol}$ is large, [Co(III)] is small and, hence, the reaction develops according to equations (M) and (N) with increasing rate (the appearance of $\dot{B}z$ increases the rate of formation of $Bz\dot{O}_2$). When the rate of O_2 absorption becomes greater than the rate of O_2 transport to solution, the rates of reactions (O) and (P) increase and the process shifts to the second stage (see Figures 6.41 and 6.42), at which Co(III) concentration rapidly drops and $[O_2]$ in solution gradually increases. Since oxygen in the first stage is absorbed, BzH can be oxidized by Br_2 formed via the reaction

$$\text{(Q)} \quad \mathrm{BzOOH + 2Br^- + 2H^+ \longrightarrow BzOH + Br_2 + H_2O};$$

$$\text{(R)} \quad \mathrm{Br_2 + BzH + H_2O \longrightarrow BzOH + 2Br^- + 2H^+}.$$

Thus, Br^- catalyzes the decomposition of peroxy acid to benzoic acid:

$$\mathrm{BzOOH + BzH \longrightarrow 2BzOH}.$$

For the explanation of oscillations, reactions (Q) and (R) are evidently less important than processes involving Co(III). A simplified four-step mechanism explaining the oscillations, which includes one reversible and three irreversible steps and is described by three independent variables, was developed by Noyes with co-workers (see [24]).

The existence of critical concentrations of catalysts (metal salts or metal ions M^{n+}), which determine the transition from the region of oxidation acceleration to inhibition, was well known in the liquid-phase oxidation of hydrocarbons. These critical parameters were explained in the analysis of two-dimensional bistable model reactions in a continuous-flow stirred tank reactor [80, 185]. It was demonstrated that small variations in the initial conditions ($[ROOH]_0$, $[M^{n+}]_0$) can lead to switching of the acceleration and inhibition regimes. The dependence of the critical concentration $[M^{n+}]_{crit}$ on the initial concentration of ROOH was also justified.

6.4.3 Oxidative carbonylation of alkynes in solutions of palladium complexes

As was noted above, a new class of catalytic oscillators in gas–liquid heterophase systems was discovered in 1985 [26, 27] as a result of investigations of the oxidative carbonylation of acetylene in alcohol–organic solvent (DMF [186], acetone [187], DMSO [27]) mixtures (see also [188, 189]). In particular, it was found that in the $PdBr_2$–HBr–PPh_3–C_4H_9OH–DMSO system in a continuous-gas-flow (C_2H_2, CO) complete-mix (stirred tank) reactor, the formation of dibutyl esters (DBEs) of maleic and fumaric acids and the oxidation of CO to CO_2 by DMSO via reaction

$$CO + (CH_3)_2SO \longrightarrow CO_2 + (CH_3)_2S \tag{6.248}$$

is accompanied by stable oscillations in the platinum electrode potential, solution color (from yellow-orange to greenish-brown), and rate of absorption of the gas ($CO + C_2H_2$) mixture (Figure 6.44).

Since the rate of DBE formation is virtually independent of the DMSO molar fraction (N_{DMSO}) in the BuOH–DMSO system, the main role of DMSO in this process probably consists of participating as oxidant in reaction (6.248), the rate of which increases with N_{DMSO} (starting at about $N_{DMSO} = 0.2$). Oscillations in the rate of gas uptake begin at about $N_{DMSO} \cong 0.35$. The rates of formation of CO_2 and dimethyl sulfide (DMS) are approximately the same. The volume of absorbed gases was measured past the cryogenic trap where DMS was condensed (at about −70 °C). Therefore, the fact that $V_{abs} \approx 0$ in the course of oscillations in V implies the absence of the interaction of CO with acetylene and alcohol

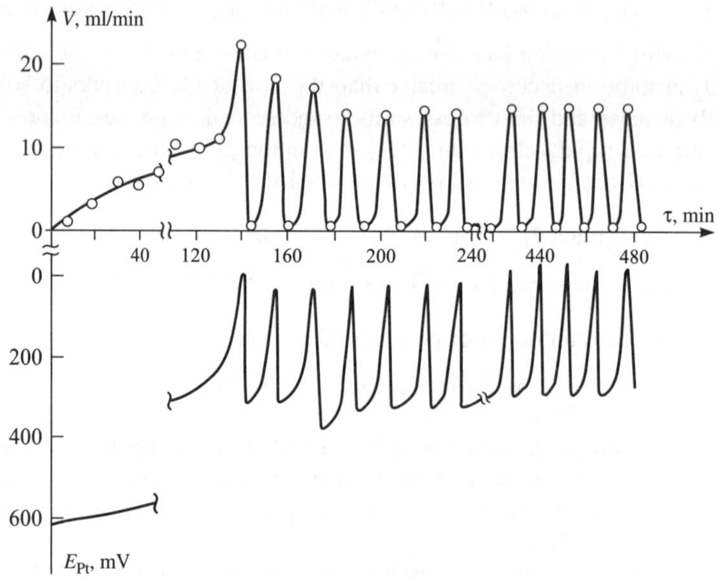

Figure 6.44 Kinetic curves of gas absorption (V) and platinum electrode potential (E_{Pt}) changes in $PdBr_2$ – HBr – PPh_3 – ROH – DMSO system for [HBr] = 0.3 M, α = [L]/[Pd] = 1, V_{sp} = 1.1 min^{-1}, and a catalyst volume of 30 ml

and the termination of DBE formation. At $N_{\mathrm{DMSO}} = 1$, oscillations in the absorption rate and electrode potential ceased, but the absorption of CO : $C_2H_2 = 1{:}\ 1$ gaseous mixture continued at a significant rate (0.4–0.45 M/h), probably with the formation of a copolymerization product (the rate of CO_2 formation was about half that of CO absorption). The oxidative synthesis of DBE was ensured by the reduction of acetylene oligomers, since the oxidation balance ($\varphi = (r_{\mathrm{DBE}} + r_{\mathrm{CO_2}})/r_{\mathrm{DMS}}$) amounted in some cases up to $\varphi = 2$–3.

The mechanism of reaction in DMSO was not studied, but the results of observations [27] and data of other investigations of the oxidative carbonylation of alkynes [190–192] allowed making some judgments. The process of acetylene carbonylation in this case, as well as in the PdI_2–KI–ROH, $PdBr_2$–HBr–PPh_3–ROH, and $PdBr_2$–acetone systems [192], begins after an induction period during which Pd(II) is reduced to Pd(I). Palladium hydride complexes (HPdBr) formed in the reaction

$$\mathrm{Pd_2Br_2 + C_2H_2 + 2CO + 2BuOH \longrightarrow DBE + 2HPdBr} \qquad (6.249)$$

are oxidized by acetylene oligomers as

$$\mathrm{HPdBr} + {}^1\!/_n(\mathrm{CH{=}CH})_n + \mathrm{HBr} \longrightarrow \mathrm{PdBr_2} + {}^1\!/_n(\mathrm{CH_2CH_2})_n \qquad (6.250)$$

and partly by DMSO as

$$\mathrm{HPdBr + (CH_3)_2SO + HBr \longrightarrow PdBr_2 + (CH_3)_2S + H_2O.} \qquad (6.251)$$

Step (6.251) is also involved in the oxidation of CO to CO_2 by palladium bromide:

$$\mathrm{PdBr_2 + CO + H_2O \longrightarrow HPdBr + CO_2 + HBr,} \qquad (6.252)$$

although some other mechanisms of CO oxidation can be operative as well. The formation of DMS and its complexes with C Pd(II) and Pd(I) probably plays a significant role in the appearance of oscillations in the system under consideration. Since it was established that characteristics of the carbonylation process in DMSO depend but little on the concentration of PPh_3, the dependence of oscillation regimes on the concentrations of solution components and reactants was mostly studied in the $PdBr_2$–KBr–HBr–DMSO–ROH system [27].

Oscillatory regimes in the oxidative carbonylation of alkynes were also studied in more detail in the PdX_2–KX–MeOH [20, 28, 193, 194]. In particular, the oxidative carbonylation of phenylacetylene (PA) in the iodide system (PdI_2–KI–MeOH) proceeds in the regime of self-oscillations in a system closed with respect to PA, under conditions of intense stirring of a gas–liquid mixture (in either a gas-flow or completely closed reactor). This is accompanied by oscillations in the platinum electrode potential, pH (glass electrode), and reaction rate as manifested by a stepwise variation of the absorbed gas ($CO + O_2$) volume (Figure 6.45) [28]. This reaction is described by the following equation,

$$\mathrm{PhC{\equiv}CH + 2CO + 2MeOH + {}^1\!/_2O_2 \longrightarrow}$$

I (70%) **II** (4%) (6.253)

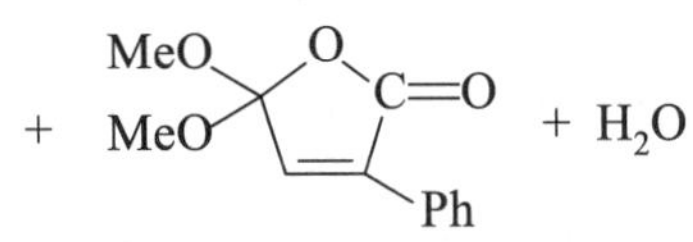

III (15%)

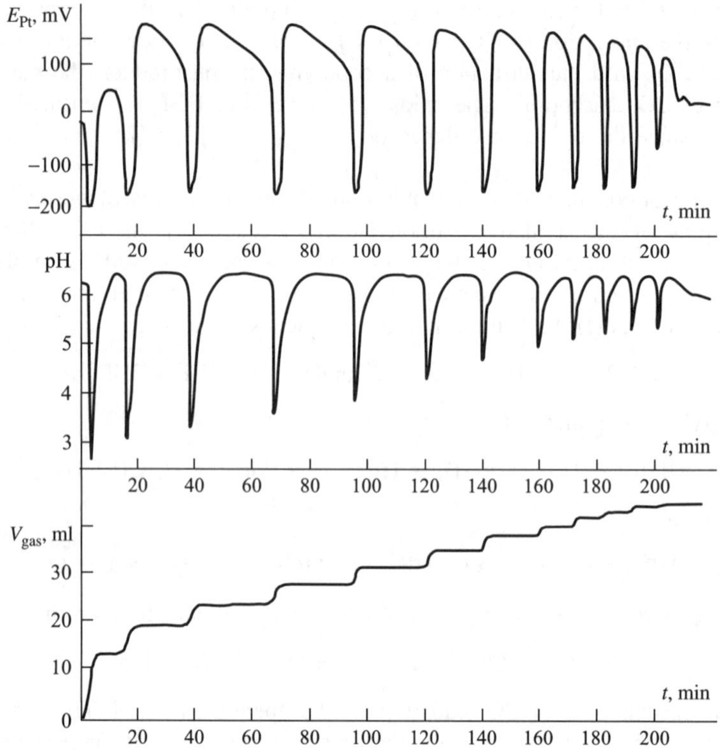

Figure 6.45 Oscillations of E_{Pt}, pH, gas ($CO + O_2$) absorption (V_{gas}) in PdI_2 (0.01 M) – KI (0.4 M) – NaOAc (0.0024 M) – phenylacetylene (0.1 M) system (reproduced with permission from the American Chemical Society © 1997)

and the parallel process of CO oxidation to CO_2. The amount of formed water approximately corresponds to the sum of products (**I** + **II** + **III**). The curves presented in Figure 6.45 were obtained in a system with sodium acetate addition (0.0024 M), which significantly influences the initial stage of the process, i.e., the time of attaining the regime of regular oscillations. It was shown that an oscillatory regime is also attained in the absence of a platinum electrode.

It was also established that the characteristics of oscillations and even the very fact of their appearance strongly depend on the purity of the PA employed [194]. Highly reproducible results are obtained with freshly distilled PA of no less than 98 % purity. It is also possible to use distilled PA that was stored at 0 °C, but for a period of time not exceeding about one month. If these requirements are satisfied, the process characteristics observed in the gas–liquid mixture (in the absence of NaOAc) depend neither on the stirrer rotation speed (within 250–900 rpm, see [182]) nor on the intensity of illumination in the visible spectral range. Figure 6.46 shows the typical pattern of oscillations in the PdI_2–KI–MeOH system [194]. During the period of time between solution preparation and the onset of reactor purging with a $CO + O_2$ mixture, the solution pH typically establishes on a level of 7–8, while the platinum electrode potential is within $E_{Pt} = 50–100$ mV. After the onset of gas purging through the reactor with a solution, pH decreases to 5–6 and E_{Pt} falls between 0 and −20 mV. After PA introduction (see Figure 6.46), E_{Pt} decreases further to −(70–100) mV, pH drops to 2–2.5, and the gas absorption begins.

Regular oscillations begin after a certain induction period (10–15 min) and proceed with gradual decay within about two hours. It was established that, in the case of using Pd(I) iodide complexes instead of

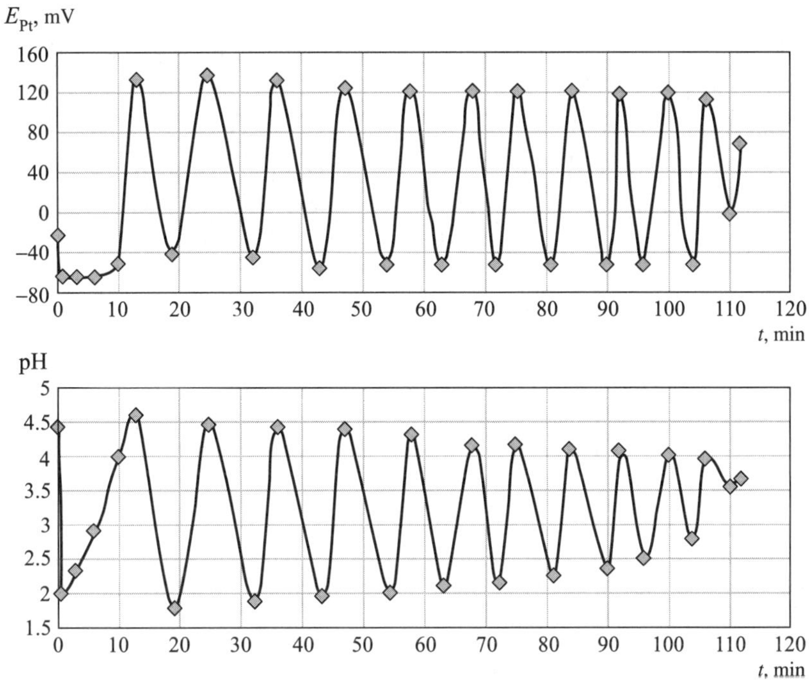

Figure 6.46 Variation of characteristics (E_{Pt}, pH) during oxidative carbonylation of phenylacetylene (PA) under standard process conditions (stirrer rotation speed, $\omega = 700$ rpm; $[KI]_0 = 0.4$ M, $[PdI_2]_0 = 0.01$ M, $[PA]_0 = 0.1$ M, and $[CO]_0$: $[O_2]_0 = 3:2$ (reproduced with permission from editorial board of Izv. Akad. Nauk, Ser. Khim. journal © 2003 [20])

PdI_2 [194], the induction period disappears, the self-oscillations begin immediately after the introduction of PA, and the system occurs initially in a strongly reduced state ($E_{Pt} = -50$ mV) (Figure 6.47). In this experiment, E_{Pt} in the course of oscillations does not exceed zero level, which is indicative of a relatively low concentration of Pd(II) in this system.

It was established that the process can be inhibited by KI, HI and I_2 and that there is a certain critical pH level below which self-oscillations do not appear and the carbonylation process does not start. The concentration of I_2 must not exceed $8 \cdot 10^{-4}$ M. The damping of oscillations can be related to both the depletion of PA and the binding of Pd in stable organometallic compounds. Adding PA brings back an oscillatory regime, but the restored regime is different from the initial one. An oscillatory regime in this system was also reproduced under somewhat different conditions [200–203], where a stepwise heat evolution synchronous with pH oscillations was detected [200].

Assuming, in agreement with the aforementioned results and the other data [186–192], especially the results of investigation of the PdI_2–KI–HCl–BuOH system [190, 191], that active species in the oxidative carbonylation of alkynes at the triple bond are Pd (I) complexes (e.g., $Pd_2I_4{}^{2-}$), the process of Pd(I) formation from Pd(II) can be represented by the following reactions:

$$\text{(G1)} \quad PdI_2 + CO + H_2O \longrightarrow HPdI + CO_2 + HI;$$

$$\text{(G2)} \quad PdI_2 + HPdI \longrightarrow Pd_2I_2 + HI.$$

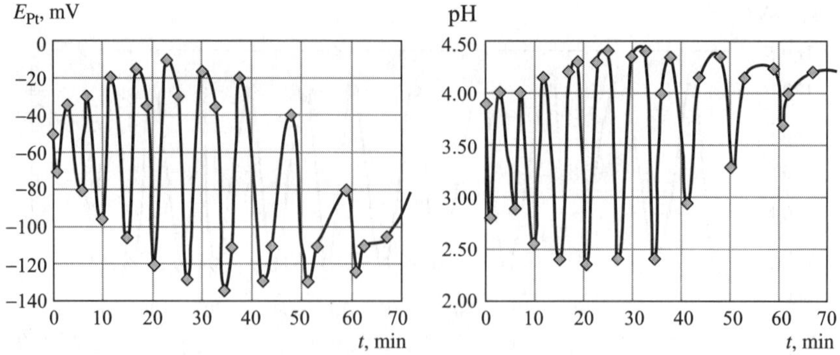

Figure 6.47 Variation of characteristics (E_{Pt}, pH) during oxidative carbonylation of phenylacetylene (PA) in a system with Pd(I) potassium iodide complex $[\kappa Pd_4I_5 \cdot 12H_2O]$ ($\omega = 700$ rpm, $[KI]_0 = 0.4$ M, $[Pd(I)]_0 = 0.02$ M, $[PA]_0 = 0.1$ M, and $[CO]_0 : [O_2]_0 = 3:2$ (reproduced with permission from editorial board of *Kinet. Katal.* journal © 2001 [194])

In the case of reaction in methanol, the content of water in the used solvent was about 0.05 M. In this case, the reduced state of palladium corresponded to low pH values. It is also not excluded that HPdI is formed in the first step (after PA introduction):

$$PdI_2 + CO + PhC{\equiv}CH + MeOH \longrightarrow \text{Products} + HPdI + HI$$

Reaction (G1) is undoubtedly a block of elementary and pseudo-elementary steps:

$$\text{(G1a)} \quad PdI_2 + CO + H_2O \rightleftharpoons IPdCOOH + HI;$$

$$\text{(G1b)} \quad IPdCOOH \longrightarrow HPdI + CO_2$$

Writing this block in the form of (G1) also admits a variant where reaction (G1a) is irreversible.

It was demonstrated [191] that Pd(I) iodide complexes did not disappear until very large P_{O_2} values were attained (20–30 % in the gas phase), but these complexes are readily oxidized by I_2 (provided that it is present in a stoichiometric ratio of $Pd(1){-}Pd_2^{2-} : I_2 \cong 1:1$):

$$Pd_2I_4^{2-} + I_2 \longrightarrow 2PdI_3^-$$

The formation of carbonylation products with the participation of Pd(I) under conditions of low concentration of HI (at pH 2–4.5) must be accompanied by the formation of Pd(II) hydride complexes, which was convincingly demonstrated for the synthesis of succinic anhydride from acetylene in bromide systems in acetone and acetonitrile [195]. For this reason, the overall reaction of formation of diesters (P) can be presented as follows:

$$\text{(G3)} \quad Pd_2I_2 + PhC \equiv CH + 2CO + 2MeOH \longrightarrow P + 2HPdI.$$

Evidently, reactions (G2) and (G3) provide eventually an autocatalytic process of PdI_2 reduction:

$$PdI_2 + HPdI + PhC \equiv CH + 2CO + 2MeOH \longrightarrow P + 2HPdI + HI.$$

If the first step of Pd_2I_2 and PA interaction in block (G3) is assumed to be a slow reaction, then the rate of block (G3) can depend only on $[Pd_2I_2]$ and [PA]:

$$W_3 = k_3[Pd_2I_2][PA]. \qquad (6.254)$$

The accumulation of a relatively large HPdI concentration at the stage of reduction (accompanied by the drop in E_{Pt} and pH), also makes possible the following reaction:

$$\text{(G4)} \quad \text{HPdI} \rightleftharpoons \text{Pd(0)} + \text{HI}.$$

However, Pd^0 is probably rapidly oxidized by one of the oxidant species that are present (PdI_2, O_2) or formed (I_2, H_2O_2) in the system. Apparently, for the completion of a catalytic cycle, it is necessary to oxidize HPdI to PdI_2 or Pd_2I_2. There are many possibilities for this process, since both O_2, I_2 and H_2O_2 can play the role of oxidants. Simple titration of the solution with thiosulfate in various phases of the oscillatory process showed that an oxidant is present in concentrations that are much greater than is possible in the case of I_2. Thus, the most probable oxidant of Pd(I) is H_2O_2.

The oxidation of Pd(II) hydride complexes by oxygen was studied by Stahl et al. [196, 197]. The mechanism of HPdI oxidation by oxygen can be represented by the following reactions

$$\text{(G5)} \quad \text{HPdI} + \text{O}_2 \overset{K_5}{\rightleftharpoons} \text{HOOPdI},$$

$$\text{(G6)} \quad \text{HOOPdI} + \text{HI} \xrightarrow{k_6} \text{H}_2\text{O}_2 + \text{PdI}_2$$

which corresponds to

$$W_6 = k_6 K_5 [\text{HPdI}][\text{HI}][\text{O}_2]. \tag{6.255}$$

Alternatively, step (G6) can be replaced by the following reaction:

$$\text{(G7)} \quad \text{HOOPdI} + \text{HPdI} \xrightarrow{k_7} \text{H}_2\text{O}_2 + \text{Pd}_2\text{I}_2,$$

with the rate expressed as

$$W_7 = k_7 K_5 [\text{HPdI}]^2 [\text{O}_2]. \tag{6.256}$$

In one of the first models, reactions (G5) and (G7) were replaced by a simple block of steps that took into account the disproportionation of Pd_2I_2,

$$\text{(G8)} \quad \text{Pd}_2\text{I}_2 \rightleftharpoons \text{PdI}_2 + \text{Pd(0)},$$

or the dissociation of HPdI (G4).

The block of steps for the formation of H_2O_2 was described as the oxidation process:

$$\text{(G9)} \quad 2\text{HPdI} + \text{O}_2 \longrightarrow \text{H}_2\text{O}_2 + \text{PdI}_2 + \text{Pd(0)},$$

the rate of which corresponded to the rate of steps described by Eq. (6.255).

The oxidation of the aforementioned hydride complex and Pd(0) complex (probably, PdI_2^{2-}) can be performed by iodine and/or hydrogen peroxide, e.g., as

$$\text{(G10)} \quad \text{HPdI} + \text{I}_2 \longrightarrow \text{PdI}_2 + \text{HI},$$

$$\text{(G11)} \quad \text{Pd(0)} + \text{I}_2 \longrightarrow \text{PdI}_2$$

while the formation of I_2 can be represented by a block of steps as

$$\text{(G12)} \quad 2\text{HI} + \text{H}_2\text{O}_2 \longrightarrow \text{I}_2 + 2\text{H}_2\text{O}$$

with the limiting first step rate

$$W_{12} = k_{12}[\text{HI}][\text{H}_2\text{O}_2]. \tag{6.257}$$

Now, let us write the complete mechanism consisting of steps (G1–G3), (G10, G11), and blocks of steps (G9) and (G12) and include the concentrations of substances that weakly vary in the course of reactions into the rate constants (kinetic model I):

$$
\begin{array}{llll}
(1) & \text{X}_1 \xrightarrow{k_1} \text{X}_2 + \text{X}_3 & (W_1 = k_1[\text{X}_1]); & (\text{G1}) \\
(2) & \text{X}_1 + \text{X}_2 \xrightarrow{k_2} \text{X}_4 + \text{X}_3 & (W_2 = k_2[\text{X}_1][\text{X}_2)]; & (\text{G2}) \\
(3) & \text{X}_4 + \text{A} \xrightarrow{k_3} 2\text{X}_2 & (W_3 = k_3[\text{X}_4][\text{A}]); & (\text{G3}) \\
(4) & 2\text{X}_2 \xrightarrow{k_4} \text{X}_1 + \text{X}_5 + \text{X}_6 & (W_4 = k_4[\text{X}_2]^2); & (\text{G9}) \\
(5) & \text{X}_2 + \text{X}_7 \xrightarrow{k_5} \text{X}_1 + \text{X}_3 & (W_5 = k_5[\text{X}_2][\text{X}_7]); & (\text{G10}) \\
(6) & \text{X}_5 + \text{X}_7 \xrightarrow{k_6} \text{X}_1 & (W_6 = k_6[\text{X}_5][\text{X}_7]); & (\text{G11}) \\
(7) & 2\text{X}_3 + \text{X}_6 \xrightarrow{k_7} \text{X}_7 & (W_7 = k_7[\text{X}_3][\text{X}_6]), & (\text{G12})
\end{array}
$$

where $X_1 = PdI_2$; $X_2 = HPdI$; $X_3 = HI$; $X_4 = Pd_2I_2$; $X_5 = Pd(0)$; $X_6 = H_2O_2$; $X_7 = I_2$; and A = PA (the PA concentration in some variants of the models was set constant and then the temporal variation of [A] was ignored).

The constants for kinetic model I were determined in a series of trials, which were aimed at finding a region of self-oscillations and obtaining the shapes of oscillations E_{Pt} and pH close to experimental patterns. The lower limit for k_3 was evaluated from the experimental rate of CO absorption. The interrelations of rate constants for fast reactions, including autocatalysis (steps (2) and (3)) and the origination (step (1)) and loss (steps (4) and (5)) of autocatalyst HPdI, were estimated using an analogy of these steps to those in the Oregonator model [24].

The system of differential equations for kinetic model I was numerically integrated using the STEP program package [198] for the following set of rate constants:

$$
\begin{aligned}
&k_1 = 10^{-3}\text{s}^{-1},\ k_2 = 680\,\text{M}^{-1}\cdot\text{s}^{-1},\ k_3 = 10^2\,\text{M}^{-1}\cdot\text{s}^{-1},\ k_4 = 600\,\text{s}^{-1},\\
&k_5 = 2\cdot 10^4\,\text{M}^{-1}\cdot\text{s}^{-1},\ k_6 = 0.6\,\text{M}^{-1}\cdot\text{s}^{-1},\ k_7 = 0.15\,\text{M}^{-1}\cdot\text{s}^{-1},
\end{aligned}
$$

and initial conditions (in mole/liter):

$$
\begin{aligned}
&[\text{A}]_0 = 1.0; \quad [\text{X}_1]_0 = 0.03; \quad [\text{X}_3]_0 = 0.25\cdot 10^{-3},\\
&[\text{X}_7]_0 = 10^{-3}; \quad [\text{X}_2]_0 = [\text{X}_4]_0 = [\text{X}_5]_0 = [\text{X}_6]_0 = 10^{-8}.
\end{aligned}
$$

The results of numerical calculations for a closed system show the presence of non-decaying oscillations (Figure 6.48) and relaxation oscillations with the shapes close to those observed in experiment (Figures 6.49 and 6.50).

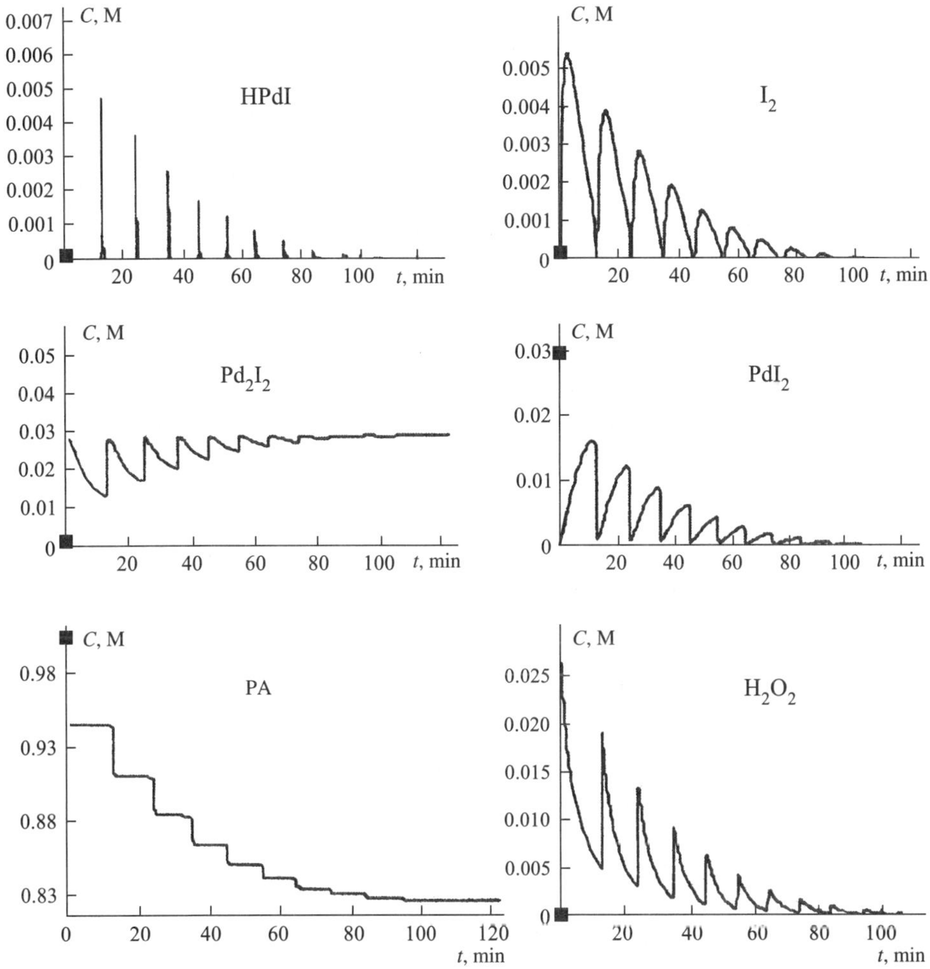

Figure 6.48 Temporal variation of concentrations of reactants and intermediates calculated according to kinetic model I (reproduced with permission from editorial board of *Izv. Akad. Nauk, Ser. Khim.* journal © 2003 [20])

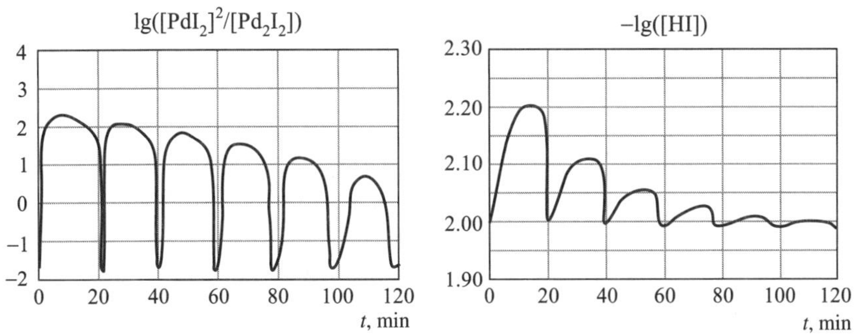

Figure 6.49 Results of mathematical simulation of oxidative carbonylation of phenylacetylene (PA) based on a solution of the system of differential according to kinetic model I [193]

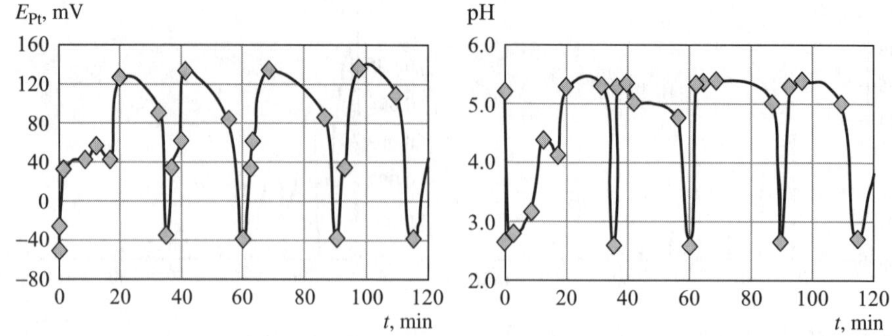

Figure 6.50 Oscillations of platinum electrode potential E_{Pt} and pH during oxidative carbonylation of phenylacetylene (PA) according to kinetic model I at $[KI]_0 = 0.4$ M, $[PdI_2]_0 = 0.005$ M, $[PA]_0 = 0.1$ M, and $[CO]_0$: $[O_2]_0 = 3: 2$ (reproduced with permission from editorial board of *Kinet. Katal.* journal © 2001 [194])

Let us also consider kinetic model II, which includes essentially the same steps (1)–(7) but with a different set of constants and without allowance for the participation of H_2O_2 in the process. It can be suggested that hydrogen peroxide exhibits rapid catalytic decomposition in the presence of I^- ions,

$$2H_2O_2 \xrightarrow{I^-} 2H_2O + O_2, \tag{6.258}$$

and does not participate in reaction (G12) of HI oxidation to I_2. The formation of I_2 in kinetic model 2 was described by the following reaction:

$$(G13) \quad 2HI + O_2 \longrightarrow I_2 + (H_2O_2).$$

Accordingly, kinetic model II is written as follows:

$$(1) \quad X_1 \xrightarrow{k_1} X_2 + X_3;$$

$$(2) \quad X_1 + X_2 \xrightarrow{k_2} X_4 + X_3;$$

$$(3) \quad X_4 + A \xrightarrow{k_3} 2X_2;$$

$$(4) \quad 2X_2 \xrightarrow{k_4} X_1 + X_5 + (H_2O_2);$$

$$(5) \quad X_2 + X_7 \xrightarrow{k_5} X_1 + X_3;$$

$$(6) \quad X_5 + X_7 \xrightarrow{k_6} X_1;$$

$$(7) \quad 2X_3 \xrightarrow{k_7} X_7 + (H_2O_2)(W_7 = k_7[X_3]).$$

where notation is the same as in kinetic model I. The numerical calculations for kinetic model II were performed for the following rate constants:

$$k_1 = 5 \cdot 10^{-4} s^{-1},\ k_2 = 34\,M^{-1} \cdot s^{-1},\ k_3 = 50\,M^{-1} \cdot s^{-1},\ k_4 = 3 \cdot 10^3\,M^{-1} \cdot s^{-1},$$
$$k_5 = 2.4 \cdot 10^6\,M^{-1} \cdot s^{-1},\ k_6 = 30\,M^{-1} \cdot s^{-1},\ k_7 = 0.02\,s^{-1}$$

and the following initial conditions (in mole/liter):

$$[A]_0 = 1.0;\ [X_1]_0 = 0.03;\ [X_3]_0 = 10^{-3};$$
$$[X_7]_0 = 10^{-8};\ [X_2]_0 = [X_4]_0 = [X_5]_0 = 10^{-8}.$$

The results of numerical calculations show the presence of non-decaying oscillations over a period of 80 min with a stepwise variation of [PA] (Figure 6.51). As can be seen from Figures 6.51e and 6.52 (constructed using the data from Figure 6.51), palladium in this system predominantly occurs in the oxidized state.

Thus, within the framework of both kinetic models, the concentrations of all intermediates exhibit relaxation oscillations with the shapes close to those observed in experiment (Figure 6.50). These results present only the first steps in an analysis of kinetic models I and II and other hypothetical mechanisms. Since both models include two constant balance relations ($[Pd]_0 = const$, $[I]_0 = const$), the basis set of kinetic model I consists of five differential equations, while that of model II includes four differential equations. The concentration of Pd(0) in both models, as well as $[H_2O_2]$ in kinetic model I can be quasi-stationary. Future tasks are to analyze the steady-state solutions of other systems of differential equations and the corresponding parametric portraits in a broad range of rate constants and initial values of variables (A, X_1, and X_3).

The possibility of generating relatively simple mechanisms (consisting of 7-8 steps) based on conditionally maximal reaction network (32 reactions, including blocks of steps and elementary steps) was considered in [20]. On this basis, about 350 kinetic models were proposed [193], the discrimination of which presents a complicated specific problem.

Investigations of the reactions of diester synthesis from alkynes RC ≡CH (where R=Ph, Me, $HOCH_2$, $HOC(Me)_2$) in various alcohols and citraconic anhydride synthesis from methylacetylene (MA) showed that the oscillatory behavior is a characteristic feature of these reactions [20, 199]. Self-oscillations were also observed for oxidative carbonylation reactions in the $PdBr_2$–KBr–MeOH system [199] with various ratios of CO and O_2 [20]. The oscillation shapes, periods, and amplitudes strongly depend on the solvent (cf. Figures 6.46, 6.50, 6.53 and 6.54).

In all cases, the reaction of oxidative carbonylation proceeds initially with the participation of palladium complexes in a reduced state, but for a very short period of time. Most of the time, the system occurs in an oxidized state (most probably, Pd(II)]. The transition takes place on attaining a certain critical concentration of HPdI, which is necessary for the formation of Pd_2I_2 catalyst via the reduction of PdI_2 and HI. This is followed by an increase in the concentration of oxidants (H_2O_2 and I_2 for kinetic model 1 and I_2 for model 2) and a growth in E_{Pt} and pH, after which HPdI species disappear and the catalyst passes into an oxidized state. The reduction process corresponds to the following overall reaction:

$$\text{(A)}\quad 2PdI_2 + [CO, RC \equiv CH, H_2O, MeOH] \longrightarrow 2HPdI + 2HI + CO_2 + P.$$

The oxidation process can be described by two reactions:

$$\text{(B)}\quad 5HPdI + 2O_2 + H_2O_2 + 5HI \longrightarrow 5PdI_2 + 6H_2O$$

(for the accumulation of PdI_2) or

$$\text{(C)}\quad 5HPdI + 2O_2 + I_2 + 3HI \longrightarrow 5PdI_2 + 4H_2O$$

(for the accumulation of I_2).

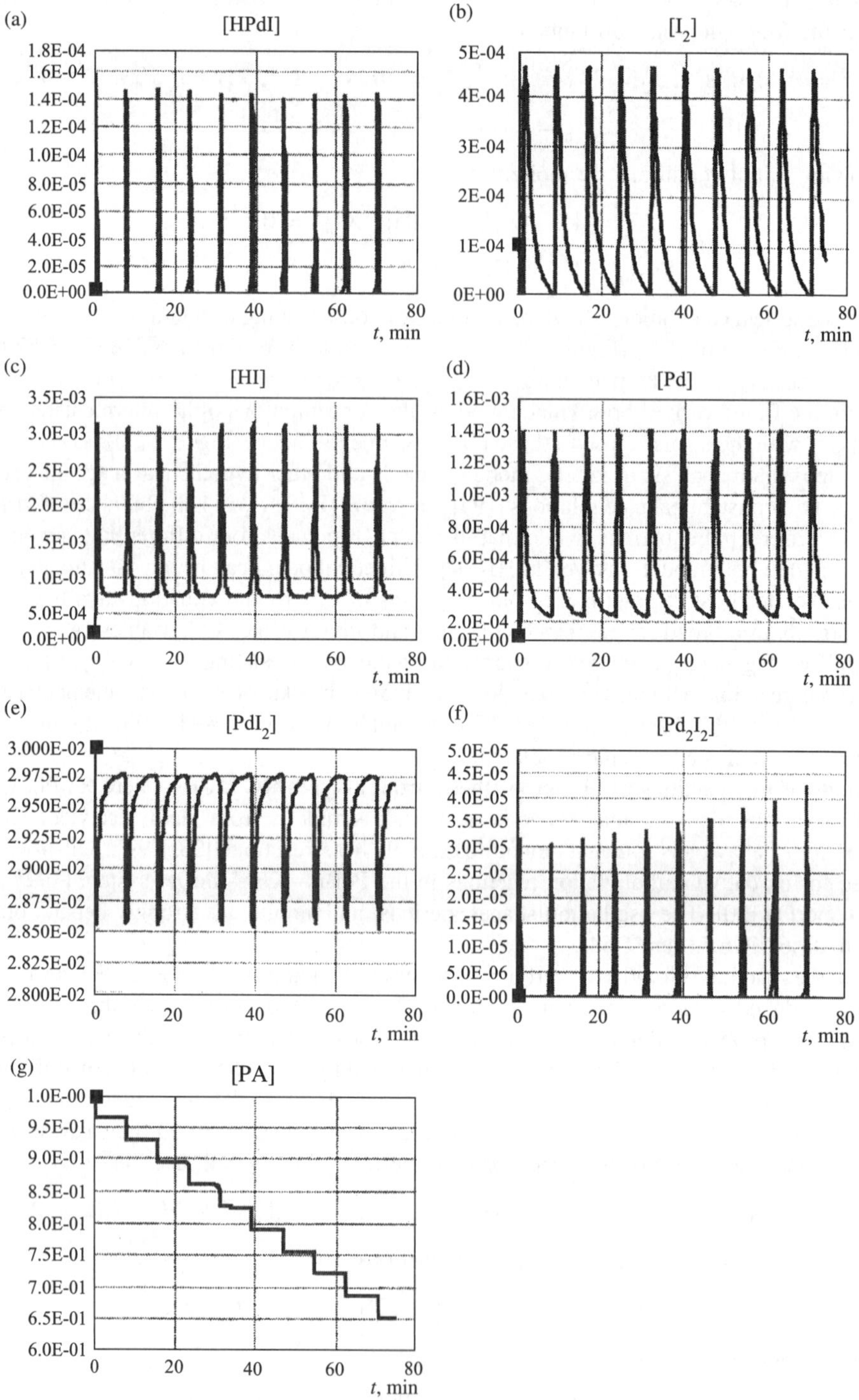

Figure 6.51 Temporal variation of concentrations of reactants and intermediates calculated according to kinetic model II [199]

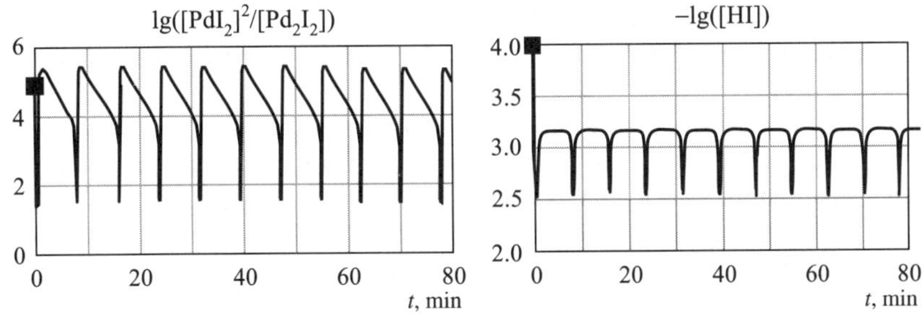

Figure 6.52 Results of mathematical simulation of oxidative carbonylation of phenylacetylene (PA) based on a solution of the system of differential according to kinetic model 2 (reproduced with permission from editorial board of *Kinet. Katal.* journal © 2001 [194])

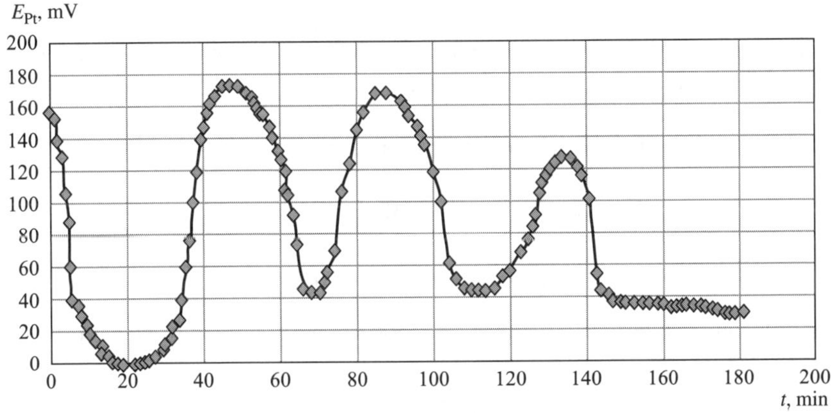

Figure 6.53 Variation of platinum electrode potential E_{Pt} during oxidative carbonylation of methylacetylene (MA) at $[LiBr]_0 = 0.2$ M, $[PdBr_2]_0 = 0.0109$ M, $V_{(CH_3)_2CO} = 10$ ml, $[MA]:[CO]:[O_2] = 5:3:2$, $[H_2O]_0 = 0.73$ M [199]

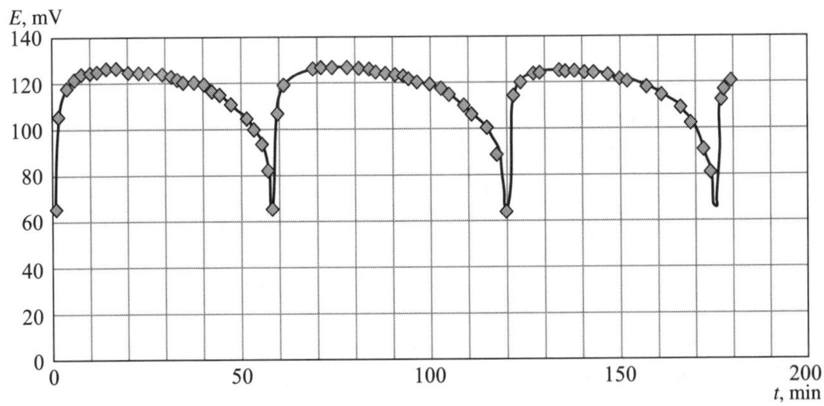

Figure 6.54 Variation of platinum electrode potential E_{Pt} during oxidative carbonylation of dimethylethinylcarbinol (DMEC) at $[KI]_0 = 0.4$ M, $[PdI_2]_0 = 0.0033$ M, $[DMEC]_0 = 0.1$ M, $[CO]_0$: $[O_2]_0 = 3:2$ [199]

It was noted that, for the appearance of self-oscillations, it is necessary in some cases to introduce NaOAc or Et_3N in order to increase pH of the system. Apparently, the process of autocatalyst origination can be inhibited by an acid.

Investigations of the kinetics of oxidative carbonylation of acetylene in a gas-flow reactor revealed the phenomenon of bistability with the transition from one P_{O_2} region, where the rate of succinic anhydride formation is greater than that of maleic anhydride formation ($r_{SA} > r_{MA}$), to another P_{O_2} region, in which $r_{MA} \gg r_{SA}$, upon a very small increase in P_{O_2} (Figure 6.55) [29]. These experiments demonstrated that a control over the steady states in systems with multiple steady states opens ways to the control of selectivity. For example, the selectivity of the maleic anhydride synthesis increases from 33 to 93 % when the reaction trajectory passes through a critical point [29]. The transition to an oscillatory regime in the PdI_2–KI system significantly increases the selectivity of phenylacetylene carbonylation with respect to the yield of oxidative carbonylation products [203]. Based on the experience gained in constructing models of the oxidative carbonylation of alkynes, the analysis of the hysteresis phenomena in heterogeneous reactions involving so-called chemical triggers (see Section 6.3) [14], and the investigations of mechanisms of the synthesis of succinic and maleic anhydrides (in particular, H–D exchange processes and kinetic isotope effects) [195], it is possible to formulate a reasonable hypothesis that would provide a basis for understanding the phenomena of bistability and self-oscillatory regimes.

Reactions of the formation of Pd_2Br_2 species (binuclear carbonyl bromide complexes, see Section 5.3.7) in acetone can be modeled as analogous to mechanism I in the synthesis of diesters:

$$\text{(T1)}\quad PdBr_2 \xrightarrow{CO,\ H_2O} HPdBr + HBr + CO_2$$

$$\text{(T2)}\quad PdBr_2 + HPdBr \longrightarrow Pd_2Br_2 + HBr$$

The synthesis of maleic anhydrides (MA) can be represented by a block of steps for the formation of a dimeric hydride complex:

$$\text{(T3)}\quad Pd_2Br_2 \xrightarrow{2CO,\ C_2H_2, H_2O} (HPdBr)_2(MA),$$

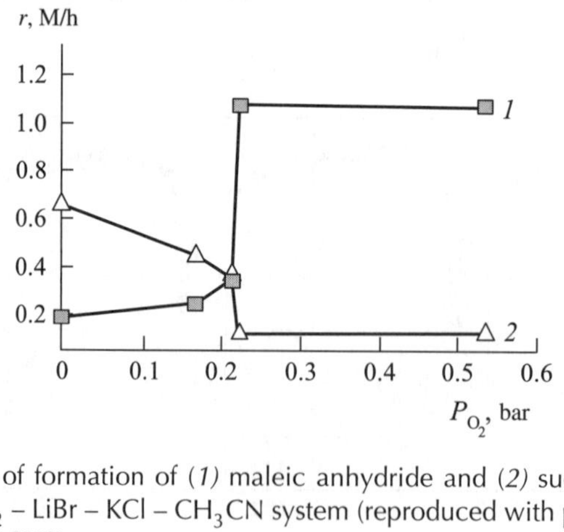

Figure 6.55 Plots of the rate of formation of (*1*) maleic anhydride and (*2*) succinic anhydride versus partial pressure of oxygen in the $PdBr_2$ – LiBr – KCl – CH_3CN system (reproduced with permission from editorial board of *Kinet. Katal.* journal © 1998 [29])

which is converted via three pathways as follows [195]:

$$\text{(T4)}\quad (HPdBr)_2(MA) \rightleftharpoons 2HPdBr + MA$$

$$\text{(T5)}\quad (HPdBr)_2(MA) \xrightarrow{O_2} Pd_2Br_2 + MA + (H_2O_2).$$

$$\text{(T6)}\quad (HPdBr)_2(MA) \longrightarrow Pd_2Br_2 + SA$$

The reactions of chain termination for the autocatalyst (HPdBr) and the catalyst of MA and SA formation (Pd_2Br_2) are as follows:

$$\text{(T7)}\quad 2HPdBr \xrightarrow{O_2} Pd_2Br_2 + (H_2O_2)$$

$$\text{(T8)}\quad Pd_2Br_2 + 2HBr \xrightarrow{O_2} 2PdBr_2 + (H_2O_2)$$

Under the condition of rapid decomposition of H_2O_2 (catalyzed by Br^- and, probably, Pd_2Br_2), the participation of hydrogen peroxide as an intermediate can be ignored to the first approximation. It is believed that a numerical analysis of model system (T1)–(T8), and kinetic models I and II in the synthesis of diesters can reveal various dynamic behavior of these systems.

In concluding Section 6.4 and all of Chapter 6, it is expedient to emphasize the importance of studying the dynamic behavior of kinetic models and critical phenomena in catalytic chemistry. A "nonlinear view" on the kinetics and mechanisms of catalytic reactions is important in many respects and opens the following possibilities:

(i) deeper insight into the mechanisms of complex catalytic processes and the interrelationship between structure of reaction mechanisms and their kinetic and dynamic behavior;
(ii) control over the selectivity of catalytic processes;
(iii) solving problems of the stability of reactors and developing methods for the control of reactor efficiency;
(iv) using critical phenomena for the discrimination of hypotheses concerning reaction mechanisms;
(v) developing dynamical nonlinear chemical models of processes in living cells and biological objects.

The author was greatly impressed by his experience of dealing with nonlinear systems and getting acquainted with theoretical investigations presented in this chapter, especially in the field of deterministic chaos. These impressions can well be expressed by ironical lines of V.G. Einstein, late professor of the M.V. Lomonosov Moscow Institute of Fine Chemical Technology (MITKhT):

The Scholar found the Truth, one can't endure.
He stares at his Formulae in Trance:
Laws rule the Universe de-jure,
De-facto – Randomness and Chance!

V. Einstein
The Open Date (Collection of Poetry)
MITKhT Publishing Center, Moscow, 2001
Translated from Russian by L. Davydova

References

1. D. A. Frank-Kamenetskii, *Diffuziya i teploperedacha v khimicheskoi kinetike* (Diffusion and Heat Transfer in Chemical Kinetics, 2nd. ed.), Nauka, Moscow, 1967.
2. R. Aris, *Introduction to the Analysis of Chemical Reactions*, Prentice-Hall, Englewood Cliffs, NJ, 1965.
3. B. V. Vol'ter and I. E. Sal'nikov, *Ustoichivost' rezhimov raboty khimicheskikh reaktorov* (Stability of Operation Regimes in Chemical Reactors), Khimiya, Moscow, 1972.
4. A. M. Zhabotinskii, in *Kolebatel'nye protsessy v biologicheskikh i khimicheskikh sistemakh* (Oscillatory Processes in Biological and Chemical Systems), G. M. Frank (Ed.), Nauka, Moscow, 1967, p. 252.
5. A. M. Zhabotinskii, *Kontsentratsionnye avtokolebaniya* (Concentration Self-Oscillations), Nauka, Moscow, 1974.
6. P. Glansdorff and I. Prigogine, *Thermodynamic Theory of Structure, Stability and Fluctuations*. Wiley-Interscience. 1970.
7. G. Nicolis and I. Prigogine, *Self-Organization in Non-Equilibrium Systems*, John Wiley & Sons, New York, 1997.
8. D. Kondepudi and I. Prigogine, *Modern Thermodynamics: From Heat Engines to Dissipative Structures*, John Wiley & Sons, Chichester, 1998.
9. R. J. Field and M. Burger (Eds), *Oscillations and Traveling Waves in Chemical Systems*, Wiley-Interscience, New York, 1985.
10. J. J. Tayson, *The Belousov–Zhabotinskii Reaction*, Springer-Verlag, Berlin, 1976.
11. G. R. Ivanitskii, V. I. Krinskii and E. E. Sel'kov, *Matematicheskaya biofizika kletki* (Mathematical Biophysics of Cells), Nauka, Moscow, 1978.
12. G. S. Yablonsky, V. I. Bykov and V. I. Elokhin, *Kinetika model'nykh reaktsii geterogennogo kataliza* (Kinetics of Model Reactions of Heterogeneous Synthesis), Nauka, Novosibirsk, 1984; G. S. Yablonsky and V. I. Elokhin, *Kinetic Models of Heterogeneous Catalysis*, in *Perspectives in Catalysis*, J. M. Thomas and K. I. Zamaraev (Eds), Blackwell Science Publishers, London, 1992, pp. 191–249.
13. V. I. Bykov, *Modelirovanie kriticheskikh yavlenii v khimicheskoi kinetike* (Modeling Critical Phenomena in Chemical Physics), Nauka, Moscow, 1988.
14. V. I. Bykov, *Modelirovanie kriticheskikh yavlenii v khimicheskoi kinetike* (Series "Sinergetika: ot proshlogo k budushchemu" [Modeling Critical Phenomena in Chemical Kinetics, 2nd ed. (Series "Synergetics: From Past to Furure"], KomKniga, Moscow, 2006.
15. M. G. Slin'ko and M. M. Slin'ko *Usp. Khim.*, **49**(4), 561 (1980).
16. M. M. Slin'ko and N. I. Jaeger *Oscillating Heterogeneous Catalytic Systems*, Elsevier, Amsterdam, 1994.
17. P. Gray and S. K. Scott, *Chemical Oscillations and Instabilities: Non-Linear Chemical Kinetics*, Oxford University Press, Oxford, 1994.
18. O. N. Temkin, A. V. Zeigarnik, R. E. Valdes-Perez and L. G. Bruk, *Kinet. Katal.*, **41** (3), 331 (2000).
19. P. Ellison, M. Feinberg, M. H. Yue and H. Saltsburg, *J. Mol. Catal., A: Chem.*, **154**, 169 (2000).
20. S. N. Gorodskii, E. S. Kalenova, L. G. Bruk and O. N. Temkin, *Izv. Akad. Nauk, Ser. Khim.*, No. 7, 1452 (2003).
21. M. G. Slin'ko, in *Iz istorii kataliza* (From the History of Catalysis), V. D. Kal'ner (Ed.), Kalvis, Moscow, 2005, pp. 103–170.
22. S. E. Shnol', *Vopr. Med. Khim.*, No. 4, 443 (1958).
23. C. Walter, *Enzyme Kinetics*, Ronald Press, New York, 1966.
24. R. M. Noyes, *J. Phys. Chem.*, **94** (11), 4404 (1990).
25. M. L. Kaliya, *Razrabotka mednykh i serebryanykh katalizatorov okisleniya izoamilovogo spirta i kineticheskoi modeli protsessa* (Development and Investigation of Copper and Silver Catalysts of Isoamyl Alcohol Oxidation and the Kinetic Model of this Process), Candidate Dissertation in Chemistry, MITKhT (Moscow Institute of Fine Chemical Technology), Moscow, 1982.
26. G. M. Shulyakovskii and O. N. Temkin, Tezisy dokladov 6 Vsesoyuznoi konferentsii "Kataliticheskie reaktsii v zhidkoi faze" (Abstracts of Papers of the 6th All-Union Conf. "Catalytic Reactions in the Gas Phase"), Alma-Ata, 1983, Vol. 2, p. 8.

27. G. M. Shulyakovskii, O. N. Temkin, N. V. Bykanova and A. N. Nyrkova, in *Khimicheskaya kinetika i kataliz: Kineticheskie modeli zhidkofaznykh reaktsii* (Chemical Kinetics in Catalysis: Kinetic Models of Liquid-Phase Reactions), S. L. Kiperman (Ed.), Chernogolovka, 1985, p. 112.
28. A. V. Malashkevich, L. G. Bruk and O. N. Temkin, *J. Phys. Chem. A.*, **101**, 9825 (1997).
29. L. G. Bruk, I. V. Oshanina, A. S. Zakieva and O. N. Temkin, *Kinet. Katal.*, **39**(2), 183 (1998).
30. B. V. Vol'ter, *Khimiya i Zhizn'*, No. 2, (1985); No. 5, (1986).
31. B. V. Vol'ter, *Znanie - Sila*, No. 12, (1994).
32. H. Landolt, *Ber.*, **19**, 1317 (1886); **20**, 745 (1887).
33. J. Eggert, *Helv. Chim. Acta*, **32** (III), 692 (1949).
34. J. Eggert and Th. Rohr, *Helv. Chim. Acta*, **36**, 855 (1953).
35. Gy. Rabai, Gy. Bazsa and M. T. Beck, *J. Am. Chem. Soc.*, **101** (22), 6746 (1979).
36. J. Liesegang, *Naturwiss. Wochschr.*, **11**, 353 (1896).
37. J. S. Morgan, *J. Chem. Soc., Trans.*, **109**, 274 (1916).
38. A. Lotka, *J. Phys. Chem.*, **14**, 271 (1910).
39. A. Lotka, *J. Am. Chem. Soc.*, **42**, 1595 (1920).
40. (a) W. C. Bray, *J. Am. Chem. Soc.*, **43**, 1262 (1921); (b) W. C. Bray and H. A. Liebhafsky, *J. Am. Chem. Soc.*, **53**, 38 (1931); (c) H. A. Liebhafsky, R. Furuichi and G. M. Roe, *J. Am. Chem. Soc.*, **103**, 51 (1981).
41. A. A. Andronov, *Compt. Rend.* Paris, **18** (15), 559 (1929).
42. D. I. Trubetskov, *Kolebaniya i volny dlya gumanitariev* (Oscillations and Waves for Students in Humanities), Izdat. GUNTs "Kolledzh", Saratov. Gos.Univ., Saratov, 1997.
43. A. A. Andronov, A. A. Vitt and S. E. Khaikin, *Teoriya kolebanii* (Theory of Oscillations, 3rd ed.), Nauka, Moscow, 1981.
44. F. M. Shemyakin, P. F. Mikhalev, *Fiziko-khimicheskie periodicheskie protsessy* (Physicochemical Periodic Processes), Izdat. AN SSSR, Moscow–Leningrad, 1983.
45. B. P. Belousov, in *Avtovolnovye protsessy v sistemakh s diffuziei (Autowave Processes in Systems with Diffusion)* Gor'kovsk. Gos. Univ, Gorki, 1951, p. 76 (Reproduced as additional chapter in Russian translation of monograph [9]: *Kolebaniya i begushchie volny v khimicheskikh sistemakh*, Mir, Moscow, 1988).
46. B. P. Belousov, in *Sbornik referatov po radiatsionnoi meditsine za 1958g* (1958 *Collection of Abstracts in Radiation Medicine*), Medgiz, Moscow, 1959, p. 145.
47. A. N. Zaikin and A. M. Zhabotinsky, *Nature*, **225**, 535 (1970).
48. A. M. Turing, *Phil. Trans. Roy. Soc. London Ser. B*, **237**, 37 (1952).
49. A. M. Zhabotinskii, *Zh. Fiz. Khim.*, **42**, 1649 (1968).
50. W. Ebeling, *Strukturbildung bei irreversiblen prozessen*, Teubner, Leipzig, 1976.
51. K. B. Yatsimirsii and L. P. Tikhonov, *Coord. Chem. Rev.*, **63**, 241 (1985).
52. T. S. Briggs and W. C. Rausher *J. Chem. Educ.*, **50**, 496 (1973).
53. R. A. Schmitz, K. R. Graziani and J. L. Hudson, *J. Chem. Phys.*, **67**, 3040 (1977).
54. P. Hugo, *Ber. Bunsenges., Phys. Chem.*, **74**, 121 (1970).
55. P. Hugo and M. Jakubith, *Chem.-Ing.-Techn.*, **44**, 383 (1972).
56. H. Beusch, P. Fieguth and E. Wicke, *Adv. Chem. Ser.*, **109**, 615 (1972).
57. V. D. Belyaev, M. M. Slin'ko, V. I. Timoshenko and M. G. Slin'ko, *Kinet. Katal.*, **14** (3), 810 (1973).
58. M. M. Slin'ko, *Kinet. Katal.*, **36** (6), 930 (1995).
59. J. Ogata and T. Morimoto, *J. Chem. Soc. B*, 74 (1969).
60. J. H. Jensen, *J. Am. Chem. Soc.*, **105**, 2639 (1983).
61. M. G. Roelofs, E. Wasserman and J. H. Jensen, *J. Am. Chem. Soc.*, **109**, 4207 (1987).
62. A. B. Rovinskii and A. M. Zhabotinskii, *Teor. Eksp. Khim.*, **14**, 183 (1978); **15**, 25, 343 (1979).
63. J. Maselko, *Chem., Phys. Lett.*, **73**, 194 (1980).
64. P. Sevcik and L. U. Adamcikova, *Collect. Czech. Chem. Comm.*, **47**, 891 (1982).
65. Q. Ouyang, W. Y. Tam, P. De Kepper et al., *J., Phys. Chem.*, **91**, 2181 (1987).
66. H.-P. Lin and J.-J. Jwo, *J. Phys. Chem.*, **99**, 6897 (1995).
67. L. Kuhnert and K. W. Pehl, *Chem., Phys. Lett.*, **84**(1), 155 (1981).
68. L. Kuhnert and K. W. Pehl, *Chem., Phys. Lett.*, **84**(1), 159 (1981).

69. M. Handlirova and A. Tockstein, *Collect. Czech. Chem. Comm.*, **45**(10), 2621 (1980).
70. K. B. Yatsimirskii, L. P. Tikhonova, L. N. Zakrevskaya and Ya. D. Lampeka, *Dokl. Akad. Nauk SSSR*, **261**(3), 647 (1981).
71. L. P. Tikhonova and V. Ya. Zayats *Teor. Eksp. Khim.*, **16**(4), 548 (1980).
72. F. Bolletta and V. Balzani, *J. Am. Chem. Soc.*, **104**(15), 4250 (1982).
73. R. P. Rastogi, K. Varma, R. D. Yadava and K. S. Singh, *J. Chem. Soc., Faraday Trans. I*, **81**, 751 (1985).
74. N. Ganapathisubramanian and R. M. Noyes, *J. Phys. Chem.*, **85**, 1103 (1981).
75. M. Orban, *J. Am. Chem. Soc.*, **108**, 6893 (1986).
76. M. Orban and I. R. Epstein, *J. Am. Chem. Soc.*, **109** (1), 101 (1987).
77. R. M. Noyes and S. D. Furrow, *J. Am. Chem. Soc.*, **104** (1), 45 (1982).
78. M. Orban and I. R. Epstein, *J. Am. Chem. Soc.*, **111**, 2891 (1989).
79. Q. Quyang and P. De Kepper, *J. Phys. Chem.*, **91**, 6040 (1987).
80. V. M. Trainin and A. S. Zaitsev, *Kinet. Katal.*, **30** (3), 713 (1989).
81. J. Young, B. Franzus and T. T. S. Huang *Int. J. Chem. Kinet.*, **14**, 749 (1982).
82. M. G. Slin'ko and G. S. Yablonsky, in *Nestatsionarnye i neravnovesnye protsessy v geterogennom katalize*, Series: *Problemy kinetiki i kataliza*. (Nonstationary and Nonequilibrium Processes in Heterogeneous Catalysis, Series: Problems in Kinetics and Catalysis), Nauka, Moscow, 1978, Vol. 17, p. 154.
83. Yu. I. Pyatnitskii, N. V. Pavlenko and N. I. Il'chenko, *Vvedenie v nelineinuyu kinetiku geterogennogo kataliza* (Introduction to Nonlinear Kinetics of Heterogeneous Catalysis), Natl. Acad. Sci. Ukraine, Kiev, 1998.
84. I. I. Ioffe, L. M. Pis'men, *Inzhenernaya khimiya geterogennogo kataliza* (Chemical Engineering of Heterogeneous Catalysis), Khimiya, Leningrad, 1972.
85. 85. T. I. Khomenko *Issledovanie mekhanizma avtokataliza v sinteze uglevodov iz formal'degida* (Investigation of Mechanism of Autocatalysis in Synthesis of Carbohydrates from Formaldehyde), Candidate Dissertation in Chemistry, Institute of Chemical Physics, USSR Acad. Sci., Moscow, 1977; T. I. Khomenko, M. M. Sakharov, and O. A. Golovina, *Usp. Khim.*, **49** (6), 1079 (1980).
86. A. N. Simonov, O. P. Pestunova, L. G. Matvienko and V. N. Parmon, *Kinet. Katal.*, **48**(2), 261 (2007).
87. H. B. Kagan and T. O. Luukas, in *Comprehensive Asymmetric Catalysis*, E. N. Jacobsen, A. Pfaltz and H. Yamamoto (Eds.), Springer-Verlag, Berlin, 1999, Ch. 4, p. 102.
88. D. Gurel and O. Gurel, *Oscillating Chemical Reactions*, Springer-Verlag, Berlin–Heidelberg, 1983.
89. M. Eigen and P. Schuster, *Hypercycle: A Principle of Natural Self-Organization*, Springer Berlin–Heidelberg, 1979.
90. N. M. Bazhin, V. A. Ivanchenko and V. N. Parmon, *Termodinamika dlya khimikov* (Thermodynamics for Chemists), Khimiya-Koloss, Moscow, 2004; V. Parmon, *Thermodynamics of Non-Equilibrium Processes for Chemists with Particular Application to Catalysis*, Elsevier Sci., New York, 2009.
91. E. M. Kol'tsova, Yu. D. Tret'yakov, L. S. Gordeev and A. A. Vertegel, *Nelineinaya dinamika i termodinamika neobratimykh protsessov v khimii i khimicheskoi tekhnologii* (Nonlinear Dynamics and Thermodynamics of Irreversible Processes in Chemistry and Chemical Technology), Khimiya, Moscow, 2001.
92. M. I. Temkin, *Kinet. Katal.*, **25**(2), 299 (1984).
93. P. De Kepper and Zh. Buassonad, in *Oscillations and Traveling Waves in Chemical Systems*, R. J. Field and M. Burger (Eds), Wiley-Interscience Publ., New York, 1985, p. 249.
94. (a) F. F. Seelig, *Proc. Ann., New York Acad. Sci.*, **316**(2), 339 (1979). N. N. Kol'tsov, V. Kh. Fedotov and B. V. Alekseev, *Zh. Fiz. Khim.*, **63**(7), 1966 (1989).
95. 95. (a) G. B. Sergeev, I. A. Leenson and L. A. Tyurina, *Dokl. Akad. Nauk SSSR*, **240**(6), 1390 (1978); *Zh. Fiz. Khim.*, **55**, 2209 (1980); (b) G. B. Sergeev and T. N. Rostovshchikova, *Usp. Khim.*, **52**, 455 (1983).
96. J. D. Murray, *Nonlinear Differential Equation Model in Biology*, Clarendon Press, Oxford. 1977.
97. G. Iooss and D. D. Joseph, *Elementary Stability and Bifurcation Theory*, Springer Verlag, New York, 1980.
98. N. N. Bautin and E. A. Leontovich, *Metody i priemy kachestvennogo issledovaniya dinamicheskikh sistem na ploskosti* (Methods and Techniques of Qualitative Investigation of Dynamical Systems on Plane), Nauka, Moscow, 1990.
99. L. P. Shil'nikov, A. L. Shil'nikov, D. V. Turaev and L. Chua, *Metody kachestvennoi teorii v nelineinoi dinamike* (Methods of Qualitative Theory in Nonlinear Dynamics), IKS, Moskva–Izhevsk, 2004, Part. 1.

100. V. N. Arnol'd, *Obyknovennye differentsial'nye uravneniya* (Ordinary Differential Equations), Nauka, Moscow, 1971.
101. M. Holodniok, F. Klic, M. Kubicek and M. Marek, *Methods of Analysis of Nonlinear Dynamical Models*, Academia, Prague, 1986.
102. Yu. A. Danilov, *Lektsii po nelineinoi dinamike* (Lectures on Nonlinear Dynamics), Postmarket, Moscow, 2001.
103. B. L. Clarke, *J. Chem, Phys.*, **60** (4), 1481 (1974).
104. B. L. Clarke, *J. Chem, Phys.*, **60** (4), 1493 (1974).
105. B. L. Clarke, *J. Chem. Phys.*, **62** (2), 773 (1975).
106. B. L. Clarke, *J. Chem. Phys.*, **62** (9), 3726 (1975).
107. B. L. Clarke, *J. Chem. Phys.*, **64** (10), 4165 (1976).
108. B. L. Clarke, *Adv. in Chem. Phys.*, **43**, 7 (1980).
109. B. L. Clarke, *J. Chem. Phys.*, **75**, 4970 (1981).
110. B. L. Clarke and C. Jefferies, *J. Chem. Phys.*, **82**, 3107 (1985).
111. A. N. Ivanova, *Kinet. Katal.*, **20**(4), 1019 (1979).
112. A. N. Ivanova, *Kinet. Katal.*, **20**(4), 1024 (1979).
113. A. N. Ivanova and B. L. Tarnopol'skii, *Kinet. Katal.*, **20**(6), 1541 (1979).
114. A. N. Ivanova, G. A. Furman, V. I. Bykov and G. S. Yablonsky, *Dokl. Akad. Nauk SSSR*, **242**(4), 872 (1978).
115. A. N. Ivanova and G. A. Furman *Khim. Fiz.*, No. 1, 931984.
116. G. S. Yablonsky, V. N. Bykov and A. N. Gorban', Kin*eticheskie modeli kataliticheskikh reaktsii* (Kinetic Models of Catalytic Reactions), Nauka, Novosibirsk, 1983; G. S. Yablonskii, V. I. Bykov, A. N. Gorban' and V. I. Elokhin, in *Comprehensive Chemical, Kinetics*, Elsevier, Amsterdam–Oxford, 1991, Vol. 32.
117. B. V. Alekseev, V. Kh. Fedotov and N. I. Kol'tsov *Dokl. Akad. Nauk SSSR*, **306**(4), 884 (1989).
118. N. I. Kol'tsov, B. V. Alekseev and V. Kh. Fedotov *Dokl. Akad. Nauk*, **337**(6), 761 (1994).
119. F. Horn and R. Jackson, *Arch. Rat. Mech. Anal.*, **47**, 81 (1972).
120. M. Feinberg, *Arch. Rat. Mech. Anal.*, **49**, 187 (1972).
121. M. Feinberg and F. Horn, *Chem. Eng. Sci.*, **29** (3), 775 (1974).
122. M. Feinberg, in *Dynamic and Modeling of Reactive Systems*, W. Stewart, W. H. Ray and C. Conley (Eds), Academic Press, 1980, pp. 59–130.
123. M. Feinberg, *Chem. Eng. Sci.*, **72** (10), 2229 (1987).
124. M. Feinberg, in *Patterns and Dynamics in Reactive Media*, R. Aris et al. (Eds), Series: *IMA Volumes in Mathematics and Its Applications*, Berlin,Springer, 1991, Vol. 37, pp. 43–70.
125. P. M. Schlosser and M. Feinberg *Chem. Eng. Sci.*, **49**(11), 1749 (1994).
126. L. F. Razon and R. A. Schmitz, *Chem. Eng. Sci.*, **42**(5), 1005 (1987).
127. E. E Sel'kov, *Issledovanie uslovii vozniknoveniya periodicheskikh kolebanii v sistemakh s obratnoi svyaz'yu, in: Kolebatel'nye protsessy v biologicheskikh i khimicheskikh sistemakh* (Oscillatory Processes in Biological and Chemical Systems), G. M. Frank (Ed.), Nauka, Moscow, 1967, p. 81.
128. J. A. Christiansen, *Adv. Enzymol.*, **23**, 83 (1961).
129. V. I. Bykov and G. S. Yablonsky, *Kinet. Katal.*, **18**, 1305 (1977).
130. V. I. Bykov, G. S. Yablonsky and V. I. Elokhin, *Kinet. Katal.*, **20**(4), 1029 (1979).
131. V. I. Bykov, G. S. Yablonsky and V. I. Elokhin, *Kinet. Katal.*, **20**(4), 1033 (1979).
132. G. S. Yablonsky and M. Z. Lazman, *Teor. Eksp. Khim.*, **32**(3), 153 (1996).
133. V. I. Bykov, G. S. Yablonsky and V. F. Kim, *Dokl. Akad. Nauk SSSR*, **242**(3), 637 (1978).
134. A. I. Khibnik, V. I. Bykov and G. S. Yablonsky, *Zh. Fiz. Khim.*, **61**(5), 1388 (1987).
135. V. N. Parmon, *Dokl. Akad. Nauk*, **377**(4), 510 (2001).
136. V. N. Parmon, *Zh. Fiz. Khim.*, **76**(1), 142 (2002).
137. M. D. Korzukhin and A. M. Zhabotinskii, in *Molekulyarnaya biofizika* (Molecular Biophysics), Nauka, Moscow, 1965, p. 52.
138. P. Gray and S. K. Scott, *J. Phys. Chem.*, **89**, 22 (1985).
139. F. B. Cook, P. Gray, D. G. Knaap and S. K. Scott, *J. Phys. Chem.*, **93**, 2749 (1989).
140. R. Lefever, *J. Chem. Phys.*, **49**, 4977 (1968).
141. E. E. Selkov, *Eur. J. Biochem.*, **4**, 79 (1968).

142. R. Lefever and G. Nicolis, *J. Theor. Biol*., **30**, 267 (1971).
143. J. J. Tyson, *J. Phys. Chem*., **58**(9), 3919 (1973).
144. J. J. Tyson and J. C. Light, *J. Phys. Chem*., **59**(8), 4164 (1973).
145. E. N. Lorenz, *J. Atmos. Sci*., **20**, 130 (1963).
146. C. T. Sparrow, *The Lorenz Equation, Bifurcations, Chaos and Strange Attractors*, Springer-Verlag, Berlin, 1982.
147. S. K. Scott, *Chemical Chaos (Int. Ser. Monographs on Chemistry)*, Oxford University Press, Oxford, 1993.
148. S. H. Strogatz, *Nonlinear Dynamics and Chaos*, Westview Press, 2001.
149. D. Kaplan and L. Glass, *Understanding Nonlinear Dynamics*, Springer, Berlin, 1997.
150. A. M. Zhabotinskii and M. D. Korzukhin, in *Kolebatel'nye protsessy v biologicheskikh i khimicheskikh sistemakh* (Oscillatory Processes in Biological and Chemical Systems), G. M. Frank (Ed.), Nauka, Moscow, 1967, p. 223.
151. M. D. Korzukhin, in *Kolebatel'nye protsessy v biologicheskikh i khimicheskikh sistemakh* (Oscillatory Processes in Biological and Chemical Systems), G. M. Frank (Ed.), Nauka, Moscow, 1967, pp. 231, 242.
152. P. De Kepper, A. Ressi and A. Pacault, *Comt. Rend. Acad. Sci*., Ser. *C*., **283**, 371 (1976).
153. M. Marek and E. Svobodova, *Biophys. Chem*., **3**, 263 (1975).
154. K. R. Graziani, J. L. Hudson and R. A. Schmitz, *Chem. Eng. J*., **12**, 9 (1976).
155. J. L. Hudson, M. Hart and D. Marinko, *J. Chem., Phys*., **71**, 1601 (1979).
156. J. L. Hudson, J. C. Mankin, J. McCullough and P. Lambda, in *Nonlinear Phenomena in Chemical Dynamics*, Springer-Verlag, Berlin, 1981, p. 44.
157. O. E. Rossler and K. Wegmann, *Nature*, **271**, 89 (1978).
158. L. F. Olsen and H. Degn, *Nature*, **267**, 177 (1977).
159. R. J. Field, E. Koros and R. M. Noyes, *J. Amer. Chem. Soc*., **94**, 8649 (1972).
160. R. J. Field and R. M. Noyes, *J. Chem. Phys*., **60**, 1877 (1974).
161. L. Gyorgyi, T. Turanyi and R. J. Field, *J. Chem. Phys*., **94**, 7162 (1990).
162. T. Turanyi, L. Gyorgyi and R. J. Field, *J. Chem. Phys*., **97**, 1931 (1993).
163. K. Showalter, R. M. Noyes and K. Bar-Eli, *J. Chem. Phys*., **69**, 2514 (1978).
164. R. D. Janz, D. J. Vanacek and R. J. Field, *J. Chem. Phys*., **73**, 3132 (1980).
165. J. C. Roux, R. H. Simoyi and H. L. Swinney, *Physica D*., **8**, 257 (1983).
166. F. Argoul, A. Arneodo, P. Richetti and J. C. Roux, *J. Chem. Phys*., **86**(6), 3325 (1987).
167. P. Richetti, J. C. Roux, F. Argoul and A. Arneodo, *J. Chem. Phys*., **86**(6), 3339 (1987).
168. F. Argoul, A. Arneodo and P. Richetti, *Physics Letters*, A, **120**, 269 (1987).
169. J. L. Hudson and O. E. Rossler, in *Modeling of Patterns in Space and Time*, W. Jager and J. Marray (Ed.), Springer, Berlin, 1984 (cited from [17]).
170. W. Geiseler and K. Bar-Eli, *J. Phys. Chem*., **85**, 908 (1981).
171. W. Geiseler, *J. Phys. Chem*., **86**, 4394 (1982).
172. K. Bar-Eli, *J. Phys. Chem*., **89**, 2855 (1985).
173. P. Ruoff and R. M. Noyes, *J. Phys. Chem*., **93**, 7394 (1989).
174. M. Menzinger, M. Boukalouch, P. De Kepper et al., *J. Phys. Chem*., **90**, 313 (1986).
175. M. Menzinger and P. Jankowski, *J. Phys. Chem*., **90**, 1217 (1986).
176. R.-S. Li and J. Li, *Chem. Phys. Letters*, 144 (1), 96 (1988).
177. P. Sevcik and I. Adamcikova, *Chem. Phys. Letters*, **146** (5), 419 (1988).
178. L. Lopez-Tomas and F. Sagues, *J. Phys. Chem*., **95**, 701 (1991).
179. V. K. Vanag and D. P. Melikhov, *J. Phys. Chem*., **99**, 17372 (1995).
180. V. K. Vanag, in *Rossiiskaya nauka: Vystoyat' i vozrodit'sya* (Russian Science: Withstand and Survive), Nauka, Moscow, 1997, p. 207.
181. S. Di Lorenzo and F. D'Alba, *J. Chem. Soc., Farad. Trans*., **81**, 421 (1985).
182. V. K. Vanag and M. V. Alfimov, *J. Phys. Chem*., **97**, 1878 (1993).
183. S. C. Muller, T. Plesser and B. Hess, *Science*, **230** (8), 661 (1985).
184. M. G. Roelofs, E. Wasserman, J. H. Jensen and A. E. Nader, *J. Am. Chem. Soc*., **105**, 6329 (1983).
185. V. M. Trainin and A. S. Zaitsev, *Kinet. Katal*., **31**(1), 51 (1990).
186. G. M. Shulyakovskii, A. N. Nyrkova, O. N. Temkin et al., *Kinet. Katal*., **23** (3), 618 (1982).

187. G. R. Emel'yanova, *Kataliticheskii sintez yantarnoi kisloty i ee efirov karbonilirovaniem atsetilena* (Catalytic Synthesis of Succinic Acid and its Esters via Acetylene Carbonylation), Candidate Dissertation in Chemistry, MITKhT (Moscow Institute of Fine Chemical Technology), Moscow, 1984.
188. O. N. Temkin, L. G. Bruk and G. M. Shulyakovskii, *Khim. Prom-st*, No. 5, 22 (1983).
189. O. N. Temkin, G. K. Shestakov and Yu. A. Treger, *Atsetilen: khimiya, mekhanizmy reaktsii, tekhnologiya* (Acetylene: Chemistry, Reaction Mechanisms, Technology), Khimiya, Moscow, 1991.
190. N. G. Mekhryakova, L. G. Bruk, O. L. Kaliya, O. N. Temkin and A. Yu. Prudnikov, *Kinet. Katal*., **20**, 629 (1979).
191. L. G. Bruk, *Metody vydvizheniya i diskriminatsii gipoteticheskikh mekhanizmov kataliticheskikh reaktsii kak osnova razrabotki tekhnologii karbonilirovaniya alkinov* (Methods of Generation and Discrimination of Hypothetical Mechanisms of Catalytic Reactions: Basis for Developing Alkyne Carbonylation Technology), Doctoral Dissertation in Chemistry, MITKnΓ (Moscow Institute of Fine Chemical Technology), Moscow, 1996.
192. O. N. Temkin and L. G. Bruk, *Kinet. Katal*., **44** (5), 661 (2003).
193. Gorodskii S. N. *Razrabotka kineticheskoi modeli kataliziruemogo komleksami palladiya avtokolebatel'nogo protsessa okislitel'nogo karbonilirovaniya alkinov* (Developing a Kinetic Model of Self-Oscillatory Process of Alkyne Oxidative Carbonylation Catalyzed by Palladium Complexes), Candidate Dissertation in Chemistry, MITKhT (Moscow Institute of Fine Chemical Technology), Moscow, 2001.
194. S. N. Gorodskii, A. N. Zakharov, A. V. Kulik, L. G. Bruk and O. N. Temkin, *Kinet. Katal*., **42**, 280 (2001).
195. L. G. Bruk, I. V. Oshanina, A. P. Kozlova, E. V. Vorontsov and O. N. Temkin, *J. Mol. Catal A: Chem*., **104**, 9 (1995).
196. M. M. Konnick, B. A. Gandhi, I. Guzei and S. S. Stahl, *Angew. Chem. Int. Ed*., **45**, 2904 (2006).
197. S. S. Stahl, *Angew. Chem. Int. Ed*., **43**, 3400 (2004).
198. S. I. Fadeev, I. A. Gainova and A. Yu. Berezin, *Primenenie paketa "STEP" dlya chislennogo issledovaniya sistem nelineinykh uravnenii avtonomnykh sistem obshchego vida* (Applying STEP Program Package to Numerical Investigation of Systems of Nonlinear Equations of Autonomous Dynamical Systems of General Type), Institute of Mathematics, Novosibirsk, 1995.
199. S. N. Gorodskii (2006), unpublished results [private communication].
200. K. Novakovic, C. Grosjean, S. K. Scott, A. Whiting, M. J. Willis and A. R. Wright, *Chem., Phys. Lett*., **435**, 142 (2007).
201. K. Novakovic, C. Grosjean, S. K. Scott, A. Whiting, M. J. Willis and A. R. Wright, *Phys. Chem. Chem., Phys*., **10**, 749 (2008).
202. C. Grosjean, K. Novakovic, S. K. Scott, A. Whiting, M. J. Willis and A. R. Wright, *J. Mol. Catalysis A: Chem*., **284**, 33 (2008).
203. K. Novakovic, A. Mukherjee, M. J. Willis, A. R. Wright and S. K. Scott, *Phys. Chem. Chem., Phys*., **11**, 9044 (2009).
204. S. R. De Groot and P. Mazur, *Non-Equilibrium Thermodynamics*, North-Holland Publ. Amsterdam, 1962.

7

Rational Strategy for Designing Kinetic Models and Studying Complex Reaction Mechanisms

Preceding chapters considered kinetic aspects of homogeneous catalysis with metal complexes in order to illustrate the main features of catalytic systems involving such complexes, which must be taken into consideration when designing kinetic models of catalytic reactions. These features include the multicomponent character of reaction media (and the related complexity of catalysts and reactants), polynuclear structure of complexes, multi-route (in particular, chain) mechanisms, linear and nonlinear steps (including critical phenomena), and kinetic polyfunctionality in the case of multicomponent catalytic systems.

In this chapter, both traditional and new approaches to the investigation of reaction mechanisms are analyzed from the standpoint of the role of the kinetic method. A rational strategy of using the formal kinetic approach in elucidating the mechanisms of numerous practically important but yet insufficiently studied complex reactions and designing their theoretically justified kinetic models [1, 2] is also put forward. Naturally, the choice of a rational strategy is important for all types of catalytic reactions and systems studied using the kinetic method.

By the end of the 1960s, a strategy (referred to below as *traditional*) had been developed based on the predominant use of a kinetic experiment as the source of information concerning the step schemes of reactions (schemes of reaction pathways). Indeed, the kinetic equations were used to "derive" the reaction schemes involving the symbols of intermediates. Some other methods of studying reaction mechanisms were also used, but mostly in order to verify the schemes that followed from formal kinetic equations.

Drawbacks of the traditional approach have long been known by specialists in computational mathematics and mathematical statistics engaged in this process, who showed that, for a number of reasons, the only actually reliable way to construct kinetic models and establish reaction mechanisms was from generated hypotheses via justified design and active implementation of kinetic experiments to a discrimination of these hypotheses [3–8]. Nevertheless, the traditional approach to constructing kinetic models was extensively used and claimed to be effective [9–12]. Let us consider these issues in more detail.

Homogeneous Catalysis with Metal Complexes: Kinetic Aspects and Mechanisms, First Edition. O. N. Temkin.

7.1 Stages in the development of chemical kinetics and methodological aspects of the strategy of studying complex reaction mechanisms

The role of chemical kinetics as a tool for the investigation of reaction mechanisms in the 20th century has been critically assessed, depending on the achievements of this approach. In the middle of that century, some excitement was related to a breakthrough in the fields of chain reactions [13], acid catalysis [14, 15], enzyme reaction kinetics [16, 17], and application of the transition state theory to heterogeneous processes, including those on nonuniform surfaces [18, 19]. It might seem that a key to chemical reaction mechanisms had been found. Chemical kinetics was referred to as the science of reaction mechanisms. Those hopes and approaches are well illustrated by the title of an article by S.L. Kiperman: "From Kinetic Models – to the Mechanism of Catalytic Reactions" [11]. However, the third quarter of the 20th century showed a clear tendency towards a drop in interest in kinetic investigations. What were the reasons for that trend?

According to the opinion of J.C. Bond, a well-known specialist in the mechanisms of heterogeneous catalytic reactions, the kinetic method has been less frequently used since chemists realized that, in practice, it is almost always possible to propose more than one mechanism consistent with the observed kinetics [20]. It also became clear that the same mechanism can lead, for various assumptions concerning the relative rates of steps, to manifold kinetic equations [18].

Subsequent achievements, including the appearance of new physical methods for the investigation of intermediate species in gas-phase reactions [21] and on the surface of solid catalysts [22], the development of IR spectroscopy and NMR techniques for *in situ* investigations, and the perfection of X-ray diffraction methods for analysis of the structure of complexes and organometallic compounds, gave grounds to hope that, once the intermediates are known, the reaction mechanism can also be unambiguously established. Naturally, the opposite occurred. An analysis of the results of investigations into a large number of compounds found on the surface of solid catalysts, and assumed to be intermediates, showed [23] that few of these compounds were checked for their correspondence to the kinetic criteria. As a possible criterion, it was suggested [24, 25] to use spectrokinetic methods to compare the rates of consumption of surface intermediate species measured *in situ* to the rates of formation of the reaction products. Should these rates be close, it would provide convincing evidence in favor of the proposed mechanism. However, it is expedient to recall the opinion (already cited in the Introduction) that no one statement about a reaction mechanism can be accepted unless it can be demonstrated that it agrees with the observed kinetics [20], that is, with all of the observed kinetic relationships.

During the same period of development of the kinetic approach, it became clear that the inverse problem of chemical kinetics (i.e., evaluation of the rate constants) and the task of statistical identification of a kinetic model by the results of kinetic experiments cannot be solved without certain knowledge of the mechanism, i.e., notions about the process steps, proposed structure of the kinetic mechanism, and preliminary estimations of the rate constants [1, 6–8].

Some doubts concerning possibilities of the formal kinetic approach were also related to the appearance of a large number of kinetic models for the same reaction carried out under quite close conditions. This situation was especially frequently encountered in the field of heterogeneous catalysis. For example, approximately ten kinetic equations were proposed for the synthesis of vinyl acetate from acetylene on a $Zn(OAc)_2/C$ catalyst [26, 27]. By the beginning of the 1990s, at least 17 kinetic equations and mechanisms existed for the synthesis of methanol from CO and H_2, most of which referred to the same catalyst [28].

The development of the theory of routes and the investigation of numerous multi-route reactions inspired chemists engaged in kinetics to recognize the incorrectness of the traditional approach (from experiment to mechanistic scheme) [1], which fails to reconstruct the scheme of a multi-route reaction directly from experimental data. It also became clear that the validity of any (hypothetical) mechanism cannot be proved experimentally. According to this concept, which had been formulated many years previously by specialists in the theory of experiment [29] and mathematical chemistry [30b] and in a more general form by philosophers [30a], the correct approach consists of rejecting hypotheses that are not working properly.

The approach to obtaining a positive result from investigations in the form of a set of working hypotheses was justified in 1890 by American geologist T.C. Chamberlin [31]. According to this concept, a researcher's efforts should be devoted to generating a set of possible hypotheses concerning the phenomenon under consideration, followed by their discrimination with a reduction to a residual set of working hypotheses. In formulating a set of hypotheses, the researcher should avoid giving his subjective preference to any one of them. Unfortunately, for various objective and subjective reasons Chamberlin's ideas were not developed until the last quarter of the 20th century. Since then, this approach has been successfully applied, based on the hypotheses of maximal mechanism and complete reaction networks [32–34], to radical chain processes and now finds increasing use (see below) in homogeneous catalysis with metal complexes and heterogeneous catalysis.

Deeper insight into all features of the kinetic method, important methodological aspects, and limitations of the traditional formal kinetic approach led to revision [1, 32, 33] of the traditional strategy of investigations in this field [11, 12] and also stimulated a growing interest into the investigation of complex reaction mechanisms using more modern approaches.

It would be expedient to mention several important objective factors that caused the revision of strategy and stimulated the activity of research in chemical kinetics by the end of the 20th century.

(i) For the preceding half a century, the development of the theory of reaction mechanisms favored the understanding of an important natural phenomenon, whereby there existed multiple mechanisms of the same chemical reaction [1, 27, 33, 35–37]. For example, a change in the degree of oxidation of a metal catalyst in the sequence Pd(0) → Pd(I) → Pd(II) leads to three principally different mechanisms of alkyne carbonylation in the same catalytic system [38–40]. In the Heck reaction of olefin arylation by aryl halides, there are coexisting routes involving PdL_2 and PdX_4^{2-} complexes as well as colloidal and metallic palladium [41]. In other words, each reaction system can be described by a certain set of elementary steps that form a reaction network, various blocks of which can work depending on the catalyst nature, conditions, substituents in the substrate, and degrees of metal oxidation in different active complexes. This concept of the mechanism of complex reactions was justified by Rafer-De Poorter [42] in application to the Fischer–Tropsch synthesis. In investigations of the mechanism of formation of five products in a conjugate process of alkyne hydroalkoxycarbonylation, a set of 1344 possible mechanisms was analyzed and those consistent with the kinetic experiments were found [1, 39].

(ii) The development of basic chemistry led to significant changes in our notions about the object of investigation in chemical kinetics. It is now evident that the mechanisms of most reactions (in particular, catalytic) represent highly complicated combinations of steps [1]. Even a "simple" gas-phase reaction such as hydrogen oxidation has a reaction network consisting of 30 elementary steps [34]. A reaction network proposed for the Fischer–Tropsch synthesis comprised 55 steps [42]. The homogeneous process of methane oxidation was analyzed using a network consisting of 92 steps [43]. A description of the observed kinetics of heterogeneous/homogeneous oxidative coupling of methane with the formation of C_2 hydrocarbons required the use of 200 homogeneous elementary steps [44], while the homogeneous oxidative coupling of methane with an allowance for the formation and conversion of C_3 hydrocarbons required 500 homogeneous elementary steps [45].

Highly complicated reaction networks were used for the description of mechanisms of the catalytic conversion of methane by carbon dioxide [46] and some other heterogeneous reactions. For example, an analysis of the process of methane conversion into synthesis gas involved 277 homogeneous and 129 heterogeneous steps [47a], while the process of oxidation of a mixture of C_3–C_6 paraffins was modeled using a kinetic scheme involving 150 intermediates and 3000 steps [47b]. The state of the problem of kinetic modeling of multistep processes of heterogeneous/homogeneous oxidation of C_1-C_4 alkanes was recently analyzed in [48]. Some examples of more complicated mechanisms and large reaction networks in the catalysis with metal complexes were also given in Chapter 3.

(iii) Computer facilities have been developed that are capable of solving large systems of differential equations and discriminating hypothetical schemes with preset combinations of rate constants [43] and special routines have been written [36] for generating hypotheses (see below).

(iv) A tendency appeared to combine the physicochemical and formal kinetic approaches in the study of reaction mechanisms. A concept was formulated, according to which a reaction mechanism should be considered as a unity of topological (structural) and chemical components [49] (see also Section 1.1.1). It is important to note that both these components are equal. In general, it is impossible first to unambiguously establish the structure of a mechanism (i.e., a scheme of steps [1]) based on a formal kinetic description and then fill it with a "chemical content". In particular cases of relatively simple, single-route linear mechanisms, this goal can be achieved as demonstrated in numerous examples given in Chapters 2 and 3. However, in general, it is difficult (or even impossible) to extract the information necessary for the correct identification of a mechanistic scheme from kinetic experiments unless the possible schemes are outlined and the corresponding tasks for the formal kinetic analysis and other methods of investigation are formulated.

The understanding of the importance of the topological component of the mechanism (i.e., interrelationship of the intermediates and reaction routes) was helped by the development of methods of graph theory in application to the chemical kinetics (see, e.g., [37], Section 1.4 and Chapter 4).

This concept of the reaction mechanism implies the need for using a strategy (mentioned above and referred to below as *rational*) in establishing mechanisms and designing kinetic models. Within the framework of this strategy, the formal kinetic description plays a different role as compared to that in the traditional approach and becomes one of the necessary tools for the discrimination of hypotheses.

(v) The traditional strategy assumed the existence of a linear mechanism that should be studied under steady-state or quasi-steady-state conditions, thus admitting the possibility of describing kinetic dependences in terms of fractional-rational equations. At the same time, the experience gained in both heterogeneous and homogeneous catalysis gave more and more reasons to use nonlinear mechanisms. Indeed, in homogeneous catalysis with metal complexes, most of the elementary steps are essentially nonlinear and involve ligands, hydrogen ions, and/or other species in addition to metal-containing intermediates (see Chapter 6). As was noted above, in some cases it is possible to obtain fractional-rational equations by using an artificial linearization of steps, but this approach does not offer a general solution to the problem of nonlinear steps and nonlinear kinetic models.

In the last quarter of the 20th century, numerous investigations were devoted to the phenomena of self-sustained oscillations in both heterogeneous and homogeneous catalysis (see Chapter 6) and the multiplicity of steady states. These so-called critical kinetic phenomena are in many cases indicative of the presence of various nonlinearities in real catalytic systems. Naturally, the traditional approach to designing kinetic models was absolutely inapplicable for the analysis of nonlinear mechanisms and nonlinear models.

It is interesting to note that, at the initial stage of investigations of the kinetics and mechanisms of oscillatory reactions, the generation and discrimination of hypothetical reaction mechanisms was considered ineffective because of a lack of data on the elementary steps of these processes (e.g., BZ reactions) [50]. It was believed that an expedient approach consisted of describing the experiment by a simple, strongly idealized model (set of differential equations), followed by adding variables (intermediates) and steps such that only terms corresponding to the law of mass action would be retained in the right-hand parts of equations. This approach was called the inversed method of quasi-steady-state concentrations [50]. Korzukhin [51] proved a theorem according to which any empirical system can be reduced to this form (see Section 6.2). Later, it was understood [52] that this approach is ineffective and provides little knowledge about a real mechanism, since there are numerous ways of reducing an empirical kinetic model to a chemically (and kinetically) reasonable form. Now let us consider the possible strategies of using the kinetic method.

7.2 Alternative strategies for studying complex reaction mechanisms and designing kinetic models

There are two approaches or two trends that are characteristic of all fields of natural sciences, and which can also be traced in the science of chemical reaction mechanisms and dynamic behaviors of reaction systems. The first approach is based upon a concept that is rejected in more developed fields of science [53], according to which a theory (e.g., mechanism of reaction) can be constructed based on empirical observations. The second approach, which employs experiment for discriminating theoretical predictions (hypotheses), now slowly but firmly enters into the arsenal of chemists engaged in the study of chemical reaction mechanisms. Moreover, our experience gained in this field as analyzed in Section 7.1 shows that there is little hope for an effective use of the empirical approach. In designing an experiment, a researcher poses questions which, to a considerable extent, contain answers. In the most formalized form, these questions have the form of hypotheses and the experiment offers a method for their discrimination.

7.2.1 Traditional strategy

In its pure form, the traditional strategy is now rarely encountered, but certain approaches are conventional and the concepts of *kinetic model* and *reaction mechanism* are commonly accepted [10, 12]. In particular, the traditional strategy involves a mathematical description based on a kinetic experiment (empirical model), which is compared to a scheme of possible reaction steps. This analysis yields a physical model [12] (in Kiperman's term, a kinetic model). Naturally, it is recommended to also use the data of other physicochemical methods in order to justify the reaction scheme and the mechanism formulated on its basis. Thus we have a sequence of procedures (operations) that can be represented by the following scheme:

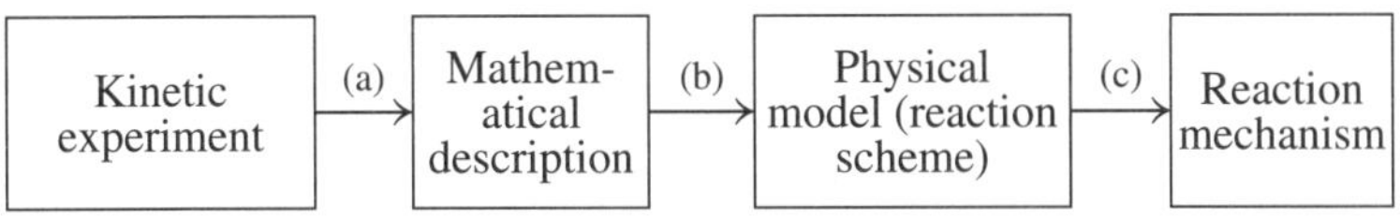

(7.1)

Having obtained a mathematical equation that is "adequate to experiment" at stage (a) and selecting a consistent (non-controversial) scheme of steps at stage (b), the researcher arrives at a single kinetically "proved" reaction mechanism. As noted above, this approach is methodologically incorrect, since the proposed mechanism cannot be proved. However, all other stages in strategy (7.1) also meet with serious objection. The main drawback is the absence of general algorithms for single-valued accomplishing of each of the three stages in the traditional approach. Indeed, in order to pass from the primary kinetic data to a mathematical description at stage (a), the researcher must first select a certain type of mathematical function – in other words, use a hypothesis concerning a mechanistic scheme (structure of mechanism) or the form of a proposed kinetic equation. The search for the possible mechanistic schemes and/or equations is usually tried after finishing the experiment, which is an evident disadvantage. An algorithm of passage (b) from the mathematical description to reaction scheme (i.e., to a physically justified model) in strategy (7.1) was developed only for single-route reactions with relatively simple linear mechanisms [1, 19]. No commonly accepted algorithms were also proposed for stage (c). Thus, all stages in the traditional procedure are unavoidably subjective, driven by the natural wish to find at least one equation (and select the corresponding "mechanistic scheme") not contradicting the experiment. As a rule, discrimination between a representative set of hypotheses is not even mentioned.

An illustrative example of the possibilities of the traditional strategy was offered by the results of a meeting organized by the University of Akron (USA) and Union Carbide Co. (USA), which were reported at the Annual AIChE Conference in 1984 (Denver, USA) [54]. Preliminary, one possible mechanism of methanol synthesis from CO and H_2 (comprising four reversible reactions without a rate-controlling step) was chosen, a kinetic model with realistic rate constants was formulated, and the experimental results in a continuous flow stirred tank reactor were simulated (in the form of the reaction rate as a function of the partial pressures of reactants and the temperature). The results of this multi-factor experiment (with a preset random error) were provided to participants of the meeting (which represented universities and companies from 19 countries) with a task of deriving a kinetic model and describing the reactor behavior. The obtained results allowed the initiators of that comparative test to extract much useful information, but one result is especially noteworthy. Among the 20 kinetic models proposed, not one coincided with the preset mechanism, even in general features. Almost all kinetic equations poorly modeled the behavior of a fixed-bed 12-m-high reactor as described by the primary model.

The author has frequently met situations where the traditional strategy was implemented under conditions of a single-factor experiment and led to a kinetic equation (i.e., a satisfactory mathematical description), for which a mechanistic scheme and a mechanism of the reaction were proposed that exhibited no physical and chemical sense in the course of subsequent investigations. For example, a study of the kinetics of trichloroethylene synthesis from acetylene in the $CuCl_2$–CuCl–HCl–NH_4Cl system in a steady-state regime (anodic oxidation of Cu(I)) led to the following kinetic equation (for $P_{C_2H_2} = \text{const}$):

$$r = k_3 \left[\mathrm{Cu^{II}}\right]_\Sigma + \frac{\left[\mathrm{Cu^{II}}\right]_\Sigma^2 \left[\mathrm{Cu^{I}}\right]_\Sigma}{k_4 \left[\mathrm{Cu^{II}}\right]_\Sigma^2 + \left(k_5' \left[\mathrm{Cu^{II}}\right]_\Sigma + k_6'\right) \left[\mathrm{H_3O^+}\right]}, \tag{7.2}$$

which satisfactorily described the process [55]. It was even suggested that one of the intermediates formed in the irreversible step is monochloroacetylene (CA). Subsequent investigations (see Examples 3.17 and 4.8) showed that CA in fact appears as an intermediate and subsequently converts into di- and trichloroethylene according to the following scheme:

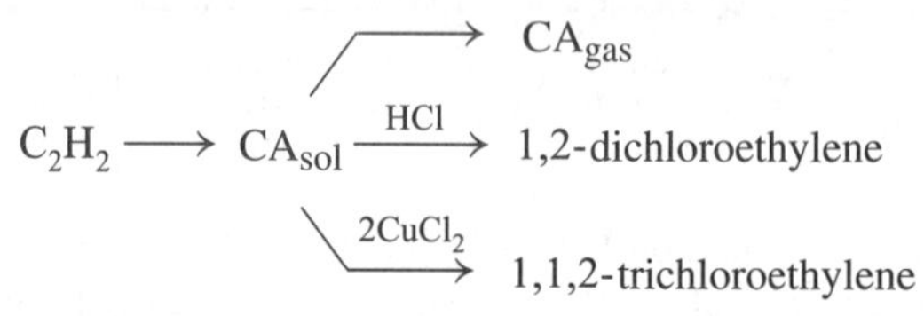

(7.3)

However, measurements of the kinetics of CA formation and conversion do not lead to Eq. (7.2), which therefore only provides an empirical mathematical description of the overall process. Scheme 7.3 also suggests that the steady-state concentration of CA in solution must depend on the rate of its passage from solution to the gas phase.

At the same time, while critically assessing possibilities of the traditional strategy in designing kinetic models and modeling reaction mechanisms, it must also be fairly noted that many well-known equations and mechanisms (including those described in Chapters 2, 3 and 4) have been obtained within the framework of this approach. The formal kinetic approach played a very important role in the creation of a theory of reactivity in organic chemistry [14], in the theory of radical chain reactions, and in acid catalysis [15], catalysis with metal complexes [27, 56–58], and enzymatic catalysis [9, 17, 59, 60]. The author also successfully applied this strategy for more than two decades in his investigations in the field of catalysis with metal complexes.

On the other hand, it should be noted that most of the well-justified equations and mechanisms were obtained using of a modified approach that may be called *iterative* strategy. This was essentially the traditional strategy, which proceeded from a kinetic experiment, but it was supplemented with iterative steps as follows:

$$\text{Kinetic experiment} \longrightarrow \text{Hypotheses} \longrightarrow \text{Refining experiment} \longrightarrow \text{Hypotheses} \longrightarrow \quad (7.4)$$

and employed a wide spectrum of complementary nonkinetic methods of investigation.

At the beginning of the 1980s, the Laboratory of Kinetics and Catalysis at the M.V. Lomonosov Moscow State Institute of Fine Chemical Technology carried out a comparative study of the traditional (iterative) approach and that based on the preliminary generation of hypotheses about the mechanism or complete reaction network (*rational* strategy). The investigations were performed for a reaction involving the formation of five products under the conditions of conjugate homogeneous hydroalkoxycarbonylation of acetylene [39, 61]. These experiments convincingly demonstrated that the traditional procedure is ineffective in investigations of the mechanisms of multi-route catalytic reactions.

7.2.2 Rational strategy

A rational strategy that corresponds to Chamberlin's ideas offers a methodologically consistent scheme of investigation, which is supported by modern computer facilities and effective routines. This procedure includes the following stages:

- generation of hypotheses;
- analysis of hypotheses (derivation of kinetic equations under various assumptions and determination of critical factors for each hypothesis);
- hypotheses discrimination planning;
- experimental verification (all kinds of experiment);
- selection of retained working hypotheses.

Generated hypotheses are discriminated using a combination of theoretical, chemical, and physicochemical methods. As a result, the initial set is reduced to a minimum, and the retained hypotheses are subjected to the formal kinetic description (with estimation of rate constants) and statistical identification. This stage includes the discrimination of steps on a theoretical level, chemical discrimination of steps (or blocks of steps), spectrokinetic [25] and isotope-kinetic discrimination of intermediates and steps, discrimination of conjugation nodes for multi-route reactions and branched reaction networks (see Section 3.3), and discrimination of retained complete kinetic models.

This sequence of stages offers a clear and logical scheme of hypothetical-deductive investigation. The value of gained information naturally depends on the number of generated and analyzed hypotheses. Evidently, the role of a hypothetical reaction mechanism in the rational strategy is no less important than the role of the physical model (7.1), since the structure of the mechanism (i.e., reaction scheme) in the general case cannot be determined without an analysis of the hypotheses about the reaction mechanism.

The above algorithm of the rational strategy emphasizes the important role of hypotheses (on any level of complexity and in any number) that can be used in arbitrary combination with experiments. Let us consider separate stages of this procedure in more detail.

Generation of hypotheses. Presently, the volume of our knowledge makes possible the formulation of numerous and very complicated hypotheses for any homogeneous and heterogeneous reactions. For example, analysis of the aforementioned process of acetylene carbonylation in solutions of palladium complexes in 1980 allowed us to formulate reaction networks comprising six sequential steps for the synthesis of acrylate, four steps for maleate, eight steps for succinate, and seven steps for propionate [1, 39, 62]. More recently, in constructing hypothetical schemes for the formation of C_1 products from CO and H_2 on metal catalysts, we obtained reaction networks made up of 130 to 1600 steps [63]. Naturally, for the generation of hypotheses, it is better to use systematic approaches and formalized methods, which help to eliminate a fraction of subjectivism. Anyhow, the author believes that the generation of hypotheses by any means is a basis for the correct strategy in investigating reaction mechanisms and designing kinetic models [6–8].

An analysis of the available a priori information about features of the process under consideration precedes the generation of hypotheses and allows requirements to be formulated [8] that must be obeyed by all hypothetical mechanisms. Naturally, the greater the volume of a priori information available, the smaller the number of hypotheses subject to discrimination. Depending on the type of processes studied, their specific features, and special interests of a researcher, we can speak of *four groups of hypotheses*. Each group can serve as a basis for the planning and conducting of experiments used to discriminate generated hypotheses.

For catalytic processes, the first and very important group includes the *hypotheses about a catalyst*. In the case of homogeneous catalysis with metal complexes, this group refers to issues such as the mechanism of active complex formation (see Chapter 5), state of the catalyst in solution, equilibrium complex formation, specific role of the solvent, degree of solution nonideality, effect of solution components on the solubility of substrates in heterophase reactions (see Chapter 8) [55, 27, 64, 65], and effect of reaction medium on the metal complexes (see Chapter 5) [66]. In the case of heterogeneous catalysis, this group includes hypotheses concerning the composition and nature of the surface, phase composition of a catalyst, degree and character of nonuniformity of the surface, adsorption of reactants and products, and conversions of surface centers during their interaction with the reaction medium [67]

The second group includes the *hypotheses about steps of a mechanism*. Hypothetical mechanisms can be discriminated on the level of separate steps. In this case, it is also necessary to provide the most complete set of hypotheses for their subsequent experimental discrimination. This approach has been most completely and consistently described by Rozovskii and Lin [28]. According to them, theoretically justified models are generated using independent information about the mechanism, even including data provided partly by the formal kinetic method. It is evident that, in the case of complex multi-route reactions, studying only separate steps without a kinetic analysis of the overall process cannot lead to establishing the topological structure of a mechanism and the form of a kinetic model. Nevertheless, this method is advantageous to the traditional kinetic approach.

In cases of highly complicated, stoichiometrically ambiguous reactions, it is very useful to employ the third group that contains the *hypotheses about conjugation nodes in the mechanistic scheme* (see Section 3.3) [37, 49, 62]. An analysis of the conjugation nodes is more expedient if there are hypotheses about the mechanism (for planning the discrimination experiments). However, this analysis is also useful in the absence of a general scheme (see Section 3.3) [47], although the formulation of a kinetic model in this case encounters problems that were mentioned above (see also Section 7.1).

The fourth group includes the *complete hypotheses about reaction mechanism*. These hypotheses provide a basis for the planning of experiments, including those for the discrimination of separate steps, intermediates, conjugation nodes, and the entire mechanism. In the absence of formalized methods for generating these hypotheses, significant aid in the formulation of sets of single- and multi-route mechanisms based on the available set of steps can be provided by the topological approaches and methods of graph theory [68, 69].

Hypotheses discrimination planning. Once a set of hypotheses is generated and a system of a priori requirements to all hypothetical mechanisms is known, it is possible to formulate criteria for distinguishing the hypotheses [8] and to plan the chemical, physicochemical, and kinetic discrimination experiments. For experiments of the latter type, it is also necessary to know the initial estimations of parameters of the kinetic model (in addition to the proposed structure and mechanism of this model) [1, 3–8].

Hypotheses discrimination based on results of experiments. The author's experience gained from investigations of the kinetics and mechanisms of complex reactions leads to a conclusion (in compete agreement with the opinion of other researchers [8, 28]) that the role of quantitative kinetic data and solutions of the inverse kinetic problem in the procedure of hypotheses discrimination must be minimized.

The main priority in the system of hypotheses discrimination must be given to the results of chemical and physicochemical experiments and qualitative kinetic data (analysis of conjugation nodes and character of kinetic relations) rather than to attempts to discriminate kinetic models by the kinetic method (i.e., by solving the inverse kinetic problem for all hypotheses). For example, in the aforementioned investigation of acetylene carbonylation [39, 62], the discrimination of 1344 hypotheses by simple kinetic and chemical experiments retained 90 mechanisms, but an analysis of the conjugation nodes showed that no one of these mechanisms could ensure qualitative coincidence of the $r_{P_k}/r_{P_j} = f(P_{CO}, P_{C_2H_2}, C_{HCl})$ dependences with experimental data for all nodes. An additional assumption that some products (acrylate) could simultaneously be formed via two different routes revealed four mechanisms with the behavior of conjugation nodes not contradicting experiment, which could therefore be used for solving the inverse kinetic problem [1]. As was pointed out in Chapter 3, an artificial modification of the topological structure of a mechanism by the directed creation of new conjugation nodes can provide a very effective discrimination of hypotheses.

The above consideration of stages of the rational strategy leads to a concept of the "theoretically justified kinetic model". This can be defined as the model based on a mechanism obtained upon the discrimination of an initial set of hypotheses by the planned chemical, physicochemical, and kinetic methods, the parameters of which were determined with allowance for the requirements to solutions of the inverse kinetic problem. In homogeneous catalysis with metal complexes, these models can be referred to as molecular kinetic models.

The examples of investigation into the reaction mechanisms and construction of the kinetic models, which were considered in the preceding chapters, cannot be always strictly classified as employing exactly one of the strategies under consideration, since researchers rarely give explicit formulations of their hypotheses (even if these existed prior to the beginning of kinetic experiments). Proceeding from the measurements of kinetics and attempting to interpret their results, these researchers rarely try to explain their choice of one or another mathematical model, which must undoubtedly be justified, at least on the intuitive level. Nevertheless, it would be useful to classify, albeit rather conditionally, the above examples into several groups with respect to the strategy employed.

The traditional strategy was mostly used to study single-route reactions, those with predominant contribution from one route, or reactions that were considered (without discussion) as single-route ones (Table 7.1).

The iterative strategy (procedure) was also successfully used in the case of two-route mechanisms. Many of the kinetic models and mechanisms obtained using this approach were important for the theory and practice of homogeneous catalysis with metal complexes (Table 7.2).

Examples of using the rational strategy in application to homogeneous catalysis with metal complexes are not very numerous (Table 7.3). This strategy was more frequently employed in investigations of the mechanisms of heterogeneous catalysis (see below).

Let us consider the procedure of hypothesis generation in more detail.

Table 7.1 Traditional strategy for studying mechanisms of reactions involving metal complexes

No	Reaction	Example no.
1	Reductive elimination of H_2	2.2
2	Protolysis of Ar–Pd bond	2.3
3	Catalytic hydrolysis of Ar–Si bond	2.4
4	Oxidative addition of R_3SiH to Ir(I)	2.6
5	Hydrocarbon oxidation in Co(II)–Br^- catalyst solutions	2.17
6	Catalytic decomposition of ethylbenzene hydroperoxide in Mo(VI) complex solutions	2.20
7	Catalytic epoxidation of propylene by organic hydroperoxides	2.21
8	Oxidation of styrene by *p*-benzoquinone in Pd(II) chloride solutions	2.23
9	Oxidation and oxidative dimerization of thiophene in Pd(II) complex solutions	2.24
10	Conjugated iodoplatination of acetylene	2.26
11	Hydrogenation of olefins in Rh(I) complex solutions	2.27
12	Hydrogenation of maleic acid in Ru(II) complex solutions	2.28
13	Oxidation of η^3-allyl Pd(II) complex by *p*-benzoquinone	2.30
14	Catalytic decomposition of H_2O_2 to triplet O_2 in Ni(II) complex solutions	2.32
15	Epoxidation of olefins by organic hydroperoxides in Mo(VI) complex solutions	2.35
16	Decomposition of organic hydroperoxides in Co complex solutions	2.35
17	Isomerization of dichlorobutenes catalyzed by $FeCl_3$	2.36
18	Oxidation of Cr(III) by cerium(IV)	2.37
19	Oxidation of H_2 by Cr(VI) catalyzed by Cu(II)	2.38
20	Catalytic oxidative decarboxylation of acetic acid by cerium(IV)	2.39
21	Hydrogenation of cyclohexene in Ti complex solutions	2.40
22	Hydrogenation of olefins in Ru(II) complex solutions	2.41
23	Arylation of olefins by benzene in Pd(II) complex solutions	2.42
24	Arylation of olefins by benzene	2.43
25	Arylation of styrene by phenylmercuric acetate	2.44
26	Oxidative acetoxylation of propylene in solutions of giant Pd clusters	2.45
27	Isomerization of 4-chloro-1,2-butadiene in CuCl solutions	3.1
28	Co-cyclization of acetylene and nitriles in Co(I) complex solutions	3.5
29	Cyclodimerization of 1,4-norbornadiene in Ni(0) complex solutions	3.13
30	Cyclopropanation of styrene by diazaethyl acetate in Co(II) phthalocyaninate complex solutions	3.14
31	Oxidation of 1,3-diene by Cu(II) complexes in alcohols and glycols	4.4
32	Oxidatve dehydrocondensation of alkynes in Cu(II) and Cu(I) complex solutions	4.7
33	Oxidation of ethylene to acetaldehyde (Eqs (4.23), (4.24)), acetaldehyde dimethyl acetal [Eqs (4.25)–(4.27)], and 1,2-oxidation products (Eqs (4.33), (4.34), (4.36), (4.37))	Section 4.1
34	Chlorination of ArH in $SbCl_5$–S_2Cl_2 solutions (Eqs (4.181), (4.182), (4.186))	Section 4.2.3
35	Oxidative chlorocarboxylation of acetylene in $PdCl_2$ and $HgCl_2$ solutions	4.17

7.3 Hypothesis generation methods and examples

Generating hypotheses is a creative task and, hence, neither now nor in the future can it be fully formalized. Nevertheless, some elements of this procedure are able to be formalized. Moreover, it is a researcher's intuition, supported by computer facilities, that provides a basis for successfully solving the task of generating hypotheses about the mechanisms of catalytic reactions.

Table 7.2 Iterative strategy for studying mechanisms of reactions involving metal complexes

No	Reaction	Example no.
1	Oxidative addition of CH_3I to Pt(0) complex	2.5
2	Oxidative addition of H_2 to Rh(I) complex	2.7
3	Oxidation of H_2 by Pd(II) complexes	2.8
4	Isotope exchange in RH–D_2O system in Pt(II) complex solutions	2.11
5	Isotope exchange in RH–D_2O system and alkane oxidation in Pt(II) complex solutions	2.12
6	Methylation of Pt(II) complexes by dimethyl sulfate	2.13
7	Alkylation of Pt(II) complexes by ethylene oxide	2.14
8	Oxidation of PcM by benzoyl peroxide and catalytic oxidation of diphenylpicrylhydrazine	2.15
9	Cyclopalladation of N,N-dimethylbenzylamines	2.16
10	Oxidation of glycolate anion by Ir(IV) complexes	2.18
11	Hydroperoxide decomposition catalyzed by Cu(II)	2.19
12	Oxidation of ethylene by *p*-benzoquinone in Pd(II) complex solutions	2.22
13	Epoxidation of olefins by hydrogen peroxide in Pt(II) complex solutions	2.25
14	Hydrocarbalkoxylation of acetylene in Pd(II) complex solutions	2.29
15	Catalytic decomposition of H_2O_2 to singlet oxygen by molybdate anion	2.31
16	Oxidative dimerization of styrene in Pd(II) complex solutions	2.33
17	Oxidative coupling of ArH by Tl(III) catalyzed by Pd(II) complexes	2.34
18	Oxidative acetoxylation of olefins [*p*-benzoquinone, Pd(II)]	2.45
19	Catalytic hydrogenation and isomerization of olefins catalyzed by *closo*-rhodacarborans	2.46
20	Asymmetric hydrogenation of olefins in Rh(I) complex solutions	3.3
21	Hydrogenation of olefins in Rh(I) complex solutions	3.7
22	Hydrocarboxylation of olefins in Pd(II) complex solutions	3.10
23	Hydrocarbalkoxylation of olefins in Pd(II) complex solutions	3.11
24	Hydrocarbalkoxylation of styrene in Pd(0) complex solutions	3.12
25	Hydrochlorination and oxidative chlorination of acetylene in $HgCl_2$ and $CuCl_2$ solutions	3.15
26	Conjugated hydrochlorination and oxidative chlorination of chloroacetylene in Cu(I)–Cu(II) system	3.17
27	Hydroiodination of acetylene in Pt(IV) complex solutions	3.20
28	Conjugated hydroiodination and reduction of acetylene to ethylene in Pt(II) complex solutions	3.26
29	Substitutional oxidative chlorination of acetylene and chloroacetylene in Cu(I)–Cu(II) system	4.8
30	Hydration of methylacetylene to acetone and propionaldehyde in CuCl–RSH–HCl system	4.20
31	Catalytic synthesis of 1-chloro-1,3-butadiene in $PdCl_2$–$HgCl_2$ system	4.21
32	Oxidation of ethylene in ethanol solutions of $PdCl_2$ and $CuCl_2$	4.1

Hypotheses are generated based on the accumulated and generalized knowledge about reaction mechanisms, active species (intermediates), and the reactions (steps) of their formation and decomposition. Enhanced by the creative imagination of researchers, this knowledge is a source of new hypotheses. In the past three decades, numerous algorithms and programs have been developed that use the accumulated knowledge with the aid of computers to generate hypotheses (hypothetical reaction mechanisms). Well-known programs were reviewed in [1, 36].

Table 7.3 Rational strategy for studying mechanisms of reactions involving metal complexes

No	Reaction	Example no.
1	Reductive elimination of CH_3I from Pt(IV) complexes	2.8
2	Hydrochlorination of vinylacetylene in Cu(I) complex solutions	3.2
3	Catalytic synthesis of vinyl chloride, 1,2-dichloroethylene, vinylidene chloride in Cu(I)–Cu(II) solutions	3.16
4	Reductive elimination of ethane from Pt(IV) *cis*-dimethyl complex	3.18
5	Reductive elimination of CH_3Cl from Pt(IV) complex and isotope exchange in CH_4–D_2O system in Pt(II) complex solutions	3.19
6	Double cross-metathesis of olefins and cycloolefins in Mo complex solutions	3.21
7	Arylation of ethylene and deuteroethylene in Pd(II) complex solutions	3.22
8	Oxidative dimerization of alkynes in Cu(I)–Cu(II) system	3.23
9	Oxidation of alkylbenzenes by potassium permanganate in acid media	3.24
10	Oxidation of hydrocarbons and perfluorocarbons by hydrogen peroxide in vanadium(V) complex solutions	3.25
11	Oxidative carbonylation of alkynes at ≡C–H bonds in $PdCl_2$–$CuCl_2$ system	4.18

There are two kinds of programs for the generation of hypotheses–empirical and formal-logical [70, 71]. Programs of the first type employ the libraries of *generalized reactions* compiled by experts. A generalized reaction is the description of a chemical reaction deprived of specific details that distinguish each particular reaction from other, generally similar, processes. These details can include atoms or molecular fragments that are not immediately involved in the redistribution of electrons and/or formal rearrangement of chemical bonds. In writing these generalized reactions, the names of atoms can be replaced by symbols, for example:

$$\begin{aligned} &\mathrm{M} + \mathrm{HX} \longrightarrow \mathrm{HMX}; \\ &\mathrm{C{-}M{-}C'} \longrightarrow \mathrm{M} + \mathrm{C{-}C'}. \end{aligned} \tag{7.5}$$

The degree of generalization can vary strongly, but particular reactions are usually not considered, since they can only lead to trivial hypotheses, all of which have already been described. Generalized reactions act as chemical operators which, being applied to a molecular fragment, transform it into a new fragment. Once the initial molecule is known, the entire elementary reaction (step) is established.

Thus, a program algorithm is reduced to a search for molecular fragments to which certain preset generalized reactions can be applied. Besides various initial restrictions set by a user, the quality of hypotheses can be determined by factors such as the degree of generality of a selected operator (chemical transformation).

Programs of the second (formal-logical) type construct a reaction using the rules of logics (stoichiometric relations, trials of all possible bond redistributions, etc.). These programs typically possess a highly interactive character and pose high requirements on the choice of initial restrictions and selection criteria [36].

One of the most widely used methods for the development of formal-logical programs intended to generate hypothetical reaction mechanisms is based on the mathematical model of Dugunji and Ugi [72]. This approach was implemented in a program package developed within the framework of the Munich Project [73]. The Dugunji–Ugi model was also used to a greater or lesser extent in the development of other routines such as RAIN [36, 73–75], GRACE [76], the Dozmorov program [77], NetGen [78, 79], and KOMSICAT [80–84]. The latter program was ideologically close to GRACE, but employed only preset group matrices [36] and hence should be more correctly classified to the empirical type (expert selection of the characteristic structural types of the elementary transformations with the participation of a catalyst). The first type also includes the MECHEM program that was developed by Valdes-Perez [85–94]. This program generates mechanisms using the principles of stoichiometry and employs a combinatorial trial of various sets of conversions of reactants, catalysts, and intermediates, which is guided by certain logical rules (restrictions). First, the "framework" mechanism of a reaction is formed and subsequently the elementary steps and particular substances are specified [36].

As was noted above, a feature in common for all programs belonging to the first type is the use of generalized reactions (transforms). The method used for the description of these transforms is rather arbitrary. The general algorithm consists of recognizing molecular fragments to which the preset generalized reactions (reaction templates) can be applied with the subsequent generation of particular reactions and chemical structures [36].

One of the first programs of the empirical type was CAMEO [95, 96]. This program is oriented toward the generation of mechanisms of organic reactions, in which the most probable sites of attack by active species can be determined and the relative reactivity can be evaluated.

The TAMREAC program [97–99] is oriented toward investigations in the field of catalysis with metal complexes (organometallic compounds). This program can generate elementary reactions that involve one metal atom. The TAMREAC program employs a large systematized database of elementary reactions, which is organized as a library of generalized transforms. This library includes the reactions of metal oxidation and reduction, metal complex formation (leading to an increase in the metal coordination number) and dissociation (with the formation of coordination vacancies), reactions without a change in the degree of oxidation and/or metal coordination number, radical reactions, and electron transfer reactions. Each file of the TAMREAC library comprises several generalized reactions described as a structural transformation of a certain reactive fragment. A detailed description of this program is presented in review [36] and the original investigations [98, 99].

The ChemNet program [1, 36, 100] is oriented toward the generation of catalytic reactions with the participation of metal complexes, transition metals, and their oxides. This program employs transformations (templates of elementary steps), initial substances, and restrictions. The transformations represent descriptions of elementary steps with various degrees of generalization, which are proposed by the user or taken from the database. The ChemNet database for reactions of metal complexes or metal atoms on the surface of metal catalysts contains about 2500 elementary steps. The program generates reaction networks, from which one can "cut" particular mechanisms [2].

The system of empirical methods also includes the generation of hypotheses using sets of elementary steps, combinatorial algorithms, and kinetic graphs (for linear mechanisms) [68, 69].

Let us consider examples of the application of some programs to homogeneous and heterogeneous catalytic processes.

7.4 Hypothesis generation programs: Application examples and related problems

Below we consider examples of the application of simple combinatorial programs, including those using kinetic graphs, and some programs of higher level such as MECHEM, ChemGen, TAMREAC, and ChemNet. Special approaches were developed that made possible the use of large reaction networks for the generation of kinetic models.

7.4.1 Combinatorics on kinetic graphs

Combinatorial algorithms are used to generate sets of routes [26, 101, 102] based on elementary steps (library of steps, see [102]), which were preliminarily selected by experts for the generation of reaction

Table 7.4 Possible types of complexes between catalyst, reactants, and products (reproduced with permission from Elsevier Ltd. © 1992 [68])

No.	Notation	Complex	No.	Notation	Complex
1	X_1	X_0[ROOH]	8	X_8	$X_0[ROH]_2$
2	X_2	X_0[Ol]	9	X_9	X_0[ROOH][Ol]
3	X_3	X_0[Ep]	10	X_{10}	X_0[ROOH][Ep]
4	X_4	X_0[ROH]	11	X_{11}	X_0[ROOH][ROH]
5	X_5	$X_0[ROOH]_2$	12	X_{12}	X_0[Ol][Ep]
6	X_6	$X_0[Ol]_2$	13	X_{13}	X_0[Ol][ROH]
7	X_7	$X_0[Ep]_2$	14	X_{14}	X_0[Ep][ROH]

networks [39]. In the case of linear mechanisms involving large numbers of equilibrium steps that determine the complexity of intermediates (pendant vertices of kinetic graphs), the kinetic graphs (KGs) are used to generate complete sets of hypotheses from a limited number of elementary steps [37, 68, 69]. This approach provides a justified solution of the problem of selecting hypothetical mechanisms for their subsequent discrimination with allowance for a particular system and reaction conditions.

Example 7.1 [68]

► An analysis of published data on the mechanism of olefin epoxidation by hydroperoxides in solutions of metal (M = Mo, V) compounds,

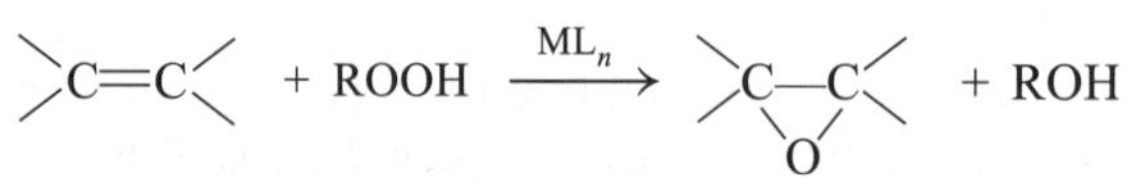

(7.6)

allowed a series of metal complexes containing ROOH, olefin (Ol), epoxide (Ep), and ROH to be selected, which can exist under the epoxidation reaction conditions, e.g., in Mo(VI) complex solutions [68]. The formation of complexes **1–6** and **8** (see Table 7.4) has physicochemical justification, while the other complexes can be constructed based on general considerations with allowance for two coordination vacancies in the metal-catalyst complex (X_0).

The possible stages of the reaction mechanism, which are listed in Table. 7.5, are not linearly independent. For example, the sum of steps 1 + 9 is equivalent to the sum of steps 6 + 16. By restricting the consideration to four steps in single-route epoxidation mechanisms with intermediates X_1, X_2, X_9, and X_{14} (steps 17–21), we obtain six types of the mechanisms that are presented (with the corresponding KGs) in Table 7.6. These mechanisms are denoted by triple codes, where the first digit indicates the number of routes, the second shows the number of intermediates in the cycle, and the third enumerates the mechanisms with the given number of intermediates.

Table 7.7 shows the KGs of single-route mechanisms with the maximum numbers of pendant vertices (see Table 7.5). Using these graphs, it is possible to construct many variants of potential mechanisms by eliminating a certain number of pendant vertices. The total number of variants, which can be obtained by varying the number of pendant vertices from 1 to 12, amounts to 2047.

Using single-route KGs (Table 7.6), it is possible to obtain 13 two-route mechanisms with common vertex X_0 (class B), common step 1, 2, or 19, or two common steps 20 and 21 (class C_2). The completeness of generating two-route (and more complicated) mechanisms can be checked by comparing them to the available standard KG tables [37]. If necessary, this procedure of hypothesis generation can be continued with the three- and four-route mechanisms. For example, the combination of mechanisms 1.3.1 and 1.4.2

Table 7.5 Possible steps in epoxidation mechanism (reproduced with permission from Elsevier Ltd. © 1992 [68])

No	Step		
1	X_0 + ROOH	$\rightleftharpoons$	X_1
2	X_0 + Ol	$\rightleftharpoons$	X_2
3	X_0 + Ep	$\rightleftharpoons$	X_3
4	X_0 + ROH	$\rightleftharpoons$	X_4
5	X_0+2ROOH	$\rightleftharpoons$	X_5
6	X_0 + 2Ol	$\rightleftharpoons$	X_6
7	X_0 + 2Ep	$\rightleftharpoons$	X_7
8	X_0 + 2ROH	$\rightleftharpoons$	X_8
9	X_1 + Ol	$\rightleftharpoons$	X_9
10	X_0 + ROOH + Ep	$\rightleftharpoons$	X_{10}
11	X_0 + ROOH + ROH	$\rightleftharpoons$	X_{11}
12	X_2 + ROOH	$\rightleftharpoons$	X_9
13	X_0 + Ol + Ep	$\rightleftharpoons$	X_{12}
14	X_0 + Ol + ROH	$\rightleftharpoons$	X_{13}
15	X_0 + Ep + ROH	$\rightleftharpoons$	X_{14}
16	X_6 + ROOH	$\rightleftharpoons$	X_9 + Ol
17	X_1 + Ol	$\longrightarrow$	X_0 + Ep + ROH
18	X_2 + ROOH	$\longrightarrow$	X_0 + Ep + ROH
19	X_9	$\longrightarrow$	X_0 + Ep + ROH
20	X_9	$\longrightarrow$	X_{14}
21	X_{14}	$\longrightarrow$	X_0 + Ep + ROH

yields the following three-route mechanism (3 5 1):

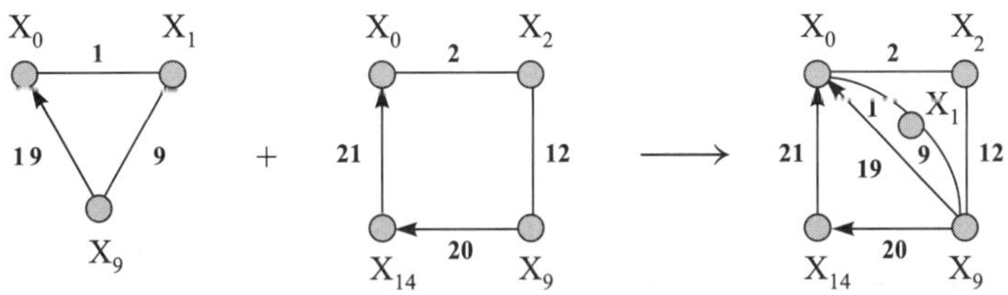

Considering all the reversible steps among the total of 19 one- and two-route mechanisms as quasi-equilibrium, on can obtain 19 kinetic equations for KGs with maximum numbers of pendant vertices. These equations take into account the complexity of a catalyst with respect to all reactants and products (see Table 7.5) [68]. The obtained models can be used for the discrimination of hypotheses. The most complicated models provide a convenient basis for any justified simplification. ■

Example 7.2 [69]

► In studying the mechanism and constructing a model for the synthesis of vinyl acetate (VA) on a supported zinc acetate catalyst according to the following reaction,

$$C_2H_2 + AcOH \longrightarrow CH_2{=}CHOAc, \tag{7.7}$$

researchers compared from nine [103] to 19 [26] hypotheses. This process was modeled by generating hypotheses on KGs and bipartite graphs [69].

Table 7.6 Six possible types of single-route mechanisms and corresponding kinetic graphs (reproduced with permission from Elsevier Ltd. © 1992 [68])

Code	Mechanism	Kinetic graph (KG)
1.2.1	1 $X_0 + ROOH \rightleftharpoons X_1$ 17 $X_1 + Ol \longrightarrow X_0 + Ep + ROH$	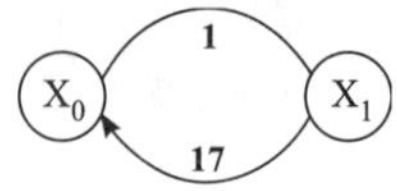
1.2.2	2 $X_0 + Ol \rightleftharpoons X_2$ 18 $X_2 + ROOH \longrightarrow X_0 + Ep + ROH$	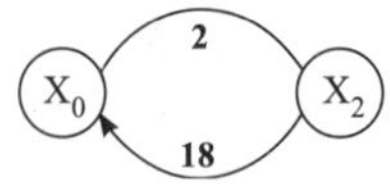
1.3.1	1 $X_0 + ROOH \rightleftharpoons X_1$ 9 $X_1 + Ol \rightleftharpoons X_9$ 19 $X_9 \longrightarrow X_0 + Ep + ROH$	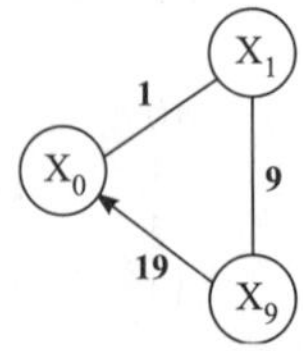
1.3.2	2 $X_0 + Ol \rightleftharpoons X_2$ 12 $X_2 + ROOH \rightleftharpoons X_9$ 19 $X_9 \longrightarrow X_0 + Ep + ROH$	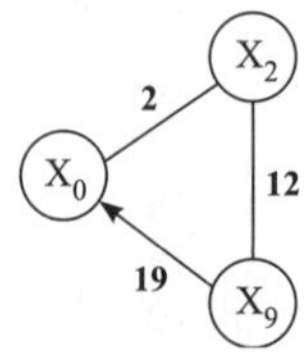
1.4.1	1 $X_0 + ROOH \rightleftharpoons X_1$ 9 $X_1 + Ol \rightleftharpoons X_9$ 20 $X_9 \longrightarrow X_{14}$ 21 $X_{14} \longrightarrow X_0 + Ep + ROH$	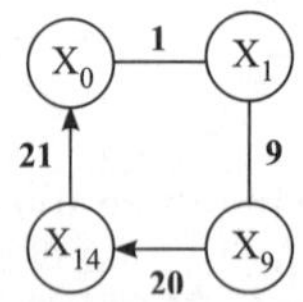
1.4.2	2 $X_0 + Ol \rightleftharpoons X_2$ 12 $X_2 + ROOH \rightleftharpoons X_9$ 20 $X_9 \longrightarrow X_{14}$ 21 $X_{14} \longrightarrow X_0 + Ep + ROH$	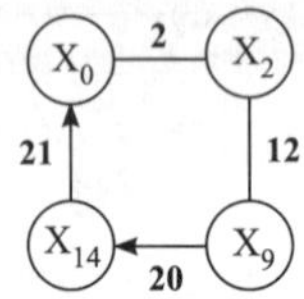

An analysis of published data on the VA synthesis and the mechanisms of other reactions involving the addition of HX molecule (with the acid hydrogen atom) to alkynes (acetylene, methylacetylene, etc.) [27] allows the set of potential surface compounds and reactions of their formation and conversion to be restricted to variants that are listed in Tables 7.8 and 7.9. The list of intermediates (Table 7.8) includes organometallic compounds (Z′A, Z′AB) formed as a result of acetate anion addition to acetylene π complexes ZA and ZAB, and the product of interaction of ZA and ZB in a proposed nonlinear acetate anion transfer to acetylene (ZAZB). The formation of intermediate ions [i.e., ionized molecules on the surface such as, e.g., $(AcO)_2Zn^{-}CH{=}CH^{+}$] was ignored. At the same time, the formation of ZnO via the reaction of zinc acetate decomposition, which is possible at elevated temperatures (230–250 °C), was also taken into consideration in selecting the surface steps (Table 7.9, step 23):

$$Zn(OAc)_2 \longrightarrow ZnO + Ac_2O. \tag{7.8}$$

Table 7.7 Single-route mechanisms and corresponding kinetic graphs (KGs) with pendant vertices (reproduced with permission from Elsevier Ltd. © 1992 [68])

Code	KG	Code	KG
1.2.1	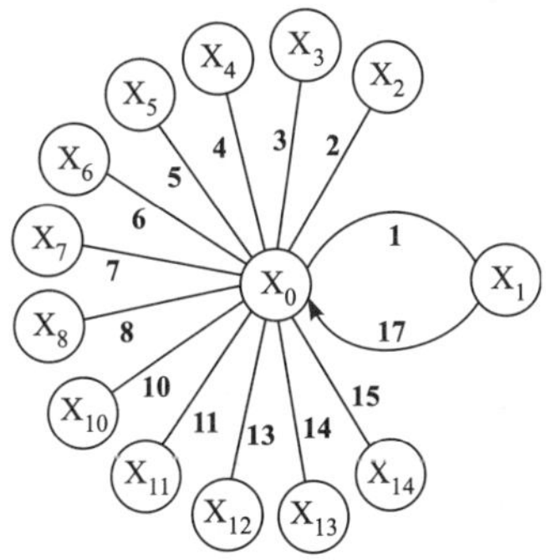	1.3.2	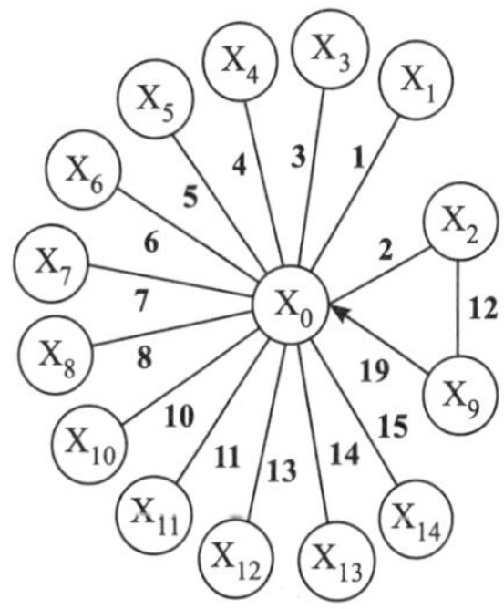
1.2.2	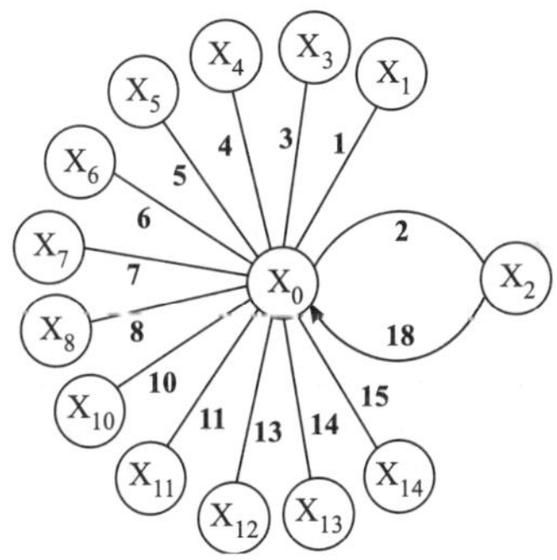	1.4.1	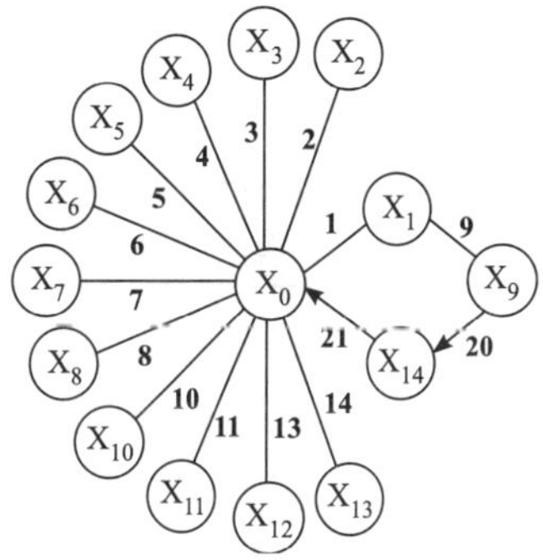
1.3.1	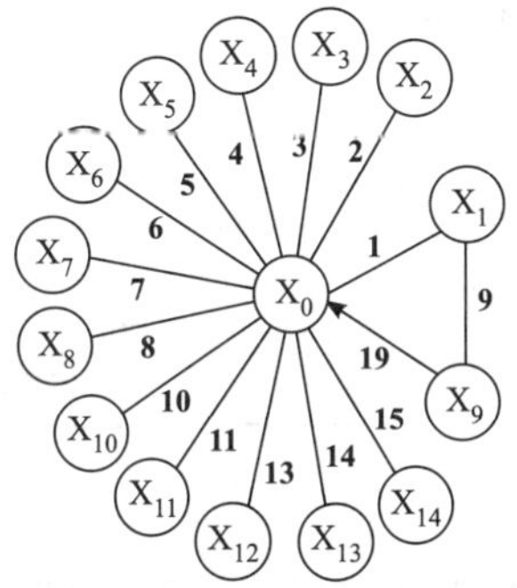	1.4.2	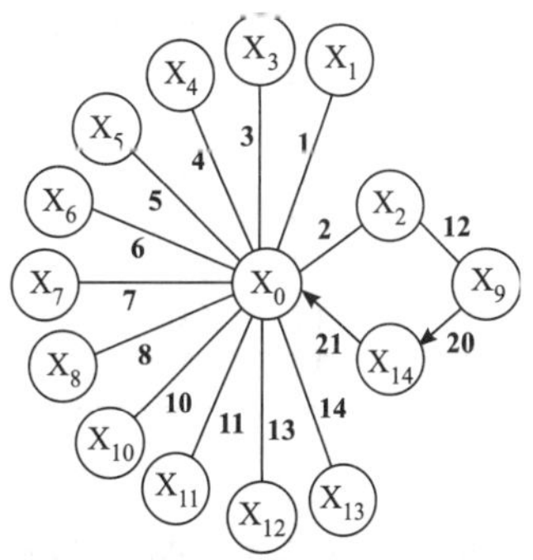

The analysis also took into consideration the possibility of zinc acetate regeneration (step 24), which can proceed either as an elementary act or as a sequence of two irreversible steps. Since the ratio of rates of reaction (7.7) in the forward and reverse directions

$$W^+/W^- = \exp[-\Delta G_T/(RT)]$$

under experimental conditions (175–230 °C, $P_A/P_B = 3$–15) amounted to 10^5–10^3, the reverse process could be ignored (i.e., steps 13–22 are irreversible). Since the fraction of side products did not exceed

Table 7.8 Possible intermediates in vinyl acetate synthesis (reproduced with permission from AIPC Nauka and editorial board of *Kinet. Katal.* journal © 2000 [69])

Compound*	Proposed chemical structure*	Notation
Z	$Zn(OAc)_2$ or $[Zn(OAc)_2]_2$	Z_0
ZA	$Zn(OAc)_2(C_2H_2)$	Z_1
Z′A	$(OAc)Zn(C_2H_2OAc)$	Z_2
ZB	$H[Zn(OAc)_3]$	Z_3
ZB2	$H_2[Zn(OAc)_4]$	Z_4
ZP	$Zn(OAc)_2 \cdot (CH_2CHOAc)$	Z_5
ZAB	$H[Zn(OAc)_3(C_2H_2)]$	Z_6
Z′AB	$H[(OAc)_2Zn(C_2H_2OAc)]$	Z_7
ZBP	$H[Zn(OAc)_3(CH_2CHOAc)]$	Z_8
ZAZB	$[Zn(OAc)_2(C_2H_2)Zn(OAc)_3H]$	Z_9
–	ZnO	Z_{10}

*Notes: Z = catalyst, A = acetylene, B = acetic acid.

Table 7.9 Possible steps in vinyl acetate synthesis (reproduced with permission from AIPC Nauka and editorial board of *Kinet. Katal.* journal © 2000 [69])

No.	Step			No.	Step		
1	Z + A	$\rightleftharpoons$	ZA	**13**	ZA + B	$\longrightarrow$	Z + P
2	Z + B	$\rightleftharpoons$	ZB	**14**	ZA + B	$\longrightarrow$	ZP
3	Z + 2B	$\rightleftharpoons$	ZB_2	**15**	ZA + ZB	$\longrightarrow$	ZAZB
4	Z + P	$\rightleftharpoons$	ZP	**16**	Z′A + B	$\longrightarrow$	Z + P
5	ZA + B	$\rightleftharpoons$	ZAB	**17**	ZB + A	$\longrightarrow$	Z + P
6	ZB + A	$\rightleftharpoons$	ZA + B	**18**	ZAB	$\longrightarrow$	ZP
7	ZB + A	$\rightleftharpoons$	ZAB	**19**	ZAB	$\longrightarrow$	Z + P
8	ZB + B	$\rightleftharpoons$	ZB_2	**20**	ZAB + B	$\longrightarrow$	ZBP
9	ZB + P	$\rightleftharpoons$	ZBP	**21**	Z′AB + B	$\longrightarrow$	ZB + P
10	ZP + B	$\rightleftharpoons$	ZB + P	**22**	ZAZB	$\longrightarrow$	2Z + P
11	ZAB	$\rightleftharpoons$	Z′AB	**23**	Z	$\longrightarrow$	$ZnO + Ac_2O$
12	ZA	$\rightleftharpoons$	Z′A	**24**	ZnO + 2B	$\longrightarrow$	$Z + H_2O$

3–5 % of the VA yield, only routes leading to VA formation via reaction (7.7) and the products of steps 23 and 24 (only appearing in insignificant amounts at 230 °C) were taken into consideration. Finally, in addition to the steps of adsorption and desorption (steps 1–5, 7–9), the possibility of adsorption substitution (steps 6, 10) was also taken into account [28]. The resulting most probable reaction network for VA synthesis (Table 7.9) included 10 surface compounds and allowed the set of possible pathways (mechanisms) leading from initial reactants to the target product (VA) to be determined.

The hypotheses were generated using two- and three-step sequences (routes) of steps, which led to the formation of VA or $(CH_3CO)_2O$, and various combinations of the adsorption steps that led to filling of the catalyst surface with reactants (A, B) and the product (P) [69]. The pendant vertices in this analysis were not formally arranged, but placed with allowance for the chemical nature of intermediates involved in a cycle, their conversion steps, and the following restrictions:

(i) Pendant vertices do not repeat (double) intermediates of a cycle.
(ii) The composition of transition states in a cyclic sequence is not repeated (doubled) by a pendant vertex. For example, pendant vertex ZAB (Z_6) does not enter into the maximal graph 1 (Table 7.10):

$$ZA + B \longrightarrow Z + P;$$

$$ZA + B \rightleftharpoons ZAB.$$

(iii) An intermediate (vertex) that is involved in a fast irreversible step (following the limiting step) is not connected to a pendant vertex. For example, vertex Z′A (Z_2) in the cyclic sequence is not connected to the pendant vertex in maximal graph 4.
(iv) An intermediate connected to a pendant vertex is also not repeated (doubled) by another pendant vertex. For example, molecule B in maximal graph 9 is not connected to intermediate ZP (Z_5), since product ZBP is already represented by vertex Z_8:

$$ZB + P \rightleftharpoons ZBP;$$

$$ZP + B \rightleftharpoons ZBP.$$

Let us consider the algorithm for constructing graphs of a hypothetical mechanism using the aforementioned rules and restrictions by example of graphs 1 and 2 from Table 7.10 (Figure 7.1).

1. Select one of the steps that lead to the VA formation (e.g., step 13) and determine pathways to (and, if necessary, from) the intermediates in this step.
2. Admit the reversibility of step 1 (in general, steps 1–10) and consider the direction of this step. The first (in the given cycle) irreversible step is assumed to be rate-controlling (subgraphs in Figures 7.1a and 7.1b).
3. Consider the adsorption of reactants (A, B) and product (P) on intermediates of the cycle, i.e., make assumptions concerning the natures of pendant vertices. For example, in route of steps 1 and 13, it is natural to assume the formation of products ZB (Z_3), ZP (Z_5), and ZB_2 (Z_4) on the active surface center Z (Z_0) (Figure 7.1c). The adsorption of B on ZA is not considered, since the interaction of B and ZA is reflected by step 13 and leads to the formation of BA. Similarly, the adsorption of A on Z_0 is also not considered, since the formation of ZA that is different from Z_1 and not involved in the route of VA formation is of low probability. It can be assumed that the adsorption of VA on free Z_0 is more probable (Z_5) than the adsorption of VA on zinc acetate that is already bound to A.
4. Superimpose the subgraph of Figure 7.1c onto the subgraphs of Figures 7.1a and 7.1b. This yields the corresponding maximal graphs of mechanisms (graphs 1 and 2 in Table 7.10).
5. Determine particular variants of hypothetical mechanisms as all possible combinations of the selected pendant vertices on the given cyclic graph. Table 7.11 lists the possible combinations (mechanisms) for the reaction under consideration and gives their codes. For maxima graphs 1 and 2, these are mechanisms 1.1–1.8 and 2.1–2.8 (Table 7.11).

In this example, graphs 1–10 refer to the mechanisms involving active centers Z_0 (Z), while graphs 11–17 refer to the mechanisms in which the active center of VA synthesis is Z_3 (ZB). The mechanisms described by graphs 18–20 also admit the steps of zinc acetate decomposition (23 and 24). Variants

Table 7.10 Maximal graphs of possible mechanisms of vinyl acetate synthesis (reproduced with permission from AIPC Nauka and editorial board of *Kinet. Katal.* journal © 2000 [69])

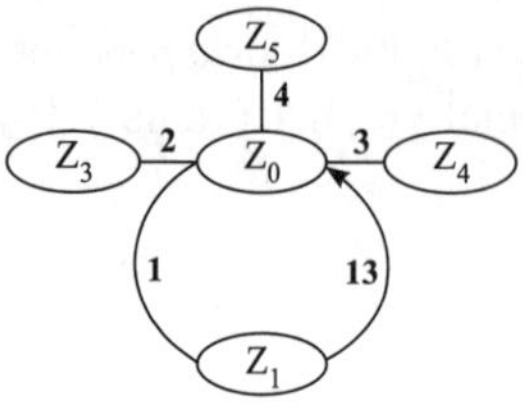

Graph 1

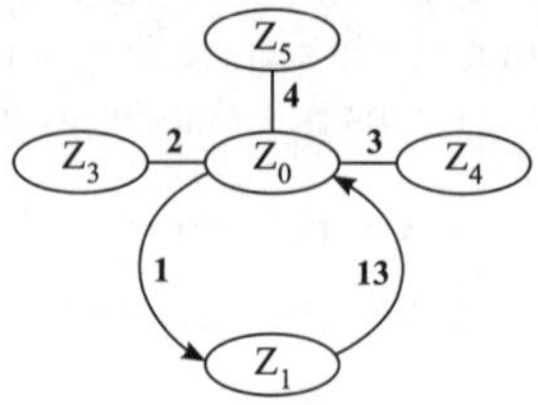

Graph 2

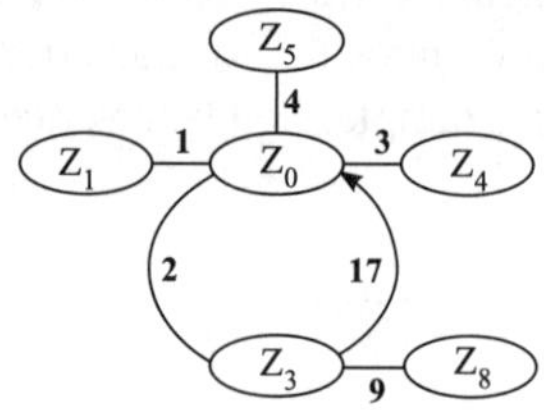

Graph 3

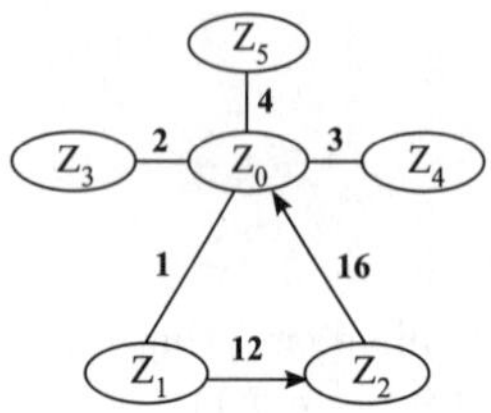

Graph 4

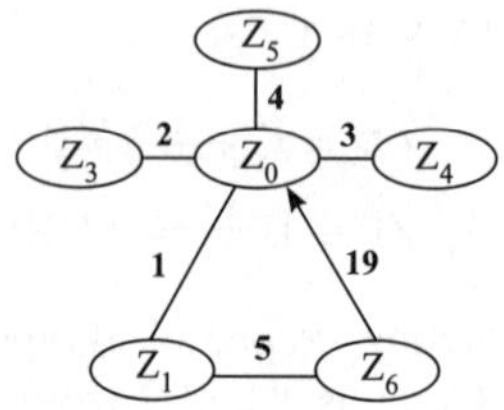

Graph 5

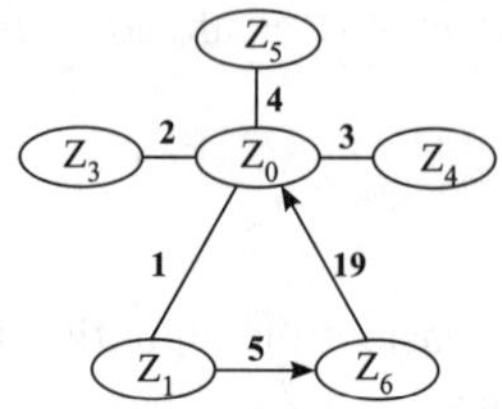

Graph 6

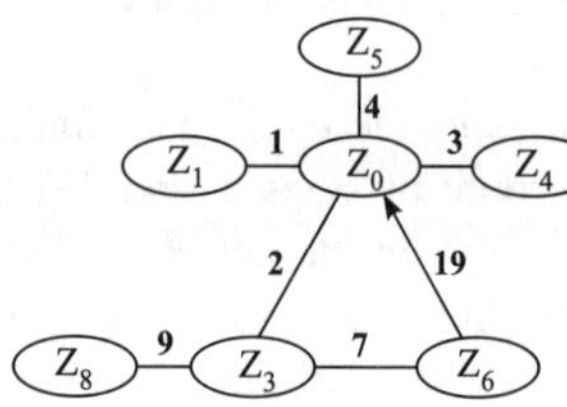

Graph 7

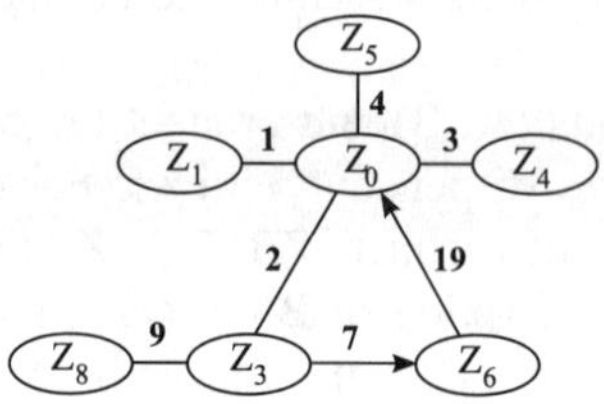

Graph 8

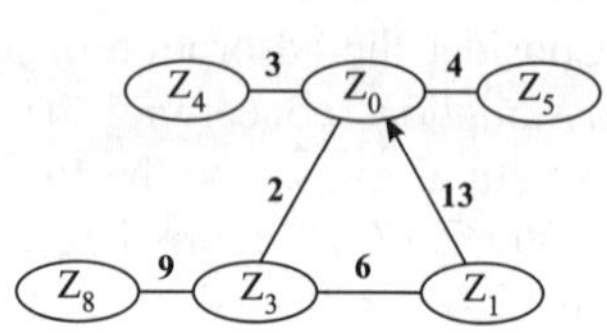

Graph 9

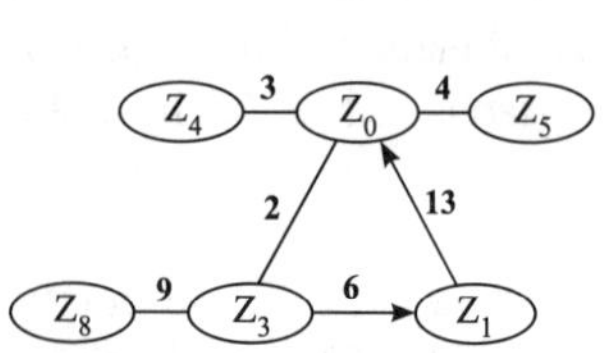

Graph 10

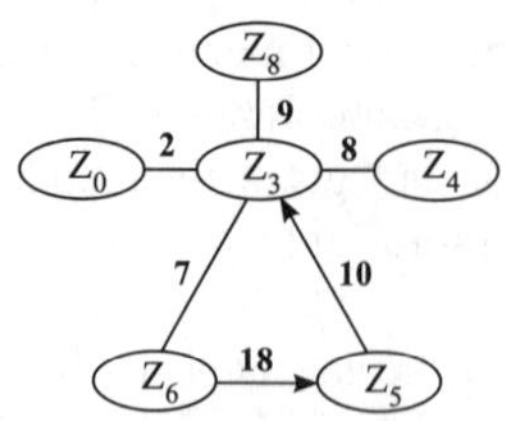

Graph 11

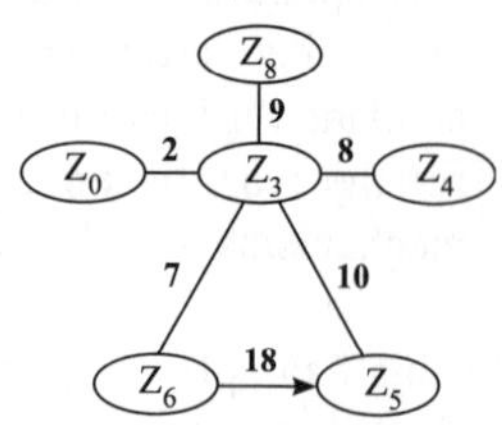

Graph 12

Table 7.10 *(continued)*

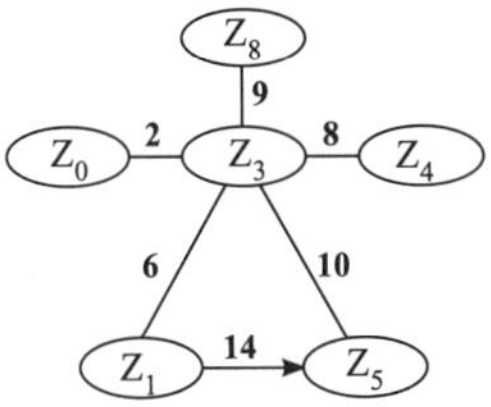

Graph 13

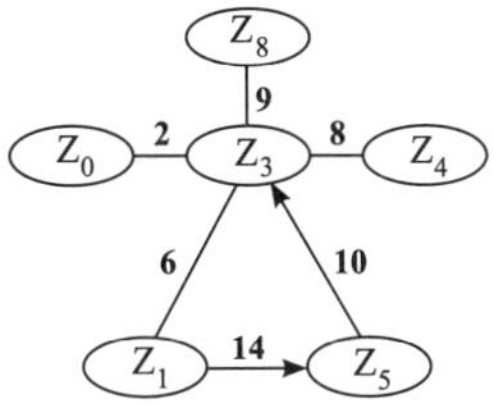

Graph 14

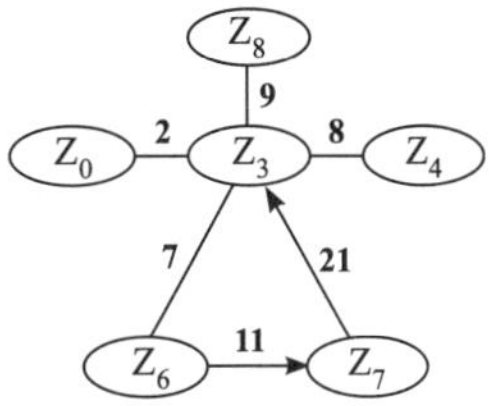

Graph 15

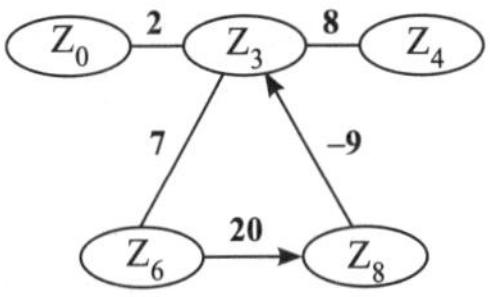

Graph 16

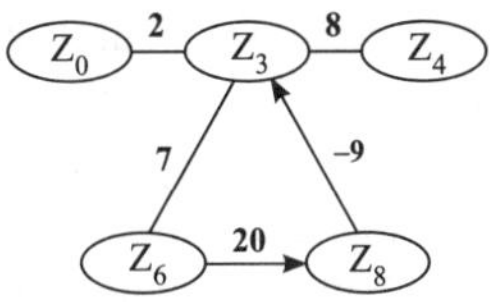

Graph 17

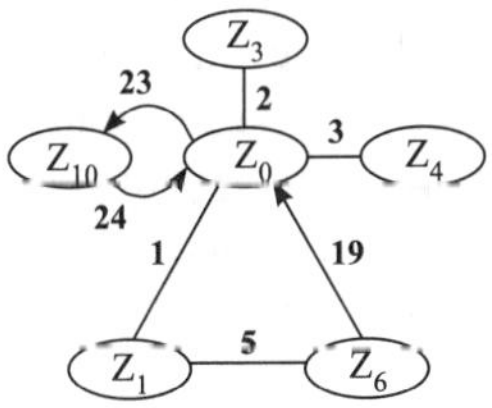

Graph 18

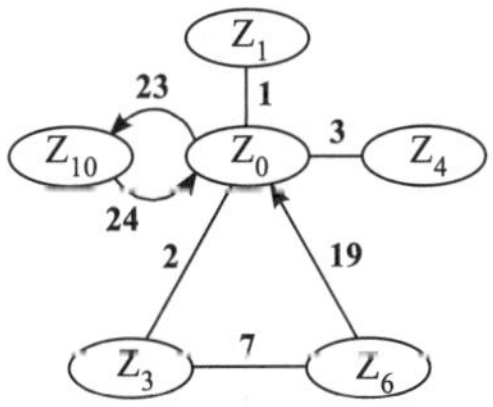

Graph 19

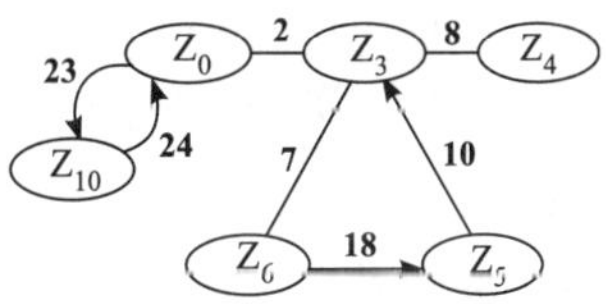

Graph 20

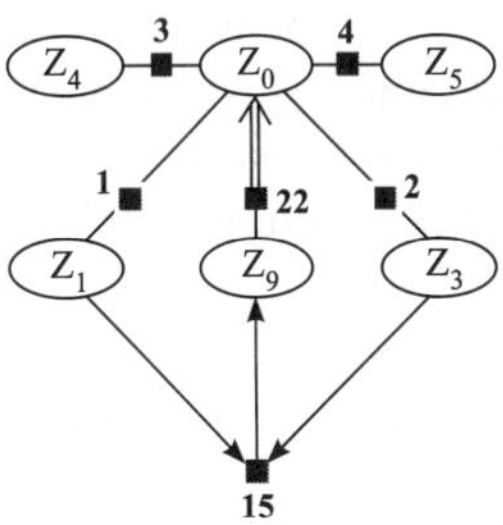

Graph 21

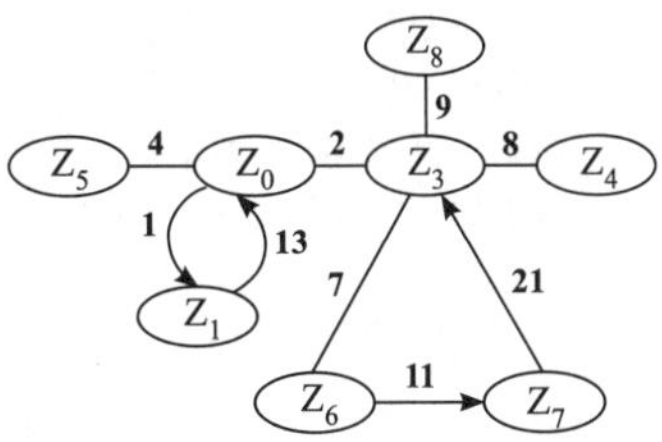

Graph 22

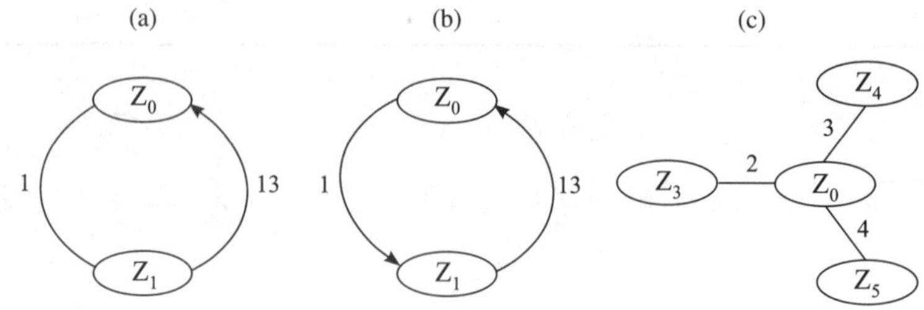

Figure 7.1 Constructing graphs 1 and 2 for Table 7.10 (see text for explanations)

21.1–21.4 (graph 21) belong to nonlinear mechanisms and, accordingly, the corresponding maximal graph is represented by a bipartite graph with vertices of two kinds (Table 7.10, graph 21), which are depicted by ovals that correspond to surface compounds and by black squares representing steps, so that graph 21 has two nonlinear steps (15 and 22) and four linear steps (1–4). Graph 22 represents the maximal graph of the simplest two-route mechanism of VA synthesis via cycles of steps 1, 13 and 7, 11, 21. Note that a variant of graph 3 with limiting step 2 was not considered because it leads to zero or negative reaction orders with respect to acetylene, which contradicts all available experimental data. Graphs 18–20 were simplified by rejecting variants with pendant vertices containing product P [69].

The kinetic equations (models) were obtained from the mechanisms presented in Tables 7.10 and 7.11, with allowance for the quasi-equilibrium character of reversible steps in the cyclic sequences of routes and with an assumption of a uniform catalyst surface. Equations were also derived using some additional assumptions concerning the surface coverage. For a total of 150 mechanisms, which were "cut" from the reaction network, 92 equations were obtained and discriminated based on the results of kinetic experiments performed in a non-gradient reactor for two different samples of $Zn(OAc)_2/C$ catalyst [69]. In the course of model discrimination by statistical techniques, it was established that, depending on the method of preparation of catalysts with the same composition, significantly different kinetic equations and, hence, substantially different hypotheses are obtained with respect to the compositions of surface compounds and reaction mechanisms. ■

7.4.2 ChemComb (Comb 1) program

Single-route mechanisms can be generated using a simple combinatorial algorithm that employs the Comb 1 program and a library of transforms (generalized reactions) to form sets of combined steps, which contain all the mechanisms that are possible with allowance for preset restrictions. This combinatorial algorithm can also be implemented in a non-computer-aided variant [102]. Let us consider the application of the Comb 1 program to an analysis of the mechanism of succinic anhydride synthesis [104].

Example 7.3 [104]

► It was preliminarily established that the following reaction proceeds in the $PdBr_2$–LiBr–acetone (acetonitrile) system at 40 °C and yields succinic anhydride (SA) at ~70 % selectivity with respect to acetylene ($[H_2O] = 0.3$ M) [105]:

$$C_2H_2 + 2CO + H_2O \longrightarrow OC{-}O{-}CO \quad (SA) \qquad (7.9)$$

Table 7.11 Variants (codes) of mechanisms of vinyl acetate synthesis (reproduced with permission from AIPC Nauka and editorial board of *Kinet. Katal.* journal © 2000 [69])

Code	Pendant vertices	Code	Pendant vertices	Code	Pendant vertices	Code	Pendant vertices
1.1	–	4.6	Z_3, Z_5	8.3	Z_4	13.1	–
1.2	Z_3	4.7	Z_4, Z_5	8.4	Z_5	13.2	Z_4
1.3	Z_4	4.8	Z_3, Z_4, Z_5	8.5	Z_8	13.3	Z_8
1.4	Z_5	5.1	–	8.6	Z_1, Z_4	13.4	Z_4, Z_8
1.5	Z_3, Z_4	5.2	Z_3	8.7	Z_1, Z_5	14.1	–
1.6	Z_3, Z_5	5.3	Z_4	8.8	Z_1, Z_8	14.2	Z_4
1.7	Z_4, Z_5	5.4	Z_5	8.9	Z_4, Z_5	14.3	Z_8
1.8	Z_3, Z_4, Z_5	5.5	Z_3, Z_4	8.10	Z_4, Z_8	14.4	Z_4, Z_8
2.1	–	5.6	Z_3, Z_5	8.11	Z_5, Z_8	15.1	–
2.2	Z_3	5.7	Z_4, Z_5	8.12	Z_1, Z_4, Z_5	15.2	Z_4
2.3	Z_4	5.8	Z_3, Z_4, Z_5	8.13	Z_1, Z_4, Z_8	15.3	Z_8
2.4	Z_5	6.1	–	8.14	Z_1, Z_5, Z_8	15.4	Z_4, Z_8
2.5	Z_3, Z_4	6.2	Z_3	8.15	Z_4, Z_5, Z_8	16.1	–
2.6	Z_3, Z_5	6.3	Z_4	8.16	Z_1, Z_4, Z_5, Z_8	16.2	Z_4
2.7	Z_4, Z_5	6.4	Z_5	9.1	–	17.1	–
2.8	Z_3, Z_4, Z_5	6.5	Z_3, Z_4	9.2	Z_4	17.2	Z_4
3.1	–	6.6	Z_3, Z_5	9.3	Z_5	18.1	–
3.2	Z_1	6.7	Z_4, Z_5	9.4	Z_8	18.2	Z_3
3.3	Z_4	6.8	Z_3, Z_4, Z_5	9.5	Z_4, Z_5	18.3	Z_4
3.4	Z_5	7.1		9.6	Z_4, Z_8	18.4	Z_3, Z_4
3.5	Z_8	7.2	Z_1	9.7	Z_5, Z_8	19.1	–
3.6	Z_1, Z_4	7.3	Z_4	9.8	Z_4, Z_5, Z_8	19.2	Z_1
3.7	Z_1, Z_5	7.4	Z_5	10.1	–	19.3	Z_4
3.8	Z_1, Z_8	7.5	Z_8	10.2	Z_4	19.4	Z_1, Z_4
3.9	Z_4, Z_5	7.6	Z_1, Z_4	10.3	Z_5	20.1	–
3.10	Z_4, Z_8	7.7	Z_1, Z_5	10.4	Z_8	20.2	Z_4
3.11	Z_5, Z_8	7.8	Z_1, Z_8	10.5	Z_4, Z_5	21.1	–
3.12	Z_1, Z_4, Z_5	7.9	Z_4, Z_5	10.6	Z_4, Z_8	21.2	Z_4
3.13	Z_1, Z_4, Z_8	7.10	Z_4, Z_8	10.7	Z_5, Z_8	21.3	Z_5
3.14	Z_1, Z_5, Z_8	7.11	Z_5, Z_8	10.8	Z_4, Z_5, Z_8	21.4	Z_4, Z_5
3.15	Z_4, Z_5, Z_8	7.12	Z_1, Z_4, Z_5	11.1	–	22.1	–
3.16	Z_1, Z_4, Z_5, Z_8	7.13	Z_1, Z_4, Z_8	11.2	Z_4	22.2	Z_4
4.1	–	7.14	Z_1, Z_5, Z_8	11.3	Z_8	22.3	Z_5
4.2	Z_3	7.15	Z_4, Z_5, Z_8	11.4	Z_4, Z_8	22.4	Z_8
4.3	Z_4	7.16	Z_1, Z_4, Z_5, Z_8	12.1	–	22.5	Z_4, Z_5
4.4	Z_5	8.1	–	12.2	Z_4	22.6	Z_4, Z_8
4.5	Z_3, Z_4	8.2	Z_1	12.3	Z_8	22.7	Z_5, Z_8
				12.4	Z_4, Z_8	22.8	Z_4, Z_5, Z_8

In addition to SA, this reaction yields maleic anhydride (MA, ~7 %), succinic and maleic acids (~12 %), acrylic acid (AA, ~7 %) and significantly smaller amounts of ethylene, propionic acid, and 2(5H)-furanone. The results of investigations of the influence of oxidants and the data of potentiometric measurements and IR spectroscopy led to a conclusion that catalytically active species in the SA and MA synthesis are Pd(I) complexes [40, 104, 106].

The procedure of generating hypotheses about the mechanisms of SA and MA synthesis via reaction (7.9) included the following stages:

1. Determination of a set of chemical bond transformations necessary for the formation of SA from C_2H_2, CO, and H_2O, that is, selection of the conversion formulas. A simple variant of the conversion formula includes the following acts:

   - cleavage of C≡C bond into C=C (A);
   - cleavage of C=C bond into C–C (B);
   - conversion of C≡O bond into >C=O (C);
   - cleavage of O–H bond (D);
   - formation of C–C bond (E);
   - formation of C–H bond (F);
   - formation of C–O bond (G).

   Then, allowance for the number of the bond transformations C, D, E, F, and G yields the conversion formula $ABC^2D^2C^2F^2G^2$.
2. Choice of transformations (generalized reactions) that involve conversions A–G from a bank (database) of steps (library of transforms) with participation of metal complexes and organometallic compounds [36, 102, 107]. A total of 20 transforms were selected, nine of which (2, 3, 5, 7, 9, 10, 11, 18, and 19) involve the simultaneous scission and formation of two bonds. For example, transform 2 involves the scission of C=C bond (B) and the formation of C–H bond (F). This set of transforms is as follows (M denotes a metal atom that enters into the metal-containing compound):

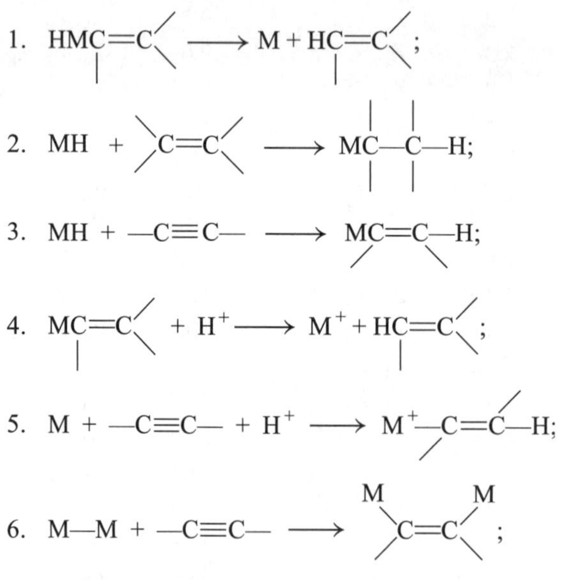

7. MOH + CO ⟶ MCOOH;

8. 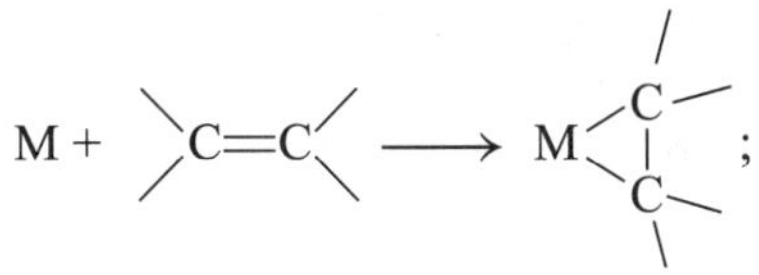

9. 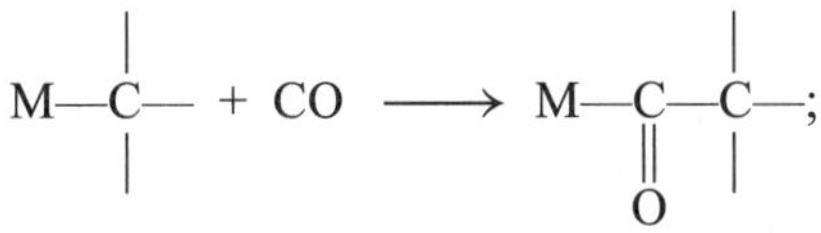

10. 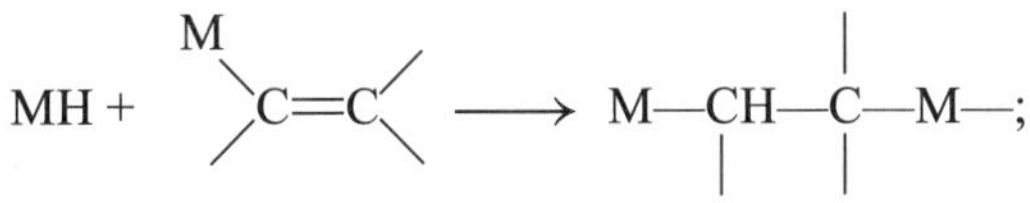

11. 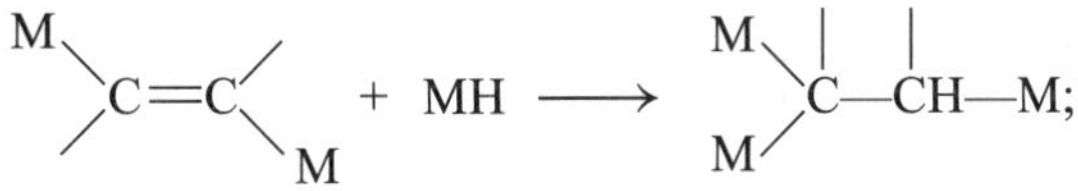

12. 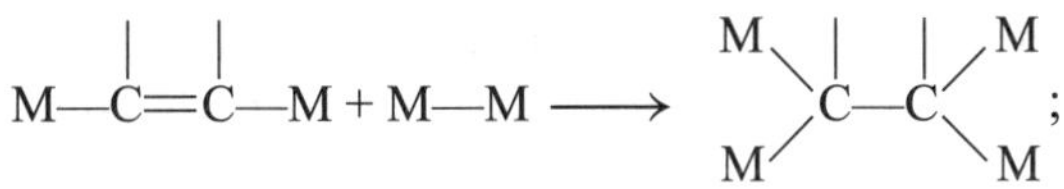

13. 

14. $M + H_2O \longrightarrow HMOH;$

15. $M^{2+} + H_2O \longrightarrow M^{+}OH + H^{+};$

16. $M{-}M + H_2O \longrightarrow MH + HOM;$

17. 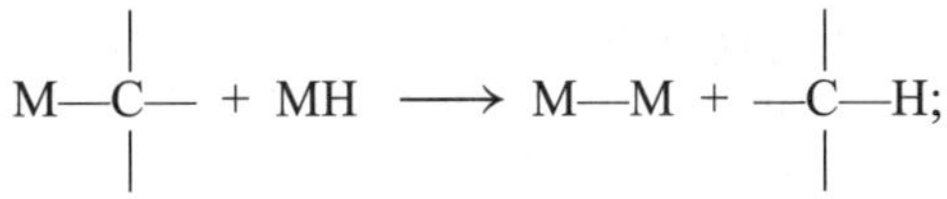

18. 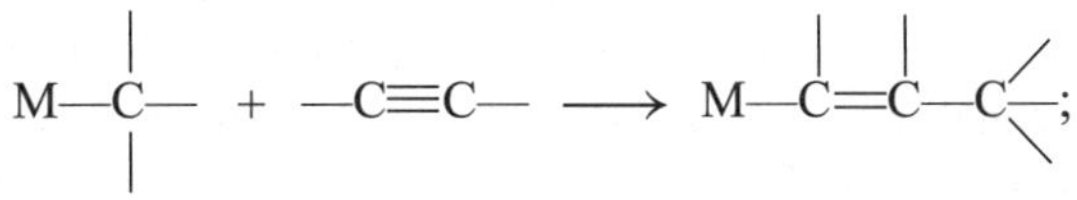

19. 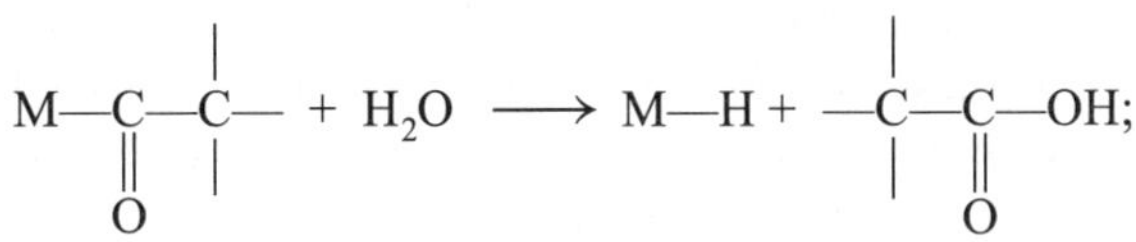

20. 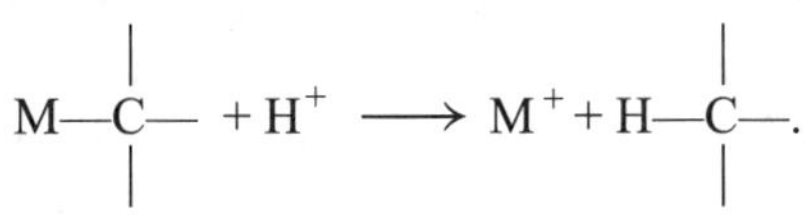

3. Combinatorial enumeration of encoded transforms in accordance with the conversion formulas with the aid of the Comb 1 program (or manually). This procedure yielded 500 combinations of steps.
4. Analysis of the obtained combinations in order to reveal those possessing chemical sense [102]. The analysis performed in [104] gave 46 reasonable combinations, i.e., hypothetical mechanisms. These hypotheses were divided into four groups with respect to the type and presence of organic intermediates in the SA synthesis:
   (i) organic intermediate: ethylene (six mechanisms);
   (ii) organic intermediate: acrylic acid (22 mechanisms);
   (iii) organic intermediate: maleic anhydride (eight mechanisms);
   (iv) no organic intermediates (10 mechanisms).

At the first stage of hypotheses discrimination, mechanisms were excluded in which the role of a catalyst (i.e., a complex that interacted with C_2H_2 or CO) was played by Pd(0) and Pd(II) complexes. Complexes HPdBr formed from $PdBr_2$ were considered as the catalysts of C_2H_2 carbonylation. After this stage, 23 hypotheses were left including those of groups I (3), II (8), III (4), and IV (8). Not all of the sequences of steps, which are derived using the conversion formula, contain a complete catalytic cycle (i.e., involve elimination of the initial catalyst species). The necessary steps of formation and regeneration of the active centers can be added at the stage of mechanism specification for the kinetic model construction. For example, it was assumed that HPdBr, which is formed in the course of $PdBr_2$ reduction by carbon monoxide

$$PdBr_2 + CO + H_2O \longrightarrow CO_2 + HPdBr + HBr$$

rapidly reduces $PdBr_2$ to Pd_2Br_2 [or the corresponding Pd(I) complex]:

$$PdBr_2 + HPdBr \longrightarrow Pd_2Br_2 + HBr.$$

At the next stage of selection of the working hypotheses, the roles of ethylene, acrylic acid, and maleic anhydride as potential intermediates were elucidated. For this purpose, each substance was introduced into the reaction system at an amount exceeding the maximum concentration observed during acetylene carbonylation. As a result of these experiments, it was established that ethylene and acrylic acid are not intermediates in the process under consideration, while maleic anhydride is a possible intermediate. Measurements of the kinetic isotope effect for H_2O replaced by D_2O allowed two mechanisms (14 and 15) to be excluded from group III and two other mechanisms (21 and 22) to be excluded from group IV. In six of the remaining eight mechanisms, an important role is played by the step of protodemetallation of organometallic intermediates. Since the initial system does not contain HBr, the appearance of hydrogen bromide is only possible at the expense of $PdBr_2$ hydrolysis and Pd(II) reduction to Pd(I), that is, in amounts not exceeding the initial values of $[Pd]_\Sigma$. The investigation of analogous acetonitrile solutions showed a virtually complete absence of HBr, which rapidly interacts with CH_3CN to form various adducts. One of these adducts was isolated and studied by X-ray diffraction [108], which showed that its composition corresponds to $(MeCN)_2(HBr)_3$ (see Chapter 8). Thus, hypotheses 12, 17, 18, 19, and 23 were also excluded from consideration.

In order to discriminate between the remaining hypotheses 13 (group III) and 16 (group IV), a single-factor kinetic experiment was carried out. Estimations obtained from statistical descriptions of this experiment in terms of mechanisms 13 and 16 (for their detailed variants 13.1, 13.2, 16.1, and 16.2) turned out to be rather close. Nevertheless, the results of an analysis of the distribution of deuterium between the reaction products and the products of MA exchange with D_2O, as well as a sharp increase in the rate of MA formation upon the introduction of O_2, allowed mechanism 13 supplemented by the steps of formation of inactive complexes of Pd(I) with CO and C_2H_2 (mechanism 13.2) to be preferred [104]:

$$(1)\ \ Pd_2Br_2(CO)_2 + C_2H_2 \underset{k_{-1}}{\overset{k_1}{\rightleftharpoons}} \underset{(X_1)}{Pd_2Br_2(CO)(C_2H_2)} + CO;$$

$$(2)\ \ X_1 + CO \xrightarrow{k_2} \underset{(X_2)}{BrPdCOCH{=}CHCOPdBr};$$

$$(3)\ \ X_2 + H_2O \longrightarrow \underset{(X_3)}{(MA)Pd_2Br_2(H)_2};$$

$$(4)\ \ X_3 + C_2H_2 \xrightarrow{k_3} MA + \underset{(X_4)}{(C_2H_2)Pd_2Br_2(H)_2};$$

$$(5)\ \ X_3 + CO \underset{k_{-4}}{\overset{k_4}{\rightleftharpoons}}$$

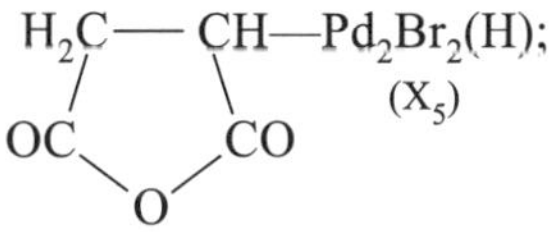

$$(6)\ \ X_5 \xrightarrow[CO]{k_5} SA + Pd_2Br_2(CO)_2;$$

$$(7)\ \ X_4 \longrightarrow X_6 \longrightarrow \text{Acrylic acid};$$

$$Pd_2Br_2(CO)_2 + CO \overset{K_{CO}}{\rightleftharpoons} Pd_2Br_2(CO)_3;$$

$$Pd_2Br_2(CO)(C_2H_2) + C_2H_2 \overset{K_{C_2H_2}}{\rightleftharpoons} Pd_2Br_2(CO)(C_2H_2)_2.$$

Here, step (4) was assumed to be irreversible at small MA concentrations in the beginning of the process. The oxidation of a hydride ligand in X_3 by oxygen (or some other oxidants) leads to a decrease in the rate of SA formation and an increase in the MA yield from reaction

$$C_2H_2 + 2CO + {}^1\!/_2\, O_2 \longrightarrow MA. \tag{7.10}$$

■

7.4.3 MECHEM program

This program generates a set of simple, single-route mechanisms of catalytic reactions for a given set of reactants, products, and justified intermediates (i.e., a minimum number of hypothetical intermediates) with allowance for certain restrictions imposed by the user [36, 85–94]. Then the user can proceed with the search for mechanisms at the next level of complexity by (i) demanding that the program generate mechanisms involving new steps and substances that were absent in the previously formulated mechanisms and/or (ii) introducing new restrictions not obeyed by the already formulated mechanisms. MECHEM program can employ up to 100 restrictions that accumulate the knowledge gained previously on the mechanisms of catalytic reactions. Latest versions of this program allow a chemist to seek for reaction mechanisms in an interactive regime [109–111].

The typical procedure and restrictions were described [109, 110] by example of the generation of mechanisms for the reaction of ethane hydrogenolysis on metal catalysts (see also [90]). One of the mechanisms generated by MECHEM program is as follows:

$$MM + H_2 \xrightarrow{(1)} 2HM;$$

$$C_2H_6 + MM \xrightarrow{(1)} HM + C_2H_5M;$$

$$C_2H_5M + MM \xrightarrow{(1)} HM + MCH_2CH_2M;$$

$$MCH_2CH_2M \xrightarrow{(1)} HM + CH_2{=}CHM;$$

$$CH_2{=}CHM + MM \xrightarrow{(1)} MCH{=}CHM + HM;$$

$$MCH{=}CHM \xrightarrow{(1)} 2HC{\equiv}M;$$

$$HC{\equiv}M + HM \xrightarrow{(2)} M + CH_2{=}M;$$

$$CH_2{=}M + HM \xrightarrow{(2)} M + CH_3M;$$

$$CH_3M + HM \xrightarrow{(2)} MM + CH_4;$$

where the reversibility of steps is not specified and figures at the arrows indicate the stoichiometric numbers of steps for the following overall equation:

$$CH_3CH_3 + H_2 \longrightarrow 2CH_4.$$

The use of interactive regimes by a chemist generating hypotheses on a computer with the aid of MECHEM program was demonstrated [111] for a model reaction of ethylene hydrogenation on transition metals, which is one of the most thoroughly studied processes in heterogeneous catalysis. The intermediate discrimination of hypotheses was based on a huge volume of experimental data accumulated over about seven decades of investigations of the mechanism of this reaction. At the stage of hypotheses generation, all mechanisms previously described in the literature were obtained, even exotic variants such as a mechanism involving ethylidyne intermediates ($M_3C{-}CH_3$) and a mechanism with the C–C bond rupture and ethane formation from two $M{-}CH_3$ intermediate species [111].

Eleven mechanisms of ethylene hydrogenation generated by MECHEM program so as to include only reversible steps were investigated [112] with the aim to find variants that admitted the multiplicity of steady states (MSS) discovered by Ellison et al. [112]. It was establish that four of the analyzed mechanisms admitted the MSS phenomenon. According to the Feinberg deficiency-one and deficiency-two theorems (see Section 6.2) [124], these mechanisms are characterized by a certain set of positive rate constants of steps, which can ensure the observed number of steady states and the corresponding values of concentrations and reaction rates in a continuous-flow stirred-tank reactor [113].

An analysis of 80 single- and two-route mechanisms of ethylene hydrogenation (13 of which were previously described in the literature) for their correspondence to two steady states observed for a catalytic process on the surface of rhodium showed that only nine mechanisms admit the existence of these observed multiple states [113]. Thus, the existence of critical phenomena (e.g., MSS, self-oscillations) and their analysis provide a powerful factor [112] for discriminating between a large number of working hypotheses (see also Chapter 6).

Based on the MECHEM program, a method was developed [115] for predicting the distribution of isotope labels in products and reactants, which can be used for the discrimination of hypotheses and the isotope experiment planning.

Example 7.4 [115]

▶ Illustrative examples for assessing the hypothesis discrimination ability of experiments with labeled atoms in the framework of MECHEM program are offered by the mechanisms of propylene hydrogenation on Pt(111) [116], which differ only by the positions of M–C bonds in the last precursor of propane. For the sake of simplicity, propylidene intermediate was excluded from these mechanistic schemes.

Mechanism 1:

$$\text{(1) } H_2 + 2M \rightleftharpoons 2MH;$$
$$\text{(2) } C_3H_6 + 2M \rightleftharpoons MCH(CH_3){-}CH_2M;$$
$$\text{(3) } MH + MCH(CH_3){-}CH_2M \rightleftharpoons 2M + MCH_2{-}CH_2{-}CH_3;$$
$$\text{(4) } MH + MCH_2{-}CH_2{-}CH_3 \longrightarrow 2M + CH_3{-}CH_2{-}CH_3.$$

Mechanism 2:

$$\text{(1) } H_2 + 2M \rightleftharpoons 2MH;$$
$$\text{(2) } C_3H_6 + 2M \rightleftharpoons MCH(CH_3){-}CH_2M;$$
$$\text{(3) } MH + MCH(CH_3){-}CH_2M \rightleftharpoons 2M + M{-}CH(CH_3)_2;$$
$$\text{(4) } MH + M{-}CH(CH_3)_2 \longrightarrow 2M + CH_3{-}CH_2{-}CH_3;$$

Mechanism 3 (1 + 2):

$$\text{(1) } H_2 + 2M \rightleftharpoons 2MH;$$
$$\text{(2) } C_3H_6 + 2M \rightleftharpoons MCH(CH_3){=}CH_2M;$$
$$\text{(3) } MH + MCH(CH_3){-}CH_2M \rightleftharpoons 2M + M{-}CH(CH_3)_2;$$
$$\text{(4) } MH + MCH(CH_3){-}CH_2M \rightleftharpoons 2M + MCH_2{-}CH_2{-}CH_3;$$
$$\text{(5) } MH + M{-}CH(CH_3)_2 \longrightarrow 2M + CH_3{-}CH_2{-}CH_3;$$
$$\text{(6) } MH + MCH_2{-}CH_2{-}CH_3 \longrightarrow 2M + CH_3{-}CH_2{-}CH_3.$$

Computer simulations were performed for two series of experiments

$$\text{(I) } CH_2{=}CHCH_3 + D_2;$$
$$\text{(II) } CH_2{=}CHCD_3 + H_2,$$

where all steps (except for the irreversible) of the mechanisms under consideration were assumed to be quasi-equilibrium and the H/D exchange in propane was ignored. The results of computer experiments showed that the number of steps in the reaction network involving labeled compounds significantly increases. Indeed, the number of steps in the series of mechanisms 1, 2, and 1 + 2 for experiment I amount to 24, 98, and 358, while those for experiment II are 7, 98, and 358, respectively. Evidently, the whole set of labeled compounds can hardly be predicted without using computers.

What conclusions can be made based on the obtained data array? Several inferences (rules) concerning the discrimination of hypotheses are as follows:

(i) Experiment II provides more information than experiment I, since the latter experiment yielded 11 substances in common for mechanisms 1 and 2, while the former gave only five such compounds.

(ii) Compounds C_3D_8, $CD_2HCDHCD_2H$, $CDH_2CD_2CD_3$, $CH_3CD_2CD_3$, $CDH_2CD_2CD_2H$, and $CDH_2CD_2CDH_2$ are not formed according to mechanisms 1 or 2.
(iii) Mechanism 3 (1 + 2) yields all the possible deuterated and partly deuterated products (a total of 56 compounds).
(iv) If propane C_3H_8 is observed in experiment II, the isotope experiment entirely corresponds to mechanism 2.
(v) If both experiments yield the same set of deuterated products, then mechanism 2 is valid.
(vi) If a deuterium atom is bound to carbon atom in position 2 of propane in experiment (II), mechanism I must be rejected.
(vii) Propanes $d_8 - d_4$ are not formed in the framework of mechanism 1.

It was shown [115] how the MECHEM program could be used for an analysis of the possible pathways of n-heptane dehydrocyclization to toluene with four positions of ^{14}C label in the initial heptane structure (Figure 7.2). ■

7.4.4 NetGen program

Hypothesis generation program NetGen [78] is based on the use of a BE (bond–electron) matrix apparatus developed by Dugunji and Ugi [72–74]. With various modifications (see, e.g. [117, 118]), this program was successfully applied to constructing kinetic models for very complicated reactions such as the hydroisomerization of paraffins [79] and Fischer–Tropsch synthesis [119]. An effective approach was provided by using a small number of reaction BE matrices to generate large reaction networks consisting of repeated steps (e.g., acid catalysis, radical processes, hydrocarbon chain propagation in the Fischer–Tropsch synthesis).

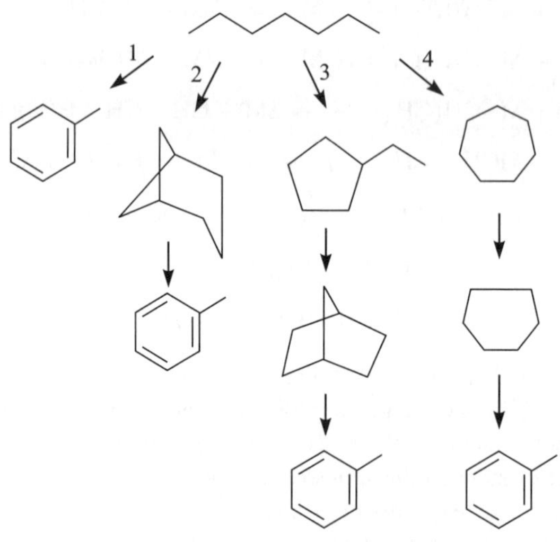

Figure 7.2 Pathways of heptane dehydrocyclization with participation of organic intermediates (reproduced with permission from John Wiley& Sons © 1998)

Example 7.5 [79]

► Let us consider a procedure used to construct a kinetic model for n-hexadecane (C_{16}) hydroisomerization on bifunctional supported Pt/alumosilicate catalysts. By this reaction, normal paraffins are converted into branched, less waxy paraffins, almost without the formation of cracking products. The construction of a kinetic model [79] for this process consisted of three stages.

Stage I: generation of a reaction network by means of NetGen and its modifications.
Stage II: automated transformation of this reaction network into a system of algebraic and differential equations (OdeGen routine);
Stage III: solution of obtained equations (with allowance for the reactor type) and comparison to experiment.

In stage I, a scheme of processes on the bifunctional Pt/alumosilicate catalyst was constructed using reactions of eight types representing standard conversions of the reactive centers of reactants to the products of elementary steps:

- dehydrogenation of alkanes into alkenes on Pt (considered as elementary process):

$$-CH_2-CH_2- \xrightarrow{Pt} -CH{=}CH- + H_2$$

- protonation of alkenes on acid centers of the catalyst:

$$-CH{=}CH- + H^+ \longrightarrow -\overset{+}{C}HCH_2-;$$

- hydride shift:

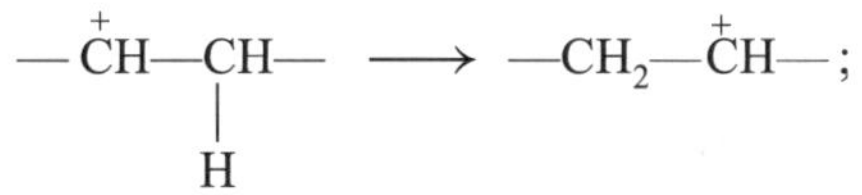

- methyl shift:

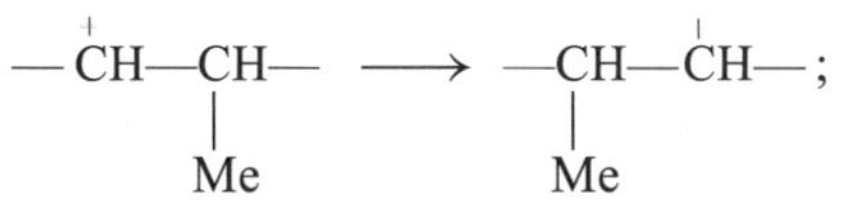

- skeletal isomerization via protonated cyclopropane (SICP):

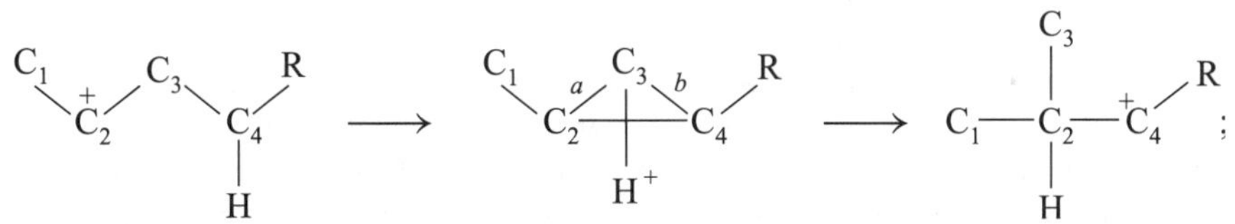

- cleavage of C – C bond (β-elimination of carbenium ion):

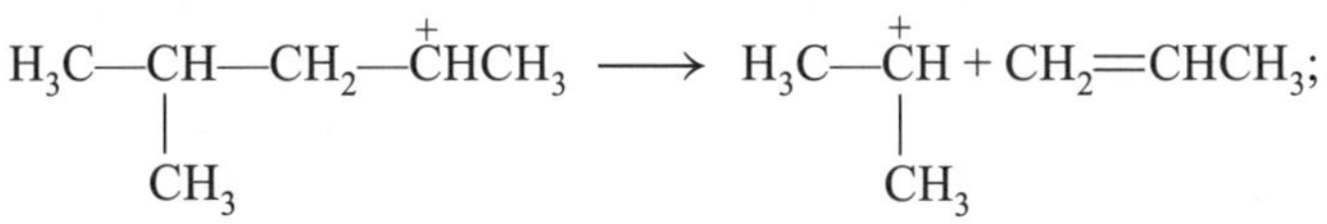

- deprotonation of carbenium ion:

$$-\overset{+}{C}HCH_2- \longrightarrow -CH{=}CH- + H^+;$$

- hydrogenation of alkenes:

$$-CH{=}CH- + H_2 \xrightarrow{Pt} -CH_2CH_2-.$$

In the case of initial C_{16} paraffin that is converted into molecules with smaller numbers of carbon atoms, about 50 000 isomeric paraffins and about 500 000 isomeric olefins (including stereoisomers) can appear. In order to obtain a kinetic model of a "reasonable" dimension, it is necessary use some rules (restrictions) based on the generalized knowledge gained about these eight reactions. Thirteen rules that helped to restrict the reaction network complication were as follows [79]:

1. All carbenium ions below C_{16} are deprotonated and hydrogenated.
2. Primary carbenium ions are not formed.
3. Tertiary carbenium ions are stable and not subject to SICP.
4. SICP increases the number of branching sites and/or length of side chains.
5. SICP does not produce geminal branching (experimentally, quaternary carbon atoms are seldom found in isomers).
6. SICP does not allow α–branching relative to existing branching sites.
7. Side-chain length does not exceed two carbon atoms.
8. A maximum of three branching sites is admitted.
9. Dehydrogenation is possible at any site of linear paraffin and only in β position relative to branching in isoparaffins.
10. Secondary cracking (β-cleavage of C–C bonds) is prohibited.
11. C-type of β-cleavage, in which secondary carbenium ion forms another secondary ion, is forbidden (in some models).
12. Hydride ion transfer is not considered (except for hydride 1, 2-shift).
13. The number of allowed SICPs and methyl and hydride shifts is determined as a function of the number of branching sites in carbenium ion-reactant.

These rules provide the entire spectrum of necessary products, while retaining the model dimension on a level acceptable for solving kinetic problems.

Application of the above approach with various restrictions yielded 25 models, which were analyzed with respect to their dimensionality and spectrum of products. The "optimum" model 10.13 involves 799 substances (393 olefins, 345 ions) and 2207 reactions (steps). The corresponding set of paraffins includes 30 isomers of C_{16} and 29 products of their cracking (C_3–C_{13}). The obtained reaction network was characterized by only eight rate constants (for the aforementioned eight transformations), which were evaluated and refined in comparison to experiment. This model successfully described the experimental values of hexadecane conversion and molar yields of products, including the group yields for monobranched C_{16}, polybranched C_{16}, and cracking products, as functions of the contact time in a flow reactor. However, the application of model 10.13 required several weeks for its optimization because of the large volume of calculations.

In order to reduce the number of reactions and substances while retaining the spectrum of products (C_{16} paraffins and cracking products in the absence of olefins), the transformations of substances were rated [120]. According to this approach, a rank is introduced that determines the order of appearance of a given substance on proceeding from the initial reactant (whose rank is zero). Thus, each type of transformation can be assigned a limiting rank, on attaining which this reaction is terminated. For example, dehydrogenation does not proceed for a substance of rank ≥ 3. The maximum rank determines the level of complexity of a model. The rank-based criterion of termination of the network generation was also used in the case of radical reactions [120]. Model 10.13 had a maximum rank of 10. A simplified model (10.13 PRR) with reduced ranks for all reactions involves 496 substances and 1443 steps. The time necessary for a numerical solution of the corresponding system of equations decreased 25 times, while the quality of description of the experimental data was retained. A very good description of the experiment was obtained using a variant with three different catalysts and three sets of rate constants. ■

The above example convincingly demonstrates the possibilities of using large model reaction networks. It should be noted that the use of ranking in order to restrict the growth of reaction networks can hardly be formalized. The task of choosing ranks is no less difficult than that of operating with large networks.

7.4.5 TAMREAC program

This program was oriented to study the mechanisms of organometallic catalysis [97–99]. The TAMREAC program developers formulated a system of selection rules that are used – together with data on the type of chemical transformation (reductive elimination, insertion, metallacycle formation, etc.), metal coordination number, number of valence electrons, and geometry of the metal complex – to obtain an expert conclusion concerning the possibility of a given reaction. All these data, represented in the form of selection matrices, are very useful for a researcher. However, the volume of information used by the TAMREAC program to derive a conclusion is by no means exhaustive and, hence, the obtained judgment can only be taken into consideration, while the possibility of applying it to particular investigations is very restricted. In studying particular systems, a researcher has the right to decide on whether to take the available expert recommendations into account. Therefore, in a more acceptable variant, a researcher (program user) should have the possibility to formulate their own selection rules within various schemes stipulated by the program.

The TAMREAC program was tested on a large number of reactions and in some cases gave very interesting results. For example, the program was applied to modeling a reaction network for ethylene hydroformylation in a catalytic system with $HRh(CO)_2L_2 (L = PPh_3)$ [98]. The initial substances were ethylene, H_2, CO, and $HRh(CO)_2L_2$. As can be seen from the results presented in Tables 7.12 and 7.13, the program "synthesized" rather interesting side products. A search for these in a real catalytic system can help to accept or reject some of the proposed reaction [36].

7.4.6 ChemNet program

This program is based on libraries of elementary steps that are selected by experts (or by a program user) and written in the form of generalized transforms [36, 100, 102]. The ChemNet program was used to model both heterogeneous [63, 100] and homogeneous [100, 104, 110, 121] catalytic processes. Several illustrative application examples are given below.

Example 7.6 [100]

▶ The process of ethane hydrogenolysis via reaction

$$C_2H_6 + H_2 \longrightarrow 2CH_4 \tag{7.11}$$

was also used to test the MECHEM program [90, 109]. A procedure of generating hypotheses about the mechanism of this reaction with the aid of the ChemNet program was described in [107, 100]. Preliminary analysis of the literature (see, e.g., [122–127]) led to the following conclusions [107] based on experimental and/or theoretical grounds.

Table 7.12 List of substances involved in reaction network generated by the TAMREAC program for ethylene hydroformylation [98]

No.	Chemical formula	No.	Chemical formula
1	$HRhL_2(CO)_2$	21	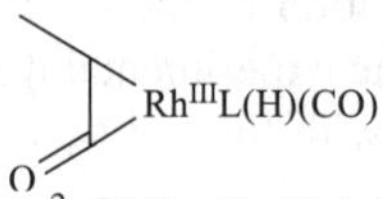
2	$HRhL_2(CO)_2(\eta^2\text{-}C_2H_4)$	22	$[\eta^2\text{-}C(O){=}CHCH_3]RhL(H)(CO)$
3	$HRhL(CO)_2$	23	$O{=}CH{-}CH(CH_3){-}Rh^{III}L_2(H)_2CO$
4	$C_2H_5RhL_2(CO)_2$	24	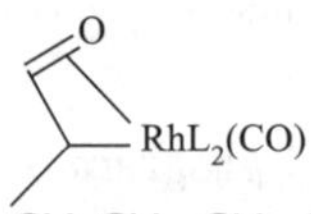
5	$C_2H_5CORhL_2(CO)$	25	$CH_3CH{=}CH{-}O{-}RhL_2(CO)$
6	$C_2H_5CORhL_2(H)_2(CO)$	26	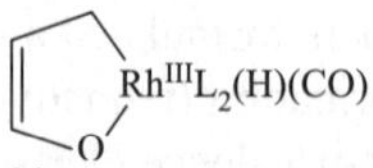
7	$HRhL_2(CO)$	27	H_2
8	$CH_3CH{=}RhL_2(H)(CO)_2$	28	$HO{-}CH{=}CHCH_2{-}RhL_2(CO)$
9	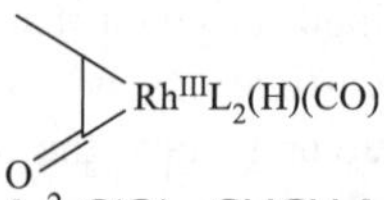	29	$HO{-}CH{=}CHCH_2{-}Rh^{III}L_2(H)_2(CO)$
10	$[\eta^2\text{-}C(O){=}CHCH_3]{-}RhL_2(H)CO$	30	$CH_3CH{=}CH{-}O{-}Rh^{III}L_2(H)_2(CO)$
11	$O{=}CH{-}CH(CH_3){-}RhL_2CO$	31	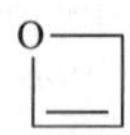
12	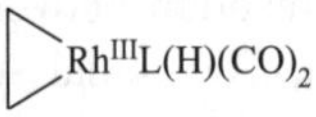	32	$[\eta^2\text{-}CH_2{=}CH{-}C(O)]RhL_2(H)(CO)$
13	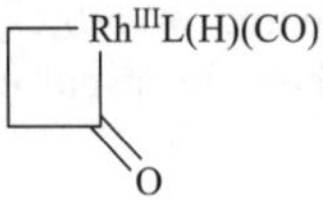	33	$CH_3CH{=}CH{-}OH$
14	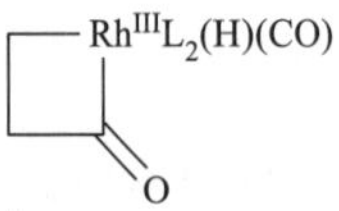	34	$O{=}CHCH_2CH_2{-}RhL_2(CO)$
15	L	35	$O{=}CHCH_2CH_2{-}Rh^{III}L_2(H)_2(CO)$
16	$HRhL(CO)_2(\eta^2\text{-}C_2H_4)$	36	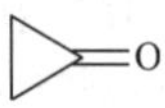
17	$C_2H_5RhL(CO)_2$	37	C_2H_4
18	$C_2H_5Rh^{III}L(H)_2(CO)_2$	38	C_2H_6
19	CO	39	$CH_2{=}CH{-}CH{=}O$
20	$CH_3CH{=}RhL(H)(CO)_2$	40	C_2H_5CHO

Table 7.13 Reaction network generated by the TAMREAC program for ethylene hydroformylation [98]

No.	Reaction *	No.	Reaction *	No.	Reaction *	No.	Reaction *
1	$1 \to 3 + 15$	23	$7 + 39 \to 32$	45	$13 \to 12$	67	$23 \to 7 + 40$
2	$1 \to 7 + 19$	24	$7 + 40 \to 6$	46	$13 + 15 \to 14$	68	$23 \to 11 + 27$
3	$1 + 37 \to 2$	25	$7 + 40 \to 23$	47	$14 \to 7 + 36$	69	$24 \to 11$
4	$2 \to 1 + 37$	26	$8 \to 4$	48	$14 \to 13 + 15$	70	$24 \to 25$
5	$2 \to 4$	27	$8 \to 9$	49	$15 + 16 \to 2$	71	$25 + 27 \to 30$
6	$2 \to 16 + 15$	28	$8 \to 20 + 15$	50	$16 \to 3 + 37$	72	$25 \to 24$
7	$3 + 15 \to 1$	29	$9 \to 10$	51	$16 \to 9$	73	$25 \to 26$
8	$3 + 37 \to 16$	30	$9 \to 11$	52	$16 \to 12$	74	$26 \to 7 + 31$
9	$4 \to 2$	31	$9 \to 16$	53	$16 \to 20$	75	$26 \to 25$
10	$4 \to 5$	32	$9 \to 21 + 15$	54	$17 + 15 \to 4$	76	$26 \to 28$
11	$4 \to 8$	33	$10 \to 5$	55	$17 \to 20$	77	$28 \to 26$
12	$4 \to 17 + 15$	34	$10 \to 9$	56	$17 + 27 \to 18$	78	$28 + 27 \to 29$
13	$5 \to 4$	35	$10 \to 11$	57	$18 \to 3 + 38$	79	$29 \to 7 + 33$
14	$5 \to 10$	36	$10 \to 22 + 15$	58	$18 \to 17 + 27$	80	$29 \to 28 + 27$
15	$5 + 27 \to 6$	37	$11 \to 9$	59	$20 + 15 \to 8$	81	$30 \to 7 + 33$
16	$6 \to 5 + 27$	38	$11 \to 10$	60	$20 \to 16$	82	$30 \to 25 + 27$
17	$6 \to 7 + 40$	39	$11 \to 24$	61	$20 \to 17$	83	$32 \to 7 + 39$
18	$7 + 19 \to 1$	40	$11 \to 32$	62	$20 \to 21$	84	$32 \to 11$
19	$7 + 31 \to 26$	41	$11 + 27 \to 23$	63	$21 + 15 \to 9$	85	$32 \to 34$
20	$7 + 33 \to 30$	42	$12 \to 9$	64	$21 \to 22$	86	$34 + 27 \to 35$
21	$7 + 33 \to 29$	43	$12 \to 13$	65	$22 + 15 \to 10$	87	$34 \to 32$
22	$7 + 36 \to 14$	44	$12 \to 16$	66	$22 \to 21$	88	$35 \to 7 + 40$
						89	$35 \to 34 + 27$

*Numbers of compounds according to Table 7.12.

1. Reaction (7.11) is always accompanied by the appearance of hydrogen atoms adsorbed on the surface of a catalyst metal. The possible ways to the formation of HM intermediates always include the dissociative adsorption of hydrogen molecules, which is due to a higher reactivity of H_2 compared to alkanes.
2. Ethane can be adsorbed on the metal surface only at the expense of C–H bond rupture. At the same time, the C–C bind is retained. In subsequent steps, the C–C bond can also be cleaved when some of the ethane hydrogen atoms are replaced by metal atoms.
3. The process of hydrogenolysis involves the formation of M–C bonds.
4. The cleavage of C–C bond leads to the formation of surface C_1 compounds.
5. Hydrogenation of the surface C_1 compounds leads to the formation of methane.

Based on these conclusions, five generalized reactions (transforms) have been selected. These transforms, together with a condition that active centers on the metal surface are monovalent, imply that carbene and carbyne species are bound to different centers on the metal catalyst surface. Accordingly, fragments of the M≡C– type are denoted by $M_3{\equiv}C-$ and those of the M=C< type, by $M_2{=}C<$. The selected five transforms involved the following substitutions:

1. $M{-}M + R_1{-}H \longrightarrow M{-}H + M{-}R_1 \quad (R_1 = C \text{ or } H);$
2. $M{-}H + M{-}R_1 \longrightarrow M{-}M + R_1{-}H \quad (R_1 = C \text{ or } H);$
3. $M{-}R_2 + C{-}H \longrightarrow M{-}C + H{-}R_2 \quad (R_2 = C, H \text{ or } M);$

$$4.\ \text{M–C} + \text{H–R}_2 \longrightarrow \text{M–R}_2 + \text{C–H} \quad (\text{R}_2 = \text{C, H or M});$$

$$5.\ \text{M–CR}_3\text{R}_4\text{–CR}_5\text{R}_6\text{R}_7 + \text{M–R}_8 \longrightarrow \text{M–CR}_3\text{R}_4\text{–M} + \text{CR}_5\text{R}_6\text{R}_7\text{R}_8$$
$$[\text{R}_i = \text{H or M}(i = 3 \div 8)].$$

On this basis, the ChemNet program generated a reaction network [107, 100] that consisted of 150 reversible steps (bidirectional reactions were also reduced to reversible steps) and 24 irreversible elementary steps, and involved six substances (participants). According to imposed restrictions, the number of atoms in molecules of a substance could not exceed eight, and a molecule could not contain more than two carbons, six metal atoms, and six hydrogen atoms. Each metal atom M was considered to be an active surface center with a unity coordination number. The reaction network generated by the ChemNet program contained all the known mechanisms reported in the literature. For example, it contained a mechanism proposed by Mahaffy and Hansen [124] based on the study of ethane hydrogenolysis on iridium films at 80–205 °C in a broad range of partial pressures of all reactants. This mechanism consists of the following steps:

$$\text{M–M} + \text{Me–Me} \rightleftharpoons \text{Me–CH}_2\text{M} + \text{MH};$$
$$\text{M–M} + \text{H}_2 \rightleftharpoons 2\text{MH};$$
$$\text{Me–CH}_2\text{M} + \text{M–M} \rightleftharpoons \text{MH} + \text{Me–CHM}_2;$$
$$\text{MCH}_2\text{–Me} + \text{M–H} \longrightarrow 2\text{M–Me};$$
$$\text{M–Me} + \text{MH} \rightleftharpoons \text{M–M} + \text{CH}_4.$$

and it is assumed that the C–C bond in the ethyl complex is ruptured under the action of MH hydride and the resulting methyl complexes on the surface are hydrogenated to methane.

Gudkov et al. [125] studied the reaction on Pt/SiO_2 and PtFe/SiO_2 catalysts and proposed two hypotheses about the mechanisms, which were believed to be valid under different reaction conditions. The first mechanism includes the following steps:

$$\text{M–M} + \text{Me–Me} \rightleftharpoons \text{Me–CH}_2\text{M} + \text{MH};$$
$$\text{M–M} + \text{H}_2 \rightleftharpoons 2\text{MH};$$
$$\text{MCH}_2\text{–Me} + \text{M–M} \longrightarrow \text{M}_2\text{CH}_2 + \text{M–Me};$$
$$\text{M}_2\text{CH}_2 + \text{MH} \rightleftharpoons \text{M–M} + \text{M–Me};$$
$$\text{M–Me} + \text{MH} \rightleftharpoons \text{M–M} + \text{CH}_4.$$

where the C–C is also only ruptured in the ethyl complex, but under the action of active centers of the catalyst surface. The second mechanism consists of the following steps:

$$\text{M–M} + \text{Me–Me} \rightleftharpoons \text{Me–CH}_2\text{M} + \text{MH};$$
$$\text{M–M} + \text{H}_2 \rightleftharpoons 2\text{MH};$$
$$\text{Me–CH}_2\text{M} + \text{M–M} \rightleftharpoons \text{MH} + \text{MCH}_2\text{–CH}_2\text{M};$$
$$\text{MCH}_2\text{–CH}_2\text{M} + \text{M–M} \rightleftharpoons \text{MH} + \text{MCH}_2\text{–CHM}_2;$$
$$\text{MCH}_2\text{–CHM}_2 + \text{M–M} \rightleftharpoons \text{MH} + \text{M}_2\text{CH–CHM}_2;$$

$$M_2CH\text{–}CHM_2 + M\text{–}M \longrightarrow 2M_3CH;$$
$$M_3CH + MH \rightleftharpoons M\text{–}M + M_2CH_2;$$
$$M_2CH_2 + MH \rightleftharpoons M\text{–}M + MMe;$$
$$MMe + MH \rightleftharpoons M\text{–}M + CH_4.$$

according to which the C–C bond is cleaved in a symmetric surface complex $CHM_2\text{–}CHM_2$ [125, 100]. Both these mechanism are contained in the reaction network generated by ChemNet program.

Goddard et al. [126] studied the mechanism of ethane hydrogenolysis on Pt, Pd, Ir, and Co catalysts supported on SiO_2, and proposed the following general mechanism:

$$(1)\ M\text{–}M + Me\text{–}Me \rightleftharpoons Me\text{–}CH_2M + MH;$$
$$(2)\ M\text{–}M + H_2 \rightleftharpoons 2MH;$$
$$(3)\ Me\text{–}CH_2M + M\text{–}M \rightleftharpoons MH + MCH_2\text{–}CH_2M;$$
$$(4)\ MCH_2\text{–}CH_2M + M\text{–}M \rightleftharpoons MH + M_2CH\text{–}CH_2M;$$
$$(5)\ M_2CH\text{–}CH_2M + M\text{–}M \rightleftharpoons MH + M_2CH\text{–}CHM_2;$$
$$(6)\ MCH_2\text{–}CH_2M + M\text{–}M \longrightarrow 2M_2CH_2;$$
$$(7)\ M_2CH\text{–}CH_2M + M\text{–}M \longrightarrow M_3CH + M_2CH_2;$$
$$(8)\ M_2CH\text{–}CHM_2 + M\text{–}M \longrightarrow 2M_3CH;$$
$$(9)\ M_3CH + MH \rightleftharpoons M\text{–}M + M_2CH_2;$$
$$(10)\ M_2CH_2 + MH \rightleftharpoons M\text{–}M + MMe;$$
$$(11)\ MMe + MH \rightleftharpoons M\text{–}M + CH_4.$$

Based on this scheme, three reaction routes were constructed that consisted of steps 1–3, 6, 10, and 11 (route *A*), steps 1–3, 4, 7, and 9–11 route *B*), and steps 1–3, 4, 5, 8, and 9–11 (route *C*). It was eventually concluded that route *C* (via the formation of M_2CHCHM_2 and subsequent C–C bond rupture) did not agree with the experiment, although this conclusion was at variance with the results obtained in [125]. Route *A* or *B* can predominate depending on the particular catalyst and reaction conditions.

Zaera and Somorjai [127] suggested that the hydrogenolysis on Pt(111) surface proceeds with the participation of ethylidyne compounds and proposed the following mechanism:

$$M\text{–}M + Me\text{–}Me \rightleftharpoons Me\text{–}CH_2M + MH;$$
$$Me\text{–}CH_2M + M\text{–}M \rightleftharpoons MH + Me\text{–}CHM_2;$$
$$Me\text{–}CHM_2 + M\text{–}M \rightleftharpoons MMe + M_3CH;$$
$$M_3CH + MH \rightleftharpoons M\text{–}M + M_2CH_2;$$
$$M_2CH_2 + MH \rightleftharpoons M\text{–}M + MMe;$$
$$MMe + MH \rightleftharpoons M\text{–}M + CH_4;$$
$$MeCHM_2 + 3M\text{–}M \longrightarrow 2M_3CH + 2MH.$$

All the steps of this mechanism, except the last one, are contained in the reaction network generated by the ChemNet program. This last step is not elementary, but it can be represented as a combination of elementary steps.

The reaction network obtained in [107, 100] using the ChemNet program contains many steps that are never discussed by most researchers engaged in the conversions of hydrocarbons. In particular, these are steps in which a hydrogen atom bound to carbon is transferred between two organometallic intermediates. For example, a key step in the mechanism of ethane hydrogenolysis according to Zaera and Somorjai [127] is the following reaction:

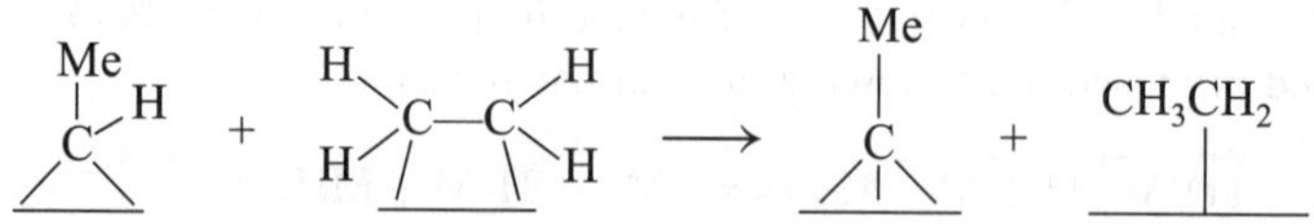
(7.12)

Steps like this were also obtained in the mechanisms of ethylene hydrogenation generated by the MECHEM program [111]. The presence of these steps significantly increases the dimensions of reaction networks (ChemNet) or the number of separate mechanisms (MECHEM). For this reason, an experimental or quantum-mechanical investigation of the probability of existence of these steps with transitions states of the A and B types [100, 111],

X = H, R,

A B

would be very useful. The possibility of excluding such steps could significantly reduce the number of generated hypothetical mechanisms. ■

Example 7.7 [110]

► This study demonstrated the application of the ChemNet program to reaction network generation for acetylene hydrocarboxylation

$$C_2H_2 + CO + H_2O \longrightarrow CH_2{=}CHCOOH, \qquad (7.13)$$

catalyzed by Pd(II) and Pd(0) complexes and showed the possibilities of the MECHEM program in the search for simple single-route mechanisms on the manifold steps and intermediates provided by the obtained reaction network. The knowledge gained previously for the mechanisms of close reactions was summarized in the form of 11 generalized transforms:

1. $Pd^{II}{-}OR' + C{\equiv}O \longrightarrow Pd^{II}C({=}O)R'$ $(R' = H, C)$;
2. $Pd^{II}{-}C({=}O){-} + HOH \longrightarrow Pd^{II}H + HO{-}C({=}O){-}$;
3. $Pd^{II}Br + HOH \longrightarrow HBr + Pd^{II}OH$;
4. $Pd^{II}{-}C({=}O){-} + C{\equiv}C \longrightarrow Pd^{II}{-}C{=}C{-}C{=}O$;

5. $Pd^{II}\text{—}C + HBr \longrightarrow H\text{—}C + Pd^{II}\text{—}Br;$

6. $H\text{—}C\text{—}Pd^{II} + C{\equiv}O \longrightarrow Pd^{II}\text{—}\underset{\underset{O}{\|}}{C}\text{—}C\text{—}H;$

7. $H\text{—}Pd^{II} + C{\equiv}C \longrightarrow Pd^{II}\text{—}C{=}C\text{—}H;$

8. $BrPd^{II}\text{—}H \longrightarrow H\text{—}Br + Pd^{0};$

9. $H\text{—}Br + Pd^{0} \longrightarrow BrPd^{II}\text{—}H;$

10. $Pd^{0} + HC{\equiv}CH \longrightarrow$ 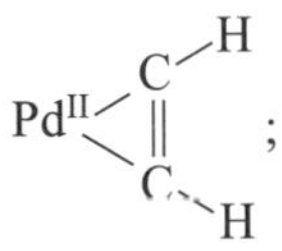 ;

11. $C\text{—}Pd^{II}\text{—}C \longrightarrow C\text{—}C + Pd^{0}.$

Mechanisms involving Pd(I) complexes [35, 38–40 104–106] were not employed in the computer experiment under consideration. The preselected set of transforms included (i) steps that were previously proposed for the synthesis of acrylic acid (AA) and its esters (**1–7, 10**), and (ii) steps characteristic for the catalysis with metal complexes, which undoubtedly exist and can be involved in the process under consideration (**8, 9, 11**). The maximum degree of palladium oxidation in all intermediates corresponded to Pd(II). In order to take into account differences in the properties of CO molecules and >C=O carbonyl groups, a higher bond order for CO is used as in reactions **1** and **6**. Reaction **6** allows the insertion of C≡O into the bond of palladium to hydrogen-bonded (C–H) carbon, while the insertion of CO into the bond of palladium to carbon that is bound to any other atom (e.g., Pd–C=O) is prohibited.

The level of generality used in writing the above transforms significantly influences the results obtained using a reaction-network-generating program. For example, transforms **8** and **11** describe the reactions of reductive elimination. If these transforms were replaced by a single more general transform

$$R_1Pd^{II}R_2 \longrightarrow R_1\ R_2 + Pd^{0}(R_1 = H, C; R_2 = Br, C),$$

the higher level of generality would actually be equivalent to an increase in the number of transforms in the initial set. Indeed, the addition of steps such as

$$HPd^{II}\text{-}C \longrightarrow H\text{-}C + Pd^{0};$$
$$C\text{-}Pd^{II}\text{-}Br \longrightarrow C\text{-}Br + Pd^{0},$$

to reactions **8** and **11** significantly increases the dimension of the generated reaction network.

Initial limitations to reaction network generation included (i) restriction of the maximum number of atoms in intermediates to 15, (ii) restriction of the maximum numbers of C, O, and Pd atoms in the intermediates to 6, 4, and 1, respectively, and (iii) restriction of the maximum coordination number of palladium in intermediates to 2.

The operation of the ChemNet program with the above restrictions yielded a reaction network that contained 160 elementary steps and involved 77 substances, which included 20 organic substances that could potentially be the side product of carbonylation and H_2CO_3.

The application of the MECHEM program to the obtained network revealed 35 simple mechanisms, each containing four to eight steps and involving four to nine intermediates. The program was stopped when all three types of mechanism reported in the available literature were found, including the hydroxycarbonyl type (20 mechanisms), hydride type (10 mechanisms), and metallacycle type (2 mechanisms). Three mechanisms had a hybrid character, representing hydride-hydroxycarbonyl (2 mechanisms) and hydride-metallacycle (1 mechanism) mixed types.

All mechanisms with the number of steps above four contained the steps of active catalyst formation. The stoichiometric numbers of all steps were selected with allowance for the Horiuti rule so as to eventually obtain the overall stoichiometric equation (7.13). Using the generated reaction network, it is also possible to obtain multi-route mechanisms that would lead to AA and any preset combination of side products. Examples of simple mechanisms are as follows (numbers in angle brackets refer to the total list of mechanisms [110]).

< 1 > Hydroxycarbonyl mechanism:

$$(1)\ H_2O + PdBr_2 \xrightarrow{(1)} HBr + BrPdOH;$$

$$(2)\ BrPdOH + CO \xrightarrow{(1)} BrPdCOOH;$$

$$(3)\ BrPdCOOH + C_2H_2 \xrightarrow{(1)} BrPdCH{=}CHCOOH;$$

$$(4)\ BrPdCH{=}CHCOOH + HBr \xrightarrow{(1)} AA + PdBr_2.$$

< 7 > Hydride mechanism:

$$(1)\ H_2O + PdBr_2 \xrightarrow{(0)} HBr + BrPdOH;$$

$$(2)\ BrPdOH + CO \xrightarrow{(0)} BrPdCOOH;$$

$$(3)\ BrPdCOOH + H_2O \xrightarrow{(0)} HPdBr + H_2CO_3;$$

$$(4)\ HPdBr + C_2H_2 \xrightarrow{(1)} CH_2{=}CHPdBr;$$

$$(5)\ CH_2{=}CHPdBr + CO \xrightarrow{(1)} CH_2{=}CHCOPdBr;$$

$$(6)\ CH_2{=}CHCOPdBr + H_2O \xrightarrow{(1)} HPdBr + AA.$$

< 27 > Simple hybrid hydride-hydroxycarbonyl mechanism:

$$(1)\ H_2O + PdBr_2 \xrightarrow{(0)} HBr + BrPdOH;$$

$$(2)\ BrPdOH + CO \xrightarrow{(0)} BrPdCOOH;$$

$$(3)\ BrPdCOOH + H_2O \xrightarrow{(0)} HPdBr + H_2CO_3;$$

$$(4)\ HPdBr + H_2O \xrightarrow{(1)} HBr + HPdOH;$$

$$(5)\ HPdOH + CO \xrightarrow{(1)} HPdCOOH;$$

$$(6)\ HPdCOOH + C_2H_2 \xrightarrow{(1)} CH_2{=}CHPdCOOH;$$

$$(7)\ CH_2{=}CHPdCOOH \xrightarrow{(1)} AA + Pd^0;$$

$$(8)\ HBr + Pd^0 \longrightarrow HPdBr.$$

< 33 > Metallacycle mechanism:

$$(1)\ H_2O + PdBr_2 \xrightarrow{(0)} HBr + BrPdOH;$$

$$(2)\ BrPdOH + CO \xrightarrow{(0)} BrPdCOOH;$$

$$(3)\ BrPdCOOH + H_2O \xrightarrow{(0)} HPdBr + H_2CO_3;$$

$$(4)\ HPdBr \xrightarrow{(1)} Pd^0 + HBr;$$

(5) 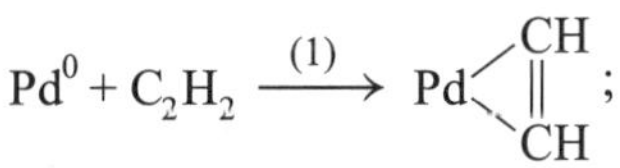

(6) 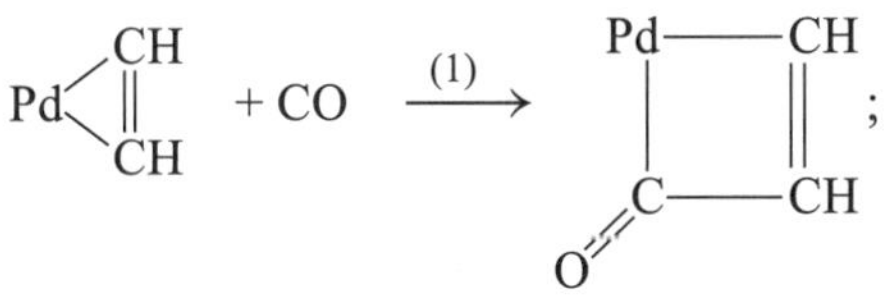

(7) 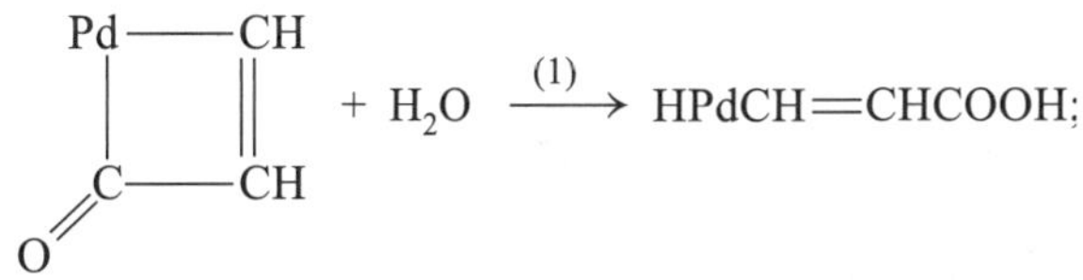

$$(8)\ HPdCH{=}CHCOOH + HBr \xrightarrow{(1)} AA + HPdBr.$$

The entire program-generated reaction network was presented in [110], where specific features of all mechanisms were discussed and the corresponding arguments *pro et contra* were considered. ■

Example 7.8 [121, 129]

► Based on experimental investigations of the mechanism of reaction

$$PhC{\equiv}CH + CO + MeOH + 2CuCl_2 + 2NaOAc \xrightarrow{PdCl_2} \\ \longrightarrow PhC{\equiv}CCOOMe + 2AcOH + 2NaCl + 2CuCl \tag{7.14}$$

and an analysis of the available literature [130, 131], five hypothetical mechanisms were proposed. A discrimination of these hypotheses left a single mechanism (M5) that did not contradict the experiment:

$$\mathrm{PhC{\equiv}CH + CuCl \xrightarrow[-HCl]{} CuC{\equiv}CPh \xrightarrow{PdCl_2} ClPdC{\equiv}CPh \xrightarrow{CO}}$$

$$\mathrm{\longrightarrow ClPdCOC{\equiv}CPh \xrightarrow{MeOH} PhC{\equiv}CCOOMe + HCl + Pd^0;}$$

$$\mathrm{Pd^0 + 2CuCl_2 \longrightarrow PdCl_2 + 2CuCl}$$

In order to obtain a more complete set of hypotheses using a formalized approach, the possible hypothetical mechanisms of reaction (7.14) were generated by the ChemNet and MECHEM programs [121, 129], after which the working hypotheses were selected using both known results and new experiments. For this purpose, the following set of 13 generalized transforms was adopted:

$$\begin{array}{ll}
\mathrm{M{-}Cl + H{-}C{\equiv}C \longrightarrow M{-}C{\equiv}C + H{-}Cl} & \mathrm{M = Cu(I);} \\
\mathrm{M_1{-}C + M_2{-}Cl \longrightarrow M_2{-}C + M_1{-}Cl} & \mathrm{M_1 = Cu(I),\ M_2 = Pd(II);} \\
\mathrm{M{-}C{\equiv}C + C{\equiv}O \longrightarrow M{-}C({=}O){-}C{\equiv}C} & \mathrm{M = Pd(II);} \\
\mathrm{M{-}Cl + MeO{-}H \longrightarrow MeO{-}M + H{-}Cl} & \mathrm{M = Pd(II),\ Cu(I);} \\
\mathrm{M{-}O + C{\equiv}O \longrightarrow M{-}C({=}O){-}O} & \mathrm{M = Pd(II);} \\
\mathrm{M{-}C{=}O + MeO{-}H \longrightarrow M{-}H + MeO{-}C{=}O} & \mathrm{M = Pd(II);} \\
\mathrm{M_1 + Cl{-}M_2{-}Cl \longrightarrow M_1{-}Cl + M_2{-}Cl} & \mathrm{M_1 = Pd(0) \longrightarrow Pd(I)\ or} \\
 & \mathrm{\quad Pd(I) \longrightarrow Pd(II);\ M_2 = Cu(II);} \\
\mathrm{M_1{-}OMe + M_2{-}Cl \longrightarrow M_1{-}Cl + M_2{-}OMe} & \mathrm{M_1 = Cu(I),\ M_2 = Pd(II);} \\
\mathrm{X{-}M{-}C \longrightarrow X{-}C + M} & \mathrm{M_1 = Pd(II) \longrightarrow Pd(0);\ X = C,\ O;} \\
\mathrm{M_1{-}Cl + M_2{-}Y \longrightarrow M_1{-}Y + M_2{-}Cl} & \mathrm{M_1 = M_2 = Pd(II),\ Y = H,\ C,\ O;} \\
\mathrm{M{-}C{=}O + C{\equiv}C \longrightarrow M{-}C{=}C{-}C{=}O} & \mathrm{M = Pd(II);} \\
\mathrm{M{-}C{=}C{-}H \longrightarrow M{-}H + C{\equiv}C} & \mathrm{M = Pd(II);} \\
\mathrm{M{-}C{\equiv}C + O{-}C({=}O){-}O \longrightarrow M{-}O + C{\equiv}C{-}C({=}O){-}O} & \mathrm{M = Pd(II).}
\end{array}$$

This set was supplemented by the following additional restrictions:

- initial reactants: PhC≡CH, CO, MeOH, $PdCl_2$, $CuCl_2$, CuCl;
- maximum number of atoms in hypothetical intermediates is 20 (Ph and Me groups are considered as single unchanged pseudo-atoms);
- maximum number of carbon atoms in these intermediates is three (Ph and Me groups not included);
- each intermediate may contain no more than one metal atom;
- each intermediate may contain no more than one Ph group and no more than three oxygen atoms;
- allowed degrees of oxidation corresponded to Cu(I) and Cu(II) for copper and Pd(0), Pd(I), Pd(II) for palladium;
- maximum coordination number for Cu and Pd in intermediates was restricted to 2.

The operation of the ChemNet program with the above restrictions yielded a reaction network that contained 233 elementary steps and involved 34 substances. On this network, the MECHEM program revealed 41 simple hypothetical mechanisms that corresponded to the stoichiometry of the following simplified reaction:

$$\mathrm{PhC{\equiv}CH + CO + MeOH + 2CuCl_2 \longrightarrow PhC{\equiv}CCOOMe + 2HCl + 2CuCl.} \tag{7.15}$$

Among these hypotheses, there were eight mechanisms with six intermediates and seven steps, 13 mechanisms with seven intermediates and eight steps, and 20 mechanisms with six intermediates and nine steps. The previously proposed mechanisms [130] involving MeOCOCl (M3) and PhC≡CCl (M4) were prohibited at the stage of selecting transforms and did not enter into the resulting set of 41 hypotheses. The other three established mechanisms – M1 with participation of ClPd(Ph)C=CHCOOMe (Heck's hypothesis), M2 with $(MeO)_2CO$, and M5 – were represented in the obtained set of hypotheses.

The results of experiments carried out previously showed evidence for the presence of an induction period on the curves of formation of the main (target) and side products (PhC≡CC≡CPh, PhC≡CCl, CO_2), decrease in this period upon the introduction of CuCl, autocatalysis with the participation of CuCl, and some other facts [130, 131]. Allowance for these results allowed four hypotheses corresponding to mechanisms M1 and M2 to be excluded from the consideration. The remaining 37 mechanisms, which involved the step of formation of a Pd(II) ethynyl compound, could not be discriminated based on the previously known data.

An analysis of these 37 hypotheses suggested that an additional discriminating factor could be provided by measurements of the kinetic isotope effect (KIE) in the CH_3OH/CH_3OD pair. In cases where methanol molecule (or MeO group) is included into the mechanism after the step of ethynyl group transfer from CuC≡CPh to Pd(II) (which is most probably the rate-limiting step), the KIE value must be close to unity. Then, if the rupture of the O–H bond in methanol takes place before or at the limiting stage, the KIE value can reach a level of 3–4 [56]. The experiments showed that KIE was 0.88 ± 0.17 (i.e., close to unity), which implied that all mechanisms except five could be rejected. Among the remaining eight-step hypotheses, mechanism 13 included a low-probable step of the formation of methoxyacyl intermediate

$$\mathrm{MeOPdC{\equiv}CPh + ClPdCOC{\equiv}CPh \longrightarrow ClPdC{\equiv}CPh + MeOPdCOC{\equiv}CPh},$$

and, hence, could also be discarded.

The remaining four hypotheses differed insignificantly and involved identical initial steps,

$$\text{(1) } \mathrm{CuCl_2 + Pd \longrightarrow CuCl + PdCl;}$$

$$\text{(2) } \mathrm{CuCl_2 + PdCl \longrightarrow CuCl + PdCl_2;}$$

$$\text{(3) } \mathrm{PhC{\equiv}CH + CuCl \longrightarrow PhC{\equiv}CCu + HCl;}$$

$$\text{(5) } \mathrm{PhC{\equiv}CCu + PdCl_2 \longrightarrow CuCl + ClPdC{\equiv}CPh.}$$

after which these mechanisms began to differ from each other.

Mechanism 5:

$$\text{(13) } \mathrm{ClPdC{\equiv}CPh + MeOH \longrightarrow MeOPdC{\equiv}CPh + HCl;}$$

$$\text{(11) } \mathrm{MeOPdC{\equiv}CPh + CO \longrightarrow MeOPdCOC{\equiv}CPh;}$$

$$\text{(7) } \mathrm{MeOPdCOC{\equiv}CPh \longrightarrow Pd + MeOCOC{\equiv}CPh.}$$

Mechanism 6:

$$\text{(13) } \mathrm{ClPdC{\equiv}CPh + MeOH \longrightarrow MeOPdC{\equiv}CPh + HCl;}$$

$$\text{(10) } \mathrm{MeOPdC{\equiv}CPh + CO \longrightarrow MeOCOPdC{\equiv}CPh;}$$

$$\text{(6) } \mathrm{MeOCOPdC{\equiv}CPh \longrightarrow Pd + MeOCOC{\equiv}CPh.}$$

Mechanism 7:

$$\text{(14) ClPdC}\equiv\text{CPh} + \text{CO} \longrightarrow \text{ClPdCOC}\equiv\text{CPh;}$$

$$\text{(18) ClPdCOC}\equiv\text{CPh} + \text{MeOH} \longrightarrow \text{HPdCl} + \text{MeOCOC}\equiv\text{CPh;}$$

$$\text{(12) HPdCl} \longrightarrow \text{Pd} + \text{HCl.}$$

Mechanism 8:

$$\text{(14) ClPdC}\equiv\text{CPh} + \text{CO} \longrightarrow \text{ClPdCOC}\equiv\text{CPh;}$$

$$\text{(19) ClPdCOC}\equiv\text{CPh} + \text{MeOH} \longrightarrow \text{MeOPdCOC}\equiv\text{CPh} + \text{HCl;}$$

$$\text{(7) MeOPdCOC}\equiv\text{CPh} \longrightarrow \text{Pd} + \text{MeOCOC}\equiv\text{CPh.}$$

Mechanisms 5–8 differ from each other by the following features:

- inner-sphere (mechanism 5) versus outer-sphere (mechanisms 7 and 8) attack of nucleophile on carbon atom of carbonyl group;
- insertion of CO at the Pd–C (mechanisms 5 and 8) or Pd–OMe (mechanism 6) bond;
- demetallation of the last intermediate preceding the product, with the formation of product and HPdCl (mechanism 7) or Pd (mechanisms 5, 6, and 8).

The further choice between mechanisms 5–8 presents a difficult physicochemical task, since the kinetic method cannot discriminate these hypotheses. For example, upon writing mechanism 7 with steps (1)–(3), (5), (14), (18), and (12) and taking into account the formation of side product PhC≡CCl (see Section 4.2.1) as

$$\text{(3) PhC}\equiv\text{CH} + \text{CuCl} \underset{k_{-1}}{\overset{k_1}{\rightleftharpoons}} \text{PhC}\equiv\text{CCu} + \text{H}^+ + \text{Cl}^-;$$

$$\text{PhC}\equiv\text{CCu} + \text{CuCl}_2 \underset{k_{-2}}{\overset{k_2}{\rightleftharpoons}} \text{X}_1;$$

$$\text{X}_1 + \text{CuCl}_2 \xrightarrow{k_3} \text{PhC}\equiv\text{CCl} + 3\text{CuCl};$$

$$\text{(5) PhC}\equiv\text{CCu} + \text{PdCl}_2 \xrightarrow{k_4} \text{PhC}\equiv\text{CPdCl} + \text{CuCl};$$

$$\text{(14) PhC}\equiv\text{CPdCl} + \text{CO} \overset{K_5}{\rightleftharpoons} \text{PhC}\equiv\text{CCOPdCl};$$

$$\text{(18) PhC}\equiv\text{CCOPdCl} + \text{MeOH} \xrightarrow{k_6} \text{HPdCl} + \text{PhC}\equiv\text{CCOOMe};$$

$$\text{HPdCl} + 2\text{CuCl}_2 \longrightarrow \text{PdCl}_2 + 2\text{CuCl} + \text{HCl},$$

we obtain kinetic equations that agree with the experimental results.

The last step can be represented as the decomposition of HPdCl,

$$\text{HPdCl} \longrightarrow \text{HCl} + \text{Pd}$$

together with steps (1) and (2). Steps (14), (18), (1), and (2) do not participate in the kinetic equation. Under conditions of $[\text{LiCl}]_0 = 3$ M (i.e., $[\text{LiCl}]_0 \gg [\text{CuCl}]_0$), the concentrations of chloride ions are constant and [CuCl], [$CuCl_2$], and [$PdCl_2$] are proportional to the corresponding initial (total) values of $[CuCl]_0$, $[CuCl_2]_0$, and $[PdCl_2]_0$. The other mechanisms (5, 6, and 8) with appropriate assumptions yield the same kinetic equations as that following from mechanism 7. ■

In order to generate hypothetical reaction networks for a highly complicated synthesis of hydrocarbons and oxygen-containing compounds on metal catalysts according to the Fischer–Tropsch process, the probable surface compounds and reactions were thoroughly analyzed using the ChemNet program so as to select theoretically and experimentally justified transforms and reaction pathways that lead to the formation of C_1 and C_2 products [63].

Example 7.9 [63]

► Using modern notions about the chemistry of metal surfaces and the nature of surface organic compounds [22, 23, 132, 133] and the presently available computer programs, it is possible to generate reaction networks with dimensions significantly exceeding the 55-step network that was proposed in 1981 for the Fischer–Tropsch synthesis by Rafer-De Poorter [42]. Let us consider in more detail the possible approaches to constructing reaction networks of the new generation. It was suggested [63] to generate reaction networks using the following sequence of stages.

Analysis of possible surface compounds formed in reactions of CO and H_2

The synthesis of hydrocarbons and oxygen-containing compounds from CO and H_2 includes the bond rupture in CO molecules and the formation of C–C, C–H, and O–H bonds:

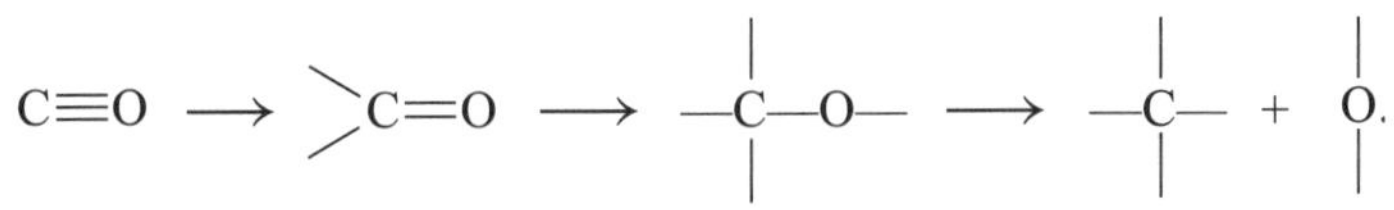

In addition, the formation of hydrocarbons and oxygen-containing compounds is accompanied by the rupture of C–H and O–H bonds. Evidently, all these processes must proceed on the surface of a metal with the participation of intermediates containing M–H, M–C, and M–O bonds. In order to find a form of the representation of intermediates that is convenient for the computer-aided generation of reaction networks, the possible types of adsorbed species and molecules and all intermediates were considered, which can take part in the process according to the results of investigations of the mechanism of the Fischer–Tropsch synthesis and related surface compounds. The adopted description is convenient for use with ChemNet program. Figures 7.3–7.6 show the selected structures of molecules and molecular fragments adsorbed on a metal surface.

Choice of elementary steps and transforms

Determination of the transforms that would ensure the formation and conversion of preselected intermediate species begins with an analysis of the main types of elementary steps and related topology identifiers (see Section 1.4.4). This analysis shows that most of the elementary steps not involving the formation of free radicals, both in homogeneous catalysis with metal complexes and in heterogeneous catalysis on metals, proceed with the simultaneous (consistent) cleavage and formation of two pairs of two-electron bonds (including π bonds). These steps are described by a four-term cyclic topology identifier (G_1) of the following type:

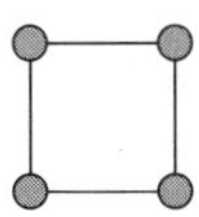

G_1

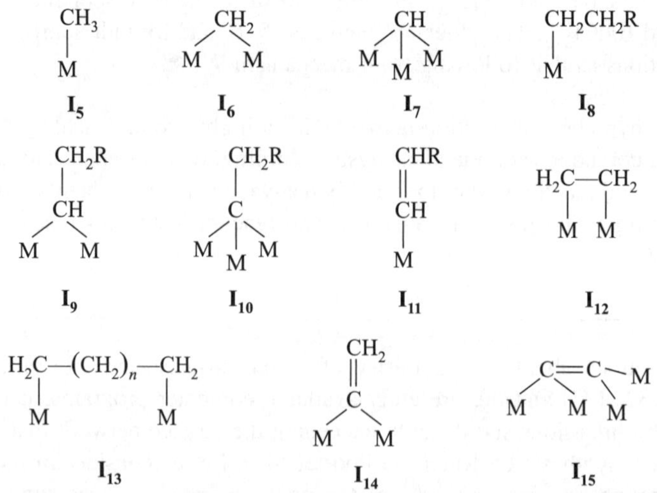

Figure 7.3 Intermediates with hydrocarbon groups and C_2 fragments (reproduced with permission from editorial board of *Izv. Akad. Nauk, Ser. Khim.* journal © 2002 [63])

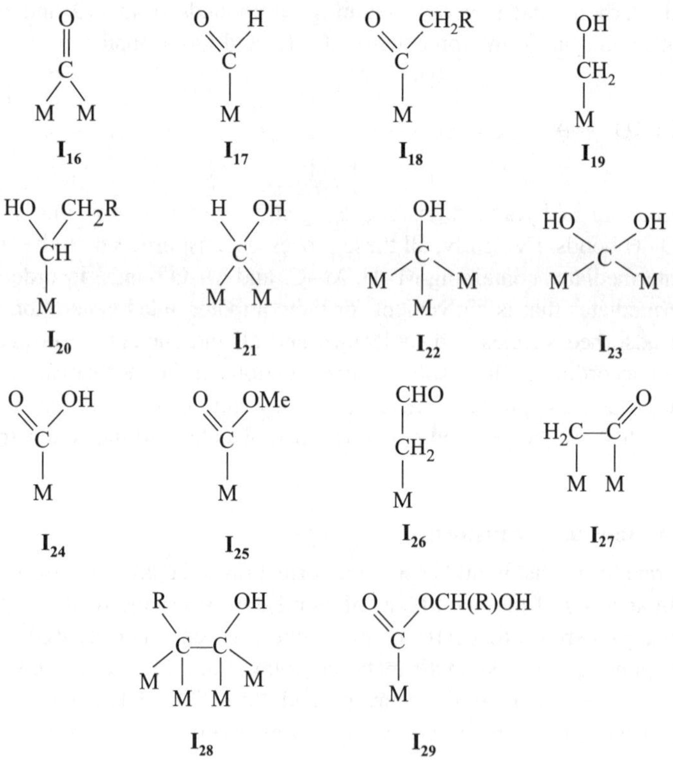

Figure 7.4 Intermediates with oxygen-containing fragments and M–C bonds (reproduced with permission from editorial board of *Izv. Akad. Nauk, Ser. Khim.* journal © 2002 [63])

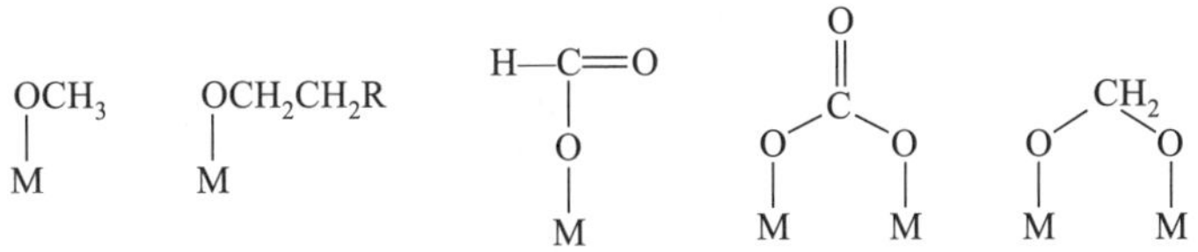

I_{35} I_{36} I_{37} I_{38} I_{39}

Figure 7.5 Oxygen-containing intermediates with M–O bonds

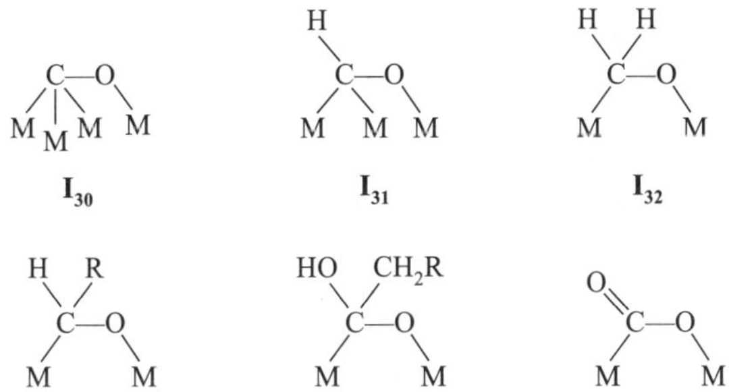

I_{30} I_{31} I_{32} I_{33} I_{33a} I_{34}

Figure 7.6 C- and O-coupled bridging intermediates

Other types of topology identifiers are also possible, but are rarely encountered.

If the vertices of graph G_1 are occupied by a combination of four elements (C, H, O, and M for metal) involved in the elementary steps of the Fischer–Tropsch synthesis, then we obtain a labeled (colored) topology identifier (LTI), which corresponds to a graph with vertices given definite labels (or painted in different colors). Each LTI may correspond to a certain transform. For example, LTI 3 (Figure 7.7) corresponds to the following generalized transform

$$\begin{array}{c} \text{M—C} \\ + \\ \text{H—H} \end{array} \rightleftharpoons \begin{array}{c} \text{M} \\ | \\ \text{H} \end{array} + \begin{array}{c} \text{C} \\ | \\ \text{H} \end{array}, \qquad (7.16)$$

that can be either further specified by defining substituents at C atom or employed in the general form. In order to select a set of transforms that can be adopted to generate a reaction network of syntheses involving CO and H_2, it is expedient to enlist (enumerate) the possible LTIs so as to define all types of elementary steps in the most general form and then choose the transforms that are provided with proper theoretical and/or experimental justification.

Since four elements (M, C, H, O) can be variously arranged at the vertices of graph G_1, the problem reduces to finding permutations $M_nC_mH_xO_y$ of these elements at four vertices of this cyclic graph with allowance for the following restrictions:

$$n + m + x + y = 4;$$

$$n = 1, 2;\; m = 0, 1, 2;\; x = 0, 1, 2;\; y = 0, 1, 2.$$

Using the well-known Pólya's theorem (enumeration theorem) for painting the vertices of a square in four colors, one can obtain 27 LTIs, including 12 with $n = 2$ and 15 with $n = 1$. From the former 12 LTIs with $n = 2$, six graphs should be rejected because they correspond to the reactions of exchange with

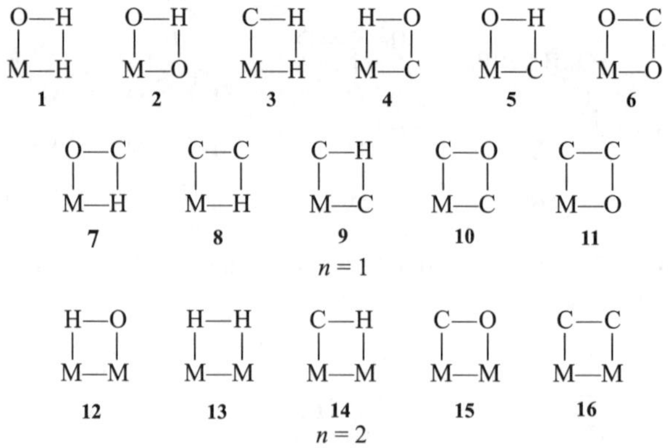

Figure 7.7 Labeled topology identifiers for steps of the Fischer–Tropsch synthesis (reproduced with permission from editorial board of *Izv. Akad. Nauk, Ser. Khim.* journal © 2002 [63])

adsorbed particles. In addition, three LTIs with O–O edge are rejected since no peroxide formation takes place in the syntheses from CO and H_2. Finally, two graphs with a diagonal arrangement of H atoms are rejected in the group of LTIs with $n = 1$, which describe symmetric reactions of hydrogen exchange. The remaining 16 LTIs are depicted in Figure 7.7.

The obtained set of LTIs 1–16 (Figure 7.7) can be supplemented by a graph (LTI 17) with $n = 1$ and $m = 3$ and a graph (LTI 18) corresponding to the possible three-term reaction cycle:

17 18

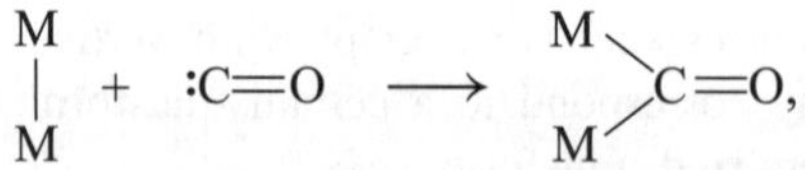

(7.17)

Thus we obtain a set of topology identifiers that is sufficient to describe the steps of any product formation in the Fischer–Tropsch synthesis with allowance for the adopted restrictions. In the framework of the method of valence schemes adopted in [63], the addition of CO at the M–M bond according to reaction (7.17) is considered as the addition of singlet carbene ($:CH_2$) or isonitrile at this bond.

The selected LTIs were used for an analysis of the possible elementary steps with the participation of intermediates I_1–I_{39} and the related generalized transforms. Each of 17 four-term cyclic LTIs was assigned the corresponding generalized transform T1–T17, for example, as

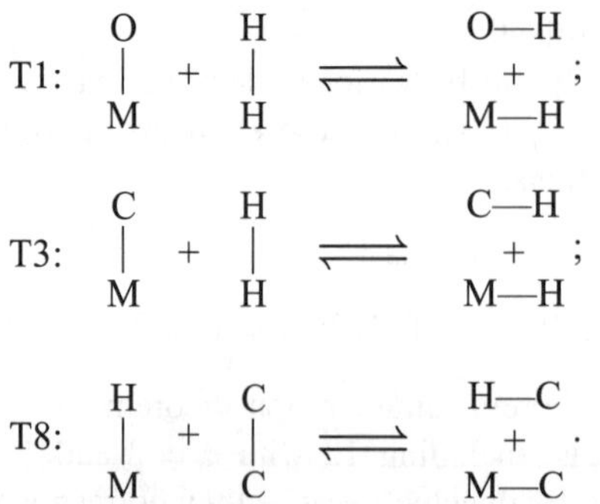

Since the topology identifiers include only bonds to be cleaved and formed, the transforms refer to the corresponding σ and π bonds such as M–O in T1 (M–OR, M=O) and C–C in T8 (C=C and C–C in the case of σ bond metathesis).

Reaction network generation

Based on the above analysis of steps and possible transforms, sets of transforms are selected for the generation of reaction networks of the formation C_1 products (Table 7.14) and C_2 products (Table 7.15) from CO and H_2. Table 7.16 presents some less frequently encountered transforms, the use of which, together with the transforms in Table 7.14, allowed the upper limiting size of the reaction network to be determined for C_1 products (with preset restrictions). Computer-aided experiments in nine variants (I–VIII, IVa) with various sets of transforms yielded the reaction networks containing from 132 (VIII) to 1647 (IV) elementary steps. All these networks included the identical sets of 27 intermediates and led to six C_1 products and water molecule: CH_4 (P_1), CH_3OH (P_2), CO_2 (P_3), H_2O (P_4), HCOOH (P_5), CH_2O (P_6) and $CH_2(OH)_2$ (P_7). For thermodynamic reasons, product CH_2O can exist only in an adsorbed state or as a hydrate (P_7).

The maximum number of steps in a reaction network (not involving the transforms from Table 7.16) is 574. Most of the obtained intermediates (21 from a total of 27) are known from published data (I_1–I_7, I_{16}, I_{17}, I_{19}, I_{21}–I_{24}, I_{30}–I_{35}, I_{37}, I_{39}), while six intermediates were previously not reported. All these new intermediates (Figure 7.8) and compound I_{39} are in fact the products of hydrogen replacement by metal atoms in methylene hydrate $CH_2(OH)_2$. Carbonate intermediates are not obtained because of a restriction imposed on the total number of oxygen atoms ($y \leq 2$).

Analysis of mechanisms of C_1 product formation

Detailed analysis was performed for a reaction network of variant III, which contained 220 steps, but included 20 more or less generalized transforms (Table 7.14) with some restrictions in transforms 3.1, 4.3, 5.1, and 6.1. This network is illustrated here by some fragments (depicted in abbreviated form), which include all pathways leading to the formation of methane (Figure 7.9) and methanol (Figure 7.10). These "cut" reaction networks contain all the known mechanisms of methane and methanol formation as well as some new routes. Mechanisms of conjugate reactions of the formation of CH_4/CO_2 and CH_4/HCOOH were also considered [63]. The obtained reaction network offers various modes of conjugation for the processes of hydrocarbon and oxygen-containing product formation and can be used to discriminate hypotheses with allowance for experimental data. ■

7.4.7 Large reaction networks and problems in discrimination of hypotheses and construction of compact kinetic models

The examples presented above show that large reaction networks can be successfully managed in cases where the rate constants of elementary steps (gas- and liquid-phase free-radical reactions, acid-catalyzed processes) can be reliably evaluated. At the same time, even when generating reaction networks for these reactions, it may become necessary to restrict the network size by excluding low-probability intermediates and steps. These networks of reasonable dimensions must simplify the discrimination of hypotheses and provide models that ensure qualitatively adequate descriptions of the formation of main products (groups of products) without losing significant chemical details.

The most natural and simple factor that can be used to limit the expansion of reaction networks is a restriction on the number of heavy atoms (C, O, N, and metal catalyst). This number is used as a criterion in all programs of hypothesis and network generation. A useful approach to the program of restricting the combinatorial growth in the number of substances and steps consists of rating (ranking) the substances

Table 7.14 Transformations used to generate reaction network for Fischer–Tropsch synthesis with formation of C_1 products (reproduced with permission from editorial board of *Izv. Acad. Nauk, Ser. Khim.* journal © 2002 [63])

Topology identifier	Transformation no.	Transformation	Application variant
O—H / M—H **1**	1.1	M–O + H–H → M–H + O–H	**I –IVa**
	1.2	M–H + XO–H → M–OX + H–H	**I –IVa**(**III** : X≠M)
C—H / M—H **3**	3.1	M–C + H–H → M–H + C–H	**I–IVa, VIII** (**III**: sp^3-hybridization for C)
	3.2	M–H + C–H → M–C + H–H	**I, II, IV, IVa**
H—O / M—C **4**	4.1	M–H + C=O → M–C–O–H	**I, II, IV–VIII**
	4.2	M–C(XY)–O–H → M–H + (XY)C=O	**I–VIII**(**VIII**: X=Y=H)
	4.3	M–CX + O–H → M–H + O–CX	**I–IVa, VI** (**I–III**: X=O)
O—H / M—C **5**	5.1	M–CX + YO–H → M–OY + XC–H	**I–IV** (**II**: X≠O, Y≠M; **III**: sp^3- hybridization for C, X≠O, M, CH_x, Y≠M)
O—C / M—O **6**	6.1	M–OX + C=O → M–O–C–OX	**I–IV** (**I, III**: X=H)
	6.2	M–O–C–OH → M–OH + C=O	**I, II, IV, IVa** (**I**: X=H)
O—C / M—H **7**	7.1	M–H + XC–OY → M–OY + XC–H	**I, II, IV, IVa, VI–VIII** (**VIII**: X=H_3, Y=M)
	7.2	M–O + C–H → M–H + C–O	**I, II, IV, IVa**
	7.3	M–H + XC=O → M–O–CX–H	**I–IVa, VI–VIII** (**VIII**:X=H)
	7.4	M–O–C(XY)–H → M–H + (XY)C=O	**I–IVa, VI–VIII** (**VIII**: X=H, Y=M)
H—O / M—M **12**	12.1	M–H + M–O → M–M + O–H	**I–VIII**
	12.2	M–M + O–H → M–H + M–O	**I–VIII**
H—H / M—M **13**	13.1	M–M + H–H → M–H + M–H	**I–VIII**
	13.2	M–H + M–H → M–M + H–H	**I–IVa, VI–VIII**

Table 7.14 *(continued)*

Topology identifier	Transformation no.	Transformation	Application variant
 **14**	14.1	M–C + M–H → M–M + C–H	**I–VIII**
	14.2	M–M + C–H → M–C + M–H	**I–VIII**
 **15**	15.1	M–M + (XY)C–OZ → M–C(XY) + M–OZ	**I–VIII** (**VIII**: X=Z=M, Y=O)
	15.2	M–C + M–O → M–M + C–O	**I, II, IV–VIII**
	15.3	M–M + XC=O → M–CX–O–M	**I–V, VII, VIII** (**VIII**: X=O)
	15.4	M–C(XY)–O–M → M–M + (XY)X=O	**I–V, VII, VIII** (**VIII**: X=Y=M)
 **18**	18.1	M–M + C≡O → M–C(O)–M	**I–VIII**
	18.2	M–C(O)–M → M–M + C≡O	**I–VIII**

Table 7.15 Transformations used to generate reaction network for Fischer–Tropsch synthesis with formation of C_2 products containing C–C bonds (reproduced with permission from editorial board of *Izv. Acad. Nauk, Ser. Khim.* journal © 2002 [63])

Topology identifier	Transformation no.	Transformation
 **8**	8.1	M–H + C=C → M–C–C–H
	8.2	M–C–C–H → M–H + C=C
 **11**	11.1	M–O + C=C → M–C–C–O
	11.2	M–C–C–O → M–O + C=C
	11.3[a]	M–R + C=O → M–O–C–R
 **16**	16.1	M–C + M–C → M–M + C–C
	16.2	M–M + C=C → M–C–C–M
	16.3	M–C–C–M → M–M + C=C
 **17**	17.1[b]	M–CH_3 + C–CR → M–CR + CH_3–C

Notes: [a]R = alkyl; [b]R≠H.

Table 7.16 Additional transformations used to generate reaction network for Fischer–Tropsch synthesis with formation of C_1 products (reproduced with permission from editorial board of *Izv. Acad. Nauk, Ser. Khim.* journal © 2002 [63])

Topology identifier	Transformation no.	Transformation	Application variant
O—H M—O **2**	2.1	M–OX + YO–H → M–OY + XO–H	**IVa, VI, VII** (**VI**: X=C, Y=H)
H—O M—C **4**	4.4	M–H + O–C → M–C + O–H	**IV, IVa**
O—H M—C **5**	5.2	M–O + C–H → M–C + O–H	**IV**
O—C M—O **6**	6.3	M–O + C–O → M–O + C–O	**IV, IVa**
C—H M—C **9**	9.1	M–C + XC–H → M–CX + C–H	**IV, VI, VII** (**VI, VII**: X=M)
C—O M—C **10**	10.1	M–C + C–O → M–C + C–O	**IV, VI**

[79, 120, 134]. However, as was noted above, the task of assigning ranks in a large reaction network is not that simple.

At present, three strategies are actively being developed for decreasing the dimensionality (order) of a model, i.e., reducing the order of a related system of differential equations, while retaining key information about the system dynamics [134–136]. The obtained system of decreased order must admit a simpler mathematical treatment, require lower computational facilities, and contain a smaller number of unknowns than the initial complete model.

The first strategy is called *lumping* and consists of combining substances with similar behavior (reactivity) into groups of pseudo-substances. This lumping (grouping) procedure leads to a decrease in the dimension of the vector of concentrations c, which is converted into a vector (c') of reduced dimension:

$$c' = h(c)$$

The second strategy consists of an analysis of the sensitivity of reaction and is aimed at determining and eliminating steps depending on their influence on the behavior of important intermediates and yield of target products.

The third strategy is based on an analysis of the characteristic time scales for various reactions (determination of fast and slow subsystems) and the use of the quasi-steady-state and quasi-equilibrium (pre-equilibrium) approximations. As a result, a model can be simplified by converting some differential

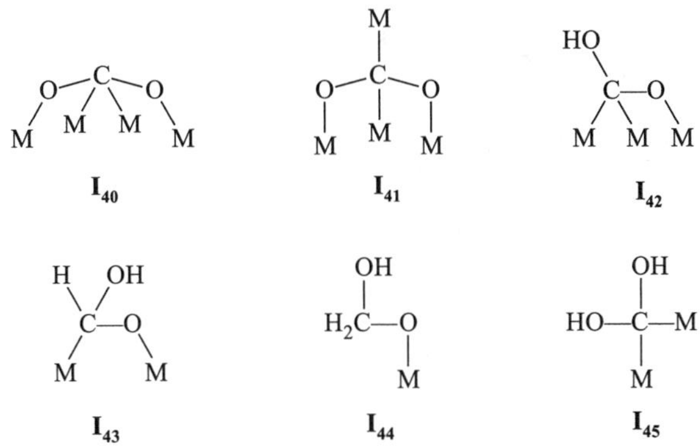

Figure 7.8 New intermediates for reaction network of C_1 product formation

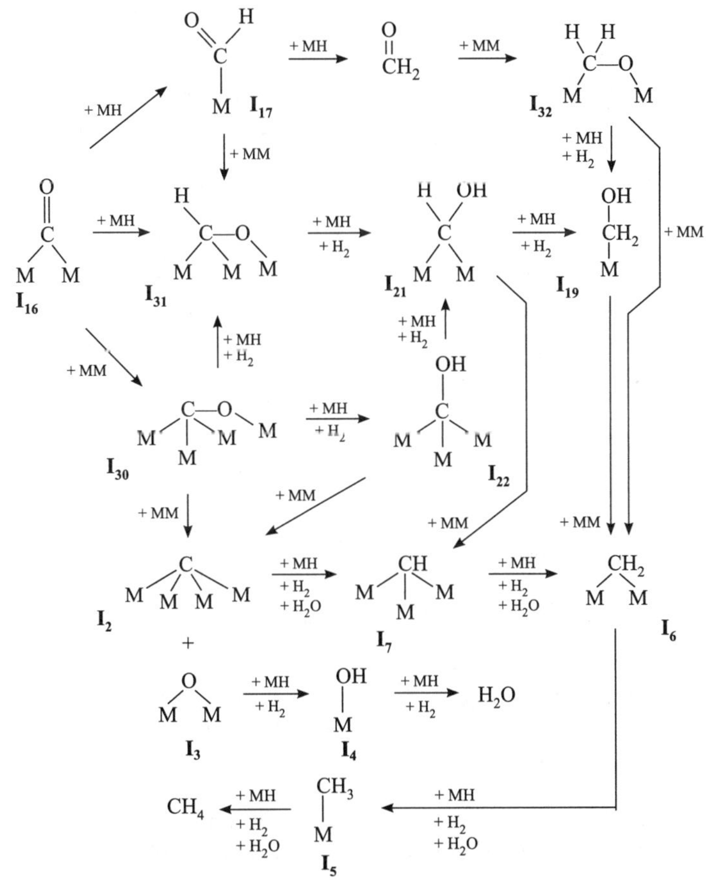

Figure 7.9 Reaction network for methane formation from CO and H_2 (reproduced with permission from editorial board of *Izv. Akad. Nauk, Ser. Khim.* journal © 2002 [63])

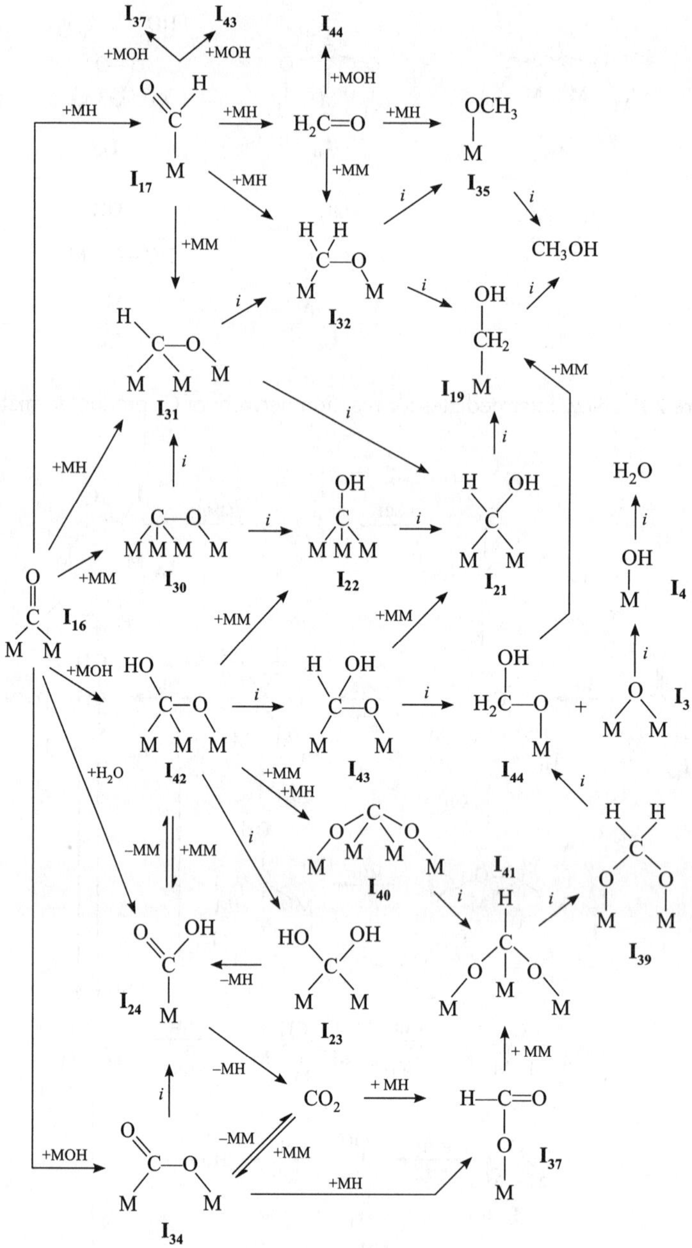

Figure 7.10 Reaction network for methanol formation from CO and H_2 (i : $+H_2$, +MH) (reproduced with permission from editorial board of *Izv. Akad. Nauk, Ser. Khim.* journal © 2002 [63])

equations into algebraic, but this approach is only acceptable for fairly small, compact models. Mathematical principles of the three strategies have been discussed in [136]. The time-scale analysis of separate steps can be used for the so-called regional analysis of models and the formulation of kinetic models that adequately describe reactions on a given time scale [135].

The implementation of each strategy requires knowledge of rate constants of separate steps. For this reason, methods of determining or evaluating the kinetic parameters (activation energies E_{act}) and thermochemical characteristics (enthalpies ΔH) of steps are given special attention. Evidently, if it were possible to estimate the probabilities of steps and/or the formation of particular intermediates in the course of reaction network generation, this would allow the network dimension to be significantly reduced and justified compact kinetic models to be obtained. However, methods and software for the computer-aided generation of reaction networks were developed much faster than the simple and convenient techniques for the rapid evaluation of ΔH and E_{act}. At present, new empirical and phenomenological approaches have been extensively developed, which are successfully applied to homogeneous gas-phase reactions and heterogeneous catalytic processes on metal crystal faces. Let us consider some of these approaches to evaluation of the probabilities of steps and mechanisms in more detail.

- **Using enthalpies of steps or heats of atomization of intermediates for discriminating steps**

By setting an interval of permissible ΔH_T° values, i.e., introducing critical enthalpies ($\Delta H_{T\ \mathrm{cr}}^\circ$) for endothermal steps, it is possible to exclude steps with $\Delta H_T^\circ > \Delta H_{T\ cr}^\circ$. This approach naturally retains exothermal steps that might be kinetically hindered. The same condition can also be formulated in terms of a change in the free energy (ΔG_T°). For example, Borchardt et al. [137] studied the mechanism of diborane oxidation by atomic oxygen O (^3p) in a continuous-flow tubular reactor and considered 15 substances with relatively small molecular weights consisting of B, O, and H atoms, and nine O- and H-containing particles. Enumeration of all mono- and bimolecular reactions involving these species gave over 1000 steps, which were reduced to a network of 670 steps by rejecting all reactions with $\Delta G_{300}^\circ > -4.0\,\mathrm{kcal/mole}$ (i.e., with more positive values). The final set of 40 steps, which was selected upon a comparison to experiment, adequately described the behavior of HO• radicals in the course of reaction over the entire tube length. It was pointed out in Sections 1.4.2 and 3.4.2 that the use of ΔG_T° values as a criterion for the probability of steps for consequent reactions or multi-route schemes is inexpedient, since a kinetic conjugation can make even a thermodynamically unfavorable process ($\Delta G_T^\circ > 0$) quite realizable, with a positive chemical affinity. Apparently, it is only in the case of a large positive ΔG_T° value (above 10–15 kcal/mole) that the subsequent conversion of an intermediate cannot change the sign of chemical affinity and impart an endoergonic step a positive rate in the desired direction ($W^+/W^- > 1$).

The energy criterion of selection was implemented in the COMSICAT program [36, 80–84] both for discriminating hypotheses and selecting potential catalysts of CO hydrogenation. In order to assess the most stable particle ensembles (intermediates), Shtokolo et al. [80] calculated the energy of atomization by the method of interacting bonds using semiempirical parameters of bonding between atoms present in the ensemble. In discriminating mechanisms of reactions on a well-known metal, the most probable process is that possessing a smoother distribution of the intermediate ensemble atomization energies over steps (i.e., without sharp jumps and depressions in the atomization energy distribution). In a similar manner, a most effective catalyst can be selected for a chosen mechanism by varying the energies of M–X bonds (X = C, H, O), finding the smoothest profile of atomization energies, and comparing it to the M–X values known for various metals [36]. The choice in favor of the criterion of uniform distribution of atomization energies (and, hence, reaction enthalpies) over steps of a route is not strictly justified, since a high stability of an intermediate with respect to other participants of the step does not necessarily imply its low reactivity at the next step.

- **The Brønsted–Polanyi equation and Bell–Evans–Polanyi principle of linearity**

Attempts at using thermodynamic characteristics of reactions (steps) for estimating $\lg k, \Delta G^{\neq}$, and E_a (activation energy) values have a long history (see, e.g., 7.13–7.15). In 1914, Taylor discovered a correlation between catalytic activity and strength of an acid-catalyst. In 1924, Brønsted and Pedersen [138] expanded Taylor's observations to a broad field of acid catalysis and demonstrated a correlation between the rate constant k of acid-catalyzed reaction and equilibrium dissociation constants of HA molecules (K_{HA}), which can be described by the following equation:

$$K = GK_{\mathrm{HA}}^{\alpha} \tag{7.18}$$

that is valid in a definite interval of K_{HA} values. Although this equation was originally related to different reactions, it provided a basis for the principle of free energy linearity that is written as

$$\Delta G^{\neq} = \beta + \alpha \Delta G^{\circ}. \tag{7.19}$$

Then, Horiuti and Polanyi [139] suggested that E_a for any series of steps of the same type can also be related to the corresponding free energy change (ΔG°) by the following linear relationship (Polanyi equation):

$$E_a = E_a^{\circ'} + \alpha \Delta G^{\circ}. \tag{7.20}$$

Later, Evans and Polanyi [140] suggested the following equation:

$$E_a = E_a^{\circ} + \alpha \Delta H^{\circ}. \tag{7.21}$$

Here, α is the so-called transfer coefficient and $E_a^{\circ'}$ and E_a° are the internal barriers of reaction (for $\Delta H^{\circ} = 0, E_a = E_a^{\circ}$). Eventually, Horiuti and Polanyi [139] and Bell [15] almost simultaneously proposed a theoretical justification of linear relationships (7.20) and (7.21).

Following Bell [15], let us describe the motion of proton along the coordinate of reaction

$$\mathrm{XH + Y \longrightarrow X + HY}$$

by a simplified scheme presented in Figure 7.11a, where potential curves **I** and **II** refer to the reactants and products, respectively. If base Y is changed, the potential curve of products shifts by $\delta \Delta H$ to position **III**, while the position of energy minimum with respect to the reaction coordinate is assumed to remain the same. Approximating potential curves in the vicinity of their intersection by straight lines and using simple geometric considerations, we can obtain a relationship between δE_a and $\delta \Delta H$. Indeed, expressing the length of segment $\mathrm{AE}_1^{\neq}$ in triangle (Figure 7.11b) via other sides b and a and the absolute values of slopes $S_1 = \mathrm{tg}\Theta_1$ and $S_2 = \mathrm{tg}\Theta_2$, we eventually obtain the relation

$$\mathrm{AB} + \mathrm{BE}_1^{\neq} = \mathrm{AE}_1^{\neq} \quad \text{or} \quad \frac{a}{S_1} + \frac{a}{S_2} = \frac{b}{S_1},$$

which yields

$$a = b\left(\frac{S_2}{S_1 + S_2}\right).$$

Taking into account that $a = E_2^{\neq} - E_1^{\neq} = E_{a2} - E_{a1} = \delta E_a$ and $b = \Delta H_2^{\circ} - \Delta H_1^{\circ} = \delta \Delta H^{\circ}$, we obtain

$$\delta E_a = \alpha \delta \Delta H^{\circ} \tag{7.22}$$

or $E_a = E_a^{\circ} + \alpha \Delta H^{\circ}$, provided that the ratio of slopes S_2/S_1 remains unchanged when ΔH° is varied.

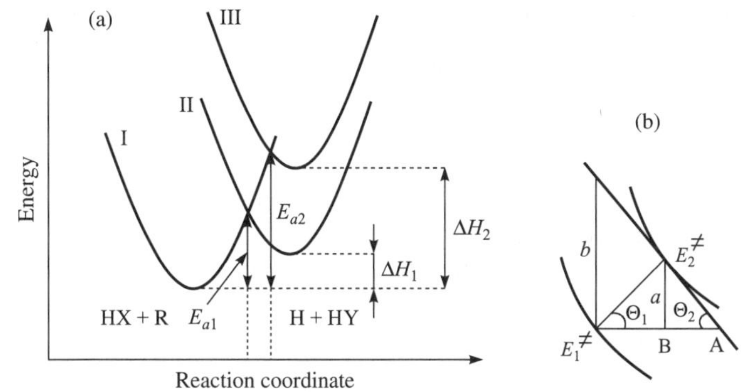

Figure 7.11 Geometric interpretation of Polanyi equation and Bell Evans – Polanyi linearity principle (see text for explanations)

The above linear relationship is valid in a limited interval of ΔH° values and corresponds to very rough approximations (linearity of potential curves in the vicinity of their intersection, unchanged position of the energy minimum for various Y, absence of X . . . Y repulsion, etc.), which was clearly understood by Polanyi. Despite this fact, the Polanyi equation as well as the general principle of linearity (Brønsted equation), was widely used in homogeneous acid catalysis, description of reactions between radicals and molecules (Polanyi–Semenov equation), organic chemistry (Hammett and Taft equations [14]), and even in heterogeneous catalysis.

In the case of endothermal reactions, it is evident that $E_a \geq \Delta H\,(\Delta H > 0)$, which implies that the interval of ΔH values is limited by $E_a^\circ/(1-\alpha)$ and, hence, $\Delta H^\circ \leq E_a^\circ/(1-\alpha)$. For example, at $\alpha = 0.75$ and $E_a^\circ = 10.0$ kcal/mole, we have $\Delta H^\circ \leq 4E_a^\circ$, (i.e., $\Delta H^\circ \leq 40.0$ kcal/mole). For exothermal reactions,

$$E_a = E_a^\circ - \alpha|\Delta H|,$$

so that $E_a^\circ \geq \alpha|\Delta H|$ and $|\Delta H| \leq E_a^\circ/\alpha$. At $\alpha = 0.25$, we have $|\Delta H| \leq 4E_a^\circ$.

The coefficients of Polanyi equation for radical–molecule reactions were determined by Semenov [13], according to which

$$E_a = 11.5 + 0.75\Delta H^\circ \tag{7.23}$$

in the region of $\Delta H^\circ > 0$, and

$$E_a = 11.5 - 0.25|\Delta H^\circ| \tag{7.24}$$

in the region of $\Delta H^\circ < 0$.

Westley [141] refined the values of E_a° based on an analysis of 482 reactions of hydrogen atom transfer. Although the dependence of E_a on ΔH° in a wide range of ΔH° has a clear parabolic character, linear approximations quite satisfactorily agree with experiment for $E_a^\circ = 9.0$ kcal/mole in the intervals of $0 < \Delta H^\circ < 36.0$ kcal/mole ($\alpha = 0.75$) and $0 > \Delta H^\circ > -36.0$ kcal/mole ($\alpha = 0.25$). The mean-square deviation for E_a over the entire data array amounts to ~4 kcal/mole. More precise E_a values can be obtained for series of reactants with close nature and then refined for each set of E_a° and α values.

In the field of heterogeneous catalysis, the Brønsted equation was apparently used for the first time by M.I. Temkin [142] in order to describe the dependence of the rate constants of elementary surface reactions on ΔG° of the adsorption steps. The estimation of activation energies for the adsorption step and the conversion of surface intermediates using linear correlations with the corresponding heats of adsorption and surface reactions in the form of Polanyi equations is still used for constructing kinetic models in heterogeneous catalysis [119, 143–150].

- **The Marcus equations**

An extensive investigation of numerous deviations from the linear relationships for free energies [15, 138] began after the appearance of a theory of electron transfer in polar media (Marcus [151], Levich and Dogonadze [152]), a theory of proton transfer (Dogonadze et al. [153, 154]), and the transfer of proton and other atoms and species (Marcus [155], Kreevoy and Konasewich [156], etc.). The following parabolic equation was obtained:

$$\Delta G^{\neq} = \frac{(E_r + \Delta G^\circ)^2}{4E_r}, \tag{7.25}$$

where E_r is the energy of reorganization for the solvent and other heavy particles (for the classical particle motion over a barrier). Equation (7.25) and the expression for the activation energy

$$E_a = E_a^\circ \left(1 + \frac{\Delta H^\circ}{4E_a^\circ}\right)^2 \tag{7.26}$$

(where E_a° is the internal energy barrier) are called the Marcus equations.

Let us derive Eq. (7.26) following Marcus [151] and Masel [157]. The activation energy for the transfer of atom B in the reaction

$$\mathrm{A + BC \longrightarrow AB + C}$$

is considered as a sum of the work (W_r') performed for bringing reactants into contact (without their significant distortion) and the energy (E_a') of atom B transfer to atom A:

$$E_a = E_a' + W_r'. \tag{7.27}$$

The energies of bonds BC and AB in the vicinity of the intersection of their potential energy curves are approximated by the parabolas E_{AB} and E_{BC} (Figure 7.12) [157], which yields

$$E_{AB}(r_x) = a_1(r_x - r_1)^2 + E_1; \tag{7.28}$$

$$E_{BC}(r_x) = a_2(r_x - r_2)^2 + \Delta H^\circ + E_2. \tag{7.29}$$

Assuming for the sake of simplicity that $a_1 = a_2$ and $E_1 = E_2$ and taking into account that $E_{AB} = E_{BC}$ at the point of intersection of parabolas ($r_x = r^{\neq}$), we obtain

$$r^{\neq} = \frac{r_1 + r_2}{2} + \frac{\Delta H^\circ}{2a_1(r_2 - r_1)}. \tag{7.30}$$

Substituting expression (7.30) for $r^{\neq}$ into Eq. (7.28) for the barrier coordinate ($r_x = r^{\neq}$), we obtain

$$E_a' = E_{AB}(r^{\neq}) = a_1 \cdot \left[\frac{r_2 - r_1}{2} + \frac{\Delta H^0}{2a_1 \cdot (r_2 - r_1)}\right]^2 + E_1. \tag{7.31}$$

At $a_1(r_2 - r_1)^2/4 = E_a^\circ$, Eq. (7.31) simplifies to

$$E_a' = E_a{}^0 \left(1 + \frac{\Delta H^0}{4E_a{}^0}\right) + E_1, \tag{7.32}$$

and then another assumption ($W_r' = 0$ and $E_1 = 0$) leads us to Eq. (7.26).

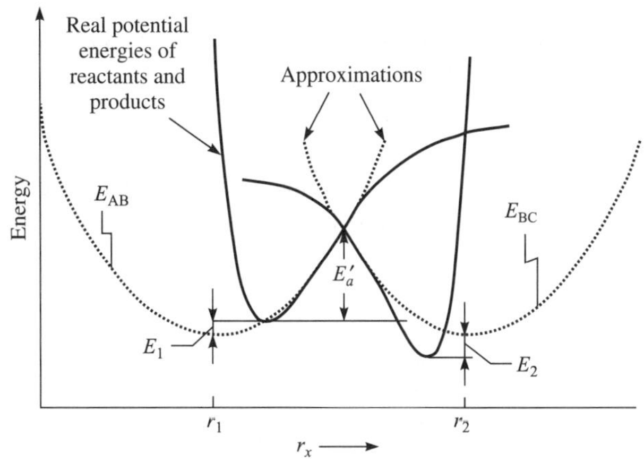

Figure 7.12 Illustration of the derivation of the Marcus equation (reproduced with permission from John Wiley & Sons © 2001 [157])

The Marcus equation well describes experimental data on the transfer of electrons, protons, atoms, and radicals [157], but only until $|\Delta H^0| = 4E_a{}^0$ for $\Delta H^\circ < 0$. At more negative values of $\Delta H^\circ(|\Delta H| < 4E_a{}^0)$ the activation energies E_a begin to increase (Marcus inverted region) and the Marcus equation fails to adequately describe experimental data on the transfer of atoms and radicals. For radical reactions, some other variants of parabolic relationships between E_a and ΔH° were also proposed [64, 158]. However, in contrast to these, the Marcus equations [153, 154] are attractive by containing a single empirical parameter (E_a°). In simple cases of radical substitution, E_a° can be estimated from data on the symmetric reactions with $\Delta H^\circ = 0$ or $\Delta G^\circ = 0$ [157]. For example, in order to estimate E_a for reactions of the type

$$\mathrm{X + R\text{–}Y \longrightarrow X\text{–}R + Y}$$

it is possible to use the values of $E_a = E_a^\circ$ for symmetric reactions

$$\mathrm{X + R\text{–}X \longrightarrow X\text{–}R + X} \quad (E_a^\circ(xx));$$

$$\mathrm{Y + R\text{–}Y \longrightarrow Y\text{–}R + Y} \quad (E_a^\circ(yy))$$

and calculate $E_{a(xy)}^\circ$ as a mean of these:

$$E_{a(xy)}^\circ = \frac{1}{2}\left(E_{a(xx)}^\circ + E_{a(yy)}^\circ\right). \tag{7.33}$$

Nonlinear relationships between E_a and ΔH° have been also observed in heterogeneous catalysis. For example, a nonlinear correlation was found between the activation energy and ΔH° for the dissociative adsorption of O_2 on crystal faces of 11 metals [159]. The E_a values were estimated based on the results of comparative quantum-chemical calculations of the energies of initial and final states in the classical and quantum nonadiabatic approach and a classical adiabatic approximation [160–162].

- **The Blowers–Masel model and equation**

Suggesting that the main defect in the Marcus model is related to the assumption that $r_2 - r_1$ distances remain constant when substituents are varied, Blowers, Masel et al. [157, 163–165] attempted to take into account the influence of electron repulsion (Pauli's effect) on $r_2 - r_1$ or, in other words, to consider an increase in the barrier height as a result of the appearance of electrons on an antibonding molecular orbital (MO) in a transition state in reactions of the type

$$H\cdot + CH_3CH_3 \longrightarrow CH_4 + \cdot CH_3.$$

The potential energy profile along the reaction coordinate was approximated by a sum of three terms, including the energy of extension of the C–C bond, the energy of formation of the C–H bond, and the energy of MO distortion caused by the Pauli electron repulsion:

$$E(l_{\text{C-C}}, l_{\text{C-P}}) = E_{\text{C-C}}(l_{\text{C-C}}) + E_{\text{C-H}}(l_{\text{C-H}}) + E_{\text{Pauli}}. \tag{7.34}$$

The $E_{\text{C-C}}$ and $E_{\text{C-H}}$ potential curves were approximated by the Morse potentials:

$$E_{\text{C-C}}(l_{\text{C-C}}) = W_{\text{C-C}}\left(\left[\exp\left(-\alpha_{\text{C-C}}(l_{\text{C-C}} - l^{eq}_{\text{C-C}})\right) - 1\right]^2 - 1\right) \tag{7.35}$$

$$E_{\text{C-H}}(l_{\text{C-H}}) = W_{\text{C-H}}\left(\left[\exp\left(-\alpha_{\text{C-H}}(l_{\text{C-H}} - l^{eq}_{\text{C-H}})\right) - 1\right]^2 - 1\right) \tag{7.36}$$

where $W_{\text{C-C}}, W_{\text{C-H}}$ are the corresponding bond dissociation energies; $\alpha_{\text{C-C}}, \alpha_{\text{C-H}}$ are the constant factors; and $l^{\text{eq}}_{\text{C-C}}, l^{\text{eq}}_{\text{C-H}}$ are the equilibrium bond lengths. The electron repulsion energy was approximated by a simple exponential function of the bond length:

$$E_{\text{Pauli}} = a_0 \exp(-\beta_{\text{C-C}} l_{\text{C-C}} - \beta_{\text{C-H}} l_{\text{C-H}}). \tag{7.37}$$

Denoting exponents in Eqs (7.35) and (7.36) by $n_{\text{C-C}}$ and $n_{\text{C-H}}$ (bond orders according to Pauling),

$$n_{\text{C-C}} = \exp\left(-\alpha_{\text{C-C}}(l_{\text{C-C}} - l^{\mathbf{eq}}_{\text{C-C}})\right); \tag{7.38}$$

$$n_{\text{C-H}} = \exp\left(-\alpha_{\text{C-H}}(l_{\text{C-H}} - l^{\text{eq}}_{\text{C-H}})\right), \tag{7.39}$$

we eventually obtain the following expression:

$$\begin{aligned} &E(n_{\text{C-C}}, n_{\text{C-H}}) = \\ &= W_{\text{C-C}}\left([n_{\text{C-C}} - 1]^2 - 1\right) + W_{\text{C-H}}\left([n_{\text{C-H}} - 1]^2 - 1\right) + a_p\, n_{\text{C-C}}^{q_{\text{C-C}}}\, n_{\text{C-H}}^{q_{\text{C-H}}}, \end{aligned} \tag{7.40}$$

where

$$q_{\text{C-C}} = \beta_{\text{C-C}}/\alpha_{\text{C-C}}, \quad q_{\text{C-H}} = \beta_{\text{C-H}}/\alpha_{\text{C-H}};$$

$$a_p = a_0 \exp(-\beta_{\text{C-C}} l^{\text{eq}}_{\text{C-C}} - \beta_{\text{C-H}} l^{\text{eq}}_{\text{C-H}}).$$

Equating the derivatives of the energy with respect to the bond order in the transition state to zero, ($\partial E/\partial n_{\text{C-C}} = 0$ and $\partial E/\partial n_{\text{C-H}} = 0$), determining the $n^{\neq}_{\text{C-C}}$ and $n^{\neq}_{\text{C-H}}$ values for the transition states from the obtained equations, and substituting these values into Eq. (7.40), we arrive at an expression for estimating E_a at $q_{\text{C-C}} = q_{\text{C-H}} = 1$ (Blowers–Masel equation):

$$E_a = \frac{(W_0 + 0.5\Delta H^0)(a_p - 2W_0 + \Delta H^0)^2}{a_p{}^2 - 4W_0^2 + (\Delta H^0)^2} \tag{7.41}$$

where $W_0 = (W_{\text{C-C}} + W_{\text{C-H}})/2$ and a_p is a single parameter related to the internal barrier E_a° (at $\Delta H^\circ = 0$) as

$$a_p = 2W_0 \frac{W_0 + E_a^\circ}{W_0 - E_a^\circ}. \tag{7.42}$$

Table 7.17 Estimations of internal barriers E_a° (reproduced with permission from Wiley International © 2001 [157])

Reaction	Interval of E_a°, kcal/mole	Best estimate of E_a°, kcal/mole
Hydrogen atom transfer $RH + X\bullet \rightarrow R\bullet + HX$	7–15	10
Radical substitution at carbon atom (transfer of R group to H•) $H\bullet + R{-}X \rightarrow H{-}R + X\bullet$	40–50	45
Transfer of other groups	40–60	50

Note that both the numerator and denominator in Eq. (7.41) are polynomials in powers of W_0 and ΔH°. At $\Delta H/(4E_a^\circ) > 1$ we have $E_a \cong \Delta H^\circ$, at $\Delta H/(4E_a^\circ) < -1$ we have $E_a \cong 0$, and in the interval $-1 \leq \Delta H/(4E_a^\circ) \leq 1$ the activation energy E_a is described by Eq. (7.41).

According to Masel [157], the adequacy of experimental data description by Eq. (7.41) in the intermediate region of ΔH° is close to the quality of results provided by the Marcus equation, but Eq. (7.41) also well describes the experiment in the Marcus inverted region. An analysis of the obtained E_a values showed that the results of calculations depend, but not strongly, on W_0. Table 7.17 presents the E_a° values recommended by Masel [157] for elementary steps in the gas phase.

Masel [157] believed that Eq. (7.41) could also be used for estimating E_a of reactions on the surface of metals. The following algorithm for selecting active metal catalysts was proposed, which is based on the generation of hypotheses about the mechanisms of catalytic reactions and the estimation of E_a for separate steps using the Blowers–Masel equation (7.41):

1. Estimate permissible E_a values for a required rate of the overall process. For example, if the anticipated process turnover frequency (TOF) is 1 s^{-1} in a temperature interval of 100 300 °C (so that $r/[M] = k = 1\,s^{-1}$, where [M] is the concentration of active centers), then it can be shown that $E_a \cong 34$ kcal/mole at 573 K. Accordingly, it is accepted that all steps must have $E_a \leq 34$ kcal/mole. It should be noted that this approach is correct provided that the observed activation energy of the overall process is determined by that of the limiting step (i.e., without any significant contribution of the enthalpies of quasi-equilibrium steps) and the kinetic equation has a simple form.
2. Generate possible mechanisms of the reaction under consideration on the surface of metals.
3. Select a group of metals capable of participating in the given reaction in the framework of generated hypotheses. For example, the synthesis of ammonia according to a dissociative mechanism implies the use of metals (Fe, Mn, Cr, Mo, W, Os, Ru, etc.) on which the dissociative adsorption of N_2 can take place.
4. Evaluate ΔH° using available experimental data for the heats of adsorption, energies of M–element bonds, or contributions of these bonds (M–H, M–C, M–N, M–O) to the thermal effect of reaction.
5. Calculate a_p by formula Eq. (7.42) using E_a° estimations from Table 7.18 for reactions of various types on the surface of metals.
6. Estimate E_a using Eq. (7.41).
7. Compare the obtained E_a values for each step for metals with the given permissible E_a value and select a group of metals that may contain the possible active catalysts.

Table 7.18 Approximate E_a° values for reactions on transition metal surfaces (reproduced with permission from Wiley International © 2001 [157])

Reaction type	E_a°, kcal/mole
Hydrogen atom transfer	10
Rupture of C–C- and C–O bonds	40
Rupture of multiple bonds	50

It is naturally expected that the obtained estimations are rather rough, but the results of calculations according to the Masel scheme showed that, for example (in the framework of the dissociative mechanism), the surface of Fe can and that of Pt cannot be a catalyst for the ammonia synthesis at the given temperature and TOF. Calculations of the activation energy of the dissociative adsorption of O_2 on Ni (111) gave E_a values within 0.24–0.85 kcal/mole [159], which are close (within experimental error) to $E_a = 5.3$ kcal/mole calculated by Eq. (7.41) for the process

$$O_2 + 4z \longrightarrow 2O_{ads}.$$

- **The Shustorovich (UBI-QEP) method**

Among the known phenomenological approaches to calculations of the enthalpies and activation energies for the chemisorption and reactions on crystalline surfaces of the transition and post-transition (Cu, Ag, Au) metals, the most developed scheme is offered by the unity bond index–quadratic exponential potential (UBI–QEP) method [166–168]. This approach was proposed by Shustorovich [169, 170] and initially called the bond order conservation–Morse potential (BOC–MP) method. Calculations according to this method involve quite a small number of observable parameters, namely, the atomic heats of adsorption, total bond energies of molecules in the gas phase, numbers of atoms per unit cell on the crystal surface, and M–M distances. The main assumptions of the UBI–QEP method are as follows:

(i) The two-center interaction M–A depends only on the M–A distance (r) and this quasi-spherical interaction is described by the following quadratic exponential potential (QEP):

$$E(x) = -Q_0[2x - x^2], \tag{7.43}$$

where $Q_0 = -E_{min}$ for the equilibrium M–A bond energy and $x = x(r)$ is expressed as

$$x(r) = \exp[-(r - r_0)/a]. \tag{7.44}$$

Function (7.44), which is called the bond index (BI), is analogous to the Pauling bond order (cf. Eq. (7.38)) and $x(r_0) = 1$ at $r = r_0$.

(ii) An additive scheme can be used to express the potential of multicenter interaction M_n–A and, hence, the binding energy Q_n, via the corresponding two-center binding energy values Q_0 for M–A. Restrictions are imposed on the BI values x_i for two-center bonds M_i–A($i = 1, 2, \ldots, n$) so that the total BI for the M_n–A interaction, which is a sum of BIs over all two-center contributions M_i–A, remains constant and equal to unity:

$$x_a = \sum x_{iA} = 1. \tag{7.45}$$

This restriction imposed on the total BI is called the unity bond index (UBI) principle.

(iii) The value of n is limited by the size of a regular unit cell on the metal surface with $n = 3$ (equilateral triangle) and $n = 4$ (square), so that $x_i = 1/n$.

Within the framework of the UBI–QEP method, special approaches were developed for taking into account the influence of adjacent groups bound to atom A, the adsorption of polynuclear molecules, the effect of surface coverage, and the surface diffusion of adsorbed species [168]. This method provides simple formulas for the barrier height. In particular, for the surface dissociation reaction

$$AB_{ads} \longrightarrow A_{ads} + B_{ads} \tag{7.46}$$

we have

$$E_{AB(ads)} = \frac{1}{2}\left(\Delta H^{\circ} + \frac{Q_A Q_B}{Q_A + Q_B}\right), \tag{7.47}$$

$$\Delta H^{\circ} = Q_{AB} + D_{AB} - Q_A - Q_B, \tag{7.48}$$

where ΔH° is the change in the enthalpy of reaction (7.46); Q_{AB}, Q_A, Q_B are the heats of binding (chemisorption) of molecules AB and atoms A and B on the surface; and D_{AB} is the energy of molecule AB dissociation in the gas phase.

The activation energy for the dissociation of molecule AB adsorbed from the gas phase is described by the following equation:

$$E_{AB(gas)} = \frac{1}{2}\left(D_{AB} - Q_{AB} - (Q_A + Q_B) + \frac{Q_A Q_B}{Q_A + Q_B}\right). \tag{7.49}$$

Considering the case of a diatomic molecule A_2 and reasonably assuming that $Q_{A_2} \ll Q_A + Q_B$, we obtain from Eq. (7.48) the flowing relationship that is analogous to the Polanyi equation and reflects the Bell–Evans–Polanyi principle:

$$E_{A_2(\mathbf{gas})} = 0.5D_{A_2} - 0.75Q_A. \tag{7.50}$$

Developments of the UBI–QEP method and its numerous applications to particular problems in heterogeneous catalysis have been reviewed in [168a, 168b]. The results of calculations of the parameters of chemisorption processes and activation energies of steps and an analysis of all the kinetic models based on the UBI–QEP method are usually compared to experimental data and quantum-chemical estimations. Since the beginning of the 1990s, much effort has been spent in the development and application of methods of the density functional theory (DFT) [171–173]. Although the DFT calculations of chemisorption were not as successful as those for the gas-phase and, sometimes, even liquid-phase processes, an increasing number of researchers use DFT estimations for a qualitative comparison to experiment [168c].

Let us briefly consider some results pertaining to particular surface reactions and chemisorption processes.

Various hypotheses for the mechanism of ethane hydrogenolysis on the surface of various metals were analyzed by the UBI-QEP method [174–179], which has also been applied to the processes of hydrocarbon dehydrogenation and cracking [174, 180]. The results of a kinetic analysis of the systems of differential equations for processes in a continuous plug flow reactor for a 20-step Sinfelt–Taylor mechanism of ethane hydrogenolysis [179–181] showed that the hydrogenolysis on (111) crystal faces of three metals (Pd, Pt, and Ni) at 478 K, $P_{H_2} = 20.26$ kPa, and $P_{C_2H_6} = 3.04$ kPa proceeds predominantly via the formation of adsorbed ethyl groups in which the C–C bond rupture step takes place as

$$CH_3CH_{2ads} + 2z \rightleftharpoons CH_{3ads} + CH_{2ads}. \tag{7.51}$$

Table 7.19 Observed reaction orders with respect to ethane and hydrogen and activation energies for ethane hydrogenolysis reactions in various systems (in comparison to results of calculations)* (reproduced with permission from editorial board of *Kinet. Katal.* journal © 2004 [179])

System	T, K	P_{H_2}, bar	$P_{C_2H_6}$, bar	n_{H_2}	$n_{C_2H_6}$	E_a°, kcal/mole	Ref.
Ni/SiO_2	450–492	0.2	0.03	−2.4	1	40.6	[122, 123]
Ni (111)	520–600	0.13	0.0013	–	–	*46.1*	[182]
Ni (111)	478–650	0.1–1	0.01–0.1	*−1.9*	*1*	*52.4*	[179]
Pd/SiO_2	616–650	0.2	0.03	−2.5	0.9	58	[122, 123]
Pd/SiO_2	478	0.20	0.01–0.10	−2.6	1	56.7	[183]
Pd (111)	478–650	0.1–1	0.01–0.1	*−1.8*	*1*	*55.5*	[179]
Pt/SiO_2	344–385	0.2	0.03	−2.5	0.9	54	[122, 123]
Pt (111)	550–640	0.13	0.0013		1.2	*36.6*	[182]
	570–625	0.13	0.0013			*43.5*	[182]
	473–623	0.13	0.013	−0.55		34	[127]
Pt/SiO_2	478	0.20	0.01–0.10	*−2.5*	*1*	55.3	[183]
Pt (111)	478–650	0.1–1	0.01–0.1	*−2*	*1*	*59*	[179]

Note: *Calculated data are italicized.

The activation energies of step (7.51) in the forward direction amount to 26.8, 33.4 and 24.4 kcal/mole for Pd, Pt, and Ni, respectively. Table 7.19 presents the activation energies and reaction orders with respect to P_{H_2} and $P_{C_2H_6}$ for the rates of ethane hydrogenolysis on three metals according to the experimental data and results of UBI–QEP calculations [179]. As can be seen from these comparative data, the coincidence of experiment and theory is quite good, the more so if we take into account that the calculation method employs a rough approximation and the comparison is rather incorrect because of differences in the surface nature of supported (M/SiO_2) and single-crystalline metals.

The UBI–QEP method was also used for an analysis of the mechanisms of CO hydrogenation to hydrocarbons in the Fischer–Tropsch synthesis [180, 184–186], to methane and methanol [187, 188] and to methanol [189–191]. Fishtik et al. [191] used ΔH° and E_a values of reactions on the (111) copper surface [180, 189] for the comparison and discrimination of the mechanisms of methanol synthesis and water-gas conversion.

With a view to predicting the activity of single-crystalline metal surfaces in the reaction

$$CO + H_2O \rightleftharpoons CO_2 + H_2 \qquad (7.52)$$

this process was studied in detail in [192], where the hypothetical mechanisms were generated by the MECHEM program and the activation energies and enthalpies of 17 selected steps were estimated using the UBI–QEP method. The results of kinetic modeling in a continuous-flow stirred-tank reactor were used to calculate the degrees of CO conversion at various temperatures. Using the estimations obtained for 510 K, the following order of catalytic activities of the metals was proposed:

$$Cu \gg Ni > Fe > Pd, Pt \gg Ag, Au.$$

The order of activities according to the experimental data for metals supported on Al_2O_3, was somewhat different:

$$Cu \gg Ni \gg Pt > Fe \sim Pd,$$

although, as noted above, a comparison between single-crystalline and supported metal surfaces is not completely correct.

Liao and Zhang [193] used DFT and UBI–QEP methods to study the steps of methane dissociation on the surface of metals, which are important for the processes of methane oxidative and steam conversion and reactions of methane with carbon dioxide. The (111)-oriented metal surfaces were modeled by clusters containing 7, 10, and 13 metal atoms. The obtained values of the heat of adsorption of various CH_{4-x} species ($x = 0, 1, 2, 3, 4$) and the energies of C–H bonds in gas-phase CH_{4-x} molecules ($x < 4$) were used to calculate the values of ΔH_S° (subscript S refers to surface) and E_a for 10 metals. The activation energies increase in the following order:

$$\mathrm{Ru} \approx \mathrm{Rh} \approx \mathrm{Ni} < \mathrm{Ir} \approx \mathrm{Pd} \approx \mathrm{Pt} < \mathrm{Cu} < \mathrm{Ag} \sim \mathrm{Au},$$

whereas experimental estimations of the activity of metals in the process of methane conversion are better correlated with the series of $\Delta H_{S(\mathrm{tot})}^\circ$ values for M_{10} clusters in the reaction

$$\mathrm{CH}_{4S} \longrightarrow \mathrm{C}_S + 4\mathrm{H}_S. \tag{7.53}$$

for which

	Ni(M_7)	< Rh	< Ru	< Ir	< Pt	< Pd	≪ Cu	< Au	< Ag
$\Delta H_{S(\mathrm{tot})}^\circ$, (kcal/mole) :	−18.56	−15.35	0.23	6.41	23.1	33.0	147.5	183.7	230.9

It was also shown that an oxygen atom adsorbed over a metal atom on the surface reduces the enthalpy of methane dissociation in the reaction

$$\mathrm{CH}_{4S} + \mathrm{O}_S \longrightarrow \mathrm{CH}_{3S} + \mathrm{OH}_S, \tag{7.54}$$

on all metals, while an oxygen atom adsorbed over a "hole" between neighboring metal atoms on the surface decreases ΔH_S only for Pt, Cu, Au, and Ag.

Hei et al. [194] carried out a detailed analysis of steps in the mechanism of methane conversion by carbon dioxide in reaction

$$\mathrm{CH}_4 + \mathrm{CO}_2 \longrightarrow 2\mathrm{CO} + 2\mathrm{H}_2 \tag{7.55}$$

and determined the E_a and ΔH° values of reactions in a network consisting of 84 steps on Cu(111), Ni(111), Pd(111), Pt(111), Rh(111), Ru(001), Ir(111), and Fe(110) surfaces. The results of calculations showed that both "difficult" steps of methane and CO_2 dissociation are slow processes, but the dissociation of methane with the formation of H_S facilitates the dissociation of CO_2 as

$$\mathrm{CO}_{2S} + \mathrm{H}_S \longrightarrow \mathrm{CO}_S + \mathrm{OH}_S,$$

while the appearance of O_S upon the dissociation of CO_2 facilitates the dissociation of methane as

$$\mathrm{CH}_{4S} + \mathrm{O}_S \longrightarrow \mathrm{CH}_{3S} + \mathrm{OH}_S.$$

Based on the estimated barriers for the slow steps, the following order of metal activity in the process of methane reforming with carbon dioxide was proposed:

$$\mathrm{Fe} > \mathrm{Ni} > \mathrm{Rh} > \mathrm{Ru} > \mathrm{Ir} > \mathrm{Pd} > \mathrm{Pt} > \mathrm{Cu}.$$

The UBI–QEP method was also successfully applied to an analysis of the mechanisms of ammonia synthesis and decomposition [195] and to simulation of the adsorption of various molecules by the Monte

Carlo technique [196–198] with allowance for the dependence of the adsorption energy characteristics on the degree of surface coverage by the adsorbed species. It should be also noted that a modified formalism of the UBI–QEP method [199] makes possible the estimation of the binding energies of polyatomic symmetric (C_2H_2, $CH_2{=}CH_2$, NH=NH, $NH_2{-}NH_2$) and asymmetric (N_2O) molecules without knowledge of the distribution of the total energy over the bonds.

- **Quantum-chemical estimations of enthalpies and E_a of steps in catalysis with metal complexes**

Considerable progress in quantum-chemical calculations of the stability and reactivity of heavy metal complexes was related to the development of new theoretical methods (mostly on the DFT basis) [171–173 200–204] and the availability of powerful computational facilities. For example, only the calculations for compounds and complexes of Pt and Pd in 1990–2000 were reported in more than 400 publications [205]. In the framework of investigations of the catalytic processes in solutions of transition and post-transition (*d*-block) metal complexes, it is possible to separate four main problems that are worthy of special attention [204]:

(i) Correct description of the molecular and electron structure of precursors of active species, active complexes, and intermediates,
(ii) Exact estimation of the energy characteristics of steps and activation barriers.
(iii) Modeling of the potential energy surface (and, in particular, reaction trajectories) by the method of internal reaction coordinates (IRCs); determining the C–H–M bond angles and C–H bond lengths as functions of the intramolecular distance r_{IRC} (reaction coordinate) for the intramolecular oxidative addition of C–H to a metal in complexes

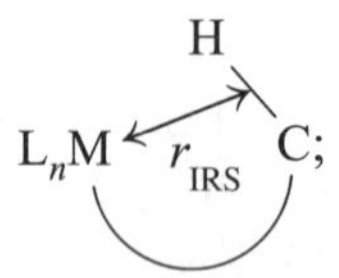

(iv) Calculation of the IR absorption spectra of intermediates in a catalytic cycle.

Most calculations of the energy characteristics of elementary steps were performed for the gas phase (vacuum), although the solvent effect is also taken into account in an increasing number of investigations. In these calculations, the solvent molecules are usually introduced into the coordination sphere of a metal atom (in the presence of coordination vacancies). The role of solvent in the stabilization of intermediates or transition states is described in the framework of a continuum model (as solvation energy) [206] or a discrete model, by introducing a certain number of solvent molecules into the metal complex environment and optimizing the entire ensemble geometry with respect to the reaction coordinate.

A large proportion of quantum-chemical investigations have been devoted to typical elementary steps involved in the catalysis with metal complexes [201–203, 205], including the oxidative addition to (and reductive elimination from) metal complexes (at H–H, C–H, C–C, O–H, Si–H, Si–X, Si–Si, and B–B bonds), migratory insertion of olefins, alkynes, and CO into M–H, M–C bonds (for metals of the second row of the transition series), β-H elimination, metathesis of σ bonds, nucleophilic addition to π ligands in π complexes, and some other. In particular, Dedieu [205, 207] obtained estimates for all the main intermediates and transition states in some catalytic cycles and rather complete fragments of reaction mechanisms that determine the process rates.

Data obtained by Cui et al. [208] provided an explanation of the experimentally observed catalysis of the reaction of alkyne *cis*-diboration by $Pt(PR_3)_2$ and the absence of catalysis by analogous Pd(0) complexes.

The model reaction

$$C_2H_2 + (HO)_2B{-}B(OH)_2 \longrightarrow (HO)_2BCH{=}CHB(OH)_2 \tag{7.56}$$

with $Pt(PH_3)_2$ and $Pd(PH_3)_2$ catalysts was studied by DFT techniques using a B3LYP exchange-correlation functional. The most pronounced difference between the two reaction mechanisms is related to the energy characteristics of the step of oxidative addition of a B–B bond to metal. This step has a greater barrier (14.0 kcal/mole) for Pt(0), but the obtained compound is stable (the process is exothermal). The barrier on Pd(0) is smaller (8.6 kcal/mole), but the corresponding compound is unstable (reverse barrier is ~0) and the reaction of oxidative addition is endothermal. These factors determine differences between the energy profiles observed for two metals and between the slow steps on Pt and Pd. Differences in the energy characteristics of steps of the oxidative addition on the two metals are explained by the dissimilar ground states of Pd(0) (d^{10}) and Pt(0) (d^9s^1). Indeed, the excited state of Pd(0) (d^9s^1) is lying 21.9 kcal/mole above the ground state, and this excited state participates in the steps of oxidative addition on palladium.

Analogous investigation [209] of the reaction of acetylene thioboration [$RS{-}B(OR)_2$] showed that, in agreement with experiment, Pd(0) complexes must catalyze this reaction. It was suggested that Pt(0) complexes, in contrast to the case of diboration reaction, would not catalyze the thioboration process because of the step of reductive elimination

$$\underset{\displaystyle \overset{|}{SH}}{L_2Pt}CH{=}CHB(OH)_2 \longrightarrow L_2Pt + HSCH{=}CHB(OH)_2. \tag{7.57}$$

In order that the quantum-chemical calculations of the rates of steps would be useful to experimenters as a criterion for discriminating hypotheses and selecting more realistic mechanistic schemes, one should not only be correct and lucky in choosing the computational schemes, but also correctly formulate questions and select well justified molecular models for estimating the reactivity of intermediates. In this context, let us consider an investigation of mechanisms of the reaction of ethylene oxidation by Pd(II) acetate in aqueous acetic acid. Siegbahn [211–213] studied this reaction in an aqueous chloride system (Wacker process) by DFT methods with allowance for the solvent effect. It was established that the *trans*-addition of water molecule to ethylene in a π complex is preferred (in comparison to the inner-sphere *cis*-insertion of ethylene into the Pd–OH bond). It was also shown that a weak exothermicity (4.5 kcal/mole) and a small barrier (5.7 kcal/mole) can result from the addition of a chain of three hydrogen-bonded water molecules:

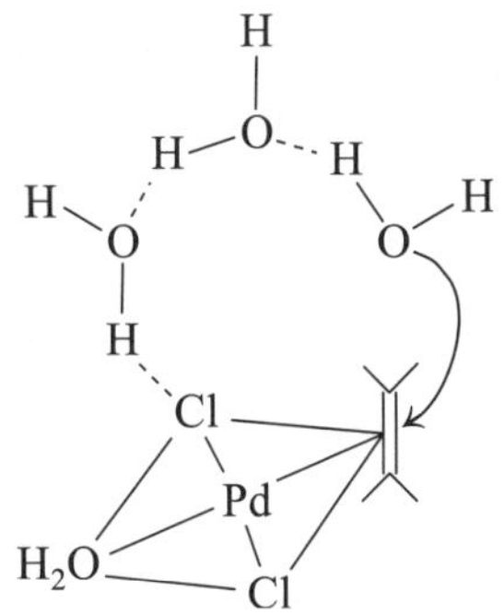

The formation of ~ $PdCH_2CH_2OH$ was not considered as a limiting step.

Based on an analysis [214] of the mechanism of vinyl acetate synthesis via the oxidative esterification of ethylene (Moiseev reaction, see Section 2.2), Kragten et al. [210] studied the Wacker process under analogous conditions, but in the presence of water. In this case, both the setting of a computational task and the choice of elementary steps raise several questions that are listed below with some comments:

(i) The given system was not studied by kinetic methods and, hence, no comparison of the results of quantum-chemical calculations to experiment was possible.
(ii) The state of $Pd(OAc)_2$ in aqueous acetic acid is unknown and, hence, the choice of a neutral dimer complex $(H_2O)_2Pd(\mu{-}HCOO)_2Pd(OCOCH_3)_2$, cannot be justified.
(iii) Comparison of the inner-sphere process

$$L{=}Pd(OAc)_2 + C_2H_4 \rightleftharpoons L{=}Pd(OAc)(C_2H_4) + OAc^-;$$

$$L{=}Pd(OAc)(C_2H_4) + H_2O \rightleftharpoons L{=}Pd(OH)(C_2H_4) + HOAc;$$

$$L{=}Pd(OH)(C_2H_4) \longrightarrow L{=}Pd\overset{CH_2CH_2}{\longleftarrow}OH,$$

(where L is $(H_2O)_2Pd(\mu\text{-}HCOO)_2$) to the outer-sphere attack of π-complex by the OH^- ion

$$L{=}Pd(C_2H_4)(OAc) + OH^- \longrightarrow L{=}\overset{\ominus}{Pd}(OAc)CH_2CH_2OH \qquad (7.58)$$

can hardly be justified, since it was demonstrated long ago [56, 215] that the step of OH^- ion attack in an aqueous acid medium (e.g., in aqueous acetic acid) cannot be brought into correspondence with the kinetic data.
(iv) In the proposed (more favorable) outer-sphere process, all catalyst is rapidly converted (by the consequent steps of H atom transfer) from intermediate (24) into intermediate (26) [$L = Pd(OAc)(CH(OH)CH_3)$ with $\Delta H = -28.5$ kcal/mole]

$$\underset{(24)}{\sim PdCH_2CH_2OH} \rightleftharpoons \underset{(25)}{\sim PdH(CH_2{=}CHOH)} \rightleftharpoons \underset{(26)}{\sim PdCH(OH)CH_3}; \qquad (7.59)$$

$$\underset{(26)}{\sim PdCH(OH)CH_3} \xrightarrow{B} \sim Pd + BH^+ + CH_3CHO. \qquad (7.60)$$

the conversion of which (with a barrier of 14.6 kcal/mole) is the only rate-limiting step (7.60) of the oxidation process.

It is rather surprising that the elimination of proton from an α–hydroxy derivative by OAc^- ion or any other base (e.g., H_2O) can be a slow step, the more so that the reduction of Pd(II) to Pd(0) in this process can be facilitated by the adjacent Pd(II) with the formation of a Pd–Pd fragment:

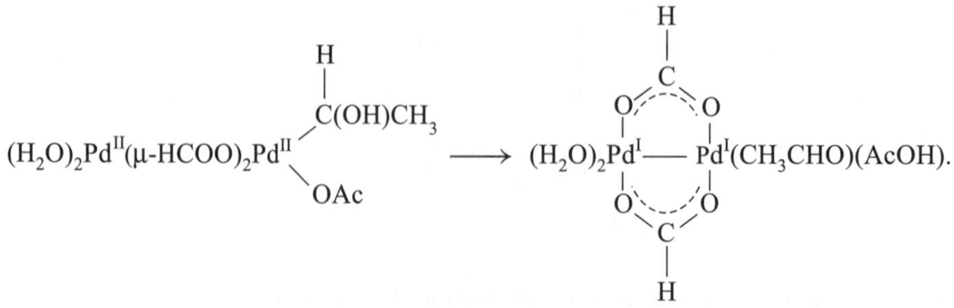

(7.61)

It should be noted that some other theoretical investigations were also devoted to separate steps in the Wacker process, including the conversion of intermediates of type (26) (elimination of palladium hydride or reductive deprotonation) [216–219], *syn*-addition of Pd-bound OH nucleophile to ethylene in π complex and *syn*-addition of OH under the action of water on *cis*-$PdCl_2(H_2O)(C_2H_4)$ with elimination of H_3O^+ [217], and anti-addition of water molecule to π complex *cis*-$PdCl_2(OH)(C_2H_4)^-$ [218]. The study by Keith et al. [216] revealed an important problem in theoretical calculations, whereby an apparently reliable methodology may incorrectly describe the energy characteristics of a proton eliminated in the transition state even in the presence of a counter-ion with a certain number of water molecules. The error margin in these calculations for ion pairs can reach 15 kcal/mole [217b].

Despite evident progress achieved in theoretical analysis of the reactivity of complicated transition metal complexes, quantum-chemical calculations still cannot be a general method for estimating ΔH and E_a of steps in large reaction networks encountered in the homogeneous catalysis with metal complexes. The choice between variants in a computational method (e.g., DFT techniques) and a basis set of wave functions, making allowance for the electron correlations and relativistic effects, and taking into account the solvent effect are still not formalized procedures that depend on the experience of researchers and require much time for iterations.

It should be noted that essentially semiempirical DFT and hybrid DFT methods (B3LYP variant) give quite satisfactory results in calculations of the ground-state energies of heavy metal complexes, but are much less suited for calculations of the transition-state energies, especially in solution. Making allowance for the solvation (especially, nonspecific) in the case of charged complexes is still a procedure that involves big uncertainties.

The development of effective approaches to calculations is, in turn, hindered by the lack of data on the energies of M–L, M–H, and M–R bonds as dependent on the nature of metal M, ligand environment, complex geometry, and degree of metal oxidation [220–222] and by a weakly developed database on the kinetics of elementary steps [223]. The same factors apparently hinder the development of phenomenological (semiempirical) approaches (analogous, e.g., to the UBI–QEP method for reactions on metal surfaces) to estimation of the energy characteristics of elementary steps in reactions of metal complexes.

Rapid progress in the quantum-chemical approaches to an analysis of large reaction networks with participation of metal complexes in solution, including the estimation of reaction rates from overall equations of a catalytic reaction, provides optimism in waiting for the development of practically acceptable methods of discriminating steps and whole mechanisms in investigations of large reaction networks. This will make it possible to use all advantages of the rational strategy in studying reaction mechanisms and constructing "molecular" kinetic models.

References

1. O. N. Temkin, L. G. Bruk and A. V. Zeigarnik, *Kinet. Katal*., **34** (3), 445 (1993).
2. O. N. Temkin, *Zh. Ross. Khim. o-va im. D.I. Mendeleeva*, **44** (4), 58 (2000).
3. V. N. Pisarenko and A. G. Pogorelov, *Planirovanie kineticheskikh eksperimentov* (Planning of Kinetic Experiments), Nauka, Moscow, 1969.
4. L. S. Polak (Ed.), *Primenenie vychislitel'noi matematiki v khimicheskoi i fizicheskoi kinetike* (Application of Computational Mathematics in Chemical and Physical Kinetics), Nauka, Moscow, 1969.
5. B. S. Snagovskii and G. M. Ostrovskii, *Modelirovanie kinetiki geterogennykh kataliticheskikh reaktsii* (Modeling the Kinetics of Heterogeneous Catalytic Reactions), Khimiya, Moscow, 1976.
6. L. S. Polak, M. Ya. Gol'denberg and A. A. Levitskii, *Vychislitel'nye metody v khimicheskoi kinetike* (Computational Methods in Chemical Kinetics), Nauka, Moscow, 1984.
7. V. G. Gorskii, *Planirovanie kineticheskogo eksperimenta* (Kinetic Experiment Planning), Nauka, Moscow, 1984.
8. E. F. Brin, *Usp. Khim*., **54** (3), 428 (1987).
9. I. V. Berezin and A. A. Klesov, *Prakticheskii kurs khimicheskoi i fermentativnoi kinetiki* (Practical Course on Chemical and Enzyme Kinetics), MGU, Moscow, 1976.

10. S. L. Kiperman, *Kinetika geterogennykh reaktsii. Metodicheskie voprosy kinetiki* (Kinetics of Heterogeneous Reactions. Methodological Problems of Kinetics), Inst. Org. Khim.–Inst. Khim. Fiz. AN SSSR, Chernogolovka, 1983, p. 5.
11. S. L. Kiperman, in *Mekhanizm kataliza, Chast' 1, Priroda kataliticheskogo deistviya* (Mechanism of Catalysis, Part 1: Nature of Catalytic Action), Nauka, Novosibirsk, 1984, p. 159.
12. R. Schmid and V. N. Sapunov, *Non-Formal, Kinetics*, Verlag-Chemie, Weinheim, 1982 (*Neformal'naya, Kinetika*, Mir, Moscow, 1985).
13. N. N. Semenov, *Tsepnye reaktsii* (Chain Reactions), 2nd ed., Nauka, Moscow, 1985.
14. L. P. Hammett, *Physical Organic Chemistry*, 2nd ed., McGraw-Hill, NewYork, 1970.
15. (a) R. P. Bell, *The Proton in Chemistry*, 2nd ed., Chapman and Hall, London, 1973; (b) R. Bell, *Proton v Khimii* (The Proton in Chemistry), R. R. Dogonadze (Ed.), Mir, Moscow, 1977 [additional chapter by E. D. German and R. R. Dogonadze], pp. 350–376.
16. M. Dixon and E. C. Webb, *Enzymes*, 2nd ed., Academic Press, New York, 1964.
17. C. Walter, *Enzyme, Kinetics*, Ronald Press, New York, 1966.
18. S. L. Kiperman, *Vvedenie v kinetiku geterogennykh kataliticheskikh reaktsii* (Introduction to Kinetics of Heterogeneous Catalytic Reactions), Nauka, Moscow, 1964.
19. M. I. Temkin, *Adv. Catal.*, **28**, 173 (1979).
20. J. C. Bond and R. H. Cunningham, *J. Catal.*, **166**, 172 (1997).
21. S. H. Bauer, *Int. J. Chem. Kinetics*, **22**, 113 (1990).
22. G. A. Somorjai, *Introduction to Surface Chemistry and Catalysis*, John Wiley & Sons, New York, 1994.
23. O. V. Krylov and V. A. Matyshak, *Promezhutochnye soedineniya v geterogennom katalize* (Intermediates in Heterogeneous Catalysis), Nauka, Moscow, 1996.
24. K. Tamaru, *Dynamic Heterogeneous Catalysis*, Academic Press, London, 1978, pp. 96–135.
25. V. A. Matyshak, *Kinet. Katal.*, **30** (1), 168 (1989).
26. K. B. S. Prasad, K. B. S. Saibabu and R. Vaidyeswaran, *J. Catal.*, **72**, 338 (1981).
27. O. N. Temkin, G. K. Shestakov and Yu. A. Treger, *Atsetilen: khimiya, mekhanizmy reaktsii, tekhnologiya* (Acetylene: Chemistry, Reaction Mechanisms, Technology), Khimiya, Moscow, 1991.
28. A. Ya. Rozovskii and G. I. Lin, *Teoreticheskie osnovy sinteza metanola* (Theoretical Backgrounds of Methanol Synthesis, Khimiya, Moscow, 1990.
29. V. V. Nalimov, *Teoriya eksperimenta* (Theory of Experiment), Nauka, Moscow, 1971.
30. K. R. Popper, *Objective Knowledge*, Oxford University Press, Oxford, 1972; (b) N. Trinajstic, in *Studies in Physical and Theoretical Chemistry*, Elsevier Sci. Publ., Amsterdam, 1989, Vol. 63, p. 557.
31. T. C. Chamberlin, *Science* (old series), **15**, 92 (1890); reprinted in *Science*, **148**, 754 (1965).
32. D. V. Gal, in *Problemy khimicheskoi kinetiki* (Problems in Chemical Kinetics), Nauka, Moscow, 1979, p. 139.
33. I. M. Emanuel' and D. Gal, *Okislenie etilbenzola (mekhanizmy reaktsii)* (Oxidation of Ethylbenzene (Reaction Mechanisms), Nauka, Moscow, 1984.
34. V. I. Dimitrov, *Prostaya kinetika* (Simple Kinetics), Nauka, Novosibirsk, 1982.
35. O. N. Temkin and L. G. Bruk, *Usp. Khim.*, **52**, 206 (1983).
36. A. V. Zeigarnik, L. G. Bruk, O. N. Temkin, V. A. Likholobov and L. I. Maier, *Usp. Khim.*, **65**, 125 (1996) [*Russ. Chem. Rev.*, **65**, 117 (1996)].
37. O. N. Temkin, A. V. Zeigarnik and D. G. Bonchev, *Chemical Reaction Networks: A Graph-Theoretical Approach*, CRC Press, Boca Raton, FL, 1996.
38. O. N. Temkin, O. L. Kaliya, L. N. Zhir-Lebed', L. G. Bruk, and N. G. Mekhryakova, in *Gomogennoe okislenie (Trudy Instituta organicheskogo kataliza i elektrokhimii AN Kaz. SSR)* [Homogeneous Oxidation (Scientific Works of Institute of Organic Catalysis and Electrochemistry, Acad. Sci. Kazakh SSR)], Nauka, Alma-Ata, 1978, Vol. 17, pp. 1–33.
39. L. G. Bruk, *Metody vydvizheniya i diskriminatsii gipoteticheskikh mekhanizmov kataliticheskikh reaktsii kak osnova razrabotki tekhnologii karbonilirovaniya alkinov* (Methods of Generation and Discrimination of Hypothetical Mechanisms of Catalytic Reactions: Basis for Developing Alkyne Carbonylation Technology), Doctoral Dissertation in Chemistry, MITKhT (Moscow Institute of Fine Chemical Technology), Moscow, 1996.
40. O. N. Temkin and L. G. Bruk, *Kinet. Katal.*, **44** (5), 661 (2003).

41. A. F. Shmidt, V. V. Smirnov, O. V. Starikova and A. V. Elaev, *Kinet. Katal.*, **42** (2), 223 (2001); *Kinet. Catal.*, **42**, 199 (2001).
42. C. K. Rafer-De Poorter, *Chem. Rev.*, **81**, 447 (1981).
43. V. I. Vedeneev et al., *Kinet. Katal.*, **29**, 7, 14, 1291, 1297 (1988); **37**, 20 (1996).
44. O. V. Krylov and V. S. Arutyunov, *Okislitel'nye prevrashcheniya metana* (Oxidative Transformations of Methane), Nauka, Moscow, 1999.
45. C. A. Mims, R. Mauti, A. M. Dean and K. D. Roze, *J. Phys. Chem.*, 98, 13357 (1994).
46. M. C. J. Bradford and M. A. Vannice, *Catal. Rev. Sci. Eng.*, **41**, 1 (1999).
47. (a) T. Faravelli, A. Goldaniga, E. Ranzi, et al., in *Natural Gas Conversion (Studies in Surface Science and Catalysis*, Vol. 119), B. Delmon and J. T. Yates (Eds), Elsevier, 1998, p. 575; (b) E. Ranzi, T. Faravelli, P. Gafuri and A. Sogaro, *Combust. Flame*, 108, 24 (1997).
48. M. Sinev, V. Arutyunov and A. Romanets, *Adv. Chem. Eng.*, **32**, 169 (2007).
49. O. N. Temkin, L. G. Bruk and D. Bonchev, *Teor. Eksp. Khim.*, **24** (3), 282 (1988).
50. A. M. Zhabotinskii and M. D. Korzukhin, in *Kolebatel'nye protsessy v biologicheskikh i khimicheskikh sistemakh* (Oscillatory Processes in Biological and Chemical Systems), G. M. Frank (Ed.), Nauka, Moscow, 1967, p. 223.
51. M. D. Korzukhin, *Kolebatel'nye protsessy v biologicheskikh i khimicheskikh sistemakh* (Oscillatory Processes in Biological and Chemical Systems), G. M. Frank (Ed.), Nauka, Moscow, 1967, pp. 231, 241.
52. A. M. Zhabotinskii, in *Oscillations and Traveling Waves in Chemical Systems*, R. J. Field and M. Burger (Eds), Wiley-Interscience Publ., New York, 1985; Mir, Moscow, 1988, p. 20.
53. L. B. Bazhenov and P. Kh. Samorodnitskii, *Vopr. Filos.*, No. 6, 93 (1976).
54. J. M. Berty, S. Lee, F. Szeifert and J. B. Crooply, *AIChE Meeting (August 1984, Denver)*, p. 37.
55. S. M. Brailovskii, O. N. Temkin, M. S. Karpova and R. M. Flid, *Kinet. Katal.*, **16** (2), 408 (1975).
56. I. I. Moiseev, *π-Kompleksy v zhidkofaznom okislenii olefinov* (π Complexes in Liquid-Phase Oxidation of Olefins), Nauka, Moscow, 1970.
57. P. M. Henry, *Palladium Catalyzed Oxidation of Hydrocarbons*, Reidel, Dordrecht–Boston, 1980.
58. C. A. Tolman and J. W. Faller, *Mechanistic Studies of Catalytic Reactions Using Spectroscopic and Kinetic Techniques*, in *Homogeneous Catalysis with Metal Phosphine Complexes*, L. H. Pignolet (Ed.), New York, Plenum Press, 1983, pp. 13–110.
59. S. D. Varfolomeev and S. V. Zaitsev, *Kineticheskie metody v biokhimicheskikh issledovaniyakh* (Kinetic Methods in Biochemical Investigations), MGU, Moscow, 1982.
60. S. D. Varfolomeev and K. G. Gurevich, *Biokinetika: Prakticheskii kurs* (Biokinetics: Practical Course), FAIR-Press, Moscow, 1999.
61. O. N. Temkin, S. M. Brailovskii, L. G. Bruk et al., in *Khimicheskaya kinetika v katalize:*, *Kineticheskie modeli zhidkofaznykh reaktsii* (Chemical Kinetics in Catalysis: Kinetic Models of Liquid-Phase Reactions), Akad. Nauk SSSR, Chernogolovka, 1985, pp. 59, 120.
62. O. N. Temkin and L. G. Bruk, *Materialy 3 Vses. konf. "Mekhanizmy kataliticheskikh reaktsii"* (Proceedings of the 3rd All-Union Conf. "Mechanisms of Catalytic Reactions"), Nauka, Novosibirsk, 1982, Vol. 2, p. 101.
63. O. N. Temkin, A. V. Zeigarnik, A. E. Kuz'min, L. G. Bruk and E. V. Slivinskii, *Izv. Akad. Nauk, Ser. Khim.*, No. 1, 1 (2002) [*Russ. Chem. Bull.*, **51**, 1 (2002)].
64. E. T. Denisov, O. M. Sarkisov and G. I. Likhtenshtein, *Khimicheskaya kinetika* (Chemical, Kinetics), Khimiya, Moscow, 2000.
65. E. S. Rudakov, *Reaktsii alkanov s okislitelyami, metallokompleksami i radikalami v rastvorakh* (Reactions of Alkanes with Oxidants, Metal Complexes, and Radicals in Solutions), Naukova Dumka, Kiev, 1985.
66. O. N. Temkin, S. M. Brailovskii and L. G. Bruk, *Tezisy dokladov Vsesoyuznoi konferentsii "Mekhanizmy kataliticheskikh reaktsii"* (Proceedings of the All-Union Conf. "Mechanisms of Catalytic Reactions"), Nauka, Moscow, 1978, p. 74.
67. A. Ya. Rozovskii, *Katalizator i reaktsionnaya sreda* (Catalyst and Reaction Medium), Nauka, Moscow, 1988.
68. D. I. Kamenskii, D. G. Bonchev and O. N. Temkin, *Appl. Catal. A*, **88**, 1 (1992).
69. H. I. Abanto-Chavez, K. B. Hoang and O. N. Temkin, *Kinet. Katal.*, **41**, 701 (2000).
70. N. S. Zefirov and E. V. Gordeeva, *Usp. Khim.*, **56**, 1753 (1987).

71. R. Barone and M. Chanon, in *Computer Aids to Chemistry*, G. Vernin and M. Chanon (Eds), Ellis Horwood, Chichester, 1986, p. 19.
72. J. Dugunji and I. Ugi, *Top. Curr. Chem.*, **39**, 19 (1973).
73. I. Ugi, J. Bauer, K. Bley et al., *Angew. Chem. Int. Ed.*, **32**, 201 (1993).
74. I. Ugi, J. Bauer, C. Blomberg et al., *J. Chem. Inf. Comput. Sci.*, **34**, 3 (1994).
75. E. Fontain and K. Reimtsa, *J. Chem. Inf. Comput. Sci.*, **31**, 97 (1991).
76. Y. Yoneda, *Bull. Chem. Soc. Jpn.*, **52**, 8 (1979).
77. S. V. Dozmorov, *React. Kinet. Catal. Lett.*, **19**, 289 (1982).
78. L. J. Broadbelt, S. M. Stark and M. T. Klein, *Ind. Eng. Chem. Res.*, **33**, 790 (1994).
79. T. I. Mizan and M. T. Klein, *Catal. Today*, **50**, 159 (1999).
80. L. I. Shtokolo, L. G. Shteingauer, V. A. Likholobov and A. V. Fedotov, *React., Kinet. Catal. Lett.*, **26**, 227 (1984).
81. L. I. Mayer, L. G. Shteingauer, A. V. Fedotov, and V. A. Likholobov, in: *Khimicheskii sintez na osnove odnouglerodnykh molekul* (Proceedings of the All-Union Conference "Chemical Synthesis Based on One-Carbon Molecules," December 3-7, 1984, Moscow), p. 8.
82. V. A. Likholobov, L. I. Mayer, N. N. Bulgakov, A. V. Fedotov and L. G. Shteingauer, in: *Homogeneous and Heterogeneous Catalysis*, Yu. Yermakov and V. Likholobov (Eds.), VNU Science Press BV, Utrecht, 1986, p. 229.
83. L. G. Shteingauer, L. I. Mayer, N. N. Bulgakov et al., *React. Kinet. Catal. Lett.*, **36**, 139 (1988).
84. L. I. Maier, *Prognozirovanie kataliticheskikh svoistv atomov metallov s ispol'zovaniem metodov komp'yuternogo sinteza na primere protsessa gidrirovaniya okisi ugleroda* (Predicting Catalytic Properties of Metal Atoms with the Aid of Computer Synthesis Methods in Application to Carbon Oxide Hydrogenation Process), Candidate Dissertation in Chemistry, Institute of Catalysis, Novosibirsk, 1990.
85. R. E. Valdes-Perez, PhD Thesis, Carnegie Mellon University, School of Computer Science, CMU-CS-90-191, 1990.
86. R. E. Valdes-Perez, *Tetrahedron Comput. Meth.*, **3**, 277 (1990).
87. R. E. Valdes-Perez, *J. Comput. Chem.*, **13**, 1079 (1992).
88. R. E. Valdes-Perez, *J. Phys. Chem.*, **96**, 2394 (1992).
89. R. E. Valdes-Perez, *J. Comput. Chem.*, **14**, 1454 (1993).
90. R. E. Valdes-Perez, *Catal. Lett.*, **28**, 79 (1994).
91. R. E. Valdes-Perez, *J. Chem. Inf. Comput. Sci.*, **34**, 976 (1994).
92. R. E. Valdes-Perez, *J. Comput. Chem.*, **15**, 1266 (1994).
93. R. E. Valdes-Perez, *J. Chem. Inf. Comput. Sci.*, **31**, 554 (1991).
94. R. E. Valdes-Perez, *Artif. Intell.*, **65**, 247 (1994).
95. T. D. Salatin and W. L. Jorgensen, *J. Org. Chem.*, **45**, 2043 (1980).
96. W. L. Jorgensen, E. R. Laird, A. J. Gushurts et al., *Pure Appl. Chem.*, **62**, 1921 (1990).
97. R. Barone, M. Chanon and M. L. H. Green, *J. Organomet. Chem.*, **185**, 85 (1980).
98. I. Theodosiou, R. Barone and M. Chanon, *J. Mol. Catal.*, **32**, 27 (1985).
99. I. Theodosiou, R. Barone and M. Chanon, *Adv. Organomet. Chem.*, **26**, 165 (1986).
100. S. I. Shalgunov, A. V. Zeigarnik, L. G. Bruk, and O. N. Temkin, *Izv. Akad. Nauk, Ser. Khim.*, No. 10, 1891 (1999) [*Russ. Chem. Bull.*, **48**, 1867 (1999)].
101. G. H. Graaf, E. J. Stamhus and A. Beenackers, *Chem. Eng. Sci.*, **43**, 3185 (1988).
102. L. G. Bruk, A. V. Zeigarnik, O. N. Temkin and R. E. Val'des-Peres *Metody vydvizheniya gipotez o mekhanizmakh slozhnykh reaktsii, Uchebnoe posobie* (Methods for Generating Hypotheses about Mechanisms of Complicated Reactions: A Learning Guide), MITKhT, Moscow, 1999.
103. H. B. Goyal, A. P. Kudchadker and R. D. Srivastava, *J. Appl. Chem. Biotechnol.*, **28**, 547 (1978).
104. L. G. Bruk, I. V. Oshanina, A. P. Kozlova, O. N. Temkin and K. Yu. Odintsov, *Izv. Akad. Nauk, Ser. Khim.*, No. 6, 1104 (1998) [*Russ. Chem. Bull.*, **47**, 1071 (1998)].
105. O. N. Temkin and L. G. Bruk, *Proceedings of the 9th Soviet-Japanese Seminar on Catalysis*, Novosibirsk, 1990, p. 83.

106. L. G. Bruk, I. V. Oshanina, A. P. Kozlova, E. V. Vorontsov and O. N. Temkin, *J. Mol. Catal. A.: Chem*., **104**, 9 (1995).
107. A. V. Zeigarnik, *Modelirovanie mekhanizmov kataliticheskikh reaktsii s ispol'zovaniem metodov teorii grafov i kombinatornykh algoritmov* (Modeling Mechanisms of Catalytic Reactions Using Graph Theory Methods and Combinatorial Algorithms), Candidate Dissertation in Chemistry, MITKhT (Moscow Institute of Fine Chemical Technology), Moscow, 1996.
108. I. V. Oshanina, N. A. Kokoreva, L. G. Bruk, I. L. Eremenko and S. E. Nefedov, *Izv. Akad. Nauk, Ser. Khim*., No. 11, 2274 (1998).
109. R. E. Valdes-Perez and A. V. Zeigarnik, *J. Mol. Catal. A: Chem*., **119**, 405 (1997).
110. A. V. Zeigarnik, R. E. Valdes-Perez, O. N. Temkin, L. G. Bruk and S. I. Shalgunov, *Organometallics*, **16**, 3114 (1997).
111. A. V. Zeigarnik, R. E. Valdes-Perez, and O. N. Temkin, *Langmuir*, **14**, 4510 (1998).
112. O. N. Temkin, A. V. Zeigarnik, R. E. Valdes-Perez and L. G. Bruk, *Kinet. Katal*., **41** (3), 331 (2000).
113. P. Ellison, M. Feinberg, M. H. Yue and H. Saltsburg, *J. Mol. Catal. A: Chem*., **154**, 169 (2000).
114. P. Ellison and M. Feinberg, *J. Mol. Catal. A: Chem*., **154**, 155 (2000).
115. A. V. Zeigarnik and R. E. Valdes-Perez, *J. Comput. Chem*., **19**, 741 (1998).
116. P. S. Cremer, X. Su, Y. R. Shen and G. A. Somorjai, *J. Phys. Chem*., **100**, 16302 (1996).
117. L. J. Broadbelt, S. M. Stark and M. T. Klein, *Chem. Eng. Sci*., **49**, 4991 (1994).
118. L. J. Broadbelt, S. M. Stark and M. T. Klein, *Chem. Eng. Sci*., **20**, 113 (1996).
119. D. J. Klinke and L. J. Broadbelt, *Chem. Eng. Sci*., **54**, 3379 (1999).
120. L. J. Broadbelt, S. M. Stark and M. T. Klein, *Ind. Eng. Chem. Res*., **34**, 2566 (1995).
121. L. G. Bruk, S. N. Gorodskii, A. V. Zeigarnik, R. E. Valdes-Perez and O. N. Temkin, *J. Mol. Catal. A: Chem*., **130**, 29 (1998).
122. J. Sinfelt, *Catal. Rev*., **3**, 175 (1969).
123. J. Sinfelt, *Adv. Catal*., **23**, 91 (1973).
124. P. Mahaffy and R. S. Hansen, *J. Chem. Phys*., **71**, 1853 (1979).
125. B. S. Gudkov, L. Guczi and P. Tetenyi, *J. Catal*., **74**, 207 (1982).
126. A. S. Goddard, M. D. Amiridis, J. E. Rekovske et al., *J. Catal*., **117**, 155 (1989).
127. F. Zaera and G. A. Somorjai, *J. Phys. Chem*., **89**, 3211 (1985).
128. F. Zaera and G. A. Somorjai, *J. Am. Chem. Soc*., **106**, 2288 (1984).
129. L. G. Bruk, S. N. Gorodskii, A. V. Zeigarnik, R. E. Valdes-Perez and O. N. Temkin, *Izv. Akad. Nauk, Ser. Khim*., 882 (1999) [*Russ. Chem. Bull*., **48**, 873 (1999)].
130. T. T. Zung, L. G. Bruk and O. N. Temkin, *Izv. Akad. Nauk, Ser. Khim*., 1860 (1993).
131. T. T. Zung, L. G. Bruk and O. L. Temkin, *Mendeleev Comm*., 2 (1994).
132. F. Zaera, *Chem. Rev*., **95**, 2651 (1995).
133. B. E. Bent, *Chem. Rev*., **96**, 1361 (1996).
134. D. J. Klinke and L. J. Broadbelt, *AIChE J*., **43** (7), 1828 (1997).
135. R. G. Forsyth and M. L. Mavrovouniotis, *J. Chem. Inf. Comput. Sci*., **37**, 258 (1997).
136. M. S. Okino and M. L. Mavrovouniotis, *Chem. Rev*., **98** (2), 391 (1998).
137. D. B. Borchardt,, J.-G. Choi, K. Suzuki and S. H. Bauer, *J. Chem. Phys*., **88**, 6282 (1988).
138. J. N. Brønsted and K. J. Pedersen, *Z. Phys. Chem*., **108**, 185 (1924); J. N. Brønsted, *Chem. Rev*., **5**, 231 (1928).
139. J. Horiuti and M. Polanyi, *Acta Physicochim. URSS*, **2**, 505 (1935).
140. M. G. Evans and M. Polanyi, *Trans. Faraday Soc*., **34**, 11 (1938).
141. F. Westley, *Table of Recommended Rate Constants for Chemical Reactions Occurring in Combustion*, NBS, Washington, DC, 1980.
142. M. I. Temkin, *Zh. Fiz. Khim*., **15** (3), 296 (1941).
143. I. Mochida and Y. Yoneda, *J. Catal*. **7**, 386 (1967).
144. M. Neurock and M. T. Klein, *Polycyclic Arom. Compds*., **3**, 231 (1993).
145. G. Yaluris, J. E. Rekoske, L. M. Aparicio et al., *J. Catal*., **153**, 54 (1995).
146. V. Pallasana and M. Neurock, *J. Catal*., **191**, 301 (2000).

147. Z. P. Liu and P. Hu, *J. Chem. Phys.*, **114**, 8244 (2001).
148. A. Logadotter, T. H. Rod, J. K. Norskov et al., *J. Catal.*, **197**, 229 (2001).
149. J. K. Norskov, *J. Catal.*, **209**, 275 (2002).
150. T. Bligaard, J. K. Norskov, S. Dahl et al., *J. Catal.*, **224**, 206 (2004).
151. R. A. Marcus, *J. Chem. Phys.*, **24**, 966, 979 (1956).
152. V. G. Levich and R. R. Dogonadze, *Dokl. Akad. Nauk SSSR*, **124**, 123 (1960).
153. R. R. Dogonadze, A. M. Kuznetsov and V. G. Levich, *Electrochim. Acta*, **13**, 1025 (1968).
154. E. D. German and R. R. Dogonadze, *Quantum-Mechanical Theory of Proton Transfer Reaction Kinetics*, in *Proton v Khimii* (The Proton in Chemistry), R. R. Dogonadze (Ed.), Mir, Moscow, 1977 [additional chapter by E. D. German and R. R. Dogonadze], pp. 350–376.
155. R. A. Marcus, *J. Phys. Chem.*, **72**, 891, 4249 (1968).
156. M. M. Kreevoy and D. E. Konasewich, *Adv. Chem. Phys.*, **21**, 243 (1971).
157. R. I. Masel, *Chemical, Kinetics and Catalysis*, Wiley International, 2001.
158. E. T. Denisov, *Usp. Khim.*, 69 (2), 166 (2000); E. T. Denisov, *Models for Abstraction and Addition Reactions of Free Radicals*, in *General Aspects of the Chemistry of Radicals*, Z. B. Alfassi (Ed.), John Wiley & Sons, Chichester, 1999, pp. 79–137.
159. E. D. German and M. Sheintuch, *J. Phys. Chem. A*, **109**, 7957 (2005).
160. E. D. German and A. M. Kuznetsov, *J. Mol. Struct.*, **153**, 671 (2004).
161. E. D. German, A. M. Kuznetsov and M. Sheintuch, *Surf. Sci.*, **159**, 554 (2004).
162. E. D. German, A. M. Kuznetsov and M. Sheintuch, *Surf. Sci.*, **170**, 554 (2004).
163. P. Blowers, L. Ford and R. I. Masel, *J. Phys. Chem.*, **102**, 9267 (1998).
164. P. Blowers and R. I. Masel, *AIChE J.*, **45**, 1794 (1999).
165. P. Blowers and R. I. Masel, *J. Phys. Chem. A*, **103**, 7047 (1999).
166. E. Shustorovich and H. Sellers, *Surf. Sci. Rep.*, **31**, 1 (1998).
167. H. Sellers and E. Shustorovich, *Surf. Sci.*, **504**, 167 (2002).
168. (a) E. M. Shustorovich and A. V. Zeigarnik, *Zh. Fiz. Khim.*, **80** (1), 8 (2006) [*Russ. J. Phys. Chem. A*, **80** (1), 4 (2006)]; (b) A. V. Zeigarnik and E. Shustorovich, *Russ. J. Phys. Chem. B*, **1** (4), 330 (2007); (c) E. Shustorovich, *Russ. J. Phys. Chem. B*, **1** (4), 307 (2007).
169. E. Shustorovich, *Adv. Catal.*, **37**, 101 (1990).
170. E. Shustorovich, *Accounts Chem. Res.*, **21**, 183 (1988).
171. D. E. Ellis (Ed.), *Density Functional Theory of Molecules, Clusters, and Solids*, Kluwer Academic, Dordrecht, 1995.
172. J. F. Dobson, G. Vignali and M. P. Das, *Electronic Density Functional Theory: Recent Progress and New Directions*, Plenum Press, New York, 1998.
173. R. A. Van Santen and M. Neurock, *Catal. Rev. Sci. Eng.*, **37**, 557 (1995).
174. E. Shustorovich and A. Bell, *Surf. Sci.*, **205**, 492 (1988).
175. A. V. Zeigarnik, R. E. Valdes-Peres and O. N. Myatkovskaya, *J. Phys. Chem. B*, **104**, 10578 (2000).
176. A. V. Zeigarnik, R. E. Valdes-Peres and N. I. Egorova, *Surf. Sci.*, **487**, 146 (2001).
177. A. V. Zeigarnik and O. N. Myatkovskaya, *Kinet. Katal.*, **42**, 464 (2001).
178. S. R. Deshmukh, A. B. Mhadshwar, D. G. Vlachos, *J. Phys. Chem. B*, **107**, 6153 (2003).
179. I. V. Kuz'min and A. V. Zeigarnik, *Kinet. Katal.*, **45**, 595 (2004).
180. A. Bell, *Relationship of Reaction Energetics to the Mechanism and Kinetics of Heterogeneously Catalyzed Reactions*, in *Metal-Surface Reaction Energetics*, E. Shustorovich (Ed.), VCH Publ., New York, 1991, p. 191.
181. G. C. Bond, A. D. Hooper, J. C. Slaa and A. O. Taylor, *J. Catal.*, **163**, 319 (1996).
182. J. A. Rodriguez and D. W. Goodman, *Surf. Sci. Rep.*, **14** (1–2), 1 (1991).
183. J. A. Dumesic, D. F. Rudd, L. M. Aparicio et al., *The Microkinetics of Heterogeneous Catalysis*, Am. Chem. Soc., Washington, DC, 1993.
184. E. Shustorovich, *Catal. Lett.*, **7**, 107 (1990).
185. E. Shustorovich and A. Bell, *Surf. Sci.*, **248**, 359 (1991).
186. J. B. Benziger, in *Metal-Surface Reaction Energetics: Theory and Applications in Heterogeneous Catalysis, Chemisorption and Surface Diffusion*, E. Shustorovich (Ed.), VCH, New York, 1991, p. 53.

187. E. Shustorovich and A. Bell, *J. Catal*., **113**, 341 (1988).
188. A. Bell and E. Shustorovich, *J. Catal*., **121**, 1 (1990).
189. E. Shustorovich and A. Bell, *Surf. Sci*., **253**, 386 (1991).
190. P. Paredas Olivera, E. M. Patrito and H. Sellers, *Surf. Sci*., **327**, 330 (1995).
191. I. Fishtik, A. Alexander and R. Datta, *Surf. Sci*., **430** (1–3), 1 (1999).
192. A. V. Zeigarnik, K. Kellakhem, R. Datta et al., *Kinet. Katal*., **46** (4), 509 (2005).
193. M.-S. Liao and Q.-E. Zhang, *J. Mol. Catal. A: Chem*., **136**, 185 (1998).
194. M. J. Hei, H. B. Chen, J. Yi et al, *Surf. Sci*., **417**, 82 (1998).
195. E. Shustorovich and A. Bell, *Surf. Sci*., **259**, L791 (1991).
196. A. V. Zeigarnik, L. A. Abramova and S. P. Baranov, *Kinet. Katal*., **44** (1), 152 (2003).
197. S. P. Baranov, L. A. Abramova, A. V. Zeigarnik and E. Shustorovich, *Surf. Sci*., **555**, 20 (2004).
198. L. A. Abramova, A. V. Zeigarnik, S. P. Baranov and E. Shustorovich, *Surf. Sci*., **565**, 45 (2004).
199. E. Shustorovich and A. V. Zeigarnik, *Surf. Sci*., **527**, 137 (2003).
200. T. R. Cundari, *Coord. Chem. Rev*., **147**, 87 (1996).
201. T. Ziegler, *Can. J. Chem*., **73**, 743 (1995).
202. D. Boyd and K. Lipkowski (Eds), *Reviews in Computational Chemistry*, VCH, New York, 1996, Vol. 8.
203. (a) P. W. N. M. van Leeuwen, K. Morokuma and J. H. van Lenthe (Eds.), *Theoretical Aspects of Homogeneous Catalysis*, Kluwer Acad. Press., Dordrecht, 1995; (b) G. Frenking (Ed.), *Theoretical Aspects of Transition Metal Catalysis* (Topics in Organometallic Chemistry, Vol. 12), Springer-Verlag, Berlin, 2005.
204. T. R. Cundari, J. Deng, W. Fu et al., *J. Chem. Inf. Comp. Sci*., **38**, 941 (1998).
205. A. Dedieu, *Chem. Rev*., **100**, 543 (2000).
206. J. Tomasi and M. Persico, *Chem. Rev*., **94**, 2027 (1994).
207. A. Dedieu, in *Theoretical Aspects of Homogeneous Catalysis*, P. W. N. M. van Leeuwen, K. Morokuma, and J. H. van Lenthe (Eds), Kluwer Acad. Press., Dordrecht, 1995, p. 167.
208. Q. Cui, D. G. Musaev and K. Morokuma, *Organometallics*, **17**, 742 (1998).
209. Q. Cui, D. G. Musaev and K. Morokuma, *Organometallics*, **17**, 1383 (1998).
210. D. D. Kragten, R. A. Van Santen and J. J. Lerou, *J. Phys. Chem. A*, **103**, 80 (1999).
211. Per E. M. Siegbahn, *Struct. Chem*., **6** (4/5), 271 (1995).
212. Per E. M. Siegbahn, *J. Am. Chem. Soc*., **117**, 5409 (1995).
213. Per E. M. Siegbahn, *J. Phys. Chem*., **100**, 14672 (1996).
214. D. D. Kragten, R. A. Van Santen, M. Neurock and J. J. Lerou, *J. Phys. Chem. A*, **103** (15), 2756 (1999).
215. P. M. Henry, *J. Am. Chem. Soc*., **86**, 3246 (1964).
216. J. A Keith., J. Oxgaard and W. A. Goddard, *J. Am. Chem. Soc*., **128**, 3132 (2006).
217. (a) J. A. Keith, R. J. Nielsen, J. Oxgaard and W. A. Goddard, *J. Am. Chem. Soc*., **129**, 12342 (2007); (b), J. A. Keith, R. J. Nielsen, J. Oxgaard, W. A. Goddard III and P. M. Henry, *Organometallics*, **28**, 1618 (2009).
218. S. A. Beyramabadi, H. Eshtiagh-Hosseini, M. R. Housaindokht and A. Morsali, *Organometallics*, **27**, 72 (2008).
219. S. A. Beyramabadi, H. Eshtiagh-Hosseini, M. R. Housaindokht and A. Morsali, *J. Mol. Struct. (TheoChem)*, **903**, 108 (2009).
220. Ch. Elschenbroich and A. Selzer, *Organometallics*, VCH, Weinheim, 1989.
221. I. B. Rabinovich, V. P. Nistratov, V. I. Tel'noi and M. S. Sheiman, *Termodinamika metalloorganicheskikh soedinenii* (Thermodynamics of Organometallic Compounds), Izdat. Nizhegorodsk. Gos. Univ, Nizhni Novgorod, 1996.
222. J. A. M. Simoes and J. L. Beauchamp, *Chem. Rev*., **90**, 629 (1990).
223. J. Collman, L. Hegedus, J. Norton and R. Finke, *Principles and Applications of Organotransition Metal Chemistry* (*Graduate Textbook*), University Science Books, Mill Valley, CA, 1987, Vol. 1.

8

Effect of Medium on Reaction Rates in Homogeneous Catalysis with Metal Complexes

Kinetic experiments that are planned for the discrimination of mechanistic hypotheses must take into account all features of the catalytic reactions with participation of metal complexes which were considered in the preceding chapters:

- nature of active complexes and mechanisms of their formation;
- complexity functions of metal catalysts;
- probabilities of multi-route mechanisms;
- existence of route-conjugation nodes;
- presence of nonlinear steps and probability of appearance of nonlinear models.

In order to obtain the correct experimental kinetic data required to discriminate the generated hypothetical mechanisms, it is also necessary to select an "ideal" reactor in which the heat and mass transfer in homogeneous heterophase processes does not influence kinetic behavior [1–6]. It is also no less important to take into account the possible nonideality of a medium in which a given reaction takes place, that is, the possibility of changes in the rate constants and equilibrium constants depending on the variable concentrations of all participants of the reaction, including solvents, electrolytes, reactants, and metal complexes. In other words, it is necessary to solve the task of creating the "ideal" medium that does not influence the character of kinetic relationships, i.e., providing conditions for the study of kinetics under which an a priori nonideal medium would behave as the ideal.

Another related problem that was originally formulated by Brønsted and then studied by Hammett [7] and other researchers consists of developing approaches to a quantitative description of the influence of a medium on the rate of a chemical reaction [8–10]. As an application to the catalysis with metal complexes, the problem of ideality of the reaction medium was studied in [11] and treated in a learning guide [12].

Chapters 2–4 presented various approaches to the problem of eliminating the influence of a medium (mostly for aqueous electrolyte solutions) with allowance for the effects of association in solvent–reactant and solvent–catalyst systems. The role of a solvent in the formation of catalytically active complexes was analyzed in Section 5.2.

Homogeneous Catalysis with Metal Complexes: Kinetic Aspects and Mechanisms, First Edition. O. N. Temkin.

Problems pertaining to the medium influencing the reaction rate and equilibrium in both aqueous and nonaqueous solutions of acids, salts, and metal complexes and the thermodynamics of solvation and ion association processes have long ago constituted a special field of physical chemistry [6–10, 13–31]. However, despite considerable achievements in understanding of the nature of salt effects [19, 23, 26, 28–31] and the influence of various solvent properties on the reaction rates [13, 27] complex-formation equilibria [26, 31], it is still difficult to imagine the scope of a theory that would be capable of predicting the effect of the ionic medium or the solvent nature on a multi-step process involving metal complexes. Most probably, any correlations between the reaction rates over routes in a multi-step mechanism and the characteristics of a medium are almost impossible. Apparently, sufficiently effective quantum-chemical approaches will be developed in the future, since this direction is already extensively explored (see Section 7.4). At the same time, correlation dependences for separate stages of a catalytic process can be useful in both planning kinetic experiments and interpreting their results.

This chapter addresses some issues important for the planning of kinetic experiments in aqueous, aqueous–organic, and organic systems. The presentation employs the minimum mathematical apparatus. Speaking of the use of aqueous and nonaqueous electrolyte solutions in kinetic experiments, we must bear in mind two important circumstances:

- The necessity of providing conditions for the constancy (stabilization) of the activity coefficients of all reaction participants, including the transition states of elementary steps, in order to determine the rate constants and equilibrium constants using the law of mass action.
- The necessity of interpreting and predicting salt effects, at least for comparatively simple mechanisms.

The effect of a reaction medium on the rate constants and equilibrium constants of complex formation and other processes was considered in Chapter 1, where two basic equations were introduced that described the rate constants of elementary steps and complexity functions of the catalyst and/or substrate. From the standpoint of the theory of transition state (theory of absolute reaction rates), which led to a new interpretation of the concepts formulated by Brønsted, the equilibrium constants $K_j^{\neq}$ of the formation of activated complexes at each step of a complicated reaction depend on the ionic medium as described by the following equation:

$$k' = \frac{kT}{h}K_j^{\neq} = \frac{kT}{h}K_{j0}^{\neq}\frac{\prod_i \gamma_{ij}^{\alpha_i}}{\gamma_j^{\neq}}, \tag{8.1}$$

where k' is the rate constant of an elementary step, $K_{j0}^{\neq}$ is the thermodynamic equilibrium constant of the formation of activated complexes in an infinitely diluted solution, and γ_{ij} and $\gamma_j^{\neq}$ are the coefficients of activity of the ith reactant and activated complex, respectively, at the jth step. Quasi-equilibrium steps also enter into the kinetic equation via the thermodynamic equilibrium constants and γ_i values of all participants involved in these steps.

Note that, in the case of catalysis with metal complexes, the coefficients of activity of all complex species not participating in the process (represented by pendant vertices on kinetic graphs) also enter into the kinetic equation via the complexity function

$$F_{\mathrm{M}} = 1 + \sum_i K_i[L]^n, \tag{8.2}$$

where $K_i = K_{i0}\gamma_{\mathrm{M}}\gamma_{\mathrm{L}}^n/\gamma_{\mathrm{ML}_n}$.

In investigations of a catalytic process in a concentrated solution, the activity of a solvent will also enter into the kinetic equation via Eqs (8.1) and (8.2) (see below). A change of the solvent will naturally lead to the replacement of all γ, but the passage from one solvent to another may also involve a change in the thermodynamic constants $K^{\neq}_{j0}$ and K_{i0}.

8.1 Effect of electrolytes on the activity coefficients of reaction medium components

In 1921, G.N. Lewis and M. Randall proposed an empirical rule of *constant ionic strength* (defined as $I = \frac{1}{2}\sum_i C_i z_i^2$), according to which the electrolyte activity coefficients are constant for all solutions with $I = \text{const}$. Somewhat later, Debye and Hückel established a relationship between γ_i and I. The so-called Debye–Hückel limiting law, written as

$$\lim_{I\to 0} \frac{\partial \lg\gamma_i}{\partial I^{1/2}} = -Az_i^2, \tag{8.3}$$

very well describes the behavior of electrolyte solutions [7] and the theoretical value of $A = 0.5$ is in most cases valid for aqueous solutions at 25 °C. However, beginning with electrolyte concentrations on the order of 0.01 M (for multiply charged electrolytes, at still lower concentrations), the relationship between γ_i and $I^{1/2}$ is no longer described by the formula $\lg \gamma_i = -Az_i^2\sqrt{I}$ that corresponds to the limiting law expressed by Eq. (8.3). In the region of ionic strengths on the order of several tenths of mole per liter, the solutions are in many cases adequately described by the following equation:

$$\lg \gamma_\pm = -Az_i^2 I^{1/2} + BI - CI^{3/2}, \tag{8.4}$$

where the coefficient at $I^{1/2}$ is predicted by the theory, while coefficients B and C are empirical constants. In a broader range of concentrations, the electrolyte solutions are described by the equation

$$\lg \gamma_i = -\frac{Az_i^2\sqrt{I}}{1 + b\sqrt{I}} + \beta I, \tag{8.5}$$

where both coefficients b and β vary on the passage from one substance to another [31]. In some cases, $\lg \gamma_\pm$ can be described by a simpler function of the concentration of an inert electrolyte. In particular, for acid solutions with salt concentrations above $0.5\,m$, a linear relationship between $\lg \gamma_\pm$ and molality $\tilde{m}_S$ is observed [32] in a range of up to $10-15\,\tilde{m}_S$:

$$\lg \gamma_\pm = \lg \gamma_\pm^0 + K\tilde{m}_S. \tag{8.6}$$

In the region of high electrolyte concentrations, the values of $\gamma_\pm$ and some functions related to the activity (e.g., Hammett acidity h_0) most frequently exhibit exponential dependence on the ionic strength or concentration of electrolytes. For example, the acidity of strong acids as a function of the ionic strength in a broad interval of electrolyte concentrations can be described by the Moiseev–Flid equation (see Section 2.3)

$$-H_0 = \lg C_{H_3O^+} + L\Delta I - \lg a_{H_2O}, \tag{8.7}$$

where $C_{H_3O^+}$ is H_3O^+ concentration (in mole/liter); L is the coefficient weakly dependent on the salt nature, acid type, and temperature; and ΔI is a change in the ionic strength due to neutral salts [33, 34]. The equation of type (8.7) with the additional (KI^2) term,

$$-H_0 = \lg C_{H_3O^+} + LI + KI^2 - \lg a_{H_2O},$$

well describes the acidity functions for aqueous H_2SO_4, $HClO_4$, HCl, and HBr solutions up to concentrations of 8–11 M [34]. For 1–1 acid electrolytes, the acid concentration is equal to ionic strength I. Linear dependences of increments δH_0 on the concentrations of inorganic salts were also obtained by Long and Paul [35–37].

It was established [7] that the validity of the constant ionic strength rule is restricted to dilute solutions of 1–1 electrolytes. For this reason, it is a common practice to use the Brønsted method, according to which a large excess of an inert electrolyte is added in order to ensure that concentrations of reactants and variable components in solution are much lower than the background electrolyte concentration. This approach was later called the *constant ionic medium* principle and developed by Leden, Biderman and Sillen [38, 39] (see also [16]).

Numerous models are currently available to calculate the activity coefficients [40], including both models that represent developments of the Debye–Hückel theory and alternative models such as the mean spherical approximation (MSA) that takes into account ions of various dimensions [40]. Apparently, an approach that is most acceptable in practice is offered by the principle of constant high-ionic medium created by the electrolytes that are indifferent with respect to reaction participants and other components of the reaction medium. It was suggested that the background ion concentration must account for 95 % of the total content of various ion species in solution. Experience gained up to now [29, 31] in investigations of the thermodynamics of ion reactions in nonaqueous media (including systems with low values of the dielectric permittivity ε) shows that, irrespective of the presence of a background electrolyte in the associated state in these media, its introduction into the reaction medium leads to the constancy (stabilization) of the activity coefficients of reactants.

Let us summarize the possible approaches to maintaining constant activity coefficients and compositions of outer-sphere complexes in kinetic experiments, which were considered in preceding chapters:

1. The first approach is to maintain a large constant excess of an electrolyte with a formally constant ionic strength (I = const). The content of charged complexes of the metal catalyst must not exceed ~5 % of the background electrolyte concentration. Examples 2.10, 2.22, 2.23, 2.24, 3.1, 3.20, 3.23, 3.26, and 4.7 correspond to $I = 1$–5 M (electrolyte concentration up to 5 M). For example, the rate of molecular hydrogen (H_2) oxidation catalyzed by Pd(II) chloride complexes passes through a maximum with respect to $[Cl^-]$ at $I = 1$ M (Example 2.10). It is interesting to note that the rate of $PdCl_2$ reduction by hydrogen at a variable concentration of electrolyte (LiCl, $LiClO_4$, $HClO_4$) within 0–6 M without ionic strength conservation passes through similar maxima [41]. In this case, γ_i, $\gamma^{\neq}$, $[Cl^-]$, and H_2 solubility are varied (see below).
2. It is possible to maintain a high constant concentration of specific non-complex-forming anions or cations (high anionic or cationic background). It was demonstrated long ago by Brønsted that, in reactions with ions of the same sign (especially in multiply charged electrolytes), the reaction rate depends on the concentration of oppositely charged ions rather than on I (see, e.g. [7]). Later, with allowance for the probability of ion association and the formation of outer-sphere complex compounds, the study of anionic complexes at a constant cationic background was suggested [16]. Electrolytes conventionally used to maintain a constant cationic background are $LiClO_4$ and $NaClO_4$. It was established that, at a cationic background ≥4 M, the concentration of anionic complexes can be varied within 10–15 % of this level. Systems with a cationic background above 5 M are considered in Examples 3.1, 3.2 (12 M), 4.8 (8 M NH_4Cl), 4.20 (12 M NH_4Cl) and in Section 2.3.

The process of ion association in these concentrated solution leads to the appearance of complicated aggregates in the medium. For example, Pd(II) complexes in chloride solutions can be represented as $PdCl_q(Li)_m(ClO_4)_n(H_2O)_p^{2+m-q-n}$ associates, so that

$$[PdCl_2]_\Sigma = \sum_q\sum_m\sum_n\sum_p [PdCl_q(Li)_m(ClO_4)_n(H_2O)_p].$$

As was noted above, the most important factor for anionic complexes is their association with cations, while the n values are usually close to zero (in some cases, ClO_4^- ions can also participate in the complex formation and outer-sphere associates). The concentration stability constants of $PdCl_3^-$ and $PdCl_4^{2-}$ complexes determined by conventional methods can be expressed by the following equation (under conditions of constant activity coefficients of all components):

$$\beta_{PdCl_q} = \sum_m \sum_p K_q [Li^+]^m \; [H_2O]^p. \tag{8.8}$$

These values enter into the kinetic equation, e.g., via the complexity function. Naturally, the constancy of β_{PdCl_2} can only be ensured provided that $[Li^+] = \text{const}$ and $a_{H_2O} = \text{const}$.

In selecting an inert (indifferent) electrolyte, it is also necessary to take into account specific features of the system studied. For example, in investigations of Tl(I) complexes, potassium fluoride proved to be a convenient background electrolyte because the stability constant of $TlClO_4$ amounts to 0.32 ± 0.04 for $I = 1$ M at 30 °C and no formation of fluoride complexes under these conditions was detected [16]. Thus, it is expedient to compare the rate constants (or equilibrium constants) determined in solutions with the same ionic medium rather than with the same ionic strength (for $I \geq 1$ M).

It should be noted that the formation of outer-sphere complexes [16, 28, 29] plays an important role in the kinetics of chemical reactions and this factor must be taken into consideration. This is illustrated by the influence of the nature of salt cations in aqueous solution on the rate of the following reaction [42]:

$$Fe(CN)_6^{3-} + KI \longrightarrow Fe(CN)_6^{4-} + {}^1\!/_2 I_2 + K^+,$$

the mechanism of which involves the formation of $KFe(CN)_6^{2-}$ outer-sphere complex with $K_{stab} = 3.9 \cdot 10^2$. The rate of this reaction is described by the following equation,

$$r = k[K^+][Fe(CN)_6^{3-}][I^-]^2, \tag{8.9}$$

and depends on the salt cation nature, decreasing in the following order for various cations:

$$Cs^+ > NH_4^+ > K^+ > Na^+ > Li^+ > Et_4N^+.$$

In order to determine the degree of influence of a solution component (e.g., a metal catalyst complex) on the ionic strength and the activity coefficients of all species present in solution, it is convenient to use the aforementioned Moiseev–Flid equation (8.7) (see Section 2.3). Then, by monitoring the variation of H_0 using the indicator technique in a solution of electrolytes, metal complexes, and small amounts of acid, it is possible to judge any change in the ionic strength and on the role of outer-sphere association in this system. For example, it was found that, in the $CuCl–NH_4Cl–H_2O$ system containing a large number of polynuclear and multiply charged $Cu_pCl_q^{(q-p)-}$ anions at high concentrations of NH_4^+, the value of H_0 was almost independent of the concentration of CuCl. Taking into account the variation of a_{H_2O} and using Eq. (8.7), it can be shown that I varies no more than by 5–10 %. This behavior can be related to the formation of associates of the $Cu_pCl_q(NH_4^+)_m(H_2O)_n$ type, the charge on which is naturally reduced and does not contribute to a change in the ionic strength (ΔI). In contrast, the dissolution of salt ($HgSO_4$) in the $H_2SO_4–H_2O$ system led to an increase in the acidity ($-H_0$), which allowed the concentration of $Hg(SO_4)_2^{2-}$ sulfate complex to be evaluated [33].

In the $HgCl_2–HCl–H_2O$ system, an increase in $[HgCl_2]_0$ also led to a sharp increase in the acidity ($-H_0$) at various concentrations of HCl (Figure 8.1) [43, 44], while the activity of water in the $HCl–H_2O$ system does not change in a broad interval of $HgCl_2$ concentrations [45]. Assuming that coefficient L in

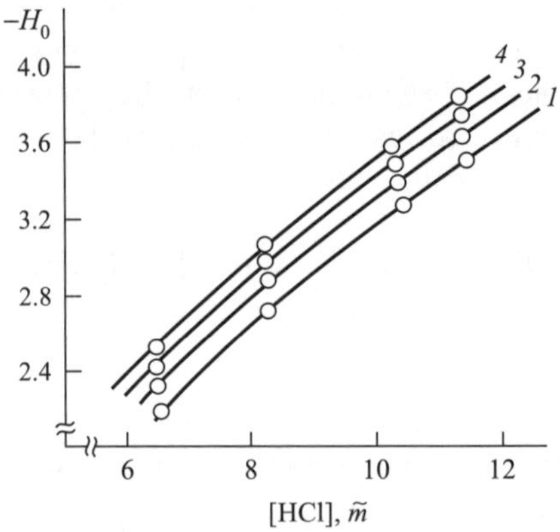

Figure 8.1 Plots of acidity function H_0 versus HCl concentration in $HCl–H_2O–HgCl_2$ solutions at 60 °C for $[HgCl_2] = 0$ (1); 1 $\tilde{m}$ (2); 2 $\tilde{m}$ (3); 3 $\tilde{m}$ (4) (reproduced with permission from editorial board of *Kinet. Katal.* journal © 1970 [44])

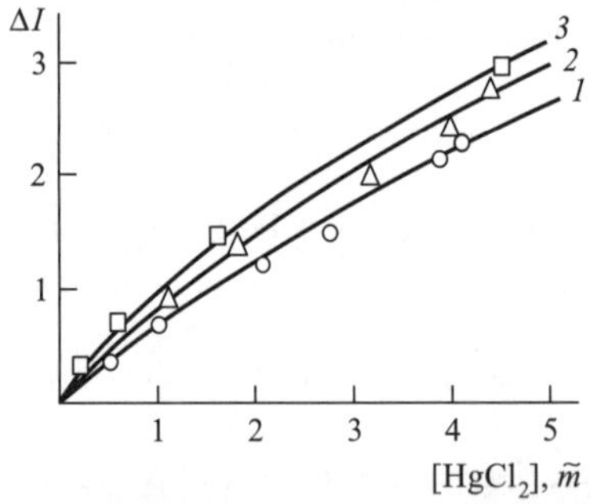

Figure 8.2 Plots of ΔI versus $[HgCl_2]_\Sigma$ at 60 °C for [HCl] = 6.5 $\tilde{m}$ (1); 8.2 $\tilde{m}$ (2); 10.2 $\tilde{m}$ (3) (reproduced with permission from editorial board of *Kinet. Katal.* journal © 1970 [44])

Eq. (8.7) is independent of the mercuric chloride content, a change in H_0 can be used to determine the ion strength increment ΔI, i.e., the concentration of $HgCl_4{}^{2-}$ complexes:

$$-(H_0' - H_0) = L\Delta I; \quad \Delta I = [HgCl_4{}^{2-}].$$

Since ΔI increases with [HCl] at $[HgCl_2]_0$ = const (Figure 8.2), there is no significant formation of outer-sphere complexes (or ion pairs) of the H_3O^+ $[HgCl_4{}^{2-}]$ type that decrease the anion charge. Any growth in ΔI is probably due to an increase in the fraction of $HgCl_4{}^{2-}$ complexes as compared to $HgCl_2(H_2O)_4$ and $HgCl_3(H_2O)^-$, which takes place with decreasing activity of water. In 8–10 M aqueous HCl solutions at 60 °C, the concentration of $HgCl_4{}^{2-}$ amounts to 85–100 % at $[HgCl_2] = 1$ M and decreases to 60–65 % at $[HgCl_2] = 5$ M. These solutions are used for the catalytic hydrochlorination of acetylene [43, 45–47].

3. Maintaining a constant concentration of an inert electrolyte also favors retention of the activity coefficients γ_S of uncharged reactant and solvent molecules during variation of the concentrations of charged complexes, acid reactants, and charged nucleophilic and electrophilic participants. As is well known, a

growth in the ionic strength of a solution is accompanied by an increase in the concentration constants of dissociation of weak uncharged acids and the concentration of H^+_{sol} ions [7]. In cases where a change in the ionic strength is unavoidable, it is necessary to take into account the effect of the medium on the solubility of nonelectrolytes (in particular, gases, hydrocarbons, and halogenated hydrocarbons) in a polar ionic medium (see Section 8.2).

4. In order to determine the form of a kinetic equation, it is expedient in some cases to use relationships of the type $r = f(a_i)$, where a_i are the activities of reactants and component solutions that enter into the kinetic equation (or quantities that are directly proportional to these activities) [11, 48, 49] (see Section 2.3). This approach does not solve all problems, since the constants of a kinetic equation will unavoidably also contain the activity coefficients $\gamma_j^{\neq}$ of all non-quasi-equilibrium steps. However, the use of activities instead of concentrations would make possible a significant expansion of the interval of possible reactant concentrations since, at very small concentrations of activated complexes ($C_j^{\neq}$), the variation of solution composition at a constant ionic strength must rather insignificantly influence the $\gamma_j^{\neq}$ values. In some systems, the concentrations of free reactants cannot either be measured or expressed via complexity because of the formation of polynuclear complexes. An example is offered by the $CuCl{-}NH_4Cl{-}H_2O$. The only approach that allowed the form of kinetic equations for some processes in this system to be determined (see Section 2.3 and Chapters 3 and 4) was based on the evaluation of activities of Cu^+ and Cl^- ions by measuring electromotive force (emf) in the corresponding charge transfer chains. Once the a_i values are known, it is possible to maintain them constant during determination of the kinetic equation. Instead of the $a_{H_3O^+}$, which is difficult to measure in concentrated solutions, it is convenient to use the product of $h_0 a_{H_2O}$, which is proportional to the activity $a_{H_3O^+}$, therefore $a_{H_3O^+} = b h_0 a_{H_2O}$ (see Section 2.3).
This approach should not be used when the kinetic equations involving the complexity function (F_M) of a metal catalyst are determined in a broad range of electrolyte concentrations without maintaining constant ionic strength, since F_M in this case will also include the activity coefficients of all complexes. In addition, the analysis of $r = f(a_i)$ relationships cannot be performed in terms of the average-ion activity coefficients $\gamma_\pm$ of salts or acids, since the $\gamma_\pm$ values can significantly differ from γ_+ and γ_-. An illustrative example is offered by the results of a spectrophotometric investigation of plutonium(IV) nitrate complexes [50]. It was established that the optical density of a reaction solution (proportional to the concentration of $Pu(NO_3)_n^{(n-4)-}$), which was studied as a function of the product $C_{NO_3^-} \gamma^{\pm}_{NO_3^-}$, strongly depends on the cation nature, which implies that the average-ion activity of nitrates of NH_4^+, Li^+, Na^+, and Ca^{2+} does not correctly reflect the activity of NO_3^-. In addition, this system is probably characterized by the outer-sphere association and the variation of $\gamma_{Pu(NO_3)_n^{(n-4)-}}$ and the activity of water. Schwabe [32] reported the following values of molar activity coefficients f_{H^+} and f_{Cl^-} for 0.01 M HCl solutions with different concentrations of $NaClO_4$:

$NaClO_4$, mole/kg	f_{H^+}, kg/mole	f_{Cl^-}, kg/mole
1	1.38	0.457
16	198.0	0.479

The results of kinetic investigations for a large number of reactions can be satisfactorily described by ignoring the variations of γ_i for all complexes in solution and using average-ion activity coefficients in a broad range of electrolyte concentrations [51]. However, results of this processing can hardly be considered as evidence in favor of the adopted mechanistic hypotheses and proposed metal complexity structures. As was shown above, there are approaches that can be used that take into account or eliminate the influence of medium composition (salt effects) on the rate of catalytic processes. However, these methods are mostly applicable only to aqueous electrolyte solutions.

8.2 Effect of electrolytes on the solubility of nonelectrolytes (gases and organic compounds)

The effect of electrolytes on the solubility of gases (He, H_2, N_2, Ar, CO, O_2) and/or organic substances in aqueous solutions plays an important role in kinetic investigations. The salt-induced decrease (salting-out) or increase (salting-in) in the solubility of nonelectrolytes has been also observed in organic solvents with variable concentrations of electrolytes (strong acids) [52]. Extremal dependences of the distribution constants $\alpha = C_{gas}/C_{sol}$ under the conditions of $C_{sol} = 10^{-6}–10^{-4}$ M were reported for H_2, CH_4, isobutane, and He in aqueous H_2SO_4 solutions at concentrations within 0–100 % [5, 53] and for C_2H_2 in aqueous HCl at various concentrations [43].

The salting-out effect in the region of low electrolyte concentrations in an aqueous solution is described by the Sechenov equation:

$$\lg \frac{C_{H_2O}}{C} = Bm_c, \tag{8.10}$$

where B is the salting-out coefficient, m_c is the electrolyte concentration, and C is the concentration of a given substance in solution (C_{sol}). This equation can also be written in alternative forms as

$$\lg \frac{\alpha}{\alpha_{H_2O}} = Bm_c; \tag{8.11}$$

$$\lg \gamma = Bm_c, \tag{8.12}$$

where γ is the activity coefficient of the dissolved substance.

For example, in the region of solution concentrations within 80–100 % H_2SO_4, the value of lg $(\alpha/\alpha_{H_2SO_4})$ for methane, cyclopentane, and isobutane linearly increases with the concentration of electrolyte ($HSO_4^- \cdot H_3O^+$) on the passage from 100 % H_2SO_4 to 80 % H_2SO_4 solution, which implies that the solubility of hydrocarbons decreases in this direction (being the maximum in concentrated H_2SO_4) [53].

In order to describe the entire range of electrolyte concentrations (i.e., both salting-out and -in effects), Rudakov et al. [53] studied the model of a multi-component (two-component) solvent, for which a reference solution was a mixture of two solvents (water and acid) rather than pure water or acid alone, with the coefficient α^* dependent on the composition. For this system, the generalized Sechenov equation is as follows [53]:

$$\lg \frac{\alpha}{\alpha^*} = B^*c, \tag{8.13}$$

where $\lg \alpha^* = \sum_j x_j \lg \alpha_j$, and x_j and α_j ($j = 1, 2$) are the molar fractions of components and Henry's distribution constants for pure components, respectively.

For the $H_2O–H_2SO_4$ system, we obtain

$$\lg \frac{\alpha}{\alpha^*} = \lg \alpha - x_{H_2O} \lg \alpha_{H_2O} - x_{H_2SO_4} \lg \alpha_{H_2SO_4} = B^*c \tag{8.14}$$

or alternatively

$$\lg \frac{\alpha}{\alpha_{H_2O}} = x_{H_2SO_4} \lg \frac{\alpha_{H_2O}}{\alpha_{H_2SO_4}} + B^*c. \tag{8.15}$$

Equation (8.15) explains why the Sechenov equation written as (8.11) fails to describe the entire range of sulfuric acid concentrations. There are only two nonelectrolytes (H_2 and He) for which $\alpha_{H_2O} = \alpha_{H_2SO_4}$ and, hence, the entire range of sulfuric acid concentrations is described by a single straight line.

In the coordinates of Eq. (8.13), all the nonelectrolytes studied (H_2, He, N_2, O_2, CH_4, isobutane, cyclohexane, isopentane, and cyclopentane) fit to the same straight line (Figure 8.3) [5, 53] in the regions of H_2SO_4 solution concentrations both within 0–80 %, where

$$c = \frac{1}{2}([HSO_4^-] + [SO_4^{2-}] + [H_5O_2^+] + [H_3O^+]),$$

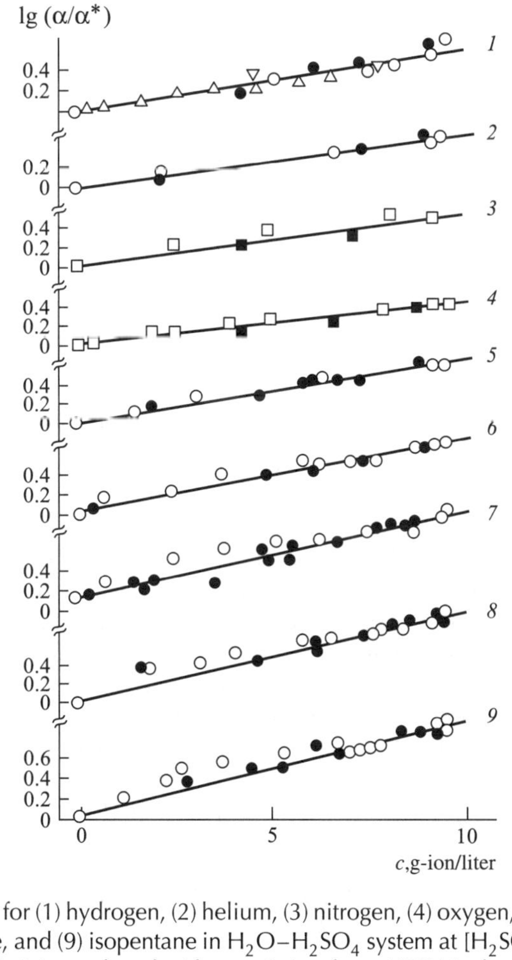

Figure 8.3 Plots of Eq. (8.13) for (1) hydrogen, (2) helium, (3) nitrogen, (4) oxygen, (5) methane, (6) cyclopentane, (7) isobutane, (8) cyclohexane, and (9) isopentane in $H_2O–H_2SO_4$ system at $[H_2SO_4]$ = 0–80 wt % (open circles and 80–100 wt % (black circles) (reproduced with permission from AIPC Nauka © 1987)

and within 80–100 %, where

$$c = \frac{1}{2}([H_3O^+] + [HSO_4^-]) = [HSO_4^-].$$

Since the ion composition of a medium is not always known, it is possible to use a simpler relationship for $\lg(\alpha/\alpha^*)$. It was shown that this value is proportional to a deviation from the additive molar volume (i.e., to a relative compression or electrostriction of the medium interacting with the electrolyte) [53]:

$$\lg \frac{\alpha}{\alpha^*} = \lg \alpha - x_A \lg \alpha_A - x_B \lg \alpha_B = L\left(\frac{V_E}{V}\right), \tag{8.16}$$

where V is the molar volume of the mixture, $V_E = x_A V_A - x_B V_B$, and L is a coefficient dependent on the nonelectrolyte. It was suggested [53] that this model can also be used for estimating the solubility of nonelectrolytes in a mixture of nonelectrolytic solvents.

The entire interval of concentrations of the solvent–electrolyte system can also be described using another interpretation of the Sechenov equation, which can also explain the appearance of extrema in α values for nonelectrolytes [43, 54]. According to this approach, the dissolution process is treated as a chemical reaction between nonelectrolyte molecule S and solvent (H_2O) molecules:

$$S_{gas} + nH_2O \rightleftharpoons S_{aq}, \tag{8.17}$$

for which the thermodynamic distribution constant can be written as

$$K_a = \frac{C_{gas} a^n_{H_2O}}{a_S}. \tag{8.18}$$

In particular, for pure water we have

$$K_a = \frac{C^0_{gas} (a^0_{H_2O})^n}{a_{S0}}, \tag{8.19}$$

where a_S and a_{S0} are the activities of a dissolved substance (solute) in the electrolyte solution and pure water, respectively. For $C_{gas} = C^0_{gas}$ and $a^0_{H_2O} = 1$, we obtain the Sechenov-Flid equation:

$$\lg \frac{c_{S0}}{c_S} = \lg \frac{\gamma_S}{\gamma_{S0}} - n \lg a_{H_2O}. \tag{8.20}$$

As can be seen from this relationship, the right-hand side of the Sechenov equation (Bm_c) describes the electrolyte effect on the activity coefficient γ_S of solute (S) and the activity of solvent (H_2O). Since a_{H_2O} in an electrolyte solution is always smaller than unity, the term $-n\lg a_{H_2O}$ is positive and c_S decreases. The activity of nonelectrolyte S can both increase and decrease with increasing m_c. Thus, the interplay of two contributions to Eq. (8.20) explains both the salting-out and -in effects of electrolyte.

Figure 8.4 shows the typical extremal plot of lg (C_{S0}/C_S) versus lg a_{H_2O} for acetylene in the HCl–H_2O system (up to 7 $\tilde{m}$ HCl) at 60 °C [43]. The value of n for Eq. (8.20) can be estimated as the derivative $d(\lg C_{S0}/C_S)/d(\lg a_{H_2O})$ at [HCl] → 0, which yields $n = 1.14$. Once n is known, relationship (8.20) with $\gamma_{S0} = 1$ can be used to calculate γ_S. The activity of acetylene in HCl solution for $n = 1$ is described by the following equation:

$$a_{C_2H_2} = \frac{P_{C_2H_2} a_{H_2O}}{K_a RT} = \frac{P_{C_2H_2} a_{H_2O}}{K_{H0}}, \tag{8.21}$$

where K_{H0} is the Henry constant at infinite dilution ([HCl] → 0).

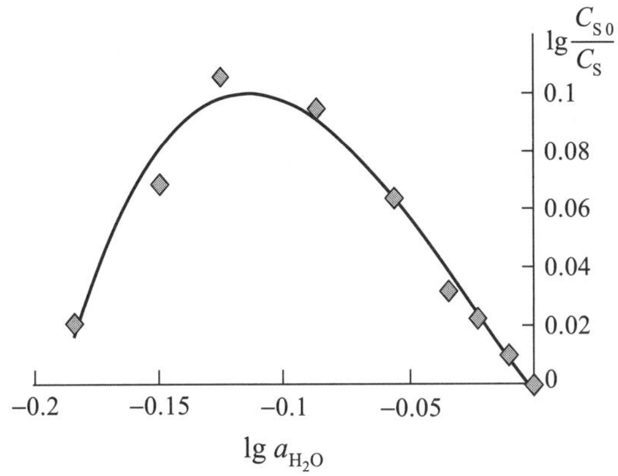

Figure 8.4 Plot of acetylene solubility $\lg(C_{S0}/C_S)$ versus $\lg a_{H_2O}$ in HCl–H_2O system at 60 °C [43]

It should be noted that writing the equilibrium constants for the reaction of nonelectrolyte MA dissociation into M^+ and A^- ions or the rate constants of alkylhalide hydrolysis with allowance for the molecules of solvent (H_2O) – in the form of so-called "complete" equilibrium and rate constants – leads to interesting effects [55]. In particular, the values of

$$K_0 = K/[H_2O]^n \quad (8.22)$$

$$k_0 = k/[H_2O]^{\pm n} \quad (8.23)$$

for some processes depend only on the temperature and are independent of the pressure in a range of 1–3000 bar [55]. The value of power index n for 1–1 electrolytes in water is amounts to ~6.5, while that for 2–2 electrolytes reaches $n \approx 10$. In the reaction of CH_3Br hydrolysis, $n = 10$–13. The experimental values of equilibrium and kinetic constants in this model remain unchanged without recourse to volatility or activity coefficients up to the region of supercritical conditions.

The influence of HCl concentration (or P_{HCl}) on the solubility of C_2H_2 in the C_2H_5OH–HCl system at 40 °C was described [52, 56] by an experimental dependence of the following type:

$$[C_2H_2] = \frac{aP_{HCl}P_{C_2H_2}}{1 + bP_{HCl}}. \quad (8.24)$$

This relationship was taken into account in determining the reaction orders with respect to [HCl] and $[C_2H_2]$ in the C_2H_5OH–HCl–H_2PtCl_4 system in investigation [52, 56] of the acetylene hydrochlorination reaction (see below).

Makitra et al. [57] obtained a useful semiempirical relationship based on the principle of linearity of free energies, which takes into account the solvent effect on the solubility of gases. According to this, the molar fraction N of a dissolved gas obeys the following equation:

$$\lg N = A_0 + A_1\frac{n^2 - 1}{n^2 + 2} + A_2\frac{E - 1}{2E + 1} + A_3\delta^2 + A_4B + A_5E, \quad (8.25)$$

where A_1, A_2, and A_3 are the coefficients of nonspecific interactions; A_4, A_5 are the coefficients of specific interactions; B is the basicity; E is the electrophilicity; and δ^2 is the density of cohesive energy (see Section 8.5.1). Using this equation, a satisfactory description of solubility was obtained [57] for the experimental data on seven gases (He, H_2, N_2, O_2, CO, CH_4, C_2H_4) in various (from 9 to 20) solvents [58]. Alternative variants of correlation equations for description of the solute–solvent interaction could be obtained within the framework of the Linear Solvation Energy Relations (LSER) theory [59].

Xu et al. [60] described and theoretically justified molecular-thermodynamic models for calculating the Henry constants in nonpolar and poplar solvents, including 1–1 electrolyte solutions, in a broad range of temperatures and concentrations, which were applied to 78 systems and 11 gases in aqueous solutions of 20 electrolytes.

8.3 Effect of electrolytes on the rates of elementary reactions between ions and uncharged substrates

The influence of electrolytes (salt effects) in reactions between charged particles is usually described in terms of the Debye–Hückel theory by writing proper polynomials in powers of I according to Eqs (8.4) [7]. However, as was noted above for salt effects in these reactions, the rate constants of reactions between charges of the same sign even at relatively small ionic strengths ($I \sim 0.5$) are determined by the concentration of ions of the opposite sign rather than by I (Olson–Simonson rule) [7]. These observations essentially provided a basis for the rule (principle) of constant high-ionic medium (high cationic or anionic background). Evidently, when studying a broad range of electrolyte concentrations, one should also take into account changes in the activity of the solvent in addition to variations in the activity coefficients of charged reactant species and transition states.

In the case of uncharged substrates, approaches based on the Debye–Hückel theory are inapplicable even in a rough (limiting) approximation, since $Z_{M^{n+}}Z_S = 0$ and, hence, the rate constants must not depend on the electrolyte concentration. These reactions should be approached from the standpoint of salting-out effects as described by the Sechenov equation [5, 60–62]. In investigations of the hydrolysis of alkylhalides (A) catalyzed by heavy metal ions, a limiting step is the interaction of molecule A with metal ion M^z:

$$A + M^z \longrightarrow [A\text{---}M^z]^{\neq} \longrightarrow \cdots. \tag{8.26}$$

At electrolyte concentrations within 1–5 M, the salt effects can be described by the following relation that has the form analogous to the Sechenov equation:

$$\lg \frac{k}{k_0} = (B - B^{\neq})m_c, \tag{8.27}$$

where $k = k_0$ at $m_c = 0$; $B > 0$ is the salting-out coefficient from Eq. (8.10); and $B^{\neq}$ is the salting-out coefficient for a transition state with respect to M^z ion [62]. Equation (8.27) proved to be useful in extrapolating the rate constants to zero ionic strength [60, 61]. It was established that $B^{\neq}$ depended on the activity of water ($a_{H_2O} = a_W$) and apparently reflected a change in the degree of dehydration of the transition state as compared to that of M^z ion.

Parker [63, 64] suggested modeling the influence of solvation effects on $\Delta G^{\neq}$ (i.e., modeling various degrees of solvation of the transition state and initial reactants) by a real chemical equilibrium. In the framework of this approach, the formation of a transition state at the limiting step (8.26) can be written as an equilibrium with allowance for the dehydration of this transition state [65]:

$$A + M^z(H_2O)_j \overset{K^{\neq}}{\rightleftharpoons} [A\text{---}M^z(H_2O)_{j-i}]^{\neq} + iH_2O. \tag{8.28}$$

Then, the rate of step (8.26) can be expressed as

$$W = \frac{kT}{h} K^{\neq} \frac{\gamma_A \gamma_{M^z}}{\gamma^{\neq} a_W^i} [A][M^z(H_2O)_j], \quad (8.29)$$

where a_W is the activity of water. Then, adopting that $\gamma_{M^z}/\gamma^{\neq} = 1$ for species of the same sign, we obtain

$$k = k_0 \gamma_A / a_W^i. \quad (8.30)$$

Finally, by assuming that γ_A is only determined by the ratio distribution constants as

$$\gamma_A = \alpha/\alpha_0, \quad (8.31)$$

and using Eqs (8.30) and (8.31), the following equation can be obtained:

$$\lg \frac{k}{\alpha} = \lg \frac{k_0}{\alpha_0} - i \lg a_W, \quad (8.32)$$

that takes into account the effect of electrolytes on the rate constant via their influence on α and a_W [62, 64]. Since $a_W < 1$, the rate constant (more precisely, the ratio k/α) will increase as the activity of water decreases with increasing electrolyte concentration.

Equation (8.32) was verified for the reaction of H_2 oxidation by Pd^{2+} ions in three systems, $PdSO_4$–H_2SO_4, $Pd(ClO_4)_2$–$HClO_4$, and $PdSO_4$–H_2SO_4 (2 M)–$Mg(ClO_4)_2$, in the interval of $\Delta \lg a_W = 3.5$ ($i = 0.6$). This equation was also valid for the reactions of hydrocarbon oxidation in the $PdSO_4$–H_2SO_4 system in the same interval of a_W, ($i = 1 - 1.6$), as well as for the M^{+n}–S_N1 reactions of alkylhalide hydrolysis (with Tl^{3+}, Hg^{2+}, and Al^{3+} ions) for $i = 6$–19 [62]. The i value strongly depends on the nature of the ion, type of reactant, and mechanism of reaction. For example, this coefficient is negative ($i = -0.7$) in the H_2–Ag^+ system [64] and vanishes ($i = 0$) in the $H_2(RH) - MnO_4^-$ system [62, 64]. For the reactions of H_2 and RH oxidation with Pd^{2+} and Pt^{2+} ions, the i value shows approximately linear growth with increasing molar volume of the substrate [62].

It is possible to propose a modified interpretation of the nature of coefficient i in Eq. (8.32), which takes into account that (i) dehydration upon the formation of a transition state can be related not only with M^z ion, but also with substrate A, and (ii) coefficient γ_A can also depend on a_W according to hypothetical Eq. (8.20). In this case, we have

$$k = k_0 \gamma_A / a_W^m; \quad (8.33)$$

$$\gamma_A = (\alpha/\alpha_0) a_W^n, \quad (8.34)$$

where m is a coefficient that characterizes the number of H_2O molecules liberated upon the transition state formation, so that $m - n = i$. Then,

$$k = k_0 \frac{\alpha}{\alpha_0} a_W^{n-m}; \quad (8.35)$$

$$\lg \frac{k}{\alpha} = \lg \frac{k_0}{\alpha_0} - (m - n) \lg a_W, \quad (8.36)$$

and the sign of i is determined by the relation between m and $n : m > n$ yields $i > 0$; $m < n$ corresponds to $i < 0$; and $m = n$ implies that $i = 0$.

Different variants of the approach that takes into account a change in the degree of hydration of the transition state at the limiting step were also used to describe the acid catalysis in concentrated solutions of strong acids [19].

8.4 Kinetics of catalytic reactions in concentrated aqueous electrolyte (HCl) solutions

Preceding chapters have given numerous examples of constructing kinetic models for catalytic reactions in concentrated aqueous solutions of Cu(I), Cu(II), and Hg(II) complexes under the conditions of constant ionic strength or cationic background. Here we consider an attempt to construct a kinetic model of the process in which a reactant (HCl) is the electrolyte with a concentration variable within broad limits.

The hydrochlorination of acetylene in the $HgCl_2$–HCl–H_2O system at 50–80 °C and a pressure of $P = 1$ bar according to the reaction

$$C_2H_2 + HCl \xrightarrow{HgCl_2} CH_2{=}CHCl \tag{8.37}$$

proceeds at a significant rate in the intervals of component concentrations within 8–14 $\tilde{m}$ HCl and 0.1–5 $\tilde{m}$ $HgCl_2$ [43–47]. As was demonstrated above, the state of mercuric chloride in HCl solutions is determined by the ratio of $HgCl_3^-$ and $HgCl_4^{2-}$ species, so that an increase in $[HgCl_2]_0$ leads to a growth in h_0 at almost constant a_{H_2O} and a significant decrease in P_{HCl} [43].The reaction order with respect to $P_{C_2H_2}$ changes from 1 to ~0.4 depending on the conditions (Figure 8.5). It was found that the total activity of mercuric chloride, a_{HgCl_2}, calculated from the results of potentiometric measurements *in situ* decreases with increasing $P_{C_2H_2}$ [43, 46].[1] In some experimental series, the observed rate constant

$$k_{obs} = \frac{W_{VC}}{P_{C_2H_2} a_{HgCl_2}} \tag{8.38}$$

remained constant, while in some other series ([HCl] = 8.2 $\tilde{m}$, $[HgCl_2] = 1$ M, 80 °C, $P_{HCl} = 0.04$ bar) this value exhibited some growth with increasing $P_{C_2H_2}$. These observations show evidence for the binding of $HgCl_2$ in intermediate compounds and a possible variation of activities a_{Cl^-} and $a_{H_3O^+}$ (h_0).

Since the increase in $[HgCl_2]_0$ (Figure 8.6) leads to a growth in the ionic strength of solution, which is accompanied by a change in the coefficients of activity of all charged and uncharged particles and a decrease in the concentration of Cl^-, the kinetic curves cannot be described using an expression for the complexity function, even if it includes the activities of Cl^- and $HgCl_2$, for example, as

$$F_{Hg} = \frac{[HgCl_2]_0}{a_{HgCl_2}} = \frac{1}{\gamma_{HgCl_2}} + \frac{K_1 a_{Cl^-}}{\gamma_{HgCl_3^-}} + \frac{K_2 a_{Cl^-}^2}{\gamma_{HgCl_4^{2-}}} + \frac{K_3 a_{C_2H_2}}{\gamma_3} + \frac{K_4 a_{Cl^-} a_{C_2H_2}}{\gamma_4}, \tag{8.39}$$

where K_1, K_2, K_3, and K_4 are the thermodynamic equilibrium constants of the formation of complexes $HgCl_3^-$, $HgCl_4^{2-}$, $HgCl_2(C_2H_2)$, and $HgCl_3(C_2H_2)^-$, respectively, from mercuric chloride, Cl^- ions, and acetylene. Evidently, a change in the HCl concentration will lead to variations in the activity coefficients of all complexes, as well as the activities of C_2H_2, H_3O^+, Cl^-, and $HgCl_2$. For this reason, the reaction rate was considered as a function of these activities and the temperature T:

$$W = f(a_{C_2H_2}, a_{HgCl_2}, a_{H_3O^+}, a_{Cl^-}, a_{H_2O}, T). \tag{8.40}$$

The activity of acetylene was calculated using Eq. (8.21), a_{HgCl_2} was determined potentiometrically, and a_{H_2O} was experimentally measured as described in [43, 45]. The activity a_{HgCl_2} of mercuric chloride was evaluated from a difference between the potentials (all determined relative to the standard calomel electrode, s.c.e.) of a platinum electrode E_{Pt} in the $HgCl_2$–$Hg_2Cl_{2(solid)}$–HCl–H_2O system (*in situ*) and a mercury

[1]The value of $\tilde{a}_{HgCl_2}$ was almost proportional to the activity a_{HgCl_2} determined in series of experiments with various $P_{C_2H_2}$ (see below).

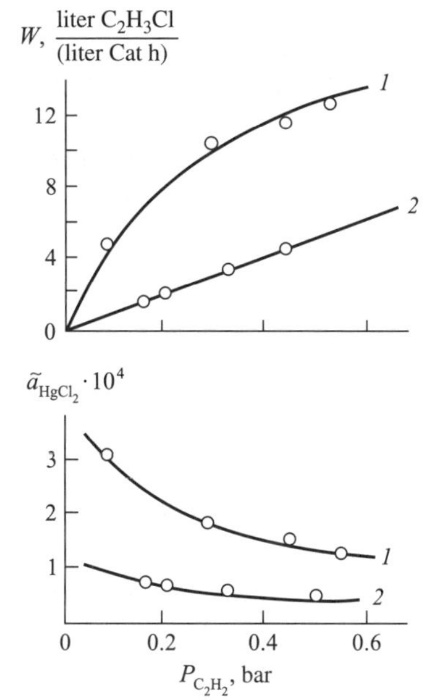

Figure 8.5 Plots of reaction rate W and activity of $HgCl_2$ versus acetylene saturated vapor pressure at P_{HCl} = const, $T = 60\,°C$, and $[HgCl_2]_\Sigma = 2$ M (1) and 1 M (2) (reproduced with permission from editorial board of *Kinet. Katal.* journal © 1970 [46])

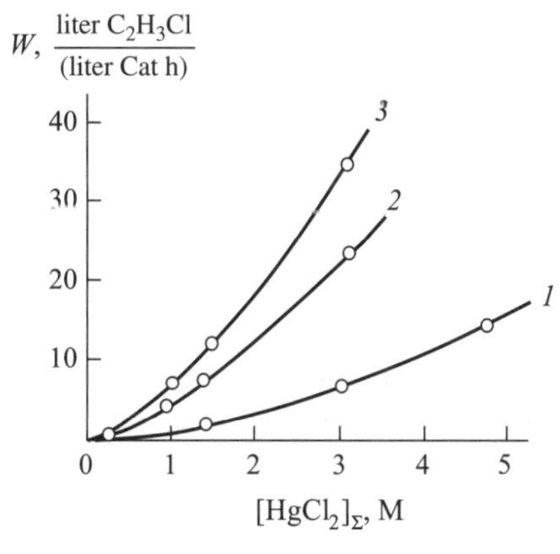

Figure 8.6 Plots of reaction rate W versus $[HgCl_2]_\Sigma$ at P_{HCl} = const, $P_{C_2H_2}$ = const, and various temperatures $T = 50\,°C$ (1); $60\,°C$ (2); $70\,°C$ (3) (reproduced with permission from editorial board of *Kinet. Katal.* journal © 1970 [46])

electrode E_{Hg} in the $Hg_2Cl_{2(solid)}$–HCl–H_2O system at the same HCl concentrations and temperatures. Under these conditions, the difference $E_{Pt} - E_{Hg}$ does not include the diffusion potentials and, hence,

$$a_{HgCl_2} = \exp[(E_{Pt} - E_{Hg}) - (E^{\circ}_{Pt} - E^{\circ}_{Hg})]\frac{F}{RT}\left(\frac{a'_{Cl^-}}{a''_{Cl^-}}\right), \tag{8.41}$$

where a''_{Cl} is the activity of Cl^- in the system free of $HgCl_2$, a'_{Cl} is the activity of Cl^- in a catalytic solution, and $\tilde{a}_{\mathrm{HgCl_2}} = a_{\mathrm{HgCl_2}} a''_{\mathrm{Cl^-}}/a'_{\mathrm{Cl^-}}$. The most difficult task in determining $a_{\mathrm{HgCl_2}}$ is related to the measurement of $a'_{\mathrm{Cl^-}}$ and $a''_{\mathrm{Cl^-}}$, that is, the activities of Cl^- in a catalytic system under the reaction conditions and in a model solution with the same concentration of HCl (in the presence of $Hg_2Cl_{2(\mathrm{solid})}$).

The values of $a_{\mathrm{H_3O^+}}$ and $a_{\mathrm{Cl^-}}$ [and, hence, the form of function (8.40)] were determined from the Hammett acidity for aqueous acid solutions which is defined as

$$h_0 = \frac{a_{\mathrm{H_3O^+}} f_{\mathrm{B}}}{a_{\mathrm{H_2O}} f_{\mathrm{BH^+}}}, \tag{8.42}$$

or in a modern variant as [66]

$$h_0 = \frac{a_{\mathrm{H_5O_2^+}} f_{\mathrm{B}}}{a_{\mathrm{H_2O}} f_{\mathrm{BH(H_2O)^+}}}. \tag{8.43}$$

Determination of the activities of neutral indicators B (using the solubility and distribution techniques) [67, 68] showed that f_{B} values differently and nonmonotonically change for various B depending on the acid concentration. Since h_0 exhibited monotonic growth with increasing [HX], Hammett hypothesis was valid for the acidity of various neutral indicators B and C

$$\frac{f_{\mathrm{B}}}{f_{\mathrm{BH^+}}} = \frac{f_{\mathrm{C}}}{f_{\mathrm{CH^+}}} \tag{8.44}$$

and for the protonation of their charged forms:

$$\frac{f_{\mathrm{B}}}{f_{\mathrm{BH^+}}} = \frac{f_{\mathrm{CH^+}}}{f_{\mathrm{CH_2^+}}} = \frac{f_{\mathrm{A^-}}}{f_{\mathrm{AH}}}. \tag{8.45}$$

Based on this, Vinnik [68] suggested that the relationship $f_{\mathrm{B}}/f^{+}_{\mathrm{BH^+}} = 1$ was valid irrespective of the nature of base B. In this case, the acidity of aqueous acid solutions is equal to the thermodynamic activity of proton ($h_0 = a^{+}_{\mathrm{H}}$ in Hammett's terms), or

$$a_{\mathrm{H_3O^+}} = h_0 a_{\mathrm{H_2O}}; \tag{8.46}$$

$$a_{\mathrm{H_5O_2^+}} = h_0 a_{\mathrm{H_2O}}. \tag{8.47}$$

Based on the Vinnik hypothesis and Eq. (8.46), it is possible to determine the activity $a_{\mathrm{H_3O^+}}$ in the course of the catalytic process. Then, using this $a_{\mathrm{H_3O^+}}$ value and a thermodynamic equilibrium constant of HCl dissolution (or the inverse value) for the reaction

$$\mathrm{H_3O^+_{aq} + Cl^-_{aq} \xrightleftharpoons{K_{HCl}} HCl_{gas} + H_2O_{liquid}}, \tag{8.48}$$

it is possible to calculate the required activities of Cl^- ($a'_{\mathrm{Cl^-}}$ and $a''_{\mathrm{Cl^-}}$):

$$a_{\mathrm{Cl^-}} = \frac{P_{\mathrm{HCl}}}{K_{\mathrm{HCl}} h_0}.$$

Thus, using the above approaches, all variables in (8.40) were determined (Figure 8.7).

Investigations of the catalytic reactions in concentrated solutions of copper(I) salts (see Section 2.3) showed that, in the case of using the activities of solution components and reactants, the effective rate constants remain unchanged when the background cation concentrations are varied within broad limits

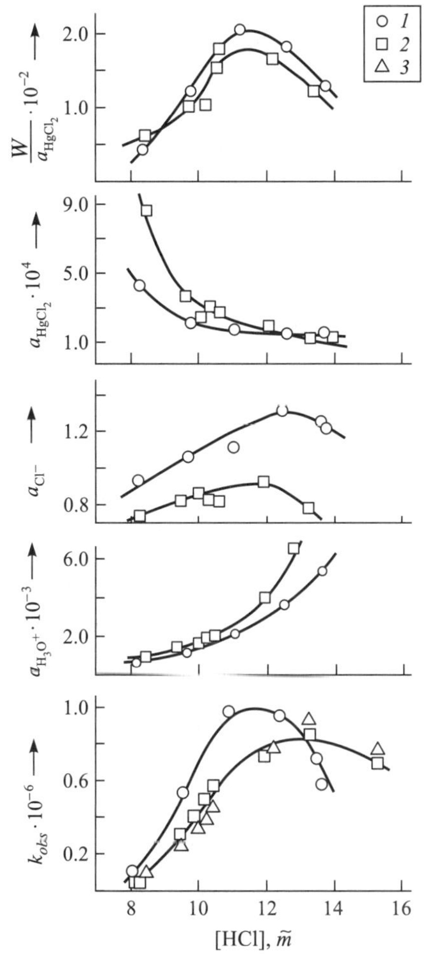

Figure 8.7 Plots of various process parameters versus [HCl]: (1, 2) experimental data for $[HgCl_2] = 2$ M and $T = 60\,°C$ and $70\,°C$, respectively; (3) calculated values of k_{obs} for $T = 70\,°C$ (reproduced with permission from AIPC Nauka © 1971)

(from 7 to $12\,\tilde{m}$) and the ionic medium is changed by adding $MgCl_2$ or $CaCl_2$. If the ratio of acitivity coefficients f_B/f_{BH^+} in the $HgCl_2$–HCl–H_2O system remains constant when the acid concentration [HCl] is varied from $8–14\,\tilde{m}$, then the values of $a_{H_3O^+}$ and a_{Cl^-} will also be determined to within a constant factor of $f_B/f_{BH^+} = \text{const} = b$ even if the Vinnik hypothesis is not valid. Therefore, it is natural to suggest that all activity coefficients $\gamma_i^{\neq}$ of transition states will also remain constant in the given interval of [HCl]. The ionic strength in the catalytic system with $8–12\,\tilde{m}$ HCl at 60 °C varies between $I = 9.6$ M and $I = 13.6$ M. For these large values of I, the values of $\gamma_i^{\neq}$ and b would most probably not change significantly.

The form of Eq. (8.40) was determined within the framework of assumptions made in the above considerations. The observed changes in E_{Pt} (sat. Hg_2Cl_2) in the $HgCl_2$–NH_4Cl–H_2O system with acetylene passed at various temperatures led to a conclusion that all organomercuric intermediates (including π complexes and β-chlorovinyl mercuric compounds) are formed via reversible steps (as indicated by a reversible E_{Pt} drop within 10–80 mV) [43, 47]. It was also established that the product of Cl^- addition to a π complex represents a *trans*-β-chlorovinyl mercury(II) derivative [47]. All these observations led to the following hypothetical mechanism of the reaction of vinyl chloride synthesis [47]:

$$\begin{aligned}
&(1)\ HgCl_n^{(n-2)-} + C_2H_{2\,aq} \overset{K_1}{\rightleftharpoons} (HgCl_n \cdot C_2H_2)_{aq}^{(n-2)-};\\
&(2)\ (HgCl_n \cdot C_2H_2)^{(n-2)-} + Cl^- \underset{k_{-2}}{\overset{k_2}{\rightleftharpoons}} \underset{(\sigma)}{Cl_{n-m}HgCH{=}CHCl^{(n-m-1)-}} + mCl^-;\\
&(3)\ \sigma + H_3O^+ \xrightarrow{k_3} CH_2{=}CHCl + HgCl_{(n-m)}^{(n-m-2)-}.
\end{aligned} \tag{8.49}$$

The reaction rate according to the mechanism with steps (1)–(3) under steady-state conditions can be expressed (with allowance for the above assumptions) by the following equation with coefficients that are independent of the medium composition:

$$W = \frac{k_1 a_{C_2H_2} a_{HgCl_n^{(n-2)-}} a_{H_3O^+} a_{Cl^-}}{k_3 a_{H_3O^+} + k_2 a_{Cl^-}^m}, \tag{8.50}$$

where $k_1 = K_1 k_2 k_3$. Let us replace the activity $a_{C_2H_2}$ by the partial pressure of acetylene (Eq. (8.21)) and express the activity $a_{HgCl_n^{(n-2)-}}$ using the value of a_{HgCl_2} that yields $a_{HgCl_n^{(n-2)}} = K_n a_{HgCl_2} a_{Cl^-}^{n-2}$, where K_n is the equilibrium constant of reaction

$$HgCl_2 + (n-2)Cl^- \overset{K_n}{\rightleftharpoons} HgCl_n^{(n-2)-} \quad (n = 3, 4). \tag{8.51}$$

As a result, we obtain the following formula for the reaction rate:

$$W = \frac{\alpha P_{C_2H_2} a_{HgCl_2} a_{H_3O^+} a_{Cl^-}^{(n-1)} a_{H_2O}}{\beta a_{H_3O^+} + a_{Cl^-}^m}, \tag{8.52}$$

where $\alpha = K_1 K_n k_2 k_3 / K_H$ and $\beta = k_3 / k_{-2}$.

The statistical processing of results for various combinations of $n = 2$, 3 and $m = 0$, 1, 2 for $k_{obs} = f(a_{H_3O^+}, a_{Cl^-})$ expressed as

$$k_{obs} = \frac{W}{P_{C_2H_2} a_{HgCl_2} a_{H_2O}} = \frac{\alpha a_{Cl^-}^{(n-1)}}{\beta a_{H_3O^+} + a_{Cl^-}^m} \tag{8.53}$$

gave the following most probable combinations: (I) $n = 3$, $m = 2$ (at 50 and 60 °C) and (II) $n = 3$, $m = 1$ (at 70 °C). Apparently, there is a change in the contributions of routes I and II with increasing reaction temperature. Figure 8.7 shows a comparison of the calculated rate constants and experimental values of k_{obs} for 70 °C [47].

Equation (8.52) contains $\gamma_3^{\neq}$ (entering into coefficient α) and ratio of the coefficients of activity of transition states (entering into coefficient β) for steps (−2) and (3):

$$\beta = \frac{k_{30}}{k_{-20}} \frac{\gamma_{-2}^{\neq}}{\gamma_3^{\neq}}. \tag{8.54}$$

As was noted above, the coefficients α and β must remain constant in the presence of a large excess of background electrolyte (8–14 $\tilde{m}$ HCl) by analogy with the case of concentrated NH_4Cl solutions (see Section 2.3, Figure 2.18). Should the coefficient $b = f_B/f_{BH}{}^+$ vary in the indicated interval of [HCl], this circumstance would not change the form of the relationship $k_{obs} = f(a_{H_3O^+})$ since, according to the Hammett hypothesis, the weight of step (3) in scheme (8.49) can be expressed via h_0 ($\omega_3 = k_3 h_0$). However, in this situation, the values of m and n in Eq. (8.53) can change.

Another (more empirical) approach was used in investigations of the kinetics of monovinylacetylene (MVA) hydrochlorination in the HCl–H_2O system at 10 °C [69]. The reaction kinetics was studied in a closed reactor with a two-phase HCl–H_2O–hexane mixture, where MVA was dissolved in hexane. The products of reactions

$$C_4H_{4\,aq} + HCl_{aq} \longrightarrow \underset{\text{(2-CBD)}}{CH_2{=}CH{-}C(Cl){=}CH_2}$$

$$C_4H_{4\,aq} + HCl_{aq} \longrightarrow \underset{\text{(4-CBD)}}{CH_2{=}C{=}CH{-}CH_2Cl}$$

in the form of 2-chloro-1,3-butadiene (2-CBD) and 4-chloro-1,2-butadiene (4-CBD) at a ratio of ~1 : 2 were also determined in hexane. A significant rate of hydrochlorination was only observed at an acid concentration of 8–12 M HCl (in water). In experiments devoted to the effect of [HCl] on the rate of MVA consumption (r_Σ), the initial MVA concentrations in hexane were the same. Calculations of the MVA solubility in hydrochloric acid at 10 °C according to the Sechenov equation showed that the solubility in the entire interval of [HCl] = 8–12 M decreases by ~7 %, which falls within the limits of experimental error. For this reason, the form of relationship $[C_4H_4] = f[HCl]$ was not studied.

For the adopted assumptions, the reaction rate can be described by the following equation:

$$r_\Sigma = k[MVA][HCl]^1 \tag{8.55}$$

(for [HCl] = 8–12 M). Assuming the relationship $[HCl] = f(h_0)$ and complete dissociation of HCl, this equation can be rewritten as follows:

$$r_\Sigma = \frac{k[MVA]h_0[Cl^-]}{1 + Kh_0}. \tag{8.56}$$

Equation (8.56) is characteristic of the acid catalysis under conditions of a significant protonation of the substrate:

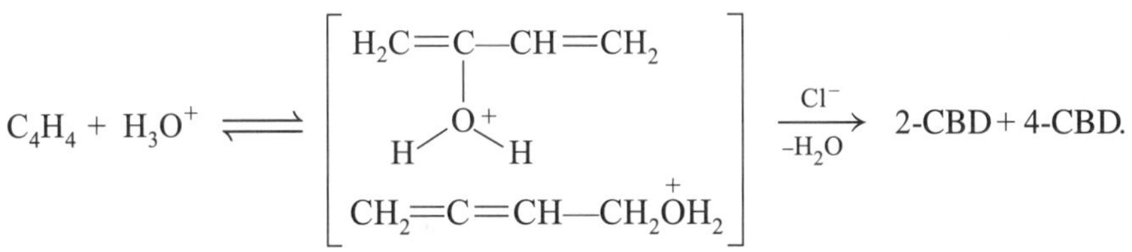

As the temperature is increased, the rate of 2-CBD formation grows more rapidly than that of 4-CBD and, at $T = 50$ °C, 2-CBD becomes the main product. In addition, the products contain a small amount of hydrochlorinated 2-CBD (2,4-dichloro-2-butene). Upon decreasing the temperature and conducting the reaction in liquid HCl ($T = -80$ °C), the formation of the usual hydrochlorination products is accompanied by the appearance of cyclic products of the azulene series (IR and UV spectroscopy data) [69].

The addition of LiCl and benzo-15-crown-5 (crown ether, CE) to the aqueous phase at 40 °C produces a significant (30- to 40-fold) growth in the reaction rate, which increases with both [CE] and [LiCl]. At [LiCl] = 2 M, the reaction rate is described by a two-term equation representing Eq. (8.55) with an additional term:

$$r_\Sigma = k[MVA][HCl]^4 + k_1[MVA][CE][HCl]^n, \tag{8.57}$$

where $n = 1.5$. The reaction via the second route entering into (8.57) leads to a predominant formation of 4-CBD, so that the selectivity with respect to this allene compound reaches 70 %. It was suggested [69] that an active role in this route is played by a complex (ion pair) of $[(CE)Li_{aq}]^+ HCl_2^-$.

In investigations of the kinetics of acetylene hydrochlorination in hydrochloric acid with $H[Pt(C_2H_4)Cl_3]$ catalyst (Zeise's acid) at 30 °C in an interval of [HCl] 4–12 M [52], it was taken into account that, according to [43] and preselected conditions in a continuous-flow stirred tank reactor, the solubility of C_2H_2 varies by no more than 10 % and, hence, these changes can be ignored. Under these assumptions, the reaction rate can be expressed as follows:

$$r_{VC} = k[Pt]_0[C_2H_2][HCl]^2, \tag{8.58}$$

or alternatively [68] as

$$r_{VC} = k_1[Pt]_0[C_2H_2]\frac{h_0}{1 + Ah_0}. \tag{8.59}$$

In terms of $P_{C_2H_2}$ and P_{HCl} (in an interval of $P_{HCl} = 10$–80 kPa), the reaction rate can also be described by the following equation:

$$r_{VC} = k_2[Pt]_0 P_{C_2H_2} P_{HCl}^{0.25}, \tag{8.60}$$

where $k_2 = 1.2\,h^{-1} \cdot kPa^{-1.25}$ ($T = 30$ °C).

8.5 Organic solvents in homogeneous catalysis with metal complexes

The possibility of using correlation equations for calculating the solubility of gases in organic solvents, including electrolyte solutions, has already been mentioned in Section 8.2. Now let us consider in more detail problems pertaining to the influence of the reaction medium, including salt effects, on the thermodynamics and kinetics of catalytic reactions with participation of metal complexes in nonaqueous media. The consideration will be based on monographs [7, 9, 13, 19, 23, 25–27, 29–31].

8.5.1 Main physical and chemical properties of solvents

Processes involved in the solvation of nonpolar reactants, electrolytes, metal complexes, and transition states are determined by the ability of a solvent to interact with solute, the state of solvent molecules (presence of associates and molecular complexes), and the possible formation of structured fragments in the bulk of the solvent. In addition, a solvent can behave like a liquid medium with the properties producing a nonspecific influence on the course of chemical reactions.

Evidently, various properties of substantially dissimilar liquid media (solvents) such as, e.g., water (H_2O), dimethylformamide (DMF), hexane, CCl_4, and R_4NX_{liquid}, differently influence the kinetics of chemical reactions. Monograph [27] summarized data on the eight most important properties in a set of 11 frequently used solvents.

The static *dielectric permittivity* ε of a solvent determines the energies of the ion–ion, ion–dipole, and dipole–dipole electrostatic (Coulomb's) interactions and characterizes a decrease in the intensity of interaction between charged particles in solution as compared to that in vacuum. It is a common practice to use the value of $1/\varepsilon$ that is proportional to the free energy ΔG_{el} of the electrostatic interaction (Table 8.1).

Table 8.1 Free energies of electrostatic interactions [27]

Interaction type	ΔG_{el}, kJ/mole		
	$\varepsilon = 60–170$	$\varepsilon = 10–60$	$\varepsilon = 2–10$
Ion–ion (i-i)	10	15	100
Ion–dipole (i-d)	(1.0)	(1.5)	20
Dipole–dipole (d-d)	(1.1)	(0.2)	5

Note: ΔG_{el} values in parentheses are comparable with the energy of thermal motion of solvent molecules at 300 K.

It is also possible to use other related dielectric characteristics such as the Kirkwood function X_a defined (for $\varepsilon = 10–60$) as

$$X_a = \frac{(\varepsilon - 1)}{(2\varepsilon + 1)}. \tag{8.61}$$

In chemistry, the ε value is conventionally treated as characterizing the polarity of a solvent, i.e., its ability to stabilize ions in solution, although this treatment is not as strict since the term "polarity" refers to a dipole moment (μ) of the solvent. The theory establishes a generally parallel variation of the permittivity ε and dipole moment μ [$\mu = f(\varepsilon, n^2)$], but sometimes this parallelism is violated. For example, alcohols possess moderate dipole moment, but high permittivities due to increased dipole moments of linear molecular associates. The association of molecules with large μ values can also lead to a decrease in the resulting dipole moment, together with a decrease in the permittivity. Melts of $R_4N^+A^-$ (ionic liquids) possess low permittivities ($\varepsilon = 4–6$) at comparatively large $\mu \cong 15\,D$. There are alternative methods for evaluating the polarity of solvents. In particular, the energies of transition to an excited state (manifested in the electron spectra) for some ion pairs (betaines) characterize a transition from the ionic state to a nonequilibrium uncharged state, thus making it possible to define a polarity scale (E_T) or determine the Dimroth–Reichardt $E_T(30)$ parameter [30]. The E_T value can be related to the aforementioned Kirkwood function X_a (i.e., to the permittivity) by various correlation equations.

Another important characteristic of solvents is *polarizability*, which characterizes the dipole moment induced in a solvent molecule by an electrolyte ion and the exchange (orbital) interactions between the electrolyte ion and solvent molecule. The polarizability increases with the molecular volume and can be estimated from the refractive index n_i of a given solvent as

$$\alpha_i = \frac{M_{ri}(n_i^2 - 1)}{d(n_i^2 + 2)}, \tag{8.62}$$

where M_{ri} and d are the molecular mass and density, respectively, of the solvent. In correlation equations, the polarizability is usually represented by a parameter P that is defined as follows:

$$P = \frac{n^2 - 1}{n^2 + 2}. \tag{8.63}$$

In the framework of the quantum theory of chemical reactions in polar media [14, 15], permittivity is treated as a quantity that varies as a result of the oscillations of solvent molecules and, thus, being a function of some coordinates and frequencies. Therefore, an important role is played by the dynamic properties (polarization fluctuations) of a medium. According to this model of a solvent, it is possible to consider a potential energy surface of the reacting system as a function of both the coordinates of reactant molecules and the coordinates that characterize the state of the dynamic polarization of the solvent.

Other important physical parameters of a solvent are the molar volume (V_M), viscosity (η), electric conductivity, and cohesive energy density (which is a measure of the energy that is necessary to form a cavity in the solvent which would accommodate a molecule of solute). The cohesive energy density can be defined in different ways, in particular as

$$Q_{vap}/V_M, \quad \sigma/V_M^{1/3}, \quad E_\eta/V_M,$$

where Q_{vap} is the heat of vaporization, σ is the surface tension, and E_η is the viscous flow activation energy.

The ability of a solvent to participate in the chemical interaction with ions and/or metal complexes is conventionally characterized in terms of various scales of basicity (for electron-donor properties) or acidity (for electron-acceptor properties) of solvents, the most popular of which are the Gutmann scales of the donor (DN) and acceptor (AN) numbers of solvents. The DN is defined as the enthalpy of mixing of $SbCl_5$ and a given solvent S in dichloroethane ($DN = 0$) with the formation of (S)$SbCl_5$ complexes [13, 27]. With respect to the Gutmann DN scale, some well-known solvents are arranged in the following sequence:

$$\text{HMPTA (38.8)} > \text{Py (33.1)} > \text{DMSO (29.8)} > \text{DMF (26.6)} > \text{THF (20)} > CH_3CN(14.1).$$

The AN values are determined from the ^{31}P NMR chemical shifts for $Et_3P=O$ in various solvents, whereby an acceptor interacts with the electron pair on the oxygen atom. The value of $AN = 0$ is assigned to hexane and the scale unit is defined as 1/100 of the ^{31}P NMR chemical shift of $Et_3P=O \cdot SbCl_5$ adduct in dichloroethane [13]. The DN and AN values have been employed in the correlation analysis for organic reactions and reactions of metal complexes with simple mechanisms, e.g., in the following relationships:

$$\lg k = a_1(DN) + a_2(AN) + \lg k_0,$$

where k_0 is the rate constant of reaction in a noncoordinating solvent [13]. Numerous correlations have been reported between the Gutmann DN and AN values and other characteristics of solvents as well as between these and other scales proposed for the donor and acceptor properties of solvents [13, 27, 30]. Different variants of using these parameters and issues pertaining to the interrelation of donor–acceptor properties and permittivity, the existence of the phenomenon of nonspecific solvation, and the reasons of deviations from correlations with DN and AN, which are known for solvents "structured" by hydrogen bonds (H_2O, ROH, RCOOH, $HCONH_2$) were considered in monographs [13, 30].

The physical and chemical characteristics of solvents were used for their numerous classifications. In particular, according to Parker [63], organic solvents are conveniently divided into three groups as follows:

(i) Protonic (protic) solvents (ArOH, ROH, RCOOH), with $\mu = 1.4–2.3\,D$ and a broad range of ε (6–30).
(ii) Aprotic nonpolar solvents ($\varepsilon < 15$, $\mu < 2.0\,D$), in which electrolytes form ion pairs and associates.
(iii) Aprotic polar solvents ($\varepsilon > 15$, $\mu < 2.0\,D$), featuring dissociation of electrolytes and weak association of ions (DMF, DMSO, HMPTA, CH_3CN, etc.).

Naturally, much more detailed classifications also exist, e.g., with respect to the type of hydrogen bonding (two- and three-dimensional H-bond networks), proximity of physical characteristics, etc. [27, 30, 19].

8.5.2 Association of solvents and formation of molecular complexes

Nonelectrolyte solutions are called associated, provided there exist molecular associates with lifetimes in a range of $10^{-11}-10^{-3}$ s and the enthalpy of formation within $\sim$1–50 kJ/mole [70]. Typical examples are offered by systems with hydrogen bonds. Situations encountered in various solutions are as follows:

(i) Ketones, R_3N, and $CHCl_3$ do not form homoassociates. In heteroassociates, they form single bonds with other components (molecular complexes).
(ii) One of the solution components is capable of forming two bonds (e.g., ROH, R_2NH, RCOOH). Then, chains of homo- and heteroassociates are formed in solutions.
(iii) Solution components are capable of forming three or more bonds. Then the associates possess a more complicated topology than linear chains and appear as two-dimensional (2D) grids or 3D structures (H_2O, $HCONH_2$, ethylene glycol, polyatomic acids). Among these, the structure of H_2O and aqueous organic solvents have been the most thoroughly studied [19, 30, 31].

In liquid water, H_2O molecules exist in at least two forms. The first of these appears as loose 3D formations (clusters) that retain to a considerable extent the structure of ice known as "ice I", with a tetrahedral coordination of oxygen. The second form comprises closely packed monomeric H_2O molecules [19]. The dimensions and shapes of clusters can rapidly vary and these are called "twinkling". Most water molecules in a cluster are linked to four other H_2O molecules that occupy vertices of a tetrahedron. It is believed that voids of clusters may contain monomeric water molecules (clathrate hydrates) [19]. An alternative point of view is that water has a unified, structurally and dynamically inhomogeneous 3D network formed due to hydrogen bonds (H-bonds) [71a].

It has been established that the dissolution of salts in water and aqueous organic media is accompanied by the unique phenomenon of so-called negative hydration of ions [71]. Thermodynamically, this phenomenon is characterized by positive values of ΔH° and ΔS° in a short-range (SR) hydration domain. For example, ΔH°_{SR} for K^+, Rb^+, Cs^+, Cl^-, Br^-, and I^- ions ranges from 5 to 90 kJ/mole. Negative values of ΔH°_{SR} and ΔS°_{SR} have been reported for Li^+, F^-, Ba^{2+}, Sc^{2+}, Ca^{2+}, and Mg^{2+} ions [72].

A quantitative measure for separating ions into two groups, with positive and negative hydration, can be provided by the ΔS_{II} value that characterizes a total change in the entropy of water related to the dissolution of ions [21]. According to this, negative hydration reflects a prevailing effect of water structure disordering over its ordering in the course of ion hydration.

Kinetically, a difference between the two types of hydration is related to the different lifetimes of a water molecule in the hydration shell and in the bulk of water. In the case of negative hydration, the lifetime of a water molecule in the hydration shell is shorter than that in usual structures, so that the H_2O molecule is driven out into the surrounding medium. In particular, the lifetime of water molecules in the hydration shell of negatively hydrated ions is below 10^{-12} s, while those in the hydration shell of positively hydrated ions reach 10^{-4} s (Ni^{2+}), 1 s (Al^{3+}), and even 10^6 s (Rh^{3+}). A water molecule occurs at the site of a quasi-crystalline lattice on the average for about $3 \cdot 10^{-12}$ s and performs several dozen oscillations. H-bonds exist over about the same period of time [73].

The introduction of an organic solvent of about 0.3 molar fractions into water leads to the breakage of its structure, which is manifested by the appearance of anomalous physical properties of these mixed systems and extrema on the composition dependences of $\Delta H^{\neq}$ and $\Delta S^{\neq}$ in reactions of RX solvolysis [19]. In water–alcohol systems with increasing x_2 (molar fraction of ROH), minima are observed on the isotherms of the heat of mixing, excess entropy of mixing, partial molar volume (V_{ROH}), activity coefficient of water, chemical shift of proton in NMR spectra of OH groups and chloroform, and in other properties. For example, significant minima were observed on the plots of $\Delta H^{\neq}$ and $\Delta S^{\neq}$ versus solvent composition for the solvolysis of *tert*-BuCl [19] and transition metal complexes [74]. These anomalous effects were observed for ethanol, 1- and 2-propanol, *tert*-butanol, tetrahydrofuran (THF), dioxane, acetone, and dimethylsulfoxide (DMSO). Apparently, the contribution due to new structures induced by the added organic solvent reaches maximum at the points of extrema in the properties. As x_2 increases further, these structures are broken. It was suggested that, at small x_2, clathrates are formed and the molecules of solute (S) occur in an environment of 24–28 water molecules that form a polyhedron. The formation of these shells is completed at a solvent-to-solute ratio of $H_2O:S \cong 28:1$. Upon attaining a state where all polyhedra are sharing edges (that takes place at $H_2O:S = 17:1$), the structure is rapidly destroyed.

In aqueous dioxane solutions, clusters with H-bonded water appear only at a water content above 0.6 molar fractions [26]. It is interesting to note that the dipole moment of water molecules in dioxane is greater than (μ_{H_2O}) in the gas phase. This increase is related to the polarization of water by dioxane (D):

$$HO\cdot\cdot H\cdot\cdot D$$

In alcohols, the molecules can form linear and cyclic associates $(ROH)_n$ with $n = 2–7$. In contrast to the case of aqueous solutions, the dissolution of substances in alcohols does not lead to the formation or breakage of these structures. The role of alcohol associates in the kinetics of catalytic reactions and the corresponding kinetic mechanisms were considered in Section 8.2. N-alkyl amides, similar to alcohols, also form chains as a result of H-bonds, which significantly increase their dielectric permittivity.

In polar aprotic solvents, some ordering takes place due to the dipole–dipole interactions, but the lifetimes of these formations are 10–100 times shorter than those of associates with same dimensions in alcohols. It is believed that molecular chains exist in both DMSO and DMF, but the role of these associates is by no means comparable with the role of association in proton-containing solvents.

Let us briefly consider specific features of some mixed organic solvents [70].

DMF–$CHCl_3(x)$ system. At $0 < x < 0.3$, complexes with compositions 1:1 and 1:2 are formed, for which the association constants at 298 K amount to $35 \cdot 10^{-2}$ and $6.2 \cdot 10^{-2}$, respectively (on a molar concentration scale).

DMF–EtOH system. In this mixture, alcohol associates coexist with molecular complexes. The enthalpy of formation of O–H··O bonds (EtOH–DMF) is $\Delta H^\circ = -11.1$ kJ/mole.

DMF – acetone system. This system exhibits no specific interactions.

$CH_3CN–CHCl_3$ system. Acetonitrile associates (dimers) are broken and molecular complexes $CH_3C{\equiv}N\cdots H–CCl_3$ are formed with K_{ass} (298 K) = 0.38. Analogous weak associates also exist in the $CHCl_3$ – acetone and $CHCl_3$–EtOH systems.

The interest in mixed binary and ternary solvents in the catalysis with metal complexes is related to the thermomorphic properties of these systems, that is, their ability to modify the mutual solubility depending on the temperature with the transition from a homogeneous single-phase to a two-phase state. The thermomorphic properties can be used to solve the tasks of isolating the catalytic reaction product and recycling a catalyst. For example, the 40 % N-methylpyrrolidone – (1–2 %) H_2O – α-olefin system is homogeneous at 110 °C. A solution of Rh complex with monosulfated triphenylphosphine ligand in this system catalyzes the homogeneous hydroformylation of higher olefins [75, 76]. Then, the cooling from 110 °C to 25 °C leads to phase separation with the passage of Rh complexes into the N-methylpyrrolidone phase.

In recent years, much attention has been devoted to systems with fluororganic solvents and ligands, which possess thermomorphic properties [77–79]. These properties are also inherent in poly(ethylene glycol)–organic solvent systems with phosphine ligands containing poly(ethylene oxide) fragments [80]. An example of three-component thermomorphic system is offered by the propylene carbonate–dodecane–*N*-octylpyrrolidone system that was used in the hydroformylation of higher olefins. At 125 °C, the region of phase separation significantly decreases and a catalyst occurs in a homogeneous three-component system.

At 28 °C, the system passes to a two-phase state and the catalyst occurs in the propylene carbonate–*N*-octylpyrrolidone phase, while the product is retained in the dodecane phase [81, 82].

In recent years, ionic solvents or ionic liquids of the B^+X^- type have also became popular in the catalytic organic chemistry ([19], see also Section 8.7). This is related to the presence of a high concentration of non-associated and non-solvated ions in these media (representing melts of organic salts) and their compatibility with various organic compounds. Due to strong electrostatic interactions, salts in the melt retain a short-range order inherent in the solid. The heats of melting for R_4NBr type salts amount to 16–42 kJ/mole [19]. The ionic liquids also admit solving of the task of separating products from homogeneous catalysts.

At present, making allowance for all equilibria in organic solvents, their mixtures, electrolyte–organic solvent systems, and ionic liquids is still impossible, but some attempts to develop chemical approaches to solving the problems of dissolution and calculating the properties of these liquid systems have been undertaken. In particular, Usanovich [83] and his followers convincingly demonstrated that, in some solutions with two non-associated components, allowance for the formation of stoichiometric complexes

$$m\mathrm{A} + n\mathrm{B} \rightleftharpoons \mathrm{A}_m\mathrm{B}_n$$

makes it possible to describe the liquid–vapor equilibrium in terms of the thermodynamics of ideal gases and solutions and to explain deviations from Raoult's law and Schröder–Le Chatellier law for solubility without recourse to the theory of activities [83]. Even the properties of solutions such as viscosity and electric conductivity can be described by considering chemical equilibria in the system [84–86]. Accordingly, the equilibrium constants and thermodynamic properties of solutions and melts can be determined using viscometric techniques [86].

Models have been proposed that make it possible to calculate the physicochemical and thermodynamic characteristics of mixed solvents as functions of the composition with allowance for the formation of associates [31, 87]. It has been shown that data on the influence of the composition of mixed solvents on their macroscopic properties and thermodynamic characteristics of processes (including complex formation) are frequently interpreted without due allowance for a change in the number of moles per unit mass or unit volume of solution [87]. For example, the stability constant (K_s) expressed in molar concentration units for Co(II) monoamine complex in a water–methanol system increases by a factor of 1.65 when the molar fraction of methanol is increased to 0.55, while the rational constant (K_w) expressed in molar fractions remains constant.

8.5.3 Metal complexes in organic and aqueous-organic solvents

In coordination chemistry, a solvent is almost always treated as representing inner-sphere ligands that fully or partly form the coordination sphere, outer-sphere ligands, or a solvating agent that influences both specific and nonspecific solvation of all participants in the formation and transformation of complexes involved in the catalytic process.

The phenomenon of inner-sphere coordination of solvent (S) molecules is not only characteristic of the complexes of transition and post-transition metals, but generally typical of all metals in groups I, II, and III of the Periodic Table. The solvation of Li^+, Na^+, and other alkali metal ions by polar solvent (THF, ethers, dioxane) is successfully explained in terms of the coordination of solvent, by considering a donor–acceptor mechanism of solvate formation with a maximum coordination number (4 for Li and Na) independent of the nature of a polar solvent. A specific solvation of cations in organometallic compounds of alkali metals explains the effect of small additives of a donor substance to MR solutions in hydrocarbon solvents on the rate of polymerization of α-olefins and dienes and on the structure of polybutadiene [23, 88]. The formation of heavy metal complexes with solvent molecules is natural for solvents such as pyridine, amines, nitriles, amides, and DMSO, which behave as strong ligands (L) and form strong M–L bonds.

However, solvents such as alcohols are also involved in the complex formation processes. For example, the dissolution of $CuBr_2$ in n-butanol is accompanied by the formation of a strong octahedral complex cation $CuBr(BuOH)_5{}^+$. Even low-soluble salts are readily dissolved in coordinating solvents, for example, AgI in NH_3 and CuCl in nitriles. In the latter case, a 40 wt % solution of CuCl can be obtained in adiponitrile (see Chapter 5). The coordination with an organic solvent can ensure stabilization of oxidized metals states that are unstable in water. For example, stable complexes of $CuClO_4$ and $CuNO_3$ ($Cu(RCN)_4ClO_4$ and $Cu(RCN)_4NO_3$, respectively) are formed in nitriles. The coordination can also lead to a change in the structure and electron configuration of the initial complex. For example, flat square diamagnetic Ni(II) complexes are transformed in pyridine into paramagnetic octahedral complexes containing two pyridine molecules in the inner sphere.

The reactions of complex formation with solvent molecules are essentially the exchange of ligands and solvent molecules in the metal coordination sphere. Specific effects of organic solvents are usually revealed by a comparison of the equilibrium constants of complex formation in water, aqueous-organic, and organic solvent. Depending on the nature of a solvent (i.e., its donor and acceptor properties), the nature of metal and ligands, and coordination number n (in ML_n), the stability constants (K_{stab}) can vary with increasing molar fraction of an organic solvent added to water, in particular, these constants can linearly grow, exhibit minima and maxima, or remain almost unchanged [26, 89].

On a qualitative level, the data on K_{stab} in various media are interpreted by comparing the donor numbers of solvents and the character of ligand solvation (acceptor numbers, H-bonds, etc.). The available results can be summarized as follows [26, 31].

(i) The stability of metal chloride complexes in CH_3CN is higher than in water, which is explained by a lower *DN* of acetonitrile and less significant stabilization of Cl^- ions by acetonitrile.

(ii) The solvation of ligands (L), especially X^- anions, is a very important factor among those determining the stability of MX_n complexes. In aqueous solutions, the stability of Cd^{2+} halide complexes (K_1) increases, in accordance with the Ahrland–Chatt rule and the Pearson hard and soft acid and base principle (see Section 1.4), in the following order:

$$Cl^- < Br^- < I^-.$$

This series of stability is also satisfactorily explained by a stronger solvation of Cl^- ions. In DMSO (which is a stronger donor), the solvation of X^- is weak and the stability (K_{stab}) series is consistent with the order of Lewis basicity of the anions:

$$Cl^-(16\,000) > Br^-(850) > I^-(150).$$

Here, despite a stronger Cd^{2+} (S) binding, no energy is spent to desolvate X^- anion. It should be also noted that, in both water and DMSO, various steps of coordination involve a change in the stereochemistry of complexes with a transition from $Cd(S)_6^{2+}$ octahedra to $CdX_n(S)_{4-n}$ tetrahedra. The solvation of chloride ions in DMF is also weaker than in water. For this reason, K_{stab} values for $CuCl_n^{(n-2)-}$ complexes in DMF are many orders higher than those in water. Equilibrium constants are also known for the formation of Cl^- monosolvates with MeOH and ethylene glycol (EG). While the values of permittivity ε are relatively close, the solvation constants differ rather significantly: $K_{Cl^-(EG)} = 0.39$ against $K_{Cl^-(MeOH)} = 0.03$ [26]. It is interesting to note that HCl molecules in EG are almost completely dissociated, while the dissociation constant in methanol is 0.17. In DMSO, where the solvation of Cl^- is very weak, the dissociation of HCl molecules is still less pronounced ($K_{diss} \cong 0.05$) despite a relatively high ε value (46.7) and leads to the formation of complex anions $HCl_2{}^-$. In THF, where the solvation of Cl^- is even lower than in DMSO, the dissociation constant for

HCl is $1.1 \cdot 10^{-6}$, while the constant of formation of $HCl^-_{2(S)}$ from $HCl_{(S)}$ and $Cl^-_{(S)}$ is $2.6 \cdot 10^7$. Poor solvation of Cl^- and H^+ in THF even leads to the formation of dimers $(HCl)_2$ with $K = 1.6 \cdot 10^8$. The problem of anion solvation was considered in much detail in [90, 91].

(iii) Quite reliable correlations have been established between K_{stab} and DN.

(iv) The results of determining K_{stab} in various media are interpreted in terms of a change $\Delta_t G_r^\circ$ in the Gibbs energy for the complex formation process upon the transition from water to mixed solvent and the Gibbs energies of reactant solvation

$$M + L \rightleftharpoons ML \quad (8.64)$$

upon the transfer of reactants from water into mixed solvent:

$$\Delta_t G_r^\circ = (\Delta_t G_{ML}^\circ - \Delta_t G_M^\circ) - \Delta_t G_L^\circ. \quad (8.65)$$

An increase in the stability ($\Delta_t G_r^\circ < 0$) is observed in cases where the destabilization of a ligand ($\Delta_t G_L^\circ > 0$) exceeds the difference in $\Delta_t G^0$ between ML and M. Figure 8.8 shows a plot of $\Delta_t G_i^\circ$ versus molar fraction of DMSO in water for a monoethylenediamine complex of Cu^{2+} [89]. In most cases it was found that $(\Delta_t G_{ML}^\circ - \Delta_t G_M^\circ) \leq \Delta_t G_L^\circ$.

(v) A specific solvation of ligand (L, X^-) and complex also naturally influences the rate of ligand replacement by solvent molecules. In flat square complexes of Pt(II), Pd(II), Ni(II), Au(III), etc., the rate of reaction

$$MX_4 + Y \longrightarrow MX_3Y + X \quad (8.66)$$

is usually described by the following equation:

$$r = k_1[MX_4], \quad (8.67)$$

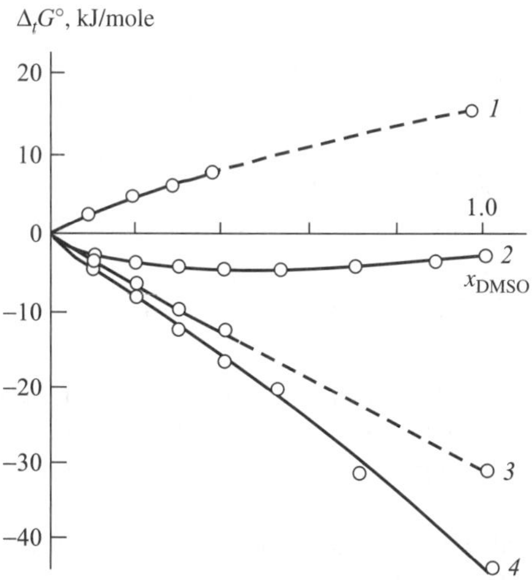

Figure 8.8 Solvation contributions of reactants to changes in Gibbs energy for the formation of monoethylenediamine complexes of Cu^{2+} in H_2O–DMSO system: (1) $\Delta_t G_L^\circ$; (2) $\Delta_t G_r^\circ$; (3) $\Delta_t G_{ML}^\circ$; (4) $\Delta_t G_M^\circ$ (reproduced with permission from AIPC Nauka © 1998)

where $k_1 = k_S[S] + k_Y[Y]$ and k_S is the rate constant for a route involving solvent molecule (S) as a reactant. Here, the first term ($k_S[S]$) reflects the solvent-catalyzed replacement of ligand X. In solvents whose molecules are capable of coordinating to metal, the term with k_S predominates, whereas in solvents not coordinated by metal the main contribution is due to the term with k_Y. For example, in the reaction of Cl^- exchange in *trans*-$PtCl_2Py_2$ complex ($k_S \gg k_{Cl^-}[Cl^-]$), the value of $k_S \cdot 10^5$ [s^{-1}] decreases in the series DMSO, H_2O, C_2H_5OH as 380, 3.5, 1.4, respectively. In poorly coordinating solvents ($k_Y[Y] \gg k_S$) [50], k_Y can vary within broad limits due to a specific solvation of Cl^- anions and nonspecific effects that can modify both the transition and initial states. In the same reaction, k_Y [$m^{-1}\,s^{-1}$] varies within seven orders of magnitude in the series CCl_4 (10^4), C_6H_6 (10^2), DMF (10^{-3}).

The role of specific donor and acceptor properties of solvents in ligand substitution processes was studied in more detail by carrying out the same reaction

$$CoCl_3S^- + Cl^- \underset{k_2}{\overset{k_1}{\rightleftharpoons}} CoCl_4{}^{2-} + S \tag{8.68}$$

in six solvents [13]. It was established that the rate constant k_1 depends primarily on the donor number of a solvent and, as *DN* grows, k_1 decreases in the series

nitromethane > benzonitrile > acetonitrile > acetone > DMF > dimethylacetamide.

This correlation is described by the following equation:

$$\ln k_1 = 0.002(AN) - 0.37(DN) + 12.2. \tag{8.69}$$

Weakening of the Co–S bonds in the transition state is more significant than the formation of bonds with Cl^- ions, which are stabilized by the acceptor solvent. In contrast, the rate constant k_2 increases predominantly due an increase in the acceptor number (AN) as follows:

$$\ln k_2 = 0.31(AN) + 0.015(DN) - 5.97. \tag{8.70}$$

In this case, the weakening of Co–Cl bonds under the action of an acceptor solvent is also more important than the formation of Co–S bonds in the transition state.

The use of mixed solvents encounters the problem of selective (preferential) solvation [26, 92, 93], which is additionally complicated by the lack of detailed information about the composition of solvates and associates in binary solvents. An illustrative example is offered by the influence of the composition of a water–dioxane mixture on the rate constant of metathesis of σ bonds Hg–X

$$HgBr_2 + Hg(CN)_2 \underset{k_2}{\overset{k_1}{\rightleftharpoons}} 2Hg(CN)Br. \tag{8.71}$$

Figure 8.9 shows the dependences of k_1 and k_2 for this reaction on the molar fraction (x_2) of dioxane in the solvent mixture [26]. Despite the existence of several maxima and minima on these curves, the value of $K_{eq} = k_1/k_2$ varies slightly, for example, $K_{eq} = 1.13$ and 1.3 for the points of maxima at $x_2 = 0.2$ and 0.7, respectively.

The specific solvation in mixed solvents also determines the solubility of metal complexes [30]. For example, chloro-oxalato-tris-pyridine Rh(III) dissolves in a pyridine (Py)–H_2O mixture but not in pure components.

The passage from aqueous to nonaqueous solutions leads to a very strong change in the standard electrode potentials E_0 of H^+ and metal ions [26, 27, 90]. This is related both to a change in the constants of stability

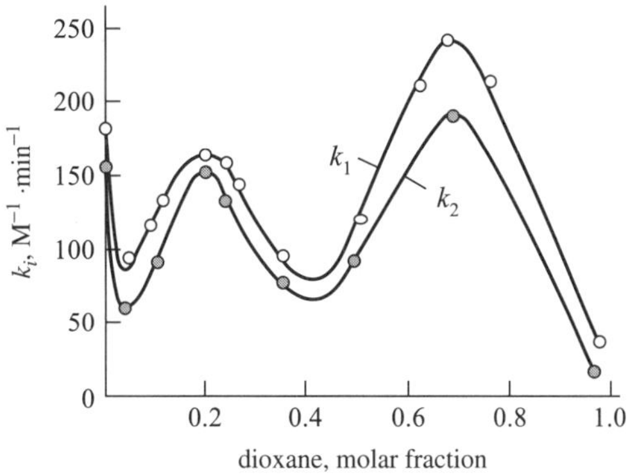

Figure 8.9 Plots of the rate constants of formation and dissociation of mixed Hg(II) cyanide–bromide complex versus molar fraction of dioxane in mixture with water (reproduced from Academiai Kiado Zrt. (Budapest) © 1982 [26])

(for E_0 of ML_n complexes) and to the specific solvation (coordination) of oxidized and reduced forms of metal ions. The modified standard electrode potentials are calculated using the E_0 values in aqueous solutions and a change ΔG_t° in the Gibbs energy during the passage from aqueous to nonaqueous solutions:

$$\Delta G_t^\circ = nF(E_S^\circ - E_{aq}^\circ); \tag{8.72}$$

$$E_S^\circ = \frac{\Delta G_t^\circ}{nF} + E_{aq}^\circ. \tag{8.73}$$

Table 8.2 presents the values of E_S° relative to E° of standard hydrogen electrode in water for various redox pairs and solvents. In the case of Cu^+/Cu^{2+}, it is clearly seen that the oxidation potential for the electrode reaction

$$Cu^{2+} + e^- \longrightarrow Cu^+$$

significantly increases in the series

$$\text{DMSO} < H_2O < \text{Py} < CH_3CN.$$

In liquid NH_3, $E^\circ_{Cu^+/Cu^{2+}}$ is negative and Cu^+ ion becomes a strong reductant, the properties of which are close to those of Zn° (per electron). Another strong reductant in this medium is Hg°. Other E° scales also exist, in which the electromotive force (emf) is determined relative to the hydrogen electrode in water with allowance for a change in the solvation energy on the passage from one solvent to another [27].

The nature and stability of the outer-sphere complexes involving solvent molecules are still not sufficiently studied. Using the results of investigations of the Raman spectra of tetrahedral complexes CdX_n, $Cd(CN)_n$, HgX_n, and $Hg(CN)_n$ in six solvents [26] showed evidence for the existence of both inner-sphere complexes MX_3S^- and outer-sphere complexes $MX_4(S)^{2-}$. In the case of anionic complexes and proton-containing solvents, it is also necessary to take into account the formation of hydrogen bonds $M{-}X \cdots H$ and the electrostatic interaction between ions and solvent dipoles. Many of the extraction methods for separating ions are based on the formation of their outer-sphere complexes that are very soluble in an organic solvent [50]. For example, complex $Co(NH_3)_6{}^{3+}$ adds four salicylate anions and a molecule of cyclohexanone. The data of ^1H and ^{13}C NMR revealed the existence of the outer-sphere adducts of cobalt(II) bis-(acetylacetonate)-bis-(pyridinate) [94] and some other octahedral 1 : 1 complexes of Co(II)

Table 8.2 Standard electrode potentials (V) at 25 °C [90]

	H_2O	CH_3OH	CH_3CN	DMSO	Py	THT	$NH_{3(liq)}$
$H_{2(gas)}/H^+$	0.00	+0.12	+0.48	−0.20	−0.18		−1.00
Na_{solid}/Na^+	−2.71	−2.61	−2.56	−2.85	−2.54	−2.41	−2.89
K_{solid}/K^+	−2.92	−2.82	−2.83	−3.06	−2.86		−3.04
Rb_{solid}/Rb^+	−2.93	−2.83	−2.86	−3.04	−2.69		−3.06
Cs_{solid}/Cs^+	−2.92	−2.83	−2.87	−3.01	−2.61		−3.08
Cu_{solid}/Cu^+	+0.52		−0.02	+0.09	−0.33	+0.21	−0.64
Cu_{solid}/Cu^{2+}	+0.34		+0.06	+0.08	+0.08		−0.60
Cu^+/Cu^{2+}	+0.16		+1.21	+0.07	+0.49		−0.56
Ag_{solid}/Ag^+	+0.80	+0.87	+0.58	+0.45	+0.25	+0.27	−0.24
Au_{solid}/Au^+	+1.83		+1.51	+1.21	+0.80		
Zn_{solid}/Zn^{2+}	−0.77	−0.62	−0.41	−1.01			−1.54
Cd_{solid}/Cd^{2+}	−0.41	−0.24	−0.19	−0.69			−1.20
	+0.80			+0.46	+0.46		
Hg_{liq}/Hg^{2+}	+0.85			+0.50	+0.44		−0.33
	+0.91			+0.54	+0.41		
Tl_{solid}/Tl^+	−0.34	−0.29	−0.23	−0.56	−0.35	−0.25	
$Cl^-/Cl_{2(gas)}$	1.36	+1.22	+0.93	+0.95	+1.02	+0.84	+0.91
$Br^-/Br_{2(liq)}$	+1.07	+0.95	+0.74	+0.80	+0.86	+0.80	+0.73
$I^-/I_{2(solid)}$	+0.54	+0.46	+0.34	+0.45	+0.34		+0.28

Note: DMSO = dimethylsulfoxide, Py = pyridine, THT = tetrahydrothiophene.

with $CHCl_3$, CH_2Cl_2, H_2O, and benzene, the stability constants of which vary within $K_{stab} = 10^{-2}–1\,M^{-1}$ (25 °C) depending on the ligand and solvent. In these complexes, the distance from the metal to hydrogen atom of chloroform in the plane of chelate ligands amounts to ~3.0 Å. The formation of these complexes should be taken into account in an analysis of the solvent effect on the rate of catalytic reactions.

Another specific feature in the behavior of metal complexes in aqueous organic solvents is related to the ability of a solvent to induce redox processes in the coordination spheres of complexes. For example, $CoS_6^{2+}[Co(CO)_4{}^-]_2$ complexes are formed in aqueous alcohols and DMF from $Co_2(CO)_8$ (see Section 5.2). An analogous reaction takes place with Fe(0) pentacarbonyl [26]:

$$4Fe(CO)_5 + 6Py \xrightarrow[-9CO]{} Fe(Py)_6{}^{2+} + Fe_3(CO)_{11}{}^{2-}. \qquad (8.74)$$

In concluding this section, it should be emphasized that researchers must pay attention to the presence of even minor amounts of water (below 0.1 M) in organic solvents. As is known, the addition of a small amount of water to solutions of AgX_n^{n-1} in DMSO and propylene carbonate leads to a significant decrease in K_{stab} [50]. The formation of *cis*-complexes $PdBr_2[P(OPh)_3]_2$ in butanol in the presence of 0.1 M H_2O was discussed in Section 5.3.1. The difficulty of studying the effect of small amounts of water on the equilibrium and rates of catalytic reactions is related to the fact that the conventionally used "dry" solvents already contain no less than $10^{-2}–10^{-3}$ M H_2O, which is comparable to the typical concentration of a metal complex catalyst.)

8.5.4 Ion association, ion pairs, and specific salt effect in organic solvents

The concept of ion pair (IP) as a chemically individual entity was introduced by Bjerrum in 1926. The energy of dissociation of ion pair M^+X^- into free ions is determined primarily by the energy of Coulomb's interaction in the pair that, in turn, depends on the radii of ions and dielectric permittivity ε of solvent:

$$E_{\text{diss}} = \frac{Ne^2}{(r_1 + r_2)\varepsilon}. \qquad (8.75)$$

The IP pair stability in a medium also depend on the ion–dipole interaction and the dispersion forces. In 1954, R.M. Fuoss and S. Winstein proposed the consideration of two types of IPs:

- tight (contact, intimate) ion pairs;
- loose (solvate-separated) ion pairs.

The process of dissociation of a salt (ionophore) in a solvent begins with the solvation of cation and anion and the appearance of a tight IP. Then, the solvent molecules enter into the space between ions (by means of cation solvation) and the IP becomes solvate-separated. The subsequent solvation of cation (and, to a lesser extent, of anion in aprotic solvents) leads to the IP dissociation into free ions. For example, Na^+R^- ion pairs in hydrocarbons separate upon adding THF according to the following scheme [23]:

$$\text{NaR} \underset{(1)}{\overset{\text{THF}}{\rightleftharpoons}} (\text{THF})_n\text{Na}^+\text{R}^{\bullet -}(\text{THF})_x \underset{(2)}{\overset{K_{\text{tight}}}{\rightleftharpoons}} \text{Na(THF)}_4{}^+\text{R}^{\bullet -}(\text{THF})_y \underset{(3)}{\overset{K_{\text{sep}}}{\rightleftharpoons}}$$
$$\rightleftharpoons \text{Na(THF)}^+_{4+m} + \text{R}^{\bullet -}(\text{THF})_z$$

(where $R^{\bullet -}$ is radical anion of naphthalene) that consists of two ionization stages [(1) and (2)] and dissociation stage (3).

As can be seen, an important factor in the conversion of a polar (but not ionic) compound into an IP is the ability of a solvent (i) to coordinate to a cation (donor properties, *DN*) with the formation of a primary solvation shell (coordination sphere) and (ii) to interact with unshared electron pairs of anion X^- (acceptor properties, *AN*). Solvate-separated IPs are denoted $M^+//X^-$. Notions about IPs and free ions are useful for considering dilute solutions of ionic compounds (ionophores) in media with relatively low permittivities ($2 < \varepsilon < 15$). However, these notions lose sense in concentrated solutions and ionic liquids (featuring more complicated associates) and media with high ε values (IPs are not formed).

Concentrated aqueous solutions feature the appearance of ion triads (e.g., $M^+X^-M^+$ and $X^-M^+X^-$) and more complicated aggregates, the formation of which in the media with low permittivities ($\varepsilon < 10$) is characterized by large association constants. In these media, salts of the $R_4N^+X^-$ type form aggregates involving 5–25 molecules (in benzene). The association constants for the formation of Bu_4N^+ (picrate)$^-$ dimers and trimers in benzene amount approximately to $10\,M^{-1}$ and $30\,000\,M^{-2}$, respectively [19]. Even inorganic salts (e.g., LiBr, LiNCS, LiCl) can form dimers $(LiX)_2$ and tetramers $(LiX)_4$ in aprotic solvents (for example, $(LiBr)_4$ in Et_2O) [95, 96]. Another well-known example is offered by oligomers of organometallic compounds of the $(LiR)_n$ type [23] (see Chapter 2). Aggregates of I–I electrolytes are capable of catalyzing various organic reactions [95, 97]. For example, aggregates of $LiClO_4$ in Et_2O apparently catalyze organic reactions [95, 98].

Table 8.3 Equilibrium constants of formation (K_{ass}) and concentrations of ion pairs ([IP]) in I–I electrolyte [19]

Solvent	K_{ass}	[IP], mol%
H_2O	1.3	5
Propylene carbonate	1.8	8
DMSO	3.3	12
DAS*	7.1	15
CH_3OH	9.9	17
CH_3COCH_3	75	35
CH_3COOH	32 000	90
C_6H_6, dioxane	1019	–

Note: *DAS – dipolar aprotic solvent [(CH_3CN, DMF, nitromethane, $PhNO_2$ ($\varepsilon \sim 36$)].

Table 8.3 presents estimations of the equilibrium constants (K_{ass}) of IP formation and IP concentrations for a I–I electrolyte (0.1 M) at 25 °C and a distance of $a = 5\,\text{Å}$ between the nuclei. According to these data, the logarithmic association constant of $(iso\text{-}C_5H_{11})_4N^+N_3^-$ in aqueous dioxane varies within ~15 units with an increase in the dioxane content and a decrease in the solvent permittivity ε according to the following formula [19]:

$$\lg K_{acc} = A/\varepsilon. \tag{8.76}$$

A large volume of reference data on the ion association is summarized in monograph [19].

Transition metal compounds consisting of ionic components, which were referred to above as outer-sphere complexes, essentially represent loose IPs in which the cations and anions are separated by ligands. These ligands may also include solvent molecules, e.g., in complexes such as $Cr(H_2O)_6^{3+}\,Cl^-$, $Fe(CN)_6^{4-}\,K^+$, $Fe(CO)_4^{2-}\,Na^+$.

The salt effects in both aprotic and protonic (with low ε) solvents are predominantly related to a specific participation of cations and anions of a salt in the elementary steps of a complicated complex formation process or in a limiting steps of an organic reaction rather than to changes in the activity coefficients (γ_i, γ_{ML}, or $\gamma^{\neq}$) as in the case of aqueous media. As was demonstrated above, the salt effects can also be manifested in aqueous media provided that an electrolyte concentration exceeds 3–4 M. According to Loupy and Tchoubar [29], specific salt effects are classified as follows:

(a) acid–base interactions;
(b) exchange reactions between IPs;
(c) effect on equilibrium dissociation of IPs;
(d) effect on structure of transition states;
(e) association with solvent molecules containing OH groups (drying effect).

In organic solvents, it can also be expected that salts (electrolytes) will influence the solubility of gases and organic substrates possessing limited solubilities.

Acid–base electrolyte effects. The acid–base properties of electrolyte M^+X^- depend on the nature of both M^+ and X^-. The harder the anion X^- (F^-) and the softer the cation (Cs^+, R_4N^+) that constitute a salt, the more basic are the properties of this salt. If the cation becomes harder (Li^+) and/or the anion becomes softer (I^-, ClO_4^-, PF_6^-), the salt acquires more acidic properties.

The interaction of a substrate with Li^+ cations changes the nucleophilic and electrophilic properties of this substrate. For example, $(ROH)Li^+$ becomes a more acidic and less nucleophilic reactant, while nitrile $RC{\equiv}NLi^+$ is characterized by increased electrophilicity. The interaction with a hard anion X^- increases the nucleophilicity of alcohol $\overset{-\delta}{RO}\cdots H\cdots X^-$. If the substitution reaction in RY in a medium with low ε is determined by the ionization step $RY \rightarrow R^+Y^-$, then the reaction rate significantly increases in the presence of an acid salt as follows:

$$R{-}Y + M^+X^- \longrightarrow \left| \overset{+\delta}{R}\cdot\cdot\overset{-\delta}{Y}\cdot\cdot M^+X^- \right|^{\neq} \longrightarrow R^+Y^- + M^+X^-.$$

The salt effect increases as the ability of a medium to solvate cation M^+ decreases. For example, the ionization of $CH_3OC_6H_4C(CH_3)_2CH_2OTs$ (where Ts is tosylate) in the presence of $LiClO_4$ in diethyl ether proceeds $3 \cdot 10^5$ times faster than in the absence of this salt [95]. The effect of $LiClO_4$ increases with decreasing donor properties of the solvent (Et_2O < THF < DMF < DMSO). A system of 5 M $LiClO_4$–Et_2O possesses unique properties [98] and, in particular, significantly increases the stereoselectivity and rate (compared to the case without the salt) of the Diels–Adler reactions, allyl and 1,3-sigmatropic rearrangements, and some other processes.

Exchange reactions. The exchange between IPs of reactants and IPs of an added electrolyte can lead to the formation of new, more reactive IP species:

$$M_1{}^+Nu^- + M^+X^- \rightleftharpoons M^+Nu^- + M_1{}^+X^-. \tag{8.77}$$

Exchange reactions have been suggested to explain the appearance of a "specific" salt effect that accompanies the solvolysis of RY. In particular, the acceleration of transition $RY \rightarrow R^+//Y^-$ (i.e., the "normal" salt effect) is manifested at relatively large $LiClO_4$ concentrations. At small [$LiClO_4$] values, an exchange reaction proceeds with the formation of solvate-separated $R^+//ClO_4{}^-$ ion pair, which is more reactive with respect to subsequent transformations and does not relax back to RY. Exchange reactions with the formation of more reactive loose IPs (e.g., $R_4N^+//Nu^-$) play an important role in interphase catalysis.

Equilibrium of IP dissociation. Numerous examples demonstrate that the reactivity of a free anion (nucleophile) is much higher than that of the same anion in a tight IP [29]. In the anionic polymerization (e.g., with LiR), the rate constants of chain propagation on free carbanions are higher by four orders of magnitude than those on IPs [23]. The salt effects also significantly influence the degree of IP dissociation. The ion dissociation constant of electrolytes in aprotic polar solvents increases with the size of cation in the series $Li^+ \ll Cs^+ \ll R_4N^+$. This mechanism of salt action is essentially also related to the exchange process (8.77).

Effect on structure of transition states. In cases where it is possible to choose between the active centers of a substrate, the interaction with the cation of a salt can determine the direction of an electrophilic reaction. For example, in the case of a keto–enole equilibrium, the binding of an enolate anion to Li^+ ion leads to the predominant O-alkylation. The proper arrangement of reactants around a cation favors the formation of macrocycles. In reactions of organometallic compounds, the nature of cation in the electrolyte salt and permittivity (ε) of the solvent can change the process direction. For example, the keto–enole equilibrium (or mesomerism of two structures)

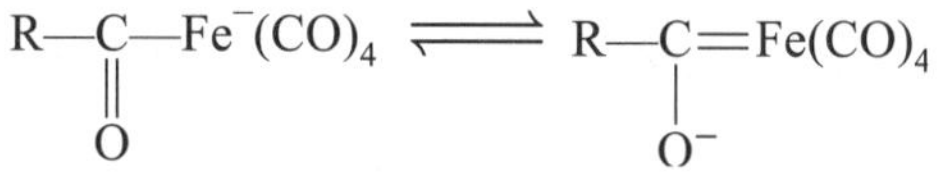

depends on the nature of cation. In weakly polar solvents (CH_2Cl_2, Et_2O), Li^+ and Na^+ cations stabilize the structure

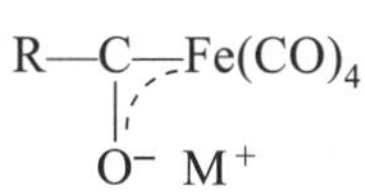

and direct the anion alkylation by alkyl halide R′X toward oxygen atom:

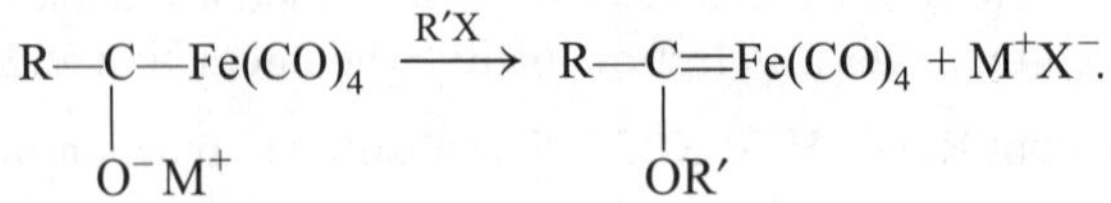

If hard cations are replaced by soft species R_4N^+, the direction of the addition is changed and alkylation proceeds at the metal as follows [99]:

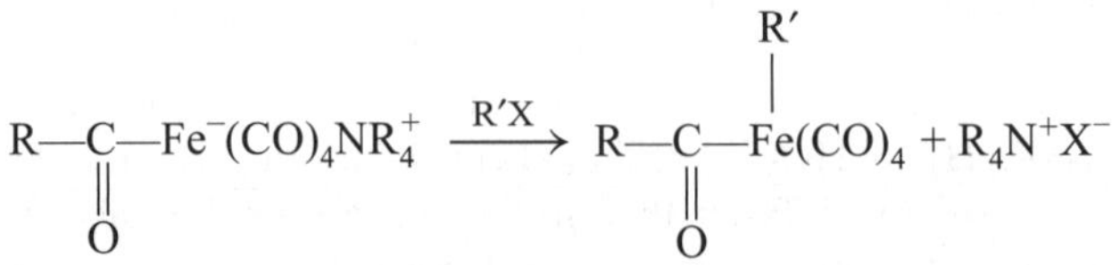

Desolvation of reactants. This phenomenon is most clearly manifested in aqueous and aqueous-organic media and is also known as the "drying effect" [29]. A decrease in the degree of reactant solvation in the presence of salts leads to an increase in the reactivity. For example, the acidity and electrophilicity of protons decrease in the following order of $H^+A^- > H_3O^+ > H(H_2O)_2{}^+ > H(H_2O)_6{}^+$. The desolvation also significantly affects the nucleophilicity (basicity) of anions such as OH^- and CH_3O^-. In addition, the concentration of ions also changes as a result of the binding of solvent molecules and a decrease in the amount of moles of free solvent molecules, e.g. for CH_3OH as

$$RH + CH_3O^- \rightleftharpoons R^- + CH_3OH \qquad (8.78)$$

For example, $(CH_3)_4N^+OH^-$ (0.01 M) in a DMSO–H_2O (50 mol %) mixture is 10^4 times stronger base than in water. The effect of OH^- dehydration is also manifested by an increase in γ_{OH^-}. Similarly, an increase in h_0 of the acid by adding electrolytes (see above) is explained by a decrease in a_{H_2O} and a growth in the proton activity ($\gamma_{H_3O^+}$). The well-known increase in the acidity of mineral acids in CH_3COOH upon adding electrolytes is also attributed to the drying effect of proton desolvation [29].

The aforementioned five mechanisms of manifestation of the salt effects in organic solvents and organic reactions are frequently operative in various combinations, so that it is only possible to speak of a prevailing role of some mechanism. The role of electrolytes is even more diverse in the catalysis with metal complexes and reactions of metal-containing coordination compounds, where soft anions play the role of ligands and hard ones stabilize the cationic forms of complexes by means of exchange reactions

$$[M]Cl + Li^+BF_4{}^- \rightleftharpoons [M]^+BF_4{}^- + Li^+Cl^-. \qquad (8.79)$$

The conversion of an anionic complex of Rh(I) into dianionic significantly accelerates the reaction of oxidative addition of CH_3I to Rh(I) in acetic acid due to an increase in the nucleophilicity of Rh(I) [29]:

$$Rh(CO)_2I_2{}^- + M^+X^- \rightleftharpoons Rh(CO)_2I_2X^{2-} + M^+;$$

$$Rh(CO)_2I_2X_2{}^- + CH_3I \longrightarrow CH_3Rh(CO)_2I_2X^{2-} + I^-$$

$$(X = I, OAc; M = Li, RN^+).$$

(X–I, OAc; M–Li, R_4N).

The addition of ArX to PdL_n is accelerated by LiCl, $R_4N^+Cl^-$, and LiOAc. A large number of other interesting examples, not yet fully justified, are presented in monograph [29].

The effect of electrolytes on the reactions kinetics, which is related to changes in the structure of aqueous–organic solvents, was also clearly demonstrated in the study of isocyanate hydrolysis to carbamic acid in a water–dioxane mixture [100]:

$$RNCO + H_2O \longrightarrow RNHCOOH \tag{8.80}$$

It was found that additives of an electrolyte (KI, negative hydration) that partly decomposes the structure of H-bonded water clusters lead to an increase in the observed reaction order with respect to $[H_2O]$ in a 3–20 M H_2O interval. The same effect of increasing temperature (12–50 °C) is also related to the partial breakage of the structure of H-bonded clusters. The additives of Et_4NCl (positive hydration) stabilize the H-bond lattice structure in the ice-like structure of water and thus decrease the reaction order with respect to $[H_2O]$.

The results were interpreted using an approach that treats associated solvents as pseudo-polymeric systems, which was developed by Tiger, Tarasov, Entelis et al. [101–103]. According to the scaling invariance concept, all long-chain polymers exhibit, to the first approximation, similar properties in solution, provided there is a properly selected scale of comparison. Here, the role of a parameter determining the properties (e.g., specific viscosity η) is played by the length of macromolecules (rather than the chemical composition of chain units). The scaling index ν in the relationships such as, e.g., $\eta \sim P^{\nu}$ (where P is the degree of polymerization), characterizes the degree of filling of the volume of macromolecules (coils, globules) by their folded units and is determined by the properties of a particular solvent. By analogy with polymer molecules, it is accepted that a relationship between the size r of a cluster of H-bonded molecules (H_2O, ROH) and their concentration C (at $C > C_{crtit}$) must possess a power character [101, 104]:

$$r \sim (C/C_{crit})^{-n}. \tag{8.81}$$

Since the reaction of molecule A with molecules of a cluster must proceed when reactant A crosses the cluster surface, the observed pseudo-first order reaction rate constant k_{obs} must be proportional to the cluster surface area ($\sim r^2$) and concentration ($\sim 1/r^3$). Then, using relation (8.81), we obtain

$$k_{obs} \sim 1/r \sim (C/C_{crit})^{n}. \tag{8.82}$$

where the power index n formally corresponds to the observed reaction order with respect to $[H_2O]$. In the study of reactions between isocyanates and alcohols [102, 103], it was established that the passage to solvents capable of breaking the H-bonded alcohol cluster structure leads to an increase in the n value. The interaction of an associated solvent with aprotic acids is accompanied by breakage of the associated structure and a change in the solvent properties [26].

8.6 Strong protonic acids in organic solvents and kinetics of catalytic reactions with metal complexes in these media

Hydrogen chloride solutions in aqueous–organic and organic media are used as catalysts for organic reactions or media for catalysis with metal complexes. For example, organic reactions were catalyzed by DMF–HCl complexes in DMF [105] and by strong protonic acids in aqueous–organic solutions [19, 30, 68, 106]. Hydrochlorination of acetylene to vinyl chloride is catalyzed by Cu(I) [107] and Pt(II) [108, 109] complexes in the DMF–HCl system, Hg(II), Pt(II), Rh(III), and Ir(III) complexes in the C_2H_5OH–HCl and C_2H_5OH–CH_3CN–HCl systems [108–110], and Rh(III) and Ru(IV) complexes in the N-methylpyrrolidone –HCl system [111].

Hydrochlorination of monovinylacetylene in ethanol, DMF, and *N*-methylpyrrolidone (NMP) is catalyzed by $RhCl_3$ [69]. The addition of HCl to methylacetylene (MA) is catalyzed by $RhCl_3$, $RuCl_3$, and $HAuCl_4$ complexes in DMF [112] and ethanol [110]. In contrast to Cu(I), Hg(II), and Pt(II) complexes, these compounds also catalyze the anti-Markovnikov addition of HCl to MA with the formation of *cis*-1-chloroprene ($CH_3CH{=}CHCl$). The anti-Markovnikov product ($ClCH{=}CH{-}CH{=}CH_2$) can also be obtained from vinylacetylene in $RhCl_3$ solutions in DMF and NMP [69].

Systems such as HCl–H_2O–ethanol, HCl–ethanol, HCl–DMF, HCl–NMP, and some others, which contain strong acid HA and organic solvent (S), are very interesting subjects for investigation, since the acid-to-solvent molar ratio reaches HA : S = 2 and above. Let us consider data available on the structure and acidity of these systems.

8.6.1 Structure and properties of strong acid solutions in organic solvents

The results of IR spectroscopic investigations using the frustrated multiple internal reflection (FMIR) technique in aqueous, aqueous–organic and organic solutions [66, 113–122] led to a conclusion [66] that the simplest stable protonic solvates were ions with strong symmetric H-bonds involving two solvent molecules (HS_2^+). In water, these are dihydrates ($H_5O_2^+$), in absolute (anhydrous) ethanol, they are $(EtOH)_2H^+$ ions, and in aqueous ethanol solutions, $(EtOH)H(H_2O)^+$ and $H_5O_2^+$ species appear that can exist in the form of free ions and/or ion pairs (IPs). The results of potentiometric investigations into HCl solutions in 95 vol % aqueous ethanol at temperatures within 5–50 °C [123] showed that, even in dilute HCl solutions (0.1 $\tilde{m}$), about half of HCl molecules occur in the form of IPs, most probably such as $(EtOH)_2H^+Cl^-$, $(EtOH)(H_2O)H^+Cl^-$, and $H_5O_2^+Cl^-$. The IP dissociation constant in 92.4 wt % ethanol at 25 °C was about 0.0325 (in molar concentration units).

The values of entropy and enthalpy of the HCl transfer from water to an aqueous–ethanol solution amount to $\Delta S_t^\circ = -100.0$ J/(mole · K) and $\Delta H_t^\circ = -19\,400$ J/mole. It was found that, as the ethanol concentration decreased to 40 wt %, the ΔS_t° and ΔH_t° values became positive. According to Vinnik [106, 124], the acidity of aqueous–ethanol HCl solutions exhibited no anomalies as the ethanol content was reduced to 34 wt %.

A comparison of the HCl acidity functions in low-water ethanol solutions (below 1 % water) to H_0 of aqueous solutions showed that the entire range of HCl concentrations can be subdivided into two groups. In dilute HCl solutions, the acidity grows faster with HCl content in anhydrous ethanol systems, whereas for concentrated solutions it grows faster in aqueous HCl [125].

Nahlovsky and Chvalovsky [126] studied the acidity of HCl in absolute ethanol (below 0.05 vol % water) and in solutions with 5 vol % (or 6.3 wt %) H_2O at 25 °C. The results were compared to the acidity function of HCl–ethanol–H_2O (48 vol %) [127] and HCl–H_2O systems. The pK_B values of indicators were references to pK_B in water. The results (Figure 8.10) showed that $(H_0)_I$ depended in a complicated manner on the composition of the medium. The $(H_0)_I$ functions in absolute ethanol and water almost coincide (for the adopted reference). The values of acidity (h_0) in the HCl–ethanol system are higher in the region of low HCl concentrations than in 6.3 % aqueous HCl solution, but the acidity of the latter solutions at [HCl] > 4 M exceeds that of the anhydrous system. The acidity of low-water ethanol solutions also increases (by ~ 0.35 units of H_0) on the passage from 1–3 vol % to 5 vol % H_2O at the same molar content of HCl if pK_{BH^+} of indicators are references to pK_B in the medium studied (Table 8.4) [125].

Solutions in the DMF–HCl and DMF–H_2O–HCl systems have been studied in even more detail. The solubility of HCl in DMF reaches nearly stoichiometric concentrations (2 mole HCl per mole DMF) [107]. At a 1 : 1 molar ratio, the solution slowly converts into a solid complex with $T_m = 37$ °C. As [HCl] increases further, the system becomes liquid and P_{HCl} over the solution sharply increases. The IR spectroscopic (FMIR) study [115] of these solutions showed that all HCl at 25 °C in a concentration range of 0–5 M

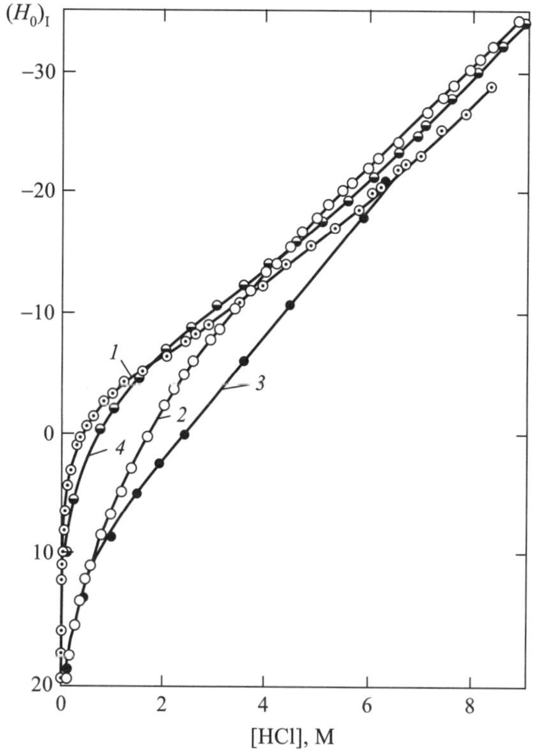

Figure 8.10 Plots of acidity functions $(H_0)_I$ versus [HCl] in (1) absolute ethanol; (2) 5.0 % H_2O; (3) 48 % H_2O; (4) 100 % H_2O (reproduced from editorial board of *Collect. Czech. Chem. Commun.* journal © 1968 [126])

Table 8.4 Values of acidity function H_0^S in dilute aqueous HCl solutions at 25 °C [125]

HCl, M	1.09 % H_2O	HCl, M	3.0 % H_2O	HCl, M	5.0 % H_2O
0.03	0.71	0.03	0.8	0.03	–
0.31	−0.84	0.315	–	–	–
1.25	−1.84	1.26	−1.56	1.26	−1.62
1.5	−1.99	1.6	−1.99	1.65	−2.09
2.1	−2.19	2.2	−2.14	2.3	−2.30
3.0	−2.46	3.1	−2.51	3.2	−2.75
4.6	−2.98	4.7	−3.09	4.9	−3.45
6.5	−3.67	6.7	−3.74	6.9	−4.25
7.7	−4.15	7.9	−4.18	8.25	−4.69
8.9	−4.58	6.2	−4.58	9.5	−5.04

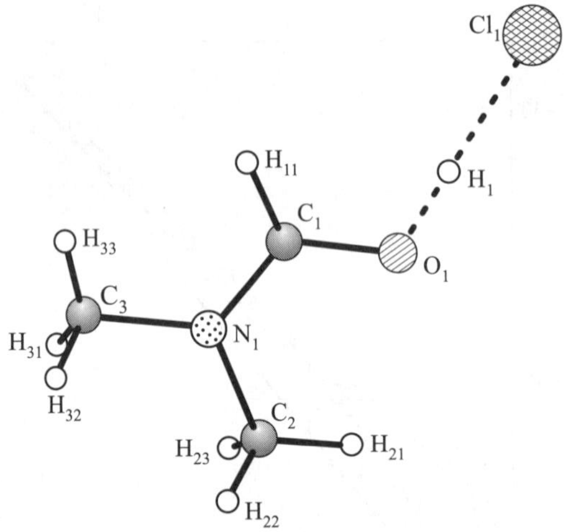

Figure 8.11 Structure of complex K2 (DMF·HCl) (reproduced with permission from AIPC Nauka and editorial board of *Dokl. Akad. Nauk* journal © 1998 [129])

is bound in K1 complex with a composition of 2DMF · HCl. In 5–10 M HCl solutions, K1 complexes coexist with DMF · HCl (K2), while at [HCl] = 10.5 M only K2 complexes are retained. It was established that two DMF molecules in the K1 complex are not identical and the DMF–HCl system contains no ions with highly symmetric H-bond in S_2H^+. At [HCl] > 10.5 M, K3 complex (DMF · 2HCl) appears.

The K2 complex represents adduct with proton in an intermediate position ($Me_2N-CH=O \cdots H \cdots Cl$) [128]. The Raman spectra of this system [118] display three absorption bands assigned to complexes K1 (1670 cm^{-1}), K2 (1690 cm^{-1}), and K3 (1720 cm^{-1}). Data of polarized Raman spectroscopy showed that the acid–base interaction in these complexes is mediated by amide oxygen, which is protonated in K3 with the formation of ion pairs $Me_2N^+{=}CHOH \cdot HCl_2{}^-$.

The ability of DMF to effective solvation of protons was also confirmed by the IR spectroscopy of DMF–HCl–H_2O solutions [116, 122]. It was established that DMF molecules even at a large excess of water replace H_2O molecules in $(H_2O)_2H^+$. This results in the formation of $[H_2O \cdots H \cdots OCHNMe_2]^+$ ions with a quasi-symmetric H-bond (in contrast to the situation in aqueous–ethanol solutions of HCl, where ethanol is completely displaced from disolvate under the same conditions). At a ratio of $H_2O : HCl \geq 3$, only mixed proton solvates are formed irrespective of [DMF].

The above assumptions concerning the structure of K2 species were confirmed by the X-ray diffraction analysis of single crystals (Figure 8.11) [129], which showed that the complex represents a quasi-ion pair with incomplete proton transfer to oxygen:

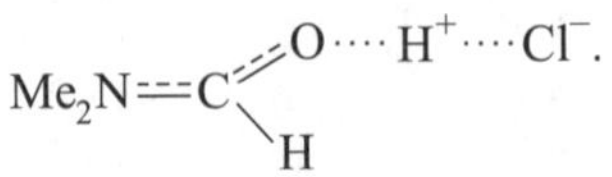

Here, the H–Cl bond length [1.723(15) Å] is longer than the usual values in hydrogen-chloride-coordinated chlorine ions $Cl^- \cdots H-Cl$ (1.649 and 1.471 Å, respectively), but shorter than in protonated dimethylacetamide $Cl^-[HOC(Me)NMe_2]^+$ (1.983 Å) [129]. Close data on the stability and structure of $(DMF)_m(HCl)_n$ complexes were obtained in quantum-chemical *ab initio* calculations [130].

The acidity of DMF–HCl [131] and DMF–HCl – 1,1,2,2-tetrachloroethane [132, 133] systems was studied using the indicator technique. Table 8.5 presents the H_0 values calculated from K_{BH^+} for the

Table 8.5 Values of acidity function H_0 at various HCl concentrations in DMF at 25 °C [131]

HCl, M	$-H_0$	Hcl, M	$-H_0$	Hcl, M	$-H_0$
0.025	−1.78	4.0	1.05	10.6	2.69
0.04	−1.49	4.5	1.15	10.8	2.80
0.07	−1.16	5.0	1.24	11.0	2.91
0.1	−0.98	5.5	1.34	11.2	3.04
0.2	−0.64	6.0	1.43	11.4	3.17
0.3	−0.44	6.5	1.53	11.6	3.30
0.4	−0.32	7.0	1.63	11.8	3.44
0.6	−0.10	7.5	1.74	12.0	3.56
0.8	0.07	8.0	1.85	12.2	3.65
1.0	0.19	8.5	1.96	12.4	3.73
1.2	0.29	9.0	2.08	12.6	3.81
1.4	0.38	9.2	2.14	12.8	3.88
1.6	0.45	9.4	2.19	13.0	3.95
1.8	0.53	9.6	2.26	13.5	4.11
2.0	0.59	9.8	2.32	14.0	4.23
2.5	0.73	10.0	2.40	14.5	4.35
3.0	0.84	10.2	2.49	15.0	4.46
3.5	0.94	10.4	2.59		

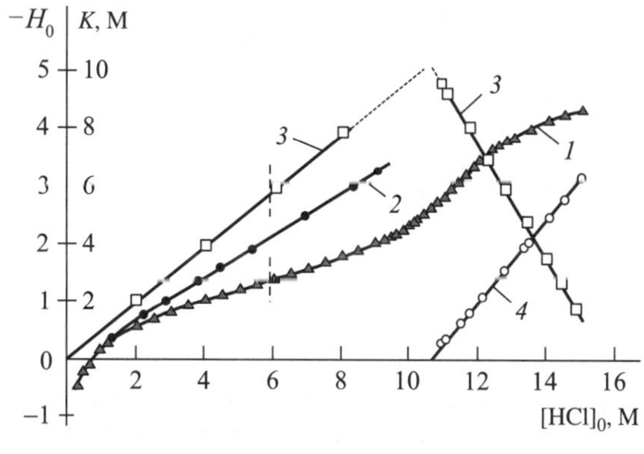

Figure 8.12 Plots of (1, 2) acidity function H_0 for HCl–DMF and HCl–H_2O systems, respectively, and (3, 4) concentration K of complexes K2 and K3, respectively, versus initial HCl concentration (reproduced with permission from editorial board of *Izv. Akad. Nauk, Ser. Khim.* journal © 1994 [131])

DMF–HCl system. It was found that the results of measuring H_0 as a function of $[HCl]_0$ at 25 and 39.5 °C almost coincided. The $H_0 = f[HCl]_0$ curves (Figure 8.12) obtained in the intervals of $[HCl]_0$ up to 1.6 M for water and DMF were also close. Above 2 M HCl, the acidity of HCl solutions in water was significantly higher than that of DMF solutions. The acidity of HCl in DMF solutions exhibited faster growth at $[HCl]_0 > 10.5$ M, which was related to an increase in the concentration of K3 complexes (Figure 8.12). In the indicated interval of [HCl], the concentration of K2 complexes passed through a maximum.

The addition of tetrachloroethane (TCE) in equimolar ratio to DMF makes the system homogeneous in the entire interval of HCl concentrations. In the presence of TCE, the acidity of solutions significantly

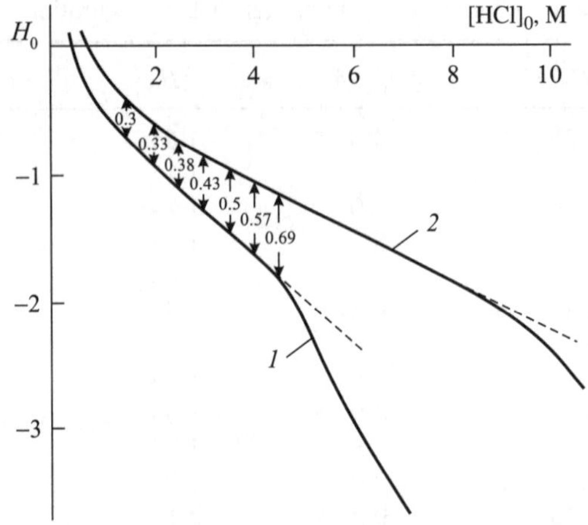

Figure 8.13 Plots of acidity function H_0 versus initial HCl concentration for (1) HCl–DMF–TCE and (2) HCl–DMF systems (reproduced with permission from editorial board of *Izv. Akad. Nauk, Ser. Khim.* journal © 1996 [132])

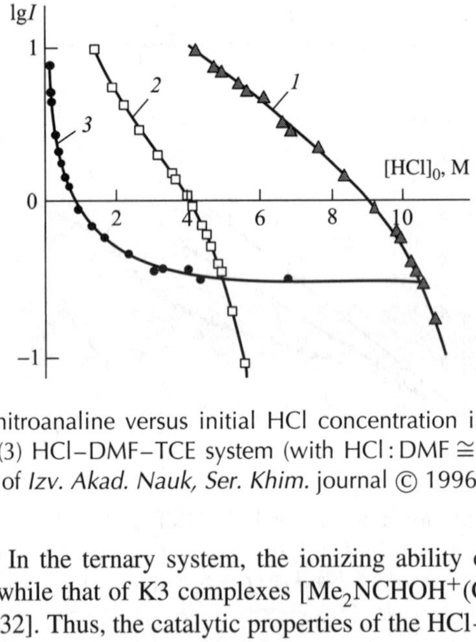

Figure 8.14 Plots of lg*I* of 2-nitroanaline versus initial HCl concentration in (1) DMF [131], (2) equimolar DMF–TCE mixture [132], and (3) HCl–DMF–TCE system (with HCl : DMF ≅ 1 : 1) at 25 °C (reproduced with permission from editorial board of *Izv. Akad. Nauk, Ser. Khim.* journal © 1996 [133])

increases (Figure 8.13) [132]. In the ternary system, the ionizing ability of K1 and K2 complexes with quasi-ion structure decreases, while that of K3 complexes [$Me_2NCHOH^+(ClHCl)^-$ ion pairs] increases as compared to the HCl–DMF [132]. Thus, the catalytic properties of the HCl–DMF system can be improved by selecting a proper solvent at the same HCl concentrations. Figure 8.14 shows the dependence of the indicator ratio $I = C_B/C_{BH^+}$ on $[HCl]_0$ for 2-nitroaniline. The behavior of lg*I* in HCl–DMF (curve *1*) [131] and HCl–DMF–TCE with equimolar ratio of components (curve *2*) [132] is significantly different. Curve *3*, which presents lg*I* values at various TCE concentrations and the same ratio of HCl : DMF = 1 (corresponding to K2 complex), clearly reveals that the ionizing ability of the medium (i.e., K2 complexes) increases with decreasing [TCE] until obtaining a two-component HCl–DMF system at $[HCl]_0 = 10.5$ M (intersection of curves *1* and *3*) [133].

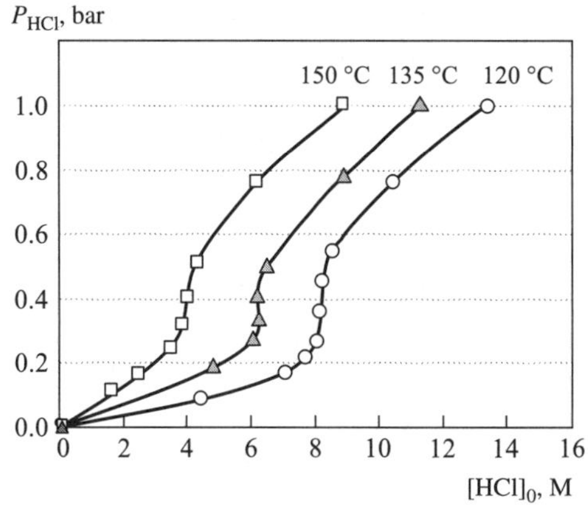

Figure 8.15 Plots of P_{HCl} versus initial HCl concentration in HCl–NMP system [111]

The HCl–NMP system at 25 °C is liquid within two narrow intervals of the acid concentration: 0–8 % HCl and 37.5–43.4 % HCl. For this reason, the interaction in this system were studied by the IR spectroscopy (FMIR) in equimolar HCl–NMP–TCE mixture at [TCE] = 62.83 wt %. Similar to DMF solutions, NMP also forms complexes K4 (2NMP · HCl), K5 (NMP · HCl), and K6 (NMP · 2HCl) [117] with close characteristics and analogous behavior. The melting temperature of K5 complex is about 88 °C [107]. The formation of these complexes is also manifested on the dependence of P_{HCl} on [HCl] in a TCE-free system that is homogeneous at 120–150 °C (Figure 8.15) [111].

Interesting information on the composition of the products of interaction of methanesulfonic acid CH_3SO_3H (MSA) with DMF [118], acetonitrile [120], DMSO [134] and ethyl acetate [135] and the influence of TCE on the properties of molecular and ion complexes of MSA with aprotic solvents was obtained from the results of IR spectroscopy measurements [119]. Investigations were also performed for the trifluoroacetic acid – acetonitrile [136] and MSA–H_2O [137, 138] systems.

The MSA–DMF system features the formation of 1 : 1 and 2 : 1 complexes in the interval of MSA concentrations within 0–67 mol %. Complexes of the 2 : 1 type also appear in concentrated MSA solutions (>67 mol %) [118] in the form of $SH^+A_2H^-$ ion pairs, where $A = CH_3SO_3^-$. Anions species of $[A\cdots H\cdots A]^-$ are also formed in the MSA–H_2O and MSA–DMSO systems, but the mechanisms of solvent ionization are different [118]. Indeed, in the case of H_2O and DMSO, cations with highly symmetric H-bonds ($H_5O_2^+$, $(DMSO)_2\ H^+$) are formed, while the ionization of DMF by MSA is accompanied by the complete proton transfer to amide oxygen. If the acid in solution is in excess, $SH^+A_2H^-$ complexes are transformed from the tight to solvate-separated IP type [118].

In the case of solvents possessing lower basicity than that of DMF (e.g., ethyl acetate [135], acetonitrile [120]), molecular complexes of the AH·S type are formed that do not belong to compounds with quasi-ionic H-bonds [139]. Acetonitrile is almost never protonated even in concentrated H_2SO_4 solutions [120]. At MSA concentrations in acetonitrile not exceeding 50 mol %, there exists a molecular complex

$$CH_3SO_2OH\cdots N{\equiv}CCH_3(K7).$$

At higher MSA concentrations, the K7 complex is solvated by the acid molecules that leads to the formation of associates (K8) with a structure different (IR data) from that of K7 complex.

The HCl–CH_3CN system has not been studied in much detail, but the available data [140] show that hydrogen halides (HCl, HBr) can form various compounds with nitriles, including complexes with immonium halide cations,

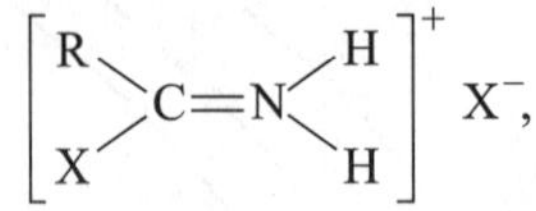

and condensation products with C–N and C–Br bonds. The reactions with HBr (and, probably, HCl) with the formation of stable adducts proceed very rapidly, so that free HBr content in acetonitrile cannot be measured. One of these adducts with a composition of $(CH_3CN)_2(HBr)_3$ was isolated and its structure was determined by X-ray diffraction (Figure 8.16) [140].

Measurements of the acidity functions in the MSA – ethyl acetate (EA) system [135] showed that this system possesses a much higher ability to protonate bases, which implies that molecular complex EA·MSA is a stronger acid than K2 and K3 complexes formed in DMF and is comparable in this respect to $H_5O_2{}^+$ ion [135]. In order to compare the three systems, their acidity functions were normalized to the same standard state (100 % MSA, $H_0 = -7.74$). Figure 8.17 shows the obtained curves of $H_0 = f[\mathrm{MSA}]_0$. In MSA solutions with DMF, 2-pyrrolidone, and ethyl acetate, the introduction of TCE retains the structure of ion pairs with AHA^- anions, while not changing the strength of H-bonds in molecular complexes [119].

Kislina et al. [141] also studied the effect of acid anions (MSA and H_2SO_4 in 2-pyrrolidone) and SH^+ counterions (MSA in DMF and 2-pyrrolidone) on the ionizing ability of $(AHA)^-$ ions with strong symmetric bonds. It was established that, under comparable conditions, the replacement of 2-pyrrolidone molecule by DMF in SH^+ cation leads to an insignificant decrease in the acidity. In H_2SO_4 solutions, $(AHA)^-$ ions possess a much greater ionizing ability than in MSA solutions.

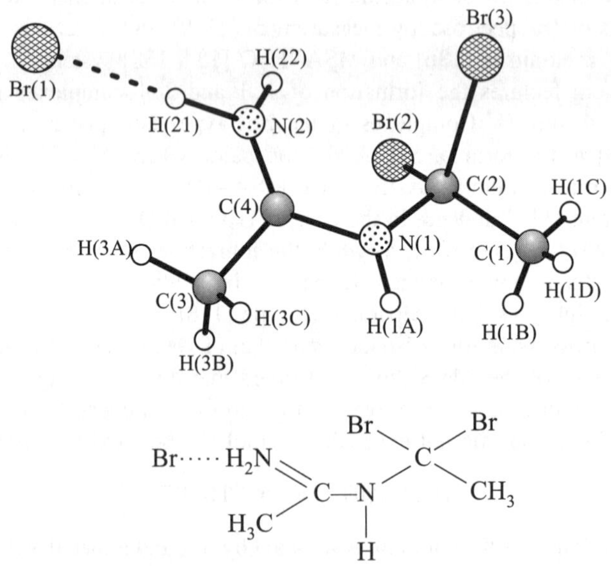

Figure 8.16 Structure of $(CH_3CN)_2(HBr)_3$ adduct (reproduced with permission from editorial board of *Izv. Akad. Nauk, Ser. Khim.* journal © 1998 [140])

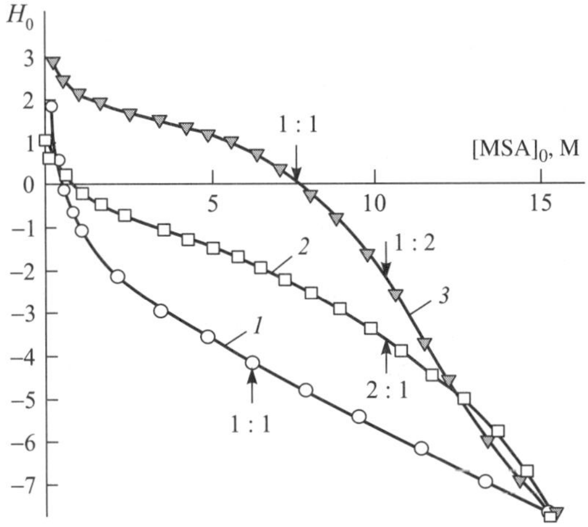

Figure 8.17 Plots of acidity functions H_0 versus [MSA] concentration in (1) ethyl acetate, (2) H_2O, and (3) DMF (reproduced with permission from editorial board of *Izv. Akad. Nauk, Ser. Khim.* journal © 1985 [137])

8.6.2 Kinetics of catalytic reactions in HCl–NMP, HCl–C_2H_5OH, and HCl–C_2H_5OH–CH_3CN systems

Investigation of the kinetics of acetylene hydrochlorination catalyzed by metal complexes in HCl – organic solvent (S) systems was originally performed by L. Gasparyan with co-workers for the HCl–DMF–CuCl and HCl–NMP–CuCl systems [107, 142]. Both these systems exhibited an interesting phenomenon, whereby the dissolution of CuCl in liquid phases containing S · 2HCl (according to the solution composition) was accompanied by an increase in the HCl pressure above the liquid. In the case of DMF solution, an increase in [CuCl] from 0 to 6.5 M leads to a growth in P_{HCl} from 0.7 to 2.0 bar, while in NMP solution, an change in [CuCl] from 0 to 8 M increases this pressure from 0.68 to 2.2 bar. Thus, every mole of CuCl displaces about one mole of HCl. Taking into account the nature of these complexes (containing 2HCl per mole solvent), the observed phenomenon can be explained by the displacement of HCl from HCl_2^- anion:

$$SH^+HCl_2^- + CuCl \longrightarrow SH^+CuCl_2^- + HCl\uparrow . \qquad (8.83)$$

At large concentrations of CuCl (>4 M), less than one mole of HCl is displaced per mole of dissolved CuCl, which is indicative of the possible formation of polynuclear anions such as $Cu_2Cl_3^-$. It was also established that an increase in the concentration of HCl in DMF significantly decreases the solubility of acetylene, but this fact was not taken into account in constructing kinetic models of the process.

The interest in nonaqueous system was significantly stimulated by the discovery of the catalysis of acetylene hydrochlorination reaction by Pt(II) complexes in absolute ethanol [107] and by Rh(III) and Ru(III) complexes in aprotic solvents.[2] Let us briefly summarize the results of investigations of the kinetics of acetylene hydrochlorination catalyzed by Rh(III) and Pt(II) complexes in nonaqueous media.

HCl–NMP–$RhCl_3$ system [111]. The kinetics of vinyl chloride (VC) synthesis was studied in a gradient-free flow reactor in a temperature interval of 110–180 °C, where the system was homogeneous. The

[2] Unpublished data of L.A. Gasparyan by private communication.

reaction rate exhibited a sharp maximum at $T = 170\,°C$. The main body of data was obtained at temperatures within 110–150 °C. The observed variable reaction order with respect to $P_{C_2H_2}$ at various P_{HCl} and T was considered as a result of the formation of π complex or β-chlorovinyl derivative of Rh(III):

$$[Rh] + C_2H_{2\ solution} \rightleftharpoons [Rh](C_2H_2). \tag{8.84}$$

The effect of [HCl] was studied in the region of existence of NMP · HCl and NMP · 2HCl complexes ($P_{HCl} = 0.2$–0.6 bar). In this interval (see Figure 8.15), the concentration of HCl changes is rather slightly (from 7.3 to 9 M) at 120 °C, but exhibits an almost two-fold growth (from 3 to 6 M) at 135 °C and (from 2 to 4 M) at 150 °C. The effect of [HCl] on the solubility of C_2H_2 was ignored, while the influence on the reaction rate was approximated by the following equation:

$$r_{VC} = \frac{aP_{HCl}}{1 + bP_{HCl}}. \tag{8.85}$$

The entire body of experimental results was satisfactorily described by the following equation

$$r_{VC} = \frac{kP_{C_2H_2}P_{HCl}[Rh]_\Sigma}{1 + k_1 P_{HCl}P_{C_2H_2}}, \tag{8.86}$$

which is valid at 100–150 °C with $k = (1.77 \pm 0.1) \cdot 10^4 \exp(-3000/T)$ $[h^{-1} \cdot bar^{-2}]$ and $k_1 = (0.7 \pm 0.1)\ \exp(1086/T)$ $[bar^{-2}]$. In order to interpret this equation, it is also necessary to obtain data for the effect of [HCl] on the solubility of acetylene and for the variation of NMP · HCl and NMP · 2HCl complex concentrations with increasing P_{HCl} at $T = 100$–$150\,°C$. The process of acetylene hydrochlorination takes place essentially in a melt of these hydrochlorides, i.e., in a typical ionic liquid. It is interesting to note that, by applying the HCl–NMP–$RhCl_3$ system onto activated carbon, it is possible to obtain supported liquid-phase catalysts NMP–$RhCl_3$/C [111, 143] characterized by high activity at a satisfactory stability, with turnover frequencies up to TOF = 7900 h^{-1} at 135 °C and 12 000 h^{-1} at 150 °C. In this "heterogenized" system called *supported liquid phase catalyst* (SLPC) [144]), the reaction rate in a differential reactor is described by the following equation [143]:

$$r_{VC} = \frac{k_5 P_{C_2H_2}P_{HCl}[Rh]_\Sigma}{1 + bP_{HCl}}. \tag{8.87}$$

It was rather surprising that, despite evident differences between Eqs (8.86) and (8.87), the dependences of TOF = $r_{VC}/[Rh]_\Sigma$ on the temperature for homogeneous catalysts and SLPCs turned out to be very close (with $T_{max} \approx 170\,°C$) [111].

HCl–C_2H_5OH–Pt(II) and HCl–C_2H_5OH–CH_3CN–Pt(II) systems. Investigations of the catalytic activity of $M[Pt(C_2H_4)Cl_3]$ (M = H, Na, K) complexes in the reaction of acetylene hydrochlorination in the HCl–H_2O and HCl–DMF systems [109, 145] showed that the catalysts exhibited maximum activity and stability in ethanol and an ethanol – acetonitrile (1 : 1, v/v.) mixture.

The solubility of acetylene in both systems at 40 °C obeys the Henry law, but significantly decreases with increasing [HCl] [52, 145]. At the same time, the solubility of HCl in both systems does not obey the Henry law. In order to describe the behavior of [HCl] = $f(P_{HCl})$, it was suggested that a monosolvate was formed that could be considered in concentrated HCl solutions as an ion pair of the SH^+Cl^- type. According to data of Pal'm and Kyrisaar [146], there was 1 mole of ethanol in solvate per mole HCl in solution in the entire range of HCl concentrations studied. Janz [147] reported on the isolation of a solid adduct with the composition C_2H_5OH · HCl at–65 °C. Then, the equilibrium dissolution of HCl in ethanol (system I) was described by the following equation

$$HCl_{gas} + C_2H_5OH \overset{K_1}{\rightleftharpoons} C_2H_5OH \cdot HCl \tag{8.88}$$

and the material balances with respect to C_{HCl} and $C_{C_2H_5OH}$ were written as

$$C^{\Sigma}_{HCl} = C_{C_2H_5OH \cdot HCl} = K_1 P_{HCl} C_{C_2H_5OH};$$

$$C^{\Sigma}_{C_2H_5OH} = C_{C_2H_5OH \cdot HCl} + C_{C_2H_5OH}.$$

From these equations, the dependence of C^{Σ}_{HCl} on P_{HCl} can be expressed as

$$C^{\Sigma}_{HCl} = \frac{K_1 P_{HCl}}{1 + K_1 P_{HCl}} C^{\Sigma}_{C_2H_5OH}, \tag{8.89}$$

where $K_1 = 3.24\ \text{bar}^{-1}$ at 40 °C.

The use of other solvates ($2C_2H_5OH \cdot HCl$ and $C_2H_5OH \cdot 2HCl$) for the description of the equilibrium dissolution of HCl did not improve the model expressed by Eq. (8.89) [145]. The dependence of $C_{C_2H_2}$ on C_{HCl} (P_{HCl}) was described using two hypothetical equilibria as follows:

$$C_2H_{2\ gas} + C_2H_5OH \overset{K_2}{\rightleftharpoons} \underset{(X_1)}{C_2H_5OH \cdot C_2H_2}; \tag{8.90}$$

$$C_2H_{2\ gas} + C_2H_5OH \cdot HCl \overset{K_3}{\rightleftharpoons} \underset{(X_2)}{C_2H_5OH \cdot HCl \cdot C_2H_2}. \tag{8.91}$$

An assumption that the contribution of complex X_2 to the balance with respect to alcohol was small led to the following equation

$$C^{\Sigma}_{C_2H_2} = \frac{(K_2 + K_1 K_3 P_{HCl}) P_{C_2H_2}}{1 + K_1 P_{HCl}} C^{\Sigma}_{C_2H_5OH} \tag{8.92}$$

that satisfactorily described the experiment for $K_2 = 3.56 \cdot 10^{-2}\ \text{bar}^{-1}$ and $K_3 = 1.4 \cdot 10^{-2}$ bar. Allowance for a dependence of the density of solutions on P_{HCl} as

$$\rho = 0.9478 P^{0.03}_{HCl}\ \text{(kg/liter)}, \tag{8.93}$$

led to the following dependences of $C^{\Sigma}_{C_2H_5OH}$, C^{Σ}_{HCl}, and $C^{\Sigma}_{C_2H_2}$ (mol/l) on P_{HCl} [52, 56, 145]:

$$C^{\Sigma}_{C_2H_5OH} = \frac{20.6(1 + K_1 P_{HCl}) P^{0.03}_{HCl}}{1 + 5.81 P_{HCl}}; \tag{8.94}$$

$$C^{\Sigma}_{HCl} = \frac{66.78 P^{0.03}_{HCl}}{1 + 5.81 P_{HCl}}; \tag{8.95}$$

$$C^{\Sigma}_{C_2H_2} = \frac{0.73 P^{0.03}_{HCl} P_{C_2H_2} (1 + 1.274 P_{HCl})}{1 + 5.81 P_{HCl}}. \tag{8.96}$$

Since the concentration of acetylene in solution exceeds C_{Pt} by more than one or even two orders of magnitude and $C^{\Sigma}_{HCl} \gg C_{Pt}$, it can be assumed that $C^{\Sigma}_{HCl} = C_{HCl}$ and $C^{\Sigma}_{C_2H_2} = C_{C_2H_5}$. Analogous dependences of $C_{HCl} = f(P_{HCl})$ and $C_{C_2H_2} = f(P_{HCl})$ were obtained in ethanol–acetonitrile solutions (system II) [52].

Investigations of the kinetic laws governing systems I and II showed that the reaction rate was described by an equation of the first order with respect to $[Pt(II)]_\Sigma$ ($HPt(C_2H_4)Cl_3$) and first order with respect to $P_{C_2H_2}$ in the entire interval of C_{HCl} (from 3.5 to 9 M). In the region where $C_{HCl} < 30$ wt %, the dependences on C_{HCl} were determined in a closed system because P_{HCl} in this region varied within about 10 Torr, while C_{HCl} in the course of kinetic experiments changed by about 2% (large excess of HCl).

The dependence of the reaction rate on C_{HCl} was determined with allowance for the first order with respect to $[Pt]_\Sigma$ and $C_{C_2H_2}$. As a result, the following equation was obtained for both systems I and II:

$$r_{VC} = kC_{Pt}C_{C_2H_2}C_{HCl}^3, \tag{8.97}$$

where (at 40 °C) $k = 5.0\,M^{-4}\,h^{-1}$ for system I and $k = 3.5\,M^{-4}\,h^{-1}$ for system II.

According to published data [126], the dependence of Hammett acidity h_0 on C_{HCl} in absolute ethanol is satisfactorily described by the following equation:

$$h_0 = BC_{HCl}^3. \tag{8.98}$$

Then, the reaction rate in system I can be expressed as

$$r_{VC} = kC_{Pt}C_{C_2H_2}h_0. \tag{8.99}$$

With this approach, the transition state in a rate-limiting step certainly includes a Pt(II) complex with acetylene and one proton, while there are presently no grounds to make allowance for the participation of associates $(HCl)_n$ involving two or three molecules. Thus, the best description of processes in system I is provided by the following empirical equation [which is analogous to kinetic Eq. (8.87) for the NMP–$RhCl_3$/C system]:

$$r_{VC} = \frac{kC_{Pt}C_{C_2H_2}C_{HCl}}{1 + KP_{HCl}}, \tag{8.100}$$

where (at 40 °C) $k = 5.9\,h^{-1}\,kPa^{-2}$, $K = 3.12\,KPa^{-1}$ [52, 56].

System II, in which the activity of a catalyst remains unchanged for about 40 h, is characterized by a much higher stability of operation than system I. Judging from the obtained results, the role of CH_3CN in the stabilization of a catalyst is related to the drying of solution from traces of water formed in the C_2H_5OH–HCl system. The acid catalysis of CH_3CN hydrolysis is accompanied by the formation of small amounts of NH_4Cl (precipitate) and CH_3COOH.

The analysis carried out in this section gives hopes that the kinetic investigations of metal complex catalysts operating in complicated systems based on concentrated solutions of strong acids in nonaqueous solvents become accessible and give promising results. Definite achievements in understanding the mechanisms of acid catalysis and the catalysis with metal complexes in these systems would be ensured by establishing a relationship between the nature of acid solvates and their reactivity in various processes.

At a ratio of HA : S = 2, systems HCl–DMF, HCl–NMP, CH_3SO_3H–DMF, and CH_3SO_3H–NMP considered in this section represent protonic ionic liquids (melts of $SH^+HA_2^-$ salts). Of considerable interest to both researchers and technologists are aprotic ionic liquids that constitute a new class of solvents potentially capable of replacing organic solvents in both basic synthetic chemistry and the chemical industry.

Specific features of the kinetics of catalytic reactions in ionic liquids (ILs) and adequate approaches to the investigation and description of reactions in these media in the catalysis with metal complexes are still not completely clear. At the same time, evidently good prospects make it expedient to provide a brief description of the specific features and properties of ILs and the examples of catalytic processes in these unique solvents.

8.7 Ionic liquids in catalytic chemistry

Salts with organic (in particular, organoelemental) cations, possessing melting temperatures below 100 °C, are called ionic liquids (ILs). The lower limit of the melting temperature known presently is −96 °C. Examples of cations encountered in ILs are as follows: [148–152]:

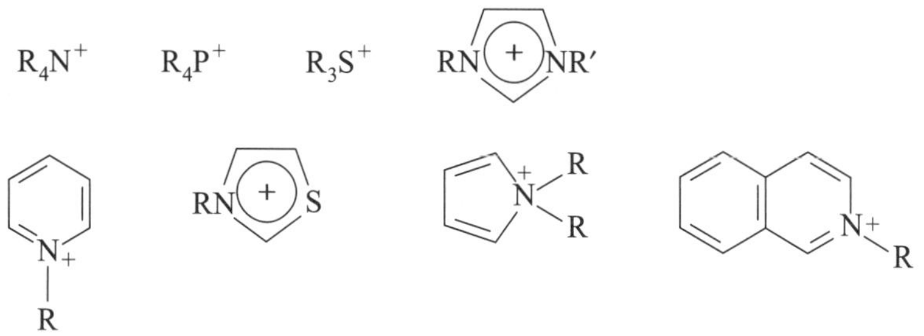

The possible anions for ILs are

Cl^-, $AlCl_4^-(Al_2Cl_7^-)$, PF_6^-, BF_4^-, NO_3^-, SO_4^{2-}, $CF_3SO_3^-$(OTf), SbF_6^-, $ROSO_3^-$, $SnCl_3^-$, $ZnCl_3^-$ [8.152].

ILs of the first generation (appearing at the beginning of the 1960s) contained quaternary salts R_4NCl and $AlCl_3$ or $R_4N^+AlCl_4^-$ and $R_4N^+Al_2Cl_7^-$ [152]. Later (at the beginning of 1990), ILs stable with respect to O_2 and moisture were synthesized based on dialkylimdazolium cations and BF_4^- and PF_6^- anions. More recently, stable salts with $(CF_3SO_2)_2N$ and $(CF_3SO_2)_3C$ anions were obtained. The stability of anions with respect to oxidation (including anodic) increases in the following order:

$$Cl^-,\ F^-,\ Br^- < AlCl_4^-,\ Al_2Cl_7^- \le BF_4^-,$$
$$PF_6^- \le CF_3SO_3^-, (CF_3SO_2)_2N^-,\ (CF_3SO_2)_3C^-.$$

The results of using spectroscopic and diffraction methods for the investigation of ILs showed that the structures of these salts in the liquid and solid phases are very close [152, p. 127]. Some ILs with long alkyl groups (C_n with $n \ge 12$) exhibit mesomorphism of the liquid-crystal type and exist either as true ILs or as mesophases of the smectic and other types.

The ILs are polar solvents with the polarity characteristics on an intermediate level between water and chloro-organic solvents. The polarity of ILs is close to that of lower alcohols and aprotic solvents (DMF, DMSO). There were attempts to apply the Dimroth–Reichardt $E_T(30)$ criterion for evaluating the polarity of ILs [149]. It was pointed put that the polar properties of ILs significantly depend on the nature of a dissolved substance [153]. In contrast to typical polar solvents, ILs are not coordinated to metal ions and complexes. In systems with coordinating anions (Hal^-, $SnCl_3^-$), metal complexes are localized in the anion (such as $AlCl_3$).

The melting temperature (T_m) of ILs is determined by the length of alkyl groups in the cation and the nature of the cation and anion. In the case of chloroaluminates, T_m depends on the molar fraction of $AlCl_3$ in the R^+Cl^-–$AlCl_3$ system. For example, in the case of 1-ethyl-3-methylinidazolium chloride, $[emim]^+Cl^-$, the melting temperature varies, depending on the molar fraction of $AlCl_3$, from −96 to

200 °C. The T_m of $[emim]^+ X^-$ with the series of anions decreases from 87 °C to −14 °C in the following order [149]:

$$Cl^- > PF_6^- > NO_2^- > NO_3^- > AlCl_4^- \geq BF_4^- > NTf_2^- > CF_3SO_3^- > CF_3CO_2^-.$$

because the process of IL solidification (crystallization or glass formation) is very slow and is accompanied by the appearance of supercooled liquid phases [152].

ILs are characterized by the absence of measurable vapor pressure (i.e. that above 10^{-2} Torr). The absence of any significant volatility of an IL solvent is one of the most characteristic and positive features of ILs, especially in technological stages of reaction product isolation and catalyst recycling. Although aprotic ILs are significantly more stable than proton-containing ones ($R_3NH^+X^-$, $R_3PH^+X^-$), the upper limit of their thermal stability is commonly accepted to be about 150 °C, although there are examples of higher decomposition temperatures in particular for $[emim]BF_4$ (~300 °C) and $[emim][(CF_3SO_2)_2N]$ (above 400 °C) [149]. ILs possess not high densities (1.15–1.4 kg/cm^3), but rather high viscosities (14–373 cP) that are 10–100 times greater than the viscosities of traditional solvents (viscosity of ethylene glycol is 16.1 cP).

The choice of anion is a key factor that determines the solubility of water, ether, and other organic solvents in ILs. The mutual solubility (miscibility) of ILs and other solvents varies from completely miscible to almost immiscible hydrophobic systems. For example, the content of water in IL with butylimidazolium cation and different anions varies as follows:

$[bmim]^+$	Content, wt %
$[bmim][NTf_2]$	1.4
$[bmim][PF_6]$	1.17
$[bmim][BF_4]$	Complete mixing
$[bmim][CF_3SO_3]$	Complete mixing

An increase in the alkyl length of R_4N^+ can produce a significant (1000-fold) increase in the solubility of nonpolar substances (e.g., 1-octene) in ILs [149, 152]. ILs can dissolve fullerenes C_{60} and numerous polymers, but not alkanes and nonpolar organic solvents. Aromatic compounds are dissolved somewhat better. For example, the solubility of benzene in chloroaluminate ILs reaches up to 50 vol %. A linear correlation was established between the logarithmic distribution constants of organic substances in the IL–H_2O and octanol – H_2O systems for IL = $[bmim]PF_6$ [152].

The solubility of gases in ILs has not yet been sufficiently studied, but some data indicate [150] that the solubility of gases in $[bmim]PF_6$ decreases in the following order:

$$CO_2 > C_2H_4 > C_2H_6 > CH_4 > O_2.$$

It was established that the solubility of gases (such as O_2, CO, H_2) in ILs is much lower than in traditional solvents. For this reason, mass transfer plays a significant role in the heterophase processes in ILs [152]. It should be noted that CO_2 in a supercritical state (sc-CO_2) is very well dissolved in $[bmim][PF_6]$. At 80 bar CO_2 in a two-phase sc-CO_2/$[bmim][PF_6]$ system, the IL contains up to 60 vol % of dissolved CO_2 [149].

Chloroaluminate ILs are capable of exhibiting both base and acid properties. The basicity is manifested at a molar fraction of $AlCl_3$ below 0.5. The system is neutral at $x_{AlCl_3} = 0.5$, and the acid properties (Lewis' acidity) are manifested at $x_{AlCl_3} > 0.5$, where $Al_2Cl_7^-$ and $Al_3Cl_{10}^-$ anions appear. The acid properties are manifested in the ability of ILs to catalyze the reactions of acylation and alkylation of

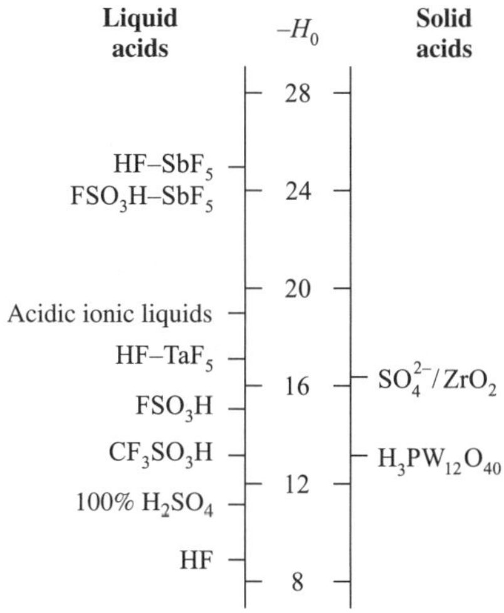

Figure 8.18 Acidity scale for liquid and solid acids (reproduced with permission from Wiley-VCH Verlag, GmbH & Co. KGaA © 2000 [149])

aromatic compounds according to the Friedel–Crafts reaction. Here, it should be noted that small amounts of water (400–4500 ppm) that are present even in dried ILs can provide conditions for the manifestation of a protonic acidity. A protonic acid in IL can possess a high ionizing ability. It was established [149, 154] that the dissolution of a gaseous HCl in acidic [emim]Cl/AlCl$_3$ ionic liquid leads to a superacid system, in which the acidity is about seven orders of magnitude higher than that of 100 % H_2SO_4 and triflic acid ($H_0 \cong -18.6$) (Figure 8.18). It was suggested that this system features an exceptionally weakly solvated proton ("bare" proton in the ion pair) with a high reactivity that is formed via the reaction

$$[\text{bmim}]^+\text{Al}_2\text{Cl}_7^- + \text{HCl}_{\text{gas}} \longrightarrow [\text{bmim}]^+\text{AlCl}_4^- + \text{H}^+[\text{AlCl}_4^-].$$

Evidently, these IL-based superacids are much safer and more convenient for use in catalytic processes such as cracking and isomerization of paraffins. Chloroaluminate ILs even without HCl catalyze the cracking of polyethylene at 90 °C [148] with the formation of lower paraffins and cycloalkanes (and without the formation of aromatics and alkenes) and the dimerization and oligomerization of olefins.

The results of neutron diffraction in the case of a binary [emim]Cl–HCl mixture showed the existence of HCl_2^- species with asymmetric H-bonds, which interacted with C–H-group hydrogen of imidazolium cation [152]. The "neutral" ILs based on dialkylimidazolium cation and BF_4^-, PF_6^- anions are ideal solvents for metal complex catalysts. Organic salts with Cl^- anion dissolve most transition and nontransition metal halides with the formation of metallate anions such as [emim]$_2$PdCl$_4$ and [emim]AuCl$_4$. Many ionic and neutral complexes are also effectively dissolved, the latter due to the process of complex formation with IL components [155]. It was noted that imidazolium cations can be deprotonated in the presence of platinum-group metals with the formation of carbene ligands,

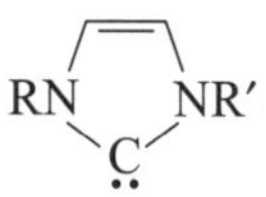

the complexes of which exhibit unusual catalytic properties [149]. It is also possible that alkylimidazolium cations would be dealkylated with the formation of N–M bonds. For example, [bmin]$_2$PdCl$_4^{2-}$ in

[bmim]BF_4 in the presence of H_2O converts into a complex

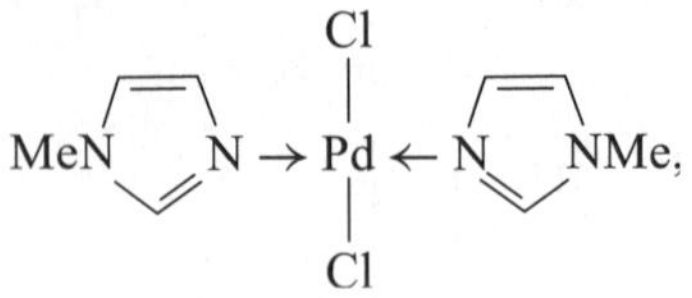

that is active in the hydrodimerization of 1,3-butadiene [149].

The best immobilization of metal complexes ML_n in ILs (i.e., the maximum retaining ability of ILs with respect to the metal catalyst passage to the organic phase) is achieved using ionic forms of complexes such as $RhL(PPh_3)$ PF_6, $RhL_2(BF_4)_2$, and $H_4Ru_4(ArH)_4(BF_4)_2$ [152], or salt-like ligands such as monosulfated or trisulfated phosphines [75, 156], $Co(\eta^5\text{-}C_5H_4PPh_2)_2{}^+ PF_6{}^-$, etc. [157–160].

Promising ligands are offered by the phosphine derivatives of imidazolium [161–163]

as well as phosphine derivatives of quaternary ammonium salts such as [164]

$$(Bu^t)_2PCH_2CH_2N^+Me_3Cl^-$$

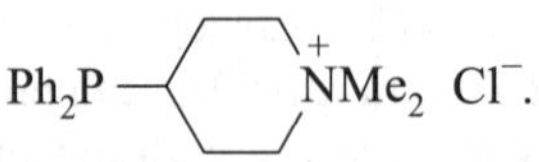

Apparently, the first attempt to use ILs in the catalysis with metal complexes was undertaken by Parshal [155] with respect to the hydrogenation and hydroformylation of alkenes. A $PtCl_2(PR_3)_2$ catalyst complex was dissolved in $Et_4N^+Sn_2Cl_5{}^-$ melt ($T_m = 78\,°C$) known to satisfactorily dissolve olefins. Probably, the resulting complex $(PR_3)_2PtCl(SnCl_3)$ [149] was dissolved in the melt of initial salt and, on cooling, a two-phase system appeared that consisted of a solid salt and a liquid phase of organic products.

Numerous examples of catalysis with transition metal complexes in ILs have been presented in reviews [149, 152, 164]. The investigated reactions include: hydrogenation (including asymmetric catalysis); oxidation by oxygen, hydrogen peroxide, and alkyl hydroperoxides; hydroformylation (Pt, Ru, Co, Rh); hydroalkoxycarbonylation, carbonylation of arylhalides; reactions of Heck, Suzuki, Stille, and Glyazer–Zalkind; olefin dimerization (Difasol-process, IFP); oligomerization; hydrodimerization; telomerization; nucleophilic allyl substitution (Trost–Tsuji), diene ring-closing metathesis, and other.

A low coordinating ability of ILs makes possible the coexistence of cationic active centers with coordination vacancies, which is extremely important for the catalysis. This possibility was demonstrated in the investigation of ethylene copolymerization with CO in $(bipy)Pd(CH_3)^+SbF_6{}^-$ complex solutions in IL ($C_6Py^+NTf_2{}^-$) [165, 166].

Using the functionalization of alkyl groups in imidazolium cations, it is possible to obtain modified ILs intended to perform special tasks. For this purpose, alcohol-, acid-, and NH_2 groups can be introduced. This direction has been extensively developed in applications of ILs (see, e.g., [167]).

Although, as noted above, the kinetic investigations of processes in ILs have not yet received proper attention, there is some research activity. For example, it was reported that the kinetics of oxidative addition of CH_3I to Vaska's complex and ligand substitution in Rh(I) complexes were studied in IL ($C_6Py^+NTf_2{}^-$) [168].

In concluding this section, it should be emphasized that especially interesting technological possibilities were revealed by combining supercritical CO_2 fluid (sc-CO_2) and IL for the homogeneous catalytic hydrogenation and hydroformylation of alkenes [152]. In this system, sc-CO_2 dissolves in IL and favors the dissolution of H_2, CO, and alkene. The reaction products pass to the sc-CO_2 phase in the two-phase sc-CO_2–IL system, leaving the catalyst in the IL phase. The use of supercritical fluids (CO_2, H_2O, alkanes) in the homogeneous catalysis with metal complexes and enzymatic catalysis demonstrated interesting possibilities of controlling the activity of catalysts and selectivity of reactions, besides evident technological and ecological advantages of these new catalytic systems [152b, 169–175].

References

1. R. Aris, *Introduction to the Analysis of Chemical Reactions*, Prentice-Hall, Englewood Cliffs, NJ, 1965.
2. N. N. Lebedev, M. N. Manakov and V. F. Shvets, *Teoriya khimicheskikh protsessov osnovnogo organicheskogo i neftekhimicheskogo sinteza* (Theory of Chemical Processes of Basic Organic and Petrochemical Synthesis), Khimiya, Moscow, 1984.
3. G. V. Odabash'yan and V. F. Shvets, *Laboratornyi praktikum po khimii i tekhnologii osnovnogo organicheskogo i neftekhimicheskogo sinteza* (Laboratory Guide in Chemistry and Technology of Basic Organic and Petrochemical Synthesis), Khimiya, Moscow, 1992.
4. G. Astarita, *Mass Transfer with Chemical Reaction*, Elsevier Publ. Co., Amsterdam, 1967; Dzh. Astarita, *Massoperedacha s khimicheskoi reaktsiei*, Russian translation ed. by L. A. Serafimov (additional chapter by L. A. Serafimov and M. I. Balashov), Khimiya, Moscow, 1971.
5. E. S. Rudakov, *Reaktsii alkanov s okislitelyami, metallokompleksami i radikalami v rastvorakh* (Reactions of Alkanes with Oxidants, Metal Complexes, and Radicals in Solutions), Naukova Dumka, Kiev, 1985.
6. E. T. Denisov, O. M. Sarkisov and G. I. Likhtenshtein, *Khimicheskaya kinetika* (Chemical Kinetics), Khimiya, Moscow, 2000.
7. L. P. Hammett, *Physical Organic Chemistry*, 2nd ed., McGraw-Hill, New York, 1970.
8. E. S. Amis, *Solvent Effects on Reaction Rates and Mechanisms*, Academic Press, New York, 1966.
9. S. G. Entelis and R. P. Tiger, *Kinetika reaktsii v zhidkoi faze* (Kinetics of Liquid-Phase Reactions), Khimiya, Moscow, 1973.
10. E. A. Moelwyn-Hughes, *The Chemical Statics and Kinetics of Solutions*, Academic Press, New York, 1971.
11. O. N. Temkin, *Kinetika i mekhanizm kataliticheskikh prevrashchenii atsetilena v rastvorakh kompleksov metallov* (Kinetics and Mechanism of Catalytic Transformations of Acetylene in Solutions of Metal Complexes), Doctoral Dissertation in Chemistry, MITKhT (Moscow Institute of Fine Chemical Technology), Moscow, 1972.
12. O. N. Temkin, *Kinetika kataliticheskikh reaktsii v rastvorakh kompleksov metallov* (Kinetics of Catalytic Reactions in Metal Complex Solutions), MITKhT (Moscow Institute of Fine Chemical Technology), Moscow, 1980.
13. R. Schmid and V. N. Sapunov, *Non-Formal Kinetics*, Verlag-Chemie, Weinheim, 1982.
14. R. R. Dogonadze, *Kvantovaya teoriya khimicheskikh reaktsii v polyarnoi zhidkosti* (Quantum Theory of Chemical Reactions in Polar Liquids), Znanie, Moscow, 1973.
15. R. R. Dogonadze and A. M. Kuznetsov, Kinetics of Chemical Reactions in Polar Media, in *Itogi Nauki i Tekhniki, Ser. Fiz. Khim., Kinetika* (Advances in Science and Technology, Ser., Physical Chemistry: Kinetics), VINITI, Moscow, 1973, Vol. 2.
16. F. Ya. Kul'ba and V. E. Mironov, *Khimiya talliya* (Chemistry of Thallium), Goskhimizdat, Moscow, 1963.
17. A. V. Storonkin and L. S. Lilich (Eds), *Khimiya i termodinamika rastvorov* (Chemistry and Thermodynamics of Solutions), Izdat. MGU, Moscow, 1964.
18. V. I. Belevantsev and B. I. Peshchevitskii, *Issledovanie slozhnykh ravnovesii v rastvore* (Investigation of Complex Equilibria in Solutions), Nauka, Novosibirsk, 1978.
19. J. E. Gordon, *The Organic Chemistry of Electrolyte Solutions*, Wiley-Interscience, New York, 1975.
20. Yu. Ya. Fialkov, A. N. Zhitomirskii and Yu. A. Tarasenko, *Fizicheskaya khimiya nevodnykh rastvorov* (Physical Chemistry of Non-Aqueous Solutions], Khimiya, Leningrad, 1973.
21. G. A. Krestov, *Termodinamika ionnykh protsessov v rastvorakh* (Thermodynamics of Ion Processes in Solutions), Khimiya, Leningrad, 1984.
22. T. I. Mikulin (Ed.), *Voprosy fizicheskoi khimii rastvorov elektrolitov* (Problems in Physical Chemistry of Electrolyte Solutions), Khimiya, Leningrad, 1968.

23. M. Szwarc, *Carbanions: Living Polymers and Electron Transfer Processes*, Wiley-Interscience, New York, 1968.
24. M. Szwarc (Ed.), *Ions and Ion Pairs in Organic Reactions*, Wiley-Interscience, New York, 1972.
25. G. A. Krestov, V. N. Afanas'ev, A. V. Agafonov et al. (Eds), *Kompleksoobrazovanie v nevodnykh sredakh*, Series: *Problemy khimii rastvorov* (Complex Formation in Non-Aqueous Media, Series: Problems in Chemistry of Solutions), Nauka, Moscow, 1989.
26. K. Burger, *Solvation, Ionic and Complex Formation Reactions in Non-Aqueous Solvents: Experimental Methods for their Investigation*, Elsevier, Amsterdam, 1983.
27. Yu. Ya. Fialkov, *Rastvoritel' kak sredstvo upravleniya khimicheskim protsessom* (Solvent as a Means of Controlling Chemical Process), Khimiya, Leningrad, 1990.
28. V. E. Mironov and I. D. Isaev, *Vvedenie v khimiyu vneshnesfernykh kompleksnykh soedinenii metallov v rastvorakh* (Introduction to Chemistry of Outer-Sphere Complex Compounds of Metals in Solutions), KrGU, Krasnoyarsk, 1986.
29. A. Loupy and B. Tchoubar, *Effets de Sels en Chimie Organique et Organometallique*, Dunod–Bordas, Paris, 1988.
30. C. Reichardt, *Solvents and Solvent Effects in Organic Chemistry*, 2nd ed., VCH, Weinheim, 1988.
31. A. M. Kutepov (Ed.), *Dostizheniya i problemy teorii sol'vatatsii: Strukturno-termodinamicheskie aspekty, Ser. Problemy khimii rastvorov* (Advances and Problems in the Theory of Solvation: Structural and Dynamic Aspects, Series: Problems in Chemistry of Solutions), Nauka, Moscow, 1998.
32. K. Schwabe, *Trudy 2 Vsesoyuznoi konferentsii po teorii rastvorov* (Proceedings of the 2nd All-Union Conf. on the Theory of Solutions (Alma-Ata, 1968)). Nauka, Alma-Ata, 1971, p. 19.
33. R. M. Flid and I. I. Moiseev, *Zh. Prikl. Khim.*, **27**, 1145 (1954).
34. I. I. Moiseev, *Issledovanie v oblasti mekhanizma zhidkofaznoi gidratatsii atsetilena* (Investigation into Mechanism and Kinetics of Liquid-Phase Hydration of Acetylene), Candidate Dissertation in Chemistry, MITKhT (Moscow Institute of Fine Chemical Technology), Moscow, 1958.
35. F. A. Long and D. McInture, *J. Am. Chem. Soc.*, **76**, 3240, 3243 (1954).
36. M. A. Paul, *J. Am. Chem. Soc.*, **76**, 3236 (1954).
37. F. A. Long and M. A. Paul, *Chem. Rev.*, **57**, 935 (1957).
38. J. Leden, *Z. Phys. Chem. A*, **188**, 160 (1941).
39. G. Biderman and L. G. Sillen, *Arkiv Kemi*, **5**, 425 (1953).
40. L. P. Safonov, A. M. Kolker and A. N. Kinchin, in: *Dostizheniya i problemy teorii sol'vatatsii: Strukturno-termodinamicheskie aspekty, Ser. Problemy khimii rastvorov* (Advances and Problems in Theory of Solvation: Structural and Dynamic Aspects, Series: Problems in Chemistry of Solutions), A. M. Kutepov (Ed.), Nauka, Moscow, 1998, pp. 63–105.
41. A. B. Fasman, V. A. Golodov and L. M. Pustyl'nikov, *Izv. Sib. Otd. Akad. Nauk SSSR*, **36**, 144 (1968).
42. G. Majed and K. E. Hawlett, *J. Chem. Soc.* A., 679 (1968).
43. N. F. Alekseeva, *Issledovaniya v oblasti kataliticheskogo gidrokhlorirovaniya atsetilena v solyanokislykh rastvorakh khlornoi rtuti* (Investigations in the Field of Catalytic Hydrochlorination of Acetylene in Hydrochloric Solutions of Mercuric Chloride), Candidate Dissertation in Chemistry, MITKhT (Moscow Institute of Fine Chemical Technology), Moscow, 1971.
44. N. F. Alekseeva, O. N. Temkin and R. M. Flid, *Kinet. Katal.*, **11** (6), p. 1580 (1970).
45. N. F. Alekseeva, O. N. Temkin and R. M. Flid, *Zh. Fiz. Khim.*, **45**, 74 (1971).
46. N. F. Alekseeva, O. N. Temkin and R. M. Flid, *Kinet. Katal.*, **11** (4), 1071 (1970).
47. O. N. Temkin, N. F. Alekseeva, R. M. Flid, O. L. Kaliya, G. K. Shestakov and T. I. Dolgina, *Dokl. Akad. Nauk SSSR*, **196**, 836 (1971).
48. P. V. Danckwerts, *Gas-Liquid Reactions*, McGraw Hill, New York, 1970.
49. O. N. Temkin, G. K. Shestakov and Yu. A. Treger, *Atsetilen: khimiya, mekhanizmy reaktsii, tekhnologiya* (Acetylene: Chemistry, Reaction Mechanisms, Technology), Khimiya, Moscow, 1991.
50. M. T. Beck, *Chemistry of Complex Equilibria*, Van Nostrand Reinhold, London, 1970.
51. D. V. Sokol'skii, Ya. A. Dorfman and T. L. Rakitskaya, *Protonno-aprotonnyi kataliz*, (Protonic–Aprotic Catalysis), Nauka, Alma-Ata, 1975.

52. L. A. Sil'chenko, S. A. Panova, G. K. Shestakov and O. N. Temkin, *Kinet. Katal.*, **39** (1), 28 (1998).
53. E. S. Rudakov, A. I. Lutsyk and S. Yu. Suikov, *Zh. Fiz. Khim.*, **61** (5), 1153 (1987).
54. R. M. Flid, 1968, unpublished results [private communication].
55. W. L. Marshall, *J. Phys. Chem.*, **74**, 346 (1970).
56. D. Kamenski, S. D. Dimitrov, L. A. Sil'chenko, G. K. Shestakov and O. N. Temkin, *Appl. Catal.*, **67**, 159 (1990).
57. R. G. Makitra, Ya. N. Pirig and T. I. Politanskaya, *Zh. Obshch. Khim.*, **1**, 54 (1981).
58. E. Wilhelm and R. Battino, *Chem. Rev.*, **73**, 1 (1973).
59. M. J. Kamlet, R. M. Doherty, G. R. Famini and R. W. Taft, *Acta Chem. Scand., Ser. B*, **41**, 589 (1987).
60. Y. Xu, G. Liu and Y. Hu, *J. Chem. Ind. Eng. (China)*, **3** (2), 163 (1988).
61. (a) E. S. Rudakov, *Reaktsionnaya sposobnost' organicheskikh soedinenii* (Reactivity of Organic Compounds), Tartu State Univ., Tartu, **7** (3), 779 (1970); (b) V. V. Zamashchikov, S. L. Litvinenko and E. S. Rudakov, *Reaktsionnaya sposobnost' organicheskikh soedinenii* (Reactivity of Organic Compounds), Tartu State Univ., Tartu, **16** (1), 95 (1979).
62. E. S. Rudakov, *Dokl. Akad. Nauk SSSR*, **275** (3), 666 (1984).
63. A. J. Parker, *Chem. Rev.*, **69**, 11 (969).
64. A. J. Parker, U. Mayer, R. Schmid and V. Gutmann, *J. Org. Chem.*, **43**, 1843 (1978).
65. E. S. Rudakov, A. P. Yaroshenko and V. V. Zamashchikov, *React., Kinet. Catal. Lett.*, **25** (3–4), 207 (1984).
66. G. V. Yukhnevich, E. G. Tarakanova, V. D. Maiorova and N. B. Librovich, *Usp. Khim.*, **64** (10), 963 (1995).
67. L. B. Librovich and M. I. Vinnik, *Dokl. Akad. Nauk SSSR*, **166**, 647 (1966).
68. M. I. Vinnik, *Issledovanie kineticheskikh zakonomernostei khimicheskikh protsessov v vodnykh rastvorakh sil'nykh kislot i osnovanii* (Studying Kinetic Laws of Chemical Processes in Aqueous Solutions of Strong Acids and Bases), Doctoral Dissertation in Chemistry, Inst. Chem. Phys., USSR Akad. Sci., Moscow, 1966.
69. I. V. Bozhko, *Gomogenno-kataliticheskoe godrokhlorirovanie vinilatsetilena i khloroprena* (Homogeneous Catalytic Hydrochlorination of Vinylacetylene and Chloroprene), Candidate Dissertation in Chemistry, MITKhT (Moscow Institute of Fine Chemical Technology), Moscow, 1989.
70. G. A. Al'per and M. Yu. Nikiforov, in *Dostizheniya i problemy teorii sol'vatatsii: Strukturno-termodinamicheskie aspekty, Ser. Problemy khimii rastvorov* (Advances and Problems in Theory of Solvation: Structural and Dynamic Aspects, Series: Problems in Chemistry of Solutions), A. M. Kutepov (Ed.), Nauka, Moscow, 1998.
71. (a) G. G. Malenkov, *Zh. Strukt. Khim.*, **47** (Suppl), S5 (2006); (b) O. Ya. Samoilov, *Struktura vodnykh rastvorov elektrolitov i sol'vatatsiya ionov* (Structure of Aqueous Electrolyte Solutions and Solvation of Ions), Izdat. Akad. Nauk SSSR, Moscow, 1957.
72. V. K. Abrosimov and Al. G. Krestov, in *Dostizheniya i problemy teorii sol'vatatsii: Strukturno-termodinamicheskie aspekty, Ser. Problemy khimii rastvorov* (Advances and Problems in Theory of Solvation: Structural and Dynamic Aspects, Series: Problems in Chemistry of Solutions), A. M. Kutepov (Ed.), Nauka, Moscow, 1998, p. 5.
73. A. N. Glebov, A. R. Budanov, *Soros. Obrazov. Zh.*, **9**, 72 (1996).
74. A. E. Eid and C. F. Wells, *J. Chem. Soc., Farad. Trans.*, **81**, 1401 (1985).
75. H. W. Bohnen and B. Cornils, *Adv. Catal.*, **42**, 1 (2002).
76. Patent EP 035082V1 (1989), Union Carbide Co.
77. P. Bhattacharyya, B. Croxtall, J. Fawcett et al., *J. Fluor. Chem.*, **101**, 247 (2000).
78. D. F. Foster, D. J. Adams, D. Gudmansen et al., *J. Chem. Soc., Chem. Comm.*, 722 (2002).
79. E. Perperi, Y. Huang, G. Manos et al., *J. Chem. Soc., Dalton Trans.*, 2062 (2004).
80. Y. Wang, C. Feng, J. Jiang and Z. Jin, *Proceedings of the 14th Int. Symp. on Homogeneous Catalysis (Munich, Germany, 2004)*, Book of Abstracts, p. 251.
81. C. Fangewisch and A. Behr, *Chem. Eng. Technology*, **25**, 143 (2002).
82. A. Behr, D. Obst, C. Schutte and T. Schosser, *J. Mol. Catal.* A, **206**, 179 (2003).
83. M. I. Usanovich, *Issledovaniya v oblasti teorii rastvorov i teorii kislot i osnovanii* (Investigations into the Theory of Solutions, Acids, and Bases), Nauka, Alma-Ata, 1970.
84. T. N. Sumarokova, *Zh. Obshch. Khim.*, **61** (6), 1385 (1991).
85. T. N. Sumarokova, *Zh. Obshch. Khim.*, **59** (7), 1484 (1989).

86. T. N. Sumarokova and E. N. Berger, *Zh. Prikl. Khim*., **70** (9), 1432 (1997).
87. G. A. Krestov, *Pure Appl. Chem*., **59** (9), 1203 (1987).
88. A. R. Gantmakher and R. V. Basova, in *Metody elementoorganicheskoi khimii: Li, K, Na, Rb i Cs* (Methods of Organoelemental Chemistry: Li, K, Na, Rb and Cs), Nauka, 1971, Vol. 2, p. 915.
89. V. A. Shormanov and V. A. Sharnin, in *Dostizheniya i problemy teorii sol'vatatsii: Strukturno-termodinamicheskie aspekty, Ser. Problemy khimii rastvorov* (Advances and Problems in the Theory of Solvation: Structural and Dynamic Aspects, Series: Problems in Chemistry of Solutions), A. M. Kutepov (Ed.), Nauka, Moscow, 1998, p. 172.
90. I. Persson, *Pure Appl. Chem*., **58** (8), 1153 (1986).
91. A. G. Sharpe, *J. Chem. Educ*., **67** (4), 309 (1990).
92. M. J. Blandamer and J. Burgess, *Pure Appl. Chem*., **62** (1), 9 (1990).
93. Y. Marcus, *J. Chem. Soc. Farad. Trans*., **86** (12), 2215 (1990).
94. K. I. Zamaraev *Proceedings of the 15th Int. Conf. Coordination Chemistry (Moscow, 1973)*, Vol. 1, p. 113.
95. Y. Pocker and J. C. Ciola, *J. Am. Chem. Soc*., **110**, 2904 (1988).
96. M. Chabanel, *Pure Appl. Chem*., **62** (1), 35 (1990).
97. D. Seebach, *Angew. Chem. Int. Ed*., **27**, 1624 (1988).
98. H. Waldmann, *Angew. Chem*., **103** (10), 1335 (1991).
99. M. F. Semmelhack and R. Tamura, *J. Am. Chem. Soc*., **105**, 4099 (1983).
100. R. P. Tiger, M. A. Levina, S. G. Entelis and M. A. Andreev, *Kinet. Katal*., **43** (5), 709 (2002).
101. R. P. Tiger, D. N. Tarasov and S. G. Entelis, *Khim. Fizika*, **15** (11), 11 (1996).
102. D. N. Tarasov, R. P. Tiger, S. G. Entelis et al., *Kinet. Katal*., **40** (1), 32 (1999).
103. D. N. Tarasov, R. P. Tiger and S. G. Entelis, Associated Solutions as Pseudo-Polymer Systems: New Approach to Liquid Phase Kinetics and Allowance for Medium Effect, in: *Polimery 2000* [Polymers 2000 (A Collection of Scientific Works)], IKhF RAN, Moscow, 2000, Vol. 2, p. 59.
104. P.-G. de Gennes, *Scaling Concepts in Polymer Physics*, Cornell University Press, Ithaca, New York, 1979.
105. L. F. De Buyck, *Jansen Chim. Acta*, **6** (3), 29 (1988).
106. M. I. Vinnik, *Kinet. Katal*., **28** (1), 100 (1987).
107. L. A. Gasparyan, *Gidrokhlorirovanie atsetilenovykh soedinenii v rastvorakh khloristoi medi* (Hydrochlorination of Acetylene Compounds in Mercuric Chloride Solutions), Candidate Dissertation in Chemistry, VNII Polimer, Yerevan, 1975.
108. N. Dan and S. D. Khue, in: *Fundamental Research in Homogeneous Catalysis*, A. E. Shilov (Ed.), Gordon & Breach, Philadelphia, 1986, Vol. 2, p. 675.
109. L. A. Sil'chenko, S. A. Panova, G. K. Shestakov and O. N. Temkin, *Kinet. Katal*., **38** (6), 861 (1997).
110. G. K. Shestakov, L. A. Sil'chenko, R. S. Korobko and S. A. Panova, *Tezisy dokladov Vsesoyuznoi konferentsii "Primenenie metallokompleksnogo kataliza v organicheskom sinteze"* (Abstracts of Papers of All-Union Conf. "Application of Catalysis with Metal Complexes in Organic Synthesis"), Ufa, 1989, p. 106.
111. S. A. Panova, *Fiziko-khimicheskie osnovy gidrokhlorirovaniya atsetilena na rodievykh katalizatorakh* (Physicochemical Principles of Acetylene Hydrochlorination on Rhodium Catalysts), Candidate Dissertation in Chemistry, MITKhT (Moscow Institute of Fine Chemical Technology), Moscow, 1990.
112. O. N. Temkin, O. Yu. Antonova, S. A. Kezin et al., *Proceedings of the 12th Int. Symp. on Homogeneous Catalysis (Stockholm, 2000)*, Book of Abstracts, p. 224.
113. G. V. Yukhnevich, E. G. Kokhanova, A. I. Pavlyuchko and V. V. Volkov, *J. Mol. Struct*., **122**, 1 (1985).
114. G. V. Yukhnevich, E. G. Tarakanova, V. D. Mayorov and N. B. Librovich, *J. Mol. Struct*., **265**, 237 (1992).
115. V. D. Maiorov, S. G. Sysoeva, O. N. Temkin and I. S. Kislina *Izv. Akad. Nauk, Ser. Khim*., No. 9, 1577 (1993).
116. I. S. Kislina, V. D. Maiorov and S. G. Sysoeva, *Khim. Fiz*., **13** (2), 92 (1994).
117. V. D. Maiorov, I. S. Kislina, S. G. Sysoeva and O. N. Temkin, *Izv. Akad. Nauk, Ser. Khim*., No. 9, 1757 (1995).
118. V. V. Burdin, I. S. Kislina, V. D. Maiorov et al., *Izv. Akad. Nauk, Ser. Khim*., No. 12, 2484 (1998).
119. I. S. Kislina, V. D. Maiorov and S. G. Sysoeva, *Kinet. Katal*., **43** (5), 714 (2002).
120. I. S. Kislina, V. D. Maiorov, S. G. Sysoeva and N. B. Librovich, *Kinet. Katal*., **45** (3), 393 (2004).
121. N. B. Librovich, *Khim. Fiz*., **11**, 627 (1992).
122. I. S. Kislina, N. B. Librovich and V. D. Maiorov, *Khim. Fiz*, **26** 25 (2007).

123. K. H. Pool and R. G. Bates, *J. Chem. Thermodynam.*, **1**, 21 (1969).
124. S. G. Sysoeva, I. S. Kislina and M. I. Vinnik, *Izv. Akad. Nauk SSSR, Ser. Khim.*, 286 (1987).
125. I. S. Kislina, M. Yu. Kulikova, S. G. Sysoeva and O. N. Temkin, *Izv. Akad. Nauk, Ser. Khim.*, **10**, 1722 (1994).
126. B. Nahlovsky and V. Chvalovsky *Coll. Czech. Chem. Comm.*, **33**, 3122 (1968).
127. D. P. N. Satchell, *J. Chem. Soc.*, 3524 (1957).
128. A. A. Pankov, V. Yu. Borovkov and V. B. Kazanskii, *Khim. Fiz.*, **4**, 249 (1985).
129. I. S. Kislina, S. G. Sysoeva and N. B. Librovich, *Dokl. Akad. Nauk*, **360** (5), 649 (1998).
130. E. G. Tarakanova, O. Yu. Tsoi, G. V. Yukhnevich et al., *Kinet. Katal.*, **45** (3), 385 (2004).
131. I. S. Kislina, S. G. Sysoeva and O. N. Temkin, *Izv. Akad. Nauk, Ser. Khim.*, No. 6, 1025 (1994).
132. I. S. Kislina, S. G. Sysoeva and O. N. Temkin, *Izv. Akad. Nauk, Ser. Khim.*, No. 1, 77 (1996).
133. I. S. Kislina, S. G. Sysoeva and O. N. Temkin, *Izv. Akad. Nauk, Ser. Khim.*, 1446 (1996).
134. A. P. Kirilova, V. D. Maiorov, A. I. Serebryanskaya et al., *Izv. Akad. Nauk, Ser. Khim.*, 2435 (1986).
135. I. S. Kislina and S. G. Sysoeva, *Izv. Akad. Nauk, Ser. Khim.*, 961 (2001).
136. I. S. Perelygin and A. N. Afanas'eva, *Zh. Prikl. Spektrosk.*, **19** (3), 500 (1973)
137. A. P. Kirilova, V. D. Maiorov, A. I. Serebryanskaya et al., *Izv. Akad. Nauk, Ser. Khim.*, 1493 (1985).
138. A. P. Kirilova, V. D. Maiorov, A. I. Serebryanskaya et al., *Izv. Akad. Nauk, Ser. Khim.*, 2724 (1987).
139. V. V. Burdin, V. D. Maiorov and N. B. Librovich, *Izv. Akad. Nauk, Ser. Khim.*, No. 2, 292 (2000).
140. I. V. Oshanina, N. A. Kokoreva, L. G. Bruk, O. N. Temkin, I. L. Eremenko and S. E. Nefedov, *Izv. Akad. Nauk, Ser. Khim.*, 2274 (1998).
141. I. S. Kislina, S. G. Sysoeva and N. B. Librovich, *Kinet. Katal.*, **46** (1), 44 (2005).
142. L. A. Gasparyan, T. K. Manukyan, S. S. Kazazyan and I. I. Oganesyan, *Arm. Khim. Zh.*, **24**, 727 (1971).
143. S. A. Panova, G. K. Shestakov and O. N. Temkin, *J. Chem. Soc., Chem. Comm.*, 977 (1994).
144. B. Gronewold and D. Hesse, *Hung. J. Ind. Chem.*, **19**, 259 (1991).
145. L. A. Sil'chenko, *Razrabotka fiziko-khimicheskikh osnov tekhnologii sinteza vinilkhlorida iz atsetilena na nertutnykh katalizatorakh* (Development of Physicochemical Principles of Vinyl Chloride Synthesis from Acetylene on Nonmercuric Catalysts), Candidate Dissertation in Chemistry, MITKhT (Moscow Institute of Fine Chemical Technology), Moscow, 1987.
146. V. A. Pal'm and A. O. Kyrisaar, *Reaktsionnaya sposobnost' organicheskikh soedinenii* (Reactivity of Organic Compounds), Tartu State Univ., Tartu, 1974, Issue 11 (2), p. 519.
147. G. Janz, *Chem. Rev.*, **60**, 209 (1960).
148. M. J. Earle and K. R. Seddon, *Pure Appl. Chem.*, **72**, 1391 (2000).
149. P. Wasserscheid and W. Keim, *Angew. Chem. Int. Ed.*, **39**, 3772 (2000).
150. R. D. Rogers and K. R. Seddon (Eds), *Ionic Liquids: Industrial Applications to Green Chemistry*, American Chemical Society, Washington, 2002.
151. R. D Rogers and K. R. Seddon (Eds.), *Ionic Liquids as Green Solvents*, American Chemical Society, Washington, 2003.
152. (a) P. Wasserscheid and T. Welton (Eds.), *Ionic Liquids in Synthesis*, Wiley-VCH, Weinheim, 2003; (b) D. J. Cole-Hamilton and R. P. Tooze (Eds.), *Catalyst Separation, Recovery and Recycling. Chemistry and Process Design* (Series: Catalysis by Metal Complexes, Vol. 30), Springer, Dordrecht, 2006.
153. D. W. Armstrong, L. He and Y.-S. Liu *Anal. Chem.*, **71**, 3873 (1999).
154. C. L. Hussey, *Adv. Molten Salt Chem.*, **5**, 185 (1983).
155. G. N. Parshall, *J. Am. Chem. Soc.*, **91**, 8716 (1972).
156. C. Bellefon, E. Pollet and P. Grenouillet, *J. Mol. Catal.*, **145**, 121 (1999).
157. C. C. Brasse, U. Englert, A. Saltzer et al., *Organometalics*, **19**, 3818 (2000).
158. P. Wasserscheid and H. Waffenschmidt, *J. Mol. Catal. A*, **164**, 61 (2000).
159. H. Waffenschmidt, *Zur Hydroformylierung in Ionischen Flussigkeiten*, Dissertation, RWTH, Aachen, 2000.
160. S. G. Lee, Y. J. Zhang, J. Y. Piao et al., *J. Chem. Soc., Chem. Comm.*, 2624 (2003).
161. S. G. Lee, Y. J. Zhang, J. H. Park et al., *Proceedings of the 14th Int. Symp. on Homogeneous Catalysis (Munich, Germany, 2004)*, Book of Abstracts, p. 40.
162. M. Sußner, S. Zehfuß and H. Plenio, *Proceedings of the 14th Int. Symp. on Homogeneous Catalysis (Munich, Germany, 2004)*, Book of Abstracts, p. 208.

163. K. H. Shaughnessy, R. S. Booth and L. R. Moore *Proceedings of the 14th Int. Symp. on Homogeneous Catalysis (Munich, Germany, 2004)*, Book of Abstracts, p. 169.
164. V. I. Parvulescu and C. Hardacre, *Chem. Rev.*, **107**, 2615 (2007).
165. M. A. Klingshim, G. A. Broker, J. D. Holbrey et al., *J. Chem. Soc., Chem. Comm.*, 1394 (2002).
166. C. Hardacre, J. D. Holbrey, S. P. Katadare and K. R. Seddon, *Green Chem.*, **4**, 143 (2002).
167. J. H. Davis, *Chem. Lett.*, **33** (9), 1072 (2004).
168. K. H. Shaughnessy, S. J. P'Pool and M. D. Sliger, *Proceedings of the 14th Int. Symp. on Homogeneous Catalysis (Munich, Germany, 2004)*, Book of Abstracts, 168.
169. M. Poliakoff, S. M. Howdle and S. G. Kazarian, *Angew. Chem. Int. Ed.*, **34**, 1275 (1995).
170. P. E. Savage, S. Gopalan, T. I. Mizan et al., *AIChE Journal*, **41** (7), 1723–1778 (1995).
171. C. A. Eckert, B. L. Knutson and P. E. Debenedetti, *Nature*, **383**, 313 (1996).
172. P. G. Jessop, T. Ikariya and R. Noyori, *Chem. Rev.*, **99**, 475 (1999).
173. J. R. Hyde, P. Licence, D. Carter and M. Poliakoff, *Appl. Catal.*, A, **222**, 119 (2001).
174. P. E. Savage, *Chem. Rev.*, **99**, 603 (1999).
175. A. J. Mesiano, E. J. Beckman and A. J Russell, *Chem. Rev.*, **99**, 623 (1999).

Conclusion

Eight chapters of this monograph present achievements in the theory of mechanisms of homogeneous reactions catalyzed by metal complexes, which are based primarily on applications of the kinetic method. Kinetic models of most commercial processes and reactions of potential interest for low-scale production of chemicals and of importance for synthetic organic chemistry have been analyzed.

The author hopes that he has managed to convince readers that, both methodically (technically) and methodologically, the most rational approach to studying mechanisms and constructing theoretically justified (molecular) kinetic models consists of proceeding from hypotheses via their discrimination to working kinetic models. Within the framework of this strategy, the use of the kinetic method is a necessary stage in the procedure of discriminating hypotheses. It should be emphasized that the greater (within reasonable limits) is the number of formulated hypotheses and the more justified are the assumptions used in the generation of hypotheses, the more appropriate to real mechanisms will be the resulting kinetic models. On the other hand, the availability of even a minimum number of hypotheses formulated using either formal or non-formal methods makes the experimental stage of investigation (and the planning of discriminating experiments) a meaningful, logically justified process.

The development and application of the kinetic method in homogeneous catalysis with metal complexes revealed some unsolved problems relating to both the theory of complex reaction mechanisms and the kinetic method proper, as well as problems in the coordination chemistry adjacent to catalysis and the chemistry of solutions, which are important for effectively using the kinetic method. Let us briefly discuss these problems.

A large volume of information has been amassed on the types of elementary steps involving inorganic and organometallic coordination compounds, but the amount of quantitative data on the rate constants of elementary steps (k_{ES}) for a large number of metals (in various oxidation states) and numerous ligands is still very restricted. Almost no quantitative data are available for the dependences of k_{ES} on the donor–acceptor properties and sterical characteristics of ligands for elementary steps of different types with various metal complexes. Very few quantitative experimental data were reported on the enthalpies (ΔH°), activation energies (E_{act}), and kinetic isotope effects of elementary steps.

In an analysis of large reaction networks, it might be possible to carry out a preliminary discrimination of hypothetical elementary steps using theoretical notions based on the methods of quantum chemistry (ΔH°, E_{act}), but use of these methods in application to reactions with participation of metal complexes in solutions is still very laborious and not very effective. The development of effective semi-empirical approaches capable of taking into account the solvent effects while being as simple as the UBI–QEP method in heterogeneous catalysis (see Chapter 7) would be very useful. Should these methods appear, the procedure of generating hypotheses will become a stage of the scientifically justified approach to selecting effective metal complex catalysts, thus making true the dream of chemists engaged in catalysis.

While all the restrictions and difficulties encountered in solving the inverse problem of chemical kinetics in the case of linear mechanisms for both steady- and non-steady-state kinetic models are quite clear,

Homogeneous Catalysis with Metal Complexes: Kinetic Aspects and Mechanisms, First Edition. O. N. Temkin.

questions concerning the choice of initial approximations for k_{ES}, the identifiability of parameters, and the evaluation of constants from the results of measurements of concentrations and reaction rates for the processes with nonlinear mechanisms and complicated dynamic behavior (oscillatory reactions) are not as clear. Solving the inverse kinetic problem in this case is still very difficult, and no effective algorithm is available for this purpose as yet.

Since both the generation and analysis (discrimination) of mechanistic hypotheses for catalytic processes with nonlinear dynamics are extremely laborious procedures, a very good aid for investigations in this field of chemical kinetics would be provided by the development of a program package with convenient interface for a chemist-user, for a computer-aided analysis of the hypotheses about reaction mechanisms with the aim to reveal potential poly-steady-state character of processes and the possible existence of self-oscillatory regimes. Judging from the available publications, it can be suggested that algorithms for such programs can be based on the graph-theoretical approaches developed by Clarke, Vol'pert, Ivanova, and Feinberg, as well as on the methods employing stoichiometric matrices of mechanisms developed by Kol'tsov with co-workers.

The existing approach to constructing kinetic models for catalysis with metal complexes still does not devote proper attention to the steps of formation and loss of active reaction centers. Elucidating the nature of active species and the mechanisms of their formation from precursors in solution is still among the most difficult tasks. These issues have become even more topical since evidence appeared for catalysis by colloidal metals, nanoclusters, giant clusters, and other nanodimensional particles, which can either behave as inactive species and reservoirs of active ones (sleeping catalysts) or serve as microheterogeneous catalysts and/or participants of homogeneous-heterogeneous catalytic reaction cycles.

For elucidating the topological structure of linear or artificially linearized multi-route mechanisms and discriminating between related hypotheses, it is possible to use an analysis of the conjugation nodes (or relative selectivities). For nonlinear multi-route mechanisms, establishing a structure of the mechanism based on simple kinetic dependences is very difficult, if at all possible. In this case, the procedure of generating and discriminating hypotheses is the only effective approach.

Another complicated task is the investigation of reaction kinetics in solutions of polynuclear complexes, especially in cases where the processes of association and clusterization accompany a given catalytic reaction, involve reactants, and are kinetically irreversible. A solution to this task can be provided by measuring *in situ* the concentrations of complexes under steady-state or quasi-steady-state conditions using spectroscopic and/or electrochemical methods or by determining *in situ* the concentrations of metal ions and ligands as demonstrated for $Cu_mCl_n^{(n-m)-}$ complexes (see Section 2.3). In this context, it is necessary to mention an evident deficit of data on the thermodynamic characteristics of polynuclear metal complexes in solution.

In the case of aqueous solutions of metal complexes and electrolytes (salts and strong acids), the methods of maintaining constant activity coefficients of all reaction components in solution while varying the concentrations of reaction participants and monitoring the activity of water are generally clear. In contrast, the question of how an ionic medium influences the activity coefficients of transition states in all elementary steps still remains open. The validity of an assumption about the absence of this medium effect on $\gamma_j^{\neq}$ at relatively high concentrations of electrolytes requires additional investigation.

Serious investigation is also necessary with respect to the influence of electrolytes on k_{ES} in nonaqueous media under conditions of the existence of ion pairs, outer-sphere complexes, and associates and the variation of solvent activity. This is especially important in view of various thermodynamic and kinetic effects related to ion pairs and the possible influence of electrolyte concentrations on the coefficients of activity of outer-sphere complexes, free ions and intermediates in the form of ion pairs. Evidently, some "white spots" in the theory of nonaqueous electrolyte solutions account for the related problems in kinetic investigations.

On the other hand, an allowance for all equilibria of the formation of ion pairs, metal–solvent complexes, outer-sphere complexes, and homo- and heteroassociates formed by the solvent can make it possible in the future to exclude the concept of activity from the thermodynamics of solution and chemical reaction kinetics. Grounds for these expectations are provided by the works by M. I. Usanovich and his followers, and S. G. Entelis and R P. Tiger.

An analysis of the dynamic behavior of various kinetic models of homogeneous catalytic reactions and the necessary chemical conditions (related to the mechanism) for the appearance of critical phenomena such as self-oscillations showed that processes with the mechanisms including steps that are nonlinear with respect to intermediates, nonlinear material balances with respect to the catalyst, substrates, and products, and the steps of formation and loss of active centers (chain mechanisms) can be considered as systems admitting the manifestation (under certain conditions) of nonlinear dynamic phenomena. There are many potential systems for investigating the phenomena of nonlinear dynamics in catalysis with metal complexes. These are, in particular, the reactions of oxidation under conditions of degenerate chain branching, reactions of oxidative and reductive carbonylation, conjugated oxidation of CO and hydrocarboxylation of olefins, cross-coupling in polyfunctional catalytic systems, oxosynthesis in cobalt-containing systems, and hydroalcoxycarbonylation of olefins and epoxides in cobalt carbonyl – pyridine – imidazole systems.

Let these suggestions and expectations be the end of a journey into the world of mechanisms of catalysis with metal complexes as studied by the methods of chemical kinetics and other chemical and physical methods, the journey that the author believes to be fascinating.

It should be emphasized that, despite the aforementioned problems, difficulties, and unanswered questions, a kinetic investigation correctly carried out within the framework of the rational strategy of kinetic model construction, with an understanding of all possibilities and limitations of the kinetic method, is the necessary stage in studying a mechanism of any reaction. The author must ascertain regretfully that the number of publications devoted to the mechanisms of chemical reactions in general and the reaction kinetics in particular has dramatically decreased over the last 10 to15 years. A shift of research interests from the basic problems of chemical science toward the search for new ligands, catalysts, materials, "smart" molecules, and molecular devices is quite natural and explainable, but this is hoped to be only a temporary stage in the oscillatory development of chemistry.

In the course of writing this book, the author had to fight against a permanent desire to go beyond the limits that he initially adopted by deciding to consider only the kinetic aspects of catalysis with metal complexes, while declining from considering questions concerning the reactivity of intermediates, identification of their structures, influence of ligands, etc. Of course, sometimes these boundaries were crossed, especially as far as the author's own investigations were concerned. At the same time, setting the aforementioned limitations was evidently correct. This allowed us to concentrate on one side of the general problem, which was expedient, the more so that separate intermediates and particular steps are considered in a number of excellent monographs. Moreover, in the author's opinion, the whole material of this book claims against simple correlations (given much attention in modern experimental investigations) of the activity of catalysts and selectivity of complicated processes, on the one hand, and simple sterical and electronic characteristics of ligands, on the other hand. The problem of increasing the activity of a catalyst should presently be solved using methods of combinatorial catalysis and/or theoretical approaches based on hypothetical mechanisms of a given reaction.

While working on this book, the author remembered and traced once again the entire history of the development of catalysis with metal complexes in the second half of the 20th century, of which he was happy to be a modest participant together with his colleagues and friends. In this context, the author would like to share with readers two conclusions he arrived at during the beginning of 1980s, which are related to the content of this monograph. The first conclusion, which might presently seem to be rather trivial, is that the chemistry is a united science both conceptually and objectively. Indeed, notions about the nature of chemical bonding, approaches to the problem of chemical reactivity, and even basically important structures of compounds are essentially the same in organic, organoelemental, organometallic, and even inorganic chemistry. An especially important role in justifying this conclusion was played by the development of organometallic chemistry, chemistry of metal polyhedral framework clusters, and coordination chemistry, as well as by the works of R. Hoffman, in particular, those devoted to the isolobal analogy concept. The author believes that the absence of "unifying" chemical courses in chemical universities hinders the formation of a consistent chemist's world outlook.

The second conclusion is related to the author's being not indifferent to catalysis with metal complexes and catalytic chemistry in general. In the entire progress of chemical knowledge and the creation of theoretical chemistry, there were two peaks, two remarkable achievements based on the generalization of chemical investigations. These are the Periodic Law (D. I. Mendeleev) and the Classical Theory of Chemical Structure (A. N. Butlerov, A. Werner, and many others). The author believes that Catalytic Chemistry is a third peak in the evolution of chemistry that deserves attention, since this science has united the numerous notions, rules and principles gained in various fields of chemistry for several hundred years.[1] Already in the 1950s, academician A. A. Balandin frequently used the term "catalytic chemistry" and considered it to be an apex of the chemical knowledge. At the beginning of the 21st century, this statement is especially true and important. It should be noted that achievements in the chemistry of enzymes also significantly contributed to the development of catalytic chemistry.

When a metal-containing complex with a chiral fluoroorganic ligand in solution converts organic and inorganic molecules in the course of a regio- and stereoselective reaction via extremely complicated intermediates into target products and returns in one of the steps into the initial state – what else can be a phenomenon that would better unify the various fields of chemistry?

It is a remarkable fact that traditional synthetic organic chemistry gradually ceases to be "traditional" and also transforms into catalytic chemistry based on catalysis with metal complexes. Thus, the phenomenon of catalysis not only ensured (according to one hypothesis) the appearance of life on the Earth, but also led to the top generalization of chemical knowledge – catalytic chemistry, which is a powerful tool in any chemist's hands.

[1] O.N. Temkin, Catalytic Reactions, in *Sovremennoe estestvoznanie (Entsyklopediya)* (Encyclopedia of Modern Natural Sciences), G. F. Voronin (Ed.), Nauka-Flinta, Moscow, 1999, Vol. 1, pp. 137–145.

Subject Index

Homogeneous Catalysis with Metal Complexes: Kinetic Aspects and Mechanisms, First Edition. O. N. Temkin.

Index of Metals

Homogeneous Catalysis with Metal Complexes: Kinetic Aspects and Mechanisms, First Edition. O. N. Temkin.

Index of Reactions

Homogeneous Catalysis with Metal Complexes: Kinetic Aspects and Mechanisms, First Edition. O. N. Temkin.